Lecture Notes in Computer Science 3173

Commenced Publication in 1973
Founding and Former Series Editors:
Gerhard Goos, Juris Hartmanis, and Jan van Leeuwen

Editorial Board

David Hutchison
  Lancaster University, UK
Takeo Kanade
  Carnegie Mellon University, Pittsburgh, PA, USA
Josef Kittler
  University of Surrey, Guildford, UK
Jon M. Kleinberg
  Cornell University, Ithaca, NY, USA
Friedemann Mattern
  ETH Zurich, Switzerland
John C. Mitchell
  Stanford University, CA, USA
Moni Naor
  Weizmann Institute of Science, Rehovot, Israel
Oscar Nierstrasz
  University of Bern, Switzerland
C. Pandu Rangan
  Indian Institute of Technology, Madras, India
Bernhard Steffen
  University of Dortmund, Germany
Madhu Sudan
  Massachusetts Institute of Technology, MA, USA
Demetri Terzopoulos
  New York University, NY, USA
Doug Tygar
  University of California, Berkeley, CA, USA
Moshe Y. Vardi
  Rice University, Houston, TX, USA
Gerhard Weikum
  Max-Planck Institute of Computer Science, Saarbruecken, Germany

Fuliang Yin   Jun Wang   Chengan Guo (Eds.)

Advances in Neural Networks – ISNN 2004

International Symposium on Neural Networks
Dalian, China, August 19-21, 2004
Proceedings, Part I

 Springer

Volume Editors

Fuliang Yin
Chengan Guo
Dalian University of Technology
School of Electronic and Information Engineering
Dalian, Liaoning, China
E-mail: {flyin, cguo}@dlut.edu.cn

Jun Wang
The Chinese University of Hong Kong
Department of Automation and Computer-Aided Engineering
Shatin, New Territories, Hong Kong
E-mail: jwang@acae.cuhk.edu.hk

Library of Congress Control Number: 2004095623

CR Subject Classification (1998): F.1, F.2, D.1, G.2, I.2, C.2

ISSN 0302-9743
ISBN 3-540-22841-1 Springer Berlin Heidelberg New York

This work is subject to copyright. All rights are reserved, whether the whole or part of the material is concerned, specifically the rights of translation, reprinting, re-use of illustrations, recitation, broadcasting, reproduction on microfilms or in any other way, and storage in data banks. Duplication of this publication or parts thereof is permitted only under the provisions of the German Copyright Law of September 9, 1965, in its current version, and permission for use must always be obtained from Springer. Violations are liable to prosecution under the German Copyright Law.

Springer is a part of Springer Science+Business Media

springeronline.com

© Springer-Verlag Berlin Heidelberg 2004
Printed in Germany

Typesetting: Camera-ready by author, data conversion by PTP-Berlin, Protago-TeX-Production GmbH
Printed on acid-free paper   SPIN: 11312390   06/3142   5 4 3 2 1 0

Preface

This book constitutes the proceedings of the International Symposium on Neural Networks (ISNN 2004) held in Dalian, Liaoning, China during August 19–21, 2004. ISNN 2004 received over 800 submissions from authors in five continents (Asia, Europe, North America, South America, and Oceania), and 23 countries and regions (mainland China, Hong Kong, Taiwan, South Korea, Japan, Singapore, India, Iran, Israel, Turkey, Hungary, Poland, Germany, France, Belgium, Spain, UK, USA, Canada, Mexico, Venezuela, Chile, and Australia). Based on reviews, the Program Committee selected 329 high-quality papers for presentation at ISNN 2004 and publication in the proceedings. The papers are organized into many topical sections under 11 major categories (theoretical analysis; learning and optimization; support vector machines; blind source separation, independent component analysis, and principal component analysis; clustering and classification; robotics and control; telecommunications; signal, image and time series processing; detection, diagnostics, and computer security; biomedical applications; and other applications) covering the whole spectrum of the recent neural network research and development. In addition to the numerous contributed papers, five distinguished scholars were invited to give plenary speeches at ISNN 2004.

ISNN 2004 was an inaugural event. It brought together a few hundred researchers, educators, scientists, and practitioners to the beautiful coastal city Dalian in northeastern China. It provided an international forum for the participants to present new results, to discuss the state of the art, and to exchange information on emerging areas and future trends of neural network research. It also created a nice opportunity for the participants to meet colleagues and make friends who share similar research interests.

The organizers of ISNN 2004 made great efforts to ensure the success of this event. We would like to thank Dalian University of Technology for the sponsorship, various IEEE organizations (especially the IEEE Circuits and Systems Society) for the technical co-sponsorship, and the members of the ISNN 2004 Advisory Committee for their spiritual support. We would also like to thank the members of the Program Committee and additional referees for reviewing the papers and the Publication Committee for checking and compiling the papers in a very short period of time. We would also like to thank the publisher, Springer-Verlag, for their agreement and cooperation to publish the proceedings as a volume of the Lecture Notes in Computer Science. Finally, we would like to thank all the authors for contributing their papers. Without their high-quality papers, this symposium would not have been possible.

August 2004

Fuliang Yin
Jun Wang
Chengan Guo

ISNN 2004 Organization

ISNN 2004 was organized and sponsored by Dalian University of Technology in cooperation with the Chinese University of Hong Kong. It was technically cosponsored by the IEEE Circuits and Systems Society, IEEE Computational Intelligence Society (Beijing, Hong Kong, and Singapore Chapters), and IEEE Control Systems Society and Robotics and Automation Society (Hong Kong Joint Chapter).

Committees

General Co-chairs *Fuliang Yin*, Dalian, China
Jun Wang, Hong Kong

Advisory Committee Co-chairs *Gengdong Cheng*, Dalian, China
Yixin Zhong, Beijing, China
Jacek M. Zurada, Louisville, USA

Advisory Committee Members

Shun-ichi Amari, Tokyo, Japan
Zheng Bao, Xi'an, China
Guoliang Chen, Hefei, China
Ruwei Dai, Beijing, China
Anthony Kuh, Honolulu, USA
Chunbo Feng, Nanjing, China
Toshio Fukuda, Nagoya, Japan
Zhenya He, Nanjing, China
Kararo Hirasawa, Fukuoka, Japan

Frank L. Lewis, Fort Worth, USA
Yanda Li, Beijing, China
Erkki Oja, Helsinki, Finland
Zong Sha, Beijing, China
Tzyh-Jong Tarn, St. Louis, USA
Shoujue Wang, Beijing, China
Zhongtuo Wang, Dalian, China
Youshou Wu, Beijing, China
Bo Zhang, Beijing, China

Program Committee Co-chairs *Chengan Guo*, Dalian, China
Andrzej Cichocki, Tokyo, Japan
Mingsheng Zhao, Beijing, China

Program Committee Members

Sabri Arik (Istanbul, Turkey), *Amit Bhaya* (Rio de Janeiro, Brazil), *Jinde Cao* (Nanjing, China), *Yijia Cao* (Hangzhou, China), *Laiwan Chan* (Hong Kong), *Ke Chen* (Manchester, UK), *Luonan Chen* (Osaka, Japan), *Tianping Chen* (Shanghai, China), *Yiu Ming Cheung* (Hong Kong), *Chuanyin Dang* (Hong Kong), *Wlodzislaw Duch* (Torun, Poland), *Mauro Forti* (Siena, Italy), *Jun Gao* (Hefei, China), *Shuzhi Sam Ge* (Singapore), *Xinping Guan* (Qinhuangdao, China), *Dewen Hu* (Changsha, China), *DeShuang Huang* (Hefei, China), *Donald L. Hung* (San Jose, USA), *Danchi Jiang* (Canberra, Australia), *Licheng Jiao* (Xi'an, China), *H.K. Kwan* (Windsor, Canada), *Xiaoli Li* (Birmingham, UK), *Yuanqing Li* (Tokyo, Japan), *Xue-Bin Liang* (Baton Rouge, USA), *Lizhi Liao*

(Hong Kong), *Xiaofeng Liao* (Chongqing, China), *Chin-Teng Lin* (Hsingchu, Taiwan), *Derong Liu* (Chicago, USA), *Baoliang Lu* (Shanghai, China), *Hongtao Lu* (Shanghai, China), *Fa-Long Luo* (San Jose, USA), *Qing Ma* (Kyoto, Japan), *Zongyuan Mao* (Guangzhou, China), *Xuemei Ren* (Beijing, China), *Rudy Setiono* (Singapore), *Peter Sincak* (Kosice, Slovakia), *Jianbo Su* (Shanghai, China), *Fuchun Sun* (Beijing, China), *Johan Suykens* (Leuven, Belgium), *Ying Tan* (Hefei, China), *Dan Wang* (Singapore), *Lipo Wang* (Singapore), *Wei Wu* (Dalian, China), *Yousheng Xia* (Hong Kong), *Zhongben Xu* (Xi'an, China), *Simon X. Yang* (Guelph, Canada), *Hujun Yin* (Manchester, UK), *Jianwei Zhang* (Hamburg, Germany), *Liming Zhang* (Shanghai, China), *Liqing Zhang* (Shanghai, China), *Yi Zhang* (Chengdu, China), *Weixin Zheng* (Sydney, Australia)

Organizing Committee Chair *Min Han* (Dalian, China)

Publication Co-chairs *Hujun Yin* (Manchester, UK)
Tianshuang Qiu (Dalian, China)

Publicity Co-chairs *Tianshuang Qiu* (Dalian, China)
Derong Liu (Chicago, USA)
Meng Joo Er (Singapore)

Table of Contents, Part I

Part I Theoretical Analysis

Approximation Bounds by Neural Networks in L^p_ω 1
    Jianjun Wang, ZongBen Xu, Weijun Xu

Geometric Interpretation of Nonlinear Approximation Capability
for Feedforward Neural Networks 7
    Bao-Gang Hu, Hong-Jie Xing, Yu-Jiu Yang

Mutual Information and Topology 1: Asymmetric Neural Network 14
    *David Dominguez, Kostadin Koroutchev, Eduardo Serrano,
    Francisco B. Rodríguez*

Mutual Information and Topology 2: Symmetric Network 20
    *Kostadin Koroutchev, David Dominguez, Eduardo Serrano,
    Francisco B. Rodríguez*

Simplified PCNN and Its Periodic Solutions 26
    Xiaodong Gu, Liming Zhang, Daoheng Yu

On Robust Periodicity of Delayed Dynamical Systems with
Time-Varying Parameters 32
    Changyin Sun, Xunming Li, Chun-Bo Feng

Delay-Dependent Criteria for Global Stability of Delayed Neural
Network System 38
    Wenlian Lu, Tianping Chen

Stability Analysis of Uncertain Neural Networks with Delay 44
    Zhongsheng Wang, Hanlin He, Xiaoxin Liao

A New Method for Robust Stability Analysis of a Class of
Recurrent Neural Networks with Time Delays 49
    Huaguang Zhang, Gang Wang

Criteria for Stability in Neural Network Models with Iterative Maps 55
    Carlos Aguirre, Doris Campos, Pedro Pascual, Eduardo Serrano

Exponential Stability Analysis for Neural Network
with Parameter Fluctuations 61
    Haoyang Tang, Chuandong Li, Xiaofeng Liao

Local Stability and Bifurcation in a Model of Delayed Neural Network 67
Yiping Lin, Zengrong Liu

On the Asymptotic Stability of Non-autonomous Delayed Neural Networks 72
Qiang Zhang, Haijun Wang, Dongsheng Zhou, Xiaopeng Wei

Global Exponential Stability of Cohen-Grossberg Neural Networks
with Multiple Time-Varying Delays 78
Kun Yuan, Jinde Cao

Stability of Stochastic Cohen-Grossberg Neural Networks 84
Lin Wang

A Novel Approach to Exponential Stability Analysis of
Cohen-Grossberg Neural Networks 90
Anhua Wan, Weihua Mao, Chun Zhao

Analysis for Global Robust Stability of Cohen-Grossberg Neural
Networks with Multiple Delays 96
Ce Ji, Huaguang Zhang, Huanxin Guan

On Robust Stability of BAM Neural Networks with Constant Delays 102
Chuandong Li, Xiaofeng Liao, Yong Chen

Absolutely Exponential Stability of BAM Neural Networks
with Distributed Delays 108
Wenjun Xiong, Qiuhao Jiang

Stability Analysis of Discrete-Time Cellular Neural Networks 114
Zhigang Zeng, De-Shuang Huang, Zengfu Wang

Novel Exponential Stability Criteria for Fuzzy Cellular Neural
Networks with Time-Varying Delay 120
Yong Chen, Xiaofeng Liao

Stability of Discrete Hopfield Networks with Delay in Serial Mode 126
Runnian Ma, Youmin Xi, Hangshan Gao

Further Results for an Estimation of Upperbound of Delays
for Delayed Neural Networks 132
Xueming Li, Xiaofeng Liao, Tao Xiang

Synchronization in Two Uncoupled Chaotic Neurons 138
Ying Wu, Jianxue Xu, Daihai He, Wuyin Jin, Mi He

Robust Synchronization of Coupled Delayed Recurrent Neural Networks 144
Jin Zhou, Tianping Chen, Xiang Lan

Part II Learning and Optimization

Self-Optimizing Neural Networks 150
    Adrian Horzyk, Ryszard Tadeusiewicz

Genetically Optimized Self-Organizing Neural Networks Based
on PNs and FPNs 156
    Ho-Sung Park, Sung-Kwun Oh, Witold Pedrycz, Hyun-Ki Kim

A New Approach to Self-Organizing Hybrid Fuzzy Polynomial Neural
Networks: Synthesis of Computational Intelligence Technologies 162
    Ho-Sung Park, Sung-Kwun Oh, Witold Pedrycz, Daehee Park, Yongkab Kim

On Soft Learning Vector Quantization Based on Reformulation 168
    Jian Yu, Pengwei Hao

A New Approach to Self-Organizing Polynomial Neural Networks
by Means of Genetic Algorithms 174
    Sung-Kwun Oh, Byoung-Jun Park, Witold Pedrycz, Yong-Soo Kim

Fuzzy-Kernel Learning Vector Quantization 180
    Daoqiang Zhang, Songcan Chen, Zhi-Hua Zhou

Genetic Generation of High-Degree-of-Freedom Feed-Forward
Neural Networks 186
    Yen-Wei Chen, Sulistiyo, Zensho Nakao

Self-Organizing Feature Map Based Data Mining 193
    Shangming Yang, Yi Zhang

Diffusion and Growing Self-Organizing Map: A Nitric Oxide Based
Neural Model 199
    Shuang Chen, Zongtan Zhou, Dewen Hu

A New Adaptive Self-Organizing Map 205
    Shifeng Weng, Fai Wong, Changshui Zhang

Evolving Flexible Neural Networks Using Ant Programming
and PSO Algorithm 211
    Yuehui Chen, Bo Yang, Jiwen Dong

Surrogating Neurons in an Associative Chaotic Neural Network 217
    Masaharu Adachi

Ensembles of RBFs Trained by Gradient Descent 223
    Carlos Hernández-Espinosa, Mercedes Fernández-Redondo,
    Joaquín Torres-Sospedra

Gradient Descent Training of Radial Basis Functions 229
  Mercedes Fernández-Redondo, Carlos Hernández-Espinosa,
  Mamen Ortiz-Gómez, Joaquín Torres-Sospedra

Recent Developments on Convergence of Online Gradient Methods
for Neural Network Training 235
  Wei Wu, Zhengxue Li, Guori Feng, Naimin Zhang, Dong Nan,
  Zhiqiong Shao, Jie Yang, Liqing Zhang, Yuesheng Xu

A Regularized Line Search Tunneling
for Efficient Neural Network Learning 239
  Dae-Won Lee, Hyung-Jun Choi, Jaewook Lee

Transductive Learning Machine Based on the Affinity-Rule
for Semi-supervised Problems and Its Algorithm 244
  Weijiang Long, Wenxiu Zhang

A Learning Algorithm with Gaussian Regularizer for Kernel Neuron 252
  Jianhua Xu, Xuegong Zhang

An Effective Learning Algorithm of Synergetic Neural Network 258
  Xiuli Ma, Licheng Jiao

Sparse Bayesian Learning Based on an Efficient Subset Selection 264
  Liefeng Bo, Ling Wang, Licheng Jiao

A Novel Fuzzy Neural Network with Fast Training
and Accurate Generalization 270
  Lipo Wang, Bing Liu, Chunru Wan

Tuning Neuro-fuzzy Function Approximator by Tabu Search 276
  Guangyuan Liu, Yonghui Fang, Xufei Zheng, Yuhui Qiu

Finite Convergence of MRI Neural Network for Linearly Separable
Training Patterns 282
  Lijun Liu, Wei Wu

A Rapid Two-Step Learning Algorithm for Spline Activation
Function Neural Networks with the Application on Biped Gait Recognition 286
  Lingyun Hu, Zengqi Sun

An Online Feature Learning Algorithm Using HCI-Based
Reinforcement Learning 293
  Fang Liu, Jianbo Su

Optimizing the Weights of Neural Networks Based on Antibody
Clonal Simulated Annealing Algorithm 299
  Xiaoyi Jin, Haifeng Du, Wuhong He, Licheng Jiao

Backpropagation Analysis of the Limited Precision on High-Order
Function Neural Networks 305
    Minghu Jiang, Georges Gielen

LMS Adaptive Notch Filter Design Based on Immune Algorithm........... 311
    Xiaoping Chen, Jianfeng Gao

Training Radial Basis Function Networks with Particle Swarms.............. 317
    Yu Liu, Qin Zheng, Zhewen Shi, Junying Chen

Optimizing Weights by Genetic Algorithm for Neural Network Ensemble 323
    Zhang-Quan Shen, Fan-Sheng Kong

Survival Density Particle Swarm Optimization
for Neural Network Training 332
    Hongbo Liu, Bo Li, Xiukun Wang, Ye Ji, Yiyuan Tang

Modified Error Function with Added Terms
for the Backpropagation Algorithm 338
    Weixing Bi, Xugang Wang, Ziliang Zong, Zheng Tang

Robust Constrained-LMS Algorithm 344
    Xin Song, Jinkuan Wang, Han Wang

A Novel Three-Phase Algorithm for RBF Neural Network Center Selection 350
    Dae-Won Lee, Jaewook Lee

Editing Training Data for kNN Classifiers with Neural Network Ensemble 356
    Yuan Jiang, Zhi-Hua Zhou

Learning Long-Term Dependencies in Segmented Memory Recurrent
Neural Networks 362
    Jinmiao Chen, Narendra S. Chaudhari

An Optimal Neural-Network Model for Learning Posterior
Probability Functions from Observations 370
    Chengan Guo, Anthony Kuh

The Layered Feed-Forward Neural Networks and Its Rule Extraction 377
    Ray Tsaih, Chih-Chung Lin

Analysing Contributions of Components and Factors to Pork Odour
Using Structural Learning with Forgetting Method 383
    Leilei Pan, Simon X. Yang, Fengchun Tian, Lambert Otten,
    Roger Hacker

On Multivariate Calibration Problems 389
    Guo-Zheng Li, Jie Yang, Jun Lu, Wen-Cong Lu, Nian-Yi Chen

Control of Associative Chaotic Neural Networks Using a
Reinforcement Learning 395
    Norihisa Sato, Masaharu Adachi, Makoto Kotani

A Method to Improve the Transiently Chaotic Neural Network 401
    *Xinshun Xu, Jiahai Wang, Zheng Tang, Xiaoming Chen, Yong Li,
Guangpu Xia*

A New Neural Network for Nonlinear Constrained
Optimization Problems 406
    Zhiqing Meng, Chuangyin Dang, Gengui Zhou, Yihua Zhu, Min Jiang

Delay PCNN and Its Application for Optimization 413
    Xiaodong Gu, Liming Zhang, Daoheng Yu

A New Parallel Improvement Algorithm for Maximum Cut Problem 419
    *Guangpu Xia, Zheng Tang, Jiahai Wang, Ronglong Wang, Yong Li,
Guang'an Xia*

A New Neural Network Algorithm for Clique Vertex-Partition Problem 425
    Jiahai Wang, Xinshun Xu, Zheng Tang, Weixing Bi, Xiaoming Chen, Yong Li

An Algorithm Based on Hopfield Network Learning for Minimum
Vertex Cover Problem 430
    *Xiaoming Chen, Zheng Tang, Xinshun Xu, Songsong Li, Guangpu Xia,
Jiahai Wang*

A Subgraph Isomorphism Algorithm Based on Hopfield Neural Network 436
    Ensheng Yu, Xicheng Wang

A Positively Self-Feedbacked Hopfield Neural Network
for N-Queens Problem 442
    Yong Li, Zheng Tang, Ronglong Wang, Guangpu Xia, Jiahai Wang

NN-Based GA for Engineering Optimization 448
    Ling Wang, Fang Tang

Applying GENET to the JSSCSOP 454
    Xin Feng, Lixin Tang, Hofung Leung

Part III   Support Vector Machines

Improvements to Bennett's Nearest Point Algorithm for Support
Vector Machines 462
    Jianmin Li, Jianwei Zhang, Bo Zhang, Fuzong Lin

Distance-Based Selection of Potential Support Vectors by Kernel Matrix 468
    Baoqing Li

A Learning Method for Robust Support Vector Machines 474
   Jun Guo, Norikazu Takahashi, Tetsuo Nishi

A Cascade Method for Reducing Training Time and the Number
of Support Vectors 480
   Yi-Min Wen, Bao-Liang Lu

Minimal Enclosing Sphere Estimation and Its Application to SVMs
Model Selection 487
   Huaqing Li, Shaoyu Wang, Feihu Qi

Constructing Support Vector Classifiers with Unlabeled Data 494
   Tao Wu, Han-Qing Zhao

Nested Buffer SMO Algorithm for Training Support Vector Classifiers 500
   Xiang Wu, Wenkai Lu

Support Vector Classifier with a Fuzzy-Value Class Label 506
   Chan-Yun Yang

RBF Kernel Based Support Vector Machine with Universal
Approximation and Its Application 512
   Junping Wang, Quanshi Chen, Yong Chen

A Practical Parameters Selection Method for SVM 518
   Yongsheng Zhu, Chunhung Li, Youyun Zhang

Ho–Kashyap with Early Stopping Versus Soft Margin SVM for Linear
Classifiers – An Application 524
   *Fabien Lauer, Mohamed Bentoumi, Gérard Bloch, Gilles Millerioux,
   Patrice Aknin*

Radar HRR Profiles Recognition Based on SVM with
Power-Transformed-Correlation Kernel 531
   Hongwei Liu, Zheng Bao

Hydrocarbon Reservoir Prediction Using Support Vector Machines 537
   Kaifeng Yao, Wenkai Lu, Shanwen Zhang, Huanqin Xiao, Yanda Li

Toxic Vapor Classification and Concentration Estimation for Space
Shuttle and International Space Station 543
   Tao Qian, Roger Xu, Chiman Kwan, Bruce Linnell, Rebecca Young

Optimal Watermark Detection Based on Support Vector Machines 552
   Yonggang Fu, Ruimin Shen, Hongtao Lu

Online LS-SVM Learning for Classification Problems Based
on Incremental Chunk 558
   *Zhifeng Hao, Shu Yu, Xiaowei Yang, Feng Zhao, Rong Hu,
   Yanchun Liang*

A Novel Approach to Clustering Analysis Based
on Support Vector Machine 565
  Zhonghua Li, ShaoBai Chen, Rirong Zheng, Jianping Wu, Zongyuan Mao

Application of Support Vector Machine in Queuing System 571
  Gensheng Hu, Feiqi Deng

Modelling of Chaotic Systems with Novel Weighted Recurrent Least
Squares Support Vector Machines 578
  Jiancheng Sun, Taiyi Zhang, Haiyuan Liu

Nonlinear System Identification Based on an Improved Support
Vector Regression Estimator 586
  Li Zhang, Yugeng Xi

Anomaly Detection Using Support Vector Machines 592
  Shengfeng Tian, Jian Yu, Chuanhuan Yin

Power Plant Boiler Air Preheater Hot Spots Detection System Based
on Least Square Support Vector Machines 598
  Liu Han, Liu Ding, Jin Yu, Qi Li, Yanming Liang

Support Vector Machine Multiuser Detector for TD-SCDMA
Communication System in Multipath Channels 605
  Yonggang Wang, Licheng Jiao, Dongfang Zhao

Eyes Location by Hierarchical SVM Classifiers 611
  Yunfeng Li, Zongying Ou

Classification of Stellar Spectral Data Using SVM 616
  Fei Xing, Ping Guo

Iris Recognition Using Support Vector Machines 622
  Yong Wang, Jiuqiang Han

Heuristic Genetic Algorithm-Based Support Vector Classifier
for Recognition of Remote Sensing Images 629
  Chunhong Zheng, Guiwen Zheng, Licheng Jiao

Landmine Feature Extraction and Classification of GPR Data Based
on SVM Method 636
  Jing Zhang, Qun Liu, Baikunth Nath

Occupant Classification for Smart Airbag Using Stereovision and
Support Vector Machines 642
  Hyun-Gu Lee, Yong-Guk Kim, Min-Soo Jang, Sang-Jun Kim, Soek-Joo Lee,
  Gwi-Tae Park

Support Vector Machine Committee for Classification 648
  Bing-Yu Sun, De-Shuang Huang, Lin Guo, Zhong-Qiu Zhao

Automatic Modulation Classification by Support Vector Machines 654
  Zhijin Zhao, Yunshui Zhou, Fei Mei, Jiandong Li

Part IV Blind Source Separation, Independent Component Analysis and Principal Component Analysis

Blind Source Separation Using for Time-Delay Direction Finding 660
  Gaoming Huang, Luxi Yang, Zhenya He

Frequency-Domain Separation Algorithms for Instantaneous Mixtures 666
  Tiemin Mei, Fuliang Yin

Cumulant-Based Blind Separation of Convolutive Mixtures 672
  Tiemin Mei, Fuliang Yin, Jiangtao Xi, Joe F. Chicharo

A New Blind Source Separation Method Based on Fractional Lower
Order Statistics and Neural Network 678
  Daifeng Zha, Tianshuang Qiu, Hong Tang, Yongmei Sun, Sen Li, Lixin Shen

A Clustering Approach for Blind Source Separation with More
Sources than Mixtures. 684
  Zhenwei Shi, Huanwen Tang, Yiyuan Tang

A Blind Source Separation Algorithm with Linear Prediction Filters 690
  Zhijin Zhao, Fei Mei, Jiandong Li

A Blind Source Separation Based Micro Gas Sensor Array
Modeling Method 696
  Guangfen Wei, Zhenan Tang, Philip C.H. Chan, Jun Yu

A Novel Denoising Algorithm Based on Feedforward Multilayer
Blind Separation 702
  Xiefeng Cheng, Ju Liu, Jianping Qiao, Yewei Tao

The Existence of Spurious Equilibrium in FastICA 708
  Gang Wang, Dewen Hu

Hardware Implementation of Pulsed Neural Networks Based
on Delta-Sigma Modulator for ICA 714
  Hirohisa Hotta, Yoshimitsu Murahashi, Shinji Doki, Shigeru Okuma

Study on Object Recognition Based on Independent Component Analysis 720
  Xuming Huang, Chengming Liu, Liming Zhang

Application of ICA Method for Detecting Functional
MRI Activation Data 726
  Minfen Shen, Weiling Xu, Jinyao Yang, Patch Beadle

Single-Trial Estimation of Multi-channel VEP Signals Using
Independent Component Analysis 732
  Lisha Sun, Minfen Shen, Weiling Xu, Francis Chan

Data Hiding in Independent Components of Video 738
  Jiande Sun, Ju Liu, Huibo Hu

Multisensor Data Fusion Based on Independent Component Analysis
for Fault Diagnosis of Rotor 744
  Xiaojiang Ma, Zhihua Hao

Substructural Damage Detection Using Neural Networks and ICA 750
  Fuzheng Qu, Dali Zou, Xin Wang

Speech Segregation Using Constrained ICA 755
  Qiu-Hua Lin, Yong-Rui Zheng, Fuliang Yin, Hua-Lou Liang

Adaptive RLS Implementation of Non-negative PCA Algorithm
for Blind Source Separation 761
  Xiao-Long Zhu, Xian-Da Zhang, Ying Jia

Progressive Principal Component Analysis 768
  Jun Liu, Songcan Chen, Zhi-Hua Zhou

Extracting Target Information in Multispectral Images Using a
Modified KPCA Approach 774
  Zhan-Li Sun, De-Shuang Huang

Use PCA Neural Network to Extract the PN Sequence in Lower SNR
DS/SS Signals 780
  Tianqi Zhang, Xiaokang Lin, Zhengzhong Zhou

Hierarchical PCA-NN for Retrieving the Optical Properties
of Two-Layer Tissue Model 786
  Yaqin Chen, Ling Lin, Gang Li, Jianming Gao, Qilian Yu

Principal Component Analysis Neural Network Based Probabilistic
Tracking of Unpaved Road 792
  Qing Li, Nannig Zheng, Lin Ma, Hong Cheng

Chemical Separation Process Monitoring Based on Nonlinear
Principal Component Analysis 798
  Fei Liu, Zhonggai Zhao

An Adjusted Gaussian Skin-Color Model Based on Principal
Component Analysis 804
  Zhi-Gang Fan, Bao-Liang Lu

Convergence Analysis for Oja+ MCA Learning Algorithm 810
  Jiancheng Lv, Mao Ye, Zhang Yi

On the Discrete Time Dynamics of the MCA Neural Networks 815
  Mao Ye, Zhang Yi

The Cook Projection Index Estimation Using the Wavelet Kernel Function 822
  Wei Lin, Tian Zheng, Fan He, Xian-bin Wen

Part V Clustering and Classification

Automatic Cluster Number Determination via BYY Harmony Learning 828
  Xuelei Hu, Lei Xu

Unsupervised Learning for Hierarchical Clustering Using
Statistical Information.. 834
  Masaru Okamoto, Nan Bu, Toshio Tsuji

Document Clustering Algorithm Based on Tree-Structured Growing
Self-Organizing Feature Map 840
  Xiaoshen Zheng, Wenling Liu, Pilian He, Weidi Dai

Improved SOM Clustering for Software Component Catalogue 846
  Zhuo Wang, Daxin Liu, Xiaoning Feng

Classification by Multilayer Feedforward Ensembles 852
  Mercedes Fernández-Redondo, Carlos Hernández-Espinosa,
  Joaquín Torres-Sospedra

Robust Face Recognition from a Single Training Image per Person
with Kernel-Based SOM-Face 858
  Xiaoyang Tan, Songcan Chen, Zhi-Hua Zhou, Fuyan Zhang

Unsupervised Feature Selection for Multi-class Object Detection
Using Convolutional Neural Networks..................................... 864
  Masakazu Matsugu, Pierre Cardon

Recurrent Network as a Nonlinear Line Attractor for Skin Color
Association 870
  Ming-Jung Seow, Vijayan K. Asari

A Bayesian Classifier by Using the Adaptive Construct Algorithm
of the RBF Networks 876
  Minghu Jiang, Dafan Liu, Beixing Deng, Georges Gielen

An Improved Analytical Center Machine for Classification 882
  FanZi Zeng, DongSheng Li, ZhengDing Qiu

Analysis of Fault Tolerance of a Combining Classifier 888
  Hai Zhao, Bao-Liang Lu

Training Multilayer Perceptron with Multiple Classifier Systems 894
Hui Zhu, Jiafeng Liu, Xianglong Tang, Jianhuan Huang

Learning the Supervised NLDR Mapping for Classification 900
Zhonglin Lin, Shifeng Weng, Changshui Zhang, Naijiang Lu, Zhimin Xia

Comparison with Two Classification Algorithms of Remote Sensing
Image Based on Neural Network 906
Yumin Chen, Youchuan Wan, Jianya Gong, Jin Chen

Some Experiments with Ensembles of Neural Networks
for Classification of Hyperspectral Images............................. 912
Carlos Hernández-Espinosa, Mercedes Fernández-Redondo, Joaquín Torres-Sospedra

SAR Image Recognition Using Synergetic Neural Networks Based
on Immune Clonal Programming 918
Shuiping Gou, Licheng Jiao

Speaker Identification Using Reduced RBF Networks Array................. 924
Han Lian, Zheng Wang, Jianjun Wang, Liming Zhang

Underwater Acoustic Targets Classification Using Welch Spectrum
Estimation and Neural Networks 930
Chunyu Kang, Xinhua Zhang, Anqing Zhang, Hongwen Lin

Automatic Digital Modulation Recognition Using Wavelet Transform
and Neural Networks 936
Zhilu Wu, Guanghui Ren, Xuexia Wang, Yaqin Zhao

Local Face Recognition Based on the Combination of ICA and NFL 941
Yisong Ye, Yan Wu, Mingliang Sun, Mingxi Jin

Facial Expression Recognition Using Kernel Discriminant Plane 947
Wenming Zheng, Xiaoyan Zhou, Cairong Zou, Li Zhao

Application of CMAC-Based Networks on Medical Image Classification 953
Weidong Xu, Shunren Xia, Hua Xie

Seismic Pattern Recognition of Nuclear Explosion Based on
Generalization Learning Algorithm of BP Network and Genetic Algorithm 959
Daizhi Liu, Renming Wang, Xihai Li, Zhigang Liu

A Remote Sensing Image Classification Method Using Color
and Texture Feature... 965
Wen Cao, Tian-Qiang Peng, Bi-Cheng Li

Remote Sensing Image Classification Based on Evidence Theory
and Neural Networks 971
  Gang Chen, Bi-Cheng Li, Zhi-Gang Guo

Neural Networks in Detection and Identification of Littoral Oil
Pollution by Remote Sensing 977
  Bin Lin, Jubai An, Carl Brown, Hande Zhang

Identifying Pronunciation-Translated Names from Chinese Texts
Based on Support Vector Machines 983
  Lishuang Li, Chunrong Chen, Degen Huang, Yuansheng Yang

FPGA Implementation of Feature Extraction and Neural Network
Classifier for Handwritten Digit Recognition 988
  Dongsheng Shen, Lianwen Jin, Xiaobin Ma

Replicator Neural Networks for Outlier Modeling
in Segmental Speech Recognition 996
  László Tóth, Gábor Gosztolya

A Novel Approach to Stellar Recognition by Combining EKF and RBF Net 1002
  Ling Bai, Ping Guo

Pattern Recognition Based on Stability of Discrete Time Cellular
Neural Networks 1008
  Zhigang Zeng, De-Shuang Huang, Zengfu Wang

The Recognition of the Vehicle License Plate Based
on Modular Networks 1015
  Guangrong Ji, Hongjie Yi, Bo Qin, Hua Xu

Neural Network Associator for Images and Their Chinese Characters 1021
  Zhao Zhang

Exploring Various Features to Optimize Hot Topic Retrieval on WEB 1025
  Lan You, Xuanjing Huang, Lide Wu, Hao Yu, Jun Wang,
  Fumihito Nishino

Author Index 1033

Table of Contents, Part II

Part VI Robotics and Control

Application of RBFNN for Humanoid Robot Real Time Optimal
Trajectory Generation in Running 1
   Xusheng Lei, Jianbo Su

Full-DOF Calibration-Free Robotic Hand-Eye Coordination
Based on Fuzzy Neural Network 7
   Jianbo Su, Qielu Pan, Zhiwei Luo

Neuro-Fuzzy Hybrid Position/Force Control for a Space Robot
with Flexible Dual-Arms 13
   Fuchun Sun, Hao Zhang, Hao Wu

Fuzzy Neural Networks Observer for Robotic Manipulators
Based on H_∞ Approach 19
   Hong-bin Wang, Chun-di Jiang, Hong-rui Wang

Mobile Robot Path-Tracking
Using an Adaptive Critic Learning PD Controller 25
   Xin Xu, Xuening Wang, Dewen Hu

Reinforcement Learning and ART2 Neural Network Based Collision
Avoidance System of Mobile Robot 35
   Jian Fan, GengFeng Wu, Fei Ma, Jian Liu

FEL-Based Adaptive Dynamic Inverse Control
for Flexible Spacecraft Attitude Maneuver 42
   Yaqiu Liu, Guangfu Ma, Qinglei Hu

Multivariable Generalized Minimum Variance Control
Based on Artificial Neural Networks and Gaussian Process Models 52
   Daniel Sbarbaro, Roderick Murray-Smith, Arturo Valdes

A Neural Network Based Method for Solving
Discrete-Time Nonlinear Output Regulation Problem in Sampled-Data Systems . 59
   Dan Wang, Jie Huang

The Design of Fuzzy Controller
by Means of CI Technologies-Based Estimation Technique 65
   Sung-Kwun Oh, Seok-Beom Roh, Dong-Yoon Lee, Sung-Whan Jang

A Neural Network Adaptive Controller
for Explicit Congestion Control with Time Delay 71
   Bo Yang, Xinping Guan

Robust Adaptive Control Using Neural Networks and Projection 77
   Xiaoou Li, Wen Yu

Design of PID Controllers Using Genetic Algorithms Approach
for Low Damping, Slow Response Plants 83
   PenChen Chou, TsenJar Hwang

Neural Network Based Fault Tolerant Control of a Class
of Nonlinear Systems with Input Time Delay 91
   Ming Liu, Peng Liu, Donghua Zhou

Run-to-Run Iterative Optimization Control of Batch Processes 97
   Zhihua Xiong, Jie Zhang, Xiong Wang, Yongmao Xu

Time-Delay Recurrent Neural Networks for Dynamic Systems Control 104
   Xu Xu, Yinghua Lu, Yanchun Liang

Feedforward-Feedback Combined Control System
Based on Neural Network 110
   Weidong Zhang, Fanming Zeng, Guojun Cheng, Shengguang Gong

Online Learning CMAC Neural Network Control Scheme
for Nonlinear Systems 117
   Yuman Yuan, Wenjin Gu, Jinyong Yu

Pole Placement Control for Nonlinear Systems via Neural Networks 123
   Fei Liu

RBF NN-Based Backstepping Control
for Strict Feedback Block Nonlinear System and Its Application 129
   Yunan Hu, Yuqiang Jin, Pingyuan Cui

Model Reference Control Based on SVM 138
   Junfeng He, Zengke Zhang

PID Controller Based on the Artificial Neural Network 144
   Jianhua Yang, Wei Lu, Wenqi Liu

Fuzzy Predictive Control Based on PEMFC Stack 150
   Xi Li, Xiao-wei Fu, Guang-yi Cao, Xin-jian Zhu

Adaptive Control for Induction Servo Motor
Based on Wavelet Neural Networks 156
   Qinghui Wu, Yi Liu, Dianjun Zhang, Yonghui Zhang

The Application of Single Neuron Adaptive PID Controller
in Control System of Triaxial and Torsional Shear Apparatus 163
    *Muguo Li, Zhendong Liu, Jing Wang, Qun Zhang, Hairong Jiang,
Hai Du*

Ram Velocity Control in Plastic Injection Molding Machines
with Neural Network Learning Control 169
    *Gaoxiang Ouyang, Xiaoli Li, Xinping Guan, Zhiqiang Zhang,
Xiuling Zhang, Ruxu Du*

Multiple Models Neural Network Decoupling Controller
for a Nonlinear System 175
    Xin Wang, Shaoyuan Li, Zhongjie Wang, Heng Yue

Feedback-Assisted Iterative Learning Control
for Batch Polymerization Reactor.................................. 181
    Shuchen Li, Xinhe Xu, Ping Li

Recent Developments on Applications of Neural Networks
to Power Systems Operation and Control: An Overview 188
    Chuangxin Guo, Quanyuan Jiang, Xiu Cao, Yijia Cao

A Novel Fermentation Control Method Based on Neural Networks 194
    Xuhua Yang, Zonghai Sun, Youxian Sun

Modeling Dynamic System by Recurrent Neural Network
with State Variables... 200
    Min Han, Zhiwei Shi, Wei Wang

Robust Friction Compensation for Servo System
Based on LuGre Model with Uncertain Static Parameters 206
    Lixin Wei, Xia Wang, Hongrui Wang

System Identification Using Adjustable RBF Neural Network
with Stable Learning Algorithms 212
    Wen Yu, Xiaoou Li

A System Identification Method Based on Multi-layer Perception
and Model Extraction 218
    Chang Hu, Li Cao

Complex Model Identification Based on RBF Neural Network 224
    Yibin Song, Peijin Wang, Kaili Li

Part VII Telecommunications

A Noisy Chaotic Neural Network Approach to Topological Optimization
of a Communication Network with Reliability Constraints 230
    Lipo Wang, Haixiang Shi

Space-Time Multiuser Detection Combined
with Adaptive Wavelet Networks over Multipath Channels 236
    Ling Wang, Licheng Jiao, Haihong Tao, Fang Liu

Optimizing Sensor Node Distribution with Genetic Algorithm
in Wireless Sensor Network 242
    Jianli Zhao, Yingyou Wen, Ruiqiang Shang, Guangxing Wang

Fast De-hopping and Frequency Hopping Pattern (FHP) Estimation
for DS/FHSS Using Neural Networks 248
    Tarek Elhabian, Bo Zhang, Dingrong Shao

Autoregressive and Neural Network Model Based Predictions
for Downlink Beamforming 254
    Halil Yigit, Adnan Kavak, Metin Ertunc

Forecast and Control of Anode Shape in Electrochemical Machining
Using Neural Network 262
    Guibing Pang, Wenji Xu, Xiaobing Zhai, Jinjin Zhou

A Hybrid Neural Network and Genetic Algorithm Approach
for Multicast QoS Routing 269
    Daru Pan, Minghui Du, Yukun Wang, Yanbo Yuan

Performance Analysis of Recurrent Neural Networks Based
Blind Adaptive Multiuser Detection in Asynchronous DS-CDMA Systems 275
    Ling Wang, Haihong Tao, Licheng Jiao, Fang Liu

Neural Direct Sequence Spread Spectrum Acquisition 281
    Tarek Elhabian, Bo Zhang, Dingrong Shao

Multi-stage Neural Networks for Channel Assignment
in Cellular Radio Networks 287
    Hyuk-Soon Lee, Dae-Won Lee,, Jaewook Lee

Experimental Spread Spectrum Communication System Based on CNN 293
    Jianye Zhao, Shide Guo, Daoheng Yu

Neural Congestion Control Algorithm in ATM Networks
with Multiple Node 299
    Ruijun Zhu, Fuliang Yin, Tianshuang Qiu

Neural Compensation of Linear Distortion in Digital Communications 305
   Hong Zhou

On the Performance of Space-Time Block Coding
Based on ICA Neural Networks 311
   Ju Liu, Hongji Xu, Yong Wan

ICA-Based Beam Space-Time Block Coding
with Transmit Antenna Array Selection 317
   Hongji Xu, Ju Liu

Nonlinear Dynamic Method to Suppress Reverberation
Based on RBF Neural Networks 324
   Bing Deng, Ran Tao

Part VIII Signal, Image, and Time Series Processing

A New Scheme for Detection and Classification
of Subpixel Spectral Signatures in Multispectral Data 331
   Hao Zhou, Bin Wang, Liming Zhang

A Rough-Set-Based Fuzzy-Neural-Network System
for Taste Signal Identification 337
   Yan-Xin Huang, Chun-Guang Zhou, Shu-Xue Zou, Yan Wang,
   Yan-Chun Liang

A Novel Signal Detection Subsystem of Radar Based on HA-CNN 344
   Zhangliang Xiong, Xiangquan Shi

Real-Time Detection of Signal in the Noise Based
on the RBF Neural Network and Its Application 350
   Minfen Shen, Yuzheng Zhang, Zhancheng Li, Jinyao Yang,
   Patch Beadle

Classification of EEG Signals Under Different Brain Functional States
Using RBF Neural Network 356
   Zhancheng Li, Minfen Shen, Patch Beadle

Application of a Wavelet Adaptive Filter Based on Neural Network
to Minimize Distortion of the Pulsatile Spectrum 362
   Xiaoxia Li, Gang Li, Ling Lin, Yuliang Liu, Yan Wang,
   Yunfeng Zhang

Spectral Analysis and Recognition Using Multi-scale Features
and Neural Networks 369
   YuGang Jiang, Ping Guo

A Novel Fuzzy Filter for Impulse Noise Removal 375
 Chang-Shing Lee, Shu-Mei Guo, Chin-Yuan Hsu

Neural Network Aided Adaptive Kalman Filter
for Multi-sensors Integrated Navigation 381
 Lin Chai, Jianping Yuan, Qun Fang, Zhiyu Kang, Liangwei Huang

Solely Excitatory Oscillator Network for Color Image Segmentation 387
 Chung Lam Li, Shu Tak Lee

Image De-noising Using Cross-Validation Method
with RBF Network Representation 393
 Ping Guo, Hongzhai Li

Ultrasonic C-scan Image Restoration Using Radial Basis Function Network 399
 Zongjie Cao, Huaidong Chen, Jin Xue, Yuwen Wang

Automatic Image Segmentation
Based on a Simplified Pulse Coupled Neural Network 405
 Yingwei Bi, Tianshuang Qiu, Xiaobing Li, Ying Guo

Face Pose Estimation Based on Eigenspace Analysis
and Fuzzy Clustering 411
 Cheng Du, Guangda Su

Chaotic Time Series Prediction
Based on Local-Region Multi-steps Forecasting Model 418
 Minglun Cai, Feng Cai, Aiguo Shi, Bo Zhou, Yongsheng Zhang

Nonlinear Prediction Model Identification and Robust Prediction
of Chaotic Time Series 424
 Yuexian Hou, Weidi Dai, Pilian He

Wavelet Neural Networks for Nonlinear Time Series Analysis 430
 Bo Zhou, Aiguo Shi, Feng Cai, Yongsheng Zhang

Part IX Biomedical Applications

Neural Networks Determining Injury by Salt Water
for Distribution Lines 436
 Lixin Ma, Hiromi Miyajima, Noritaka Shigei, Shuma Kawabata

EEG Source Localization Using Independent Residual Analysis 442
 Gang Tan, Liqing Zhang

Classifying G-protein Coupled Receptors with Support Vector Machine 448
 Ying Huang, Yanda Li

A Novel Individual Blood Glucose Control Model Based on Mixture
of Experts Neural Networks 453
   Wei Wang, Zheng-Zhong Bian, Lan-Feng Yan, Jing Su

Tracking the Amplitude Variation of Evoked Potential by ICA and WT 459
   Haiyan Ding, Datian Ye

A Novel Method for Gene Selection and Cancer Classification 465
   Huajun Yan, Zhang Yi

Nonnegative Matrix Factorization for EEG Signal Classification 470
   Weixiang Liu, Nanning Zheng, Xi Li

A Novel Clustering Analysis Based on PCA and SOMs
for Gene Expression Patterns 476
   Hong-Qiang Wang, De-Shuang Huang, Xing-Ming Zhao, Xin Huang

Feedback Selective Visual Attention Model
Based on Feature Integration Theory 482
   Lianwei Zhao, Siwei Luo

Realtime Monitoring of Vascular Conditions
Using a Probabilistic Neural Network 488
   Akira Sakane, Toshio Tsuji, Yoshiyuki Tanaka, Kenji Shiba, Noboru Saeki, Masashi Kawamoto

Capturing Long-Term Dependencies
for Protein Secondary Structure Prediction 494
   Jinmiao Chen, Narendra S. Chaudhari

A Method for Fast Estimation of Evoked Potentials
Based on Independent Component Analysis 501
   Ting Li, Tianshuang Qiu, Xuxiu Zhang, Anqing Zhang, Wenhong Liu

Estimation of Pulmonary Elastance Based on RBF Expression 507
   Shunshoku Kanae, Zi-Jiang Yang, Kiyoshi Wada

Binary Input Encoding Strategy Based Neural Network
for Globulin Protein Inter-residue Contacts Map Prediction 513
   GuangZheng Zhang, DeShuang Huang, Xin Huang

Genetic Regulatory Systems Modeled by Recurrent Neural Network 519
   Xianhua Dai

A New Computational Model of Biological Vision for Stereopsis 525
   Baoquan Song, Zongtan Zhou, Dewen Hu, Zhengzhi Wang

A Reconstruction Approach to CT with Cauchy RBFs Network 531
   Jianzhong Zhang, Haiyang Li

Part X Detection, Diagnostics, and Computer Security

Fault Diagnosis on Satellite Attitude Control
with Dynamic Neural Network 537
    HanYong Hao, ZengQi Sun, Yu Zhang

A New Strategy for Fault Detection of Nonlinear Systems
Based on Neural Networks 543
    Linglai Li, Donghua Zhou

Non-stationary Fault Diagnosis Based on Local-Wave Neural Network 549
    Zhen Wang, Ji Li, Zijia Ding, Yanhui Song

Hybrid Neural Network Based Gray-Box Approach to Fault Detection
of Hybrid Systems 555
    Wenhui Wang, Dexi An, Donghua Zhou

Application of Enhanced Independent Component Analysis
to Leak Detection in Transport Pipelines 561
    Zhengwei Zhang, Hao Ye, Rong Hu

Internet-Based Remote Monitoring and Fault Diagnosis System 567
    Xing Wu, Jin Chen, Ruqiang Li, Fucai Li

Rough Sets and Partially-Linearized Neural Network
for Structural Fault Diagnosis of Rotating Machinery 574
    Peng Chen, Xinying Liang, Takayoshi Yamamoto

Transient Stability Assessment Using Radial Basis Function Networks 581
    Yutian Liu, Xiaodong Chu, Yuanyuan Sun, Li Li

An Optimized Shunt Hybrid Power Quality Conditioner
Based on an Adaptive Neural Network for Power Quality Improvement
in Power Distribution Network 587
    Ming Zhang, Hui Sun, Jiyan Zou, Hang Su

Cyclic Statistics Based Neural Network for Early Fault Diagnosis
of Rolling Element Bearings 595
    Fuchang Zhou, Jin Chen, Jun He, Guo Bi, Guicai Zhang, Fucai Li

Application of BP Neural Network for the Abnormity Monitoring
in Slab Continuous Casting 601
    Xudong Wang, Man Yao, Xingfu Chen

A TVAR Parametric Model Applying for Detecting
Anti-electric-Corona Discharge 607
    Zhe Chen, Hongyu Wang, Tianshuang Qiu

The Study on Crack Diagnosis Based on Fuzzy Neural Networks
Using Different Indexes 613
  Jingfen Zhang, Guang Meng, Deyou Zhao

Blind Fault Diagnosis Algorithm for Integrated Circuit
Based on the CPN Neural Networks........................... 619
  Daqi Zhu, Yongqing Yang, Wuzhao Li

A Chaotic-Neural-Network-Based Encryption Algorithm
for JPEG2000 Encoded Images 627
  Shiguo Lian, Guanrong Chen, Albert Cheung, Zhiquan Wang

A Combined Hash and Encryption Scheme by Chaotic Neural Network 633
  Di Xiao, Xiaofeng Liao

A Novel Symmetric Cryptography Based on Chaotic Signal Generator
and a Clipped Neural Network 639
  Tsing Zhou, Xiaofeng Liao, Yong Chen

A Neural Network Based Blind Watermarking Scheme for Digital Images...... 645
  Guoping Tang, Xiaofeng Liao

Color Image Watermarking Based on Neural Networks................ 651
  Wei Lu, Hongtao Lu, Ruiming Shen

A Novel Intrusion Detection Method
Based on Principle Component Analysis in Computer Security 657
  Wei Wang, Xiaohong Guan, Xiangliang Zhang

A Novel Wavelet Image Watermarking Scheme Combined
with Chaos Sequence and Neural Network 663
  Jian Zhao, Mingquan Zhou, Hongmei Xie, Jinye Peng, Xin Zhou

Quantum Neural Network for Image Watermarking 669
  Shiyan Hu

An NN-Based Malicious Executables Detection Algorithm
Based on Immune Principles 675
  Zhenhe Guo, Zhengkai Liu, Ying Tan

The Algorithm for Detecting Hiding Information Based on SVM............. 681
  JiFeng Huang, JiaJun Lin, XiaoFu He, Meng Dai

An E-mail Filtering Approach Using Neural Network................... 688
  Yukun Cao, Xiaofeng Liao, Yunfeng Li

Part XI Other Applications

Topography-Enhanced BMU Search in Self-Organizing Maps 695
  James S. Kirk, Jacek M. Zurada

Nonlinear Optimal Estimation for Neural Networks Data Fusion 701
  Ye Ma, Xiao Tong Wang, Bo Li, JianGuo Fu

Improvement of Data Visualization Based on SOM 707
  Chao Shao, Houkuan Huang

Research on Neural Network Method in GPS Data Transformation 713
  Min Han, Xue Tian, Shiguo Xu

Dynamic File Allocation in Storage Area Networks
with Neural Network Prediction 719
  Weitao Sun, Jiwu Shu, Weimin Zheng

Competitive Algorithms for Online Leasing Problem
in Probabilistic Environments 725
  Yinfeng Xu, Weijun Xu

Wavelet Network for Nonlinear Regression Using Probabilistic Framework 731
  Shu-Fai Wong, Kwan-Yee Kenneth Wong

A Neural Network Approach for Indirect Shape from Shading 737
  Ping Hao, Dongming Guo, Renke Kang

A Hybrid Radial Basis Function Neural Network
for Dimensional Error Prediction in End Milling 743
  Xiaoli Li, Xinping Guan, Yan Li

Bayesian Neural Networks for Life Modeling and Prediction
of Dynamically Tuned Gyroscopes 749
  Chunling Fan, Feng Gao, Zhihua Jin

Furnace Temperature Modeling for Continuous Annealing Process
Based on Generalized Growing and Pruning RBF Neural Network 755
  Qing Chen, Shaoyuan Li, Yugeng Xi, Guangbin Huang

A Learning-Based Contact Model in Belt-Grinding Processes 761
  Xiang Zhang, Bernd Kuhlenkötter, Klaus Kneupner

Application of General Regression Neural Network
to Vibration Trend Prediction of Rotating Machinery 767
  Zhipeng Feng, Fulei Chu, Xigeng Song

A Novel Nonlinear Projection to Latent Structures Algorithm 773
  Shi Jian Zhao, Yong Mao Xu, Jie Zhang

Inversing Reinforced Concrete Beams Flexural Load Rating
Using ANN and GA Hybrid Algorithm 779
   Zeying Yang, Chengkui Huang, Jianbo Qu

Determining of the Delay Time for a Heating Ventilating
and Air-Conditioning Plant Using Two Weighted Neural Network Approach 786
   Mengdi Hu, Wenming Cao, Shoujue Wang

Intelligent Forecast Procedures for Slope Stability
with Evolutionary Artificial Neural Network 792
   Shouju Li, Yingxi Liu

Structural Reliability Analysis via Global Response Surface Method
of BP Neural Network 799
   Jinsong Gui, Hequan Sun, Haigui Kang

Modeling Temperature Drift of FOG by Improved BP Algorithm
and by Gauss-Newton Algorithm 805
   Xiyuan Chen

A Novel Chaotic Neural Network for Automatic Material Ratio System 813
   Lidan Wang, Shukai Duan

Finite Element Analysis of Structures
Based on Linear Saturated System Model 820
   Hai-Bin Li, Hong-Zhong Huang, Ming-Yang Zhao

Neural Network Based Fatigue Cracks Evolution 826
   Chunsheng Liu, Weidong Wu, Daoheng Sun

Density Prediction of Selective Laser Sintering Parts
Based on Artificial Neural Network 832
   Xianfeng Shen, Jin Yao, Yang Wang, Jialin Yang

Structure Optimization of Pneumatic Tire Using an Artificial Neural Network ... 841
   XuChun Ren, ZhenHan Yao

A Multiple RBF NN Modeling Approach to BOF Endpoint Estimation
in Steelmaking Process 848
   Xin Wang, Shaoyuan Li, Zhongjie Wang, Jun Tao, Jinxin Liu

Application of Neural Network on Wave Impact Force Prediction 854
   Hongyu Zhang, Yongxue Wang, Bing Ren

RBF Neural Networks-Based Software Sensor
for Aluminum Powder Granularity Distribution Measurement 860
   Yonghui Zhang, Cheng Shao, Qinghui Wu

A Technological Parameter Optimization Approach
in Crude Oil Distillation Process Based on Neural Network 866
  Hao Tang, Quanyi Fan, Bowen Xu, Jianming Wen

A Neural Network Modeling Method for Batch Process 874
  Yi Liu, XianHui Yang, Jie Zhang

Modelling the Supercritical Fluid Extraction of Lycopene
from Tomato Paste Waste Using Neuro-Fuzzy Approaches 880
  Simon X. Yang, Weiren Shi, Jin Zeng

Detection of Weak Targets with Wavelet and Neural Network 886
  Changwen Qu, You He, Feng Su, Yong Huang

An Artificial Olfactory System Based on Gas Sensor Array
and Back-Propagation Neural Network 892
  Huiling Tai, Guangzhong Xie, Yadong Jiang

Consumer Oriented Design of Product Forms 898
  Yang-Cheng Lin, Hsin-Hsi Lai, Chung-Hsing Yeh

An Intelligent Simulation Method Based on Artificial Neural Network
for Container Yard Operation 904
  Chun Jin, Xinlu Liu, Peng Gao

A Freeway Traffic Incident Detection Algorithm
Based on Neural Networks 912
  Xuhua Yang, Zonghai Sun, Youxian Sun

Radial Basis Function Network for Chaos Series Prediction 920
  Wei Chi, Bo Zhou, Aiguo Shi, Feng Cai, Yongsheng Zhang

Feature Extraction and Identification of Underground Nuclear Explosion
and Natural Earthquake Based on FM^mlet Transform
and BP Neural Network 925
  Xihai Li, Ke Zhao, Daizhi Liu, Bin Zhang

A Boosting-Based Framework for Self-Similar
and Non-linear Internet Traffic Prediction 931
  Hanghang Tong, Chongrong Li, Jingrui He

Traffic Flow Forecasting Based on Parallel Neural Network 937
  Guozhen Tan, Wenjiang Yuan

A High Precision Prediction Method by Using Combination
of ELMAN and SOM Neural Networks 943
  Jie Wang, Dongwei Yan

Short-Term Traffic Flow Forecasting Using Expanded Bayesian
Network for Incomplete Data 950
    Changshui Zhang, Shiliang Sun, Guoqiang Yu

Highway Traffic Flow Model Using FCM-RBF Neural Network 956
    Jian-Mei Xiao, Xi-Huai Wang

Earthquake Prediction by RBF Neural Network Ensemble 962
    Yue Liu, Yuan Wang, Yuan Li, Bofeng Zhang,
    Gengfeng Wu

Rainfall-Runoff Correlation
with Particle Swarm Optimization Algorithm 970
    Kwokwing Chau

A Study of Portfolio Investment Decision Method
Based on Neural Network... 976
    Yongqing Yang, Jinde Cao, Daqi Zhu

Portfolio Optimization for Multi-stage Capital Investment
with Neural Networks.. 982
    Yanglan Zhang, Yu Hua

Efficient Option Pricing via a Globally Regularized Neural Network 988
    Hyung-Jun Choi, Hyo-Seok Lee, Gyu-Sik Han, Jaewook Lee

Effectiveness of Neural Networks for Prediction
of Corporate Financial Distress in China 994
    Ji-Gang Xie, Jing Wang, Zheng-Ding Qiu

A Neural Network Model on Solving
Multiobjective Conditional Value-at-Risk 1000
    Min Jiang, Zhiqing Meng, Qiying Hu

DSP Structure Optimizations –
A Multirate Signal Flow Graph Approach 1007
    Ronggang Qi, Zifeng Li, Qing Ma

Author Index... 1013

Approximation Bounds by Neural Networks in L_ω^p

JianJun Wang[1], ZongBen Xu[1], and Weijun Xu[2]

[1] Institute for Information and System Science, Faculty of Science,
Xi'an Jiaotong University, Xi'an, Shaan'xi, 710049, P.R. China
wjj761229@163.com
[2] School of Management, Xi'an Jiaotong University,
Xi'an, Shaan'xi, 710049, P.R. China
xuweijun75@163.com

Abstract. We consider approximation of multidimensional functions by feedforward neural networks with one hidden layer of Sigmoidal units and a linear output. Under the Orthogonal polynomials basis and certain assumptions of activation functions in the neural network, the upper bounds on the degree of approximation are obtained in the class of functions considered in this paper. The order of approximation $O(n^{-\frac{r}{d}})$, d being dimension, n the number of hidden neurons, and r the natural number.

1 Introduction

Universal approximation capabilities for a broad range of neural network topologies have been established by researchers like cybenko [1], Ito [2], and T.P.Chen [3]. Their work concentrated on the question of denseness. But from the point of application, we are concerned about the degree of approximation by neural networks. For any approximation problem, the establishment of performance bounds is an inevitable but very difficult issues. As we know, feedforward neural networks (FNNS) have been shown to be capable of approximating general class of functions, including continuous and integrable ones. Recently, several researchers have been derive approximation error bounds for various functional classes (see, for example, [4–9]) approximated by neural networks.

While many open issues remain concerning approximation degree, we stress in this paper on the issue of approximation of functions defined over $[-1, 1]^d$ by FNNS. In [6], the researcher took some basics tools from the theory of weighted polynomial approximation of multivariate functions (The weight function is $\omega(x) = exp(-Q(x))$), under certain assumptions on the smoothness of functions being approximated and on the activation functions in the neural network, the authors present upper bounds on the degree of approximation achieved over the domain R^d. In this paper, using the Chebyshev Orthogonal series from the approximation theory and moduli of continuity, we obtain upper bounds on the degree of approximation in $[-1, 1]^d$. We take advantage of the properties of the Chebyshev polynomial and the methods of paper [6], we yield the desired results, which can be periodically extend to the space R^d.

2 Approximation of Multi-polynomials in L_ω^p

Before introducing the main results, we firstly introduce some basic results on Chebyshev polynomials from the approximation theory. For convenience, we introduce a weighted norm of a function f given by

$$\|f\|_{p,\omega} = \left(\int_{[-1,1]^d} \omega(\mathbf{x}) | f(\mathbf{x}) |^p \, d\mathbf{x}\right)^{\frac{1}{p}}, \tag{1}$$

where $1 \leq p < \infty$, $\omega(\mathbf{x}) = \Pi_{i=1}^d \omega(x_i)$, $\omega(x_i) = (1-x_i^2)^{-\frac{1}{2}}$, $\mathbf{x} = (x_1, x_2, \ldots, x_d) \in \mathbf{R}^d$, $\mathbf{m} = (m_1, m_2, \ldots, m_d) \in \mathbf{Z}^d$, $d\mathbf{x} = dx_1 dx_2 \cdots dx_d$. We denote the class of functions for which $\|f\|_{p,\omega}$ is finite by L_ω^p.

For function $f : \mathbf{R}^d \longrightarrow R$, the class of functions we wish to approximate in this work is defined as follows.

$$\Psi_{p,\omega}^{r,d} = \{f : \|f^{(\lambda)}\|_{p,\omega} \leq M, |\lambda| \leq r\}, \tag{2}$$

where $\lambda = (\lambda_1, \lambda_2, \ldots, \lambda_d)$, $|\lambda| = \lambda_1 + \lambda_2 + \cdots + \lambda_d$, $f^{(\lambda)} = \frac{\partial^{|\lambda|} f}{\partial x_1^{\lambda_1} \cdots \partial x_d^{\lambda_d}}$, r is a natural number, and $M < \infty$.

As we known, Chebyshev polynomial of a single real variable is a very important polynomial in approximation theory. Using the notation of (1), we introducing multivariate Chebyshev polynomials: $T_\mathbf{0}(\mathbf{x}) = \frac{1}{\sqrt{\pi}}$, $T_\mathbf{n}(\mathbf{x}) = \prod_{i=1}^d T_{n_i}(x_i)$, $T_{n_i}(x_i) = \frac{2}{\sqrt{\pi}} \cos(n_i \arccos x_i)$. Evidently, for any $\mathbf{m}, \mathbf{l} \in Z^d$, we have

$$\int_{[-1,1]^d} T_\mathbf{m}(\mathbf{x}) T_\mathbf{l}(\mathbf{x}) \omega(\mathbf{x}) d\mathbf{x} = \begin{cases} 1 & \mathbf{m} = \mathbf{l}, \\ 0 & \mathbf{m} \neq \mathbf{l}. \end{cases}$$

For $f \in L_\omega^p$, $\mathbf{m} \in Z^k$, let $\widehat{f}(m) = \int_{[-1,1]^d} f(\mathbf{x}) T_\mathbf{m}(\mathbf{x}) \omega(\mathbf{x}) d\mathbf{x}$, then we have the orthogonal expansion $f(\mathbf{x}) \sim \sum_{m=0}^\infty \widehat{f}(m) T_\mathbf{m}(\mathbf{x})$, $x \in [-1,1]^d$.

For one-dimension, degree of approximation of a function g by polynomials of degree m as follows.

$$E_m(g, \mathbf{P_m}, L_\omega^p) = \inf_{P \in P_m} \|g - P\|_{p,\omega}, \tag{3}$$

where $\mathbf{P_m}$ stands for the class of degree-m algebraic polynomials. From [10], we have a simple relationship which we will be used in the following. Let g be differentiable, and then we have

$$E(g, P_m, L_\omega^p) \leq M_1 m^{-1} E(g', P_m, L_\omega^p), \tag{4}$$

$$E(g, P_m, L_\omega^p) \leq \|g\|_{p,\omega}. \tag{5}$$

Let $S_n(f,t) = \sum_{k=1}^{n-1} \widehat{f}(k) T_\mathbf{k}(\mathbf{x})$, and the de la Valle Poussin Operators is defined by

$$V_n(f,t) = \frac{1}{n+1} \sum_{m=\frac{n+3}{2}}^{n+1} S_n(f,t).$$

Furthermore, we can simplify $V_n(f,t)$ as

$$V_n(f,t) = \sum_{k=1}^{n} \xi_k \widehat{f}(k) T_k(t), \qquad (6)$$

where

$$\xi_k = \begin{cases} \frac{m-1}{2(m+1)} & \text{if } 0 \leq k \leq \frac{m+3}{2}, \\ \frac{m-k}{m+1} & \text{if } \frac{m+3}{2} \leq k \leq m. \end{cases}$$

A basic result concerning Valle Poussin Operators $V_m(f,t)$ is

$$E_{2m}(f, P_{2m}, L_\omega^p) \leq \|f - V_m f\|_{p,\omega} \leq E_m(f, P_m, L_\omega^p).$$

Now we consider a class of multivariate polynomials defined as follows.

$$\mathbf{P}_m = \{P : P(x) = \sum_{0 \leq |i| \leq |m|} b_{i_1, i_2, \ldots, i_d} x_1^{i_1} \cdots x_d^{i_d}, b_{i_1, i_2, \ldots, i_d} \in R, \forall i_1, \ldots, i_d \}.$$

Hence, we have the following theorem.
Theorem 1. For $1 \leq p < \infty$, let $f \in \Psi_{p,\omega}^{r,d}$. Then for any $\mathbf{m} = (m_1, m_2, \ldots, m_d)$, $m_i \leq m$, we have

$$\inf_{p \in \mathbf{P}_m} \| f - p \|_{p,\omega} \leq C m^{-r}.$$

Proof. We consider the Chebyshev orthogonal polynomials $T_\mathbf{m}(\mathbf{x})$, and obtain the following equality from (6)

$$V_{i,m_i}(f) = \sum_{s=1}^{m_i} \xi_s \widehat{f}_{s,i} T_s(x_i),$$

where $\widehat{f}_{s,i} = \int_{[-1,1]^d} f(\mathbf{x}) T_s(x_i) \omega(x_i) dx_i$. Hence, we define the following operators

$$\begin{aligned} V(f) &= V_{1,m_1} V_{2,m_2} \cdots V_{d,m_d} f \\ &= \sum_{s_1=1}^{m_1} \cdots \sum_{s_d=1}^{m_d} \xi_{s_1} \cdots \xi_{s_d} f_{s_1,\ldots,s_d} T_{s_1}(x_1) \cdots T_{s_d}(x_d), \end{aligned} \qquad (7)$$

where $f_{s_1,\ldots,s_d} = \int_{[-1,1]^d} (\prod_{i=1}^{d} \omega(x_i) T_{s_i}(x_i)) f(\mathbf{x}) d\mathbf{x}$. Then we have

$$\begin{aligned} \| f - V(f) \|_{p,\omega} &= \| f - V_{1,m_1}(f) + V_{1,m_1}(f) - V_{1,m_1} V_{2,m_2}(f) \\ &\quad + V_{1,m_1} V_{2,m_2}(f) - \cdots - V(f) \|_{p,\omega} \\ &\leq \sum_{i=1}^{d} \| V_0 \cdots V_{i-1,m_i-1} f - V_0 \cdots V_{i,m_i} f \|_{p,\omega}, \end{aligned} \qquad (8)$$

where V_0 is the identity operator. Let $g = V_0 \cdots V_{i-1,m_i-1}f$, then $V_{i,m_i}g = V_0 \cdots V_{i,m_i}f$, $g^{r_i}(\mathbf{x}) = V_0 \cdots V_{i-1,m_i-1}D^{r_i}f(\mathbf{x})$. We view $V_{i,m_i}g$ as a one-dimensional function x_i. Using (3), (4), (5), we have

$$\| g - V_{i,m_i}g \|_{p,\omega} \leq C_1 E_{m_i}(g, \mathbf{P}_{\mathbf{m_i}}, L^p_\omega)$$
$$\leq C_1 M_1^{r_i}(\frac{1}{m_i}) \cdots (\frac{1}{m_i - r_i + 1}) E_{m_i - r_i}(g^{r_i}, \mathbf{P}_{\mathbf{m_i - r_i}}, L^p_\omega)$$
$$\leq C_1 M_1^{r_i}(\frac{1}{m_i}) \cdots (\frac{1}{m_i - r_i + 1}) \|g^{r_i}\|_{p,\omega} \qquad (9)$$
$$= C_{r_i}(\frac{1}{m_i}) \cdots (\frac{1}{m_i - r_i + 1}) \|V_0 \cdots V_{i-1,m_i-1}D^{r_i}f\|_{p,\omega}.$$

Letting $r_i = r, m_i = m, i = 1, \ldots, d$, if $m > r(r-1)$, we get from (7), (8), (9) and the inequality $\prod_{i=1}^{n}(1+a_i) \geq 1 + \sum_{i=1}^{n}a_i, (a_i \geq -1)$,

$$\| f - V(f) \|_{p,\omega} \leq C_r \sum_{i=1}^{d}(\frac{1}{m}) \cdots (\frac{1}{m-r+1}) \|D^r f\|_{p,\omega}$$
$$\leq C_r dM(\frac{1}{m}) \cdots (\frac{1}{m-r+1}) \qquad (10)$$
$$= C_r dM m^{-r}(1-\frac{1}{m})^{-1}(1-\frac{2}{m})^{-1} \cdots (1-\frac{r-1}{m})^{-1}$$
$$\leq C_r dM m^{-r}(1-\frac{r(r-1)}{2m})^{-1} \leq 2dC_r M m^{-r}.$$

In order to obtain a bound valid for all m, for $0 < m \leq r(r-1)$. Using the property of $V(f)$ and letting $C = max\{2dC_r dM, 2^{-1}M_2(r(r-1))^r\}$, we conclude an inequality of the desired type for every m.

3 Approximation by Sigmoidal Neural Networks

We consider the approximation of functions by feedforward neural networks with a ridge functions. We use the approximating function class in [6] composed of a single hidden layer feedforward neural network with n hidden units. The class of function is

$$\mathbf{F_n} = \{f : f(\mathbf{x}) = \sum_{k=1}^{n} d_k \phi(a_k \cdot \mathbf{x} + b_k); a_k \in \mathbf{R}^d, b_k, d_k \in R, k = 1, 2, ..., n\},$$
$$(11)$$

where $\phi(x)$ satisfy the following assumptions:
(1). There is a constant C_ϕ such that $| \phi^{(k)}(x) | \geq C_\phi > 0, k = 0, 1, \ldots$
(2). For each finite k, there is a finite constant l_k such that $| \phi^{(k)}(x) | \leq l_k$.
We define the distance from F to G as

$$dist(F, G, L^p_\omega) = \sup_{f \in F} \inf_{g \in G} \|f - g\|_{p,\omega},$$

where F, G are two sets in L^p_ω. We have the following results.

Theorem 2. Let condition (1) and (2) hold for the activation function $\phi(x)$. Then for every $0 < L < \infty$, $\mathbf{m} = (m_1, m_2, \ldots, m_d) \in Z_+^d$, $m_i \leq m, \epsilon > 0$ and $n > (m+1)^d$, we have

$$dist(\mathbf{BP_m}(L), \mathbf{F_n}, L_\omega^p) \leq \epsilon,$$

where

$$\mathbf{BP_m}(L) = \{P : p(x) = \sum_{0 \leq s \leq m} a_s \mathbf{x^s}; \max_{0 \leq s \leq m} |a_s| \leq L\}.$$

Proof. Firstly, we consider the partial derivative

$$\phi^{(s)}(\mathbf{w} \cdot \mathbf{x} + b) = \frac{\partial^{(|s|)}}{\partial_{w_1}^{s_1} \ldots \partial_{w_d}^{s_d}} (\phi(\mathbf{w} \cdot \mathbf{x} + b)) = \mathbf{x^s} \phi^{|s|}(\mathbf{w} \cdot \mathbf{x} + b), \quad (12)$$

where $|s| = s_1 + \ldots + s_d$, and $\mathbf{x^s} = \prod_{i=1}^d x_i^{s_i}$. Thus $\phi^{(s)}(b) = \mathbf{x^s} \phi^{|s|}(b)$.

For any fixed b and $|\mathbf{x}| < \infty$, we consider a finite difference of orders

$$\triangle_{h,\mathbf{x}}^\mathbf{s} \phi(b) = \sum_{0 \leq \mathbf{l} \leq \mathbf{s}} (-1)^{|\mathbf{l}|} C_\mathbf{s}^\mathbf{l} \phi(h\mathbf{l} \cdot \mathbf{x} + b)$$

$$= \mathbf{x^s} \int_0^h \cdots \int_0^h \phi^{(|s|)}[((a_1 + \cdots + a_{s_1})x_1 + \cdots \quad (13)$$

$$+ (a_{|s|-s_d+1} + \cdots + a_{|s|})x_d) + b] da_1 \cdots da_{|s|}$$

$$\doteq \mathbf{x^s} A_h^{|s|} \phi(x),$$

where $C_\mathbf{s}^\mathbf{l} = \prod_{i=1}^d C_{s_i}^{l_i}$, $\triangle_{h,\mathbf{x}}^\mathbf{s} \phi(b) \in \mathbf{F_n}$ with $n = \prod_{i=1}^d (1 + s_i)$. Thus

$$|\phi^\mathbf{s}(b) - h^{-|s|} \triangle_{h,\mathbf{x}}^\mathbf{s} \phi(b)| = |\mathbf{x^s}(\phi^{|s|}(b) - h^{-|s|} A_h^\mathbf{s} \phi^{|s|}(\mathbf{x}))|$$

$$= |\mathbf{x^s}(\phi^{|s|}(b) - \phi^{|s|}(b + \eta))| \quad (14)$$

$$\leq C_s \omega(\phi^{|s|}, h),$$

where we derive (14) by using (13), the mean value theorem of integral(i.e. there is a $\eta \in [0, h|\mathbf{s} \cdot \mathbf{x}|]$, such that $A_h^\mathbf{s} \phi^{|s|}(\mathbf{x}) = h^{|s|} \phi^{|s|}(\mathbf{b} + \eta))$, and the moduli of continuity $\omega(g, h) = \sup_{|t| \leq h} |f(x+t) - f(x)|$.

From the definition of $dist(F, G, L_\omega^p)$ and (14), we have

$$dist(\mathbf{BP_m}(L), \mathbf{F_n}, L_\omega^p)^p \leq \|\sum_{0 \leq s \leq m} a_s \mathbf{x^s} - \sum_{0 \leq s \leq m} a_s \frac{\triangle_{h,\mathbf{x}}^\mathbf{s} \phi(b)}{h^{|s|} \phi^{|s|}(b)} \|_{p,\omega}^p$$

$$\leq (m+1)^d \max_{0 \leq s \leq m} \{|a_s| \|\mathbf{x^s} - \frac{\triangle_{h,x}^\mathbf{s} \phi(b)}{h^{|s|} \phi^{|s|}(b)} \|_{p,\omega}^p\}$$

$$\leq (m+1)^d L \max_{0 \leq s \leq m} \{\phi^{|s|}(b)\}^{-p} \omega(\phi^{|s|}, h)$$

$$\leq (m+1)^d L C_\phi^{-p} \omega(\phi^{|s|}, h) < \epsilon.$$

The last step $\omega(\phi^{|s|}, h)$ can be made arbitrarily small by letting $h \to 0$.

Using the Theorem 1 and Theorem 2, we can easily establish our final result.

Theorem 3. For $1 \leq p < \infty$, we have

$$dist(\Psi_{p,\omega}^{r,d}, F_n, L_\omega^p) \leq Cn^{-\frac{r}{d}}.$$

References

1. Cybenko, G.: Approximation by Superpositions of a Sigmoidal Function. Math. Contr. Signals. Syst.. **2** (1989) 303–314
2. Ito, Y.: Approximation of Continuous Functions on R^d by Linear Combination of Shifted Rotations of Sigmoid Function with and without Scaling Neural Networks. Neural Networks. **5** (1992) 105–115
3. Chen, T.P., Chen, H.: Approximation Capability to Functions of Several Variables, Nonlinear Functions, and Operators by Radial Function Neural Networks. IEEE Trans. Neural Networks. **6** (1995) 904–910
4. Barron, A.R.: Universal Approximation Bound for Superpositions of a Sigmoidal Function. IEEE Trans. Inform. Theory. **39** (1993) 930–945
5. Mhaskar, H.N.: Neural Networks for Optimal Approximation for Smooth and Analytic Functions. Neural Comput.. **8** (1996) 164–177
6. Maiorov, V., Meir, R.S.: Approximation Bounds for Smooth Functions in $C(R^d)$ by Neural and Mixture Networks. IEEE Trans. Neural Networks. **9** (1998) 969–978
7. Burger, M., Neubauer, A.: Error Bounds for Approximation with Neurnal Networks. J. Approx. Theory. **112** (2001) 235–250
8. Kurkova, V., Sanguineti, M.: Comparison of Worst Case Errors in Linear and Neural Network Approximation. IEEE Trans. Inform. Theory. **48** (2002) 264–275
9. Wang, J.L., Sheng, B.H., Zhou, S.P.: On Approximation by Non-periodic Neural and Translation Networks in L_ω^p Spaces. ACTA Mathematica Sinica (in chinese). **46** (2003) 65–74
10. Timan, A.F.: Theory of Approximation of Functions of a Real Variable. New York: Macmillan. (1963)

Geometric Interpretation of Nonlinear Approximation Capability for Feedforward Neural Networks*

Bao-Gang Hu[1,2], Hong-Jie Xing[1,2], and Yu-Jiu Yang[1,2]

[1] National Laboratory of Pattern Recognition
Institute of Automation, Chinese Academy of Sciences
[2] Beijing Graduate School, Chinese Academy of Sciences
P.O. Box 2728, Beijing, 100080 China
{hubg,hjxing,yjyang}@nlpr.ia.ac.cn

Abstract. This paper presents a preliminary study on the nonlinear approximation capability of feedforward neural networks (FNNs) *via* a geometric approach. Three simplest FNNs with at most four free parameters are defined and investigated. By approximations on one-dimensional functions, we observe that the Chebyshev-polynomials, Gaussian, and sigmoidal FNNs are ranked in order of providing more varieties of non-linearities. If neglecting the compactness feature inherited by Gaussian neural networks, we consider that the Chebyshev-polynomial-based neural networks will be the best among three types of FNNs in an efficient use of free parameters.

1 Introduction

Machine learning through input-output data from examples can be considered as approximations of unknown functions. Two cases can be found in the nonlinear approximations. One is having a certain degree of knowledge about the nonlinear functions investigated. The other is in the case that *a priori* information is unavailable in regards to the degree of nonlinearity of the problem. The last case presents more difficulty in handling, but often occurs in real world problems. In this work, we will investigate feedforward neural networks (or **FNNs**) as the nonlinear approximators for the last case.

Significant studies have been reported on that FNNs are universal approximators with various basis (or activation) functions [1-3]. However, the fundamental question still remains: *Among the various basis functions, which one provides the most efficiency in approximations of arbitrary functions?* This efficiency can be evaluated by the approximation accuracy over a given number of free parameters employed by its associated FNN. Some numerical investigations have shown that the radial basis functions usually afford the better efficiency than the sigmoidal

---

* This work is supported by Natural Science of Foundation of China (#60275025, #60121302).

functions [4]. Hence, further question arises: *What are natural reasons for some basis function exhibiting better efficiency than the others?*

In this paper, we attempt to answer the two basic questions above *via* a geometric approach. The interpretations from the approach seem simply and preliminary at this stage, but we believe that a geometric approach does provides a unique tool for understanding the nature insights of nonlinear approximation capabilities of universal approximators. This paper is organized as follows. Section 2 proposes a new methodology. Three simplest FNNs are defined and examined in Section 3. A nonlinearity domain analysis is made with respect to their availability of nonlinearity components for the three FNNs in Section 4. Finaly, some remarks are given in Section 5.

2 Proposed Methodology

In the studies of approximation capability, the conventional methodology used in FNNs is generally based on the performance evaluations from approximation errors. Two common methods are employed in the selections of basis functions. One is on the estimation of error bonds, and the other is on the examination of numerical errors to the specific problems. Few studies related the basis functions to the approximation capability using a geometric approach. In this work, we propose a new methodology from the following aspects.

2.1 Nonlinearity Domain Analysis

Nonlinearity domain analysis is a novel concept. There is no existing and explicit theory for such subject. We propose this concept in order to characterize a given nonlinear function by its nonlinearity components similar to a frequency domain analysis. Here are some definitions:

Definition 1. *Free parameters, linear parameters, and nonlinear parameters.*

Any nonlinear function can be represented in a form as:

$$\mathbf{y} = f(\mathbf{x}, \theta) = f(\mathbf{x}, \theta_L, \theta_{NL}), \quad (1)$$

where $\mathbf{x} \in R^N$ and $\mathbf{y} \in R^M$ are input and output variables, respectively; θ, θ_L and θ_{NL} are free parameter set, linear and nonlinear parameter sets respectively. The behaviors and properties of nonlinear functions are controlled by free parameters. If it can change the shape or orientation of nonlinear function, this parameter will fall into a nonlinear parameter set. Otherwise, it is a linear parameter (also called location parameter).

Definition 2. *Nonlinearity domain, nonlinearity components and nonlinearity variation range.*

Nonlinearity domain is a two dimensional space used for characterization of nonlinearity of functions. Nonlinearity components are a set of discrete points with an infinite number along the horizontal axis of nonlinearity domain. The vertical axis represents the variable of nonlinearity variation range. The plot in Fig. 1 can be called *"Nonlinearity Spectrum"*, which reveals two sets of information. First, for a given function, how many nonlinearity components could be generated by changing the free parameters. Each component represents a unique class of nonlinear functions, say, NC_j for the jth nonlinearity component, which could be an "S-type curve". Second, for each included component, what is its associated nonlinearity variation range. This range exhibits the admissible range realized by the given function. A complete range is normalized within $[0,1]$. If $NV_j = 0.5$, it indicates that the given function can only span a half space in the jth nonlinearity component.

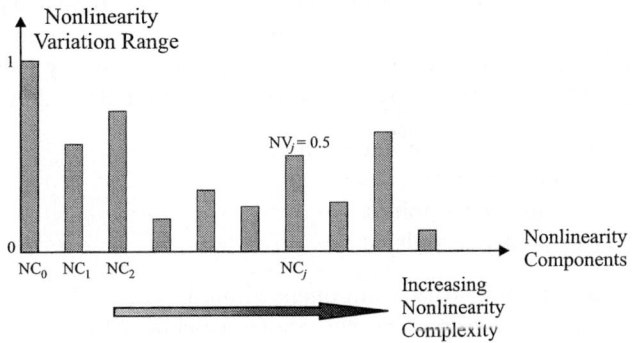

Fig. 1. Nonlinearity Domain Analysis

2.2 Definition of the Simplest-Nonlinear FNNs

The general form of FNNs is a nonlinear mapping: $f : R^N \to R^M$. In the nonlinearity domain analysis, it will be a complex task if the high dimensionality of FNNs is involved. In order to explore the nature insights of FNNs, one has to make necessary simplifications, or assumptions. In this work, we define the simplest-nonlinear FNNs for the nonlinearity domain analysis.

Definition 3. *Simplest-nonlinear FNNs.*

The simplest-nonlinear FNNs present the following features in their architectures: I. A single hidden layer. II. A single hidden node (but more for polynomial-based FNNs). III. A single-input-single-output nonlinear mapping, $f_s : R \to R$. IV. Governed by at most four free parameters:

$$y = f_s(x, \theta) = f_s(x, a, b, c, d), \qquad (2)$$

where $a, b, c, d \in R$. Further classification of the four parameters is depending on the basis function applied. We will give discussions later about the reason of choosing four free parameters, and call FNNs in eq (2) the simplest FNNs.

When we define the simplest FNNs, one principle should be followed. The conclusions or findings derived from the simplest nonlinear FNNs can be extended directly to judge approximation capability of the general FNNs. For example, we only study the nonlinearity of curves. However, the nature insights obtained from this study can also be effective to the FNNs that construct hypersurfaces.

2.3 Classification for Nonlinearity Components

In the nonlinearity domain analysis, all nonlinearity components are classified according to the geometric properties from nonlinear functions. However, there exist various geometric features for classification. These include continuity, monotonicity, symmetry, periodicity, compactness, boundness, singularity, *etc.* In this work, we restrict the studies within the smooth functions. Therefore, the geometric properties in related to the continuity and monotonicity features will be used in the classification. In this work, we consider the following aspects:

- G1. Monotonic increasing or decreasing.
- G2. Convexity or concavity.
- G3. Number of inflection points.
- G4. Number of peaks or valleys.

Therefore, each nonlinearity component should represent a unique class of nonlinear functions with respect to the above aspects. After the classification, we usually arrange the components, NC_j, along the axis in an order of increasing nonlinearity complexity. In this work, we only consider the one-dimensional nonlinear functions. Then, we immediately set $\{NC_0 : y = c\}$ and $\{NC_1 : y = ax+c\}$ to be constant and linear components, respectively. Although these two components are special cases for the zero degree of nonlinearity, both of them cannot be missed for the completeness of nonlinearity components. The next will start from simple nonlinear curves, say, "C-type" and "S-type" curves. We will give more detailed classification examples later.

3 Examination of Parameters on Three Simplest FNNs

In this section, we will examine the parameters on the simplest FNNs with three different basis functions, *i.e.*, sigmoidal, Gaussian, and Chebyshev-polynomials. Their mathematic representations are given in the following forms.

The simplest sigmoidal FNNs:

$$y = \frac{a}{1 + \exp(bx + c)} + d \qquad (3)$$

The simplest Gaussian FNNs:

$$y = a \exp\left[-\frac{(x-c)^2}{b^2}\right] + d \qquad (4)$$

The simplest Chebyshev-polynomial FNNs:

$$y = aT_3(x) + bT_2(x) + cT_1(x) + d \qquad (5)$$

where $T_i(x)$ are the Chebyshev polynomials (see [5] and their architectures of FNNs).

Both simplest sigmoidal and Gaussian FFNs apply at most four free parameters. In order to make a fair comparison, we set the cubic forms for the simplest Chebyshev-polynomial FNNs. The first analysis of the three types of FNNs is the classification of linear and nonlinear parameters. We catalogue two sets of parameters for the reason that linear parameters do not change the geometric properties (say, G1-G4 in Section 2.3) of nonlinear functions. Table 1 presents the two parameter sets for the three simplest FNNs. All linear parameters play a "shifting" role to the curves; but only nonlinear parameters can change the shape, or curvatures, of functions. We conclude that the Chebyshev shows better features over the others on the following aspects:

I. The Chebyshev presents a bigger set of nonlinear parameters, which indicates that it can form a larger space for nonlinearity variations.
II. The nonlinear parameters in the Chebyshev can be changed into linear parameters. This flexibility feature is not shared by the others.
III. The nonlinear parameters in both sigmoidal and Gaussian can produce a "scaling" effect to the curves. This will add a dependency feature to the nonlinear parameters and thus reduce the approximation capability if similarity is considered. The Chebyshev, in general, does not suffer this degeneration problem.

Table 1. Comparisons of linear and nonlinear parameters for the three simplest FNNs

	Linear parameters	Nonlinear parameters
Sigmoidal	c, d	a, b
Gaussian	c, d	a, b
Chebyshev	d	a, b, c

4 Nonlinearity Domain Analysis on Three Simplest FNNs

In this section we will conduct nonlinearity domain analysis on the three simplest FNNs. Without losing the generality, a smooth function in a compact interval, $\{f(x); x \in [0,1]\}$, will be approximated. Therefore, any function can be approximated by the linear combinations of several simplest FNNs. The approximation will allocate the proper segmentation range (by using linear parameters) from the given basis function (by using nonlinear parameters for the proper shapes). Therefore, each simplest FNN can provide the different nonlinear curves (see Fig. 2 for the sigmoidal function).

In this work, we summarize the nonlinearity components in their availability for the three simplest FNNs in Table 2, in which each NC_j is given graphically in Fig. 3. All NC_j represent unique classes of nonlinear, but smooth, functions according to the geometric properties. One can observe that the Chebyshev is

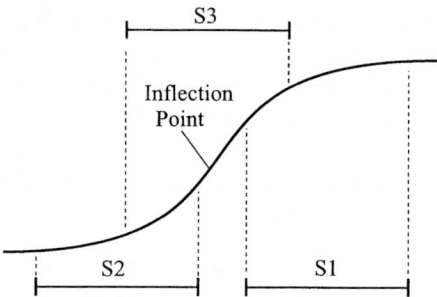

Fig. 2. Segmentation of a sigmoidal function and different nonlinear curves. (S1, S2 and S3 correspond to the C-, Inverse C- and S-curves, respectively.)

the best again in producing the most nonlinearity components. However, the nonlinearity variations range for each FNNs is not given and will be a future work.

Table 2. Comparisons of nonlinearity components in their availability for the three simplest FNNs. (The sign "√" indicates the availiabilty of its current component, otherwise it is empty).

	NC_0	NC_1	NC_2	NC_3	NC_4	NC_5	NC_6	NC_7	NC_8	NC_9	NC_{10}
Sigmoidal	√	√	√	√	√	√	√		√		
Gaussian	√	√	√	√	√	√	√		√		√
Chebyshev	√	√	√	√	√	√	√	√	√	√	√

5 Final Remarks

In this work, we investigate the FNNs with three commonly used basis functions. A geometric approach is used for interpretation of the nature in FNNs. We conclude that the Chebyshev-polynomial FNNs are the best type in comparing with the sigmoidal and Gaussian FNNs by including more nonlinearity components. However, a systematic study on the nonlinearity domain analysis

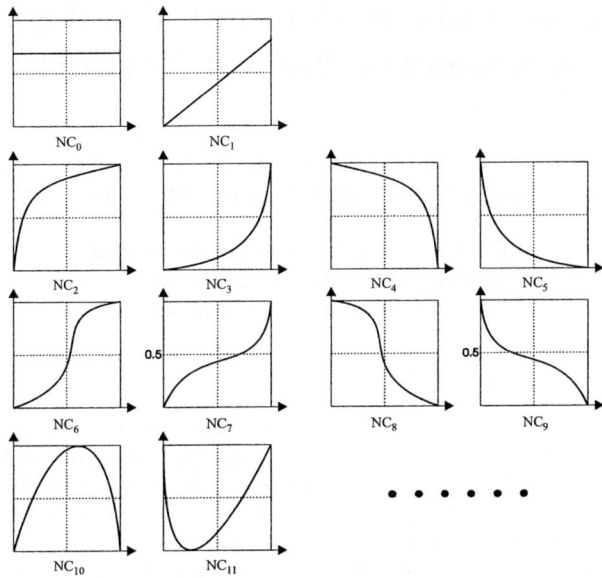

Fig. 3. Nonlinear curves and their associated nonlinearity components NC_j

is necessary to reach overall conclusions for each type of FNNs. For example, the compactness feature of the Gaussian is more efficiency in approximation of a local behavior of nonlinear functions. On the other hand, we believe that both performance-based and function-based evaluation approaches [6] will provide a complete study to ease the difficulty of *"trial and error"* in designs of universal approximators, such as fuzzy systems and neural networks.

References

1. Hornik, K., Stinchcombe, M., White, H.: Multilayer Feedforward Networks are Universal Approximators. Neural Networks. **2** (1989) 359–366
2. Park, J., Sandberg, J.W.: Universal Approximation Using Radial Basis Functions Network. Neural Computation. **3** (1991) 246-257
3. Chen, T-P., Chen, H.: Universal Approximation to Nonlinear Operators by Neural Networks with Arbitrary Activation Functions and Its Application to Dynamical Systems. IEEE Trans. on Neural Networks. **6** (1995) 911-917
4. Park, J.-W., Harley, R. G., Venayagamoorthy, G. K.: Indirect Adaptive Control for Synchronous Generator: Comparison of MLP/RBF Neural Networks Approach With Lyapunov Stability Analysis. IEEE Trans. on Neural Networks. **15** (2004) 460-464
5. Lee, T.-T., Jeng, J.-T.: The Cyebyshev-Polynomials-Based Unified Model Neural Net-works for Function Approximation. IEEE Trans. on Sys. Man Cyber. **28B** (1998) 925-935
6. Hu, B.-G., Mann, G.K.I., Gosine, R.G.: A Systematic Study of Fuzzy PID Controllers—Function-based Evaluation Approach. IEEE Trans. on Fuzzy Systems. **9** (2001) 699-712

Mutual Information and Topology 1: Asymmetric Neural Network

David Dominguez*,**, Kostadin Koroutchev**,
Eduardo Serrano**, and Francisco B. Rodríguez**

EPS, Universidad Autonoma de Madrid,
Cantoblanco, Madrid, 28049, Spain
david.dominguez@ii.uam.es

Abstract. An infinite range neural network works as an associative memory device if both the learning storage and attractor abilities are large enough. This work deals with the search of an optimal topology, varying the (small-world) parameters: the average connectivity γ ranges from the fully linked to a extremely diluted network; the randomness ω ranges from purely neighbor links to a completely random network. The network capacity is measured by the mutual information, MI, between patterns and retrieval states. It is found that MI is optimized at a certain value γ_o for a given $0 < \omega < 1$ if the network is asymmetric.

1 Introduction

The collective properties of attractor neural networks (ANN), such as the ability to perform as an associative memory, has been a subject of intensive research in the last couple of decades[1], dealing mainly with fully-connected topologies. More recently, a renewed interest on ANN has been brought by the study of more realistic architectures, such as small-world or scale-free[2][3] models. The storage capacity α and the overlap m with the memorized patterns are the most used measures of the retrieval ability for the Hopfield-like networks[4][5]. Comparatively less attention has been paid to the study of the mutual information MI between stored patterns and the neural states[6][7].

A reason for this few interest is twofold: first, while m is a global parameter running only over one site, MI is a function of the conditional probability, which depends on all states, at input and output; second, the load α is enough to measure the information if the overlap is close to $m \sim 1$, since in this case the information carried by any single (binary, uniform) neuron is almost 1 bit. The first difficulty can be solved for infinite-range connections, because this so called *mean-field networks* satisfies the conditions for the law of large numbers, and MI is a function only of the macroscopic parameters (m) and of the load rate ($\alpha = P/K$, where P is the number of uncorrelated patterns, and K is the neuron connectivity).

---
* DD thanks a Ramon y Cajal grant from MCyT.
** Supported by MCyT-Spain BFI-2003-07276 and TIC 2002-572-C02.

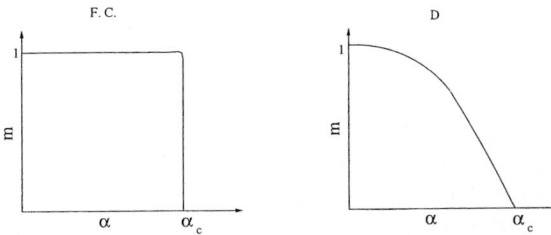

Fig. 1. The overlap m vs α for fully-connected (FC) and extremely diluted (D) networks

The second reason holds for a fully-connected network, for which the critical $\alpha_c \sim 0.14$[4], with $m \sim 0.97$ (it crashes to $m \to 0$ for larger α): in this case, the information rate is about $i \sim 0.13$. Nevertheless, in the case of diluted networks, (the connectivity of each neuron is kept large, but much smaller than N), the transition is smooth. In particular, the extremely diluted random network has load capacity $\alpha_c \sim 0.64$[8] but the overlap falls continuously to 0, which yields null information at the transition. Such indetermination shows that one must search for the value of α corresponding to the maximal MI, instead of α_c. It is seen in Fig.1. Previous works[3] studied only the overlap $m(\alpha)$.

Our main goal in this work is to solve the following question: how does the maximal $MI_m(\gamma,\omega) \equiv MI(\alpha_m, m; \gamma, \omega)$ behaves with respect to the network topology? We will show that, for larger values of the randomness ω, the extremely-diluted network performs the best. However, with asymmetric connections, smaller values of ω lead to an optimal $MI_o(\gamma) \equiv MI_m(\gamma_o, \omega)$ for intermediate levels of connectivity $0 < \omega < 1$.

2 The Information Measures

2.1 The Neural Channel

The network state in a given time t is defined by a vector of binary neurons, $\boldsymbol{\sigma}^t = \{\sigma_i^t \in \{\pm 1\}, i = 1, ..., N\}$. Accordingly, each pattern, $\boldsymbol{\xi}^\mu = \{\xi_i^\mu \in \{\pm 1\}, i = 1, ..., N\}$, are site-independent random variables, binary and uniformly distributed:

$$p(\xi_i^\mu) = \frac{1}{2}\delta(\xi_i^\mu - 1) + \frac{1}{2}\delta(\xi_i^\mu + 1). \qquad (1)$$

The network learns a set of independent patterns $\{\boldsymbol{\xi}^\mu, \mu = 1, ..., P\}$.

The task of the neural channel is to retrieve a pattern (say, $\boldsymbol{\xi}^\mu$) starting from a neuron state which is inside its attractor basin, $B(\boldsymbol{\xi}^\mu)$: $\boldsymbol{\sigma}^0 \in B(\boldsymbol{\xi}^\mu) \to \boldsymbol{\sigma}^\infty \approx \boldsymbol{\xi}^\mu$. This is achieved through a network dynamics, which couples neurons σ_i, σ_j by the *synaptic matrix* $\mathbf{J} \equiv \{J_{ij}\}$ with cardinality $\#\mathbf{J} = N \times K$.

2.2 The Overlap

For the usual binary non-biased neurons model, the relevant order parameter is the *overlap* between the neural states and a given pattern:

$$m_N^{\mu t} \equiv \frac{1}{N}\sum_i \xi_i^\mu \sigma_i^t, \qquad (2)$$

in the time step t. Note that both positive and negative pattern, $-\boldsymbol{\xi}$, carries the same information, so the absolute value of the overlap is the measure of the retrieval quality: $|m| \sim 1$ means a good retrieval.

Alternatively, one can measure the error in retrieving using the (square) Hamming distance: $E_N^{\mu t} \equiv \frac{1}{N}\sum_i |\xi_i^\mu - \sigma_i^t|^2 = 2(1 - m_N^{\mu t})$. Together with the overlap, one needs a measure of the load capacity, which is the rate of pattern bits per synapses used to store them. Since the synapses and patterns are independent, the load is given by $\alpha = \#\{\boldsymbol{\xi}^\mu\}/\#\mathbf{J} = (PN)/(NK) = P/K$.

We require our network has long-range interactions. That means, it is a mean-field network (MFN), the distribution of the states is site-independent, so every spatial correlation such as $\langle \sigma_i \sigma_j \rangle - \langle \sigma_i \rangle \langle \sigma_j \rangle$ can be neglected, which is reasonable in the limit $N \to \infty$. Hence the condition of the law of large numbers, are fulfilled. At a given time step of the dynamical process, the network state can be described by one particular overlap, let say $m_N^t \equiv m_N^{\mu t}$. The order parameters can thus be written in the asymptotic limit, $\lim N \to \infty$, as $m^t = \langle \sigma^t \xi \rangle_{\sigma,\xi}$. The brackets represent average over the joint distribution of σ, ξ for a single neuron (we can drop the index i). These are the macroscopic variables describing the state of the network at a given, unspecified, time step t of the dynamics.

2.3 Mutual Information

For this long-range system, it is enough to observe the distribution of a single neuron in order to know the global distribution. This is given by the conditional probability of having the neuron in a state σ, (we also drop the time index t), given that in the same site the pattern being retrieved is ξ. For the binary network we are considering, $p(\sigma|\xi) = (1 + m\sigma\xi)\delta(\sigma^2 - 1)$[9] where the overlap is $m = \langle\langle\sigma\rangle_{\sigma|\xi}\xi\rangle_\xi$, using $p(\sigma,\xi) = p(\sigma|\xi)p(\xi)$.

The joint distribution of $p(\sigma,\xi)$ is interpreted as an ensemble distribution for the neuron states $\{\sigma_i\}$ and inputs $\{\xi_i\}$. In the conditional probability, $p(\sigma|\xi)$, all type of noise in the retrieval process of the input pattern through the network (both from environment and over the dynamical process itself) is enclosed.

With the above expressions and $p(\sigma) \equiv \sum_\xi p(\xi)p(\sigma|\xi) = \delta(\sigma^2 - 1)$, we can calculate the *Mutual Information MI* [7], a quantity used to measure the prediction that the observer of the output (σ^t) can do about the input ($\boldsymbol{\xi}^\mu$) at each time step t. It reads $MI[\sigma;\xi] = S[\sigma] - \langle S[\sigma|\xi]\rangle_\xi$, where $S[\sigma]$ is the entropy and $S[\sigma|\xi]$ is the conditional entropy. We use binary logarithms, $\log \equiv \log_2$. The entropies are[9]:

$$\langle S[\sigma|\xi]\rangle_\xi = -\frac{1+m}{2}\log\frac{1+m}{2} - \frac{1-m}{2}\log\frac{1-m}{2}, \quad S[\sigma] = 1[bit]. \qquad (3)$$

When the network approaches its saturation limit α_c, the states can not remain close to the patterns, then m_c is usually small. We avoid this writing the information rate as

$$i(\alpha, m) = MI[\{\sigma\}|\{\boldsymbol{\xi}\}]/\#\mathbf{J} \equiv \sum_{i\mu} MI[\sigma_i|\xi_i^\mu]/(KN) = \alpha MI[\sigma;\xi]). \quad (4)$$

The information $i(\alpha, m)$ is a non-monotonic function of the overlap, which reaches its maximum value $i_m = i(\alpha_m, m)$ at some value of the load α.

3 The Model

3.1 The Network Topology

The synaptic couplings are $J_{ij} \equiv c_{ij}K_{ij}$, where the connectivity matrix has a regular and a random parts, $\{c_{ij} = c_{ij}^n + c_{ij}^r\}$. The regular part connects the K_n nearest neighbors, $c_{ij}^n = \sum_{k \in V} \delta(i - j - k)$, or $V = \{1, ..., K_n\}$ in the asymmetric case, in a closed one-dimensional lattice. The random part consists of independent random variables $\{c_{ij}^r\}$, distributed as

$$p(c_{ij}^r) = c_r \delta(c_{ij}^r - 1) + (1 - c_r)\delta(c_{ij}^r), \quad (5)$$

where $c_r = K_r/N$, with $K_r =$ the mean number of random connections of a single neuron. Hence, the neuron connectivity is $K = K_n + K_r$. The network topology is then characterized by two parameters: the $average-connectivity$, defined as $\gamma = K/N$, and the $randomness$ ratio, $\omega = K_r/K$, besides its symmetry constraints. The ω plays the role of the probability of rewiring in the $small-world$ model (SW)[2]. Our analysis of the present topology shows the same qualitative behavior for the clustering and mean-length-path of the original SW. The average-connectivity is normalized with the same scale as the information, since the load rate is $\alpha = P/K$.

The learning algorithm, K_{ij}, is given by the Hebb rule

$$K_{ij}^\mu = K_{ij}^{\mu-1} + \frac{1}{K}\xi_i^\mu \xi_j^\mu. \quad (6)$$

The network start in $K_{ij} = 0$, and after $\mu = P = \alpha K$ learning steps, it reaches a value $K_{ij} = \frac{1}{K}\sum_\mu^P \xi_i^\mu \xi_j^\mu$. The learning stage is a slow dynamics, being stationary-like in the time scale of the much faster retrieval stage.

3.2 The Neural Dynamics

The neurons states, $\sigma_i^t \in \{\pm 1\}$, are updated according to the stochastic dynamics:

$$\sigma_i^{t+1} = \text{sign}(h_i^t + Tx), \; i = 1...N, \; h_i^t \equiv \sum_j J_{ij}\sigma_j^t, \quad (7)$$

where x is a random variable and T is the temperature-like environmental noise. In the case of symmetric synaptic couplings, $J_{ij} = J_{ji}$ it can be defined an energy function $H_s = -\sum_{(i,j)} J_{ij}\sigma_i\sigma_j$ whose minima are the stable states of the dynamics Eq.(7).

In the present paper, we work out the asymmetric network (no constraints $J_{ij} = J_{ji}$), for which no energy function can be written. We restrict our analysis also for the deterministic dynamics ($T = 0$). The only stochastic effects comes from the large number of learned patterns ($P = \alpha K$, cross-talk noise).

4 Results

We have studied the behavior of the network varying the range of connectivity γ and randomness ω. In all cases we used the parallel dynamics, Eq.(7). The simulation was carried out with $N \times K = 25 \cdot 10^6$ synapses, storing K_{ij} and c_{ij}, instead of J_{ij}, i.e., we use adjacency list as data structure. For instance, with $\gamma \equiv K/N = 0.01$, we used $K = 500, N = 5 \cdot 10^4$. In [3] the authors use $K = 50, N = 5 \cdot 10^3$, which is far from asymptotic limit.

We studied the asymmetric network by searching for the stationary states of the network dynamics; we suppose $t = 20$ parallel (all neurons) updates are enough for convergence, which is true in mostly cases. We look at the behavior of fixed-point of the overlap m^t, with the memory loading $P = \alpha K$. We averaged over a window in the axis of P, usually $\delta P = 25$.

In first place, we checked for the stability properties of the network: the neuron states start precisely at a given pattern $\boldsymbol{\xi}^\mu$ (which changes at each learned step μ). The initial overlap is $m_0^\mu = 1$, so, after $t \leq 20$ time steps in retrieving, the information $i(\alpha, m; \gamma, \omega)$ for final overlap is calculated. We plot it as a function of α, and its maximum $i_m \equiv i(\alpha_m, m; \gamma, \omega)$ is evaluated. This is repeated for various values of the average-connectivity γ and randomness ω parameters. The results are in the upper part of Fig.2.

Second, we checked for the retrieval properties: the neuron states start far from a learned pattern, but inside its basin of attraction, $\boldsymbol{\sigma}^0 \in B(\boldsymbol{\xi}^\mu)$. The initial configuration is chosen with distribution: $p(\sigma^0|\xi) = (1 + m^0)/2\delta(\sigma^0 - \xi) + (1 - m^0)/2\delta(\sigma^0 + \xi)$, for all neurons (so we avoid a bias between regular/random neighbors). The initial overlap is now $m^0 = 0.1$, and after $t \leq 20$ steps, the final information $i(\alpha, m; \gamma, \omega)$ is calculated. The results are in the lower part of Fig.2. The first observation in both parts is that the maximal information $i_{max}(\gamma; \omega)$ increases with dilution (smaller γ) if the network is more regular, $\omega \simeq 1$, while it decreases with dilution if the network is more random, $\omega \simeq 0$.

The comparison between stability and retrieval properties, looking at upper and lower parts of Fig.2, shows clearly that the relation between dilution and randomness is powered respect to the basins of attraction. Random topologies have very robust attractors only if the network is diluted enough; otherwise, they are strongly damaged by bad initial conditions (while regular topologies almost lost their retrieval abilities with extreme dilution).

Each maxima of $i_{max}(\gamma; \omega)$ in Fig.2 is plotted in Fig.3. We see that, for intermediate values of the randomness parameter $0 < \omega < 1$ there is an optimal information respect to the dilution γ. We observe that the optimal $i_o \equiv i_m(\gamma_o; \omega)$ is shifted to the left (stronger dilution) when the randomness ω of the network increases. For instance, with $\omega = 0.1$, the optimal is at $\gamma \sim 0.02$ while with $\omega = 0.3$, it is $\gamma \sim 0.005$. This result does not change qualitatively with the initial condition, but respect to the attractors ($m_0 = 0.1$), even the regular topology presents an optimum at $\gamma \sim 0.1$.

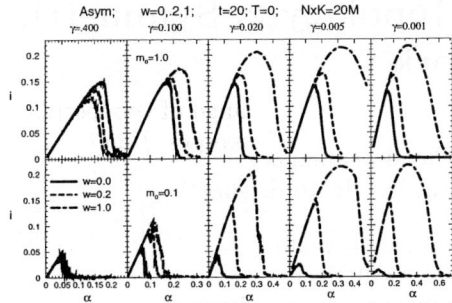 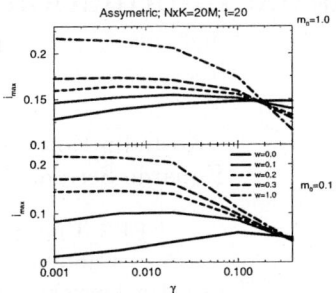

Fig. 2. The information $i \times \alpha$ with $\omega = 0, 0.2, 1$ and several γ; simulation with $N \times K = 20M$, $t \leq 20$. Upper panel: initial overlap $m^0 = 1.0$. Lower panel: $m^0 = 0.1$.

Fig. 3. The maximal information $i_{max} \times \gamma$ with $\omega = 0, .1, .2, .3, 1$; simulation. The initial overlap $m_0 = 1.0$ (upper) and $m_0 = 0.1$ (lower).

5 Conclusions

In this paper we have studied the dependence of the information capacity with the topology for an attractor neural network. We calculated the mutual information for a Hebbian model, for storing binary patterns, varying the connectivity and randomness parameters, and obtained the maxima respect to α, $i_m(\gamma, \omega)$. Finally we also presented results for the basins of attraction.

We found there is an optimal γ_o for which an optimized topology, in the sense of the information, $i_m(\gamma_o)$, holds. We believe that the maximization of information respect to the topology could be a biological criterion to build real neural networks. We expect that the same dependence should happens for more structured networks and learning rules.

References

1. Hertz, J., Krogh, J., Palmer, R.: Introduction to the Theory of Neural Computation. Addison-Wesley, Boston (1991)
2. Strogatz, D., Watts, S.: Nature **393** (1998) 440
3. McGraw, P., Menzinger, M.: Phys. Rev. E **68** (2004) 047102
4. Amit, D., Gutfreund, H., Sompolinsky, H.: Phys. Rev. A **35** (1987) 2293
5. Okada, M.: Neural Network **9/8** (1996) 1429
6. Perez-Vicente, C., Amit, D.: J. Phys. A, **22** (1989) 559
7. Dominguez, D., Bolle, D.: Phys. Rev. Lett **80** (1998) 2961
8. Derrida, B., Gardner, E., Zippelius, A.: Europhys. Lett. **4** (1987) 167
9. Bolle, D., Dominguez, D., Amari, S.: Neural Networks **13** (2000) 455

Mutual Information and Topology 2: Symmetric Network

Kostadin Koroutchev **, David Dominguez *,**
Eduardo Serrano **, and Francisco B. Rodríguez**

EPS, Universidad Autonoma de Madrid,
Cantoblanco, Madrid, 28049, Spain
{kostadin.koruchev,david.dominguez,eduardo.serrano,
francisco.rodriguez}@ii.uam.es

Abstract. Following Gardner [1], we calculate the information capacity and other phase transition related parameters for a symmetric Hebb network with small word topology in mean-field approximation. It was found that the topology dependence can be described by very small number of parameters, namely the probability of existence of loops with given length. In the case of small world topology, closed algebraic set of equations with only three parameters was found that is easily to be solved.

1 Introduction

There are 10^{11} neurons in a human brain, each neuron is connected with some 10^3 other neurons and the mean radius of connection of each neuron is about 1 mm, although some neurons has extent of decimeters [2]. It seems that, at least in the cortex, there is no critical finite-size subnetwork of neurons. From this considerations it is clear that the topology of this real neural network is far from fully connected, uniformly sparse connected or scale free.

All neural networks (NN) process information and therefore the description of the network in terms of information capacity seems to be the most adequate one. For a living system it is clear that better information processing provides evolutionary advantage.

One can expect, that in order to survive, at least to the some extend, the biological systems are optimal in sense of usage of their resources. Fixing the number of neurons, the main characteristic that can vary in a real NN is the topology of the network, that give rise to the intriguing question:

How does the topology of the network interfere on the information characteristics of the network? The first part of this communication [3] investigates asymmetric small world networks via simulation. Also more complete resume of the existing publications is given. This article tries to answer partially to the previous question, using as a model symmetrical Hebb NN with small world (SW) topology [4] at zero temperature limit.

---
* DD thanks a Ramon y Cajal grant from MCyT
** Supported by the Spanish MCyT BFI-2003-07276 and TIC 2002-572-C02.

If one considers a network of N neurons, there exists a huge number of $2^{N(N-1)/2}$ different topologies. If all neurons are connected in the same manner, the number of possible topologies is still of order of $2^{(N-1)}$. That makes impossible the exact topology description of the network because the parameters to describe it are too many. Here we show that actually, the dependence of the topology of a symmetric NN can be described by only few real numbers, with clear topological meaning.

The paper is organized as follows: In the following Section, the system is defined and replica-trick solution is given. A zero temperature limit is performed. The topological dependence is found to be expressed by topological coefficients a_k, that are the probabilities of cycle with given length $(k+2)$. Monte-Carlo procedure of finding the topological dependence is described, that makes possible to calculate the dependence on the topology for virtually any topology. It is shown that the method converges fast enough. In Section 3, a small world topology is explored. It is shown that the dependence on the SW topology parameters is trivial. A resume of the results is given in the last section.

2 Replica Trick Equations

We consider a symmetrical Hebb network with N binary neurons each one in state σ_i, $i \in \{1...N\}$. The topology of the network is described by its connectivity matrix $\mathcal{C} = \{c_{ij} | c_{ij} \in \{0,1\}, c_{ij} = c_{ji}, c_{ii} = 0\}$.

The static thermodynamics of symmetric Hebb neural network in thermodynamic limit $(N \to \infty)$ is exactly solvable for arbitrary topology, provided that each neuron is equally connected to the others, and site independent solution for the fixed point exists. Small world topology [4], the case studied in details here, is of that type. The energy of the symmetric network in state $\boldsymbol{\sigma} \equiv \{\sigma_i\}$:

$$H_s = -\frac{1}{2} \sum_{i,j} c_{ij} K_{ij} \sigma_i \sigma_j; \quad K_{ij} \equiv \frac{1}{\gamma N} \sum_{\mu=1}^{P} \xi_i^\mu \xi_j^\mu \qquad (1)$$

defines completely the equilibrium thermodynamics of the system. In the above equation P is the number of memorized patterns $\boldsymbol{\xi} \equiv \xi_i^\mu$ and γ is the average connectivity of the network, that is the fraction of existing links or the largest eigenvalue of $\mathcal{C}/N$. K_{ij} is defined in the the complementary work [3].

Following Gardner [1], using replica-symmetry trick, in thermodynamic limit $(N \to \infty)$ one obtains:

$$m^\mu = \left\langle\!\left\langle \int \frac{dz}{\sqrt{2\pi}} e^{-z^2/2} \xi^\mu \tanh \beta\gamma(\sqrt{\alpha r} z + \mathbf{m}.\boldsymbol{\xi}) \right\rangle\!\right\rangle_\xi \qquad (2)$$

$$q = \left\langle\!\left\langle \int \frac{dz}{\sqrt{2\pi}} e^{-z^2/2} \tanh^2 \beta\gamma(\sqrt{\alpha r} z + \mathbf{m}.\boldsymbol{\xi}) \right\rangle\!\right\rangle_\xi \qquad (3)$$

$$r = \mathrm{Tr}_{ij} \left[\frac{q}{\gamma} \left(1 - \frac{\beta \mathcal{C}(1-q)}{N}\right)^{-2} \left(\frac{\mathcal{C}}{N}\right)^2 \right], \qquad (4)$$

where $\alpha \equiv P/(\gamma N)$, $\beta = 1/T$ is the inverse temperature, m^μ are the order parameters, q is the mean overlap of σ_i between the replicas and r is the mean overlap of m^μ between the replicas.

The small world topology can be defined as :

$$Prob(c_{ij}=1) \equiv \omega\gamma + (1-\omega)\theta(\gamma - ((i-j+N+\gamma N/2) \bmod N)/N)), \quad (5)$$

where ω is the small world parameter defined as p in [4] and $\theta(.)$ is the θ-Heaviside function.

At $T \to 0$, assuming that only one $m^\mu \equiv m$ differs from zero, that is the system is in ferromagnetic state, and keeping the quantity $G \equiv \gamma\beta(1-q)$ finite, in the equation (2) the tanh(.) converge to sign(.), in the next equation $\tanh^2(.)$ converges to $1 - \delta(.)$ and in (4) the expression can be expanded in series by G, giving:

$$m = \mathrm{erf}(m/\sqrt{2r\alpha}) \quad (6)$$

$$G = \sqrt{2/(\pi r\alpha)}e^{-m^2/2r\alpha} \quad (7)$$

$$r = \sum_{k=0}^{\infty}(k+1)a_k G^k, \quad (8)$$

where $a_k \equiv \gamma \mathrm{Tr}((C/\gamma N)^{k+2})$. Note that a_k is the probability of existence of cycle of length $k+2$ in the connectivity graph.

The only equation explicitly dependent on the topology of the system is the equation (8) and the only topology describing characteristics important for the network equilibrium are a_k. As $k \to \infty$, a_k tends to γ. Actually, for the small world topology $\forall k > 40$, $|a_k - \gamma| < 10^{-2}$, except for extremely small γ, ω.

The coefficients a_k can be calculated using a modification of the Monte-Carlo method proposed by [1], see Fig. 2,*Left*. Namely, let us consider k as a number of iteration and let us regard a particle that at each iteration changes its position among the network connectivity graph with probabilities defined by the topology of the graph (small world). If the particle at step k is in node $j(k)$, we say that the coordinate of the particle is $x(k) \equiv j(k)/N$. When $N \to \infty$, $x(k)$ is continuous variable. If n is uniformly distributed random variable between -1/2 and 1/2, then in step $k+1$ the coordinate of the particle changes according to the rule:

$x(k+1) = x(k) + \gamma n \bmod 1$ with probability $1 - \omega$ and
$x(k+1) = x(k) + 1n \bmod 1$ with probability ω.

If the particle starts at moment $k=0$ form $x(0)=0$ and is in position $x(k)$ at step k then the probability to have a loop of length $k+1$, according to the small world topology is (assuming $x(k) \in [-1/2, 1/2]$):

$$\hat{a}_{k-1} = \theta(\gamma/2 - |x(k)|)(1-\omega) + \omega\gamma \quad (9)$$

One can estimate easily the speed of the convergence of $\langle \hat{a}_k \rangle$ to a_k. As a_k decrease with k, it is sufficient to estimate the error when k is large. But when k is large, the coordinate at the step k is uniformly distributed in $[0,1)$ and therefore the process of selecting $\hat{a}_k$ is Bernoulli process that with probability γ gives value

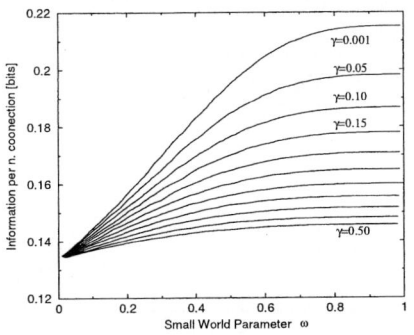
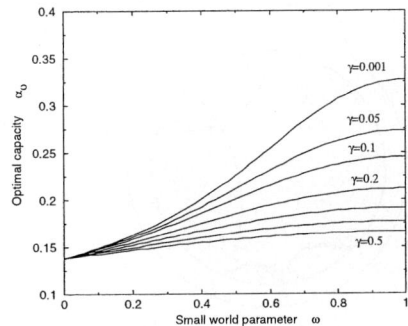

Fig. 1. *Left:* The optimal information capacity per connection $i(\alpha_o; \omega, \gamma)$ as a function of the network topology. Small world parameter ω for different dilutions γ. *Right:* The optimal capacity α_o. No minima nor maxima are observed.

of $(1 - \omega + \omega\gamma)$ and with probability $(1 - \gamma)$ gives value $\omega\gamma$. Therefore, the error of the calculation behaves as $1/\sqrt{MCtrials}$. As a practical consequence, using Monte Carlo method as few as 10000 steps provide good enough (up to 2 digits) precision for $a_k \approx \langle \hat{a}_k \rangle_{MCtrials}$.

The calculation of $a_k = a_k(\mathcal{C})$ closes the system of Eqs.(6-8) giving the possibility to solve them in respect of (G, m) for every (ω, γ, α).

The mutual information per neural connection (information capacity) is given by:

$$i(m, \alpha) \equiv \alpha \left(1 + \frac{1-m}{2} \log_2 \frac{1-m}{2} + \frac{1+m}{2} \log_2 \frac{1+m}{2}\right) \text{ [bits]}. \quad (10)$$

The interesting values of m, α are those of the phase transition $\alpha_c : m(\omega, \gamma, \alpha_c) = 0; \forall \alpha < \alpha_c \ m(\omega, \gamma, \alpha) > 0$ and those with maximal information capacity $\alpha_o : (di(m, \alpha)/d\alpha)(\alpha_o) = 0, m \neq 0$. The mean field theory provides good approximation far from the phase transition point. If $\alpha_c \approx \alpha_o$ then significant differences of the mean-field prediction can be expected.

3 Small World Topology

Fig. 1 show $i(\alpha_o)$ and α_o. as a function of the topology parameters (γ, ω). In all the cases monotonicity and smooth behavior in respect to the topological parameters is observed. Therefore, a small world topology has no significant impact on the network performance. Particularly, there is no maxima nor minima of the capacity $i(\alpha_o)$ or α_o as a function of the network topology for any intermediate values of (γ, ω) (except the trivial ones $c \to 0$, $\omega \to 1$). In contrast, using simulations it was shown that in the asymmetrical case there exists maxima of the information capacity [3].

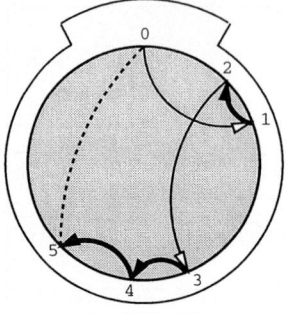 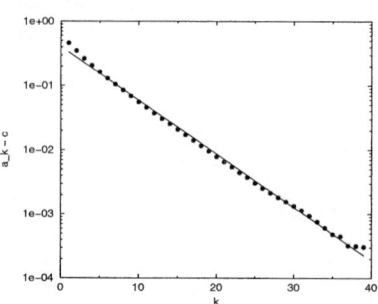

Fig. 2. *Left:* Calculus of a_k by random walk in SW topology. The network with its connection is represented by the gray circle. The height of the surrounding circle represents the probability of connection of the neuron at position 0. The "particle" moves at distance shorter then $\gamma/2$ with high probability (bold lines) and at large distances with smaller probability (thin lines). After 5 steps, the probability of existence of circle with length 6 (a_4) is given by the probability of existence of connection between x_5 and x_0 (dotted line). *Right:* A typical behavior of the topology dependent coefficients $a_k(k) - \gamma$ on the index k, for SW topology with $\gamma = 0.001$ and $\omega = 0.15$ together with the regression fit.

The mean field approximation is valid, because for all values of the parameters of SW, except the case of fully connected network, it was calculated that $1 - \alpha_o/\alpha_c > 2.5\%$.

A typical behavior of the coefficients a_k, $k > 0$ in the equation (8) is shown in Fig. 2,*Right*. One can see that a_k is roughly exponentially falling with the index k toward γ, that is

$$a_k \approx (a_1 - \gamma)e^{-(k-1)\eta} + \gamma. \qquad (11)$$

By definition, a_1 is the probability of having 3-cycle, e.g. the clusterization index of the graph.

From this observation it follows that in very good approximation the behavior of the network is described by only 3 numbers: the average connectivity γ, the clustering coefficient a_1 and η.

Substituting the values of a_k in the expression of r (8), one obtains:

$$r = 1 + \gamma[(1-G)^{-2} - 1] + (a_1 - \gamma)[(1 - e^{-\eta}G)^{-2} - 1], \qquad (12)$$

According to the numerical fitting, the parameters a_1 and η depend on γ and ω as:

$$\eta = -0.078 + 0.322\gamma - 0.985\omega + 2.07\gamma^2 - 1.65\gamma\omega - 0.73\omega^2 \qquad (13)$$
$$a_1 - \gamma = \exp(-0.413 - 1.083\gamma - 0.982\omega + 0.31\gamma^2 - 6.43\gamma\omega - 3.27\omega^2). \qquad (14)$$

Of course, one can find more precise numerical approximations for $\eta, a_1 - \gamma$, using more sophisticated approximations. Although the approximation is very rough, the results for the area of SW behavior of the graph are exact of up to 2 digits.

The equations (6, 7, 12, 13) provide closed system that solves the problem of small world topology network. This system can be solved by successive approximations.

4 Summary and Future Work

The effect of the topology on the performance of a symmetrical Hebb NN at zero temperature was found to be dependent only on small number of real parameters a_k, that are the probabilities of having cycle of length $k+2$ in the connectivity graph. The first a_k, $a_0 = 1$, because the network is symmetric, the second a_1 is the clustering coefficient of the network and the limit value of a_k by large k is the average connectivity of the network γ. Monte-Carlo procedure was designed to find this probabilities for arbitrary topology.

Empirical study of a_k shows that for a small world topology these coefficients converge as exponent to its limit value, that makes possible to express analytically the behavior of SW NN using directly its definition parameters.

Although the replica-symmetry solution for SW is stable, it is not clear if it is so for different topologies. Curiously, although lacking theoretical background, the comparison between the simple solution given here and the simulation for a wide variety of topologies, even asymmetric, is almost perfect. It would be interesting to find similar, theoretically grounded, framework for calculating the behavior of general connectivity distribution and asymmetrical networks.

The particular choice of small world topology seems to have little impact on the behavior (11) of a_k. It seams that the this behavior is universal with large number of rapidly dropping with the distance connectivities. If the mean interaction radius of each neuron is much smaller then N it is easy to show that the equation (11) holds exactly.

However, the exact behavior of η and the range of the validity of the approximation (11) deserves future attention. It would be interesting to find how the parameters just described and the mean minimal path length are connected. It is clear that η increases with the drop of the mean minimal path length.

References

1. Canning, A. and Gardner, E. *Partially Connected Models of Neural Networks*, J. Phys. A, **21**,3275-3284, 1988
2. Braitengerg, V. and Schuez, A. Statistics and Geometry of Neural Connectivity, Springer, 1998
3. Dominguez, D., Korutchev, K.,Serrano, E. and Rodriguez F.B. *Mutual Information and Neural Network Topology*. This Publication
4. Watts, D.J. and Strogatz, S.H. *Collective Dynamics of Small-World Networks*, Nature, **393**:440-442, 1998

Simplified PCNN and Its Periodic Solutions*

Xiaodong Gu[1,2], Liming Zhang[1], and Daoheng Yu[2]

[1]Department of Electronic Engineering, Fudan University, Shanghai, 200433, P.R. China
`guxiaodong@263.net`
[2]Department of Electronics, Peking University, Beijing, 100871,P.R. China

Abstract. PCNN-Pulse Coupled Neural Network, a new artificial neural network based on biological experimental results, can be widely used for image processing. The complexities of the PCNN's structure and its dynamical behaviors limit its application so simplification of PCNN is necessary. We have used simplified PCNNs to efficiently process image. In this paper dynamical behaviors of simplified PCNNs under certain conditions are analyzed in detail and we obtain the conclusion that under these conditions, simplified PCNNs have periodic solutions, i.e. their dynamical behaviors are periodical.

1 Introduction

1990, according to the phenomena of synchronous pulse bursts in the cat visual cortex) Eckhorn introduced the linking field network[1], which is different from traditional neural networks. 1993,Johnson developed PCNN(pulse coupled neural network)[2] based on the linking model. Lately, some researchers paid more attention to it. Today the research of PCNN is still at the beginning and more work need to do further. PCNN can be applied in many fields, such as image processing, object detection, optimization [3-8]. However, Pulse Coupled Neurons (PCN) constructing PCNNs are more complex than traditional artificial neurons. The complexities of the structures of PCNs and dynamical behaviors of PCNNs limit PCNNs' application. Therefore simplification of PCNN is necessary. We have used the simplified PCNNs to efficiently process image, such as remove image noise, segment image, detect image edge, and thin image[7,9]. In this paper, under certain conditions, dynamical behaviors of a simplified PCNN composed of two neurons are analyzed, so are dynamical behaviors of a simplified PCNN composed of many neurons, which are used in image processing. By analyzing in detail we obtain the conclusion that under these conditions, the dynamical behaviors of simplified PCNN are periodical. Meanwhile, the condition guaranteeing that each neuron only fires once during each periodicity is given. Researches on dynamical behaviors of simplified PCNNs are necessary for they are the bases of deepening simplified PCNNs' theories and

---

* This research was supported by China Postdoctoral Science Foundation (No.2003034282) and National Natural Science Foundation of China (No.60171036 and No.30370392).

expanding their applications. Therefore doing research on dynamical behaviors of simplified PCNNs has the significance and value in theories and applications.

In Section 2, simplified PCNN is introduced. In Section 3, under certain conditions, dynamical behaviors of PCNNs composed of two neurons and many neurons are analyzed in detail respectively. Meanwhile, the condition guaranteeing that each neuron only fires once during each periodicity is given. The last Section is conclusions.

2   Simplified PCNN

The complexities of the Pulse Coupled Neurons (PCN) limit applications of PCNNs and make it difficult to analysis the dynamical behaviors of PCNNs. Therefore, it is necessary to simplify PCNs with retaining their main characteristics.

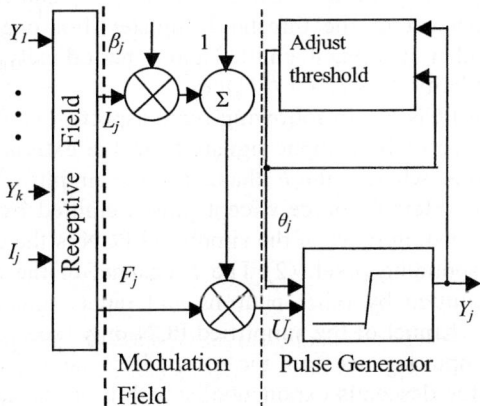

Fig. 1. A simplified PCNN neuron model

$$F_j = I_j \tag{1}$$

$$L_j = \sum_k L_{kj} = V_j^L \sum_k [W_{kj} \exp(-\alpha_{kj}^L t)] \otimes Y_k(t) \tag{2}$$

$$U_j = F_j(1 + \beta_j L_j) \tag{3}$$

$$\frac{d\theta_j(t)}{dt} = -\alpha_j^T + V_j^T Y_j(t), \quad \begin{array}{l}\textit{the lower limit of integration is just}\\ \textit{before the last firing in solution}\end{array} \tag{4}$$

$$Y_j = \text{Step}(U_j - \theta_j) = \begin{cases} 1, & \text{if } U_j > \theta_j \\ 0, & \text{else} \end{cases} \tag{5}$$

Fig.1 illustrates a simplified PCN j. The simplified PCN j consists of three parts: the receptive field, the modulation field, and the pulse generator. It has two channels. One channel is feeding input (F_j); the other is linking input (L_j). The neuron receives input signals through the receptive field. $I_j, Y_1, \ldots, Y_k$ are input signals of neuron j. Y_j is the output pulse of neuron j. I_j, an input signal from the external source, only inputs to F channel of j (see Eq.(1)). $Y_1, \ldots, Y_k$, which are output pulses emitted by other neurons connecting with j, only input to L channel of j (see Eq.(2)). In Eq.(2), L_{kj} indicates the linking response of j arising from the output pulse emitted by k. W_{kj} is synaptic gain strength between j and k. α_{kj}^L is the time constant of the linking channel of j between j and k. V_j^L is the amplitude gain of the linking channel of j. The linking input L_j is added a constant positive bias firstly. Then it is multiplied by the feeding input and the bias is taken to be unity (see Eq.(3)). β_j is the linking strength. The total internal activity U_j is the result of modulation and it is inputted to the pulse generator. If U_j is greater than the threshold θ_j, the neuron output Y_j turns into 1 (namely the neuron j fires, see Eq.(5)). Then Y_j feedbacks to make θ_j rises over U_j immediately so that Y_j turns into 0. Therefore, when U_j is greater than θ_j, neuron j outputs a pulse. On one hand, θ_j changes with the variation of the neuron's output pulse. On the other hand, θ_j drops lineally with time increasing. In Eq.(4) V_j^T and α_j^T are the amplitude gain and the time constant of the threshold adjuster. For one neuron that is not connected with any other neuron, it emitted pulse periodically. This periodicity is called its natural periodicity. It is $T_j = (V_j^T - U_j) / \alpha_j^T$.

Simplifications of a PCN are in following three aspects. (1) The F channel of the non-simplified PCN receive both input signals from the external source and pulses emitted by other neurons, whereas the F channel of the simplified PCN only received input signals from the external source except pulses emitted by other neurons. For image processing, the feeding input of the simplified PCN is the constant equal to the intensity of the corresponding pixel. (2) The L channel of the non-simplified PCN receive both pulses emitted by other neurons and input signals from the external source, whereas the L channel of the simplified PCN only received pulses emitted by other neurons except input signals from the external source. (3) For a non-simplified PCN, its threshold value descends exponentially with time increasing, whereas for a non-simplified PCN, its threshold descends linearly with time increasing.

For some applications, such as image edge detection, image thinning, we can use Unit-linking PCNs ($L_j = \text{Step}(\sum_{k \in N(j)} Y_k)$, $N(j)$ is the neighbor field of neuron j) that are obtained by making the linking inputs of simplified PCNs uniform. Unit-linking PCNs is a special kind of simplified PCNs. Connecting the neurons one another, one of which is show in Fig.1, a simplified PCNN appears.

3 Periodic Solutions of Simplified PCNN Under Certain Conditions

In this paper, under certain conditions, analyze dynamical behaviors of the simplified PCNN used for image processing. Used for image processing, PCNN is a single layer two-dimensional array of laterally linked neurons. Each neuron's feeding input is the constant equal to the corresponding pixel's intensity. Meanwhile, each neuron is connected with neurons in the nearest-neighbor field by L channel. Under certain conditions, simplified PCNN has periodic solutions.

3.1 Periodic Solutions of Simplified PCNN Composed of Two Neurons

Neuron i is connected with neuron j to construct a two-neuron system. Now analyze dynamical behaviors of this system under two following conditions.

Initial condition: these two neurons fire at the same time at the beginning.

Parameter choosing condition: For each neuron, the time constant of the linking channel is less than that of the threshold greatly ($a_{ji}^L \ll a_i^T$, $a_{ij}^L \ll a_j^T$).

Suppose F_i, the constant feeding input of neuron i, is greater than F_j, the constant feeding input of neuron j. T_i, T_j are natural periodicities of i and j respectively. According to the initial condition, at the beginning ($t=0$), these two neurons fire to emit pulses at the same time and then their thresholds increase to V_i^T, V_j^T respectively. Next thresholds θ_i, θ_j start to descend and their descending rates are decided by a_i^T, a_j^T respectively. Meanwhile these two neuron's linking inputs (L_i, L_j) increase because of receiving the pulses emitted by each other so that their total internal activities($U_i=F_i(1+\beta_i L_i)$, $U_j=F_j(1+\beta_j L_j)$) rise accordingly. Now U_i, U_j all are less than θ_i, θ_j respectively and θ_i, θ_j are equal to V_i^T, V_j^T respectively. Then U_i, U_j start to descend and their descending rates are decided by a_{ji}^L, a_{ij}^L respectively. Because $a_{ji}^L \ll a_i^T$, $a_{ij}^L \ll a_j^T$ (see parameter choosing condition), U_i, U_j delay faster than θ_i, θ_j greatly. Therefore, when $t>0$, the firing state of each neuron are independent of the previous pulses emitted by the other. It means that the interactivity between two neurons is almost instantaneous, i.e. if a neuron fires to emit a pulse and but this pulse does not excite the other neuron to fire synchronically, as to the latter, the effect of this pulse can be ignored completely in the future.

The firing frequency of neuron i (f_i) is larger than that of neuron j (f_j) for F_i is larger than F_j, so i fires at the second time before j to emit a pulse to input to the L channel of j and to make L_j increase when $t=T_i$. Then U_j increases accordingly because of the modulation ($U_j=F_j(1+\beta_j L_j)$). At the moment, if U_j is larger than θ_j, j fires too and it is captured by i so that these two neurons start to fire synchronically for ever and the synchronic frequency is f_i.

If when $t=T_i$, U_j is not larger than θ_j, j does not be captured by i. Then U_j begins to descend exponentially. Because $a_{ij}^L \ll a_j^T$, U_j descends faster than θ_j. In this situation, when $t>T_j$, U_j does not effected by the pulse emitted by i at $t=T_i$. Next if at $t=T_j$, j fires before i fires at the third time and i is captured by j, they start to fire synchronically for ever and the firing frequency is f_j ($1/T_j$). In this situation the pulse-emission pattern of this system is periodic and the periodicity is T_j. If j does not fire before i fires at the third time, i fires thirdly at $t=2T_i$. At the moment, if the pulse emitted by i makes $U_j>\theta_j$, j is captured by i so that they fire synchronically for ever and the firing frequency is $f_i/2$ ($1/2T_i$); or I is not captured by j.

Under the initial condition and the parameter choosing condition mentioned above, in a two-neuron system, a neuron can capture the other certainly. If a neuron capture the other at $t=C$, the inner state of the system at $t=C$ is as the same as that at $t=0$. Therefore the dynamical behaviors between $t=0$ and $t=C$ will be repeated between $t=C$ and $t=2C$, between $t=2C$ and $t=3C$, $t=3C$ and $t=4C$, and so on, namely solutions of this system is periodic and the periodicity is C.

3.2 Periodic Solutions of Simplified PCNN Composed of Many Neurons

Now analyze dynamical behaviors of the simplified PCNN composed of many neurons and used for image processing.

Initial condition: all neurons fire at the same time at the beginning.

Parameter choosing condition: for each neuron, the time constant of the linking channel is less than that of the threshold greatly.

If the maximal intensity and the minimal intensity of the image is F_{max} and F_{min} respectively, for all neurons, the maximal and the minimal feeding inputs are F_{max} and F_{min} respectively. T_{max} and T_{min} indicate natural periodicities of the neuron corresponding to F_{max} and the neuron corresponding to F_{min}. At $t=T_{max}$, the neurons whose feeding inputs are equal to F_{max} fire and neurons captured by they fire synchronically. The feeding inputs of these captured neurons should satisfy some condition. If j is captured by the neuron corresponding to F_{max}, inequality $F_j(1+\beta_j L_j(T_{max}))>\theta_j(T_{max})$ should be satisfied. Due to $\theta_j(T_{max})=F_{max}$, have $F_j>F_{max}/(1+\beta_j L_j(T_{max}))$. Therefore only when F_j belongs to $(F_{max}/(1+\beta_j L_j(T_{max})), F_{max})$, j is captured by the neuron corresponding to F_{max}. This is the necessary condition of being captured. When j is captured by the neuron corresponding to F_{max}, the former may not connected with the latter directly and the pulse emitted by the latter can reach the former through other neurons instantaneously and capture the j. These are phenomena of pulse emission, and synchronizing of simplified PCNN.

At $t=T_{min}$, every neuron in this system has fired once at least. If at $t=T_{min}$, set all neurons fire together, the inner state of the system is as the same as that at $t=0$ so that the dynamical behaviors between 0 and T_{min} is as the same as those between T_{min} and $2T_{min}$. If under the initial condition and the parameter choosing condition mentioned above, set all neurons fire at $t=nT_{min}$ (n is an integer), solutions of this system are periodic and the periodicity is T_{min}.

In order to be convenient for controlling and applications, we wish every neuron in this system only fire once during $[0, T_{min}]$. It can be realized by choosing suitable threshold amplitude gain. Here assume that every neuron has the same threshold amplitude gain V^T and the same threshold time constant α^T.

The neuron corresponding to F_{max} fires first at $t=T_{max}$. At $t=T_{min}$ its threshold is $\theta_{max}(T_{min})=V^T-\alpha^T(T_{min}-T_{max})$.

In order to make every neuron only fire once, inequality $\theta_{max}(T_{min})>U_{max}(T_{min})$ should be satisfied. Namely, $V^T-\alpha^T(T_{min}-T_{max})>U_{max}(T_{min})$ should be satisfied.

Due to $T_{min}=(V^T-F_{min})/\alpha^T$, $T_{max}=(V^T-F_{max})/\alpha^T$, $U_{max}(T_{min})=F_{max}(1+\beta_{max}L_{max}(T_{min}))$, have

$$V^T-(F_{max}-F_{min})>F_{max}[1+\beta_{max}L_{max}(T_{min})] \tag{6}$$

Assume that L_{max}^{max} indicates the possible maximal value of $L_{max}(T_{min})$ obtained when all neurons in neighbor field fire at the same time. Rewrite (6) as

$$V^T>F_{max}(2+L_{max}^{max})-F_{min} \tag{7}$$

When V^T satisfies inequality (7), every neuron in this system only fires once during $[0, T_{min}]$.

4 Conclusions

For a simplified PCNN composed of two neurons, solutions of this system are periodic when two following conditions are satisfied. (1) These two neurons fire at the same time at the beginning.(2) For each neuron, the time constant of the linking channel is less than that of the threshold greatly.

For a simplified PCNN composed of many neurons used for image processing, which is a single layer two-dimensional array of laterally linked, solutions of this system are periodic when three following conditions are satisfied. (1) These two neurons fire at the same time at the beginning. (2)For each neuron, the time constant of the linking channel is less than that of the threshold greatly. (3)Set all neurons fire at $t = nT_{min}$(n is an integer). In addition, choosing suitable V^T can make every neuron fire only once during one periodicity so as to be convenient for controlling and applications.

References

1. Eckhorn R., Reitboeck H.J., Arndt, M., et al: Feature Linking via Synchronization among Distributed Assemblies: Simulation of Results from Cat Cortex. Neural Computation 2 (1990) 293-307
2. Johnson J.L., Ritter D.: Observation of Periodic Waves in a Pulse-coupled Neural Network. Opt.Lett.18 (1993) 1253-1255
3. Johnson J.L., Padgett M.L.: PCNN Models and Applications, IEEE Trans. on Neural Networks 10 (1999) 480-498
4. Kuntimad G., Ranganath H.S.: Perfect Image Segmentation Using Pulse Coupled Neural Networks, IEEE Trans. on Neural Networks 10 (1999) 591-598
5. Ranganath H.S., Kuntimad G.: Object Detection Using Pulse Coupled Neural Networks. IEEE Trans. on Neural Networks 10 (1999) 615-620
6. Gu X.D., Guo S.D., Yu D.H.: New Approach for Noise Reducing of Image Based on PCNN. Journal of Electronics and Information Technology 24 (2002) 1304-1309
7. Gu X.D., Guo S.D., Yu D.H.: A New Approach for Automated Image Segmentation Based on Unit-linking PCNN. The First International Conference on Machine Learning and Cybernetics, Beijing, China (2002) 175-178
8. Caufield H.J., Kinser J.M.: Finding Shortest Path in the Shortest Time Using PCNN's. IEEE Trans.on Neural Networks 10 (1999) 604-606
9. Gu X.D., Yu D.H., Zhang L.M.: Image Thinning Using Pulse Coupled Neural Network, Pattern Recognition Letters, In Press

On Robust Periodicity of Delayed Dynamical Systems with Time-Varying Parameters

Changyin Sun[1,2,3], Xunming Li[1], and Chun-Bo Feng[2]

[1] College of Electrical Engineering, Hohai University, Nanjing 210098, China
[2] Research Institute of Automation, Southeast University, Nanjing 210096, China
[3] Departmen of Computer Science and engineering, The Chinese University of Hong Kong, Shatin, Hong Kong
cysun@ieee.org

Abstract. In this paper, robust exponential periodicity of a class of dynamical systems with time-varying parameters is introduced. Novel robust criteria to ensuring existence and uniqueness of periodic solution for a general class of neural systems are proposed without assuming the smoothness and boundedness of the activation functions. Which one is the best in previous results is addressed.

1 Introduction

Recently, there are many important results focusing mainly on equilibrium point of neural systems [1-5],[10-17]. This is because the properties of equilibrium points of neural systems play an important role in some practical problems, such as optimization solvers, associative memories, image compression, processing of moving images, and pattern classification. It is well known that an equilibrium point can be viewed as a special periodic solution of continuous-time neural systems with arbitrary period [10],[14,[16]. In this sense the analysis of periodic solutions of neural systems may be considered to be more general sense than that of equilibrium points. In addition, the existence of periodic solutions of continuous-time neural networks is an interesting dynamic behavior. It has been found applications in learning theory [9], which is motivated by the fact that learning process usually requires repetition. Among the most previous results of the dynamical analysis of continuous-time neural systems, a frequent assumption is that the activation functions are differentiable and bounded, such as the usual sigmoid-type functions in conventional neural networks [7],[8]. But, in some application, the smoothness and boundedness of the activation function are not satisfied [5],[6],[17]. Since the boundedness and smoothness assumption is not always practical, it is necessary and important to investigate the dynamical properties of the continuous-time neural systems in both theory and applications without assuming the smoothness and boundedness of the activation functions.

On the other hand, the dynamical characteristics of neural networks may often be destroyed by its unavoidable uncertainty due to the existence of modeling external disturbance and parameter fluctuation during the implementation on very-large-scale-integration (VLSI) chips. From the view point of reality, it should also naturally take into account evolutionary processes of the biological systems as well as disturbances of external influence. The deviations and perturbations of the neuron charging time constants and the weights of interconnections are bounded in general. These motivate the present investigation of dynamical properties of delayed neural networks with varying external stimuli, network parameters and intervalize the above mentioned quantities. Thus, it is important to investigate the periodicity and robustness of the network against such deviations and fluctuation. In order to overcome this difficulty, Arik [3] and Sun [12],[13] have extended the model of delayed Hopfield neural networks to interval-delayed neural systems. Mohamad [15] and Zhang [16] have investigated the dynamical properties of delayed neural networks under a periodically varying environment.

In this paper, a new concept of robust exponential periodicity is introduced. And borrowing the technique in [12],[13,[16] we will further derive some sufficient conditions to guarantee robust exponential periodicity of delayed neural networks with time-varying parameters without assuming the smoothness and boundedness of the activation functions.

2   Neural Network Model and Preliminaries

Consider the model of neural network with delays described by the following functional differential equations

$$\dot{x}_i(t) = -d_i(t)x_i(t) + \sum_{j=1}^{n} a_{ij}(t)g_j(x_j(t)) + \sum_{j=1}^{n} b_{ij}(t)f_j(x_j(t-\tau_j)) + I_i(t), \ t \geq 0 \qquad (1)$$

$$x_i(t) = \phi_i(t), \ -\tau \leq t \leq 0, i = 1, 2, \cdots, n,$$

where $0 \leq \tau_j \leq \tau$ is the transmission time delay of the j th unit, $x_i(t)$ is the state vector of the i th unit at time t, $d_i(t)$ is self-inhibition, and $a_{ij}(t)$, $b_{ij}(t)$ are interconnection weights. $A(t) = (a_{ij}(t))$ and $B(t) = (b_{ij}(t))$ are $n \times n$ interconnection matrices. $g(x) = (g_1(x_1), g_2(x_2), \ldots, g_n(x_n))^T$: $R^n \to R^n$ and $f(x) = (f_1(x_1), f_2(x_2), \ldots, f_n(x_n))^T$: $R^n \to R^n$ are nonlinear vector-valued activation functions. $I(t) = (I_1(t), I_2(t), \ldots, I_n(t))^T \in R^n$ is an input periodic continuous vector function with period ω, i.e., there exists a constant $\omega > 0$ such that $I_i(t+\omega) = I_i(t)$ ($i = 1, 2, \cdots, n$) for all $t \geq 0$, and $x = (x_1, x_2, \ldots, x_n)^T \in R^n$. Suppose further the following assumptions are satisfied.

(A₁) g is globally Lipschitz continuous (GLC) and strictly monotone increasing activation function; that is, for each $j \in \{1,2,\cdots,n\}$, there exist L_j^* and L_j such that $0 < L_j^* \le \dfrac{g_j(u)-g_j(v)}{u-v} \le L_j$ for any $u, v \in \mathbb{R}$.

(A₂) $f \in$ GLC in $\mathbb{R}^n$, for each $j \in \{1,2,\cdots,n\}$, $f_j : \mathbb{R} \to \mathbb{R}$ is globally Lipschitz continuous with Lipschitz constant M_j, i.e. $|f_j(u)-f_j(v)| \le M_j |u-v|$ for any $u, v \in \mathbb{R}$.

As a special case of neural system (1), the delayed neural networks, with constant input vector $I = (I_1, I_2, \ldots, I_n)^T \in \mathbb{R}^n$, self-inhibition $d_i(t)$ and interconnection weights $a_{ij}(t)$, $b_{ij}(t)$ being constants, have been studied widely by many researchers [1-8],[16-23],[26]. This system is described by the following functional differential equations

$$\dot{x}_i(t) = -d_i x_i(t) + \sum_{j=1}^{n} a_{ij} g_j(x_j(t)) + \sum_{j=1}^{n} b_{ij} f_j(x_j(t-\tau_j)) + I_i, \quad t \ge 0 \tag{2}$$

$$x_i(t) = \phi_i(t), \quad -\tau \le t \le 0, \, i = 1, 2, \cdots, n,$$

However, in practice implementation of neural networks (2), the value of the constants d_i and connection weight coefficients a_{ij} and b_{ij} depend on certain resistance and capacitance value which are subject to uncertainties. This may lead to some deviations in the value of d_i, a_{ij} and b_{ij}. From the view point of reality, it should also naturally take into account evolutionary processes of the biological systems as well as disturbances of external influence. The deviations and perturbations of the neuron charging time constants and the weights of interconnections are bounded in general. These motivate the present investigation of dynamical properties of delayed neural networks with varying external stimuli, network parameters and intervalize the above mentioned quantities. For example, [15],[16] have investigated the dynamical properties of (1) under a periodically varying environment; [3],[12,[131] addressed robust properties of interval neural networks with delays. In the following and throughout this paper, we further correspondingly assume self-inhibition $d_i(t)$ and interconnection weights $a_{ij}(t)$, $b_{ij}(t)$ are continuous functions with $d_i(t) \ge d_i > 0$, $a_{ii}^* = \sup_{\{0<t<\infty\}} a_{ii}(t)$, $|a_{ij}^*| = \sup_{\{0<t<\infty\}} |a_{ij}(t)|$, $|b_{ij}^*| = \sup_{\{0<t<\infty\}} |b_{ij}(t)|$ where $i, j = 1, 2, \cdots, n$.

3 Main Results

Theorem 1. Suppose (A$_1$)-(A$_2$) holds. $a_{jj} > 0$ $j = 1,2,\cdots,n$. If there exist constants $\lambda_j > 0$, α_{ij}, β_{ij}, $i,j = 1,2,\cdots,n$, such that

$$-d_i\lambda_i + L_i\lambda_i a_{ii}^* + \frac{1}{r}L_i\sum_{j=1,j\neq i}^n \lambda_j |a_{ji}^*|^{\alpha_{ji}r} + \frac{r-1}{r}\sum_{j=1,j\neq i}^n \lambda_i |a_{ij}^*|^{(1-\alpha_{ij})\frac{r}{r-1}} L_j$$

$$+\frac{1}{r}M_i\sum_{j=1}^n \lambda_j |b_{ji}^*|^{\beta_{ji}r} + \frac{r-1}{r}\lambda_i \sum_{j=1}^n |b_{ij}^*|^{(1-\beta_{ij})\frac{r}{r-1}} M_j < 0, \quad i = 1,2,\cdots,n \quad (3)$$

where $r \geq 1$, then the delayed neural system (1) is robustly exponentially periodic.

Theorem 2. Suppose (A$_1$)-(A$_2$) holds. $a_{jj} < 0$, $j = 1,2,\cdots,n$. If there exist constants $\lambda_j > 0$, α_{ij}, β_{ij}, $i,j = 1,2,\cdots,n$, such that

$$-d_i\lambda_i + L_i^*\lambda_i a_{ii}^* + \frac{1}{r}L_i\sum_{j=1,j\neq i}^n \lambda_j |a_{ji}^*|^{\alpha_{ji}r} + \frac{r-1}{r}\sum_{j=1,j\neq i}^n \lambda_i |a_{ij}^*|^{(1-\alpha_{ij})\frac{r}{r-1}} L_j$$

$$+\frac{1}{r}M_i\sum_{j=1}^n \lambda_j |b_{ji}^*|^{\beta_{ji}r} + \frac{r-1}{r}\lambda_i \sum_{j=1}^n |b_{ij}^*|^{(1-\beta_{ij})\frac{r}{r-1}} M_j < 0, \quad i = 1,2,\cdots,n \quad (4)$$

where $r \geq 1$, then the delayed neural system (1) is robustly exponentially periodic.

Remark 1. In many previous papers, various stability or periodicity criteria are obtained. It is an important issue to answer the question which one is the best and how to choose r to verify the sufficient condition in (3) or (4). Maybe, they are equivalent. The authors in [16] first address this issue. Similar to some results in [16], we can assert the condition (3) or (4) when $r = 1$ are enough in practice. Therefore, we can verify the following corollaries in practice which are equivalent to verify (3) or (4) when $r > 1$ respectively.

Corollary 1. Suppose (A$_1$)-(A$_2$) holds. $a_{jj} > 0$ $j = 1,2,\cdots,n$. If there exist constants $\lambda_j > 0$, $j = 1,2,\cdots,n$, such that

$$-d_i\lambda_i + L_i\lambda_i a_{ii}^* + L_i\sum_{j=1,j\neq i}^n \lambda_j |a_{ji}^*| + M_i\sum_{j=1}^n \lambda_j |b_{ji}^*| < 0, \quad i = 1,2,\cdots,n \quad (5)$$

then the delayed neural system (1) is robustly exponentially periodic.

Corollary 2. Suppose (A_1)-(A_2) holds. $a_{jj} < 0$, $j = 1,2,\cdots,n$. If there exist constants $\lambda_j > 0$, $j = 1,2,\cdots,n$, such that

$$-d_i\lambda_i + L_i^*\lambda_i a_{ii}^* + L_i \sum_{j=1,j\neq i}^{n} \lambda_j |a_{ji}^*| + M_i \sum_{j=1}^{n} \lambda_j |b_{ji}^*| < 0, \ i = 1,2,\cdots,n \qquad (6)$$

then the delayed neural system (1) is robustly exponentially periodic.

Remark 2. Our focus is on robust periodicity of delayed neural networks with time-varying parameters. The neural systems (1) can be looked as weak-nonlinear systems restricted by a linear system which is implied in (A_1) and (A_2). From the view point of systme, For linear systems, the output can preserve some characteristics of the input. The results obtained are consistent with the exponential stability results that were recently reported in [16] and with the robust stability results that were previously given in [12], [14]. Especially, when $I = (I_1, I_2, \ldots, I_n)^T \in R^n$, we can easily obtain the main results there in Reference [1],[2],[3]. In particular, we improved the main results of Reference [13]. Therefore, this work gives some improvements to the earlier results.

4 Conclusion

In this paper, robust exponential periodicity of delayed neural networks has been introduced. Without assuming the boundedness and differentiability of the activation functions, the easily checked conditions ensuring the robust exponential periodicity of neural systems with time-varying parameters are proposed. In addition, the results are applicable to neural networks with both symmetric and nonsymmetric interconnection matrices.

Acknowledgements. This work was supported by the National Nature Science Foundation of China under Grant 69934010 and Hohai University Startup Foundation of China.

References

1. Driessche, P.V., Zou, X.: Global Attactivity in Delayed Hopfield Neural Network Models. SIAM Journal of Applied Mathematics. 58 (1998) 1878-1890
2. Liao, X., Wang, J.: Algebraic Criteria for Global Exponential Stability of Cellular Neural Networks with Multiple Time Delays. IEEE Trans. Circuits Systems I, 50 (2003) 268-275
3. Arik, S., Tavsanoglu, V.: Global Robust Stability of Delayed Neural Networks. IEEE Transactions on Circuits and Systems I. 50 (2003) 156–160

4. Forti, M., Tesi, A.: New Conditions for Global Stability of Neural Networks with Application to Linear and Quadratic Programming Problems, IEEE Trans. Circuits and Systems I. 42 (1995) 354-366
5. Sudharsanan, S., Sundareshan, M.: Exponential Stability and a Systematic Synthesis of a Neural Network for Quadratic Minimization. Neural Networks. 4 (1991) 599-613
6. Morita, M.: Associative Memory with Non-monntone Dynamics, Neural networks. 6 (1993) 115-126
7. Tank, D.W., Hopfield, J.: Simple "Neural" Optimization Networks: an A/D Converter, Signal Decision Circuit, and a Linear Programming Circuit, IEEE Trans. Circuits & Systems. 33 (1986) 533-541
8. Kennedy, M., Chua, L.: Neural Networks for Linear and Nonlinear Programming, IEEE Trans. Circuits & Systems. 35 (1988) 554-562
9. Townley, S., Ilchmann, A., Weiss, M., et al.: Existence and Learning of Oscillations in Recurrent Neural Networks. IEEE Trans. Neural Networks. 11 (2000) 205-214
10. Sun, C., Feng, C.B.: Exponential Periodicity of Continuous-time and Discrete-time Neural Networks with Delays. Neural Processing Letters. 19 (2004) 131-146
11. Sun, C., Zhang, K., Fei, S., et al.: On Exponential Stability of Delayed Neural Networks with a General Class of Activation Functions. Physics Letters A. 298 (2002) 122-132
12. Sun, C., Feng, C.B.: Global Robust Exponential Stability of Interval Neural Networks with Delays. Neural Processing Letters. 17 (2003) 107-115
13. Sun, C., Feng, C.B.: On Robust Exponential Periodicity of Interval Neural Networks with Delays. Neural Processing Letters. 20 (2004) 1-10
14. Sun, C., Feng, C.B.: Exponential Periodicity and Stability of Delayed Neural Networks. Mathematics and Computers in Simulation. 66 (2004) 1-9
15. Mohamad, S., Gopalsamy, K.: Neuronal Dynamics in Time Varying Environments: Continuous and Discrete Time Models. Discrete and Continuous Dynamical Systems. 6 (2000) 841-860
16. Zhang, Y., Chen, T.: Global Exponential Stability of Delayed Periodic Dynamical System. Physics Letters A. 322 (2004) 344-355
17. Liang, X., Wang, J.: Absolute Exponential Stability of Neural Networks with a General Class of Activation Functions. IEEE Transaction on Circuits and Systems: I. 47 (2000) 1258-1263

Delay-Dependent Criteria for Global Stability of Delayed Neural Network System

Wenlian Lu[1] and Tianping Chen[2]*

Laboratory of Nonlinear Science, Institute of Mathematics,
Fudan University, Shanghai, 200433, P.R. China
tchen@fudan.edu.cn

Abstract. In this paper, we discuss delayed neural networks, investigating the global exponential stability of their equilibria. Delay- dependent criteria ensuring global stability are given. A numerical example illustrating the dependence of stability on the delays is presented.

1 Introduction

Delayed recurrently connected neural networks can be described by the following delayed dynamical system:

$$\frac{du_i(t)}{dt} = -d_i u_i(t) + \sum_{j=1}^{n} a_{ij} g_j(u_j(t)) + \sum_{j=1}^{n} b_{ij} f_j(u_j(t-\tau_{ij})) + I_i \quad i=1,\cdots,n \quad (1)$$

where activation functions $g_j(\cdot)$ and $f_j(\cdot)$ satisfy certain conditions. The initial conditions are $u_i(s) = \phi_i(s)$, $s \in [-\tau, 0]$, where $\tau = \max_{i,j} \tau_{ij}$, and $\phi_i(s) \in C([-\tau,0])$, $i = 1, \cdots, n$.

Recently, there are quite a few papers investigating dynamical behaviors of delayed systems (see[1,2,3,4,5,6,7]). The criteria obtained in these papers for global stability are all independent of time delays. Delays only affect convergence rate. However, there are cases that delays play the key role in discussion of whether or not a system is stable. For example, consider the following one-dimensional delayed system:

$$\dot{x} = -x(t-\tau) \quad (2)$$

Its characteristic equation is $\lambda + e^{-\lambda \tau} = 0$. Therefore, the system (2) is globally asymptotically stable if and only if $\tau < \frac{\pi}{2}$. Moreover, it was reported in [8], when minimal time lags between inputs and corresponding teacher signal are too long, learning algorithm will take too much time or even not work at all. Hence, it is needed to analyze how time delays affect stability. In [9,10], the authors investigated some special forms of system (1) and obtained some delay-independent criteria for global stability.

---

* Corresponding author

In this paper, we give some new delay-dependent criteria for global exponential stability of delayed neural networks (1). Firstly, we assume that activation functions satisfy

$$H: \quad 0 \leq \frac{g_j(\eta) - g_j(\zeta)}{\eta - \zeta} \leq G_j \text{ and } 0 \leq \frac{f_j(\eta) - f_j(\zeta)}{\eta - \zeta} \leq F_j \text{ for any } \eta \neq \zeta$$

where G_j, F_j are positive constants, $j = 1, \cdots, n$. We also define $\|x\|_{\{\xi,1\}} = \sum_{i=1}^{n} \xi|x_i|$, where $\xi > 0, i = 1, \cdots, n$ and $a^+ = \max\{0, a\}$.

2 Main Results

Firstly, we give a lemma for the existence of the equilibrium of system (1).

Lemma 1. *If $g_j(\cdot)$, $f_j(\cdot)$ satisfy H, $j = 1, \cdots, n$, and there exist $\xi_i > 0$, $i = 1, \cdots, n$ such that*

$$-d_i\xi_i + [a_{ii}\xi_i + \sum_{j \neq i}|a_{ji}|\xi_j]^+ G_i + [b_{ii}\xi_i + \sum_{j \neq i}|b_{ji}|\xi_j]^+ F_i < 0 \quad i = 1, \cdots, n \quad (3)$$

then there exists a unique equilibrium $u^* = [u_1^*, \cdots, u_n^*]^T$ of system (1).

The proof is omitted here due to the restriction of space.

Now, we establish the following main theorem.

Theorem 1. *If $g_j(\cdot)$, $f_j(\cdot)$ satisfy H, $j = 1, \cdots, n$, and there exist $\xi_i > 0$, $p_{ij} > 0$, $q_{ij} > 0$, $i, j = 1, \cdots, n$, and $B^k = (b_{ij}^k) \in R^{n,n}$, $k = 1, 2, 3, 4$ such that*

1) $-\xi_i d_i + \left[a_{ii}\xi_i - (\sum_{k=1}^{n} q_{ki}\tau_{ki})a_{ii} + \sum_{j \neq i}|a_{ji}|\xi_j + \sum_{j=1}^{n}\sum_{k \neq i} q_{jk}\tau_{jk}|a_{ki}|\right]^+ G_i$

$+ \left[b_{ii}^1\xi_i - (\sum_{k=1}^{n} q_{ki}\tau_{ki})b_{ii}^3 + \sum_{j \neq i}|b_{ji}^1|\xi_j + \sum_{j=1}^{n}\sum_{k \neq i} q_{jk}\tau_{jk}|b_{ki}^3|\right]^+ F_i + \sum_{j=1}^{n} p_{ji} < 0$

2) $p_{ij} > |b_{ij}^2|F_j\xi_i + (\sum_{k=1}^{n} q_{ki}\tau_{ki})|b_{ij}^4|F_j$

3) $q_{ij} > |b_{ij}^1|F_j\xi_i + (\sum_{k=1}^{n} q_{ki}\tau_{ki})|b_{ij}^3|F_j$

4) $b_{ij}^1 + b_{ij}^2 = b_{ij}^3 + b_{ij}^4 = b_{ij} \quad i, j = 1, \cdots, n$ (4)

Then equilibrium u^* is globally exponentially asymptotically stable.

Proof. From condition (4), there exists $\alpha > 0$ such that let

$$\beta_i = -\xi_i(d_i - \alpha) + \left[a_{ii}\xi_i - (\sum_{k=1}^{n} q_{ki}\eta_{ki})a_{ii} + \sum_{j \neq i}|a_{ji}|\xi_j + \sum_{j=1}^{n}\sum_{k \neq i} q_{jk}\eta_{jk}|a_{ki}|\right]^+ G_i$$

$$+ \left[b_{ii}^1 \xi_i - (\sum_{k=1}^n q_{ki}\eta_{ki})b_{ii}^3 + \sum_{j\neq i}^n |b_{ji}^1|\xi_j + \sum_{j=1}^n\sum_{k\neq i}^n q_{jk}\eta_{jk}|b_{ki}^3| \right]^+ F_i + \sum_{j=1}^n p_{ji}e^{\alpha\tau_{ji}} \leq 0$$

$$\gamma_{ij} = -p_{ij} + |b_{ij}^2|F_j\xi_i + (\sum_{k=1}^n q_{ki}\eta_{ki})|b_{ij}^4|F_j \leq 0$$

$$\delta_{ij} = -q_{ij} + |b_{ij}^1|F_j\xi_i + (\sum_{k=1}^n q_{ki}\eta_{ki})|b_{ij}^3|F_j \leq 0 \qquad (5)$$

Let $x_i(t) = u_i(t) - u_i^*$, $g_i^*(x_i(t)) = g_i(x_i(t) + u_i^*) - g_i(u_i)$, and $f_i^*(x_i(t)) = f_i(x_i(t) + u_i^*) - f_i(u_i)$, $i = 1, \cdots, n$. Then system (1) can be rewritten as follows:

$$\frac{dx_i(t)}{dt} = -d_i x_i(t) + \sum_{j=1}^n a_{ij} g_j^*(x_j(t)) + \sum_{j=1}^n b_{ij} f_j^*(x_j(t-\tau_{ij})) \qquad (6)$$

Let

$$L_1(t) = \sum_{i=1}^n |x_i(t)|e^{\alpha t} \quad L_2(t) = \sum_{i,j=1}^n p_{ij} \int_{t-\tau_{ij}}^t |x_j(s)|e^{\alpha(s+\tau_{ij})}ds$$

$$L_3(t) = \sum_{i,j=1}^n q_{ij} \int_{t-\tau_{ij}}^t e^{\alpha(s+\tau_{ij})} \int_s^t |\dot{x}_j(\theta)|d\theta ds \quad L(t) = L_1(t) + L_2(t) + L_3(t)$$

Noticing system (6) can be rewritten as follows:

$$\frac{dx_i(t)}{dt} = -d_i x_i(t) + \sum_{j=1}^n a_{ij} g_j^*(x_j(t)) + \sum_{j=1}^n b_{ij}^1 f_j^*(x_j(t)) + \sum_{j=1}^n b_{ij}^2 f_j^*(x_j(t-\tau_{ij}))$$

$$+ \sum_{j=1}^n b_{ij}^1 \left[f_j^*(x_j(t-\tau_{ij})) - f_j^*(x_j(t)) \right] \quad i = 1, \cdots, n$$

and differentiating $L_1(t)$, we have

$$\frac{dL_1(t)}{dt} \leq -\sum_{i=1}^n \xi_i(d_i - \alpha)e^{\alpha t}|x_i(t)| + e^{\alpha t}\sum_{i=1}^n |g_i^*(x_i(t))|\left(a_{ii}\xi_i + \sum_{j\neq i}|a_{ji}|\xi_j\right)$$

$$+ e^{\alpha t}|f_i^*(x_i(t))|\left(b_{ii}^1\xi_i + \sum_{j\neq i}|b_{ji}^1|\xi_j\right) + e^{\alpha t}\sum_{i,j=1}^n |b_{ij}^2|F_j\xi_i|x_j(t-\tau_{ij})|$$

$$+ e^{\alpha t}\sum_{i,j=1}^n |b_{ij}^1|F_j\xi_i \int_{t-\tau_{ij}}^t |\dot{(x)}_j(\theta)|d\theta$$

Differentiating $L_2(t)$, we have

$$\frac{dL_2(t)}{dt} = \sum_{i,j=1}^n p_{ij}|x_j(t)|e^{\alpha(t+\tau_{ij})} - \sum_{i,j=1}^n p_{ij}|x_j(t-\tau_{ij})|e^{\alpha t}$$

Noticing system (6) can also be rewritten as follows:

$$\frac{dx_i(t)}{dt} = -d_i x_i(t) + \sum_{j=1}^{n} a_{ij} g_j^*(x_j(t)) + \sum_{j=1}^{n} b_{ij}^3 f_j^*(x_j(t)) + \sum_{j=1}^{n} b_{ij}^4 f_j^*(x_j(t-\tau_{ij}))$$

$$+ \sum_{j=1}^{n} b_{ij}^3 \left[f_j^*(x_j(t-\tau_{ij})) - f_j^*(x_j(t)) \right] \quad i=1,\cdots,n$$

and differentiating $L_3(t)$, we have

$$\frac{dL_3(t)}{dt} \leq -\sum_{i,j=1}^{n} q_{ij} e^{\alpha t} \int_{t-\tau_{ij}}^{t} |\dot{x}_j(\theta)| d\theta + e^{\alpha t} \sum_{i=1}^{n} |g_i^*(x_i(t))| \bigg[-a_{ii} (\sum_{k=1}^{n} q_{ki} \eta_{ki})$$

$$+ \sum_{j=1}^{n} \sum_{k \neq i} |a_{ki}| q_{jk} \eta_{jk} \bigg] + e^{\alpha t} \sum_{i=1}^{n} |f_i^*(x_i(t))| \bigg[-b_{ii}^3 (\sum_{k=1}^{n} q_{ki} \eta_{ki})$$

$$+ \sum_{j=1}^{n} \sum_{k \neq i} |b_{ki}^3| q_{jk} \eta_{jk} \bigg] + e^{\alpha t} \sum_{i,j=1}^{n} |x_j(t-\tau_{ij})| b_{ij}^4 | F_j (\sum_{k=1}^{n} q_{ki} \eta_{ki})$$

$$+ e^{\alpha t} \sum_{i,j=1}^{n} \int_{t-\tau_{ij}}^{t} |\dot{x}_j(\theta)| d\theta |b_{ij}^3| F_j (\sum_{k=1}^{n} q_{ki} \eta_{ki})$$

It can be seen that if condition (5) are satisfied, then

$$\frac{dL(t)}{dt} = \frac{dL_1(t)}{dt} + \frac{dL_2(t)}{dt} + \frac{dL_3(t)}{dt}$$

$$\leq \sum_{i=1}^{n} \beta_i |x_i(t)| e^{\alpha t} + \sum_{i,j=1}^{n} \gamma_{ij} |x_j(t-\tau_{ij})| e^{\alpha t} + e^{\alpha t} \sum_{i,j=1}^{n} \delta_{ij} \int_{t-\tau_{ij}}^{t} |\dot{x}_j(\theta)| d\theta$$

$$\leq 0$$

from which it can be concluded that $L(t) \leq L(0)$. Therefore, $\|x(t)\|_{\{\xi,1\}} = O(e^{-\alpha t})$. The Theorem is proved. $\square$

Remark 1. From Theorem 1, it can be seen that the condition (4) depend on time delay τ_{ij}. For instance, picking $B^3 = B^2 = 0$, we have the following:

Corollary 1. If $g_j(\cdot)$, $f_j(\cdot)$ satisfy H, $j=1,\cdots,n$, and there exists $\xi_i > 0$, $i=1,\cdots,n$, such that

$$-\xi_i d_i + \bigg[a_{ii} \xi_i + \sum_{j \neq i} |a_{ji}| \xi_j + \tau \sum_{k,j=1}^{n} |b_{jk}| F_j \xi_i |a_{ki}| \bigg]^+ G_i + \bigg[b_{ii} \xi_i + \sum_{j \neq i} |b_{ji}| \xi_j \bigg]^+ F_i$$

$$+ \tau \sum_{k,j=1}^{n} |b_{kj}| F_j \xi_k |b_{ji}| F_i < 0 \quad i=1,\cdots,n$$

where $\tau = \max_{i,j} \tau_{ij}$. Then u^* is globally exponentially asymptotically stable.

From Corollary 1, if condition (3) is satisfied and τ is small enough, the system is stable.

Picking $B^1 = 0$ and $B^3 = 0$, we have

Corollary 2. *If $g_j(\cdot)$, $f_j(\cdot)$ satisfy H, $j = 1, \cdots, n$, and there exist $\xi_i > 0$ such that*

$$-d_i\xi_i + \left[a_{ii}\xi_i + \sum_{j\neq i}|a_{ji}|\xi_j\right]^+ G_i + \sum_{j=1}^{n}|b_{ji}|\xi_j F_i < 0 \quad i = 1, \cdots, n \quad (7)$$

Then equilibrium u^* is globally exponentially asymptotically stable.

Remark 2. It can be seen that condition (7) is independent of delays, and results of [1,2,3,4,5] are all certain forms of Corollary 2.

3 Numerical Illustration

In this section, we present a delayed system, which is stable for small delays. Instead, it is unstable for large delays.

Consider following two-dimensional delayed system:

$$\dot{u} = -u(t) + Ag(u(t)) + cAf(u(t-\tau)) + I \quad (8)$$

where $u = [u_1, u_2]^T \in R^2$, $g(u) = [g(u_1), g(u_2)]^T$, $g(s) = (s + tanh(s))/2$, $f(u) = [f(u_1), f(u_2)]^T$, $f(s) = (s + arctan(s))/2$, $F_i = G_i = d_i = 1$, for $i = 1, 2$, $I = [-5, 10]^T$, $A = \begin{bmatrix} -1 & 2 \\ 1 & -2 \end{bmatrix}$, and c is a positive number.

It can be verified that if $c > \frac{2}{7}$, the condition (7) of Corollary 2 is not satisfied. All the criteria given in [1,2,3,4,5] fail to determine whether the system (8) is stable. Instead, the condition(3) given in Lemma 1 are satisfied for any constant $c > 0$. Now, picking $c = 2$ as an example, we get the unique equilibrium $u^* = [2.1522, 2.8478]^T$ for the system (8). By Theorem 1, using Matlab Optimization Toolbox, we conclude that if $\tau < \tau^* = 0.0203$, u^* is globally exponentially asymptotically stable. Picking $\tau = 0.02$, The figure on the left in $Fig.1$ indicates the dynamical behaviors of u_1 and u_2 through time with different initial data. One can see that $u(t)$ converges to u^* no matter whatever the initial data are.

Furthermore, we examine the dependence of the stability on the delay τ. Assume that initial data are $\phi_1(s) = 4$, $\phi_2(s) = -7$, $s \in [-\tau, 0]$. Let $u(t, \tau)$ denote the solution of the system (8) with delay τ and $y(\tau) = \|u(100, \tau) - u^*\|$. The figure on the right in $Fig.1$ indicates how $y(\tau)$ changes as τ increases. It can be seen that when τ is large enough (> 1.0), u^* is unstable. It is an open problem how to estimate the precise upper bound of τ^* such that the system is stable if $\tau < \tau^*$.

Acknowledgements. This work is supported by National Science Foundation of China 69982003 and 60074005, and also supported by Graduate Innovation Foundation of Fudan University.

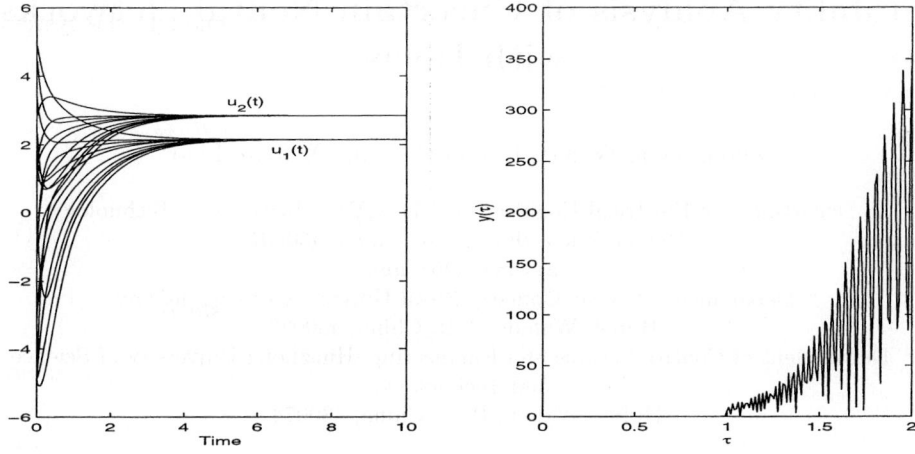

Fig. 1. The dynamics of $u_1(t)$ and $u_2(t)$ and the change of $y(\tau)$ depending on τ

References

1. Liao, X. X. and Wang, J : Algebraic Criteria for Global Exponential Stability of Cellular Neural Networks with Multiple Time Delays. IEEE Trans. CAS-1 **50** (1995), 268-275
2. Cao, J. : Global Asymptotical Stability of Delayed Bi-directional Associative Memory Neural Networks. Appl. Mtah. Comp., **142** (2003), 333-339
3. Zhang, J : Global Stability Analysis in Delayed Cellular Neural Networks. Comp. Math. Appl., **45** (2003), 1707-1720
4. Zhang, Q., et.al. : On the Global Stability of Delayed Neural Networks. IEEE Trans. Auto. Cont. **48** (2003), 794-797
5. Chen, T. P. : Global Exponential Stability of Delayed Hopfield Neural Networks. Neural Networks, **14** (2001), 977-980
6. Sabri Arik : Global Asymptotical Stability of a Large Class of Neural Networks with Constant time Delay. Phys. Lett. A, **311** (2003), 504-511
7. Lu, W. L., Rong, L. B., and Chen, T. P. : Global Convergence of DElayed Neural Network Systems. Inter. J. Neural Syst., **13** (2003), 193-204
8. Hochreiter, S. and Schmidhuber, J. : Long Short-term Memory. Neural Computation, **9** (1997), 1735-1780
9. Li, X. M., Huang, L. H., and Wu, J. H. : Further Results on the Stability of Delayed Cellular Neural Networks. IEEE Trans. CAS-1, **50** (2003), 1239-1242
10. Zhang, Q., Wei. X. P., and Xu. J. : Global Asymptotical Stability of Hopfield Neural Networks with Transmission Delays. Phys. Lett. A, **318** (2003), 399-405

Stability Analysis of Uncertain Neural Networks with Delay

Zhongsheng Wang[1], Hanlin He[2], and Xiaoxin Liao[3]

[1] Department of Electrical Engineering, ZhongYuan Institute of Technology,
Henan, Zhengzhou, P. R. China, 450007,
zswang3@263.net
[2] Department of Basic Courses, Naval University of Engineering,
Hubei, Wuhan, P. R. China, 430033
[3] Department of Control Science and Engineering, Huazhong University of Science and Technology,
Hubei, Wuhan, P. R. China, 430074.

Abstract. The globally uniformly asymptotic stability of uncertain neural networks with time delay has been discussed in this paper. Using the Razumikhin-type theory and matrix analysis method, A sufficient criterion about globally asymptotic stability of the neural networks is obtained.

1 Introduction

Neural networks has attracted the attention of the scientists, due to their promising potential for the tasks of classification, associate memory and parallel computation,etc., and various results were reported (see [1-9]). However, in hardware implementation, uncertainty and time delays occur due to disturbance between the electric components and finite switching speeds of the amplifiers, can affect the stability of a network by creating oscillatory and unstable characteristics. It is important to investigate the dynamics of uncertain neural networks with time delays. Most papers discussed the the stability of the neural networks with time delays and obtained some sufficient conditions for globally stability exponential stability, but there are few results of stability for the uncertain neural networks. In this paper,we will discuss the asymptotic behavior of uncertain neural networks with time delay,and achieve a sufficient criterion for globally asymptotic stability of the neural networks by unifying the Razumikhin-type theory and matrix analysis method.

2 System Description and Lemma

We consider the uncertain neural networks with time delay described by the differential-difference equation of the form

$$\begin{cases} \dot{x}(t) = [A + \Delta A(t)]x(t) + Bf(x(t-h(t))) \\ x(t) = \phi(t), t \in [-d, 0] \end{cases} \quad (1)$$

Where $x(t) \in R^n$ is the state of the neuron ; A, B is real constant matrices with appropriate dimension ; $h(t)$ is time-varying delay, satisfying $0 \leq h(t) \leq d$; $\phi(t)$ is a continuous vector-valued initial function defined on $[-d, 0]$; $f \in C(R^n, R^n)$ is bound function, the boundedness is l, that is $\|f(x)\| \leq l\|x\|$ and $f(0)) = 0$; uncertain parameters $\Delta A(t)$ satisfys the following

$$|\Delta A(t)| \prec A_1 \qquad (2)$$

where $|\Delta| \prec E$ means that $|\delta_{ij}| \leq e_{ij}$, δ_{ij} and e_{ij} denote, respectively, the ij-entries of Δ and E.

Lemma 1[10] : For $\alpha \in R^+$, define the $2n \times 2n$ Hamiltonia matrix as

$$H = \begin{bmatrix} A_0 & LL^T \\ -(\alpha+1) & -A_0^T \end{bmatrix} \qquad (3)$$

assume that (i) A_0 is a stable matrix and (ii) H has no eigenvalues on the imaginary axis, the Algebraic Ricatti Equation (ARE)

$$A_0^T P + P A_0 + P L L^T P + (\alpha+1) I_n = 0 \qquad (4)$$

has a positive definite solution P.

Lemma 2: for $E \in R^{n \times n}$, if $|E| \prec F$, then $EE^T \leq \Omega(F)$. Where $\Omega(F) = ndiag(FF^T)$, $A \leq B$ means that $A - B$ is semi-negative definite matrix.

Proof: Let $E = (e_{ij})_{n \times n}$, $F = (f_{ij})_{n \times n}$, for any $x = (x_1, x_2, \cdots, x_n)$,

$$x^T E E^T x = \sum_{k=1}^{n} (e_{1k} x_1 + e_{2k} x_2 + \cdots + e_{nk} x_n)^2$$

$$= \sum_{k=1}^{n} \sum_{i=1}^{n} e_{ik}^2 x_i^2 + \sum_{k=1}^{n} \sum_{i=1}^{n-1} \sum_{j=i+1}^{n} (e_{ik}^2 x_i^2 + e_{jk}^2 x_j^2)$$

$$\leq \sum_{k=1}^{n} \sum_{i=1}^{n} f_{ik}^2 x_i^2 + \sum_{k=1}^{n} \sum_{i=1}^{n-1} \sum_{j=i+1}^{n} (f_{ik}^2 x_i^2 + f_{jk}^2 x_j^2)$$

$$= \sum_{k=1}^{n} [\sum_{i=1}^{n} f_{ik}^2 x_i^2 + \sum_{i=1}^{n-1} \sum_{j=i+1}^{n} (f_{ik}^2 x_i^2 + f_{jk}^2 x_j^2)]$$

$$= n \sum_{i=1}^{n} \sum_{k=1}^{n} f_{ik}^2 x_i^2 = x^T ndiag(FF^T) x$$

that is $EE^T \leq \Omega(F)$.

3 Main Result

Let $A_0 = A, L = (\Omega^{1/2}(A_1), lB)$

Remark: $\Omega(A_1)$ can be decomposed to $\Omega^{1/2}(A_1) \Omega^{1/2}(A_1)$ due to $\Omega(A_1)$ being positive definite or semi-positive definite.

Theorem 1: The zero solution of system (1) is globally uniformly asymptotically stable if (i) A is stable and (ii) there exists $\alpha > 1$, H defined by (3) has no eigenvalues on the imaginary axis.

Proof: From the conditions (i) and (ii) of theorem 1, a positive definite matrix P can be obtained by solving Algebraic Ricatti Equation (4). Define a Lyapunov candidate

$$V(x(t)) = x^T P x$$

Taking the derivative along the trajectories of neural network (1)

$$\begin{aligned}
\dot{V}(x(t)) &= \dot{x}^T P x + x^T P \dot{x} \\
&= x^T(t)[PA + A^T P]x(t) + 2x^T(t)P\Delta A(t)x(t) + 2x^T P f(x(t-h(t))) \\
&\leq x^T(t)[PA + A^T P]x(t) + 2\left\|x^T(t)P\Delta A(t)\right\| \|x(t)\| \\
&\quad + 2\left\|x^T PB\right\| \|f(x(t-h(t)))\|
\end{aligned} \quad (5)$$

where $\|x\|$ denote the Euclid-norm of vector x.

Applying Lemma 2 and $2ab \leq a^2 + b^2$ (for any a and b) yields

$$\begin{aligned}
2\left\|x^T(t)P\Delta A(t)\right\| \|x(t)\| &\leq x^T P \Delta A(t) \Delta^T A(t) P x + \|x(t)\|^2 \\
&\leq x^T P \Omega(A_1) P x + \|x(t)\|^2
\end{aligned} \quad (6)$$

$$2\left\|x^T PB\right\| \|f(x(t-h(t)))\| \leq x^T P l^2 BB^T P x + \|x(t-h(t))\|^2 \quad (7)$$

Substituting equ.(6)-(7) into (5) yields

$$\begin{aligned}
\dot{V}(x(t)) &\leq x^T(t)[PA + A^T P + PLL^T P]x(t) \\
&\quad + \|x(t)\|^2 + \|x(t-h(t))\|^2.
\end{aligned}$$

Applying Lemma 1 and theorem A1 in Appendix [11] assume

$$\|x(t-h(t))\| \leq q \|x(t)\|, q > 1$$

then we can get the following bound on $\dot{V}(x(t))$

$$\dot{V}(x(t)) \leq -\omega \|x(t)\|^2$$

where $\omega = \alpha - q^2$, a sufficient small $q > 1$ exists such that $\omega > 0$. Thus, according to theorem A1 in Appendix, the zero solution of the system (1) is globally uniformly asymptotically stable.

4   Illustrative Example

Example: Consider the uncertain neural networks with time delay

$$\dot{x}(t) = \begin{bmatrix} -3 & r_1 \\ 0 & -4 \end{bmatrix} x(t) + \begin{bmatrix} 1 & 1 \\ 0 & 1 \end{bmatrix} \begin{bmatrix} \sin x_2(t-h(t)) & 0 \\ 0 & \sqrt{1-\cos x_1(t-h(t))} \end{bmatrix} \quad (8)$$

where r_1 is the uncertainty, $|r_1| \leq 0.3$, $h(t)$ is variable with time, $0 \leq h(t) \leq d$.

is stable,

$$A = \begin{bmatrix} -3 & 0 \\ 0 & -4 \end{bmatrix}$$

$$f(x(t-h(t))) = \begin{bmatrix} sinx_2(t-h(t)) & 0 \\ 0 & \sqrt{1-cosx_1(t-h(t))} \end{bmatrix}$$

$$\Delta A(t) = \begin{bmatrix} 0 & r_1 \\ 0 & 0 \end{bmatrix}$$

it is easy to know that $l = 1$,

$$\Omega(A_1) = \begin{bmatrix} 0.18 & 0 \\ 0 & 0 \end{bmatrix},$$

taking $\alpha = -1.1$ then

$$H = \begin{bmatrix} -3 & 0 & 2.18 & 1 \\ 0 & 4 & 1 & 1 \\ -2.1 & 0 & 3 & 0 \\ 0 & -2.1 & 0 & 4 \end{bmatrix}$$

has eigenvalues ± 1.9944, ± 3.7874, according to theorem 1, the zero solution of uncertain neural networks (8) is globally uniformly asymptotically stable.

Acknowledgement. This work is supported by National Natural Science Foundation (60274007) of China.

References

1. Liang, X. B. and Wu, L. D.: Globally Exponential Stability of Hopfield Neural Networks and Its Applications. Sci. China (series A), **25** (1995) 523-532
2. Liang, X. B. and Wu, L. D.: Globally Exponential Stability of A Class of Neural Circuits. IEEE Trans. on Circ. and Sys. I: Fundanmetntal Theory and Application, **46** (1999) 748-751
3. Cao, J. D. and Li, Q.: On the Exponential Stability and Periodic Solution of Delayed Cellular Neural Networks. J. Math. a Anal. and Appl., **252** (2000) 50-64
4. Liao, X. X. and Xiao, D. M.: Globally Exponential Stability of Hopfield Neural Networks with Time-varying Delays. ACTA Electronica Sinica, **28** (2000) 1-4
5. Arik, S. and Tavsanoglu, V.: On the Global Asymptotic Stability of Delayed Cellular Neural Networks. IEEE Trans. Circuits Syst. I, **47** (2000) 571-574
6. Liao, T. L. and Wang, F. C.: Global Stability for Cellular Neural Networks with Time Delay. IEEE Trans. Neural Networks, **11** (2000) 1481-1484
7. Zhang, Y., Heng, P. A. and Leung, K. S.: Convergence Analysis of Cellular Neural Networks with Unbounded Delay. IEEE Trans. Circuits Syst. I, **48** (2001) 680
8. Li, S. Y. and Xu, D. Y.: Exponential Attractions Domain and Exponential Convergent Rate of Associative Memory Neural Networks with Delays. Control Theory and Applications, **19** (2002) 442-444

9. Zeng, Z. G., Fu, C. J. and Liao, X. X.: Stability Analysis of Neural Networks with Infinite Time-varying Delay. J. of Math.(PRC), **22** (2002) 391-396
10. Doyle, J. C., Glover, K., Khargonekar, P. P. and Francis B. A.: State-space Solution to Standard H_2 and H_∞ Control Problem. IEEE Transaction on Automatic Control, **34** (1989) 831-846
11. Xu, B. G. and Liu, Y. Q.: An Improved Razumikhin Type Theorem and Its Applications. IEEE Transaction on Automatic Control, **39** (1994) 839-841

Appendix

An improved Razumikhin Type Theorem in [11].

Let R^n denote an n-dimensional linear vector space over the reals with the norm $\|\cdot\|$, $R = (-\infty, +\infty)$, $R^+ = [0, +\infty)$, $J = [\sigma, +\infty)$, and $J_1 = [\sigma - r, +\infty)$ for $r > 0$, where $h(t) \in R^+$. Let $C_n = C([-r, 0], R^n)$ be the Banach space of continuous functions mapping the interval $[-r, 0]$ into R^n with the topology of uniform convergence. For given $\phi \in C_n$, we define $\|\phi\| = \sup_{-r \leq \theta \leq 0} \|\phi(\theta)\|$, $\phi(\theta) \in R^n$. Let $x_t \in C_n$ be defined by $x_t(\theta) = x(t+\theta), \theta \in [-r, 0]$. Consider the initial-value problem for retarded functional differential equation(RFD)

$$\dot{x}(t) = f(t, x_t), t \geq t_0 \in J, \sigma \in R \quad (A1)$$

where $f : J \times C_n^H \to R^n$ is completely continuous, $C_n^H = \{\phi \in C_n, \|\phi\| \leq H\}, H > 0$, and $f(t, 0) = 0$ for all $t \in J$. We also suppose that for any $\phi \in C_n^H$, and for $t_0 \in J$ system (A1) possesses a unique solution with $x_t = \phi$ and the value of $x_t(t_0, \phi)$ at t is denoted by $x(t) = x(t_0, \phi)(t)$. When global results are considered, we always suppose that $C_n^H = C_n$.

Theorem A1 : If there exists a continuous function $V : J_1 \times R^n \to R^+$ and positive numbers α_1, α_2 and α_3 such that
1) $\alpha_1 \|x\|^2 \leq V(t, x) \leq \alpha_2 \|x\|^2$
for all $x \in R^n, t \in [\sigma - r, +\infty)$.
2) there exists a positive $q > 1$ such that

$$\dot{V}(t, x) \leq -\alpha_3 \|x\|^2$$

for any $t_0 \in J$ and for the class of solution $x(t)$ satisfying

$$\|x(t+\theta)\| \leq q \|x(t)\|, \theta \in [-r, 0], t \geq t_0$$

then, the zero solution of (A1) is globally uniformly asymptotically stable

A New Method for Robust Stability Analysis of a Class of Recurrent Neural Networks with Time Delays

Huaguang Zhang and Gang Wang

Institute of Information Science and Engineering,
Northeastern University, Shenyang, Liaoning, 110004, P. R. China
Hg_zhang@21cn.com   Erelong@sohu.com

Abstract. In this paper, based on the state equations of neural networks and existing theorems, we discuss a class of recurrent neural networks with time delays and investigate their robust stability of the equilibrium point for this system. A new method to analyze the robust stability under disturbance of structure of recurrent neural networks with time delays is presented, and a sufficient condition guaranteeing the robust stability is derived, which provides the theoretical basis for designing a system.

1 Introduction

In recent years, neural networks have attracted considerable attention due to their promising applications in image processing, optimization, and associative memories, ect [1,2,3,4]. In practice, time delays are often encountered in the interaction between the neurons, which would affect the stability of the system. Meantime, there are some uncertainties such as perturbations and component variations, which might lead to very complex dynamical behaviors. Thus it is very important to investigate the robust stability of this system against the modelling errors and parameter fluctuation.

A neural network with time delays can be described by the state equation

$$\frac{dx_i(t)}{dt} = -d_i x_i(t) + \sum_{j=1}^{n} a_{ij} f_j(x_j(t)) + \sum_{j=1}^{n} b_{ij} g_j(x_j(t-\tau_j)) + I_i, \quad i=1,2,...,n. \quad (1)$$

The general method of analyzing the robust stability is to construct Lyapunov function, obtain the global asymptotic stability of the equilibrium point of (1), and get the global robust stability conditions [5].

It is mentioned that most of existing papers concerning with the robust stability of neural network with time delays were done by intervalizing the coefficients d_i and connection matrices as follows [6,7]

$$d_i := \{\underline{d_i} \leq d_i \leq \overline{d_i}, \quad i=1,2,...,n\},$$
$$a_{ij} := \{\underline{a_{ij}} \leq a_{ij} \leq \overline{a_{ij}}, \quad i,j=1,2,...,n\}, \quad (2)$$
$$b_{ij} := \{\underline{b_{ij}} \leq b_{ij} \leq \overline{b_{ij}}, \quad i,j=1,2,...,n\}.$$

In this paper, based on the state equations of neural networks and existing theorems, we discuss a class of recurrent neural networks with time delays and investigate their robust stability of the equilibrium point for this system. A new method to analyze the robust stability under disturbance of structure of recurrent neural networks with time delays is presented, and a sufficient condition guaranteeing the robust stability is derived without use of the intervalizing (2). These conditions are in the form of matrix.

2 The Model Description

In this paper, we consider a class of recurrent neural networks with time delays and Lipschitz continuous activation functions described by (1). When convenient, we will also use the matrix format of equation (1) as follows

$$\frac{dx(t)}{dt} = -Dx(t) + Af(x(t)) + Bg(x(t-\tau)) + I, \qquad (3)$$

where $D = diag(d_1, d_2, ..., d_n)$ represents the linear self-feedback coefficient and $d_i > 0$, $A = (a_{ij})_{n \times n}$ and $B = (b_{ij})_{n \times n}$ are the connection weight matrix and delayed connection weight matrix, respectively, $I = (I_1, I_2, ..., I_n)^T$ is the external bias, $f(x) = (f_1(x_1), f_2(x_2), ..., f_n(x_n))^T$ and $g(x(t-\tau)) = (g_1(x_1(t-\tau_1)), g_2(x_2(t-\tau_2)), ..., g_n(x_n(t-\tau_n)))^T$ denote the activation functions. It is also assumed that there exists a T such that $0 \leq \tau_j \leq T$, $j = 1, 2, ...n$. The activation function are assumed to be bounded on R and are Lipschitz continuous, i.e., there exists $\mu_i > 0$ such that

$$\begin{aligned} |f_i(u) - f_i(v)| &\leq \mu_i |u-v|, \\ |g_i(u) - g_i(v)| &\leq \mu_i |u-v|, \end{aligned} \qquad (4)$$

for every $u, v \in R$. We denote $K = L = diag\{\mu_1, \mu_2, ..., \mu_n\}$. It is noted that the activation functions commonly used in the literature, such as the sigmoidal functions, the hyperbolic tangent function, and the piece-wise linear saturation function, satisfy these conditions.

Remark 1. It is obvious that the cellular neural network, presented by Chua L.O. and Yang L. [1], is a special case of equation (3). Therefore, the results obtained from recurrent neural networks with time delays can be applied on cellular neural networks.

To proceed, we give the following lemmas.

Lemma 1 [8]. Assume (4) is satisfied, the equilibrium point of the neural network model (3) is global asymptotically stable if $G + G^T$ is a nonsingular M-matrix, where $G := D - |A|K - |B|L$.

Now, assume that the recurrent neural network (3) has an equilibrium at x^*, i.e., x^* satisfies

$$-Dx^* + Af(x^*) + Bg(x^*) + I = 0.$$

For robust stability analysis of recurrent neural networks (3), it is often convenient to rewrite (3) as

$$\frac{dz(t)}{dt} = -Dz(t) + AF(z(t)) + BG(z(t-\tau)), \qquad (5)$$

where $z = x - x^*$, $F(z) = f(z) - f(x^*)$, and $G(z(t-\tau)) = (g_1(x_1(t-\tau_1)) - g_1(x_1^*), g_2(x_2(t-\tau_2)) - g_2(x_2^*), \ldots, g_n(x_n(t-\tau_n)) - g_n(x_n^*))^T$. Functions $F(\cdot)$ and $G(\cdot)$ defined here are bounded and satisfy the conditions given in (4).

Recurrent neural networks described by (5) are equivalent to neural networks (1) and (3). The robust stability analysis of the equilibrium point x^* of (3) can now be transformed to the robust stability analysis of the equilibrium $z = 0$ of (5).

3 Robust Stability

For non-disturbed system described by

$$\frac{dz}{dt} = f(z,t), \qquad (6)$$

where $f \in C[G_H, R^n], z \in R^n, f(z,t) \equiv z$ iff $z = 0$, let $G_H = \{(z,t), t \geq t_0, \|z\| < H\}$, and consider disturbed system as follow

$$\frac{dz}{dt} = f(z,t) + g(y,t), \qquad (7)$$

where $g \in C[G_H, R^n]$, g is essentially unknown, and bounded.

Lemma 2 [9]. If there exists a scalar function $V(z,t) \in C^1[G_H, R]$, which satisfies
1) $\varphi_1(\|z\|) \leq V(z,t) \leq \varphi_2(\|z\|)$ where $\varphi_1(\|z\|)$ and $\varphi_2(\|z\|)$ are continuous non-decreasing function.
2) $D^+V(z,t)|_{(6)} \leq -cV(z,t), \ c > 0$
3) $|V(z,t) - V(y,t)| \leq k|z-y|, \ \forall z, y \in G_H$

then the null solution of equation (6) is robust stable under structure disturbance.

We now establish our robust stability theorem.

Theorem 1. Assume (4) is satisfied, the equilibrium point $z = 0$ of the recurrent neural network model (5) is robust stable under structure disturbance if $G + G^T$ is a nonsingular M-matrix, where $G := D - |A|K - |B|L$.

Proof. It suffices to show that the equilibrium point $z = 0$ of (5) is robust stable.

Since $D - |A|K - |B|L$ is a nonsingular M-matrix, i.e., there exists constants $r_i > 0$ such that

$$r_i d_i > \mu_i \sum_{j=1}^{n} r_j |a_{ji}| + \mu_i \sum_{j=1}^{n} r_j |b_{ji}|, \quad i=1,2,\ldots,n. \tag{8}$$

(8) implies that we can always choose small constant, $c > 0$ and ε, such that

$$r_i(\varepsilon - d_i + c) + \mu_i \sum_{j=1}^{n} r_j |a_{ji}| + e^{\varepsilon\tau} \mu_i (1 + \frac{c}{\varepsilon}) \sum_{j=1}^{n} r_j |b_{ji}| \leq 0, \tag{9}$$

where $\tau = \max_{1 \leq j \leq n} \tau_j$.

We define the following scalar function $V(z,t)$,

$$V(z,t) = \sum_{i=1}^{n} r_i [|z_i(t)| e^{\varepsilon t} + \sum_{j=1}^{n} |b_{ij}| \mu_j \int_{t-\tau_j}^{t} |z_j(s)| e^{\varepsilon(s+\tau_j)} ds]. \tag{10}$$

Firstly, we will prove that the condition 2) of **lemma 2**, i.e., $D^+V(z,t)|_{(5)} \leq -cV(z,t)$, $c > 0$, is hold.

Calculating the upper right derivative $V(z,t)$ along the solutions of the recurrent neural network (5), one can derive that

$$\nabla_{(5)}(z(t),t) = \sum_{i=1}^{n} r_i \left[|z_i(t)| \varepsilon e^{\varepsilon t} + |z_i(t)|' e^{\varepsilon t} + \sum_{j=1}^{n} |b_{ij}| \mu_j (|z_j(t)| e^{\varepsilon t} e^{\varepsilon \tau_j} - |z_j(t-\tau_j)| e^{\varepsilon t}) \right]. \tag{11}$$

Because

$$|z_i(t)|' = \frac{z_i(t)}{|z_i(t)|}(-d_i z_i(t) + \sum_{j=1}^{n} a_{ij} F_j(z_j(t)) + \sum_{j=1}^{n} b_{ij} G_j(z_j(t-\tau_j))). \tag{12}$$

Substitute (12) to (11), so

$$\nabla_{(5)}(z(t),t) \leq e^{\varepsilon t} \sum_{i=1}^{n} r_i [(\varepsilon - d_i)|z_i(t)| + \sum_{j=1}^{n} |a_{ij}| \|F_j(z_j(t))\| + \sum_{j=1}^{n} |b_{ij}| \|G_j(z_j(t-\tau_j))\| + \sum_{j=1}^{n} |b_{ij}| \mu_j |z_j(t)| e^{\varepsilon \tau_j} - \sum_{j=1}^{n} |b_{ij}| \mu_j |z_j(t-\tau_j)|]. \tag{13}$$

According to (4), or in this case, $|F_j(z_j)| \leq \mu_j |z_j|$ and $|G_j(z_j)| \leq \mu_j |z_j|$, then

$$\nabla_{(5)}(z(t),t) \leq e^{\varepsilon t} \sum_{i=1}^{n} r_i [(\varepsilon - d_i)|z_i(t)| + \sum_{j=1}^{n} |a_{ij}| \mu_j |z_j(t)| + \sum_{j=1}^{n} |b_{ij}| \mu_j |z_j(t)| e^{\varepsilon \tau}]$$
$$\leq e^{\varepsilon t} [\sum_{i=1}^{n} r_i (\varepsilon - d_i)|z_i(t)| + \sum_{i=1}^{n}\sum_{j=1}^{n} r_i |a_{ij}| \mu_j |z_j(t)| + \sum_{i=1}^{n}\sum_{j=1}^{n} r_i |b_{ij}| \mu_j |z_j(t)| e^{\varepsilon \tau}]. \tag{14}$$

$$\leq e^{\varepsilon t}[\sum_{i=1}^{n}r_i(\varepsilon-d_i)|z_i(t)|+\sum_{i=1}^{n}\mu_i|z_i(t)|\sum_{j=1}^{n}r_j|a_{ji}|+\sum_{i=1}^{n}\mu_i|z_i(t)|\sum_{j=1}^{n}r_j|b_{ji}|e^{\varepsilon\tau}].$$

i.e.

$$\nabla_{(5)}(z(t),t)\leq e^{\varepsilon t}\sum_{i=1}^{n}[r_i(\varepsilon-d_i)+\mu_i\sum_{j=1}^{n}r_j|a_{ji}|+\mu_i\sum_{j=1}^{n}r_j|b_{ji}|e^{\varepsilon\tau}]|z_i(t)|. \tag{15}$$

Add $cV(z,t)$ to the two side of (15),

$$\nabla_{(5)}(z(t),t)+cV(z,t)\leq e^{\varepsilon t}\sum_{i=1}^{n}[r_i(\varepsilon-d_i+c)$$
$$+\mu_i\sum_{j=1}^{n}r_j|a_{ji}|+\mu_i\sum_{j=1}^{n}r_j|b_{ji}|e^{\varepsilon\tau}]|z_i(t)|+c\sum_{j=1}^{n}\mu_j\int_{t-\tau_j}^{t}|z_j(s)|e^{\varepsilon(s+\tau_j)}ds\sum_{i=1}^{n}r_i|b_{ij}|. \tag{16}$$

Next, we discuss the definite integral term of (16)

$$\int_{t-\tau_j}^{t}|z_j(s)|e^{\varepsilon(s+\tau_j)}ds=\frac{1}{\varepsilon}[|z_j(s)|e^{\varepsilon(s+\tau_j)}|_{t-\tau_j}^{t}-\int_{t-\tau_j}^{t}e^{\varepsilon(s+\tau_j)}d|z_j(s)|]$$
$$\leq\frac{1}{\varepsilon}[|z_j(t)|e^{\varepsilon(t+\tau_j)}-|z_j(t-\tau_j)|e^{\varepsilon t}]\leq\frac{1}{\varepsilon}|z_j(t)|e^{\varepsilon(t+\tau)}. \tag{17}$$

According to (17), then

$$\nabla_{(5)}(z(t),t)+cV(z,t)\leq e^{\varepsilon t}\sum_{i=1}^{n}[r_i(\varepsilon-d_i+c)$$
$$+\mu_i\sum_{j=1}^{n}r_j|a_{ji}|+\mu_i\sum_{j=1}^{n}r_j|b_{ji}|e^{\varepsilon\tau}]|z_i(t)|+c\sum_{j=1}^{n}\mu_j\frac{1}{\varepsilon}|z_j(t)|e^{\varepsilon(t+\tau)}\sum_{i=1}^{n}r_i|b_{ij}| \tag{18}$$
$$=e^{\varepsilon t}\sum_{i=1}^{n}[r_i(\varepsilon-d_i+c)+\mu_i\sum_{j=1}^{n}r_j|a_{ji}|+e^{\varepsilon\tau}\mu_i(1+\frac{c}{\varepsilon})\sum_{j=1}^{n}r_j|b_{ji}|]|z_i(t)|.$$

According to (9), then

$$\nabla_{(5)}(z(t),t)+cV(x,t)\leq 0. \tag{19}$$

Secondly, let

$$\alpha(\|z\|)=\sum_{i=1}^{n}r_i|z_i(t)|e^{\varepsilon t},\forall t\geq t_0, \tag{20}$$

$$\beta(\|z\|)=\sum_{i=1}^{n}r_i[|z_i(t)|e^{\varepsilon t}+\sum_{j=1}^{n}|b_{ij}|\mu_j q\tau_j],\forall t\geq t_0, \tag{21}$$

where $q=\max\{|z_j(t)|e^{\varepsilon(t+\tau_j)},|z_j(t-\tau_j)|e^{\varepsilon t}\}$.

It is obvious that they are monotone non-decreasing function of $|z_i(t)|$ and

$$\alpha(\|z\|) \leq V(z,t) \leq \beta(\|z\|). \tag{22}$$

Thirdly, $V(z,t)$ satisfies (4), i.e., $\forall z, y \in G_H$,

$$|V(z,t) - V(y,t)| \leq k|z-y|. \tag{23}$$

The equilibrium point $z = 0$ of recurrent neural network described by equation (5) is robust stable under structure disturbance.

If we discuss the cellular neural network with time delays, that is, let $\mu_i = 1, f_i(x) = g_i(x) = 0.5(|x+1| - |x-1|)$. Hence we have the following corollary.

Corollary 1. Assume (4) is satisfied, the equilibrium point x^* of the cellular neural network model (3) is robust stable under structure disturbance if $G + G^T$ is a nonsingular M-matrix, where $G := D - |A| - |B|$.

4 Conclusions

This paper discusses the robust stability under disturbance of structure of recurrent neural networks with time delays. A sufficient condition for robust stability is presented in the form of matrix, which is different from those results given in previous paper [6,7]. The sufficient condition for robust stability is independent of time delays. The results and methods of analysis and solutions proposed in this paper could be extended to other neural networks.

References

1. Chua, L.O., Yang, L.: Cellular Neural Networks: Theory. IEEE Trans. on Circuits and System, Vol. 35. (1988) 1257-1272
2. Chua, L.O., Yang, L.: Cellular Neural Networks: Application. IEEE Trans. on Circuits and System, Vol. 35. (1998) 1273-1290
3. Matsumoto, T., Chua, L.O., Yokohama, T.: Image Thinning with a Cellular Neural Network. IEEE Trans. on Circuits and System, Vol. 37. (1990) 638-640
4. Michel, A.N., Liu, D.R.: Qualitative Analysis and Synthesis of Recurrent Neural Network. Mercel Dekker, Inc, 270 Madison Avenue, New York, NY 10016 (2002)
5. Singh, V.: Robust Stability of Cellular Neural Networks with Delay: Linear Matrix Inequality Approach. IEE Proc.-Control Theory Appl., Vol. 151. (2004) 125-129
6. Liao, X., Yu, J.: Robust Stability for Interval Hopfield Neural Networks with Time Delays. IEEE Trans. Neural Networks, Vol. 9. (1998) 1042-1045
7. Arik, S.: Global Robust Stability of Delayed Neural Networks. IEEE Trans. on Circuits and Systems, Vol. 50. (2003) 156-160
8. Cao, J., Wang, J.: Global Asymptotic Stability of a General Class of Recurrent Neural Networks with Time-varying Delays. IEEE Trans. on Circuits and Systems, Vol. 50. (2003) 34-44
9. Liao, X.X.: Theory and Application of Stability for Dynamical Systems. National Defence Industry Press, No23, Haidian District, Beijing (2001)

Criteria for Stability in Neural Network Models with Iterative Maps

Carlos Aguirre, Doris Campos, Pedro Pascual, and Eduardo Serrano

GNB, Escuela Politécnica Superior, Universidad Autonoma de Madrid,
28049 Madrid, Spain
{Carlos.Aguirre,Doris.Campos,Pedro.Pascual,Eduardo.Serrano}@ii.uam.es

Abstract. Models of neurons based on iterative maps allows the simulation of big networks of coupled neurons without loss of biophisical properties such as spiking, bursting or tonic bursting and with an affordable computational effort. In this work we explore by means of the use of Lyapunov "energy" functions the asymptotic behavior of a set of coupled neurons where each neuron is modelled by an iterative map. The method here developed allows to establish conditions on the parameters of the system to achieve asymptotic stability and can be applied to different models both of neurons and network topologies.

1 Introduction

During the last few years there has been an increasing interest in the global behavior of ensembles of spiking neurons. Most of these studies consider models of neurons based on differential equations, see [1] for a complete review of these models. The models based on differential equations need a high computational effort to reproduce neuronal behavior such as spiking or bursting. Well known differential models such as the Hodking-Huxley (HH) model [2], or the the Hindmarsh-Rose (HR) model [3] require a number of floating point operations that range from 1200 operations in the case of the HH model to 70 operations in the HR model to simulate a single neuron in the network for 1 ms. This means that the simulation of the behavior of a neural network composed for thousand of neurons for even not very long periods of time is computationally inviable.

Recently, some models have solved this drawback of the differential models [9,10]. These new models are based on iterative maps that can present the same neuro–computational properties that the differential models with a very low computational effort that makes possible the simulation of big ensembles of coupled neurons during relatively long periods of time. Besides some of these models like [9] are biophysically meaningful in such a way that the parameters of the model can be selected in order to obtain a characteristic neuron behavior, such as, spiking, tonic spiking, bursting, etc.

Another important issue in the study of network dynamics is the connection topology of the network. There are some well known biological neural networks that present a clear clustering in their neurons but have small distances between each pair of neurons. These kind of highly clustered, highly interconnected sparse networks are known as Small-World (SW) networks [4]. As initial substrate for the generation of SW, the use of undirected and unweighted ring-lattices or grids is usually proposed [4]. They are used because these graphs are connected, there are not specific nodes on them, present a good transition from regular to random, which is goberned by the probability p of rewiring each edge of a original regular substrate and model a high number of real networks. Biological or artificial neural networks are not accurately represented by these models as neural networks present a clear directionality and a different coupling in their connections. In [7] a biologically inspired SW model is presented, this model includes important characteristics of biological networks such as directionality and weight in the neuronal connections.

The stability of networks where each neuron is modelled by a differential equation is an important issue and has been studied on several works [5,6]. In [5] the stability of an ensemble of differential neurons is studied when the connection topology follows the model of connectivity presented in [7]. In this work we explore by means of the use of Lyapunov "energy" functions the asymptotic behavior of a set of coupled neurons where each neuron is modelled by an iterative map. The method here developed allows to establish conditions on the parameters of the system to achieve asymptotic stability on the network and can be applied to different models both of neurons and network topologies.

2 Neuron Models

The dynamics of each neuron in the network is described by an iterative map of the form:

$$\boldsymbol{x}(t+1) = A\boldsymbol{x}(t) + F(W, \boldsymbol{x}(t), S) \quad (1)$$

where $\boldsymbol{x}(t) = (x_1(t), x_2(t), \cdots, x_N(t))$ is the network state, $A = (a_1, a_2, \cdots, a_N)$ is a diagonal positive matrix verifiyng $a_i < 1$, $S = (S_1, S_2, \cdots, S_N)$ is a constant input vector, F is a function representing the neuron input provinient from other neurons in the network and, finally $W = w_{ij}$ is the connection matrix of the network and defines the connection topology on the network. There are several models of neural networks in the bibliography that follows the previous scheme. The simplest model is the discrete integrate and fire model defined by the one dimensional map for each cell in the network in the following way:

$$x_i(n+1) = a_i x_i(n) + \sum_j w_{ij} g_j(x_j(n)) + S_i \quad (2)$$

With g an activation function that verifies $g(0) = 0$ and verifying a sector condition of the form

$$0 \leq \frac{g_i(b) - g_i(a)}{b - a} \leq k, \quad \forall\, a, b \in R, i = 1, \cdots, N \quad (3)$$

for some constant k. In particular

$$0 \leq \frac{g_i(x_i(n+1)) - g_i(x_i(n))}{x_i(n+1) - x_i(n)} \leq k, \quad i = 1, \cdots, N \qquad (4)$$

for some constant k

The stability of the differential form of this model has been widely studied in [5]. Some results for this discrete model about encoding rhythms for different network topologies can be found at [8].

3 Criteria for Stability

In the following we shift the equilibrium x^e of the network to the origin. For example in the model given by eq. (2) it is sufficient to define $y(t) = x(t) - x^e$. With this change of variable eq. (2) becomes:

$$x_i(n+1) = a_i x_i(n) + \sum_j w_{ij} f_j(x_j(n)), \text{ with } i \neq j \qquad (5)$$

where $f_i(x_j) = g_i(x_j + x_j^e) - g_i(x_j^e)$.
As in [5] we select a Lyapunov function of the form

$$V(x) = \sum_i \int_0^{x_i} f_i(s) ds \qquad (6)$$

Following [11] it can be shown that V is a Lyapunov function. If we now define a function

$$G(u) = \min_i \left(\min \left(\int_0^u f_i(s) ds, \int_0^{-u} f_i(s) ds \right) \right) \qquad (7)$$

we have $G(0) = 0$, $G(|u|) = G(u)$, $G(r) > 0$ if $r > 0$ and $G(r) \to \infty$ when $r \to \infty$.

As $V(x) = \sum_i \int_0^{x_i} f_i(s) ds$ we have that

$$V(x) \geq \sum_i G(|x_i|) \geq G(|x|). \qquad (8)$$

Where $|u|$ represents $\sum_i |u_i|$. It is possible also to show [11]

$$G(|x|) \leq V(|x|) \leq qk\|x\|^2 \text{ with } q > 1 \qquad (9)$$

If we now calculate increment of $V(x)$ along the orbits of Eq. (5) and using that f also satisfies a sector condition of the form

$$f_i(x_i)[f_i(x_i - kx_i)] \leq 0, \quad i = 1, 2, \cdots N \qquad (10)$$

we have

$$\Delta V(n) \leq f^T(x(n))[(A - I)/k + W]f(x(n)) \qquad (11)$$

where the index n represents the discrete time index and I represents the identity matrix. Therefore

$$\Delta V(n) \leq \lambda_{max}((A-I)/k + W) \| f(\boldsymbol{x}(n)) \|^2 \qquad (12)$$

where $\lambda_{max}((A-I)/k+W)$ is the largest eigenvalue of matrix $(A-I)/k+W$.

From the previous result we can say that the network defined by equation (5) is asymptotically stable around the origin if $\lambda_{max}((A-I)/k + W) \leq 0$ or equivalently $\lambda_{max}(W) < 1/k - min_i(a_i/k)$, as $(A-I)$ is a negative diagonal matrix. For the [9,10] discrete models a similar construction is out of the scope of this paper but the previous method can be followed, provided that a good Lyapunov function candidate V is found.

4 Stability in Small–World Networks

The connection matrix W determines the topology of the network. An interesting model is the Small-World (SW) model. The original SW model is described in [4], and can be resumed as follows. Take a one–dimensional undirected unweighted lattice with N nodes, where each node is connected to the k closest neighbors. Then rewire each connection in the lattice with a probability p, avoiding self connections or double connections between nodes. Here we use the model of SW presented in [7]. In this model, each edge has a direction and a positive excitatory random weigh. A variation of the previous model is also considered; in this variation both positive and negative weights are allowed, therefore introducing an inhibitory effect in the network. Another adventage of this model is that it does not introduce new edges in the graph as p grows.

We study the behavior of the largest eigenvalue of W for excitatory and excitatory–inhibitory networks as a function of the probability p and as a function of the number of nodes N. The average degree of each node is taken constant with a value of 8. Each connection weight $w_{i,j}$ is generated following a random uniform ditribution between $[0, 2*\delta]$ in the case of excitatory networks and $[-\delta, \delta]$ in the case of excitatory–inhibitory networks. Disconneted graphs or graphs that does not provide a complete set of real eigenvalues are discarded. Each plot is the average of 50 experiments.

5 Results

Figures (1) and (2) show the numerical values of the largest eigenvalue of W. In our model the stability depends more clearly on the number of nodes in the network N than in the probability p. The largest eigenvalue λ_{max} grows almost linearly with the number of nodes N. This means that for a fixed value of p there exists a value N' that verifies that if $0 \leq N \leq N'$ the network is asimptotically stable but for $N > N'$ the network is not guaranteed to be stable. This result seems to contradict the results obtained in [5] for the continuum model. The difference is in the use of different models for the construction of the SW network. In [5], the model of SW selected is a model proposed by Newman and Watts [12]. In the Newman model, for each pair of nodes, an edge is added with prability p, this model has the advantage of avoiding disconnected graphs but has the drawack that the number of edges grows linearly with p. The Newman model is similar to the SW model only when $p \ll 1$; for $p \approx 1$ we obtain a densely connected graph instead of a random graph meanwhile it does not changes significantly with the change of p.

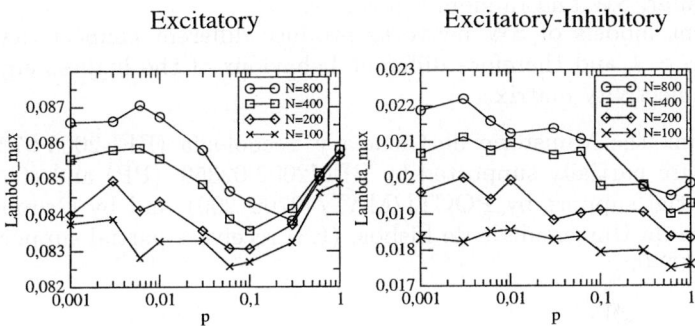

Fig. 1. Largest eigenvalue of W as function of p for only excitatory (left) and excitatory–inhibitory (right) networks, $\delta = .01$.

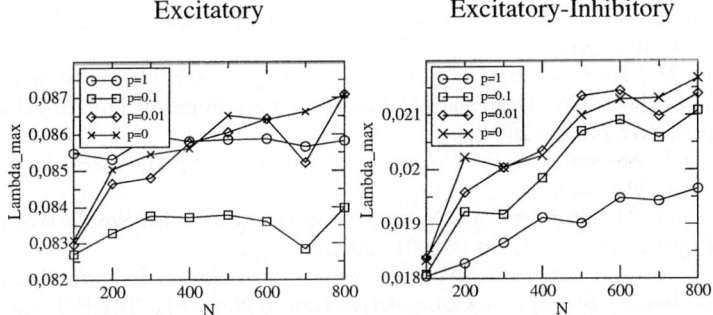

Fig. 2. Largest eigenvalue of W as function of N for only excitatory (left) and excitatory–inhibitory (right) networks, $\delta = .01$.

In our model the stability on the network it is not clearly determined by the value of p, as it can be seen in figure (1), the value of λ_{max} does not grow with the value of p, in fact it has a minimum close to the SW area $p \approx .1$ in the case of the excitatoty networks or has a minimum in the random area $p \approx 1$ in the case of excitatory–inhibitory networks. This again seems to contradict the results of [5]; the reason of this contradiction is the different meaning of the parameter p. In our experiments p only controls the "randomness" of the network, mantaining constant the other parameters of the model.

6 Conclusions

The previous results allow us to establish the following conclusions.

- The stability of a network of neurons modelled by an integrate and fire iterative map is determined by the largest eigenvalue of the connectivity matrix.
- The number of neurons in the network determines the stability of the network for regular, SW and random topologies.
- Different models of SW networks produce different connectivity matrixes when $p \approx 1$ and therefore different behaviour of the highest eigenvalue of the connectivity matrix.

We thank the Ministerio de Ciencia y Tecnología (BFI-2003-07276). (PP) and (CA) are partially supported by BFM2002-02359. (PP) and (CA) also receive a partial support by POCTI/MAT/40706/2001 and by Grupo de Fisica Matematica da Universidade de Lisboa. (ES) receive a partial support by (TIC 2002-572-C02-02).

References

1. Izhikevich, E. M.: Neural Excitability, Spiking and Bursting. International Journal of Bifurcation and Chaos **10** (2000) 1171–1266.
2. Hodgkin, A. L., Huxley, A. F.: A Quantitative Description of Membrane Current and Application to Conduction and Excitation in Nerve. Journal of Physiology **117** (1954) 165–181.
3. Rose, R. M., Hindmarsh, J. L.: The Assembly of Ionic Currents in a Thalamic Neuron, I The Three Dimensional Model. Proceedings of The Royal Society of London B, **237** (1989) 267–288.
4. Watts, D.J., Strogatz, S. H.: Collective Dynamics of Small–World Networks. Nature. **393** (1998) 440.
5. Li, C., Chen, G.: Stability of a Neural Network Model with Small–World Connections. Physical Review E **68** 052901 (2003).
6. Cohen, M. A., Grossberg S.: Absolute Stability of Global Pattern Formation and Parallel Memory Storage by Competitive Neural Networks. IEEE Transactions on Syst. Man. and Cybern. **13** (1983) 815–826.
7. Aguirre C. , Huerta R. , Corbacho F., Pascual P.: Analysis of Biologically Inspired Small–World Networks. Artificial Networks-ICANN 2002, Lecture Notes in Computer Science, Springer (2002) 27–32.
8. Aguirre C. , Campos, D. , Pascual P., Serrano, E.: Effects of Different Connectivity Patterns in a Model of Cortical Circuits. IWANN 2003, LNCS, Springer (2003).
9. Rulkov, N. F.: Modeling of Spiking–bursting Neural Behavior Using Two–dimensional Map. Physical Review E **65** 041922 (2002).
10. Izhikevich, E. M.: Simple Model of Spiking Neurons. IEEE Trans. on Neural Networks **68** 052901 (2003).
11. Liao, X. , Chen, G., Sanchez, E. N.: LMI-based Approach for Asymptotically Stability Analysis of Delayed Neural Networks. IEEE Transactions on Circuits and Systems I: Fundamental Theory and Applications, **49** 7 (2002) 1033–1049.
12. Newman, M. E. J., Watts, D. J.: Scaling and Percolation in the Small–World Network Model. Phys. Rev. E, **60** 6 (1999) 7332–7342.

Exponential Stability Analysis for Neural Network with Parameter Fluctuations

Haoyang Tang[1], Chuandong Li[2], and Xiaofeng Liao[2]

[1] College of Business Administration, Chongqing University 400030, China
[2] Department of Computer Science and Engineering, Chongqing University 400030, China
cd_licqu@163.com

Abstract. The stability of a neural network model may often be destroyed by the parameter deviations during the implementation. However, few results (if any) for the stability of such system with a certain deviation rate have been reported in the literature. In this paper, we present a simple delayed neural network model, in which each parameter deviates the reference point with a rate, and further investigate the robust exponential stability of this model and illustrate the relationship between the permissible fluctuation rate and the exponential convergence rate.

1 Introduction

During recent decades, several neural network models with or without delays have been extensively studied, particularly, regarding their stability analysis [1-10]. However, It is well known that the stability of a given system may often be destroyed by its unavoidable uncertainty due to the existence of modeling error, external disturbance and parameter fluctuation during the implementation. Recently, several robust stability criteria for interval neural networks with constant and/or time-varying delays have been proposed [11-14]) since the pioneering work of Liao and Yu [11]. In most existing literature about the uncertain neural network model, the maximum permissible deviation rate of the system parameters has not been investigated and estimated. However, this is an unavoidable and important factor to be considered during the implementation. In this paper, we present a delayed neural network model with parameter fluctuations, and further study its robust exponential. The relationship between the permissible fluctuation rate and the exponential convergence rate is also investigated.

2 Problem Formulation

Consider a delayed neural network with parameter fluctuations described by

$$\dot{x}(t) = -Ax(t) + Wf(x(t-\tau)), \tag{1}$$

where $x(t) = [x_1(t), \cdots, x_n(t)]^T$ is the neuron state vector. $A = \mathrm{diag}(a_1, a_2, \cdots, a_n)$ is a positive diagonal matrix, $W = (w_{ij})_{n \times n}$ is the connect-weighting matrix, $\tau > 0$ is transmi-ssion delay. $f(x) = [f_1(x_1), \cdots, f_n(x_n)]^T$ denotes the neuron activation functions satisfying the following assumption:

(H) f_i ($i = 1, 2, \cdots, n$) are bounded with $f(0) = 0$ and there exists constants $\sigma_i > 0$, such that

$$0 \leq |f_i(x) - f_i(y)| \leq \sigma_i |x - y|, \text{ for any } x, y \in R, i = 1, 2, \cdots, n.$$

Here, let

$$A = A_0 + \Delta A, \quad W = W_0 + \Delta W, \tag{2}$$

where $A_0 = \mathrm{diag}(a_i^{(0)})_{n \times n} > 0$ (throughout this paper, we use $\Omega > 0$ to denote the symme-trical positive matrix Ω) and $W_0 = (w_{ij}^{(0)})_{n \times n}$ are called the reference matrices of A and W, respectively. $\Delta A = \mathrm{diag}(\Delta a_i)_{n \times n} > 0$ and $\Delta W = (\Delta w_{ij})_{n \times n}$ are the fluctuation strengths, which is assumed to satisfy the following condition: There exist constant scalars $\alpha_i > 0$ and $\beta_{ij} > 0$ ($i, j = 1, 2, \cdots, n$) such that

$$|\Delta a_i| \leq \alpha_i a_i^{(0)}, \quad |\Delta w_{ij}| \leq \beta_{ij} |w_{ij}^{(0)}|. \tag{3}$$

Let

$$\underline{A} = \mathrm{diag}(\underline{a}_i)_{n \times n}, \quad \overline{W} = (\overline{w}_{ij})_{n \times n}, \quad \underline{a}_i = (1 - \alpha_i) a_i^{(0)}, \quad \overline{w}_{ij} = (1 + \beta_{ij}) |w_{ij}^{(0)}| \tag{4}$$

From (2), (3) and (4), we have that, for any $i, j = 1, 2, \cdots, n$,

$$a_i \geq \underline{a}_i, \quad |w_{ij}| \leq \overline{w}_{ij} \tag{5}$$

Moreover, by using the Gersgorin theorem, we obtain the following lemma, which is useful to derive our main result.

Lemma 1 Suppose that $W = (w_{ij})_{n \times n}$ satisfy (2)-(5), then for any $x \in R^n$, the following inequality holds:

$$x^T W W^T x \leq x^T C x \tag{6}$$

where $C = \mathrm{diag}(c_i)$ with $c_i = \sum_{k=1}^{n} \left(\overline{w}_{ik} \sum_{j=1}^{n} \overline{w}_{jk} \right)$, $i = 1, 2, \cdots, n$.

3 Main Results

Theorem 1 Suppose that there exist symmetrical and positive definite matrix Q and scalars $s > 0, k > 0$ such that

$$\begin{bmatrix} QA + AQ - 2kQ - \frac{1}{s}e^{2k\tau}C & \sqrt{s}QM \\ \sqrt{s}MQ & E_n \end{bmatrix} > 0 \qquad (7)$$

where E_n is an n-order identity matrix, $M = diag(\sigma_i)_{n \times n}$, and the symbols $\underline{A}, C$ are defined in (4) and (6), respectively. Then system (1) is globally robustly exponentially stable. Moreover,

$$\|x(t)\| \leq \left[\frac{\lambda_M(Q)}{\lambda_m(Q)} + \frac{n \times s \times \lambda_M(Q)}{2k} \max_{1 \leq i \leq n} \{M_i^2\} \right]^{\frac{1}{2}} \|\phi\| e^{-kt} \qquad (8)$$

where $\|\phi\| = \sup\limits_{\theta \in [-\bar{\tau}, 0]} \|x(\theta)\|$, $\lambda_M(Q)$ and $\lambda_m(Q)$ denote the maximum and minimum eigenvalues of a square matrix Q, respectively.

Proof Obviously, the origin is an equilibrium point of the system (1). Suppose that $x^* \neq (0, 0, \cdots, 0)^T$, which implies $f(x^*) \neq (0, 0, \cdots, 0)^T$, is also an equilibrium point of model (1), i.e.,

$$Ax^* - Vf(x^*) = 0 \qquad (9)$$

Multiplying both sides of (9) by $2x^{*T}P$, where $P = Q^{-1}$, we obtain

$$2x^{*T}PAx^* - 2x^{*T}PWf(x^*) = x^{*T}PAx^* + x^{*T}APx^* - \left(x^{*T}PWf(x^*) + x^{*T}f^T(x^*)W^TPx^*\right) = 0 \qquad (10)$$

Since

$$x^{*T}PWf(x^*) + x^{*T}f^T(x^*)W^TPx^* \leq \frac{1}{s}x^{*T}PWW^TPx^* + sf^T(x^*)f(x^*) \leq \frac{1}{s}x^{*T}PCPx^* + sx^{*T}M^2x^* \qquad (11)$$

Introducing (11) into (10) yields

$$x^{*T}P\left[AQ + QA - \frac{1}{s}C - sQM^2Q\right]Px^* \leq 0$$

Hence,

$$AQ + QA - \frac{e^{2k\tau}}{s}C - sQM^2Q \leq AQ + QA - \frac{1}{s}C - sQM^2Q \leq 0 \qquad (12)$$

It leads to a contradiction with the condition (7) in Theorem 1. Thus, there exists only unique equilibrium point of the system (1). To prove the robust exponential stability, we consider the following Lyapunov-Krasovskii functional:

$$V(x(t)) = e^{2kt}x^T(t)Px(t) + s\sum_{j=1}^{n}\int_{t-\tau}^{t}e^{2k\xi}f_j^2(x_j(\xi))d\xi \qquad (13)$$

where $P = Q^{-1}$ is a symmetric positive definite matrix and $s > 0$ is a constant scalar. The Dini's time derivative of $V(x(t))$ along the trajectories of the system (1) is

$$\dot{V}(x(t)) \leq e^{2kt}x^T(t)\left[2kP - AP - PA + \frac{1}{s}e^{2k\tau}PWW^TP + sM^2\right]x(t)$$

$$= -e^{2kt}x^T(t)P\left[QA + AQ - 2kQ - \frac{1}{s}e^{2k\tau}WW^T - sQM^2Q\right]Px(t)$$

$$\leq -e^{2kt}x^T(t)P\left[Q\underline{A} + \underline{A}Q - 2kQ - \frac{1}{s}e^{2k\tau}C - sQM^2Q\right]Px(t)$$

$$\equiv -e^{2kt}x^T(t)P\Omega Px(t)$$

where $\Omega = Q\underline{A} + \underline{A}Q - 2kQ - \frac{1}{s}e^{2k\tau}C - sQM^2Q$. Hence, it is easy to see that $\dot{V}(x(t)) < 0$ if $\Omega > 0$ and $x(t) \neq 0$. Also, $\Omega > 0$ if and only if Eq. (7) holds. Therefore, we have

$$e^{2kt}\lambda_m(P)\|x(t)\|^2 \leq V(x(t)) \leq V(x(0)) \qquad (14)$$

Note that

$$V(x(0)) = x^T(0)Px^T(0) + s\sum_{i=1}^{n}\int_{-\tau}^{0}e^{2k\xi}f_i^2(x_i(\xi))$$

$$\leq \lambda_M(P)\|\phi\|^2 + s\max_{1\leq i\leq n}\{M_i^2\}\|\phi\|^2\sum_{i=1}^{n}\int_{-\tau}^{0}e^{2k\xi}d\xi$$

$$\leq \left[\lambda_M(P) + s\max_{1\leq i\leq n}\{M_i^2\}\frac{n}{2k}\right]\|\phi\|^2$$

From (14) and (15), we easily obtain the inequality (8). The proof is thus completed.
Remark 1. The constant scalar k>0 is called the exponential convergence rate. According to the LMI (7), we can estimate it by solving the following optimization problem:

$$\begin{cases} \max\ k \\ s.t.\ \text{condition (7) is satisfied.} \end{cases} \qquad (16)$$

Let

$$\alpha_i = \alpha,\ \beta_{ij} = \beta,\ i,j = 1,2,\cdots,n. \qquad (17)$$

Then, we have that

$$\underline{A} = (1-\alpha)A_0, \quad C = (1+\beta)^2 \operatorname{diag}\left(\sum_{k=1}^{n}\left|a_{ik}^{(0)}\right|\sum_{j=1}^{n}\left|a_{jk}^{(0)}\right|\right) \tag{18}$$

Therefore, we have the following corollary.

Corollary 1 Suppose that (17) is satisfied and $M = E_n$. If there exist symmetrical and positive definite matrix Q and scalars $k > 0$ such that

$$\begin{bmatrix} (1-\alpha)QA_0 + (1-\alpha)A_0Q - 2kQ - e^{2k\tau}C & Q \\ Q & E_n \end{bmatrix} > 0 \tag{19}$$

where E_n is an n-order identity matrix, and C is defined in (18). Then system (1) is globally robustly exponentially stable.

Remark 2. Eq. (19) formulates the relationship among the fluctuation rates α, β, the time delay τ and global exponential convergence rate k under the condition of the global expo-nential stability of model (1). Therefore, given a reference model

$$\dot{x}(t) = -A_0 x(t) + W_0 f(x(t-\tau)). \tag{20}$$

We can estimate the maximum permissive fluctuation rates if k and τ are known. If α, β are prior, contrariwise, the relationship between k and τ can be estimated.

4 Conclusions

In summary, we have presented a simple delayed neural network with parameter fluctua-tions and analyzed its robust stability. In particular, we estimated the maximum permissible fluctuation rates of different parameter matrices and the effect of the time delay on the globally exponential convergence rate using the proposed stability conditions. Because the issue investigated in this paper may often arise during the implementation of neural networks, the results here are practical and should be useful for further studies of this kind of neural network model.

Acknowledgement. The work described in this paper was supported by the NNSF of China (grant no. 60271019), the Doctorate Foundation Grants from the National Education Committee of China (grant no. 20020611007), and the Applied Basic Research Grants Committee of Science and Technology of Chongqing (grant no. 7370).

References

1. Zhang Y., Heng, P.-H.: Vadakkepat, P.: Absolute Periodicity and Absolute Stability of Delayed Neural Networks. IEEE Transactions on CAS-I 49 (2002) 256-261.
2. He H, Cao J., Wang J.: Global Exponential Stability and Periodic Solutions of Recurrent Neural Networks with Delays. Physics Lett. A 198 (2002) 393-404.
3. Chen T.: Global Exponential Stability of Delayed Hopfield Neural Networks. Neural Networks 14 (2001): 977-980.

4. Liao X.: Stability of Hopfield-Type Neural Networks (I). Science in China (Scientia Sinica) Series A 14 (1995) 407-418.
5. Zhao H.: Global Stability of Neural Networks with Distributed Delays. Phys. Rev. E 68 (2003) 051909.
6. Li C., Chen G.: Stability of a Neural Network Model with Small-World Connections. Phys. Rev. E 68 (2003) 052901.
7. Guo S., Huang L.: Stability Analysis of a Delayed Hopfield Neural Network. Phys. Rev. E 67 (2003) 061902.
8. Liao X., Wong K.-W: Global Exponential Stability of Hybrid Bidirectional Associative Memory Neural Networks with Discrete Delays. Phys. Rev. E 67 (2003) 042901.
9. Zhang Y., TanK. K.: Dynamic Stability Conditions for Lotka-Volterra Recurrent Neural Networks with Delays. Phys. Rev. E 66 (2002) 011910.
10. Liao X., Yu J.: Robust Stability for Interval Hopfield Neural Networks with Time Delay. IEEE Trans. Neural Networks 9 (1998) 1042–1046.
11. Liao X., Wong K.-W., Wu Z.and Chen G.: Novel Robust Stability Criteria for Interval-Delayed Hopfield Neural Networks", IEEE Trans. Circuits Syst. I 48 (2001) 1355-1359.
12. Liao X., Wang J.: Global and Robust Stability of Interval Hopfield Neural Networks with Time-Varying Delays. Int. J. neural syst. 13 (2003) 171-182.
13. Arik S.: Global Robust Stability of Delayed Neural Networks. IEEE Trans. Circuits Syst. I 50 (2003) 156-160.

Local Stability and Bifurcation in a Model of Delayed Neural Network*

Yiping Lin[1,2] and Zengrong Liu[1]

[1] Department of Mathematics, Shanghai University,
Shanghai 200436, P.R.China
[2] Department of Applied Mathematics, Kunming University of
Science and Technology, Kunming 650093, P.R.China***
lin_yiping@hotmail.com

Abstract. A system of n-units neural network with coupled cells is investigated, the local stability of null solution is considered, and the parameter values of the periodic solution bifurcation are given.

1 Introduction

Recently, theoretical and applied studies of the neural networks have become an area of intense research activity. In 1984, Hopfield [5] introduced a continuous version of a circuit equation for a network of n saturating voltage amplifiers (neurons). The system of ordinary differential equations describes the time evolution of the voltage on input of each neuron. Updating and propagation are assumed to occur instantaneously. In recent years ,the processing in each neuron was incorporated into Hopfield's equations. The resulting delay-differential equation system has been the starting point of several recent investigations (see for example, [1], [2], [3], [6]). In this paper, we consider the following Hopfield equation

$$C_i \dot{u}_i(t) = -\frac{u_i(t)}{R_i} + \sum_{j=1}^{n} T_{ij} f_j[u_j(t-\tau_j)], \quad 1 \leq i \leq n, \qquad (1)$$

where u_i denotes the voltage on the input of the ith neuron. Each neuron is characterized by an input capacitance C_i and a transfer function f_j, T_{ij} are the elements of the connection matrix, $f_j(0) = 0$ and R_i refers to output resistance. Considering the case of identical units, so that $C_i = C, f_j = f$, and $\tau_j = \tau$, assuming that each unit is connected to the same total output resistance $R_i = R$, scaling the time and delay by RC, and letting $a_{ij} = RT_{ij}$, we obtain the normalized system

---

* This research is supported by the National Natural Science Foundation of China (No.10161007) and the Science Foundation of Education Department of Yunnan, China.
*** Corresponding address

$$\dot{x}_i(t) = -x_i(t) + \sum_{j=1}^{n} a_{ij} f[x_j(t-\tau)], \quad 1 \le i \le n . \tag{2}$$

This equation has been studied by Marcus and Westervelt [6], they state a number of conjectures, supported by heuristic and numerical computations. Bélair [1] considered the conditions of the local stability and the global stability of this equation use analytical method.

In this paper, we give a new criterion for local stability of the null solution of Equation (2) from the point of view of the bifurcation theory, and obtain the parameter values of the periodic solution bifurcation. Our result can be regarded as a complement of the paper of J. Bélair [1].

2 Criterion of Local Stability

Consider equation

$$\dot{x}_i(t) = -x_i(t) + \sum_{j=1}^{n} a_{ij} f[x_j(t-\tau)], \quad 1 \le i \le n ,$$

and let $f'(0) = \beta$. Linearizing Equation (2), we obtain, in vectorized notation, the equation

$$\dot{X}(t) = -X(t) + \beta A X(t-\tau), \tag{3}$$

where $A = (a_{ij})_{n \times n}$ is a real matrix. The characteristic roots of Equation (3) are obtained

$$\det(\lambda I + I - \beta A e^{-\lambda \tau}) = 0 . \tag{4}$$

In this paper, we only consider $\tau > 0$.

Lemma 1. *If λ is a root of Equation (4), then there is an eigenvalue d of the matrix A, such that*

$$d = \frac{1}{\beta}(1+\lambda)e^{\lambda \tau} . \tag{5}$$

Conversely, for any eigenvalue d of the matrix A, any solution λ of the equation (5) will be a characteristic root of (4).

The prove of Lemma 1 is simple and we omit it. Write (5) as

$$(\lambda \tau + \tau)e^{\lambda \tau} - \beta d \tau = 0, \tag{6}$$

and let $d_j, j = 1, 2, \cdots, n$ be the eigenvalues of A, we obtain the following theorem.

Theorem 1. *Suppose that $d_j, j = 1, 2, \cdots, n$ are all real, and let*

$$\beta_1 = \max \left\{ \frac{1}{d_{\min}^+ \tau}(\tau \cos \zeta - \zeta \sin \zeta), \frac{1}{d_{\min}^-} \right\} ,$$

$$\beta_2 = \min\left\{\frac{1}{d_{\max}^+}, \frac{1}{d_{\max}^-\tau}(\tau\cos\zeta - \zeta\sin\zeta)\right\}.$$

Then the null solution of Equation (2) is locally asymptotically stable if and only if

$$\beta_1 < \beta < \beta_2, \tag{7}$$

where ζ is the root of $\zeta = -\tau\tan\zeta$, $0 < \zeta < \pi$. $d_{\min}^+, d_{\max}^+$ are the minimum and the maximum of the positive eigenvalues of A, and $d_{\min}^-, d_{\max}^-$ are the minimum and maximum of the negative eigenvalues of A respectively.

If A has only positive eigenvalues or negative eigenvalues, we only keep the corresponding terms.

Proof. Using Lemma 1, we can reduce the study of Equation (4) to the investigation of the n scalar equations

$$(\lambda\tau + \tau)e^{\lambda\tau} - \beta d_j\tau = 0, \quad j = 1, 2, \cdots, n. \tag{8}$$

From J. K. Hale [4] we know that all roots of equation $(z+a)e^z + b = 0$ have negative real parts if and only if the following conditions hold

$$a > 0, \quad a + b > 0, \quad b < \zeta\sin\zeta - a\cos\zeta,$$

where ζ satisfies that when $a = 0$, then $\zeta = \frac{\pi}{2}$, and when $a \neq 0$, ζ is the root of the equation $\zeta = -\tau\tan\zeta$, $0 < \zeta < \pi$.

Let $\tau = a, -\beta d_j\tau = b$ in (8), then all $\lambda\tau$ thus every eigenvalue λ have negative real part if and only if

$$\frac{1}{d_j\tau}(\tau\cos\zeta - \zeta\sin\zeta) < \beta < \frac{1}{d_j}, \text{ when } d_j > 0, \tag{9}$$

$$\frac{1}{d_j} < \beta < \frac{1}{d_j\tau}(\tau\cos\zeta - \zeta\sin\zeta), \text{ when } d_j < 0, \tag{10}$$

where ζ is the root of the equation $\zeta = -\tau\tan\zeta, 0 < \zeta < \pi$.

Sum up all above conditions, we see that the null solution of Equation (2) is asymptotically stable if and only if (7) is satisfied.

If A has $2k$ complex eigenvalues $d_j, j = 1, 2, \cdots, 2k = n$. Let $d_j = R_j e^{i\varphi_j}$, we might as well let $0 < \varphi_l < \pi, l = 1, 2, \cdots, k$, and

$$\beta_3 = \min_{1 \leq l \leq k}\left\{\frac{2\varphi_l - \pi}{2R_l\tau}\right\},$$

then we obtain following theorem.

Theorem 2. *Assume that all eigenvalues $d_j, j = 1, 2, \cdots, 2k$ of A are complex, then the null solution of Equation (2) is locally asymptotically stable if and only if $\beta < \beta_3$.*

Proof. Let $d_j = R_j e^{i\varphi_j}, j = 1, 2, \cdots, 2k$, and let $\lambda+1 = r+is = \rho e^{i\theta}$, substituting in (6) yields

$$\rho e^{r\tau + i(\theta + s\tau)} = \beta R_j e^{\tau + i\varphi_j}, \tag{11}$$

then

$$\begin{cases} \rho e^{r\tau} = \beta R_j e^{\tau} \\ \varphi_j = \theta + s\tau + 2\pi m, \end{cases}$$

where m is an integer chosen so that $0 < \varphi_j < 2\pi$ and $0 < \theta < 2\pi$, we might as well set $m = 0$. Then

$$\varphi_j = \theta + \beta R_j \tau e^{(1-r)\tau} \sin \theta . \tag{12}$$

Since $r = \rho \cos \theta, \rho > 0$, then if $r \geq 1$, it means $\text{Re}\lambda \geq 0$, the value of θ is restricted to the range $\left(0, \frac{\pi}{2}\right) \cup \left(\frac{3\pi}{2}, 2\pi\right)$. From (12), we see that when φ_j satisfies

$$\frac{\pi}{2} + \beta R_j \tau < \varphi_j < \frac{3\pi}{2} - \beta R_j \tau,$$

then the eigenvalue λ of Equation (4) corresponding to d_j has negative real part, that is β satisfies

$$\begin{cases} \beta < \dfrac{2\varphi_j - \pi}{2R_j\tau}, & \text{when } 0 < \varphi_j < \pi, \\ \beta < \dfrac{3\pi - 2\varphi_j}{2R_j\tau}, & \text{when } \pi < \varphi_j < 2\pi . \end{cases} \tag{13}$$

Since A is real, the complex eigenvalues appear by pairs. If $d_1 = Re^{i\varphi_1}$ with $\varphi_1 \in (0, \pi)$, the another eigenvalue is $d_2 = Re^{i\varphi_2}$ with $\varphi_2 = 2\pi - \varphi_1 \in (\pi, 2\pi)$. Since

$$\frac{3\pi - 2\varphi_2}{2R\tau} = \frac{3\pi - 4\pi + 2\varphi_1}{2R\tau} = \frac{2\varphi_1 - \pi}{2R\tau},$$

we see that all eigenvalues of Equation (4) have negative real part if and only if $\beta < \beta_3$, and the null solution of Equation (2) is asymptotically stable.

If A has both real roots and complex roots, it can be discussed analogously by consider the criterion in Theorem 1 and Theorem 2 simultaneously.

Lemma 2. *Let d be a real eigenvalue of A. When*

$$\beta = \frac{1}{d\tau}(\tau \cos \zeta - \zeta \sin \zeta),$$

Equation (4) has a pair of pure imaginary roots $\lambda = \pm i \dfrac{\zeta}{\tau}$, where ζ satisfies $\zeta = -\tau \tan \zeta$, $0 < \zeta < \pi$. If $d = Re^{i\varphi}$, $0 < \varphi < \pi$ is a complex eigenvalue of A, then when $\beta = \dfrac{2\varphi - \pi}{2R\tau}$, Equation (4) has a pair of pure imaginary roots.

From Lemma 2 we see that if the eigenvalues of A are all real, and $\beta_1 = \dfrac{1}{d^+_{\min}\tau}(\tau \cos \zeta - \zeta \sin \zeta)$ or $\beta_2 = \dfrac{1}{d^-_{\max}\tau}(\tau \cos \zeta - \zeta \sin \zeta)$ in Theorem 1, then

Equation (4) has a pair of pure imaginary roots when $\beta = \beta_1$ or $\beta = \beta_2$. If the eigenvalues of A are all complex, then when $\beta = \beta_3$, Equation (4) has a pair of pure imaginary roots.

Suppose that when $\beta = \beta_0$, Equation (4) has a pair of pure imaginary roots $\lambda = \pm i\omega$, and all other eigenvalues have negative real parts, d stands for the eigenvalue of A corresponding to β_0. Then we have following theorem.

Theorem 3. *If $\alpha'(\beta_0) \neq 0$, then β_0 is a Hopf bifurcation value of Equation (2). Where*

$$\alpha'(\beta_0) = \frac{dRe(\lambda)}{d\beta}\bigg|_{\beta=\beta_0}$$

$$= \begin{cases} \dfrac{d}{(1+\tau^2)(\omega\tau)^2}(1+d\tau\beta_0), & \text{when } d \text{ is real;} \\ \dfrac{R}{(1+\tau^2)(\omega\tau)^2}[(1+\tau)(\cos(\varphi+\omega\tau)+\omega\tau\sin(\varphi+\omega\tau)], & \text{when } d = Re^{i\varphi} \\ & \text{is complex.} \end{cases}$$

3 Conclusion

In this paper, we consider a general n-units neural network with delay. We do not assume any restriction for connection matrix $(a_{ij})_{n\times n}$, we use only one parameter β which is the first-order coefficient of the transfer function f to obtain the stability of the null solution of the system. For this reason we can adjust β to obtain the stabilities expected. In addition, if β is a bifurcation value, the null solution will change its stability and a small periodic solution appears when the parameter passing through. These properties can be applied to the design of the delayed cellular neural networks (DCNN). DCNN possess highly important significance for some applied fields, and we provide a new method for the studies. The stability of the small bifurcation periodic solution is also a important problem, we leave it to the future investigation.

References

1. Bélair, J.: Stability in a Model of a Delayed Neural Network. J. of Dynamics and Differential Equations **5** (1993) 607–623
2. Burton, T. A.: Averaged Neural Networks. Neural Network **6** (1993) 677–680
3. Gopalsamy, K., He, X.: Stability in Asymmetric Hopfield Net with Transmission Delays. Phys. D **76** (1994) 344–358
4. Hale, J.K., Lunel, S.V.: Introduction to Functional Differential Equations. Applied Mathematical Sciences, Vol. 99. Springer-Verlag, New York Berlin Heidelberg (1993)
5. Hopfield, J.J.: Neurons with Graded Response Have Collective Computational Properties like Those of Tow-State Neurons. Proc. Natl. Acad. Sci. **81** (1984) 3088–3092
6. Marcus, C.M., Westervelt, R.M.: Stability of analog neural networks with delay. Phys. Rev. A **39** (1989) 347–359

On the Asymptotic Stability of Non-autonomous Delayed Neural Networks*

Qiang Zhang[1,2], Haijun Wang[2], Dongsheng Zhou[2], and Xiaopeng Wei[1]

[1] Center for Advanced Design Technology, Dalian University, Dalian, 116622, China
[2] School of Mechanical Engineering, Dalian University of Technology, Dalian, 116024, China
zhangq26@163.com

Abstract. The global asymptotic stability of non-autonomous delayed neural networks is discussed in this paper. By utilizing a delay differential inequality, we present several sufficient conditions which guarantee the asymptotic stability. Since these conditions do not impose differentiability on delay function, they are less conservative than some established in the earlier references. The results show that this approach is more straightforward and effective for stability analysis compared with the method of Lyapunov functionals which has been successfully applied the in the literature.

1 Introduction

The stability properties of autonomous delayed neural networks have been extensively studied in the past decades and many important results on the global asymptotic stability and global exponential stability of one unique equilibrium point have been presented, see, for example,[1]-[13] and references cited therein. However, to the best of our knowledge, few studies have considered dynamics for non-autonomous delayed neural networks [14]. In this paper, by using differential inequalities, we discuss the global asymptotic stability of non-autonomous delayed neural networks and obtain several new sufficient conditions. We do not require the delay to be differentiable.

2 Preliminaries

The dynamic behavior of a continuous time non-autonomous delayed neural networks can be described by the following state equations:

$$x_i'(t) = -c_i(t)x_i(t) + \sum_{j=1}^{n} a_{ij}(t)f_j(x_j(t)) + \sum_{j=1}^{n} b_{ij}(t)f_j(x_j(t - \tau_j(t))) + I_i(t). \tag{1}$$

---

* The project supported by the National Natural Science Foundation of China (grant nos. 60174037 and 50275013.) and China Postdoctoral Science Foundation (grant no.200303448)

where n corresponds to the number of units in a neural networks; $x_i(t)$ corresponds to the state vector at time t; $f(x(t)) = [f_1(x_1(t)), \cdots, f_n(x_n(t))]^T \in R^n$ denotes the activation function of the neurons; $A(t) = [a_{ij}(t)]_{n \times n}$ is referred to as the feedback matrix, $B(t) = [b_{ij}(t)]_{n \times n}$ represents the delayed feedback matrix, while $I_i(t)$ is a external bias vector at time t, $\tau_j(t)$ is the transmission delay along the axon of the jth unit and satisfies $0 \le \tau_i(t) \le \tau$.

Throughout this paper, we will assume that the real valued functions $c_i(t) > 0, a_{ij}(t), b_{ij}(t), I_i(t)$ are continuous functions. The activation functions $f_i, i = 1, 2, \cdots, n$ are assumed to satisfy the following hypothesis

$$|f_i(\xi_1) - f_i(\xi_2)| \le L_i |\xi_1 - \xi_2|, \forall \xi_1, \xi_2. \tag{2}$$

This type of activation functions is clearly more general than both the usual sigmoid activation functions and the piecewise linear function (PWL): $f_i(x) = \frac{1}{2}(|x+1| - |x-1|)$ which is used in [3].

The initial conditions associated with system (1) are of the form

$$x_i(s) = \phi_i(s), \ s \in [-\tau, 0], \ \tau = \max_{1 \le i \le n}\{\tau_i^+\} \tag{3}$$

in which $\phi_i(s)$ are continuous for $s \in [-\tau, 0]$.

Lemma 1. *Assume $k_1(t)$ and $k_2(t)$ are nonnegative continuous functions. Let $x(t)$ be a continuous nonnegative function on $t \ge t_0 - \tau$ satisfying inequality (4) for $t \ge t_0$.*

$$x'(t) \le -k_1(t)x(t) + k_2(t)\bar{x}(t) \tag{4}$$

where $\bar{x}(t) = \sup_{t-\tau \le s \le t}\{x(s)\}$. If the following conditions hold

$$\begin{array}{l} 1) \int_0^\infty k_1(s)ds = +\infty \\ 2) \int_{t_0}^t k_2(s) e^{-\int_s^t k_1(u)du} ds \le \delta < 1. \end{array} \tag{5}$$

then, we have $\lim_{t \to \infty} x(t) = 0$.

Proof. It follows from (4) that

$$x(t) \le x(t_0) e^{-\int_{t_0}^t k_1(s)ds} + \int_{t_0}^t k_2(s) e^{-\int_s^t k_1(u)du} \bar{x}(s)ds, \ t \ge t_0 \tag{6}$$

For $t \ge t_0$, let $y(t) = x(t)$, and for $t_0 - \tau \le t \le t_0$, $y(t) = \sup_{t_0 - \tau \le \theta \le t_0}[x(\theta)]$. From (6), we obtain

$$x(t) \le x(t_0) + \delta \sup_{t_0 - \tau \le \theta \le t}[x(\theta)], \ t \ge t_0 \tag{7}$$

then, we can get

$$y(t) \le x(t_0) + \delta \sup_{t_0 - \tau \le \theta \le t}[y(\theta)], \ t \ge t_0 - \tau \tag{8}$$

Since the right hand of (8) is nondecreasing, we have

$$\sup_{t_0-\tau\leq\theta\leq t} [y(\theta)] \leq x(t_0) + \delta \sup_{t_0-\tau\leq\theta\leq t} [y(\theta)], \ t \geq t_0 - \tau \tag{9}$$

and

$$x(t) = y(t) \leq \frac{x(t_0)}{1-\delta}, \ t \geq t_0 \tag{10}$$

By condition 1), we know that $\lim_{t\to\infty} \sup x(t) = x^*$ exists. Hence, for each $\varepsilon > 0$, there exists a constant $T > t_0$ such that

$$x(t) < x^* + \varepsilon, \ t \geq T \tag{11}$$

From (6) combining with (11), we have

$$\begin{aligned}x(t) &\leq x(T)e^{-\int_T^t k_1(s)ds} + \int_T^t k_2(s)e^{-\int_s^t k_1(u)du}\bar{x}(s)ds \\ &\leq x(T)e^{-\int_T^t k_1(s)ds} + \delta(x^* + \varepsilon), \ t \geq T\end{aligned} \tag{12}$$

On the other hand, there exists another constant $T_1 > T$ such that

$$\begin{aligned}x^* - \varepsilon &< x(T_1) \\ e^{-\int_T^{T_1} k_1(u)du} &\leq \varepsilon\end{aligned} \tag{13}$$

therefore,

$$x^* - \varepsilon < x(T_1) \leq x(T)\varepsilon + \delta(x^* + \varepsilon) \tag{14}$$

Let $\varepsilon \to 0^+$, we obtain

$$0 \leq x^* \leq \delta x^* \tag{15}$$

this implies $x^* = 0$. This completes the proof.

3 Global Asymptotic Stability Analysis

In this section, we will use the above Lemma to establish the asymptotic stability of system (1). Consider two solutions $x(t)$ and $z(t)$ of system (1) for $t > 0$ corresponding to arbitrary initial values $x(s) = \phi(s)$ and $z(s) = \varphi(s)$ for $s \in [-\tau, 0]$. Let $y_i(t) = x_i(t) - z_i(t)$, then we have

$$\begin{aligned}y_i'(t) = &-c_i(t)y_i(t) + \sum_{j=1}^n a_{ij}(t)\left(f_j(x_j(t)) - f_j(z_j(t))\right) \\ &+ \sum_{j=1}^n b_{ij}(t)\left(f_j(x_j(t-\tau_j(t))) - f_j(z_j(t-\tau_j(t)))\right)\end{aligned} \tag{16}$$

Set $g_j(y_j(t)) = f_j(y_j(t) + z_j(t)) - f_j(z_j(t))$, one can rewrite Eq.(16) as

$$y_i'(t) = -c_i(t)y_i(t) + \sum_{j=1}^n a_{ij}(t)g_j(y_j(t)) + \sum_{j=1}^n b_{ij}(t)g_j(y_j(t-\tau_j(t))) \tag{17}$$

Note that the functions f_j satisfy the hypothesis (2), that is,

$$|g_i(\xi_1) - g_i(\xi_2)| \leq L_i|\xi_1 - \xi_2|, \forall \xi_1, \xi_2. \tag{18}$$
$$g_i(0) = 0$$

By Eq.(17), we have

$$D^+|y_i(t)| \leq -c_i(t)|y_i(t)| + \sum_{j=1}^n |a_{ij}(t)|L_j|y_j(t)| + \sum_{j=1}^n |b_{ij}(t)|L_j|y_j(t-\tau_j(t))| \tag{19}$$

Theorem 1. Let $k_1(t) = \min_i \sum_{j=1}^n \left\{ 3c_i(t)\delta_{ij} - |a_{ij}(t)|^{3\alpha_{ij}^1} L_j^{3\beta_{ij}^1} - |a_{ij}(t)|^{3\alpha_{ij}^2} L_j^{3\beta_{ij}^2} \right.$

$\left. - |a_{ji}(t)|^{3\alpha_{ji}^3} L_i^{3\beta_{ji}^3} - |b_{ij}(t)|^{3\zeta_{ij}^1} L_j^{3\eta_{ij}^1} - |b_{ij}(t)|^{3\zeta_{ij}^2} L_j^{3\eta_{ij}^2} \right\} > 0,$

$k_2(t) = \max_i \left\{ \sum_{j=1}^n |b_{ij}(t)|^{3\zeta_{ij}^3} L_j^{3\eta_{ij}^3} \right\}$, where $\delta_{ij} = \begin{cases} 1, & i=j \\ 0, & i \neq j \end{cases}$,

$\alpha_{ij}^1, \alpha_{ij}^2, \alpha_{ij}^3, \beta_{ij}^1,$
$\beta_{ij}^2, \beta_{ij}^3, \zeta_{ij}^1, \zeta_{ij}^2, \zeta_{ij}^3, \eta_{ij}^1, \eta_{ij}^2, \eta_{ij}^3$ are real constants and satisfy $\alpha_{ij}^1 + \alpha_{ij}^2 + \alpha_{ij}^3 = \beta_{ij}^1 + \beta_{ij}^2 + \beta_{ij}^3 = \zeta_{ij}^1 + \zeta_{ij}^2 + \zeta_{ij}^3 = \eta_{ij}^1 + \eta_{ij}^2 + \eta_{ij}^3 = 1$, then Eq.(1) is globally asymptotically stable if

$$\begin{aligned} &1)\ \int_0^\infty k_1(s)ds = +\infty \\ &2)\ \int_{t_0}^t k_2(s) e^{-\int_s^t k_1(u)du} ds \leq \delta < 1 \end{aligned} \tag{20}$$

Proof. Let $z(t) = \frac{1}{3}\sum_{i=1}^n |y_i(t)|^3$. Calculate the upper right derivative of $z(t)$ along the solutions of Eq.(17), we have

$$D^+z(t) \leq \sum_{i=1}^n |y_i(t)|^2 D^+|y_i(t)|$$

$$\leq \sum_{i=1}^n |y_i(t)|^2 \left\{ -c_i(t)|y_i(t)| + \sum_{j=1}^n |a_{ij}(t)|L_j|y_j(t)| \right.$$

$$\left. + \sum_{j=1}^n |b_{ij}(t)|L_j|y_j(t-\tau_j(t))| \right\}$$

$$= -\sum_{i=1}^n c_i(t)|y_i(t)|^3 + \sum_{i=1}^n \sum_{j=1}^n |a_{ij}(t)|L_j|y_i(t)|^2|y_j(t)|$$

$$+ \sum_{i=1}^n \sum_{j=1}^n |b_{ij}(t)|L_j|y_i(t)|^2|y_j(t-\tau_j(t))|$$

$$= -\sum_{i=1}^n c_i(t)|y_i(t)|^3$$

$$+ \sum_{i=1}^{n}\sum_{j=1}^{n} \left(|a_{ij}(t)|^{\alpha_{ij}^1} L_j^{\beta_{ij}^1} |y_i(t)|\right) \times \left(|a_{ij}(t)|^{\alpha_{ij}^2} L_j^{\beta_{ij}^2} |y_i(t)|\right)$$

$$\times \left(|a_{ij}(t)|^{\alpha_{ij}^3} L_j^{\beta_{ij}^3} |y_j(t)|\right)$$

$$+ \sum_{i=1}^{n}\sum_{j=1}^{n} \left(|b_{ij}(t)|^{\zeta_{ij}^1} L_j^{\eta_{ij}^1} |y_i(t)|\right) \times \left(|b_{ij}(t)|^{\zeta_{ij}^2} L_j^{\eta_{ij}^2} |y_i(t)|\right)$$

$$\times \left(|b_{ij}(t)|^{\zeta_{ij}^3} L_j^{\eta_{ij}^3} |y_j(t - \tau_j(t))|\right) \quad (21)$$

Employing the element inequality $3abc \leq a^3 + b^3 + c^3$ $(a, b, c \geq 0)$ in (21), we get

$$D^+ z(t) \leq -\sum_{i=1}^{n} c_i(t)|y_i(t)|^3 + \frac{1}{3}\sum_{i=1}^{n}\sum_{j=1}^{n} \left\{|a_{ij}(t)|^{3\alpha_{ij}^1} L_j^{3\beta_{ij}^1} |y_i(t)|^3 \right.$$

$$\left. + |a_{ij}(t)|^{3\alpha_{ij}^2} L_j^{3\beta_{ij}^2} |y_i(t)|^3 + |a_{ij}(t)|^{3\alpha_{ij}^3} L_j^{3\beta_{ij}^3} |y_j(t)|^3 \right\}$$

$$+ \frac{1}{3}\sum_{i=1}^{n}\sum_{j=1}^{n} \left\{|b_{ij}(t)|^{3\zeta_{ij}^1} L_j^{3\eta_{ij}^1} |y_i(t)|^3 + |b_{ij}(t)|^{3\zeta_{ij}^2} L_j^{3\eta_{ij}^2} |y_i(t)|^3 \right.$$

$$\left. + |b_{ij}(t)|^{3\zeta_{ij}^3} L_j^{3\eta_{ij}^3} |y_j(t - \tau_j(t))|^3 \right\}$$

$$= -\frac{1}{3}\sum_{i=1}^{n}\sum_{j=1}^{n} \left\{ 3c_i(t)\delta_{ij} - |a_{ij}(t)|^{3\alpha_{ij}^1} L_j^{3\beta_{ij}^1} \right.$$

$$- |a_{ij}(t)|^{3\alpha_{ij}^2} L_j^{3\beta_{ij}^2} - |a_{ji}(t)|^{3\alpha_{ji}^3} L_i^{3\beta_{ji}^3}$$

$$\left. - |b_{ij}(t)|^{3\zeta_{ij}^1} L_j^{3\eta_{ij}^1} - |b_{ij}(t)|^{3\zeta_{ij}^2} L_j^{3\eta_{ij}^2} \right\} |y_i(t)|^3$$

$$+ \frac{1}{3}\sum_{i=1}^{n}\sum_{j=1}^{n} |b_{ij}(t)|^{3\zeta_{ij}^3} L_j^{3\eta_{ij}^3} |y_j(t - \tau_j(t))|^3$$

$$\leq -k_1(t) z(t) + k_2(t) \bar{z}(t) \quad (22)$$

By using the above Lemma, we know that if the conditions 1) and 2) are satisfied, we have $\lim_{t\to\infty} z(t) = \lim_{t\to\infty} \frac{1}{3}\sum_{i=1}^{n} |y_i(t)|^3 = 0$, this implies that $\lim_{t\to\infty} y_i(t) = 0$. The proof is completed.

Theorem 1 possesses many free adjustable parameters. By taking different parameter values, we can obtain a set of independent sufficient conditions. For example, taking $\alpha_{ij}^1 = \alpha_{ij}^2 = \alpha_{ij}^3 = \beta_{ij}^1 = \beta_{ij}^2 = \beta_{ij}^3 = \zeta_{ij}^1 = \zeta_{ij}^2 = \zeta_{ij}^3 = \eta_{ij}^1 = \eta_{ij}^2 = \eta_{ij}^3 = \frac{1}{3}$ in Theorem 1, we can easily derive the following Corollary 1.

Corollary 1. Let
$$k_1(t) = \min_i \sum_{j=1}^{n} \{3c_i(t)\delta_{ij} - 2|a_{ij}(t)|L_j - |a_{ji}(t)|L_i - 2|b_{ij}(t)|L_j\} > 0$$

$k_2(t) = \max_i \left\{ \sum_{j=1}^{n} |b_{ij}(t)| L_j \right\}$. Eq.(1) is globally asymptotically stable if

$$\begin{array}{l} 1)\ \int_0^\infty k_1(s)ds = +\infty \\ 2)\ \int_{t_0}^t k_2(s) e^{-\int_s^t k_1(u)du} ds \le \delta < 1. \end{array} \quad (23)$$

Remark 1. Note that the criteria obtained here are independent of delay and the coefficients $c_i(t), a_{ij}(t)$ and $b_{ij}(t)$ may be unbounded. For this reason, our results improve and extend those earlier established in [2], [4], [7], [13], [14].

References

1. Arik, S.: An Improved Global Stability Result for Delayed Cellular Neural Networks. IEEE Trans.Circuits Syst.I **49** (2002) 1211–1214
2. Cao, J., Wang, J.: Global Asymptotic Stability of a General Class of Recurrent Neural Networks with Time-Varying Delays. IEEE Trans.Circuits Syst.I **50** (2003) 34–44
3. Chua, L.O., Yang, L.: Cellular Neural Networks: Theory and Applications. IEEE Trans.Circuits Syst.I **35** (1988) 1257–1290
4. Huang, H., Cao, J.: On Global Asymptotic Stability of Recurrent Neural Networks with Time-Varying Delays. Appl.Math.Comput. **142** (2003) 143–154
5. Liao, X., Chen, G., Sanchez, E.N.: LMI-Based Approach for Asymptotically Stability Analysis of Delayed Neural Networks. IEEE Trans.Circuits Syst.I **49** (2002) 1033–1039
6. Liao, X.X., Wang, J.: Algebraic Criteria for Global Exponential Stability of Cellular Neural Networks with Multiple Time Delays. IEEE Trans.Circuits Syst.I **50** (2003) 268–274
7. Mohamad, S., Gopalsamy, K.: Exponential Stability of Continuous-Time and Discrete-Time Cellular Neural Networks with Delays. Appl.Math.Comput. **135** (2003) 17–38
8. Zeng, Z., Wang, J., Liao, X.: Global Exponential Stability of a General Class of Recurrent Neural Networks with Time-Varying Delays. IEEE Trans.Circuits Syst.I **50** (2003) 1353–1358
9. Zhang, J.: Globally Exponential Stability of Neural Networks with Variable Delays. IEEE Trans.Circuits Syst.I **50** (2003) 288–290
10. Zhang, Q., Ma, R., Xu, J.: Stability of Cellular Neural Networks with Delay. Electron. Lett. **37** (2001) 575–576
11. Zhang, Q., Ma, R., Wang, C., Xu, J.: On the Global Stability of Delayed Neural Networks. IEEE Trans.Automatic Control **48** (2003) 794–797
12. Zhang, Q., Wei, X.P. Xu, J.: Global Exponential Convergence Analysis of Delayed Neural Networks with Time-Varying Delays. Phys.Lett.A **318** (2003) 537–544
13. Zhou, D., Cao, J.: Globally Exponential Stability Conditions for Cellular Neural Networks with Time-Varying Delays. Appl.Math.Comput. **131** (2002) 487–496
14. Jiang, H., Li, Z., Teng, Z.: Boundedness and Stability for Nonautonomous Cellular Neural Networks with Delay. Phys.Lett.A **306** 313–325

Global Exponential Stability of Cohen-Grossberg Neural Networks with Multiple Time-Varying Delays*

Kun Yuan and Jinde Cao

Department of Mathematics, Southeast University, Nanjing 210096, China
jdcao@seu.edu.cn

Abstract. A new sufficient condition is presented ensuring the global exponential stability of Cohen-Grossberg neural networks with multiple time-varying delays by using an approach based on the Halanay inequality combing with Young inequality. Furthermore, a more subtle estimate is also given for the exponential decay rate.

1 Introduction

Cohen-Grossberg Neural Network (CGNNs) was proposed by Cohen and Grossberg in [9]. This model has been extensively studied in [1][7-10][11] due to its promising potential applications in classification, parallel computing, associative memory, especially in solving some optimization problems [11]. In general, time delays are often involved in neural networks due to various reasons, for example, the unavoidable finite switching speed of amplifiers in circuit implementation of a neural network, or deliberately introduced to achieve tasks of dealing with motion-related problems. Thus, the authors in [4] investigated Cohen-Grossberg neural networks (CGNNs) with multiple delays. To date, most works on CGNNs have been restricted to simple cases of constant delays, but few authors considered the stability of CGNNs with multiple variable delays. However, absolute constant delays for a process of dynamics change is almost not existent in practice, constant delay is only an ideal approach of variable delay. Therefore, the studies of the systems with variable delays have more important significance than ones of systems with constant delays. The aim of this paper is to investigate the exponential stability of CGNNs with multiple time-varying delays and give the exponential decay rate.
The organization of this paper is as follows. In section 2, we will derive new sufficient conditions for checking global exponential stability of the equilibrium point of CGNNs with multiple variable delays, an illustrate example is given in section 3, and conclusion follows in section 4.

---

* This work was jointly supported by the National Natural Science Foundation of China under Grant 60373067, the Natural Science Foundation of Jiangsu Province, China under Grants BK2003053 and BK2003001, Qing-Lan Engineering Project of Jiangsu Province and the Foundation of Southeast University, Nanjing, China under grant XJ030714.

2   Global Exponential Stability Analysis

The Cohen-Grossberg neural network with multiple time-varying delays can be described by the following state equations:

$$\dot{x}_i(t) = -d_i(x_i(t))(c_i(x_i(t)) - \sum_{j=1}^{n} a_{ij} f_j(x_j(t)) - \sum_{k=1}^{K}\sum_{j=1}^{n} b_{ij}^k f_j(x_j(t-\tau_{ij}^k(t))) + J_i), \quad (1)$$

where $n \geq 2$ is the number of neurons, x_i denotes the state variable associated with the ith neuron, $d_i(x_i)$ represents an amplification function and $c_i(x_i)$ is behaved function. The feedback matrix $A = (a_{ij})_{n\times n}$ indicates the strength of the neuron interconnections within the network, while the delayed feedback matrix $B^k = (b_{ij}^k)_{n\times n}$ indicates the strength of the neuron interconnections within the network which is associated with time-varying delay parameter $\tau_{ij}^k(t)$. Finally, the activation function f_j describes the manner in which the neurons respond to each other. Typically, $\tau_{ij}^k(t) > 0$ represents the delay parameter and it is assumed that $\tau = \max(\tau_{ij}^k(t))$, $1 \leq i,j \leq n$, $1 \leq k \leq K$. Furthermore, J_i is the external constant input from outside of the system.
To establish the exponential stability of the model (1), it is first necessary to make the following assumptions regarding functions d_i, c_i and f_i:
(A_1) : Each function $d_i(x_i)$, where $i = 1, 2, \cdots, n$, is bounded, positive and continuous. Furthermore, $0 < m_i \leq d_i(x_i) \leq M_i < \infty$ for all $x_i \in R$;
(A_2) : Functions $c_i(x_i)$, where $i = 1, 2, \cdots, n$, are derivative. Moreover, $c_i'(x_i) > \gamma_i > 0$ for all $x_i \in R$;
(A_3) : Each function $f_i : R \to R$, $i = 1, 2, \cdots, n$ is bounded and satisfies $|f_i(\xi_1) - f_i(\xi_2)| \leq L_i|\xi_1 - \xi_2|$ for each $\xi_1, \xi_2 \in R$, $\xi_1 \neq \xi_2$.
Suppose that the system (1) is supplemented with initial conditions of the form

$$x_i(s) = \phi_i(s), \quad s \in [-\tau, 0], \quad i = 1, 2, \cdots, n,$$

in which $\phi_i(s)$ is continuous for $s \in [-\tau, 0]$ and Eq.(1) has an equilibrium point $x^* = (x_1^*, \cdots, x_n^*)$. We denote

$$\|\phi - x^*\| = \sup_{-\tau \leq s \leq 0} [\sum_{j=1}^{n} |\phi_j(s) - x_j^*|^r]^{\frac{1}{r}}.$$

We can rewrite the system (1) as follows through the transformation $z_i(t) = x_i(t) - x_i^*$,

$$\dot{z}_i(t) = -d_i(z_i + x_i^*)[c_i(z_i + x_i^*) - c_i(x_i^*) - \sum_{j=1}^{n} a_{ij}(f_j(z_j(t) + x_j^*) - f_j(x_j^*))$$

$$-\sum_{k=1}^{K}\sum_{j=1}^{n} b_{ij}^k(f_j(z_j(t-\tau_{ij}^k(t)) + x_j^*) - f_j(x_j^*))]. \quad (2)$$

Definition 1. We say that an equilibrium point $x^* = (x_1^*, \cdots, x_n^*)$ is globally exponentially stable if there exist constants $\epsilon > 0$ and $M \geq 1$ such that

$$\|x(t) - x^*\| \leq M\|\phi - x^*\|e^{-\epsilon t}, \quad t \leq 0.$$

In order to simplify the proofs, we first present some lemmas as follows.

Lemma 1. *(Young inequality [3,5])Assume that $a > 0$, $b > 0$, $p \geq 1$, $\frac{1}{p} + \frac{1}{q} = 1$, then the following inequality*

$$ab \leq \frac{1}{p}a^p + \frac{1}{q}b^q$$

holds.

Lemma 2. *(Halanay inequality [2,6])Let α and β be constants with $0 < \beta < \alpha$. Let $x(t)$ be a continuous nonnegative function on $t \geq t_0 - \tau$ satisfying $x'(t) \leq -\alpha x(t) + \beta \bar{x}(t)$ for $t \geq t_0$, where $\bar{x}(t) = \sup_{t-\tau \leq s \leq t} x(s)$. Then*

$$x(t) \leq \bar{x}(t_0)e^{-\delta(t-t_0)},$$

where δ is bound on the exponential convergence rate and is the unique positive solution of

$$\delta = \alpha - \beta e^{\delta \tau}.$$

Theorem 1. *Assume that there exist real constants ξ_{ij}, η_{ij} and positive constants $\lambda_i > 0$, $p > 1$, $i, j = 1, \cdots, n$ such that*

$$\alpha = \min_{1 \leq i \leq n} m_i \{p\gamma_i - \sum_{j=1}^{n}[\frac{p}{q}L_j|a_{ij}|^{(1-\xi_{ij})q} + \sum_{k=1}^{K}\frac{p}{q}L_j|b_{ij}^k|^{(1-\eta_{ij})q} + \frac{\lambda_j}{\lambda_i}L_i|a_{ji}|^{p\xi_{ji}}]\}$$

$$> \max_{1 \leq i \leq n} m_i \{\sum_{k=1}^{K}\sum_{j=1}^{n}\frac{\lambda_j}{\lambda_i}L_i|b_{ji}^k|^{p\eta_{ji}}\} = \beta, \quad (3)$$

where $\frac{1}{p} + \frac{1}{q} = 1$. Then the equilibrium point x^* of system (1) is globally exponentially stable.

Proof. Consider the function

$$V(t) = \sum_{i=1}^{n}\lambda_i|z_i(t)|^p,$$

where $z_i(t) = x_i(t) - x_i^*$.
Calculating and estimating the upper right derivative D^+V of V along the solution of (2) as follows,

$$D^+V(t) = \sum_{i=1}^{n}\lambda_i p|z_i(t)|^{p-1}\text{sign}(z_i(t))\dot{z}_i(t)$$

$$= \sum_{i=1}^{n} \lambda_i p |z_i(t)|^{p-1} sign(z_i(t))\{-d_i(z_i + x_i^*)[c_i(z_i + x_i^*) - c_i(x_i^*)]$$

$$- \sum_{j=1}^{n} a_{ij}(f_j(z_j(t) + x_j^*) - f_j(x_j^*))$$

$$- \sum_{k=1}^{K} \sum_{j=1}^{n} b_{ij}^k (f_j(z_j(t - \tau_{ij}^k(t)) + x_j^*) - f_j(x_j^*))]\}$$

$$\leq \sum_{i=1}^{n} \lambda_i m_i p [-\gamma_i |z_i(t)|^p + \sum_{j=1}^{n} |a_{ij}||L_j||z_i(t)|^{p-1}|z_j(t)|$$

$$+ \sum_{k=1}^{K} \sum_{j=1}^{n} |b_{ij}^k||L_j||z_i(t)|^{p-1}|z_j(t - \tau_{ij}^k(t))|]. \quad (4)$$

By Lemma 1, we have

$$|a_{ij}||z_i(t)|^{p-1}|z_j(t)| \leq \frac{1}{p}|a_{ij}|^{p\xi_{ij}}|z_j(t)|^p + \frac{1}{q}|a_{ij}|^{(1-\xi_{ij})q}|z_i(t)|^p, \quad (5)$$

$$|b_{ij}^k||z_i(t)|^{p-1}|z_j(t-\tau_{ij}^k(t))| \leq \frac{1}{p}|b_{ij}^k|^{p\eta_{ij}}|z_j(t-\tau_{ij}^k(t))|^p + \frac{1}{q}|b_{ij}^k|^{(1-\eta_{ij})q}|z_i(t)|^p. \quad (6)$$

Substituting inequalities (5) and (6) into (4), we obtain

$$D^+V(t) \leq \sum_{i=1}^{n} \lambda_i m_i p \{-\gamma_i |z_i(t)|^p + \sum_{j=1}^{n} \frac{1}{p} L_j |a_{ij}|^{p\xi_{ij}}|z_j(t)|^p$$

$$+ \sum_{j=1}^{n} \frac{1}{q} L_j |a_{ij}|^{(1-\xi_{ij})q}|z_i(t)|^p + \sum_{k=1}^{K} \sum_{j=1}^{n} \frac{1}{p} L_j |b_{ij}^k|^{p\eta_{ij}}|z_j(t-\tau_{ij}^k(t))|^p$$

$$+ \sum_{k=1}^{K} \sum_{j=1}^{n} \frac{1}{q} L_j |b_{ij}^k|^{(1-\eta_{ij})q}|z_i(t)|^p\}$$

$$= \sum_{i=1}^{n} \lambda_i m_i \{-p\gamma_i + \sum_{j=1}^{n} \frac{\lambda_j}{\lambda_i} L_i |a_{ji}|^{p\xi_{ji}} + \sum_{j=1}^{n} \frac{p}{q} L_j |a_{ij}|^{(1-\xi_{ij})q}$$

$$+ \sum_{k=1}^{K} \sum_{j=1}^{n} \frac{p}{q} L_j |b_{ij}^k|^{(1-\eta_{ij})q}\}|z_i(t)|^p$$

$$+ \sum_{i=1}^{n} \lambda_i m_i [\sum_{k=1}^{K} \sum_{j=1}^{n} \frac{\lambda_j}{\lambda_i} L_i |b_{ji}^k|^{p\eta_{ji}}]|z_i(t-\tau_{ji}^k(t))|^p$$

$$\leq - \min_{1 \leq i \leq n} m_i \{p\gamma_i - \sum_{j=1}^{n} [\frac{p}{q} L_j |a_{ij}|^{(1-\xi_{ij})q}$$

$$+ \sum_{k=1}^{K} \frac{p}{q} L_j |b_{ij}^k|^{(1-\eta_{ij})q} + \frac{\lambda_j}{\lambda_i} L_i |a_{ji}|^{p\xi_{ji}}]\} V(t)$$

$$+ \max_{1\leq i\leq n} m_i \{\sum_{k=1}^{K}\sum_{j=1}^{n} \frac{\lambda_j}{\lambda_i} L_i |b_{ji}^k|^{p\eta_{ji}}\} \overline{V}(t). \tag{5}$$

According to Lemma 2, we obtain

$$V(t) \leq \overline{V}(t_0) e^{-\delta(t-t_0)}.$$

Furthermore, we have

$$\|x(t) - x^*\| \leq (\frac{\lambda_{max}}{\lambda_{min}})^{\frac{1}{p}} \|\phi - x^*\| e^{-\frac{\delta}{p}t},$$

where $\delta = \alpha - \beta e^{\delta \tau}$.
Therefore, the proof is completed.

Corollary 1. *If the following inequality holds*

$$\min_{1\leq i\leq n} m_i \{2\gamma_i - \sum_{j=1}^{n}[L_j(|a_{ij}| + \sum_{k=1}^{K}|b_{ji}^k|) + L_i|a_{ji}|]\}$$

$$> \max_{1\leq i\leq n} m_i \{\sum_{k=1}^{K}\sum_{j=1}^{n} L_i |b_{ji}^k|\},$$

then the equilibrium point x^* is globally exponentially stable.

Proof. It is easy to check that the inequality (3) is satisfied by taking $p = q = 2$, $\lambda_i = 1$, $\xi_{ij} = \eta_{ij} = 0.5$, for $i, j = 1, 2, \cdots, n$, and hence the Theorem 1. implies Corollary 1.

3 An Illustrate Example

Consider the following Cohen-Grossberg neural network with multiple time-varying delays

$$\frac{dx_1}{dt} = -(2 + cosx_1))[2x_1 - f_1(x_1) - 0.5 f_2(x_2)$$
$$-0.09 f_1(x_1(t - |\sin t|)) + 0.08 f_2(x_2(t - |\sin t|))$$
$$-0.05 f_1(x_1(t - |\cos t|)) - 0.1 f_2(x_2(t - |\cos t|)) + 1],$$

$$\frac{dx_2}{dt} = -(2 + sinx_2))[2x_2 + 0.5 f_1(x_1) - f_2(x_2)$$
$$+0.05 f_1(x_1(t - |\sin t|)) - 0.15 f_2(x_2(t - |\sin t|))$$
$$+0.02 f_1(x_1(t - |\cos t|)) - 0.04 f_2(x_2(t - |\cos t|)) + 2],$$

where $f_i(x_i) = \tanh(x_i)$, $i = 1, 2$,

$$A = \begin{bmatrix} 1 & 0.5 \\ -0.5 & 1 \end{bmatrix}, \quad B^1 = \begin{bmatrix} 0.09 & -0.08 \\ -0.05 & 0.15 \end{bmatrix}, \quad B^2 = \begin{bmatrix} 0.05 & 0.1 \\ -0.02 & 0.04 \end{bmatrix}.$$

Obviously, $m_i = 1$, $\gamma_i = 2$ and $L_i = 1$ for $i = 1, 2$. By calculating, we obtain

$$\min_{1 \leq i \leq 2} m_i \{2\gamma_i - \sum_{j=1}^{2}[L_j(|a_{ij}| + \sum_{k=1}^{2}|b_{ji}^k|) + L_i|a_{ji}|]\} = 0.5$$

$$\max_{1 \leq i \leq 2} m_i \{\sum_{k=1}^{2}\sum_{j=1}^{2} L_i|b_{ji}^k|\} = 0.32 < 0.5.$$

From the Corollary 1., we deduce the equilibrium point is globally exponentially stable.

4 Conclusions

During the implementation process of neural networks, time-varying delays are inevitable. Aiming at the cases, in this paper, we have investigated the global exponential stability of a class of Cohen-Grossberg neural networks with multiple time-varying delays via Halanay inequality. We yield conditions that guarantee global exponential stability of the equilibrium point. Moreover, the obtained result shows explicitly a more subtle evaluation of exponential decay rate and is hence of practical interest in designing a fast and stable neural network.

References

1. Hwang, C.C., Cheng, C.J., Liao, T.L.: Globally Exponential Stability of Generalized Cohen-Grossberg Neural Networks with Delays. Phy. Lett. A **319** (2003) 157-166
2. Cao, J., Wang, J.: Absolute Exponential Stability of Recurrent Neural Networks with Lipschitz-continuous Activation Functions and Time Delays, Neural Networks, **17** (2004) 379-390.
3. Hardy, G.H., Littlewood, J.E., Polya, G.: Inequalities, second ed. Cambridge University Press, London (1952)
4. Ye, H., Michel, A.N., Wang, K.: Analysis of Cohen-Grossberg Neural Networks with Multiple Delays. Phy. Rev. E **51** (1998) 2611-2618
5. Cao, J.: New Results Concerning Exponential Stability and Periodic Solutions of Delayed Cellular Neural Networks. Phy. Lett. A **307** (2003) 136-147
6. Gopalsamy, K.: Stability and Oscillations in Delay Differential Equations of Populations Dynamics. Kluuner Academic, dordrecht (1992)
7. Wang, L., Zou, X.F.: Exponential Stability of Cohen-Grossberg Neural Networks. Neural Networks **15** (2002) 415-422
8. Wang, L., Zou, X.F.: Harmless Delays in Cohen-Grossberg Neural Networks. Phys. D **170** (2002) 162-173
9. Cohen, M., Grossbeerg, S.: Absolute Stability of Global Pattern Formation and Parallel Memory Storage by Competitive Neural Networks. IEEE Trans. Systems, Man and Cybernetics **13**(1983) 815-826
10. Chen, T.P., Rong, L.B.: Delay-independent Stability Analysis of Cohen-Grossberg Neural Networks. Phy. Lett. A **317** (2003) 436-449
11. Takahashi, Y.: Solving Optimization Problems with Variable-constraint by an Extend Cohen-Grossberg Model. Theoretical Computer Science **158** (1996) 279-341

Stability of Stochastic Cohen-Grossberg Neural Networks

Lin Wang

Department of Mathematics and Statistics
McMaster University
Hamilton, ON, Canada L8S 4K1

Abstract. Almost sure stability and instability of stochastic Cohen–Grossberg neural networks are addressed in this paper. Our results can be used as theoretic guidance to stabilize neural networks in practical applications when stochastic noise is take into consideration.

1 Introduction

The dynamics of neural networks have been extensively investigated in the past two decades. See, for example, [2], [3], [4], [6], [11], [14] and the reference therein. Most neural network models proposed and discussed in the literature are deterministic. However, in real nervous systems and in the implementation of artificial neural networks, noise is unavoidable [5] and should be taken into consideration in modeling. Therefore it is of practical importance to study the stochastic neural networks (see [1], [8] and [9]).

In this paper, we consider a stochastic Cohen-Grossberg neural network model, described by the following stochastic differential equations

$$du(t) = -A(u(t))[b(u(t)) - Wg(u(t))]dt + \sigma(u(t))dB(t), \ t \geq 0, \quad (1)$$

where $u(t) = (u_1(t), \ldots, u_n(t))^T$ is the neuron states vector; $A(u(t)) = diag(a_i(u_i(t)))$; $b(u(t)) = (b_1(u(t)), \cdots, b_n(u(t)))^T$; $W = (w_{ij})_{n \times n}$ is the connection matrix; $g(u) = (g_1(u_1), \cdots, g_n(u_n))^T$ is the activation functions vector; $\sigma = (\sigma_{ij})_{n \times n}$ is the diffusion coefficient matrix and $B(t) = (B_1(t), \ldots, B_n(t))^T$ is an n-dimensional Brownian motion.

It follows from a standard textbook [10] on stochastic differential equation that if $a_i(u), g_i(u), \sigma_{ij}(u)$ are locally Lipschitz and satisfy the linear growth condition, then, for any given initial data u_0, there is a unique solution for system (1). We denote it by $u(t; u_0)$.

In what follows, we assume that $b_i(0) = 0, g_i(0) = 0$ for $i \in N(1, n)$ and $\sigma(0) = \mathbf{0}$ so that $u = 0$ is a trivial (equilibrium) solution of (1). For any undefined concept appears in this paper, we refer to [10].

Let $C^{2,1}(\mathbb{R}^n \times \mathbb{R}_+; \mathbb{R}_+)$ denote the family of all nonnegative functions $V(u;t)$ on $\mathbb{R}^n \times \mathbb{R}_+$ which are twice differentiable in x and once in t. For each such $V(u;t)$, we define an operator $\mathcal{L}V$ associated with (1) as

$$\mathcal{L}V(u;t) = V_t(u;t) + V_u(u;t)(-A(u(t))[B(u(t)) - Wg(u(t))]) \\ + \tfrac{1}{2}trace[\sigma^T(u(t))V_{uu}(u;t)\sigma(u(t))], \quad (2)$$

where

$$V_t(u;t) = \frac{\partial V(u;t)}{\partial t}, \quad V_u = \left(\frac{\partial V(u;t)}{\partial u_1}, \cdots, \frac{\partial V(u;t)}{\partial u_n}\right), \quad V_{uu} = \left(\frac{\partial^2 V(u;t)}{\partial u_i \partial u_j}\right)_{n \times n}.$$

Let $N(1,n) = \{1, 2, \ldots, n\}$ and $|u| = (\sum_{i=1}^n u_i^2)^{1/2}$ for $u = (u_1, u_2, \ldots, u_n)^T \in \mathbb{R}^n$. We give some assumptions which will be used later.

(H1) for each $i \in N(1,n)$, there is $\gamma_i > 0$ such that $ub_i(u) \geq \gamma_i u^2$;
(H2) for each $i \in N(1,n)$, $|g_i(u)| \leq L_i |u|$;
(H3) for each $i \in N(1,n)$, $ug_i(u) > 0$ for $u \neq 0$;
(H4) for each $i \in N(1,n)$, $\alpha_i \leq a_i(u) \leq \bar{a}_i$;
(H5) $|\sigma(u)|^2 \leq k|u|^2$;
(H6) for each $i \in N(1,n)$, $0 \leq \frac{b_i(u) - b_i(v)}{u - v} \leq \beta_i \; \forall u, v \in \mathbb{R}$ with $u \neq v$.

Our main results concerning stability and instability of stochastic Cohen–Grossberg neural networks (1) are given in Section 2.

It is Cohen and Grossberg [3] who first proposed the Cohen–Grossberg neural network model. For other papers dealing with dynamics of continuous–time generalized Cohen-Grossberg neural networks, we refer to [12] and [13].

2 Main Results

First, we give the following lemma which plays a crucial role in establishing our main results.

Lemma 1. *Assume that there exist a symmetric positive definite matrix Q and two real numbers $\mu \in \mathbb{R}$ and $\rho \geq 0$ such that*

$$2u^T Q[-A(u)b(u) + A(u)Wg(u)] + trace(\sigma^T Q \sigma) \leq \mu u^T Q u$$

and

$$|u^T Q \sigma(u)|^2 \geq \rho (u^T Q u)^2, \quad \forall u \in \mathbb{R}^n. \tag{3}$$

Then we have

$$\limsup_{t \to \infty} \frac{1}{t} \log(|u(t; u_0)|) \leq -(\rho - \frac{\mu}{2}) \quad a.s.$$

whenever $u_0 \neq 0$. If $\rho > \frac{\mu}{2}$, then the trivial solution of (1) is almost surely exponentially stable.

Proof. The proof follows from Theorem 4.3.3 [10] by letting $V(u;t) = u^T Q u$.

Based on this lemma, we establish our main results as follows.

Theorem 1. *Suppose that (H1), (H2), (H4) and (H5) hold. Assume also that there are a matrix $Q = diag(q_1, q_2, \ldots, q_n) > 0$, a real number $\rho \geq 0$ and some positive constants ξ_i, $i \in N(1,n)$ such that (3) holds and*

$$\lambda := \min_{i \in N(1,n)} \left\{ 2\alpha_i q_i \gamma_i - \bar{a}_i q_i \sum_{j=1}^n |w_{ij}| L_j \xi_j - \frac{L_i}{\xi_i} \sum_{j=1}^n \bar{a}_j q_j |w_{ji}| \right\} > 0.$$

If $\rho > \frac{\mu}{2}$ with

$$\mu := k \frac{\max_{i \in N(1,n)} q_i}{\min_{i \in N(1,n)} q_i} - \frac{\lambda}{\max_{i \in N(1,n)} q_i},$$

then the trivial solution of (1) is almost surely exponentially stable, i.e.,

$$\limsup_{t \to \infty} \frac{1}{t} \log(|u(t; u_0)|) \leq -\left(\rho - \frac{\mu}{2}\right) \quad a.s.$$

whenever $u_0 \neq 0$.

Proof. Let $V(u,t)$ be $V(u,t) = u^T(t)Qu(t)$. Then, by (2),

$$\mathcal{L}V = 2u^T Q[-A(u)b(u) + A(u)Wg(u)] + trace(\sigma^T Q \sigma)$$

$$= -2 \sum_{i=1}^{n} a_i(u_i) q_i u_i b_i(u_i) + 2 \sum_{j=1}^{n} a_i(u_i) q_i u_i \sum_{j=1}^{n} w_{ij} g_j(u_j)$$

$$+ trace(\sigma^T(u) Q \sigma(u))$$

Following from the assumptions (H1), (H2), (H4) and (H5), we have

$$\mathcal{L}V \leq -2 \sum_{i=1}^{n} \alpha_i q_i \gamma_i u_i^2 + 2 \sum_{j=1}^{n} \bar{a}_i q_i u_i \sum_{j=1}^{n} |w_{ij}| L_j |u_j|$$

$$+ trace(\sigma^T(u) Q \sigma(u))$$

$$\leq -2 \sum_{i=1}^{n} \alpha_i q_i \gamma_i u_i^2 + \sum_{j=1}^{n} \bar{a}_i q_i \sum_{j=1}^{n} |w_{ij}| L_j (\xi_j u_i^2 + \frac{1}{\xi_j} u_j^2)$$

$$+ \max_{i \in N(1,n)} q_i |\sigma(u)|^2$$

$$= -\sum_{i=1}^{n} \left[2\alpha_i q_i \gamma_i - \bar{a}_i q_i \sum_{j=1}^{n} |w_{ij}| L_j \xi_j - \frac{L_i}{\xi_i} \sum_{j=1}^{n} \bar{a}_j q_j |w_{ji}| \right] u_i^2$$

$$+ \max_{i \in N(1,n)} q_i |\sigma(u)|^2$$

$$\leq -\lambda \sum_{i=1}^{n} u_i^2 + k \max_{i \in N(1,n)} q_i \sum_{i=1}^{n} u_i^2$$

$$\leq \left(-\frac{\lambda}{\max_{i \in N(1,n)} q_i} + k \frac{\max_{i \in N(1,n)} q_i}{\min_{i \in N(1,n)} q_i} \right) \sum_{i=1}^{n} q_i u_i^2$$

$$= \mu u^T Q u$$

The rest of the proof is a consequence of Lemma 1.

Note that if $\mu < 0$, then one can take $\rho = 0$. Thus, we have

Corollary 1. *Suppose that* (H1), (H2), (H4) *and* (H5) *hold. Assume also that there are a matrix* $Q = diag(q_1, q_2, \ldots, q_n) > 0$ *and some positive constants* ξ_i, $i \in N(1,n)$ *such that*

$$\lambda := \min_{i \in N(1,n)} \left\{ 2\alpha_i q_i \gamma_i - \bar{a}_i q_i \sum_{j=1}^{n} |w_{ij}| L_j \xi_j - \frac{L_i}{\xi_i} \sum_{j=1}^{n} \bar{a}_j q_j |w_{ji}| \right\} > 0.$$

If
$$\mu := k \frac{\max_{i \in N(1,n)} q_i}{\min_{i \in N(1,n)} q_i} - \frac{\lambda}{\max_{i \in N(1,n)} q_i} < 0,$$

then the trivial solution of (1) is almost surely exponentially stable, i.e.,
$$\limsup_{t \to \infty} \frac{1}{t} \log(|u(t; u_0)|) \leq \frac{\mu}{2} \quad a.s.$$

whenever $u_0 \neq 0$.

If we denote $[w]^+ = \max\{0, w\}$, then the same argument, together with assumption (H3), gives the following

Theorem 2. *Assume that that (H1)-(H5) hold. Assume also that there are a matrix $Q = diag(q_1, q_2, \ldots, q_n) > 0$ and a real number $\rho \geq 0$ such that (3) holds and*

$$\hat{\lambda} := \min_{i \in N(1,n)} \left\{ 2\alpha_i q_i \gamma_i - 2\bar{a}_i q_i L_i [w_{ii}]^+ - \bar{a}_i q_i \sum_{j=1, j \neq i}^{n} |w_{ij}| L_j - L_i \sum_{j=1, j \neq i}^{n} \bar{a}_j q_j |w_{ji}| \right\} > 0.$$

If $\rho > \frac{\mu_1}{2}$ with
$$\mu_1 := k \frac{\max_{i \in N(1,n)} q_i}{\min_{i \in N(1,n)} q_i} - \frac{\hat{\lambda}}{\max_{i \in N(1,n)} q_i},$$
then the trivial solution of (1) is almost surely exponentially stable, i.e.,
$$\limsup_{t \to \infty} \frac{1}{t} \log(|u(t; u_0)|) \leq -\left(\rho - \frac{\mu_1}{2}\right) \quad a.s.$$

whenever $u_0 \neq 0$.

Corollary 2. *Suppose that (H1)-(H5) hold. Assume also that there is a matrix $Q = diag(q_1, q_2, \ldots, q_n) > 0$ such that*

$$\hat{\lambda} := \min_{i \in N(1,n)} \left\{ 2\alpha_i q_i \gamma_i - 2\bar{a}_i q_i L_i [w_{ii}]^+ - \bar{a}_i q_i \sum_{j=1, j \neq i}^{n} |w_{ij}| L_j - L_i \sum_{j=1, j \neq i}^{n} \bar{a}_j q_j |w_{ji}| \right\} > 0.$$

If
$$\mu_1 := k \frac{\max_{i \in N(1,n)} q_i}{\min_{i \in N(1,n)} q_i} - \frac{\hat{\lambda}}{\max_{i \in N(1,n)} q_i} < 0,$$

then the trivial solution of (1) is almost sure exponential stable, i.e.,
$$\limsup_{t \to \infty} \frac{1}{t} \log(|u(t; u_0)|) \leq \frac{\mu_1}{2} \quad a.s.$$

whenever $u_0 \neq 0$.

Next we present an instability result.

Theorem 3. *Assume that (H2)-(H4) and (H6) hold. If there are a matrix $Q = diag(q_1, q_2, \ldots, q_n) > 0$ and a real number $\hat{\rho} > 0$ such that*

$$|u^T Q\sigma(u)|^2 \leq \hat{\rho}(u^T Q u)^2 \tag{4}$$

holds for all $u \in \mathbb{R}^n$ and $|\sigma(u)|^2 \geq \bar{k}|u|^2$, then the solution of (1) satisfies

$$\liminf_{t \to \infty} \frac{1}{t} \log(|u(t; u_0)|) \geq \frac{\hat{\mu}}{2} - \hat{\rho} \quad a.s.$$

whenever $u_0 \neq 0$ and $\hat{\mu}$ will be specified later in the proof. Particularly if $\frac{\hat{\mu}}{2} - \hat{\rho} > 0$, then (1) is almost surely exponentially unstable.

Proof. Let $V = u^T Q u = \sum_{i=1}^{n} q_i u_i^2$, then

$$\mathcal{L}V = -2u^T QA(u)b(u) + 2u^T QA(u)Wg(u) + \mathrm{trace}(\sigma^T(u)Q\sigma(u))$$

$$\geq -2\sum_{i=1}^{n} \bar{a}_i \beta_i q_i u_i^2 + 2\sum_{i=1}^{n} g_i(u_i) \sum_{j=1}^{n} a_j q_j w_{ji} u_j$$

$$+ \min\{q_i, i \in N(1,n)\}|\sigma(u)|^2$$

$$\geq -2\sum_{i=1}^{n} \bar{a}_i \beta_i q_i u_i^2 + 2\sum_{i=1}^{n} [w_{ii}]^- L_i \bar{a}_i q_i u_i^2$$

$$- \sum_{i=1}^{n} \left(\frac{L_i}{q_i} \sum_{j=1, j\neq i}^{n} \bar{a}_j q_j |w_{ji}| + \bar{a}_i \sum_{j=1, j\neq i}^{n} L_j |w_{ij}| \right) q_i u_i^2$$

$$+ \min\{q_i, i \in N(1,n)\}|\sigma(u)|^2$$

$$\geq -\nu \sum_{i=1}^{n} q_i u_i^2 + \min\{q_i, i \in N(1,n)\}\bar{k}|u|^2,$$

where

$$\nu := \max_{i \in N(1,n)} \left\{ 2\bar{a}_i \beta_i + \frac{L_i}{q_i} \sum_{j=1, j\neq i}^{n} \bar{a}_j q_j |w_{ji}| + \bar{a}_i \sum_{j=1, j\neq i}^{n} L_j |w_{ij}| - 2[w_{ii}]^- \bar{a}_i L_i \right\}$$

and $[w_{ii}]^- = \min(0, w_{ii})$. Then we have

$$\mathcal{L}V \geq \hat{\mu} \sum_{i=1}^{n} q_i u_i^2,$$

where

$$\hat{\mu} := \bar{k} \frac{\min\{q_i, i \in N(1,n)\}}{\max\{q_i, i \in N(1,n)\}} - \nu.$$

The rest of the proof follows from Theorem 4.3.5. of [10].

References

1. Blythe, S., Mao, X., Liao, X.: Stability of stochastic delay neural networks, J. the Franklin Institute, **338**(2001), 481–495.
2. Carpenter, G.: Neural network Models for pattern recognition and associative memory, Neural Networks, **2**(1989), 243–257.
3. Cohen, M., S. Grossberg, S.: Absolute stability and global pattern formation and parallel memory storage by competitive neural networks, IEEE Trans. Syst. Man. Cybern. SMC, **13**(1983), 815–825.
4. Forti, M.: On global asymptotic stability of a class of nonlinear systems arising in neural network theory, J. Differential Equations **113**(1994), 246–264.
5. Haykin, S.: Neural Networks, Prentice-Hall, NJ, 1994.
6. Hopfield, J.: Neurons with graded response have collective computational properties like those of two-stage neurons, Proc. Nat. Acad. Sci. U.S.A. **81**(1984), 3088–3092.
7. Liao, X., Chen, G. Sanchez, E.: Delay-dependent exponential stability analysis of delayed neural networks: an LMI approach, Neural Networks, **15**(2002), 855–866.
8. Liao, X., Mao, X.: Exponential stability and instability of stochastic neural networks, Stochast. Anal. Appl., **14**(1996), 165–185.
9. Liao, X., Mao, X.: Stability of stochastic neural networks, Neural, Parallel Sci. Comput, **4**(1996), 205–114.
10. Mao, X.: Stochastic Differential Equations & Applications, Horwood Publishing, Chichester, 1997.
11. Tank, D. Hopfield, J.: Simple neural optimization networks: An A/D converter, signal decision circuit and a linear programming circult, IEEE Trans. Circuits Systems, **33**(1986), 533–541.
12. Wang, L., Zou, X.: Harmless delays in Cohen-Grossberg neural networks, Physica D. **170**(2002), 162–173.
13. Wang, L., Zou, X.: Exponential stability of Cohen–Grossberg neural networks, Neural Networks, **15** (2002): 415-422.
14. Wang, J.: Recurrent neural networks for computing pseudoinverses of rank-deficient matrices, SIAM J. Sci. Comput. **18**(1997), 1479–1493.

A Novel Approach to Exponential Stability Analysis of Cohen-Grossberg Neural Networks

Anhua Wan[1,2], Weihua Mao[2], and Chun Zhao[1,3]

[1] Faculty of Science, Xi'an Jiaotong University, Xi'an 710049, P. R. China
anhuawan@tom.com, ah_one@sohu.com
[2] Department of Applied Mathematics, Faculty of Science, South China Agricultural University, Guangzhou 510642, P. R. China
mwh_eastport@sohu.com
[3] School of Mathematics and Computer Sciences, Ningxia University, Yinchuan 750021, P. R. China
zhaochun@mailst.xjtu.edu.cn

Abstract. The stability of neural networks is fundamental for successful applications of the networks. In this paper, the exponential stability of Cohen-Grossberg Neural Networks is investigated. To avoid the difficulty of the construction of a proper Lyapunov function, a new concept named generalized relative nonlinear measure is introduced, and thus a novel approach to stability analysis of neural networks is developed. With this new approach, sufficient conditions for the existence, uniqueness of the equilibrium and the exponential stability of the Cohen-Grossberg neural networks are presented. Meanwhile, the exponential convergence rate of the networks to stable equilibrium point is estimated, and the attraction region of local stable equilibrium point is also characterized.

1 Introduction

The stability of neural networks is of great significance for both practical and theoretical purposes and has become an important area of research on neural networks in recent years (see, for example, [1],[2],[3],[4] and references therein).

We consider Cohen-Grossberg neural networks model of the form (see [5],[6])

$$\frac{\mathrm{d}}{\mathrm{d}t}u_i(t) = -a_i(u_i(t))\left[b_i(u_i(t)) - \sum_{j=1}^{n} w_{ij} f_j(u_j(t)) + I_i\right], \quad i = 1, 2, \cdots, n, \quad (1)$$

where $u_i(t)$ denotes the state variable associated with the i-th neuron, a_i denotes an amplification function, b_i is an appropriately behaved function, $W = (w_{ij})$ is the $n \times n$ connection weight matrix, f_i denotes the activation function, I_i denotes a constant external input, and $n \geq 2$ is the number of neurons in the networks.

The stability of Cohen-Grossberg neural networks(CGNNs) is of considerable importance and has attracted much attention (e.g. [6],[7],[8],[9]), since it includes many models in population biology, neurobiology, evolutionary theory etc., and it also takes the famous Hopfield-type neural networks(HNNs) as its special case([10]). The main difficulty for stability analysis of (1) results from the nonlinearity of a_i, b_i and f_i. Almost all the existing works have been performed under some special assumptions on a_i, b_i and f_i such as differentiability, boundedness and monotonicity. Removing these usual assumptions, in this paper, we only assume that each f_i is Lipschitz continuous (i.e., for each $i = 1, 2, \cdots, n$, there exists a constant $M_i > 0$ such that $|f_i(s_1) - f_i(s_2)| \leq M_i|s_1 - s_2|$ ($\forall s_1, s_2 \in \mathbf{R}$)), and $m_i = \sup\limits_{s_1,s_2 \in R, s_1 \neq s_2} \frac{|f_i(s_1) - f_i(s_2)|}{|s_1 - s_2|}$ is the minimal Lipschitz constant of f_i. Besides, we assume that each a_i is bounded, positive (i.e., $\forall u_i \in R, 0 < \grave{\alpha}_i \leq a_i(u_i) \leq \acute{\alpha}_i$) and Lipschitz continuous, and each b_i is Lipschitz continuous. We do not make any supposition on the connection matrix $W = (w_{ij})_{n \times n}$.

It is well known that most of the existing approaches to stability analysis of neural networks are based on the Lyapunov function approach. However, the construction of a valid Lyapunov function is usually rather difficult. In Qiao et al. [11], based on a new concept named nonlinear measure of nonlinear operator, a novel approach to stability analysis of nonlinear systems in R^n is proposed and successfully applied to Hopfield-type neural networks model. The purpose of the present paper is to further develop an effective approach to stability analysis for nonlinear systems in Banach space by introducing the substantial generalization of the nonlinear measure concept, and then apply the new approach to the stability analysis of the Cohen-Grossberg neural networks. By this novel approach, we not only derive some new criteria for the exponential stability of neural networks (1), but also characterize the exponential decay estimation of solution of (1) and the attraction region of local stable equilibrium point.

2 Stability Analysis via Generalized Nonlinear Measure

In this section, we first generalize the notion of nonlinear measure in [11], and then develop a novel approach to stability analysis of nonlinear systems. By means of the approach, we will examine stability of neural networks model (1).

Assume X is a Banach space with endowed norm $\|\cdot\|$, Ω is an open subset of X, F is a nonlinear operator defined on Ω, and $x(t) \in \Omega$. Consider the following system
$$\frac{dx(t)}{dt} = F(x(t)), \ t \geq t_0 . \tag{2}$$

Definition 1. (*Qiao, Peng & Xu [11]*) *Suppose R^n is the n-dimensional real vector space endowed with 1-norm $\|\cdot\|_1$ ($\|x\|_1 = \sum_{i=1}^n |x_i|$), Ω is an open subset of R^n, $F : \Omega \to R^n$ is an operator, and x^0 is any fixed point in Ω. The constant $m_\Omega(F) = \sup\limits_{x,y \in \Omega, x \neq y} \frac{\langle F(x) - F(y), sign(x-y) \rangle}{\|x-y\|_1}$ is called the nonlinear measure of F on Ω; The constant $m_\Omega(F, x^0) = \sup\limits_{x,x^0 \in \Omega, x \neq x^0} \frac{\langle F(x) - F(x^0), sign(x-x^0) \rangle}{\|x-x^0\|_1}$ is called the relative nonlinear measure of F at x^0. Here, $\langle \cdot, \cdot \rangle$ denotes the inner product of vectors in R^n, and $sign(x) = (sign(x_1), sign(x_2), \cdots, sign(x_n))^T$ denotes the sign vector of $x \in R^n$, where $sign(x_i)$ is the usual sign function of each $x_i \in \mathbf{R}$.*

The nonlinear measure plays an important role in characterizing stability of system (2) when $X = R^n$. In this special case, if $m_\Omega(F) < 0$, then (2) has a unique equilibrium point in Ω and the equilibrium point is exponentially stable ([11]). We expect that there may exist some quantity which can characterize the stability of general nonlinear system (2) in any Banach space. Peng & Xu [12] introduced the following concept of general nonlinear operators to analyze the asymptotic behavior of nonlinear semigroup of Lipschitz operators.

Definition 2. (Peng & Xu [12]) *Suppose Ω is an open subset of Banach space X, and $F : \Omega \to X$ is an operator. $\alpha_\Omega(F) = \sup\limits_{x,y\in\Omega, x\neq y} \frac{1}{\|x-y\|} \lim\limits_{r\to+\infty} [\|(F+rI)x - (F+rI)y\| - r\|x-y\|]$ is called the generalized nonlinear measure of F on Ω.*

Inspired by the above definitions, we further introduce a new notion for general nonlinear operators.

Definition 3. *Suppose Ω is an open subset of Banach space X, $F : \Omega \to X$ is an operator, and $x^0 \in \Omega$ is any fixed point. The constant $\alpha_\Omega(F, x^0) = \sup\limits_{x\in\Omega, x\neq x^0} \frac{1}{\|x-x^0\|} \lim\limits_{r\to+\infty} [\|(F+rI)x - (F+rI)x^0\| - r\|x-x^0\|]$ is called the generalized relative nonlinear measure of F at x^0.*

Similar to nonlinear measure and the generalized nonlinear measure, we can readily verify that the generalized relative nonlinear measure satisfies: (1) $\alpha_\Omega(G+H, x^0) \leq \alpha_\Omega(G, x^0) + \alpha_\Omega(H, x^0)$; (2) $\alpha_\Omega(kG, x^0) = k\alpha_\Omega(G, x^0), \forall k \geq 0$; (3) $\alpha_\Omega(G + lI, x^0) = \alpha_\Omega(G, x^0) + l, \forall l \in \mathbf{R}$. Moreover, it is clear that $\forall x^0 \in \Omega$, $\alpha_\Omega(F, x^0) \leq \alpha_\Omega(F)$ holds, and it is worth noting that this inequality may hold strictly, which can be clearly shown by the following example.

Example 1. Consider nonlinear system (2) where $F(x) = -\frac{3}{5}x + \sin^2 x, \forall x \in \mathbf{R}$. By Definitions 2 and 3, we immediately deduce that $\alpha_\mathbf{R}(F) = -\frac{3}{5} + 1 = \frac{2}{5} > 0$ and $\alpha_\mathbf{R}(F, 0) = -\frac{3}{5} + \sup\limits_{x\neq 0} \frac{\sin^2 x}{x} < 0$, thus $\alpha_\mathbf{R}(F, 0) < \alpha_\mathbf{R}(F)$ holds.

We will show that the generalized (relative) nonlinear measure can characterize the uniqueness of equilibrium point of nonlinear systems in Banach space.

Lemma 1. *If $\alpha_\Omega(F) < 0$, then F is a one-to-one mapping on Ω.*

Proof. Suppose there exist $x_1, x_2 \in \Omega (x_1 \neq x_2)$ satisfy $F(x_1) = F(x_2)$. Then, in view of Definition 2, we have $\alpha_\Omega(F) \geq \frac{1}{\|x_1-x_2\|} \lim\limits_{r\to+\infty} [\|(F+rI)x_1 - (F+rI)x_2\| - r\|x_1-x_2\|] = 0$, which contradicts to $\alpha_\Omega(F) < 0$. Thus, F is one-to-one on Ω.

Lemma 2. *If $x^* \in \Omega$ is an equilibrium point of system (2) and $\alpha_\Omega(F, x^*) < 0$, then there is no other equilibrium point of (2) in Ω than x^*.*

Proof. Otherwise, let $\hat{x} \in \Omega(\hat{x} \neq x^*)$ be any other equilibrium point of (2). Then $F(\hat{x}) = F(x^*) = 0$. By Definition 3, we infer $\alpha_\Omega(F, x^*) \geq \frac{1}{\|\hat{x}-x^*\|} \lim\limits_{r\to+\infty} [\|(F+rI)\hat{x} - (F+rI)x^*\| - r\|\hat{x}-x^*\|] = 0$, which contradicts to $\alpha_\Omega(F, x^*) < 0$. Thus $\hat{x} = x^*$, and therefore we assert the equilibrium point of (2) is unique in Ω.

In the following, we will justify that the generalized (relative) nonlinear measure can also characterize the stability of nonlinear systems in any Banach space.

Theorem 1. *If the operator F in the system (2) satisfies $\alpha_\Omega(F) < 0$, then there is at most one equilibrium point of (2) in Ω. Moreover, any two solutions $x(t)$ and $y(t)$ respectively initiated from $x(t_0) = x_0 \in \Omega$ and $y(t_0) = y_0 \in \Omega$ satisfy*

$$\|x(t) - y(t)\| \leq e^{\alpha_\Omega(F)(t-t_0)} \|x_0 - y_0\|, \quad \forall t \geq t_0. \tag{3}$$

Proof. Since $\alpha_\Omega(F) < 0$, it follows from Lemma 1 that F is one-to-one in Ω, therefore there is at most one point $u \in \Omega$ such that $F(u) = 0$, i.e., there is at most one equilibrium point of system (2) in Ω.

Assume $x(t)$ and $y(t)$ are the solutions of (2) respectively initiated from the initial values $x(t_0) = x_0 \in \Omega$ and $y(t_0) = y_0 \in \Omega$. Thus we have $(e^{rt}x(t))'_t = re^{rt}x(t) + e^{rt}Fx(t) = e^{rt}(F+rI)x(t)$ for all $t \geq t_0$ and $r \in \mathbf{R}$.

$\forall x_0, y_0 \in \Omega, r > 0, t > s \geq t_0, e^{rt}[x(t) - y(t)] = e^{rs}[x(s) - y(s)] + \int_s^t e^{ru}[(F + rI)x(u) - (F + rI)y(u)]du$, then $e^{rt}\|x(t) - y(t)\| - e^{rs}\|[x(s) - y(s)]\| \leq \int_s^t e^{ru}\|(F + rI)x(u) - (F + rI)y(u)\|du$. Then for almost all $t \geq t_0$, we infer that $(e^{rt}\|x(t) - y(t)\|)'_t \leq e^{rt}\|(F + rI)x(t) - (F + rI)y(t)\|$. Therefore, we have $\|x(t) - y(t)\|'_t \leq \|(F + rI)x(t) - (F + rI)y(t)\| - r\|x(t) - y(t)\|$. Let $r \to +\infty$, then we deduce $\|x(t) - y(t)\|'_t \leq \alpha_\Omega(F)\|x(t) - y(t)\|$. Integrating the inequality over interval $[t_0, t]$ yields the expected estimation (3).

Theorem 2. *Suppose $x^* \in \Omega$ is an equilibrium point of the system (2) and $\Gamma \subset \Omega$ is a neighborhood of x^*. If $\alpha_\Gamma(F, x^*) < 0$, then x^* is the unique equilibrium point of (2) in Γ, x^* is exponentially stable, and furthermore, the exponential decay estimation of any solution $x(t)$ initiated from $x(t_0) = x_0 \in \Gamma$ satisfies*

$$\|x(t) - x^*\| \leq e^{\alpha_\Gamma(F, x^*)(t-t_0)}\|x_0 - x^*\|, \forall t \geq t_0. \tag{4}$$

Proof. It follows from Lemma 2 that x^* is the unique equilibrium point of system (2) in Γ. Assume $x(t)$ is any solution of system (2) initiated from $x(t_0) = x_0 \in \Gamma$.

Analogous to the proof of Theorem 1, we can readily justify that $\|x(t) - x^*\|'_t \leq \alpha_\Gamma(F, x^*)\|x(t) - x^*\|$, and integration of the inequality from t_0 to t immediately yields $\|x(t) - x^*\| \leq e^{\alpha_\Gamma(F, x^*)(t-t_0)}\|x_0 - x^*\|$. Therefore, it follows from $\alpha_\Gamma(F, x^*) < 0$ that the equilibrium point x^* is exponentially stable in Γ, and the exponential decay estimation of solution $x(t)$ is governed by the inequality (4).

Remark 1. Theorem 2 provides new criterion for the exponential stability of (2). Since $\alpha_\Gamma(F, x^*) \leq \alpha_\Gamma(F)$, the condition $\alpha_\Gamma(F) < 0$ can sufficiently ensure the exponential stability of (2). However, it should be emphasized that generalized relative nonlinear measure sometimes is more desirable than generalized nonlinear measure in quantifying exponential stability of equilibrium point of nonlinear system. For example, 0 is obviously an equilibrium point in Example 1, we have $\alpha_\mathbf{R}(F) > 0$ while $\alpha_\mathbf{R}(F,0) < 0$, then we infer by Theorem 2 that 0 is the unique equilibrium point and it is globally exponentially stable in real number space.

Theorems 1 and 2 lead to a new approach to stability analysis for nonlinear systems, now we apply them to the investigation into the exponential stability of the Cohen-Grossberg neural networks (1).

Theorem 3. *Suppose that there exists $\lambda_i > 0$ such that $(u_i - v_i)[b_i(u_i) - b_i(v_i)] \geq \lambda_i(u_i - v_i)^2$ for any $u_i, v_i \in R$, $i = 1, 2, \cdots, n$, and*

$$\|W\|_1 \cdot \max_{1 \leq j \leq n}(\acute{\alpha}_j m_j) < \min_{1 \leq j \leq n}(\grave{\alpha}_j \lambda_j), \quad j = 1, 2, \cdots, n, \tag{5}$$

where $\|W\|_1 = \max_{1 \leq j \leq n} \sum_{i=1}^{n} |w_{ij}|$ is the matrix norm of the connection matrix W. Then for each group of inputs I_i, neural networks (1) is globally exponential stable, and the exponential decay estimation is governed by

$$\|u(t) - u^*\|_1 \leq e^{-b(t-t_0)} \|u_0 - u^*\|_1, \quad \forall t \geq t_0, \tag{6}$$

where $b = \min_{1 \leq j \leq n}(\grave{\alpha}_j \lambda_j) - \max_{1 \leq j \leq n}(\acute{\alpha}_j m_j \sum_{i=1}^{n} |w_{ij}|)$, $u(t)$ is any solution of neural networks (1) initiated from u_0, and u^* is the unique equilibrium point of (1).

Proof. Define the operator F as $F(u) = (F_1(u), F_2(u), \cdots, F_n(u))^T$, $\forall u = (u_1, u_2, \cdots, u_n)^T \in R^n$, where $F_i(u) = -a_i(u_i)b_i(u_i) + a_i(u_i)\left[\sum_{j=1}^{n} w_{ij} f_j(u_j) - I_i\right]$, $i = 1, 2, \cdots, n$. Then neural networks (1) is reduced to the following system

$$\frac{du(t)}{dt} = F(u(t)), \quad u(t) \in R^n, t \geq t_0. \tag{7}$$

In view of the inequality (5), we immediately deduce $\alpha_{R^n}(F) \leq -b < 0$. Then we infer by Lemma 1 and Theorem 1 that system (7) has a unique equilibrium point u^*, which is globally exponentially stable, and the exponential decay estimation of any solution $u(t)$ of (7) initiated from u_0 is governed by $\|u(t) - u^*\|_1 \leq e^{-b(t-t_0)} \|u_0 - u^*\|_1$, $\forall t \geq t_0$. Therefore the results in Theorem 3 holds naturally.

Theorem 4. *Suppose $u^* = (u_1^*, u_2^*, \cdots, u_n^*)^T$ is an equilibrium point of neural networks (1), $\Gamma \subset R^n$ is a neighborhood of u^*, and Γ_j is the projection of Γ on the j-th axis of R^n. If there exists $\lambda_i(\Gamma) > 0$ such that $(u_i - v_i)[b_i(u_i) - b_i(v_i)] \geq \lambda_i(\Gamma)(u_i - v_i)^2$ for any $u_i, v_i \in \Gamma_i, i = 1, 2, \cdots, n$, and $\min_{1 \leq j \leq n}(\grave{\alpha}_j \lambda_j(\Gamma)) > \max_{1 \leq j \leq n}(\acute{\alpha}_j m_j(\Gamma) \sum_{i=1}^{n} |w_{ij}|)$, $j = 1, 2, \cdots, n$, where $m_j(\Gamma) = \sup_{v \in \Gamma_j, v \neq u_j^*} \frac{|f_j(v) - f_j(u_j^*)|}{|v - u_j^*|}$, then networks (1) is exponentially stable on Γ, and the exponential decay estimation obeys $\|u(t) - u^*\|_1 \leq e^{-b(t-t_0)} \|u_0 - u^*\|_1, \forall t \geq t_0$, where $b = \min_{1 \leq j \leq n}(\grave{\alpha}_j \lambda_j(\Gamma)) - \max_{1 \leq j \leq n}(\acute{\alpha}_j m_j(\Gamma) \sum_{i=1}^{n} |w_{ij}|)$ and $u(t)$ is any solution of (1) initiated from $u_0 \in \Gamma$.*

Proof. Theorem 2 and proof similar to that of Theorem 3 yield the results.

3 Conclusions

By employing the new concept of generalized relative nonlinear measure of a nonlinear operator, we developed a novel approach to stability analysis for neural networks. By means of this new approach, we derived sufficient conditions for the exponential stability of Cohen-Grossberg neural networks, which generalize some known results. Moreover, we proposed the decay estimation of the solution of the neural networks to stable equilibrium point, and characterized the attraction region of local stable equilibrium point. The new approach can also be applicable to stability analysis of other general nonlinear systems besides neural networks.

References

1. Fang, Y., Kincaid, T.G.: Stability Analysis of Dynamical Neural Networks. IEEE Transactions on Neural Networks **7**(4) (1996) 996–1006
2. Forti, M., Manetti, S., Marini, M.: Necessary and Sufficient Condition for Absolute Stability of Neural Networks. IEEE Transactions on Circuits and Systems I: Fundamental Theory and Applications **41**(7) (1994) 491–494
3. Truccolo, W.A., Rangarajan, G., Chen, Y.H., Ding, M.Z.: Analyzing Stability of Equilibrium Points in Neural Networks: A General Approach. Neural Networks **16**(10) (2003) 1453–1460
4. Qiao, H., Peng, J.G., Xu, Z.B., Zhang, B.: A Reference Model Approach to Stability Analysis of Neural Networks. IEEE Transactions on Systems, Man and Cybernetics, Part B **33**(6) (2003) 925–936
5. Cohen, M.A., Grossberg, S.: Absolute Stability of Global Pattern Formation and Parallel Memory Storage by Competitive Neural Networks. IEEE Transactions on Systems, Man and Cybernetics **SMC-13**(5) (1983) 815–821
6. Wang, L., Zou, X.F.: Exponential Stability of Cohen-Grossberg Neural Networks. Neural Networks **15** (2002) 415–422
7. Chen, T.P., Rong, L.B.: Robust Global Exponential Stability of Cohen-Grossberg Neural Networks with Time Delays. IEEE Transactions on Neural Networks **15**(1) (2004) 203–206
8. Ye, H., Michel, A.N., Wang, K.: Qualitative Analysis of Cohen-Grossberg Neural Network with Multiple Delays. Physical Review E **51** (1995) 2611–2618
9. Hwang, C.C., Cheng C.J., Liao, T.L.: Globally Exponential Stability of Generalized Cohen-Grossberg Neural Networks with Delays. Physics Letters A **319**(1-2) (2003) 157–166
10. Tank, D.W., Hopfield, J.J.: Simple "Neural" Optimization Networks: An A/D Converter, Signal Decision Circuit, and a Linear Programming Circuit. IEEE Transactions on Circuits and Systems **33**(5) (1986) 533–541
11. Qiao, H., Peng, J.G., Xu, Z.B.: Nonlinear Measures: A New Approach to Exponential Stability Analysis for Hopfield-Type Neural Networks. IEEE Transactions of Neural Networks **12**(2) (2001) 360–370
12. Peng, J.G., Xu, Z.B.: On Asymptotic Behaviours of Nonlinear Semigroup of Lipschitz Operators with Applications. Acta Mathematica Sinica **45**(6) (2002) 1099–1106
13. Söderlind, G.: Bounds on Nonlinear Operators in Finite-Dimensional Banach Spaces. Numerische Mathematik **50** (1986) 27–44

Analysis for Global Robust Stability of Cohen-Grossberg Neural Networks with Multiple Delays

Ce Ji, Huaguang Zhang, and Huanxin Guan

Institute of Information Science and Engineering, Northeastern University,
Shenyang 110004, P.R. China
hanfj119@163.net

Abstract. Global Robust stability of a class of Cohen-Grossberg neural networks with multiple delays and parameter perturbations is analyzed. The sufficient conditions for the globally asymptotic stability of equilibrium point are given by way of constructing a suitable Lyapunov functional. Combined with the linear matrix inequality (LMI) technique, a practical corollary is derived. All results are established without assuming any symmetry of the interconnecting matrix, and the differentiability and monotonicity of activation functions.

1 Introduction

Since Cohen and Grossberg proposed a class of neural networks in 1983 [1], this model has received increasing interest. For the Cohen-Grossberg neural networks, many sufficient conditions for asymptotic stability, or exponential stability have been obtained. By Lyapunov functional, [2], [3] established some criteria for the globally asymptotic stability of this model. In [4], the global stability of Cohen-Grossberg neural networks was derived under the assumption that the interconnecting matrix T was symmetric. However, parameter perturbations are unavoidably encountered during the implementation of neural networks and in general result in the disappearance of stable memories, or the change of interconnecting structure. Hence, it is very difficult to realize absolute symmetry of interconnecting structure. On the other hand, the finite switching speed of amplifiers and the inherent communication time of neurons inevitably induce time delays, which may bring oscillation or network instability. Thus, [5], [6] investigated Cohen-Grossberg neural networks with multiple delays and obtained several exponential stability criteria.

In recent years, considerable efforts have been devoted to the robust stability analysis for Hopfield neural networks [7], [8]. However, there are few existing results on the robust stability for Cohen-Grossberg neural networks. Aiming at this case, we will analyze the global robust stability of Cohen-Grossberg neural networks with multiple delays and perturbations of interconnecting weights in this paper. The paper is organized as follows. In Section 2, the network model will be described as the basis of later sections. In Section 3, we will establish the sufficient conditions for the global robust stability of this model. Finally, Section 4 will present the future work.

2 Model Description and Preliminaries

The Cohen-Grossberg neural network model with multiple delays can be described by equation

$$\dot{u}(t) = -\alpha(u(t))\left[\beta(u(t)) - T_0 G(u(t)) - \sum_{k=1}^{K} T_k G(u(t-\tau_k)) + I\right], \qquad (1)$$

where $u = (u_1, \cdots, u_n)^T \in \Re^n$, $\alpha(u) = diag[\alpha_1(u_1), \cdots, \alpha_n(u_n)]$, $\beta(u) = [\beta_1(u_1), \cdots, \beta_n(u_n)]^T$, $G(x) = [g_1(u_1), \cdots, g_n(u_n)]^T$, and $I = diag(I_1, \cdots, I_n)$. Amplification function $\alpha_i(\cdot)$ is positive, continuous and bounded, behaved function $\beta_i(\cdot)$ is continuous, $g_i(\cdot)$ is the activation function representing the *ith* neuron, I_i denotes the external input. $T_0 = [t_{0ij}] \in \Re^{n \times n}$ denotes that part of the interconnecting structure which is not associated with delay, $T_k = [t_{kij}] \in \Re^{n \times n}$ denotes that part of the interconnecting structure which is associated with delay τ_k, where τ_k denotes *kth* delay, $k = 1, \cdots, K$ and $0 < \tau_1 < \cdots < \tau_K < +\infty$.

Considering the influences of parameter perturbations for model (1), then we can describe (1) as

$$\dot{u}(t) = -\alpha(u(t))\left[\beta(u(t)) - (T_0 + \Delta T_0)G(u(t)) - \sum_{k=1}^{K}(T_k + \Delta T_k)G(u(t-\tau_k)) + I\right], \qquad (2)$$

where $\Delta T_0 = \left[\Delta t_{0ij}\right] \in \Re^{n \times n}$, $\Delta T_k = \left[\Delta t_{kij}\right] \in \Re^{n \times n}$, $k = 1, \cdots, K$. Now, the interconnecting matrix $T = T_0 + \Delta T_0 + \sum_{k=1}^{K}(T_k + \Delta T_k)$ is nonsymmetric due to the perturbation terms $\Delta T_0, \Delta T_k$.

The equilibrium point of system (1) is said to be globally robustly stable with respect to the perturbation ΔT_0 and ΔT_k, if the equilibrium point of system (2) is globally asymptotically stable.

Throughout this paper, we assume that

(H$_1$) For $i = 1, \cdots, n$, $0 < \alpha_i^m \leq \alpha_i(u_i) \leq \alpha_i^M$.

(H$_2$) For $i = 1, \cdots, n$, $\beta_i(u_i)$ is continuous and differentiable, and satisfies

$$\beta_i'(u_i) \geq \beta_i^m > 0. \qquad (3)$$

(H$_3$) For $i = 1, \cdots, n$, $g_i(u_i)$ is bounded and satisfies

$$0 \leq \frac{g_i(x) - g_i(y)}{x - y} \leq \sigma_i^M, \quad \forall x, y \in \Re \text{ with } x \neq y. \qquad (4)$$

By the Brouwer's fixed-point theorem [9], we can prove that if (H$_1$), (H$_2$) hold, and $g_i(u_i)$ is bounded, then for every input I, there exists one equilibrium point for (2).

If we suppose $x_e = (x_{e1}, \cdots, x_{en})^T$ is an equilibrium point of network, then by means of coordinate translation $x = u - x_e$, we can obtain the new description of the neural networks (2)

$$\dot{x}(t) = -A(x(t))\left[B(x(t)) - (T_0 + \Delta T_0)S(x(t)) - \sum_{k=1}^{K}(T_k + \Delta T_k)S(x(t - \tau_k))\right], \quad (5)$$

where $x = (x_1, \cdots, x_n)^T \in \Re^n$,

$$A(x) = diag[a_1(x_1), \cdots, a_n(x_n)], \ a_i(x_i) = \alpha_i(x_i + x_{ei}), \quad (6)$$

$$B(x) = [b_1(x_1), \cdots, b_n(x_n)]^T, \ b_i(x_i) = \beta_i(x_i + x_{ei}) - \beta_i(x_{ei}), \quad (7)$$

$$S(x) = [s_1(x_1), \cdots, s_n(x_n)]^T, \ s_i(x_i) = g_i(x_i + x_{ei}) - g_i(x_{ei}). \quad (8)$$

From the above assumptions (H$_1$), (H$_2$) and (H$_3$), we have

$$0 < \alpha_i^m \leq a_i(x_i) \leq \alpha_i^M, \quad (9)$$

$$b_i(x_i)/x_i \geq \beta_i^m > 0, \quad (10)$$

$$0 \leq s_i(x_i)/x_i \leq \sigma_i^M, \quad (11)$$

for $i = 1, \cdots, n$. Thus, we shift the equilibrium x_e to the origin. In order to study the global robust stability of the equilibrium point for (1), it suffices to investigate the globally asymptotic stability of equilibrium point $x = 0$ of system (5).

3 Robust Stability

Theorem. For any bounded delay τ_k, $k = 1, \cdots, K$, the equilibrium point $x = 0$ of system (5) is globally asymptotically stable, if there exist a positive definite matrix P and positive diagonal matrixes $\Lambda_k = diag[\lambda_{k1}, \cdots, \lambda_{kn}]$, where $\lambda_{ki} > 0$, $i = 1, \cdots, n$, $k = 0, \cdots, K$, such that

$$S^M = -(B^m A^m P + P A^m B^m) + \sum_{k=0}^{K} \Lambda_k + \sum_{k=0}^{K} P A^M (T_k + \Delta T_k) E^M \Lambda_k^{-1} E^M (T_k + \Delta T_k)^T A^M P \quad (12)$$

is negative definite, where $A^M = diag[\alpha_1^M, \cdots, \alpha_n^M]$, $B^m = diag[\beta_1^m, \cdots, \beta_n^m]$, $E^M = [\sigma_1^M, \cdots, \sigma_n^M]$.

Proof. We can rewrite (5) as

$$\dot{x}(t) = -A(x(t))\left[B(x(t)) - (T_0 + \Delta T_0)E(x(t))x(t) - \sum_{k=1}^{K}(T_k + \Delta T_k)E(x(t-\tau_k))x(t-\tau_k)\right], \quad (13)$$

where

$$E(x) = diag[\varepsilon_1(x_1), \cdots, \varepsilon_n(x_n)], \quad \varepsilon_i(x_i) = s_i(x_i)/x_i, \quad i = 1, \cdots, n. \quad (14)$$

Then from (11), we have $\varepsilon_i(x_i) \in [0, \sigma_i^M]$.

Here, we introduce the following Lyapunov functional

$$V(x_t) = x^T(t)Px(t) + \sum_{k=1}^{K}\int_{t-\tau_k}^{t} x_t^T(\theta)\Lambda_k x_t(\theta)d\theta. \quad (15)$$

Clearly, we have $\lambda_{\min}(P)\|x_t(0)\|^2 \leq V(x_t) \leq \left(\lambda_{\max}(P) + \sum_{k=1}^{K}\tau_k\lambda_{\max}(\Lambda_k)\right)|x_t|^2$.

The derivative of $V(x_t)$ with respect to t along any trajectory of system (13) is given by

$$\dot{V}(x_t) = \dot{x}^T(t)Px(t) + x^T(t)P\dot{x}(t) + \sum_{k=1}^{K}x^T(t)\Lambda_k x(t) - \sum_{k=1}^{K}x^T(t-\tau_k)\Lambda_k x(t-\tau_k)$$

$$= -x^T(t)\overline{B}^T A^T Px(t) - x^T(t)PA\overline{B}x(t) + \sum_{k=1}^{K}x^T(t)\Lambda_k x(t) + x^T(t)\Lambda_0 x(t)$$

$$+ x^T(t)PA(T_0 + \Delta T_0)E(x(t))\Lambda_0^{-1}E^T(x(t))(T_0 + \Delta T_0)^T A^T Px(t)$$

$$- \left[\Lambda_0^{\frac{1}{2}}x(t) - \Lambda_0^{-\frac{1}{2}}E^T(x(t))(T_0 + \Delta T_0)^T A^T Px(t)\right]^T$$

$$\times \left[\Lambda_0^{\frac{1}{2}}x(t) - \Lambda_0^{-\frac{1}{2}}E^T(x(t))(T_0 + \Delta T_0)^T A^T Px(t)\right]$$

$$+ \sum_{k=1}^{K}x^T(t)PA(T_k + \Delta T_k)E(x(t-\tau_k))\Lambda_k^{-1}E^T(x(t-\tau_k))(T_k + \Delta T_k)^T A^T Px(t)$$

$$- \sum_{k=1}^{K}\left[\Lambda_k^{\frac{1}{2}}x(t-\tau_k) - \Lambda_k^{-\frac{1}{2}}E^T(x(t-\tau_k))(T_k + \Delta T_k)^T A^T Px(t)\right]^T$$

$$\times \left[\Lambda_k^{\frac{1}{2}}x(t-\tau_k) - \Lambda_k^{-\frac{1}{2}}E^T(x(t-\tau_k))(T_k + \Delta T_k)^T A^T Px(t)\right]$$

$$\leq -x^T(t)(\overline{B}^T A^T P + PA\overline{B})x(t) + \sum_{k=0}^{K}x^T(t)\Lambda_k x(t)$$

$$+ x^T(t)PA(T_0 + \Delta T_0)E(x(t))\Lambda_0^{-1}E^T(x(t))(T_0 + \Delta T_0)^T A^T Px(t)$$

$$+\sum_{k=1}^{K} x^T(t)PA(T_k+\Delta T_k)E(x(t-\tau_k))\Lambda_k^{-1}E^T(x(t-\tau_k))(T_k+\Delta T_k)^T A^T Px(t), \quad (16)$$

where $\bar{B} = diag[b_1(x_1)/x_1, \cdots, b_n(x_n)/x_n]$. For convenience, we denote $A(x(t))$ and $B(x(t))$ as A and B, respectively.

For any given t, if we let $y_k^T(t) = (y_{k1}, \cdots, y_{kn}) = x^T(t)P(T_k+\Delta T_k)$, then by (11), the last term of (16) assumes the form

$$\sum_{k=1}^{K} y_k^T(t)E(x(t-\tau_k))\Lambda_k^{-1}E^T(x(t-\tau_k))y_k(t) = \sum_{k=1}^{K}\sum_{i=1}^{n} y_{ki}^2 \lambda_{ki}^{-1} \varepsilon_i^2(x_i(t-\tau_k)) \leq \sum_{k=1}^{K}\sum_{i=1}^{n} y_{ki}^2 \lambda_{ki}^{-1}(\sigma_i^M)^2$$

$$= \sum_{k=1}^{K} x^T(t)PA(T_k+\Delta T_k)E^M \Lambda_k^{-1} E^M (T_k+\Delta T_k)^T A^T Px(t), \quad (17)$$

Similarly, we have

$$x^T(t)PA(T_0+\Delta T_0)E(x(t))\Lambda_0^{-1}E^T(x(t))(T_0+\Delta T_0)^T A^T Px(t)$$

$$\leq x^T(t)PA(T_0+\Delta T_0)E^M \Lambda_0^{-1} E^M (T_0+\Delta T_0)^T A^T Px(t), \quad (18)$$

Since $A^M, \bar{B}, A^m, B^m$ are all positive diagonal matrixes and P is a positive definite matrix, then by (9), (10) and the relevant matrix theory, we can prove easily

$$x^T(t)\bar{B}^T A^T Px(t) \geq x^T(t)B^m A^m Px(t) > 0, \quad (19)$$

$$x^T(t)PA\bar{B}x(t) \geq x^T(t)PA^m B^m x(t) > 0, \quad (20)$$

$$\sum_{k=1}^{K} x^T(t)PA(T+\Delta T_k)E^M \Lambda_k^{-1} E^M (T+\Delta T_k)^T A^T Px(t)$$

$$\leq \sum_{k=1}^{K} x^T(t)PA^M (T+\Delta T_k)E^M \Lambda_k^{-1} E^M (T+\Delta T_k)^T A^M Px(t), \quad (21)$$

$$x^T(t)PA(T_0+\Delta T_0)E^M \Lambda_0^{-1} E^M (T_0+\Delta T_0)^T A^T Px(t)$$

$$\leq x^T(t)PA^M (T_0+\Delta T_0)E^M \Lambda_0^{-1} E^M (T_0+\Delta T_0)^T A^M Px(t). \quad (22)$$

Considering (17)~(22), we can express (16) as

$$\dot{V}(x_t) \leq x^T(t)\left\{-(B^m A^m P + PA^m B^m) + \sum_{k=0}^{K}\Lambda_k + \sum_{k=0}^{K} PA^M(T_k+\Delta T_k)E^M \Lambda_k^{-1} E^M (T_k+\Delta T_k)^T A^M P\right\}x(t)$$

$$= x^T(t)S^M x(t). \quad (23)$$

Thus, $\dot{V}(x_t) < 0$ if S^M is negative definite. By the functional differential equations theory, for any bounded delay $\tau_k > 0$, $k=1,\cdots,K$, the equilibrium point $x=0$ of system (5) is globally asymptotically stable.

According to the well-known Schur complement, we can obtain the following corollary expressed by a linear matrix inequality (LMI).

Corollary. For any bounded delay τ_k, $k = 1,\cdots,K$, the equilibrium point $x = 0$ of system (5) is globally asymptotically stable, if

$$\begin{bmatrix} -\left(B^m A^m P + PA^m B^m\right) + \sum_{k=0}^{K} \Lambda_k & PA^M(T_K + \Delta T_K)E^M & \cdots & \cdots & \cdots & PA^M(T_0 + \Delta T_0)E^M \\ E^M(T_K + \Delta T_K)^T A^M P & -\Lambda_K & 0 & 0 & 0 & 0 \\ \vdots & 0 & \ddots & 0 & 0 & 0 \\ \vdots & 0 & 0 & \ddots & 0 & 0 \\ \vdots & 0 & 0 & 0 & \ddots & 0 \\ E^M(T_0 + \Delta T_0)^T A^M P & 0 & 0 & 0 & 0 & -\Lambda_0 \end{bmatrix} < 0. \quad (24)$$

The condition is in the form of linear matrix inequality and hence is convenient to verify.

4 Future Work and Acknowledgements

In applications, the bound of time delays is frequently not very large and is usually known. Therefore, the next research work is to discuss further whether we can obtain the sufficient conditions for the global robust stability of equilibrium point, which depend on time delays τ_k, $k = 1,\cdots,K$.

This work is supported by the National Natural Science Foundation of China (60244017, 60325311).

References

1. Cohen, M., Grossberg, S.: Absolute Stability of Global Pattern Formation and Parallel Memory Storage by Competitive Neural Networks. IEEE Transactions on Systems, Man and Cybernetics, Vol. 13. (1983) 815–826
2. Wang, L., Zou, X.F.: Harmless Delays in Cohen-Grossberg Neural Networks. Physica D, Vol. 170. (2002) 162–173
3. Chen, T.P., Rong, L.B.: Delay-Independent Stability Analysis of Cohen-Grossberg Neural Networks. Physics Letters A, Vol. 317. (2003) 436–449
4. Ye, H., Michel, A.N., Wang, K.: Analysis of Cohen-Grossberg Neural Networks with Multiple Delays. Physical Review E, Vol. 51. (1995) 2611–2618
5. Hwang, C.C., Cheng, C.J., Liao, T.L.: Globally Exponential Stability of Generalized Cohen-Grossberg Neural Networks with Delays. Physics Letters A, Vol. 319. (2003) 157–166
6. Wang, L., Zou, X.F.: Exponential Stability of Cohen-Grossberg Neural Networks. Neural Networks, Vol. 15. (2002) 415–422
7. Liao, X.X., Wu, K.Z., Chen, G.: Novel Robust Stability Criteria for Interval-Delayed Hopfield Neural Networks. IEEE Transactions on Circuits and Systems-I, Vol. 48. (2001) 1355–1358
8. Arik, S.: Global Robust Stability of Delayed Neural Networks. IEEE Transactions on Circuits and Systems-I, Vol. 50. (2003) 156–160
9. Gopalsamy, K.: Stability and Oscillations in Delay Differential Equations of Population Dynamics. Kluwer Academic Publishers, Dordrecht (1992)

On Robust Stability of BAM Neural Networks with Constant Delays

Chuandong Li, Xiaofeng Liao, and Yong Chen

College of computer science and engineering, Chongqing University 400030, China
cd_licqu@163.com

Abstract. The problems of determining the robust stability of bidirectional associative memory neural networks with delays are investigated in this paper. An approach combining the Lyapunov-Krasovskii stability theorem with the linear matrix inequality (LMI) technique is taken to study the problems, which provide bounds on the interconnection matrix and the activation functions. Some criteria for the robust stability, which give information on the delay-independence property, are derived. The results obtained in this paper provide one more set of easily verified guidelines for determining the robust stability of delayed BAM (DBAM) neural networks, which are less conservative and less restrictive than the ones reported recently in the literature. Some typical examples are presented to show the effectiveness of results.

1 Introduction

Bi-directional associative memory (BAM) is an important network model with the ability of information memory and information association, which is crucial for application in pattern recognition and automatic control engineering (see [1-3] and the references cited therein). BAM neural network is described originally by the set of ordinary differential equations [1-3]. Recently, the time delays have been introduced into this model [4-10], which is described by

$$\dot{u}(t) = -Au(t) + W_1 f(v(t-\tau)) + I \tag{1a}$$

$$\dot{v}(t) = -Bv(t) + W_1 g(u(t-\sigma)) + J \tag{1b}$$

where $A = \text{diag}(a_1, a_2, \ldots, a_l) > 0$ and $B = \text{diag}(b_1, b_2, \ldots, b_m) > 0$ denote the neuron charging time constants and passive decay rates, respectively. $W_1 = \left(W_{ij}^{(1)}\right)_{l \times m}$ and $W_2 = \left(W_{ij}^{(2)}\right)_{m \times l}$ are the synaptic connection strengths; $I = (I_1, I_2, \ldots, I_l)$ and $J = (J_1, J_2, \ldots, J_l)$ indicate the exogenous inputs. Time delays τ and σ correspond to the finite speeds of the axonal transmission of signals. f and g represent the activation functions of the neurons and the propagational signal functions with $f(v(t-\tau)) = (f_1(v_1(t-\tau_1)), \cdots; f_m(v_m(t-\tau_m)))^T$ and $g(u(t-\sigma)) = (g_1(u_1(t-\sigma_1)), \cdots; g_l(u_l(t-\sigma_l)))^T$, respectively. Moreover, we assume that, throughout this paper, the activation functions satisfy

(H) f_i and g_j are bounded with $f_i(0) = 0$, $g_j(0) = 0$ and there exist $L_i > 0$ and $M_j > 0$ such that, for any $x, y \in R$,

$$|f_i(x) - f_i(y)| \leq L_i |x - y|, \ i = 1, 2, \cdots, m,$$

$$|g_j(x) - g_j(y)| \leq M_i |x - y|, j = 1, 2, \cdots, n.$$

Arguing similarly with the analysis in [14-17], in this paper, we assume that

$$A \in A_I = \{A = diag(a_i)_{l \times l} | \ \underline{a}_i \leq a_i \leq \overline{a}_i, i = 1, 2, \cdots, l\}$$

$$B \in B_I = \{B = diag(b_i)_{m \times m} | \ \underline{b}_i \leq b_i \leq \overline{b}_i, i = 1, 2, \cdots, m\} \quad (2)$$

$$W_1 \in W_{1I} = \left\{W_1 = \left(w_{ij}^{(1)}\right)_{l \times m} \middle| \ \underline{w}_{ij}^{(1)} \leq w_{ij}^{(1)} \leq \overline{w}_{ij}^{(1)}, i = 1, 2, \cdots, l; j = 1, 2, \cdots, m\right\}$$

$$W_2 \in W_{2I} = \left\{W_2 = \left(w_{ij}^{(2)}\right)_{m \times l} \middle| \ \underline{w}_{ij}^{(2)} \leq w_{ij}^{(2)} \leq \overline{w}_{ij}^{(2)}, i = 1, 2, \cdots, m; j = 1, 2, \cdots, l\right\} \quad (3)$$

Moreover, for notational convenience, we define, for $i = 1, 2, \cdots, l$ and $j = 1, 2, \cdots, m$,

$$\underline{A} = diag(\underline{a}_i)_{l \times l}, \underline{B} = diag(\underline{b}_j)_{m \times m}, w_{ij}^{(1)*} = max\left\{\left|\underline{w}_{ij}^{(1)}\right|, \left|\overline{w}_{ij}^{(1)}\right|\right\}, w_{ji}^{(2)*} = max\left\{\left|\underline{w}_{ji}^{(2)}\right|, \left|\overline{w}_{ji}^{(2)}\right|\right\} \quad (4)$$

and,

$$c_i = \sum_{j=1}^{m}\left(w_{ij}^{(1)*} \sum_{k=1}^{l} w_{kj}^{(1)*}\right), \ C = diag(c_i), \ i = 1, 2, \cdots, l.$$

$$d_i = \sum_{j=1}^{l}\left(w_{ij}^{(2)*} \sum_{k=1}^{m} w_{kj}^{(2)*}\right), \ D = diag(d_i), \ i = 1, 2, \cdots, m. \quad (6)$$

It is well known that bounded activation functions always guarantee the existence of an equilibrium point for system (1). For notational convenience, we will always shift an intended equilibrium point (u^*, v^*) of system (1) to the origin by letting $x(t) = u(t) - u^*$ and $y(t) = v(t) - v^*$, which yields the following system:

$$\dot{x}(t) = -Ax(t) + W_1 F(y(t - \tau))$$

$$\dot{y}(t) = -By(t) + W_2 G(x(t - \sigma)) \quad (7)$$

where $x(t) = [x_1(t), x_2(t), \cdots, x_n(t)]^T$ is the state vector of the transformed system; and $F(y) = [F_1(y_1), F_2(y_2), \cdots, F_m(y_m)]^T$ and $G(x) = [G_1(x_1), G_2(x_2), \cdots, G_l(x_l)]^T$ with

$F_i(y_i(t-\tau_i)) = f_i(y_i(t-\tau_i) + v_i^*) - f_i(v_i^*)$, and $G_i(x_i(t-\sigma_i)) = g_i(x_i(t-\sigma_i) + u_i^*) - g_i(u_i^*)$.

Obviously, the equilibrium point u^* of system (1) is globally robustly stable if and only if the origin of system (7) is globally robustly stable. Thus in the sequel, we only consider system (7).

2 Main Results

Theorem 1 Suppose that the assumption (H) is satisfied and there exist symmetrical positive definite matrices $P > 0$, $Q > 0$ and constant scalars $\alpha > 0$, $\beta > 0$ such that the following LMIs hold.

$$\begin{bmatrix} \underline{A}P + P\underline{A} - \frac{1}{\alpha}C & \sqrt{\beta}\Sigma_M P \\ \sqrt{\beta}P\Sigma_M & E_x \end{bmatrix} > 0, \tag{8}$$

and

$$\begin{bmatrix} \underline{B}Q + Q\underline{B} - \frac{1}{\beta}D & \sqrt{\alpha}\Sigma_L Q \\ \sqrt{\alpha}Q\Sigma_L & E_y \end{bmatrix} > 0, \tag{9}$$

where C and D are defined by (5) and (6), respectively, $\Sigma_M = diag(M_1, M_2, \cdots, M_l)$, E_x and E_y denote the identity matrices with appropriate dimension, $\Sigma_L = diag(L_1, L_2, \cdots, L_m)$. Then the equilibrium point, i.e., the origin, of system (7) is unique and globally robust asymptotically stable.

Before presenting the proof of the theorem above, we first state a lemma, which is used below.

Lemma 1 For any given real matrix $W = (w_{ij})_{l \times m}$, suppose that

$$W = (w_{ij})_{l \times m} \in \{W = (w_{ij})_{l \times m} | \underline{w}_{ij} \le w_{ij} \le \overline{w}_{ij}, i = 1, 2, \cdots, l; j = 1, 2, \cdots, m\}.$$

Let

$$w_{ij}^* = max\{|\underline{w}_{ij}|, |\overline{w}_{ij}|\}, \quad \widetilde{W} = diag(\widetilde{w}_i), \quad \widetilde{w}_i = \sum_{j=1}^{m}\left(w_{ij}^* \sum_{k=1}^{l} w_{kj}^*\right).$$

Then the inequality $x^T W W^T x \le x^T \widetilde{W} x$ holds for any real vector $x \in R^l$.

Proof Let $H = (h_{ij})_{l \times l} \equiv W W^T$. Then $h_{ij} = \sum_{k=1}^{m} w_{ik} w_{jk}$ and

$$\widetilde{w}_i - h_{ii} = \sum_{j=1}^{m}\left(w_{ij}^* \sum_{k=1}^{l} w_{kj}^*\right) - \sum_{j=1}^{m}(w_{ij}^2) \ge \sum_{j=1}^{m}\left(w_{ij}^* \sum_{k=1}^{l} w_{kj}^*\right) - \sum_{j=1}^{m}(w_{ij}^{*2}) \ge 0 \tag{10}$$

Furthermore, note that

$$\tilde{w}_i - h_{ii} - \sum_{j \neq i}^{m} |h_{ij}| \geq \sum_{j=1}^{m} \left(w_{ij}^* \sum_{k=1}^{l} w_{kj}^* \right) - \sum_{j=1}^{m} \left(|w_{ij}| \sum_{k=1}^{l} |w_{kj}| \right) \geq 0 \quad (11)$$

From (10) and (11), we can see that $\tilde{W} - WW^T$ is a semi-positive definite matrix, which completes the proof.

Proof of Theorem 1 It is worth noting that the LMI (8) is equivalent to

$$\Omega_x = \underline{A}P + P\underline{A} - \frac{1}{\alpha}C - \beta P \Sigma_M \Sigma_M P > 0,$$

and LMI (9) equivalent to

$$\Omega_y = \underline{B}Q + Q\underline{B} - \frac{1}{\beta}D - \alpha Q \Sigma_L \Sigma_L Q > 0.$$

Obviously, the origin is the equilibrium point of (7). To show the uniqueness, we assume that there exists $(x^*, y^*) \neq (0, 0)$ such that

$$-Ax^* + W_1 F(y^*) = 0, \quad (12)$$

and,

$$-By^* + W_2 G(x^*) = 0. \quad (13)$$

Multiplying both sides of (12) and (13) by $2x^{*T}P^{-1}$ and $2y^{*T}Q^{-1}$, respectively, we have

$$0 = 2x^{*T}P^{-1}\left(-Ax^* + W_1 F(y^*)\right)$$

$$\leq -x^{*T}\left(P^{-1}\underline{A} + \underline{A}P^{-1}\right)x^* + \frac{1}{\alpha}x^{*T}P^{-1}DP^{-1}x^* + \alpha y^{*T}\Sigma_M \Sigma_M y^*,$$

$$0 = 2y^{*T}Q^{-1}\left(-By^* + W_2 G(x^*)\right)$$

$$\leq -y^{*T}\left(Q^{-1}B + BQ^{-1}\right)y^* + \frac{1}{\beta}y^{*T}Q^{-1}DQ^{-1}y^* + \beta x^{*T}\Sigma_L \Sigma_L x^*$$

Therefore,

$$0 \leq -x^{*T}P^{-1}\Omega_x P^{-1}x^* - y^{*T}Q^{-1}\Omega_y Q^{-1}y^*,$$

which contradicts $\Omega_x > 0$ and $\Omega_y > 0$. Hence, the origin is the unique equilibrium of (7). To complete the proof, we construct a Lyapunov-Krasovskii functional [11] as follows:

$$V(x(t),y(t)) = x^T(t)P^{-1}x(t) + y^T(t)Q^{-1}y(t)$$

$$+ \alpha \sum_{i=1}^{m} \int_{t-\tau_i}^{t} F_i^2(y_i(s))ds + \beta \sum_{i=1}^{m} \int_{t-\sigma_i}^{t} G_i^2(x_i(s))ds$$

By Lemma 1, the time derivative of (14) along the solution of system (7) is

$$\dot{V}(x(t),y(t)) \leq -x^T(t)P^{-1}\left[\underline{A}P + P\underline{A} - \frac{1}{\alpha}C - \beta P\Sigma_L \Sigma_L P\right]P^{-1}x(t)$$

$$- y^T(t)Q^{-1}\left[\underline{B}Q + Q\underline{B} - \frac{1}{\beta}D - \alpha Q\Sigma_M \Sigma_M Q\right]Q^{-1}y(t)$$

$$= -x^T(t)P^{-1}\Omega_x P^{-1}x(t) - y^T(t)Q^{-1}\Omega_y Q^{-1}y(t).$$

Hence, $\dot{V}(x(t),y(t)) < 0$ when $\Omega_x > 0$ and $\Omega_y > 0$, i.e., LMI (8) and LMI (9) hold. The proof is thus completed.

3 Conclusions

In this paper, we have proposed a sufficient condition for determining the robust asymptotical stability of a general class of interval BAM models based on Lyapunov-Krasovskii stability theory and LMI technique. Our results have been shown to be less restrictive and be easier to be verified than those reported recently in the literature.

Acknowledgement. The work described in this paper was partially supported by the National Natural Science Foundation of China (Grant No.60271019), the Doctorate Foundation of the Ministry of Education of China (Grant NO.20020611007), the Applied Basic Research Grants Committee of Science and Technology of Chongqing (Grant NO.7370).

References

1. Kosko B.: Bi-Directional Associative Memory. IEEE trans. on Systems, Man, and Cybernetics 18 (1998) 49-60.
2. Mathai C., Upadhyaya B. C.: Performance Analysis and Application of the Bi-Directional Associative Memory to Industrial Spectral Signatures. Proceeding of IJCNN.1(1989) 33-37.
3. Elsen I., Kraiaiss K.F. and Krumbiegel D.: Pixel Based 3D Object Recognition with Bi-Directional Associative Memory. International Conference on neural networks 3 (1997) 1679-1684.
4. Liao X, Yu J. and Chen G.: Novel Stability Criteria for Bi-Directional Associative Memory Neural Networks with Time Delays. Int. J. Circ. Theor. Appl. 30 (2002) 519-546.

5. Liao X, Yu J.: Qualitative Analysis of bi-Directional Associative Memory Networks with Time Delays. Int. J. Cir. and App. 26 (1998) 219–229.
6. Liao X, Liu G, Yu J.: BAM Network with axonal Signal transmission Delay. Journal of Electronics 19 (1997) 439-444.
7. Liao X, Mu W, Yu J.: Stability Analysis of Bi-Directional Association Memory with Axonal Signal Transmission Delay. Proceedings of the third International Conference on Signal Processing, 2 (1996) 1457-1460.
8. Liao X, Liu G, YU J.: Qualitative Analysis of BAM Networks. Journal of circuits and systems 1 (1996) 13-18.
9. Gopalsamy, K., He, X.: Delay-Dependent Stability in Bi-Directional Associative Memory Networks. IEEE transaction on neural networks 5 (1994) 998-1002.
10. Gopalsamy, K., He, X. Z.: Delay-Dependent Stability in Bi-Directional Associative Memory Networks. IEEE Trans. on Neural Networks 5 (1994) 998–1002.
11. Hale, J. K., Lunel, S. M. V.: Introduction to the Theory of Functional Differential Equations. Applied mathematical sciences, New York: Springer (1991)
12. Boyd, S., Ghaoui, L. EI, Feron, E. Balakrishnan, V.: Linear matrix inequalities in systems and control theory. SIAM, Philadephia PA (1994).
13. Nesterov, Y., & Nemirovsky, A.: Interior point polynomial methods in convex Programming. SIAM, Philadephia PA (1994).
14. Liao X., Yu, J.: Robust Stability for Interval Hopfield Neural Networks with Time Delay. IEEE Trans. Neural Networks, 9(1998) 1042–1046.
15. Liao X., Wong, K.-W., Wu Z., Chen G.: Novel Robust Stability Criteria for Interval-Delayed Hopfield Neural Networks. IEEE Trans. Circuits Syst. I 48 (2001) 1355-1359.
16. Liao X., Wang,J.: Global and Robust Stability of Interval Hopfield Neural Networks with Time-Varying Delays. Int. J. neural syst. 13 (2003) 171-182.
17. Arik, S.: Global Robust Stability of Delayed Neural Networks. IEEE Trans. Circuits Syst. I 50(2003) 156-160.

Absolutely Exponential Stability of BAM Neural Networks with Distributed Delays*

Wenjun Xiong[1] and Qiuhao Jiang[1,2]

[1] Department of Mathematics, Southeast University, Nanjing 210096, China
[2] Department of Mathematics, China Pharmaceutical University, Nanjing 210009, China.
xiongwenjun@126.com

Abstract. Some novel criteria are obtained for checking the absolute stability of the equilibrium point for bidirectional associative memory networks with distributed delays, where the activation functions only need to be partially Lipschitz continuous, but not bounded or differentiable.

1 Introduction

The bidirectional associative memory (BAM) neural networks were first proposed by Kosko [2], and have been widely applied to some practical fields. In [1,4,5,8,9,10], the various stabilities are discussed for different recurrent neural networks, where the activation functions need to be bounded or differentiable. Recently, it is noted that, in [3,7], the authors discussed the absolutely exponential stability of several recurrent neural networks. However, in applications, neural networks usually have a spatial extent due to the presence of an amount of parallel pathways with a variety of axon sizes and lengths. Motivated by above discussions, we shall consider the absolutely exponential stability of BAM neural networks with distributed delays, where the activation functions only require to be partially Lipschitz continuous.

2 Preliminaries

Consider the following BAM neural networks:

$$\frac{dx_i}{dt} = -\alpha_i x_i(t) + \sum_{j=1}^{m} a_{ji} f_j(y_j(t - \tau_{ji})) + \sum_{j=1}^{m} p_{ji} \int_0^\infty H_{ji}(s) f_j(y_j(t - s)) ds + I_i,$$

$$\frac{dy_j}{dt} = -\beta_j y_j(t) + \sum_{i=1}^{n} b_{ij} f_i(x_i(t - \sigma_{ij})) + \sum_{j=1}^{n} q_{ij} \int_0^\infty K_{ij}(s) f_i(x_i(t - s)) ds + J_j,$$

(2.1)

---

* This work was jointly supported by the National Natural Science Foundation of China, the Natural Science Foundation of Jiangsu Province, China.

where $i = 1, 2, \cdots, n$, $j = 1, 2, \cdots, m$, n and m correspond to the number of neurons in X-layer and Y-layer respectively, $x_i(t)$ and $y_j(t)$ are the activations of the ith neuron and the jth neuron, respectively. $\alpha_i > 0$ and $\beta_j > 0$, $a_{ji}, b_{ij}, p_{ji}, q_{ij}$ are the connection weights, and H_{ji} and K_{ij} denote the refractoriness of the ith neuron and jth neuron after they have responded. $\tau_{ji} > 0$, $\sigma_{ij} > 0$ denote synapse time delays, $I = (I_1, I_2, \cdots, I_n)^T$ and $J = (J_1, J_2, \cdots, J_m)^T$ are the external constant inputs.

Throughout this paper, we assume that
(H_1) The delay kernels $H_{ji}, K_{ij} : [0, \infty) \to [0, \infty)$ are continuous and satisfy:

$$\int_0^\infty H_{ji}(s) = \int_0^\infty K_{ij}(s) = 1 > 0, \quad \int_0^\infty H_{ji} e^{\mu s} ds < \infty, \quad \int_0^\infty K_{ij} e^{\mu s} ds < \infty,$$

where $i = 1, 2, \cdots, n$, $j = 1, 2, \cdots, m$, and $\mu > 0$ is a constant.

The initial values associated with (2.1) are assumed to be of the forms:

$$x_i(t) = \phi_i(t), \quad y_j(t) = \varphi_j(t), \quad t \in (-\infty, 0], \tag{2.2}$$

where $\phi_i(t), \varphi_j(t) : (-\infty, 0] \to R$ are continuous.

Define $B = \{z(t) = (x(t), y(t))^T | x(t) = (x_1(t), x_2(t), \cdots, x_n(t))^T, y(t) = (y_1(t), y_2(t), \cdots, y_m(t))^T\}$. For $\forall z \in B$, we define the norm $||z(t)|| = \sum_{i=1}^n |x_i(t)| + \sum_{j=1}^m |y_j(t)|$. Clearly, B is a Banach space.

Suppose $z^* = (x^*, y^*)^T$ be the equilibrium of systems (2.1), we denote

$$||(\phi, \varphi)^T - (x^*, y^*)^T|| = \sum_{i=1}^n \sup_{t \leq 0} |\phi_i(t) - x_i^*| + \sum_{j=1}^m \sup_{t \leq 0} |\varphi_j(t) - y_j^*|.$$

Definition 1: A function $h(\rho) : R \to R$ is said to be partially Lipschitz continuous in R if for any $\rho \in R$ there exists a positive number l_ρ such that

$$|h(\theta) - h(\rho)| \leq l_\rho |\theta - \rho|, \quad \forall \theta \in R.$$

Definition 2: The activation functions f_i is said to belong to the class PLI (denotes $f_i \in PLI$), if for each $f_i : R \to R$ is a partially Lipschitz continuous and monotone nondecreasing function.

Definition 3: The equilibrium point z^* of Eqs. (2.1) is said to be globally exponentially stable (GES), if there exist constants $\lambda > 0$ and $M \geq 1$ such that

$$\sum_{i=1}^n |x_i(t) - x_i^*| + \sum_{j=1}^m |y_j(t) - y_j^*| \leq M ||(\phi, \varphi)^T - (x^*, y^*)^T|| e^{-\lambda t}, \quad t \geq 0.$$

Definition 4: The neural networks (2.1) are said to be absolutely exponentially stable (AEST), if it possesses a unique and GES equilibrium point for every function $f_i \in PLI$, and every input I with J.

3 Main Results

3.1 Existence and Uniqueness of the Equilibrium Point

Due to $f_i \in PLI$, there exists a positive number $l_i > 0$ such that $0 \leq \frac{f_i(z(t)) - f_i(0)}{z(t)} \leq l_i$ and $0 \leq \frac{f_i(z(t)) - f_i(z^*)}{z(t) - z^*} \leq l_i$.

Theorem 1: Under the assumption (H_1), for every $f_i \in PLI$ and the external inputs I, J, if $\max\left(\max_{1 \leq j \leq m}(\sum_{i=1}^{n} \frac{l_j}{\alpha_i}|a_{ji} + p_{ji}|), \max_{1 \leq i \leq n}(\sum_{j=1}^{m} \frac{l_i}{\beta_j}|b_{ij} + q_{ij}|)\right) < 1$, then systems (2.1) have a unique equilibrium point.

Proof: Obviously, the existence of the equilibrium of Eqs. (2.1) is equivalent to the existence of the solution of the following systems:

$$x_i = \frac{1}{\alpha_i}\sum_{j=1}^{m}(a_{ji} + p_{ji})f_j(y_j) + \frac{1}{\alpha_i}I_i, \quad y_j = \frac{1}{\beta_j}\sum_{i=1}^{n}(b_{ij} + q_{ij})f_i(x_i) + \frac{1}{\beta_j}J_j. \quad (3.1)$$

And we define a subset $S = \{(x,y)^T | \sum_{i=1}^{n}|x_i| + \sum_{j=1}^{m}|y_j| \leq r\} \subseteq R^{n+m}$, where

$$r = \frac{\sum_{i=1}^{n}\frac{1}{\alpha_i}|I_i| + \sum_{j=1}^{m}\frac{1}{\beta_j}|J_j| + \sum_{i=1}^{n}\sum_{j=1}^{m}\left(\frac{1}{\alpha_i}|a_{ji} + p_{ji}||f_j(0)| + \frac{1}{\beta_j}|b_{ij} + q_{ij}||f_i(0)|\right)}{1 - \max\left(\max_{1 \leq j \leq m}(\sum_{i=1}^{n}\frac{l_j}{\alpha_i}|a_{ji} + p_{ji}|), \max_{1 \leq i \leq n}(\sum_{j=1}^{m}\frac{l_i}{\beta_j}|b_{ij} + q_{ij}|)\right)}.$$

Clearly, S is a close bounded convex subset of the Banach space R^{n+m}. Define the continuous mapping $H: S \to R^{n+m}$ is as follows:

$$H(x,y) = (H_1(x,y)^T, H_2(x,y)^T)^T, \quad H_1(x,y) = (H_{11}(x,y), \cdots, H_{1n}(x,y))^T,$$

$$H_2(x,y) = (H_{21}(x,y), H_{22}(x,y), \cdots, H_{2m}(x,y))^T, \quad \text{where}$$

$$H_{1i}(x,y) = \frac{1}{\alpha_i}(\sum_{j=1}^{m}(a_{ji} + p_{ji})f_j(y_j) + I_i), \quad H_{2j}(x,y) = \frac{1}{\beta_j}(\sum_{i=1}^{n}(b_{ij} + q_{ij})f_i(x_i) + J_j).$$

Then we have

$$\sum_{i=1}^{n}|H_{1i}(x,y)| \leq \sum_{i=1}^{n}\frac{1}{\alpha_i}I_i + \sum_{i=1}^{n}\frac{1}{\alpha_i}\sum_{j=1}^{m}|a_{ji} + p_{ji}|(|f_j(y_j) - f_j(0)| + |f_j(0)|)$$

$$\leq \sum_{i=1}^{n}\frac{1}{\alpha_i}I_i + \sum_{i=1}^{n}\frac{1}{\alpha_i}\sum_{j=1}^{m}|a_{ji} + p_{ji}||f_j(0)| + \max_{1 \leq j \leq m}(\sum_{i=1}^{n}\frac{1}{\alpha_i}l_j|a_{ji} + p_{ji}|)\sum_{j=1}^{m}|y_j|,$$

similarly,

$$\sum_{j=1}^{m}|H_{2j}(x,y)|$$

$$\leq \sum_{j=1}^{m}\frac{1}{\beta_j}J_j + \sum_{j=1}^{m}\frac{1}{\beta_j}\sum_{i=1}^{n}|b_{ij} + q_{ij}||f_i(0)| + \max_{1 \leq i \leq n}(\sum_{j=1}^{m}\frac{1}{\beta_j}l_i|b_{ij} + q_{ij}|)\sum_{i=1}^{n}|x_i|.$$

Therefore,

$$\|H(x,y)\| = \sum_{i=1}^{n} |H_{1i}(x,y)^T| + \sum_{j=1}^{m} |H_{2j}(x,y)^T|$$

$$\leq \sum_{i=1}^{n} \frac{1}{\alpha_i} I_i + \sum_{j=1}^{m} \frac{1}{\beta_j} J_j + \sum_{i=1}^{n} \sum_{j=1}^{m} \left(\frac{1}{\alpha_i}|a_{ji}+p_{ji}||f_j(0)| + \frac{1}{\beta_j}|b_{ij}+q_{ij}||f_i(0)|\right)$$

$$+ \max\left(\max_{1\leq j\leq m}(\sum_{i=1}^{n}\frac{1}{\alpha_i}l_j|a_{ji}+p_{ji}|), \max_{1\leq i\leq n}(\sum_{j=1}^{m}\frac{1}{\beta_j}l_i|b_{ij}+q_{ij}|)\right)r = r,$$

which implies H maps the close bounded convex set S into S. According to Brouwer fixed point theorem ([6]), systems (3.1) have a solution $(x^*, y^*)^T \in S$ such that $H(x^*, y^*)^T = (x^*, y^*)^T$. Obviously, $(x^*, y^*)^T$ is the equilibrium point of neural networks (2.1).

In the following, we will prove the solution of Eqs. (3.1) is unique. Suppose $(\widetilde{x}, \widetilde{y})^T$ is another solution of Eqs. (3.1), and $(\widetilde{x}, \widetilde{y})^T \neq (x^*, y^*)^T$. Due to $f_j \in$ PLI, we obtain $\sum_{i=1}^{n} |x_i^* - \widetilde{x}_i| \leq \max_{1\leq j\leq m}(\sum_{i=1}^{n} \frac{l_j}{\alpha_i}|a_{ji}+p_{ji}|)\sum_{j=1}^{m} |y_j^* - \widetilde{y}_j|$ and $\sum_{j=1}^{m} |y_j^* - \widetilde{y}_j| \leq \max_{1\leq i\leq n}(\sum_{j=1}^{m} \frac{l_i}{\beta_j}|b_{ij}+q_{ij}|)\sum_{i=1}^{n} |x_i^* - \widetilde{x}_i|$, then

$$\sum_{i=1}^{n} |x_i^* - \widetilde{x}_i| + \sum_{j=1}^{m} |y_j^* - \widetilde{y}_j|$$

$$\leq \max\left(\max_{1\leq j\leq m}(\sum_{i=1}^{n}\frac{l_j}{\alpha_i}|a_{ji}+p_{ji}|), \max_{1\leq i\leq n}(\sum_{j=1}^{m}\frac{l_i}{\beta_j}|b_{ij}+q_{ij}|)\right)$$

$$\times \left(\sum_{i=1}^{n} |x_i^* - \widetilde{x}_i| + \sum_{j=1}^{m} |y_j^* - \widetilde{y}_j|\right) < \sum_{i=1}^{n} |x_i^* - \widetilde{x}_i| + \sum_{j=1}^{m} |y_j^* - \widetilde{y}_j|,$$

which is a contradiction. Therefore this completes the proof.

3.2 Absolutely Exponential Stability of BAM Neural Networks

Let z^* be the equilibrium of systems (2.1), and $z(t)$ be an arbitrary solution of Eqs. (2.1). Then the Eqs. (2.1) can be rewritten

$$\frac{du_i}{dt} = -\alpha_i u_i(t) + \sum_{j=1}^{m} a_{ji} F_j(v_j(t-\tau_{ji})) + \sum_{j=1}^{m} p_{ji} \int_0^\infty H_{ji}(s) F_j(v_j(t-s)) ds,$$

$$\frac{dv_j}{dt} = -\beta_j v_j(t) + \sum_{i=1}^{n} b_{ij} F_i(u_i(t-\sigma_{ij})) + \sum_{i=1}^{n} q_{ij} \int_0^\infty K_{ij}(s) F_i(u_i(t-s)) ds, \quad (3.2)$$

where $u_i(t) = x_i(t) - x_i^*$, $v_j(t) = y_j(t) - y_j^*$, $F_j(v_j) = f_j(v_j + y_j^*) - f_j(y_j^*)$, and $F_i(u_i) = f_i(u_i + x_i^*) - f_i(x_i^*)$.

Theorem 2: Under the assumptions of Theorem 1, if $-\alpha_i + \sum_{j=1}^{m} |a_{ji}| l_j + \sum_{j=1}^{m} |p_{ji}| l_j < 0$, and $-\beta_j + \sum_{i=1}^{n} |b_{ij}| l_i + \sum_{i=1}^{n} |q_{ij}| l_i < 0$, then systems (2.1) are AEST.

Proof: Let

$$F_i(\bar{\mu}) = -\alpha_i + \bar{\mu} + \sum_{j=1}^{m} |a_{ji}| l_j e^{\bar{\mu}\tau_{ji}} + \sum_{j=1}^{m} |p_{ji}| l_j \int_0^\infty H_{ji}(s) e^{\bar{\mu}s} ds,$$

$$G_j(\bar{\mu}) = -\beta_j + \bar{\mu} + \sum_{i=1}^{n} |b_{ij}| l_i e^{\bar{\mu}\sigma_{ij}} + \sum_{i=1}^{n} |q_{ij}| l_i \int_0^\infty K_{ij}(s) e^{\bar{\mu}s} ds.$$

According to (H_1) and the conditions, we can get $F_i(0) < 0$, $G_j(0) < 0$. Obviously, $F_i(\bar{\mu})$ and $G_j(\bar{\mu})$ are continuous functions on R. Therefore, there exists $\mu > 0$ such that

$$F_i(\mu) < 0, \quad G_j(\mu) < 0. \tag{3.3}$$

Consider functions $U_i(t)$ and $V_j(t)$: $U_i(t) = e^{\mu t}|u_i(t)|$, $V_j(t) = e^{\mu t}|v_j(t)|$, $t \in R$. From (3.2) and $f_i \in PLI$, we have

$D^+ U_i(t)$
$\leq (\mu - \alpha_i) U_i(t) + \sum_{j=1}^{m} \left(|a_{ji}| l_j V_j(t - \tau_{ji}) e^{\tau_{ji} \mu} + |p_{ji}| l_j \int_0^\infty H_{ji}(s) e^{\mu s} V_j(t-s) ds \right),$

$D^+ V_j(t)$
$\leq (\mu - \beta_j) V_j(t) + \sum_{i=1}^{n} \left(|b_{ij}| l_i U_i(t - \sigma_{ij}) e^{\sigma_{ij} \mu} + |q_{ij}| l_i \int_0^\infty K_{ij}(s) e^{\mu s} U_i(t-s) ds \right),$

where D^+ denotes the upper right derivative and $t > 0$.

Let $D > 1$ denotes an arbitrary real number and

$$P = \|(\phi, \varphi)^T - (x^*, y^*)^T\| = \sum_{i=1}^{n} \sup_{t \leq 0} |u_i(t)| + \sum_{j=1}^{m} \sup_{t \leq 0} |v_j(t)| > 0. \tag{3.4}$$

Obviously, we can get $U_i(t) < DP$, $V_j(t) < DP$, $t \in (-\infty, 0]$. Then we claim that $U_i(t) < DP$, $V_j(t) < DP$, $t > 0$. In fact, if it is not true, there exist two possible cases:

Case 1: There exist some c ($1 \leq c \leq n$) and $t_1 > 0$ such that

$$U_i(t) < DP, \quad V_j(t) < DP, \quad for \ t \in (-\infty, t_1],$$
$$U_c(t) < DP, \quad U_c(t_1) = DP, \quad D^+ U_c(t_1) \geq 0, \ t \in (-\infty, t_1), \ i \neq c, \tag{3.5}$$

where $i = 1, 2, \cdots, n$, $j = 1, 2, \cdots, m$. Using (3.5), we have

$$0 \leq D^+U_c(t_1) \leq \left(\mu - \alpha_i + \sum_{j=1}^{m}|a_{ji}|l_j e^{\mu\tau_{ji}} + \sum_{j=1}^{m}|p_{ji}|l_j \int_0^\infty H_{ji}(s)e^{\mu s}ds\right)DP,$$

according to (3.3), it follows that $0 \leq D^+U_c(t_1) < 0$, which is a contradiction.
Case 2: There exist some l $(1 \leq l \leq m)$ and $t_2 > 0$ such that

$$U_i(t) < DP, \quad V_j(t) < DP, \quad for \quad t \in (-\infty, t_2],$$

$$V_l(t) < DP, \quad V_l(t_2) = DP, \quad D^+V_l(t_2) \geq 0, \quad t \in (-\infty, t_2), \quad j \neq l, \qquad (3.6)$$

where $i = 1, 2, \cdots, n$, $j = 1, 2, \cdots, m$. Similarly, we obtain $0 \leq D^+V_l(t_2) < 0$, which is also a contradiction.

Therefore, we obtain $U_i(t) < DP$, $V_j(t) < DP$, $t > 0$. Then we have

$$\sum_{i=1}^{n}|x_i - x_i^*| + \sum_{j=1}^{m}|y_j - y_j^*| \leq e^{-\mu t}M\|(\phi,\varphi)^T - (x^*, y^*)^T\|, \quad t > 0,$$

where $M = (n+m)D \geq 1$. This completes the proof.

4 Conclusion

Some sufficient conditions have been obtained to ensure the AEST of the considered model based fixed point method and new analysis techniques. Several previous results are improved and extended. Moreover, our criteria can be easily checked in practice.

References

1. Anping, Ch., Jinde, C., Lihong, H.: Exponential Stability of BAM Neural Networks with Transmission Delays. Neurocomputing, **57** (2004) 435-454.
2. Kosko, B.: Bi-directional Associative Memories. IEEE Trans. Systems Man Cybernet, **18** (1) (1988) 49-60.
3. Jinde, C., Jun, W.: Absolute Exponential Stability of Recurrent Neural Networks with Lipschitz-continuous Activation Functions and Time Delays. Neural Networks, **17** (2004) 379-390.
4. Jinde, C., Lin, W.: Exponential Stability and Periodic Oscillatory Solution in BAM Networks with Delays. IEEE Trans. Neural Networks, **13**(2) (2002) 457-463.
5. Jinde, C., Meifang, D.: Exponential Stability of Delayed Bidirectional Associative Memory Networks. Appl. Math. Comput., **135**(1) (2003) 105-112.
6. Schwartz, J.: Nonlinear Functional Analysis. Gordon and Breach, New York, 1969.
7. Jiye, Zh., Yoshihiro, S., Takashi, I.: Absolutely Exponential Stability of a Class of Neural Networks with Unbounded Delay. Neural Networks, **17** (2004) 391-397.
8. Jinde, C.: Global Asymptotic Stability of Delayed Bidirectional Associative Memory Neural Networks. Appl. Math. Comput., **142** (2-3)(2003) 333-339.
9. Jinde, C.: Global Stability Conditions for Delayed CNNs. IEEE Trans. Circuits Syst.-I, **48** (11)(2001) 1330-1333.
10. Jinde, C.: A Set of Stability Criteria for Delayed Cellular Neural Networks. IEEE Trans. Circuits Syst.-I, **48** (4)(2001) 494-498.

Stability Analysis of Discrete-Time Cellular Neural Networks

Zhigang Zeng[1,2], De Shuang Huang[1], and Zengfu Wang[2]

[1] Intelligent Computing Lab, Hefei Institute of Intelligent Machines,
Chinese Academy of Sciences, P.O.Box 1130, Hefei Anhui 230031, China
{zhigangzeng}@iim.ac.cn
[2] Department of Automation, University of Science and Technology of China
Hefei, Anhui, 230026, China

Abstract. Discrete-time cellular neural networks (DTCNNs) are formulated and studied in this paper. Several sufficient conditions are obtained to ensure the global stability of DTCNNs with delays based on comparison methods (not based on the well-known Liapunov methods). Finally, the simulating results demonstrate the validity and feasibility of our proposed approach.

1 Introduction

In recent years, several methods have been proposed for designing associative memories using both continuous-time neural networks (CNNs) and discrete-time cellular neural networks (DTCNNs) [1]-[3]. There have been active investigations recently into the dynamics and applications of CNNs introduced by Chua and Yang [4]. A cellular neural network is a non-linear dynamic circuit consisting of many processing units called cells arranged in a two- or three-dimensional array. The structure of a cellular neural network is similar to that of a cellular automata in which each cell is connected only to its neighbouring cells. In a circuit of cellular neural network, each cell contains elements, typically a capacitor and a resistor, piecewise linear or non-linear output functions and an external input source introduced from outside the network; thus CNNs can be regarded as dissipative dynamical systems. For a circuit assembly of CNNs, one can refer to Chua and Yang [4]. Cellular neural networks have been found useful in areas of signal processing, image processing, associative memories, pattern classification (see for instance [1]) and in this respect it is important for the network to be completely stable in the sense that the network has a unique equilibrium point and every neural state trajectory approaches the equilibrium point. Most of the studies mentioned above deal with cellular neural networks in which the dynamical processing and transmission of signals are assumed to be instantaneous. Such networks are described by systems of coupled ordinary differential equations. In most circuits however, it is usually expected that time delays exist during the processing and transmission of signals [5]. Roska and Chua [6] considered CNNs wherein discrete delays are incorporated in the processing parts of the network architectures and they have also found applications of such networks in image processing and pattern classification. It is essential therefore that delayed cellular neural networks be globally asymptotically stable. In addition, we usually need to consider the discrete-time systems in practice such as computer simulation, etc [7,8]. Motivated by the above discussions, our aim in this paper is to consider the global stability of DTCNNs.

This paper consists of the following sections. Section 2 describes some preliminaries. The main results are stated in Sections 3. Simulation results of one illustrative example are given in Section 4. Finally, concluding remarks are made in Section 5.

2 Preliminaries

In this paper, we always assume that $t \in \mathcal{N}$, where $\mathcal{N}$ is the set of all natural number.

Consider a two-dimensional DTCNNs described by a space-invariant template where the cells are arranged on a rectangular array composed of N rows and M columns. The dynamics of such a DTCNNs are governed by the following normalized equations:

$$\Delta x_{ij}(t) = -c_{ij}x_{ij}(t) + \sum_{k=k_1(i,r_1)}^{k_2(i,r_1)} \sum_{l=l_1(i,r_2)}^{l_2(i,r_2)} a_{k,l}f(x_{i+k,j+l}(t))$$

$$+ \sum_{k=k_1(i,r_1)}^{k_2(i,r_1)} \sum_{l=l_1(i,r_2)}^{l_2(i,r_2)} b_{k,l}f(x_{i+k,j+l}(t-\kappa_{ij})) + u_{ij}, \quad (1)$$

where $\Delta x_i(t) = x_i(t+1) - x_i(t)$, $i \in \{1, 2, \cdots, N\}$, $j \in \{1, 2, \cdots, M\}$, $k_1(i, r_1) = \max\{1-i, -r_1\}$, $k_2(i, r_1) = \min\{N-i, r_1\}$, $l_1(j, r_2) = \max\{1-j, -r_2\}$, $l_2(j, r_2) = \min\{M - j, r_2\}$, x_{ij} denotes the state of the cell located at the crossing between the i-th row and j-th column of the network, r_1 and r_2 denote neighborhood radius, and r_1 and r_2 are positive integer, $A = (a_{kl})_{(2r_1+1)\times(2r_2+1)}$ is the feedback cloning template defined by a $(2r_1 + 1) \times (2r_2 + 1)$ real matrix, $B = (b_{kl})_{(2r_1+1)\times(2r_2+1)}$ is the delay feedback cloning template defined by a $(2r_1 + 1) \times (2r_2 + 1)$ real matrix, $c_{ij} > 0$, $f(\cdot)$ is the activation function defined by

$$f(x_{ij}(t)) = (|x_{ij}(t) + 1| - |x_{ij}(t) - 1|)/2, \quad (2)$$

$\kappa_{ij} \in \mathcal{N}$ are the time delays upper-bounded by a natural number κ, and $u = (u_{11}, u_{12}, \cdots, u_{NM})^T \in \Re^{N \times M}$ is an external input.

For $k = 0, 1, \cdots, (N-1)$ and $p = 1, 2, \cdots, M$, let $\hat{c}_{(k \times M + p)(k \times M + p)} = c_{(k+1)p}$. Denote $\hat{C} = diag\{\hat{c}_{11}, \hat{c}_{22}, \cdots, \hat{c}_{NM \times NM}\}$. An alternative expression for the state equation of DTCNNs (1) is obtained by ordering the cells in some way (e.g., by rows or by columns) and by cascading the state variables into a state vector $y = (y_1, y_2, \cdots, y_{N \times M})^T \hat{=} (x_{11}, x_{12}, \cdots, x_{1M}, x_{21}, \cdots, x_{2M}, \cdots, x_{NM})^T$.

The following compact form is then obtained:

$$\Delta y(t) = -\hat{C}y(t) + \hat{A}f(y(t)) + \hat{B}f^{(d)}(y(t)) + \hat{u}, \qquad (3)$$

where the coefficient matrices $\hat{A}$ and $\hat{B}$ are obtained through the templates A and B, the input vector $\hat{u}$ is obtained through u_{ij}, vector-valued activation function $f(y(t)) = (f(y_1(t)), f(y_2(t)), \cdots, f(y_{N\times M}(t)))^T$, $f^{(d)}(y(t)) = (f(y_1(t-\hat{\kappa}_1)), f(y_2(t-\hat{\kappa}_2)), \cdots, f(y_{N\times M}(t-\hat{\kappa}_{N\times M})))^T$, the delay $\hat{\kappa}_i$ is obtained through κ_{ij}. Denote $n = N \times M$, $\hat{A} = (\hat{a}_{ij})_{n\times n}$; $\hat{B} = (\hat{b}_{ij})_{n\times n}$.

3 Asymptotic Stability

The initial value problem for DTCNNs (1) requires the knowledge of initial data $\{x(-\kappa), \cdots, x(0)\}$. This vector is called initial string. For every initial string, there exists a unique solution $\{x(t)\}_{t \geq -\kappa}$ of (1) that can be calculated by the explicit recurrence formula

$$x_{ij}(t+1) = (1-c_{ij})x_{ij}(t) + \sum_{k=k_1(i,r_1)}^{k_2(i,r_1)} \sum_{l=l_1(i,r_2)}^{l_2(i,r_2)} a_{k,l} f(x_{i+k,j+l}(t))$$

$$+ \sum_{k=k_1(i,r_1)}^{k_2(i,r_1)} \sum_{l=l_1(i,r_2)}^{l_2(i,r_2)} b_{k,l} f(x_{i+k,j+l}(t-\kappa_{ij})) + u_{ij}, \quad t \geq 1. \quad (4)$$

In general it is difficult to investigate the asymptotic behaviour of the solutions using formula (4). The next result gives an asymptotic estimate by comparison methods.

For $i, j \in \{1, 2, \cdots, n\}$, let

$$T_{ij} = \begin{cases} \hat{c}_{ii} - |\hat{a}_{ii}| - |\hat{b}_{ii}|, & i=j, \\ -|\hat{a}_{ij}| - |\hat{b}_{ij}|, & i \neq j, \end{cases} \qquad \tilde{T}_{ij} = \begin{cases} 2 - \hat{c}_{ii} - |\hat{a}_{ii}| - |\hat{b}_{ii}|, & i=j, \\ -|\hat{a}_{ij}| - |\hat{b}_{ij}|, & i \neq j. \end{cases}$$

Denote matrices $T_1 = (T_{ij})_{n\times n}, T_2 = (\tilde{T}_{ij})_{n\times n}$.

Theorem 3.1. If $\{y(t)\}_{t \geq -\kappa}$ be a sequence of real number vectors satisfying (3), and for $\forall i \in \{1, 2, \cdots, n\}$, $\hat{c}_{ii} \in (0, 1)$, T_1 is a nonsingular M-matrix, then DTCNNs (3) have a unique equilibrium point $y^* = (y_1^*, \cdots, y_n^*)^T$, and there exist positive constants β and $\lambda_0 \in (0, 1)$ such that for $\forall i \in \{1, 2, \cdots, n\}$, $|y_i(t) - y_i^*| \leq \beta \lambda_0^t$, $t \geq 1$; i.e., DTCNNs (3) are globally asymptotical stable.

Proof. Let $H(y) = \hat{C}^{-1}[\hat{A}f(y) + \hat{B}f(y) + \hat{u}]$. Since $f(y)$ is a bounded vector function, using Schauder fixed point theorem, there exists $y^* = (y_1^*, \cdots, y_n^*)^T$ such that $H(y^*) = y^*$, i.e., DTCNNs (3) have an equilibrium point.

Since T_1 is a nonsingular M-matrix, there exist positive constants $\gamma_1, \gamma_2, \cdots, \gamma_n$ such that for $\forall i \in \{1, 2, \cdots, n\}$,

$$\gamma_i \hat{c}_{ii} - \sum_{j=1}^{n} \gamma_j |\hat{a}_{ij}| - \sum_{j=1}^{n} \gamma_j |\hat{b}_{ij}| > 0. \qquad (5)$$

Let $\eta_i(\lambda) = \gamma_i \lambda^{\kappa+1} - (\gamma_i(1-\hat{c}_{ii}) + \sum_{j=1}^n \gamma_j|\hat{a}_{ij}|)\lambda^\kappa - \sum_{j=1}^n \gamma_j|\hat{b}_{ij}|$, then $\eta_i(0) = -\sum_{j=1}^n \gamma_j|\hat{b}_{ij}| \leq 0$, $\eta_i(1) > 0$. Hence, there exists $\lambda_{0i} \in (0,1)$ such that $\eta_i(\lambda_{0i}) = 0$, and $\eta_i(\lambda) \geq 0$, $\lambda \in [\lambda_{0i}, 1)$.

In fact, if $\eta_i(0) \neq 0$, we can choose the largest value of $\lambda \in (0,1)$ satisfying $\eta_i(\lambda_{0i}) = 0$, since $\eta_i(\lambda)$ is a polynomial and it has at most $\kappa + 1$ real roots; if $\eta_i(0) = 0$, we can choose $\lambda_{0i} = 1 - \hat{c}_{ii} + (\sum_{j=1}^n \gamma_j|\hat{a}_{ij}|)/\gamma_i$. (5) implies $1 - \hat{c}_{ii} + (\sum_{j=1}^n \gamma_j|\hat{a}_{ij}|)/\gamma_i < 1$; $-\hat{c}_{ii} \in (-1, 0)$ implies $1 - \hat{c}_{ii} + (\sum_{j=1}^n \gamma_j|\hat{a}_{ij}|)/\gamma_i > 0$.

Choose $\lambda_0 = \max_{1 \leq i \leq n}\{\lambda_{0i}\}$, then for $\forall j \in \{1, 2, \cdots, n\}$,

$$\eta_j(\lambda_0) \geq 0. \tag{6}$$

Let $z_i(t) = (y_i(t) - y_i^*)/\gamma_i$, then according to (3),

$$\Delta z_i(t) = -\hat{c}_{ii} z_i(t) + [\sum_{j=1}^n \hat{a}_{ij}\tilde{f}(z_j(t)) + \sum_{j=1}^n \hat{b}_{ij}\tilde{f}(z_j(t - \hat{\kappa}_j))]/\gamma_i, \tag{7}$$

where $\tilde{f}(z_j(t)) = f(\gamma_j z_j(t) + y_j^*) - f(y_j^*)$. Let $\bar{\Upsilon} = \max_{1 \leq i \leq n}\{\max\{|z_i(0)|, |z_i(-1)|, \cdots, |z_i(-\hat{\kappa}_i)|\}\}$, then for natural number t, $|z_i(t)| \leq \bar{\Upsilon}\lambda_0^t$. Otherwise, there exist $p \in \{1, 2, \cdots, n\}$ and natural number $q \geq 1$ such that $|z_p(q)| > \bar{\Upsilon}\lambda_0^q$, and for all $j \neq p, j \in \{1, 2, \cdots, n\}$, $|z_j(s)| \leq \bar{\Upsilon}\lambda_0^s$, $-\kappa \leq s \leq q$; $|z_p(s)| \leq \bar{\Upsilon}\lambda_0^s$, $-\kappa \leq s < q$.

If $z_p(q) > \bar{\Upsilon}\lambda_0^q$, since $1 - \hat{c}_{pp} \geq 0$, from (7),

$$\bar{\Upsilon}\lambda_0^q < z_p(q) \leq \bar{\Upsilon}\lambda_0^{q-1}\{(1 - \hat{c}_{pp}) + [\sum_{j=1}^n \gamma_j|\hat{a}_{pj}| + \sum_{j=1}^n \gamma_j|\hat{b}_{pj}|\lambda_0^{-\kappa}]/\gamma_p\};$$

i.e., $\gamma_p \lambda_0^{\kappa+1} < [\gamma_p(1-\hat{c}_{pp}) + \sum_{j=1}^n \gamma_j|\hat{a}_{pj}|]\lambda_0^\kappa + \sum_{j=1}^n \gamma_j|\hat{b}_{pj}|$, this contradicts (6).

It is similar to proof that $z_p(q) \geq -\bar{\Upsilon}\lambda_0^q$.

Hence for natural number $t \geq 1$, $|y_i(t) - y_i^*| = \gamma_i|z_i(t)| \leq \gamma_i \bar{\Upsilon}\lambda_0^t$. Choose $\beta = \max_{1 \leq i \leq n}\{\gamma_i \bar{\Upsilon}\}$, the result of Theorem 3.1 holds.

Theorem 3.2. If $\{y(t)\}_{t \geq -\kappa}$ be a sequence of real number vectors satisfying (3), and for $\forall i \in \{1, 2, \cdots, n\}$, $\hat{c}_{ii} \in [1, 2)$, T_2 is a nonsingular M-matrix, then DTCNNs (3) have a unique equilibrium point $y^* = (y_1^*, \cdots, y_n^*)^T$, and there exist positive constants β and $\lambda_0 \in (0, 1)$ such that for $\forall i \in \{1, 2, \cdots, n\}$, $|y_i(t) - y_i^*| \leq \beta \lambda_0^t$, $t \geq 1$; i.e., DTCNNs (3) are globally asymptotical stable.

Proof. It is similar to the proof of Theorem 3.1 that DTCNNs (3) have an equilibrium point y^*.

Since T_2 is a nonsingular M-matrix, it is similar to the proof of Theorem 3.1 that there exist positive constants $\gamma_1, \gamma_2, \cdots, \gamma_n$ and $\lambda_0 \in (0, 1)$ such that for $\forall i \in \{1, 2, \cdots, n\}$,

$$\gamma_i \lambda^{\kappa+1} - \gamma_i(\hat{c}_{ii} - 1)\lambda^\kappa - \sum_{j=1}^n \gamma_j|\hat{a}_{ij}|\lambda^\kappa - \sum_{j=1}^n \gamma_j|\hat{b}_{ij}| \geq 0. \tag{8}$$

Let $z_i(t) = (y_i(t) - y_i^*)/\gamma_i$, $\bar{\Upsilon} = \max_{1 \leq i \leq n}\{\max\{|z_i(0)|, |z_i(-1)|, \cdots, |z_i(-\hat{\kappa}_i)|\}\}$. Hence, for natural number t, $|z_i(t)| \leq \bar{\Upsilon}\lambda_0^t$. Otherwise, there exist $p \in$

$\{1, 2, \cdots, n\}$ and natural number $q \geq 1$ such that $|z_p(q)| > \bar{\varUpsilon}\lambda_0^q$, and for all $j \neq p, j \in \{1, 2, \cdots, n\}, |z_j(s)| \leq \bar{\varUpsilon}\lambda_0^s, -\kappa \leq s \leq q; \ |z_p(s)| \leq \bar{\varUpsilon}\lambda_0^s, -\kappa \leq s < q$.
If $z_p(q) > \bar{\varUpsilon}\lambda_0^q$, since $\hat{c}_{pp} - 1 \geq 0$, from (7),

$$\bar{\varUpsilon}\lambda_0^q < z_p(q) \leq \bar{\varUpsilon}\lambda_0^{q-1}\{(\hat{c}_{pp} - 1) + [\sum_{j=1}^{n}\gamma_j|\hat{a}_{pj}| + \sum_{j=1}^{n}\gamma_j|\hat{b}_{pj}|\lambda_0^{-\kappa}]/\gamma_p\},$$

i.e., $\gamma_p\lambda_0^{\kappa+1} < [\gamma_p(\hat{c}_{pp}-1)+\sum_{j=1}^{n}\gamma_j|\hat{a}_{pj}|]\lambda_0^\kappa + \sum_{j=1}^{n}\gamma_j|\hat{b}_{pj}|$, this contradicts (8).
It is similar to proof that $z_p(q) \geq -\bar{\varUpsilon}\lambda_0^q$.
Hence, the result of Theorem 3.2 holds.
Let $N_1 \bigcup N_2 = \{1, 2, \cdots, n\}, N_1 \bigcap N_2$ is empty.

Theorem 3.3. If $\{y(t)\}_{t \geq -\kappa}$ be a sequence of real number vectors satisfying (3), and for $\forall i \in N_1$, $\hat{c}_{ii} \in (0, 1)$, $\hat{c}_{ii} - \sum_{p=1}^{n}|\hat{a}_{ip}| - \sum_{p=1}^{n}|\hat{b}_{ip}| > 0$; for $\forall j \in N_2$, $\hat{c}_{jj} \in [1, 2)$, $2 - \hat{c}_{jj} - \sum_{p=1}^{n}|\hat{a}_{jp}| - \sum_{p=1}^{n}|\hat{b}_{jp}| > 0$, then DTCNNs (3) have a unique equilibrium point $y^* = (y_1^*, \cdots, y_n^*)^T$, and there exist positive constants β and $\lambda_0 \in (0, 1)$ such that for $\forall i \in \{1, 2, \cdots, n\}, |y_i(t) - y_i^*| \leq \beta\lambda_0^t, \ t \geq 1$; i.e., DTCNNs (3) are globally asymptotical stable.

Proof. When $i \in N_1$, it is similar to the proof of Theorem 3.1 that $y_i(t) \to y_i^*, t \to +\infty$. When $j \in N_2$, it is similar to the proof of Theorem 3.2 that $y_i(t) \to y_i^*, t \to +\infty$. Hence, DTCNNs (3) are globally asymptotical stable.

4 Simulation Result

In this section, we give one example to illustrate the new results.

Example 1. Consider a two dimensional space invariant CNN with a neighborhood radius $r = 1$, its delay cloning template B is a 3×3 real matrix; i.e.,

$$\begin{bmatrix} b_{-1,-1} & b_{-1,0} & b_{-1,1} \\ b_{0,-1} & b_{0,0} & b_{0,1} \\ b_{1,-1} & b_{1,0} & b_{1,1} \end{bmatrix} = \begin{bmatrix} 0.4 & 0.3 & 0 \\ 0.1 & 0.1 & 0.1 \\ 0.4 & 0.1 & 0.4 \end{bmatrix}.$$

Assume the cloning template $A = 0$ and $N = M = 2$, $\hat{c}_{ii} = 0.9$, then from (3),

$$\begin{cases} \Delta x_{11}(t) = -0.9x_{11}(t) + 0.1f_1(t) + 0.1f_2(t) + 0.1f_3(t) + 0.4f_4(t) \\ \Delta x_{12}(t) = -0.9x_{12}(t) + 0.1f_1(t) + 0.1f_2(t) + 0.4f_3(t) + 0.1f_4(t) \\ \Delta x_{21}(t) = -0.9x_{21}(t) + 0.3f_1(t) + 0.1f_3(t) + 0.1f_4(t) \\ \Delta x_{22}(t) = -0.9x_{22}(t) + 0.4f_1(t) + 0.3f_2(t) + 0.1f_3(t) + 0.1f_4(t), \end{cases} \quad (9)$$

where $f_1(t) = f(x_{11}(t-1)), f_2(t) = f(x_{12}(t-2)), f_3(t) = f(x_{21}(t-3)), f_4(t) = f(x_{22}(t-4))$. According to Theorem 3.1, (9) is globally asymptotical stable. Simulation results with 36 random initial strings are depicted in Figure 1.

5 Concluding Remarks

In this paper, DTCNNs are formulated and studied. Three theorems are obtained to ensure the global stability of DTCNNs with delays based on comparison methods (not based on the well-known Liapunov methods). They are shown

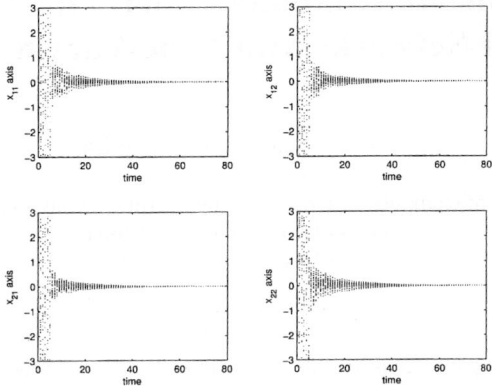

Fig. 1. The relation between t and x_{ij}

that all trajectories of DTCNNs (1) (i.e., DTCNNs (3)) converge to an (unique) equilibrium point when some sufficient conditions on weight matrices are satisfied. Conditions of these results can be directly derived from the parameters of DTCNNs, are very easy to verified. Hence, it is very convenience in application. Finally, the simulating results demonstrate the validity and feasibility of our proposed approach.

References

1. Grassi, G.: On Discrete-time Cellular Neural Networks for Associative Memories. IEEE Trans. Circuits Syst. I. **48** (2001) 107-111
2. Liu, D., Michel, A. N.: Sparsely Interconnected Neural Networks for Associative Memories with Applications to Cellular Neural Networks. IEEE Trans. Circ. Syst. II, **41** (1994) 295-307
3. Michel, A. N., Wang, K., Liu, D., Ye H.: Qualitative Limitations Incurred in Implementations of Recurrent Neural Networks. IEEE Trans. Contr. Syst. Technol. **15** (1995) 52-65
4. Chua, L. O., Yang, L.: Cellular Neural Networks: Theory. IEEE Trans. Circuits Syst. **35** (1988) 1257-1272
5. Liao, X. X., Wang, J.: Algebraic Criteria for Global Exponential Stability of Cellular Neural Networks with Multiple Time Delays. IEEE Trans. Circuits and Systems I. **50** (2003) 268-275
6. Roska, T., Chua, L. O.: Cellular Neural Networks with Nonlinear and Delay-type Template. Int. J. Circuit Theor. Appl. **20** (1992) 469-481
7. Mohamad, S., Gopalsamy, K.: Exponential Stability of Continuous-time and Discrete-time Cellular Neural Networks with Delays. Applied Mathematics and Computation, **135** (2003) 17-38
8. Barabanov, N. E., Prokhorov, D. V.: Stability Analysis of Discrete-Time Recurrent Neural Networks. IEEE Trans. Neural Networks, **13** (2002) 292-303

Novel Exponential Stability Criteria for Fuzzy Cellular Neural Networks with Time-Varying Delay

Yong Chen and Xiaofeng Liao

Department of Computer Science and Engineering, Chongqing University,
400044 Chongqing, P. R. China
xfliao@cqu.edu.cn

Abstract. This paper considers the problem of the global exponential stability of fuzzy cellular neural networks with time-varying delays. By employing Halanay-type inequalities, some novel delay-independent sufficient conditions under which the neural networks converge exponentially to the equilibria are derived. Our results are easier verifiable and less restrictive than previously known results.

1 Introduction

Fuzzy cellular neural networks (FCNN), introduced by T. Yang and L. B. Yang [1], which combined fuzzy logic and cellular neural networks (CNN) into one system. The studies [2] are shown its potential in image processing and pattern recognition. Usually, it is important for the network to be stable in the sense that the network has a unique and global convergence equilibrium point. In [3], the authors studied FCNN without time delays. But in practical application, time delays exist due to the processing and transmission of signals. It is well known that time delays can affect the dynamic behavior of neural networks [4]. Hence, the investigation of exponential stability of FCNN with time delay is very significant.

In this paper, with Hanaley-type inequalities the sufficient conditions for global exponential stability of FCNN with time-varying delays are discussed. Moreover, the exponential convergence rate is given for the convenience of estimating the speed of convergence.

2 Exponential Stability of FCCN with Time-Varying Delays

The differential equations of FCNN without time delays are proposed in [1]. To simulate the interneuron signal transmission delays, the differential equations of FCNN with time-varying delay is suggested as (1)

The suffices u, x, I_i denote input, state and bias of cell, respectively. $a_{ij}^{and}, a_{ij}^{or}, b_{ij}^{and}$ and b_{ij}^{or} are elements of fuzzy feedback MIN template, fuzzy feedback

MAX template, fuzzy feed-forward MIN template and fuzzy feed-forward MAX template, respectively. a_{ij}, b_{ij} are elements of feedback template and feed-forward template, respectively. $\wedge$ and $\vee$ denote fuzzy AND, fuzzy OR respectively.

$$\begin{cases} \dot{x}_i(t) = -r_i x_i(t) + \sum_{j=1}^{n} a_{ij} f_j(x_j(t)) + \sum_{j=1}^{n} b_{ij} u_j + I_i + \bigwedge_{j=1}^{n} a_{ij}^{and} f_j(x_j(t-\tau_{ij}(t))) \\ + \bigvee_{j=1}^{n} a_{ij}^{or} f_j(x_j(t-\tau_{ij}(t))) + \bigwedge_{j=1}^{n} b_{ij}^{and} u_j + \bigvee_{j=1}^{n} b_{ij}^{or} u_j, \quad i=1,2,\cdots,n \end{cases} \quad (1)$$

The time-varying delays $\tau_{ij}(t)$ are non-negative and $\tau_{ij}(t) \leq \tau$ for all $t > 0, 1 \leq i, j \leq n$, where τ is a positive constant. The initial function space is given as

$$\psi = \{\psi(s): -\tau \leq s \leq 0\} \in C([-\tau, 0]; R^n) \quad (2)$$

In this paper, we assume $f_j(\cdot)$ satisfy following requirements:

(H1) $\quad f_j: R \to R$

$$|f_j(u) - f_j(v)| \leq L|u-v|, \; j=1,2,\ldots n \quad (3)$$

for all $u, v \in R$, where $L > 0$ denote the Lipschitz constant.

Before we investigate the existence and the exponential stability of the equilibria of FCNN with time-varying delays, some lemmas are introduced.

Lemma 1. Suppose x and x' are two states of system (1), then we have

1) $\left| \bigwedge_{j=1}^{n} a_{ij}^{and} f_j(x_j(t)) - \bigwedge_{j=1}^{n} a_{ij}^{and} f_j(x'_j(t)) \right| \leq L \max_{1 \leq j \leq n} \{|a_{ij}^{and}|\} \max_{1 \leq j \leq n} \{|x_j - x'_j|\} \quad (4)$

2) $\left| \bigvee_{j=1}^{n} a_{ij}^{or} f_j(x_j(t)) - \bigvee_{j=1}^{n} a_{ij}^{or} f_j(x'_j(t)) \right| \leq L \max_{1 \leq j \leq n} \{|a_{ij}^{or}|\} \max_{1 \leq j \leq n} \{|x_j - x'_j|\} \quad (5)$

where L is given as (3).

Proof: 1) Suppose there exist k and l such that

$\bigwedge_{j=1}^{n} a_{ij}^{and} f_j(x_j) = a_{ik}^{and} f_k(x_k)$, $\bigwedge_{j=1}^{n} a_{ij}^{or} f_j(x'_j) = a_{il}^{or} f_l(x'_l)$. Then we have

$$\left| \bigwedge_{j=1}^{n} a_{ij}^{and} f_j(x_j) - \bigwedge_{j=1}^{n} a_{ij}^{and} f_j(x'_j) \right| \leq \max\{|a_{ik}^{and}||f_k(x_k) - f_k(x'_k)|,$$
$$|a_{il}^{and}||f_l(x_l) - f_l(x'_l)|\}$$

$$\leq \max_{1 \leq j \leq n}\{|a_{ij}^{and}||f_j(x_j) - f_j(x'_j)|\} \leq L \max_{1 \leq j \leq n}\{|a_{ij}^{and}|\} \max_{1 \leq j \leq n}\{|x_j - x'_j|\}$$

2) The proof is similar to that of 1). The proof is completed.

We introduce the Halanay-type inequality [5] in a scalar version.
Lemma 2. (Halanay inequality). Let $v(t) > 0$ for $t \in R$, suppose that

$$v' \leq -av(t) + b\left(\sup_{-\tau \leq s \leq t} v(s)\right) \quad \text{for } t > t_0 \qquad (6)$$

If $a > b > 0$, there exist $\gamma > 0$ and $k > 0$ such that $v(t) \leq ke^{-\gamma(t-t_0)}$ for $t > t_0$.

In the following, we denote the equilibrium of system (1) as $x^* = [x_1^*, x_2^*, \ldots, x_n^*]^T$.

Theorem 1. Assume $\tau \geq \tau_{ij}(t) \geq 0$, for all $t > 0$, $i, j = 1, 2, \ldots, n$, where τ is a positive constant. If the following inequalities hold:

$$\sum_{j=1}^{n}\left(|a_{ij}|\right) + \max_{1 \leq j \leq n}\left\{|a_{ij}^{and}|\right\} + \max_{1 \leq j \leq n}\left\{|a_{ij}^{or}|\right\} < \frac{r_i}{L}, \quad i = 1, 2, \ldots n. \qquad (7)$$

Then, for every input vector u, system (1) has a unique equilibrium x^* and there exit constants $\alpha > 0$ and $\beta > 0$ such that

$$\sum_{i=1}^{n}|x_i(t) - x_i^*| \leq \beta e^{-\alpha t} \qquad (8)$$

Proof. The proof for the existence of the equilibrium is similar to [3]. In the following, we will proof the uniqueness and the exponentially stability of the equilibrium point. Without loss of generality, suppose x^*, x^{**} be any two equilibrium points of system (1), without losing generality, suppose $|x_s^* - x_s^{**}| = \max_{1 \leq j \leq n}\{|x_j^* - x_j^{**}|\}$, where $s \in \{1, 2, \ldots n\}$. Then with (3),(4),(5) and (7), from system (1) we have

$$r_s |x_s^* - x_s^{**}| \leq \left|\sum_{j=1}^{n} a_{sj}\left(f_j(x_j^*) - f_j(x_j^{**})\right)\right| + \left|\bigwedge_{j=1}^{n} a_{sj}^{and} f_j(x_j^*) - \bigwedge_{j=1}^{n} a_{sj}^{and} f_j(x_j^{**})\right|$$

$$+ \left|\bigvee_{j=1}^{n} a_{ij}^{or} f_j(x_j^*) - \bigvee_{j=1}^{n} a_{ij}^{or} f_j(x_j^{**})\right|$$

$$\leq L\left(\sum_{j=1}^{n}|a_{sj}||x_j^* - x_j^{**}| + \left[\max_{1 \leq j \leq n}\{|a_{sj}^{and}|\} + \max_{1 \leq j \leq n}\{|a_{sj}^{or}|\}\right]\max_{1 \leq j \leq n}\{|x_j^* - x_j^{**}|\}\right)$$

$$\leq L\left(\sum_{j=1}^{n}|a_{sj}| + \max_{1 \leq j \leq n}\{a_{sj}^{and}\} + \max_{1 \leq j \leq n}\{|a_{ij}^{or}|\}\right)|x_s^* - x_s^{**}|$$

$$0 \leq \left(-\frac{r_s}{L} + \sum_{j=1}^{n}|a_{sj}| + \max_{1 \leq j \leq n}\{|a_{sj}^{and}|\} + \max_{1 \leq j \leq n}\{|a_{sj}^{or}|\}\right)|x_s^* - x_s^{**}| \leq 0$$

then we get $\left|x_s^* - x_s^{**}\right| = 0$, with the supposition $\left|x_s^* - x_s^{**}\right| = \max\limits_{1 \le j \le n}\left\{\left|x_j^* - x_j^{**}\right|\right\}$, we have $\left|x_j^* - x_j^{**}\right| = 0, j = 1,2,\ldots,n$. It implies that $x^* = x^{**}$, the uniqueness of the equilibrium is proved.

Now we prove the global exponential stability of x^* with the constrains of (7). We use $x(t)$ to denote an arbitrary solution of system (1). We first let

$$\beta_1 = \max_{1 \le j \le n}\left\{\sup_{-\tau \le s \le 0}\left|x_j(s) - x_j^*\right|\right\} \tag{9}$$

and define functions $P_i(.), i = 1,2,\ldots,n$ as

$$P_i(\mu_i) = \frac{r_i - \mu_i}{L} - \left(\sum_{j=1}^{n}\left(\left|a_{ij}\right|\right) + \max_{1 \le j \le n}\left\{\left|a_{ij}^{and}\right|\right\} + \max_{1 \le j \le n}\left\{\left|a_{ij}^{or}\right|\right\}\right)e^{\tau\mu_i} \tag{10}$$

where $\mu_i \in [0,+\infty)$. With condition (7), we have $P_i(0) > 0$ and, $P_i(\infty) \to -\infty$ as $\mu_i \to +\infty, i \in \{1,2,\ldots,n\}$, then for every $i \in \{1,2,\ldots,n\}$, there exist a constant $\alpha_i > 0$ which satisfy $P_i(u_i) > 0$ when $u_i < \alpha_i$ and $P_i(\alpha_i) = 0$. We let $\alpha = \min\limits_{1 \le i \le n}\{\alpha_i\}$, for $i \in \{1,2,\ldots,n\}$, we have

$$P_i(\alpha) \ge \frac{r_i - \alpha}{L} - \left(\sum_{j=1}^{n}\left(\left|a_{ij}\right|\right) + \max_{1 \le j \le n}\left\{\left|a_{ij}^{and}\right|\right\} + \max_{1 \le j \le n}\left\{\left|a_{ij}^{or}\right|\right\}\right)e^{\tau\alpha} \ge 0 \tag{11}$$

We define functions $X_i(.), i \in \{1,2,\ldots,n\}$, as

$$X_i(t) = e^{\alpha t}\left|x_i(t) - x_i^*\right| \tag{12}$$

From (12), it is obvious that $X_i(t) \le \beta_1$ for all $i \in \{1,2,\ldots,n\}, t \in [-\tau,0]$, and β_1 is given by (9). Now we claim that

$$X_i(t) \le \beta_1, \quad \text{for all } i \in \{1,2,\ldots,n\}, t > 0. \tag{13}$$

Suppose that (13) dose not hold, then there is one component among $X_i(.)$ (say $X_k(.)$) and a first time $t_1 > 0$ such that

$$X_k(t) \le \beta_1, t \in [-\tau,t_1], X_k(t_1) = \beta_1, \frac{d^+ X_k(t_1)}{dt} > 0 \tag{14}$$

while $X_i(t) \le \beta_1, i \ne k, t \in [-\tau,t_1]$
with (3),(4) and (5),we obtain from (1) the following inequalities:

$$\frac{d^+}{dt}\left|x_i(t)-x_i^*\right| \leq -r_i\left|x_i(t)-x_i^*\right| + L\sum_{j=1}^n |a_{ij}|\left|x_j(t-\tau_{ij}(t))-x_j^*\right|$$
$$+ L\left(\max_{1\leq j\leq n}\{|a_{ij}^{and}|\} + \max_{1\leq j\leq n}\{|a_{ij}^{or}|\}\right)\max_{1\leq j\leq n}\{|x_j(t-\tau_{ij}(t))-x_j^*|\} \tag{15}$$

then with (12) and (15) a system of Halanay-type inequalities is derived as follows

$$\begin{aligned}\frac{d^+ X_i(t)}{dt} &\leq -(r_i-\alpha)X_i(t) + L\sum_{j=1}^n |a_{ij}|\left|x_j(t-\tau_{ij}(t))-x_j^*\right|e^{\alpha t}\\ &+ L\left(\max_{1\leq j\leq n}\{|a_{ij}^{and}|\} + \max_{1\leq j\leq n}\{|a_{ij}^{or}|\}\right)\max_{1\leq j\leq n}\{|x_j(t-\tau_{ij}(t))-x_j^*|\}e^{\alpha t}\\ &\leq -(r_i-\alpha)X_i(t) + L\sum_{j=1}^n |a_{ij}|e^{\tau\alpha}\sup_{-\tau\leq s\leq t}(X_j(s))\\ &+ L\left(\max_{1\leq j\leq n}\{|a_{ij}^{and}|\} + \max_{1\leq j\leq n}\{|a_{ij}^{or}|\}\right)e^{\tau\alpha}\max_{1\leq j\leq n}\left\{\sup_{-\tau\leq s\leq t}(X_j(s))\right\}\\ &\leq -(r_i-\alpha)X_i(t) + L\left(\sum_{j=1}^n |a_{ij}| + \left[\max_{1\leq j\leq n}\{|a_{ij}^{and}|\}\right]\right.\\ &\left.+ \max_{1\leq j\leq n}\{|a_{ij}^{or}|\}\right)e^{\tau\alpha}\max_{1\leq j\leq n}\left\{\sup_{-\tau\leq s\leq t}(X_j(s))\right\}\end{aligned} \tag{16}$$

Substituting (14) into (16), with conditions (7) we have

$$0 < \frac{d^+ X_k(t_1)}{dt} \leq -\left(\frac{r_k-\alpha}{L} - \sum_{j=1}^n |a_{kj}| - \left(\max_{1\leq j\leq n}\{|a_{kj}^{and}|\} + \max_{1\leq j\leq n}\{|a_{kj}^{or}|\}\right)e^{\tau\alpha}\right)L\beta_1 \leq 0 \tag{17}$$

which contradicts to (14). Consequently, the inequality (13) must hold. Now, we put (12) into (13), then the following inequality is obtained

$$\left|x_i(t)-x_i^*\right| \leq \beta_1 e^{-\alpha t}, \quad i=1,2,\ldots,n$$

Let $\beta = n\beta_1$, then we have

$$\sum_{i=1}^n \left|x_i(t)-x_i^*\right| \leq \beta e^{-\alpha t},$$

which asserts the global exponential stability of x^* and the convergence rate α of system (1). The proof is completed.

One example is given to show that our results are less restrictive than previously known results in [3] and easier verifiable. Consider the following cell template matrixes $A = \begin{pmatrix} 0.4 & 0.2 \\ 0.1 & 0.4 \end{pmatrix}$, $A^{and} = \begin{pmatrix} 0.25 & 0.1 \\ 0.1 & 0.25 \end{pmatrix}$, $A^{or} = \begin{pmatrix} 0.15 & 0.2 \\ 0.2 & 0.15 \end{pmatrix}$, where we let

$f_j(x_j(\cdot)) = \frac{1}{2}(|x_j(\cdot)+1| - |x_j(\cdot)-1|)$. Let $R_x = \mathrm{diag}([r_1, r_2, ..., r_n])$, in this example, we set $R_x = I$, where I means unit matrix. We note that the matrix

$$|A| + |A^{and}| + |A^{or}| = \begin{pmatrix} 0.8 & 0.5 \\ 0.4 & 0.8 \end{pmatrix},$$ where $|A|$ means $\{|a_{ij}|\}_{n \times n}$. Then we have

$$S = I - R_x(|A| + |A^{and}| + |A^{or}|) = \begin{pmatrix} 0.2 & -0.5 \\ -0.4 & 0.2 \end{pmatrix}$$

It is obvious that the matrix S is not a nonsingular M-matrix, thereby the condition given in [3] does not hold. However, it can be easily verified that the conditions in Theorem 1 of this paper are satisfied.

3 Conclusions

In this paper, the easily verifiable sufficient conditions for the exponential convergence of FCNN are derived. They are less restrictive and robust to time-varying delays and lead to the efficient design of FCNN.

Acknowledgments. The work was supported by the National Natural Science Foundation of China under grant No. 60271019.

References

1. Yang. T., Yang. L. B., Wu. C. W., Chua. L. O.: Fuzzy cellular neural networks: theory. Proceedings of IEEE International Workshop on Cellular Neural networks and Applications, (1996) 181-186
2. Yang. T., Yang. L. B., Wu. C. W., Chua. L. O.: Fuzzy cellular neural networks: applications. Proceedings of IEEE International Workshop on Cellular Neural networks and Applications, (1996) 225-230
3. Yang. T., Yang. L. B., The Global stability of Fuzzy Cellular Neural Network. IEEE Trans. Circuist Syst_ I, Vol 43. No. 10. (1996) 880-884,
4. Civalleri. P. P., Gilli. M., On stability of cellular neural networks with delays. IEEE Trans. Circuist Syst_ I, Vol 40. No. 3. (1993) 157-164
5. Gopalsamy. K., Stability and oscillations in delay differential equations of population dynamics. Kluwer, Dordrecht. (1992).

Stability of Discrete Hopfield Networks with Delay in Serial Mode

Runnian Ma[1], Youmin Xi[1], and Hangshan Gao[2]

[1] School of Management, Xi'an Jiaotong University, Xi'an, 710049, China
m314@163.com
[2] Department of Engineering Mechanics, Northwester Plytechnical University, Xi'an, 710072, China
gaomountain@sohu.com

Abstract. Discrete Hopfield neural networks with delay are extension of discrete Hopfield neural networks. The stability of the networks is known to be bases of successful applications of the networks. The stability of discrete Hopfield neural networks with delay is mainly investigated in serial mode. Several new sufficient conditions for the networks with delay converging towards a stable state are obtained. The obtained results here generalize the existing results on stability of both discrete Hopfield neural networks without delay and with delay in serial mode.

1 Introduction

The discrete Hopfield neural network (DHNN) is one of the famous neural networks with a wide range of applications, such as content addressable memory, pattern recognition, and combinatorial optimization. Such applications heavily depend on the dynamic behavior of the networks. Therefore, the researches on the dynamic behavior are a necessary step for the design of the networks. Because the stability of DHNN is not only the foundation of the network's applications, but also the most basic and important problem, the researches on stability of the DHNN have attracted considerable interest[1-5]. Recently, the discrete Hopfield neural network with delay (DHNND) is presented. The DHNND is an extension of the DHNN. Also, the stability of DHNND is an important problem. The stability of DHNND is investigated in serial mode and some results on serial stability of DHNND are given[6-8]. However, all previous researches on the networks assume that interconnection matrix W^0 is symmetric or quasi-symmetric and interconnection matrix W^1 is row-diagonally dominant. In this paper, we improve the previous stability conditions of DHNND and obtain some new stability conditions for the DHNND converging towards a stable state.

This paper is organized as follows. Section two introduces some notations and definitions used in the paper. Section three investigates the stability of the DHNND and gives some new conditions for the DHNND converging to an equilibrium point.

2 Notations and Preliminaries

The DHNND with n neurons can be determined by two $n \times n$ real matrices $W^0 = (w_{ij}^0)_{n \times n}$, $W^1 = (w_{ij}^1)_{n \times n}$, and an n-dimensional column vector $\theta = (\theta_1, \cdots, \theta_n)^T$, denoted by $N = (W^0 \oplus W^1, \theta)$. There are two possible values for the state of each neuron: 1 or -1. Denote the state of neuron i at time $t \in \{0,1,2,\cdots\}$ as $x_i(t)$, the vector $X(t) = (x_1(t), \cdots, x_n(t))^T$ is the state of the whole neurons at time t.

The updating mode of the DHNND is determined by the following equations

$$x_i(t+1) = \text{sgn}(\sum_{j=1}^n w_{ij}^0 x_j(t) + \sum_{j=1}^n w_{ij}^1 x_j(t-1) + \theta_i), i \in I = \{1,2,\cdots,n\} \quad (1)$$

where $t \in \{0,1,2,\cdots\}$, and the sign function is defined as follows

$$\text{sgn}(u) = \begin{cases} 1, & \text{if } u \geq 0 \\ -1, & \text{if } u < 0 \end{cases}.$$

We rewrite equation (1) in the compact form

$$X(t+1) = \text{sgn}(W^0 X(t) + W^1 X(t-1) + \theta).$$

If the state X^* satisfies the following condition

$$X^* = \text{sgn}(W^0 X^* + W^1 X^* + \theta)$$

then the state X^* is called a stable state (or an equilibrium point).

Let $N = (W^0 \oplus W^1, \theta)$ starting from any initial states $X(0) = X(1)$. For $t \geq 2$, if there exists a neuron $i \in I$ such that $x_i(t) \neq x_i(t-1)$, then choose neuron i to update according to Eq.(1), else choose a neuron randomly to update by the Eq.(1). This mode of operations is called a serial mode.

For any initial states $X(0) = X(1)$, if there exists $t_1 \in \{0,1,2,\cdots\}$ such that every updating sequence $X(0), X(1), X(2), X(3), \cdots$ satisfies $X(t) = X(t_1)$ for all $t \geq t_1$, then we call that the initial states $X(0) = X(1)$ converge towards a stable state, and the DHNND is called serial stability.

A matrix $A = (a_{ij})_{i,j \in I}$ is called to be row--diagonally dominant or column-diagonally dominant, if the matrix A satisfies the following conditions

$$a_{ii} \geq \sum_{j \in I(j \neq i)} |a_{ij}|, \text{ or } a_{ii} \geq \sum_{j \in I(j \neq i)} |a_{ji}|, i \in I = \{1,2,\cdots,n\}.$$

3 The Stability in Serial Mode

In order to prove the main results in this paper, we give one lemma as follows.

Lemma [8]. Let matrix W^0 with $w_{ii}^0 \geq 0$. For any initial states $X(0) = X(1)$, if $X(t-1) \neq X(t)$, then $X(t) = X(t+1)$, $X(t-1) = X(t)$.

Theorem. Let $W^0 = A^0 + B^0 + C^0$, $W^1 = A^1 + B^1 + C^1$, $A^0 = (a_{ij}^0)_{i,j \in I}$, $A^1 = (a_{ij}^1)_{i,j \in I}$, $B^0 = (b_{ij}^0)_{i,j \in I}$, $B^1 = (b_{ij}^1)_{i,j \in I}$, $C^0 = (c_{ij}^0)_{i,j \in I}$, $C^1 = (c_{ij}^1)_{i,j \in I}$. If matrix $B^0 + B^1$ is column-diagonally dominant, $C^0 + C^1$ is row-diagonally dominant, and

$$w_{ii}^0 \geq 0, \ a_{ii}^0 + a_{ii}^1 \geq \frac{1}{2} \sum_{j \in I} \left| a_{ij}^0 + a_{ij}^1 - a_{ji}^0 - a_{ji}^1 \right|, \ \forall i \in I \quad (2)$$

then the DHNND $N = (W^0 \oplus W^1, \theta)$ converges towards a stable state in serial mode.

Proof. Let

$$\varepsilon_i = \max\left\{ \sum_{j=1}^n w_{ij}^0 x_j^0 + \sum_{j=1}^n w_{ij}^1 x_j^1 + \theta_i \left| \sum_{j=1}^n w_{ij}^0 x_j^0 + \sum_{j=1}^n w_{ij}^1 x_j^1 + \theta_i < 0, x_j^0, x_j^1 \in \{-1,1\}, j \in I \right. \right\}.$$

If there are no $x_j^0, x_j^1 \in \{-1,1\}, j \in I$ such that $\sum_{j=1}^n w_{ij}^0 x_j^0 + \sum_{j=1}^n w_{ij}^1 x_j^1 + \theta_i < 0$, then ε_i can be chosen as any negative number. Set $\overline{\theta}_i = \theta_i - \frac{\varepsilon_i}{2}, t \in \{0,1,2,\cdots\}$, then

$$x_i(t+1) = \text{sgn}(\sum_{j=1}^n w_{ij}^0 x_j(t) + \sum_{j=1}^n w_{ij}^1 x_j(t-1) + \theta_i)$$

$$= \text{sgn}(\sum_{j=1}^n w_{ij}^0 x_j(t) + \sum_{j=1}^n w_{ij}^1 x_j(t-1) + \overline{\theta}_i) \quad (3)$$

The stability of the DHNND (1) is equivalent to the stability of the DHNND (3). Obviously, the DHNND (3) is strict, i.e.,

$$\sum_{j=1}^n w_{ij}^0 x_j(t) + \sum_{j=1}^n w_{ij}^1 x_j(t-1) + \overline{\theta}_i \neq 0$$

for any $X(t-1)$ and $X(t)$. By the definition of sign function, we can prove that, if $x_i(t+1) = 1$ or -1, we have

$$x_i(t+1)(\sum_{j \in I} w_{ij}^0 x_j(t) + \sum_{j \in I} w_{ij}^1 x_j(t-1) + \overline{\theta}_i) > 0 \quad (4)$$

In this paper, the energy function (Lyapunov function) of the DHNND (3) is used as follows

$$E(t) = E(X(t)) = -\frac{1}{2} X^T(t)(A^0 + A^1)X(t) - X^T(t)(B^0 + B^1)X(t) - X^T(t)\overline{\theta}$$

where $\overline{\theta} = (\overline{\theta}_1, \overline{\theta}_2, \cdots, \overline{\theta}_n)^T$. Then

$$\Delta E(t) = E(t+1) - E(t)$$

$$= -\frac{1}{2} \Delta x_i(t) \sum_{j=1}^n (a_{ij}^0 + a_{ij}^1) x_j(t) - \frac{1}{2} \Delta x_i(t) \sum_{j=1}^n (a_{ji}^0 + a_{ji}^1) x_j(t)$$

$$-\frac{1}{2}\Delta x_i(t)\sum_{j=1}^{n}(a_{ij}^0+a_{ij}^1)\Delta x_j(t)$$

$$-\Delta x_i(t)(\sum_{j=1}^{n}w_{ij}^0 x_j(t)+\sum_{j=1}^{n}w_{ij}^1 x_j(t-1)+\overline{\theta}_i)$$

$$+\Delta x_i(t)\sum_{j=1}^{n}(w_{ij}^0 x_j(t)+w_{ij}^1 x_j(t-1))-\Delta x_i(t)\sum_{j=1}^{n}(b_{ij}^0+b_{ij}^1)x_j(t)$$

$$-\Delta x_i(t)\sum_{j=1}^{n}(b_{ji}^0+b_{ji}^1)x_j(t+1) \quad (5)$$

where $\Delta x_i(t) = x_i(t+1) - x_i(t)$. We consider two cases in the following.

Case 1 If $\Delta X(t) = 0$, then $\Delta E(t) = 0$.

Case 2 If $\Delta X(t) \neq 0$, then $X(t-1) = X(t)$ by the lemma. Therefore, we replace $X(t-1)$ by $X(t)$, and then $\Delta E(t)$ can be rewritten as

$$\Delta E(t) = \frac{1}{2}\Delta x_i(t)\sum_{j=1}^{n}(a_{ij}^0+a_{ij}^1)x_j(t)-\frac{1}{2}\Delta x_i(t)\sum_{j=1}^{n}(a_{ji}^0+a_{ji}^1)x_j(t)$$

$$-\frac{1}{2}\Delta x_i(t)\sum_{j=1}^{n}(a_{ij}^0+a_{ij}^1)\Delta x_j(t)$$

$$-\Delta x_i(t)(\sum_{j=1}^{n}w_{ij}^0 x_j(t)+\sum_{j=1}^{n}w_{ij}^1 x_j(t-1)+\overline{\theta}_i)+\Delta x_i(t)\sum_{j=1}^{n}(c_{ij}^0+c_{ij}^1)x_j(t)$$

$$-\Delta x_i(t)\sum_{j=1}^{n}(b_{ji}^0+b_{ji}^1)x_j(t+1)$$

$$= \alpha(t)+\beta(t)+\gamma(t) \quad (6)$$

where

$$\alpha(t) = \frac{1}{2}\Delta x_i(t)\sum_{j=1}^{n}(a_{ij}^0+a_{ij}^1)x_j(t)-\frac{1}{2}\Delta x_i(t)\sum_{j=1}^{n}(a_{ji}^0+a_{ji}^1)x_j(t)$$

$$-\frac{1}{2}\Delta x_i(t)\sum_{j=1}^{n}(a_{ij}^0+a_{ij}^1)\Delta x_j(t)$$

$$\leq -\frac{1}{2}|\Delta x_i(t)|^2 (a_{ii}^0 + a_{ii}^1) + \frac{1}{2}|\Delta x_i(t)| \sum_{j\in I}|a_{ij}^0 + a_{ij}^1 - a_{ji}^0 - a_{ji}^1|$$

$$\leq -2(a_{ii}^0 + a_{ii}^1 - \frac{1}{2}\sum_{j\in I}|a_{ij}^0 + a_{ij}^1 - a_{ji}^0 - a_{ji}^1|) \leq 0 \tag{7}$$

$$\beta(t) = -\Delta x_i(t)(\sum_{j=1}^{n} w_{ij}^0 x_j(t) + \sum_{j=1}^{n} w_{ij}^1 x_j(t-1) + \overline{\theta}_i)$$

$$= -2x_i(t+1)(\sum_{j=1}^{n} w_{ij}^0 x_j(t) + \sum_{j=1}^{n} w_{ij}^1 x_j(t-1) + \overline{\theta}_i) < 0 \tag{8}$$

$$\gamma(t) = \Delta x_i(t) \sum_{j=1}^{n}(c_{ij}^0 + c_{ij}^1)x_j(t) - \Delta x_i(t) \sum_{j=1}^{n}(b_{ji}^0 + b_{ji}^1)x_j(t)$$

$$= -2x_i(t) \sum_{j=1}^{n}(c_{ij}^0 + c_{ij}^1)x_j(t) - 2x_i(t+1) \sum_{j=1}^{n}(b_{ji}^0 + b_{ji}^1)x_j(t+1)$$

$$= -2(c_{ii}^0 + c_{ii}^1 + \sum_{j=1(j\neq i)}^{n}(c_{ij}^0 + c_{ij}^1)x_i(t)x_j(t))$$

$$- 2(b_{ii}^0 + b_{ii}^1 + \sum_{j=1(j\neq i)}^{n}(b_{ji}^0 + b_{ji}^1)x_i(t+1)x_j(t+1))$$

$$\leq -2(c_{ii}^0 + c_{ii}^1 - \sum_{j=1(j\neq i)}^{n}|c_{ij}^0 + c_{ij}^1|) - 2(b_{ii}^0 + b_{ii}^1 - \sum_{j=1(j\neq i)}^{n}|b_{ji}^0 + b_{ji}^1|) \leq 0 \tag{9}$$

Based respectively on (2) and (4), we know that above inequalities (7) and (8) are both true. By matrix $B^0 + B^1$ being column-diagonally dominant and matrix $C^0 + C^1$ being row-diagonally dominant, we obtain that above inequality (9) holds.

As proved above, this implies that, if $\Delta X(t) = 0$, then $\Delta E(t) = 0$, if $\Delta X(t) \neq 0$, then $\Delta E(t) < 0$. Therefore, the energy function is strict. The proof is completed.

Corollary. Let $W^0 = A^0 + B^0 + C^0$, $W^1 = A^1 + B^1 + C^1$, matrix B^0 and B^1 be column-diagonally dominant, matrix C^0 and C^1 be row-diagonally dominant. If one of the following conditions is satisfied, then the DHNND $N = (W^0 \oplus W^1, \theta)$ converges towards a stable state in serial mode.

1) Matrices A^0 and A^1 with

$$a_{ii}^0 \geq \frac{1}{2}\sum_{j\in I}|a_{ij}^0 - a_{ji}^0|, \quad a_{ii}^1 \geq \frac{1}{2}\sum_{j\in I}|a_{ij}^1 - a_{ji}^1|, \quad \forall i \in I$$

2) Matrix $A^0 + A^1$ is symmetric with $a_{ii}^0 \geq 0$ and $a_{ii}^0 + b_{ii}^0 \geq 0$ for each $i \in I$.

Remark 1. In the theorem 1, if matrices $B^0 = C^0 = A^1 = B^1 = 0$, then the result is one result of reference [8], especially, if matrix A^1 is symmetric too, the result is one result in reference [6,7]. In fact, many results on serial stability in references can be deemed to some direct results of the theorem and the corollary in this paper.

Remark 2. If there exists a positive diagonal matrix $M = diag(m_1, \cdots, m_n)$ such that matrices MA^0, MB^0, MC^0, MA^1, MB^1, and MC^1 satisfy the conditions in theorem or corollary, the corresponding results also hold.

Example. Let $N = (W^0 \oplus W^1, \theta)$, where

$$W^0 = \begin{pmatrix} 1 & 0 \\ -3 & 2 \end{pmatrix} = \begin{pmatrix} 0 & -1 \\ -1 & 0 \end{pmatrix} + \begin{pmatrix} 1 & 1 \\ -2 & 2 \end{pmatrix} = A^0 + C^0$$

$$W^1 = \begin{pmatrix} 0 & -4 \\ -3 & 1 \end{pmatrix} = \begin{pmatrix} 0 & -3 \\ -3 & 0 \end{pmatrix} + \begin{pmatrix} 0 & -1 \\ 0 & 1 \end{pmatrix} = A^1 + B^1$$

The stability in serial mode can not be validated since the previous stability conditions references are not satisfied. However, the conditions 2) of the corollary are satisfied, therefore, the network converges towards a stable state in serial mode.

4 Conclusion

This paper mainly studies the stability of DHNND and obtains some results. The conditions for the DHNND converging towards a stable state in serial mode are given. These results generalize the existing results. If matrix W^1=0, the DHNND is the same as DHNN. So, the results also generalize some results on stability of the DHNN.

References

1. Hopfield, J.J.: Neural networks and physical systems emergent collective computational abilities. Proc.Nat.Acad.Sci.USA, Vol.79(1982) 2554-2558
2. Bruck, J: On the convergence properties of the Hopfield model. Proceedings IEEE, Vol.78 (1990) 1579-1585
3. Zongben Xu and C.P.Kwong: Global convergence and asymptotic stability of asymmetrical Hopfield neural networks. J. Mathematical Analysis and Applications, Vol.191(1995) 405-426
4. Donqliang Lee: New stability conditions for Hopfield neural networks in partial simultaneous update mode. IEEE,Trans.Neural Networks, Vol.10(1999) 975-978
5. Runnian Ma, Qiang Zhang and Jin Xu, Convergence of discrete-time cellular neural networks. Chinese J. Electronics, Vol.11(2002) 352-356
6. Shenshan Qiu, Eric C.C.Tang , Daniel S.Yeung: Stability of discrete Hopfield neural networks with time-delay. Proceedings of IEEE International Conference on system, Man Cybernetics(2000) 2546-2552
7. Shenshan Qiu, Xiaofei Xu, Mingzhou Liu et al: Convergence of discrete Hopfield-type neural network with delay in serial mode. J. Computer Research and Development, Vol.36(1999) 546-552
8. Runnian Ma, Qiang Zhang and Jin Xu: Convergence of discrete Hopfield neural network with delay in serial mode. J. Xidian University, Vol.28(2001) 598-602

Further Results for an Estimation of Upperbound of Delays for Delayed Neural Networks

Xueming Li, Xiaofeng Liao, and Xiang Tao

Department of Computer Science and Engineering
Chongqing University, Chongqing, 400044 P. R. China
xfliao@cqu.edu.cn

Abstract. In this paper, we have studied the global stability of the equilibrium of Hopfield neural networks with discrete delays. Criteria for global stability are derived by means of Lyapunov functionals and the estimates on the allowable sizes of delays are also given. Our results are better than those given in the existing literature.

1 Introduction

In recent years, the stability problem of time-delay neural networks has been extensively investigated [3-15] due to theoretical interest as well as application considerations. In particular, stability analysis is a power tool for practical systems and associative memory [1-15] since delays are often encountered in various neural systems.

However, the effects of delays on the stability are more complicated. Depending on whether the stability criterion itself contains the delay argument as a parameter, the stability criteria for time-lag systems can be classified into two categories, namely delay-independent criteria and delay-dependent criteria. In the past, most of the investigations on neural systems with various types of delays are confined to the delay-independent analysis [3,4-14]. For the most general delayed neural systems such as the Hopfield model (see the following Eq. (1)), it is believed that small delays are also harmless for the global asymptotic stability, but this has not been proved yet [15].

In this paper, we studied the problem of the effects of delays on the stability. Our main motivation is to provide an upperbound of delays for the delay-dependent global asymptotic stability of the equilibrium of Hopfield neural networks with discrete delays.

2 Main Results

In this section, we consider the following Hopfield neural networks with discrete delays

$$\frac{dx_i(t)}{dt} = -b_i x_i(t) + \sum_{j=1}^{n} a_{ij} f_j(x_j(t - \tau_{ij})) + I_i \tag{1}$$

in which $i = 1, 2, \cdots, n$, b_i represents the rate with which the ith unit will reset its potential to the resting state in isolation when disconnected from the network and external inputs, aij denotes the strength of the jth unit on the ith unit at time t, corresponds to the transmission delay of the ith unit along the axon of the jth unit at time t, Ii denotes the external bias or clamped input form outside the network to the ith unit, xi corresponds to the membrane potential of the ith at time t, fj(xj) denotes the conversion of the membrane potential of the jth unit into its firing rate. Throughout this section, we assume that $b_i > 0, \tau_{ij} > 0, a_{ij} \in R$ and $I_i \in R$ are constants.

In this paper, we consider solution of (1) with the initial values of the type

$$x_i(t) = \varphi_i(t), \quad t \in [t_0 - \tau, t_0], i = 1, 2, \cdots, n, \tag{2}$$

where $\tau = \max_{1 \le i, j \le n} \{\tau_{ij}\}$ and φ_i is a bounded continuous function on $(-\infty, t_0]$. It follows step by step that every solution of (1) exists in the future. Our aim is to give an estimation of the upperbound of delays τ_{ij} for the global asymptotic stability of the delayed Hopfield neural network model (1). We prove rigorously that the unique equilibrium of system (1) is globally asymptotic stable under two simple assumptions when the delays are smaller than the bound. Such a result has important significance in both theory and application for the design of the globally stable networks.

Our results are expressed as follows.

Theorem 1. Assume that the function fj(x) satisfies:
(H1) fj(x) is a bounded function for $x \in R$ and $j = 1, 2, \cdots, n$;
(H$_2$) There exists a constant $\mu_j > 0$ such that

$$|f_j(x) - f_j(y)| \le \mu_j |x - y| \text{ for } x, y \in R \text{ j=1,2,\&,n}.$$

Moreover, assume that

$$2b_i - \sum_{j=1}^{n} (|a_{ij}|\mu_j + |a_{ji}|\mu_i) \ge 0 \text{ for all } i = 1, 2, \cdots, n.$$

Then the system (1) has a unique equilibrium that is globally stable if the delays τ_{ij} $(i, j = 1, 2, \cdots, n)$ are less than or equal to the value τ^*, i.e.,

$$\tau_{ij} \le \tau^* = \min_{1\le i \le n}\left\{\frac{2b_i - \sum_{j=1}^{n}\left(|a_{ij}|\mu_j + |a_{ji}|\mu_i\right)}{\sum_{j=1}^{n}\sum_{k=1}^{n}\mu_j\left(\mu_k|a_{ik}||a_{jk}| + \mu_i|a_{ji}||a_{kj}|\right)}\right\} \qquad (3)$$

Proof. By Lemma[7,8], the system (1) exists an equilibrium, say $x^* = (x_1^*, x_2^*, \cdots, x_n^*)^T$. Let $u_i(t) = x_i(t) - x_i^*$, then the system (1) reduces to

$$\frac{du_i(t)}{dt} = -b_i u_i(t) + \sum_{j=1}^{n} a_{ij}\left[f_j(u_j(t-\tau_{ij}) + x_j^*) - f_j(x_j^*)\right] \qquad (4)$$

Hence, we have

$$\frac{d^+|u_i(t)|}{dt} \le -b_i|u_i(t)| + \sum_{j=1}^{n}|a_{ij}|\mu_j|u_j(t-\tau_{ij})| \qquad (5)$$

If let $y_i(t) = |u_i(t)|$, then

$$\frac{d^+ y_i(t)}{dt} \le -b_i y_i(t) + \sum_{j=1}^{n}|a_{ij}|\mu_j y_j(t-\tau_{ij}) \qquad (6)$$

Clearly, the globally asymptotic stability of (1) implies the uniqueness of equilibrium of (4). Therefore, it is sufficient to prove that all solutions of (4) satisfying $\lim_{t\to\infty} y_i(t) = 0$.

We define the following Lyapunov functionals

$$V(y_t) = \sum_{i=1}^{n}\left(V_{i1}(y_t) + V_{i2}(y_t)\right)$$

where,

$$V_{i1}(y_t) = \frac{1}{2}y_i^2(t)$$

and

$$V_{i2}(y_t) = \frac{1}{2}\sum_{j=1}^{n}|a_{ij}|\mu_j \int_{t-\tau_{ij}-\tau_{jk}}^{t-\tau_{jk}}\left(\sum_{k=1}^{n}|a_{jk}|\mu_k y_k^2(s)\right)ds$$

Calculate the upper Dini derivative of $V_{i1}(y_t)$ and apply inequality $2ab \leq a^2 + b^2$, we obtain

$$\frac{d^+V_{i1}(y_t)}{dt} \leq y_i(t)[-b_i y_i(t) + \sum_{j=1}^{n}|a_{ij}||\mu_j|y_j(t-\tau_{ij})]$$

$$= -b_i y_i^2(t) + \sum_{j=1}^{n}|a_{ij}||\mu_j|y_i(t)y_j(t) - y_i(t)\sum_{j=1}^{n}|a_{ij}||\mu_j|\int_{-\tau_{ij}}^{} \dot{y}_j(s)ds$$

$$\leq -b_i y_i^2(t) + \sum_{j=1}^{n}|a_{ij}||\mu_j|y_i(t)y_j(t) + \sum_{j=1}^{n}|a_{ij}||\mu_j|\int_{-\tau_{ij}}^{}\left(\sum_{k=1}^{n}|a_{jk}||\mu_k||y_i(t)||y_k(s-\tau_{jk})|\right)ds$$

$$\leq -b_i y_i^2(t) + \frac{1}{2}\sum_{j=1}^{n}|a_{ij}||\mu_j|y_i^2(t) + \frac{1}{2}\sum_{j=1}^{n}|a_{ij}||\mu_j|y_j^2(t)$$

$$+ \frac{1}{2}\sum_{j=1}^{n}|a_{ij}||\mu_j|\int_{-\tau_{ij}}^{}\sum_{k=1}^{n}|a_{jk}||\mu_k|[y_i^2(t) + y_k^2(s-\tau_{jk})]ds$$

$$\leq -b_i y_i^2(t) + \frac{1}{2}\sum_{j=1}^{n}|a_{ij}||\mu_j|y_i^2(t) + \frac{1}{2}\sum_{j=1}^{n}|a_{ij}||\mu_j|y_j^2(t)$$

$$+ \frac{1}{2}\sum_{j=1}^{n}[|a_{ij}||\mu_j|\left(\sum_{k=1}^{n}|a_{jk}||\mu_k|\tau\right)]y_i^2(t) + \frac{1}{2}\sum_{j=1}^{n}|a_{ij}||\mu_j|\int_{-\tau_{ij}}^{}\sum_{k=1}^{n}|a_{jk}||\mu_k|y_k^2(s-\tau_{jk})ds$$

$$\leq -b_i y_i^2(t) + \frac{1}{2}\sum_{j=1}^{n}|a_{ij}||\mu_j|y_i^2(t) + \frac{1}{2}\sum_{j=1}^{n}|a_{ij}||\mu_j|y_j^2(t)$$

$$+ \frac{1}{2}\sum_{j=1}^{n}[|a_{ij}||\mu_j|\left(\sum_{k=1}^{n}|a_{jk}||\mu_k|\tau\right)]y_i^2(t) + \frac{1}{2}\sum_{j=1}^{n}|a_{ij}||\mu_j|\int_{-\tau_{ij}-\tau_{jk}}^{-\tau_{jk}}\left(\sum_{k=1}^{n}|a_{jk}||\mu_k|y_k^2(s)\right)ds$$

We can also calculate

$$\frac{d^+V_{i2}(y_t)}{dt} = \frac{1}{2}\sum_{j=1}^{n}|a_{ij}||\mu_j|\int_{-\tau_{ij}-\tau_{jk}}^{-\tau_{jk}}\left(\sum_{k=1}^{n}|a_{jk}||\mu_k|y_k^2(t)\right)ds$$

$$+ \frac{1}{2}\sum_{j=1}^{n}|a_{ij}||\mu_j|\int_{-\tau_{jk}}^{}\left(\sum_{k=1}^{n}|a_{jk}||\mu_k|y_k^2(s)\right)ds - \frac{1}{2}\sum_{j=1}^{n}|a_{ij}||\mu_j|\int_{-\tau_{ij}-\tau_{jk}}^{}\left(\sum_{k=1}^{n}|a_{jk}||\mu_k|y_k^2(s)\right)ds$$

$$= \frac{1}{2}\sum_{j=1}^{n}|a_{ij}||\mu_j|\left(\sum_{k=1}^{n}|a_{jk}||\mu_k|y_k^2(t)\right)\tau - \frac{1}{2}\sum_{j=1}^{n}|a_{ij}||\mu_j|\int_{-\tau_{ij}-\tau_{jk}}^{}\left(\sum_{k=1}^{n}|a_{jk}||\mu_k|y_k^2(s)\right)ds$$

Hence,

$$\frac{dV(y_t)}{dt} \leq \frac{1}{2}\sum_{i=1}^{n}\{-2b_i + \sum_{j=1}^{n}(|a_{ij}||\mu_j| + |a_{ji}||\mu_i|) + \tau\sum_{j=1}^{n}[|a_{ij}||\mu_j|(\sum_{k=1}^{n}|a_{jk}||\mu_k|)]\}y_i^2(t)$$

$$+ \frac{\tau}{2}\sum_{i=1}^{n}\sum_{j=1}^{n}|a_{ij}||\mu_j|(\sum_{k=1}^{n}|a_{jk}||\mu_k|)y_k^2(t) \quad (7)$$

$$\leq \frac{1}{2}\sum_{i=1}^{n}\{-2b_i + \sum_{j=1}^{n}(|a_{ij}||\mu_j| + |a_{ji}||\mu_i|) + \tau\sum_{j=1}^{n}\sum_{i=1}^{n}\mu_j(\mu_k|a_{ij}||a_{jk}| + \mu_i|a_{kj}||a_{ji}|)\}y_i^2(t)$$

$$< 0$$

By integrating (7) from t_0 to t, we obtain

$$V(t)+\sum_{i=1}^{n}\beta_i \int_{t_0}^{t} y_i^2(s)ds \leq V(t_0) \tag{8}$$

This, together with $V(t) \geq 0$, implies that $V(t)$ is bounded on $(t_0, +\infty)$, moreover,

$$\limsup_{t \to +\infty} \sum_{i=1}^{n}\beta_i \int_{t_0}^{t} y_i^2(s)ds \leq V(t_0) < +\infty \tag{9}$$

Again, from (1), (H$_1$) and Lemma[7,8], it follows that $x_i(t)$ and dx_i/dt are bounded for $t>t_0$. Hence, $y_i(t)$ is a uniformly continuous function for $t>t_0$. The uniform continuity of $y_i(t)$ on $(t_0, +\infty)$ together with (9) implies $\lim_{t \to +\infty} y_i(t) = 0$. The proof is complete.

Example. We consider an example given in [15]. It is easy to calculate the value τ^* when the values of b_i and a_{ij} are given for all i,j=1,2. The calculated results are given in Table 1.

Table 1. Comparison for our results with [15]

τ^*	$\tau^*_{[15]}$	b_1	b_2	a_{11}	a_{12}	a_{21}	a_{22}
1	0.3638	0.7	0.7	0.1	0.1	0.3	0.3
1.5	0.4688	0.7	0.7	0.1	0.1	0.5	0.1
0.5263	0.2198	0.8	1.8	0.5	0.1	0.1	0.7
1.875	0.6818	0.7	0.7	0.3	0.1	0.1	0.3
1.6666	0.6838	0.8	5.3	0.1	0.5	0.1	0.1
1.4705	0.0279	0.8	5.3	0.1	0.3	0.9	0.1

In Table 1, τ^* is calculated by means of Theorem 1. However, $\tau^*_{[15]}$ is the result of [15]. Then, from Table 1, we can easily see that our results are better than those given in [15].

3 Conclusions

Usually, there are general results for stability independent of delays, one may expect sharper, delay-dependent stability conditions. This is because the robustness of independent of delay properties is of course counterbalanced by very conservative conditions. In engineering practice, information on the delay range is generally available and delay-dependent stability criteria are likely to give better performances. In this paper, we have shown that time delays are not negligible for the global stability of the delayed Hopfield networks. Some criteria are also derived by means of Lyapunov functionals.

Acknowledgments. The work described in this paper was supported by a grant from the National Natural Science Foundation of China (No. 60271019) and the Doctorate Foundation of the Ministry of Education of China (No. 20020611007).

References

1. Hopfield, J. J.: Neuronal networks and physical systems with emergent collective computational abilities. Proceedings of the National Academy of Sciences, 79, (1982) 2554-2558
2. Hopfield, J. J.: Neurons with graded response have collective computational properties like those of two-state neurons. Proceeding of the National Academy of Sciences, 81, (1984) 3058-3092
3. van der Driessche, P., Zou, X.: Global attractivity in delayed Hopfield neural network models. SIAM Journal of on Applied Mathematics, 58, (1998) 1878-1890
4. Marcus, C. M., Westervelt, R. M.: Stability of analog neural networks with delay. Physical Review A, 39, (1989) 347-359
5. Gopalsamy, K., He, X.: Stability in asymmetric Hopfield nets with transmission delays. Physica D, 76, (1994a) 344-358
6. Gopalsamy, K., He, X.: Delay-independent stability in bi-directional associative memory networks. IEEE Transactions on Neural Networks, 5, (1994b) 998-1002
7. Liao, X. F., Yu, J. B.: Robust stability for interval Hopfield neural networks with time delays", IEEE Transactions on Neural Networks, 9, (1998) 1042-1046
8. Liao, X. F., Wong, K. W., Wu, Z. F., Chen, G.: Novel robust stability criteria for interval delayed Hopfield neural networks. IEEE Transactions on CAS-I, 48, (2001) 1355-1359
9. Liao, X. F., Yu, J. B.: Qualitative analysis of bi-directional associative memory networks with time delays. Int. J. Circuit Theory and Applicat., 26, (1998) 219-229
10. Liao, X. F., Yu, J. B., Chen, G.: Novel stability criteria for bi-directional associative memory neural networks with time delays. Int. J. Circuit Theory and Applicat., 30, (2002) 519-546
11. Liao, X. F., Wong, K. W., Yu, J. B.: Novel stability conditions for cellular neural networks with time delay. Int. J. Bifur. And Chaos, 11, (2001) 1853-1864
12. Liao, X. F., Wu, Z. F., Yu, J. B.: Stability analyses for cellular neural networks with continuous delay. Journal of Computational and Applied Math., 143, (2002) 29-47
13. Liao, X. F., Chen, G. Sanchez, E. N.: LMI-based approach for asymptotically stability analysis of delayed neural networks. IEEE Transactions on CAS-I, 49, (2002a) 1033-1039
14. Liao, X. F., Chen, G. Sanchez, E. N.: Delay-dependent exponential stability analysis of delayed neural networks: an LMI approach. Neural Networks, 15, (2002b) 855-866
15. Chen A., Cao J., and Huang L.: An estimation of upperbound of delays for global asymptotic stability of delayed Hopfield neural networks, IEEE Transactions on CAS-I, Vol. 49, No. 7, (2002) 1028-1032

Synchronization in Two Uncoupled Chaotic Neurons*

Ying Wu[1], Jianxue Xu[1], Daihai He[2], Wuyin Jin[1], and Mi He[1]

[1] School of Architectural Engineering and Mechanics,
Xi'an Jiaotong University, Xi'an 710049
Wying36@163.com
[2] Department of Mathematics and Statistics,
McMaster University, Hamilton L8S4K1, Canada

Abstract. Using the membrane potential of a chaotic neuron as stimulation signal to synchronize two uncoupled Hindmarsh-Rose (HR) neurons under different initial conditions is discussed. Modulating the corresponding parameters of two uncoupled identical HR neurons, full synchronization is realized when the largest condition lyapunov exponent (LCLE) becomes negative at the threshold of stimulation strength. Computing the interspiks interval (ISI) sequence shows synchronized chaotic response of modulated neurons is different from stimulation signal. Modulating input currents of two uncoupled HR neurons with parameters mismatch, phase synchronization is obtained when the LCLEs of two systems change to both negative, and synchronized response of two systems is in phase synchronization with stimulation signal.

1 Introduction

Since the concept of chaos synchronization was reported by Pecora and Carroll[1] in 1990, it has been studied widely in fields such as communications [2], and neural science[3,4]. The concept of synchronization has been generalized, such as phase synchronization (PS)[5], which means the entrainment of phases of chaotic oscillators, whereas their amplitudes remain chaotic, and PS has been observed in nonlinear neural systems [6].

Experiments of brain activity show that synchronous response to same stimul-ation may appear in different neurons arrays separated from each other in same region of the brain, even appear in two different region of brain [7]. It means that the synchronization doesn't result from feedback connection only.

The Hindmarsh-Rose (HR) model [8] exhibits a multi-time-scaled burst-rest behavior. In the Ref. [3], Phase synchronization in two coupled HR neurons is discussed. In the Ref. [4], noise-induced synchronization is studied numerically between two uncoupled identical HR neurons.

---

* Project supported by the National Natural Science Foundation of China (Grant Nos. 10172067 and 30030040).

In this paper, the membrane potential of chaotic neuron is used to modulate the corresponding parameters of two uncoupled HR models to realize synchronization. Analyzing the change of the largest condition lyapunov exponent (LCLE) with stimulation strength k, the result shows that the LCLEs becomes negative at the onset of the synchronization, which is different from the result of two coupled HR neurons reported in Ref. [3]. The interspiks interval (ISI) sequences have relation with processing and encoding information [9,10], and synchronized response of two modulated neurons is discussed by computing the ISI sequence as well.

2 Hindmarsh-Rose Model and Synchronization Method

HR model has the following equations:

$$\dot{x} = y - ax^3 + bx^2 - z + I$$
$$\dot{y} = c - dx^2 - y \qquad (1)$$
$$\dot{z} = r[S(x-\chi) - z]$$

The HR neuron is characterized by three time-dependent variables: the membrane potential x, the recovery variable y, and a slow adaptation current z. In the simulation, let $a = 1.0, b = 3.0, c = 1.0, d = 5.0, S = 4.0, r = 0.006, \chi = -1.56$, and $I = 3.0$, I denotes the input current. Because the HR neuron model has multiple time-scale dynamics [8], the burst of action potentials often consists of a group of spikes. The projection of the attractor in the $z - x$ plane is given in Fig.1 (a). The double precision fourth-order Runge kutta method with integration time step 0.01 was used, the initial condition is (0.1,1.0,0.2).

Two uncoupled HR models equations are listed as follow:

$$\dot{x}_{1,2} = y_{1,2} - a_{1,2}x_{1,2}^3 + bx_{1,2}^2 - z_{1,2} + I_{1,2}$$
$$\dot{y}_{1,2} = c_{1,2} - d_{1,2}x_{1,2}^2 - y_{1,2} \qquad (2)$$
$$\dot{z}_{1,2} = r_{1,2}[S_{1,2}(x_{1,2} - \chi_{1,2}) - z_{1,2}]$$

where (x_1, y_1, z_1) and (x_2, y_2, z_2) denote the state vectors of two HR models, $(a_1, b_1, c_1, d_1, r_1, S_1, I_1, \chi_1)$ and $(a_2, b_2, c_2, d_2, r_2, S_2, I_2, \chi_2)$ denote parameters vectors.

To synchronize two uncoupled HR neurons in Eq.(2) under different initial conditions, The membrane potential of HR neuron in Eq. (1) (denoted by $x_s(t)$) is used as stimulation signal to modulate the corresponding parameters of two neurons, such as that the parameters $b_{1,2}$ are modulated, then:

$$b_{1,2}(t) = b_{1,2} + kx_s(t) \qquad (3)$$

where k is stimulation strength. Synchronization error is as follow:

$$\varepsilon(t) = \sqrt{(x_2 - x_1)^2 + (y_2 - y_1)^2 + (z_2 - z_1)^2} \qquad (4)$$

if $\varepsilon(t) \to 0$ as $t \to \infty$, we can say that two systems attain full synchronization.

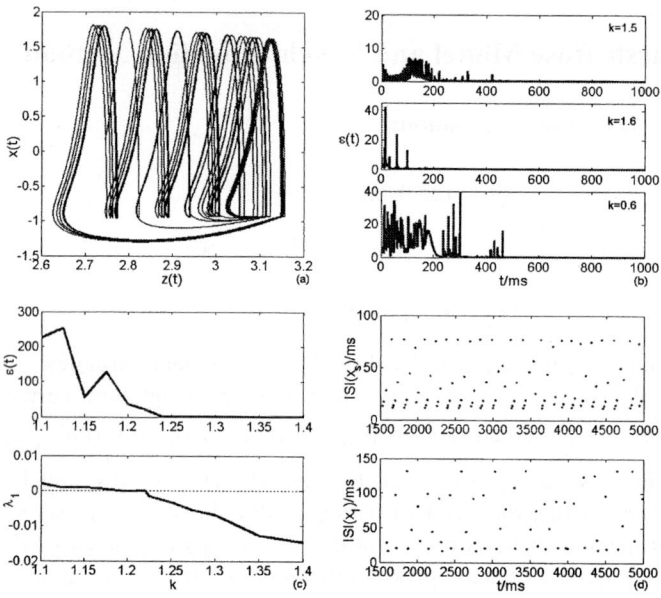

Fig. 1. (a) The projection of the attractor of Hindmarsh-Rose model in $z - x$ plane (b) The synchronization error $\varepsilon(t)$ courses, with parameters d, b and a being modulated, (from top to bottom) respectively; (c) The synchronization error $\varepsilon(t)$ and the largest condition Lyapunov exponent λ_1 via stimulation strength k, with parameter d being modulated; (d) ISI(x_s) course of the stimulation signal and ISI(x_1) course of the modulated neuron at the onset of synchronization. The initial conditions are (1.0,0.2,0.2) and (-1.0,0.8,0.3) for two modulated neurons respectively.

3 Synchronizing Two Uncoupled HR Neurons

3.1 Full Synchronization in Two Identical HR Neurons

The corresponding parameters of two HR neurons in Eq. (2) are the same as that in section 2.

When the corresponding parameters d, b and a of two HR neurons in Eq. (2) are modulated, the synchronization error $\varepsilon(t)$ courses are given in Fig. 1(b). The result shows that two neurons can attain full synchronization at threshold of stimulation strength.

Analyzing the largest condition lyapunov exponent (LCLE) λ_1 shows the LCLE becomes negative at the threshold of stimulation strength k, and the synchronization will be obtained. Fig.1(c) gives the LCLE and the synchronization error $\varepsilon(t)$ versus stimulation strength k when the parameters d of two neurons are modulated, it can be seen that the LCLE becomes negative at $k \approx 1.23$ and the synchronization error $\varepsilon(t)$ vanishes. It is known that the largest lyapunov exponent didn't change to negative when two coupled HR neurons achieve synchronization [3], which is quite different from the result of this paper. Fig.1(d) shows ISI courses of stimulation signal x_s and state variable x_1 of modulated neuron at the onset of synchronization. Obviously, ISI(x_1) is different from ISI(x_s), which means that two systems is synchronized to a chaotic response different from stimulation signal.

3.2 Phase Synchronization in Two Different HR Neurons

Now let's discuss two HR neurons in Eq. (2) with parameters mismatch, due to which full synchronization will become phase synchronization (PS). In numerical stimulation, let $a_1 = 1.1, a_2 = 0.9, c_1 = 0.9, c_2 = 1.1$, and other parameters remain the same as that in section 2.

HR model exhibits a multi-time-scaled burst-rest behavior, and consists of infinite rotation centers. We can find a projection plane of normal rotating trajectory in which the trajectory has a single rotation center, and then the phase can be computed in this projection plane [3]. In this paper, the $x(t) - x(t-\tau)$ plane is used, and $\tau = 0.5$ is time delay, that can be confirmed by mutual information method.

The phase can be introduced conveniently as:

$$\theta(t) = \arctan[x(t-\tau)/x(t)] \quad (5)$$

$\theta(t)$ is computed in four different quadrants,

$$\phi(t) = \theta(t) + 2\pi m \quad (6)$$

$$m(t+1) = \begin{cases} m(t)+1 & \phi(t) \text{ change from the forth quadrant to first one} \\ m(t) & \end{cases}$$

$$m(1) = 0 \quad (7)$$

Phase synchronization occurs when the phases of two oscillators have the continual relationship $\phi_1 \approx \phi_2$ with time [3].

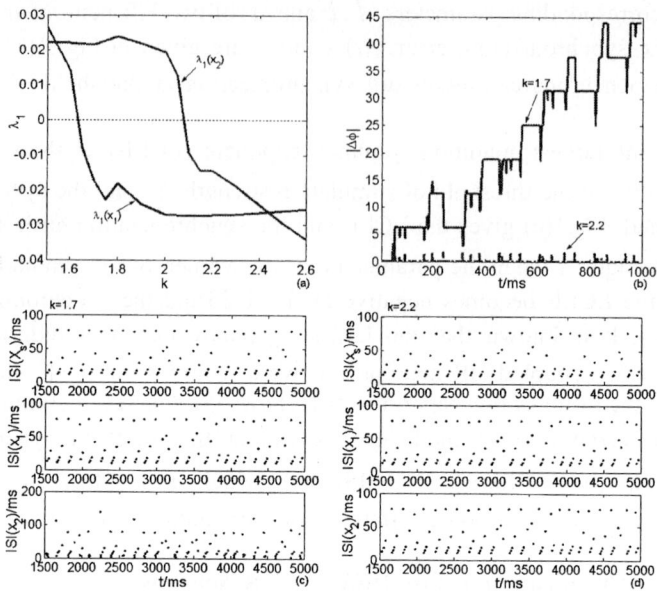

Fig. 2. Input current I_1, I_2 are modulated, (a) the largest condition Lyapunov exponent λ_1 of two different neurons via stimulation strength k ; (b) the absolute phase difference $|\Delta\phi|$ time course of two different neurons for $k=1.7$ and $k=2.2$, respectively; (c), (d) ISI(x_s) course of the stimulation signal, and ISI(x_1), ISI(x_2) course of modulated neurons for $k=1.7$ and $k=2.2$, respectively. The initial conditions are (1.0,0.2,0.2) and (-1.0,0.8,0.3) for two modulated neurons respectively.

Numerical results show that two uncoupled neurons with parameters mismatch will become phase synchronization by modulating input current I_1, I_2 when the LCLEs of two systems are both negative at the threshold of stimulation strength k. Figure 2(a) shows the LCLE plots of two neurons versus stimulation strength k. It can be seen that the LCLE of the first neuron changes to negative at $k \approx 1.7$ and that of the second neuron becomes negative at $k \approx 2.2$. Two neurons obtain PS at $k=2.2$, which can be seen in Fig.2 (b), which gives the absolute phase differences curves for $k=1.7$ and $k=2.2$, respectively.

Analyzing ISIs of neurons shows that ISI(x_1) is the same as ISI(x_s) at $k=1.7$ (such as Fig.2(c)), implying that the first neuron achieves PS with stimulating signal[3] when the LCLE $\lambda_1(x_1)$ becomes negative. When the two modulated neurons obtain PS at $k=2.2$, the ISIs of them are the same as the ISI(x_s) (such as Fig.2(d)), implying that synchronized response of two neurons is in phase synchronization with stimulating signal.

4 Conclusion

In summary, two uncoupled identical neurons can fully synchronize to a chaos response different from stimulating signal by using the membrane potential of a chaos neuron to modulate corresponding parameters of two systems, when stimulation strength attains the threshold at which the largest condition lyapunov exponent (LCLE) becomes negative. As for two uncoupled neurons with parameters mismatch, they can achieve phase synchronization by using the membrane potential of chaos neuron to modulate input currents of two systems, when the LCLEs of two systems become both negative, and the synchronized response is in phase with the stimulating signal.

This paper proposes reasonable experiment methods for synchronizing two neuron systems, such as modulating physiology environment of neuron cell or input current, verifies that two uncoupled neurons can achieve synchronous response to same external stimulation, and provide a foundation for studying complex neuron systems.

References

1. Pecora, L.M., Carroll, T.L.: Synchronization in Chaotic Systems. Phys. Rev. Lett. 64 (1990) 821-824
2. Cuomo, K.M., Oppenheim, A.V.: Circuit Implementation of Synchronized Chaos with Applications to Communications. Phys. Rev. Lett. 71(1993) 65-68
3. Shuai, J.W., Durand, D.M.: Phase Synchronization in two coupled Chaotic Neurons. Phys. Lett. A 264 (1999) 289-296
4. He, D.H., Shi, P.l., Stone, L.: Noise-induced Synchronization in Realistic Models. Phys. Rev. E 67 (2003) 027201
5. Rosenblum, M.G., Pikovsky, A.S., Kurths, J.: Phase Synchronization of Chaotic Oscillators. Phys. Rev. lett. 76 (1996) 1804-1807
6. Schader, C., Rosenblum, M.G., Kurths, J., Abel, H.: Heartbeat Synchronized with ventilation. Nature 392 (1998) 239-240
7. Gray, C.M., Koening, P., Engel, A.K., et al: Oscillatory response in cat visual cortex exhibit inter_columnar synchronization which reflects global stimulus properties. Nature 338 (1989) 334-337
8. Hindmarsh, J.L., Rose, R.M.: A Model of Neuronal Bursting Using Three Coupled First Order Differential Equation. Proc.R .Soc London B. 87 (1984) 221-225
9. Jin, W.Y., Xu, J.X., Wu, Y., et al: Rate of Afferent Stimulus Dependent Synchronization and Coding in Coupled Neurons System. Chaos Soliton & Fractals 21 (2004) 1221-1229
10. Xu, J.X., Gong, Y.F., Ren, W.: Propagation of periodic and Chaotic Potential trains along Nerve Fibers. Physica D 100 (1997) 212-224

Robust Synchronization of Coupled Delayed Recurrent Neural Networks

Jin Zhou[1,2], Tianping Chen[1], and Xiang Lan[2]

[1] Laboratory of Nonlinear Science, Institute of Mathematics,
Fudan University, Shanghai, 200433, P. R. China
{Jinzhou,Tchen}@fudan.edu.cn
[2] Department of Applied Mathematics and Physics,
Hebei University of Technology, Tianjin, 300130, P. R. China
Xianglan.htu@eyou.com.cn

Abstract. This paper investigates synchronization dynamics of a system of linearly and diffusively coupled identical delayed recurrent neural networks. A simple yet generic criterion for robust synchronization of such coupled recurrent neural networks is given. Furthermore, the theoretical result is applied to a typical coupled chaotic delayed Hopfied neural networks, and is also illustrated by numerical simulations.

1 Introduction

Synchronization of coupled systems is a topic of intensive research in the past decade. Because they can exhibit many interesting phenomena such as spatiotemporal chaos, auto waves, spiral waves etc. In addition, there are many synchronous systems existing in the real world where synchronization takes place among massive cells or nodes with fully or almost fully connections, examples of this kind include biological neural networks, coupled lattices of VLSIs and CNNs, and various complex small-world and scale-free networks (see [1,2,3]).

Recently, there has been increasing interest in applications of the dynamical properties of delayed recurrent neural networks such as delayed Hopfied neural networks and delayed cellular neural networks (CNN) in signal processing, pattern recognition, optimization and associative memories, and pattern classification. Most of previous studies are predominantly concentrated on the stability analysis and periodic oscillations of such kind of networks [4,5,6,7]. However, it has been shown that such networks can exhibit some complicated dynamics and even chaotic behaviors, there are strange attractors even in first-order continuous-time autonomous delayed Hopfied neural networks. On the other hand, experimental and theoretical studies have revealed that a mammalian brain not only can display in its dynamical behavior strange attractors and other transient characteristics for its associative memories, but also can modulate oscillatory neuronal synchronization by selective visual attention [8, 9]. Therefore, investigation of synchronization dynamics of such coupled neural networks is indispensable for practical design and applications of delayed neural networks.

This paper focuses on synchronization dynamics in a system composing of identical delayed recurrent neural networks. Based on the Lyapunov functional methods, a simple but less-conservative criterion is derived for robust synchronization of such coupled delayed neural networks. It is shown that robust synchronization of coupled delayed neural networks is ensured by a suitable design of the coupling matrix and the inner linking matrix. To this end, the theoretical results will be illustrated by numerical simulations on an coupled chaotic delayed Hopfied neural networks.

2 Model Descriptions

First, we consider an isolate delayed recurrent neural networks, which is described by the following set of differential equations with delays [4,5,6,7]:

$$\dot{x}_r(t) = -c_r h_r(x_r(t)) + \sum_{s=1}^{n} a_{rs}^0 f_s(x_s(t)) + \sum_{s=1}^{n} a_{rs}^\tau g_s(x_s(t-\tau_{rs})) + u_r(t), r = 1, 2, \cdots, n. \tag{1}$$

or, in a compact form,

$$\dot{x}(t) = -Ch(x(t)) + Af(x(t)) + A^\tau g(x(t-\tau)) + u(t), \tag{1}'$$

where $x(t) = (x_1(t), \cdots, x_n(t))^\top \in R^n$ is the state vector of the neural network, $C = \mathrm{diag}(c_1, \ldots, c_n)$ is a diagonal matrix with $c_r > 0$, $r = 1, 2, \cdots, n$, $A = (a_{rs}^0)_{n \times n}$ is a weight matrix, $A^\tau = (a_{rs}^\tau)_{n \times n}$ is the delayed weight matrix, $u(t) = (u_1(t), \cdots, u_n(t))^\top \in R^n$ is the input vector function, $\tau(r) = (\tau_{rs})$ with the delays $\tau_{rs} \geq 0$, $r, s = 1, 2, \cdots, n$, $h(x(t)) = [h_1(x_1(t)), \cdots, h_n(x_n(t))]^\top$, $f(x(t)) = [f_1(x_1(t)), \cdots, f_n(x_n(t))]^\top$ and $g(x(t)) = [g_1(x_1(t)), \cdots, g_n(x_n(t))]^\top$.

Clearly, (1) or (1)' is generalization of some additive delayed neural networks such as delayed Hopfied neural networks and delayed cellular neural networks (CNNs).

Next, we list some assumptions which will be used in the main results of this paper [4,5,6,7]:

(A_0) $h_r : R \to R$ is differentiable and $\eta_r = \inf_{x \in R} h'_r(x) > 0$, $h_r(0) = 0$, where $h'_r(x)$ represents the derivative of $h_r(x)$, $r = 1, 2, \cdots, n$.

Each of the activation functions in both $f_r(x)$ and $g_r(x)$ is globally Lipschitz continuous, i.e., either (A_1) or (A_2) is satisfied:

(A_1) There exist constants $k_r > 0, l_r > 0$, $r = 1, 2, \cdots, n$, for any two different $x_1, x_2 \in R$, such that

$$0 \leq \frac{f_r(x_1) - f_r(x_2)}{x_1 - x_2} \leq k_r, \quad |g_r(x_1) - g_r(x_2)| \leq l_r |x_1 - x_2|, \quad r = 1, 2, \cdots, n.$$

(A_2) There exist constants $k_r > 0, l_r > 0$, $r = 1, 2, \cdots, n$, for any two different $x_1, x_2 \in R$, such that

$$|f_r(x_1) - f_r(x_2)| \leq k_r |x_1 - x_2|, \quad |g_r(x_1) - g_r(x_2)| \leq l_r |x_1 - x_2|, \quad r = 1, 2, \cdots, n.$$

Obviously, Condition (A_2) is less conservative than (A_1), and some general activation functions in conventional neural networks, such as the standard sigmoidal functions and piecewise-linear function, satisfy Conditions (A_1) or (A_2).

Finally, a configuration of the coupled delayed recurrent neural networks is formulated. Consider a dynamical system consisting of N linearly and diffusively coupled identical delayed recurrent neural networks. The state equations of this system are

$$\dot{x}_i(t) = -Ch(x_i(t)) + Af(x_i(t)) + A^\tau g(x_i(t-\tau)) + u(t) + \sum_{j=1}^{N} b_{ij} \Gamma x_j(t), i = 1, 2 \cdots, N. \quad (2)$$

where $x_i(t) = (x_{i1}(t), \cdots, x_{in}(t))^\top \in R^n$ are the state variables of the ith delayed recurrent neural networks, $\Gamma = \mathrm{diag}(\gamma_1, \cdots, \gamma_n) \in R^{n \times n}$ is a diagonal matrix, which means that if $\gamma_i \neq 0$ then two coupled networks are linked through their ith state variables, and the coupling matrix $B = (b_{ij}) \in R^{N \times N}$ represents the coupling configuration of the coupled networks. For simplicity, assume that the coupling is symmetric, i.e. $b_{ij} = b_{ji}$ and $b_{ij} \geq 0$ $(i \neq j)$ but not all zero, with *the diffusive coupling connections* [2,3]: $\sum_{j=1}^{N} b_{ij} = \sum_{j=1}^{N} b_{ji} = 0$, $i = 1, 2 \cdots, N$. This implies that the coupling matrix B is a symmetric irreducible matrix. In this case, it can be shown that zero is an eigenvalue of B with multiplicity 1 and all the other eigenvalues of B are strictly negative [2,9].

3 Main Results

The delayed dynamical networks (2) is said to be *robust synchronization (strong synchronization)* if *the synchronous manifold* $\Lambda = \{x_1, x_2, \cdots, x_N \in R^n \mid x_i = x_j : i, j = 1, 2, \cdots, N.\}$ is uniformly globally exponentially asymptotically stable. Clearly, the diffusive coupling conditions ensure that the synchronization state $x_1(t) = x_2(t) = \cdots = x_N(t) \in \Lambda$ be a solution $s(t) \in R^n$ of an individual delayed neural network $(1)'$, namely,

$$\dot{s}(t) = -Ch(s(t)) + Af(s(t)) + A^\tau g(s(t-\tau)) + u(t). \quad (3)$$

Here, $s(t)$ can be an equilibrium point, a periodic orbit, or a chaotic attractor. Clearly, stability of synchronization state $s(t)$ is determined by the dynamics of an isolate delayed neural networks (3), the coupling matrix B and the inner linking matrix Γ.

For the definitions of exponential stability of synchronization manifold Λ and the related stability theory of retarded dynamical systems, see also [4,5,6,7,9,10]. Base on the Lyapunov functional method and Hermitian matrices theory, the following sufficient criterion for robust synchronization of the coupled network (2) is established.

Theorem 1 *Consider the delayed dynamical networks (2). Let the eigenvalues of its coupling matrix B be ordered as $0 = \lambda_1 > \lambda_2 \geq \lambda_3 \geq \cdots, \lambda_N$. If one of* the following conditions is satisfied, then the delayed dynamical system (2) will achieve robust synchronization.

1) Assume that (A_0) and (A_1) hold, and there exist n positive numbers $p_1, \cdots, p_n$, and two numbers $r_1 \in [0,1], r_2 \in [0,1]$, with $\lambda(\gamma_i) = \begin{cases} \lambda_2, & \text{if } \gamma_i > 0, \\ 0, & \text{if } \gamma_i = 0, \\ \lambda_N, & \text{if } \gamma_i < 0, \end{cases}$

and

$$\alpha_i \stackrel{\text{def}}{=} -c_i \eta_i p_i + \frac{1}{2} \sum_{\substack{j=1 \\ j \neq i}}^{n} \left(p_i |a_{ij}^0| k_j^{2r_1} + p_j |a_{ji}^0| k_i^{2(1-r_1)} \right)$$

$$+ \frac{1}{2} \sum_{j=1}^{n} \left(p_i |a_{ij}^\tau| l_j^{2r_2} + p_j |a_{ji}^\tau| l_i^{2(1-r_2)} \right),$$

such that for all $i = 1, 2, \cdots, n$,

$$(a_{ii}^0)^+ p_i k_i + \alpha_i + p_i \gamma_i \lambda(\gamma_i) < 0, \tag{4}$$

where $(a_{ii}^0)^+ = \max\{a_{ii}^0, 0\}$.

2) Assume that (A_0) and (A_2) hold, and for all $i = 1, 2, \cdots, n$,

$$|a_{ii}^0| p_i k_i + \alpha_i + p_i \gamma_i \lambda(\gamma_i) < 0. \tag{5}$$

Brief Proof: 1) Since B is a symmetric and irreducible matrix, by the Lemma 6 of [3] (Wu & Chua, 1995), there exists a $L \times N$ matrix $M = (m_{ij})_{L \times N} \in M_2^N(1)$, such that $B = -M^\top M$, where $M \in M_2^N(1)$ is a set of matrices, which is composed of $L \times N$ matrices. Each row (for instance, the i-th row) of $M \in M_2^N(1)$ has exactly one entry β_i and one $-\beta_i$, where $\beta_i \neq 0$. All other entries are zeroes. Moreover, for any pair of indices i and j, there exist indices $j_1, \cdots, j_l$, where $j_1 = i$ and $j_l = j$, and $p_1, \cdots, p_{l-1}$, such that the entries $m_{p_q i_q} \neq 0$, and $m_{p_q i_{q+1}} \neq 0$ for all $1 \leq q < l$. Let $\mathbf{C} = (I_N \otimes C), \mathbf{A} = (I_N \otimes A), \mathbf{A}^\tau = (I_N \otimes A^\tau), \mathbf{\Gamma} = (I_N \otimes \tau), \mathbf{B} = (B \otimes \Gamma), \mathbf{M} = (M \otimes I_N)$, where the notation $\otimes$ indicates the Kronecker product of both matrices, I_N stands for N order identity matrix. Define $\mathbf{u}(t) = (u^\top(t), \cdots, u^\top(t))^\top, x(t) = (x_1^\top(t), \cdots, x_N^\top(t))^\top, \mathbf{h}(x) = (h(x_1), \cdots, h(x_N))^\top, \mathbf{f}(x) = (f(x_1), \cdots, f(x_N))^\top, \mathbf{g}(x) = (g(x_1), \cdots, g(x_N))^\top$, one can rewrite the system (2) as follows:

$$\dot{x}(t) = -\mathbf{C}\mathbf{h}(x(t)) + \mathbf{A}\mathbf{f}(x(t)) + \mathbf{A}^\tau \mathbf{g}(x(t - \mathbf{\Gamma})) + \mathbf{u}(t) + \mathbf{B}x(t). \tag{6}$$

Let $y(t) = \mathbf{M}x(t) = (y_1(t), \cdots, y_L(t))^\top, y_i(t) = (y_{i1}(t), \cdots, y_{in}(t))^\top, i = 1, 2, \cdots, L$. One also write $\overline{y}_j(t) = (\overline{y}_{1j}(t), \cdots, \overline{y}_{Lj}(t))^\top, \overline{x}_j(t) = (\overline{x}_{1j}(t), \cdots, \overline{x}_{Nj}(t))^\top$, hence $\overline{y}_j(t) = M\overline{x}_j(t)$ for $j = 1, 2, \cdots, n$. According to the structure of $\mathbf{M}$ and $y(t) = \mathbf{M}x(t)$, one can use the $|y| = |\mathbf{M}x|$ to measure the the distance from x to the synchronous manifold Λ.

Let $P = \text{diag}(p_1, p_2, \cdots, p_n), \mathbf{P} = I_L \otimes P$. Then one can construct a Lyapunov functional of the following form with respect to system (6)

$$V = \frac{1}{2} \left\{ x^\top(t) \mathbf{M}^\top \mathbf{P} \mathbf{M} x(t) e^{\varepsilon t} + \sum_{i=1}^{L} \sum_{r=1}^{n} \sum_{s=1}^{n} p_r |a_{rs}^\tau|^{2(1-r_2)} \int_{t-\tau_{rs}}^{t} y_{is}^2(u) e^{\varepsilon(u + \tau_{rs})} du \right\}$$

$$= \frac{1}{2}\sum_{i=1}^{L}\sum_{r=1}^{n} p_r \left\{ y_{ir}^2(t)e^{\varepsilon t} + \sum_{s=1}^{n} |a_{rs}^\tau| l_s^{2(1-r_2)} \int_{t-\tau_{rs}}^{t} y_{is}^2(u) e^{\varepsilon(u+\tau_{rs})} \, du \right\} \quad (7)$$

Differentiating V with respect to time along the solution of (6), from (A_0), (A_1) and (4), then by some elementary but tedious computations, one obtains

$$\frac{dV}{dt}\Big|_{(6)} \leq -\sum_{i=1}^{n} \bar{x}_i^\top(t) B \Big(p_i \gamma_i B + ((a_{ii}^0)^+ p_i k_i + \alpha_i') I_n \Big) \bar{x}_i(t) e^{\varepsilon t} < 0, \quad (8)$$

from which it can be concluded that $V(t) \leq V(0)$, for all $t \geq 0$. Therefore, let $Q = \max\left\{ 1, \left[\left(\min_{1 \leq r \leq n} \{p_r\} \right)^{-1} \left(\max_{1 \leq s \leq n} \left\{ p_s + \sum_{r=1}^{n} p_r |a_{rs}^\tau| l_s^{2(1-r_2)} e^{\varepsilon \tau_{rs}} \tau \right\} \right) \right]^{\frac{1}{2}} \right\}$, from the construction of the Lyapunov functional (7), it can be derived that $|y(t)| = \left(\sum_{i=1}^{L} y_i^2(t) \right)^{\frac{1}{2}} \leq Q \left(\sum_{i=1}^{L} \|\phi_i\|^2 \right)^{\frac{1}{2}} e^{-\frac{\varepsilon}{2} t} = Q \|\phi\| e^{-\frac{\varepsilon}{2} t}$, where $\tau = \max_{1 \leq r, s \leq n} \{\tau_{rs}\}$. This completes the proof of 1) in Theorem 1. The proof of 2) is precisely the same as that for 1) in Theorem 1 if $(a_{ii}{}^0)^+$ is replaced by $|a_{ii}^0|$ in (8) from (A_0) and (A_2), and hence it is omitted.

Remark 1. It can be seen that if $h(x) = x$ and $f(x) = g(x)$ in (2), Theorem 1 is actually a generalization of important result associated with the same problem obtained in [9] (see [9. Theorem 1]).

Example 1. Consider a dynamical system consisting of 3 linearly and diffusively coupled identical delayed recurrent neural networks model (2), in which $x_i(t) = (x_{i1}(t), x_{i2}(t))^\top$, $h(x_i(t)) = (x_{i1}(t), x_{i2}(t))^\top$, $f(x_i(t)) = g(x_i(t)) = ($tanh$(x_{i1}(t)),$ tanh$(x_{i2}(t)))^\top$, $u(t) = (0,0)^\top, \tau = (1), i = 1,2,3$. $C = \begin{bmatrix} 1 & 0 \\ 0 & 1 \end{bmatrix}$, $A = \begin{bmatrix} 2.0 & -0.1 \\ -5.0 & 3.0 \end{bmatrix}$ and $A^\tau = \begin{bmatrix} -1.5 & -0.1 \\ -0.2 & -2.5 \end{bmatrix}$. It should be noted that the isolate neural networks $\dot{x}(t) = -Ch(x(t)) + Af(x(t)) + A^\tau g(x(t-\tau)) + u(t)$ is actually a chaotic delayed Hopfied neural networks [8] (see Fig. 1). Let the coupling matrix $B = \begin{bmatrix} -8 & 2 & 6 \\ 2 & -4 & 2 \\ 6 & 2 & -8 \end{bmatrix}$ and $\Gamma = \text{diag}(\gamma_1, \gamma_2)$. The matrix B has eigenvalues $0, -6$ and -14. By taking $\eta_r = k_r = l_r = 1$ $(r = 1, 2)$, $r_1 = r_2 = \frac{1}{2}$, $p_1 = p_2 = 1$, it is easy to verify that if $\gamma_1 > 0.86, \gamma_2 > 1.2$. then the conditions of Theorem 1 are satisfied. Hence, the delayed dynamical networks (2) will achieve robust synchronization. This numerical simulation is done by using the Delay Differential Equations (DDEs) Solvers in Matlab Simulink Toolbox (see Fig. 2).

4 Conclusions

In this paper, a general model of a system of linearly and diffusively coupled identical delayed recurrent neural networks has been formulated and its synchronization dynamics have been studied. In particularly, a simple sufficient

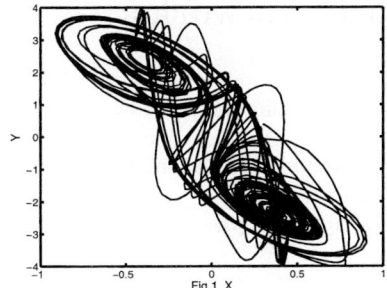
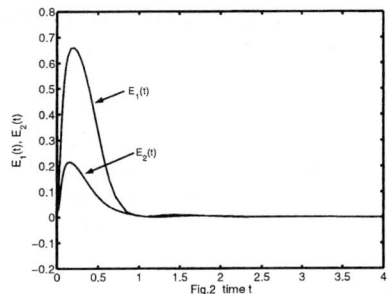

Fig. 1. A fully developed double-scroll-like chaotic attractors of the isolate delayed Hopfied neural networks.

Fig. 2. Robust synchronization of the coupled networks (2) with the parameters $\gamma_1 = \gamma_2 = 1.2$, where $E_k(t) = \sum_{i,j \in \{1,2,3\}} |x_{ik}(t) - x_{jk}(t)|$, $k = 1, 2$

criterion for robust synchronization of such coupled networks has been derived analytically. It is shown that the theoretical results can be applied to some typical chaotic recurrent neural networks such as delayed Hopfied neural networks and delayed cellular neural networks (CNN).

Acknowledgements. This work was supported by the National Science Foundation of China 10171061 and the Science Foundation of Education Commission of Hebei Province 2003013.

References

1. Chen, G., Dong, X.: From Chaos to Order: Methodologies, Perspectives, and Applications, World Scientific Pub. Co, Singapore (1998)
2. Wang, X. F., Chen, G.: Synchronization in Scale-Free Dynamical Networks: Robustness and Fragility. IEEE Trans. CAS-1. 49 (2002) 54-62
3. Wu, C. W., Chua, L. O.: Synchronization in an Array Linearly Coupled Dynamical System. IEEE Trans. CAS-1. 42 (1995) 430-447
4. Chen, T.: Global Exponential Stability of Delayed Hopfield Neural networks. Neural Networks. 14 (2001) 977-980
5. Huang, H., Cao, J., Wang, J.: Global Exponential Stability and Periodic Solutions of Recurrent Neural Networks with Delays. Phys. Lett. A. 298 (2002) 393-404
6. Zhou, J., Liu., Z., Chen, G.: Dynamics of Periodic Delayed Neural Networks. Neural Networks. 17 (2004) 87-101
7. Zhou, J., Liu., Z., Chen, G.: Global Dynamics of Periodic Delayed Neural Networks Models. Dyns. Dis. Conti. Impul. Ser B. 10 (2004) In Press
8. Lu, H.: Chaotic Attractors in Delayed Neural Networks. Phys. Lett. A. 298 (2002) 109–116
9. Chen, G., Zhou, J., Liu, Z.: Global Synchronization of Coupled Delayed Neural Networks and Applications to Chaotic CNN Model. Int. J. Bifur. Chaos. 14 (2004) In Press
10. Hale, J. K.: Introduction to Functional Differential Equations, Springer-Verlag, Berlin Heidelberg New York (1977)

Self-Optimizing Neural Networks

Adrian Horzyk and Ryszard Tadeusiewicz

University of Mining and Metallurgy, Department of Automatics
Mickiewicza Av. 30, 30-059 Cracow, Poland
horzyk@agh.edu.pl, rtad@ia.agh.edu.pl

Abstract. The paper is concentrated on two essential problems: neural networks topology optimization and weights parameters computation that are often solved separately. This paper describes new solution of solving both selected problems together. According to proposed methodology a special kind of multilayer ontogenic neural networks called Self-Optimizing Neural Networks (SONNs) can simultaneously develop its structure for given training data and compute all weights in the deterministic way based on some statistical computations that are incomparably faster then many other training methods. The described network optimization process (both structural and parametrical) finds out a good compromise between a minimal topology able to correctly classify training data and generalization capability of the neural network. The fully automatic self-adapting mechanism of SONN does not use any *a priori* configuration parameters and is free from different training problems.

1 Introduction

Neural networks (NNs) are very modern computational tools, but many practical problems with fixing different configuration parameters makes them difficult to use. Discovering an appropriate NN topology determines the success of the training process, but takes into account the arbitrary factors like number of layers, number of neurones and connections template. Many's the time computation of optimal weights is problematic for the given NN structure because of many convergence problems of the training algorithm, e.g. local minima. Moreover, the good result of training does not automatically determine good result of generalization – let us recall only the overfiting problem. Furthermore, large input training vectors produce the problem of dimensionality [1,2]. These difficulties lead many researchers to the decision to use other alternative methods that are more deterministic and their functionality are easier to foresee.

NNs still remain attractive and popular. The ability of NNs to generalize the knowledge collected in the training process is determined by proper computation of many NNs parameters especially topology, weights and activation functions. The natural brain collects the knowledge and trains habits from the beginning of its embryological growth and develops its structure to reflect these knowledge and habits. On the other hand, the structure and type of artificial NN is usually matched experimentally for given training data [1,2]. Some other methods can be sometimes used to create NN topology, e.i. genetic or evolutional algorithms.

This paper describes a special kind of supervised ontogenic NNs called Self-Optimizing Neural Networks (SONNs) that have characteristic multilayer NN topology with partially inter- and supralayer connections (Fig. 1). Such a topology is achieved as a result of adaptation process based on given training data. The topology is built up progressively in the self-adapting process after some probabilistic estimation of training data features [4,5]. The SONN adaptation process is capable to extract the most general and discriminating features of the data and create only the major (the most representative) connections and neurons. Furthermore, the data dimension of the input training vectors is automatically reduced if possible avoiding the curse of dimensionality problem. Moreover, the described SONN self-adapting process can successfully replace the usually used training process and is incomparably faster in comparison to it. The SONN automatically and simultaneously adapts the topology of the NN and all weights of it in the very fast deterministic process.

2   SONN – Theoretical Background

The training data $U = \{(u^1, C^{m_1}), \ldots, (u^N, C^{m_N})\}$ usually consist of the pairs: the input vector $u^n = [u_1^n, \ldots, u_K^n]$ (in this paper defined as $u_k^n \in \{-1, 0, +1\}$ with following interpretation: false (-1), don't known or undefined (0), true $(+1)$)) and the adequate class $C^{m_n} \in \{C^1, \ldots, C^M\}$. The importance of the features $1, \ldots, K$ of any input vector $u^n \in C^{m_n}$ can be estimated in view of the classification process. In order to extract the most important features of the patterns that characterize each class and discriminate it from others the definition of the following three coefficients will be useful:

1. Coefficient $p_k^n \in [0; 1]$ is a measure of representativeness of feature k of pattern n in its class m and is computed as the Laplace probability [3] of correct classification of the given training pattern n after the feature k:

$$\forall_{n \in \{1, \ldots, N\}} \forall_{k \in \{1, \ldots, K\}} \; p_k^n = \begin{cases} \frac{P_k^m}{P_k^m + N_k^m} & \text{if } u_k^n > 0 \; \& \; u^n \in C^m \\ 0 & \text{if } u_k^n = 0 \\ \frac{N_k^m}{P_k^m + N_k^m} & \text{if } u_k^n < 0 \; \& \; u^n \in C^m \end{cases} \quad (1)$$

where $\forall_{m \in \{1, \ldots, M\}} \forall_{k \in \{1, \ldots, K\}} \; P_k^m = \sum_{u_k^n \in \mathcal{P}^m} u_k^n \; \wedge \; N_k^m = \sum_{u_k^n \in \mathcal{N}^m} -u_k^n$,

$\mathcal{P}^m = \{u_k^n : u_k^n > 0 \& u^n \in U \cup C^m \& n \in \{1, \ldots, N\}\}$,
$\mathcal{N}^m = \{u_k^n : u_k^n < 0 \& u^n \in U \cup C^m \& n \in \{1, \ldots, N\}\}$.

2. Coefficient $q_k^n \in [0; 1]$ measures how much is the value of the feature k of the training pattern n representative for class m in view of the whole data set. This coefficient makes the SONN insensitive for differences in quantity of training patterns representing classes in the training data set. The coefficient is computed as the Laplace probability [3] of correct classification of the training pattern n from class m after the feature k:

$$\forall_{n \in \{1, \ldots, N\}} \forall_{k \in \{1, \ldots, K\}} \; q_k^n = \begin{cases} \frac{P_k^m}{P_k} & \text{if } u_k^n > 0 \; \& \; u^n \in C^m \\ 0 & \text{if } u_k^n = 0 \\ \frac{N_k^m}{N_k} & \text{if } u_k^n < 0 \; \& \; u^n \in C^m \end{cases} \quad (2)$$

where $\forall_{k\in\{1,...,K\}} P_k = \sum\limits_{u_k^n \in \mathcal{P}} u_k^n, \mathcal{P} = \{u_k^n : u_k^n > 0 \& u^n \in U \& n \in \{1,...,N\}\}$ and
$\forall_{k\in\{1,...,K\}} N_k = \sum\limits_{u_k^n \in \mathcal{N}} -u_k^n, \mathcal{N} = \{u_k^n : u_k^n < 0 \& u^n \in U \& n \in \{1,...,N\}\}$

3. Coefficient $r_k^n \in [0;1]$ measures how much is the value of the feature k of the training pattern n rare in view of the whole data set and is computed as follows:

$$\forall_{N>1} \forall_{n\in\{1,...,N\}} \forall_{k\in\{1,...,K\}} \quad r_k^n = \begin{cases} \frac{N_k}{P_k+N_k-1} & \text{if } u_k^n > 0 \\ 0 & \text{if } u_k^n = 0 \\ \frac{P_k}{P_k+N_k-1} & \text{if } u_k^n < 0 \end{cases} \quad (3)$$

The coefficients (1,2,3) are used to define the fundamental coefficient of discrimination $d_k^n \in [0;1]$ that defines how well the feature k discriminate between the training pattern n of the class m and the training patterns of the other classes:

$$\begin{array}{l} \forall_{n\in\{1,...,N\}} \forall_{k\in\{1,...,K\}} \\ d_k^n = p_k^n \cdot q_k^n \cdot r_k^n = \end{array} \begin{cases} \frac{(P_k^m)^2 \cdot N_k}{(P_k^m+N_k^m)\cdot P_k \cdot (P_k+N_k-1)} & \text{if } u_k^n > 0 \ \& \ u^n \in C^m \\ 0 & \text{if } u_k^n = 0 \\ \frac{(N_k^m)^2 \cdot P_k}{(P_k^m+N_k^m)\cdot N_k \cdot (P_k+N_k-1)} & \text{if } u_k^n < 0 \ \& \ u^n \in C^m \end{cases} \quad (4)$$

The discrimination coefficient is insensitive for quantitative representation of classes and concerns statistical differences of features quantity after their values in classes and the training data set. The bigger discrimination coefficient the more important and discriminative is the feature in view of classification and *vice versa*. The following architecture optimization process starts from the selection of the most important and discriminative features (called major features in the Fig. 1) for each training pattern.

The major features are selected after the maximal values of the corresponding discrimination coefficients and used to construct an ontogenic NN. The construction process of the NN aggregates the features of the training patterns of the same class after the sign of their values and the quantity of their same values in such a way to minimize the quantity of connections in the constructed SONN. The fundamental connections of the NN structure form a tree for each class (Fig. 1) completed with the input connections. After each construction period the NN is evaluated for all training patterns and the answers of the NN are checked up on univocal classification. If the answers are not univocal the incorrectly classified training patterns are marked and represented by more major features in the next development period in order to achieve their discrimination from other patterns of different classes. The ontogenic construction process stops just after all training patterns are univocally classified. The correctness of training patterns classification is built in the construction process and is always fully performed. The resultant NN correctly discriminates all training patterns of different classes. The SONN adapts simultaneously structure and weights. The weights values are computed for all major features $i \in J$ (figure 1) as follows:

$$\forall_{i \in J} \ w_i = \frac{u_i^n \cdot d_i}{\sum\limits_{j \in J} d_j} \quad (5)$$

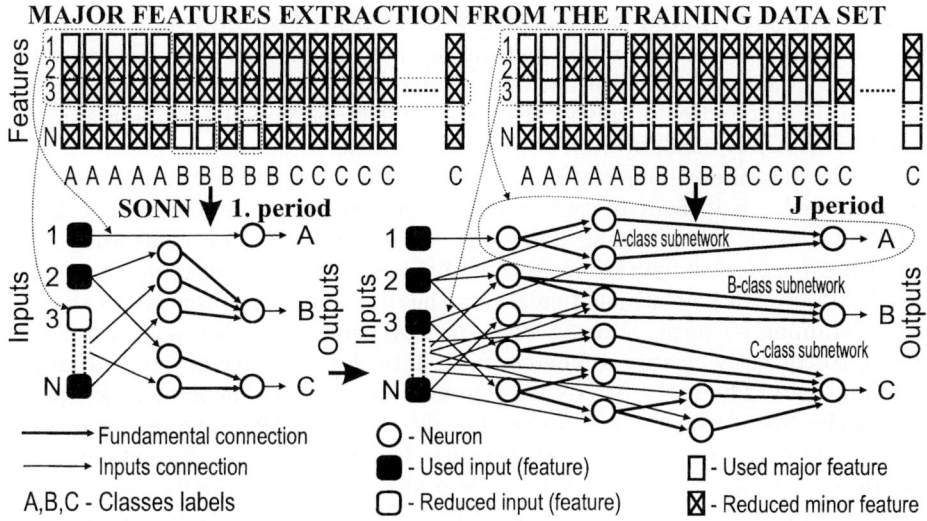

Fig. 1. The SONN development process based on the most discriminative features

where $\forall_{n \in \{1,...,N\}} u_0^n = 1$ and $d_0 = \sum_{l \in L} d_l$ for each neuron except the input neurons, $\forall_{n \in \{1,...,N\}} u_0^n = 0$ and $d_0 = 0$ for each input neuron, $J = A \cup \{0\}$ is the set of features that are transformed to connections of the neuron, $A \subset \{1, 2, ..., K\}$ is the set of accordant already not used (in previous layers) features related with training patterns concerned with the neuron, $L \subset \{1, 2, ..., K\} \cup \{0\}$ is the set of features of preceding neuron that are already transformed into connections.

The activation function of neurons except the output ones is defined as

$$y = f(u) = \sum_{j \in J} w_j \cdot u_j \qquad (6)$$

The activation function of output neuron m is defined as

$$y^m = f_{SGMMAX}(Y) = \begin{cases} y_{MIN} & if \ |y_{MAX}| < |y_{MIN}| \\ y_{MAX} & if \ |y_{MAX}| \geq |y_{MIN}| \end{cases} \qquad (7)$$

where $Y = \{y^1, \cdot, y^T\}$, y^t is the output of some neuron of a previous layer, $y_{MIN} = \min\{y_1, \cdot, y_T\}$, $y_{MAX} = \max\{y_1, \cdot, y_T\}$. The output neurons return the values in the rage $[-1; +1]$. The SONN outputs give information about the similarities of input vector to each class. The maximum output value appoints the class to which the input vector is the most similar.

Final classification is a simple process of SONN evaluation. The input vector can consist of bipolar $\{-1; +1\}$, trivalent $\{-1; 0; +1\}$ or continuous values in the rage $[-1; +1]$.

The described SONN development resembles the process of natural brain embryo development. The quantity of the training data inputs (data dimension) is automatically reduced as a side-effect of the described process. Some inputs are reduced if the corresponding features are not included into the NN construction process as major features for all training data. The curse of dimensionality problem is automatically solved in view of SONN methodology. The maximum number of SONN configuration periods is equal the dimension of given training data. The described configuration process does not stop if the training data are contradicted, e.g. if there are two or more same training input vectors that define different classes. The training data have to be checked up on contradiction in order not to loop configuration process of the SONN. The development process of the SONN can be continued enlarging the discrimination property of the NN for all training patterns. Such an extra development of SONN is sometimes necessary to correctly classify difficult testing data that are very different from the training data. The extra SONN development can increase generalization property of the NN as well as the NN size. The described SONN development process is very fast even for big data bases because: Discrimination coefficients are ones computed for all training patterns. The major features are simply selected after the maximum discrimination coefficients. The SONN does not require any *a priori* coefficients. All necessary computations are fully automatic for given training data! The SONN structure increases only if necessary. Sometimes the quantities of the NN elements (neurons, connections, layers) can even be smaller than in previous periods. This is a result of optimization process that takes place in each development period of the SONN. The SONN optimization algorithm minimizes the NN elements for the selected major features that change from period to period. The features are aggregated and transformed into the NN if only possible. Even not exactly defined input vectors can be used as training patterns with no essential harm to the described configuration process. The described self-optimization process of NNs configuration works only on input training vectors consisting of trivalent values $\{-1, 0, +1\}$. In order to use SONN to continuous data, the data have to be transformed to binaries. The binary transformation can be performed by intervals or other methods. If there are more then one maximal output values the result of classification is not univocal. In some cases such answer is an advantage that can help human expert to decide or to perform other action or research. The negative values of the output neuron suggest that the input vector is more similar to the inverse patterns of the appropriate class. This feature of SONN can be a great advantage when recognizing and classifying images that can be inverse. In the other cases the negative values of output neurons means hardly any similarity to these classes.

The SONN has been successfully implemented and adapted to many classification and recognition tasks (OCR, medical and pharmacological diagnosing, classical benchmarks). It has been tested also on continuous and discrete data after the simple binary transformation by intervals: Iris data (150 training vectors) features have been transformed to binaries by different intervals producing about 43 binary features. The SONN created for the iris data in 13 adaptation periods consists of 29 neurons, 102 connections, 5 layers, 25 from 43 input dimension. Mushrooms data (8124 training vectors) features have been transformed to binaries by intervals producing 121 binary features. The SONN created for the mushrooms data in 59 adaptation periods consists of 137 neurons, 632 connections, 8 layers, 64 from 121 input dimension. In result 8124 training vectors have been correctly classifies with 100% similarity to the appropriate classes.

3 Conclusions

The described methodology of the SONN for simultaneous configuration and adaptation of classifying NN builds up the topology optimized for the given training data and computes optimal weighs for them. It does not need to use any *a priori* given configuration parameters and is free from learning convergence problems. The SONN always gradually adapts its topology and weights for the given training data in the deterministic way. Small changes of training data almost always result in non-trivial changes of optimized SONN topology and weights parameters. The methodology of SONN building can be compared to the processes that take place during natural brain embryo development. Moreover, the SONN can increase its discrimination ability - if necessary. Such extra discrimination improvement influences generalization properties of the NN even after finishing of the fundamental SONN development. Such an additional development produces larger topologies. The created SONN always correctly classifies all training data and discriminates all training patterns from the patterns of other classes. The SONN offers very good generalization properties, because it is created after the most representative and discriminating features of the training data producing the well-representative model of any given classification problem. The SONN can also automatically recognize and classify inverse patterns of defined classes. Finally, the computational time of the described SONN is incomparable faster then in many classifying or NNs training methods.

Acknowledgements. Support from research funds 10.10.120.493 and the Polish Committee for Scientific Research is gratefully acknowledged.

References

1. Duch, W., Korbicz, J., Rutkowski, L., Tadeusiewicz, R. (eds): Biocybernetics and Biomedical Engineering. Neural Networks. EXIT Warsaw **6** (2000) 257–566
2. Fiesler, E., Beale, R. (eds): Handbook of Neural Computation. IOP Publishing Ltd and Oxford University Press, Bristol & New York (1997) B3–C1
3. Hellwig, Z.: Elements of Calculus of Probability and Mathematical Statistics PWN Warsaw (1993) 40–50
4. Horzyk, A.: New Efficient Ontogenic Neural Networks for Classification Tasks. Advanced Computer Systems, Soldek J., Drobiazbiewicz L. (eds) INFORMA Szczecin (2003) 466–473
5. Horzyk, A.: New Ontogenic Neural Network Classificator Based on Probabilistic Reasoning. Advances in Soft Computing. Neural Networks and Soft Computing, Rutkowski L., Kacprzyk J. (eds) Physica Verlag, Springer-Verlag Company, Heidelberg (2003) 188–193

Genetically Optimized Self-Organizing Neural Networks Based on PNs and FPNs

Ho-Sung Park[1], Sung-Kwun Oh[1], Witold Pedrycz[2], and Hyun-Ki Kim[3]

[1] Department of Electrical Electronic and Information Engineering, Wonkwang University,
344-2, Shinyong-Dong, Iksan, Chon-Buk, 570-749, South Korea
ohsk@wonkwang.ac.kr
http://autosys.wonkwang.ac.kr
[2] Department of Electrical and Computer Engineering, University of Alberta, Edmonton,
AB T6G 2G6, Canada
and Systems Research Institute, Polish Academy of Sciences, Warsaw, Poland
pedrycz@ee.ualberta.ca
[3] Department of Electrical Engineering, University of Suwon, South Korea
hkkim@suwon.ac.kr

Abstract. In this study, we introduce and investigate a class of neural architectures of self-organizing neural networks (SONN) that is based on a genetically optimized multilayer perceptron with polynomial neurons (PNs) or fuzzy polynomial neurons (FPNs), develop a comprehensive design methodology involving mechanisms of genetic optimization and carry out a series of numeric experiments. We distinguish between two kinds of SONN architectures, that is, (a) Polynomial Neuron (PN) based and (b) Fuzzy Polynomial Neuron (FPN) based self-organizing neural networks. The GA-based design procedure being applied at each layer of SONN leads to the selection of preferred nodes with specific local characteristics (such as the number of input variables, the order of the polynomial, and a collection of the specific subset of input variables) available within the network.

1 Introduction

When the dimensionality of the model goes up, so do the difficulties. To help alleviate the problems, one of the first approaches along the line of a systematic design of non-linear relationships between system's inputs and outputs comes under the name of a Group Method of Data Handling (GMDH) [1]. While providing with a systematic design procedure, GMDH comes with some drawbacks. To alleviate the problems associated with the GMDH, Self-Organizing Neural Networks (SONN) (viz. polynomial neuron (PN)-based SONN and fuzzy polynomial neuron (FPN)-based SONN, or called SOPNN/FPNN) were introduced by Oh and Pedrycz [2-4] as a new category of neural networks or neuro-fuzzy networks. Although the SONN has a flexible architecture whose potential can be fully utilized through a systematic design, it is difficult to obtain the structurally and parametrically optimized network because of the limited design of the nodes (viz. PNs or FPNs) located in each layer of the SONN. In order to generate a structurally and parametrically optimized network, such parameters need to be optimal.

In this study, we introduce a new genetic design approach; as a consequence we will be referring to these networks as genetically optimized SONN (gSONN). The determination of the optimal values of the parameters available within an individual PN or FPN (viz. the number of input variables, the order of the polynomial, and a collection of preferred nodes) leads to a structurally and parametrically optimized network.

2 The Architecture and Development of the Self-Organizing Neural Networks (SONN)

2.1 Polynomial Neuron (PN) Based SONN and Its Topology

By choosing the most significant input variables and an order of the polynomial among various types of forms available, we can obtain the best one – it comes under a name of a partial description (PD). The detailed PN involving a certain regression polynomial is shown in Table 1. The choice of the number of input variables, the polynomial order, and input variables available within each node itself helps select the best model with respect to the characteristics of the data, model design strategy, non-linearity and predictive capabilities.

Table 1. Different forms of regression polynomial building a PN and FPN

Order of the polynomial	No. of inputs		1	2	3
Order	FPN	PN			
0	Type 1		Constant	Constant	Constant
1	Type 2	Type 1	Linear	Bilinear	Trilinear
2	Type 3	Type 2	Quadratic	Biquadratic-1	Triquadratic-1
	Type 4	Type 3		Biquadratic-2	Triquadratic-2

1: Basic type, 2: Modified type

2.2 Fuzzy Polynomial Neuron (FPN) Based SONN and Its Topology

The FPN consists of two basic functional modules. The first one, labeled by **F**, is a collection of fuzzy sets that form an interface between the input numeric variables and the processing part realized by the neuron. The second module (denoted here by **P**) is about the function – based nonlinear (polynomial) processing. Proceeding with the FPN-based SONN architecture essential design decisions have to be made with regard to the number of input variables and the order of the polynomial forming the conclusion part of the rules as well as a collection of the specific subset of input variables.

Table 2. Polynomial type according to the number of input variables in the conclusion part of fuzzy rules

Type of the consequence polynomial \ Input vector	Selected input variables in the premise part	Selected input variables in the consequence part	Entire system input variables
Type T	A	A	B
Type T*	A	B	B

Where notation **A**: Vector of the selected input variables $(x_1, x_2, \ldots, x_i)$, **B**: Vector of the entire system input variables $(x_1, x_2, \ldots x_i, x_j \ldots)$, Type T: $f(A)=f(x_1, x_2, \ldots, x_i)$ - type of a polynomial function standing in the consequence part of the fuzzy rules, Type T*: $f(B)=f(x_1, x_2, \ldots x_i, x_j \ldots)$ - type of a polynomial function occurring in the consequence part of the fuzzy rules

3 Genetic Optimization of SONN

GAs have been theoretically and empirically demonstrated to provide robust search capabilities in complex spaces thus offering a valid solution strategy to problems requiring efficient and effective searching. To retain the best individual and carry it over to the next generation, we use elitist strategy [7]. As mentioned, when we construct PNs or FPNs of each layer in the conventional SONN, such parameters as the number of input variables (nodes), the order of polynomial, and input variables available within a PN or a FPN are fixed (selected) in advance by the designer. This could have frequently contributed to the difficulties in the design of the optimal network. To overcome this apparent drawback, we resort ourselves to the genetic optimization for more detailed flow of the development activities.

4 The Algorithm and Design Procedure of Genetically Optimized SONN (gSONN)

Overall, the framework of the design procedure of the SONN based on genetically optimized multi-layer perceptron architecture comprises the following steps.
[Step 1] *Determine system's input variables.*
[Step 2] *Form a training and testing data.*
[Step 3] *Decide initial information for constructing the SONN structure.*
[Step 4] *Decide a structure of the PN or FPN based SONN using genetic design.*
[Step 5] *Estimate the coefficient parameters of the polynomial in the selected node(PNs or FPNs).*
[Step 6] *Select nodes (PNs or FPNs) with the best predictive capability and construct their corresponding layer.*
[Step 7] *Check the termination criterion.*
[Step 8] *Determine new input variables for the next layer.*
The SONN algorithm is carried out by repeating steps 4-8 of the algorithm.

5 Experimental Studies

We illustrate the performance of the network and elaborate on its development by experimenting with data coming from the gas furnace process [2-6, 8-12]. The delayed terms of methane gas flow rate, $u(t)$ and carbon dioxide density, $y(t)$ are used as 6 system input variables with vector format, [$(t-3)$, $u(t-2)$, $u(t-1)$, $y(t-3)$, $y(t-2)$, and $y(t-1)$]. The output variable is $y(t)$.

Table 3. Computational aspects of the genetic optimization of PN-based and FPN-based SONN

	Parameters	1st layer	2nd to 5th layer
GA	Maximum generation	100	100
	Total population size	60	60
	Selected population size (W)	30	30
	Crossover rate	0.65	0.65
	Mutation rate	0.1	0.1
	String length	3+3+30	3+3+30
PN-based SONN	Maximal no.(Max) of inputs to be selected	$1 \leq l \leq$Max(2~5)	$1 \leq l \leq$Max(2~5)
	Polynomial type (Type T)(#)	$1 \leq T \leq 3$	$1 \leq T \leq 3$
FPN-based SONN	Maximal no.(Max) of inputs to be selected	$1 \leq l \leq$Max(2~5)	$1 \leq l \leq$Max(2~5)
	Polynomial type (Type T) of the consequent part of fuzzy rules(##)	$1 \leq T \leq 4$	$1 \leq T \leq 4$
	Consequent input type to be used for Type T (###)	Type T*	Type T
	Membership Function (MF) type	Triangular Gaussian	Triangular Gaussian
	No. of MFs per input	2	2

l, T, Max: integers, #, ## and ### : refer to Tables 1-2 respectively.

In case of PN-based gSONN, Fig. 1 illustrates the detailed optimal topologies of the network with 1 or 2 layers.

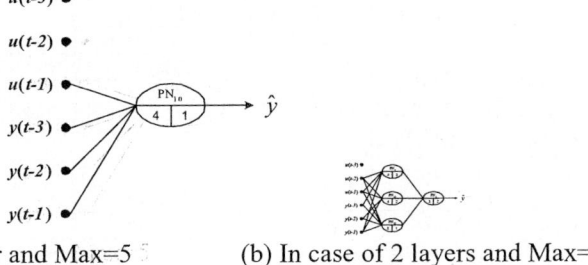

(a) In case of 1 layer and Max=5   (b) In case of 2 layers and Max=5

Fig. 1. PN–based genetically optimized SONN (gSONN) architecture

In case of FPN-based gSONN, Fig. 2 illustrate the detailed optimal topologies of the network with 1 layer. As shown in Fig. 2, the genetic design procedure at each stage (layer) of SONN leads to the selection of the preferred nodes (or FPNs) with optimal local characteristics.

160 H.-S. Park et al.

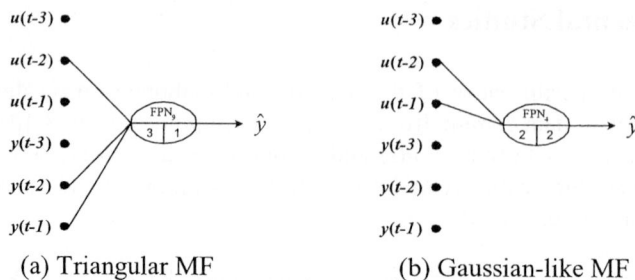

(a) Triangular MF   (b) Gaussian-like MF

Fig. 2. FPN-based gSONN architecture in case of using entire system input vector format (Type T*)

Table 4. Comparative analysis of the performance of the network; considered are models reported in the literature

Model					PI	PI$_s$	EPI$_s$
Box and Jenkin's model [8]					0.710		
Sugeno and Yasukawa's model [9]					0190		
Pedrycz's model [5]					0.320		
Chen's model [10]					0.268		
Oh and Pedrycz's model [6]					0.123	0.020	0.271
Kim et al.'s model [11]						0.034	0.244
Lin and Cunningham's model [12]						0.071	0.261
Generic SOPNN [2]		Basic SOPNN (SI=4, 5th layer)			0.027	0.021	0.085
		Modified SOPNN (SI=4, 5th layer)			0.035	0.017	0.095
Advanced SOPNN [3]		Basic SOPNN (SI=4, 5th layer)				0.020	0.119
		Modified SOPNN (SI=4, 5th layer)				0.018	0.118
SONN*[4]	Type I (SI=2)	Basic SONN	Case 1(5th layer)			0.016	0.266
			Case 2(5th layer)			0.016	0.265
		Modified SONN	Case 1(5th layer)			0.013	0.267
			Case 2(5th layer)			0.013	0.272
	Type II (SI=4)	Basic SONN	Case 1(5th layer)			0.016	0.116
			Case 2(5th layer)			0.016	0.128
		Modified SONN	Case 1(5th layer)			0.016	0.133
			Case 2(5th layer)			0.018	0.131
Proposed gSONN	PN-based	Max=4		1st layer		0.035	0.125
				5th layer		0.014	0.100
		Max=5		1st layer		0.035	0.125
				5th layer		0.012	0.091
	FPN-based	Max=4 (Type T)	Triangular	1st layer		0.021	0.136
				5th layer		0.015	0.106
		Max=5 (Type T)	Gaussian-like	1st layer		0.016	0.147
				5th layer		0.013	0.096
		Max=4 (Type T*)	Triangular	1st layer		0.021	0.136
				5th layer		0.011	0.106
		Max=4 (Type T*)	Gaussian-like	1st layer		0.012	0.145
				5th layer		0.008	0.093

*: denotes "conventional optimized FPN-based SONN".

6 Concluding Remarks

In this study, we introduced a class of genetically optimized self-organizing neural networks, discussed their topologies, came up with a detailed genetic design procedure, and used these networks to nonlinear system modeling. The key features of this approach can be enumerated as follows: **1)** The depth (layer size) and width (node size of each layer) of the gSONN can be selected as a result of a tradeoff between accuracy and complexity of the overall model. **2)** The structure of the network is not predetermined (as in most of the existing neural networks) but becomes dynamically adjusted and optimized during the development process. **3)** With a properly selected type of membership functions and the organization of the layers, FPN based gSONN performs better than other fuzzy and neurofuzzy models. **4)** The gSONN comes with a diversity of local neuron characteristics such as PNs or FPNs that are useful in copying with various nonlinear characteristics of the nonlinear systems.

Acknowledgements. This work has been supported by KESRI(R-2003-B-274), which is funded by MOCIE(Ministry of commerce, industry and energy)

References

1. A. G. Ivakhnenko.: Polynomial theory of complex systems. IEEE Trans. on Systems, Man and Cybernetics. SMC-1 (1971) 364-378
2. S.-K. Oh and W. Pedrycz.: The design of self-organizing Polynomial Neural Networks. Information Science. 141 (2002) 237-258
3. S.-K. Oh, W. Pedrycz and B.-J. Park.: Polynomial Neural Networks Architecture: Analysis and Design. Computers and Electrical Engineering. 29 (2003) 703-725
4. S.-K. Oh and W. Pedrycz.: Fuzzy Polynomial Neuron-Based Self-Organizing Neural Networks. Int. J. of General Systems. 32 (2003) 237-250
5. W. Pedrycz.: An identification algorithm in fuzzy relational system. Fuzzy Sets and Systems. 13 (1984) 153-167
6. S.-K. Oh and W. Pedrycz.: Identification of Fuzzy Systems by means of an Auto-Tuning Algorithm and Its Application to Nonlinear Systems. Fuzzy sets and Systems. 115 (2000) 205-230
7. D. Jong, K. A.: Are Genetic Algorithms Function Optimizers?. Parallel Problem Solving from Nature 2, Manner, R. and Manderick, B. eds., North-Holland, Amsterdam.
8. D. E. Box and G. M. Jenkins.: Time Series Analysis, Forcasting and Control, California. Holden Day. (1976)
9. M. Sugeno and T. Yasukawa.: A Fuzzy-Logic-Based Approach to Qualitative Modeling. IEEE Trans. Fuzzy Systems. 1 (1993) 7-31
10. J. Q. Chen, Y. G. Xi, and Z.J. Zhang.: A clustering algorithm for fuzzy model identification. Fuzzy Sets and Systems. 98 (1998) 319-329
11. E.-T. Kim, et al.: A simple identified Sugeno-type fuzzy model via double clustering. Information Science. 110 (1998) 25-39
12. Y. Lin, G. A. Cunningham III.: A new approach to fuzzy-neural modeling. IEEE Trans. Fuzzy Systems. 3 (1995) 190-197

A New Approach to Self-Organizing Hybrid Fuzzy Polynomial Neural Networks: Synthesis of Computational Intelligence Technologies

Hosung Park[1], Sungkwun Oh[1], Witold Pedrycz[2], Daehee Park[1], and Yongkab Kim[1]

[1] Department of Electrical Electronic and Information Engineering, Wonkwang University,
344-2, Shinyong-Dong, Iksan, Chon-Buk, 570-749, South Korea
{neuron, ohsk, parkdh, ykim}@wonkwang.ac.kr
http://autosys.wonkwang.ac.kr

[2] Department of Electrical and Computer Engineering, University of Alberta, Edmonton,
AB T6G 2G6, Canada
and Systems Research Institute, Polish Academy of Sciences, Warsaw, Poland
pedrycz@ee.ualberta.ca

Abstract. We introduce a new category of fuzzy-neural networks-Hybrid Fuzzy Polynomial Neural Networks (HFPNN). These networks are based on a genetically optimized multi-layer perceptron with fuzzy polynomial neurons (FPNs) and polynomial neurons (PNs). The augmented genetically optimized HFPNN (namely gHFPNN) results in a structurally optimized structure and comes with a higher level of flexibility in comparison to the one we encounter in the conventional HFPNN. The GA-based design procedure being applied at each layer of HFPNN leads to the selection of preferred nodes (FPNs or PNs) available within the HFPNN. In the sequel, two general optimization mechanisms are explored. First, the structural optimization is realized via GAs whereas the ensuing detailed parametric optimization is carried out in the setting of a standard least square method-based learning.

1 Introduction

In particular, when dealing with high-order nonlinear and multivariable equations of the model, we require a vast amount of data for estimating all its parameters. The Group Method of Data Handling (GMDH)[1] introduced by A.G. Ivakhnenko is one of the approaches that help alleviate the problem. But, GMDH has some drawbacks. In alleviating the problems of the GMDH algorithms, Polynomial Neural Networks(PNN)[2] was introduced as a new class of networks. Combination of neural networks and fuzzy systems has been recognized as a powerful alternative approach to develop fuzzy systems. In the sequel, as a new category of neuro-fuzzy networks, Hybrid Fuzzy Polynomial Neural Networks (HFPNN) [3] as well as Fuzzy Polynomial Neural Networks (FPNN) [4] was proposed. Although the HFPNN has flexible architecture whose potential can be fully utilized through a systematic design, it is difficult to obtain the structurally and parametrically optimized network because of the limited design of the nodes (viz. FPNs and PNs) located in each layer of the network.

In this paper, we study a genetic optimization-driven new neurofuzzy topology, called genetically optimized Hybrid Fuzzy Polynomial Neural Networks (gHFPNN) and discuss a comprehensive design methodology supporting their development. Each node of the first layer of gHFPNN, that is a fuzzy polynomial neuron (FPN) operates as a compact fuzzy inference system. The networks of the second and higher layers of the gHFPNN come with a high level of flexibility as each node (processing element forming a PN). The determination of the optimal values of the parameters available within an individual PN and FPN (viz. the number of input variables, the order of the polynomial, and a collection of preferred nodes) leads to a structurally and parametrically optimized network.

2 The Architecture and Development of the Hybrid Fuzzy Polynomial Neural Networks

2.1 The Architecture of Fuzzy Polynomial Neurons (FPN) Based Layer of gHFPNN

This neuron, regarded as a generic type of the processing unit, dwells on the concept of fuzzy sets. The FPN consists of two basic functional modules. The first one, labeled by **F**, is a collection of fuzzy sets that form an interface between the input numeric variables and the processing part realized by the neuron. The second module (denoted here by **P**) concentrates on the function – based nonlinear (polynomial) processing.

Table 1. Different forms of the regression polynomials building PN and FPN

No. of inputs Order of the polynomial			1	2	3
Order	FPN	PN			
0	Type 1		Constant	Constant	Constant
1	Type 2	Type 1	Linear	Bilinear	Trilinear
2	Type 3	Type 2	Quadratic	Biquadratic-1	Triquadratic-1
	Type 4	Type 3		Biquadratic-2	Triquadratic-2

1: Basic type, 2: Modified type

Proceeding with the FPN-based layer of gHFPNN architecture essential design decisions have to be made with regard to the number of input variables and the order of the polynomial forming the conclusion part of the rules as well as a collection of the specific subset of input variables. Especially for the consequence part, we consider two kinds of input vector formats in the conclusion part of the fuzzy as shown in Table 2.

Table 2. Polynomial type according to the number of input variables in the conclusion part of fuzzy rules

Type of the consequence polynomial \ Input vector	Selected input variables in the premise part	Selected input variables in the consequence part	Entire system input variables
Type T	A	A	B
Type T*	A	B	B

Where notation **A**: Vector of the selected input variables $(x_1, x_2,..., x_i)$, **B**: Vector of the entire system input variables$(x_1, x_2, ...x_i, x_j ...)$, Type T: $f(A)=f(x_1, x_2,..., x_i)$ - type of a polynomial function standing in the consequence part of the fuzzy rules, Type T*: $f(B)=f(x_1, x_2, ...x_i, x_j ...)$ - type of a polynomial function occurring in the consequence part of the fuzzy rules

2.2 The Architecture of the Polynomial Neuron(PN) Based Layer of gHFPNN

As underlined, the PNN algorithm in the PN based layer of gHFPNN is based on the GMDH method and utilizes a class of polynomials such as linear, quadratic, modified quadratic, etc. to describe basic processing realized there.
The estimated output $\hat{y}$ reads as

$$\hat{y} = c_0 + \sum_{i=1}^{N} c_i x_i + \sum_{i=1}^{N}\sum_{j=1}^{N} c_{ij} x_i x_j + \sum_{i=1}^{N}\sum_{j=1}^{N}\sum_{k=1}^{N} c_{ijk} x_i x_j x_k \cdots \quad (1)$$

The detailed PN involving a certain regression polynomial is shown in Table 1.

3 Genetic Optimization of gHFPNN

Genetic algorithms (GAs) are optimization techniques based on the principles of natural evolution. To retain the best individual and carry it over to the next generation, we use elitist strategy [5]. As mentioned, when we construct PNs and FPNs of each layer in the conventional HFPNN, such parameters as the number of input variables (nodes), the order of polynomial, and input variables available within a PN and a FPN are fixed (selected) in advance by the designer. This could have frequently contributed to the difficulties in the design of the optimal network. To overcome this apparent drawback, we resort ourselves to the genetic optimization.

4 The Algorithm and Design Procedure of Genetically Optimized HFPNN (gHFPNN)

Overall, the framework of the design procedure of the HFPNN based on genetically optimized multi-layer perceptron architecture comprises the following steps.

[Step 1] *Determine system's input variables.*
[Step 2] *Form a training and testing data.*
[Step 3] *Decide initial information for constructing the gHFPNN structure.*
[Step 4] *Decide a structure of the PN and FPN based layer of gHFPNN using genetic design.*
[Step 5] *Estimate the coefficient parameters of the polynomial in the selected node (PN or FPN).*
[Step 6] *Select nodes (PNs or FPNs) with the best predictive capability and construct their corresponding layer.*
[Step 7] *Check the termination criterion.*
[Step 8] *Determine new input variables for the next layer.*
The HFPNN algorithm is carried out by repeating steps 4-8 of the algorithm.

5 Simulation Study

We demonstrate how the gHFPNN can be utilized to predict future values of a chaotic Mackey-Glass time series [3, 6-12]. From the Mackey-Glass time series $x(t)$, we extracted 1000 input-output data pairs in the following format:
$$[x(t-18), x(t-12), x(t-6), x(t) ; x(t+6)]$$
where, t=118 to 1117. The first 500 pairs were used as the training data set while the remaining 500 pairs formed the testing data set. To come up with a quantitative evaluation of the network, we use the standard RMSE performance. Table 3 summarizes the list of parameters used in the genetic optimization of the network.

Table 3. Summary of the parameters of the genetic optimization

	Parameters	1st layer	2nd to 5th layer
GA	Maximum generation	100	100
	Total population size	60	60
	Selected population size (W)	30	30
	Crossover rate	0.65	0.65
	Mutation rate	0.1	0.1
	String length	3+3+30	3+3+30
HFPNN	Maximal no.(Max) of inputs to be selected	1≤*l*≤Max(2~5)	1≤*l*≤Max(2~5)
	Polynomial type (Type T) of the consequent part of fuzzy rules	1≤T≤4	1≤T≤4
	Consequent input type to be used for Type T (*)	Type T*	Type T
	Membership Function (MF) type	Triangular Gaussian	Triangular Gaussian
	No. of MFs per input	2	2

l, T, Max: integers, T* means that entire system inputs are used for the polynomial in the conclusion part of the rules.

Fig. 1 illustrate the detailed optimal topologies of gHFPNN for 1 layer and Max=3 in case of Type T*: those are quantified as PI=4.6e-4, EPI=4.4e-4 for triangular MF, and PI=3.6e-5, EPI=4.5e-5 for Gaussian-like MF.

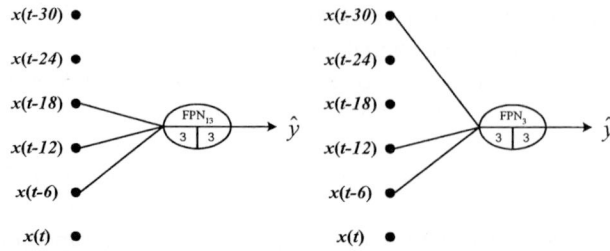

(a) Triangular MF(1 layer and Max=3)   (b) Gaussian-like MF(1 layer and Max=3)

Fig. 1. gHFPNN architecture in case of using entire system input vector format (Type T*)

Table 4 gives a comparative summary of the network with other models.

Table 4. Comparative analysis of the performance of the network; considered are models reported in the literature

Model				PI	PI_s	EPI_s	NDEI*
Wang's model [6]				0.044			
				0.013			
				0.010			
Backpropagation MLP [7]							0.02
ANFIS [8]					0.0016	0.0015	0.007
FNN model [9]					0.014	0.009	
Recurrent neural network [10]				0.0138			
SONN** [11]	Basic (5th layer)		Case 1		0.0011	0.0011	0.005
			Case 2		0.0027	0.0028	0.011
	Modified (5th layer)		Case 1		0.0012	0.0011	0.005
			Case 2		0.0038	0.0038	0.016
HFPNN [3]	Triangular		5th layer		7.0e-4	6.0e-4	
	Gaussian		5th layer		4.8e-5	7.1e-5	
Proposed gHFPNN	Max=5 (Type T)	Triangular	1st layer		4.8e-5	1.2e-4	
			2nd layer		4.5e-5	9.9e-5	
	Max=5 (Type T)	Gaussian	1st layer		6.1e-5	1.0e-4	
			2nd layer		5.1e-5	7.2e-5	
	Max=5 (Type T*)	Triangular	1st layer		7.7e-5	1.6e-4	
	Max=5 (Type T*)	Gaussian	1st layer		3.6e-5	4.5e-5	

*Non-dimensional error index (NDEI) as used in [12] is defined as the root mean square errors divided by the standard deviation of the target series. ** is called "conventional optimized FPNN".

6 Concluding Remarks

In this study, the GA-based design procedure of Hybrid Fuzzy Polynomial Neural Networks (HFPNN) along with its architectural considerations has been investigated. Through the consecutive generation of a layer through a growth process (iteration) of the gHFPNN, the depth (layer size) and width (node size of each layer) of the network could be flexibly selected based on a diversity of local characteristics of these preferred FPNs and PNs (such as the number of input variables, the order of the consequent polynomial of rules/the polynomial order, and a collection of specific subset of input variables) available within HFPNN. The design methodology comes as a hybrid structural optimization (based on GMDH method and genetic optimization) and parametric learning being viewed as two fundamental phases of the design process. Through the proposed framework of genetic optimization we can efficiently search for the optimal network architecture (structurally and parametrically optimized network) and this becomes crucial in improving the performance of the resulting model.

Acknowledgements. This paper was supported by Wonkwang University in 2003.

References

1. A.G. Ivahnenko.: Polynomial theory of complex systems. IEEE Trans. on Systems, Man and Cybernetics. SMC-12 (1971) 364-378
2. S.-K. Oh and W. Pedrycz.: The design of self-organizing Polynomial Neural Networks. Information Science. 141 (2002) 237-258
3. S.-K. Oh, W. Pedrycz, and D.-W. Kim.: Hybrid Fuzzy Polynomial Neural Networks. Int. J. of Uncertainty, Fuzziness and Knowledge-Based Systems. 10 (2002) 257-280
4. B.-J. Park, W. Pedrycz, and S.-K. Oh.: Fuzzy Polynomial Neural Networks : Hybrid Architectures of Fuzzy Modeling. IEEE Transaction on Fuzzy Systems, 10 (2002) 607-621.
5. D. Jong, K. A.: Are Genetic Algorithms Function Optimizers?. Parallel Problem Solving from Nature 2, Manner, R. and Manderick, B. eds., North-Holland, Amsterdam
6. L. X. Wang, J. M. Mendel.: Generating fuzzy rules from numerical data with applications. IEEE Trans. Systems, Man, Cybern. 22 (1992) 1414-1427
7. R. S. Crowder III.: Predicting the Mackey-Glass time series with cascade-correlation learning. In: D. Touretzky, G. Hinton, and T. Sejnowski, editors, Proceedings of the 1990 Connectionist Models Summer School, Carnegic Mellon University, (1990) 117-123
8. J. S. R. Jang.: ANFIS: Adaptive-Network-Based Fuzzy Inference System. IEEE Trans. System, Man, and Cybern. 23 (1993) 665-685
9. L. P. Maguire, B. Roche, T. M. McGinnity, L. J. McDaid.: Predicting a chaotic time series using a fuzzy neural network. Information Sciences. 112(1998) 125-136
10. C. James Li, T. -Y. Huang.: Automatic structure and parameter training methods for modeling of mechanical systems by recurrent neural networks. Applied Mathematical Modeling. 23 (1999) 933-944
11. S.-K. Oh, W. Pedrycz, T.-C. Ahn.: Self-organizing neural networks with fuzzy polynomial neurons. Applied Soft Computing. 2 (2002) 1-10
12. A. S. Lapedes, R. Farber.: Non-linear Signal Processing Using Neural Networks: Prediction and System Modeling. Technical Report LA-UR-87-2662, Los Alamos National Laboratory, Los Alamos, New Mexico 87545. (1987)
13. B.-J. Park, D.-Y. Lee, S.-K. Oh: Rule-Based Fuzzy Polynomial Neural Networks in Modeling Software Process Data. International Journal of Control, Automation and Systems, 1(3) (2003) 321-331
14. H.-S. Park, S.-K. Oh: Rule-based Fuzzy-Neural Networks Using the Identification Algorithm of GA hybrid Scheme. International Journal of Control, Automation and Systems, 1(1) (2003) 101-110

On Soft Learning Vector Quantization Based on Reformulation*

Jian Yu[1] and Pengwei Hao[2,3]

[1] Dept. of Computer Science, Beijing Jiaotong University
Beijing, 100044, P.R .China
jianyu@center.njtu.edu.cn
[2] Center of Information Science, Peking University
Beijing, 100871, P.R. China
phao@cis.pku.edu.cn
[3] Dept. of Computer Science, Queen Mary, University of London,
London, E1 4NS, UK
phao@dcs.qmul.ac.uk

Abstract. The complex admissibility conditions for reformulated function in Karayiannis model is obtained based on the three axioms of radial basis function neural network. In this paper, we present an easier understandable assumption about vector quantization and radial basis function neural network. Under this assumption, we have obtained a simple but equivalent criterion for admissible reformulation function in Karayiannis model. We have also discovered that Karayiannis model for vector quantization has a trivial fixed point. Such results are useful for developing new vector quantization algorithms.

1 Introduction

It is well known that vector quantization is a data compression method, which encodes a set of data points into a reduced set of reference vectors. In the literature, many vector quantization strategies are proposed. One common way to design vector quantizer is based on cluster analysis. Therefore, many researchers studied cluster analysis and vector quantization together, e.g. in [1]. In [2], Karayiannis proposed an axiomatic approach to soft learning vector quantization and clustering based on reformulation. In this paper, this model is called Karayiannis model. As indicated in [2], Karayiannis model leads to abroad family of soft learning vector quantization and clustering algorithms, including FCM [3], fuzzy learning vector quantization (FLVQ) [4], entropy-constrained learning vector quantization (ECLVQ) [5], and so on.

Based on Karayiannis model, many variations for radial basis neural networks are proposed, see [6]. In the previous research, reformulated function plays a pivotal role

---

* This work was partially supported by the Key Scientific Research Project of MoE, China under Grant No.02031, the National Natural Science Foundation of China under Grant No.60303014, and the Foundation for the Authors of National Excellent Doctoral Dissertation, China, under Grant 200038.

in constructing numerous reformulated radial basis function neural networks (RRBF). The admissibility condition for reformulated function is obtained based on the three axioms of radial basis function neural network (RBF). However, such three axiomatic requirements are not easily understood, especially in mathematical form. In this paper, we obtain an efficient criterion for admissible reformulation function in Karayiannis model based on a novel assumption of RBF and vector quantization, which is proved to be equivalent to the complex admissibility conditions for reformulated function in Karayiannis model.

The remainder of this paper is organized as follows: In Section 2, RBF, vector quantization and Karayiannis model are briefly related. In Section 3, a novel and intuitive assumption about RBF and vector quantization is proposed, which leads to a necessary condition for Karayiannis model to perform well. Moreover, we prove that such a condition is equivalent to the complex admissibility conditions for reformulated function in Karayiannis's model. In the final section, we draw the conclusion.

2 RBF, Vector Quantization, and Karayiannis Model

It is well known that the number and the centers of radial basis functions play an important role in the performance of RBF neural networks. Similarly, the size of codebook and the reference vectors are the key to vector quantization. Therefore, it is natural to establish a connection between vector quantization (in particular, LVQ) and RBF, see [2][6]. As noted in [6], a mapping can be the basis for constructing LVQ algorithms and RBF networks, more details can be found in [6]. In the following, we will describe this mapping.

Let $X=\{x_1, x_2, ..., x_n\}$ be a s-dimensional data set, $v=\{v_1, v_2, ..., v_c\}$ is the codebook, the centers (or prototype). Karayiannis defines a mapping $R^s \to R$ as below:

$$y = f\left(w_0 + \sum_{i=1}^{c} w_i g\left(\|x - v_i\|^2\right)\right) \tag{1}$$

where $f(x)$ and $g(x)$ are everywhere differentiable functions of $x \in (0, +\infty)$ and $\|\ \|$ denotes the Euclidean norm. Obviously, selecting appropriate $f(x)$ and $g(x)$ is very important for the performance of LVQ and RBF.

In order to tackle this issue, Karayiannis made the following assumption about RBF in [6]: an RBF is a composition of localized receptive fields, and the locations of these receptive fields are determined by the centers. And he also observed the three facts about RBF as following:

1. *The response of RBF's to any input is positive.*
2. *When the centers are considered as the centers of receptive fields, it is expected that the response of any RBF becomes stronger if an input approaches its corresponding center.*
3. *The response of any RBF is more sensitive to an input as this input approaches its corresponding center.*

Based on the above three observations, the three axioms of Karayiannis model are presented. In the following, we briefly introduce Karayiannis model.

By observing that the FCM and ECFC algorithms correspond to reformulation functions of the same basic form, Karayiannis [2] proposed a family of functions in the general form as follows:

$$R = \frac{1}{n}\sum_{k=1}^{n} f(S_k), \quad S_k = \frac{1}{c}\sum_{i=1}^{c} g(d_{ik}), d_{ik} = \|x_k - v_i\|^2, \quad f(g(x)) = x \tag{2}$$

Obviously, (1) is a basis to develop (2). As Karayiannis noted, minimization of admissible reformulation functions (2) using gradient descent can produce a variety of batch LVQ algorithms. The gradient of R with respect to v_i is:

$$\frac{\partial R}{\partial v_i} = -\frac{2}{nc}\sum_{k=1}^{n} f'(S_k) g'(\|x_k - v_i\|^2)(x_k - v_i) \tag{3}$$

Hence, the update equation for the centers can be obtained as:

$$\Delta v_i = -\eta_i \frac{\partial R}{\partial v_i} = \eta_i \sum_{k=1}^{n} \alpha_{ik}(x_k - v_i),$$ where η_i is the learning rate for the prototype v_i, and $\alpha_{ik} = f'(S_k) g'(\|x_k - v_i\|^2)$ is the competition function.

Then the LVQ algorithms can be implemented as follows:

Set the generator $f(x)$, the initial centers $\{v_{i,0} | 1 \le i \le c\}$, the termination limit ε, the maximum number of iteration L, and $t=1$

Step 1. $v_{i,t} = v_{i,t-1} + \eta_{i,t} \sum_{k=1}^{n} \alpha_{ik,t}(x_k - v_{i,t-1})$, $1 \le i \le c$ where $\{v_{i,t-1} | 1 \le i \le c\}$ is the set centers obtained after the $(t-1)$-th iteration, $\eta_{i,t}$ is the learning rate at iteration t and $\alpha_{ik,t} = f'(S_{k,t-1}) g'(\|x_k - v_{i,t-1}\|^2)$ with $S_{k,t-1} = c^{-1}\sum_{i=1}^{c} g(\|x_k - v_{i,t-1}\|^2)$.

Step 2. If $\max_i \|v_{i,t} - v_{i,t-1}\| < \varepsilon$, or $l > L$, then stop; else $l=l+1$ and go to step 1.

where $\eta_{i,t-1}$, are learning rates between 0 and 1.

According to the above three axioms, Karayiannis found that the properties of admissible competition functions should satisfy the following three axioms:

Axiom 1: if $c=1$, then $\alpha_{1k} = 1$, $1 \le k \le n$;

Axiom 2: $\alpha_{ik} \ge 0$, $1 \le i \le c$, $1 \le k \le n$;

Axiom 3: If $\|x_k - v_p\| > \|x_k - v_q\| > 0$, then $\alpha_{pk} < \alpha_{qk}$, $\forall p \ne q$.

According to Axiom 1-3, Karayiannis proved the following theorem.

Theorem 1 (Karayiannis, [2]): Let $X = \{x_1, x_2, \ldots, x_n\} \subset R^s$ be a finite set of feature vectors which are represented by the set of $c<n$ prototypes $V = \{v_1, v_2, \ldots, v_c\} \subset R^s$. Then the function R defined by (2) is admissible reformulation function of the first (second) kind in accordance with the axiomatic requirements 1-3 if $f(x)$ and $g(x)$ are differentiable everywhere functions of $x \in (0, +\infty)$ satisfying $f(g(x)) = x$, $f(x)$ and $g(x)$ are both monotonically decreasing (increasing) functions of x, and $g'(x)$ is a monotonically increasing (decreasing) function of $x \in (0, +\infty)$.

Obviously, the condition of admissible functions for Karayiannis model is too complex. In other words, it is not easy to judge whether or not a function is admissible for Karayiannis model, and Karayiannis himself also made many efforts to study more specific admissible reformulation function for Karayiannis model, see [2], [7], [8].

In Section 3, we present a simple assumption about vector quantization and RBF, which leads to a more efficient criterion for selecting function $f(x)$.

3 A Simple Assumption About RBF and Vector Quantization

When encoding or vector quantization, it is expected that all reference vectors are different. For example, Buhmann & Kuhnel thought that configurations with degenerate reference vectors are inadmissible in [9]. Similarly, the centers of receptive fields of RBF networks are not expected to be coincidental. However, many soft learning vector quantization algorithms (like ECLVQ) can output degenerate reference vectors, i.e. coincidental reference vectors. Therefore, it is naturally supposed that degenerate reference vectors are not the stable solution of vector quantization algorithms. The similar assumption has been used for clustering algorithms, see [10], [11], and [12].

For brevity, let Ω denote all fixed points of Karayiannis model, then it includes the saddle points and attractive points. A saddle point is unstable, i.e. not able to stand sufficiently small disturbing. All attractive points are stable. Therefore, the degenerate reference vectors are stable if they are the attractive points of Karayiannis model. It is better if Karayiannis model has no degenerate reference vectors, but unfortunately, it is easy to prove that $\forall 1 \leq i \leq c, v_i = \bar{x}$ is a fixed point of Karayiannis model, where $\bar{x} = \sum_{k=1}^{n} x_k / n$. Therefore, the output of Karayiannis model will be $\bar{x}$ with a great probability if $\forall 1 \leq i \leq c, v_i = \bar{x}$ is stable. Obviously, it is not the case we hope to face. What measure can we take to avoid such happening? First, we need to do is to find a criterion to judge whether a fixed point is stable for Karayiannis model. Theorem 2 offers such a criterion.

Theorem 2: If $\forall x, f''(x) < 0$ then $\forall 1 \leq i \leq c, v_i = \bar{x}$ is a stable fixed point of Karayiannis model, i.e. $\forall 1 \leq i \leq c, v_i = \bar{x}$ is a strict local minimum of R.

Proof: See Appendix A.

As analyzed above, it is unacceptable for $\forall 1 \leq i \leq c, v_i = \bar{x}$ is a stable fixed point of Karayiannis model. Therefore, it is a natural requirement for Karayiannis model that $\forall x, f''(x) \geq 0$. It seems inconsistent with Theorem 1, but we indeed have Theorem 3.

Theorem 3: $f(x)$ and $g(x)$ are differentiable everywhere functions of $x \in (0, +\infty)$ satisfying $f(g(x)) = x$, and both monotonically decreasing (increasing) functions of x, and $g'(x)$ is a monotonically increasing (decreasing) function of $x \in (0, +\infty)$ if and only if $f(x)$ and $g(x)$ are differentiable everywhere functions of $x \in (0, +\infty)$ satisfying $f(g(x)) = x$ and $f''(x) > 0$.

Proof: See Appendix B.

Apparently, Theorem 3 verifies that the new criterion is simple but equivalent to the complex condition in Theorem 1. Moreover, Theorem 2 tells us that our criterion has its own meaning totally different from three observations on radial basis functions. It also opens a new door to understand the properties of radial basis function networks. As a matter of fact, the optimality test for Karayiannis model is obtained, too.

4 Conclusions and Discussions

In [2], Karayiannis proposed a soft learning vector quantization based on reformulation. The complex admissibility conditions for reformulated function in Karayiannis model is obtained based on the three axioms of radial basis function neural network. Based on an easier understood assumption about vector quantization and radial basis function neural network, we obtained a simple but equivalent criterion about admissible reformulation function in Karayiannis's model. Moreover, we have found the optimality test for Karayiannis model, and discovered that Karayiannis model for vector quantization has a trivial fixed point. Such results are also useful for constructing new vector quantization algorithms.

References

1. Equitz, W.H.: A New Vector Quantization Clustering Algorithm. IEEE Transactions on Acoustics, Speech, and Signal Processing, 37(10) (1989) 1568-1575
2. Karayiannis, N.B.: An Axiomatic Approach to Soft Learning Vector Approach Quantization and Clustering. IEEE Transactions on Neural Networks, 10(5) (1999) 1153-1165
3. Bezdek, J.C.: Cluster Validity with Fuzzy Sets. J.Cybernt., 3(3) (1974) 58-72
4. Karayiannis, N.B., Bezdek, J.C.: An Integrated Approach to Fuzzy Learning Vector Quantization and Fuzzy C-means Clustering. IEEE Transactions on Fuzzy Systems, vol.5 (1997) 622-628
5. Karayiannis, N.B.: Entropy-Constrained Learning Vector Quantization Algorithms and Their Application in Image Compression. SPIE Proc., vol.3030 (1997) 2-13
6. Karayiannis, N.B.: Reformulated Radial Basis Neural Networks Trained by Gradient Descent. IEEE Transactions on Neural Networks, 10(3) (1999) 657 - 671
7. Karayiannis, N.B., Randolph-Gips, M.M.: Soft Learning Vector Quantization and Clustering Algorithms Based on Non-Euclidean Norms: Multinorm Algorithms. IEEE Transactions on Neural Networks, 14 (1) (2003) 89-102
8. Karayiannis, N.B.: Soft Learning Vector Quantization and Clustering Algorithms Based on Ordered Weighted Aggregation Operators. IEEE Transactions on Neural Networks, 11(5) (2003) 1093 - 1105
9. Buhmann, J., Kuhnel, H.: Vector Quantization with Complexity Costs. IEEE Trans. On Information Theory, 39(4) (1993) 1133-1145
10. Yu, J.: General c-means Clustering Model and Its Applications. CVPR2003, v.2 (2003) 122-127
11. Yu, J., Yang, M.S.: A Study on Generalized FCM. RSFDGrC'2003, Lecture Notes in Computer Science, no.2639, Springer-Verlag Heidelberg (2003) 390-393
12. Yu, J., Cheng, Q.S., Huang, H.K.: Analysis of the Weighting Exponent in the FCM. IEEE Transactions on Systems, Man and Cybernetics-part B: Cybernetics, 34(1) (2004) 634-639

Appendix A: Proof of Theorem 2

$$1 \le i, j \le c, \frac{\partial R^2}{\partial v_i \partial v_j} = \frac{4}{c^2 n} \sum_{k=1}^{n} f''(S_k) g'(d_{jk}) g'(d_{ik})(x_k - v_i)(x_k - v_j)^T$$

$$+ \frac{4}{nc} \delta_{ij} \sum_{k=1}^{n} f'(S_k) g''(d_{ik})(x_k - v_i)(x_k - v_i)^T + \frac{2}{nc} \delta_{ij} \sum_{k=1}^{n} f'(S_k) g'(d_{ik}) I_s \quad (4)$$

Set $\vartheta = (v_1, v_2, \cdots, v_c) = (\bar{x}, \bar{x}, \cdots, \bar{x}) \in \Omega$, and $d_{k\bar{x}} = (x_k - \bar{x})^T (x_k - \bar{x})$. Noticing $f(g(x)) = x$, we have $f'(g(x))g'(x) = 1$. Then, $\forall i, d_{ik} = d_{ck}$, $f'(S_k)g'(d_{ik}) = 1$. By straight calculation, the second term of the Taylor series expansion of R on the point $\vartheta = (v_1, v_2, \cdots, v_c) \in \Omega$ is as follows:

$$\varphi_v^T \left(\frac{\partial^2 R}{\partial v_i \partial v_j} \right) \varphi_v = \frac{4}{n} \sum_{k=1}^{n} f''(g(d_{k\bar{x}}))(g'(d_{k\bar{x}}))^2 \left(\sum_{i=1}^{c} c^{-1} \varphi_{v_i}, (x_k - \bar{x}) \right)^2$$

$$+ 2 \sum_{i=1}^{c} c^{-1} \varphi_{v_i}^T \left(I_{s \times s} + \sum_{k=1}^{n} \frac{2g''(d_{k\bar{x}})}{n \times g'(d_{k\bar{x}})} (x_k - \bar{x})(x_k - \bar{x})^T \right) \varphi_{v_i} \quad (5)$$

Noticing that $\left(\sum_{i=1}^{c} c^{-1} \varphi_{v_i}, (x_k - \bar{x}) \right)^2 \le \sum_{i=1}^{c} c^{-1} (\varphi_{v_i}, (x_k - \bar{x}))^2$, and $f(g(x)) = x$ we have $f''(g(x))(g'(x))^2 = -\frac{g''(x)}{g'(x)}$, then we can obtain (6) from (5) if $f''(x) < 0$:

$$\varphi_v^T \left(\frac{\partial^2 R}{\partial v_i \partial v_j} \right) \varphi_v \ge -\frac{4}{n} \sum_{k=1}^{n} \frac{g''(d_{k\bar{x}})}{g'(d_{k\bar{x}})} \sum_{i=1}^{c} c^{-1} (\varphi_{v_i}, (x_k - \bar{x}))^2 + 2 \sum_{i=1}^{c} c^{-1} \varphi_{v_i}^T \varphi_{v_i}$$

$$+ \sum_{k=1}^{n} \frac{4g''(d_{k\bar{x}})}{n \times g'(d_{k\bar{x}})} \sum_{i=1}^{c} c^{-1} (\varphi_{v_i}, (x_k - \bar{x}))^2 = 2 \sum_{i=1}^{c} c^{-1} \varphi_{v_i}^T \varphi_{v_i} \ge 0 \quad (6)$$

∎

Appendix B: Proof of Theorem 3

As $f(g(x)) = x$, $f'(g(x))g'(x) = 1$, obviously, we have $\forall x \ge 0$ $f'(x) \ne 0$, and $g'(x) \ne 0$. According to Darboux's Theorem, we know that $\forall x$, $f'(x) > 0$, otherwise $\forall x > 0$, $f'(x) < 0$. It is easy to prove that $f'(x) < 0$ if and only if $g'(x) < 0$ or that $f'(x) > 0$ if and only if $g'(x) > 0$. Noticing that $f''(g(x))(g'(x))^2 + f'(g(x))g''(x) = 0$, we have $f''(x) > 0$ if and only if $\frac{g''(x)}{g'(x)} < 0$.

∎

A New Approach to Self-Organizing Polynomial Neural Networks by Means of Genetic Algorithms

Sung-Kwun Oh[1], Byoung-Jun Park[1], Witold Pedrycz[2], and Yong-Soo Kim[3]

[1] Department of Electrical Electronic and Information Engineering, Wonkwang University,
344-2, Shinyong-Dong, Iksan, Chon-Buk, 570-749, South Korea
{ohsk, lcap}@wonkwang.ac.kr
[2] Department of Electrical and Computer Engineering, University of Alberta, Edmonton,
AB T6G 2G6, Canada
and Systems Research Institute, Polish Academy of Sciences, Warsaw, Poland
pedrycz@ee.ualberta.ca
[3] Division of Computer Engineering, Daejeon University, Korea
kjstj@dju.ac.kr

Abstract. In this paper, we introduce a new architecture of Genetic Algorithms (GA)-based Self-Organizing Polynomial Neural Networks (SOPNN) and discuss a comprehensive design methodology. The proposed GA-based SOPNN gives rise to a structurally optimized structure and comes with a substantial level of flexibility in comparison to the one we encounter in conventional PNNs. The design procedure applied in the construction of each layer of a PNN deals with its structural optimization involving the selection of preferred nodes (or PNs) with specific local characteristics (such as the number of input variables, the order of the polynomial, and a collection of the specific subset of input variables) and addresses specific aspects of parametric optimization. An aggregate performance index with a weighting factor is proposed in order to achieve a sound balance between approximation and generalization (predictive) abilities of the network.

1 Introduction

Computational Intelligence technologies such as neural networks, fuzzy sets and evolutionary computing have expanded and enriched a field of modeling quite immensely. In the sequel, they have also given rise to a number of new methodological issues and increased our awareness about tradeoffs one has to make in system modeling [1]. One of the first approaches along the line of a systematic design of nonlinear relationships between system's inputs and outputs comes under the name of a Group Method of Data Handling (GMDH) [2]. The GMDH algorithm generates an optimal structure of the model through successive generations of partial descriptions of data (PDs) being regarded as quadratic regression polynomials of two input variables. While providing with a systematic design procedure, GMDH comes with some drawbacks. To alleviate the problems associated with the GMDH, Polynomial Neural Networks (PNN) were introduced by Oh. et al. [3-4] as a new category of neural networks. Although the

PNN has a flexible architecture whose potential can be fully utilized through a systematic design, it is difficult to obtain the structurally and parametrically optimized network because of the limited design of the polynomial neurons (PNs) located in each layer of the PNN. In this study, in addressing the above problems with the conventional PNN as well as the GMDH algorithm, we introduce a new genetic design approach; as a consequence we will be referring to these networks as GA-based SOPNN. The determination of the optimal values of the parameters available within an individual PN (viz. the number of input variables, the order of the polynomial, and input variables) leads to a structurally and parametrically optimized network. As a result, this network is more flexible as well as exhibits simpler topology in comparison to the conventional PNNs discussed in the previous research. In the development of the network, we introduce an aggregate objective function (performance index) that deals with training data and testing data, and elaborate on its optimization to produce a meaningful balance between approximation and generalization abilities of the network [5].

2 Polynomial Neural Networks

2.1 PNN Based on Polynomial Neurons (PNs)

By choosing the most significant input variables and an order of the polynomial among various types of forms available, we can obtain the best one – it comes under a name of a partial description (PD) or polynomial neuron (PN). It is realized by selecting nodes at each layer and eventually generating additional layers until the best performance has been reached. Such methodology leads to an optimal PNN structure. The design of the PNN structure proceeds further and involves a generation of some additional layers. These layers consist of PNs (PDs) for which the number of input variables is the same across the layers. The detailed PN involving a certain regression polynomial is shown in Table 1.

Table 1. Different forms of regression polynomial building a PN

Order of the polynomial		No. of inputs	1	2	3
Order	FPN	PN			
0	Type 1		Constant	Constant	Constant
1	Type 2	Type 1	Linear	Bilinear	Trilinear
2	Type 3	Type 2	Quadratic	Biquadratic-1	Triquadratic-1
	Type 4	Type 3		Biquadratic-2	Triquadratic-2

2.2 Genetic Optimization of PNN

Genetic algorithms (GAs) are optimization techniques based on the principles of natural evolution. GAs have been theoretically and empirically demonstrated to provide robust search capabilities in complex spaces thus offering a valid solution strategy to problems requiring efficient and effective searching [6]. In this study, for the optimization of the PNN model, GA uses the serial method of binary type, roulette-wheel used in the selection process, one-point crossover in the crossover operation, and a binary inversion (complementation) operation in the mutation operator. To retain the best individual and carry it over to the next generation, we use elitist strategy [7].

3 The Algorithms and Design Procedure of GA-Based SOPNN

The optimization design method of the GA-based SOPNN is described in details.
[Step 1] *Determine system's input variables*
[Step 2] *Form training and testing data*
[Step 3] *Decide initial information for constructing the SOPNN structure*
a) the stopping criterion, b) The maximum number of input variables coming to each node in the corresponding layer, c) The total number W of nodes to be retained (selected) at the next generation of the SOPNN algorithm, d) The value of weighting factor of the aggregate objective function.
[Step 4] *Decide PN structure using genetic design*
 This concerns the selection of the number of input variables, the polynomial order, and the input variables to be assigned in each node of the corresponding layer.
The 1st sub-chromosome contains the number of input variables, the 2nd sub-chromosome involves the order of the polynomial of the node, and the 3rd sub-chromosome (remaining bits) contains input variables coming to the corresponding node (PN).
[Step 5] *Estimate the coefficients of the polynomial assigned to the selected node (PN)*
The vector of coefficients $\mathbf{C}_i$ is derived by minimizing the mean squared error between y_i (original output) and $\hat{y}_i$ (model output).
[Step 6] *Select nodes (PNs) with the best predictive capability and construct their corresponding layer*
In this Step, the objective function is used. The objective function (or cost function) is employed to decrease the error and to increase the predictability (generalization) capability of the model. Having this in mind, the objective function includes the performance index for training (*PI*), the performance index for evaluation (*EPI*) that are combined (weighted) by means of some weighting factor θ. More specifically we have

$$f(PI, EPI) = \theta \times PI + (1-\theta) \times EPI \quad (1)$$

The fitness function reads as

$$F(fitness\ function) = \frac{1}{1 + f(PI, EPI)} \quad (2)$$

[Step 7] *Check the termination criterion*
As far as the performance index is concerned (that reflects a numeric accuracy of the layers), a termination is straightforward and comes in the form,

$$F_1 \leq F_* \tag{3}$$

Where, F_1 denotes a maximal fitness value occurring at the current layer whereas F_* stands for a maximal fitness value that occurred at the previous layer. In this study, we use the Root Mean Squared Error (RMSE) as performance index.

$$E(PI_s \text{ or } EPI_s) = \sqrt{\frac{1}{N}\sum_{p=1}^{N}(y_p - \hat{y}_p)^2} \tag{4}$$

[Step 8] *Determine new input variables for the next layer*
If (3) has not been met, the model is expanded. The outputs of the preserved nodes (z_{1i}, z_{2i}, ..., z_{Wi}) serves as new inputs to the next layer (x_{1j}, x_{2j}, ..., x_{Wj})($j=i+1$).
The GA-based SOPNN algorithm is carried out by repeating steps 4-8.

4 Simulation Results

In this section, we demonstrate how the GA-based SOPNN can be utilized to predict future values of a chaotic Mackey-Glass time series. The performance of the network is also contrasted with some other models existing in the literature [8-13]. From the Mackey-Glass time series $x(t)$, we extracted 1000 input-output data pairs in the following format:

[$x(t-30)$, $x(t-24)$, $x(t-18)$, $x(t-12)$, $x(t-6)$, $x(t)$; $x(t+6)$]

where, t=118 to 1117. The first 500 pairs were used as the training data set while the remaining 500 pairs formed the testing data set.

Table 2. Summary of the parameters of the genetic optimization

	Parameters		1st ~ 5th layer
GA	Maximum generation		200
	Total population size		150
	Selected population size		100
	Crossover rate		0.65
	Mutation rate		0.1
	String length	Max=2~5	3+3+30
		Max=10	4+3+60
PNN	No. of inputs to be selected (l)		$1 \leq l \leq$ Max(2~5, 10)
	Type (T)		$1 \leq T \leq 3$
	Weighting factor (θ)		$0 \leq \theta \leq 1$

The values of performance index vis-à-vis the number of layers of the network are shown in Fig. 1. The best results for the network in the 5th layer coming with PI=0.00085 and EPI=0.00094 have been reported when using θ=0.75 with Type 2.

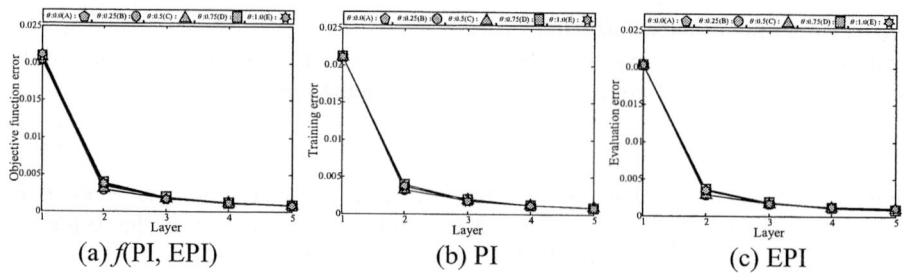

Fig. 1. Performance index of GA-based SOPNN with respect to the increase of the number of layer

Fig. 2 illustrates the optimization process by visualizing the values of the performance index obtained in successive generations of GA.

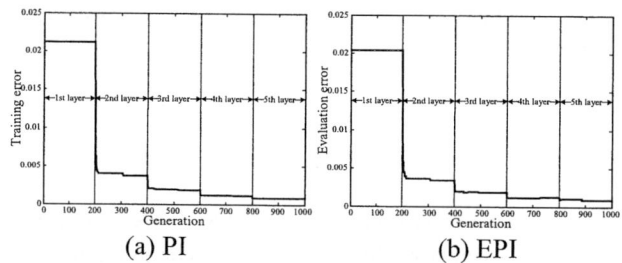

Fig. 2. The optimization process of each performance index by the GAs

Table 3 gives a comparative summary of the network with other models.

Table 3. Comparison of performance with other modeling methods

Model		PI	PI$_s$	EPI$_s$	NDEI
Wang's model[8]		0.013			
		0.010			
Cascaded-correlation NN[9]					0.06
Backpropagation MLP[9]					0.02
6th-order polynomial[9]					0.04
Recurrent neural network[10]		0.0138			
FNN model[11]				0.014	0.009
ANFIS[12]			0.0016	0.0015	0.007
Our model	θ=0.75 (5th layer)	0.0008	0.0008	0.0009	0.003

Non-dimensional error index (NDEI) as used in [13] is defined as the root mean square errors divided by the standard deviation of the target series.

5 Conclusions

In this study, the GA-based Self-Organizing Polynomial Neural Networks (SOPNN) has been investigated. The GA-based design procedure applied at each stage (layer) of the SOPNN leads to the selection of these preferred nodes (or PNs) with local characteristics (such as the number of input variables, the order of the polynomial, and input variables) available within PNN. These options contribute to the flexibility of the resulting architecture of the network. The design methodology comes as a hybrid structural optimization and parametric learning being viewed as two fundamental phases of the design process. Most importantly, through the proposed framework of genetic optimization we can efficiently search for the optimal network architecture and this becomes crucial in improving the performance of the resulting model.

Acknowledgements. This work was supported by Korea Research Foundation Grant (KRF-2003-002-D00297).

References

1. Dickerson, J. A. and Kosko, B.: Fuzzy Function Approximation with Ellipsoidal Rules. IEEE Trans. Syst., Man, Cybernetics. Vol. 26 (1996) 542-560
2. Ivakhnenko, A. G.: Polynomial Theory of Complex Systems. IEEE Trans. on Systems, Man and Cybernetics. SMC-1 (1971) 364-378
3. Oh, S.K. and Pedrycz, W.: The Design of Self-organizing Polynomial Neural Networks. Information Science. 141 (2002) 237-258
4. Oh, S.K., Pedrycz, W. and Park, B.J.: Polynomial Neural Networks Architecture: Analysis and Design. Computers and Electrical Engineering. 29 (2003) 703-725
5. Oh, S.K. and Pedrycz, W.: Identification of Fuzzy Systems by Means of an Auto-Tuning Algorithm and Its Application to Nonlinear Systems. Fuzzy Sets and Systems. 115 (2000) 205-230
6. Michalewicz, Z.: Genetic Algorithms + Data Structures = Evolution Programs. Springer-Verlag. Berlin Heidelberg New York (1996)
7. D. Jong, K. A.: Are Genetic Algorithms Function Optimizers?. Parallel Problem Solving from Nature 2, Manner, R. and Manderick, B. eds., North-Holland, Amsterdam.
8. Wang, L. X. and Mendel, J. M.: Generating Fuzzy Rules from Numerical Data with Applications. IEEE Trans. Systems, Man, Cybern. 22 (1992) 1414-1427
9. Crowder III, R. S.: Predicting the Mackey-Glass Time Series with Cascade-Correlation Learning. In: D. Touretzky, G. Hinton, and T. Sejnowski, Editors, Proceedings of the 1990 Connectionist Models Summer School, Carnegie Mellon University, (1990) 117-123
10. Li, C. J., Huang, T.Y.: Automatic Structure and Parameter Training Methods for Modeling of Mechanical Systems by Recurrent Neural Networks. Applied Mathematical Modeling. 23 (1999) 933-944
11. Maguire, L. P., Roche, B., McGinnity, T. M., McDaid, L. J.: Predicting a Chaotic Time Series using a Fuzzy Neural Network. Information Sciences. 112(1998) 125-136
12. Jang, J. S. R..: ANFIS: Adaptive-Network-Based Fuzzy Inference System. IEEE Trans. System, Man, and Cybern. 23 (1993) 665-685
13. Lapedes, A. S., Farber, R.: Non-linear Signal Processing Using Neural Networks: Prediction and System Modeling. Technical Report LA-UR-87-2662, Los Alamos National Laboratory, Los Alamos, New Mexico 87545. (1987)

Fuzzy-Kernel Learning Vector Quantization

Daoqiang Zhang[1], Songcan Chen[1], and Zhi-Hua Zhou[2]

[1] Department of Computer Science and Engineering
Nanjing University of Aeronautics and Astronautics
Nanjing 210016, China
{dqzhang, s.chen}@nuaa.edu.cn
[2] National Laboratory for Novel Software Technology
Nanjing University, Nanjing 210093, China
zhouzh@nju.edu.cn

Abstract. This paper presents an unsupervised fuzzy-kernel learning vector quantization algorithm called FKLVQ. FKLVQ is a batch type of clustering learning network by fusing the batch learning, fuzzy membership functions, and kernel-induced distance measures. We compare FKLVQ with the well-known fuzzy LVQ and the recently proposed fuzzy-soft LVQ on some artificial and real data sets. Experimental results show that FKLVQ is more accurate and needs far fewer iteration steps than the latter two algorithms. Moreover FKLVQ shows good robustness to outliers.

1 Introduction

The self-organizing map (SOM) due to Kohonen [1] is an ingenious neural network and has been widely studied and applied in various areas. The SOM network uses the neighborhood interaction set to approximate lateral neural interaction and discover the topological structure hidden in the data. The unsupervised learning vector quantization (LVQ) [2] can be seen as a special case of the SOM, where the neighborhood set contains only the winner node. Such learning rule is also called the winner-take-all principle.

LVQ has attracted a lot of attentions because of its learning simplicity and efficiency. However, LVQ suffers from several major problems when used for unsupervised clustering. Firstly, LVQ is sequential, so the final result severely depends on the order which the input patterns are presented to the network and usually a lot of numbers of iteration steps are needed for termination. Secondly, LVQ suffers the so-called prototype under-utilization problem, i.e., only the winner is updated for each input. Finally, because of adopting Euclidean distance measure, LVQ can cause bad performance when the data is non-spherical distribution, and especially contains noises or outliers.

To solve the first problem, the batch LVQ is proposed, which comes from the notion of batch SOM [1]. And fuzzy membership functions are introduced to original LVQ to overcome the second problem. For example, Yair et al. [3] proposed a soft competitive learning scheme to LVQ. To simultaneously address the above two problems some batch version of fuzzy LVQs are proposed. Bezdek et al. [2, 4] proposed the well-known fuzzy LVQ (FLVQ). Wu and Yang [5] presented a fuzzy-soft LVQ (FSLVQ). However, both FLVQ and FSLVQ use the Euclidean distance measure, and hence they are effective only when data set is spherical alike distributed. In addition, according to Huber's robust statistics [6], the Euclidean distance measure is not robust, i.e., sensitive to noises and outliers. To solve this problem, a sequential kernel SOM was proposed in one of our recent works, where a kernel-induced distance measures replaces original Euclidean one in order to improve robustness to outliers [7].

In this paper, we advance the sequential kernel SOM to the batch type of clustering learning network and call it a fuzzy-kernel LVQ (FKLVQ). Our goal aims to making FKLVQ simultaneously solve the three problems of LVQ by fusing the batch learning, fuzzy membership functions, and kernel-induced distance measures together. We made comparisons between FKLVQ and other types of batch LVQ on some artificial and real data sets.

2 Fuzzy-Kernel Learning Vector Quantization

As already mentioned in the previous section, FKLVQ consists of three main parts, i.e. batch learning, fuzzy memberships and kernel-induced distance measures. Before presenting FKLVQ algorithm, we first introduce the batch LVQ, fuzzy membership function and kernel-induced distance measures used in FKLVQ.

2.1 Batch LVQ

The LVQ for unsupervised clustering is a special case of the SOM network [5]. Suppose W_i in R^s is the weight vector of the node i and the input sample x_k in R^s is presented online at time t, sequential LVQ updates its neuron i as follows:

$$W_i(t) = W_i(t-1) + \alpha(t) h_{ik} \left(x_k - W_i(t-1) \right). \tag{1}$$

Here $\alpha(t)$ is the scalar-valued learning rate, $0<\alpha(t)<1$, and decreases monotonically with time. h_{ik} is an indicative function whose value is 1 if i is the winner node j, and 0 otherwise, where the winner node j is computed as follows:

$$\forall i, \quad \| x_k - W_j(t-1) \| \leq \| x_k - W_i(t-1) \|. \tag{2}$$

Suppose that the sample set are $X=\{x_1,\ldots,x_n\}$, where n is fixed. the sequential LVQ can be replaced by the following batch version which is significantly faster and does not require specification of any learning rate $\alpha(t)$.

Assume that the online algorithm will converge to a stationary state W_i^*, then the expectation values of $W_i(t)$ and $W_i(t-1)$ must be equal as t goes to infinity. In other words, in the stationary state we must have

$$E\left[h_{ik} \left(x_k - W_i^* \right) \right] = 0. \tag{3}$$

Applying the empirical distribution to solve the above equation, we have the batch learning formula of LVQ with [5]

$$W_i^* = \frac{\sum_k h_{ik} x_k}{\sum_k h_{ik}} . \qquad (4)$$

Since the determination of h_{ik} still depends on W_i^* according to Eq. (2), an alternate iteration between Eqs. (2) and (4) is used to approximate the explicit solution of W_i^*.

2.2 Fuzzy Membership Function

The above batch LVQ is in fact equivalent to traditional *k*-means (also called hard *c*-means) clustering algorithm, whose fuzzy extension is the widely used fuzzy *c*-means algorithm (FCM) [8]. The key point of FCM is the use of membership function which originally exists in fuzzy sets. Given $X=\{x_1,\ldots,x_n\}$, FCM obtains a fuzzy *c*-partition of X with $\{u_1,\ldots,u_c\}$, where $u_{ik} = u_i(x_k)$ takes value in the interval [0,1] such that $\sum_i u_{ik}=1$ for all k. By optimizing the objective function of FCM, one can obtain the alternate iterative equations. When the Euclidean distance is used, the update equation of the membership u_{ik} is as follows [8]:

$$u_{ik} = \left(\sum_{j=1}^{c} (\| x_k - W_i \| / \| x_k - W_j \|)^{2/(m-1)} \right)^{-1} . \qquad (5)$$

Here $m>1$ denotes the degree of fuzziness, and as m approximate to 1^+, FCM degenerate into crisp k-means algorithm.

Inspired by the success of FCM, many researchers also introduced the fuzzy membership to original batch LVQ in a similar way [2, 5, 9]. That is often achieved by representing h_{ik} with some monotone functions of u_{ik} and afterwards alternately iterating between h_{ik} and W_i, e.g. the FLVQ due to Bezdek et al. [2] and the FSLVQ proposed by Wu and Yang [5].

2.3 Kernel-Induced Distance Measures

Given input set X and a nonlinear mapping function Φ, which maps x_k from the input space X to a new space F with higher or even infinite dimensions. The kernel function is defined as the inner product in the new space F with: $K(x,y) = \Phi(x)^T \Phi(y)$, for x, y in input space X.

An important fact about kernel function is that it can be directly constructed in original input space without knowing the concrete form of Φ. That is, a kernel function implicitly defines a nonlinear mapping function. There are several typical kernel functions, e.g. the Gaussian kernel: $K(x,y)=exp(-\|x-y\|^2/\sigma^2)$, and the polynomial kernel: $K(x,y)=(x^T y + 1)^d$. From a kernel K, we have

$$d(x,y) = \| \Phi(x) - \Phi(y) \| = \sqrt{K(x,x) - 2K(x,y) + K(y,y)} . \qquad (6)$$

Here $d(x, y)$ defines a class of kernel-induced non-Euclidean distance measures with varying kernel functions. In our early works, the kernel-induced distance measures have been adopted in sequential kernel SOM [7], clustering [10, 11] and image denoising [12] respectively. And it has been proved that the measure induced by the Gaussian kernel is more robust to noises and outliers compared with original Euclidean measure [11]. Hereafter, we only discuss the Gaussian kernel in the rest of the paper. For Gaussian kernel, we have $K(x,x)=1$ for all x. Thus the distance measure in Eq. (6) can be simplified as $d(x,y) = sqrt\,(2(1-K(x,y)))$.

2.4 The Proposed FKLVQ

We are in position to propose the FKLVQ algorithm now. Define the objective function between the weight vector W_i and the input sample x_k as follows:

$$J(W_i, x_k) = K(W_i, W_i) - 2K(W_i, x_k) + K(x_k, x_k). \qquad (7)$$

Minimizing Eq. (7) by gradient descent, we obtain the update equation for W_i as:

$$W_i(t) = W_i(t-1) - \alpha(t) h_{ik} \left(\frac{\partial J(W_i, x_k)}{\partial W_i} \right). \qquad (8)$$

Especially for the Gaussian kernel $K(x,y)=exp(-\|x-y\|^2/\sigma^2)$, $K(W_i,W_i)=1$, so we have

$$W_i(t) = W_i(t-1) - \alpha(t) h_{ik} K(x_k, W_i(t-1))(x_k - W_i(t-1)) \cdot 2/\sigma^2. \qquad (9)$$

Here $\alpha(t)$ is defined before, but h_{ik} is calculated as follows:

$$h_{ik} = \left(\frac{u_{ik}}{\max_{1 \leq i \leq c} \{u_{ik}\}} \right)^{(1+\sqrt{t}/c)}, \qquad (10)$$

where u_{ik} is the fuzzy membership under kernel-induced measures. In a similar way to FCM, for the Gaussian kernel, u_{ik} can be derived as follows [10]:

$$u_{ik} = \left(\sum_{j=1}^{c} \left((1-K(x_k, W_i))/(1-K(x_k, W_j)) \right)^{1/(m-1)} \right)^{-1}. \qquad (11)$$

So far, we have derived the sequential fuzzy-kernel LVQ algorithm. Its batch version can be constructed in a similar way to that of batch LVQ. From Eq. (9), in the stationary state we must have

$$E\left[h_{ik} K(x_k, W_i)(x_k - W_i) \right] = 0. \qquad (12)$$

Applying the empirical distribution to solve the above equation, we have the batch learning formula of fuzzy-kernel LVQ with

$$W_i = \frac{\sum_k h_{ik} K(W_i, x_k) x_k}{\sum_k h_{ik} K(W_i, x_k)}. \tag{13}$$

Here h_{ik} is determined by Eq. (10), and by alternately iterating between Eqs. (11), (10) and (13), we get the FKLVQ algorithm.

3 Experimental Results

In this section, we make numerical comparison between the proposed FKLVQ and other batch algorithms such as FCM, FLVQ and FSLVQ on some artificial and real data sets. The Gaussian kernel is used for FKLVQ.

The fist example is an artificial data set which contains two clusters. Two clusters contain respectively 50 and 49 sample patterns and are separately centered at the points (0,0) and (3,0) with Gaussian distributions. Besides the two clusters, there exists an outlier at (200, 0). In this experiment, the parameters used in the algorithms are set to $m=2$, $c=2$, $\sigma =20$, and the maximum number of iterations is 50. Fig. 1 shows the comparison of four algorithms. From Fig. 1, results of FCM, FLVQ and FSLVQ are severely affected by the outlier, and the numbers of misclassified samples are all 49. However, FKLVQ successfully avoids the disturbance of the outlier and correctly classified the two clusters.

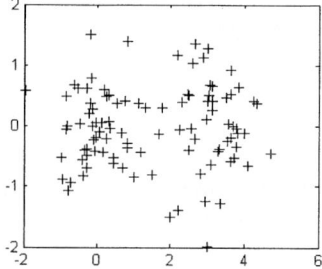 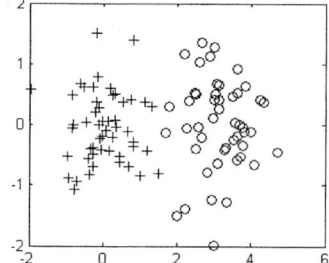

Fig. 1. Comparisons of performances of the four algorithms on artificial data set with outlier (not plotted in the figure): Left is the result by FCM, FLVQ and FSLVQ, where samples from both real clusters are classified to one group and the outlier is classified to the other group; Right is the result of FKLVQ where disturbance of the outlier is completely avoided.

The second example is the well-known Iris data set. It contains 3 clusters with 50 samples each. In this experiment, the parameters are set to $m=2$, $c=3$, $\sigma =10$, and the maximum number of iterations is 50. Table 1 gives comparison of accuracies and numbers of iterations of the four algorithms. From Table 1, FKLVQ achieves better accuracy and needs much less iteration than FLVQ and FSLVQ. For clustering the Iris data, FKLVQ only needs few tens of iterations. However, it was reported in [7] that the sequential kernel SOM typically needs hundreds to thousands of iterations for classifying the Iris data. Thus FKLVQ is much superior to kernel SOM as far as the computation efficiency is concerned.

Table 1. Comparison of accuracy and number of iterations of the four algorithms

| Algo- | Number of misclassified samples | | | | Number of |
rithms	Cluster 1	Cluster 2	Cluster 3	Total	iterations
FCM	0	3	13	16	14
FLVQ	0	3	14	17	50
FSLVQ	0	3	14	17	41
FKLVQ	0	6	5	11	13

Acknowledgements. This work was supported by the National Outstanding Youth Foundation of China under the Grant No. 60325207, the Jiangsu Science Foundation under the Grant No. BK2002092, and the Jiangsu Science Foundation Key Project.

References

1. Kohonen, T.: The Self-Organizing Map. Neurocomputing 21 (1998) 1-6
2. Bezdek, J.C., Pal, N.R.: Two Soft Relatives of Learning Vector Quantization. Neural Networks 8 (1995) 729-743
3. Yair, E., Zeger, K., Gersho, A.: Competitive Learning and Soft Competition for Vector Quantization Desigh. IEEE Trans. Signal Process. 40 (1992) 294-309
4. Tsao, E.C.K., Bezdek, J.C., Pal, N.R.: Fuzzy Kohonen Clustering Networks. Pattern Recognition 27 (1994) 757-764
5. Wu, K.L., Yang, M.S.: A Fuzzy-Soft Learning Vector Quantization. Neurocomputing 55 (2003) 681-697
6. Huber, P.J.: Robust statistics. Wiley, New York (1981)
7. Pan, Z.S., Chen, S.C., Zhang, D.Q.: A Kernel-based SOM classifier in Input Space. ACTA ELECTRONICA SINICA 32 (2004) 227-231
8. Bezdek, J.C.: Pattern Recognition with Fuzzy Objective Function Algorithms. Plenum Press, New York (1981)
9. Huntsberger, T., Ajjimarangsee, P.: Parallel Self-Organizing Feature Maps for Unsupervised Pattern Recognition. Intern. J. Gen. Systerms 16 (1990) 357-372
10. Zhang, D.Q., Chen, S.C.: Clustering Incomplete Data Using Kernel-based Fuzzy C-Means Algorithm. Neural Processing Letters 18 (2003) 155-162
11. Zhang, D.Q., Chen, S.C.: A Novel Kernelised Fuzzy C-Means Algorithm with Application in Medical Image Segmentation. Artificial Intelligence in Medicine, in press (2004)
12. Tan, K.R., Chen, S.C., Zhang, D.Q.: Robust Image Denoising Using Kernel-induced Measures. Accepted for publication in International Conference on Pattern Recognition (2004)

Genetic Generation of High-Degree-of-Freedom Feed-Forward Neural Networks

Yen-Wei Chen[1,2], Sulistiyo[3], and Zensho Nakao[3]

[1] College of Information Science and Eng., Ritsumeikan University, 525-8577, Japan
chen@is.ritsumei.ac.jp
[2] Institute of Computational Science and Eng., Ocean Univ. of China, Shandong, China
[3] Faculty of Eng.,Univ. of the Ryukyus, Okinawa 903-0213, Japan

Abstract. Appropriate definition of neural network architecture prior to data analysis is crucial for successful data mining. This is particularly important when the underlying model of the data is unknown. The proposed algorithm is intended to develop automatically an appropriate neural network (including the number of layers, the number of processing elements per layer, and types of each processing element) needed to solve the given problem. Genetic programming (GP) is used to develop the neural network's structure and the resilient-back-propagation (RPROP) will be used to train the neural network.

1 Introduction

Most of the various neural network (NN) architectures and training paradigms described in the literature [3] assume that the architecture of the neural network has been determined before the training process begins. That is, they suppose that a selection has been made for the number of layers of linear threshold processing elements, the number of processing elements in each layer, and the nature of the allowable connectivity between the processing elements. In real world, behavior of training data is generally unknown. So methods to build and to evolve neural networks automatically are in need.

In the past, the problem of automatically obtaining the topology of a neural network has been tackled by several authors. Constructive algorithms start with a network of a given minimum size and add nodes and links as appropriate, whereas destructive algorithms assume a network of a certain maximum size and remove the unneeded components. It has been shown that both methods are somewhat restrictive because they limit, in some way, the final network architectures that can be reached. Another class is to employ evolutionary algorithms (such as genetic algorithms (GAs)) and evolutionary programming (such as GP). These are population-based search methods which do not have such constraints. In GAs, both the structure and parameters of a network are represented as (usually) fixed-length strings and a population of such strings (or genotypes) is evolved, mainly by recombination. However, one problem with this approach is the sized of the required genotype bit string [4]; for a fully connected network with N neurons, there must be at least N^2 genetic components to specify the connection weights. This large genome size leads to impractically long convergence time. This problem has been addressed by Gruau [2] and Koza [5], who developed a compact cellular growth (constructive) algorithm based on symbolic S-expressions that are evolved (GP). Although this method can evolve very elaborate structures, we have observed that it takes very long to converge to an optimum, which is unsuitable for certain applications. The slowness is as a result of using GP, which is a structure optimizer, to also optimize the weights, which are numerical variables. Another method proposed by [1] separates between the evolution of the structures (by GP) and the evolution of the weights (by SA). It speeds up the convergence rate, but nevertheless, some weaknesses still exist. First, SA does not guarantee that each cycle will give a better result. Second, the weights are only located at the tree-connections made by GP. So compared to the conventional NN, with the same number of nodes, the resulting structure has less connection. Therefore to solve the same problem we need more processing nodes hence increases processing time. Our proposed method is intended to overcome this problem. The basic idea is to separate between trees evolved by GP and the related NN's structures. GP will build GP's tree structures which consist of various NN's processing nodes. Several rules are applied to build high density NN's connections based on GP's tree structure. So high degree-of-freedom as stated in the title means the resulting neural networks may consist of various processing nodes and have higher density of connections compared to previous similar projects.

Furthermore we employ RPROP as training algorithm, in which learning characteristic are dynamically determined to speed up convergence rate.

2 Evolving Neural Networks

GP is a machine learning methodology that generates computer programs to solve problems using a process that is inspired by biological evolution by natural selection. GP begins with an initial population of randomly generated computer programs, all of which are possible solutions to given problems. This step is essentially a random search or sampling of the space of all possible solutions. Trees made by GP (i.e., Fig.1) are used to determine position and status of each node and to track down member of each branch (i.e., member of branch 8 are nodes [8, 9, 10, 11]). Based on position and status of each node, new connections will be made to form NN. Steps to generate tree-structures:
- Generate a dummy node, as a marker of a-tree-top.
- Generate a random integer number to determine number(s) of each node's branch(s).
- Generate a random number for each branch to determine its related new-node's processing function.
- if Number-of-Layer < Maximal-Number-of-Layers then Generate a random number for each new node to determine whether it is a terminal node or not; otherwise assign all new nodes as terminal nodes.
- Return to step-2 until all branches are terminated.

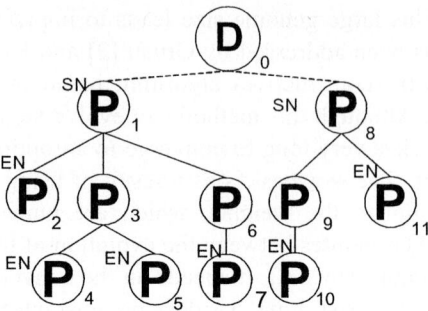

Fig. 1. Genetic Programming tree

Next, each of these trees are executed (create the related NN and proceed with NN training) and assigned a fitness that is proportional to the related NN's error and number of processing nodes after training phase. Then, the best computer programs, or solutions, are selected to undergo genetic operations based on Darwin's principle of survival of the fittest. Reproduction takes place with a subset of the best solutions, such that those solutions are directly copied into the next generation. Crossover, or recombination, takes place between another subset of solutions. This operation is used to create new computer program by combining components of two parents programs. An over all flow chart is shown in Fig.2.

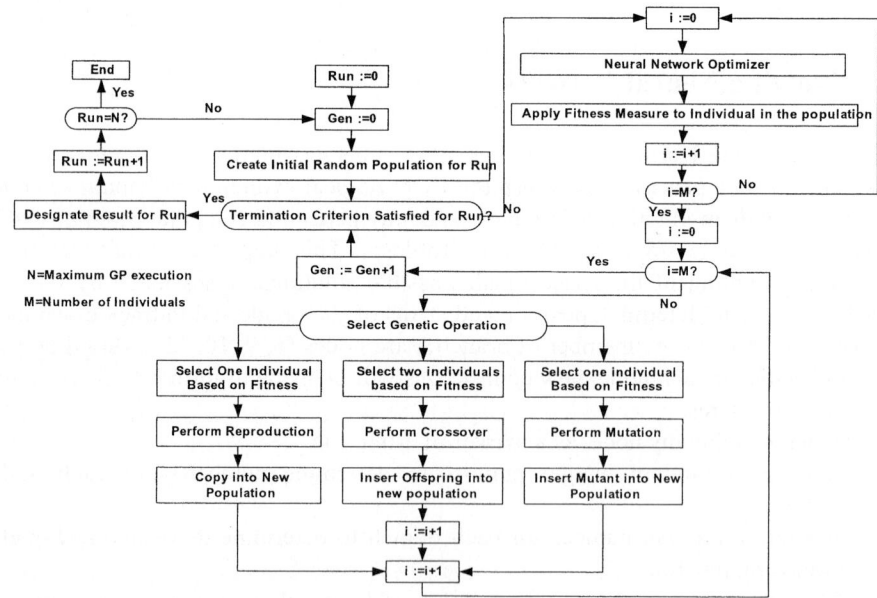

Fig. 2. A flowchart of the GPNN

Even though all processing node are the same, based on their position, those processing nodes are differentiated into 3 types: (P_{SN}), (P), and (P_{EN}). Connection existing

in each tree-structure does not represent connection of the related NN. The NN's connections will be generated based on these following rules:
- (D) is erased.
- (P_{SN})s are processing nodes which are directly connected to (D). They will be connected to all output-nodes.
- (P)s will be connected to all (P)s or (P_{SN})s of previous layer.
- (P_{EN})s, which are located at the end of each branch of GP's tree, directly process all inputs. They are also connected to all (P)s or (P_{SN})s of previous layer.

The related NN of Fig.1 is shown in Fig. 3.

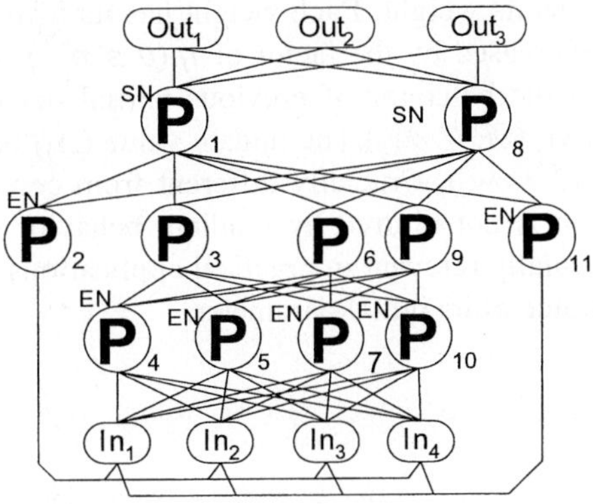

Fig. 3. NN representation based on Fig.1.

3 Measuring Fitness

Structure evolution is intended to search the most beneficial NN (NN with less processing node and computing time but able to solve the given problem with certain criteria).

Each tree structure's participation level in creating the next generation depends on its related NN's fitness. Fitness will be measured based on 2 parameters: a) accuracy (dependent on error level); b) efficiency (dependent on numbers of processing node). The formula is:

$$F_i = 1 - C_1 \left[E_i \bigg/ \sum_{k=1}^{N} E_k \right] - C_2 \left[R_i \bigg/ \sum_{k=1}^{N} R_k \right] \quad (1)$$

where C_1 and C_2 are constants ($0 \leq C_1, C_2 \leq 1$ & $C_1+C_2=1$); F_i is fitness, E_i is error, and R_i is number of processing nodes of to the-i^{th} NN of current generation.

4 Training Algorithm

GP builds NNs in seemingly random way, so the resulting NNs are unpredictable. Therefore each connection has its own optimum-learning-characteristic. Constants learning algorithm like conventional Backpropagation [7] will not achieve a fast convergence rate or a robust learning algorithm. The system needs learning algorithm with ability to automatically adjust learning rate based on behavior of each connection.

RPROP (6) is an efficient learning scheme, which performs direct adaptation of the weight step based on local gradient information. Compare to [7], the only difference is the way RPROP updates its weight. Each weight has its own individual update value (Δ_{ij}), whose value is increased by the factor of η ($0 \leq \eta^- \leq 1 \leq \eta^+$). Values of η are selected depending on the behavior of previous partial derivative ($\partial E^{(t-1)}/ \partial w_{ij}$) and current partial derivative ($\partial E^{(t)}/ \partial w_{ij}$). The update value ($\Delta_{ij}$) use the current weight or bias (w_{ij}) based on ($\partial E^{(t)}/ \partial w_{ij}$). Crucially different from constant-learning-algorithm, the effort of adaptation is not blurred by gradient behavior or whatsoever. Another important feature, especially relevant in practical applications, is the robustness of the RPROP against the choice of its initial parameter.

5 Experiments

The proposed neural network has been applied to cancer diagnosis problems. Neural networks are used to classify a tumor as either benign or malignant based on cell descriptions gathered by microscopic examination. The data set is obtained from the University of Wisconsin Hospitals, from Dr. William H. Wolberg. There are 699 training data with 458 Benign and 241 Malignant. Following 9 inputs are used for diagnosis by neural networks: Clump Thickness; Uniformity of Cell Size; Uniformity of Cell Shape; Marginal Adhesion; Single Epithelial Cell Size; Bare Nuclei; Bland Chromatin; Normal Nucleoli; Mitoses. Each input is scaled from 1 to 10. Output is classified into 2 classes: Benign and Malignant.

Experiments are done with following 4 neural networks: (A) Koza's method [5]; (B) Esparcia's method [1]; (C) Same structure with (B), but SA learning algorithm is replaced by back-propagation [9]; (D) the proposed method. Each structure in (B) is given 1000 of SA, while (C) and (D) is given 1000 training iterations. The number of population is 40. Maximum number of processing node at first iteration is 50 nodes and maximum processing time is 20 minutes. The process is terminated if the 50% of the newly generated individuals have more than 100 nodes in order to simulate real applications in which number of processing nodes is limited. The results are summarized in Fig.Fig.4. As shown in Fig. 4, the proposed method (D) is much superior to other methods in both number of nodes and errors of classification.

6 Conclusion

The proposed method is shown to have better performance compared to previous similar methods. The main reasons are [1] it has higher density of connections hence increasing variables which can be optimized, or, in other word, has higher degree of freedom; [2] it applies adaptive learning to solve the problem so that each bias or weight has its own optimum learning characteristic; hence the system has higher learning robustness; [3] it guarantees that each NNs contains all inputs so no data is neglected.

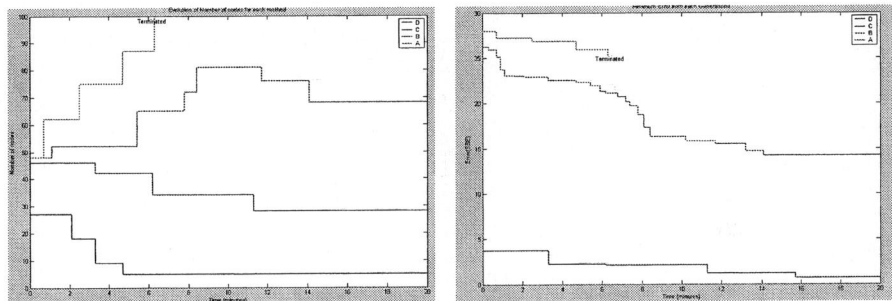

Fig. 4. Experimental results

Previous experiments are just preliminary experiments and the given problems are also toy problems. Even though it doesn't explore real ability of the proposed system yet, it is evident that the proposed method is better. In the future we will test the proposed method to solve real problems and conduct thorough measurement to show how good the proposed algorithm is.

We believe that it is just a beginning of something that will grow faster and wider, since this method provides a good automatic neural networks generator; hence no more need to make the NN manually with trial-and-error algorithm.

References

1. Esparcia Alc'azar, A. I., Sharman, K. C.: Genetic Programming Techniques That Evolve Recurrent Neural Network Architectures for Signal Processing, IEEE Signal Processing Society Workshop, Kyoto, Japan (1996) 139-148
2. Gruau, F.: Genetic Micro Programming of Neural Networks, Advance in Genetic Programming, The MIT press (1994) 495-518
3. Hinton, G. E.: Connectionist Learning Procedures, Artificial Intelligence, 40 (1989) 185-234
4. Kitano, H.: Designing Neural Network Using Genetic Algorithm with Graph Generation System, Complex System, 4 (1990) 461-476
5. Koza, J. R., Rice, J. P.: Genetic Generation of Both the Weights and Architecture for Neural Network, IEEE Press, Vol II (1991)

6. Riedmiller, M., Braun, H.: A Direct Adaptive Method for Faster Backpropagation Learning: the RPROP Algorithm, Proceeding of the IEEE International Conference on Neural Networks, San Francisco (1993) 586-591
7. Rumelhart, D. E., Hinton, G. E., Williams, R. J.: Learning Internal Representations by Error Propagation, Parallel Distributed Processing: Explorations in the Microstructure of Cognition, Vol. 1, MIT press (1986) 318-362
8. Mangasarian, O. L., Wolberg, W. H.: Cancer Diagnosis via Linear Programming, SIAM News, **23** (1980) 1-18
9. Sulistiyo: Private Communication.

Self-Organizing Feature Map Based Data Mining

Shangming Yang and Yi Zhang

University of Electronic Science and Technology of China, Chengdu, 610054, China
minn003@163.com zhangyi@uestc.edu.cn
http://cilab.uestc.edu.cn

Abstract. In data mining, Apriori algorithm for association rules mining is a traditional approach. However, it takes too much time in scanning database for finding the frequent itemsets. In this paper, based on SOM clustering, a novel algorithm is introduced. In this algorithm, each transaction is converted to an input vector, SOM is employed to train these input vectors, from which we achieve the visualization of the relationship between the items in a database. The time efficiency and the visualized map units make the proposed approach a particularly attractive alternative to current data mining algorithms.

1 Introduction

Data mining or knowledge discovery in database is to find new knowledge from database. However, the dimensionality, complexity, or amount of data is prohibitively large for manual analysis. With a huge amount of data stored in databases, it is increasingly important to develop powerful tools for mining interesting knowledge from it. In recent years, the computational efficiency of modern computer technology is making the mining fast and precise. One of the most interesting developments in this area is the application of neural computation.

The self-organizing map (SOM) [5] is an unsupervised neural network algorithm. It has been widely applied to solve problems such as pattern recognition, financial data analysis, image analysis, process monitoring, and fault diagnosis [3, 6].

The purpose of this paper is to introduce a new algorithm to make use of the SOM in data mining. Some major components of data mining are discussed and the applications of the SOM in association rules mining are proposed.

2 Data Mining Concepts and Algorithms

In general, data mining has three major components [1], they are clustering or classification, association rules mining, and sequential pattern analysis. In classification we generate a set of grouping rules which can be used to classify data in the future. Using SOM as an intermediate step makes the clustering a two-level approach [4]: the data set is first clustered using SOM, and then the SOM is clustered. One big benefit of this approach is that computational load decreases. For this reason, it is convenient to cluster a set of prototypes rather than directly the data.

Association rules mining is to mine some associated relationships among a set of objects in a database, it requires iterative scanning of large relational database which is time consuming in processing. An association rules mining algorithm Apriori has been developed for rule mining in large transaction databases by [8]. Combination of this algorithm and SOM neural networks will speed up the mining.

In sequential pattern analysis, we explore the database to discover patterns that occur in sequence. This deals with data that appear in separate transactions. For example, if a customer buys item A in the first month of the year, then s/he buys item B in the second month and item C in the third month etc.

3 Using Self-Organizing Map in Association Rules Mining

3.1 The Basic Concepts of Association Rules Mining

In association rules mining [1], a set of items is referred as itemset. An itemset that contains k items, is a k-itemset. Suppose $I = \{i_1, i_2, ..., i_m\}$ is a set of items, D is a set of database transactions, where each transaction T is a set of items such that $T \subseteq I$. Let A be set of items. A transaction set T is said to contain A if and only if $A \subseteq T$. An association rule is an implication of the form $A \Rightarrow B$, $A \subset I$, $B \subset I$ and $A \cap B = \phi$. It has the following two significant properties:

$$support : (A => B) = P(A U B)$$

The probability of itemset D contains both itemset A and B

$$confidence : (A => B) = P(A/B)$$

The probability of itemset D contains A also contains B.

Rules that satisfy both minimum support threshold and a minimum confidence threshold are called strong. Association rules mining is the rule to generate support and confidence which are greater than or equal to their minimum support and confidence. An itemset satisfies minimum support is called a frequent itemset, the set of frequent k-itemsets is commonly denoted by L_k.

3.2 SOM Clustering

Transactions as in Table 1 in a database can be modelled as data vectors, each transaction will be converted to an input vector. If it has item i in the transaction, the ith component in the vector will be 1, otherwise 0. The data modelling can be done at the time when transactions are extracted from the database.

To train the input vectors, a SOM or GHSOM [3] neural network can be initialized to generate map units. Different from the usual neural networks we

Table 1. Transaction data for Allen's Electronics Branch

TID	List of item Ids
T01	I1, I4, I9, I12
T02	I1, I4, I9
T03	I1, I4, I9, I13
T04	I1, I4, I9, I12, I13
T05	I2, I5, I6, I11
T06	I2, I5, I6, I8 I10, I11
T07	I1, I5, I6, I10
T08	I3, I5, I6, I10, I11
T09	I2, I5, I6, I11
T10	I2, I5, I6, I8, I10, I11
T11	I1, I5, I6, I10
T12	I3, I5, I6, I10, I11
T13	I3, I5, I6, I8, I10, I11
T14	I3, I5, I6, I7, I8, I11
T15	I3, I5, I6, I7, I8
T16	I3, I5, I6, I7
T17	I1, I4, I9, I12
T18	I1, I4, I9, I10, I12
T19	I2, I4, I9, I10, I12
T20	I2, I5, I6, I10

used in other pattern, in this particular training, the network has only one vector in each row, each neuron just has neighbors in different rows. From the map units, one can easily find the relationship between different items. Transactions in Table 1 are converted to SOM input samples as follows:

$\{1,0,0,1,0,0,0,0,1,0,0,1,0\}$ $\{1,0,0,1,0,0,0,0,1,0,0,0,0\}$
$\{1,0,0,1,0,0,0,0,1,0,0,0,1\}$ $\{1,0,0,1,0,0,0,0,1,0,0,1,1\}$
$\{0,1,0,0,1,1,0,0,0,0,1,0,0\}$ $\{0,1,0,0,1,1,0,1,0,1,1,0,0\}$
$\{1,0,0,0,1,1,0,0,0,1,0,0,0\}$ $\{0,0,1,0,1,1,0,0,0,1,1,0,0\}$
$\{0,1,0,0,1,1,0,0,0,0,1,0,0\}$ $\{0,1,0,0,1,1,0,1,0,1,1,0,0\}$
$\{1,0,0,0,1,1,0,0,0,1,0,0,0\}$ $\{0,0,1,0,1,1,0,0,0,1,1,0,0\}$
$\{0,0,1,0,1,1,0,1,0,1,1,0,0\}$ $\{0,0,1,0,1,1,1,1,0,0,1,0,0\}$
$\{0,0,1,0,1,1,1,1,0,0,0,0,0\}$ $\{0,0,1,0,1,1,1,0,0,0,0,0,0\}$
$\{1,0,0,1,0,0,0,0,1,0,0,1,0\}$ $\{1,0,0,1,0,0,0,0,1,1,0,1,0\}$
$\{0,1,0,1,0,0,0,0,1,1,0,1,0\}$ $\{0,1,0,0,1,1,0,0,0,1,0,0,0\}$

The SOM algorithm has two phases [5], the first is the search phase, during which each node i computes the Euclidean distance between its weight vector $\mathbf{w}_i(t)$ and the input vector $\mathbf{T}(t)$ as following:

$$d(\mathbf{T}(t), \mathbf{w}_i(t)) = ||\mathbf{T}(t) - \mathbf{w}_i(t)|| = \sqrt{\sum_{j=1}^{n}(\tau_i(t) - w_{ij}(t))^2} \qquad (1)$$

Then it chooses the closest neuron i_0 such that

$$\mathbf{w}_{i_0} = \min_i d(\mathbf{T}(t), \mathbf{w}_i(t)) \qquad (2)$$

where neuron i_0 is called the winner or winning cell.

The second phase is the update phase, during this phase, a small number of neurons within a neighborhood around the winning cell, including the winning cell itself, are updated by

$$\mathbf{w}_i(t+1) = \mathbf{w}_i(t) + \alpha(t) * \mathbf{T}(t) - \mathbf{w}_i(t) \qquad (3)$$

where $\alpha(t)$ is the learning constant.

3.3 Association Rules Mining on SOM Clusters

The purpose of using SOM to train the data set is to obtain the visualization of structure of the transactions and reduce the association rules mining time. After iterations of SOM training, if two or more neurons are in the same cluster, it means these neurons have one more similar or same components in their weight vectors, therefore transactions which have some or even exactly same items will be in same cluster. There are more transactions in one cluster; the support of itemsets in the cluster will be higher. Using the SOM trained outputs, we have the following two approaches to generate association rules for the database:

1. Obtain the association rules by observing SOM trained map units. Fig. 1 is the map units of 5000 iterations SOM trained transactions of Allen's Electronics Branch [1]. From the three classes in this figure, we have immediate conclusions about the association rules for this mining. For example, itemset {I5, I6}, {I3, I5, I6}, {I2, I5, I6}, {I5, I6, I10}, and {I1, I4, I9} and their subsets have bigger support counts than any other itemsets, these items must have some rules associated.

2. Generate association rules for a giving minimum support count. For a k-itemset, in Fig. 1, the support count for 1-itemset is the total count of 1s in each column. If the support count of an 1-itemset is less than the minimum support count, then the support count of any k-itemset which contains this 1-itemset will be less than the minimum support count. For saving time, we remove these columns if their total counts in the map units are less than the minimum support count, then use Apriori algorithm to generate strong association rules for the remained items. This will take less scanning time but generate same association rules for the database. In Fig. 1, if the minimum support count is 4, then column 7 and column 13 will be removed from the map units because their support counts are 3 and 2 respectively. Table 2 shows the remained items and their 1-item set support counts, the association rules will be generated based on these items.

```
Class 1  0 0 1 0 1 1 1 0 0 0 0 0 0
         0 0 1 0 1 1 1 1 0 0 0 0 0
         0 0 1 0 1 1 1 1 0 0 1 0 0
         0 0 1 0 1 1 0 1 0 1 1 0 0
         0 0 1 0 1 1 0 0 0 1 1 0 0
         0 0 1 0 1 1 0 0 0 1 1 0 0
         1 0 0 0 1 1 0 0 0 1 0 0 0
         1 0 0 0 1 1 0 0 0 1 0 0 0

Class 2  0 1 0 0 1 1 0 0 0 1 0 0 0
         0 1 0 0 1 1 0 1 0 1 1 0 0
         0 1 0 0 1 1 0 1 0 1 1 0 0
         0 1 0 0 1 1 0 0 0 0 1 0 0
         0 1 0 0 1 1 0 0 0 0 1 0 0

Class 3  1 0 0 1 0 0 0 0 1 0 0 1 1
         1 0 0 1 0 0 0 0 1 0 0 0 1
         1 0 0 1 0 0 0 0 1 0 0 0 0
         1 0 0 1 0 0 0 0 1 0 0 1 0
         1 0 0 1 0 0 0 0 1 0 0 1 0
         1 0 0 1 0 0 0 0 1 1 0 1 0
         0 1 0 1 0 0 0 0 1 1 0 1 0
```

Fig. 1. Outputs of 5000 iterations of SOM trained transactions

The benefit for this algorithm is, in a large database, it may generate thousands of transactions for analysis, but we are only interested in these data which appear repeatedly. Items in the transactions just appear a few times may not be important for the mining. In Apriori algorithm association rules mining, computation of frequent k-itemsets L_k ($k = 1, 2, ...$) and their candidates C_k will be very tedious, we use SOM clustering to classify the transaction data, it always organizes the data to be neighborhood if they have some similar properties (same items appear in many different transactions). Therefore the structures of those data are visualized, from the structured map we can remove some columns and choose data for association rules mining.

4 Conclusions

Many algorithms are proposed to use SOM for data mining. Apriori algorithm for data mining is a very efficient utility when database is relatively small. However, it takes too much time to repeatedly scan the large-scale database. The combination of SOM training and Apriori algorithm can take the advantage. In this algorithm, each input vector represents a transaction in a database, Kohonen self-organizing map is used to train input vectors. According to the principles of the SOM, it is a similarity of neighborhood neural network, on which we have the visualization of the relationship between items in the database. The

Table 2. Items used to generate association rules for the database

Itemset	Sup. Count
{I01}	8
{I02}	6
{I03}	6
{I04}	7
{I05}	13
{I06}	13
{I08}	5
{I09}	7
{I10}	10
{I11}	8
{I12}	6

simulation experiment shows the feature map can be formed in very short time. By reviewing the SOM outputs, one can easily determine the association rules in the database. It's clearly the proposed algorithm makes the data mining a significantly improvement.

References

1. Han, J., Kamber, M.: Data Mining – Concepts and Techniques. Higher Education Press, Beijing, China (2001)
2. Rauber, A., Merkl, D., Dittenbach, M.: The Growing Hierarchical Self-organizing Map: Exploratory Analysis of High-dimensional Data. IEEE Transactions on Neural Networks, Vol. 13, No. 6 (2002) 1331-1341
3. Kohonen, T., Kaski, S., Lagus, K., Salojarvi, J., Honkela, J., Paatero, V., Saarela, A.: Self-organization of a Massive Document Collection. IEEE Transactions on Neural Networks, Vol. 11, No. 3 (2000) 574–585
4. Vesanto, J., Alhoniemi, E.: Clustering of the Self-organizing Map. IEEE Transactions on Neural Networks, Vol. 11, No. 3 (2000) 586–600
5. Kohonen, T.: Self-organizing Maps. Springer-Verlag Berlin and Heidelberg, Germany (1995)
6. Debock, G., Kohonen, T.: Visual Exploration in Finance Using Self-organizing Maps. Springer-Verlag, London (1998)
7. Agrawal, R., Srikant, R.: Fast Algorithms for Mining Association Rules. In Proc. 1994 Int. Conf. Very Large Databases (VLDB'94) (1994)
8. Agrawal, R., Srikant, R.: Mining Sequential Patterns. Proc. of the Int'l Conference on Data Engineering (ICDE), Taipei, Taiwan (1995)

Diffusion and Growing Self-Organizing Map: A Nitric Oxide Based Neural Model*

Shuang Chen, Zongtan Zhou, and Dewen Hu

Department of Automatic Control, National University of Defense Technology,
Changsha, Hunan, 410073, P.R.C.
dhu@nudt.edu.cn

Abstract. This paper presents a new diffusion and growing neural network model called DGSOM for self-organization, where more generalized ideas of short-range competition and long-range cooperation are adopted while introducing the diffusion mechanism of intrinsic NO as its most remarkable characteristic. The new DGSOM model can compartmentalize input space rationally and efficiently, and generated topological connections among neurons can reflect the dimensionality and structure of input signals. Experiments and simulations indicated that the embedding of NO diffusion mechanism improve the performance of self-organization remarkably, especially the flexibility and rapidity of response for dynamic distribution.

1 Introduction

The ideas about self-organization were initially inspired by human's cognition ability of learning knowledge from environment automatically without teacher. The essential and goal of a self-organizing neural model is to train the neural network to compartmentalize the space of input data with the interrelation and distribution retained rationally, and to reflect intrinsic topological relations and structure of the diverse input. A well-known self-organizing model is the Self-Organizing Map(SOM) presented by T. Kohonen(see [1]). Other self-organizing models, such as neural gas, growing cell structures, growing grid, and growing neural gas(see [2]), which are mostly inspired by and based on Kohonen's SOM model, are also widely used.

In many practical cases applying SOM, little or no information about the input distribution is available, and it is hard to determine a priori the appropriate number of nodes and the topology of network. Further more, depending on the relation between inherent data dimensionality and dimensionality of the target space, some information on the topological arrangement of the input data may be lost in the mapping process(see [3]). In addition, many neural network models with growing mechanism having a fixed insertion rate policy might not always be desirable, since it may lead to unnecessary or untimely insertions.

---
* Supported by Natural Science Foundation of China (60171003), the Distinguished Young Scholars Fund of China (60225015), Ministry of Science and Technology of China(2001CCA04100) and Ministry of Education of China (TRAPOYT Project).

This paper proposes a new self-organizing model with growing mechanism called Diffusion and Growing Self-Organizing Map(DGSOM).The DGSOM model has the capability of topology learning and is adaptive to dynamic distribution. It adds neurons through competition mechanism, updates the topology of network using a Competitive Hebbian Learning(CHL) fashion, and uses diffusion mechanism of Nitric Oxide(NO) as global coordinator of self-organizing learning,which is the most remarkable characteristic of the proposed DGSOM model.

Description about growing mechanism, topology learning process and diffusion mechanism of NO, and the detailed DGSOM algorithm are presented in Sect.2. Experiments, simulations, results and some discussion concerning the novel DGSOM model are presented in Sect.3 and Sect.4.

2 Diffusion and Growing Self-Organizing Map

2.1 Background and Some Key Features

For most self-organizing phenomena in the nature, the main factor that influences the distribution of units is the competition for resources. In the design of self-organizing neural network model, the neurons and the input distribution can be regarded as competitors and resources respectively. This is the first primary idea of DGSOM model. The second point of the model concerns generating a rational topological structure corresponding to the intrinsic interrelationship of input data, which can be achieved by CHL algorithm. The algorithm generates a number of connections between the neurons of network while the positions of neurons do not change at all.

Moreover, to construct a practical self-organizing map and to make it flexible and adaptable, a rational adjustment strategy is rather important. As we know, the discovery of NO as new neurotransmitter changes scientist's ideas about interaction between neurons, and it is also heuristic to our design of self-organizing neural network model. The diffusion mechanism of NO as a global coordinator of self-organizing learning, is the last but most important aspect of the proposed DGSOM model.

Intrinsic NO, as intercellular signalling molecule in the nervous system, was first suggested by Garthwaite et al. in 1988 (see [4]). R.F. Furchgott, L.J. Ignarro and F. Murad who have won the Nobel Prize(1998) have confirmed this through experiments. NO violates some of the key tenets of classical neural signalling(see [5]). Conventional neurotransmission is essentially two-dimensional, however NO acts four-dimensionally in space and time, affecting volumes of the brain containing many neurons and synapses. The NO diffusion mechanism embedding in DGSOM is simplified by modeling NO diffusion from a point-source(see [5]):

$$C(d) = a_{\text{NO}} \cdot \exp\left(-d^2/2 \cdot \sigma_{\text{NO}}^2\right) . \tag{1}$$

Where $C(d)$ is the concentration of NO at a point with distance d from source. The values of a_{NO} and σ_{NO} are parameters depended on given application, but unvaried with time.

2.2 Configuration and Algorithm of DGSOM

The DGSOM model assumes that each node(neuron) r consists of a reference vector $\mathbf{w}_r$, a counter C_r to record the winning numbers and a set of connections with measurement called age_{rj} defining the topological neighbors of node r. While initializing, parameter t is set to 0, and there are only two nodes with reference vectors chosen randomly, with counter C_r for each node initialized to 0. Some global parameters such as threshold $WIN_{\max}$, forgetting interval T, $age_i, age_f, a_{NO}, \sigma_{NO}, \sigma_i$ and σ_f are set accordingly.

Step 1. Generate an input vector $\mathbf{u}$ conforming to specific distribution.

Step 2. Locate the two nodes s and t nearest to input vector $\mathbf{u}$, namely, the two nodes with reference vectors $\mathbf{w}_s$ and $\mathbf{w}_t$ such that $\|\mathbf{u} - \mathbf{w}_s\|$ is the smallest and $\|\mathbf{u} - \mathbf{w}_t\|$ is the second smallest.

Step 3. Increase counter C_s of the winner-node s by 1 and decide: if $C_s > WIN_{\max}$, a new node q will be inserted without any connection to any existing nodes. $\mathbf{w}_q$ and C_q of the new node are set as:

$$\mathbf{w}_q = \mathbf{u} \ . \tag{2}$$

$$C_q = C_s = C_s/2 \ . \tag{3}$$

else ($C_s \leq WIN_{\max}$) modify the reference vectors with a weight coefficient of e_r (which will be described in the next section):

$$\Delta \mathbf{w}_r = e_r \cdot [(\mathbf{u} - \mathbf{w}_s)/\|(\mathbf{u} - \mathbf{w}_s)\|], \quad \forall r \ . \tag{4}$$

Step 4. Set age_{st} of connection between s and t to 0, and increase age_{sj} of all topological neighboring connections from node s by 1.

Step 5. Determine if there is any connection with $age_{rj} > age_{\max}$ and then remove it. Parameter $age_{\max}$ is also varying with time t:

$$age_{\max}(t) = age_i(age_f/age_i)^{t/t_{\max}} \ . \tag{5}$$

Step 6. If $t = kT$, decrease the winning number C_r by 1 for all nodes r, $k = 1, 2, 3 \ldots$:

$$C_r = C_r - 1, \quad \forall r \ . \tag{6}$$

Step 7. Increase the time parameter t and return to Step 1 if $t < t_{\max}$.

2.3 Growing and Diffusion Mechanism of DGSOM

If we treat the input as resources with specific distribution, the winning counter C_r can be regarded as the resources that node r possess. If the winning number of node r is larger than threshold $WIN_{\max}$, which means that r possesses too many resources, then a new node is required to share the resources to make the resources distributed more evenly and rationally. After that, the winning counter

C_r must be reassigned according to (3). The new node does not connect with any existing nodes because its local situation is unknown, but new connections may be formed gradually in later learning.

In (4), mechanism of short-range competition and long-range cooperation is adopted for nodes adjustment. When the winner-node s is selected, and it is not time to insert a new node, the reference vector of node s will be adjusted toward $\mathbf{u}$ with a fixed weight scaling of e_s. At the same time the node s generates NO which spreads through the whole network, and the diffusion of NO will pull reference vectors of other nodes located within the diffusion range along the direction of $(\mathbf{u}-\mathbf{w}_s)$, with a weight coefficient of $D_{\mathbf{w}_r}$. On the other hand, if any node has connections with the winner-node s, they will be pushed away along $(\mathbf{u}-\mathbf{w}_s)$ with a weight of $e_{\mathbf{w}_r}$ to balance the pulling effect of NO diffusion. Weight $D_{\mathbf{w}_r}$ of NO diffusion has a simplified form as in (1), while $e_{\mathbf{w}_r}$ is an exponentially time-decaying factor:

$$D_{\mathbf{w}_r} = a_{\text{NO}} \cdot \exp(-\|\mathbf{w}_{\text{winner}} - \mathbf{w}_r\|^2 / 2 \cdot \sigma_{\text{NO}}^2) \ . \tag{7}$$

$$e_{\mathbf{w}_r}(t) = \sigma_{\text{i}} (\sigma_{\text{f}}/\sigma_{\text{i}})^{t/t_{\max}} \ . \tag{8}$$

$$e_r = D_{\mathbf{w}_r} - e_{\mathbf{w}_r} \ . \tag{9}$$

Since the reference vectors of all nodes are slowly moving around, it may cause the construction of the network to become uneven, which may bring a very obvious phenomenon that the topological connections may be intersectant. A local aging mechanism is used to remove the invalid connections, which is done in Step 4 and Step 5. In (6) the winning number C_r is decreased every T steps. This makes the influence of early input signals be forgotten and prevents the parameter C_r from increasing infinitely. Parameters $WIN_{\max}$ and T are related to node numbers and growing speed of the neural network.

3 Experiments and Simulations

Computer simulations and experiments are also carried out to test the capability and performance of DGSOM. Only two typical cases are described below to illustrate the effect of NO diffusion limiting to the length of this paper.

Experiment on Shifting Distribution of Input. Figure 1 illustrates the ability of DGSOM with NO diffusion mechanism to track a shifting data distribution. The distribution of a randomly generated 2-D input vectors is changing its range in the learning course of DGSOM as follows: first the distribution of input lies in the range of (0,0.4)...(0,0.4), then shifts to range of (0.4,0.8)...(0.3,0.7) after 5000 iterations, and finally moves to (0,0.4) ...(0.6,1) after 10000 iterations.

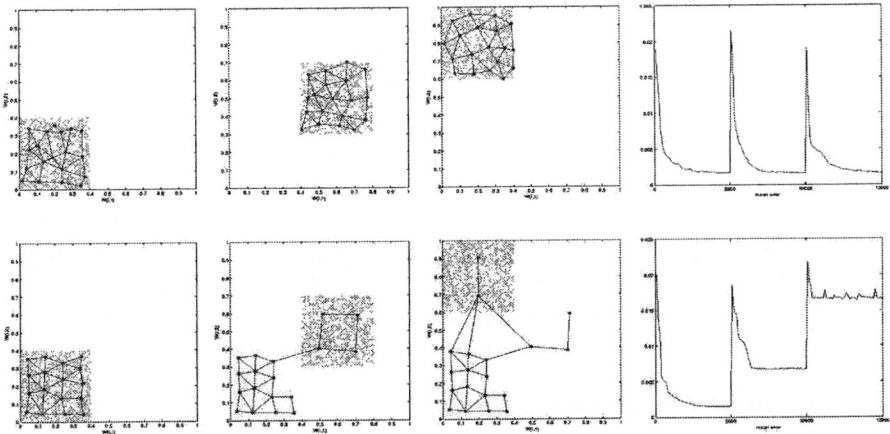

Fig. 1. The first row is the result of DGSOM and the second row is the result of the same model but without NO diffusion mechanism. The first column is the state after 5000 iterations for two cases. The second column is the state after 10000 iterations, in which the input distribution jumps for the first time. The third column is the state after 15000 iterations, in which the input distribution jumps for the second time. The fourth column shows the mean-error of two networks which is computed every 100th iteration. The three minimum errors of original DGSOM are 0.0016, 0.0016 0.0017 respectively, and the three minimum errors of diffusion-removed model are 0.0015, 0.0067 0.0164.

In the two cases described above(one with NO diffusion mechanism and another without), most of the nodes are positioned during the first 5000 iterations. In DGSOM model, almost all the nodes can track the shifting of distribution. While in the opposite case where NO diffusion is cancelled, almost all the nodes are left behind and become inactive when the distribution shifts.

Low Dimensional Distributed Input Embedded in High Dimensional Space. Another characteristic of DGSOM is that the model can reflect the intrinsic dimensionality of input signals adaptively. Generally and simply to say, for 1-D network, each node has about *two* coterminous nodes, and for 2-D network, each node has about *four* coterminous nodes etc. Thus we can estimate the essential dimensionality of the input data by computing the average number of coterminous nodes in the network after self-organizing learning. What is more important for most practical situation, is that DGSOM can reflect the essential structure of input distribution more adaptively and efficiently. Figure 2 shows an input distributed as a spirality curve embedded in three-dimensional space(unknown a priori for DGSOM learning) and the DGSOM model mapping the intrinsic 1-D distribution rationally, where the average number of coterminous nodes is 2.

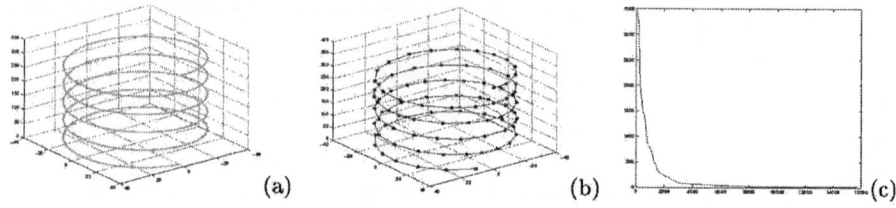

Fig. 2. (a) Distribution of the input vector; (b) Result mapping with 92 nodes. Parameters used: $WIN_{max} = 100$; $T = 100$; $age_i = 100$; $age_f = 30$; $\sigma_{NO} = 2.5$; $a_{NO} = 0.002$; $\sigma_i = 0.0001$; $\sigma_f = 0.000015$; (c) The mean local-error and the minimum value is 14.856.

4 Conclusion and Discussion

The DGSOM model presented in this paper is a simple and speedy one for most self-organizing task. It can not only track the diverse distribution of input space, but also reflect the intrinsic dimensionality and structure of the input signals. Introducing the mechanism of NO diffusion as global adaptation coordinator can make the model more preferable, stable and flexible, which was also proved by the results of experiments and simulations.

By far, the stop criterion of node growing in the network might be that, for example, the maximum iterations of the learning process or the maximum nodes number has been achieved, etc. More impersonality criterions should be considered, e.g. the residual error is small enough or the performance of a given aspect is good enough. Problems of determining appropriate and impersonal criterions will be studied in our further research work.

References

1. Kohonen, T.: Self-Organizing Maps. 2nd edn. Springer, Berlin Heidelberg New York (1997)
2. Fritzke, B.: Growing Self-Oganizing Networks - Why?. In: Verleysen, M. (ed.): European Symposium on Artificial Neural Network. D-Facto Publishers, Brussels (1996) 61–72
3. Fritzke, B.: A Growing Neural Gas Network Learns Topologies. In: Tesauro, G., Touretzky, D.S., Leen, T.K. (eds.): Advances in Neural Information Processing Systems 7. MIT Press, Cambridge MA (1995) 625–632
4. Garthwaite, J., Charles, S., Chess-Williams, R.: Endothelium-Derived Relaxing Factor Release on Activation of NMDA Receptors Suggests Role as Interneurons Messager in The Brain. Nature. **336** (1988) 385–388.
5. Philippides, A., Husbands, P., O'Shea, M.: Four-Dimensional Neuronal Signaling by Nitric Oxide: A Computational Analysis. The Journal of Neuroscience. **20** (2000) 1199–1207

A New Adaptive Self-Organizing Map

Shifeng Weng, Fai Wong, and Changshui Zhang

State Key Laboratory of Intelligent Technology and Systems,
Department of Automation, Tsinghua University, Beijing 100084, China
{wengsf00, huangh01}@mails.tsinghua.edu.cn
zcs@mail.tsinghua.edu.cn

Abstract. The self-organizing map (SOM) method developed by Kohonen is a powerful tool for visualizing high-dimensional datasets. To improve the adaptability of SOM, this paper proposes a new model based on Kohenen SOM. The new model has integrated a series of evolutionary working mechanisms of neurons. The microcosmic analysis gives the reason that those introduced mechanisms can conquer the problems of instability in competitive learning. The empirical evidences show the performance of the proposed model.

1 Introduction

The self-organizing map (SOM) method developed by Kohonen is a powerful tool for visualizing high-dimensional datasets [1]. It converts complex, nonlinear statistical relationships among the high-dimensional dataset into simple geometric relationships on a low dimensional display. There are many reports about SOM in literatures, ranging from theory analysis to engineering applications [2,3,4].

However, the ability of SOM has been limited by the fact that the mapping-range of SOM is low dimensional rigid grids (usually 1-D or 2-D). For this problem, some approaches have been proposed [5,6,7]. Among them, double SOM (DSOM) [5] and adaptive coordinate (AC) [7] methods introduce a special feature in training algorithm, not only the weights of a neuron is modified during the learning process, but also the coordinates of the neuron is updated accordingly. Visually, neurons are moving around on a low dimensional grid. Neurons that have similar weights tend to move closed to each other.

In fact, the modification of position vectors of neurons during the training processing in either AC or DSOM is an adaptive reaction against the input data. These two methods put much attention to the design of movement action of neurons, while neglected other adaptive reactions.

This paper has introduced a further working mechanism for the neurons, which includes moving, splitting, vanishing and merging under specific conditions. This algorithm is called Adaptive Self-Organizing feature Maps (ASOM). ASOM model not only inherits the features of AC and DSOM approaches, but also makes the learning algorithm more robust and adaptive. The experimented results show the performance of the proposed ASOM model.

2 Related Work

Although AC and DSOM have a common essence to preserve the cluster relation by monitoring the movement of neurons, the training algorithms they applied are different. The algorithm applied by AC approach is:

$$p_i(t+1) = p_i(t) + \triangle Dist_i(t+1) \times (p_c(t) - p_i(t)) \tag{1}$$

$$\triangle Dist_i(t+1) = \frac{Dist_i(t) - Dist_i(t+1)}{Dist_i(t)}, \tag{2}$$

where $Dist(t)$ is a table, in which the distance between the weight vectors and a randomly selected input pattern are stored at step t; $\triangle Dist_i(t+1)$ presents the relative change of neural unit i at step $t+1$; the position of neural unit i is denoted by $p_i(t)$, which actually means coordinate (x, y) in the case of a two-dimensional output plan; $p_c(t)$ is the position of the winning neuron.

While the DSOM algorithm adopts an updating rule similar to the conventional SOM updating rule to move neurons. And that is why it is called the double SOM.

$$p_i(t+1) = p_i(t) + \eta(t)h(t)[p_i(t) - p_c(t)] \tag{3}$$

$$\eta(t) = \frac{\eta_p}{1+t} exp\{-s_p(1 + \frac{t}{t_{max}} D(p_i(t), p_c(t)))\} \tag{4}$$

$$h(t) = exp\{-s_x(1 + \frac{t}{t_{max}}[D(w_i(t), x) - D(w_c(t), x)])\}, \tag{5}$$

where s_p and s_x are two predetermined scalar parameters and η_p is the initial learning rate; $\eta(t)$ gives the difference between the position of ith neural unit and the winning unit; $h(t)$ denotes the difference between weight vectors and input pattern; D is a distance function; and t_{max} is the maximum number of iterations for training.

From the above stated formulas, it is found that, the algorithms for adjusting the movement of the winning neuron in both the AC and DSOM approaches are of the same nature, and can be expressed by the following learning rule,

$$p_i(t+1) = p_i(t) + \beta(p_c(t) - p_i(t)). \tag{6}$$

The only difference is the choice of moving rate, β, in each method to have different interpretation. These two methods put much attention to the design of movement action of neurons, while neglected other adaptive reactions.

3 Working Mechanism

Let's consider three situations during the competitive learning process (See Fig.1). Note that all the situations would lead to instability of network, and they are often met in competitive learning.

In order to conquer these problems and make the network performance better, a series of evolutionary working mechanisms of neurons has been introduced in the ASOM model.

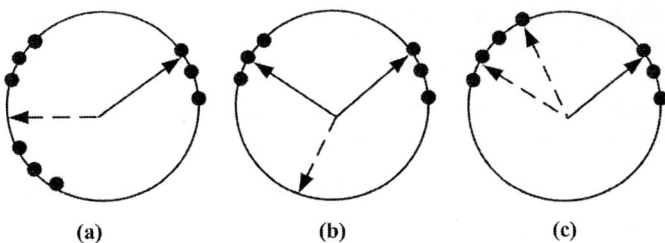

(a) (b) (c)

Fig. 1. Problems of instability in competitive learning. The spots are the input vectors, and the pointing arrows are the weight vectors. During the competitive learning process, the winning arrow tends to approach to the input vector. Dashed arrows are in problem status. (a)Oscillate. Once a weight vector reaches the middle of two clusters, the weight vector tends to oscillate between these clusters and never come to a stable state. (b)Learn nothing. Occasionally, the initial weight of a neuron is located so far from any input vectors that it never wins the competition, and therefore hardly learns during the training process. As a result, these neurons have no meaning in representative (c)Be redundant. Although having chance to win the competition, too many neurons will degrade not only the learning performance but also the visual effect

1. Moving - neurons with similar weights tend to move closed to each other.
2. Vanishing - 'dummy' neurons that rarely learn from the input patterns vanish (or are removed) from the network in the training phase.
3. Merging - neurons that have very close weight and position vectors combine into a single unit.
4. Splitting - neuron that frequently wins the competitions during the learning process cleaves into several to moderate the learning frequency.

4 Training Algorithm of ASOM

The network topology is initialized as $N \times N$ rectangular grid, where each neuron, i, is attached with a weight vector, w_i, and a position vector, p_i. During the organizing process, the learning rule for the weight vectors w_is is [5]:

$$w_i(t+1) = w_i(t) + \eta(t)g(t)[x - w_i(t)], \qquad (7)$$

where $\eta(t) = \eta_w/(1+t)$ and $g(t) = exp\{-s_w(1 + t/t_{max})D(p_i(t), p_c(t))\}$. η_w is the initial learning rate, and s_w is a parameter to regulate how fast the function $g(t)$ decreases. The learning rule for the position vectors adopts Equation (3).

Furthermore, the evolutionary actions of neurons are triggered during a learning epoch if it meets the specific conditions. A neural unit will vanish, i.e. simply be removed, if the following condition holds:

$$win_{acc} < win_{min} \times N\_Inner \times Unit_{num}, \qquad (8)$$

where win_{acc} is the winning accumulator for a neural unit, win_{min} is a scalar parameter, N_Inner is the number of inner iterations and $Unit_{num}$ is the total number of units still alive.

The splitting action is being controlled by the following condition:

$$win_{acc} > win_{max} \times N\_Inner \times Unit_{num}, \qquad (9)$$

where win_{max} is a predetermined scalar parameters. The jth generated neurons are initialized by the following equations, for $j = 1, 2...J$,

$$p_j^x = p_i^x + 2R_1(rand() - 0.5) \quad and \quad p_j^y = p_i^y + 2R_1(rand() - 0.5) \qquad (10)$$

$$w_j^k = w_i^k + 2R_2(rand() - 0.5), \quad k = 1, 2...K, \qquad (11)$$

where J is the number of neurons generated, R_1 and R_2 are the splitting radii for position vectors and weight vectors. p_j^x and p_j^y denote the x and y components of the position vector, while w_j^k is the kth component of the weight vector for the jth generated neuron. K is the dimension of weight vector. The function $rand()$ generates a rand number uniformly distributed on the interval $(0,1)$.

Neurons merging process is taken place when the following condition holds:

$$D(p_i, p_j) < \varepsilon \quad and \quad D(w_i, w_j) < \delta, \qquad (12)$$

where ε and δ define the distance threshold between two neighboring neural units i and j. The weight and position vectors for the new unit are simply the average of those of neuron i and j.

In addition, all of the described evolutionary actions are constrained by the following stochastic condition:

$$rand() < p_a, \quad p_a \in (0, 1), \qquad (13)$$

where p_a is a predetermined constant. This stochastic condition is introduced to make the neurons evolve in a more natural way. Here we give the pseudo code of the learning algorithm for this ASOM model.

Learning algorithm of ASOM

```
Initialize global parameters;
Do N_outer times {
    Do N_inner times {
        For each randomly selected input pattern vector {
            Determine the winning unit;
            Update the weight vectors;
            Update the position vectors;
            Collect the statistical information;
        }
    }
    For each neural unit
        It vanishes, if Equation (8) and (13) hold;
    For each neural unit
        It splits, if Equation (9) and (13) hold;
    For each two neural units
        They merge, if Equation (12) and (13) hold;
}
```

5 Experimental Results

In order to demonstrate the performance ASOM, Iris plants data set has been used in the experiment [8]. This is one of the best-known database to be found in the pattern recognition literature. It contains three classes of data set with 50 instances each, where each class refers to a type of iris plant. Each sample is described by 4 numeric attributes. In our experiment, DSOM is to be used for comparison. We ran ASOM and DSOM under the same initial conditions as ASOM. Fig. 2 illustrates the projection results by both methods.

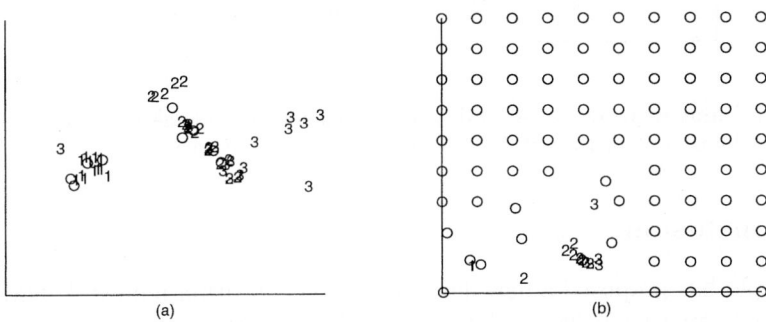

Fig. 2. Projection results of Iris Data Set produced by ASOM(a) and DSOM(b). Having finished the learning of network, all the samples are input to the network one by one. We attach the wining neural units with the class label of input sample. The neural units which haven't won are denoted by 'o's. All the parameter values used are: Neural Units, 100; N_Outer, 5; N_Inner, 3; p_a, 0.5; η_w, 0.2; s_w, 1; η_p, 0.1; s_p, 1; s_x, 3; win_{min}, 0.01; win_{max}, 0.1; R_1, 0.05; R_2, 0.05; J, 4; ϵ, 0.03; and δ, 0.03

It is obvious that DSOM retains a number of unlearned neurons in the plot. It is because the initial weight vectors of these units locate far from the input patterns that they never win the competition, and therefore hardly learn and adjust their positions in output space. On the contrary, these redundant neurons are eliminated from the network in ASOM algorithm. Also in DSOM, as the learning proceeds, neurons that have similar weight vectors are approaching to each other and finally converge at a certain point tightly. In contrast, this problem has been overcomed by the evolutionary actions in our ASOM algorithm.

From the projection result of ASOM, it is clear that class 2 and class 3 are quite closed to each other, and the class 1 is completely isolated. This result is well consistent to the prior knowledge: one class is linearly separable from the other 2; the latter are not linearly separable from each other [8].

Another advantage of ASOM is that it can give the sequential training status of the network (See Fig. 3). Initially, neurons are fixed in lattices. At the beginning of the learning process, the neurons are very active and have a relatively bigger change in the network structure (the first three frames). Then gradually, the network comes to stable as the process proceeds (the last three frames).

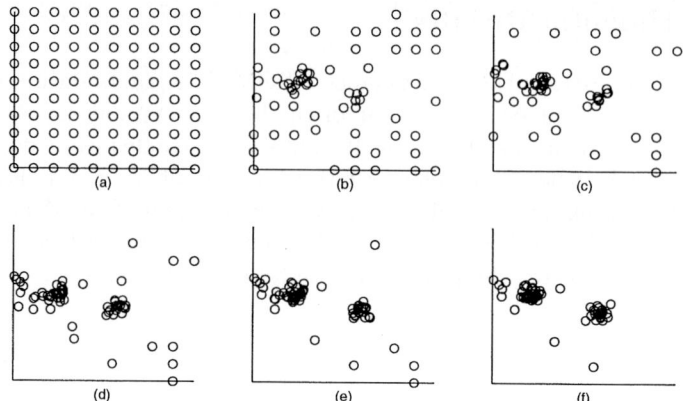

Fig. 3. Training sequences of ASOM model. Each subplot associates with a status after one outer loop in algorithm of ASOM. All the neural units are denotes by 'o'

6 Conclusion

This paper has proposed a new data projection algorithm named ASOM. It has integrated some adaptive working mechanisms of neurons, where the neurons are allowed to move, vanish, split and merge when responding to the external stimulus. The experiments showed the good performance of ASOM.

Acknowledgements. This work was supported by National High Technology Research and Development Program of China (863 Program) under contract No.2001AA114190.

References

1. Kohonen, T.: Self-Organizing Maps. 3rd edn. Springer-Verlag, Berlin (2001)
2. Kraaijveld, M.A., Mao, J., Jain, A.K.: A Nonlinear Projection Method Based an Kohonen's Topology Preserving Maps. IEEE Trans. Neural Networks. Vol. 6 (1995) 548-559
3. Rubner, J., Schlten, P.: Development of Feature Detectors by Self-Organization. Biol. Cybern. Vol. 62 (1990) 193-199
4. Kohonen, T., Oja, E., Simula, O., Visa, A., Kangas, J.: Engineering Application of the Self-Organizing Map. Proc.IEEE. Vol. 84(10) (1996) 1358-1383
5. Su, M.C., Chang, H.T.: A New Model of Self-Organizing Neural Networks and Its Application in Data Projection. IEEE Trans. Neural Networks. Vol. 12(1) (2001) 153-158
6. Murtagh, F.: Interpreting the Kohonen Self-Organizing Feature Map Using Contiguity Constrained Clustering. Pattern Recognition Lett. Vol. 16 (1995) 399-408
7. Merkl, D., Rauber, A.: Alternative Ways for Cluster Visualization in Self-Organizing Maps. Proc. Workshop on Self-Organizing Maps (1997) 106-111
8. http://www.ics.uci.edu/~mlearn/MLRepository.html (1998)

Evolving Flexible Neural Networks Using Ant Programming and PSO Algorithm

Yuehui Chen, Bo Yang, and Jiwen Dong

School of Information Science and Engineering
Jinan University, Jinan 250022, P.R.China
yhchen@ujn.edu.cn

Abstract. A flexible neural network (FNN) is a multilayer feedforward neural network with the characteristics of: (1) overlayer connections; (2) variable activation functions for different nodes and (3) sparse connections between the nodes. A new approach for designing the FNN based on neural tree encoding is proposed in this paper. The approach employs the ant programming (AP) to evolve the architecture of the FNN and the particle swarm optimization (PSO) to optimize the parameters encoded in the neural tree. The performance and effectiveness of the proposed method are evaluated using time series prediction problems and compared with the related methods.

1 Introduction

Designing of a neural networks for a given task usually suffers from the following difficulties: (1)choosing the appropriate architecture; (2)determining the connection ways between the nodes; (3)selecting the activation functions; and (4)finding a fast convergent training algorithm.

Many efforts have been made to construct a neural network for a given task in different application areas. A recent survey can be found in [8]. These approaches can be summaried as follows:

o Optimization of weight parameters, i.e., gradient descent method, evolutionary algorithm, tabu search, random search, etc.;
o Optimization of architectures, i.e., constructive and pruning algorithms, genetic algorithm, genetic programming, etc.;
o Simultaneously optimization of the architecture and parameters, i.e. EPNet [8], Neuroevolution [5] etc.;
o Optimization of learning rules.

Evolving architecture and parameters of a higher order Sigma-Pi neural network based on a sparse neural tree encoding has been proposed in [9]. Where only Σ and Π neurons are used and there are no activation functions used for the neurons. A recent approach for evolving the neural tree model based on probabilistic incremental program (PIPE) and random search algorithm has been proposed in our previous work [3].

In this paper, a new approach for evolving flexible neural network based on a neural tree encoding is proposed. The approach employs the ant programming [1] to evolve the architecture of the FNN and the particle swarm optimization to optimize the parameter encoded in the neural tree.

The paper is organized as follows: Section 2 gives the encoding method and evaluation of the FNN. A hybrid learning algorithm for evolving the FNN is given in Section 3. Section 4 presents some simulation results for the time series forecasting problems. Finally in section 5 we present some concluding remarks.

2 Encoding and Evaluation

Encoding. A tree-structural based encoding method with specific instruction set is selected for representing a FNN in this research. The reason for choosing this representation is that the tree can be created and evolved using the existing or modified tree-structure-based approaches, i.e., Genetic Programming (GP), PIPE algorithms, Ant Programming (AP).

The used instruction set for generating the FNN tree is as follows,

$$I = \{+_2, +_3, \ldots, +_N, x_1, x_2, \ldots, x_n\}, \tag{1}$$

where $+_i (i = 2, 3, \ldots, N)$ denote non-leaf nodes' instructions and taking i arguments. $x_1, x_2, \ldots, x_n$ are leaf nodes' instructions and taking no argument each. In addition, the output of each non-leaf node is calculated as a single neuron model. For this reason the non-leaf node $+_i$ is also called a i-inputs neuron instruction/operator. Fig.1 (left) shows tree structural representation of a FNN.

In the creation process of neural tree, if a nonterminal instruction, i.e., $+_i (i = 2, 3, 4, \ldots, N)$ is selected, i real values are randomly generated and used for representing the connection strength between the node $+_i$ and its children. In addition, two parameters a_i and b_i are randomly created as flexible activation function parameters and attach them to node $+_i$. In this study the used flexible activation function is described as

$$f(a_i, b_i, x) = e^{-(\frac{x-a_i}{b_i})^2}. \tag{2}$$

Evaluation. The output of a FNN can be calculated in a recursive way. For any nonterminal node, i.e., $+_i$, the total excitation is calculated as

$$net_i = \sum_{j=1}^{i} w_j * y_j \tag{3}$$

where $y_j (j = 1, 2, \ldots, i)$ are the input to node $+_i$. The output of the node $+_i$ is then calculated by

$$out_i = f(a_i, b_i, net_i) = e^{-(\frac{net_i - a_i}{b_i})^2}. \tag{4}$$

Thus, the overall output of flexible neural tree can be computed from left to right by depth-first method, recursively.

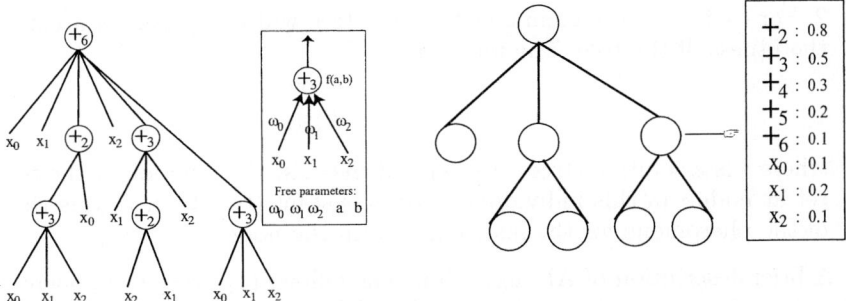

Fig. 1. Left: Tree-structural representation of an example FNN with instruction set $I = \{+_2, \ldots, +_6, x_0, x_1, x_2\}$. Right: Pheromone tree, in each node a pheromone table holds the quantity of pheromone associated with all possible instructions.

Objective function. In this work, the fitness function used for AP and PSO is given by mean square error (MSE):

$$Fit(i) = \frac{1}{P} \sum_{j=1}^{P} (y_1^j - y_2^j)^2 \qquad (5)$$

where P is the total number of training samples, y_1^j and y_2^j are the actual and model outputs of j-th sample. $Fit(i)$ denotes the fitness value of i-th individual.

3 An Approach for Evolving a FNN

Ant Programming for evolving the architecture of FNN. Ant programming is a new method which applies the principle of the ant systems to automated program synthesis [1]. In the AP algorithm, each ant will build and modify the trees according to the quantity of pheromone at each node. The pheromone table appears as a tree. Each node owns a table which memorize the rate of pheromone to various possible instructions (Fig.1 (right)).

First, a population of programs are generated randomly. The table of pheromone at each node is initialized at 0.5. This means that the probability of choosing each terminal and function is equal initially. The higher the rate of pheromone, the higher the probability to be chosen. Each program (individual) is then evaluated using a predefined objective function. The table of pheromone is update by two mechanisms:

- 1. Evaporation decreases the rate of pheromone table for every instruction on every node according to following formula :

$$P_g = (1 - \alpha) P_{g-1} \qquad (6)$$

where P_g denotes the pheromone value at the generation g, α is a constant ($\alpha = 0.15$).

- 2. For each tree, the components of the tree will be reinforced according to the fitness of the tree. The formula is:

$$P_{i,s_i} = P_{i,s_i} + \frac{\alpha}{Fit(s)} \quad (7)$$

where s is a solution (tree), $Fit(s)$ its fitness, s_i the function or the terminal set at node i in this individual, α is a constant ($\alpha = 0.1$), P_{i,s_i} is the value of the pheromone for the instruction s_i in the node i.

A brief description of AP algorithm is as follows:(1) every component of the pheromone tree is set to an average value; (2) random generation of tree based on the pheromone tree; (3) evaluation of ants using Eqn.(5); (4) update of the pheromone table according to Eqn.(6) and Eqn.(7); (5) go to step (1) unless some criteria is satisfied.

Parameter optimization with PSO. For the parameters optimization of FNN, a number of global and local search algorithms, i.e., GA, EP, gradient based learning method can be employed. The basic PSO algorithm is selected for parameter optimization due to its fast convergence and ease to implementation.

The PSO [6] conducts searches using a population of particles which correspond to individuals in evolutionary algorithm (EA). A population of particles is randomly generated initially. Each particle represents a potential solution and has a position represented by a position vector $\mathbf{x_i}$. A swarm of particles moves through the problem space, with the moving velocity of each particle represented by a velocity vector $\mathbf{v_i}$. At each time step, a function f_i (Eqn.(5) in this study) representing a quality measure is calculated by using $\mathbf{x_i}$ as input. Each particle keeps track of its own best position, which is associated with the best fitness it has achieved so far in a vector $\mathbf{p_i}$. Furthermore, the best position among all the particles obtained so far in the population is kept track of as $\mathbf{p_g}$. In addition to this global version, another version of PSO keeps track of the best position among all the topological neighbors of a particle.

At each time step t, by using the individual best position, $\mathbf{p_i(t)}$, and the global best position, $\mathbf{p_g(t)}$, a new velocity for particle i is updated by

$$\mathbf{v_i(t+1)} = \mathbf{v_i(t)} + c_1\phi_1(\mathbf{p_i(t)} - \mathbf{x_i(t)}) + c_2\phi_2(\mathbf{p_g(t)} - \mathbf{x_i(t)}) \quad (8)$$

where c_1 and c_2 are positive constant and ϕ_1 and ϕ_2 are uniformly distributed random number in $[0,1]$. The term $\mathbf{v_i}$ is limited to the range of $\pm\mathbf{v_{max}}$. If the velocity violates this limit, it is set to its proper limit. Changing velocity this way enables the particle i to search around its individual best position, $\mathbf{p_i}$, and global best position, $\mathbf{p_g}$. Based on the updated velocities, each particle changes its position according to the following equation:

$$\mathbf{x_i(t+1)} = \mathbf{x_i(t)} + \mathbf{v_i(t+1)}. \quad (9)$$

The proposed learning algorithm. The general learning procedure for the optimal design of the FNN can be described as follows.

1) Create the initial population randomly (FNNs and their corresponding parameters);

Fig. 2. Case 1: The structure of optimized FNN for prediction of Box and Jenkins data with 2-inputs (left), and with 10-inputs (right).

2) Structure optimization by AP algorithm.
3) If the better structure is found, then go to step 4), otherwise go to step 2);
4) Parameter optimization by PSO algorithm. In this stage, the tree structure is fixed, and it is the best tree taken from the end of run of the structure search. All of the parameters encoded in the best tree formulated a parameter vector to be optimized by PSO;
5) If the maximum number of PSO search is reached, or no better parameter vector is found for a significantly long time (say 100 steps for maximum 2000 steps) then go to step 6); otherwise go to step 4);
6) If satisfied solution is found, then stop; otherwise go to step 2).

4 Case Studies

Developed the FNN is applied to a time-series prediction problem: Box and Jenkins time series [2]. The used population sizes for AP and PSO are 150 and 30, respectively.

Case 1. The inputs of the prediction model are $u(t-4)$ and $y(t-1)$, and the output is $y(t)$. 200 data samples are used for training and the remaining data samples are used for testing the performance of the evolved model. The used instruction sets for creating the FNN model is $I = \{+_2, \ldots, +_8, x_0, x_1\}$. Where x_0 and x_1 denotes the input variables $u(t-4)$ and $y(t-1)$, respectively.

The results were obtained from training of the FNN with 20 different experiments. The average MSE value for training and test data sets are 0.000680 and 0.000701, respectively. The optimal structure of the evolved FNN is shown in Fig.2 (left).

Case 2. For the second simulation, 10 inputs variables are used for constructing the FNN in order to test the input-selection ability of the algorithm. The used instruction sets for creating the FNN is $I = \{+_2, \ldots, +_8, x_0, x_1, \ldots, x_9\}$. Where $x_i (i = 0, 1, \ldots, 9)$ denotes $u(t-6), u(t-5), u(t-4), u(t-3), u(t-2), u(t-1)$ and $y(t-1), y(t-2), y(t-3), y(t-4)$, respectively.

The average MSE value for training and test data sets are 0.000291 and 0.000305, respectively. The optimal structure of FNN is shown in Fig. 2 (right). From the figure, it is can be seen that the FNN model has some capability to

Table 1. Comparative results of different modelling approaches

Model name and reference	Number of inputs	MSE
ARMA [2]	5	0.71
FuNN model [7]	2	0.0051
ANFIS model [4]	2	0.0073
Case 1	2	0.00068
Case 2	10	0.00029

select the associated input variables to construct the FNN model. A comparison result of different methods for forecasting Jikens-Box data is shown in Table 1.

5 Conclusion

In this paper, a FNN model and its design method were proposed. A combined approach of AP to evolving the architecture and PSO to optimize the free parameters encoded in the neural tree was developed. Simulation results on time series prediction problem shown the feasibility and effectiveness of the proposed method. It should be noted that other tree-structure-based evolutionary algorithm and parameter optimization algorithm can also be employed to accomplish the same tasks.

References

1. Birattari, M., Di Caro, G., and Dorigo M. : Toward the formal foundation of Ant Programming. In Third International workshop, ANTS2002. LNCS 2463, 2002 188-201
2. Box, G. E. P. : Time series analysis, forecasting and control. San Francisco Holden Day (1970)
3. Chen, Y., Yang, B., Dong, J. Nonlinear systems modelling via optimal design of neural trees. International Journal of Neural ystems. **14**, (2004) 125-138
4. Jang, J.S. : (1997) Neuro-fuzzy and soft computing: a computational approach to learning and machine intelligence. Upper saddle River, NJ:prentice-Hall (1997)
5. Kenneth O. S., and Risto Miikkulainen: Evolving neural networks through augmenting topologies. Evolutionary Computation. **10**, (2002) 99-127
6. Kennedy, J. et al.,: Particle Swarm Optimization. Proc. of IEEE International Conference on Neural Networks. **IV** (1995) 1942-1948
7. Kasabov, K. et al., : FuNN/2 - A fuzzy neural network architecture for adaptive learning and knowledge acquisition. Information Science **101** (1997) 155-175
8. Yao, X.: Evolving artificial neural networks. Proceedings of the IEEE. **87**, (1999) 1423-1447
9. Zhang B.T., et al.: Evolutionary induction of sparse neural trees. Evolutionary Computation. **5**, (1997) 213-236

Surrogating Neurons in an Associative Chaotic Neural Network

Masaharu Adachi

Tokyo Denki University, Department of Electronic Engineering
2-2 Kanda-Nishiki-cho, Chiyoda-ku, Tokyo 101-8457, Japan
adachi@d.dendai.ac.jp

Abstract. A method of surrogate data, which is originally proposed for nonlinear time series analysis, is applied to an associative chaotic neural network in order to see which statistic of the deterministic chaos of the constituent neurons in the network for the dynamical association is important. The original associative network consists of 16 chaotic model neurons whose individuals may exhibit deterministic chaos by themselves. The associative network, whose synaptic weights are determined by a conventional auto–associative matrix of the three orthogonal patterns, shows chaotic retrievals of the stored patterns. The method of surrogation is applied to replace several neuronal sites by the surrogate data to see which statistic of the deterministic chaos of the constituent neurons is important to show the chaotic retrieval. The result shows that the auto–correlation of the time series of the output of the constituent neurons is important for maintaining the chaotic retrieval.

1 Introduction

A chaotic neural network model was proposed[1][2] and it has been applied to an associative memory[3] and to solving the optimization problems[4][5]. The network model consists of model neurons that exhibit deterministic chaos by themselves, namely, without connections to the other neurons in the network[1][2].

In the application of the network model to an associative memory, it was already reported that the network retrieves all the stored patterns with chaotic behavior[3], however, the role of the individual constituent neurons has not been clarified well. Therefore, in the present paper, we attempt to deal with checking the importance of the individual deterministic dynamics of the neuron in the network by applying the method of surrogate data[6] to the associative chaotic neural network. At first, run the complete associative chaotic neural network with the parameter so that the network shows chaotic retrieval of the stored patterns[3] in order to record the output time series data of the constituent neurons in the network. Next, the surrogate data of the output time series of several sites are generated by using the random shuffling or the phase randomizing method[6]. Then the network with replacing several sites with surrogate data is simulated and compared with the original network for the retrieval characteristics. Through the comparison we attempt to estimate the importance of the individual deterministic dynamics for the chaotic associative dynamics in the whole network.

2 Associative Chaotic Neural Network

The associative chaotic neural network to be investigated in the present paper consists of a chaotic neuron model[1][2] that exhibits deterministic chaos by itself. The synaptic weights of the associative chaotic neural network are determined by the conventional auto–associative matrix[7][8] of the stored pattern vectors.

The operation of the associative chaotic neural network model is represented by the following equations:

$$x_i(t+1) = f\{\eta_i(t+1) + \zeta_i(t+1)\}, \tag{1}$$

$$\eta_i(t+1) = k_f \eta_i(t) + \sum_{j=1}^{16} w_{ij} x_j(t), \tag{2}$$

$$\zeta_i(t+1) = k_r \zeta_i(t) - \alpha x_i(t) + a_i \tag{3}$$

where $x_i(t)$ denotes output of the ith neuron at discrete-time t. The variables $\eta_i(t)$ and $\zeta_i(t)$ denote internal states for feedback inputs from the constituent neurons and for the refractoriness, respectively. k_f and k_r are the decay parameters for the feedback inputs and the refractoriness, respectively. α denotes refractory scaling parameter. The parameters w_{ij} and a_i denote the synaptic weights from the jth neuron to the ith neuron and the sum of the threshold and the external inputs to the ith neuron ($a_i = a$ for every neuron in this paper), respectively. Output function of the neuron is denoted by f; in this paper, we use the logistic function represented by

$$f(y) = \frac{1}{1 + \exp(-y/\varepsilon)}, \tag{4}$$

where ε is a parameter for the steepness of the function[1][2]. We examine on the associative chaotic neural network with 16 chaotic neurons. The stored patterns for the network are three 16-dimensional binary patterns that are orthogonal with each other with average firing rate of each pattern is set to be equal to 0.5. Therefore, the synaptic weights are determined by the following equation[7][8].

$$w_{ij} = \frac{4}{3} \sum_{p=1}^{3} (x_i^{(p)} - \overline{x})(x_j^{(p)} - \overline{x}) \tag{5}$$

with $w_{ii} = 0$ where $x_i^{(p)}$ is the ith component of the pth stored pattern. $\overline{x}$ denotes spatially averaged value of the stored patterns.

It has been reported that the network exhibits chaotic sequential patterns that include the stored patterns when the parameters of the network are set to certain values[2] and it has also been reported that the network, as a whole, in such chaotic retrieval of the stored patterns shows orbital instability which implies deterministic chaos[3]. Since the constituent neuron of the network has an ability to exhibits deterministic chaos by itself, it is not clear whether such chaotic associative dynamics of the network is caused by the deterministic chaos of the constituent neurons or the mutual connections among them. Before addressing the problem, more primitive problem than it is that which statistic of the deterministic chaos of the constituent neurons is important for the chaotic retrieval. Therefore, in the present paper, we attempt to address the primitive problem by applying the method of surrogate data[6] to the analysis of the chaotic associative dynamics.

3 Surrogation Algorithms and Their Application to the Network

The method of surrogate data[6] was proposed originally for the nonlinear time series analysis. The method is useful for distinguishing time series data generated by deterministic chaos from the ones by stochastic processes. Several algorithms for generating surrogate data were proposed[6]. The following two algorithms are used for generating surrogate data in the present paper.

3.1 Random Shuffling

An algorithm called random shuffling, 'RS' for its abbreviation, is just randomly shuffling the time order of the original time series. The RS surrogate data made by the algorithm preserves empirical histogram of the original data but does not preserve the deterministic structure of the original data[6].

3.2 Phase Randomizing

The other algorithm used in the present paper is called phase randomizing, 'PR' for its abbreviation, and it is divided into the following three steps. (1) Compute power spectrum of the original time series by the Fourier transform. (2) Randomize the phase structure of the power spectrum. (3) The inverse Fourier transform is performed to generate surrogate data. The surrogate data by the algorithm preserves the power spectrum of the original data but the empirical histogram is not the same with the original one[6].

3.3 Surrogating Neurons in the Network

As it is already explained in the beginning of the section, the method of surrogate data is proposed originally for the nonlinear time series analysis. It means that the method is usually performed off line. However, in the present paper, we use the method on–line in some sense as follows. Our purpose to use the method is to see which statistic of the output of individual chaotic neuron in the associative chaotic neural networks is important. For the purpose we replace several neurons by surrogate data of the neurons' output that are stored from the original network. The procedures for the surrogation of the neurons in the network are summarized as follows.

(1) Run the original associative chaotic neural network for the predetermined duration L. In the present paper $L = 8192$ is used for every case. During the run, the outputs x_k (for $k \in S$; where S is a set of neuron index to be surrogated) of several sites are stored to make the surrogate data in the next procedure.

(2) Generate RS or PR surrogate of the stored output data in (1).

(3) Set the initial conditions as the same as in (1) and run the network with replacing the feedback input(s) x_k by the surrogate data s_k generated in (2) for several site and other neurons are preserved. The updating of the internal state η of Eq.(2) is replaced by the following equation.

$$\eta_i(t+1) = k_f \eta_i(t) + \sum_{j \in P} w_{ij} x_j(t) + \sum_{k \in S} w_{ik} s_k(t), \tag{6}$$

where, P denotes a set of neuron index to be preserved.

(4) Retrieval characteristics of the networks in (1) and in (3) are compared.

4 Results of the Surrogation

We make comparisons between the original network and the network with the surrogate neurons on the retrieval frequencies of the stored patterns. The effects of the surrogate neurons are examined for both the RS and PR surrogate algorithms. Figure 1 shows an example of the time series of the output x_1 of a neuron in the original network, the RS and PR surrogate data of x_1, respectively. Note that the RS surrogate holds the empirical histogram of the original, however, the PR surrogate does not. Therefore, we shift the PR surrogate to fit the average value of the original data.

The comparison on the retrieval frequencies among the original network and the networks with surrogate neurons are executed as follows.

(1) Count the retrieval frequency of the three stored patterns during the 8192 iterations of the original network. The frequency was 923 in the case of Fig.1.

(2) Count the retrieval frequency of the network with one to three RS surrogate neuron(s) during the 8192 iterations. For each number of surrogate neurons, 20 RS surrogate data are made and used as the output of the surrogate neurons. Then the retrieval frequencies are averaged over the 20 cases for each numbers of surrogate neurons. The retrieval of the stored pattern for the network with surrogate neurons is judged by the comparisons only on the preserved sites since we can not expect the retrieval in the surrogate site.

(3) Same experiment as (2) using PR surrogate algorithm is executed.

The results of the comparison are summarized in Table 1. The averaged retrieval frequencies and the ratio of the frequency to that of the original network(i.e., the averaged frequency divided by 923) are shown in the table. From the table we find that the network with the PR surrogate neurons keeps the retrieval better than that of RS neurons except for the case with 3 surrogate neurons. In this case the retrieval frequency of the RS neurons itself is larger than the PR, however, the network with the RS neurons converges to a periodic state with one stored pattern and it is far away from the chaotic retrieval. The result suggests that the power spectrum of the output of the constituent neurons is much important than the empirical histogram of them for maintaining the chaotic association of the network.

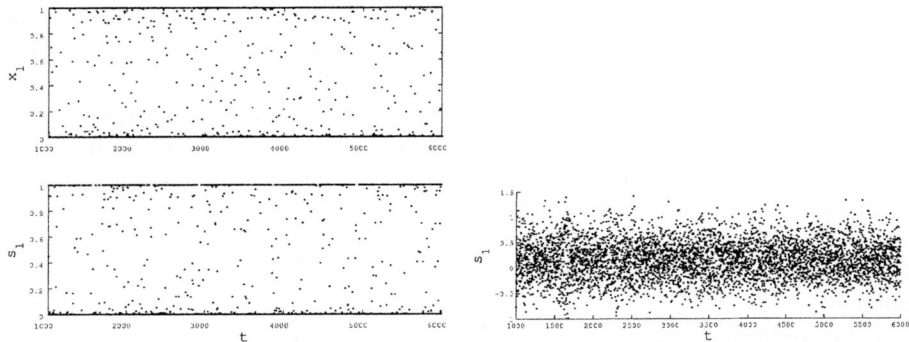

Fig. 1. Example of (top)the original time series of the output x_1 of a neuron in the network, (bottom left)the RS surrogate data of x_1 and (bottom right)the PR surrogate data.

Table 1. Comparison of the retrieval frequencies between RS and PR surrogate.

	RS surrogate		PR surrogate	
# of Surrogate Neurons	Freq.	Ratio (%)	Freq.	Ratio (%)
1	365.1	39.6	415.5	45.0
2	185.6	20.1	280.4	30.4
3	235.2	25.5	115.9	12.6

5 Conclusions and Discussions

In the present paper, a method of surrogate data, that was originally proposed for a nonlinear time series analysis, is applied to an analysis of an associative chaotic neural network. Surrogation of the neurons in the chaotic neural network that replacing some constituent neurons in the network by their surrogate data is proposed for the analysis. The retrieval frequencies of the original network and the network with two surrogate methods are compared. The results shows that the auto-correlation kept in the power spectrum with the phase randomizing is much important than the empirical histogram of the neuronal outputs of the original network.

The method of surrogate data was applied to see the effect of chaotic noise in the discrete-time Hopfield network for solving the traveling salesman problems[9]. They evaluate the solving ability of the Hopfield network with external input generated by chaotic time series and compare the ability with the network with the surrogate data of the chaotic time series. Their result shows that the auto-correlation kept in the surrogate of the chaotic time series is also important for solving the TSP.

Our surrogate data, that is generated from the data obtained form the complete network in the chaotic association, differs the case in the above investigation on the Hopfield network since they just applying chaotic time series and its surrogate data that are 'independent' from the network. However, it is worth to note that both approaches suggest the importance of the auto-correlation in the chaotic time series for both in chaotic association and in solving the optimization problems.

A future problem is applying the proposed method of surrogating neurons to the chaotic neural networks for solving the optimization problems.

References

1. Aihara, K., Takabe, T. and Toyoda M.: Chaotic Neural Networks. *Physics Letters A*, **144** (1990) 333–340
2. Aihara, K.: Chaotic Neural Networks. In: Kawakami, H. (ed.): Bifurcation Phenomena in Nonlinear Systems and Theory of Dynamical Systems. World Scientific, Singapore, (1990) 143–161
3. Adachi, M. and Aihara, K.: Associative Dynamics in a Chaotic Neural Network. Neural Networks, **10**, (1997) 83–98
4. Nozawa, H.: A Neural Network Model as a Globally Coupled Map and Applications Based on Chaos. Chaos, **2**, (1992) 377–386
5. Hasegawa, M., Ikeguchi, T. and Aihara, K.: Combination of Chaotic Neurodynamics with the 2-opt Algorithm to Solve Traveling Salesman Problems. Phys. Rev. Lett., **79**, (1997) 2344–2347
6. Theiler, J., Eubank, S., Longtin, A., Galdrikian, B. and Farmer, J.D.: Testing for Nonlinearity in Time Series: The Method of Surrogate Data. Physica D, **58** (1992) 77-94
7. e.g., Kohonen, T.: Correlation Matrix Memories. IEEE Trans., **C-21** (1972) 353–359
8. Hopfield, J.J.: Neural Networks and Physical Systems with Emergent Collective Computation Abilities. Proceedings of National Academy of Sciences, USA, **79**, (1982) 2445–2558
9. Hasegawa, M., Ikeguchi, T. and Aihara, K.: An Analysis on Additive Effects on Nonlinear Dynamics for Combinatorial Optimization. The Institute of Electronics, Information and Communication Engineers – IEICE Trans. Fundamentals, **E80-A** (1997) 206–213

Ensembles of RBFs Trained by Gradient Descent[1]

Carlos Hernández-Espinosa, Mercedes Fernández-Redondo,
and Joaquín Torres-Sospedra

Universidad Jaume I. Dept. de Ingeniería y Ciencia de los Computadores. Avda Vicente
Sos Baynat s/n. 12071 Castellon. Spain.
{espinosa, redondo}@icc.uji.es

Abstract. Building an ensemble of classifiers is an useful way to improve the performance. In the case of neural networks the bibliography has centered on the use of Multilayer Feedforward (MF). However, there are other interesting networks like Radial Basis Functions (RBF) that can be used as elements of the ensemble. Furthermore, as pointed out recently the network RBF can also be trained by gradient descent, so all the methods of constructing the ensemble designed for MF are also applicable to RBF. In this paper we present the results of using eleven methods to construct a ensemble of RBF networks. The results show that the best method is in general the *Simple Ensemble*.

1 Introduction

Probably the most important property of a neural network (NN) is the generalization capability. One method to increase this capability with respect to a single NN consist on training an ensemble of NNs, i.e., to train a set of NNs with different weight initialization or properties and combine the outputs in a suitable manner.

In the field of ensemble design, the two key factors to design an ensemble are how to train the individual networks and how to combine the different outputs. It seems clear from the bibliography that this procedure generally increases the generalization capability in the case of the NN Multilayer Feedforward (MF) [1,2].

However, in the field of NNs there are other networks besides MF, and traditionally the use of ensembles of NNs has restricted to the use of MF.

Another useful network which is quite used in applications is Radial Basis Functions (RBF). This network can also be trained by gradient descent [3]. So with a fully supervised training, it can be also an element of an ensemble.

2 Theory

In this section, first we briefly review the basic concepts of RBF networks and after that we review the different method of constructing the ensemble.

---

[1] This research was supported by the project MAPACI TIC2002-02273 of CICYT in Spain

2.1 RBF Networks with Gradient Descent Training

A RBF has two layer of networks. The first layer is composed of neurons with a Gaussian function and the second layer has neurons with a linear function. The output is:

$$F_k(x) = \sum_{q=1}^{Q} w_q^k \cdot \exp\left(-\sum_{n=1}^{N} \left(C_{q,n}^k - X_n\right)^2 \Big/ \left(\sigma_q^k\right)^2\right) \quad (1)$$

Where $C_{q,n}^k$ are the centers of the Gaussian units, σ_q^k control the width of the Gaussian functions and w_q^k are the weights among the Gaussian units and the output units.

The parameters which are changed during the training process [3] are $C_{q,n}^k$ and w_q^k. The equations for the adaptation of the weights is the following:

$$\Delta w_q^k = \eta \cdot \varepsilon_k \cdot \exp\left(-\sum_{n=1}^{N} \left(C_{q,n}^k - X_n\right)^2 \Big/ (\sigma)^2\right) \quad (2)$$

Where η is the step size and ε_k is the difference between the target and the output. And the equation for the adaptation of the centers is number 3.

$$\Delta C_q = \eta \cdot (X_k - C_q) \cdot \frac{2}{\sigma} \cdot \exp\left(-\sum_{n=1}^{N} \left(C_{q,n}^k - X_n\right)^2 \Big/ (\sigma)^2\right) \cdot \sum_{k=1}^{n_o} \varepsilon_k \cdot w_q^k \quad (3)$$

2.2 Ensemble Design Methods

Simple Ensemble. A simple ensemble can be constructed by training different networks with the same training set, but with different random initialization.

Bagging. This method is described in reference [4]. It consists on generating different datasets drawn at random with replacement from the original training set. After that, we train the different networks in the ensemble with these different datasets.

Bagging with Noise (BagNoise). It was proposed in [2]. It is a modification of Bagging, we use in this case datasets of size 10·N (N number of training patterns) generated like Bagging. Also we introduce a random noise in every selected training point.

Boosting. This ensemble method is reviewed in [4]. It is conceived for a ensemble of three networks. The first network is trained with the whole training set. After that, for the second network, we pass all patterns through the first network and we use a subset of them with 50% of patterns incorrectly classified and 50% correctly classified. Finally, the original patterns are presented to both networks. If the two networks disagree in the classification, we add the training pattern to the third training set.

CVC. It is reviewed in [1]. In k-fold cross-validation, the training set is divided into k subsets and k-1 subsets are used to train the network. Similarly, by changing the subset that is left out, we can construct k classifiers, with different training sets.

Adaboost. We have implemented the algorithm *"Adaboost.M1"* in the reference [5]. The successive networks are trained with a training set selected randomly from the original set, but the probability of selecting a pattern changes depending on the correct classification of the pattern and on the performance of the last trained network.

Decorrelated (Deco). This ensemble method was proposed in [6]. It consists on introducing a penalty term added to the error function. The penalty for network *j* is:

$$Penalty = \lambda \cdot d(i,j)(y - f_i) \cdot (y - f_j) \qquad (4)$$

Where λ is a parameter, y is the target of the pattern and f_i and f_j are the outputs of networks *i* and *j* in the ensemble. The term $d(i,j)$ is 1 for $i=j-1$ and 0 otherwise.

Decorrelated2 (Deco2). It was proposed also in reference [6]. It is basically the same method of *"Deco"* but with a different term $d(i,j)$, is 1 when $i=j-1$ and *i* is even.

Evol. This method was proposed in [7]. In each iteration (presentation of a training pattern), it is calculated the output of the ensemble by voting. If the output is correctly classified we continue with the next iteration and pattern. Otherwise, the network with an erroneous output and lower MSE is trained until the output is correct. This procedure is repeated for several nets until the vote correctly classifies the pattern.

Cels. It was proposed in [8]. This method also uses a penalty term added to error function. In this case the penalty term for network number *i* is in equation 5.

$$Penalty = \lambda \cdot (f_i - y) \cdot \sum_{j \neq i} (f_j - y) \qquad (5)$$

Ola. This ensemble method was proposed in [9]. First, several datasets are generated by using Bagging. Each network is trained with one of this datasets and with "virtual data". The virtual data is generated by selecting samples for the original training and adding a random noise. The target is calculated by the output of the ensemble.

3 Experimental Results

We have applied the eleven ensemble methods to nine problems from the UCI repository. Their names are Balance Scale (BALANCE), Cylinders Bands (BANDS), Liver Disorders (BUPA), Credit Approval (CREDIT), Glass Identification (GLASS), Heart Disease (HEART), the Monk's Problems (MONK'1, MONK'2) and Voting Records (VOTE). A full description can be found in the UCI repository (http://www.ics.uci.edu /~mlearn/ MLRepository.html).

We trained ensembles of 3 and 9 networks. We repeated this process of training an ensemble ten times for ten different partitions of data in training, cross-validation and test sets. In this way, we can obtain a mean performance of the ensemble for each database (the mean of the ten ensembles) and an error in the performance. The results are in table 1 and 2 for the case of ensembles of 3 networks and in table 3 and 4 for 9.

We have included also in table 1 and 2 the mean performance of a single network for comparison.

Table 1. Results for the ensemble of three networks

	BALANCE	BAND	BUPA	CREDIT	GLAS
Single Net.	90.2 ± 0.5	74.0 ± 1.1	70.1 ± 1.1	86.0 ± 0.8	93.0 ± 0.6
Adaboost	88.0 ± 1.2	73 ± 2	68.4 ± 0.5	84.7 ± 0.6	91.3 ± 1.3
Bagging	89.7 ± 0.8	73 ± 2	70.1 ± 1.6	87.4 ± 05.	93.8 ± 1.2
Bag_Noise	89.8 ± 0.8	73.1 ± 1.3	64 ± 2	87.1 ± 0.7	92.2 ± 0.9
Boosting	88.2 ± 0.8	70.7 ±1.8	70.6 ± 1.6	86.6 ± 0.7	92.2 ± 0.9
Cels	89.5 ± 0.8	75.3 ± 1.4	69.3 ± 1.4	86.9 ± 0.5	93.0 ± 1.0
CVC	90 ± 0.7	75.1 ± 1.4	69.9 ± 1.5	87.5 ± 0.5	92.4 ± 1.1
Decorrelated	89.8 ± 0.8	73.3 ± 1.7	71.9 ± 1.6	87.1 ± 0.5	93.2 ± 1.0
Decorrelated2	89.8 ± 0.8	73.8 ± 1.4	71.3 ± 1.4	87.2 ± 0.5	93.4 ± 1.0
Evol	89.5 ± 0.9	67.6 ± 1.3	63 ± 2	84.6 ± 1.1	88 ± 2
Ola	88.2 ± 0.9	73.1 ± 1.3	68.1 ± 1.5	86.1 ± 0.7	89.4 ± 1.8
Simple Ense	89.7 ± 0.7	73.8 ± 1.2	71.9 ± 1.1	87.1 ± 0.5	93.2 ± 1.0

Table 2. Results for the ensemble of three networks, continuation of Table 1

	HEART	MOK1	MOK2	VOTE
Single Net.	82.0 ± 1.0	98.5 ± 0.5	91.3 ± 0.7	95.4 ± 0.5
Adaboost	82.0 ± 1.4	88.0 ± 1.7	84.9 ± 1.5	95.1 ± 0.8
Bagging	83.6 ± 1.8	99.1 ± 0.6	88.0 ± 1.4	95.8 ± 0.6
Bag_Noise	83.2 ± 1.5	94.3 ± 1.3	89.5 ± 1.1	95.6 ± 0.7
Boosting	82.4 ± 1.2	94.9 ± 1.4	88.4 ± 1.0	95.9 ± 0.6
Cels	82.7 ± 1.7	94.3 ± 1.3	89.5 ± 1.3	94.9 ± 0.6
CVC	83.9 ± 1.6	95.5 ± 1.3	84.9 ± 1.7	95.9 ± 0.7
Decorrelated	84.1 ± 1.4	99.5 ± 0.4	89.8 ±1.3	96.0 ± 0.7
Decorrelated2	83.3 ± 1.4	99.4 ± 0.4	91.9 ± 1.2	96.1 ± 0.6
Evol	79 ± 2	74.8 ± 1.4	64.6 ± 1.6	92.3 ± 0.7
Ola	81.5 ± 1.2	75.1 ± 1.1	83.1 ± 1.1	96.1 ± 04.
Simple Ense	84.6 ± 1.5	99.6 ± 0.4	90.9 ± 1.1	96.4 ± 0.6

Table 3. Results for the ensemble of nine networks

	BALANCE	BAND	BUPA	CREDIT	GLAS
Single Net.	90.2 ± 0.5	74.0 ± 1.1	70.1 ± 1.1	86.0 ± 0.8	93.0 ± 0.6
Adaboost	91.8 ± 0.8	71.5 ± 1.2	69.6 ± 1.1	84.8 ± 0.8	93.1 ± 1.3
Bagging	90 ± 0.8	74.3 ± 1.4	71.0 ± 1.5	87.5 ± 0.5	93.6 ± 1.2
Bag_Noise	90 ± 0.8	73.1 ± 1.1	64.1 ± 1.9	87.1 ± 0.5	91.4 ± 0.9
Cels	89.7 ± 0.8	74.0 ± 1.2	69.4 ± 1.9	87.1 ± 0.5	92.4 ± 1.1
CVC	89.8 ± 0.8	73.6 ± 1.3	70 ± 2	87.6 ± 0.5	93.0 ± 1.1
Decorrelated	89.8 ± 0.8	73.5 ± 1.8	71.4 ± 1.4	87.2 ± 0.6	93.0 ± 1.0
Decorrelated2	89.8 ± 0.8	73.8 ± 1.8	71.6 ± 1.2	87.2 ± 0.5	93.2 ± 1.0
Evol	88.1 ± 1.1	67.6 ± 1.3	63 ± 2	83.4 ± 1.3	82.6 ± 1.8
Ola	88.5 ± 0.7	74.5 ± 1.2	69.6 ± 1.4	81.8 ± 1.1	90.6 ± 1.2
Simple Ense	89.7 ± 0.7	73.3 ± 1.4	72.4 ± 1.2	87.2 ± 0.5	93.0 ± 1.0

By comparing the results of tables 1 and 2 with the results of a single network we can see that the improvement of the ensemble is database and method dependent. Sometimes, the performance of an ensemble (as in case of Balance) is worse than the single network, the reason may be the combination method (*output averaging*) which does not exploit the performance of the individual networks. Besides that, there is one method which clearly perform worse than the single network which is *Evol*, but we obtained the same result for ensembles of Multilayer Feedforward networks.

Table 4. Results for the ensemble of nine networks, continuation of Table 3

	HEART	MOK1	MOK2	VOTE
Single Net.	82.0 ± 1.0	98.5 ± 0.5	91.3 ± 0.7	95.4 ± 0.5
Adaboost	81.0 ± 1.5	91.1 ± 1.4	87.0 ± 1.2	95.9 ± 0.6
Bagging	84.9 ± 1.2	99.4 ± 0.4	89.3 ± 1.2	95.9 ± 0.6
Bag_Noise	83.6 ± 1.6	95.5 ± 1.4	90.1 ± 1.2	96.0 ± 0.7
Cels	82.5 ± 1.4	93.3 ± 0.7	87.5 ± 1.2	94.9 ± 0.6
CVC	84.1 ± 1.3	99.4 ± 0.5	90.5 ± 1.0	96.1 ± 0.7
Decorrelated	84.1 ± 1.3	99.4 ± 0.4	90.3 ± 1.1	96.1 ± 0.7
Decorrelated2	84.7 ± 1.4	99.6 ± 0.4	91.1 ± 1.2	96.1 ± 0.6
Evol	78 ± 2	74.8 ± 1.4	64.6 ± 1.6	92.3 ± 0.7
Ola	78 ± 2	71.1 ± 1.6	82.0 ± 1.4	96.1 ± 0.4
Simple Ense	83.9 ± 1.5	99.6 ± 0.4	91.4 ± 1.2	96.3 ± 0.6

To see the results more clearly, we have also calculated the percentage of error reduction of the ensemble with respect to a single network as equation 6.

$$PorError_{reduction} = 100 \cdot \frac{PorError_{single\ network} - PorError_{ensemble}}{PorError_{single\ network}} \quad (6)$$

In this last equation, $PorError_{single\ network}$ is the error percentage of a single network (for example, 100-90.2=9.8% in the case of Balance, see table 1) and $PorError_{ensemble}$ is the error percentage in the ensemble with a particular method.

The value of this percentage ranges from 0%, where there is no improvement by the use of a ensemble method with respect to a *single network*, to 100% where the error of the ensemble is 0%. There can also be negative values, which means that the performance of the ensemble is worse than the performance of the *single network*.

This new measurement is relative and can be used to compare clearly the methods. In table 5 and 6 we have the results for the ensemble of 9 networks. Furthermore we have calculated the mean of the percentage of error reduction across all databases and is in the last column with header "Mean".

Table 5. Percentage of error reduction for the ensemble of 9 networks

	BALANCE	BAND	BUPA	CREDIT	GLAS
Adaboost	16.33	-9.62	-1.67	-8.57	1.43
Bagging	-2.04	1.15	3.01	10.71	8.57
Bag_Noise	-2.04	-3.46	-20.07	7.86	-22.86
Cels	-5.10	0	-2.34	7.86	-8.57
CVC	-4.08	-1.54	-0.33	11.43	0
Decorrelated	-4.08	-1.92	4.35	8.57	0
Decorrelated2	-4.08	-0.77	5.02	8.57	2.86
Evol	-21.42	-24.62	-23.75	-18.57	-148.57
Ola	-17.35	1.92	-1.67	-30	-34.29
Simple Ense	-5.10	-2.69	7.69	8.57	0

According to this mean measurement there are five methods which perform worse than the *single network*, they are *Adaboost, BagNoise, Cels, Evol* and *Ola*. The performance of *Evol* is clear, in all databases is worse. *Ola, Adaboost, BagNoise* and *Cels* are in general problem dependent (unstable).

The best and most regular method across all databases is the *Simple Ensemble*.

Finally, to see the influence of the number of networks in the ensemble we have obtained the results of the mean percentage of error reduction for ensembles of 3 and 9 networks, from the results is clear that in general there is an improvement in performance from the ensemble of three to the ensemble of nine networks.

Table 6. Percentage of error reduction for the ensemble of 9 networks, continuation of Table 5

	HEART	MOK1	MOK2	VOTE	MEAN
Adaboost	-5.56	-493.33	-49.43	10.87	-59.95
Bagging	16.11	60	-22.99	10.87	9.49
Bag_Noise	8.89	-200	-13.79	13.04	-25.82
Cels	2.78	-346.67	-43.68	-10.87	-45-18
CVC	11.67	60	-9.20	15.22	9.24
Decorrelated	11.67	60	-11.49	15.22	9.14
Decorrelated2	15	73.33	-2.30	15.22	12.53
Evol	-22.22	-1580	-306.90	-67.39	-245.94
Ola	-22.22	-1826.67	-106.90	15.22	-224.66
Simple Ense	10.56	73.33	1.15	19.57	12.56

4 Conclusions

In this paper we have presented results of eleven different methods to construct an ensemble of RBF networks, using nine different databases. The results showed that in general the performance is method and problem dependent, sometimes the performance of the ensemble is even worse than the *single network*, the reason can be that the combination method (*output averaging*) is not appropriate. The best and most regular performing method across all databases was the simple ensemble, i.e., the rest of methods proposed to increase the performance of MF seems not to be useful in RBF networks. Perhaps, another reason of this result in the combination method which may not be very appropriate as commented before, the future research will go in the direction of trying other combination methods with ensembles of RBF networks.

References

1. Tumer, K., Ghosh, J.: Error Correlation and Error Reduction in Ensemble Classifiers. Connection Science. **8** no. 3 & 4, (1996) 385-404
2. Raviv, Y., Intrator, N.: Bootstrapping with Noise: An Effective Regularization Technique. Connection Science. **8** no. 3 & 4, (1996) 355-372
3. Karayiannis, N.B., Randolph-Gips, M.M.: On the Construction and Training of Reformulated Radial Basis Function Neural Networks. IEEE Trans. On Neural Networks. **14** no. 4, (2003) 835-846
4. Drucker, H., Cortes, C., Jackel, D., et al.: Boosting and Other Ensemble Methods. Neural Computation. **6** (1994) 1289-1301
5. Freund, Y., Schapire, R.: Experiments with a New Boosting Algorithm. Proceedings of the Thirteenth International Conference on Machine Learning. (1996) 148-156
6. Rosen, B.: Ensemble Learning Using Decorrelated Neural Networks. Connection Science. **8** no. 3 & 4, (1996) 373-383
7. Auda, G., Kamel, M.: EVOL: Ensembles Voting On-Line. Proc. of the World Congress on Computational Intelligence. (1998) 1356-1360
8. Liu, Y., Yao, X.: A Cooperative Ensemble Learning System. Proc. of the World Congress on Computational Intelligence. (1998) 2202-2207
9. Jang, M., Cho, S.: Ensemble Learning Using Observational Learning Theory. Proceedings of the International Joint Conference on Neural Networks. **2** (1999) 1281-1286

Gradient Descent Training of Radial Basis Functions[1]

Mercedes Fernández-Redondo, Carlos Hernández-Espinosa,
Mamen Ortiz-Gómez, and Joaquin Torres-Sospedra

Universidad Jaume I, D. de Ingeniería y Ciencia de los Computadores, Avda. Vicente Sos
Baynat s/n, 12071 Castellón, Spain.
espinosa@icc.uji.es

Abstract. In this paper we present experiments comparing different training algorithms for Radial Basis Functions (RBF) neural networks. In particular we compare the classical training which consist of a unsupervised training of centers followed by a supervised training of the weights at the output, with the full supervised training by gradient descent proposed recently in same papers. We conclude that a fully supervised training performs generally better. We also compare *Batch training* with *Online training* of fully supervised training and we conclude that *Online training* suppose a reduction in the number of iterations and therefore increase the speed of convergence.

1 Introduction

A RBF has two layer of neurons. The first layer, in its traditional form, is composed of neurons with a Gaussian transfer function and the second layer has neurons with a linear transfer function. The output of a RBF can be calculated with equation 1 and 2.

$$\hat{y}_{i,k} = \mathbf{w}_i^T \cdot \mathbf{h}_k = \sum_{j=1}^{c} w_{ij} \cdot h_{j,k} \tag{1}$$

$$h_{j,k} = \exp\left(-\frac{\|\mathbf{x}_k - \mathbf{v}_j\|^2}{\sigma^2}\right) \tag{2}$$

Where v_j are the center of the Gaussian functions (GF), σ control the width of the GF and w_i are the weights among the Gaussian units (GU) and the output units.

As equations 1 and 2 show, there are three elements to design in the neural network: centers and widths of the GU and the weights among the GU and output units.

There are two different procedures to design the network. One is to train the networks in two steps. First we find the centers and widths by using same unsupervised clustering algorithm and after that we train the weights among hidden units and output units by a supervised algorithm. This process is usually fast, and the most important step is the training of centers and widths [1-4].

---

[1] This research was supported by the project MAPACI TIC2002-02273 of CICYT in Spain.

The second procedure is to train simultaneously the centers and weights in a full supervised fashion, similar to the algorithm Backpropagation (BP) for Multilayer Feedforward. This procedure was not traditionally used because it has the same drawbacks of BP, long training time and high computational cost. However, it has received quite attention recently [5-6].

In [5-6] it is used a sensitivity analysis to show that the traditional GU (called "exponential generator function") of the RBF network has low sensitivity for gradient descent training for a wide range of values of the widths, this parameter should be tuned carefully. As an alternate two different transfer functions are proposed, called in the papers "lineal generator function" and "cosine generator function" with a better sensitivity. Unfortunately, the experiments shown in the papers are performed with only two databases and the RBF networks are compared with equal number of GU. In our opinion the number of GU should be determined by trial and error and cross-validation, and it can be different for a traditional unsupervised training and a gradient descent training.

In contrast, in this paper we present more complete experiments and include in the experiments four traditional unsupervised training algorithms and a fully gradient descent training with the three transfer functions analysed in papers [5-6].

2 Theory

2.1 Training by Gradient Descent

"Exponential (EXP) Generator" Function. This RBF has the usual Gaussian transfer function described in equations 1 and 2. The equation for adapting the weights is:

$$\Delta \mathbf{w}_p = \eta \cdot \sum_{k=1}^{M} \varepsilon_{p,k}^0 \cdot \mathbf{h}_k \tag{3}$$

Where η is the learning rate, M the number of training patterns and $\varepsilon_{p,k}^0$ is the output error, the difference between the target of the pattern and the output.

The equation for adapting the centers is the following:

$$\Delta \mathbf{v}_q = \eta \cdot \sum_{k=1}^{M} \varepsilon_{p,k}^h \cdot (\mathbf{x}_k - \mathbf{v}_q) \tag{4}$$

Where η is the learning rate and $\varepsilon_{p,k}^h$ is the hidden error given by the following equations:

$$\varepsilon_{p,k}^h = \alpha_{q,k} \cdot \sum_{i=1}^{n_o} \varepsilon_{i,k}^0 \cdot w_{iq} \qquad \alpha_{q,k} = \frac{2}{\sigma^2} \cdot \exp\left(-\frac{\|\mathbf{x}_k - \mathbf{v}_q\|^2}{\sigma^2}\right) \tag{5}$$

In the above equations n_o is the number of outputs and these equation are for *Batch training*, the equations for *Online training* are evident.

"Lineal (LIN) Generator" Function. In this case the transfer function of the hidden units is the following:

$$h_{j,k} = \left(\frac{1}{\|\mathbf{x}_k - \mathbf{v}_j\|^2 + \gamma^2} \right)^{\frac{1}{m-1}} \qquad (6)$$

Where we have used $m=3$ in our experiments and γ is a parameter that should be determined by trial and error and cross-validation.

The above equations 3, 4 and 5 are the same, but in this case $\alpha_{q,k}$ is different and is given in the following equation:

$$\alpha_{q,k} = \frac{2}{m-1} \cdot \left(\|\mathbf{x}_k - \mathbf{v}_q\|^2 + \gamma^2 \right)^{\frac{m}{1-m}} \qquad (7)$$

"Cosine (COS) Generator" Function. In this case the transfer function is:

$$h_{j,k} = \frac{a_j}{\left(\|\mathbf{x}_k - \mathbf{v}_j\|^2 + a_j^2 \right)^{1/2}} \qquad (8)$$

Equations 3 and 4 are the same, but in this case the hidden error is different:

$$\varepsilon_{p,k}^h = \left(\frac{h_{j,k}^3}{a_j^2} \right) \cdot \sum_{i=1}^{no} \varepsilon_{i,k}^0 \cdot w_{iq} \qquad (9)$$

The parameter a_j is also adapted during training, the equation is the following:

$$\Delta a_j = \left(\frac{\eta}{a_j} \right) \cdot \sum_{i=1}^{no} h_{j,k} \cdot (1 - h_{j,k}^2) \cdot \varepsilon_{p,k}^h \qquad (10)$$

2.2 Training by Unsupervised Clustering

Algorithm 1. This training algorithm is the simplest one. It was proposed in [1]. It uses adaptive k-means clustering to find the centers of the GU.

After finding the centers, we should calculate the widths of the GU. For that, it is used a simple heuristic, we calculate the mean distance between one center and one of the closets neighbors, P, for example, the first ($P=1$), second ($P=2$), etc.

Algorithm 2. It is proposed in reference [2]. The GU are generated incrementally, in stages. A stage is characterized by a parameter δ that specifies the maximum radius for the hypersphere that includes the random cluster of points that is to define the GU, this parameter is successively reduced in every stage k. The GU at any stage are randomly selected, by choosing an input vector x_i from the training set and search for all other training vectors within the δ_k neighborhood of x_i. The training vector are used to define the GU (the mean is the center, and the standard deviation the width). The stages are repeated until the cross-validation error increases. The algorithm is complex and the full description can be found in the reference.

Algorithm 3. It is proposed in [3]. They use a one pass algorithm called *APC-III*, clustering the patterns class by class instead of the entire patterns at the same time.

The *APC-III* algorithm uses a constant radius to create the clusters, in the reference this radius is calculated as the mean minimum distance between training patterns multiplied by a constant α.

Algorithm 4. It is proposed in reference [4]. The GU are generated class by class, so the process is repeated for each class. In a similar way to algorithm 2 the GU are generated in stages. A stage is characterized by its majority criterion, a majority criterion of 60% implies that the cluster of the GU must have at least 60% of the patterns belonging to its class. The method will have a maximum of six stages, we begin with a majority criterion of 50% and end with 100%, by increasing 10%. The GU at any stage h are randomly selected, by picking a pattern vector x_i of class k from the training set and expand the radius of the cluster until the percentage of patterns belonging to the class falls below the majority criterion. To define the next GU another pattern x_i of class k is randomly picked from the remaining training set and the process repeated. The successive stage process is repeated until the cross-validation error increases. The algorithm is complex and the full description is in the reference.

3 Experimental Results

We have applied the training algorithms to nine different classification problems from the UCI repository of machine learning databases. They are Balance Scale (BALANCE), Cylinders Bands (BANDS), Liver Disorders (BUPA), Credit Approval (CREDIT), Glass Identification (GLASS), Heart Disease (HEART), the Monk's Problems (MONK'1, MONK'2) and Voting Records (VOTE). The complete data and a full description can be found in the UCI repository (http://www.ics.uci.edu/~mlearn/MLRepository.html).

The first step was to determine the appropriate parameters of the algorithms by trial and error and cross-validation. We have used an extensive trial and error procedure.

After that, with the final parameters we trained ten networks with different partition of the data in training set, cross-validation set and test set, also with different random initialization of centers and weights. With this procedure we can obtain a mean performance in the database (the mean of the ten networks) and an error.

These results are in Table 1, 2 and 3. We have for each database the mean percentage in the test and the mean number of clusters in the network.

Comparing the results of the same algorithm trained by gradient descent in the case of *Batch training* and *Online training*, we can see that the differences in performance are not significant. The unique two cases where there is a difference are in *EXP*, Mok1 (Batch= 94.7 and Online= 98.5) and in *LIN*, Mok2 (Batch= 82.8 and Online= 89.6). The fundamental difference between both training procedures is in the number of iterations and the value of the learning step. For example, 8000 iterations, $\eta=0.001$ in *EXP* Batch for Balance and 6000 iterations, $\eta=0.005$ in *EXP* Online.

Table 1. Performance of the different algorithms, Radial Basis Functions

DATABASE	TRAINING ALGORITHM							
	Exp Batch		Exp Online		Lineal Batch		Lineal Online	
	Perc.	Cluster	Perc.	Cluster	Perc.	Cluster	Perc.	Cluster
Balance	90.2±0.5	45	90.2±0.5	60	90.1±0.5	45	90.6±0.5	50
Band	74.1±1.1	110	74.0±1.1	40	74.5±1.1	30	73.4±1.0	35
Bupa	69.8±1.1	35	70.1±1.1	40	71.2±0.9	10	69.7±1.3	15
Credit	86.1±0.7	40	86.0±0.8	30	86.2±0.7	10	85.8±0.8	10
Glass	92.9±0.7	125	93.0±0.6	110	91.4±0.8	35	92.4±0.7	30
Heart	82.0±1.0	155	82.0±1.0	20	82.1±1.1	15	81.8±1.1	10
Monk1	94.7±1.0	60	98.5±0.5	30	93.2±0.7	15	94.5±0.7	15
Monk2	92.1±0.7	80	91.3±0.7	45	82.8±1.2	25	89.6±1.2	50
Vote	95.6±0.4	35	95.4±0.5	5	95.6±0.4	25	95.6±0.4	10

Table 2. Performance of the different algorithms, Radial Basis Functions (continuation)

DATABASE	TRAINING ALGORITHM							
	Cosine Batch		Cosine Online		UC Alg. 1		UC Alg. 2	
	Perc.	Cluster	Perc.	Cluster	Perc.	Cluster	Perc.	Cluster
Balance	89.9±0.5	25	90.0±0.7	40	88.5±0.8	30	87.6±0.9	88.5±1.6
Band	75.0±1.1	120	74.9±1.1	125	74.0±1.5	60	67±2	18.7±1.0
Bupa	69.9±1.1	15	70.2±1.1	40	59.1±1.7	10	57.6±1.9	10.3±1.5
Credit	86.1±0.8	10	86.1±0.8	25	87.3±0.7	20	87.5±0.6	95±14
Glass	93.5±0.8	105	92.6±0.9	15	89.6±1.9	100	79±2	30±2
Heart	82.1±1.0	25	81.9±1.1	15	80.8±1.5	100	80.2±1.5	26±4
Monk1	89.8±0.8	100	90.2±1.0	145	76.9±1.3	90	72±2	93±8
Monk2	87.9±0.8	125	86.6±1.1	45	71.0±1.5	90	66.4±1.7	26±4
Vote	95.6±0.4	20	95.4±0.4	10	95.1±0.6	40	93.6±0.9	53±5

Table 3. Performance of the different algorithms, Radial Basis Functions (continuation)

DATABASE	TRAINING ALGORITHM			
	UC Alg. 3		UC Alg. 4	
	Perc.	Cluster	Perc.	Cluster
Balance	88.0±0.9	94.7±0.5	87.4±0.9	45±7
Band	67±4	97.2±0.3	65.8±1.4	4.5±1.3
Bupa	60±4	106.2±0.3	47±3	11±5
Credit	87.9±0.6	161.10±0.17	86.4±0.9	32±4
Glass	82.8±1.5	59.9±0.7	81.2±1.8	22±2
Heart	72±4	71.8±0.6	78±3	10±2
Monk1	68±3	97.4±0.6	64±2	23±6
Monk2	66.5±0.8	143±0	71.6±1.5	20±2
Vote	94.1±0.8	120.30±0.15	76±5	5.0±1.1

Comparing *EXP*, *LIN* and *COS* generator functions, we can see that the general performance is quite similar and the differences are not significant except in the case Monk1 where the performance of *EXP* is clearly better. In other aspects, *EXP* and *LIN* functions need a higher number of trials for the process of trial and error to design the network (*EXP*, Band= 280 trials; *LIN*, Band= 280 trials; *COS*, Band= 112), because cosine generator functions adapt all parameters and the initialization is less important. But in contrast, the number of iterations needed to converge by *COS* functions is usually superior (*EXP*, Band= 10000 iterations; *LIN*, Band= 15.000; *COS*, Band= 75000), so globally speaking the computational cost can be considered similar.

Comparing unsupervised training algorithms among them, it seems clear that the classical algorithm 1, k-means clustering shows the better performance.

Finally, comparing unsupervised training with fully supervised training we can see that the best alternative (under the performance point of view) is supervised training, it achieves a better performance in databases Balance, Bupa, Glass, Heart, Monk1, and Monk2. In 6 of 9 databases. So, gradient descent is the best alternative.

In order to perform a further comparison, we include the results of Multilayer Feedforward with Backpropagaion in Table 4. We can see that the results of RBF are better. This is the case in all databases except Credit, Heart and Voting.

Table 4. Performance of Multilayer Feedforward with Backpropagation

DATABASE	Number of Hidden	Percentage
BALANCE	20	87.6±0.6
BANDS	23	72.4±1.0
BUPA	11	58.3±0.6
CREDIT	15	85.6±0.5
GLASS	3	78.5±0.9
HEART	2	82.0±0.9
MONK'1	6	74.3±1.1
MONK'2	20	65.9±0.5
VOTING	1	95.0±0.4

4 Conclusions

In this paper we have presented a comparison of unsupervised and fully supervised training algorithms for RBF networks. The algorithms are compared using nine databases. Our results show that the fully supervised training by gradient descent may be the best alternative under the point of view of performance. The results of RBF are also compared with the results of Multilayer Feedforward with Backpropagation, the performance of a RBF network is better. So we think it is a better alternative.

References

1. Moody, J., Darken, C.J.: Fast Learning in Networks of Locally-Tuned Processing Units. Neural Computation. **1** (1989) 281-294
2. Roy, A., Govil, S., et al.: A Neural-Network Learning Theory and Polynomial Time RBF Algorithm. IEEE Trans. on Neural Networks. **8** no. 6, (1997) 1301-1313
3. Hwang, Y., Bang, S.: An Efficient Method to Construct a Radial Basis Function Neural Network Classifier. Neural Network. **10** no. 8, (1997) 1495-1503
4. Roy, A., Govil, S., et al.: An Algorithm to Generate Radial Basis Function (RBF)-Like Nets for Classification Problems. Neural Networks. **8** no. 2, (1995) 179-201
5. Krayiannis, N.: Reformulated Radial Basis Neural Networks Trained by Gradient Descent. IEEE Trans. on Neural Networks. **10** no. 3, (1999) 657-671
6. Krayiannis, N., Randolph-Gips, M.: On the Construction and Training of Reformulated Radial Basis Functions. IEEE Trans. Neural Networks. **14** no. 4, (2003) 835-846

Recent Developments on Convergence of Online Gradient Methods for Neural Network Training*

Wei Wu[1], Zhengxue Li[1], Guori Feng[2], Naimin Zhang[1,3],
Dong Nan[1], Zhiqiong Shao[1], Jie Yang[1], Liqing Zhang[1], and Yuesheng Xu[4]

[1] Appl. Math. Dept., Dalian University of Technology, Dalian 116023, China
wuweiw@dlut.edu.cn
[2] Math. Dept., Shanghai Jiaotong University, Shanghai 200000, China
[3] Information Engineering College, Dalian University, Dalian 116622, China
[4] Syracuse University, New York State, USA

Abstract. A survey is presented on some recent developments on the convergence of online gradient methods for feedforward neural networks such as BP neural networks. Unlike most of the convergence results which are of probabilistic and non-monotone nature, the convergence results we show here have a deterministic and monotone nature. Also considered are the cases where a momentum or a penalty term is added to the error function to improve the performance of the training procedure.

Keywords: Convergence, Online gradient methods, BP Neural networks, Momentum term, Penalty term

In this paper, we consider the following feedforward neural networks

$$\zeta = g(w \cdot \xi), \ \zeta \in R, \xi \in R^N \tag{1}$$

or

$$\zeta = g(W \cdot G(V\xi)), \ \zeta \in R, \xi \in R^N \tag{2}$$

where g is a given smooth nonlinear activation function; ζ is the output of the network in response to the input ξ; $w \in R^N$, $W \in R^n$ and $V \in R^{n \times N}$ are the weights of the networks to be determined; n is the number of the hidden neurons; and

$$G(x) = (g(x_1, g(x_2), \cdots, g(x_n)) \in R^n, \forall x = (x_1, \cdots, x_n)^T \in R^n$$

In particular, the network (2) equipped with a back-propagation learning procedure is called a BP neural network ([8]).

The weights are obtained through a training iteration procedure using a given set of training samples $\{\xi^k, O^k\}_{k=1}^{J} \subset R^N \times R$. To this end, we choose an error

---

* Partly supported by the National Natural Science Foundation of China.

function $E(\omega)$ as follows, where ω denotes either w for Case (1), or W and V for Case (2).

$$E(\omega) = \frac{1}{2} \sum_{j=1}^{J} \left(O^j - \zeta^j\right)^2 \qquad (3)$$

The purpose of network training is to find a weight ω^* such that

$$E(\omega^*) = \min E(\omega) \qquad (4)$$

To solve this special optimization problem, the engineering community often prefers using the so-called *online gradient method* (OGM) (cf. [3,4,5]). So at the m-th step of the refinement of the present ω^m ($m = 1, 2, \cdots$), we choose an input example ξ^k, and accordingly define

$$\omega^{m+1} = \omega^m - \eta_m \frac{\partial((O^k - \zeta^k)^2/2)}{\partial \omega} \qquad (5)$$

and $\eta_m > 0$ is the learning rate which may depend on m. We observe that a summation in (5) over k from 1 to J will lead to the ordinary gradient method.

Usually the online gradient method chooses ξ^k from $\{\xi^k\}_{k=1}^{J}$ in a *completely stochastic* order (referred to as OGM-CS for short). For simplicity of analysis, we can choose the training examples in a *fixed* order $\{\xi^1, O^1, \xi^2, O^2, \cdots, \xi^J, O^J, \xi^1, O^1, \xi^2, O^2, \cdots\}$, and the corresponding online gradient method is referred to as OGM-F. We also consider a *partially stochastic* order (OGM-PS) where in the m-th cycle of iteration, the examples are arranged as $\{\xi^{m,1}, O^{m,1}, \xi^{m,2}, O^{m,2}, \cdots, \xi^{m,J}, O^{m,J}\}$ which is a random permutation of $\{\xi^1, O^1, \xi^2, O^2, \cdots, \xi^J, O^J\}$.

By now, most of the convergence results on OGM are of probabilistic nature (see [3,4,5]). Wu and Xu [12] gives a deterministic convergence result of OGM-F for input examples being linearly independent (so $J \leq N$). Wu, Feng and Li investigated OGM-F in [9], where the input examples are allowed to be linearly dependent, so the more important case $J > N$ is allowed. Now, we need the following assumptions.

(A1) The functions $|g(t)|$, $|g'(t)|$ and $|g''(t)|$ are uniformly bounded for $t \in R$.
(A2) The learning rate $\eta^m = \eta^{mJ+i} > 0$, for $i = 0, 1, \cdots, J-1$, is defined by

$$\frac{1}{\eta^m} = \frac{1}{\eta^{m-1}} + \beta, \; m = 1, 2, \cdots$$

where $\beta > 0$ and $\eta^0 > 0$ are suitably chosen constants (see [9]).

Theorem 1. [9] *If (A1) and (A2) are satisfied, then the online gradient method (5) for the network training of (1) is monotone and convergent in the following sense ($E_\omega(\omega)$ is the gradient of $E(\omega)$).*

$$E\left(\omega^{(m+1)J}\right) \leq E\left(\omega^{mJ}\right) \qquad (6)$$

$$\lim_{i \to \infty} \left\| E_\omega\left(\omega^i\right) \right\| = 0 \qquad (7)$$

This result is extended to OGM-PS in [6]. See [6] for the motivation for making such an extension.

Theorem 2. [6] *If (A1) and (A2) are satisfied, then OGM-PS for the network training of (1) is monotone and convergent in the sense of (6) and (7), respectively.*

[10] extends the result of Theorem 1 to BP neural networks (2), in which a new condition is needed:
(A3) The sequence $\|\omega^k\|$, $k = 1, 2, \ldots$, is bounded.

Theorem 3. ([10]) *If (A1), (A2) and (A3) are satisfied, then OGM-F for the network training of (2) is monotone and convergent in the sense of (6) and (7), respectively.*

The weights of the network may become very large in the training, causing difficulties in the implementation of the network by electronic circuits. The following error function with a penalty term is introduced in [13] to prevent this situation.

$$E(w) = \frac{1}{2}\sum_{j=1}^{J}((O^j - g(w \cdot \xi^j))^2 + \lambda(w \cdot \xi^j)^2), \quad \lambda > 0 \qquad (8)$$

Theorem 4. ([13]) *Assume that (A1) and (A2) are satisfied and that the error function (8) is used. Then OGM-F for the network training of (1) is monotone and convergent in the sense of (6) and (7) respectively, and $\|w\|$ is bounded.*

In [15], a momentum term is added in the following fashion to the updating of the weights in order to improve the stability of the learning procedure of (1):

$$w^{k+1} = w^k + \eta\sum_{j=1}^{J}(O^j - g(w^k \cdot \xi^j))g\prime(w^k \cdot \xi^j)\xi^j + \tau_k(w^k - w^{k-1}), \quad k = 1, 2, \cdots \qquad (9)$$

where the momentum factor τ_k is chosen to be

$$\tau_k = \begin{cases} \tau\|\sum_{j=1}^{J} g\prime(w^k \cdot \xi^j)\xi^j\|/\|\triangle w^k\|, & \text{if } \triangle w^k \neq 0 \\ 0, & \text{else} \end{cases} \qquad (10)$$

and τ is a suitable constant.

Theorem 5. ([15]) *Assume that (A1) is satisfied. Then the OGM-F with a momentum term as in (9) for the network training of (1) is monotone and convergent in the sense of (6) and (7), respectively.*

We mention that Theorems (4) and (5) have been extended to BP neural network (2) in our recent work ([14],[16]) under some stronger conditions.

References

1. Ellacott, S.W.: The Numerical Analysis Approach of Neural Networks. In: Taylor, J.G. (ed.): Mathematical Approaches to Neural Networks. North-Holland (1993) 103-138
2. Fine, T.L., Mukherjee, S.: Parameter Convergence and Learning Curves for Neural Networks. Neural Computation **11** (1999) 747-769
3. Finnoff, W.: Diffusion Approximations for the Constant Learning Rate Backpropagation Algorithm and Resistance to Locol Minima. Neural Computation **6** (1994) 285-295
4. Haykin, S.: Neural Networks. 2nd edn. Prentice Hall (1999)
5. Liang, Y.C. et. al.: Successive Approximation Training Algorithm for Feedforward Neural Networks. Neurocomputing **42** (2002) 11-322
6. Li, Z., Wu, W., Tian, Y.: Convergence of an Online Gradient Method for Feedforward Neural Networks with Stochastic Inputs. Journal of Computational and Applied Mathematics **163(1)** (2004) 165-176
7. Liu, W.B., Dai, Y.H.: Minimization Algorithms Based on Supervisor and Searcher Cooperation: I-Fast and Robust Gradient Algorithms for Minimization Problems with Strong Noise. Journal of Optimization Theory and Application **111** (2001) 359-379
8. Rumelhart, D.E., Hinton, G.E.: Learning Representations of Back-propagation Errors. Nature **323** (1986) 533-536
9. Wu, W., Feng, G.R., Li, X.: Training Multylayer Perceptrons via Minimization of Sum of Ridge Functions. Advances in Computational Mathematics **17** (2003) 331-347
10. Wu, W., Feng, G., Li, Z., Xu, Y.: Deterministic Convergence of an Online Gradient Method for BP Neural Networks. Accepted by IEEE Transactions on Neural Networks, 2004
11. Wu, W., Shao, Z.: Convergence of Online Gradient Methods for Continuous Perceptrons with Linearly Separable Training Patterns. Applied Mathematics Letters **16** (2003) 999-1002
12. Wu, W., Xu, Y.S.: Deterministic Convergence of an On-line Gradient Method for Neural Networks. Journal of Computational and Applied Mathematics **144**(2002) 335-347
13. Zhang, L., Wu, W.: Online Gradient Method with a Penalty Term for Feedforward Neural Networks. To appear
14. Zhang, L., Wu, W.: Online Gradient Method with a Penalty Term for BP Nneural Networks. To appear
15. Zhang, N., Wu, W., Zheng, G.: Convergence of Gradient Method with a Momentum Term for Feedforward Neural Networks. To appear
16. Zhang, N., Wu, W.: Convergence of Gradient Method with a Momentum Term for BP Neural Networks. To appear

A Regularized Line Search Tunneling for Efficient Neural Network Learning

Dae-Won Lee, Hyung-Jun Choi, and Jaewook Lee

Department of Industrial Engineering,
Pohang University of Science and Technology,
Pohang, Kyungbuk 790-784, Korea.
{woosuhan,chj,jaewookl}@postech.ac.kr

Abstract. A novel two phases training algorithm for a multilayer perceptron with regularization is proposed to solve a local minima problem for training networks and to enhance the generalization property of networks trained. The first phase is a trust region-based local search for fast training of networks. The second phase is an regularized line search tunneling for escaping local minima and moving toward a weight vector of next descent. These two phases are repeated alternatively in the weight space to achieve a goal training error. Benchmark results demonstrate a significant performance improvement of the proposed algorithm compared to other existing training algorithms.

1 Introduction

Many supervised learning algorithms of multilayer perceptrons (MLPs, for short) find their roots in nonlinear minimization algorithms. For example, error back-propagation, conjugate gradient, and Levenberg-Marquardt methods have been widely used and applied successfully to solve diverse problems such as pattern recognition, classification, robotics and automation, financial engineering, and so on [4]. These methods, however, have difficulties in finding a good solution when the error surface are very rugged since they often get trapped in poor sub-optimal solutions.

To overcome the problem of local minima and to enhance generalization capability, in this letter, we present a new efficient regularized method for MLPs and demonstrate its superior performance on some difficult benchmark neural network learning problems.

2 Proposed Method

The proposed method is based on viewing the supervised learning of a MLP as an unconstrained minimization problem with a regularization term as follows:

$$\min_{\mathbf{w}} E_\lambda(\mathbf{w}) = E_{train}(\mathbf{w}) + \lambda E_{reg}(\mathbf{w}), \tag{1}$$

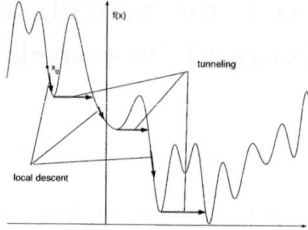

Fig. 1. Basic Idea of Tunneling Scheme

where $E_{train}(\cdot)$ is a training error cost function averaged over the training samples which is a highly nonlinear function of the synaptic weight vector $\mathbf{w}$ and $E_{reg}(\cdot)$ is a regularization term to smooth the networks (for example, $E_{reg}(\mathbf{w}) = \|\mathbf{w}\|^2$ is a weight decay term. See ([4]) for other regularization terms). The proposed training algorithm consists of two phases. The first phase employs a trust region-based local search to retain the rapid convergence rate of second-order methods in addition to the globally convergent property of gradient descent methods. The second phase employs a regularized line search tunneling to generate a sequence of weight vectors converging to a new weight vector with a lower minimum squared error (MSE), E_λ. The repeated iteration of these two phases alternatively forms a new training procedure which results in fast convergence to a goal error in the weight space. (See Figure 1.)

2.1 Phase I (Trust Region-Based Local Search)

The basic procedure of a trust region-based local search ([3]) adapted to Eq. (1) is as follows. For a given weight vector $\mathbf{w}(n)$, the quadratic approximation $\hat{E}$ is defined by the first two terms of the Taylor approximation to E_λ at $\mathbf{w}(n)$;

$$\hat{E}(\mathbf{s}) = E_\lambda(\mathbf{w}(n)) + \mathbf{g}(n)^T\mathbf{s} + \frac{1}{2}\mathbf{s}^T\mathbf{H}(n)\mathbf{s} \qquad (2)$$

where $\mathbf{g}(n)$ is the local gradient vector and $\mathbf{H}(n)$ is the local Hessian matrix. A trial step $\mathbf{s}(n)$ is then computed by minimizing (or approximately minimizing) the trust region subproblem stated by

$$\min_{\mathbf{s}} \hat{E}(\mathbf{s}) \quad \text{subject to} \quad \|\mathbf{s}\|_2 \leq \Delta_n \qquad (3)$$

where $\Delta_n > 0$ is a trust-region parameter. According to the agreement between predicted and actual reduction in the function E as measured by the ratio

$$\rho_n = \frac{E_\lambda(\mathbf{w}(n)) - E_\lambda(\mathbf{w}(n) + \mathbf{s}(n))}{\hat{E}(\mathbf{0}) - \hat{E}(\mathbf{s}(n))}, \qquad (4)$$

Δ_n is adjusted between iterations as follows:

$$\Delta_{n+1} = \begin{cases} \|\mathbf{s}(n)\|_2/4 & \text{if } \rho_n < 0.25 \\ 2\Delta_n & \text{if } \rho_n > 0.75 \text{ and } \Delta_n = \|\mathbf{s}(n)\|_2 \\ \Delta_n & \text{otherwise} \end{cases} \quad (5)$$

The decision to accept the step is then given by

$$\mathbf{w}(n+1) = \begin{cases} \mathbf{w}(n) + \mathbf{s}(n) & \text{if } \rho_n \geq 0 \\ \mathbf{w}(n) & \text{otherwise} \end{cases} \quad (6)$$

which means that the current weight vector is updated to be $\mathbf{w}(n) + \mathbf{s}(n)$ if $E_\lambda(\mathbf{w}(n) + \mathbf{s}(n)) < E_\lambda(\mathbf{w}(n))$; Otherwise, it remains unchanged and the trust region parameter Δ_n is shrunk and the trial step computation is repeated.

2.2 Phase II (Regularized Line Search Tunneling)

Despite of its rapid and global convergence properties ([3]), the trust region-based local search would get trapped at a local minimum, say $\mathbf{w}^*$. To escape from this local minimum, our proposed method, which we call a regularized line search tunneling, attempts to compute a weight vector, say $\hat{\mathbf{w}}$ of next descent, by minimizing a subproblem given by

$$\min_{t>0} E_\lambda(\mathbf{w}(t)) \quad (7)$$

where $\{\mathbf{w}(t) : t > 0\}$ is the solution trajectory of a tunneling dynamics described by

$$\frac{d\mathbf{w}(t)}{dt} = -\nabla E_{reg}(\mathbf{w}(t)), \quad \mathbf{w}(0) = \mathbf{w}^* \quad (8)$$

One distingushed feature of the proposed tunneling technique is that the obtained weight vector $\hat{\mathbf{w}}$ is located normally outside the convergent region of $\mathbf{w}^*$ with respect to the trust-region method of Phase I so that applying a trust-region local search to $\hat{\mathbf{w}}$ leads us to get another locally optimal weight vector. Another feature of the proposed method is that the value of E_{reg} becomes relatively small during the regularized line search tunneling in Eq. (8). Consequently, these features make it easier to find a new weight vector of next descent with a lower MSE, thereby enhancing generalization ability.

3 Simulation Results

To evaluate the performance of the proposed algorithm, we conducted experiments on some benchmark problems we have found in the literature. The neural network models for the applied benchmark problems (Iris, Sonar, 2D-sinc, and Mackey-Glass) are 4-6-3-1, 60-5-1, 2-15-1, and 2-20-1, respectively. Table 3 shows the performance of our proposed algorithm compared to error back-propagation

Table 1. Experimental Results results for the benchmark data

Benchmark		EBPR	DTR	BR	LMR	GA	SA	Proposed
	T	235	263	1.7	1.4	347	499	18.2
Iris	E1	1%	1%	1.2%	1.7%	4.9%	5.8%	1.1%
	E2	4%	4%	6%	4.6%	8.2%	11.2%	4%
	T	273	265	-	16.5	182.9	224.6	43.4
Sonar	E1	1.0%	0.0%	-	0.0%	7.95%	13.1%	0.0%
	E2	27.8%	27.8%	-	29.8%	36.2%	41.5%	25.0%
	T	2652	2843	11.8	10.0	1824	3966	12.9
2D-sinc	E1	0.0055	0.0052	0.0092	0.0054	0.0112	0.0287	0.0052
	E2	0.0078	0.0074	0.0120	0.0087	0.0142	0.0308	0.0073
	T	671	6365.3	156.2	3.6	4765.6	9454	15.0
Mackey-Glass	E1	0.5640	0.0664	0.483	0.6109	0.7313	0.7919	0.0418
	E2	0.5643	0.0668	0.597	0.6147	0.7235	0.7730	0.0428

based regularization (EBPR) [4], Dynamic tunneling based regularization (DTR) [5], Baysian Regularization (BR) [2], Levenberg-Marquardt based regularization (LMR) [4], genetic algorithm based network training (GA) and simulated annealing based network training (SA). Experiments were repeated a hundred times for every algorithm in order to decrease effects of randomly chosen initial weight vector. The criteria for comparison are the average time of training (T), the mean squared (or misclassification) training error(E1) and test error(E2). The results demonstrate that the new algorithm not only successfully achieves the goal training error and smaller test error for all these benchmark problems but also is substantially faster than these state-of-art methods.

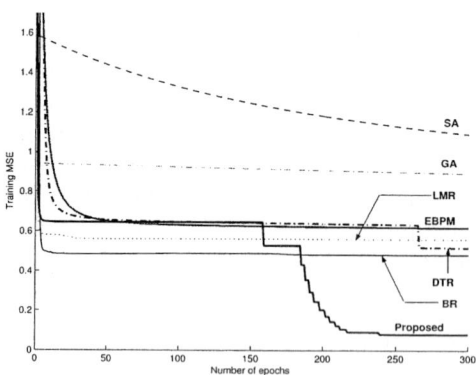

Fig. 2. Convergence curve for Mackey-Glass problem

4 Conclusion

In this paper, a new deterministic method for training a MLP has been developed. This method consists of two phases: Phase I for approaching a new local minimum in terms of a trust region-based local search and Phase II for escaping from this local minimum in terms of line search tunneling. Benchmark results demonstrate that the proposed method not only successfully achieves the goal training error but also is substantially faster than other existing training algorithms.

The proposed method has several features: First it does not require a good initial guess. Second, even in the complex network architecture it can bring appropriate tunneling directions against a trapped local minimum. Finally, weights converge relatively small value and reduce regularization error term. The robust and stable nature of the proposed method enable to apply it to various supervised learning problems. An application of the method to more large-scale benchmark problems remains to be investigated.

Acknowledgement. This work was supported by the Korea Research Foundation under grant number KRF-2003-041-D00608.

References

1. Barhen, J., Protopopescu, V., Reister, D.: TRUST: A Deterministic Algorithm for Global Optimization. Science, Vol. 276 (1997) 1094-1097
2. Foresee, F.D., Hagan, M.T.: Gauss-Newton Approximation to Bayesian Regularization. In Procedeengs, International Joint Conference on Neural Networks (1997) 1930-1935
3. Nocedal, J., Wright, S.J.: Numerical Optimization. Springer, New York (1999)
4. Haykin, S.: Neural Networks: A Comprehensive Foundation. Prentice-Hall, New York (1999)
5. Singh, Y.P., Roychowdhury, P.: Dynamic Tunneling Based Regularization in Feedforward Neural Networks. Artificial Intelligence, Vol. 131 (2001) 55-71

Transductive Learning Machine Based on the Affinity-Rule for Semi-supervised Problems and Its Algorithm

Weijiang Long and Wenxiu Zhang

Institute of Information and Systems,
Faculty of Sciences, Xi'an Jiaotong University, Xi'an 710049
wjlong@mailst.xjtu.edu.cn

Abstract. One of the central problems in machine learning is how to effectively combine unlabelled and labelled data to infer the labels of unlabelled ones. In this article, transductive learning machines are introduced based on a so-called affinity rule that if two objects are close in input space then their outputs should also be close, to obtain the solution of semi-supervised learning problems. The analytic solution for the problem and its iterated algorithm are obtained. Some simulations about pattern classification are conducted to demonstrate the validity of the proposed method in different situations. An incremental learning algorithm adapting to on-line data processing is also derived.

1 Introduction

One of the most important subjects in current data-mining research is semi-supervised learning in which some of the observations have been assigned and labelled by the supervisor, while the labels of others are not obtained for various reasons. We respectively call these two kinds of observations the labelled data and unlabelled data. The main problem to study is how to infer the proper label of the unlabelled data using relevant domain knowledge and the observations including labelled and unlabelled data. The classical method for solving this problem is so-called induction-deduction method in which the labelled data are first analyzed to find a generalized rule and regard this rule to be justified for future observations (that is, from particularity to generality), and then this general rule is applied to the unlabelled data to infer their labels (that is, from generality to particularity).

In recent years, however, the transductive method proposed by Vapnik[7], has gained much concern. The rationale of this method is that, if the problem we face is of a particular situation, there is no need for a universal rule since conclusion from particularity to particularity is not impossible. For the problem of semi-supervised learning, a general rule which is applied to both the unlabelled data and the possible other observations is indeed unnecessary in that only labelling those particular observations of unlabelled data is what we concern.

The realization of semi-supervised learning transduction method is quite skillful. It avoids solving a more difficult, general question which needs more evidence to support, but fully uses the information provided by the unlabelled data, and hence has many advantages. Up to now, there have been several examples of successful realization of transduction and experimentation on its superiority against traditional method. Chapell et al [4] implemented transductive inference by minimizing the leave-one-out error of ridge regression. Bennettet [2] introduced semi-supervised support vector machines (S^3VM) by overall risk minimization. Joachims [6] introduced transductive support vector machines (TSVM) to deal with text classification. Zhou [8] used objects programming involving norm to derive the solution of semi-supervised learning problems. However, it should be pointed out that is different from our paper. Zhou Guang-ya et al gave many affinity-measures and applying rules in [10] which can be used in our discussing on the machine learning problems.

In this paper, we realized the semi-supervised learning transduction method based on the affinity-rule by measuring the affinity of different objects in general space. The principle comes from an intuitive fact that similar objects should have similar outputs. Because of the loose assumptions, this method has wider applications, more concise solution, and more practical recursive algorithm. In fact, as the derivation originates from only the general concept of similarity comparing of the objects and affinity-rule, this method is one of the most applicable methods to various types of data, for example, to the generalized symbolic data.

2 Modelling

Let $\mathbf{X}$ be the input space and $\mathbf{Y}$ be the output space. The obtained data set is

$$\{(\mathbf{x}_1,\mathbf{y}_1),(\mathbf{x}_2,\mathbf{y}_2),\cdots,(\mathbf{x}_l,\mathbf{y}_l),\mathbf{x}_{l+1},\cdots,\mathbf{x}_{l+u}\},$$

$$\mathbf{x}_i \in \mathbf{X}, 1 \leq i \leq l+u; \mathbf{y}_i \in \mathbf{Y}, 1 \leq i \leq l, n = l+u,$$

where $L = \{(\mathbf{x}_1,\mathbf{y}_1),(\mathbf{x}_2,\mathbf{y}_2),\ldots,(\mathbf{x}_l,\mathbf{y}_l)\}$ is labelled data set and $U = \{\mathbf{x}_{l+1}, \mathbf{x}_{l+2},\ldots\mathbf{x}_{l+u}\}$ is unlabelled data set. $\{\mathbf{x}_1,\mathbf{x}_2,\ldots,\mathbf{x}_l\}$ are labelled objects and $\mathbf{y}_i$ is the label of $\mathbf{x}_i$. We aim at inferring the labels of unlabelled data. For convenience, we assume that the labels stand for the different classes and discuss the classification problem. Let the number of the classes be c and $\mathbf{e}_i$ be the vector with the i-th element being 1 and others being 0. Therefore $\mathbf{Y}$ may be taken as $\{\mathbf{e}_1,\mathbf{e}_2,...,\mathbf{e}_c\}$ in classification problem. In order to obtain the labels estimates of $\mathbf{x}_i$ for $i > l$, a general labelling variable $\mathbf{z}_i \in R^c$ is often considered, where $\mathbf{z}_i = (z_{i1}, z_{i2}, \cdots, z_{ic})^T$ is an auxiliary variable to simplify the complexity in solving this problem, and $\mathbf{z}_i$ is not necessary to hold the same form as $\mathbf{e}_i$. Then we use $\mathbf{y}_i = \mathbf{e}_{\arg\max\{z_{ik}; 1 \leq k \leq c\}}$ as the estimates of the labels. The following conditions are requested for $\mathbf{z}_i$. 1)If $\mathbf{x}_i$ is close to $\mathbf{x}_j$, then the general label $\mathbf{z}_i$ is also close to $\mathbf{z}_j$. 2)For the labelled data, the $\mathbf{z}_i$ and $\mathbf{y}_i$ are as close as possible. The measure of the closeness for two objects can be taken as various modes including the inclusion degree like that in [9], similarity (or dissimilarity, distance) etc..

The former measures have wider applications. For example, they can be used for the general symbolic data in the cases where even the symmetry does not hold. We use $s_{ij} = s(\mathbf{x}_i, \mathbf{x}_j)$ to measure the affinity between two input objects $\mathbf{x}_i$ and $\mathbf{x}_j$ where (s_{ij}) need not to be symmetric, but assume them to be positive. If the entries of (s_{ij}) are not all positive, we may consider $s_{ij}^* = s_{ij} + s, s > -\min\{s_{ij}; 1 \leq i, j \leq n\}$. Let $t_{ij} = t(\mathbf{z}_i, \mathbf{z}_j)$ be the affinity-measure between two outputs $\mathbf{z}_i$ and $\mathbf{z}_j$, and $f(v), g(v), h(v)$ be increasing functions of v. The idea based on the affinity-rule is that the greater the affinity-measure between $\mathbf{x}_i$ and $\mathbf{x}_j$, the smaller the $-f(s_{ij})$ as the dis-affinity-measure between $\mathbf{x}_i$ and $\mathbf{x}_j$, and the smaller the $g(t_{ij})$ dis-affinity-measure between $\mathbf{z}_i$ and $\mathbf{z}_j$; $t(\mathbf{z}_i, \mathbf{y}_i)$ should be as small as possible for $1 \leq i \leq l$. Therefore, we obtain a general framework as follows:

$$\max_{\mathbf{Z}_1, \mathbf{Z}_2, \ldots, \mathbf{Z}_n, \xi} -\sum_{i=1}^n \sum_{j=1}^n f(s(\mathbf{x}_i, \mathbf{x}_j)) g(t(\mathbf{z}_i, \mathbf{z}_j)) - C \sum_{i=1}^l h(\xi_i)$$
$$\text{s. t.} \quad r(t(\mathbf{z}_i, \mathbf{y}_i)) \leq \xi_i, \quad \xi_i \geq 0.$$

where C is a penalty factor which takes a tradeoff role among the multi-objects functions.

3 Methods and Properties

As a concrete realization, we take the decreasing function of the squared distance between two objects as the affinity-measure, and $\|\mathbf{z}_i - \mathbf{z}_j\|^2$ as dis-affinity. Let $f(v) = h(v) = v$, $g(t(z_i, z_j)) = r(t(\mathbf{z}_i, \mathbf{z}_j)) = \|\mathbf{z}_i - \mathbf{z}_j\|^2$, and denote $\mathbf{z}_i = (z_{i1}, z_{i2}, \ldots, z_{ic})^T, Z = (\mathbf{z}_1^T, \mathbf{z}_2^T, \ldots, \mathbf{z}_n^T)^T = (\mathbf{z}_{(1)}, \mathbf{z}_{(2)}, \ldots, \mathbf{z}_{(c)}), \mathbf{y}_i = (y_{i1}, y_{i2}, \ldots, y_{ic})^T$, $1 \leq i \leq l$; $\mathbf{y}_j = 0, j \geq l+1$ and $Y_0 = (\mathbf{y}_1^T, \mathbf{y}_2^T, \ldots, \mathbf{y}_n^T)^T = (\mathbf{Y}_{(1)}, \mathbf{Y}_{(2)}, \ldots, \mathbf{Y}_{(c)})$. We always suppose $s_{ij} > 0$ without special declaration. Then we obtain the formal expression for semi-supervised transductive learning machine based on the affinity-rule as follows:

$$\left. \begin{array}{c} \min_{\mathbf{Z}_1, \mathbf{Z}_2, \ldots, \mathbf{Z}_n, \xi} \sum_{i=1}^n \sum_{j=1}^n s_{ij} \|\mathbf{z}_i - \mathbf{z}_j\|^2 + C \sum_{i=1}^l \xi_i \\ \text{s. t.} \quad \|\mathbf{z}_i - \mathbf{y}_i\|^2 \leq \xi_i, \quad \xi_i \geq 0, \quad 1 \leq i \leq l \end{array} \right\} \quad (P)$$

This is a convex programming and there is a globally optimal solution. Its Lagrangian function is

$$L = \sum_{i=1}^n \sum_{j=1}^n s_{ij} \|\mathbf{z}_i - \mathbf{z}_j\|^2 + C \sum_{i=1}^l \xi_i + \sum_{i=1}^l \lambda_i (\|\mathbf{z}_i - \mathbf{y}_i\|^2 - \xi_i) - \sum_{i=1}^l \mu_i \xi_i,$$

$$\lambda_i \geq 0, \mu_i \geq 0.$$

Note that

$$\sum_{i=1}^n \sum_{j=1}^n s_{ij} \|\mathbf{z}_i - \mathbf{z}_j\|^2 = \sum_{k=1}^c \left[\sum_{i=1}^n \sum_{j \neq i} (s_{ij} + s_{ji}) z_{ik}^2 - 2 \sum_{i=1}^n \sum_{j<i} (s_{ij} + s_{ji}) z_{ik} z_{jk} \right].$$

Let $w_{ij} = \frac{1}{2}(s_{ij} + s_{ji}), j \neq i;$ $w_{ii} = 0,$ $W = (w_{ij})_{n \times n};$

$$d_i = \sum_{j \neq i} w_{ij}, \quad D = \text{diag}(d_1, d_2, \cdots, d_n), \quad A = D - W,$$

$$I_l = \text{diag}(1, 1, \cdots, 1, 0, \cdots, 0), \quad \Lambda = \text{diag}(\lambda_1, \lambda_2, \cdots, \lambda_l, 0, \cdots, 0),$$

and $\mu = (\mu_1, \mu_2, \cdots, \mu_l)^T$, where there are l 1's in I_l, and A is a symmetric matrix. Therefore,

$$L_P = 2\sum_{k=1}^{c} \mathbf{z}_{(k)}^T A \mathbf{z}_{(k)} + \sum_{i=1}^{l} \lambda_i \|\mathbf{z}_i - \mathbf{y}_i\|^2 + \sum_{i=1}^{l} (C - \lambda_i - \mu_i)\xi_i, \lambda_i \geq 0, \mu_i \geq 0 \quad (1)$$

To obtain the $K - T$ points of (P), we calculate the derivative:

$$\left. \begin{array}{l} \frac{\partial L}{\partial \xi_i} = C - \lambda_i - \mu_i \\ \frac{\partial L}{\partial \mathbf{z}_{(k)}} = 4A\,\mathbf{z}_{(k)} + 2\Lambda(\mathbf{z}_{(k)} - \mathbf{y}_{(k)}) \end{array} \right\}$$

When Z is the $K - T$ points of (P), then

$$\left. \begin{array}{l} C - \lambda_i - \mu_i = 0, \quad \lambda_i \geq 0, \quad \mu_i \geq 0 \\ (2A + \Lambda)\,\mathbf{z}_{(k)} = \Lambda\,\mathbf{y}_{(k)} \end{array} \right\} \quad (2)$$

and the KKT conditions are hold. That is,

$$\lambda_i(\|\mathbf{z}_i - \mathbf{y}_i\|^2 - \xi_i) = 0, \qquad \mu_i \xi_i = 0, \qquad 1 \leq i \leq l \quad (3)$$

In consideration of the limited space, we will omit the proofs of the following some lemmas and theorems.

Lemma 1. The solutions of (P) with the condition (2) are constant values when $\lambda_i \equiv 0$. If the number of the labels for the labelled data is larger than one, then there exist an i_0 such that $\lambda_{i_0} > 0$. That is, $\Lambda \geq 0$, and $\Lambda \neq 0$.

Lemma 2. When $\Lambda \geq 0$, and $\Lambda \neq 0$ we have 1) $(2A + \Lambda)$ is invertible, and 2) $\rho((D + \frac{1}{2}\Lambda)^{-1}W) < 1$.

We deal with the dual problem of (P) now. Substituting the results of (2) into (1), and using (2), we obtain $0 \leq \lambda_i \leq C$. By Lemma 2, we have $L_D = -\sum_{k=1}^{c} \mathbf{y}_{(k)}^T \Lambda (2A + \Lambda)^{-1} \Lambda \mathbf{y}_{(k)} + \sum_{k=1}^{c} \mathbf{y}_{(k)}^T \Lambda \mathbf{y}_{(k)}$. So, when $\Lambda \geq 0$, and $\Lambda \neq 0$, the dual programming is

$$\left. \begin{array}{l} \max_{\lambda} -\sum_{k=1}^{c} \mathbf{y}_{(k)}^T \Lambda (2A + \Lambda)^{-1} \Lambda \mathbf{y}_{(k)} + \sum_{k=1}^{c} \mathbf{y}_{(k)}^T \Lambda \mathbf{y}_{(k)} \\ \text{s.t.}\ \ 0 \leq \lambda_i \leq C, \qquad \Lambda \neq 0 \end{array} \right\} \quad (D)$$

Lemma 3. For $\Lambda \geq 0$, and $\Lambda \neq 0$, L_D reaches the maximum when $\Lambda = CI_l$.

Theorem 4. For the nontrivial solutions, $(Z, \xi; \Lambda, \mu) = (Z^*, \xi^*; CI_l, 0)$ are the Lagrangian saddle-point of (P), and (Z^*, ξ^*) are the globally optimal solution of (P), where $Z^* = U^{-1}Y_0 = C(2A+CI_l)^{-1}Y_0$ and $U = (\frac{2}{C}A+I_l)$. We can write

$$Z^* = (I-\Delta D^{-1}W)^{-1}(I-\Delta)Y_0, \qquad (4)$$

where

$$\Delta = \text{diag}\left(\frac{2d_1}{2d_1+C}, \cdots, \frac{2d_l}{2d_l+C}, 1, \cdots, 1\right)$$

and $\xi^* = \left(\|\mathbf{z}_1^* - \mathbf{y}_1\|^2, \cdots, \|\mathbf{z}_l^* - \mathbf{y}_l\|^2\right)^T$.

Proof. By Lemma 1 and 2, we know that $Z^* = C(2A+CI_l)^{-1}Y_0$ is well-defined and can be rewritten as $Z^* = (I-\Delta D^{-1}W)^{-1}(I-\Delta)Y_0$. According to the above these lemmas and the explanations, we obtain that $(2A+\Lambda)\mathbf{Z}^* = \Lambda Y_0, \xi^* = \left(\|\mathbf{z}_1^* - \mathbf{y}_1\|^2, \cdots, \|\mathbf{z}_l^* - \mathbf{y}_l\|^2\right)^T$, and that $\Lambda = CI_l, \mu = 0$ satisfy the conditions that (Z^*, ξ^*) are the $K-T$ points of the convex programming (P). Therefore $(Z^*, \xi^*; CI_l, 0)$ is the Lagrangian saddle-point of (P), and (Z^*, ξ^*) is the globally optimal solution of (P). This is the analytic solution of the transductive learning machine for semi-supervised problems based on the affinity-rule.

Theorem 5. Z^* is the limiting solution of the following recurrence equation

$$\left.\begin{array}{l} Z_0^* = Y_0 \\ Z_{m+1}^* = \Delta D^{-1}W Z_m^* + (I-\Delta)Y_0, \quad m \geq 1 \end{array}\right\} \qquad (5)$$

and the estimates of the error and a bound for Z^* are respectively

$$\|Z_m^* - Z^*\| \leq \frac{\|\Delta D^{-1}W\|^m}{1-\|\Delta D^{-1}W\|} \|\Delta(I-D^{-1}W)Y_0\|$$

and

$$\|Z^*\| \leq \frac{\|(I-\Delta)Y_0\|}{1-\|\Delta D^{-1}W\|}.$$

Similarly, we can obtain $U^{-1} = (\frac{2}{C}A+I_l)^{-1}$ with iterated algorithm if $I_{n\times n}$ is considered as Y_0.

4 Incremental Learning Algorithm

To discuss the problem about on-line data processing with real-time constrain, we study the situations with information increment. Suppose that the given $n = l+u$ data have been expressed with the notations in Section 3. Let $s = \frac{2}{C}$ and denote respectively by $A_n, I_l(n), Y_n$, and U_n the aforementioned A, I_l, Y_0 and U in the case of n given data where $U_n = sA_n + I_l(n), Z(n) = U_n^{-1}Y_n = (sA_n + I_l(n))^{-1}Y_n, I_l(n) = \text{diag}(1,1,\cdots,1,0,\cdots,0)$, and rearrange these n data in such a way that the labelled data are in the front of the unlabelled data. We add a star to $Z(n+1)$ to stand for the rearranging version of $Z(n+1)$ with $n+1$ data.

Now, we study the semi-supervised learning problem with information increment for the case that the $(n+1)$-th point is unlabelled. Let

$$Y_{n+1} = \begin{pmatrix} Y_n \\ 0 \end{pmatrix}, Z(n+1) = \begin{pmatrix} Z_n \\ \mathbf{z}_{n+1}^T \end{pmatrix}, I_l(n+1) = \begin{pmatrix} I_l(n) & 0 \\ 0 & 0 \end{pmatrix}_{(n+1)\times(n+1)},$$

and

$$A_{n+1} = \begin{pmatrix} A_n & A_{n,n+1} \\ A_{n+1,n} & d_{n+1} \end{pmatrix},$$

where $A_{n,n+1}$ are an affinity vector between the $(n+1)$-th point and the n data given before, that is, $A_{n,n+1} = (a_{1,n+1}, a_{2,n+1}, ..., a_{n,n+1})^T$, $A_{n+1,n} = A_{n,n+1}^T$, and $d_{n+1} = \sum_{j \neq n+1} a_{n+1,j}$. By the analytic solution of transductive learning machine based on the affinity-rule, we get

$$Z(n+1) = [sA_{n+1} + I_l(n+1)]^{-1} Y_{n+1} \;.$$

Therefore we have

$$\left. \begin{array}{l} sA_n Z_n + sA_{n,n+1}\mathbf{z}_{n+1}^T + I_l(n)(Z_n - Y_n) = 0 \\ sA_{n,n+1}^T Z_n + sd_{n+1}\mathbf{z}_{n+1}^T = 0 \end{array} \right\}.$$

By Sherman-Morrison-Woodbory formula

$$(M + BCD^T)^{-1} = M^{-1} - M^{-1}B(C^{-1} + D^T M^{-1} B)^{-1} D^T M^{-1},$$

we obtain

$$Z_n = [I + \frac{1}{(d_{n+1} - sA_{n,n+1}^T U_n^{-1} A_{n,n+1})} U_n^{-1} A_{n,n+1} A_{n,n+1}^T] Z(n).$$

Similarly, we also can consider the other cases with information increment.

Theorem 6. Suppose that the solution of transductive learning machine based on the affinity-rule for the n data has been given as $Z(n) = U_n^{-1} Y_n = [sA_n + I_l(n)]^{-1} Y_n$. If the incremental information is the $(n+1)$-th data which is an unlabelled point, then the incremental learning solution is

$$\left. \begin{array}{l} Z_n = \left[I + \frac{1}{(d_{n+1} - sA_{n,n+1}^T U_n^{-1} A_{n,n+1})} U_n^{-1} A_{n,n+1} A_{n,n+1}^T \right] Z(n) \\ \mathbf{z}_{n+1} = -\frac{1}{d_{n+1}} Z_n^T A_{n,n+1} \end{array} \right\} \quad (6)$$

and

$$Z^*(n+1) = Z(n+1) = (Z_n^T, \mathbf{z}_{n+1})^T.$$

5 Numerical Experiments

For convenience, we assume in this section that both the input space and output space are Euclidean space. It is well known from By [1,3] that the Gaussian

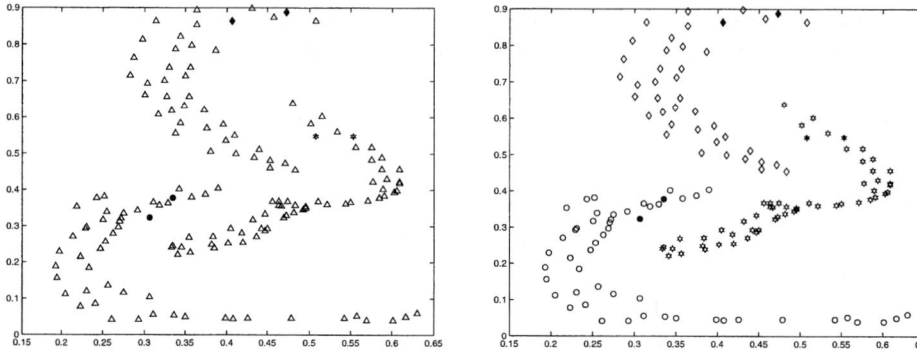

Fig. 1. The origin data plots. The triangle are unlabelled points and the others are labelled points.

Fig. 2. The plots for the results after learning. The diamond, star and circularity are '1','2' and '3' points respectively.

kernels have many good properties in support vector clustering and classification. Like the case in [5], we take Gaussian kernels in this section, and use $s(\mathbf{x}_i, \mathbf{x}_j) = \exp\left(-q \left\| z_i - z_j \right\|^2\right)$ as the affinity between two points in input space. In this case $s_{ij} > 0$ for any i and j. Two parameters are q and penalty factor C. The figures 1-2 are plotted by Matlab and the concrete steps are described as follows. (1) The original data in figures 1 are produced by randomly choosing some points in the given area, of which the points displayed by solid diamond, solid circle and star in the figure 1 are respectively labelled as different classes with different labels and those displayed by the hollow triangle are unlabelled. (2) The original data file is obtained by the coordinates of the points and the labels of the labelled points and then arranging the data in such a way that the labelled data are in front of the unlabelled data to from a vector. (3) Appropriately choose the parameters C and q. (4) The labels of the unlabelled data are inferred by the proposed method in Section 3. (5) Plot the inferred different labels of the unlabelled data with different geometrical patterns to display the results. It should be pointed out that there is not a general rite for choosing the values of the parameters C and q, which often depends on the knowledge about the semi-supervised learning problems in the studied domain. In our examples, we chose $C = \frac{1}{7}$ and $q = 6500$. The figures 1 and 2 are multi-class classifier examples. The results show that the solution of transductive learning machine performs very well, and its inferred labels can explain the manifold structure of data but not simply depend on the distances.

6 Conclusion

We have studied a semi-supervised learning problem by the transduction method in which both the labelled and the unlabelled data are used to derive the rule from particularity to particularity. The proposed transductive learning machine has realized the following goals. Firstly, similar objects get similar label measure of similarity. Secondly, the inferred labels on the labelled data match as closely as possible with the labels given by the supervisor. Our transductive learning machine based on affinity has several advantages. We efficiently utilize the information provided the unlabelled objects by using the transductive learning. This method is adaptable to general input space objects and even can be extended to the situation with default values like those in [10]. It is a quite practical method since the solution can be found by simple algorithms. The transductive learning machine is adaptive to multi-category classification naturally. In conclusion, the proposed method can make the affinity for the labels to be as consistent as possible with that for the input data, and also maintain the best consistence on the labelled data. It has a more general adaptability and provides an easier algorithm, which will make it more adaptable to on-line processes.

References

1. Ben-Hur, A. , Horn,D. , Siegelmann, H. T. , Vapnik, V.: Support Vector Clustering. Journal of Machine Learning Research , **2**, (2001) 125–137
2. Bennett, K. , Demiriz, A.: Semi-supervised Support Vector Machines. In: Kearns, M. S., Solla, S. A, Cohn, ,D. A.(eds.): Advances in Neural Information Processing Systems 11, MIT Press, Cambridge, MA (1998) 368–374
3. Burges, C. J. C.: A Tutorial on Support Vector Machines for Pattern Recognition. Data Mining and Knowledge Discovery , **2(2)**, (1998) 121–167
4. Chapelle, O. , Vapnik, V. , Weston, J.: Transductive Inference for Estimating Values of Functions. In: Solla, S. A., Leen, T. K., Müller, K. R. (eds.): Advances in Neural Information Processing Systems 12, MIT Press, Cambridge, MA (1999) 421–427
5. Chapelle, O. , Weston, J. , Schoelkopf, B.: Cluster Kernels for Semi-supervised Learning. In: Becker, S., Thrun, S., Obermayer, K. (eds.): Advances in Neural Information Processing Systems 15, MIT Press, Cambridge, MA (in press) (2003)
6. Joachims, T.: Transductive Inference for Text Classification Using Support Vector Machines. In: Bratko, I., Dzeroski, S.(eds.): Proceedings of the Sixteenth International Conferenceon Machine Learning (ICML 1999), Morgan Kaufmann Publishers, San Francisco (1999) 200–209
7. Vapnik, V.: Statistical Learning Theory. John Wiley, New York (1998)
8. Zhou, D. , Bousquet, O. , Lal, T. N. , Weston, J. , Schölkopf B.: Learning with Local and Global Consistency. In: Thrun, S., Saul, L., Schölkopf,B. (eds.): Advances in Neural Information Processing Systems 16, MIT Press , Cambridge, MA (in press)(2004)
9. Zhang, W. X., Leung, Y.: The Principle of Uncertainty Inference. Xi'an Jiaotong University Press, Xi'an (1996). (in Chinese)
10. Zhou, G. Y., Xia, L. X.: Non-metric Data Analysis and Its Applications. Science Press, Bejing (1993). (in Chinese)

A Learning Algorithm with Gaussian Regularizer for Kernel Neuron

Jianhua Xu [1] and Xuegong Zhang [2]

[1] School of Mathematical and Computer Sciences, Nanjing Normal University
Nanjing 210097, Jiangsu Province, China
xujianhua@email.njnu.edu.cn
[2] Department of Automation, Tsinghua University
Beijing 100084, China
zhangxg@mail.tsinghua.edu.cn

Abstract. In support vector machine there exist four attractive techniques: kernel idea to construct nonlinear algorithm using Mercer kernels, large margin or regularization to control generalization ability, convex objective functional to obtain unique solution, and support vectors or sparseness to reduce computation time. The kernel neuron is the nonlinear version of McCulloch-Pitts neuron based on kernels. In this paper we define a regularized risk functional including empirical risk functional and Gaussian regularizer for kernel neuron. On the basis of gradient descent method, single sample correction and momentum term, the corresponding learning algorithm is designed, which can realize four ideas in support vector machine with a simple iterative scheme and can handle the classification and regression problems effectively.

1 Introduction

In support vector machine (or SVM) for classification [1,2], there are four attractive concepts: kernel idea that can construct nonlinear algorithm in the original space through Mercer kernels, large margin or regularization that can control the generalization ability, convex objective functional that make the corresponding solution exist uniquely, and sparseness or support vectors in final discriminant function that can reduce the time consuming in the decision procedure.

Among four techniques one particularly realizes that kernel idea and kernel functions are very effective tools to construct the nonlinear algorithms for classification and regression problems. In the last several years, many kernel-based methods were presented and were referred to as kernel machines [3], for example, kernel principal component analysis [4], kernel Fisher discriminant [5], least squares support vector machines [6], and large margin kernel pocket algorithm [7].

In neural networks, the McCulloch-Pitts (M-P) neuron is the simple and basic unit [8]. In order to deal with the complicated classification and regression problems, several feedforward neural networks to realize the nonlinear mapping were constructed through neuron connection, e.g., back propagation network [9] and radial basis function network (cf. [10])). However, in [11] the nonlinear version of M-P neuron based on Mercer kernels was defined and a simple iterative procedure for single kernel neuron was designed based on the gradient descent method. In this case, a nonlinear transform can be carried out, but the kernel idea in support vector machine can be implemented only.

In the objective functional of support vector machine for classification, the key term is 2-norm square of weight vector. For the linearly separable case, minimizing it implies maximizing margin between two classes, which means that structural risk minimization principle rather than empirical risk minimization principle can be realized. However, for the nonlinearly separable case, the meaning of this term becomes vague. Simply such a term could be referred to as a regularizer or regularization term in regularization technique [12]. In neural networks, it has been found empirically that adding a proper regularizer to an objective functional in the training procedure can result in significant improvement in network generalization [13]. There mainly exist three widely used regularizers, i.e., the squared or Gaussian, absolute or Laplace, and normalized or Cauchy regularizers [14]. For the supervised learning, a detailed comparison on the three regularizers and different learning algorithms was provided, and it was concluded that the combination of the Gaussian regularizer and the second order learning algorithm can drastically improve the convergence and generalization ability [14]. Further Evgeniou et al [15] analyzed the relationship among regularization networks (i.e. feedforward neural networks with regularizer) [16], statistical learning theory and support vector machine in detail, and showed that the regularization networks also can approximately implement structural risk minimization principle when a optimal regularization parameter has been chosen. Due to regularizer and Mercer kernels, support vector machine is to solve a convex quadratic programming problem. For a proper regularization parameter, the regularized risk functional could become a convex function [3].

In support vector machine, the support vectors or sparseness occurs due to inequality constrains and Kuhn-Tucker condition. However, in regularization networks the regularizer also can be consiedered as weight decay [13,17,18], which implies that the sparseness happens through decaying connection weights and forcing small weights to be zeros (i.e., pruning the connections).

In order to realize four ideas in support vector machine with a simple approach, we define a regularized risk functional consisting of Gaussian regularizer and empirical risk functional for kernel neuron. Based on gradient descent method, single sample correction and momentum term, a corresponding learning algorithm is constructed, which can deal with the nonlinear classification and regression problems effectively.

2 M-P Neuron and Kernel Neuron

Let the training set of l i.i.d. samples be

$$\{(\mathbf{x}_1, y_1),(\mathbf{x}_2, y_2),...,(\mathbf{x}_i, y_i),...,(\mathbf{x}_l, y_l)\}, \mathbf{x}_i \in R^n, y_i \in R. \tag{1}$$

For the classification problem of binary classes (ω_1, ω_2), if $\mathbf{x}_i \in \omega_1$, then $y_i = +1$, otherwise $y_i = -1$. For the regression problem, $y_i \in R$.

The input-output relationship of classical M-P neuron is defined as

$$o(\mathbf{x}) = f((\mathbf{w} \cdot \mathbf{x}) + b), \tag{2}$$

where o is the output of neuron, $\mathbf{x}$ is the input vector, $\mathbf{w}$ and b are the weight vector and threshold respectively, and f is the transfer function, e.g., sigmoid function $f(x) = (1-e^{-x})/(1+e^{-x})$ or linear function $f(x) = x$.

The kernel neuron is the nonlinear form of M-P neuron based on Mercer kernels [11]. Its input-output relationship satisfies:

$$o(\mathbf{x}) = f(\sum_{i=1}^{l} \alpha_i k(\mathbf{x}_i, \mathbf{x}) + \beta), \tag{3}$$

where $\alpha_i \in R, i = 1, 2, \ldots l$ are coefficients corresponding to samples, β is the threshold, and $k(\mathbf{x}_i, \mathbf{x})$ is the kernel function satisfying Mercer condition, e.g., polynomial and RBF kernels [1,2]. The kernel neuron utilizes Mercer kernels to build a nonlinear transform from the original input vector space (R^n) to the real number space (R).

3 A Learning Algorithm with Gaussian Regularizer

The simplest objective functional for kernel neuron is the widely used empirical risk functional, i.e., summation of squared error between the desired output and actual output. Based on standard gradient descent scheme, a learning algorithm was constructed [11]. In this case, kernel neuron and its learning procedure only realize kernel idea in SVM. By using Gaussian regularizer, we can control generalization ability, decay the connection weights and force sparseness to happen, and design the convex objective functional and make solution exist uniquely. Thus, for kernel neuron we now define the regularized risk functional consisting of empirical risk (square error summation between the actual output and desired output) and Gaussian regularizer, that is,

$$E(\mathbf{\alpha}, \beta) = \frac{1}{2}\sum_{j=1}^{l}[y_j - o(\mathbf{x}_j)]^2 + \frac{\mu_1}{2}\sum_{j=1}^{l} \alpha_j^2, \tag{4}$$

where $\mathbf{\alpha} = [\alpha_1, \ldots, \alpha_l]^T$ and $\mu_1 \geq 0$ is the regularization parameter. For a proper μ_1, the functional (4) could become a convex function and the corresponding solution could be unique. In order to find out $\mathbf{\alpha}$ and β minimizing (4), the gradient descent scheme can be utilized. The gradient of (4) becomes

$$\frac{\partial E(\mathbf{\alpha}, \beta)}{\partial \alpha_m} = -\sum_{j=1}^{l}[y_j - f(\tilde{y}_j)]f'(\tilde{y}_j)k(\mathbf{x}_m, \mathbf{x}_j) + \mu_1 \alpha_m, \quad m = 1, \ldots l, \tag{5}$$

$$\frac{\partial E(\mathbf{a},b)}{\partial \beta} = -\sum_{j=1}^{l}\left[y_j - f(\tilde{y}_j)\right]f'(\tilde{y}_j), \tag{6}$$

where $\tilde{y}_j = \sum_{i=1}^{l}\alpha_i k(\mathbf{x}_i,\mathbf{x}_j)+\beta$, and $f'(\tilde{y}_j)$ implies the first derivative of $f(\tilde{y}_j)$.

In the iterative procedure of learning algorithm, we utilize gradient descent technique, single sample correction and momentum term. Thus, for some training sample $\mathbf{x}_j$, the iterative update formulas are

$$\begin{aligned}\alpha_m(t+1) &= \alpha_m(t)+\Delta\alpha_m(t)\\ \beta(t+1) &= \beta(t)+\Delta\beta(t)\end{aligned} \tag{7}$$

The corresponding corrections in (7) become

$$\begin{aligned}\Delta\alpha_m(t) &= \eta_1\left[y_j - f(\tilde{y}_j(t))\right]f'(\tilde{y}_j(t))k(\mathbf{x}_m,\mathbf{x}_j) - \eta_2\alpha_m + \eta_3\Delta\alpha_m(t-1)\\ \Delta\beta(t) &= \eta_1\left[y_j - f(\tilde{y}_j(t))\right]f'(\tilde{y}_j(t)) + \eta_3\Delta\beta(t-1)\end{aligned} \tag{8}$$

where η_1 indicates the learning rate, $\eta_2 = \eta_1\mu_1$, η_3 indicates the coefficient for momentum term, and $t = 1, 2, 3, \ldots$ indicates time.

To obtain a sparser discriminant function, a selective regularized risk functional could be defined to make only the connection weights decay whose absolute values are below some threshold after a learning procedure mentioned above has been finished [18]. Thus a new regularized functional for kernel neuron could be defined, i.e.,

$$E^s(\mathbf{a},\beta) = \frac{1}{2}\sum_{j=1}^{l}[y_j - o(\mathbf{x}_j)]^2 + \frac{\mu_2}{2}\sum_{\substack{j=1\\|\alpha_j|<\delta}}^{l}\alpha_j^2, \tag{9}$$

where δ is a proper threshold. Such a trick can further improve the goodness of fit of a model and decay the small connection weights.

For kernel neuron, the learning algorithm consists of two sub-procedures. By using (4) and (7), the first sub-procedure is constructed to obtain an initial solution (i.e., $\mathbf{a},\beta$). Through setting a proper threshold and using (9), the second sub-procedure is designed to force sparseness to occur. We refer to this procedure as a learning algorithm with Gaussian regularizer for kernel neuron, or simply GR-KN. Such an algorithm implements all ideas in SVM with a simple iterative procedure.

4 Experiment Results and Analysis

We use ten benchmark data sets from http://www.first.gmd.de/~raetsch to evaluate the performance of learning algorithm for classification. These sets include 100 realizations, so the average value and standard deviation can be estimated. In our experiment we utilize RBF kernel $k(\mathbf{x},\mathbf{y}) = \exp(-\|\mathbf{x}-\mathbf{y}\|^2/2\sigma^2)$ and same width parameter σ used in support vector machine [3,5] (see column 2 in Table 1). Column 3 in Table 1 shows the average error and standard deviation on the test sets using GR-KN. Column 4 implies the corresponding results from SVM [3,5]. There are seven data sets whose differences of error rate between GR-KN and SVM are less than 1%. For the diabetis data set, the largest difference (i.e., 2.13%) occurs.

Table 1. The results on ten benchmark data sets

Data Set	Width	GR-KN	SVM
Banana	0.71	11.46±0.98	11.5±0.7
Breast Cancer	5.00	26.86±5.22	26.0±4.7
Diabetis	3.16	25.63±2.30	23.5±1.7
German	5.24	25.13±2.40	23.6±2.1
Heart	7.75	17.95±3.78	16.0±3.3
Ringnorm	2.24	1.94±0.44	1.7±0.1
Thyroid	1.22	5.32±2.52	4.8±2.2
Titanic	1.00	23.24±0.73	22.4±1.0
Twonorm	4.47	3.31±0.41	3.0±0.2
Waveform	3.16	10.54±0.62	9.9±0.4

The Boston housing data set (ftp://ftp.ics.uci.com/pub/machine-learning-databases /housing) has been widely used to examine nonlinear regression methods [1,2]. It consists of 506 samples in which 13 attributes determine the median house price. In our experimet, we partition it into a training set of 401 samples, a validation set of 80 samples, and a test set of 25 samples according to the similar idea in [1,2]. This partitioning is carried out randomly 20 times. We elaborately adjust the width of RBF and other parameters for the first partitioning, where $\sigma = 0.20$. The average squared error 9.6 for 20 test sets is estimated. The average squared errors 12.4, 10.7 and 7.2 using Bagging, Boosting and SVM from 100 test sets were provided in [1,2]. Note that our partitioning is different from theirs.

These results illustrate that our learning algorithm can work well in the classification and regression problems.

5 Conclusions

For kernel neuron, in order to control generalization ability, and obtain sparseness and unique solution, the regularized risk functional is constructed, which consists of the emperical risk fuctional (squared error summation between the desired output and the actual output) and Gaussian regularizer. According to the gradient descent scheme, single sample correction and momentum term, an iterative procedure is designed. Such a learning algorithm can realize all attractive techniques in support vector machine with a simple scheme. The initial experiments on 11 benchmark data sets show that it works well on the nonlinear classification and regression problems.

Our further works will adjust paremeters (epecially width of RBF kernel) elaborately to otain the better performance on benchmark data sets for both classification and regression problems.

This work is supported by National Natural Science Foundation of China, project No. 60275007.

References

1. Vapnik, V. N.: Statistical Learning Theory. Wiley, New York (1998)
2. Vapnik, V. N.: The Nature of Statistical Learning Theory (2nd edition). Springer-Verlag, New York (1999)
3. Scholkopf, B., Smola, A. J.: Learning with Kernels - Support Vector Machines, Regularization, Optimization and Beyond. MIT Press, Cambridge CA (2002)
4. Scholkopf, B., Smola, A., Muller, K.-R.: Nonlinear Component Analysis as a Kernel Eignvalue Problem. Neural Computation. 10(5)(1998) 1299-1319
5. Mika, S., Ratsch, G., Weston, J., Scholkopf, B., Muller, K-R.: Fisher Discriminant Analysis with Kernels. In: Neural Networks for Signal Processing IX. IEEE Press, New York (1999) 41-48
6. Suykens, J. A. K., Vandewalle, J.: Least Squares Support Vector Machines. Neural Processing Letters. 9(1999) 293-300
7. Xu, J., Zhang, X., Li, Y.: Large Margin Kernel Pocket Algorithm. In: Proceedings of 2001 International Joint Conference on Neural Networks. IEEE Press, New York (2001) 1480-1485
8. McCulloch, W. S., Pitts, W.: A Logical Calculus of the Ideas Immanent in Nervous Activity. Bulletin of Mathematical Biophysics. 5(1943) 115-131
9. Rumelhart, D. E., Hinton, G. E., Williams, R. J.: Learning Representations by Back-Propagating Errors. Nature. 323(9)(1986) 533-536
10. Theodoridis, S., Koutroumbas, K.: Pattern Recognition. Academic Press, San Diego (1999)
11. Xu, J., Zhang, X., Li, Y.: Kernel Neuron and its Training Algorithm. In: Proceedings of 8th International Conference on Neural Information Processing, Vol. 2. Fudan University Press, Shanghai China (2001) 861-866
12. Tikhonov, A. N., and Arsenin, V. Y.: Solution of Ill-Posed Problem. W. H. Wanston, Washington DC (1977)
13. Pault, D. C., Nowlan, S. J., Hinton, G. E.: Experiments on Learning by Back Propagation. Technical Report. CMU-CS-86-126, Carnegie-Mallon University (1986)
14. Saito, K., Nakano, R.: Second Order Learning Algorithm with Squared Penalty Term. Neural Computation. 12(3)(2000) 709-729
15. Evgeniou, T., Pontil M., Poggio, T.: Regularization Networks and Support Vector Machines. In: Smola, A. J., Bartlett, P. L., Schölkopf, B., and Schuurmans, D. (editors), Advances in Large Margin Classifiers. MIT Press, Cambridge MA (2000) 171-204
16. Girosi, F., Jones, M., Poggio, T.: Regularization Theory and Neural Networks Architectures. Neural Computation. 7(2)(1995) 219-269
17. Reed, R.: Pruning Algorithms - A Survey. IEEE Transactions on Neural Networks. 4(5)(1993) 740-747
18. Ishikawa, M.: Structural Learning with Forgetting. Neural Networks. 9(3)(1996) 509-521

An Effective Learning Algorithm of Synergetic Neural Network

Xiuli Ma and Licheng Jiao

Institute of Intelligent Information Processing, Xidian University,
710071 Xi'an, China
Key Lab for Radar Signal Processing, Xidian University,
710071 Xi'an, China
`Lijianer@163.com, Lchjiao@Xidian.edu.cn`

Abstract. In Synergetic Neural Network (SNN), the learning problem can be reduced to how to get prototype pattern vector and adjoint vector. Here we put emphasis on the study of the former. A novel-learning algorithm of SNN is presented in this paper, which combines the self-learning ability of SNN with the global searching performance of Immunity Clonal Strategy. In comparison with learning algorithm on superposition of information (LAIS), the proposed method can avoid local extremum and improve searching efficiency. Experiments show that the new method not only overcomes the shortages of methods available but also enhances the classification accuracy rate greatly.

1 Introduction

In the late 1980s, Haken proposed to put synergetic theory into the area of pattern recognition [1]. Hence, the application of synergetics in image processing and recognition is a rising field. At the same time, its algorithms are also studied widely especially on the selection of prototype pattern vector, setting of attention parameter and its invariant properties and so on.

On the selection of prototype pattern vector, Haken proposed to select any sample from each class as prototype pattern vector. Wanger and Boebel made use of SCAP algorithm [2] by averaging training samples simply, which had been used to detect machinery parts automatically in industry. Wang *et al.* took clustering center got by C-Means clustering algorithm as prototype pattern vector. Then he proposed to a learning algorithm based on the superposition of information (LAIS) [3], which iteratively modifies the prototype pattern vector by the pattern with the highest classification inaccuracy rate. These methods improve the performance of SNN to a certain extent but they easily get into local extremum.

Considering that, a novel-learning algorithm of Synergetic Neural Network (SNN) based on Immunity Clonal Strategy (ICS) is presented in this paper. Compared with others, the new method is not easy to get into local extremum and can avoid the aimless searching at the later time of searching. The classification results show that the improved SNN can enhance the classification accuracy rate greatly.

2   Review of Synergetic Neural Network

The basic principle of SNN is that the pattern recognition procedure can be viewed as the competition process of many order parameters. Supposing the prototype pattern vectors are M and the status vectors are N, M is less than N for the sake of linear independence of prototype pattern vectors. A dynamic equation proposed by Haken, which is applicable to pattern recognition, can be given as follows:

$$\dot{q} = \sum_{k=1}^{M} \lambda_k (v_k^+ q) v_k - B \sum_{k \neq k'} v_k (v_{k'}^+ q)^2 (v_k^+ q) - Cq(q^+ q) + F(t) , \quad (1)$$

where q is the status vector of input pattern with initial value q_0, λ_k is attention parameter. If and only if λ_k is positive, patterns can be recognized. v_k is prototype pattern vector and v_k^+ is its adjoint vector that satisfies $(v_k^+, v_{k'}) = v_k^+ v_{k'} = \delta_{kk'}$. Order parameter is

$$\xi_k = (v_k^+, q) = v_k^+ q . \quad (2)$$

Corresponding dynamic equation of order parameters is

$$\dot{\xi} = \lambda_k \xi_k - B \sum_{k' \neq k} \xi_{k'}^2 \xi_k - C(\sum_{k'=1}^{M} \xi_{k'}^2) \xi_k . \quad (3)$$

Haken proved that when $\lambda_k = C > 0$, the largest initial order parameter will win and the network will then converge. In this case, the SNN can meet some real-time needs for it avoids iteration.

In SNN, learning is also the training and modification process on weight matrix. Its learning problem can be reduced to how to get prototype pattern vector and adjoint vector. In this paper, we propose a new learning algorithm, which gets prototype pattern vector by Immunity Clonal Strategy.

3   Learning Algorithm of Synergetic Neural Network Based on Immunity Clonal Strategy

3.1   Immunity Clonal Strategy

In order to enhance the diversity of the population in GA and avoid prematurity, Immunity Clonal Strategy [4] is proposed by Du *et al*. The clonal operator is an antibody random map induced by the affinity including: clone, clonal mutation and clonal selection. The state transfer of antibody population is denoted as

$$C_{MA} : A(k) \xrightarrow{clone} A'(k) \xrightarrow{mutation} A''(k) \xrightarrow{selection} A(k+1) .$$

Here antibody, antigen, the affinity between antibody and antigen are similar to the definitions of the objective function and restrictive condition, the possible solution, match between solution and the fitting function in AIS respectively. According to the affinity function $f(*)$, a point $a_i = \{x_1, x_2, \cdots, x_m\}$, $a_i(k) \in A(k)$ in the solution space will be divided into q_i different points $a_i'(k) \in A'(k)$, by using clonal operator, a new antibody population is attained after performing clonal mutation and clonal selection. It is easy to find that the essential of the clonal operator is producing a variation population around the parents according to their affinity. Then the searching area is enlarged.

Immunity Clonal Strategy is an optimizing and intelligent algorithm imitating the natural immune system. Its substance is producing a variation population around the best individual according to their affinity in the process of evolution and then enlarging the searching area. It is helpful to avoid prematurity and local extremum. The mechanism of clonal selection can increase the diversity of the antibody.

3.2 Learning Algorithm of SNN Based on Immunity Clonal Strategy

In SNN, learning is also the training and modification process on weight matrix on condition that the training samples are known. Its learning problem can be reduced to how to get prototype pattern vector and adjoint vector. We can get prototype pattern vector first and then compute adjoint vector and vice versa. The methods available to get prototype pattern vector are C-Means clustering algorithm, SCAP, SCAPAL and LAIS and so on. They are easy to get into local extremum and also search aimlessly at the later time of searching. Considering that, the new learning algorithm based on ICS is presented, which is described as follows.

Step 1: Initialize the antibody population.

Initialize the antibody population $A(0)$ randomly, which has the size of $N-1$. Every individual is a possible prototype pattern vector. At the same time, the average of the training samples in each class is calculated as prior knowledge and then is added to the initial population as an individual. As a result, the size of initial population is N.

Step 2: Calculate the affinity.

The affinity is donated by the classification accuracy rate of training samples.

Step 3: Clone.

Clone every individual in the k-th parent $A(k)$ to produce $A'(k)$. The clonal scale is a constant or determined by the affinity.

Step 4: Clonal mutation.

Mutate $A'(k)$ to get $A''(k)$ in probability of p_m.

Step 5: Calculate the affinity.

Calculate the affinity of every individual in new population $A''(k)$.

Step 6: Clonal selection.

Select the best individual with better affinity than its parent into new parent population $A(k+1)$.

Step 7: Calculate the affinity.

Calculate the affinity of every individual in new population $A(k+1)$.

Step 8: Judge halt conditions.

The halt conditions can be restricted iterative number or the classification accuracy rate of training samples. If searching algorithm reaches the halt conditions, terminate the iteration and make the best individual preserved during iterative process as the best prototype pattern vector, or else preserve the best individual in current iteration and then turn its steps to Step 3.

4 Experiments

4.1 Classification of IRIS Data

IRIS data is selected to test the performance of the new method. This data set has 150 data, which is made up of 3 classes, and each datum consists of 4 attributes. We randomly select 16 data from every class as training samples and others as testing samples.

In this experiment, the size of initial population is 10 and mutation probability is 0.1. The clonal scale is determined by the affinity. The halt condition is the average classification accuracy rate of training samples up to 100% or the iterative number up to 200. The classification results of 20 trails are shown in Table 1.

Table 1. Comparison between classification results

	Learning method based on LAIS	Proposed method based on ICS
Classification accuracy rate of training samples (%)	81.25	100
Classification accuracy rate of testing samples (%)	78.431	94.118

From Table 1 we can see that the method to get prototype pattern vector based on ICS has a much higher classification accuracy rate than on LAIS. In fact, the construction of order parameters is another problem in SNN and it can affect the classification accuracy to a certain extent.

4.2 Classification of Remote Sensing Images

By the characteristic of the remote sensing images adopted, Relative moments features are extracted. Relative moments [5] unite the computational formula of region and structure. In comparison with Hu invariant moments, it is even more universal.

From the Riemann integral, Let density function be $f(x, y)$, geometry moment m_{pq} with two dimensions and $(p+q)$ ranks, and geometry center moment u_{pq} are defined respectively as follows:

$$m_{pq} = \int_{-\infty}^{+\infty}\int_{-\infty}^{+\infty} x^p y^q f(x,y)dxdy \quad u_{pq} = \int_{-\infty}^{+\infty}\int_{-\infty}^{+\infty} (x-\bar{x})^p (y-\bar{y})^q f(x,y)dxdy, \quad (4)$$

where $\bar{x} = m_{10}/m_{00}, \bar{y} = m_{01}/m_{00}$. Let

$$\begin{aligned}
\xi_1 &= u_{20} + u_{02} \\
\xi_2 &= (u_{20} - u_{02})^2 + 4u_{11}^2 \\
\xi_3 &= (u_{30} - 3u_{12})^2 + (3u_{21} - u_{03})^2 \\
\xi_4 &= (u_{30} + u_{12})^2 + (u_{21} + u_{03})^2 \\
\xi_5 &= (u_{30} - 3u_{12})(u_{30} + u_{12})[(u_{30} + u_{12})^2 - 3(u_{21} + u_{03})^2] + (3u_{21} - u_{03})(u_{21} + u_{03})[3(u_{30} + u_{12})^2 - (u_{21} + u_{03})^2] \\
\xi_6 &= (u_{20} - u_{02})[(u_{30} + u_{12})^2 - (u_{21} + u_{03})^2] + 4u_{11}(u_{30} + u_{12})(u_{21} + u_{03})
\end{aligned} \quad (5)$$

The relative moments formulae satisfied with the invariance of structural translation, zoom and rotation are

$$R_1 = \frac{\sqrt{\xi_2}}{\xi_1} \quad R_2 = \frac{\xi_1 + \sqrt{\xi_2}}{\xi_1 - \sqrt{\xi_2}} \quad R_3 = \frac{\sqrt{\xi_3}}{\sqrt{\xi_4}} \quad R_4 = \frac{\sqrt{\xi_3}}{\sqrt[4]{|\xi_5|}} \quad R_5 = \frac{\sqrt{\xi_4}}{\sqrt[4]{|\xi_5|}}$$

$$R_6 = \frac{|\xi_6|}{\xi_1 \cdot \xi_3} \quad R_7 = \frac{|\xi_6|}{\xi_1 \cdot \sqrt{|\xi_5|}} \quad R_8 = \frac{|\xi_6|}{\xi_3 \cdot \sqrt{|\xi_2|}} \quad R_9 = \frac{|\xi_6|}{\sqrt{|\xi_5|} \cdot \xi_2} \quad R_{10} = \frac{|\xi_5|}{\xi_3 \cdot \xi_4}. \quad (6)$$

In order to test the presented method, we compared it with the algorithm based on LAIS [3]. The image set has the size of 1064 and includes boats and planes 456 and 608 respectively. The images in this set are integrated and misshapen binary images and have different rotary angles. Image set is divided into two parts: training set and testing set. In the former set, boats and planes are 120 and 160, and in the later, boats and planes are 336 and 448. Relative moments features of 10 dimensions are extracted from each image.

In this experiment, the parameters in ICS are defined as: the size of initial population is 10 and mutation probability is 0.2. The clonal scale is determined by the affinity. In EA, the size of initial population is 20, crossover probability is 0.9 and mutation probability is 0.1. The two methods have the same halt condition, which is the average classification accuracy rate of training set more than 99% or the iterative number up to 200. The statistical results of 20 trials are shown in Table 2.

From Table 2 it is well known that the learning algorithm based on ICS has a higher classification accuracy rate than on LAIS. During the training process, we observed that the method based on LAIS had a low classification accuracy rate, and if you want to get a bit improvement then the method can iterate endlessly. So the training time cannot be estimated (denoting as "—" in Table 2). Moreover, the method based on Evolutionary Algorithm has a much lower classification accuracy rate than on LAIS, which is consistent with the conclusion in paper [3].

Table 2. Comparison among classification results

		Method based on LAIS	Method based on EA	The method proposed
Training time(s)		—	17.6375	85.5076
Testing time(s)		0.0374	0.0342	0.0327
Average classification accuracy rate of training set (%)		96.33	99.297	99.33
Classification accuracy rate of testing set (%)	Boat	91.67	87.887	98.93
	Plane	96.96	75.772	98.85
Average classification accuracy rate of testing set (%)		94.63	81.099	98.89

5 Conclusions

A new learning algorithm of Synergetic Neural Network (SNN) based on Immunity Clonal Strategy (ICS) is presented in this paper, which combines the self-learning ability of SNN with the global searching performance of ICS to get prototype pattern vector. In comparison with others, the proposed method has not only a faster convergent speed but also a higher classification accuracy rate and it is not easy to get into local extremum. The tests show that the presented method is effective and reasonable. Moreover, the construction of order parameters is very important and it can affect the classification accuracy to a certain extent.

References

1. Haken, H.: Synergetic Computers and Cognition–A Top-Down Approach to Neural Nets. Springer-Verlag, Berlin (1991)
2. Wagner, T., Boebel, F.G.: Testing Synergetic Algorithms with Industrial Classification Problems. Neural Networks. 7 (1994) 1313–1321
3. Wang, H.L., Qi, F.H.: Learning Algorithm Based on The Superposition of Information. J. Infrared Millim. Waves. 19 (2000) 205–208
4. Jiao, L.C., Du, H.F.: Development and Prospect of The Artificial Immune System. Acta Electronica Sinica. 31 (2003) 1540–1548
5. Wang, B.T., Sun, J.A., Cai, A.N.: Relative Moments and Their Applications to Geometric Shape Recognition. Journal of Image and Graphics. 6 (2001) 296–300

Sparse Bayesian Learning Based on an Efficient Subset Selection

Liefeng Bo, Ling Wang, and Licheng Jiao

Institute of Intelligent Information Processing and National Key Laboratory
for Radar Signal Processing, Xidian University, Xi'an 710071, China
blf0218@163.com

Abstract. Based on rank-1 update, Sparse Bayesian Learning Algorithm (SBLA) is proposed. SBLA has the advantages of low complexity and high sparseness, being very suitable for large scale problems. Experiments on synthetic and benchmark data sets confirm the feasibility and validity of the proposed algorithm.

1 Introduction

Regression problem is one of the fundamental problems in the field of supervised learning. It can be thought of as estimating the real valued function from a samples set of noise observation. A very successful approach for regression is Support Vector Machines (SVMs) [1-2]. However, they also have some disadvantages [3]:
- To derive analytically error bars for SVMs is very difficult.
- The solution is usually not very sparse.
- Kernel function must satisfy Mercer's condition.

In order to overcome the above problems, Relevance Vector Machine (RVM) [3] is proposed, which is very elegant and obtains a high sparse solution. However, RVM needs to solve linear equations, whose cost is very expensive, and therefore not feasible for large scale problems.

Based on rank-1 update, we propose Sparse Bayesian Learning Algorithm (SBLA), which has low complexity and high sparseness, thus being very suitable for large scale problems. Experiments on synthetic and benchmark data sets confirm the feasibility and validity of SBLA.

2 Model Specification

Let $z = \{(\mathbf{x}_1, y_1), \cdots (\mathbf{x}_l, y_l)\}$ be empirical samples set drawn from

$$y_i = f(\mathbf{x}_i, \mathbf{w}) + \varepsilon_i, \quad i = 1, 2, \cdots l, \tag{1}$$

where ε_i is independent samples from some noise process which is further assumed to be mean-zero Gaussian with variance σ^2. We further assume

$$f(\mathbf{x},\mathbf{w}) = \sum_{i=1}^{l} w_i k(\mathbf{x},\mathbf{x}_i). \tag{2}$$

According to Bayesian inference, the posterior probability of $\mathbf{w}$ can be expressed as

$$P(\mathbf{w}|z) = \frac{P(z|\mathbf{w})P(\mathbf{w})}{P(z)}. \tag{3}$$

Due to the assumption of independence of z, Likelihood $P(z|\mathbf{w})$ can be written as

$$P(z|\mathbf{w}) = (2\pi\sigma^2)^{1/2} \exp\left(\frac{1}{2\sigma^2}(\mathbf{Kw}-\mathbf{y})^T(\mathbf{Kw}-\mathbf{y})\right). \tag{4}$$

If Gaussian prior $P(\mathbf{w}) = (2\pi\gamma^2)^{1/2} \exp\left(\frac{\mathbf{w}^T\mathbf{w}}{2\gamma^2}\right)$ is chosen, maximizing the log-posterior is equivalent to minimizing the following likelihood function

$$\hat{\mathbf{w}} = \arg\min\left(L(\mathbf{w},\lambda) = \left(\mathbf{w}^T\left(\mathbf{K}^T\mathbf{K}+\lambda\mathbf{I}\right)\mathbf{w} - 2\mathbf{w}\mathbf{K}^T\mathbf{y}\right)\right). \tag{5}$$

where $\lambda = \sigma^2/\gamma^2$, $\mathbf{I}$ is unit matrix.

3 Sparse Bayesian Learning Algorithm

For large datasets, the classic methods for quadratic programming such as conjugate gradient methods [4] are not feasible, due to the requisite time and memory costs. The following greedy approximation scheme can be used. Starting with an empty set $P = \emptyset$ and set $Q = \{1,2,\cdots l\}$, we select at each iteration a new basis function s from Q, and resolve the problem (5) containing the new basis function and all previously picked basis functions. The basis function is deleted when the removing criterion is satisfied and the algorithm is terminated when certain criterion is satisfied.

3.1 Adding One Observation

Let $\mathbf{H} = (\mathbf{K}^T\mathbf{K}+\lambda\mathbf{I})$ and $\mathbf{b} = \mathbf{K}^T\mathbf{y}$, then (5) can be rewritten as

$$\hat{\mathbf{w}} = \arg\min\left(L(\mathbf{w},\lambda) = \left(\mathbf{w}^T\mathbf{H}\mathbf{w} - 2\mathbf{w}\mathbf{b}\right)\right). \tag{6}$$

Assume that $P = \{p_1,\cdots,p_n\}$, $\mathbf{R}^{t-1} = (\mathbf{H}_{PP})^{-1}$, and $\mathbf{h}_s = [H_{p_1 s},\cdots H_{p_n s}]^T$ at the $(t-1)^{th}$ iteration. If the sth basis function is added in t^{th} iteration, in terms of a rank-1 update [5], we have

$$\mathbf{R}^t = \begin{bmatrix} \mathbf{R}^{t-1} & \mathbf{0} \\ \mathbf{0}^T & 0 \end{bmatrix} + \alpha \begin{bmatrix} \boldsymbol{\beta} \\ -1 \end{bmatrix} \begin{bmatrix} \boldsymbol{\beta}^T & -1 \end{bmatrix}. \tag{7}$$

where $\boldsymbol{\beta} = \mathbf{R}^{t-1}\mathbf{h}_s$, $\alpha = (H_{ss} - \mathbf{h}_s^T\boldsymbol{\beta})^{-1}$. Thus the weights can be updated by the following equations

$$\begin{bmatrix} \mathbf{w}_P^t \\ w_s^t \end{bmatrix} = \mathbf{R}^t \begin{bmatrix} \mathbf{b}_P \\ b_s \end{bmatrix} = \begin{bmatrix} \mathbf{R}^{t-1}\mathbf{b}_P \\ 0 \end{bmatrix} + \alpha \begin{bmatrix} \boldsymbol{\beta} \\ -1 \end{bmatrix} \begin{bmatrix} \boldsymbol{\beta}^T\mathbf{b}_P & -b_s \end{bmatrix}. \tag{8}$$

Together with $\mathbf{w}_P^{t-1} = \mathbf{R}^{t-1}\mathbf{b}_P$, we have

$$\begin{bmatrix} \mathbf{w}_P^t \\ w_s^t \end{bmatrix} = \begin{bmatrix} \mathbf{w}_P^{t-1} \\ 0 \end{bmatrix} + \alpha(\boldsymbol{\beta}^T\mathbf{b}_P - b_s)\begin{bmatrix} \boldsymbol{\beta} \\ -1 \end{bmatrix}. \tag{9}$$

The rest major problem is how to pick the appropriate basis function at each iteration. A natural idea is to choose s^{th} basis function for which we have the biggest decrease in the objective function, which is called pre-fitting [6]. However its cost is too expensive. Here we adopt a cheap selection criterion that is also used in numerical algebra [7]

$$s = \max_{k \in Q}\left(abs\left(g_k^t\right)\right). \tag{10}$$

where $g_k^t = \sum_{j=1}^n \mathbf{K}_{kj} w_j^t - y_k$.

3.2 Removing One Observation

Let $\mathbf{R}^{(t,s)}$ represent the matrix with the $(P_s)^{th}$ observation deleted at the tth iteration. In terms of a rank-1 update, we have

$$\left(\mathbf{R}^{(t,s)}\right)_{ij} = \left(\mathbf{R}^t\right)_{ij} - \frac{(\mathbf{R}^t)_{is}(\mathbf{R}^t)_{sj}}{(\mathbf{R}^t)_{ss}} \quad i,j \leq n; i,j \neq s. \tag{11}$$

$$\left(\mathbf{w}^{(t,s)}\right)_i = \sum_{j=1, j \neq k}^n \left((\mathbf{R}^t)_{ij} - \frac{(\mathbf{R}^t)_{is}(\mathbf{R}^t)_{sj}}{(\mathbf{R}^t)_{ss}}\right) b_{P_j} \quad j \leq n; j \neq s. \tag{12}$$

Together with $\mathbf{w}_P^t = \mathbf{R}^t\mathbf{b}_P$, (12) is simplified as

$$\left(\mathbf{w}^{(t,s)}\right)_i = w_i^t - w_s^t \frac{(\mathbf{R}^t)_{is}}{(\mathbf{R}^t)_{ss}} \quad i \leq n; i \neq s. \tag{13}$$

Then the problem is how to pick an appropriate basis function at each iteration. Let $f^{(t,k)}$ be the objective value with the $(P_k)^{th}$ observation deleted, we have

$$f^{(t,k)} = -\sum_{i,j=1, i,j \neq k}^n b_{P_i} \left(\mathbf{R}^{(t,k)}\right)_{ij} b_{P_j}. \tag{14}$$

Substituting (11) into (14), we obtain

$$f^{(t,k)} = -\sum_{i,j=1}^n b_{P_i}\left(\mathbf{R}^t\right)_{ij} b_{P_j} + \sum_{i,j=1}^n b_{P_i} \frac{(\mathbf{R}^t)_{ik}(\mathbf{R}^t)_{kj}}{(\mathbf{R}^t)_{kk}} b_{P_j} = f^t + \frac{\left(\sum_{j=1}^n (\mathbf{R}^t)_{kj} b_{P_j}\right)^2}{(\mathbf{R}^t)_{kk}} \tag{15}$$

By the virtue of $\sum_{i=1}^{n} b_{p_i}(\mathbf{R}^t)_{ik} = w_k^t$, (15) is translated into

$$\Delta f^{(t,k)} = f^{(t,k)} - f^t = \frac{(w_k^t)^2}{(\mathbf{R}^t)_{kk}}. \qquad (16)$$

Thus we can obtain s by

$$s = \arg\min_{k \in P}\left(\Delta f^{(t,k)}\right). \qquad (17)$$

If $\Delta f^{(t,s)}$ is smaller than some threshold ε, the corresponding basis function will be removed. The algorithm is terminated if the number of removed observations is larger than some threshold M.

Accordingly, Sparse Bayesian Learning Algorithm can be described as the following

---

Sparse Bayesian Learning Algorithm

1. Let $\mathbf{w}^0 = \mathbf{0}^T$, $\mathbf{g}^0 = -\mathbf{y}$, $Q = \{1, 2, 3, \cdots, l\}$, $P = \{\}$, $t = 1$;
2. $s = \arg\max_{k \in Q}\left(abs(\mathbf{g}_Q^t)\right)$;
3. Add the sth observation and update $\mathbf{R}^t$ and $\mathbf{w}^t$ according to (7) and (9);
4. $Q = Q - \{s\}$, $P = P + \{s\}$;
5. $s = \arg\min_{k \in P}\left(\Delta f^{(t,k)}\right)$;
6. If $\Delta f^{(t,s)} \leq \varepsilon$, remove the observation s and update $\mathbf{R}^t$ and $\mathbf{w}^t$ according to (11) and (13), $P = P - \{s\}$;
7. If the stop criterion is satisfied, stop.
8. $\mathbf{g}_Q^t = \mathbf{K}_{QP}\mathbf{w}_P^t$, $t = t + 1$, goto 2;

Fig. 1. Sparse Bayesian Learning Algorithm.

3.3 Parameters Selection

Let L_{min} be the minimal value of likelihood function, we have

$$(L_{min} + \mathbf{y}^T \mathbf{y})/l \approx 0. \qquad (18)$$

Then, the average contribution of one observation for L_{min} is about $\mathbf{y}^T\mathbf{y}/l$. Let $\varepsilon = \mathbf{y}^T\mathbf{y}/l/T$. If the decrease of the cost function is smaller than $\mathbf{y}^T\mathbf{y}/l/T$ after the sth observation be deleted, we will remove it. If the number of removed observations is larger than M, the algorithm is terminated. The common choice of T and M is 10^4 and $l/10$. λ can be fixed to 10^{-3}.

3.4 Complexity of Algorithms

Calculating $\mathbf{h}_s$ and updating $\mathbf{R}$ and $\mathbf{g}_Q$ are operations of cost $O(nl)$, $O(n^2)$ and $O(n(l-n))$, respectively. Therefore, the single step computational complexity of SBLA is only $O(nl)$ and successive $(n+M)$ iterations incur a computational cost of $O(n^2 l + nMl)$. Note that SBLA has low complexity since $n, M \ll l$. Besides that, the memory footprint of the algorithms is also only $O(nl)$.

4 Simulation

In order to evaluate the performance of the proposed algorithm, we performed three experiments on synthetic and benchmark data sets. Data sets for regression come from STATLOG COLLECTION. For the sake of comparison, different algorithms used the same input sequence. The elements of Gram matrix $\mathbf{K}$ were constructed using the Gaussian kernel function of the form $k(\mathbf{x}, \mathbf{y}) = \exp(\frac{\|\mathbf{x} - \mathbf{y}\|_2^2}{2\sigma^2})$. In all experiments, λ, T and M was fixed to 10^{-3}, 10^4 and $l/10$, respectively. Kernel width σ was chosen by 10-fold cross validation procedure.

The Sinc function $y = \sin(x)/x + \sigma N(0,1)$ is a popular choice of illustrating support vector machines regression. Training samples were generated from Sinc function at 100 equally-spaced x-value in [-10,10] with added Gaussian noise of standard deviation 0.1. Results were averaged over 100 random instantiations of the noise, with the error being measured over 1000 noise-free test samples in [-10,10]. The decision functions and support vectors obtained by our algorithm and SVMs are shown in Fig. 4.1.

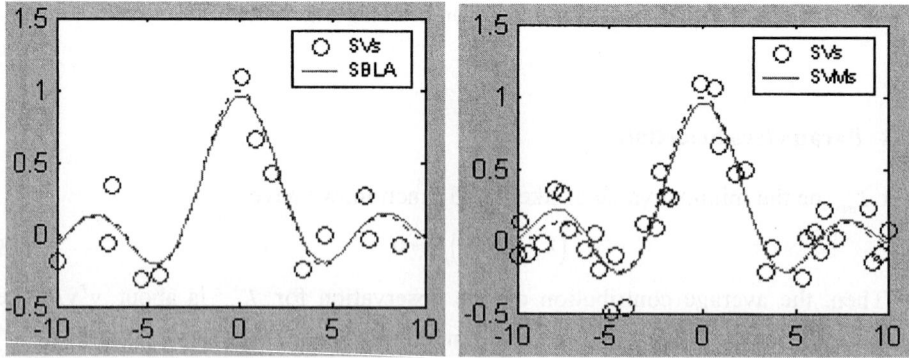

Fig. 2. Decision functions and support vectors by SBL and SVMs. Real lines denote the decision functions and circles denote support vectors.

For the Boston Housing data set, we averaged our results over 100 random split of the full dataset into 481 training samples and 25 testing samples. For the Abalone data set, we averaged our results over 10 random splits of the mother dataset into 3000 training samples and 1177 testing samples. Before experiments, we scaled all the training data in [-1,1] and then adjusted test data using the same linear transformation. The results are summarized in Table 1.

Table 1. Results obtained by SBLA and SVMs. NSV denotes the number of support vector. MSE denotes the mean squared error. For Sinc problem, unit of error is 10^{-3}.

Problem	SBLA			SVMs		
	(λ,σ)	NSV	MSE	(C,σ)	NSV	MSE
Sinc	$(10^{-3}, 2^{1.5})$	13.99 ± 1.93	0.89 ± 0.41	$(2^1, 2^{1.5})$	35.03 ± 4.87	1.24 ± 0.53
Housing	$(10^{-3}, 2^{0.5})$	62.35 ± 4.80	10.58 ± 6.91	$(2^3, 2^{0.5})$	162.40 ± 3.34	10.54 ± 8.17
Abalone	$(10^{-3}, 2^{-0.5})$	30.9 ± 4.12	4.48 ± 0.25	$(2^1, 2^{-0.5})$	1188 ± 47.12	4.48 ± 0.25

SBLA and SVMs obtained similar generalization ability; however, the solution of SBLA is much sparser than that of SVMs.

5 Conclusion

SBLA offers a simple and computationally efficient scheme for supervised learning. Its application is by no mean limited to regression problem. Work on classification problem is in progress.

Acknowledgments. This work was supported by the National Natural Science Foundation of China under grant 60372050 and 60133010 and National "863" Project grant 2002AA135080.

References

1. Vapnik, V.: The Nature of Statistical Learning Theory. Springer-Verlag, New York (1995)
2. Vapnik, V.: Statistical Learning Theory. Wiley-Interscience Publication, New York (1998)
3. Tipping, M.E.: Sparse Bayesian Learning and the Relevance Vector Machines. Journal of Machine Learning Research. 1 (2001) 211-244
4. Xing, Z.D., Cao J.R.: Matrix Numerical Analysis. Science and Technology Press, Shannxi (1999)
5. Stoer, J., Bulirsch, R.: Introduction to Numerical Analysis. Springer-Verlag, New York (1993)
6. Smola, A. J., Scholkopf, B.: Sparse Greedy Matrix Approximation for Machine Learning. Proceedings of the Seventeenth International Conference on Machine Learning. Morgan-Kaufmann, San Mateo (2000) 911-918
7. Chow, E., Saad, Y.: Approximate Inverse Preconditioners via Sparse-Sparse Iterations. SIAM Journal of Scientific and Statistical Computing. 19 (1998) 955-1023

A Novel Fuzzy Neural Network with Fast Training and Accurate Generalization

Lipo Wang[1,2], Bing Liu[2], and Chunru Wan[2]

[1] College of Information Engineering, Xiangtan University,
Xiangtan, Hunan, China,
[2] School of Electrical and Electronic Engineering, Nanyang Technology University,
Block S1, Nanyang Avenue, Singapore 639798
{elpwang, liub0002, ecrwan}@ntu.edu.sg

Abstract. For the reason of all parameters in a conventional fuzzy neural network (FNN) needed to be adjusted iteratively, learning can be very slow and may suffer from local minima. To overcome these problems, we propose a novel FNN in this paper, which shows a fast speed and accurate generalization. First we state the universal approximation theorem for an FNN with random membership function parameters (FNN-RM). Since all the membership function parameters are arbitrarily chosen, the proposed FNN-RM algorithm needs to adjust only the output weights of FNNs. Experimental results on function approximation and classification problems show that the new algorithm not only provides thousands of times of speed-up over traditional learning algorithms, but also produces better generalization performance in comparison to other FNNs.

1 Introduction

A fruitful approach of having benefits of both the artificial neural network (ANN) and fuzzy logic (FL) is to combine them into an integrated system called the fuzzy neural network (FNN)[1]. Due to its ability to approximate nonlinear functions, the FNN plays a key role in solving difficult nonlinear problems in many engineering fields [2]. In a traditional FNN, all system parameters are adjusted by using various optimization methods. However, there are two main disadvantages in these conventional methods. Firstly, these learning methods are usually far slower than required. Secondly, these methods may fall into a local minimum during the learning process. To overcome these problems, we propose a novel FNN training method called FNN-with-Random-Membership-function-parameters (FNN-RM), which arbitrarily chooses the parameters for the membership functions and analytically determine the output weights between the rule layer and the output layer. Compared with the traditional FNNs, the proposed FNN-RM can achieve a global error minimum for any rule set and shows a faster training speed and better generalization.

It has been pointed out that an FNN can be trained by fixing the parameters of membership functions first and then adjust the weights of the output layer [2]. However, in contrast to the present paper, there is no proof of universal approximation in [2] for such an FNN with fixed membership function parameters, no extensive simulations, and no specifications how to determine the parameters [2]. The present paper is also inspired by Huang et al's recent work [3], which showed that a multilayer perception neural network with randomly selected hidden neurons provides universal approximation, fast training and excellent generalization.

In this paper, we firstly state the universal approximation capability of the FNN-RM in Section 2. In Section 3, we describe the fast and accurate FNN-RM algorithm. Performance evaluation of the proposed FNN-RM is presented in Section 4. Finally, discussions and conclusions are given in Section 5.

2 Universal Approximation

We assume that the FNN is a multi-input-single-output (MISO) network $f : R^d \to R$. Results can be generalized to the case of multi-input-multi-output (MIMO) in a straightforward fashion. An FNN can be considered as a four-layer network comprised of an input layer, a membership layer, a rule layer and an output layer. We note that such an FNN is not functionally equivalent to a radial basis function network (RBFN) for the reason of all parameters in membership functions different from each other [2]. Assume the membership function is chosen as Gaussian, then the output of this fuzzy system can be described as follows:

$$f_n(x_1, x_2, ..., x_d) = \sum_{k=1}^{n} w_k g_k(x_1, x_2, ..., x_d), \quad (1)$$

where

$$g_k(x_1, x_2, ..., x_d) = \prod_{i=1}^{d} h_{ik}(x_i), \quad (2)$$

and

$$h_{ik} = \exp\left(-\frac{(x_i - a_{ik})^2}{2\sigma_{ik}^2}\right). \quad (3)$$

Here, $x_i \in R$ is the ith linguistic variable. g_k is the kth rule layer output, h_{ik} is the membership function between the kth rule node and the ith linguistic variable x_i. $a_{ik} \in R$ and $\sigma_{ik} \in R$ are, respectively, the mean and the standard deviation of the Gaussian function. w_k is the weight for the kth rule.

We now state the following: given a membership function $h(x)$ (e.g. a Gaussian function), if all the centers and standard deviations are chosen randomly, the network sequence $\{f_n\}$ can approximate any continuous function f by only adjusting the output weights between the rule layer and the output layer, with probability 1.

Theorem 1. *Given membership function* $h_{ik}(x_i, a_{ik}, \sigma_{ik}) = \exp\left(-\frac{(x_i - a_{ik})^2}{2\sigma_{ik}^2}\right)$, $i = 1, 2, ..., d$, $k = 1, 2, ..., n$, $n \geq 2$, $g_k(x_1, x_2, ..., x_d) = \prod_{i=1}^{d} h_{ik}$. *Assume* $a_k =$

$[a_{1k}, ..., a_{dk}]^T$, and when $k \neq j$, $\boldsymbol{a}_k \neq \boldsymbol{a}_j$, for any continuous function f, there exists a vector $\boldsymbol{\beta} = [\beta_1, \beta_2, ..., \beta_n]^T$ and a network sequence $f_n = \sum_{k=1}^{n} \beta_k g_k$, such that $\lim_{n \to \infty} \|f - f_n\| = 0$..

The proof for the above theorem is given in [4]. If centers and radii are arbitrarily chosen in the FNN, then with probability 1, all the centers are different. Hence, according to Theorem 1, the network will approximate any continuous function with probability 1. When other nonlinear continuous functions, such as generalized bell function [2], sigmoidal function [2], sine and cosine, are adopted as membership function, we can similarly prove the universal approximation theorem.

3 Our FNN-RM Algorithm

Assume membership functions are chosen as Gaussian, we propose the FNN-RM algorithm as follows:

Algorithm 1 *Given a training set $\varphi = \{(\boldsymbol{x}_s, \boldsymbol{t}_s) | \boldsymbol{x}_s = [x_{s1}, x_{s2}, ..., x_{sd}]^T \in R^d, \boldsymbol{t}_s = [t_{s1}, t_{s2}, ..., t_{sq}]^T \in R^q, s = 1, 2, ..., N\}$, number of rules n and membership function $h_{ik}(x, a_{ik}, \sigma_{ik})$, $i = 1, 2, ..., d$, $k = 1, 2, ..., n$, we conduct the following:*

1. *Assign arbitrary center a_{ik} and standard deviation σ_{ik}, $i = 1, 2, ..., d$, $k = 1, 2, ..., n$.*
2. *Calculate the rule layer output vector $\boldsymbol{p}_s$ for the sth input $\boldsymbol{x}_s$, where $\boldsymbol{p}_s = [p_{s1}, p_{s2}, ..., p_{sn}]^T$, $p_{sk} = g_k(x_{s1}, x_{s2}, ..., x_{sd}) = \prod_{i=1}^{d} h_{ik}(x_{si})$ is the output of the sth rule node, $s = 1, 2, ..., N$, $k = 1, 2, ..., n$.*
3. *Calculate the output weights $W = H^+ T = (H^T H)^{-1} H^T \cdot T$, where $H = [\boldsymbol{p}_1, \boldsymbol{p}_2, ..., \boldsymbol{p}_N]^T$, and $T = [\boldsymbol{t}_1, \boldsymbol{t}_2, ..., \boldsymbol{t}_N]^T$.*

According to [5], when $W = H^+ T$, we have

$$\|HW - T\|_F = \|HH^+ T - T\|_F = \min_W \|HW - T\|_F, \quad (4)$$

which means the proposed FNN-RM algorithm reaches the minimum training error and achieves a global minima for any rule set. Here $\| \cdot \|_F$ means vector $F - norms$ [5].

4 Experimental Results

We now compare the performance of our FNN-RM algorithm with conventional algorithms for FNNs. All the simulations are carried out in MATLAB 6.5 environment running in a Pentium 4, 2.4 GHZ CPU with 256 MB of RAM. We use MATLAB function ANFIS to design a tradional FNN. In our experiments, the membership functions are always chosen as Gaussian, and all input attributes are normalized to the range $[0, 1]$. 50 trials have been conducted with the FNN-RM and 5 trials with the ANFIS, since it takes much longer to train the ANFIS compared to the FNN-RM.

4.1 California Housing Data Set

The California Housing data set (www.niaad.liacc.up.pt/~ltorgo/ Regression/cal-housing.html) contains 20640 cases, 8 continuous inputs and one continuous output. In our simulation, 8000 training data and 12640 testing data are randomly selected from the entire data set for the FNN-RM algorithm, while 4000 training data, 4000 validation data and 12640 testing data are randomly selected for the ANFIS algorithm. The average results for multiple runs are shown in Table 1. Seen from Table 1, the FNN-RM algorithm obtains a much smaller error (root mean square (RMS)) and runs 52510 times faster than ANFIS does.

Table 1. Comparison in California Housing Data Set

Algorithms	Time (sec)	Training Error	Testing Error	Rule Num.
FNN-RM	0.5016	0.0741	0.0831	10
ANFIS	26339	0.190	0.263	256

4.2 Balance Scale Data Set

The Balance Scale data set [6] contains 625 cases, 4 numeric attributes and 3 classes. In our simulation, 325 training data and 300 testing data are randomly chosen for the proposed FNN-RM algorithm, while 200 training data, 125 validation data and 300 testing data are randomly chosen for the ANFIS algorithm. The average results are shown in Table 2. Seen from Table 2, the proposed FNN-RM algorithm achieves a much higher classification accuracy and runs 7929.8 times faster than ANFIS does.

4.3 Diabetes Data Set

The Diabetes data set [6] contains 768 cases, 8 attributes and 2 classes. In our simulation on the FNN-RM algorithm, 576 and 192 patterns are randomly chosen for training and testing at each trial, respectively. In the simulation on the ANFIS algorithm, 300, 276 and 192 patterns are randomly chosen for training, validation and testing, respectively. The average results for multiple runs shown in Table 2 show that our FNN-RM algorithm reaches a higher testing accuracy and runs 54617 times faster than ANFIS does. Compared with other methods' results in Table 3, the testing accuracy in our FNN-RM algorithm is the highest.

5 Discussion and Conclusion

In this paper, we first state the universal approximation capability for an FNN with arbitrary membership function parameters, with its proof given in [4]. Based

Table 2. Comparison for Diabetes and Balance Scale data sets.

Problems	Algorithms	Time (sec)	Training Accuracy	Testing Accuracy	Rule Num.
Diabetes	FNN-RM	0.0384	78.07%	77.76%	20
	ANFIS	2097.3	83.85%	64.58%	256
Balance Scale	FNN-RM	0.0443	89.72%	81.39%	10
	ANFIS	351.2886	72.62%	34.73%	256

Table 3. Comparison in accuracy for Diabetes data set.

Algorithms	Testing Accuracy
FNN-RM	77.76%
SVM [7]	76.50%
SAOCIF [8] [9]	77.32%
Cascade-Correlation [8] [9]	76.58%
AdaBoost [10]	75.60%
C4.5 [10]	71.60%
RBF [11]	76.30%
Heterogeneous RBF [11]	76.30%

on this theory, we proposed the new learning algorithm called FNN-RM. Compared with the traditional learning methods, the proposed FNN-RM method shows some surprising advantages as follows:

1. The learning speed of our FNN-RM algorithm is very fast. According to our experiments, when the number of input variables is very large (1000−10000), the learning speed of FNN-RM is at least 1000 times faster than conventional methods.
2. The proposed FNN-RM learning method can avoid being trapped in a local minimum for any rule set.
3. It is noticed from our experiments that the testing error in the FNN-RM method is always smaller than that in the traditional FNN learning methods and comparable to that obtained by support vector machines (SVMs), meaning that the proposed FNN-RM algorithm shows a excellent generalization performance.

References

1. Wang, L.X., Mendel, J.M.: Fuzzy Basis Functions, Universal Approximation, and Orthogonal Least Squares Learning. IEEE Trans. on Neural Networks, Vol. 3, No. 5 (1992) 807−814
2. Jang, J.S.R., Sun, C.T., Mizutani, E.: Neuro-fuzzy and Soft Computing. Prentice Hall International Inc. (1997)
3. Huang, G.B., Zhu, Q.Y., Siew, C.K.: Extreme Learning Machine. To Appear in 2004 International Joint Conference on Neural Networks (IJCNN'2004)(2004)

4. Liu, B., Wan, C.R., Wang, L.: A Fast and Accurate Fuzzy Neural Network with Random Membership Function Parameters. Submitted for Publication
5. Golub, G.H., Loan, C.F.V.: Matrix Computation. 3rd edn. Johns Hopkins University Press (1996)
6. Blake, C., Merz, C.: UCI Repository of Machine Learning Databases. http://www.ics.uci.edu/~mlearn/MLRepository.html, Department of Information and Computer Science, University of California, Irvine, USA (1998)
7. Ratsch, G., Onoda, T., Muller, K.R.: An Improvement of AdaBoost to Avoid Overfitting. Proceedings of the 5th International Conference on Neural Information Processing (ICONIP'1998) (1998)
8. Romero, E.: Function Approximation with Saocif: a General Sequential Method and a Particular Algorithm with Feed-forward Neural Networks. http://www.lsi.upc.es/dept/techreps/html/R01-41.html, Department de Llenguatgesi Sistemes Informatics, Universitat Politecnica de catalunya (2001)
9. Romero, E.: A New Incremental Method for Function Approximation Using Feed-forward Neural Networks. Proc. INNS-IEEE International Joint Conference on Neural Networks (IJCNN'2002) (2002) 1968–1973
10. Freund, Y., Schapire, R. E.: Experiments with a New Boosting Algorithm. International Conference on Machine Learning (1996) 148–156
11. Wilson, D.R., Martinez, T.R.: Heterogeneous Radial Basis Function Networks. Proceedings of the International Conference on Neural Networks (ICNN 96) (1996) 1263–1267

Tuning Neuro-fuzzy Function Approximator by Tabu Search*

Guangyuan Liu[1], Yonghui Fang[1,**], Xufei Zheng[2], and Yuhui Qiu[2]

[1] School of Electronic and Information Engineering, Southwest China Normal University,
Chongqing 400715, China
{liugy, fyhui}@swnu.edu.cn
[2] Faculty of Computer & Information Science, Southwest China Normal University,
Chongqing, 400715, China
{zxufei, yhqiu}@swnu.edu.cn

Abstract. Gradient techniques and genetic algorithms are currently the most widely used parameters learning methods for fuzzy neural networks. Since Gradient techniques search for local solutions and GA is easy to premature, tabu search algorithms are currently being investigated for the development of adaptive or self-tuning neuro-fuzzy approximator(NFA). By using the globe search technique, the fuzzy inference rules are built automatically. To show the effectiveness of this methodology, it has been used for modeling static nonlinear systems.

1 Introduction

In the past years, many researches were going on in the field of combining fuzzy systems and neural networks. Most of these studies took place to optimize fuzzy systems. Recently more and more researchers have used fuzzy models instead of multi-layer perceptron for approximation purposes [1~3].

The neuro-fuzzy approximator (NFA) used in this paper is based on Takagi-Sugeno linear inference rules and Radio based Function neural network (Fig. 1). It is a multi-layer network having four layers, in which the parameters are divided into nolinear parameters of the fuzzy membership functions and linear parameters of the output layer. So many researchers split the learning algorithm into two stages: the adaptation of the linear output weights in the first stage and then learning of the nonlinear parameters using backpropagation algorithm [2]. A more recent technique in implementing adaptive or self-tuning neuro-fuzzy system is by using genetic algorithms (GA) [4,5]. Since Gradient techniques search for local solutions and GA is easy to premature, finding another effective and rapid learning method is necessary.

---

* This research was supported by key project of Ministry of Education, China (104262) and fund project of Chongqing Science and technology Commission (2003-7881).
** Corresponding author.

This paper we will propose another globe optimal algorithm tabu search (TS) as the adaptive algorithms for the neuro-fuzzy approximator. Tabu search (TS) is a meta-heuristic search method originated by Glover (1986)[6,7] that has been proven very effective in solving large combinatorial optimization problem. TS is based on the use of *prohibition-based* techniques and "intelligent" schemes as a complement to basic heuristic algorithms like local search, with the purpose of guiding the basic heuristic beyond local optimum. Many researches have proven that TS has equivalence (even better) capability to GA and SA (simulated annealing). So this paper, we use TS as the parameters learning algorithms for the neuro-fuzzy network.

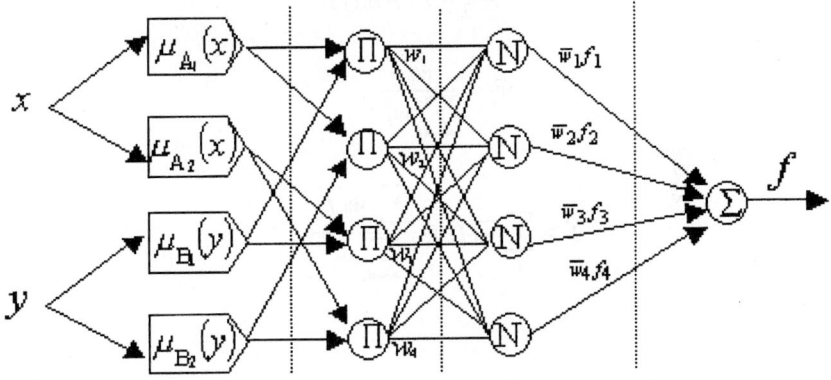

Fig. 1. The structure of the neuro-fuzzy network

2 Neuro-fuzzy Network

For simplicity, we use zero order liner function for the output of each rule. Researches have proven that zero order Takagi-Sugeno fuzzy inference system is a universal approximator[8,9]. Then the fuzzy inference rule is as follows:

$$R_j : \text{If } x \text{ is } A_i \text{ and } y \text{ is } B_k, \text{ then } f_j = r_j. \tag{1}$$

where x and y are the input variables, f is the zero order liner function.

Fig.1 shows the structure of the neuro-fuzzy approximator having two input variables. The node functions in the same layer are of the same function family. The first layer performs the fuzzification according to the membership function μ. For simplicity, we choose μ to be Gaussian function described by the relation (2). Where $\{a_i, c_i\}$ is the set of parameters. As the value of these parameters change, the shaped functions vary accordingly. Every node in the second layer is a circle node labeled Π which multiplies the incoming signals and sends the product out. Usually the T-norm operators perform generalized AND operation, shown as equation (3). The i-th node of the third layer calculates the ratio of the i-th rule's firing strength to the sum of all rule's firing strengths (equation (4)). The final layer containing only one neuron produces the result of the whole system (equation (5)).

In this paper, to tune the neuro-fuzzy network is to adapt the parameters of the preceding parameters $\{a_i, c_i\}$ and the consequent parameters $\{r_i\}$

$$\mu_{A_i}(x) = \exp\left[-\left(\frac{c_i - x}{a_i}\right)^2\right]. \tag{2}$$

$$w_i = \mu_{A_i}(x) \times \mu_{B_i}(y). \tag{3}$$

$$\overline{w}_i = \frac{w_i}{\sum_i w_i}. \tag{4}$$

$$f = \sum_i \overline{w}_i f_i = \frac{\sum_i w_i f_i}{\sum_i w_i}. \tag{5}$$

3 Tabu Search Used as Parameters Learning Algorithm for NFA

In this paper, we proposed a simultaneous tuning strategy of all of its parameters by TS, without the need to perform partial of sequential optimization. All the Gaussian input membership functions and the weights of the NFA are tuned by the TS. Furthermore, a flexible position coding strategy for the fuzzy membership functions is introduced to configure the fuzzy membership partitions in the corresponding universe of discourse. The proposed methodology is then tested on a two variables function.

3.1 Target Function and Solution Structure

The target function selected in this paper is mean square error (MSE) function having the following forms:

$$\text{MSE} = \sum_{i=1}^{P} (T(i) - O(i))^2 / P. \tag{6}$$

Where P is the number of elements in the train set. $T(i)$ is the i-th target value and $O(i)$ is the i-th output value of the neuro-fuzzy network. Using TS to tuning the parameters of neuro-fuzzy network, it is unnecessary for the energy function (or error function) to be differentiable. We also needn't divided the studying procedure into two steps but the linear and nonlinear parameters are adjusted simultaneously. We array the parameters { a_i, c_i } and { r_i } in one list. The difference from conventional heuristic methods (such as GA) is that every parameter in the list is represented by real number of itself, it's no need to encode them into binary digit of "0" and "1". The configuration of the solution is shown in Figure 2. In which, $a_N(t)$ and $c_N(t)$ are the parameters of the member function of the first layer and r_i is the consequent parameters of the output layer. When there has N input variables and every variable is divided into T fuzzy subsets, then there have T^N rules in the system. The parameter number of the neuro-fuzzy networks is $S = 2*N*T + T^N$.

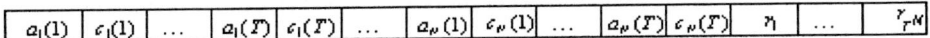

Fig. 2. Configuration of solution

3.2 The TS Algorithm Used in This Paper

According to the characters of the neuro-fuzzy network and tabu search, we design the parameters learning algorithm as follows:

Step1. Identify the number of the fuzzy set for each input variable.

Step2. Originate an initial solution x^{now} randomly, and set tabu table $H = \phi$.

Step3. Construct neighborhood $N(x^{now})$ for x^{now} through the method as follows: Produce a $\delta(i)$ ($0 \leq \delta(i) \leq 1$) randomly for every element in x^{now}. If the $\delta(i)$ is bigger than the given threshold P, then adding a little random number to the i-th element in the solution vector.

Step4. If the best solution in $N(x^{now})$ satisfies the aspiration criteria, then $x^{next} = x^{N\_best}$, go Step 6.

Step5. Construct the candidate set $Can\_N(x^{now})$ by selecting the solutions that satisfy the tabu conditions in $N(x^{now})$, Select the best solution x^{Can_best} in $Can\_N(x^{now})$, $x^{next} = x^{Can\_best}$.

Step6. $x^{now} = x^{next}$, renew the tabu table H.

Step7. repeat Step3, stop when it satisfies the stop rules.

3.3 Simulation

We have selected an approximation task to verify the availability of tabu search. The test function has the form as follows:

$$f(x,y) = \frac{\sin(x)*\sin(y)}{x*y}. \tag{7}$$

From the grid points of the range [-10,10]×[-10,10] within the input space of the above equation, 676 training data pairs were obtained first. Fig.3 presents the actual approximated shape of the function. Compared with the destination function, we can see that the output of NFA is very similar to the destination. The mean square error attached 10^{-4} Fig.4 presents the error of approximation and the search process of tabu search.

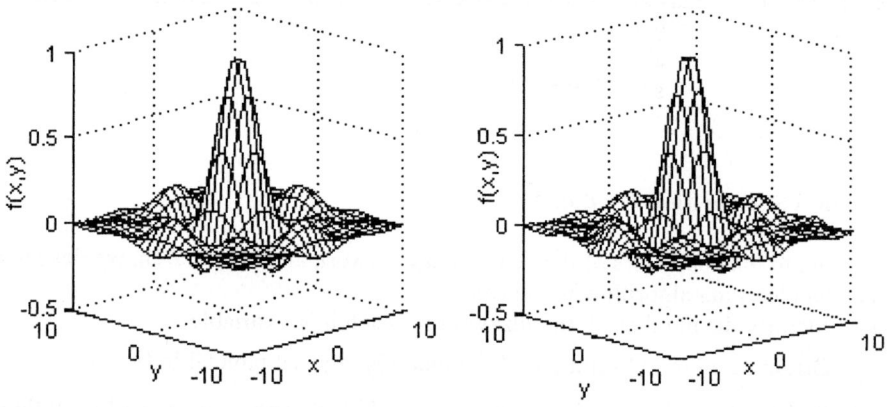

Fig. 3. The destination function for approximation(*left*) and the result of approximation(*right*)

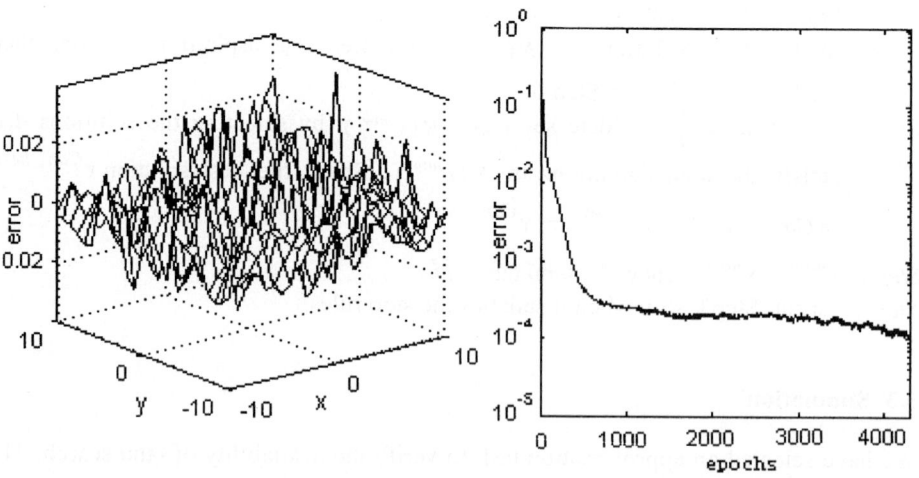

Fig. 4. Error of approximation(*left*) and the search process of tabu search(*right*)

4 Conclusion

This paper has presented an adaptive algorithm----tabu search algorithm to tuning the parameters of neuro-fuzzy approximator. The linear and nonlinear parameters of NFA are arranged in one list, all the parameters of the list present with the real value of themselves. To denote the list with the real value of the parameters, it is more clearly for the procedure, and no times is wasted to coding and decoding the parameters. Because of the good local searching ability of TS, the process always can jump the local optimum and attach the globe optimum. By using TS the problem is converted to combinatorial optimization problem, all the inference rules are built automatically. The NFA can approximate the test function perfectly after learned by tabu search algorithm. So TS is a good algorithm for tuning the neuro-fuzzy systems. Using tabu search as the learning algorithm, the process is clear and effectively. The neuro-fuzzy system based on tabu search can be used universally in many fields such as control, dynamic modeling etc.

References

1. Ishibuchi, H.: Development of fuzzy neural networks, in Fuzzy Modelling – Paradigm and Practice, ed. W.Pedrycz, Kluwer, Dordrecht (1997) 185–202
2. Tran Haoi, L., Osowski, S.: Neuro-fuzzy TSK network for approximation of static and dynamic functions. Control and Cybernetics, 31(2) (2002) 309–326
3. Jyh-Shing Roger Jang.: ANFIS: Adaptive-Network-Based Fuzzy Inference System. IEEE Transactions on Systems, Man, and Cybernetics, 23 (1993) 665–684
4. Teo Lian Seng, Marzuki Khalid, Rubiyah Yusof. Tuning of a neuro-fuzzy controller by genetic algorithms with an application to a coupled-tank liquid-level control system. International Journal of Engineering Applications on Artificial Intelligence. 11 (1998) 517–529
5. Miagkikh, V.V., Kononenko, R.N., Topchy, A.P., Melikhov, A.N.: Adaptive Genetic Search For Optimization Of Fuzzy And Neuro-Fuzzy Systems. Proc. of {IC} on Evolutionary Computation and Its Applications (1996) 245-253
6. Glover, F.: Tabu Search:part I.ORSA Journal on Computing (1989) 1:190–206
7. Glover, F.: Tabu Search:part II. ORSA Journal on Computing (1990) 2:4–3
8. Ke Zeng, Wenli Xu, Naiyao Zhang.: Universal Approximation of Special Mamdani Fuzzy Systems. Control and Decision, China 15(4) (2000)436–438
9. Kosko, B.: Fuzzy system as universal approximators. In: Proc IEEE Int Conf on Fuzzy Systems, San Diego (1992.) 1163–1170

Finite Convergence of MRI Neural Network for Linearly Separable Training Patterns*

Lijun Liu and Wei Wu**

Department of Applied Mathematics, Dalian University of Technology,
Dalian 116024, China
wuweiw@dlut.edu.cn

Abstract. MRI (Madaline Rule I) neural network has wide applications. A finite convergence for the training of MRI neural network is proved for linearly separable training patterns.

1 Introduction

A basic building block used in many neural networks is the "adaptive linear element" or Adaline, which gives the binary ±1 output through hard-limiting function $y = \text{sgn}(x)$. One of the earliest trainable layered neural networks with multiple Adalines was the MRI (Madaline Rule I) neural network of Widrow and Hoff [1],[2]. Madaline neural network has found wide applications in practice [3],[4].

Robustness analysis of Madaline neural network has been investigated in [5]–[8]. To the best of our knowledge, however, no rigorous mathematical proof has been made to the convergence of MRI. We are concerned in this paper the case where the training patterns are linearly separable. We establish a finite convergence result, that is, the training procedure will converge in finite steps of iterations.

2 The MRI Neural Network

The MRI neural network was constructed with an input layer, a hidden layer of Adalines, and a majority-vote-taker device(MAJ) as the output layer. For a given input pattern ξ^j, if the output response of the MAJ does not match the desired response O^j, the Adaline that is adapted by MRI is the one whose linear output is most close to zero. The rule of such an adaption can be either least mean square (LMS) algorithm

$$W^{new} = W^{old} + \eta(O^j - W^{old} \cdot \xi^j)\xi^j \tag{1}$$

---

* Partly supported by the National Natural Science Found of China, the Basic Research Program of the Committee of Science, Technology and Industry of National Defense of China.
** Corresponding author

or normalized least mean square (NLMS) algorithm

$$W^{new} = W^{old} + \eta \frac{(O^j - W^{old} \cdot \xi^j)\xi^j}{\|\xi^j\|^2} \quad (2)$$

where η is a positive constant.

We are supplied with a set of training pattern pairs $\{\xi^k, O^k\}_{k=1}^J \subset R^m \times \{\pm 1\}$. The usual training procedure chooses ξ^j from $\{\xi^k, O^k\}_{k=1}^J$ in a completely stochastic order, in which each pattern appears infinite times. Thus, starting from a random initial weight vector $W_i^0 \in R^m$, by the adaption procedure (1) or (2) we obtain a infinite weight sequence $\{W_i^k\}_{k=0}^\infty$ for the ith Adaline in the hidden layer.

If we set $x^k := O^k \xi^k$ then $\{\xi^k, O^k\}_{k=1}^J$ becomes $\{x^k, 1\}_{k=1}^J$. In this notation, the LMS (1) and NLMS (2) rules become respectively

$$W^{new} = W^{old} + \eta(1 - W^{old} \cdot x^j)x^j \quad (3)$$

and

$$W^{new} = W^{old} + \eta \frac{(1 - W^{old} \cdot x^j)x^j}{\|x^j\|^2} \quad (4)$$

We can see that the weight sequence $\{W_i^k\}_{k=1}^\infty$ remains unchanged under our simplification of symbols.

3 Main Results

We assume the training patterns are linearly separable, that is, there exists a vector $W^* \in R^m$, such that

$$W^* \cdot x^j > 1, \quad j = 1, 2, \cdots, J \quad (5)$$

Denote $\beta := \max_{1 \leq j \leq J} \|x^j\|$ and $\gamma := \min_{1 \leq j \leq J} W^* \cdot x^j$. For the main convergence result, we suppose that the following condition is satisfied.

$$0 < \eta < 2/\beta^2 \quad (6)$$

Theorem 1. *For the linearly separable training patterns, MRI neural network with LMS rule will converge in finite iteration steps if (6) is satisfied.*

Proof. Suppose weight sequence for the ith Adaline is $\{W_i^{i_k}\}_{k=0}^\infty$. Note that the weight must have given a wrong response to the input pattern x^{i_k} when it was updated. This means

$$W_i^{i_k} \cdot x^{i_k} < 0 \quad (7)$$

Thus according to (3) we have

$$\begin{aligned}\left\|W_i^{i_k+1} - W^*\right\|^2 &= \left\|W_i^{i_k} - W^*\right\|^2 + 2\eta\left(1 - W_i^{i_k} \cdot x^{i_k}\right)\left(W_i^{i_k} - W^*\right) \cdot x^{i_k} \\ &\quad + \eta^2\left(1 - W_i^{i_k} \cdot x^{i_k}\right)^2 \left\|x^{i_k}\right\|^2 \\ &= \left\|W_i^{i_k} - W^*\right\|^2 + \eta\left(1 - W_i^{i_k} \cdot x^{i_k}\right) \\ &\quad \left[2\left(W_i^{i_k} - W^*\right) \cdot x^{i_k} + \eta\left(1 - W_i^{i_k} \cdot x^{i_k}\right)\left\|x^{i_k}\right\|^2\right] \\ &= \left\|W_i^{i_k} - W^*\right\|^2 - \eta\left(1 - W_i^{i_k} \cdot x^{i_k}\right) \\ &\quad \left[W_i^{i_k} \cdot x^{i_k}\left(\eta\|x^{i_k}\|^2 - 2\right) + \left(2W^* \cdot x^{i_k} - \eta\|x^{i_k}\|^2\right)\right]\end{aligned} \quad (8)$$

From (5),(6) and (7), we have

$$\eta\|x^{i_k}\|^2 - 2 < 0$$

$$2W^* \cdot x^{i_k} - \eta\|x^{i_k}\|^2 > 0$$

$$1 - W_i^{i_k} \cdot x^{i_k} > 1$$

So, we have

$$\begin{aligned}\left\|W_i^{i_k+1} - W^*\right\|^2 &< \left\|W_i^{i_k} - W^*\right\|^2 - \eta\left(2W^* \cdot x^{i_k} - \eta\|x^{i_k}\|^2\right) \\ &< \left\|W_i^{i_k} - W^*\right\|^2 - \left(2\gamma - \eta\beta^2\right) \\ &= \left\|W_i^{i_k} - W^*\right\|^2 - \delta\end{aligned} \quad (9)$$

where $\delta := 2\gamma - \eta\beta^2 > 0$.

Here we can see that $\|W_i^{i_k+1} - W^*\|$ is monotonously decreasing. Applying (9) repeatedly we have

$$\left\|W_i^{i_k} - W^*\right\|^2 < \left\|W_i^{i_0} - W^*\right\|^2 - k\delta \quad (10)$$

Thus we have got an upper bound of the number of iteration steps

$$k < \|W_i^{i_0} - W^*\|^2/\delta$$

This completes the proof. □

We remark that in general $\|W_i^{i_k} - W^*\|$ will not go to zero since the iteration will stop in finite steps.

By the same technique, we can get a similar convergence result for NLMS adaption procedure (4) if

$$0 < \eta < 2 \quad (11)$$

is satisfied.

Theorem 2. *For the linearly separable training patterns, MRI neural network with NLMS rule will converge in finite iteration steps if* (11) *is satisfied.*

Proof. For the same reason, we just need to show that weights connected to the ith Adaline in the hidden layer keep unchanged after finite adaption steps. By (4) we have

$$\left\|W_i^{i_{k+1}} - W^*\right\|^2 = \|W_i^{i_k} - W^*\|^2 + 2\eta \frac{(1 - W_i^{i_k} \cdot x^{i_k})(W_i^{i_k} - W^*) \cdot x^{i_k}}{\|x^{i_k}\|^2}$$

$$+ \eta^2 \left(\frac{1 - W_i^{i_k} \cdot x^{i_k}}{\|x^{i_k}\|^2}\right)$$

$$= \|W_i^{i_k} - W^*\|^2$$

$$- \eta \frac{1 - W_i^{i_k} \cdot x^{i_k}}{\|x^{i_k}\|} \left[(\eta - 2)\frac{W_i^{i_k} \cdot x^{i_k}}{\|x^{i_k}\|} + \frac{2W^* \cdot x^{i_k} - \eta}{\|x^{i_k}\|}\right]$$

$$< \|W_i^{i_k} - W^*\|^2 - \frac{\eta}{\|x^{i_k}\|}\left(\frac{2W^* \cdot x^{i_k} - \eta}{\|x^{i_k}\|}\right) \quad (12)$$

where we have used conditions (5), (7) and (11).

To obtain our result, we define a positive constant

$$\zeta := \min_{1 \le j \le J} \frac{\eta}{\|x^j\|} \left[\frac{2W^* \cdot x^j - \eta}{\|x^j\|}\right]$$

Thus we have

$$\left\|W_i^{i_k} - W^*\right\|^2 < \|W_i^{i_0} - W^*\|^2 - k\zeta \quad (13)$$

from which we bound the iteration steps k by

$$k < \|W_i^{i_0} - W^*\|^2/\zeta \quad (14)$$

This completes the proof. □

References

1. Widrow, B., Lehr, M.A.: 30 Years of Adaptive Neural Networks: Perceptron, Madaline, and Backpropagation. Proc. of IEEE **78(9)** (1990) 1415-1441
2. Hoff, M.E.Jr.: Learning Phenomena in Networks of Adaptive Switching Circuits. Ph.D. thesis, Tech.Rep. 1554-1. Stanford Electron. Labs., Stanford, CA. (1962)
3. Widrow, B., Winter, R., Baxter, R.: Learing Phenomena in Layered Neural Networks. IEEE 1st Conf. on Neural Networks. (1987) 411-429
4. Zhu, Q.M., Tawfik, A.Y.: Quantitative Object Motion Prediction by an ART2 and Madaline Combined Neural Network: Concepts and Experiments. Artif. Intell. **8(5)** (1995) 569-578
5. Piche, S.: Robustness of Feedforward Neural Networks. IJCNN **2** (1992) 346-351
6. Oh, S., Lee, Y.: Sensitivity Analysis of Single Hidden Layer Neural Networks with Threshold Functions. Neural Networks **6(4)** (1995) 1005-1007
7. Stevenson, M., Winter, R., Widrow, B.: Sensitivity of Feedforward Neural Networks to Weight Errors. Neural Networks **1** (1990) 71-90
8. Widrow, B., Kamenetsky, M.: On the Statistical Efficiency of the LMS Family of Adaptive Algorithms. IJCNN **4** (2003) 2872-2880

A Rapid Two-Step Learning Algorithm for Spline Activation Function Neural Networks with the Application on Biped Gait Recognition

Lingyun Hu and Zengqi Sun

State Key Laboratory of Intelligent Technology and Systems, Computer Science and Technology Department, Tsinghua University, Beijing, China, 100084
huly02@mails.tsinghua.edu.cn, szq-dcs@tsinghua.edu.cn

Abstract. A fast two-stage learning algorithm is proposed to construct and optimize the weights of spline activation function neural networks (SAFNN). Feedforward network is firstly trained by back propagate (BP) algorithm, and then errors are applied to generate new neurons in hidden layers. A rapid dynamic updating algorithm is introduced to modify the new weights. Generalization capability and approximation precision are ensured by the two steps respectively. Simulation results on biped gaits demonstrate improvements in these two capabilities and of learning speed with comparison to traditional BP in SFANN and common NN.

1 Introduction

A majority of current neural networks (NN) is constructed on classical model, proposed by McCulloch and Pitts in 1943 [1] with fixed basis functions. To simplify the complexity of structure and computation in both hardware and software implementations, adaptive activation functions is proposed to provide more free parameters for representation.

A natural solution is involving adaptive sigmoid function $\frac{2a}{1+e^{-bx}} - c$ with gradient-based learning algorithm. Advantages of better data modeling of this method were proved by Chen and Chang [2]. Compared with it, a simpler solution is LUT (look-up-table) based on polynomial functions. Complexity of this method can be controlled easily with compromise of adaptation in coefficients learning and non-boundary of activation function. To overcome the first drawback, more free parameters should be introduced to smooth neuron's output, which is in accordance with structural simplicity while in contrast to computational simplicity [3]. Followed with it, Lorenzo Vecci [4] proposed the solution of spline based activation functions, whose shape can be modified through control points. Later, Stefano Guarnieri [5] presented a new structure of SAFNN with gradient-based learning algorithm.

To apply SAFNN into online biped gait recognition and planning, for its structural simplicity, a fast dynamic learning method based on BP and least-mean-square (LMS) algorithm is introduced in this paper. Besides simpler and quicker, weight connections learnt by this approach can be updated easily for new inputs given after the network has been trained. And new neurons can also be added to meet higher error and generalization criterion in this method.

2 The SAFNN

Different to traditional sigmoid activation function neurons, spline neurons adopt piecewise polynomial spline interpolation schemes to ensure a continuous first derivative and local adaptation. For given points $\{Q_1 \cdots Q_N\}$, $Q_i = [q_{xi}, q_{yi}]^T$, a general spline expression for that curve would be $F(u) = \bigcup_{i=1}^{N-3} F_i(u)$. Where $\bigcup$ is the concatenation operator on local spline basis functions and $F_i(u) = [F_{xi}(u), F_{yi}(u)]^T = \sum_{j=0}^{3} C_j(u) Q_{i+j} \cdot F_i(u)$ ($0 \leq u \leq 1$) is the i th curve span function controlled by four points $[Q_i \quad Q_{i+1} \quad Q_{i+2} \quad Q_{i+3}]^T$ and $C_j(u)$ are the Catmull-Rom (CR) cubic spline basis as (1)

$$C(u) = \begin{bmatrix} C_0(u) \\ C_1(u) \\ C_2(u) \\ C_3(u) \end{bmatrix} = \frac{1}{2}[u^3 \quad u^2 \quad u \quad 1] \times \begin{bmatrix} -1 & 3 & -3 & 1 \\ 2 & -5 & 4 & -1 \\ -1 & 0 & 1 & 0 \\ 0 & 2 & 0 & 0 \end{bmatrix} \quad (1)$$

For input s, proper local parameters of u and i can be obtained through $s = F_{xi}(u)$ and then substitute these values to $y = F_{yi}(u)$ to generate the corresponding output. To accelerate the computation of local parameters and reduce free parameters, we uniformly sample the x-axis with fixed step length $\Delta x = q_{x,i+1} - q_{x,i}$. This turned $F_{xi}(u)$ to a first degree polynomial of $F_{xi}(u) = u\Delta x + q_{x,i+1}$.

For a standard multilayer SAFNN with M layer and N_l ($l=1,\cdots M$) neurons per layer, quantities are defined as follows for later discussion. Structural details of SAFNN and spline neuron are shown in Fig1.

$x_k^l(t,p)$: Output of the k th neuron in the l th layer at the t th iteration related to the p th batch input of the NN. ($p=1,\cdots P$. $x_k^0(t)$ are inputs of the network; $x_0^l(t) = 1$ for bias computation. Indexes t and p are the same meaning in following definition.)

$w_{kj}^l(t,p)$: Weight of the kth neuron in the lth layer with respect to the jth neuron in the previous layer. (w_{0j}^l are bias weight)

$s_k^l(t,p)$: The linear combiner input of the kth neuron in the lth layer.

N: Number of control points in every neuron.

$i_k^l(t,p)$: Curve span index of the activation function for the kth neuron in the lth layer. ($1 \leq i_k^l(t,p) \leq N-3$)

$u_k^l(t,p)$: Local parameters of the $i_k^l(t)$ curve span for the kth neuron in the lth layer. ($0 \leq u_k^l(t) \leq 1$)

$q_{k,n}^l(t,p)$: Y-axis ordinate of the nth control point in the kth neuron of the lth layer. ($0 \leq n \leq N$)

$F_{k,i_k^l}^l(\cdot)$: Activation function of i_k^lth spline phase in kth neuron of lth layer.

$C_{k,m}^l(\cdot)$: mth CR polynomial function for kth neuron of lth layer. ($0 \leq m \leq 3$)

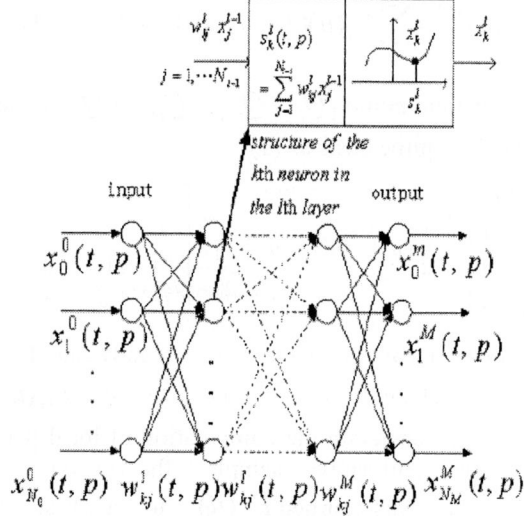

Fig. 1. Structure of SAFNN

$$i_k^l(t) = \left\lfloor \frac{s_k^l(t)}{\Delta x} + \frac{N}{2} \right\rfloor \quad (2)$$

$$u_k^l(t) = \frac{s_k^l(t)}{\Delta x} + \frac{N}{2} - i_k^l(t) \quad (3)$$

$$x_k^l(t) = F_{k,i_k^l}^l(\cdot) = \begin{cases} s_k^l(t) & l = M \\ \sum_{m=0}^{3} q_{k,(i_k^l+m)}^l C_{k,m}^l(u_k^l(t)) & l = 1, \cdots M-1 \end{cases} \quad (4)$$

Supposing $s_k^l(t, p) = \sum_{j=1}^{N_{l-1}} w_{kj}^l x_k^l$, then output of this neuron $x_k^l(t) = F_{k,i_k^l}^l(\cdot)$ can be calculated out by (2) to (4). Index parameter p is elided here for easy expression. Equation (2) and (3) implement local parameter computation as indicated before. $\lfloor \ \rfloor$ is the floor operator and the second term in equation (2) is designed to guarantee $i_k^l(t)$ be always nonnegative.

3 First Step Learning

BP algorithm is employed in the first step learning for both weights and control points in neurons. Supposing that $D_k(t)$ ($1 \le k \le N_m$) is the desired output, then $E = \sum_{p=1}^{P} E_d(t, p)$, where $E_d(t, p) = \frac{1}{2} \sum_{k=1}^{N_m} (D_k(t) - x_k^M(t))^2$ is the squared output error related to the pth desired output. Index parameter p will be also elided next for easy expression. According to BP algorithm, weights and the control points are updated with anti-gradient of error surface as (5) to (6).

$$w_{kj}^l(t+1) = w_{kj}^l(t) - \mu_w \frac{\partial E_d(t)}{\partial w_{kj}^l(t)} \quad (5)$$

$$q_{k,(i_k^l+m)}^l(t+1) = q_{k,(i_k^l+m)}^l(t) - \mu_q \frac{\partial E_d(t)}{\partial q_{k,(i_k^l+m)}^l(t)} \quad (6)$$

Details of parameters used in it can be found in [4]. This algorithm expanded BP on control points and can also be reduced to BP by setting these points not adapted.

4 Second Step Learning

Advantages of BP algorithm like global optimization and good generalization capability were retained in first step learning. To accelerate computation in BP, new neurons built to refine such areas should be dynamically added to SAFNN. Different to spline neurons initiated in first step learning, newly added N_M neurons using point pairs $[s_{N_{M-1}+j}^{M-1}(p), d_j(p) - x_j^M(p)]$ to construct $F_{N_{M-1}+j,i_k^l}^M(\cdot)$. Where $x_j^M(p)$ is the real output of network and $d_j(p)$ is the corresponding desired output for the pth batch input. The linear combiner $s_{N_{M-1}+j}^{M-1}(p,t) = \sum_{i=1}^{N_{M-2}} w_{ki}^{M-2}(p,t) x_k^{M-2}(p,t)$ $i=1,\cdots N_1, j=1,\cdots N_M$. N_M neurons were built to modify the input of corresponding ones in the Mth layer respectively. Define matrix as (7) and (8) and denote $x^{M-1}(t+1) = [x^{M-1}(t) \quad x_{N_{M-1}+1}^{M-1}(t)]$ Then the last layer can be expressed as $x^M(t) = x^{M-1}(t) w^M(t)$ and LMS solution to it is $w^M(t) = \{x^{M-1}(t)\}^+ x^M(t)$, where $\{x^{M-1}(t)\}^+$ is the pseudo inverse of $x^{M-1}(t)$ and can be calculated by (9). Parameters in it take the value as $d = \{x^{M-1}(t)\}^+ \times x_{N_{M-1}+1}^{M-1}(t)$, $b = \begin{cases} (c^+)^T \\ ((1+d^T d)^{-1} d^T \{x^{M-1}(t)\}^+)^T \end{cases}$,

$c = x_{N_{M-1}+1}^{M-1}(t) - x^{M-1}(t) \times d$.

$$w'(t) = \begin{bmatrix} w'_{1,1}(t) & \cdots & w'_{1,N_I}(t) \\ \vdots & & \vdots \\ w'_{N_I-1,1}(t) & \cdots & w'_{N_{I-1},N_I}(t) \end{bmatrix} \quad (7)$$

$$x'(t) = \begin{bmatrix} x'_1(t) & \cdots & x'_{N_I}(t) \end{bmatrix} \quad (8)$$

$$\{x^{M-1}(t+1)\}^+ = [\frac{\{x^{M-1}(t)\}^+ - db}{b}] \quad (9)$$

Finally new weights after adding neurons are $w^M(t+1) = [\frac{w^M(t) - db^T x^M(t)}{b^T x^M(t)}]$.

5 Simulation Result

Experiment is taken on pattern recognition of biped gait by SAFNN with the two step learning algorithm. (10) to (12) are equations to express cycloid character of biped gait trajectory. Result of same simulation on classical SAFNN with traditional BP algorithm is also presented for comparison. Three kinds of SISO NN as: SAFNN (1HL with 4 neurons) with two-step learning algorithm (NN1), SAFNN with BP (NN2) and common NN(1HL with 19 neurons) with BP algorithm (NN3) are designed for experiment. 28 control points were initialized by sigmoid function in each SAFNN neuron and the value of 28 was determined empirically. More neurons and larger learning rate are designed in common NN to achieve equivalent adaptable variables and better results.

$$x_7 = \frac{D_s}{\pi} \{\frac{2\pi}{N} i - \sin(\frac{2\pi}{N} i)\} + r_x \quad (10)$$

$$y_7 = 1.5 L_{fw} + r_y \qquad (11)$$

$$z_7 = \frac{L_7}{2}\{1 - \cos(\frac{2\pi}{N}i)\} + r_z \qquad (12)$$

(x_7, y_7, z_7) is the position of swing foot, (r_x, r_y, r_z) is random noises associated with (x_7, y_7, z_7). $L_2 = 50, L_3 = 40$, $L_7 = 5$ and $L_{fw} = 10$ are shank length, thigh length, ankle length and foot width respectively. $D_s = 25$ is the gait length. $i = 1, 2 \cdots N$ is sampling index and $N = 50$ is the total sampling number for one step. Fig.2 showed the training result of the three NN for (12) after 10,000 epochs. Better approximation precision and generalization capability of NN1 can be found in it easily. However, in some suitable range, more neurons in HL and training epochs would not improve the two aspects greatly in our algorithm as indicated in Fig. 3. So less neurons and epochs will be chosen under given precision to accelerate computation in gait simulation. Similar computation can be applied on hip and knee joints with the same algorithm as shown in Fig. 4.

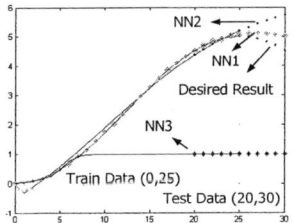 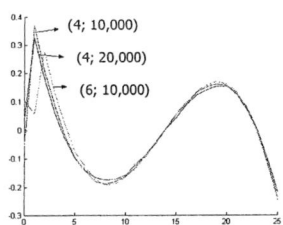 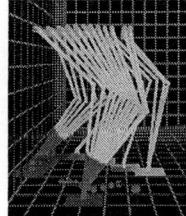

Fig. 2. Result of 3 NN trained by 0 to 25 and tested by 20 to 30.

Fig. 3. Different result with (i, j) assignment, where i: number of neurons in HL; j: training epochs.

Fig. 4. 3D simulation with trajectory learned by SAFNN

6 Conclusion

This paper focuses on two main properties of the two-step learning method for SAFNN including: 1) wider generalization capability; 2) higher approximation characteristics. Compared to classical BP algorithm, improvement in learning speed and precision should own to the dynamical neuron learning. Prove of it will be shown in another paper.

References

1. McCulloch, W. S., Pitts, W.: A Logical Calculus of the Ideas Imminent in Nervous Activity. Bull. Math. Biophys., Vol. 5, (1943) 115-133

2. Chen, C. T., Chang, W. D.: A Feedforward Neural Network with Function Shape Auto Tuning. Neural Networks, Vol. 9, (1996) 627-641
3. Piazza, F., Uncini,, A. and Zenobi, M.: Neural Networks with Digital LUT Activation Function. Proc. of the IJCNN, Beijing, China, Vol. 2, (1993) 343-349
4. Vecci, L. (ed.): Learning and Approximation Capabilities of Adaptive Spline Activation Function Neural Networks. Neural Networks, Vol. 11, (1998) 259-270
5. Guarnieri, S. (ed.): Multilayer Feedforward Networks with Adaptive Spline Activation Function. IEEE Tran. Neural Networks, Vol. 10(3), (1999) 672-683

An Online Feature Learning Algorithm Using HCI-Based Reinforcement Learning

Fang Liu and Jianbo Su

Research Center of Intelligent Robotics & Dept. of Automation
Shanghai Jiaotong University, Shanghai, Postfach 20 00 30, China
{fangliu,jbsu}@sjtu.edu.cn

Abstract. In this paper, a novel learning approach for feature selection is proposed based on the reinforcement learning algorithm. We apply reinforcement learning framework to deal with feature selection, and further add human-computer interaction (HCI) to reduce the learning complexity and speed up the convergence rate. Feature learning is formulated as an online sequential decision-making that interacts with uncertain environment. The system can learn the adaptive desired feature subset with higher discriminatory information. Experiments of detecting a specific object in different conditions are provided to verify the effectiveness of the proposed approach.

Keywords. Feature learning, reinforcement learning, HCI

1 Introduction

Feature selection is important because the choice of the features significantly affects system's performance. Most of the features obtained from image processing are irrelevant or redundant for a given task, but only a small part of the features are essential. Unfortunately, those irrelevant or redundant features not only requires more computation power to acquire in image processing, but also slow down computational speed and improve inaccuracy. Therefore, it is important to study how to quickly make out best feature set that properly matches a particular task. Feature selection is defined as finding the most compact feature set that is necessary and sufficient for a specific task. This problem is fundamental in a number of tasks like classification [3], [4], data mining [2], machine learning [1], etc. Based on the evaluation criterion, there are two strategies for feature selection. One is feature filter that evaluates the goodness of feature from the data, ignoring the classification or learning algorithm. Another is the wrapper approach that needs to measure performance to evaluate feature set, e.g., the accuracy of the classifier, cross validation. Wrapper approach finds better feature subset for a specific learning algorithm, but generally suffers from higher computational complexity [5]. Filtering approach is a faster and flexible one that can be combined with any learning algorithm. There are a number of measures (distance, dependence, information, consistency, and accuracy) to evaluate goodness of selected feature subset and remove irrelevant features.

However, these studies focus on the selection of a subset for a given classification task, but not visual context (i.e., the decision does not depend on the current visual environment). Feature selection for visual context usually adapts to novel visual environment unforeseeable by the designer. Machine learning allows the system to improve its ability in feature selection adaptively over time, based on the visual environment that it acts on. Then this feature learning procedure is incremental, interactive and task-oriented. With these characteristics, system can learn dynamically the updated desired feature subset and improve the performance over time. Moreover, machine learning can solve the feature selection problem when there is no knowledge about the visual environment and task. As an incremental, interactive, task-oriented, and model-free learning algorithm, reinforcement learning (Q-learning) is very suitable for feature selection. In this paper, reinforcement learning is used to learn feature selection strategy through interaction with visual environment, considering the probability distribution of the visual environment. Moreover, since the standard reinforcement learning suffers from average update complexity depending on the size of the state-action space. Human-computer interaction (HCI) is involved in the learning procedure to provide direct evaluation for the learning performance, which will greatly reduce learning complexity and speed up the convergence rate.

The paper is organized as follows: Section 2 presents a feature learning theory and an HCI-based feature learning scheme. Section 3 gives experimental results, followed by conclusion and future directions of our research.

2 Feature Learning

2.1 Feature Extraction and Construction of the Feature Space

Various types of low-level or mid-level features can be extracted from images, such as color, shape, motion, and texture, etc. For example, red feature and blue feature are the most commonly used features in practice as the low-level color features. Shape feature is obtained by calculating the Hough transform of an edge map for circles. For stationary camera position, the motion feature can be achieved by frame subtraction. The way to generate a feature space is by defining a feature as an attributes composition of several components. The feature attribute vector can be expressed as:

$$\mathbf{f} = (f_{color}, f_{shape}, f_{motion}, f_{texture}, \cdots), \qquad (1)$$

where $\mathbf{f}$ denotes the integration of all feature variables. f_{color}, f_{shape}, f_{motion} and $f_{texture}$ are the feature variable of color, shape, motion, and texture respectively.

2.2 Reinforcement Learning-Based Feature Learning

The feature selection procedure is designed as a model of sequential decision-making that interacts with uncertain visual environment. We define *state* as the feature attribute vector $\mathbf{f}$ in the feature space representing the properties of the image, and action as the selection of the feature subset. The *action a* at the state **f** is defined as adding, deleting, or changing a feature. Via computing a similarity function, the system receives 0 for the *reward r* if the action selects the desired feature subset, or -1 otherwise.

The general principle of reinforcement learning can be seen in [7]. In most applications, reward r of executing action a at state **f** and the transition function δ describing the change from the state $\mathbf{f}_t$ to state $\mathbf{f}_{t+1}$ when executing action a_t are initially unknown. A model-free Q-learning algorithm is developed to find optimal actions. Initially, all the values of state-action evaluation function $Q(\mathbf{f}, a)$ are initialized. Every time in one feature state, the system chooses and executes an action, and the $Q(\mathbf{f}, a)$ is updated as follows:

$$Q_{t+1}(\mathbf{f}_t, a_t) = Q_t(\mathbf{f}_t, a_t) + \alpha[r_t + \gamma \max_{a_{t+1}} Q_t(\mathbf{f}_{t+1}, a_{t+1}) - Q_t(\mathbf{f}_t, a_t)], \quad (2)$$

where $\alpha(0 < \alpha < 1)$, is the learning rate; γ is the parameter, called the discounted rate, normally we have $0 < \gamma < 1$. The value of $Q(\mathbf{f}, a)$ is the reward r of executing action a at state **f** plus the value of following the optimal policy thereafter. With each state-action pair, the system stores a value $Q(\mathbf{f}, a)$, called the action-value function for policy π. These action-values are used to compute the system's decision policy π. Watkins proved that the Q-values would converge to the optimal values given that each state-action pair is experienced infinitely and the environment is Markovian [8]. The optimal policy $\pi^*(\mathbf{f})$ can be obtained by the following definition:

$$\pi^*(\mathbf{f}) = arg \max_a Q(\mathbf{f}, a). \quad (3)$$

2.3 HCI-Based Feature Learning

The drawback of standard reinforcement learning is that the system suffers from average update complexity depending on the size of the state-action space. We propose an alternative to obtain reward r through human-computer interaction in reinforcement learning framework. The reward is defined according to different evaluation criterions that can measure distance with goal, which has the following form:

$$r_{hci} = k_1 r_1 + k_2 r_2 + \cdots + k_n r_n, \quad (4)$$

where r_i is the reward that is provided by operator based on the ith evaluation criterion, k_i is the weight of r_i, and n is the number of evaluation criteria. The system can obtain the distance measurement between current feature state and desired feature state to simplify the learning process greatly for a complex task. It constructs a interactive learning architecture that operator can give a direct feedback to learning system according to the performance of system behavior.

There are three aspects in a typical feature selection method: 1) search direction; 2) search strategies; and 3) evaluation measures [6]. Our method for feature selection differs from the conventional ones in the following properties: 1) feature selection can be interactive with uncertain visual environment; 2) searching can be in either forward or backward direction; 3) evaluation criterion is independent of the system's prior knowledge.

3 Experimental Results

3.1 The Performance of the Feature Learning

The first experiment shows that the system could learn to automatically select the desired feature subset to locate the object using reinforcement learning. Fig.1. shows the feature learning and object location. In Fig.1. (a), the original image is shown. In the scene, there are two moving objects, one is blue circle object, another one is red block. The system learns to select the feature subset that can distinguish the target (blue circle object) from the background. As can be seen, the system successfully learns to select the shape and motion features (desired feature subset) to exactly locate the target. In Fig.1. (d), the target is also detected by selecting blue and shape features, but this feature subset is not the desired one because the blue feature hampers the performance of the object location.

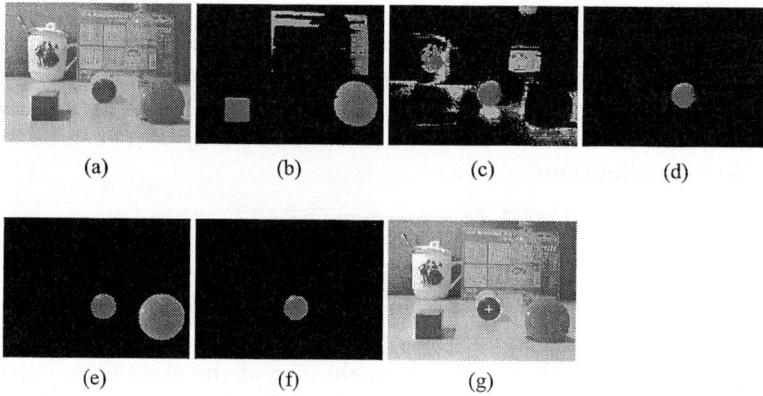

Fig. 1. The feature learning and object location. (a) the original image; (b) to (f) shows the sequence of the feature learning process. (b) red feature; (c) blue feature; (d) blue and shape (circle) features. (e) shape feature; (f) shape and motion features; (g) object location.

Fig.3. shows the learning steps per trial from the start feature subset to the desired one. The results were averaged over 200 trials. It is apparent that the number of feature selection actions decreases with the increase of the number of trials. The proposed feature learning approach is efficient to improve the performance of system.

3.2 Feature Learning Under Changed Visual Environment

In this experiment, the ability of the system to learn the adaptive desired feature subset under a changed visual environment was tested. Fig.2. shows the feature learning and object location under changed visual environment. Fig.2. (a) shows the image, where the visual environment has changed because of the disappearance of red circle object. As can be seen, the system successfully learns to select the changed desired feature subset (shape feature) to exactly locate the target. In Fig.2. (d), the target is also detected by selecting shape and motion features, but this feature subset is not the best feature subset because it is redundant for the task due to the changed visual environment. The system can learn to select changed desired feature subset by using reinforcement learning interacts with the visual environment.

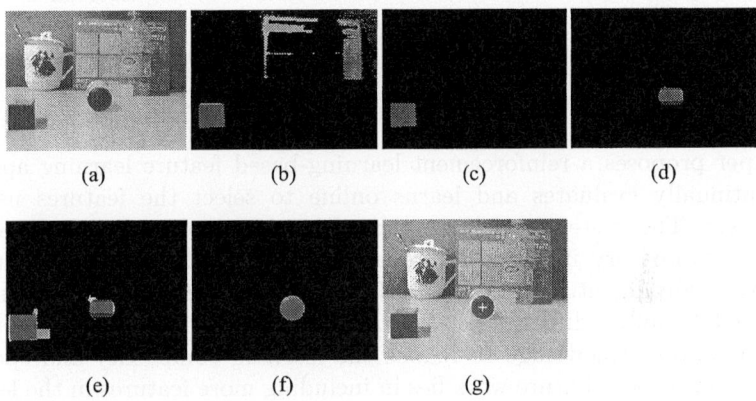

Fig. 2. Feature learning and object location under changed visual environment. (a) the original image under changed visual environment; (b) to (f) shows the sequence of the feature learning process. (b) red feature; (c) red and motion features; (d) shape (circle) and motion features; (e) motion feature; (f) shape feature; (g) object location.

Fig.4. shows the performance of feature learning under changed visual environment. The results were averaged over 200 trials. It is apparent that the number of feature selection with changed environment is less because of the simplified visual environment.

3.3 The Performance of the HCI-Based Feature Learning

In this experiment, the performance of the HCI-based feature learning has been tested. The operator should provide reward r_{hci} for reinforcement learning via HCI in order to simplify the feature learning process. Fig.5. compares the performance of feature learning and HCI-based feature learning on the number of actions per trial. The results were averaged over 200 trials. As can be seen, the number of feature selection with HCI-based feature learning is less than standard reinforcement learning, and the HCI-based feature learning is more effectively used to learn to select the desired feature subset. The experiments result shows that the HCI-based reward helps the system efficiently receive the evaluation of its action to reduce the learning complexity and speed up the convergence rate. When the feature space becomes larger, further performance improvement could be expected to reach.

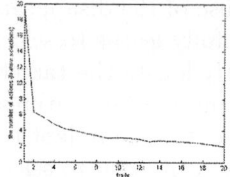 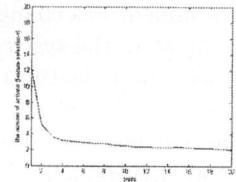 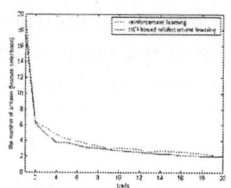

Fig. 3. Performance of feature learning

Fig. 4. Performance of feature learning under changed visual environment

Fig. 5. Comparison between two feature learning algorithms

4 Conclusions

This paper proposes a reinforcement learning-based feature learning approach that continually evaluates and learns online to select the features used for special task. The system can learn the adaptive desired feature subset with higher discriminatory information, considering the probability distribution of the visual environment. Moreover an HCI-based reinforcement learning scheme is proposed to obtain direct feedback about feature selection behavior with the help of operator's knowledge to reduce the learning complexity and speed up the convergence rate. Future work lies in including more features in the learning framework so as to extend its application to more complex tasks.

Acknowledgments. This research has been supported by NSFC, under grant 60275042.

References

1. Bell, D.A., Wang, H.: A Formalism for Relevance and its Application in Feature Subset Selection. Machine Learning, Vol.41 (2000) 175-195
2. Chen, M., Han. J., Yu, P.S.: Data Mining: an Overview from Database Perspective. IEEE Transactions on Knowledge and Data Engineering, Vol.8, No.6 (1996) 866-833
3. Dash, M., Liu, H.: Feature Selection for Classification. Intelligent Data Analysis, Vol.1, No.(1-4) (1997) 131-156
4. Jain, A.K., Duin, R.P.W., Mao, J.: Statistical Pattern Recognition: a Review. IEEE Transactions on Pattern Analysis and Machine Intelligence, Vol.22, No.1 (2000) 4-37
5. Kohavi, R., John, G.H.: Wrappers for Feature Subset Selection. Artificial Intelligence, Vol.97, No.(1-2) (1997) 273-324
6. Liu, H., Motoda, H.: Feature Selection for Knowledge Discovery and Data Mining. Kluwer Academic Publishers, Boston, (1998)
7. Sutton, R.S., Barto, A.G.: Reinforcement Learning: An Introduction. MIT Press, Cambridge, MA (1998)
8. Watkins, C.J.C.H., Dayan, P.: Technical Note: Q-Learning. Machine Learning, Vol.8 (1992) 279-292

Optimizing the Weights of Neural Networks Based on Antibody Clonal Simulated Annealing Algorithm

Xiaoyi Jin[1], Haifeng Du[2,3], Wuhong He[2], and Licheng Jiao[2]

[1] Institute for Population and Development Studies, Xi'an Jiaotong University
710049, Xi'an, China
Xiaoyijin@163.com
[2] Institute of Intelligent Information Processing, Xidian University
710071, Xi'an, China
{Haifengdu72, lchjiao1}@163.com
[3] Industry Training Center, Xi'an Jiaotong University
710049, Xi'an, China
Haifengdu72@163.com

Abstract. Based on the clonal selection theory, a new algorithm, Antibody Clone Simulated Annealing Algorithm, is put forward for optimizing the weights of neural networks. Combining the mechanism of the clonal selection and the simulated annealing, the new algorithm optimizes the weights using a population instead of single point so as to enlarge the searching range and overcome the shortcomings of the simulated annealing algorithm. The effectiveness of the method is proved by the experiments optimizing the weights of the forward neural networks.

1 Introduction

The whole weight distribution includes all the knowledge of neural networks. The conventional method obtaining the weights is to give the initial values randomly, then according to the rule (such as gradient descent) of weight changing, to make adjustment so as to obtain a better weight distribution step by step. However, the shortcoming of the algorithm and the limitation on robustness of the random initialization of neural network may influence the learning efficiency. On the other hand, the weights adjustment of neural networks, namely the learning process, is virtually a process of global search in the weight space. Therefore, the optimization strategies, such as simulated annealing, genetic algorithm etc. have been applied to neural networks learning and perform effectively [1].

Based on the mechanism of the Antibody Clone Selection Theory, this paper discusses a novel Antibody Clone Simulated Annealing Algorithm (ACSAA). The algorithm takes the energy function and the solution of problem as antigen and antibody respectively. It combines the advantage of the diversity of antibody clone with the random search ability of simulated annealing, so that the weight optimization is processed by searching not in a single point, but in a population. On the other hand, using the knowledge of explored space to direct the clone mutation and employing the local optimization strategy (such as gradient descent) to promote the antibody-antigen affinity, the algorithm becomes more effective. The relative experimental results of weight learning of neural networks also corroborate the effectiveness of the new algorithm we put forward here.

2 Simulated Annealing and Antibody Clone Selection

Based on the conception and method of statistical physics, S.Kirkpatrick etc. proposed an optimization algorithm simulating the annealing process of solid material, namely simulated annealing algorithm (SA), which is a simple and available stochastic optimization algorithm [2]. Although the simulated annealing algorithm is widely used, it still has some drawbacks. For example, due to multiple nesting cycles, it possesses high computational complexity. Meanwhile, the usage of explored space knowledge is not enough; consequently, it cannot direct the following search adequately.

Artificial immune system describes a learning technique based on the heuristic of biological immune system and the natural defense mechanism of learning outside material; it is novel and has the potential to solve the complex problems [3]. Thanks to the gene mutation of immune cells during the proliferation, the diversity of immune cells is generated, and the proliferation of these cells results in vegetative propagation system. This vegetative propagation of cells is called clone. The concept of clone has been widely applied in computer programming [4], system control [5], and interactive parallel simulation, etc. In artificial immune system, the mechanism of clone has already aroused interests of the researchers [6].

In addition, immunology indicates that the clone selection is a process, during which each antigen invading the organism can activate, differentiate and proliferate the corresponding immune cells clone, and finally remove the antibody. This mechanism can be used to the optimization search. Not trying to optimize in the overall situation, it deals with the different antibodies of antigen evolutionarily in order to improve the ability of global search, enlarge the antibody diversity and avoid falling into the local optimum to some extent.

3 Antibody Clonal Simulated Annealing Algorithm

In particular, for the problem of confirming the weights of neural networks, define the energy function as antigen E and the solution (the weight matrix or vector of the network) as antibody W. The antibody population is $C, C = \{W^1 \quad W^2 \quad \cdots \quad W^n\}$, here the corresponding energy function of $W^i(k)$ is the affinity. Additionally, considering the low measurements ability of binary code, low accuracy, length limitation of string and the problem of confirming the weight of network itself, the real number code is adopted in this paper. ACSAA can be described as Fig. 1.

ALGORITHM Antibody Clone Simulated Annealing Algorithm

Step1 Initialization: randomly set the antibody population $C(0)$, that is the weights of neural networks. Initialize temperature T_0 as high as possible, choose available annealing strategy.

Step2 Clone: According to the affinity of antibody $W^i(k)$ and antigen, clone as follow:

$$\overline{W}^i(k) = \Theta(W^i(k)) = \begin{bmatrix} W_1^i(k) & W_2^i(k) & \cdots & W_{q_i}^i(k) \end{bmatrix}^T \quad i=1,2\cdots n$$

Where, $W_j^i(k) = W^i(k) \quad j=1,2\cdots q_i$. Generally, q_i is given by:

$$q_i = \text{int}\left[N * \frac{E(W^i(k))}{\sum_{j=1}^n E(W^j(k))} \right] \quad i=1,2\cdots n$$

$N>n$ is a given integer relating to the clonal size. int (•) rounds the elements of X to the nearest integers towards infinity and int (x) returns the smallest integer bigger than x.

Step3: Clone Mutation: Mutate the antibody population after cloned

$$\overline{C}(k) = \Theta(C(k)) = \begin{bmatrix} \overline{W}^1(k) & \overline{W}^2(k) & \cdots & \overline{W}^n(k) \end{bmatrix}^T,$$

for each antibody, there is equation as follow:

$$\hat{W}_j^i(k) = \overline{W}_j^i(k) + W_j^i(k) \qquad \overline{W}_j^i(k) \in \overline{W}^i(k) \qquad i=1,2,\cdots n$$

Step4 Annealing Selection: Calculate the affinity of new antibody, and:

$$E_j^i(k) = E(\hat{W}_j^i(k)) - E(W^i(k)) \qquad j=1,2\cdots q_i$$

It is assumed that $\Delta E_{\min}(k)$ is the smallest, the energy is $E_{\min}(W_l^i(k))$. If $\Delta E_{\min}(k) < 0$, then $W_l^i(k)$ is accepted as the new situation, otherwise, it is accepted with the probability $p = \exp(\frac{-\Delta E_{\min}(k)}{kT}) > a$. If $W_l^i(k)$ is accepted, such that $W^i(k+1) = W_l^i(k)$, $E(W^i(k+1)) = E_{\min}(W_l^i(k))$, or else $W^i(k+1) = W^i(k)$, $E(W^i(k+1)) = E(W^i(k))$.

Decrease the temperature T according to the given annealing strategy.

Step5 Halt Condition: if there is at least one antibody's affinity is smaller than the best one E_{op} or no required temperature T_s in affinity $E(C(k))$, then stop, or else go to Step2.

Fig. 1. The main steps of ACSAA.

Where, k is the evolution generation, and the affinity is the square sum of errors of the neural networks. The annealing strategy is

$$T_k = \left(\frac{k-1}{k}\right)^p T_{k-1} \quad k \geq 2 \tag{1}$$

Along with the increasing iteration number of the algorithm, the decreasing speed of temperature is more and more slowly, which is corresponding with the annealing process of physical system. And it is well controlled according to an appropriate p.

Adding the item of momentum on the basis of negative gradient of the affinity realizes clone mutation, which matures the antibody-antigen affinity.

$$E_j^i(k) = \eta \frac{\partial E(W^i(k))}{\partial W^i(k)} + \delta \times E_j^i(k-1) + \xi \times \sqrt{\beta^i \times E(W^i(k)) + r^i} \times N^i(0,1) \qquad (2)$$

Where, η is the learning rate, δ is the momentum factor, $N^i(0,1)$ is a uniform random distribution with the expectation of 0 and the square error of 1, ξ is the Gaussian coefficient. Introducing the item of Gaussian mutation is to increase the diversity of population fatherly, hence ξ is very small generally. At the same time, β^i and r^i are the unknown parameters, generally set as 1 and 0.

ACSAA associates with local search strategy based on gradient descent to accelerate the mature speed, therefore it is more efficient. Compared with genetic algorithm and hybrid genetic algorithm (GA-BP)[1], ACSAA doesn't adopt crossover operator, but generates a population of mutation solution around the candidate solutions according to the affinity. Thus, the local search space is enlarged. Because the number of antibody clone is constrained by the affinity, its computational complexity is lower than that of genetic algorithm.

4 Experiments and Discussion

To prove the validity of the new approach, we take the neural network approximation problem f_1, f_2, f_3 as examples. Here, ACSAA is applied to optimize the weights of neural networks.

$$f_1(x) = \sin(5x) \qquad -1 \leq x \leq 1 \qquad (3)$$

$$f_2(x) = \frac{\sin(5x) + \log(2.5 * |x|)}{\max(abs(f_2(x)))} \qquad -1 \leq x \leq 1 \qquad (4)$$

$$f_3(x) = \frac{[10 - \frac{\sin(1/x)}{(x-0.16)^2 + 0.1}]}{\max[abs(f_3(x))]} \qquad 0 < x \leq 1 \qquad (5)$$

The multilayer forward propagation neural network is shown in Fig. 2, where the node function is Gauss function. To simplify the description, we have the network expressed uniformly as $[n_i$-n_h-$n_o]$, where n_i is the number of input variables, n_h is the number of hidden variables, and n_o is the number of output variables.

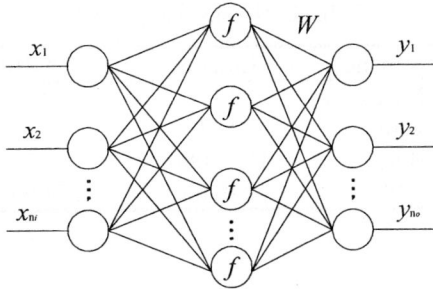

Fig. 2. Forward propagation neural network

In our experiments, $m=67$, $E_{op}=0.0001$, $N=150$, $T_0=1$, $p=2$, $Ts = 4.0 \cong 10^{-4}$, 1-20-1 is for f_1; $m=67$, $E_{op}=0.05$, $N=150$, $T_0=1$, $p=2$, $Ts = 4.0 \cong 10^{-4}$, 1-20-1 is for f_2; and $m=100$, $E_{op}=0.6$, $N=150$, $T_0=1$, $p=2$, $Ts = 4.0 \cong 10^{-4}$, 1-34-1 for f_3. Where, m is the number of input samples, and N is the antibody clone size of each iteration. Table 1 shows the statistical results of 10 independent runs for the same samples. For the comparison, Table 1 gives the numbers of the random initial training weights, the evolution generation of GA-BP and ACSAA, which have the same initial population, with the same size of 100. The approximation accuracy of neural network expressed by the sum of error square is shown in the brackets. The experimental results indicate that ACSAA can optimize the weights of neural networks more efficiently than GA-BP, especially in complex problems like approximation to f_3. Employing the results of 10 generations as the initial weights of the neural network to do further approximation to f_3, and with the same experimental conditions as above, the experimental results are shown in Table 2. It indicates ACSAA is also efficient as the initialization strategy of weights of the neural network.

The experimental results also describe that both GA-BP and ACSAA are insensitive to initial values, but the neural network learning algorithm based on the gradient descent method has the poor robustness for the initial values.

Table 1. The simulation results I

Function	Performance	Random	GA-BP	ACSAA
$f_1(0.001)$	Max	60	47	18
	Min	38	28	9
	Mean	48.4000	38.9000	14.8000
	Std	7.3666	6.1183	2.6998
$f_2(0.05)$	Max	40	20	9
	Min	18	10	5
	Mean	27	15.7000	6.9000
	Std	7.4386	3.3682	1.1005
$f_3(0.6)$	Max	70	52	17
	Min	37	16	7
	Mean	47.1000	32.6667	11.5000
	Std	11.2195	9.9499	3.5040

Table 2. The simulation results II

Function	Performance	Random	GA-BP	ACSAA
$f_3(0.5)$	Max	135	101	78
	Min	114	77	42
	Mean	125.2000	94.2000	64.8000
	Std	9.0160	15.0096	12.1454

5 Conclusion

Combining the mechanism of antibody clone and traditional strategy of simulated annealing, this paper explores a novel approach, Antibody Clone Simulated Annealing Algorithm (ACSAA). Applying the negative gradient to make the affinity mature, ACSAA overcomes the shortcoming of "random walking" of conventional SA and strengthens the characteristics of the conventional SA as a method of noisy gradient descent. Compared with the genetic algorithm and hybrid genetic algorithm applied for neural network weight optimization, ACSAA does not adopt the crossover operator, but uses the mutation operator to produce a variation around the candidate solution according to the affinity. Accordingly, it enlarges the searching area and maintains the population diversity. ACSAA is a populating search rather than single point search, and remedies the weaknesses of conventional simulated annealing algorithm. The corresponding experiments indicate that ACSAA is suitable to initialize and optimize the weights of neural networks.

References

1. Chen, G.L., Wang, X.F., Zhuang, Z.Q.: Genetic Algorithms and Its Application. People's Mail Press, Beijing (1996)
2. Wu, H.Y., Chang, B.G., Zhu, C.C.: Multi-population Parallel Genetic Algorithm Based on Simulated Annealing. Journal of Software 11 (2000) 416–420
3. Dasgupta, D., Forrest, S.: Artificial Immune Systems in Industrial Applications. In: John, A. M., Marcello, M.V. (eds.): Proceedings of the Second International Conference on Intelligent Processing and Manufacturing of Materials, Vol. 1. IEEE, Hawaii (1999) 257–267
4. Balazinska, M., Merlo, E., Dagenais, M.: Advanced Clone-Analysis to Support Object-Oriented System Refactoring. In: Cristina, C., Sun, M., Elliot, C. (eds.): Proceedings of Seventh Working Conference on Reverse Engineering, IEEE, Brisbane (2000) 98–107
5. Esmaili, N., Sammut, C., Shirazi, G.M.: Behavioral Cloning in Control of a Dynamic System. In: DeSilva, W. (ed.): IEEE International Conference on Systems, Man and Cybernetics Intelligent Systems for the 21st Century, Vol. 3. IEEE, Vancouver (1995) 2904–2909
6. Kim, J., Bentley, P.J.: Towards an Artificial Immune System for Network Intrusion Detection: An Investigation of Clonal Selection with a Negative Selection Operator. In: IEEE Neural Networks Council. (ed.): Proceedings of the 2001 Congress on Evolutionary Computation, Vol. 2. IEEE, Seoul (2001) 1244–1252

Backpropagation Analysis of the Limited Precision on High-Order Function Neural Networks

Minghu Jiang[1,2] and Georges Gielen[2]

[1]Lab of Computational Linguistics, Dept. of Chinese Language,
Tsinghua University, Beijing, 100084, China
jiang.mh@tsinghua.edu.cn
[2]Dept. of Electrical Eng., MICAS, K.U.Leuven,
Kasteelpark Arenberg 10, B-3001 Leuven-Heverlee, Belgium
georges.gielen@esat.kuleuven.ac.be

Abstract. Quantization analysis of the limited precision is widely used in the hardware realization of neural networks. Due to the most neural computations are required in the training phase, the effects of quantization are more significant in this phase. We pay attention and analyze backpropagation training and recall of the limited precision on the HOFNN, point out the potential problems and the performance sensitivity with lower-bit quantization. We compare the training performances with and without weight clipping, derive the effects of the quantization error on backpropagation for on-chip and off-chip training. Our experimental simulation results verify the presented theoretical analysis.

Keywords: Quantization, Weight Clipping, Backpropagation (BP), High Order Function Neural Networks (HOFNN)

1 Introduction

Reducing the number of quantization bits in digitally implemented neural networks affects their circuit size or response time, several weight-quantization techniques have been developed to further reduce the required accuracy without deterioration of the network performance [1]. There are two alternative approaches to weight quantization and clipping in the presence of finite precision: on-chip training and off-chip training. The on-chip updates are constrained by a finite-precision range which departs to some extent from the true gradient descent. This is not just an approximation of the gradient descent, but most of the time exhibits significantly different behavior [2], and learning appears slower when more weights reach the limits of the clipping function. During training process, many clipping methods were studied [3], [4]. A feasible approach to finding good weights is on-chip training which repeats the weight clipping at each iteration of the gradient method. Clipping during training results in neural networks with significantly better performances for recall, as this kind of training is shown to be more robust in performance for training and testing data. Because the most neural computations are required in the training phase, therefore we derive and analyze the effects of the quantization error on backpropagation for on-chip and off-chip training.

2 Quantization Analysis on Backpropagation of the HOFNN

Assuming that an N-dimensional pattern $X = \{x_1, x_2, \cdots, x_N\} \in R^N$ is nonlinearly expanded to N_w dimensions, the output of the HOFNN for node i can be written as:

$$\tilde{y}_i^{H-th} = f(y_i^{H-th}) = \tanh(\lambda y_i^{H-th} / N_w)$$

$$= \tanh(\lambda (\sum_{p_1=1}^{N} \sum_{p_2=p_1}^{N} \cdots \sum_{p_H=p_{H-1}}^{N} w_{i,p_1p_2\cdots p_H} x_{p_1} x_{p_2} \cdots x_{p_H} + \cdots + \sum_{p_1=1}^{N} \sum_{p_2=p_1}^{N} w_{i,p_1p_2} x_{p_1} x_{p_2} + \sum_{p_1=1}^{N} w_{i,p_1} x_{p_1} + w_{i,N_w}) / N_w). \tag{1}$$

Where $p_1, p_2, \cdots, p_H = 1, 2, \cdots, N$; $N_w = C_{N+H}^{H}$; H is the order of the network. λ / N_w is a normalized factor. Because most neural computations are required in the training phase, in the following part we analyze the quantization effects of the limited precisions to backpropagation training. For the sake of simplicity, we analyze firstly two kinds of pattern classifications with only an output node under the condition of off-chip training. Assume that a set of patterns in N-dimensional space $X = \{x_1, x_2, \cdots, x_N\} \in R^N$ is not linearly separable, and the extended pattern set $Z = \{z_1, z_2, \cdots, z_{N_w}\} \in R^{N_w}$ is linearly separable. The output weights are constrained within the range $\{-W_{max}, +W_{max}\}$, assume that the error energy function is E, and k is the iteration number, then the weight is updated as:

$$W(k+1) - W(k) = \eta(k) \frac{\partial E[k]}{\partial W} = \frac{\eta \lambda}{N_w} (t - \tilde{y}^{H-th}) \frac{\partial f}{\partial y^{H-th}} Z. \tag{2}$$

Where η is the learning rate, we sum up to iteration number k and obtain:

$$W(k+1) = W(k) + \frac{\eta \lambda}{N_w} (t - \tilde{y}^{H-th}) \frac{\partial f}{\partial y^{H-th}} Z = W(0) + \sum_{m=0}^{k} \frac{\eta \lambda}{N_w} (t - \tilde{y}^{H-th}) \frac{\partial f}{\partial y^{H-th}} Z. \tag{3}$$

Due to the triangular inequality, we have:

$$\|W(k+1)\|^2 \leq \|W(0)\|^2 + \sum_{m=0}^{k} [\frac{\eta \lambda}{N_w} (t - \tilde{y}^{H-th}) \frac{\partial f}{\partial y^{H-th}}]^2 \|Z\|^2 \leq \|W(0)\|^2 + 4 [\frac{\eta \lambda}{N_w}]^2 \|Z\|^2 (k+1). \tag{4}$$

Because a set of patterns in the expanded pattern space Z of the HOFNN is linearly separable, there exists a separating vector $C \in R^{N_w}$, and $\|C\| = 1$. In this case the following formulas hold:

$$C^T Z > 0 \text{ and } t - \tilde{y}^{H-th} > 0, \text{ if } Z \in class(+). \tag{5}$$

$$C^T Z < 0 \text{ and } t - \tilde{y}^{H-th} < 0, \text{ if } Z \in class(-). \tag{6}$$

Combining Eqs. (5) with (6), we know $(t - \tilde{y}^{H-th})C^T Z > 0$ always holds no matter whether $Z \in class(+)$ or $class(-)$. By multiplying vector C along two sides of Eq. (3), we obtain:

$$C^T W(k+1) = C^T W(0) + \sum_{m=0}^{k} \frac{\eta \lambda}{N_w}(t - \tilde{y}^{H-th})\frac{\partial f}{\partial y^{H-th}} C^T Z = C^T W(0) + \sum_{m=0}^{k} \frac{\eta \lambda}{N_w}|C^T Z|\frac{\partial f}{\partial y^{H-th}}|t - \tilde{y}^{H-th}|. \quad (7)$$

Where $\frac{\partial f}{\partial y^{H-th}} > 0$ always holds. Combining Eqs.(4) with (7), we have inequalities:

$$1 \geq \frac{C^T W(k+1)}{\|W(k+1)\|} \geq \frac{C^T W(0) + \sum_{m=0}^{k} \frac{\eta \lambda}{N_w}|C^T Z|\frac{\partial f}{\partial y_i^{H-th}}|t - \tilde{y}^{H-th}|}{\sqrt{\|W(0)\|^2 + 4[\frac{\eta \lambda}{N_w}]^2 \|Z\|^2 (k+1)}} \quad (8)$$

When $k \to \infty$, the denominator of the right inequality approaches $O(\sqrt{k})$, and the following condition must be valid: $\frac{\eta \lambda}{N_w}|C^T Z|\frac{\partial f}{\partial y^{H-th}}|t - \tilde{y}^{H-th}| \to 0$. For $\forall m > k$:

$$\frac{\eta \lambda}{N_w}|C^T Z|\frac{\partial f}{\partial y^{H-th}}|t - \tilde{y}^{H-th}| < \alpha. \quad (9)$$

For the output of a HOFNN, the following relation is satisfied:

$$\tilde{y}^{H-th} = f(y^{H-th}) = \tanh(\frac{\lambda}{N_w}\sum_{j=1}^{N_w} w_j z_j) < f(y_{max}^{H-th}) = \tanh(\lambda w_{max} z_{max}) < |t|. \quad (10)$$

Taking the partial derivative of both sides of Eq. (10) for y^{H-th}, we have:

$$\frac{\partial f(y^{H-th})}{\partial y^{H-th}} > \frac{\partial f(y_{max}^{H-th})}{\partial y^{H-th}} = \theta. \quad (11)$$

Combining Eqs. (9) with (11), we have:

$$|t - \tilde{y}^{H-th}| < \varepsilon = \frac{\alpha N_w}{\eta \lambda |C^T Z| \theta}. \quad (12)$$

Our aim is to try and find a set of weights which minimizes the error energy function inside the hypercube formed in the weight space. Assuming that the weight vector of node i is expressed by $W_i = [w_{i,1}, w_{i,2}, \cdots, w_{i,N_w}]$, $w_{i,j}$ is the weight which connects input node j to output node i, which are statistically independent of each other. If we use on-chip training, the weights are updated as:

$$Q[w_{i,j}(k+1) - w_{i,j}(k)] = Q[\eta Q[\delta_i] z_j]. \quad (13)$$

$$Q[\delta_i] = Q[\lambda Q[(t_i - \tilde{y}_i^{H-th})(1-(\tilde{y}_i^{H-th})^2)]/N_w]. \tag{14}$$

Where $j = 1, 2, \cdots, N_w$; $i = 1, 2, \cdots, s$; t_i and $\tilde{y}_i^{H-th}$ are the expected output and the real output of node i, respectively. We further assume that $x_i|_{max} = 1$, $i = 1, 2, \cdots, N$, meaning $z_j|_{max} = 1$. Then according to Eqs. (13) and (14) and the above analysis, when the value of the weight updates is smaller than the quantization width:

$$|w_{i,j}(k+1) - w_{i,j}(k)| = |\eta \delta_i z_j| < |\eta \delta_i z_{max}| = |\eta \delta_i| < \Delta/2. \tag{15}$$

The network training will cease, where

$$\eta \delta_i = \frac{\eta \lambda |t_i - \tilde{y}_i^{H-th}|(1-(\tilde{y}_i^{H-th})^2)}{N_w} = \frac{\eta \lambda F(W_i)}{N_w}. \tag{16}$$

The random value $F(W_i)$ in Eq. (16) can be obtained by using Anand's method [5] to our quantization analysis, the conditional expectation $E_w\{\ \}$ in the first few iterations of BP after some simple derivations is:

$$E_w\{|t_i - \tilde{y}_i^{H-th}|(1-(\tilde{y}_i^{H-th})^2)\} = [1 - \sum_{r=1}^{N_w} \frac{\Delta^2 2^{2M}}{12}(\frac{\lambda z_r}{N_w})^2]. \tag{17}$$

Where the expected value of Eq. (17) is less than 1. Further, only when the conditional expectation $E_w\{\ \}$ of Eq.(15):

$$E_w\{|w_{i,j}(k+1) - w_{i,j}(k)|\} < E_w\{|\eta \delta_i|\} = E_w\{\left|\frac{\eta \lambda F(W_i)}{N_w}\right|\} < E_w\{\left|\frac{\eta \lambda}{N_w}\right|\} < \Delta/2. \tag{18}$$

i.e., $\frac{\eta \lambda}{N_w} < \Delta/2$, which prevents the weight from changing, the network training ceases. It implies that if there is only quantization and no clipping, then when $\frac{\eta \lambda}{N_w} > \Delta/2$, the iteration can be performed. In this case, according to Eq. (12), for all p training sample sets, the error is $|t - \tilde{y}^{H-th}| < \varepsilon$ and the HOFNN will converge to minima after limited weight updates. As the effect of the gradient descent on training with finite-precision computations is difficult to measure, the statistical evaluation of the weight updates does not effectively determine the property of a network to learn [6]. Therefore, for weight clipping we just follow a qualitative analysis. During on-chip training, due to quantization and weight clipping all the contributions of a single iteration are summed in a nonlinear fashion. In addition, the weight updates may depart from the true gradient descent, especially due to nonlinear clipping at the upper and lower boundaries. During the training process, when the weights reach the border of [-W_{max}, W_{max}], and the gradient is directed out of [-W_{max}, W_{max}], the weights are updated along the projection of the gradient onto the borderline of [-W_{max}, W_{max}] instead of along the gradient itself [7]. In this case, because both quantization and

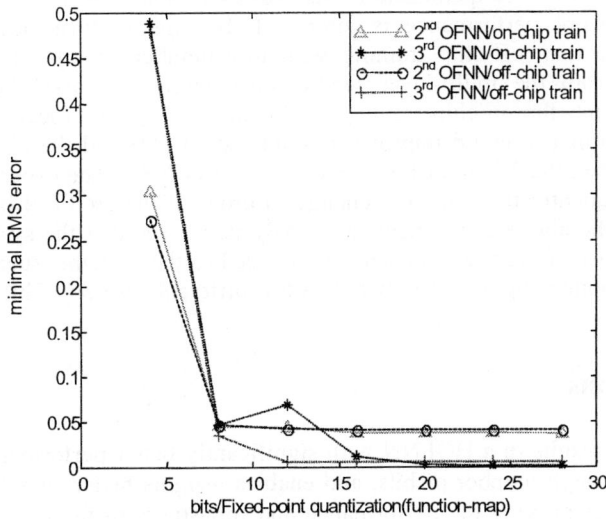

Fig. 1. Minimal RMS error for different numbers of fixed-point quantization bits

clipping errors exist, the weights are unlikely to be updated towards the direction of gradient optimization. Therefore, during the finite-precision training process, the BP algorithm is highly sensitive to weights. The training failure may be weight clipping or an insufficient weight resolution. As the number of quantization bits decreases, the weights become more quantized and the noise-to-signal ratio (NSR) increases, adversely affecting network performance.

3 Experimental Results

The experiment is the function approximation problem of the 2nd–order and the 3rd–order function neural networks (FNNs) when using different numbers of bits. The input to the network was a random number in the interval [-1, +1] and the target was the mapping to the 0.7*tanh(1.5x) curve. Training data consisted of 25 pairs of uniformly distributed data within the interval range [-1, +1], and the testing data consisted of 25 pairs of randomly selected data which differed from the training data. Fig. 1 shows a comparison of the minimal root-mean-square (RMS) error for the 2nd–order and the 3rd–order FNNs with the limited precision. The minimal RMS error decreases dramatically when going from 4 to 8 bits, to around 4-5% at 8 bits. When going further from 8 to 28 bits, the minimal RMS error decreases slowly. The 2nd–order FNN has a lower minimal RMS error than the 3rd–order FNN below 8 bits, because the latter has a higher NSR; but when the quantization is above 16 bits, the 2nd–order FNN has a higher minimal RMS error than the 3rd–order FNN, because the 3rd–order FNN has a higher nonlinear function mapping ability. The minimal RMS error of the 3rd-order FNN can effectively be decreased using more quantization bits. In this case, on-chip training has a lower minimal RMS error than off-chip training, resulting in neural networks with significantly better performance for recall. Because on-chip training with more quantization bits creates weights better suited to training and testing, network performance is improved. But the performance of on-chip training is worse than off-chip training with low number of bits of quantization (which differs from the multi-layer feedforward neural networks; because local minima are present, the on-chip training could jump out of the local minimum, in which case performance could improve to some extent [1]) and the 2nd–order FNN performs better than the 3rd–order FNN, and has a lower NSR. The experiment shows that with fewer quantization bits, the change in order of the HOFNN become more sensitive. From the above experiments and analysis, we see that the system will not work if the number of quantization states is reduced to the extreme, such as the case in which the synaptic weights can only hold a few different values (1-2bits).

4 Conclusions

On-chip training results in a HOFNN with significantly better performance than off-chip training at a high number of bits, and enables weights better suited for training and testing, thus improving the performance and robustness of the network. But the performance of on-chip training is worse than off-chip training with low number of bits of quantization. The experiments reveal that changing the order enables greater performance sensitivity with lower-bit quantization. The system will not work if the quantization bit is very low.

References

1. Jiang, M., Gielen, G.: The Effects of Quantization on Multilayer Feedforward Neural Networks. International Journal of Pattern Recognition and Artificial Intelligence. 17 (2003) 637-661
2. Marco, G., Marco, M.: Optimal Convergence of On-Line Backpropagation. IEEE Transactions on Neural Networks. 7 (1996) 251-254
3. Bayraktaroglu, I., Ogrenci, A. S., Dundar, G., et al.: ANNSyS: An Analog Neural Network Synthesis System. Neural Networks. 12 (1999) 325-338
4. Shima, T., Kimura, T., Kamatani, Y., et al: Neural Chips with On-Chip Backpropagation and / or Hebbian Learning. IEEE Journal of Solid State Circuits. 27(1992) 1868-1876
5. Anand, R., Mehrotra, K., Mohan, C. K., et al: Efficient Classification for Multiclass Problems Using Modular Neural Networks. IEEE Transactions on Neural Networks. 6 (1995) 117-123
6. Holt, J. L., Hwang, J.-N.: Finite Error Precision Analysis of Neural Network Hardware Implementations. IEEE Transactions on Computers. 42 (1993) 1380-1389
7. Jiang, M., Gielen, G.: The Effect of Quantization on the High Order Function Neural Networks. In: Neural Networks for Signal Processing - Proceedings of the IEEE Workshop, IEEE Inc. (2001) 143-152

LMS Adaptive Notch Filter Design Based on Immune Algorithm

Xiaoping Chen and Jianfeng Gao

School of Electronics & Information Engineering, Soochow University
178, GanJiang Road, Suzhou, Jiangsu, 215021, P. R. China.
{xpchen, jfgao}@suda.edu.cn

Abstract. In LMS algorithm of adaptive filter design, how to determine the learning step is an unpleasant problem. In this paper an application of Immune Algorithm (IA) to LMS Adaptive notch filter design is presented. The steps of the IA's realization and the experimental data are shown with an example. The result of the experiment has shown that the IA can find the optimal learning step of the LMS adaptive notch filter and the convergent speed of algorithm is fast.

1 Introduction

Adaptive signal process has been widely used in radar communication, sonar technology, image processing and so on. The parameters of adaptive filter can be automatically adjusted to optimal based on adaptive filter theory proposed by Widrow B. et al. A simple and actual algorithm named Least-Mean-Square algorithm (LMS) was raised by them at the meantime. However, how to determine the learning step μ is an unpleasant problem in LMS algorithm. If μ is selected too low the convergent speed of algorithm will be very slow, and if μ is selected too high the stable misadjustment (that is Mean Square Error, MSE) will be very high and the filter may be unstable. Although Widrow B. et al presented the range of learning step μ ($1 < \mu < 1/\lambda_{max}$, λ_{max} is the maximal eigenvalue of input signals) which ensures the algorithm convergent, they didn't present a way to get the optimal learning step. Thus applying IA to LMS adaptive notch filter design is introduced to determine the learning step in this paper.

IA is inspired by antigen-antibody reaction of immune system in biology. The objectives and constraints of IA are expressed as the antigen inputs. The IA antigen and antibody are equivalent to the objective and the feasible solution for a conventional optimization method[1]. The genetic operators including crossover and mutation are the following process for the production of antibodies in a feasible space. The algorithm to operate on the feasible cells will achieve very fast convergence during searching process. The concept of the information entropy is also introduced as a measure of diversity for the population to avoid falling into a local optimal solution. IA can search diverse solutions which can be candidates as optimal solutions in only one optimization procedure. By checking affinities between each antibody, several good antibodies can be obtained[2].

In this paper IA is applied to LMS adaptive notch filter design, so that the drawback that learning step μ obtained randomly can not be ensured to be optimal can be overcome. An example is provided and the experimental result is satisfactory.

This paper is organized as follows. In section 2 the LMS adaptive filter principle is introduced simply. Section 3 presents IA and its realization. Section 4 reports experimental results. Finally, some conclusions are offered in section 5.

2 LMS Adaptive Filter

The realization of adaptive filter is shown in Fig. 1. Here $s(n)$, $v(n)$, $d(n)$ and $e(n)$ are needed signal, noise, desired signal and error respectively.

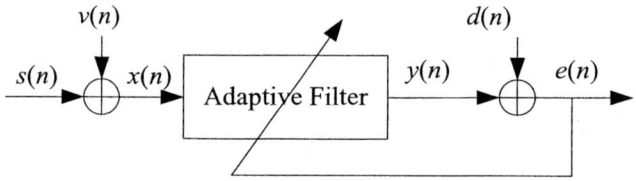

Fig. 1. Adaptive filter principle

LMS algorithm is described by the equation as follows:

$$\begin{aligned}e(n) &= d(n) - \mathbf{X}^T(n)\mathbf{W}(n) \\ \mathbf{W}(n+1) &= \mathbf{W}(n) + 2\mu e(n)\mathbf{X}(n)\end{aligned}, \quad (1)$$

where μ is learning step, $X(n)$ and $W(n)$ are input signals vector and filter weight coefficients vector respectively. The dimension of $W(n)$ is the length L of adaptive filter.

Convergent speed and stable misadjustment are two important factors to evaluate the performance of adaptive filter. Generally decreasing the learning step μ can reduce the stable misadjustment of the filter, but prolongs the time of convergent progress. When increasing μ, the convergent speed is raised, but the misadjustment of filter output is increased with bad filter effects. Many researchers presented some methods to solve the contradiction between convergent speed and stable misadjustment[3]. In fact, we can take measures to look for an optimal learning step μ to make the stable misadjustment be minimal with suitable convergent speed. IA is powerful to find the global solution to complex problems. Using IA to determine the learning step

μ in adaptive filter design is a good way which can overcome the contradiction between convergent speed and stable misadjustment.

3 Immune Algorithm and Its Realization

3.1 Immune Algorithm

In the immune system, the lymphocytes recognize the invading antigen and produce the antibodies coping with the antigen for excluding this foreign one (antigen). The number of foreign molecules that the immune system can recognize is unknown but it has been estimated to greater than 10^{16}. In spite of the diverse production of antibodies, there is the control mechanism that adjusts to produce the needed quantities in the immune system[1]. The immune system produces the diverse antibodies by recognizing the idiotypes between antigens and antibodies or between antibodies and antibodies. These combination intensities can be guessed by the affinity which is defined by using the information entropy theory.

In optimization problem the antigen, the antibody and the affinity between antigen and antibody correspond to the objective function, the solution and the combination intensity of solution respectively [4].

Although IA and GA are optimization search algorithms that mimic the process of biological evolution, they have essential discrimination in the memory drilling and the production of antibodies (individuals).

3.2 Algorithm Realization

We design a LMS adaptive notch filter in which the learning step μ is determined via IA. The algorithm realization way is to specify the repeat number of LMS algorithm 500, stable misadjustment is calculated and the optimal μ is looked for after executing LMS 500 times. The real significance is to make the misadjustment be minimal under the suitable convergent speed. In IA the other parameters are set as follows: the order L of adaptive notch filter is 32, input signal $x(n)=v(n)+s(n)$, where $s(n)$ is needed signal and $v(n)$ is noise, such as alternative current noise which should be filtered.

Encode Antibody. In this example the antibody is encoded with 16 bits binary string $d_1 d_2 ... d_{16}$, thus μ is expressed by

$$\mu = \frac{1}{\lambda_{max}} \sum_{i=1}^{16} d_i 2^{-i} . \qquad (2)$$

In the process of antibodies production there is a constraint ($1 < \mu < 1/\lambda_{max}$) for validating the legitimacy of antibodies. This constraint is satisfied naturally from Eq. (2).

Design Fitness Function. In each generation of IA, compute Mean Square Error (MSE) of every antibody μ_i. Let MSE of ith antibody be MSE_i, thus the fitness of the antibody can be defined as

$$f(\mu_i) = \frac{1/MSE_i}{\sum_{j=1}^{S} 1/MSE_j}, \qquad (3)$$

where S is the number of antibodies. The fitness probability of antibody is defined as the promotion of single antibody fitness to the sum of fitness of all antibodies,

$$p_f = f(\mu_i) \Big/ \sum_{j=1}^{S} f(\mu_j). \qquad (4)$$

Calculate Density Probability. In current population the number t of antibodies with highest density is counted. The calculation formula of density probability of these t antibodies is different from other S-t antibodies,

$$p_d = \begin{cases} \dfrac{1}{S}\left(1 - \dfrac{t}{S}\right), & t \text{ antibodies with highest density} \\ \dfrac{1}{S}\left(1 + \dfrac{t^2}{S^2 - St}\right), & \text{Other } S\text{-}t \text{ antibodies} \end{cases} \qquad (5)$$

The sum of density probability of all antibodies is equal to 1. The density probability of antibodies with high density is smaller than that of those antibodies with low density.

Promotion and Suppression of Antibodies. The selection probability of antibody is composed of its fitness probability and density probability, that is

$$p = \alpha p_f + (1 - \alpha) p_d, \qquad (6)$$

where $0 < \alpha < 1$, and $\alpha = 0.4$ in this example. From Eq. (6) it is shown that the bigger the fitness of an antibody is, the bigger its selection probability is, and the higher the density of an antibody is, the smaller its selection probability is. Thus, antibodies with high fitness are not only saved, the diversity of antibodies is but also guaranteed by the promotion and suppression between antibodies based on their density.

Crossover and Mutation. Here the crossover probability and mutation probability are set as 0.9 and 0.01 respectively. The optimal antibody is selected into next generation directly with no crossover and mutation operation that the convergence of IA is guaranteed [5].

4 Experimental Result

We design a LMS adaptive notch filter in which the learning step μ is determined via IA. Input signal is the addition of whitenoise and sine signal shown in Fig. 2. Here, $x(n)=v(n)+s(n)$, $x(n)$: input signal, $s(n)$: whitenoise, $v(n)$: sine signal. The number of samples is 500.

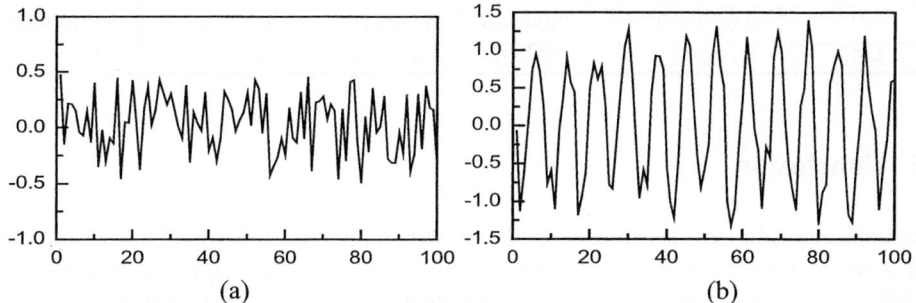

Fig. 2. Input signal: (a) white noise $s(n)$ and (b) addition signal $x(n)$

The parameters in IA are set as follows: size of population $S=60$, total generations $G=16$. MSE-μ curve shown in Fig. 3 (a) is obtained by searching enumerably to verify the validity of the algorithm in this paper. From Fig. 3 (a) we can see that when $\mu=0.0003$ MSE is minimal. So $\mu=0.0003$ can be set as the optimal μ_{opt} of LMS adaptive notch filter.

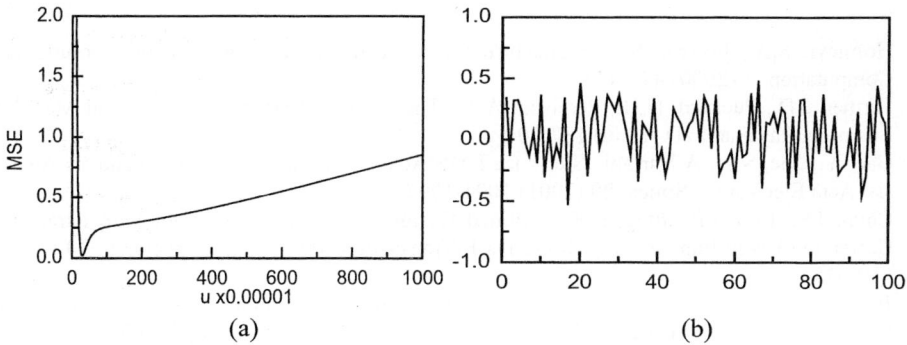

Fig. 3. (a) MSE-μ curve, (b) output signal

The next is to look for μ_{opt} by using IA. We have observed the ability for IA to find μ_{opt} within the specified evolution generations by running IA 10 times. The result in Tab. 1 shows that IA can converge the optimal learning step μ very nearly. We take the $\mu = 0.0003113$ in Tab. 1 to verify the effect of LMS adaptive notch filter. The output signal of the notch filter is shown in Fig. 3 (b) from which it is found that output signal is very similar to the original signal in Fig. 2 (a).

Table 1. Result of IA

	1	2	3	4	5
μ	0.0003113	0.0003011	0.0002987	0.0002992	0.0003666
MSE	0.02173	0.02027	0.02028	0.02027	0.05449
	6	7	8	9	10
μ	0.0003068	0.0004045	0.0003111	0.0003001	0.0003004
MSE	0.02081	0.08565	0.02167	0.02026	0.02026

5 Conclusion

Immune algorithm is a learning algorithm based on immune system in biology and one of main contents in study of artificial immune system. In this paper IA is applied to LMS adaptive notch filter design. Experimental results have shown that the adaptive notch filter designed via IA can converge the optimal learning step fast and stably and has good filter performance.

Acknowledgments. This work has been supported by the natural science fund of education office in Jiangsu Province (02KJB510001) and the open fund of the computer laboratory of Jiangsu Province (KJS02008) of China.

References

1. Hofmeyr, S.A., Forrest, S.: Architecture for an Artificial Immune System. Evolutionary Computation. 8 (2000) 443-473
2. Farmer, J.D., Packard, N.H., Perelson, A.A.: The Immune System Adaption, and Machine Learning. Physica. 22D (1986) 187-204
3. Gao, Y., Xie, S.L.: A Variable Step Size LMS Adaptive Filtering Algorithm and Its Analysis. Acta Electronica Sinica. 29 (2001) 1094-1097
4. Chun, J.S., Lim, J.P., Jung, H.K.: Optimal Design of Synchronous Motor with Parameter Correction Using Immune Algorithm. IEEE Transactions on Energy Conversion. 14 (1999) 610-615
5. Jiao, L.C., Wang, L.: A Novel Genetic Algorithm Based on Immunity. IEEE Transaction on Systems, Man, and Cybernetics-Part A: Systems and Humans. 30 (2000) 552-561

Training Radial Basis Function Networks with Particle Swarms

Yu Liu[1], Qin Zheng[1,2], Zhewen Shi[1], and Junying Chen[1]

[1] Department of Computer Science, Xian JiaoTong University, Xian 710049, P.R.China
liuyu@mailst.xjtu.edu.cn
http://www.psodream.net
[2] School of Software, Tsinghua University, Beijing 100084, P.R.China

Abstract. In this paper, Particle Swarm Optimization (PSO) algorithm, a new promising evolutionary algorithm, is proposed to train Radial Basis Function (RBF) network related to automatic configuration of network architecture. Classification tasks on data sets: Iris, Wine, Newthyroid, and Glass are conducted to measure the performance of neural networks. Compared with a standard RBF training algorithm in Matlab neural network toolbox, PSO achieves more rational architecture for RBF networks. The resulting networks hence obtain strong generalization abilities.

1 Introduction

Radial Basis Function (RBF) networks were introduced into the neural network literature by Broomhead and Lowe [1], which are motivated by observation on the local response in biologic neurons. Due to their better approximation capabilities, simpler network structures and faster learning algorithms, RBF networks have been widely applied in many science and engineering fields. RBF network is three layers feedback network, where each hidden unit implements a radial activation function and each output unit implements a weighted sum of hidden units outputs. Its training procedure is usually divided into two stages. First, the centers and widths of the hidden layer are determined by clustering algorithms such as K-means [2], vector quantizations [3], decision trees [4], and self-organizing feature maps [5]. Second, the weights connecting the hidden layer with the output layer are determined by Singular Value Decomposition (SVD) or Least Mean Squared (LMS) algorithms. The problem of selecting the appropriate number of basis functions remains a critical issue for RBF networks. The number of basis functions controls the complexity and the generalization ability of RBF networks. RBF networks with too few basis functions cannot fit the training data adequately due to limited flexibility. On the other hand, those with too many basis functions yield poor generalization abilities since they are too flexible and fit the noise in the training data. The methods mentioned above require designers to fix the structure of networks in advance according to prior knowledge. However it is difficult for designers to achieve optimal architecture. Genetic algorithms, the most popular evolutionary algorithms, have been employed to automatically evolve the structure of neural network [6]. Training technique can be formulated as an optimization problem, which includes the network structure into a set of variables that are used to minimize the prediction error. Particle Swarm Optimization (PSO) algorithm, an emerging evolutionary computation technique motivated by observations on the flocking behavior of birds, was first introduced by Eberhart and Kennedy in 1995 [7]. PSOs possess similar attractive features of genetic algorithms such as independence from gradient information of the objective function, the ability to solve complex nonlinear high dimensional problems. Furthermore, they can achieve faster convergence speed and require fewer parameters to be adjusted. In this paper, PSO algorithm is adopted to auto-configure the structure of RBF network and obtain the model parameters according to given input-output examples.

2 RBF Training Algorithm Design

The architecture of RBF network consists of three layers that have entirely different roles. The input layer, a set of source nodes, connects the network to the environment. The second layer consists of a set basis function units that carry out a nonlinear transformation from the input space to the hidden space. Usually, nonlinear transformation is based on gaussian function as follows:

$$z_i(x) = exp(-\frac{\|x - u_i\|^2}{2\sigma_i^2}) \tag{1}$$

where $\|\ldots\|$ represents Euclidean norm; u_i, σ_i, and z_i are the center, the width and the output of the i-th hidden unit, respectively. The output layer, a set of summation units, supplies the response of the network.

In this paper, we formulated the training technique as an optimization problem and employed PSO algorithm to resolve it. PSO algorithm is a population-based iterative search procedure, which simulates the social behavior of flocks of birds and schools of fishes. A group of individuals, referred to as particles, constitute a swarm. The position of each particle is a candidate solution, which is taken into fitness function to indicate the quality of the particle. The particles interact and cooperate each other to move toward better solutions. When determining the direction and step-size during the search, a particle learns from its own experience and shared information among its neighbors, by which the particle can know more about landscape and easily escape from local optima. Basing on this natural mechanism of swarm intelligence, PSO becomes a very promising global optimization method. When adopting PSO to train RBF neural networks, we must resolve two problems: encoding neural network architecture and designing fitness function.

2.1 Encoding Neural Network Architecture

The aim of training is to determine the number of hidden units, centers and widths of corresponding hidden units, and the weights that connect hidden units and output units. The choice of an efficient representation for network architecture is one of the most important issues in training. If we encode all these parameters into a particle, the length of the particle is too long and hence the search space is too large, which results in a very slow convergence rate. Since the performance of RBF networks mainly depends on the centers of hidden units, we just encode the centers into a particle for stochastic search. Then the widths and weights are determined by heuristic methods and analytic methods, respectively. PSO shows a strong ability to deal with real-valued optimization problem. Moreover, centers are also real values. In order to fully exploit potential of PSO, real-valued flags are employed, which indicate whether or not the corresponding hidden units are involved in networks. Therefore, the position of a particle is represented by concatenation of flags and centers of hidden units. Suppose the max number of hidden units is set to $hmax$, thus the structure of the particle includes two parts as follows:

Center existence array			Center vector array		
$Flag_1$	$Flag_2$	...	$Flag_{hmax}$	$Center_1$	$Center_2$	...	$Center_{hmax}$

Fig. 1. Structure of a particle

$Flag_i$ indicates whether or not the i-th hidden unit is involved into the network. If $Flag_i > 0$, the i-th hidden unit is included in the network. Otherwise the i-th hidden unit is removed from the network. In such a way, particles can be interpreted to networks with variable number of hidden units though they have a fixed length. Here, different particles correspond to networks with different number of hidden units.

2.2 Designing Fitness Function

The fitness function guides the evolution process. Here a small fitness denotes a good particle. RBF networks with good configuration should have less hidden units and high prediction accuracy. Therefore, fitness function takes into account two factors: mean squared error between network outputs and desired outputs (MSE), and the complexity of networks ($ComNN$).

$$Fitness = MSE + k \cdot ComNN \qquad (2)$$

where k is a constant which balances the impact of MSE and $ComNN$.

Calculating MSE. After a particle is interpreted into a neural network according to Section 2.1, the number of hidden units and their centers are obtained. Then the width of the i-th hidden unit is determined by the following heuristic formula:

$$\sigma_i^2 = \frac{1}{h}\sum_{j=1}^{h}\|Center_i - Center_j\|^2 \qquad (3)$$

where h is the number of hidden units that are involved in the network.

Once the centers and the widths are fixed, the task of determining weights reduces to solving a simple linear system. An analytical non-iterative Single Value Decomposition (SVD) method is adopted. The resulting network is measured on the training set where PN input-target vector pairs (patterns) are given. The MSE is calculated as the following:

$$MSE = \frac{1}{PN} \sum_{p=1}^{PN} \|t_p - o_p\|^2 \qquad (4)$$

where t_p and o_p are the desired output and network output for pattern p respectively.

Calculating *ComNN*

$$ComNN = \frac{Nhidden}{Nmaxhidden} \qquad (5)$$

where $Nhidden$ is the number of hidden units involved in networks; $Nmaxhidden$ is predefined the max number of hidden units.

3 Implementation with PSO

N particles that constitute a swarm move around in a D-dimensional space to search minimal fitness values. The i-th particle at t iteration has a position $X_i^{(t)} = (x_{i1}, x_{i2}, -, x_{iD})$ that denotes a kind of configuration of neural network as described in Section 2.1, a velocity $V_i^{(t)} = (v_{i1}, v_{i2}, -, v_{iD})$, and the best position achieved so far (*pbest*) by itself $P_i = (p_{i1}, p_{i2}, -, p_{iD})$. The best solution achieved so far by the whole swarm (*gbest*) is represented by $P_g = (p_{g1}, p_{g2}, -, p_{gD})$. The best position indicates the position where the minimal fitness is achieved. In order to rationally try various positions, at every iteration every particle adjusts its position toward two positions: *pbest* and *gbest*, which simulates the social behavior of flocks of birds and schools of fishes. The position of the i-th particle at the next iteration is calculated according to the following equations:

$$V_i^{(t+1)} = w * V_i^{(t)} + c_1 * rand() * (P_i - X_i^{(t)}) + c_2 * rand() * (P_g - X_i^{(t)}) \qquad (6)$$

$$X_i^{(t+1)} = X_i^{(t)} + V_i^{(t+1)} \qquad (7)$$

where w is inertia factor which balances the global exploitation and local exploration abilities of the swarm; c_1 and c_2 are two positive constants, called cognitive learning rate and social learning rate respectively; $rand()$ is a random function in the range [0,1]. In addition, a constant, $Vmax$, is used to limit the velocities of particles. If smaller than -$Vmax$, an element of the velocity is set equal to -$Vmax$; if greater than $Vmax$, then equal to $Vmax$. The pseudocode is as follows:

Initialize();
for $t = 1$ to the limit of iterations
    for $i = 1:N$
        $Fitness_i^{(t)} =$ EvaluationFitness($X_i^{(t)}$);
        UpdateVelocity($V_i^{(t+1)}$) according to formula (6);
        LimitVelocity($V_i^{(t+1)}$);
        UpdatePosition(($X_i^{(t+1)}$)) according to formula (7);
        if needed, update P_i and P_g;
    End
    Terminate if P_g meets problem requirements;
End

4 Experiments

The data sets used in this section were obtained from the UCI repository of Machine Learning databases [8]. Two training algorithm were compared. One was PSO algorithm whose parameters were set as follows: weight w decreasing linearly between 0.9 and 0.4, learning rate $c_1 = c_2 = 2$, and $Vmax = 5$. Max iterations were different with different data sets. The other was newrb routine that was included in Matlab neural networks toolbox as standard training algorithm for RBF neural network. The function newrb iteratively creates a radial basis network one neuron at a time. Neurons are added to the network until the sum-squared error falls beneath an error goal or the maximum number of neurons has been reached. All the experiments were conducted 30 runs. In each experiment, each data set was randomly divided into two parts: 2/3 as training set and 1/3 as test set. TrainCorrect and TestCorrect referred to mean correct classification rate averaged over 30 runs for the training and test sets, respectively. The information of data sets and results of the two algorithms were listed in Table 1.

5 Conclusions

In this paper, PSO algorithm, a population-based iterative global optimization, was implemented to train RBF networks. The method of encoding a RBF network into a particle was given, where only the centers of hidden units were encoded. In each iteration, each particle determined a kind of configuration for the centers of hidden units, according to which the widths of hidden units were calculated by heuristic methods, and connection weights between hidden layer and output layer were obtained by SVD. Consequently, a RBF network was constructed. Then it was performed on training sets to evaluate fitness for the particle. Fitness function takes into account not only MSE between network outputs and desired outputs, but also the number of hidden units, thus the resulting networks can alleviate over-fitting. Experimental results show that PSO algorithm achieves more rational architecture for RBF networks and the resulting networks hence obtain strong generalization abilities at the cost of a little longer time to train networks.

Table 1. The information of data sets, parameters and results of the two algorithms for different data sets

	Iris	Wine	New-thyroid	Glass
# of patterns	150	178	215	214
# of input units	4	13	5	9
# of output units	3	3	3	7
Max Iteration (PSO)	2000	5000	5000	5000
# of hidden (PSO)	5	15	7	12
TrainCorrect (PSO)	0.99	1	0.9650	0.8042
Test Correct (PSO)	0.98	0.9631	0.9444	0.6620
# of hidden (newrb)	9	58	26	87
TrainCorrect (newrb)	0.9850	0.9375	0.9240	0.9850
TestCorrect (newrb)	0.9560	0.6554	0.6204	0.6174

References

1. Broomhead, D., Lowe, D.: Multivariable Functional Interpolation and Adaptive Networks. Complex Systems **2** (1988) 321-355
2. Moody, J., Darken, C.: Fast Learning Networks of Locally-Tuned Processing Units. Neural Computation **3** (1991) 579-588
3. Vogt, M.: Combination of Radial Basis Function Neural Networks with Optimized Learning Vector Quantization. IEEE International Conference on Neural Networks (1993) 1841-1846
4. Kubat, M.: Decision Trees Can Initialize Radial-Basis Function Networks. IEEE Transactions on Neural Networks **9** (1998) 813-821
5. Robert, J., Howlett L.C.J.: Radial Basis Function Networks 2: New Advances in Design (2001)
6. Yao, X.: Evolving Artificial Neural Networks. Proceedings of the IEEE (1999) 87(9) 1423-1447
7. Kennedy, J., Eberhart, R.C.: Particle Swarm Optimization. Proceedings of IEEE International Conference on Neural Networks, Piscataway, NJ (1995) 1942-1948
8. Blake, C., Keogh, E., Merz, C.J.: UCI Respsitory of Machine Learning Databases (1998) www.ics.uci.edu/ mlearn/MLRepository.html

Optimizing Weights by Genetic Algorithm for Neural Network Ensemble

Zhang-Quan Shen [1,2] and Fan-Sheng Kong [2]

[1] Institute of Remote Sensing and Information System Application, Zhejiang University,
Hangzhou 310029, China
zhqshen@zju.edu.cn
[2] College of Computer Science, Zhejiang University, Hangzhou 310027, China

Abstract. Combining the outputs of several neural networks into an aggregate output often gives improved accuracy over any individual output. The set of networks is known as an ensemble. Neural network ensembles are effective techniques to improve the generalization of a neural network system. This paper presents an ensemble method for regression that has advantages over simple weighted or weighted average combining techniques. After the training of component neural networks, genetic algorithm is used to optimize the combining weights of component networks. Compared with ordinary weighted methods, the method proposed in this paper achieved high predicting accuracy on five test datasets.

1 Introduction

Neural network ensemble is a learning paradigm where a collection of a finite number of neural networks is trained for the same task. It originates from Hansen and Salamon's work, which shows that the generalization ability of a neural network system can be significantly improved through ensembling a number of neural networks [1]. Since this technology behaves remarkably well, recently it has become a very hot topic in both neural networks and machine learning communities, and has already been successfully applied to many areas such as face recognition, optical character recognition, scientific image analysis, medical diagnosis, seismic signal classification, etc [2].

In general, neural network ensemble is constructed in two steps, firstly training a number of component neural networks and then combing the component predictions [2].

As for training component neural networks, two learning approaches, boosting and bagging, have received extensive attention. Both generate by re-sampling training data sets from the original data sets to the learning algorithm which builds up a base predictor for each training data set [2].

Ensemble methods combine the outputs of the component networks. The output of an ensemble is a weighted average of the output of each network, with the ensemble weights determined as a function of the relative error of each component network determined in training. Since the optimum weights can minimize the generalization error of ensemble, it's very important to find optimum weights for individual component networks. However it's nearly impossible to find the optimum weights for the neural networks directly. The problem of optimal weights of component networks can be taken as an optimization problem. Considering that genetic algorithm has been shown as a powerful optimization tool, it's a good idea to solve the problem of optimal ensemble weights by genetic algorithm.

This paper tries to explore the advantages of using genetic algorithm to solve optimal ensemble weights. The rest of this paper is organized as follows. In section 2, the bagging and boosting algorithms are introduced. In section 3, the genetic algorithm and how to optimize the weights of component networks by genetic algorithm is described briefly. In section 4, an empirical study is reported. Finally in section 5, a short summary and some conclusions are concluded.

2 Bagging and Boosting

In this chapter, bagging and boosting algorithms are described.

The bagging algorithm (Bootstrap aggregating) uses bootstrap samples to build the base predictors. Each bootstrap sample of m instances is formed by uniformly sampling m instances from the training data set with replacement. This results that dissimilar component predictors are built with unstable learning algorithms like neural networks, and the performance of the ensemble can become better. However, bagging may slightly degrade the performance of stable algorithms. The bagging algorithm generates T bootstrap samples $B_1, B_2, ..., B_T$ and then the corresponding T base predictors $P_1, P_2, ..., P_T$. The final prediction produced by the integration of these base predictors [3].

Boosting was developed as a method for boosting the performance of any weak learning algorithm, which needs only to be a little bit better than random guessing. The AdaBoost algorithm (Adaptive Boosting) was introduced as an improvement of the initial boosting algorithm. The concept used to correspond a generation of one base predictor is a trial. AdaBoost changes the weights of the training instances after each trial based on the predicting errors made by the resulting base predictors trying to force the learning algorithm to minimize the expected error over different input distributions. Similarly, the correctly predicted instances will have a lower total weight. Thus AdaBoost generates during the T trails T training sets $S_1, S_2, ..., S_T$ where instances have weights and thus T base predictors $P_1, P_2, ..., P_T$ are built. A final prediction produced by the ensemble of these base predictors [3, 4].

The algorithm of AdaBoost for regression adopted in this paper is described as follows. The algorithm starts setting the same weights $w_i^0 = 1, i = 1, \cdots, N$ to all cases of the training sample. Then for $t=1$ to T, it executes the following steps [3]:

(1) Extract N cases with replacement from the training sample, where the probability of inclusion for the ith case is fixed to $p_i^t = w_i^t / \sum_{j=1}^{N} w_j^t$.

(2) Fit an approximating neural network $h_t(x)$ to the sample extracted.

(3) Calculate over all the examples of the training sample the relative absolute

residuals $r_i^t = \dfrac{|y_i - h_t(x_i)|}{\max |y_i - h_t(x_i)|}$ $(i = 1, \cdots, N)$, the corresponding values of a loss function $L_i^t = r_i^t$, and finally calculate the average loss $\overline{L}^t = \sum_{i=1}^{N} p_i^t L_i^t$.

(4) Calculate the coefficient for the predictor $h_t(x)$, $\beta_t = \overline{L}^t / 1 - \overline{L}^t$.

(5) Update the weights: $w_i^{t+1} = w_i^t \beta_t^{1-L_i^t}$.

To Aggregate the results obtained by the T approximating predictors, use of the weighted median with weights proportional to $\log(1/\beta_t)$.

The condition $\overline{L}^t < 0.5$ $(t = 1, \cdots, T)$ is necessary for the convergence of the algorithm, so at step 3, if the condition $\overline{L}^t < 0.5$ $(t = 1, \cdots, T)$ can not reach, then give up the tth training result, and train again.

3 Genetic Algorithms and the Problem of Optimizing the Weights of Component Networks

Suppose the task is to use an ensemble comprising N component neural networks to approximate a function $f : \Re^m \to \Re$, and the predictions of the component networks are combined through weighted averaging where a weight w_i (i=1, 2, ..., N) satisfying the conditions $0 \leq w \leq 1$ and $\sum_{i=1}^{N} w_i = 1$ is assigned to the ith component network f_i. Then we get a weight vector $w = (w_1, w_2, ..., w_N)$. Since the optimum weights should minimize the generalization error of the ensemble, the optimum weight vector w_{opt} can be expressed as [2]:

$$w_{opt} = \arg\min_w (\sum_{i=1}^{N} \sum_{j=1}^{N} w_i w_j C_{ij}) \tag{1}$$

$w_{opt.k}$, the kth (k=1, 2, ..., N) variable of w_{opt}, can be solved by *Lagrange multiplier*, which satisfies:

$$\frac{\partial (\sum_{i=1}^{N} \sum_{j=1}^{N} w_i w_j c_{ij} - 2\lambda (\sum_{i=1}^{N} w_i - 1))}{\partial w_{opt.k}} = 0 \tag{2}$$

Eq. (2) can be simplified to:

$$\sum_{\substack{j=1 \\ j \neq k}}^{N} w_{opt.k} c_{kj} = \lambda \tag{3}$$

Considering that $w_{opt.k}$ satisfies $\sum_{i=1}^{N} w_i = 1$, we get:

$$w_{opt.k} = \frac{\sum_{j=1}^{N} c_{kj}^{-1}}{\sum_{i=1}^{N} \sum_{j=1}^{N} c_{ij}^{-1}} \tag{4}$$

It seems that we can solve w_{opt} from Eq. (4). But in fact, this equation rarely works well in real-world applications. This is because when a number of neural networks are available, there are often some networks that are quite similar in performance, which makes the correlation matrix $(C_{ij})_{N \times N}$ be an irreversible or ill-conditioned matrix so that Eq. 4 can't be solved. However, although we can't solve the optimum weights of the component neural networks directly, we can try to approximate them in some way. Look at Eq. 1, we can find that it could be viewed as an optimization problem [2].

Genetic algorithms have been used to solve difficult problems with objective functions that do not possess "nice" properties such as continuity, differentiability, etc. These algorithms maintain and manipulate a family of solutions and implement a "survival of the fitness" strategy in their search for better solutions. Genetic algorithms have been shown to solve linear and nonlinear problems by exploring all regions of the state space and exponentially exploiting promising areas through mutation, crossover, and selection operations applied to individuals in the population. So the genetic algorithm has been shown as a powerful optimizing tool [5]. In this study, we try to explore the effect and potential capability of genetic algorithm applied in neural network ensembles.

4 Experiments

4.1 Data Sets

In this chapter we present experiments where the dynamic weighted ensemble algorithm is used with neural networks generated by bagging and AdaBoost. The experimental setting is described before the results of the experiments. We have used five datasets. Friedman #1, Friedman #2 and Friedman #3 have been used by Breiman in testing the performance of bagging. The Bodyfat and Abalone datasets were gotten from the UCI machine learning repository. The datasets were divided randomly into independent train and test datasets. The main characteristics of the five datasets and the functions of Friedman #1, #2 and #3 are presented in Table1 and Table 2. The table 1 includes the name of dataset, number of instances included in the train and test datasets, and the conventional statistical parameters of train and test datasets. The table 2 includes the function and the constraints on the variables for the Friedman datasets.

In our experiments, the instances contained in three Friedman data sets are generated from the functions listed in Table 2. In our experiments some noise fitted for $N(0,1)$ have been added to the train datasets of Friedman, and the test datasets are noise-free.

4.2 Experimental Methodology

In the experiments, five-fold cross validation procedure is performed on each train dataset, where ten component neural networks are trained by each compared approach in each fold, totally fifty component neural networks are generated in ensemble, and each approach is performed 20 runs for each data set.

Table 1. The characteristics of train and test data sets

Dataset	Train dataset					Test dataset				
	Instances	Mean	Min	Max	SD	Instances	Mean	Min	Max	SD
Friedman #1	4310	14.410	1.159	28.302	4.801	1090	14.653	2.444	26.676	4.634
Friedman #2	4084	478.11	29.62	1441.00	337.74	1036	475.09	30.58	1439.20	334.42
Friedman #3	4075	1.369	-2.843	5.581	1.044	1045	1.369	0.311	1.564	0.218
Bodyfat	201	18.9	0.0	47.5	8.3	51	20.0	4.1	40.1	8.8
Abalone	3364	11.4	2.5	30.5	3.2	813	11.6	5.5	28.5	3.2

Table 2. Description of Friedman datasets

Dataset	Function	Variable
Friedman #1	$y = 10\sin(\pi x_1 x_2) + 20(x_3 - 0.5)^2 + 10x_4 + 5x_5$	$x_i \sim U[0,1]$
Friedman #2	$y = \sqrt{x_1^2 + (x_2 x_3 - (\frac{1}{x_2 x_4}))^2}$	$x_1 \sim U[0,100]$, $x_2 \sim U[40\pi, 560\pi]$, $x_3 \sim U[0,1]$, $x_4 \sim U[1,11]$
Friedman #3	$y = \tan^{-1}\frac{x_2 x_3 - \frac{1}{x_2 x_4}}{x_1}$	$x_1 \sim U[0,100]$, $x_2 \sim U[40\pi, 560\pi]$, $x_3 \sim U[0,1]$, $x_4 \sim U[1,11]$

The architecture of each ANN used in this application is a multi-layer feed-forward neural network with one hidden layer constituting of 10 neurons. The activation function used in the neurons of hidden layers is the exponential-sigmoid transfer function (*logsig*), whereas for the neurons of output layer is the linear transfer function (*purelin*). The ANNs are trained by the implementation of back-propagation algorithm with adaptive learning rate (*traingda*) in $MATLAB^{®}6.5$. During the training process, the "early stopping" technique on the one of five folds of training data set is used to improve the generalization and avoid over-fitting. The maximum fail epochs is set to 20. The generalization error of each neural network is estimated in each epoch using one fold as validation data set. If the error does not decrease in 20 consecutive epochs, the training of the neural network is terminated to avoid over fitting. Before training, the data of training dataset is scaled to the range [0, 1] for high training efficiency [6].

After the training of component neural networks, the genetic algorithm employed in our study is realized by the Genetic Algorithms for Optimization Toolbox (GAOT) developed by Houck *et al.*. The genetic operators, including select, crossover, and mutation, and the system parameters, including the crossover probability, the mutation probability, are all set to the default values of GAOT. The maximum generation and the population are set to 200, and a float coding scheme that represents each weight in 64 bits is used. The evolving process finished as the maximum generation was reached.

Want to evaluate the goodness of the individuals in the evolving population; the whole training dataset is used. Let $\hat{E}_w^T$ denote the estimated generalization error of the ensemble corresponding to the individual w on the train dataset T. It is obvious that $\hat{E}_w^T$ can express the goodness of w, i.e., the smaller $\hat{E}_w^T$ is, the better w is. So, $f(w) = 1/\hat{E}_w^T$ is used as the fitness function in our study.

In order to know how well the compared ensemble methods work, for example, how significant the predicting accuracy and generalization ability are improved by utilizing these ensemble methods, in the experiments we also test the performance of single neural networks. For each dataset, fifty single neural networks are trained, and 20 runs are repeated in the same way. The training sets, architecture, parameters, and the training process of these neural networks are all crafted in the same way as that of the networks used in ensembles.

4.3 Root Mean Square Error (RMSE)

In order to evaluate the predicting accuracy of different methods, the Root Mean Square Error (RMSE) of predicted and measured values of test dataset according to:

$$RMSE = \sqrt{\frac{1}{N} \sum_{i=1}^{N} (\hat{Z}_i - Z_i)^2} \quad (5)$$

where: N = the instance number of test dataset;
$\hat{Z}_i$ = the predicted value of test dataset;
Z_i = the real value of test dataset.

4.4 Results

Fig.1 presents the predicting error (RMSE value) for different weighted ensemble methods. The label "No Weight" represents using equal weights for all component neural networks in ensemble. The label "Weight" represents using different weights for different component neural networks in ensemble, and the weight of individual component neural network is inverse proportion to its predicted error (mse, mean square error) of train dataset. The label "GAOT_W" represents keeping all component neural networks and taking the weights evolved by GAOT as their ensemble weights. The label "GAOT_S" represents, the weights evolved by GAOT are used to select the component networks and combines the predictions of the selected networks with equal weights. The label "GAOT_S_W" represents, the weights evolved by GAOT are used to select the component networks and combines the predictions of the selected networks with the normalized version of their evolved weights. The label "Single" represents the single neural network. The RMSE values of different methods are the average of 20 runs. Fig.1a lists the results of bagging and fig.1b lists the results of boosting.

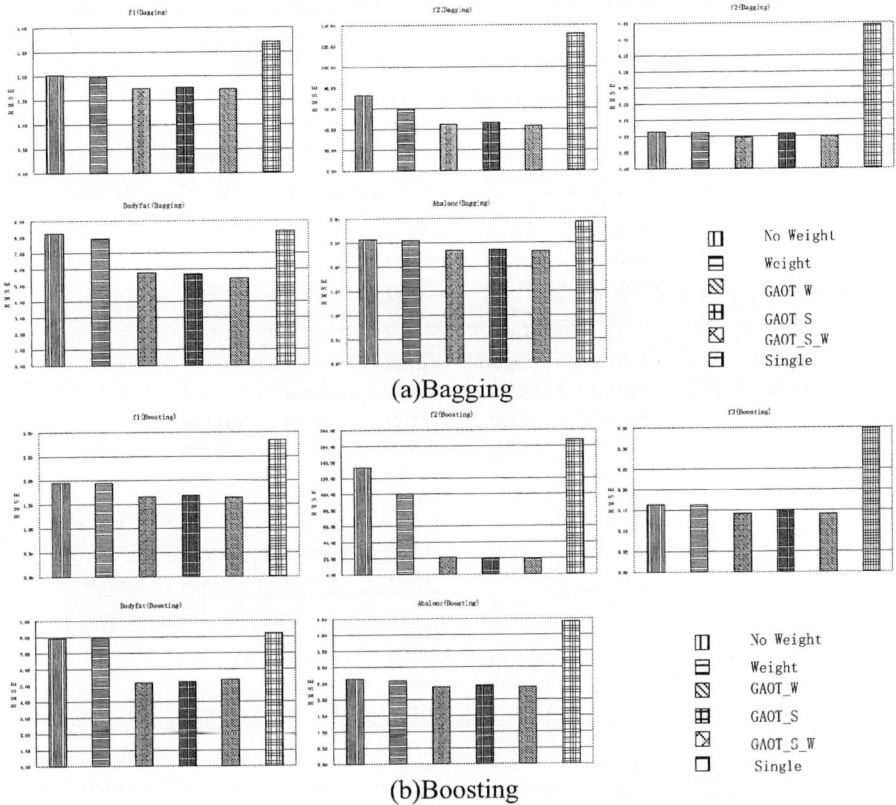

Fig. 1. Comparison of the RMSE values of test datasets for bagging and boosting with different weighted ensemble methods

Table 3 lists the pairwise one-tailed t-test results between different weight integrated methods.

Fig.1 shows that all five methods are consistently better than single neural networks in five test datasets. The pairwise one-tailed t-test results in table 3 also show this clearly. The methods using genetic algorithm are obviously better than usual ensemble methods in most test datasets. For the different methods with the same genetic algorithm, there are slightly different among them, the "GAOT_S_W" method has the lowest RMSE values except bodyfat dataset with boosting.

For Bagging, weighted methods are slightly better than simple weighted methods in all test datasets.

5 Conclusions

Ensemble methods have demonstrated spectacular success in reducing predicting error with learned predictors. These techniques form an ensemble of base predictors which produce the final prediction using simple weighted or weighted methods.

Table 3. Pairewise one-tailed t-tests for RMSE values on different methods

	Dataset		No_weight	Weight	GAOT_w	GAOT_S	GAOT_S_W
Bagging	Friedman #1	Weight	3.399E-15				
		GAOT_w	6.518E-16	5.996E-15			
		GAOT_S	1.597E-15	3.033E-14	0.0135		
		GAOT_S_W	2.459E-15	2.240E-14	2.124E-04	2.288E-04	
		Single	1.803E-28	1.099E-27	9.683E-26	1.209E-26	3.244E-25
	Friedman #2	Weight	2.100E-16				
		GAOT_w	1.066E-20	4.359E-19			
		GAOT_S	8.878E-18	4.020E-15	1.763E-04		
		GAOT_S_W	1.982E-20	8.230E-19	4.571E-06	1.859E-07	
		Single	9.102E-30	4.043E-28	5.605E-28	8.498E-26	2.800E-27
	Friedman #3	Weight	5.914E-06				
		GAOT_w	1.515E-06	6.331E-06			
		GAOT_S	9.822E-03	0.0744	3.963E-06		
		GAOT_S_W	6.503E-05	3.187E-04	8.678E-03	1.851E-04	
		Single	1.016E-25	6.184E-26	8.630E-25	8.276E-25	4.754E-25
	Bodyfat	Weight	1.937E-11				
		GAOT_w	1.954E-19	8.855E-19			
		GAOT_S	4.167E-16	6.144E-15	0.1368		
		GAOT_S_W	1.671E-17	1.054E-16	1.244E-07	4.585E-03	
		Single	5.709E-11	1.878E-11	8.107E-20	1.038E-16	5.747E-18
	Abalone	Weight	2.192E-09				
		GAOT_w	5.168E-16	3.207E-16			
		GAOT_S	2.370E-14	2.932E-14	6.825E-03		
		GAOT_S_W	9.821E-16	8.007E-16	3.649E-06	6.497E-06	
		Single	5.277E-22	2.719E-21	8.582E-21	7.173E-20	1.603E-20
Boosting	Friedman #1	Weight	6.121E-08				
		GAOT_w	5.149E-16	9.105E-16			
		GAOT_S	1.918E-15	4.105E-15	0.0358		
		GAOT_S_W	6.747E-15	1.115E-14	0.1078	6.376E-07	
		Single	7.843E-28	2.529E-28	5.613E-26	9.594E-25	4.843E-24
	Friedman #2	Weight	1.156E-17				
		GAOT_w	6.316E-19	1.027E-16			
		GAOT_S	1.462E-18	2.493E-16	6.108E-05		
		GAOT_S_W	9.188E-19	1.432E-16	1.983E-07	4.251E-03	
		Single	3.614E-15	1.660E-18	8.625E-19	1.505E-18	1.022E-18
	Friedman #3	Weight	0.0796				
		GAOT_w	7.301E-14	2.892E-15			
		GAOT_S	8.090E-08	2.985E-07	7.672E-06		
		GAOT_S_W	3.616E-13	1.063E-13	0.0416	1.465E-05	
		Single	1.442E-20	2.623E-20	9.397E-21	1.598E-20	1.177E-20
	Bodyfat	Weight	0.3368				
		GAOT_w	1.192E-21	9.296E-22			
		GAOT_S	2.639E-20	1.184E-19	0.1505		
		GAOT_S_W	1.074E-18	5.518E-18	5.217E-03	0.0756	
		Single	4.216E-14	3.904E-09	1.455E-22	5.427E-21	1.342E-19
	Abalone	Weight	3.385E-08				
		GAOT_w	1.513E-16	4.042E-15			
		GAOT_S	8.908E-12	2.707E-09	2.683E-04		
		GAOT_S_W	1.922E-17	7.489E-16	0.0142	7.777E-05	
		Single	2.054E-16	8.184E-17	2.078E-17	3.764E-17	2.575E-17

In this paper, a technique for evolving weights of component neural networks by genetic algorithm was experiment as a method instead of ordinary weighted methods with bagging and boosting. The technique is base on the optimizing ability of genetic algorithm.

The proposed optimal weighted ensemble technique was evaluated with bagging and adaboost on five datasets. The results achieved are promising and show that both bagging and boosting might give better performance with optimal weighted ensemble than with ordinary weighted methods. In three varieties of optimal weighted methods tried in our study, the method that uses the evolved weights to select the component networks and combines the predictions of the selected networks with the normalized version of their evolved weights is the best one.

For bagging, it may be necessary to consider the optimizing weight of integration; improvements are achieved by weighted or optimizing weighted methods in our study.

Acknowledgments. This research is partly supported by the National Natural Scientific Foundation of China (No. 40201021). We would like to thank the UCI machine learning repository of databases, for the two datasets used in this study.

References

1. Hansen, L.K., Salamon, P.: Neural Network Ensembles. IEEE Transactions on Pattern Analysis and Machine Intelligence, Vol. 12, No. 10 (1990) 993-1001
2. Zhou, Z.H., Wu, J.X., Tang, W.: Ensembling Neural Networks: Many Could be Better than All. Artificial Intelligence, Vol. 137 (2002) 239-263
3. Borra, S., Ciaccio, A.D.: Improving Nonparametric Regression Methods by Bagging and Boosting. Computational Statistics & Data Analysis, Vol. 38 (2002) 407-420
4. Schapire, R.E.: A Brief Introduction to Boosting. In: Thomas, D. (ed.): Proceedings of the Sixteenth International Joint Conference on Artificial Intelligence, Stockholm, Sweden (1999) 1-6
5. Houck, C.R., Joines, J.A., Kay, M.G.: A Genetic Algorithm for Function Optimization: A Matlab Implementation. Technical Report NCSU-IE-TR-95-09, North Carolina State University, Raleigh, NC (1995)
6. The Mathworks (ed.): Neural Network Toolbox User's Guide (version 4). The Mathworks, Inc., Natick, Massachussets (2001)
7. Granitto, P.M., Verdes, P.F., Navone, H.D., Ceccatto, H.A.: Aggregation Algorithms for Neural Network Ensemble Construction. In: Werner, B. (ed.): Proceedings of the VII Brazilian Symposium on Neural Networks, IEEE Computer Society, Pernambuco, Brazil (2002) 178-183

Survival Density Particle Swarm Optimization for Neural Network Training

Hongbo Liu[1,2], Bo Li[1], Xiukun Wang[1,2], Ye Ji[1,2], and Yiyuan Tang[2]

[1] Department of Computer, Dalian University of Technology,
Postfach 116023, Dalian, China
{lhb99, libo, wxk, jeyee}@tom.com
[2] Institute of Neuroinformnatics, Dalian University of Technology,
Postfach 116023, Dalian, China
brain@dlut.edu.cn

Abstract. The Particle Swarm Optimizer (PSO) has previously been used to train neural networks and generally met with success. The advantage of the PSO over many of the other optimization algorithms is its relative simplicity and quick convergence. But those particles collapse so quickly that it exits a potentially dangerous property: stagnation, which state would make it impossible to arrive at the global optimum, even a local optimum. The ecological and physical universal laws enlighten us to improve the PSO algorithm. We introduce a concept, swarm's survival density, into PSO for balancing the gravity and repulsion forces between two particles. A modified algorithm, survival density particle swarm optimization (SDPSO) is proposed for neural network training in this paper. Then it is applied to benchmark function minimization problems and neural network training for benchmark dataset classification problems. The experimental results illustrate its efficiency.

1 Introduction

Particle swarm optimization (PSO) is an evolutionary computation technique introduced by Kennedy and Eberhart in 1995 [1], inspired by social behaviours of bird flocking or fish schooling. Similar to genetic algorithms (GA), PSO is a population based optimization tool [2]. The system is initialized with a population of random solutions and searches for optima by updating generations. However, unlike GA, PSO has no evolution operators such as crossover and mutation. In PSO, the potential solutions, called particles, are "flown" through the problem space by following the current optimum particles. PSO has been successfully applied in many areas: function optimization, artificial neural network training, fuzzy system control, and other areas [3, 4]. But including the later inertia weight and constriction factor versions, most of all have a potentially dangerous property: stagnation, which state would make it impossible to arrive at the global optimum, even a local optimum. The ecological and physical universal laws enlighten us to improve the PSO algorithm. A modified algorithm, survival density particle swarm optimization (SDPSO) will be introduced for neural network training in this paper.

2 Overview of Particle Swarm Optimization

The basic PSO model consists of a swarm of particles moving in an n-dimensional search space where a certain quality measure, the fitness, can be calculated. Each particle has a position represented by a position-vector x and a velocity represented by a velocity-vector v. Each particle remembers its own best position so far in a vector p_i, i is the index of the particle and the d-th dimensional value of the vector p_i is p_{id} (i.e. the position where it achieved its best fitness). Further, a neighborhood relation is defined for the swarm. The best position-vector among all the neighbors of a particle is then stored in the particle as a vector p_g and the d-th dimensional value of the vector p_g is p_{gd}. At each iteration step the velocity is updated and the particle is moved to a new position. Firstly the update of the velocity from the previous velocity to the new velocity is (in its simplest form) determined by equation (1). And then the new position is determined by the sum of the previous position and the new velocity by equation (2), a neighborhood relation is defined for the swarm:

$$v_{id} = w*v_{id} + c_1 Rand()*(p_{id} - x_{id}) + c_2 Rand() * (p_{gd} - x_{id}) \tag{1}$$

$$x_{id} = x_{id} + v_{id} \tag{2}$$

In PSO algorithms, a particle decides where to move next, considering its own experience, which is the memory of its best past position, and the experience of its most successful neighbor. Later some actions differ from one variant of PSO to the other for improving the computational performance [5].Several researchers have analyzed it empirically [6] and theoretically [7, 8], which have shown that the particles oscillate in different sinusoidal waves and converging quickly, sometimes prematurely, especially for PSO with small inertia factor, w or constriction coefficients, c_1,c_2. From equation (1), v_{id} can become small value, but if the second term and the third term in RHS of equation (1) are both small, it cannot back to large value and lost exploration capability in some generations. This phenomenon will be referred to as *stagnation*, which state would make it impossible to arrive at the global optimum, even a local optimum.

3 Survival Density Particle Swarm Optimization (SDPSO)

The improved performance can be cast into the framework of an ecological phenomenon. Any species have their distributing laws [9]. Once the density of population in one area is too crowded and common resource is limited, the individuals have to disperse for surviving. In the other way, if the density of population in the area is too sparse, the individuals have to congregate for sharing information and fighting back natural enemy. The universal laws of gravity and repulsion tell us that a similar principle exists in the physical phenomenon [10]. In a given range, the distance between two particles exceeds the balance distance, r_0, the composition of gravity and repulsion forces makes them to get together. In the other way, it is possible to make them to apart from each other. In PSO algorithm, the third term in RHS of equation (1) could represent gravity force. But there is not repulsion force. That perhaps is just why PSO has the potentially dangerous property: premature or stagnation. We introduce a concept, swarm's survival density, into PSO for balancing the gravity and repulsion forces between each couple of particles. In our modified algorithm, survival density particle swarm optimization (SDPSO), the individuals in swarm have survival density limit. Once the density of population in one area is too crowded, the individuals would be dispersed by repulsion force for keeping on searching different regions. In the other way, if the density of population in the area is too sparse, the individuals have to congregate by gravity force for sharing information and searching a more precise solution. We add another term in RHS of equation (1), which represents repulsion force. It is inverse ratio against the distance between each couple of particles. Modifying the equation (1), a new velocity update equation is following:

$$v_{id} = w*v_{id} + c_1 Rand()*(p_{id} - x_{id}) + c_2 Rand()*(p_{gd} - x_{id}) - c_3 * r * Rand()/(p_l - x_{id})$$ (3)

$$r_{Iteration} = e^{-\alpha Iteration}$$ (4)

It is possible that $(p_l\text{-}x_{id})$ is zero, the v_{id} need to be evaluated as random value in the maximum velocity range. From the equation (4), we can see that the distance limit between two particles will cut down exponentially.

It is possible that the dimension of particles is too large to search a very good solution. An alterative method is cooperative learning [11], which can improve their performance by partitioning the vector into several subvectors. Each sub-vector is then allocated its own swarm. For example, the vector is split into N parts, each part being optimized by a swarm with M particles, leaving $N*M$ possible vectors to choose from. The simplest approach is to select the best particle from each swarm. The method involves a trade-off: increasing the number of swarms leads to a directly proportional increase in the number of error function evaluations, as one function evaluation is required for each particle in each swarm during each epoch.

4 Experiment and Result

In this section, we apply the SDPSO to benchmark function minimization problem and neural network training for benchmark dataset classification problems. The performance of the algorithms is correlative directly with the parameters selection. In our experiments, the parameter setting is following the recommendation in [6, 7, 12].

4.1 Benchmark Functions Experiment

By including a pure function minimization problem it is possible to investigate the properties of SDPSO in more detail, as more information is available on the nature of the error function as opposed to the error function of a neural network classification problem. Rastrigin function [13] is following:

$$f(x) = \sum_{i=1}^{n}(x_i^2 - 10\cos(2\pi x_i) + 10) \tag{5}$$

where $-5.12 < x_i < 5.12$, the function is continuous and multimodal; $x^*=0$, with $f(x^*)=0$.

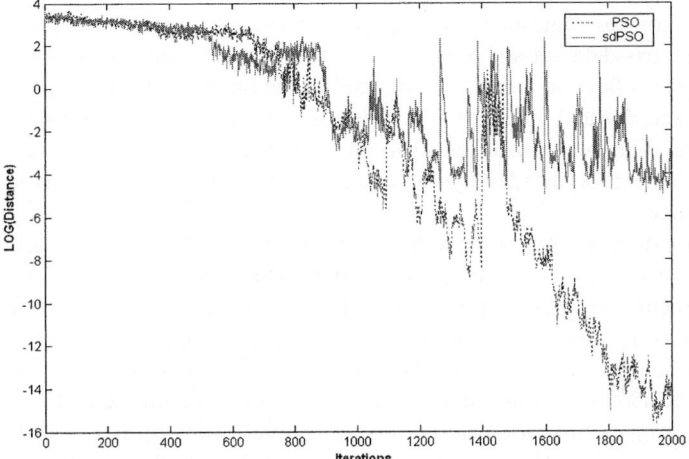

Fig. 1. The distance varying for 30-D Rastrigin function.

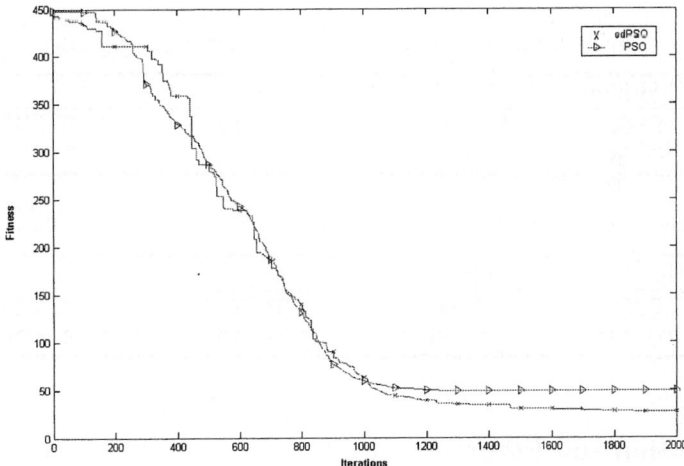

Fig. 2. Optimisation performance for 30-D Rastrigin function.

In our experiments, the global minimum (the only that achieved the goal) was situated at or near the center of the search domain. Rastrigin function has multiple local minima and has a "large scale" curvature which guides the search towards the global minimum. We compare the distances and fitness values in PSO and SDPSO for the function minimization problems in Fig.1, 2, respectively. The distances are decreased with fluctuation because of the randomcity in the algorithms. SDPSO prevents the particles from getting together too quickly. Fig.2 illustrated SDPSO algorithm efficiency, especially for more complex functions, such as Rastrigin function.

4.2 Benchmark Dataset Experiment

There are many test dataset for experiment in the UCI machine-learning repository [14]. We use Iris data set and Wisconsin Breast Cancer data for our experiment. The training set classification error, ε_t, computed as the number of misclassified patterns in the training set, divided by the total number of patterns in the training set, expressed as a percentage. The generalization error, ε_g, computed as the number of misclassified patterns in the test set, divided by the total number of patterns in the test set, expressed as a percentage.

In the iris data set, measurements of four attributes of iris flowers are provided in each data set record: sepal length, sepal width, petal length, and petal width. Fifty sets of measurements are present for each of three varieties of iris flowers, for a total of 150 records, or patterns. The neural networks can be designed with 4-3-3 architecture for the Iris classification problem. Breast cancer classification problem comprises 6 699 instances with 9 attributes and 2 classes. For the problem the neural networks can be designed with a 9-8-1 architecture. For a simpler problem, Iris classification problem, it is not obvious difference in the results of two algorithms. But for a more complex problem, breast cancer classification, SDPSO performs significantly better than PSO.

Table 1. Comparison of PSO and SDPSO for Iris Classification

Algorithm	Epochs	ε_t	ε_g
PSO	500	1.35±0.32	2.89±0.31
SDPSO	500	1.30±0.22	2.82±0.21

Table 2. Comparison of PSO and SDPSO for Breast Cancer Classification

Algorithm	Epochs	ε_t	ε_g
PSO	2000	3.08±0.26	7.38±0.24
SDPSO	2000	2.73±0.21	5.98±0.20

5 Conclusion and Future Work

The ecological and physical universal laws enlighten us that there is not only gravity but also repulsion forces between the particles. We introduced a concept, swarm's survival density, into PSO for balancing the particles' distributing. In the paper, a modified algorithm, survival density particle swarm optimization (SDPSO) was proposed for neural network training to avoid premature and eliminate stagnation in PSO. Once the density of population in one area is too crowded, the individuals would be dispersed by repulsion force for keeping on searching different regions. In the other way, if the density of population in the area is too sparse, the individuals have to congregate by gravity force for information sharing and searching a more precise solution. The SDPSO algorithm was applied to benchmark function minimization problem and neural network training for benchmark dataset classification problems. The experimental results illustrated its efficiency.

The concept, swarm's survival density, is from ecology and physics. We choose arbitrarily the density according to the particles' searching range. There would be better results if the density according to the problems. It is deserved to research how to decide the density for a swarm or multi-swarm for better efficiency and result in future work.

References

1. Kennedy, J. and Eberhart, R. C.: Particle Swarm Optimization. In: Proceedings of IEEE International Conference on Neural Networks, Perth, Australia (1995) 1942-1948
2. Eberhart, R. C., and Shi, Y.: Comparison Between Genetic Algorithms and Particle Swarm Optimization. In: Proceedings of 7th Annual Conference on Evolutionary Computation, (1998) 611-616
3. Kennedy, J., Eberhart, R.: Swarm Intelligence. Morgan Kaufmann Publishers, Inc., San Francisco, CA (2001)
4. He, Z., Wei, C., Yang, L., Gao X., Yao S., Eberhart R. and Shi Y.: Extracting Rules From Fuzzy Neural Network by Particle Swarm Optimization. In: Proceedings of IEEE International Conference on Evolutionary Computation, (1998) 4-9
5. Shi, Y. and Eberhart, R. C.: A Aodified Particle Swarm Optimizer. In: Proceedings of the International Joint Conference on Evolutionary Computation, (1998) 69-73
6. Kennedy, J.: Bare Bones Particle Swarms. In: Proceedings of IEEE Swarm Intelligence Symposium, (2003) 80-87
7. Cristian, T. I.: The Particle Swarm Optimization Algorithm: Convergence Analysis and Parameter Selection. In: Information Processing Letters, Vol. 85, 6 (2003) 317–325
8. Blackwell, T. M. and Bentley, P.: Don't Push Me! Collision-Avoiding Swarms. In: Proceeding of IEEE Conference on Evolutionary Computation, (2002) 1691-1697
9. Shizhang, Z.: Ecology: Principle, Method and Application. Fudan University Press, Shanghai, China (1994)
10. Ziqing, W.: Physics and Society. Peking University Press, Peking, China (1992)
11. Van den Bergh, F. and Engelbrecht, A.P.: Training Product Unit Networks Using Cooperative Particle Swarm Optimizers. In: Proceedings of the International Joint Conference on Neural Networks, (2001) 126-132
12. Eberhart, R. and Shi, Y.: Comparing Inertia Weights and Constriction Factors in Particle Swarm Optimization. In: Proceedings of IEEE International Conference on Evolutionary Computation, (2000) 84-88
13. Bäck, T., Rudolph, G. and Schwefel, H. P.: Evolutionary Programming and Evolution Strategies: Similarities and Differences. In: Proceedings of 2nd Annual Conference on Evolutionary Programming, San Diego, CA (1993)
14. Blake, C., Keogh, E., and Merz, C.J.: UCI Repository of Machine Learning Databases, ww.ic.uci.edu/~mlearn/MLRepository.htm (2003)

Modified Error Function with Added Terms for the Backpropagation Algorithm

Weixing Bi[1], Xugang Wang[2], Ziliang Zong[3], and Zheng Tang[1]

[1] Faculty of Engineering, Toyama University, 930-8555 Toyama, Japan
biweixing613@hotmail.com, tang@iis.toyama-u.ac.jp
[2] Intelligence Engineering Laboratory, Institute of Software, The Chinese Academy of Science, Beijing 100080, China
wxg@iel.iscas.ac.cn
[3] Faculty of Computer Science, Shandong University, 250061 Jinan, China
pipizzl@hotmail.com

Abstract. We have noted that the local minima problem in the backpropagation algorithm is usually caused by update disharmony between weights connected to the hidden layer and the output layer. To solve this problem, we propose a modified error function with added terms. By adding one term to the conventional error function, the modified error function can harmonize the update of weights connected to the hidden layer and the output layer. Thus, it can avoid the local minima problem caused by such disharmony. Moreover, some new learning parameters introduced for the added term are easy to select. Simulations on the modified XOR problem have been performed to test the validity of the modified error function.

1 Introduction

We have noted that many local minima difficulties in the backpropagation learning for feedforward neural network are closely related to the neuron saturation in the hidden layer. Once such saturation occurs, neurons in the hidden layer will lose their sensitivity to input signals, and the propagation of information is blocked severely [1]. In some cases, the network can no longer learn [2]. The same phenomenon was also observed and discussed by Andreas Hadjiprocopis [1], Christian Goerick [2], Simon Haykin [3] and Wessels et al. [4].

In this paper, we explain this phenomenon as the weights update disharmony of between hidden and output Layers and propose a robust modified error function for the backpropagation algorithm in order to avoid the local minima problem that occurs due to neuron saturation in the hidden layer. Since a three-layered network is capable of forming arbitrarily close approximation to any continuous nonlinear mapping [5], our discussion will be limited to three-layered networks. To demonstrate the efficiency of the modified error function, we apply it to the backpropagation learning and conduct the simulation on the modified XOR problem. Furthermore, we compare the results with those of the backpropagation algorithm and simulated annealing method using the conventional error function.

2 Weights Update Disharmony

The backpropagation algorithm is a gradient descent procedure used to minimize an objective function (error function) E. The most popularly used error function is the "*sum-of-squares*" that is given by

$$E = \frac{1}{2}\sum_{p=1}^{P}\sum_{j}^{J}(t_{pj} - o_{pj})^2, \qquad (1)$$

where P is the number of training patterns, t_{pj} is the target value (desired output) of the j-th component of the outputs for the pattern p, o_{pj} is the output of the j-th neuron of the actual output pattern produced by the presentation of input pattern p, and J is the number of neurons in the output layer. To minimize the error function E, the backpropagation algorithm uses the following delta rule:

$$\Delta w_{ji} = -\eta \frac{\partial E}{\partial w_{ji}}, \qquad (2)$$

where w_{ji} is the weight connected between neurons i and j and η is the learning rate. From the above equations, we can see that there is only one term related to the neuron outputs in the output layer. Weights are updated iteratively to make the neurons in the output layer approximate to their desired values. However, the conventional error function does not consider the neuron behavior in the hidden layer and what value should be produced in the hidden layer.

Usually, the activation function of a neuron is given by a sigmoid function:

$$f(x) = \frac{1}{1 + e^{-x}}. \qquad (3)$$

There are two extreme areas that are called the saturated portion of the sigmoidal curve. If weights connected to the hidden layer and the output layer are updated so inharmoniously that all the hidden neurons' outputs are driven rapidly into the extreme areas before the output neurons start to approximate to the desired signals, no weights connected to the hidden layer will be modified even though the actual outputs in the output layer are far from the desired outputs. Therefore, the local minima problem occurs.

3 Modified Error Function

To overcome such a problem, and furthermore, to avoid the local minima caused by such disharmony, the neuron outputs in the output layer and those in the hidden layer should be considered together during the iterative update procedure. Motivated by this, we have proposed modified error function in [6]. The modified error function is given by:

$$\begin{aligned}E_{new} &= \frac{1}{2}\sum_{p=1}^{P}\sum_{j}^{J}(t_{pj} - o_{pj})^2 + \frac{1}{2}\sum_{p=1}^{P}(\sum_{j}^{J}(t_{pj} - o_{pj})^2)(\sum_{j}^{H}(y_{pj} - 0.5)^2) \\ &= E_A + E_B. \end{aligned} \qquad (4)$$

where, $\sum_j^H (y_{pj} - 0.5)^2$ can be defined as the degree of saturation in the hidden layer for pattern p. y_{pj} is the output of the j-th neuron in the hidden layer, and H is the number of neurons in the hidden layer. Using the above error function as the objective function, the update rule of both the weight w_{ji} and threshold θ_j can be given as:

$$\Delta w_{ji} = -\eta_A \frac{\partial E_A}{\partial w_{ji}} - \eta_B \frac{\partial E_B}{\partial w_{ji}}, \tag{5}$$

and

$$\Delta \theta_j = -\eta_A \frac{\partial E_A}{\partial \theta_j} - \eta_B \frac{\partial E_B}{\partial \theta_j}, \tag{6}$$

where η_A and η_B are the learning rates for E_A and E_B, respectively. It has been proven that this modified error function could effectively help the network avoid some local minima problems [6]. However, the selection of learning rates of E_A and E_B often is a very difficult and is problem-dependent. That's, given a value of η_A, both an over-small and over-large value of η_B will damage the learning. The setting of the learning rates η_A and η_B is a difficult and subtle task. Thus, we propose a novel modified error function that is defined as follows:

$$E_{new} = \frac{1}{2}\sum_{p=1}^{P}\sum_{j}^{J}(t_{pj} - o_{pj})^2 + (\frac{1}{2}\sum_{p=1}^{P}\sum_{j}^{H}(y_{pj}(W) - 0.5)^2$$

$$+ \frac{1}{2}\sum_{p=1}^{P}\prod_{j}^{H}(y_{pj}(\theta_j) - 0.5)^2)$$

$$= E_A + E_B. \tag{7}$$

We can see that the new error function also consists of two terms—the first term is the conventional error function:

$$E_A = \frac{1}{2}\sum_{p=1}^{P}\sum_{j}^{J}(t_{pj} - o_{pj})^2, \tag{8}$$

and the second one is the added term concerning the hidden layer:

$$E_B = \frac{1}{2}\sum_{p=1}^{P}\sum_{j}^{H}(y_{pj}(W) - 0.5)^2 + \frac{1}{2}\sum_{p=1}^{P}\prod_{j}^{H}(y_{pj}(\theta_j) - 0.5)^2. \tag{9}$$

The added term also consists of two terms:

$$E_B = E_B(W) + E_B(\Theta), \tag{10}$$

where $E_B(W)$ is the added error function of the weights connected to the hidden layer, and $E_B(\Theta)$ is the added error function of the threshold parameters of hidden layer neurons, respectively. We redefine the added term E_B and use different forms for the weights and thresholds since we found it is excellent trade-off for both the convergence speed and global optimization.

The derivatives of the added term E_B corresponding the weights and thresholds are computed as deferent forms. Given a pattern p, for the weights connected to the hidden layer, $\frac{\partial E_B^p}{\partial w_{ji}}$ is easily obtained as follows:

$$\frac{\partial E_B^p}{\partial w_{ji}} = \frac{\partial E_B^p(W)}{\partial w_{ji}} = (y_{pj} - 0.5)\frac{\partial y_{pj}}{\partial w_{ji}} = (y_{pj} - 0.5)f'(\cdot)x_{pi}, \quad (11)$$

where x_{pi} is the i-th input for pattern p since we use the network with only one hidden layer. For the thresholds of the neurons in the hidden layer, $\frac{\partial E_B^p}{\partial \theta_j}$ can be computed as follows:

$$\frac{\partial E_B^p}{\partial \theta_{ji}} = \frac{\partial E_B^p(\Theta)}{\partial \theta_j} = (y_{pj} - 0.5)\frac{\partial y_{pj}}{\partial \theta_j}(\prod_{h \neq j}(y_{ph} - 0.5)^2)$$

$$= -(y_{pj} - 0.5)f'(\cdot)(\prod_{h \neq j}(y_{ph} - 0.5)^2). \quad (12)$$

Since this added term is used to keep the degree of saturation of the hidden layer small when E_A is large, the effect of term E_B should be diminished and will eventually become zero while the output layer approximates to the desired signals. Therefore, for training pattern p, the learning rate η_B at step $t+1$ is adapted according to the following rule:

$$\eta_B^p(t+1) = E_A^p(t)\eta_B(0), \quad (13)$$

where, $\eta_B(0)$ is the initial value of the learning rate η_B. It is set to the same value for all patterns. For the novel modified error function, the selection of learning rates η_A and η_B is much easier. Generally, if $\eta_B(0) < \eta_A$ is selected, the performance of the network is not affected too much with various η_B.

4 Simulation

In order to verify the effectiveness of the modified error function, we applied it to the backpropagation algorithm (denoted by "BP+Added-terms"). Then the modified XOR problem was used for simulation. For comparison, we also performed the backpropagation algorithm (denoted by "BP") and a global search technology—the simulated annealing method [7] (denoted by "SA") with the conventional error function. Here, weights and thresholds were initialized randomly from (-1.0, 1.0). Two aspects of the training algorithm performance, "success rate" and "training speed", were assessed for algorithm. The upper limit epochs for the BP+Added-terms and BP were set to 10,000.

The modified XOR problem is different from the classical XOR problem because one more pattern is included (that is, inputs=(0.5,0.5), teacher signal=1.0) such that a unique global minimum exists. Furthermore, several local minima exist simultaneously in this problem [8]. We used the 2-2-1 neural network to solve this problem. To show how the BP+Added-terms can avoid the local minima,

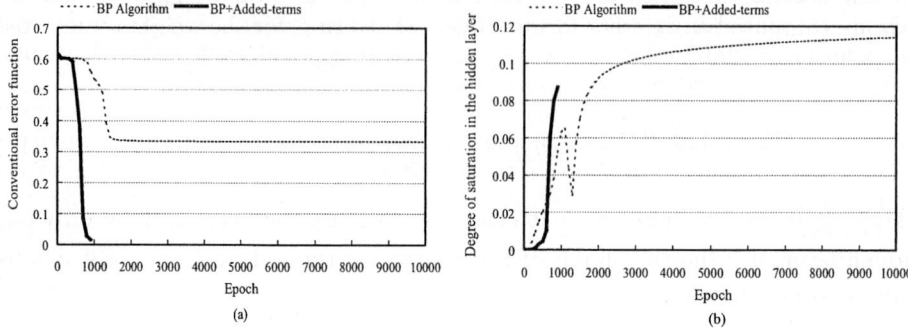

Fig. 1. Comparison of learning processes with local minima for the modified XOR problem between BP+Added-terms and BP: (a) conventional error function (E_A) vs. epochs and (b) degree of saturation in the hidden layer of all patterns vs. epochs

we compared a typical learning process of BP+Added-terms with that of BP in the case where there is the local minimum in the learning of BP. In Fig.1, the conventional error function (E_A) and the degree of saturation in the hidden layer of overall patterns are plotted as a function of epochs for both methods. We can see that the BP converged into a local minimum, while the degree of saturation in the hidden layer increased continually until it reached about 0.114. Meanwhile, the BP+Added-terms method avoided the local minimum and trained the network successfully about 900 epochs when the degree of saturation in the hidden layer was effectively neutralized by the modified error function. Table 1 shows the experimental results of the three methods based on 100 runs of this problem. For the BP, different learning rates $\eta = 0.3$, $\eta = 0.5$, and $\eta = 1.0$ were used. $\eta_B(0) = \eta_A = \eta$ was selected for the BP+Added-terms. The table shows that the backpropagation algorithm could obtain successful solutions for almost every run using the modified error function, while many failures in convergence to the global solution occurred both in the backpropagation algorithm and the simulated annealing method using the conventional error function. Although the average number of epochs of BP+Added-terms was a bit more than that of BP when $\eta = 0.3$ was used, it was almost the same as that of the BP when $\eta = 0.5$ was selected. Moreover, it was much less than those of the SA in all cases and BP in the case of $\eta = 1.0$. These results indicate that the proposed method could efficiently avoid the local minima for this problem.

5 Conclusions

In this paper, we proposed a modified error function with the added terms for the backpropagation algorithm to harmonize the update of weights connected to the hidden layer and those connected to the output layer. Therefore, local minima problems due to such disharmony could be avoided without much additional

Table 1. Experimental results for modified XOR problem

Methods		Success Rate	Average Number of Epoch
	($\eta = 0.3$)	69%	3582
BP	($\eta = 0.5$)	70%	2181
	($\eta = 1.0$)	58%	1106
	($\eta = 0.3$)	60%	7164
SA	($\eta = 0.5$)	85%	4811
	($\eta = 1.0$)	93%	3593
	($\eta_B(0) = \eta_A = 0.3$)	99%	4157
BP+Added-terms	($\eta_B(0) = \eta_A = 0.5$)	99%	2257
	($\eta_B(0) = \eta_A = 1.0$)	97%	904

computation and change in the network topology. And, the new learning parameters for the added term is not difficult to select. Finally, simulations performed on a benchmark problem demonstrated that the performance of the backpropagation algorithm was greatly improved by using the modified error function. More analysis on large problems is still required.

References

1. Hadjiprocopis, A.: Feed Forward Neural Network Entities. Ph.D. Thesis. City University London UK. (2000)
2. Goerick, C., Seelen, W.V.: On Unlearnable Problems or A Model for Premature Saturation in Backpropagation Learning. In: Proceedings of the European Symposium on Artificial Neural Networks, Brugge Belgium April 24-26 (1996) 13–18
3. Haykin, S.: Neural Networks, A Comprehensive Foundation. MacMillan Publishing New York (1994)
4. Wessels, L.F.A., Barnard, E., van Rooyen, E.: The Physical Correlates of Local Minima. In: Proceedings of the International Neural Network Conference, Paris July (1990) 985
5. Funahashi, K.: On the Approximate Realization of Continuous Mapping by Neural Networks. Neural Networks, Vol. 2. (1989) 183–192
6. Wang, X.G., Tang, Z., Tamura, H., Ishii, M.: A Modified Error Function for Backpropagation Algorithm. Neurocomputing, Vol. 57. (2004) 477–484
7. Owen, C.B., Abunawass, A.M.: Application of Simulated Annealing to the Backpropagation Model Improves Convergence. In: Proceedings of the SPIE Conference on the Science of Artificial Neural Networks II, Vol. 1966. (1993) 269–276
8. Gori, M., Tesi, A.: On the Problem of Local Minima in Backpropagation. IEEE Trans. Pattern Analysis and Machine Intelligence, Vol. 14, No. 1. (1992) 76–86

Robust Constrained-LMS Algorithm*

Xin Song, Jinkuan Wang, and Han Wang

School of Information Science and Engineering,
Northeastern University (NEU), 110004 Shenyang, China
{sxin78916,wjk,wanghan}@mail.neuq.edu.cn

Abstract. The performances of the existing adaptive array algorithms are known to degrade substantially in the presence of even slight mismatches between the actual and presumed array responses to the desired signal. Similar types of performance degradation can take place when the signal array response is known precisely but the training sample size is small. In this paper, on the basis of the constrained-LMS (CLMS) algorithm, we propose a robust constrained-LMS (RCLMS) algorithm. Our robust constrained-LMS algorithm provides excellent robustness against signal steering vector mismatches, offers fast convergence rate and makes the mean output array SINR consistently close to the optimal one. Computer simulations show better performance of our RCLMS algorithm as compared with the classical CLMS algorithm.

1 Introduction

Adaptive beamforming is a ubiquitous task in array signal processing with applications, among others, in radar, sonar, astronomy, and medical imaging [1], [2], [3], and, more recently, in wireless communications [4],[5]. In particular, the development of robust adaptive beamforming spans over two decades.

When adaptive arrays are applied to practical problems, the performance degradation of adaptive beamforming techniques may become more pronounced than in the ideal case because some of underlying assumptions on the environment, sources, or sensor array can be violated and this may cause a mismatch between the presumed and actual signal steering vectors. Adaptive array techniques are known to be very sensitive even to slight mismatches of such type that can easily occur in practical situations. In such cases, robust approaches to adaptive beamforming are required. There are several existing approaches to robust adaptive beamforming, such as linearly constrained minimum variance (LCMV) beamformer [6], diagonal loading of the sample covariance matrix [7]. But these methods cannot be expected to provide sufficient robustness improvements.

The weight vector of the CLMS algorithm is selected by minimizing the mean output power while maintaining a distortionless response toward the desired signal. Thus, the performance of the CLMS algorithm degrades in the presence of even slight mismatches between the actual and presumed array responses

---

* This work is supported by Key Program of Science and Technology from the Ministry of Education of China, under Grant no. 02085

to the desired signal. In this paper, on the basis of the CLMS algorithm, we develop a novel robust constrained-LMS (RCLMS) algorithm against the signal steering vector mismatches and small training sample size. Computer simulations demonstrate a visible performance gain of the proposed RCLMS algorithm over other traditional and robust adaptive beamforming techniques.

2 Problem Formulation

2.1 Mathematical Model

Consider a uniform linear array (ULA) with M omnidirectional sensors spaced by the distance d and D narrow-band incoherent plane waves, impinging from directions $\{\theta_0, \theta_1, \cdots, \theta_{D-1}\}$. The observation vector is given by

$$\begin{aligned}\mathbf{X}(k) &= \mathbf{s}(k) + \mathbf{i}(k) + \mathbf{n}(k) \\ &= s_0(k)\mathbf{a} + \mathbf{i}(k) + \mathbf{n}(k) \;.\end{aligned} \quad (1)$$

where $\mathbf{X}(k) = [x_1(k), x_2(k), ..., x_M(k)]^T$ is the complex vector of array observations, $\mathbf{s}(k)$, $\mathbf{i}(k)$ and $\mathbf{n}(k)$ are the desired signal, interference and noise components, respectively. Here, $s_0(k)$ is the signal waveform, $\mathbf{a}$ is the signal steering vector. The output of a narrowband beamformer is given by

$$y(k) = \mathbf{W}^H \mathbf{X}(k) \;. \quad (2)$$

where $\mathbf{W} = [w_1, w_2, ..., w_M]^T$ is the complex vector of beamformer weights and $(\cdot)^T$ and $(\cdot)^H$ stand for the transpose and Hermitian transpose, respectively. The weight vector can be found from the maximum of the signal-to-interference-plus-noise ratio (SINR)

$$\text{SINR} = \frac{\mathbf{W}^H \mathbf{R_s} \mathbf{W}}{\mathbf{W}^H \mathbf{R_{i+n}} \mathbf{W}} \;. \quad (3)$$

where $\mathbf{R_s} = E\{\mathbf{s}(k)\mathbf{s}^H(k)\}$, $\mathbf{R_{i+n}} = E\{(\mathbf{i}(k)+\mathbf{n}(k))(\mathbf{i}(k)+\mathbf{n}(k))^H\}$.

2.2 Constrained-LMS (CLMS) Algorithm

Constrained-LMS (CLMS) algorithm is a real-time constrained algorithm for determining the optimal weight vector. The optimal weight vector is the solution of the following optimization problem:

$$\min_{\mathbf{W}} \mathbf{W}^H \mathbf{R_{xx}} \mathbf{W} \qquad \text{subject to} \quad \mathbf{W}^H \mathbf{a} = 1 \;. \quad (4)$$

Optimization technique used to find $\mathbf{W}_{opt}$ will use Lagrange multiplier method, thus, the expression for $\mathbf{W}_{opt}$ becomes

$$\mathbf{W}_{opt} = \frac{\mathbf{R_{xx}^{-1}} \mathbf{a}}{\mathbf{a}^H \mathbf{R_{xx}^{-1}} \mathbf{a}} \;. \quad (5)$$

In practical applications, the sample covariance matrix $\hat{\mathbf{R}}_{xx}$ is used instead of $\mathbf{R}_{xx}$. In this case, (5) should be rewritten as

$$\hat{\mathbf{W}}_{opt} = \frac{\hat{\mathbf{R}}_{xx}^{-1}\mathbf{a}}{\mathbf{a}^H \hat{\mathbf{R}}_{xx}^{-1}\mathbf{a}} . \tag{6}$$

In an environment where complete knowledge of signal characteristics is not available and also time-vary environment, we need a "recursive algorithm". The CLMS algorithm should be described as [8]:

$$\begin{aligned}\mathbf{F} &= \mathbf{a}[\mathbf{a}^H \mathbf{a}]^{-1} , \\ \mathbf{P} &= \mathbf{I} - \mathbf{a}[\mathbf{a}^H \mathbf{a}]^{-1}\mathbf{a}^H , \\ \mathbf{W}(k+1) &= \mathbf{P}[\mathbf{W}(k) - \mu \mathbf{R}_{xx}\mathbf{W}(k)] + \mathbf{F} .\end{aligned} \tag{7}$$

The CLMS algorithm requires the knowledge of the direction-of-arrival (DOA) of the desired signal, but in practical applications, the performance degradation of the CLMS algorithm may become evident because some of underlying assumptions on the environment, sources, or sensor array can be violated and this may cause a mismatch between the presumed and actual signal steering vectors.

3 Robust Constrained-LMS (RCLMS) Algorithm

We develop a novel approach to robust adaptive beamforming that provides an improved robustness against the signal steering vector mismatches and small training sample size. Our approach is based on the constrained-LMS algorithm.

In practical applications, we assume that the norm of the steering vector distortion $\mathbf{e}$ can be bounded by some known constant $\varepsilon > 0$, $||\mathbf{e}|| \leq \varepsilon$. Then, the actual signal steering vector belongs to the set

$$\Phi(\varepsilon) = \{\mathbf{b} | \mathbf{b} = \mathbf{a} + \mathbf{e}, \ ||\mathbf{e}|| \leq \varepsilon\} . \tag{8}$$

The weight vector is selected by minimizing the mean output power while maintaining a distortionless response for the mismatched signal steering vector. Thereby the cost function of the RCLMS algorithm can be written as the following constrained minimization problem [9]:

$$\min_{\mathbf{W}} \mathbf{W}^H \mathbf{R}_{xx} \mathbf{W} \quad \text{subject to} \quad \min_{\mathbf{b} \in \Phi(\varepsilon)} |\mathbf{W}^H \mathbf{b}| \geq 1 . \tag{9}$$

We can prove that the inequality constraint in (9) is equivalent to the equality constraint. As a result, (9) can be equivalently described as

$$\min_{\mathbf{W}} \mathbf{W}^H \mathbf{R}_{xx} \mathbf{W} \quad \text{subject to} \quad \mathbf{W}^H \mathbf{a} = \varepsilon ||\mathbf{W}|| + 1 . \tag{10}$$

Use Lagrange multiplier method to obtain the optimal weight vector

$$\mathbf{W}_{rob} = -\lambda(\mathbf{R}_{xx} + \lambda \varepsilon^2 \mathbf{I} - \lambda \mathbf{a}\mathbf{a}^H)^{-1}\mathbf{a} . \tag{11}$$

The solution to (10) can be found by minimizing the function

$$H(\mathbf{W}, \lambda) = \mathbf{W}^H \mathbf{R}_{\mathbf{xx}} \mathbf{W} + \lambda(\varepsilon^2 \mathbf{W}^H \mathbf{W} - \mathbf{W}^H \mathbf{a}\mathbf{a}^H \mathbf{W} + \mathbf{W}^H \mathbf{a} + \mathbf{a}^H \mathbf{W} - 1) . \quad (12)$$

where λ is a Lagrange multiplier. Take the gradient of $H(\mathbf{W}, \lambda)$

$$\mathbf{\Gamma}(\mathbf{W}, \lambda) = (\mathbf{R}_{\mathbf{xx}} + \lambda \varepsilon^2 \mathbf{I} - \lambda \mathbf{a}\mathbf{a}^H) \mathbf{W} + \lambda \mathbf{a} . \quad (13)$$

The weight updating equation for the RCLMS algorithm becomes

$$\mathbf{W}(k+1) = \mathbf{W}(k) - \mu[(\mathbf{R}_{\mathbf{xx}} + \lambda \varepsilon^2 \mathbf{I} - \lambda \mathbf{a}\mathbf{a}^H) \mathbf{W}(k) + \lambda \mathbf{a}] . \quad (14)$$

The weight vector in (14) satisfies the constraint in (10) at every iteration.

To summarize, our proposed robust constrained-LMS (RCLMS) algorithm consists of the following steps.
Step 1) Initialize $\mathbf{W}(0)$.
Step 2) Compute the desired signal steering vector $\mathbf{a}$ and let $i = 1$.
Step 3) Compute the sample covariance matrix $\hat{\mathbf{R}}_{\mathbf{xx}}$, while updating $k = k + 1$.
Step 4) Compute λ by using (14) and the constraint in (10).
Step 5) Compute the gradient $\mathbf{\Gamma}(\mathbf{W}, \lambda)$ by using λ and (13).
Step 6) Substitute $\mathbf{\Gamma}(\mathbf{W}, \lambda)$ into (14) to update the weight vector.
Step 7) If $i = n$, stop. Otherwise, set $i = i + 1$, and go to step 3.

4 Simulations

In this section, we present some simulations to justify the performance of the proposed RCLMS algorithm. We assume a uniform linear array with $M = 10$ omnidirectional sensors spaced half a wavelength apart. For each scenario, 100 simulation runs are used to obtain each simulated point. In all examples, we assume two interfering sources with plane wavefronts and the directions of arrival (DOAs) 30° and 50°, respectively.

In the first example, we assume that the actual signal spatial signature is plane waves impinging from the DOA 5°. Fig. 1 compares two aforementioned methods (the beamformer (6), the robust beamformer (11)) in terms of the mean output array SINR versus the number of snapshots N for the fixed single-sensor $SNR = 10dB$ and $\varepsilon = 3$. Fig. 1 shows the robust beamformer (11) can be seen to outperform the beamformer (6).

In the second example, the plane-wave signal is assumed to impinge on the array from $\theta = 3°$. Fig. 2 displays the performance of the methods tested versus the number of snapshots N for the fixed $SNR = 10dB$ and $\varepsilon = 1.5$. Fig. 3 shows the performance of these algorithms versus the SNR for the fixed training data size $N = 100$. The example demonstrates that the RCLMS algorithm achieves the values of SINR that are close to the optimal one in a wide range N and SNR. Moreover, the RCLMS algorithm offers faster convergence rate.

In the last example, a scenario with the signal look direction mismatch is considered. We assume that both the presumed and actual signal spatial signatures

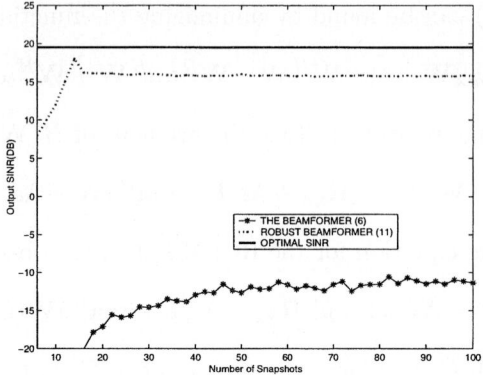

Fig. 1. Output SINR versus the training sample size N

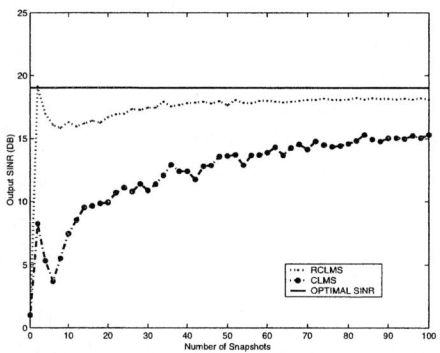

Fig. 2. Output SINR versus N

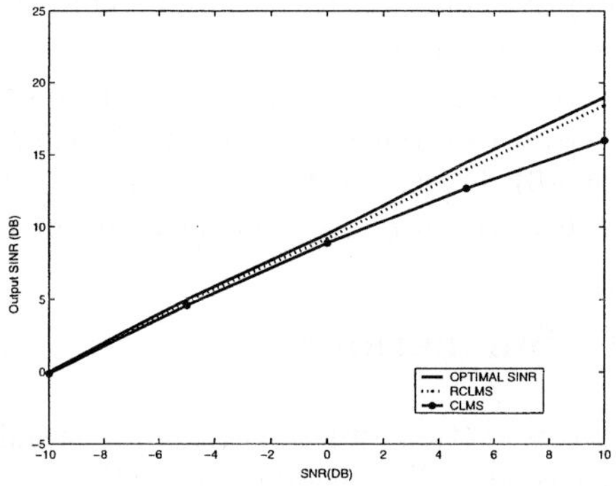

Fig. 3. Output SINR versus SNR

are plane waves impinging from the DOAs 3° and 5°, respectively. This corresponds to a 2° mismatch in the signal look direction. Fig. 4 displays the performance of the methods tested versus the number of snapshots N for $SNR = 10dB$ and $\varepsilon = 1.5$. The performance of these algorithms versus the SNR for the fixed training data size $N = 100$ is shown in Fig. 5. In this example, the CLMS algorithm is very sensitive even to slight mismatches that can easily occur in practical situations. Moreover, the CLMS algorithm shows poor performance at all values of the SNR.

5 Conclusions

The proposed RCLMS algorithm in this paper is based on the classical CLMS algorithm. The proposed RCLMS algorithm offers faster convergence rate and provides excellent robustness against signal steering vector mismatches. More-

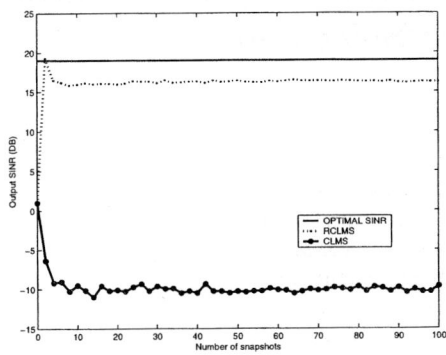 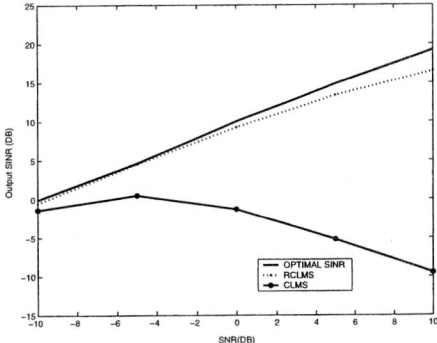

Fig. 4. Output SINR versus N **Fig. 5.** Output SINR versus SNR

over, the mean output SINR of the RCLMS algorithm is consistently close to the optimal one in a wide range of SNR and N. Our simulation figures clearly demonstrate that in all examples, the proposed RCLMS algorithm is shown to consistently enjoy a significantly improved performance.

References

1. Brennan, L.E., Mallet, J.D., Reed, I.S.: Adaptive Arrays in Airborne MTI Radar. IEEE Trans. Antennas Propagation, Vol. 24 (1976) 607-615
2. Krolik, J.L.: The Performance of Matched-Field Beamformers with Mediterranean Vertical Array Data. IEEE Trans. Signal Processing, Vol. 44 (1996) 2605-2611
3. Gorodetskaya, E.Y., Malekhanov, A.I., Sazontov, A.G., Vdovicheva, N.K.: Deep-Water Acoustic Coherence at Long Ranges: Theoretical Prediction and Effects on Large-Array Signal Processing. IEEE J. Ocean. Eng., Vol. 24 (1999) 156-171
4. Shahbazpanahi, S., Gershman, A.B., Luo, Z.Q.: Robust Adaptive Beamforming for General-Rank Signal Models. IEEE Trans. Signal Processing, Vol. 51 (2003) 2257-2269
5. Rapapport, T.S. (ed.): Smart Antennas: Adaptive Arrays, Algorithms, and Wireless Position Location. Piscataway. NJ: IEEE (1998)
6. Monzingo, R.A., Miller, T.W.: Introduction to Adaptive Arrays. New York: Wiley (1980)
7. Carlson, B.D.: Covariance Matrix Estimation Errors and Diagonal Loading in Adaptive Arrays. IEEE Trans. Aerosp. Electron. Syst., Vol. 24 (1988) 397-401
8. Godara, L.C. : Application of Antenna Arrays to Mobile Communication, Part II: Beam-Forming and Direction-of-Arrival Considerations. Proc. IEEE, Vol. 85 (1997) 1213-1216
9. Vorobyov, S.A., Gershman, A.B., Luo, Z.Q.: Robust Adaptive Beamforming Using Worst-Case Performance Optimization: A Solution to the Signal Mismatch Problem. IEEE Trans. Signal Processing, Vol. 51 (2003) 313-323

A Novel Three-Phase Algorithm for RBF Neural Network Center Selection

Dae-Won Lee and Jaewook Lee

Department of Industrial Engineering,
Pohang University of Science and Technology,
Pohang, Kyungbuk 790-784, Korea.
{woosuhan,jaewookl}@postech.ac.kr

Abstract. In this paper, we propose a new method for selecting RBF centers. The strength of our method is to determine the number and the locations of RBF centers automatically without any priori assumption about the number of centers. The proposed method consists of three phases. The first phase is to partition the input patterns into the several subsets according to their output labels. In the second and third phase, the number and the locations of RBF centers are determined using bisection algorithm and weighted mean centering. These second and third phase are iteratively repeated until to reach the goal error. The proposed method is applied to several benchmark data sets. The numerical results show that our method is robust and efficient for determining the number and the locations of centers.

1 Introduction

Radial basis function network (RBFN), due to the simplicity of its single hidden layer structure and universal property, has been widely used for nonlinear function approximation and pattern classification. Training of the RBFNs consists of selecting centers of the hidden neurons and estimating the weights that connect the hidden and the output layers. Once centers have been fixed, the network weights will be directly estimated by using the least squares algorithm.

The generalized radial basis function network(GRBFN) involves searching for a suboptimal solution in a lower-dimensional space that approximates the interpolation solution where the approximated solution $F^*(\mathbf{x})$ can be expressed as follows

$$F^*(\mathbf{x}) = \sum_{i=1}^{m} w_i \phi(\|\mathbf{x} - \mathbf{t}_i\|) \tag{1}$$

where the set of RBF centers $\{\mathbf{t}_i | i = 1, \ldots, m\}$ is to be determined.

One of the key issues in the design of RBFN specially is how to determine the number and the locations of the RBF centers. In the recent researches, a variety of ways for determining the locations of centers have been proposed. Previously reported approaches for RBF center selection include random selection from input patterns [1], and selection of centers based on clustering algorithms [2], [3]. These methods have some drawbacks in that they cannot determine the number and the locations of centers at the same time and do not consider the information of the supervised data.

In this paper, to overcome such drawbacks, the proposed method consists of three basic ingredients. Firstly, input patterns are partitioned into several subsets according to their output labels and the RBF centers are determined separately for each subset in order to reflect information of the output class label. Then these centers are combined to form a larger superset to be used in the final RBFN. Secondly, the proposed algorithm determines the number of the centers by using bi-section algorithm. Finally, the proposed algorithm determines the optimal locations of the centers by employing a weighted mean centering.

The organization of this paper is as follows : In Section 2 a new method for RBF center selection was explained. Section 3 presents an algorithm for the proposed method, and Section 4 presents experimental results applied to benchmark problems, followed by conclusions in Section 5.

2 The Proposed Method

2.1 Phase I: Partitioning Supervised Data

The proposed method is basically similar to the *selection of centers based on clustering algorithm* which is most widely used for center selection. However this kind of approaches deal with input patterns as unsupervised data during the center selection step even though supervised data are available. When the output class label is not considered, some of centers is often found to be located in the boundary between several classes, in which case these centers fail to play a role of a good feature extractor in the GRBFN.

In phase I, we partition training data into C disjoint subsets $\{\mathcal{D}_i\}_{i=1}^{C}$ according to its class to reflect the distribution of input pattern for each output class. Then, with each subset, cluster centers are found using clustering algorithm to be explained in Section 2.2 and 2.3. Finally, we use C disjoint sets of cluster centers as GRBFN centers.

2.2 Phase II: Determining the Number of Centers by Bi-section Algorithm

Phase II tries to estimate the proper number of centers automatically. The idea was originated from *Cover's theorem on the separability of patterns* [4]: the more GRBFN has hidden neurons, the more input patterns are linearly separable. It can be interpreted that a training error rate curve against the number of centers is approximately monotonic decreasing function. From this observation, in order to find the number of centers we employed the *bi-section* algorithm expressed in Fig.1. It can approximately estimate the minimum number of centers reaching the goal error rate with reducing the search space $(1 \sim N)$ in a ratio of $1/2$. One advantage of this algorithm is that it can find the number of centers within a comparatively less time because it has only $O(\log_2 N)$ complexity.

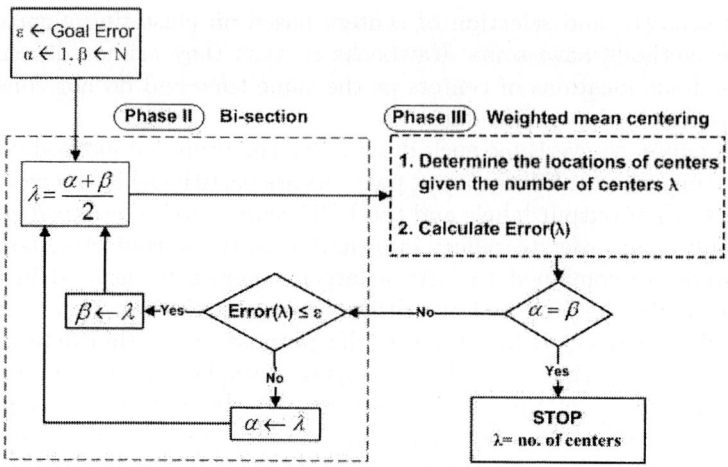

Fig. 1. Flow diagram for center selection (Phase II and III). Where Error(λ) is a misclassified rate with the λ number of centers

2.3 Phase III: Determining the Locations of Centers Using Weighted Mean Centering

Determining the locations of the RBF centers is based on optimizing the following objective function with respect to $(\mathbf{t}, \mathbf{w})$.

$$\mathcal{E} = \frac{1}{2} \sum_{i=1}^{N} \left[d_i - \sum_{j=1}^{k_i} w_j \phi(\|\mathbf{x}_i - \mathbf{t}_j\|) \right]^2 \tag{2}$$

From the necessary optimality condition, positions of centers is given by

$$\frac{\partial \mathcal{E}}{\partial \mathbf{t}_j} = \frac{1}{\sigma^2} w_j \sum_{i=1}^{N} (\mathbf{d} - \Phi \mathbf{w})_i \Phi_{ij} (\mathbf{x}_i - \mathbf{t}_j) = 0 \tag{3}$$

$$\mathbf{t}_j = \frac{\sum_{i=1}^{N} (\mathbf{d} - \Phi \mathbf{w})_i \Phi_{ij} \mathbf{x}_i}{\sum_{i=1}^{N} (\mathbf{d} - \Phi \mathbf{w})_i \Phi_{ij}} = \sum_{i=1}^{N} \left[\frac{(\mathbf{d} - \Phi \mathbf{w})_i \Phi_{ij}}{\sum_{i=1}^{N} (\mathbf{d} - \Phi \mathbf{w})_i \Phi_{ij}} \right] \mathbf{x}_i$$

$$= \sum_{i=1}^{N} \lambda_j(\mathbf{x}_i, \mathbf{t}, \mathbf{w}) \mathbf{x}_i, \quad \forall j = 1, \ldots, k_i \tag{4}$$

where $\sum_{i=1}^{N} \lambda_j(\mathbf{x}_i, \mathbf{t}, \mathbf{w}) = 1$, $\forall j = 1, \ldots, k_i$, $\Phi = [\phi(\mathbf{x}_i, \mathbf{t}_j)]_{i=1,\ldots,n_i, \ j=1,\ldots,k_i}$, $\mathbf{d} = [d_1, d_2, \ldots, d_N]^T$ and $\mathbf{w} = [w_1, w_2, \ldots, w_{m_1}]^T$. It shows that the estimate for $\mathbf{t}_j$ is merely a weighted average of the samples. However, due to the highly non-linearity of $\mathcal{E}$ with respect to the centers $\mathbf{t}_j$, it is very difficult to compute

t_j directly. Instead, we employ a so-called *weighted mean centering* scheme to implement this, which consists of two steps. In the first step, center positions t_j of Eq. (4) are approximated as a simple average of the x_i that has higher absolute value of the numerator of $\lambda_j(x_i, t, w)$. This process is repeated until no change is made. (See Section 3 for more details about this.) In the case of non-convex data set, the obtained centers in the first step are often found to be located in the regions of other classes, even though we have partitioned data sets according to their classes. To avoid this problem, in the second step, we modify RBF centers into the nearest points within the partitioned subset $\mathcal{D}_i$.

3 Algorithm

An algorithm of center selection for the proposed method is as follows.

% Phase I: Partitioning supervised data
1. Separate N training data into C disjoint subsets $\mathcal{D}_i$ containing n_i elements, according to class.

$$\{(x_i, d_i)\}_{i=1}^N \to \mathcal{D}_1 \cup \mathcal{D}_2 \cup, \ldots, \cup \mathcal{D}_C$$
$$\mathcal{D}_i = \{(x_1, i), (x_2, i), \ldots, (x_{n_i}, i)\} \quad \text{for } i = 1, \ldots, C$$

% Selecting k_i centers for each subset $\mathcal{D}_i$
2. **for** $i = 1$ **to** C **do**
   % Phase II: bi-section algorithm
   2.1. Determine the number of centers (k_i) using bi-section algorithm as explained in Fig. 1: initially, $k_i = \frac{1+n_i}{2}$.

   % Phase III: weighted mean centering
   2.2. Determine the locations of k_i centers using weighted means centering.
   2.2.1. Choose random values for the initial centers $\mathcal{T}_i = \{t_j\}_{j=1}^{k_i}$ from input space. Where t_j is the jth cluster center of subset $\mathcal{D}_i$.
   % Adjust the centers
   2.2.2. **for** $l = 1$ **to** n_i **do**
   2.2.2.1. Let $\mathcal{I}_i(x_l)$ denote the index of the best-matching center for the input vector $x_l \in \mathcal{D}_i$ as follows.

$$\mathcal{I}_i(x_l) = \arg\max_j \|(d - \Phi w)_l \Phi_{lj}\|, \quad j = 1, 2, \ldots, k_i$$

   2.2.2.2. Adjust the centers using the update rule

$$t_j \leftarrow t_j + \eta(x_l - t_j), \quad j = \mathcal{I}_i(x_l) \tag{5}$$

   where η is a learning step size.
   2.2.3. Continue the center adjusting procedure (step 2.2.2) until no change are observed in the centers $\{t_j\}_{j=1}^{k_i}$
   2.2.4. Modify centers into nearest points of partitioned subset $\mathcal{D}_i$ and determine final locations of centers.

Table 1. Benchmark data description

	Input dimension	Number of classes	Number of patterns
2-Spirals	2	2	388 [194,194]
Sonar	60	2	104 [55,49]
Heart	13	2	180 [98,82]
Vowel	10	11	528[48,48,48,48,48,48,48,48,48,48]

bracketed numbers mean the number of patterns for each class.

$\mathbf{t}_j=$ the nearest point $\mathbf{x} \in \mathcal{D}_i$ for $j = 1, 2, \ldots, k_i$

2.3. Repeat Step 2.1.~2.2. until converge to goal error.

% Complete the RBFN training

3. Combine the centers of C disjoint subsets $\{\mathcal{D}_i\}_{i=1}^{C}$ and construct generalized RBF network by using pseudo-inverse.

$$\mathcal{T} = \{\mathbf{t}_j\}_{j=1}^{K} \leftarrow \mathcal{T}_1 \cup \mathcal{T}_2 \cup, \ldots, \cup \mathcal{T}_C$$

$$\mathbf{w} = (\Phi^T \Phi + \lambda \Phi_0)^{-1} \Phi^T \mathbf{d}$$

where $\Phi = [\phi(\mathbf{x}_i, \mathbf{t}_j)]_{i=1,\ldots,N,\ j=1,\ldots,K}$, $\Phi_0 = [\phi(\mathbf{x}_i, \mathbf{t}_j)]_{i,j=1,\ldots,K}$, and $K = \sum_{i=1}^{C} k_i$.

4 Simulation Results

The algorithm described in the previous section has been simulated on four kinds of benchmark data sets (2-spiral, sonar, heart, vowel). Description of the benchmark data sets is given in Table 1.

The performance of the proposed method is compared with two widely used center selection methods. In Table 2, *KM* is the *k-means based center selection without partitioning* and *RS* is *random selection from training data without partitioning*. For these two methods, the number of centers is determined by increasing centers one by one until they achieve the goal error. For the comparison we adopted three criterion: the number of centers, the mis-classified rate, and the computing time. Simulation results are shown in Table 2. The results show that the proposed method achieves better accuracy with a slightly fewer number of RBFN centers while significantly reducing computing time.

5 Concluding Remarks

In this study, we have presented a novel three-phase algorithm for RBF center selection. The proposed method has several advantages. Firstly, it determines the number and the locations of centers automatically without any assumption about the number of centers. Secondly it selects good feature extractors by using

Table 2. Simulation results on four benchmark problems

Method	KM			RS			Proposed		
	m	E	T	m	E	T	m	E	T
2-Spiral	84	0.069	5650	88	0.080	5925	80 [35,45]	0.064	297
Sonar	37	0.096	1570	37	0.096	1112	34 [17,17]	0.096	21
Heart	129	0.97	11876	131	0.99	9723	129 [68,61]	0.093	125
Vowel	66	0.099	3658	64	0.099	2465	59 [2,5,5,2,8,7,3,2,9,2,14]	0.099	188

m is the number of centers and bracketed numbers are the number of centers for each class. E is mis-classified error rate and T is computing time to construct the GRBFN.

the information of output class label. Finally, it is robust to a data set with non-convex distribution.

Experimental results show that the proposed method is competitive with the previously reported approaches for RBF center selection. Other methods to improve efficiency of RBF center selection, such as Homotopy method [5], [6] can be also be investigated.

Acknowledgement. This work was supported by the Korea Research Foundation under grant number KRF-2003-041-D00608.

References

1. Mao, K.Z.: RBF Neural Network Center Selection Based on Fisher Ratio Class Separability Measure. IEEE Trans. Neural Networks, Vol. 13(5) (2002) 1211-1217
2. Gomm, J.B., Yu, D.L.: Selection Radial Basis Function Network Centers with Recursive Orthogonal Least Squares Training. IEEE Trans. Neural Networks, Vol. 11(2) (2000) 306-314
3. Haykin, S.: Neural Networks: A Comprehensive Doundation. Prentice Hall, New York (1999)
4. Cover, T.M.: Geometrical and Statistical Properties of Systems of Linear Inequalities with Applications in Pattern Recognition. IEEE Trans. Electronic Computers, Vol. EC-14 (1965) 326-334
5. Lee, J., Chiang, H.-D.: Constructive Homotopy Methods for Finding All or Multiple DC Operating Points of Noninear Circuits and Systems. IEEE Trans. on Circuits and Systems- Part I, Vol. 48-(1) (2001) 35-50
6. Lee, J., Chiang, H.-D.: A Singular Fixed-Point Homotopy Method to Locate the Closest Unstable Equilibrium Point for Transient Stability Region Estimate. IEEE Trans. on Circuits and Systems- Part II, Vol. 51-(4) (2004) 185-189

Editing Training Data for kNN Classifiers with Neural Network Ensemble

Yuan Jiang and Zhi-Hua Zhou

National Laboratory for Novel Software Technology,
Nanjing University, Nanjing 210093, China
{jiangyuan,zhouzh}@nju.edu.cn

Abstract. Since kNN classifiers are sensitive to outliers and noise contained in the training data set, many approaches have been proposed to edit the training data so that the performance of the classifiers can be improved. In this paper, through detaching the two schemes adopted by the Depuration algorithm, two new editing approaches are derived. Moreover, this paper proposes to use neural network ensemble to edit the training data for kNN classifiers. Experiments show that such an approach is better than the approaches derived from Depuration, while these approaches are better than or comparable to Depuration.

1 Introduction

KNN [5] is one of the most widely used lazy learning approach [1]. Given a set of n training examples, upon receiving a new instance to predict, the kNN classifier will identify k nearest neighboring training examples of the new instance and then assign the class label holding by the most number of neighbors to the new instance.

The asymptotic classification error of kNN tends to the optimal Bayes error rate as $k \to \infty$ and $k/n \to 0$ when n grows to infinity, and the error is bounded by approximately twice the Bayes error if $k = 1$ [6]. This behavior in asymptotic classification performance combining with the simplicity in concept and implementation, makes kNN a powerful classification approach capable of dealing with arbitrarily complex problems, provided there is a large training data set. However, the theoretical behavior can hardly be obtained because kNN is sensitive to outliers and noise contained in the training data set, which usually occurs in real-world applications. Therefore, it is important to eliminate outliers in the training data set and make other necessary cleaning. The approaches devoting to this purpose are referred to as *editing* approaches [6].

During the past years, many editing approaches have been proposed for kNN classifiers [7]. In this paper, the Depuration algorithm [2] is examined and two schemes adopted by it are separated so that two new editing approaches are derived. Experiments show that the effect of Depuration is very close to one new approach will worse than the other, which suggests that the Depuration algorithm has not fully exploit the schemes it adopted. Moreover, this paper proposes to use neural network ensemble to edit the training examples and obtains some success.

The rest of this paper is organized as follows. Section 2 introduces the Depuration algorithm and proposes several new editing approaches. Section 3 presents the experimental results. Section 4 concludes.

2 Editing Approaches

The Depuration algorithm was regarded as the first *prototype selection* approach [9], which consists of removing some "suspicious" training examples while changing the class labels of some other examples. Its purpose is to deal with all types of dirts in the training data set, including outliers, noise and mislabeled examples. The algorithm is based on the *generalised editing* scheme [8] where two parameters, i.e. k and k', have to be set according to $(k+1)/2 \leq k' \leq k$. When k and k' were set to 3 and 2 respectively, Sánchez et al. [9] reported that the Depuration algorithm achieved the best effect among some other editing approaches they compared.

The Depuration algorithm is summarized in Table 1, where X is the original training data set and S is the edited training data set to be returned.

Table 1. The Depuration algorithm

Let $S = X$
For each $x_i \in X$ do
  Find k nearest neighbors of x_i in $(X - \{x_i\})$
  If a class label, say c, is held by at least k' neighbors
  Then set the label of x_i in S to c
  Else remove x_i from S

Through examining Table 1, it can be found that Depuration implicitly adopts two schemes to edit the training data set. The first scheme is that if there are k' neighbors holding the same class label, then change the class label of the concerned example to the commonly agreed label; otherwise the concerned example is kept as it was. The second scheme is that if there are k' neighbors holding the same class label, then keep the concerned example as it was; otherwise the concerned example is removed. Through separating these two schemes, two new editing approaches can be derived, as shown in Tables 2 and 3. Note that similar to Depuration, both the RelabelOnly approach and the RemoveOnly approach have two parameters, i.e. k and k', have to be set according to $(k+1)/2 \leq k' \leq k$.

Neural network ensemble [12] is a learning technique where multiple neural networks are trained to solve the same problem. Since the generalization ability of a neural network ensemble is usually significantly better than that of a single

Table 2. The RelabelOnly algorithm

Let $S = X$
For each $x_i \in X$ do
    Find k nearest neighbors of x_i in $(X - x_i)$
    If a class label, say c, is held by at least k' neighbors
    Then set the label of x_i in S to c

Table 3. The RemoveOnly algorithm

Let $S = X$
For each $x_i \in X$ do
    Find k nearest neighbors of x_i in $(X - x_i)$
    If no class label is held by at least k' neighbors
    Then remove x_i from S

---

neural network, it has become a hot topic during the past years. Recently, Zhou and Jiang [10] proposed a strong rule learning algorithm through using a neural network ensemble as the preprocess of a rule inducer, and later they showed that using a neural network ensemble to preprocess the training data set could be beneficial when the training data set contains noise and has not captured the whole target distribution [11].

Inspired by these works, here a new editing approach based on neural network ensemble is proposed for kNN classifiers. As shown in Table 4, the NNEE (Neural Network Ensemble Editing) algorithm uses a popular ensemble learning algorithm, i.e. Bagging [4], to construct a neural network ensemble from the original training data set. This trained neural network ensemble is then used to classify the training examples, and the classification is used to replace the original class label of the concerned training example. The NNEE approach has two parameters to set, i.e. the number of neural networks contained in the neural network ensemble and the number of hidden units in the networks, if single-hidden-layered feedforward neural networks are used. Fortunately, our experiments show that the NNEE approach is not sensitive to the setting of these parameters.

Table 4. The NNEE algorithm

Let $S = X$
Let NNE = Bagging(X)
For each $x_i \in X$ do
    change the label of x_i in S to the label predicted by NNE

---

3 Experiments

Ten data sets from the UCI machine learning repository [3] are used in the experiments. Information on these data sets are shown in Table 5.

Table 5. Experimental data sets

Data set	Attribute		Size	Class
	Categorical	Continuous		
annealing	33	5	798	6
credit	9	6	690	2
glass	0	9	214	7
hayes-roth	4	0	132	3
iris	0	4	150	3
liver	0	6	345	2
pima	0	8	768	2
soybean	35	0	683	19
wine	0	13	178	3
zoo	16	0	101	7

On each data set, 10 runs of 10-fold cross validation is performed with random partitions. The effects of the editing approaches described in Section 2 are compared through coupling them with a 3NN classifier. The predictive accuracy of the 3NN classifiers trained on the training data sets edited by different approaches are shown in Table 6, where the values following ± are standard deviations. The parameters of Depuration, RelabelOnly and RemoveOnly are set as $k = 3$ and $k' = 2$. Therefore these algorithms are denoted as Depuration(3,2), RelabelOnly(3,2) and RemoveOnly(3,2), respectively. Five BP networks are contained in the neural network ensemble used by NNEE, and each network has one hidden layer consisting of five hidden units. Therefore here the approach is denoted as NNEE(5,5).

Table 6 shows that the NNEE approach achieves the best editing effect. In detail, it obtains the best performance on seven data sets, i.e. *annealing, credit, liver, pima, soybean, wine* and *zoo*. RemoveOnly obtains the best performance on three data sets, i.e. *glass, hayes-roth* and *wine*. It is surprising that Depuration obtains the best performance on only one data set, i.e. *iris*, as RelabelOnly does. These observations indicate that NNEE is a better editing approach than Depuration. Moreover, since the effect of Depuration is only comparable to that of RelabelOnly, it is obvious that Depuration has not exploited well the power of the two schemes it adopted, especially the scheme used by RemoveOnly. In fact, in some cases such as on *glass, hayes-roth, soybean* and *zoo*, simultaneously adopting the schemes used by RelabelOnly and RemoveOnly is even worse than adopting any of these schemes. The reason why the phenomenon appearing remains to be explored in the future.

Table 6. Predictive accuracy (%) of 3NN coupled with different editing approaches

Data set	Depuration(3,2)	RelabelOnly(3,2)	RemoveOnly(3,2)	NNEE(5,5)
annealing	89.95 ± 2.81	89.95 ± 2.81	92.76 ± 1.80	**92.81 ± 1.77**
credit	84.75 ± 3.95	84.75 ± 3.95	85.33 ± 2.50	**86.20 ± 1.97**
glass	59.90 ± 9.29	60.27 ± 9.21	**68.23 ± 5.27**	67.94 ± 6.60
hayes-roth	47.81 ± 9.09	48.34 ± 9.23	**54.31 ± 7.89**	50.50 ± 9.06
iris	**95.67 ± 4.75**	**95.67 ± 4.75**	95.20 ± 5.08	95.47 ± 3.25
liver	57.28 ± 7.25	57.28 ± 7.25	61.22 ± 5.11	**64.06 ± 5.27**
pima	72.42 ± 4.79	72.42 ± 4.79	74.35 ± 2.77	**75.57 ± 3.04**
soybean	87.76 ± 3.40	89.28 ± 3.38	89.58 ± 2.57	**90.87 ± 2.53**
wine	94.94 ± 4.26	94.94 ± 4.26	**96.05 ± 2.89**	**96.05 ± 2.89**
zoo	90.75 ± 6.79	90.95 ± 6.91	93.49 ± 3.87	**94.48 ± 4.47**

4 Conclusion

This paper proposes to use neural network ensemble to edit the training data set for kNN classifiers. In detail, a neural network ensemble is trained from the original training data set. Then, the class labels of the training examples are replaced by the labels generated by the neural network ensemble. Experiments show that such an approach could achieve better editing effect than the Depuration algorithm does.

This paper also examines the Depuration algorithm and identifies the two editing schemes it adopted. Through detaching these two schemes, this paper derives two new editing approaches from Depuration, i.e. RelabelOnly and RemoveOnly. Experiments show that the editing effect of Depuration is only comparable to that of RelabelOnly while worse than that of RemoveOnly. This discloses that the scheme of RemoveOnly does not function in the Depuration algorithm. Moreover, in some cases simultaneously using the scheme of RelabelOnly and the scheme of RemoveOnly is even worse than using either of them. Exploring the reason behind these observations is an interesting issue for future work.

Acknowledgement. This work was supported by the National Outstanding Youth Foundation of China under the Grant No. 60325207, the Fok Ying Tung Education Foundation under the Grant No. 91067, the Excellent Young Teachers Program of MOE of China, the Jiangsu Science Foundation Key Project, and the National 973 Fundamental Research Program of China under the Grant No. 2002CB312002.

References

1. Aha, D.W.: Lazy learning: special issue editorial. Artificial Intelligence Review **11** (1997) 7–10

2. Barandela, R., Gasca, E.: Decontamination of training samples for supervised pattern recognition methods. In: Ferri, F.J., Iñesta Quereda, J.M., Amin, A., Paudil, P. (eds.): Lecture Notes in Computer Science, Vol. 1876. Springer, Berlin (2000) 621–630
3. Blake, C., Keogh, E., Merz, C.J.: UCI repository of machine learning databases [http://www.ics.uci.edu/~mlearn/MLRepository.html], Department of Information and Computer Science, University of California, Irvine, CA (1998)
4. Breiman, L.: Bagging predictors. Machine Learning **24** (1996) 123–140
5. Dasarathy, B.V.: Nearest Neighbor Norms: NN Pattern Classification Techniques. IEEE Computer Society Press, Los Alamitos, CA (1991)
6. Devijver, P.A., Kittler, J.: Pattern Recognition: A Statistical Approach. Prentice Hall, Englewood Cliffs, NJ (1982)
7. Ferri, F.J., Albert, J.V., Vidal, E.: Considerations about sample-size sensitivity of a family of edited nearest-neighbor rules. IEEE Transactions on Systems, Man, and Cybernetics - Part B **29** (1999) 667–672
8. Koplowitz, J., Brown, T.A.: On the relation of performance to editing in nearest neighbor rules. Pattern Recognition **13** (1981) 251–255
9. Sánchez, J.S., Barandela, R., Marqués, A.I., Alejo, R., Badenas, J.: Analysis of new techniques to obtain quality training sets. Pattern Recognition Letters **24** (2003) 1015–1022
10. Zhou, Z.-H., Jiang, Y.: Medical diagnosis with C4.5 rule preceded by artificial neural network ensemble. IEEE Transactions on Information Technology in Biomedicine **7** (2003) 37–42
11. Zhou, Z.-H., Jiang, Y.: NeC4.5: neural ensemble based C4.5. IEEE Transactions on Knowledge and Data Engineering **16** (2004)
12. Zhou, Z.-H., Wu, J., Tang, W.: Ensembling neural networks: many could be better than all. Artificial Intelligence **137** (2002) 239–263

Learning Long-Term Dependencies in Segmented Memory Recurrent Neural Networks

Jinmiao Chen and Narendra S. Chaudhari

School of Computer Engineering, Nanyang Technological University, Singapore
pg05205549@ntu.edu.sg

Abstract. Gradient descent learning algorithms for recurrent neural networks (RNNs) perform poorly on long-term dependency problems. In this paper, we propose a novel architecture called Segmented-Memory Recurrent Neural Network (SMRNN). The SMRNN is trained using an extended real time recurrent learning algorithm, which is gradient-based. We tested the SMRNN on the standard problem of information latching. Our implementation results indicate that gradient descent learning is more effective in SMRNN than in standard RNNs.

1 Introduction

Recurrent neural networks(RNNs) use their recurrent connections to store and update context information, i.e., information computed from the past inputs and useful to produce target outputs. For many practical applications, the goal of recurrent networks is to robustly latch information. RNNs are usually trained with gradient-based algorithms such as back propagation through time(BPTT) [10] and real-time recurrent learning(RTRL) [11]. Unfortunately, the necessary conditions of robust information latching bring a problem of vanishing gradients, making the task of learning long-term dependencies hard [1]. Several approaches have been suggested to deal with the problem of vanishing gradients. Some consider alternative network architectures, such as Nonlinear Auto-Regressive models with eXogenous(NARX) recurrent neural network [6,7], Hierarchical Recurrent Neural Network [3], Long Short-Term Memory [4] and Latched Recurrent Neural Network [9]. Others try alternative optimization algorithms, such as simulated annealing algorithm [1], cellular genetic algorithm [5] and expectation-maximization algorithm [8].

To tackle the long-term dependency problems, we propose a novel recurrent neural network architecture called *Segmented-Memory Recurrent Neural Network*(SMRNN) and present experimental results showing that SMRNN outperforms conventional recurrent neural networks.

2 Segmented-Memory Recurrent Neural Networks

2.1 Architecture

As we observe, during the process of human memorization of a long sequence, people tend to break it into a few segments, whereby people memorize each segment first and then cascade them to form the final sequence. The process of memorizing a sequence in segments is illustrated in Figure 1. In Figure 1,

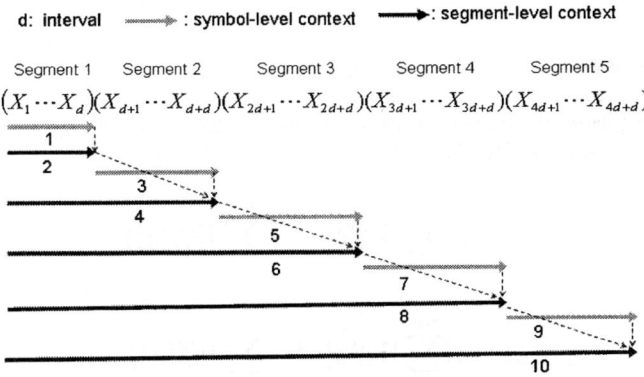

Fig. 1. Segmented memory with interval=d

the substrings in parentheses represent segments of equal length d; gray arrows indicate the update of contextual information associated to symbols and black arrows indicate the update of contextual information associated to segments; numbers under the arrows indicate the sequence of memorization.

Based on the observation on human memorization, we believe that RNNs are more capable of capturing long-term dependencies if they have segmented-memory and imitate the way of human memorization. Following this intuitive idea, we propose *Segmented-Memory Recurrent Neural Network* (SMRNN) as illustrated in Figure 2. The SMRNN has hidden layer H1 and hidden layer H2 representing symbol-level state and segment-level state respectively. Both H1 and H2 have recurrent connections among themselves. The states of H1 and H2 at the previous cycle are copied back and stored in context layer S1 and context layer S2 respectively. Most importantly, we introduce into the network a new attribute *interval*, which denotes the length of each segment.

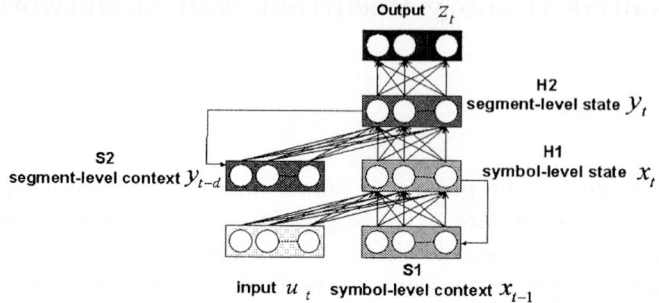

Fig. 2. Segmented-memory recurrent neural network with interval=d

2.2 Dynamics

In order to implement the segmented-memory illustrated in Figure 1, we formulate the dynamics of SMRNN with interval=d as below:

$$x_i^t = g(\sum_{j=1}^{n_X} W_{ij}^{xx} x_j^{t-1} + \sum_{j=1}^{n_U} W_{ij}^{xu} u_j^t) \qquad (1)$$

$$y_i^t = g(\sum_{j=1}^{n_Y} W_{ij}^{yy} y_j^{t-d} + \sum_{j=1}^{n_X} W_{ij}^{yx} x_j^t) \qquad (2)$$

$$z_i^t = g(\sum_{j=1}^{n_Y} W_{ij}^{zy} y_j^t) \qquad (3)$$

The variables in the above equations have the following meanings:

- n_Z, n_Y, n_X and n_U denote the numbers of neurons at output layer, hidden layer H2 (context layer S2), hidden layer H1 (context layer S1) and input layer respectively.
- W_{ij}^{zy} denotes the connection between the ith neuron at output layer and the jth neuron at hidden layer H2.
- W_{ij}^{yy} denotes the connection between the ith neuron at hidden layer H2 and the jth neuron at context layer S2.
- W_{ij}^{yx} denotes the connection between the ith neuron at hidden layer H2 and the ith neuron at hidden layer H1.
- W_{ij}^{xx} denotes the connection between the ith neuron at hidden layer H1 and the jth neuron at context layer S1.
- W_{ij}^{xu} denotes the connection between the ith neuron at hidden layer H1 and the jth neuron at input layer.
- d denotes the length of interval.
- $g(x) = 1/(1 + exp(-x))$.

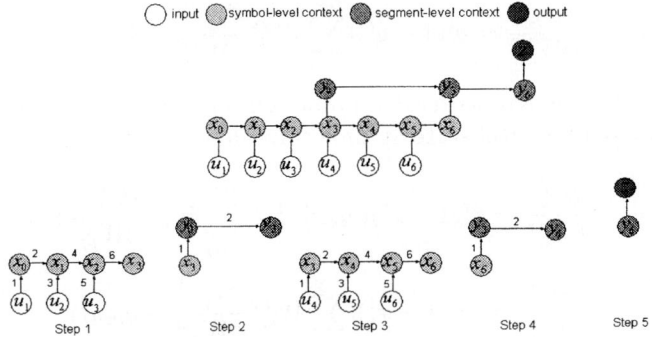

Fig. 3. Dynamics of segmented-memory recurrent neural network

We now explain the dynamics of SMRNN with an example(Figure 3). In this example, the input sequence is divided into segments with equal length 3. Then symbols in each segment are fed to hidden layer H1 to update the symbol-level context. Upon completion of each segment, the symbol-level context is forwarded to the next layer H2 to update the segment-level context. This process continues until it reaches the end of the input sequence, then the segment-level context is forwarded to the output layer to generate the final output. In other words, the network reads in one symbol per cycle; the state of H1 is updated at the coming of each single symbol, while the state of H2 is updated only after reading an entire segment and at the end of the sequence. The segment-level state layer behaves as if it cascades segments sequentially to obtain the final sequence as people often do: Every time when people finish one segment, they always go over the sequence from the beginning to the tail of the segment which is newly memorized, so as to make sure that they have remembered all the previous segments in correct order(see Figure 1).

2.3 Learning Strategy

The SMRNN is trained using an extension of the Real Time Recurrent Learning algorithm [11]. Every parameter P is initialized with small random values then updated according to gradient descent:

$$\Delta P = -\alpha \frac{\partial E^t}{\partial P} = -\alpha \frac{\partial E^t}{\partial y^t} \frac{\partial y^t}{\partial P} \quad (4)$$

where α is the learning rate and E^t is the error function at time t.

Derivatives associated to recurrent connections are calculated in a recurrent way. Derivatives of segment-level state at time t depend on derivatives at time $t - d$ where d is the length of each segment.

$$\frac{\partial y_i^t}{\partial W_{kl}^{yy}} = y_i^t(1 - y_i^t)(\delta_{ik} y_l^{t-d} + \sum_{j=1}^{n_Y} W_{ij}^{yy} \frac{\partial y_j^{t-d}}{\partial W_{kl}^{yy}}) \quad (5)$$

$$\frac{\partial y_i^t}{\partial W_{kl}^{yx}} = y_i^t(1-y_i^t)(\sum_{j=1}^{n_Y} W_{ij}^{yy} \frac{\partial y_j^{t-d}}{\partial W_{kl}^{yx}} + \delta_{ik} x_l^t) \qquad (6)$$

where δ_{ik} denotes the Kronecker delta function (δ_{ik} is 1 if $i = k$ and 0 otherwise). Derivatives of symbol-level state at time t are dependent on derivatives at time t-1.

$$\frac{\partial x_i^t}{\partial W_{kl}^{xx}} = x_i^t(1-x_i^t)(\delta_{ik} x_l^{t-1} + \sum_{j=1}^{n_X} W_{ij}^{xx} \frac{\partial x_j^{t-1}}{\partial W_{kl}^{xx}}) \qquad (7)$$

$$\frac{\partial x_i^t}{\partial W_{kl}^{xu}} = x_i^t(1-x_i^t)(\sum_{j=1}^{n_X} W_{ij}^{xx} \frac{\partial x_j^{t-1}}{\partial W_{kl}^{xu}} + \delta_{ik} u_l^t) \qquad (8)$$

3 Experimental Evaluation

3.1 The Information Latching Problem

We test the performance of SMRNN on the information latching problem. This problem is a minimal task designed by Bengio as a test that must be passed in order for a dynamic system to latch information robustly [1]. In this task, the SMRNN is trained to classify two different sets of sequences. For each sequence $X_1, X_2, \ldots, X_T$, the class $C(X_1, X_2, \ldots, X_T) \in \{0, 1\}$ depends only on the first L values of the sequence:

$$C(X_1, X_2, \ldots, X_T) = C(X_1, X_2, \ldots, X_L), \qquad (9)$$

where T is the length of the sequence. The value $X_{L+1}, \cdots, X_T$ are irrelevant for determining the class of the sequences, however, they may affect the evolution of the dynamic system and eventually erase the internally stored information about the initial values of the input. We suppose L fixed and allow sequences of arbitrary length $T \gg L$. Thus, the problem can be solved only if the network is able to store information about the initial input values for an arbitrary duration. This is the simplest form of long-term computation that one may ask a recurrent network to carry out.

In our experiments, we kept L fixed and varied T in increments of five or ten. By increasing T, we will be able to control the span of long-term dependencies, in which the output will depend on input values far in the past. For each value of T, we randomly generated two sets of sequences, one for training and the other for testing. Each set contains equal number of member sequences and nonmember sequences. Sequence $X_1, X_2, \ldots, X_T$ is classified as member if its prefix $X_1, X_2, \ldots, X_L$ is identical to a predefined string $Y_1, Y_2, \cdots, Y_L$, otherwise nonmember, i.e. $C(X_1, X_2, \cdots, X_T) = 1$ if $X_1, X_2, \ldots, X_L = Y_1, Y_2, \cdots, Y_L$ and $C(X_1, X_2, \cdots, X_T) = 0$ otherwise. Sequences in the training set are presented to the SMRNN one symbol at a time. Target output is available at the end of each sequence. At that time, the SMRNN provides an output and adjusts the weights with gradient descent. The cost function is given below:

$$E = \frac{1}{2}(z-c)^2, \qquad (10)$$

where z is the actual output and c is the target output. This cost function has a minimum when the actual output is the probability that the sequence is in the class.

3.2 Comparison Between SMRNN and Elman's Network

Researchers have proposed several architectural approaches for the long-term dependency problems. They compare their network architectures against Elman's network which is a standard architecture of RNN [6,3,4,7]. Elman's network is a two-layer backpropagation network, with the addition of a feedback connection from the output of the hidden layer to its input. In our experiments, we also take Elman's network for comparison. This feedback path allows Elman's network to learn to recognize and generate temporal patterns.

The first experiment was carried out on a SMRNN with each hidden layer having 10 units and the interval being 15. In this experiment, we kept L fixed and varied T in increments of ten. As the sequences become longer, more training samples are required to achieve a satisfactory level of generalization. Thus under circumstances whereby the network has small training error and relatively low testing accuracy, it is necessary to enlarge its training set. As illustrated in Table 1, for sequences with length 60-200, SMRNN can learn to classify the testing sequences with high accuracy. Being a comparison test, an Elman's network [2] with 10 hidden units was also trained for the same task. From the results illustrated in Table 2, we observe that Elman's network has difficulty learning to classify sequences of length 65 and the accuracy declines as the size of training

Table 1. Information latching in SMRNN.

L	T	train set size	test set size	accuracy
50	60	30	30	83.3%
50	60	50	50	96%
50	70	50	50	54%
50	70	80	80	92.5%
50	80	80	80	88.8%
50	80	100	100	97%
50	90	100	100	92%
50	90	150	150	100%
50	100	100	100	53%
50	100	150	150	98%
50	110	200	200	45.5%
50	110	300	300	96.3%
50	120	400	400	99%
50	130	500	500	99.6%
50	140	500	500	100%
50	150	600	600	100%
50	160	800	800	99.75%
50	170	1000	1000	99.9%
50	180	1200	1200	100%
50	190	1300	1300	99.9%
50	200	1400	1400	100%

Table 2. Information latching in Elman's network.

L	T	train set size	test set size	accuracy
50	55	30	30	30%
50	55	100	100	72.5%
50	55	200	200	92.5%
50	60	100	100	40%
50	60	200	200	85%
50	65	200	200	75%
50	65	300	300	60%
50	65	400	400	47.5%

set increases. This means that Elman's network has low accuracy not because of the insufficiency of training data but itself being not powerful enough. A comparison between Table 1 and Table 2 indicates that SMRNN is able to capture much longer ranges of dependencies than Elman's network.

4 Conclusion

By inserting intervals into the memory of contextual information, segmented-memory recurrent neural networks are able to perform much better on long-term dependency problem than conventional recurrent networks. But it doesn't mean that it performs poorly on tasks in which short-term dependencies are equally or even more important. Its segmented and cascading memory allows SMRNN to efficiently learn short-term dependencies, especially those within the range of a segment. Experiments show that SMRNN converges fast and generalizes well for short sequences as well.

There is a trade-off between efficient training of gradient descent and long-range information latching. For SMRNN, the ERTRL training algorithm is essentially gradient descent, hence SMRNN does not circumvent the problem of vanishing gradients. Nevertheless, SMRNN improves the performance on long-term dependency problem.

References

1. Bengio, Y., Simard, P., Frasconi, P.: Learning Long-Term Dependencies with Gradient Descent is Difficult, IEEE Transactions on Neural Networks, vol. 5, no. 2, (1994) 157-166
2. Elman, J.L.: Distributed Representations, Simple Recurrent Networks, and Grammatical Structure, Machine Learning, vol. 7, (1991) 195-226
3. Hihi, S.E., Bengio, Y.: Hierarchical Recurrent Neural Networks for Long-Term Dependencies, Advances in Neural Information Processing Systems, Perrone, M., Mozer, M., Touretzky, D.D. ed., MIT Press, (1996) 493-499
4. Hochreiter, S., Schmidhuber, J.: Long Short-Term Memory, Neural Computation, vol. 9, no. 8, (1997) 1735-1780
5. Ku, K.W.C., Mak, M.W., Siu, W.C.: A Cellular Genetic Algorithm for Training Recurrent Neural Networks, Proc. Int. Conf. Neural Networks and Signal Processing, (1995) 140-143
6. Lin, T., Horne, B.G., Tino, P., Giles, C.L.: Learning Long-Term Dependencies in NARX Recurrent Neural Networks, IEEE Trans. on Neural Networks, vol. 7, (1996) 1329-1337
7. Lin, T., Horne, B.G., Giles, C.L.: How Embedded Memory in Recurrent Neural Network Architectures Helps Learning Long-Term Temporal Dependencies, Neural Networks, vol. 11, (1998) 861-868
8. Ma, S., Ji, C.: Fast Training of Recurrent Networks Based on the EM Algorithms, IEEE Trans. on Neural Networks, vol. 9(1), (1998) 11-26
9. Branko Šter: Latched Recurrent Neural Network, Elektrotehniški vestnik, vol. 70(1-2), (2003) 46-51

10. Werbos, P.J.: Backpropagation through Time: What It Does and How to Do It, Proc. IEEE, vol. 78(10), (1990) 1550-1560
11. Williams, R.J., Zipser, D.: Gradient Based Learning Algorithms for Recurrent Connectionist Networks, Northeastern University, College of Computer Science Technical Report, NU-CCS-90-9, (1990) 433-486

An Optimal Neural-Network Model for Learning Posterior Probability Functions from Observations

Chengan Guo [1] and Anthony Kuh [2]

[1] Dalian University of Technology, Dalian, Liaoning 116023, China
cguo@dlut.edu.cn
[2] University of Hawaii at Manoa, Honolulu, HI 96822, USA
kuh@spectra.eng.hawaii.edu

Abstract. This paper presents the further results of the authors' former work [1] in which a neural-network method was proposed for sequential detection with similar performance as the optimal sequential probability ratio tests (SPRT) [2]. The analytical results presented in the paper show that the neural network is an optimal model for learning the posterior conditional probability functions, with arbitrarily small error, from the sequential observation data under the condition in which the prior probability density functions about the observation sources are not provided by the observation environment.

1 Introduction

The neural-network model proposed by [1] for learning the posterior conditional probability functions is shown in Fig. 1, in which X_t is the observation at time t that passes through the function expander to generate $\sum_{k=1}^{t} X_k^2$, $\sum_{k=1}^{t} X_k$, t and constant 1 which are then input into the one-layer neural network with two sigmoid units to generate outputs $y_0(t)$ and $y_1(t)$, respectively. The neural network is trained by a reinforcement learning method using the TD learning algorithm [3]. The reinforcement signal $R(t)$ is a binary label vector that is available only at n, the termination time of the observation sequence $\mathbf{X}_t = (X_1, \cdots, X_t)$, defined by

$$R(t) = \begin{cases} not\ available, & t < n \\ (B_0(\mathbf{X}_t), B_1(\mathbf{X}_t)), & t = n \end{cases} \quad (1)$$

where $B_i(\mathbf{X}_t)$ is the binary label information of the sequential observation data,

$$B_i(\mathbf{X}_t) = \begin{cases} not\ available, & t < n \\ 1, & \mathbf{X}_t \in H_i\ and\ t = n \\ 0, & \mathbf{X}_t \notin H_i\ and\ t = n \end{cases} \quad (2)$$

Note that since we do not have access to the statistical knowledge about the observation data, the ideal target values for the posterior probability functions are not available. Some basic questions associated with this problem that we will address in this paper are:

(i) *If the neural network model given in Fig. 1 can learn the posterior probability functions, with the binary label signal R(t) (without the probability values) as the ideal targets, from the observation sequence?*

(ii) *Does there exist an optimal solution for the neural network and how closely can the neural network approach the probability functions?*

These questions will be answered in the following sections of the paper. Analyses for the optimality of the neural network model are given in Section 2 and 3 under the same conditions as in [1]. Section 4 summarizes the conclusions of the paper.

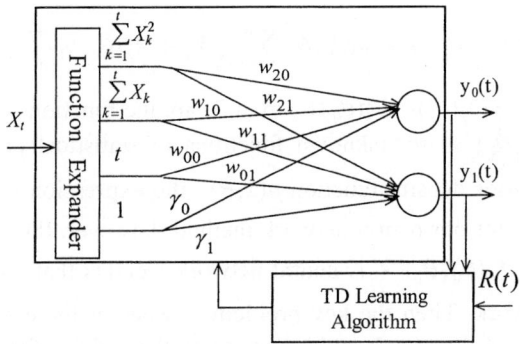

Fig. 1. Neural network model for learning posterior probability functions from observations

2 Analysis of the Approachability of the Network Model

It has been derived in [1] that, when the observation data $\{X_k, k=1,2,\cdots\}$ are independent and identically distributed sequence drawn from two hypothesis sources, the posterior conditional probability functions $Q_i(\mathbf{X}_t) = P(H_i | \mathbf{X}_t)$ are given as follows:

$$Q_1(\mathbf{X}_t) = \frac{P(H_1)p_\mathbf{X}(\mathbf{X}_t | H_1)}{P(H_1)p_\mathbf{X}(\mathbf{X}_t | H_1) + P(H_0)p_\mathbf{X}(\mathbf{X}_t | H_0)} = \frac{1}{1+e^{-(Z_t - \tau_0)}}, \quad (3)$$

$$Q_0(\mathbf{X}_t) = \frac{1}{1+e^{(Z_t - \tau_0)}} \quad (4)$$

where $\pi_i = P(H_i)$ is the prior hypothesis probability, $\tau_0 = \ln(\pi_0/\pi_1)$, $p_\mathbf{X}(\mathbf{X}_t | H_i)$ is the prior joint density function of $\mathbf{X}_t$ conditioned on the hypothesis H_i ($i=0,1$), and Z_t is the log-likelihood ratio function of the observation data $\mathbf{X}_t$,

$$Z_t = \ln\frac{p_X(X_t|H_1)}{p_X(X_t|H_0)} = \sum_{k=1}^{t}\ln\frac{p(X_k;\theta_1)}{p(X_k;\theta_0)} \tag{5}$$

where $p(x;\theta)$ is the density function of x given the parameter θ (θ is a vector in general).

The *exponential family* was considered in [1] for the density functions $p(x;\theta)$ since many commonly used random variables in engineering applications (such as Gaussian, Poisson, Exponential, and Bernoulli random variables) belong to this family in which $p(x;\theta)$ is of the form

$$p(x;\theta) = h(x)e^{C_0(\theta)+xC_1(\theta)+x^2C_2(\theta)}. \tag{6}$$

Under these conditions, the log-likelihood ratio function Z_t is given by

$$Z_t = A_0(\theta_0,\theta_1)\cdot t + A_1(\theta_0,\theta_1)\sum_{k=1}^{t}X_k + A_2(\theta_0,\theta_1)\sum_{k=1}^{t}X_k^2 \tag{7}$$

where $A_j(\theta_0,\theta_1) = C_j(\theta_1) - C_j(\theta_0)$ ($j=0,1,2$) are the functions only of θ_0 and θ_1.

Since $A_j(\theta_0,\theta_1)$'s are unknown functions of statistical parameters that are dependent of the prior density function $p(x;\theta)$, the expression (7) helps little for estimating $Q_i(X_t)$. For our neural network method, however, this expression is essential as we will take $A_j(\theta_0,\theta_1)$'s as neural network weights that may be learned through training the network. Then the key problem is whether there exist the ideal weights corresponding to $A_j(\theta_0,\theta_1)$'s and how to learn the weights for the neural network. In the following we present Theorem 1 which shows the existence of the ideal weights for the network model.

Theorem 1: For the neural network model shown in Fig. 1, when the density function $p(x;\theta)$ of the input sequence X_t ($t=1,2,\cdots$) belongs to the exponential family defined by (6), there exists a set of ideal weights, w_{ij}^0 and γ_j^0 ($i=0,1,2,3; j=0,1$) that makes the outputs of the network, $y_0(t)$ and $y_1(t)$ equal to the posterior probability functions, i. e.,

$$y_0(t,W^0) = Q_0(X_t), \quad y_1(t,W^0) = Q_1(X_t) \tag{8}$$

where W^0 represents the ideal weight vector including all w_{ij}^0 and γ_j^0 ($i=0,1,2$):

$$w_{i,1}^0 = A_i(\theta_0,\theta_1), \quad w_{i,0}^0 = -w_{i,1}^0, \text{ and } \gamma_0^0 = \ln(\pi_0/\pi_1), \quad \gamma_1^0 = -\gamma_0^0 \tag{9}$$

Proof: This theorem can directly be proved by using the results of (3), (4) and (7): note that, for the neural network shown in Fig. 1, when the weights satisfy (9) of Theorem 1, we have the output

$$y_1(t, W^0) = 1/(1+e^{-[w_{0,1}^0 \cdot t + w_{1,1}^0 \sum_{k=1}^{t} X_k + w_{2,1}^0 \sum_{k=1}^{t} X_k^2 + \gamma_1^0]})$$

$$= 1/(1+e^{-[A_0(\theta_0,\theta_1) \cdot t + A_1(\theta_0,\theta_1) \cdot \sum_{k=1}^{t} X_k + A_2(\theta_0,\theta_1) \cdot \sum_{k=1}^{t} X_k^2 - \ln\frac{\pi_0}{\pi_1}]})$$

$$= \frac{1}{1+e^{-(Z_t - \tau_0)}} = Q_1(\mathbf{X}_t) \, . \tag{10}$$

In the same way we have that

$$y_0(t, W^0) = \frac{1}{1+e^{(Z_t - \tau_0)}} = Q_0(\mathbf{X}_t) \, . \tag{11}$$

From Theorem 1, we see that this neural network is able to give the ideal target function $Q_i(\mathbf{X}_t)$ when the weights of the network are set to the ideal values W^0. This is guaranteed by the neural network architecture that is designed to have the same functional structure as $Q_i(\mathbf{X}_t)$. Furthermore, Theorem 1 implies that this neural network architecture is an optimal model. The theorem also indicates that realization of the optimum approximation capacity of the neural network depends on whether the ideal weight vector W^0 can be learned. While this depends on if an effective learning algorithm can be designed.

3 Convergence Analysis of Learning Algorithm

Now the problem we want to solve is to learn the posterior probability function $Q_i(\mathbf{X}_t)$ under the circumstance that the prior density function $p(X_k; \theta_i)$ is unknown. In practice, sample values of a probability function (such as $Q_i(\mathbf{X}_t)$) cannot be obtained by direct measurement and computation. This means that there is no "teacher" to guide the learning process for each input sequence $\mathbf{X}_t$ and therefore it is not an ordinary supervised learning problem. In order to overcome this difficulty, we assume that, during the training stage, a binary label signal $B_i(\mathbf{X}_t)$ is available at the termination time n when the whole sequence X_t ($t = 1, \cdots, n$) have been observed. We will use $B_i(\mathbf{X}_t)$ as the "desired outputs" with $\{(\mathbf{X}_t, B_i(\mathbf{X}_t))\}$ as sample data to train the network to learn $Q_i(\mathbf{X}_t)$. Then the question is, in this case, whether the learning task is accomplishable? In this section we present a convergence analysis to answer this question.

Let $\{(\mathbf{X}_n^{k,i}, B_j(\mathbf{X}_n^{k,i}))\}$ ($k = 1, 2, \cdots, K_i(n)$; $i = 0, 1$; $n = 1, 2, \cdots$) be the training set where $\mathbf{X}_n^{k,i}$ represents a training sequence of size n, the superscript "k" represents the k-th sequence with size n, "i" represents that $\mathbf{X}_n^{k,i}$ is drawn from hypothesis H_i, and $K_i(n)$ is the number of the sequences with size n and from H_i. Accordingly, the binary label function of $B_j(\mathbf{X}_n^{k,i})$ is re-defined by

$$B_j(\mathbf{X}_n^{k,i}) = \begin{cases} 1, & \text{if } i = j \\ 0, & \text{otherwise} \end{cases}. \qquad (12)$$

Define two error energy functions, $J_B(\mathbf{W})$ and $J_Q(\mathbf{W})$, respectively by

$$J_B(\mathbf{W}) = \frac{1}{N} \sum_n \sum_{i=0}^{1} \sum_{k=1}^{K_i(n)} \sum_{j=0}^{1} (y_j(n) - B_j(\mathbf{X}_n^{k,i}))^2,$$
$$J_Q(\mathbf{W}) = \sum_n p(n) \sum_{j=0}^{1} E_{\mathbf{X}_n}[(y_j(n) - Q_j(\mathbf{X}_n))^2] \qquad (13)$$

where $\mathbf{W}$ is the network weight vector to be learned, $E_{\mathbf{X}_n}[Y]$ represents the expected ensemble average over $\mathbf{X}_n$, $p(n)$ is the probability that a training sequence has size n, and N is the total number of the training sequences in the training set.

Note that the function $J_B(\mathbf{W})$ is the sample mean squared-error between the network output $y_j(n)$ and the binary signal $B_j(\mathbf{X}_n^{k,i})$ (the nominal target function), and the function $J_Q(\mathbf{W})$ is the ensemble mean squared-error between $y_j(n)$ and $Q_j(\mathbf{X}_n)$ (the ideal target function). $J_B(\mathbf{W})$ can be minimized with a LMS learning algorithm without much difficulty since its target values, $B_j(\mathbf{X}_n^{k,i})$, are available [4]. But our goal is to minimize $J_Q(\mathbf{W})$ instead of $J_B(\mathbf{W})$. It is not clear at present whether or not this is realizable since the target values for $Q_j(\mathbf{X}_n)$ are unavailable. Theorem 2 given bellow shows that $J_Q(\mathbf{W})$ can be minimized by minimizing $J_B(\mathbf{W})$.

Theorem 2: Suppose that there is a learning algorithm which is able to minimize $J_B(\mathbf{W})$ when using this algorithm to train a neural network. Then the function $J_Q(\mathbf{W})$ will automatically be minimized by the training process provided that a proper training set $\{(\mathbf{X}_n^{k,i}, B_j(\mathbf{X}_n^{k,i}))\}$ is used in the training process (i.e., the number of total training sequences of $\{(\mathbf{X}_n^{k,i}, B_j(\mathbf{X}_n^{k,i}))\}$ and the number of the training sequences with the same size n and hypothesis H_i in the set are sufficiently large).

Proof: Let

$$J_a(\mathbf{W}) = \lim_{N \to \infty} J_B(\mathbf{W}). \qquad (14)$$

Since minimizing $J_B(\mathbf{W})$ is also minimizing $J_a(\mathbf{W})$, what we need to prove is that minimizing $J_a(\mathbf{W})$ will result in minimizing $J_Q(\mathbf{W})$.

Let N_i ($i = 0,1$) be the total number of the training sequences belonging to H_i in the training set. By expanding $J_B(\mathbf{W})$ we have that

$$J_a(\mathbf{W}) = \lim_{N\to\infty} \{ \frac{N_0}{N} \sum_n \frac{K_0(n)}{N_0} \frac{1}{K_0(n)} \sum_{k=1}^{K_0(n)} [(y_0(\mathbf{X}_n^{k,0})-1)^2 + y_1^2(\mathbf{X}_n^{k,0})]$$

$$+ \frac{N_1}{N} \sum_n \frac{K_1(n)}{N_1} \frac{1}{K_1(n)} \sum_{k=1}^{K_1(n)} [(y_0^2(\mathbf{X}_n^{k,1}) + (y_1(\mathbf{X}_n^{k,1})-1)^2] \} . \quad (15)$$

Since it has been assumed that the number of total sequences in the training set is sufficiently large, by the Law of Large Numbers [5], as $N \to \infty$, we have that $N_i \to \infty, K_i(n) \to \infty$, $\lim_{N\to\infty} N_i/N = P(H_i) = \pi_i$, $\lim_{N\to\infty} K_i(n)/N_i = p_i(n)$ for $i = 0, 1$, and the two inner sample averages of (15) converge to their corresponding expected ensemble averages, respectively. Then (15) becomes

$$J_a(\mathbf{W}) = \pi_0 \sum_n p_0(n) \int_{R^n} [(y_0(\mathbf{X}_n)-1)^2 + y_1^2(\mathbf{X}_n)] p_\mathbf{X}(\mathbf{X}_n|H_0) d\mathbf{X}_n$$

$$+ \pi_1 \sum_n p_1(n) \int_{R^n} [y_0^2(\mathbf{X}_n) + (y_1(\mathbf{X}_n)-1)^2] p_\mathbf{X}(\mathbf{X}_n|H_1) d\mathbf{X}_n$$

$$= \sum_n \int_{R^n} [y_0^2(\mathbf{X}_n) + y_1^2(\mathbf{X}_n)][\pi_0 p_0(n) p_\mathbf{X}(\mathbf{X}_n|H_0) + \pi_1 p_1(n) p_\mathbf{X}(\mathbf{X}_n|H_1)] d\mathbf{X}_n$$

$$- \sum_n \int_{R^n} [2y_0(\mathbf{X}_n)\pi_0 p_0(n) p_\mathbf{X}(\mathbf{X}_n|H_0) + 2y_1(\mathbf{X}_n)\pi_1 p_1(n) p_\mathbf{X}(\mathbf{X}_n|H_1)] d\mathbf{X}_n$$

$$+ \sum_n \int_{R^n} [\pi_0 p_0(n) p_\mathbf{X}(\mathbf{X}_n|H_0) + \pi_1 p_1(n) p_\mathbf{X}(\mathbf{X}_n|H_1)] d\mathbf{X}_n . \quad (16)$$

Let $p(n)$ be the probability that a sequence $\mathbf{X}_n$ with length n is selected as training data. Then by the Law of Large Numbers [5] it follows $K_0(n) \approx N\pi_0 \, p(n) \approx N_0 \, p(n)$, $K_1(n) \approx N\pi_1 \, p(n) \approx N_1 \, p(n)$ and $p_0(n) = \lim_{N\to\infty} K_0(n)/N_0 = p_1(n) = \lim_{N\to\infty} K_1(n)/N_1 = p(n)$. Thus we have the following identities by Bayes Law for $i = 0, 1$:

$$\pi_0 p_0(n) p_\mathbf{X}(\mathbf{X}_n|H_0) + \pi_1 p_1(n) p_\mathbf{X}(\mathbf{X}_n|H_1) = p(n) p_\mathbf{X}(\mathbf{X}_n) \quad (17)$$

$$\pi_i p_i(n) p_\mathbf{X}(\mathbf{X}_n|H_i) = p(n) p_\mathbf{X}(\mathbf{X}_n) P(H_i|\mathbf{X}_n) = p(n) p_\mathbf{X}(\mathbf{X}_n) Q_i(\mathbf{X}_n) \quad (18)$$

$$\sum_n \int_{R^n} [\pi_0 p_0(n) p_\mathbf{X}(\mathbf{X}_n|H_0) + \pi_1 p_1(n) p_\mathbf{X}(\mathbf{X}_n|H_1)] d\mathbf{X}_n = 1 . \quad (19)$$

Substituting (17)~(19) into (16), we get that

$$J_a(\mathbf{W}) = \sum_n p(n) \int_{R^n} [(y_0(\mathbf{X}_n) - Q_0(\mathbf{X}_n))^2 + (y_1(\mathbf{X}_n) - Q_1(\mathbf{X}_n))^2] p_\mathbf{X}(\mathbf{X}_n) d\mathbf{X}_n$$

$$+ 1 - \sum_n p(n) \int_{R^n} [Q_0^2(\mathbf{X}_n) + Q_1^2(\mathbf{X}_n)] p_\mathbf{X}(\mathbf{X}_n) d\mathbf{X}_n$$

$$= J_Q(\mathbf{W}) + 1 - \sum_n p(n) E_{\mathbf{X}_n}[Q_0^2(\mathbf{X}_n) + Q_1^2(\mathbf{X}_n)] . \quad (20)$$

Since in (20) only $J_Q(\mathbf{W})$ is dependent on the weight $\mathbf{W}$, minimizing $J_a(\mathbf{W})$ with respect to $\mathbf{W}$ is equivalent to minimizing $J_Q(\mathbf{W})$ with respect to $\mathbf{W}$. This completes the proof of Theorem 2.

4 Conclusions

In this paper two theorems are derived to present performance analysis results for the neural-network model proposed in [1]. Theorem 1 shows that there exists a set of ideal weights that makes the outputs of the network equal to the posterior probability functions. Theorem 2 shows that the mean squared-error between the outputs and the ideal target functions can be minimized even though the sample values of the ideal target functions are not given, which solves the existence problem for learning the ideal target functions. However Theorem 2 did not give what the minimum value is. In fact, this minimum value depends on the network model and the learning algorithm used in the training process. According to the two theorems, when the network model and the learning algorithm are properly designed, such as the model shown in Section 2 and the learning algorithm given in [1], the minimum value can be made arbitrarily small. This implies the optimality of the neural-network model.

Simulation results also confirmed the validity of the performance analysis, which are omitted here due to space limitation of the paper and audiences are referred to [6] for the experimental results.

References

1. Guo, C., Kuh, A.: Temporal Difference Learning Applied to Sequential Detection. IEEE Trans. on Neural Networks, vol. 8, 3 (1997) 278-287
2. Wald, A., Wolfowitz, J.: Optimum Character of the Sequential Probability Ratio Test. Ann. Math. Statist., 19 (1948) 326-339
3. Sutton, R. S.: Learning to Predict by the Methods of Temporal Differences. Machine Learning, 3 (1988) 9-44
4. Rumelhart, D. E., McClelland, J. L., PDP Research Group: Parallel Distributed Processing. Vol. 1. Cambridge MIT Press (1986)
5. Leon-Garcia, A.: Probability and Random Processes for Electrical Engineering. Second Edition, Addison-Wesley Publishing Company, Inc. (1994)
6. Guo, C.: Applications of Reinforcement Learning in Sequential Detection and Network Routing. PhD Dissertation, University of Hawaii, USA (1999)

The Layered Feed-Forward Neural Networks and Its Rule Extraction

Ray Tsaih and Chih-Chung Lin

Department of Management Information Systems, National Chengchi University, Taipei,
Taiwan
{tsaih,m92014}@mis.nccu.edu.tw

Abstract. A mathematical study of the layered feed-forward neural networks is proposed here for identifying the rules suggested in the network. The mathematical study, not a data analysis, is proposed for identifying the premise association with each rule. It, hopefully, can be used further to deal with the predicament of ANN being a black box.

1 The Predicament of Being a Black Box

Layered feed-forward neural networks have been widely used in many fields. When the layered feed-forward neural network is used as a modeling tool, it is interesting to check if it can display some useful information. It is necessary to have a deeper analysis of the network structure, an analysis that is rather complicated mathematically. However, as Yoon, Guimaraes, and Swales in [1] argued that, after building the layered feed-forward neural networks, reading or understanding the knowledge in layered feed-forward neural networks was difficult because the knowledge was distributed over the entire network.

There are some recent studies related with extracting rules from the trained Artificial Neural Networks (ANN). For instance, in [2] and [3] try to extract rules from a trained ANN for regression problems. To identify the premise of a single rule, however, in [2] and [3] implement a data analysis on the training data set or the generated data set. No matter the data set for extracting rules is the trained one or the generated one, the amount of data instances is still finite, and the premise of a resulted rule covers merely discrete points, not an area.

Here we present a mathematical study of the layered feed-forward neural networks for identifying the rules suggested in the network. The mathematical analysis, not a data analysis, is proposed for identifying the premise associated with each rule. This paper is organized as follows. Section 2 gives the detail of the mathematical study and Section 3 offers some conclusions and future work.

2 The Study for Identifying the Rules Suggested in the Network

Assume the layered feed-forward neural network f is arranged as the one defined from equations (1) and (2) below, where $tanh(x) \equiv \dfrac{e^x - e^{-x}}{e^x + e^{-x}}$. Namely, given the cth observation $_c\mathbf{x}$, the corresponding value of the ith hidden node $_ca_i$ equals $tanh(_2w_{i0} + \sum_{j=1}^{m} {}_2w_{ij}\, _cx_j)$ and the corresponding value of $f(_c\mathbf{x})$ equals $_3w_0 + \sum_{i=1}^{p} {}_3w_i\, _ca_i$. In equations (1) and (2), m is the number of explanatory variables x_j's, p is the number of adopted hidden nodes, $_2w_{i0}$ is the bias value of the ith hidden node a_i, $_2w_{ij}$ is the weight between the jth explanatory variable x_j and the ith hidden node a_i, $_3w_0$ is the bias value, and $_3w_i$ is the weight between the ith hidden node a_i and the output node.

$$a_i(\mathbf{x}) \equiv tanh({}_2w_{i0} + \sum_{j=1}^{m} {}_2w_{ij}\, x_j), \qquad (1)$$

$$f(\mathbf{x}) \equiv {}_3w_0 + \sum_{i=1}^{p} {}_3w_i\, a_i(\mathbf{x}) = {}_3w_0 + \sum_{i=1}^{p} {}_3w_i\, tanh({}_2w_{i0} + \sum_{j=1}^{m} {}_2w_{ij}\, x_j). \qquad (2)$$

In this article, character in bold represents a column vector, a matrix or a set, and the superscript T indicates the transposition: $_2\mathbf{w}_i^T \equiv ({}_2w_{i1}, \ldots, {}_2w_{im})$, $_3\mathbf{w}^T \equiv ({}_3w_1, \ldots, {}_3w_p)$, $_2\mathbf{w}^T \equiv ({}_2\mathbf{w}_1^T, {}_2\mathbf{w}_2^T, \ldots, {}_2\mathbf{w}_p^T)$, and $\mathbf{w}^T \equiv ({}_2\mathbf{w}^T, {}_3\mathbf{w}^T)$. Here we assume $m > 1$. We also assume that $_3\mathbf{w}$ and $_2\mathbf{w}$ are non-zero vectors. Thus $_2\mathbf{W} \equiv ({}_2\mathbf{w}_1, {}_2\mathbf{w}_2, \ldots, {}_2\mathbf{w}_p)^T$ is a non-zero matrix.

The range of f is at most $({}_3w_0 - \sum_{i=1}^{p} |{}_3w_i|, {}_3w_0 + \sum_{i=1}^{p} |{}_3w_i|)$ instead of R. It is because that $-1 < a_i < 1$, thus $-\sum_{i=1}^{p} |{}_3w_i| < \sum_{i=1}^{p} {}_3w_i\, a_i < \sum_{i=1}^{p} |{}_3w_i|$ and $_3w_0 - \sum_{i=1}^{p} |{}_3w_i| < y < {}_3w_0 + \sum_{i=1}^{p} |{}_3w_i|$. y is vague if it cannot be carried out with the network system; otherwise, y is non-vague. If $y \in (-\infty, {}_3w_0 - \sum_{i=1}^{p} |{}_3w_i|) \cup ({}_3w_0 + \sum_{i=1}^{p} |{}_3w_i|, \infty)$, y is surely vague.

As Rumelhart and his collagues in [4] proposed the interesting Back-Propagation learning algorithm for training layered feed-forward neural networks, the information $\mathbf{x}$ coming to the input nodes is re-coded into $\mathbf{a} \equiv (a_1, a_2, \ldots, a_p)^T \in (-1, 1)^p$ and the output y is generated by $\mathbf{a}$ rather than the original pattern $\mathbf{x}$. In other words, f can be viewed as the composite $g \circ \mathbf{h}$ where $\mathbf{h}$ is defined by $(\mathbf{h}(\mathbf{x}))_i$, the ith component of $\mathbf{h}(\mathbf{x})$, $\equiv tanh({}_2w_{i0} + \sum_{j=1}^{m} {}_2w_{ij}\, x_j)$ and $a_i = (\mathbf{h}(\mathbf{x}))_i$ for every $i \in P \equiv \{1, 2, \ldots, p\}$, and g is defined by $g(\mathbf{a}) \equiv {}_3w_0 + \sum_{i=1}^{p} {}_3w_i\, a_i$. The hidden-layer set is $(-1, 1)^p$ because the $tanh$ activation function is used here.

For any non-vague y, let $\mathbf{X}(y) \equiv f^{-1}(y)$, the set of all elements of the input space whose images under f are y. Any input stimulus $\mathbf{x}$ in $\mathbf{X}(y)$ will result in an output value y. Thus, the following rule is suggested from the network:

Rule: If the input $\mathbf{x}$ is in the region of $\mathbf{X}(y)$, then the output value of the network is y.

$\mathbf{X}(y)$ can be viewed as the composite $\mathbf{h}^{-1} \circ \mathbf{g}^{-1}(y)$, where $\mathbf{g}^{-1}(y)$ is the set of all elements of the hidden-layer set whose images under g are y, and $\mathbf{h}^{-1}(\mathbf{a})$ is the set of all elements of the input space whose images under $\mathbf{h}$ are $\mathbf{a}$.

$\mathbf{h}^{-1}(\mathbf{a})$ equals $\{\mathbf{x}/ \sum_{j=1}^{m} {}_2w_{ij} x_j = tanh^{-1}(a_i) - {}_2w_{i0}$ for all $i \in \mathbf{P}\}$ where $tanh^{-1}(x) \equiv 0.5$ $ln(\frac{1+x}{1-x})$ is the inverse function of $tanh$. When ${}_2w_i$ is non-zero, there are parallel activation level hyperplanes, $\{\mathbf{x}/ \sum_{j=1}^{m} {}_2w_{ij} x_j = tanh^{-1}(a) - {}_2w_{i0}\}$ for all $a \in (-1, 1)$, in the input space. As stated in [5], these activation level hyperplanes make a scalar activation field in the input space. Through each point of the input space, there passes merely an activation level hyperplane that determines the associated activation value a_i of that point. All points on the level hyperplane $\{\mathbf{x}/ \sum_{j=1}^{m} {}_2w_{ij} x_j = tanh^{-1}(a_i) - {}_2w_{i0}\}$ have the same activation value a_i in the ith hidden node. p hidden nodes set up p activation fields in the input space, but these activation fields do not interfere with each other. Thus, $\mathbf{h}^{-1}(\mathbf{a})$ equals $\bigcap_{i=1}^{p} \{\mathbf{x}/ \sum_{j=1}^{m} {}_2w_{ij} x_j = tanh^{-1}(a_i) - {}_2w_{i0}\}$, which also equals $\{\mathbf{x}/ \sum_{j=1}^{m} {}_2w_{ij} x_j = tanh^{-1}(a_i) - {}_2w_{i0}$ for all $i \in \mathbf{P}\}$.

$\{\mathbf{x}/ \sum_{j=1}^{m} {}_2w_{ij} x_j = tanh^{-1}(a_i) - {}_2w_{i0}$ for all $i \in \mathbf{P}\}$ can also be represented as $\{\mathbf{x}/ {}_2\mathbf{W} \mathbf{x} = \omega(\mathbf{a})\}$. The function $\omega : (-1, 1)^p \rightarrow R^p$ is defined by $\omega(\mathbf{a}) \equiv (\omega_1(a_1), \omega_2(a_2),..., \omega_p(a_p))^T$ with $\omega_i(a_i) \equiv tanh^{-1}(a_i) - {}_2w_{i0}$ for every i. Given the vector $\mathbf{a}$, $\omega(\mathbf{a})$ is determined. Accordingly, given the vector $\mathbf{a}$, the system ${}_2\mathbf{W} \mathbf{x} = \omega(\mathbf{a})$ is a system of p linear equations in m unknowns. If $rank({}_2\mathbf{W} : \omega(\mathbf{a})) = rank({}_2\mathbf{W}) + 1$, where $rank({}_2\mathbf{W} : \omega(\mathbf{a}))$ is the rank of the augmented matrix $({}_2\mathbf{W} : \omega(\mathbf{a}))$, the system ${}_2\mathbf{W} \mathbf{x} = \omega(\mathbf{a})$ has no solution. (cf. [6], p. 108) In other words, if $rank({}_2\mathbf{W} : \omega(\mathbf{a})) = rank({}_2\mathbf{W}) + 1$, $\mathbf{h}^{-1}(\mathbf{a})$ is an empty set and the activation value $\mathbf{a}$ cannot be carried out with the neural network.

The point $\mathbf{a}$ in the hidden-layer set is vague if the activation values of $\mathbf{a}$ cannot be carried out with the neural network; otherwise, $\mathbf{a}$ is non-vague. Namely, the point $\mathbf{a}$ in the hidden-layer set is vague if $\mathbf{a}$ is mapped from nowhere in the input space. Lemma 1 gives the situation of the point $\mathbf{a}$ being vague.

Lemma 1: If $rank({}_2\mathbf{W} : \omega(\mathbf{a})) = rank({}_2\mathbf{W}) + 1$, the point $\mathbf{a}$ is vague.

The fact that $rank({}_2\mathbf{W} : \omega(\mathbf{a})) = rank({}_2\mathbf{W})$ implies $\omega(\mathbf{a})$ is in the linear hull spanned by column vectors of ${}_2\mathbf{W}$ (cf. [6], p. 92). Therefore, as stated in Lemma 2 and Lemma 3, $\{\mathbf{a}/ rank({}_2\mathbf{W} : \omega(\mathbf{a})) = rank({}_2\mathbf{W})\}$ is either the whole hidden-layer set, or a manifold[1] or a linear hull bounded by $(-1, 1)^p$. Moreover, whether there are vague points in the hidden-layer set can be determined by comparing values of $rank(_2W)$ and p. In other words, the range of **h** is $\{a/\ rank(_2W : \omega(a)) = rank(_2W)\}$ instead of $(-1, 1)^p$.

Lemma 2: If $rank(_2W) = p$, $\{a/\ rank(_2W : \omega(a)) = rank(_2W)\}$ equals $(-1, 1)^p$ and there are no vague points in the hidden-layer set.

Lemma 3: If $rank(_2W) = r$ and $r < p$, $\{a/\ rank(_2W : \omega(a)) = rank(_2W)\}$ is a r-manifold or a linear hull of dimension r bounded by $(-1, 1)^p$ and there are vague points in the hidden-layer set.

Lemma 4 states that $\mathbf{h}^{-1}(\mathbf{a})$ is a single point in the input space when **a** is a non-vague point and $rank(_2W) = m$; Lemma 5 states that $\mathbf{h}^{-1}(\mathbf{a})$ is an affine space of dimension $m - r$ in the input space when **a** is a non-vague point, $rank(_2W) = r$ and $r < m$. In other words, $\mathbf{h}^{-1}(\mathbf{a})$ is an affine space of dimension $m - rank(_2W)$ when $\mathbf{h}^{-1}(\mathbf{a})$ is non-empty.

Lemma 4: Each non-vague point **a** is mapped from a single point **x** in the input space provided that $rank(_2W) = m$.

Lemma 5: Each non-vague point **a** is mapped from any point **x** in an affine space of dimension $m - r$ in the input space provided that $rank(_2W) = r$ and $r < m$.

Conversely, $\mathbf{g}^{-1}(y)$ equals $\{a/\ \sum_{i=1}^{p} {}_3w_i\ a_i = y - {}_3w_0,\ \mathbf{a} \in (-1, 1)^p\}$, and $\mathbf{g}^{-1}(y)$ is a hyperplane bounded by $(-1, 1)^p$. When $_3\mathbf{w}$ is non-zero, there are parallel hyperplanes, $\mathbf{g}^{-1}(y)$s for all $y \in ({}_3w_0 - \sum_{i=1}^{p} /_3w_i/,\ {}_3w_0 + \sum_{i=1}^{p} /_3w_i/)$, in $(-1, 1)^p$. These hyperplanes make a scalar field (activation field) in the hidden-layer set. Through each point of the hidden-layer set, there passes merely one hyperplane that determines the associated activation value y of that point. All points on the $\mathbf{g}^{-1}(y)$ hyperplane have the same value y.

$\mathbf{g}^{-1}(y) \equiv \mathbf{A}_v(y) \cup \mathbf{A}_{nv}(y)$ where $\mathbf{A}_v(y)$ and $\mathbf{A}_{nv}(y)$ are the sets of vague and non-vague points in $\mathbf{g}^{-1}(y)$, respectively. From Lemma 1, $\mathbf{A}_{nv}(y) \equiv \{a/\ \mathbf{a} \in \mathbf{g}^{-1}(y), rank(_2W : \omega(a)) = rank(_2W)\}$. $\mathbf{A}_{nv}(y)$ can be viewed as $\{a/\ rank(_2W : \omega(a)) = rank(_2W)\} \cap \mathbf{g}^{-1}(y)$. From Lemma 6 and Lemma 7, $\mathbf{A}_{nv}(y)$ is either a hyperplane bounded by $(-1, 1)^p$ or an intersection of a hyperplane and a r-manifold or a linear hull of dimension r in $(-1, 1)^p$.

Lemma 6: $\mathbf{A}_{nv}(y)$ equals $\mathbf{g}^{-1}(y)$ provided that $rank(_2W) = p$.

Lemma 7: $\mathbf{A}_{nv}(y)$ is an intersection of a hyperplane and a r-manifold or a linear hull of dimension r in $(-1, 1)^p$ provided that $rank(_2W) = r$ and $r < p$.

Let $\mathbf{h}^{-1}(\mathbf{g}^{-1}(y)) \equiv \{\mathbf{h}^{-1}(\mathbf{a})/\ \mathbf{a} \in \mathbf{g}^{-1}(y)\}$ and $\mathbf{h}^{-1}(\mathbf{A}_{nv}(y)) \equiv \{\mathbf{h}^{-1}(\mathbf{a})/\ \mathbf{a} \in \mathbf{A}_{nv}(y)\} = \{\mathbf{x}/\ _2\mathbf{W}\ \mathbf{x} = \omega(\mathbf{a})$ with all $\mathbf{a} \in \mathbf{A}_{nv}(y)\}$. $\mathbf{h}^{-1}(\mathbf{g}^{-1}(y))$ equals $\mathbf{h}^{-1}(\mathbf{A}_{nv}(y))$ because that $\mathbf{g}^{-1}(y) \equiv \mathbf{A}_v(y) \cup \mathbf{A}_{nv}(y)$ and $\mathbf{A}_v(y)$ is mapped from nowhere in the input space. Thus $\mathbf{X}(y)$ equals $\{\mathbf{x}/\ _2\mathbf{W}\ \mathbf{x} = \omega(\mathbf{a})$ with all $\mathbf{a} \in \mathbf{A}_{nv}(y)\}$. Theorems 1 and 2 show that, when $rank(_2W) = p$, $\mathbf{X}(y)$ is either a hyperplane or a manifold in the input space and the dimension of $\mathbf{X}(y)$ is $m-1$.

Theorem 1: $\mathbf{X}(y)$ is a hyperplane in the input space provided that $m > p$, $p = 1$ and $rank(_2W) = 1$.

---

[1] The definition of manifold please refer to [7].

Theorem 2: $X(y)$ is a $(m-1)$-manifold in the input space provided that $m \geq p$, $p > 1$ and $rank(_2W) = p$.

For any non-vague y_1 and y_2 with $_3w_0 - \sum_{i=1}^{p} |_3w_i| < y_2 \leq y_1 < {_3w_0} + \sum_{i=1}^{p} |_3w_i|$, the prediction $y \leq y_2$ is activated by any **a** in $\bigcup_{y={_3w_0}-\sum_{i=1}^{p}|_3w_i|}^{y_2} A_{nv}(y)$ or any input **x** in $\bigcup_{y={_3w_0}-\sum_{i=1}^{p}|_3w_i|}^{y_2} X(y)$. In other words, there is a rule:

Rule: If the input **x** is in the region of $\bigcup_{y={_3w_0}-\sum_{i=1}^{p}|_3w_i|}^{y_2} X(y)$, then the output value of the network is less than or equals y_2. Similarly, the prediction $y_2 \leq y \leq y_1$ is activated by any **a** in $\bigcup_{y=y_2}^{y_1} A_{nv}(y)$ or any input **x** in $\bigcup_{y=y_2}^{y_1} X(y)$; the prediction $y \geq y_1$ is activated by any **a** in $\bigcup_{y=y_1}^{{_3w_0}+\sum_{i=1}^{p}|_3w_i|} A_{nv}(y)$ or any input **x** in $\bigcup_{y=y_1}^{{_3w_0}+\sum_{i=1}^{p}|_3w_i|} X(y)$.

Theorem 3 shows that, when $rank(_2W) = p$, the set $\bigcup_{y={_3w_0}-\sum_{i=1}^{p}|_3w_i|}^{y_2} A_{nv}(y)$ is an union of adjacent hyperplanes bounded by $(-1, 1)^p$. Combined with Theorem 1, the set $\bigcup_{y={_3w_0}-\sum_{i=1}^{p}|_3w_i|}^{y_2} X(y)$ is an union of adjacent hyperplanes in the input space when $m > p$, $p = 1$ and $rank(_2W) = 1$; combined with Theorem 2, the set $\bigcup_{y={_3w_0}-\sum_{i=1}^{p}|_3w_i|}^{y_2} X(y)$ is an union of adjacent $(m-1)$-manifolds in the input space when $m \geq p$, $p > 1$ and $rank(_2W) = p$.

Theorem 3: For any non-vague y_1 and y_2 with $_3w_0 - \sum_{i=1}^{p} |_3w_i| < y_1 \leq y_2 < {_3w_0} + \sum_{i=1}^{p} |_3w_i|$, sets of $\bigcup_{y={_3w_0}-\sum_{i=1}^{p}|_3w_i|}^{y_2} A_{nv}(y)$, $\bigcup_{y=y_2}^{y_1} A_{nv}(y)$ and $\bigcup_{y=y_1}^{{_3w_0}+\sum_{i=1}^{p}|_3w_i|} A_{nv}(y)$ are convex polytopes provided that $rank(_2W) = p$.

3 Summary

In conclusion, for any non-vague y_1 and y_2 with $_3w_0 - \sum_{i=1}^{p}|_3w_i| < y_1 \leq y_2 < {}_3w_0 + \sum_{i=1}^{p}|_3w_i|$, the following four rules are suggested from the trained network:

Rule 1: If the input **x** is in the region of $X(y)$, then the output value of the network is y.

Rule 2: If the input **x** is in the region of $\bigcup_{y={}_3w_0-\sum_{i=1}^{p}|_3w_i|}^{y_2} X(y)$, then the output value of the network is less than or equals y_2.

Rule 3: If the input **x** is in the region of $\bigcup_{y=y_2}^{y_1} X(y)$, then the output value of the network is within $[y_2, y_1]$.

Rule 4: If the input **x** is in the region of $\bigcup_{y=y_1}^{{}_3w_0+\sum_{i=1}^{p}|_3w_i|} X(y)$, then the output value of the network is greater than or equals y_1, where $X(y) = \{x /\ _2\mathbf{W}\,x = \omega(a)$ with all $a \in A_{nv}(y)\}$.

References

1. Yoon, Y., Guimaraes, T., Swales, G.: Integration Artificial Neural Networks with Rule-Based Expert System in Decision Support Systems. Vol. 11. (1994) 497-507
2. Setiono, R., Leow, W. K., Zurada, J. M.: Extraction of Rules from Artificial Neural Networks for Nonlinear Regression. IEEE Transactions on Neural Networks, Vol. 13. (2002) 564-577
3. Saito, K., Nakano, R.: Extracting Regression Rules from Neural Networks in Neural Networks. Vol. 15. (2002) 1279-1288
4. Rumelhart, D. E., Hinton, G. E., Williams, R.: Learning Internal Representation by Error Propagation in Parallel Distributed Processing. Vol. 1, Cambridge, MA: MIT Press (1986) 318-362
5. Tsaih, R.: An Explanation of Reasoning Neural Networks in Mathematical and Computer Modelling. Vol. 28. (1998) 37-44
6. Murty, K.: Linear Programming. John Wiley & Sons, New York (1983)
7. Munkres, J.: Topology: A First Course. Prentice-Hall, Englewood Cliffs, New Jersey (1975)

Analysing Contributions of Components and Factors to Pork Odour Using Structural Learning with Forgetting Method

Leilei Pan, Simon X. Yang, Fengchun Tian, Lambert Otten, and Roger Hacker

School of Engineering, University of Guelph, Guelph, ON N1G 2W1, Canada
syang@uoguelph.ca

Abstract. A novel neural network based approach to analysing contributions of odour components and factors to the perception of pork farm odour is proposed. A multi-component multi-factor odour analysis model is developed and learnt by an algorithm called structural learning with forgetting. Through the learning, unnecessary connections fade away and a skeletal network emerges. By analysing the resulting skeletal networks significant odour components and factors can be identified, and thus a more thorough understanding of odour model can be obtained. The proposed approach is tested with a pork farm odour database. The results demonstrate the effectiveness of the proposed approach.

1 Introduction

In pig farming, the efforts to measure and reduce pork odour intensity have been impeded by the lack of knowledge on odour generation mechanism and odour quantification techniques. Prediction of odour intensity is difficult for the fact that they are often complex mixtures of hundreds of compounds which interact with each other and a variety of factors contribute to their generation [1]. Over past decades, researchers initially attempted to determine a single-component of odour that could act as an indicator for odour intensity [2]. Later researches indicated that single-component analysis of the odour is insufficient. Janes et al. [3] proposed an analysis model of pork farm odour considering the location of measurement as a contributing source. The results of this study provide evidence that the accuracy and precision of odour intensity prediction will be improved by taking into consideration the contributing factors such as environmental conditions, etc.. However, previous multi-component multi-factor analysis did not give any idea of the relationships of odour components and factors to a complex odour. Researchers have assumed those potential odour components and factors would affect the perception of odour, but whether they would or not, and by how much, still remains unclear. An analysis of the contributions of odour components and factors to the perception of odour would allow the identification of significant odour components and major contributing factors.

Previous research on the contributions of odour components and factors is scarce, and existing researches are all based on simple statistical approaches such as principal component analysis (PCA) [4]. A disadvantage of statistical methods is that a linear relationship is forced on the system being modelled. Since the odour system exhibits non-linear behaviour, the accuracy of the statistical methods will be reduced. Artificial neural networks (ANN) have been commonly used to predict odour intensity, for their generalisation abilities, their capability to handle non-linear interactions mathematically, as well as their ability to automatically learn the relationship that exists between the inputs and outputs[5]. Though past attempts at using ANN have performed well, they were not able to identify components or factors contribute most to a complex odour.

In this paper, the main innovative aspect of the presented study relies on the capability of finding out the significant odour components and factors which contribute most to the perception of odour intensity. Various comparative studies are also presented in this paper.

2 Methods

In this section, the basic idea of structural learning with forgetting is introduced after a brief description of the multi-component multi-factor odour analysis model.

2.1 Multi-component Multi-factor Odour Analysis Model

There are many factors that are reported to affect odour intensity, including environmental conditions, type of swine facilities, etc. [6]. Many of these factors are non-numeric data which cannot be incorporated into a statistical model. The rapid development of soft computing techniques, such as neural networks, fuzzy logic, etc., allows the incorporation of these other potential contributing factors to the analysis model of pork farm odour.

2.2 Structural Learning with Forgetting

Structure learning with forgetting (SLF) is composed of three algorithms: learning with forgetting, hidden units clarification and learning with selective forgetting [7]. The second part, hidden units clarification, which is not implemented in this research, is not presented here.

Learning with Forgetting. The criterion function in learning with forgetting is given as

$$J_f = J_p + \varepsilon' \sum_{i,j} |w_{ij}| \quad (1)$$

where the first term on the right side, J_P, is the mean square error in the back propagation learning, the second term is the complexity penalty, with the relative weight ε', and total criterion J_f. During learning, the penalty term leads

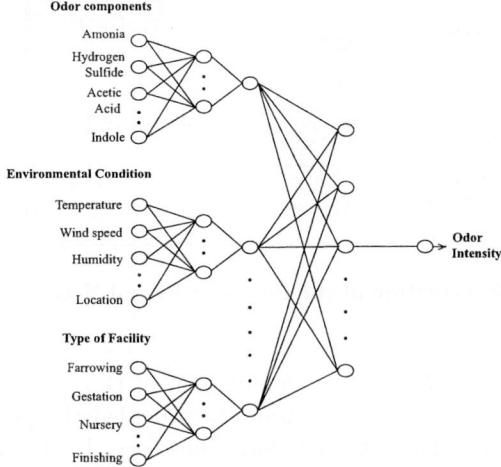

Fig. 1. Potential Multi-Component Multi-Factor Neural Network Model of Pork Odour

to a constant decay to all weights so that redundant connections and units will disappear, leaving a skeletal structure.

The weight change, Δw_{ij}, is obtained by differentiating equation (1) with respect to the connection weight w_{ij}

$$\Delta w_{ij} = -\eta \frac{\partial J_f}{\partial w_{ij}} = \Delta w'_{ij} - \varepsilon \operatorname{sgn}(w_{ij}) \quad (2)$$

where

$$\Delta w'_{ij} = -\eta \frac{\partial J_p}{\partial w_{ij}} \quad (3)$$

is the weight change due to back-propagation (BP), η is a learning rate, $\varepsilon = \eta \varepsilon'$ is the amount of decay at each weight change, and $\operatorname{sgn}(w_{ij})$ is a sign function.

Learning with Selective Forgetting. The mean square error (MSE) by learning with forgetting is larger than that by BP learning, because the former minimises the total criterion J_f instead of the quadratic criterion J. The following criterion makes only the connection weights decay whose absolute values are below a threshold θ

$$J_f = J_p + \varepsilon' \sum_{|w_{ij}| < \theta} |w_{ij}| \quad (4)$$

The penalty term makes the MSE much smaller than that in learning with forgetting, because the summation is restricted only to weak connections.

2.3 Data Sets and Proposed Odour Analysis Model

In this section, the datasets used for model development and testing are introduced and the specific neural network model proposed in this paper is developed.

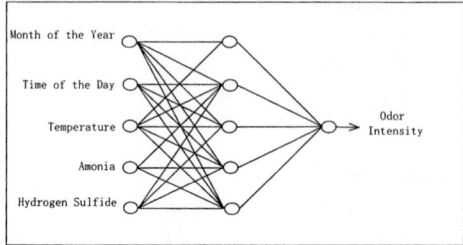

Fig. 2. Structure of the Proposed Neural Network NN1

Data Sets. Researchers from the Department of Biosystems and Agricultural Engineering at the University of Minnesota provided the pork farm data used in this paper from the Farmstead Odour Database. Each dataset has the airphase concentrations of ammonia and hydrogen sulfide with an odour intensity generated by human assessors. For each sample, the time of day and month of year, as well as outside temperature, are available.

Neural Network Architecture. To estimate the odour intensity, a neural network model (NN1) has been designed based on the pork farm data, with an hidden layer. Figure 2 illustrates the network structure. The neural network model has five input neurons and the variables include two odour components, i.e. the air concentrations of ammonia and hydrogen sulfide, and three contributing factors i.e. the time of day, month of year, and the outdoor temperature. The model starts with a fully connected network. Connections reflect the relation between each attribute and the corresponding attribute value.

3 Results and Analysis

In this section, the results of our proposed approach will be analysed. In order to ensure the validity of proposed approach, two additional neural networks are developed and analysed.

3.1 Performance Analysis

In order to analyse the effectiveness of the proposed approach, NN1 was trained both by normal BP algorithm and SLF algorithm. The performance is evaluated based on the mean square error (MSE) and standard deviation of the observed and predicted odour intensity values. The results are shown in Table 1. The MSE by SLF learning is somewhat larger than that by BP learning because the former minimises the total criterion J_f instead of the quadratic criterion J. The lower standard deviation of normal BP shows that the neural network model that was trained with BP learning provided more precise prediction than the SLF learning. But again, the difference is small.

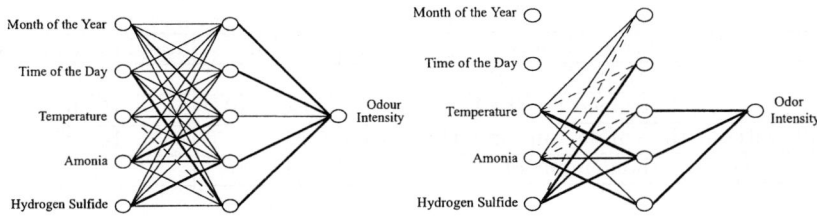

Fig. 3. Comparison of resulting structures of NN1 trained by normal BP and SLF algorithm (The dashed lines represent connections smaller than 1, and the solid lines represent connections greater than 1. The strengths of the connections are represented by the line width.)

3.2 Analysis of Potential Components and Factors

Fig. 3 shows the resulting architectures of NN1 trained by normal BP algorithm and SLF algorithm. For a connection less than 0.001, no line is drawn. Comparing the resulting structures indicates that when trained by SLF algorithm, 3 significant attributes were selected which are the concentrations of Ammonia and Hydrogen-Sulfide, and the outdoor temperature.

An examination of the resulting architectures shows that two of the factors, time of day and month of year, faded away. This result is different from what we expected before the investigation. We assumed that the time of day and month of year when the measurement was taken can generally represent such measures as farm activity, ventilation, temperature, humidity, and other physical measures. Although it certainly deserve further analysis, we believe this surprising behaviour exists because the outdoor temperature is included in the datasets as a contributing factor, thus the time of day and month of year which mostly influence temperature become less significant.

3.3 Validity of the Approach

Two additional neural network models are developed in order to ensure the validity of our proposed approach. The first one (NN2), is the same as NN1, but without the input neurons for time of day and month of year. The second one (NN3) is identical to NN1 with the exception of no hydrogen sulfide input.

The performance of NN2, and NN3 are shown in Table. 1. The MSE and the standard deviation of NN2 are both close value to the results obtained from NN1. Analysis of the results shows that time of day and month of year contribute little to the perception of pork odour.

The MSE and the standard deviation of NN3 are both significantly larger than these of NN1. Investigation of the results shows that hydrogen sulfide contributes significantly to the perception of pork odour.

The results from these two additional neural networks validates our initial results from NN1 trained by SLF.

Table 1. Performance of NN1, NN2 and NN3

type of Neural Network	MSE	Standard Deviation
NN1 (trained by BP)	5.09×10^{-3}	16.01%
NN1 (trained by SLF)	6.036×10^{-3}	17.86%
NN2 (without time of day and mouth of year)	5.13×10^{-3}	16.12%
NN3 (without input of H2S)	7.89×10^{-3}	23.1%

4 Conclusion

The analysis of contributions odour components and factors has been attracting much attention because it has potential for a more thorough and comprehensive understanding of pork odour. The novel approach proposed here is to analyse contributions of odour components and factors using neural network-based, structural learning with forgetting method. This method is based solely on data, without initial theories and preprocessing. A pork farm database has been used as an example. The results show that the proposed approach is a simple but effective method.

References

1. Zahn, J.A., DiSpirito, A.A., Do, Y.S., Brooks, B.: Correlation of human olfactory response to airbone intensity of malodourous volatile organic compounds emitted from swine effluent. Journal of Environmental Quality **30** (2001) 635–647
2. Lunn, F., Vyver, J.V.D.: Sampling and analysis of air in pig houses. Agriculture and Environment **3** (1994) 159–169
3. Janes, K., Yang, S.X., Hacker, R.: Multiple component and multiple factor analysis of poerk farm odour using neural networks. Applied Soft Computing. Submitted.
4. Doty, R.L., Smith, R., McKeown, D.A., Raj, J.: Test of human olfactory function: Principal components analysis suggests that most measure a common source of variance. Percept Psychophys **56** (1994) 701–701
5. Haykin, S.: Neural Networks: A Comprehensive Foundation. Prentice Hall, New Jersey, America (1999)
6. Zhang, Q., Feddes, J., Edeogu, I., Nyachoti, M., House, J., Small, D., Liu, C., Mann, D., Clark, G.: Odour production, evaluation and control. Manure Management Initiative Inc, North Carolina,USA (1984)
7. Ishikawa, M.: Structural learning with forgetting. Neural Networks **9** (1996) 509–521

On Multivariate Calibration Problems

Guo-Zheng Li[1], Jie Yang[1], Jun Lu[1], Wen-Cong Lu[2], and Nian-Yi Chen[2]

[1] Institute of Image Processing & Pattern Recognition,
Shanghai Jiaotong University, Shanghai, China, 200030
[2] Department of Chemistry, Shanghai University, Shanghai, China, 200436

Abstract. Multivariate calibration is a classic problem in the analytical chemistry field and frequently solved by partial least squares method in the previous work. Unfortunately there are so many redundant features in the problem, that feature selection are often performed before modeling by partial least squares method and the features not selected are usually discarded. In this paper, the redundant information is, however, reused in the learning of partial least squares method within the frame of multitask learning. Results on three multivariate calibration data sets show that multitask learning can greatly improve the accuracy of partial least squares method.

1 Introduction

Multivariate calibration[1] is a classic problem in the analytical chemistry field, which provides a convenient way to determine several components in a mixture within only one experimental step, without the tedious operation of pre-separation of these components. One of the distinct characteristics of the multivariate calibration problems is that the number of features is far more than the number of cases.

Multivariate calibration problems are most widely solved by partial least squares(PLS) method and sometimes by artificial neural networks(ANN) since ANN are powerful regression techniques [2]. Because too many redundant features will speed up the overfitting phenomena of the learning machine, feature selection methods are often employed to eliminate the redundant features [3]. However, features not selected are always discarded and not input into the later modeling any longer.

Recently, multitask learning(MTL) [4] was proposed to reuse the redundancy information, which uses some of the discarded features as extra output targets and obtain better results than that of only discarding the features not selected. Yet, accuracy improved by MTL is so slight that the researcher claimed MTL was only proper to be used in the cases that even slight improvements are needed[4].

Since there are so many redundant features in the multivariate calibration problems, intuitively, we think that MTL can help to improve the prediction accuracy of PLS or ANN on this problem. Motivated by this, we try to apply MTL on PLS method to address the multivariate calibration problems and to study whether MTL can improve the prediction accuracy of PLS. Computation will also be performed using ANN with the MTL method on the same problems, for ANN is a baseline method in the machine learning field.

The remainder of this paper is arranged as follows. Section 2 describes the multivariate calibration problems. Section 3 presents the learning methods of PLS and MTL. Section 4 demonstrates the experiments on the multivariate calibration data sets. This paper is ended up with discussions in Section 5.

2 Multivariate Calibration Problems

Multivariate calibration are very interesting problems from the analytical chemistry field, since they have two distinct characteristics, 1) There are many redundant features in the data sets; 2) The number of features are always rather more than the number of cases. Here we use three different data sets[5], all are collected using fluorescence spectrometry, the number of features, cases and target values are listed in Table 1.

Table 1. The property of the multivariate calibration data sets

Data set	Number of features	Number of Cases	Number of targets
I	211	23	3
II	141	17	2
III	116	17	2

3 Learning Methods

3.1 Partial Least Squares Method

Partial least squares(PLS) regression algorithm can build linear regression models, and it has proved to be useful in situations when the number of observed features is significantly greater than the number of observations and high multicollinearity among the features exists.

Suppose the input features $X \subset \mathbb{R}^n$ and output targets $Y \subset \mathbb{R}^m$, PLS uses a robust procedure, a nonlinear iterative partial least squares(NIPALS) algorithm [6], to solve the singular value decomposition problem of the inner product of $X^T Y$. In the NIPALS algorithm, there are two loops. The inner loop is used to extract the score vector $\mathbf{t}$ and its corresponding latent vector $\mathbf{u}$. The outer loop is to sequentially extract the latent vectors $\mathbf{t}, \mathbf{u}$ and the weight vectors $\mathbf{w}, \mathbf{c}$ from $\mathbf{X}$ and $\mathbf{Y}$ matrices in decreasing order of their corresponding singular values.

The PLS regression model can be written in matrix form as [7]

$$\mathbf{Y} = \mathbf{XB} + \mathbf{F}$$

where **B** is an $(n \times m)$ matrix of the regression coefficients and **F** is an $(N \times m)$ matrix of residuals. The matrix **B** has the form

$$\mathbf{B} = \mathbf{X}^T\mathbf{U}(\mathbf{T}^T\mathbf{X}\mathbf{X}^T\mathbf{U})^{-1}\mathbf{T}^T\mathbf{Y}$$

where the **T** and **U** are $(N \times p)$ matrices of the extracted p latent vectors, N is the number of cases.

More and more people besides the researchers from the analytical chemistry field are interested in the study of PLS method[7,8] in the machine learning community.

3.2 Multitask Learning

Multitask learning(MTL) [4] is a form of inductive transfer that is applicable to any learning method that can share part of what is learned between multiple tasks. The basic idea is to use the selected features as the input feature set and combine the target values with some of the discarded features as the target output. The terms used here is according to the previous work [4], they are arranged as in Table 2.

Table 2. The terms used in multitask learning method

Term	Explanation
Main Task	The output target values to be learned
Selected Inputs	The features selected as inputs in all experiments
Extra Features	The features selected from the discarded features
Extra Inputs	The extra features selected from the discarded features when used as inputs
Extra Outputs	The same extra features selected from the discarded features when used as outputs
STD	Standard PLS using the Selected Inputs as inputs and only the Main Task as outputs
STD+IN	Uses the Extra Features as Extra Inputs to learn the Main Task
STD+OUT	Uses the Extra Features as Extra Outputs in paralleled with the Main Task using the Selected Inputs as inputs

MTL has used many popular learning algorithms as the base learning machine such as k-nearest neighborhood, artificial neural networks, even support vector machines, etc., but obtained slight improvements [4]. Here in this work, we will apply it on the PLS method, to see if it can help to obtain better performance. We also will apply MTL on ANN in order to perform comparisons with PLS on the multivariate calibration problems. It is worth noting that ANN used in this work is weight decay based neural networks in Bayesian frame, which is insensitive to the setting of parameters to some degree[9].

Many feature selection methods have been proposed in machine learning community[10,11]. In this work, feature selection is based on Kohonen neural network which has been used as the feature selection method for multivariate calibration problems to improve the prediction accuracy[3]. In this method, data is firstly clustered according to the Euclidean distance of each feature vectors using the Kohonen neural network, then the features near the center of clusters are selected as Selected Inputs, and the other not selected features are ranked according to the Euclidean distance to the Selected Inputs. Finally, the first several features with the least distance are selected as Extra Features.

3.3 Assessment of Regression Quality

Here the leave-one-out cross validation technique is used to evaluate the above learning methods with the measure *root mean square error* (RMSE), which is defined as

$$\text{RMSE} = \sqrt{\frac{1}{mN} \sum_{j=1}^{m} \sum_{i=1}^{N} (y_{ij}^e - y_{ij})^2}$$

where $y_{ij}^{(e)}$ denotes the jth predicted real(target) value of the ith example, N denotes the number of cases and m denotes the number of target values in the Main Task of each example, which is 2 or 3 as in Table 1.

4 Results of Computation

4.1 MTL on Different Number of Selected Inputs

Data sets in section 2 have been processed. Firstly, feature selection using Kohonen neural network has been performed, then, PLS and ANN are used to perform computations for three cases of STD, STD+IN and STD+OUT. Results of RMSE are computed as the number of Selected Inputs increases, some statistical results are listed in Table 3, in which the least error in the cases of STD+OUT and the corresponding error of STD and STD+IN cases are written as RMSE_{xs}, RMSE_{xi} and RMSE_{xo} respectively.

Table 3. Statistical results using PLS and ANN within the frame of MTL

	PLS			ANN		
	RMSE_{xs}	RMSE_{xi}	RMSE_{xo}	RMSE_{xs}	RMSE_{xi}	RMSE_{xo}
Data Set I	1.0253	1.0175	0.3079	**0.2562**	0.2906	0.2798
Data Set II	0.2219	0.2188	**0.0094**	0.0303	0.0261	0.0196
Data Set III	0.2877	0.2819	**0.0244**	0.0507	0.0397	0.0502
Average	0.5116	0.5061	**0.1139**	0.1157	0.1188	0.1165

From Table 3, we can see that on the multivariate calibration problems, 1) PLS+MTL can obtain better results than ANN(+MTL); 2) Feature selection can improve the accuracy to some degree compared with the total feature subset, but the improvements are slight; 3) For the PLS method, multitask learning can greatly improve the prediction accuracy, the error of STD+OUT is about one third of that of STD or STD+IN cases on Data Set I, even one twentieth on Data Set II, and one tenth on Data Set III. However, for the ANN method, MTL only slightly improve the prediction accuracy on two data sets.

4.2 MTL on Different Number of Extra Outputs

We select the feature subset in which the number of Selected Inputs is about one third of the number of cases using Kohonen feature selection method, then we use the Selected Inputs as input and use the Extra Features plus the Main Task as outputs to build the PLS model. As the number of Extra Outputs increases in STD+OUT case results of RMSE are plotted on Fig. 1, Results of RMSE are also computed in the cases of STD and STD+IN and plotted on Fig. 1.

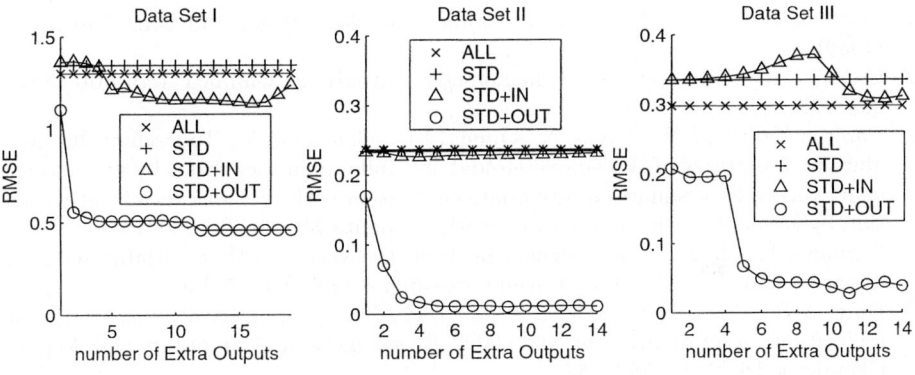

Fig. 1. Result of RMSE as the number of Extra Ouputs increases on three Data Sets

From Fig. 1, we can see that when the Extra Features are used as Extra Outputs in STD+OUT case, RMSE is rapidly decreasing as the number increases and then kept stable until 10 plus features are used. However, when the Extra Features are used as Extra Inputs in STD+IN case, slight improvements and no explicit rules on the improvements can be obtained.

5 Discussions

Beyond our imagination, on multivariate calibration problems multitask learning(MTL) using partial least squares(PLS) method can obtain so inspiring results. This owns to the special algorithm of PLS, which uses a robust nonlinear iterative partial least squares to solve the singular value decomposition of the product $X^T Y$ of the input matrix and output matrix. Therefore, when PLS method uses the Extra Features as the output target values in the multivariate calibration problems, the Extra Outputs exert constraints on the PLS regression model and depress the overfitting, then PLS can obtain more precision models and give less error prediction values on the test examples.

PLS+MTL obtain better results than ANN+MTL do on two data sets, and worse results on one data sets. We think the reason is that multivariate calibration are collinearity problems and PLS is suitable to solve them, while ANN is too powerful that it can not build the proper model. The first data sets have three outputs, it is too complex that ANN can obtain better results. Therefore, perhaps kernel PLS can be used in the modeling of the first problems to obtain better results.

Acknowledgments. This work is financed by NSF of China (50174038).

References

1. Martens, H., Ns, T.: Multivariate Calibration. John Wiley and Sons, Chichester (1989)
2. Hopke, P.K.: The evolution of chemometrics. Analytica Chimica Acta **500** (2003) 365–377
3. Capitán-Vallvey, L.F., Navas, N., Olmo, M.D., Consonni, V., Todeschini, R.: Resolution of mixtures of three nonsteroidal anti-inflammantory drugs by fluorescence using partial least squares multivariate calibration with previous wavelength selection by kohonen artificial neural networks. Talanta **52** (2000) 1069–1079
4. Caruana, R., de Sa, V.R.: Benefiting from the variables that variable selection discards. Journal of machine learning research **3** (2003) 1245–1264
5. Ding, Y.P., Chen, N.Y., Wu, Q.S., Li, G.Z., Yang, J.: Derivative spectrum simultaneous determination of NO_3^--NO_2^- by svr method. Computers and Applied Chemistry **19** (2002) 752–754
6. Wold, H.: Estimation of principal components and related models by iterative least squares. In: Multivariate Analysis. Academic Press, New York (1966) 391–420
7. Rosipal, R., Trejo, L.J.: Kernel partial least squares regression in reproducing kernel hilbert space. Journal of Machine Learning Research **2** (2001) 97–123
8. Shawe-Taylor, J., Cristianini, N.: Kernel Methods for Pattern Analysis. Cambridge University Press, Cambridge (2004)
9. Foresee, F.D., Hagan, M.T.: Gauss-newton approximation to bayesian regularization. In: Proceedings of the 1997 International Joint Conference on Neural Networks. (1997) 1930–1935
10. Kohavi, R., George, J.H.: Wrappers for feature subset selection. Artificial Intelligence **97** (1997) 273–324
11. Guyon, I., Elisseeff, A.: An introduction to variable and feature selection. Journal of machine learning research **3** (2003) 1157–1182

Control of Associative Chaotic Neural Networks Using a Reinforcement Learning

Norihisa Sato, Masaharu Adachi, and Makoto Kotani

Department of Electrical Engineering, Graduate School of Engineering, Tokyo Denki University
2-2 Kanda-Nishiki-cho, Chiyoda-ku, Tokyo 101-8457, Japan
03gdd02@ed.cck.dendai.ac.jp
adachi@d.dendai.ac.jp

Abstract. We propose a control method for associative chaotic neural networks to be chaotic using a reinforcement learning. In this method the controller gives perturbation to chaotic neural networks. The perturbation is selected by Q-learning that is a kind of the reinforcement learning. The reward of reinforcement learning is decided by the largest local divergence of network. The reward is given to the agent when the largest local divergence increases over the previous value. The states of the agent to be used for the reinforcement learning are observed by quasi-energy of the chaotic neural network. The proposed control method is useful not only for associative chaotic neural networks but also for the chaotic neural networks solving optimization problem like quadratic assignment problems or the traveling salesman problem. Because this algorithm only gives perturbations obtained for the network by the reinforcement learning

1 Introduction

At present chaotic behavior discovered in many systems. In most of the cases chaotic behavior is not favorable to the system. However, it is reported that when one attempts to solve a combinational optimization problem by chaotic neural networks[1] chaotic behavior is useful[2]. However, it is hard to find parameter values of the chaotic neural network showing chaotic behavior. Therefore, we investigate control method making chaotic neural networks be chaotic by using a reinforcement learning. We attempt to control chaotic neural network from periodic to chaotic behavior. In this paper, at first we explain the reinforcement learning used for the control. Secondly, our control algorithm by using reinforcement learning is explained. At last, the result of simulation and future problems are explained.

2 Reinfocement Learning

Reinforcement learning is one of the learning algorithms which are invented by modeling the learning with human's trial and error. Consequently, the reinforcement learning is a kind of unsupervised learning. When a control object, which is called an "agent", takes some action, the action has to be evaluated. The controller gives a reward when the agent takes a good action. On the other hand, the controller gives a penalty when the agent takes bad action[4],[5].

As a concrete reinforcement learning rule, Q-learning is adopted for the control in this paper. The feature of the Q-learning is that a state-action value function is employed to determine the action of the agent. The state-action value function has two parameters. One is the state of the agent which is expressed by v, and the other is the action of the agent which is expressed by u. Therefore, the state-action value function can be expressed by $Q(v, u)$ when the agent takes a state v and an action u. The state-action value function is defined for quantized states $\widehat{v}$'s and for candidate actions u's. At each quantized sate $\widehat{v}$, the agent selects the control action u which gives the maximum value of $Q(\widehat{v}, u)$ among u's

3 The Network to Be Controlled

Chaotic neural networks are applied to solving optimization problems[2] or to associative memory[3]. When chaotic neural networks solve those problems, behavior of chaotic neural networks must be chaotic because we intend to utilize the searching ability of chaos. However, it is hard to find the parameter values of the chaotic neural network showing chaotic behavior. Therefore, we attempt to control chaotic neural network to be chaotic using the reinforcement learning. The updating of the chaotic neural networks to be controlled is represented by Eqs.(1)–(4).

$$\eta_i(t+1) = k_f \eta_i(t) + \sum_{j=1}^{N} w_{ij} x_j(t), \qquad (1)$$

$$\zeta_i(t+1) = k_r \zeta_i(t) + \alpha x_i(t) + b_i, \qquad (2)$$

$$x_i(t+1) = f(\eta_i(t+1) + \zeta_i(t+1)), \qquad (3)$$

$$f(y) = 1/\{1 + \exp(-y/\epsilon)\}, \qquad (4)$$

where, $\eta_i(t)$ and $\zeta_i(t)$ denote internal state variables for the feedback inputs from the constituent neurons and for the refractoriness, respectively. $x_i(t)$ denotes the output of the ith neuron at discrete-time t. The parameter k_r and k_f are the decay parameters for the feedback inputs and the refractoriness of the neuron, respectively. b_i denotes temporally constant external inputs to the ith neuron. w_{ij} is the synaptic weight from the jth to the ith neuron. The synaptic weights w_{ij}'s are determined by the following equation (5) to store the pattern vectors[6],[7].

$$w_{ij} = \frac{1}{P} \sum_{\mu=1}^{P} (2x_i^{(\mu)} - 1)(2x_j^{(\mu)} - 1), \qquad (5)$$

where, P is number of stored pattern vectors. In this paper, three pattern vectors are stored to a chaotic neural network of 16 neurons. The three stored pattern vectors are orthogonal with each other.

4 Control Algorithm

4.1 State of the Agent

As it is already explained in Sec.2, quantized states are needed for the Q-learning. It is practical to represent the state of the network in one variable even a value of the variable may correspond to multiple states of the whole state vector $X = [x_1, x_2, \cdots, x_N]^T$. Because when there are many neurons in the network, it is impractical to assign the state-action value function Q for the state vector X. Therefore, a quasi-energy function E, as usually defined for the Hopfield network[7], is employed for the state of the agent in the Q-learning as follows.

$$E = -\frac{1}{2}\sum_{i=1}^{N}\sum_{j=1}^{N}w_{ij}x_i x_j + b_i x_i. \qquad (6)$$

4.2 Action of the Agent

In this algorithm, the action of the agent is a control perturbation which is a parameter perturbation to the chaotic neural networks. The agent selects an action among several candidate control perturbations according to the state-action value function Q. At a certain time t, the controller gives control perturbation $u(t)$ applied to Eq.(2) then the equation becomes $\zeta_i(t+1) = k_r \zeta_i(t) + \alpha x_i(t) + b_i + u(t)$. Where, the value of the control perturbation $u(t)$ is selected among five candidate control perturbations, i.e., $u \in \{0, 0.1, 0.3, 0.5, 0.7\}$. The network shows periodic dynamics when the perturbation is fixed to one of them. However, switching the perturbations cause the network to be chaotic as it is shown in the next section.

4.3 Reward to the Agent

The selected action is evaluated by a criterion whether the action is effective for achieving the control target. The target of the control is to make the network be chaotic. For the target, the action is evaluated whether it causes increase of the largest local divergence $\lambda_m(t)$ [8] or not, for each control iteration. We explain how to calculate the local divergence in the following. The updating of the internal states of the network represented by Eq.(1) and (2) can be seen as a nonlinear mapping of a 2N-dimensional vector composed of the internal states of η_i and ζ_i for N neurons. Therefore, the local divergence of the updating can be evaluated by calculating the eigenvalues of the Jacobian matrix of a local linearized map of the nonlinear map in each iteration. The largest eigenvalue is called the largest local divergence in this paper. When the largest local divergence increase over the previous value, the controller gives the agent a reward $r = 0$. When the largest local divergences decrease, the controller gives the agent a penalty $r < 0$.

4.4 Control Algorithm

The proposed control algorithm for making the network be chaotic in this paper is based on the algorithm that was proposed by Gadaleta et al.[9] for stabilizing unstable periodic orbits in a chaotic attractor. Our algorithm is divided into three major procedures as follows. At first, one generates quantized states for the Q-learning. Because the behavior of a network cannot be predicted at the beginning of the learning. Note that our goal of the control is generating a chaotic attractor. Secondly, Q-learning is performed for constructing the state-action function for the control. At last, the control of the chaotic neural networks is executed. The procedures of the algorithm are concretely described as follows.

i. Perform the preprocessing for the control.
   (i–a) A set of initial values is given to the chaotic neural network.
   (i–b) Prepare the candidate actions u's that are used in the Q-learning.
ii. Generate the quantized states for the Q-learning and compose the state-action value function.
   (ii–a) Update the state variables of the network.
   (ii–b) Compute the value of the quasi-energy function E.
   (ii–c) Compute the distances between E and the existing prototype of the quantized states $\widehat{E}$'s $(\widehat{E_1}, \cdots, \widehat{E_k})$ except for the beginning. If the computed distances are larger than the predetermined distance, E is registered to a new prototype state $\widehat{E}_{k+1}$. Otherwise, the prototype $\widehat{E}$ covering E is used for the state of the agent in the Q-learning.
   (ii–d) Compute the local divergence $\lambda_m(t)$ of the network at the mapping of the updating of the network.
   (ii–e) Give a reward r for the action according to the local divergence whether it increases from the previous value or not as follows.

   $$r = \begin{cases} 0 & (\lambda_m(t+1) > \lambda_m(t)) \\ -1 & (\lambda_m(t+1) \leq \lambda_m(t)) \end{cases}, \quad (7)$$

   (ii–f) Update the state-value function Q by the reward r.
   The size of the change ΔQ of the updating is computed by the following equation[5].

   $$\Delta Q(\widehat{E}, u) = \beta\{r + \gamma maxQ(\widehat{E}, u) - Q(\widehat{E}, u)\}. \quad (8)$$

   Where β and γ denote step size parameter of the learning and discount rate, respectively.
   (ii–g) For the quantized state $\widehat{E}$, select an action $u(t)$ that has largest action value function Q as follows.

   $$u(t+1) = arg \max_u Q(\widehat{E}, u). \quad (9)$$

   (ii–h) The action $u(t+1)$ is added to the bias of the network in the next iteration.

(ii–i) Stop repeating (ii–a)–(ii–h) when the number of the prototype states becomes the predetermined number N_P or when new prototype state is not generated for more than T_P iterations

(ii–j) Only perform the reinforcement learning. Repeat (ii–a)–(ii–h) when number of iteration reaches to the predetermined learning time T_Q. At this stage generating new prototypes is prohibited. In the numerical experiments that are shown in the next section, N_P, T_P and T_Q are set to be 30 000, 100 000 and 1 000 000, respectively.

iii. Perform the actual control.

(iii–a) A set of initial values is given to the chaotic neural network.

(iii–b) Update state variables of the network.

(iii–c) Compute the value of the quasi-energy function E.

(iii–d) Search the prototype of the quantized state $\widehat{E}$ that corresponds to E.

(iii–e) For the prototype of the quantized state $\widehat{E}$, select an action $u(t+1)$ that has the largest action value function Q as show in Eq.(9)

(iii–f) The action $u(t+1)$ is added to the bias of the network in the next iteration.

(iii–g) Repeat (iii–b)–(iii–d) when the number of iteration reaches to T_C. T_C is set to be 200 000 in the following numerical experiments.

5 Simulation and the Result

5.1 Conditions of the Numerical Experiment

We demonstrate efficacy of the proposed control method by computer simulation with the following conditions. We use 13 initial conditions for the Q-learning of the control. Three out of the 13 initial conditions correspond to the stored patterns. Other five initial conditions are random vectors of binary components. Another five of the initial conditions are random vectors whose components are nearly 0.5 and with a small deviation. The Q-learning is stopped at 1 000 000 iterations and the control trial is performed for 200 000 iterations after the learning. The initial conditions for the Q-learning are also used for the control trial.

5.2 Result of the Experiments

The proposed control method succeeded in making the chaotic neural network be chaotic. Namely, all the stored patterns are retrieved in the chaotic dynamics. Fig.1, shows the time course of the Hamming distance between the outputs of the network and one of the stored patterns. From the figure we confirm that the network retrieves the stored pattern and reverse of it when the distance becomes naught and 16, respectively.

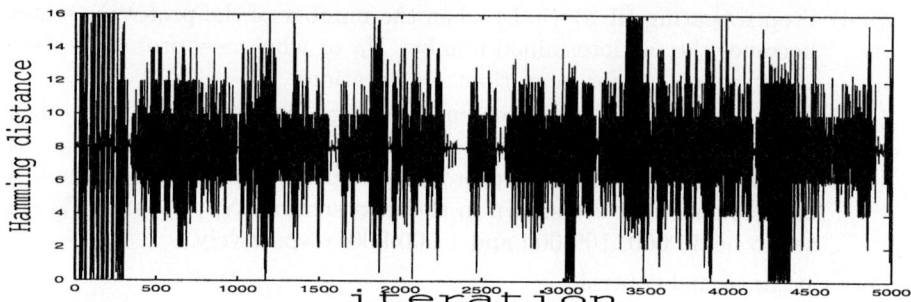

Fig. 1. Hamming distance between the output of the network and a stored pattern.

6 Conclusion

In this paper, a control method for the chaotic neural network to be chaotic is proposed. The proposed control method succeeded in controlling the chaotic neural network to be chaotic using a reinforcement learning. The action is chosen according to the state by the reinforcement learning. The chosen action causes the chaotic neural network to be aperiodic. The proposed control method is very easy to apply for the chaotic neural networks because it is just a parameter perturbation to the bias of the network. Therefore, the proposed control method can adapt other chaotic neural networks for solving optimization problems such as the traveling salesman problem or quadratic assignment problems.

References

1. Aihara, K.: Chaotic Neural Networks.: In Kawakami, H. (ed.): Bifurcation Phenomena in Nonlinear Systems and Theory of Dynamical Systems, World Scientific, Singapore, (1990) 143–161
2. e.g., Nozawa, H.: A Neural Network Model as a Globally Coupled Map and Applications Based on Chaos, Chaos, vol. 2, no. 3, (1992) 377-386
3. Adachi, M. and Aihara, K.: Associative Dynamics in a Chaotic Neural Network, Neural Netw., vol. 10, no. 1, (1997) 83-98
4. Sutton, R. S. and Barto, A. G.: Richard S. Sutton and Andrew G. Barto (eds.): Reinforcement Learning , The MIT press, (1998)
5. Watkins, C. J. C. H. and Dayan, P.: Q-learning. Machine Learning, **8**, (1992) 279-292
6. e.g., Kohonen, T.: Correlation Matrix Memories. IEEE Trans., **C-21**, (1972) 353–359
7. Hopfield, J. J.: Neural Networks and Physical Systems with Emergent Collective Computation Abilities. Proceedings of National Academy of Sciences, USA, **79**, (1982) 2445-2558
8. Shimada, I. and Nagashima, T.: A Numerical Approach to Ergodic Problem of Dissipative Dynamical systems. Prog. Theor. Phys., **61**, (1979) 1605–1616
9. Gadaleta, S. and Dangelmayr, G.: Optimal Chaos Control Through Reinfocement Learning. American Institute of Physics **9** (1999) 775-788

A Method to Improve the Transiently Chaotic Neural Network

Xinshun Xu, Jiahai Wang, Zheng Tang, Xiaoming Chen,
Yong Li, and Guangpu Xia

Faculty of Engineering, Toyama University, Toyama-shi, 930-8555, Japan
xinshun_xu@hotmail.com

Abstract. In this article, we propose a method to improve the transiently chaotic neural network by introducing several time-dependent parameters. With this method, the network processes by starting at rich chaotic dynamics, and reaches stable state for all neurons rapidly after the last bifurcation. This enables the network to have rich search ability at the beginning, and use less CPU time to reach a stable state. The simulation results on the N-queen problem confirm that this method is effective to improve TCNN in terms of both the solution quality and convergence speed.

1 Introduction

Neural networks have been shown to be a powerful tool for combinatorial optimization problems, especially for NP-hard problems. One of the well-known models among them is the Hopfield neural network[3,4]. The Hopfield-type neural network converges to a stable equilibrium point due to its gradient descent dynamics. However, it suffers from sever local minimum problems whenever applied to optimization problems. Although many methods have been presented to improve them [12, 7], the results are not always satisfactory. Recently, many artificial neural networks with chaotic dynamics have been investigated [1,9]. The chaotic neural networks have much rich and far-from equilibrium dynamics with various coexisting attractors, not only of fixed points and periodic points but also of strange attractors in spite of the simple equation. But it is usually difficult to decide when to terminate chaotic dynamics, or how to harness chaotic behavior for convergence to a stable equilibrium point corresponding to an acceptably near-optimal state [2]. In order to take advantage of both the Hopfield network's convergent dynamics and chaotic dynamics, Chen and Aihara have proposed a transiently chaotic neural network(TCNN) [2]. However, there are many parameters which can affect its convergent speed and solution quality[2]. Usually, the rich dynamics usually uses more steps for the network to get to a stable state. In this article, we present a method of introducing several time-independent parameters into the original transiently chaotic neural network model. With this method, the modified transiently chaotic neural network has relatively rich dynamics at the beginning, and converges to stable states faster after bifurcation for all neurons.

2 The Transiently Chaotic Neural Network

Chen and Aihara's transiently chaotic neural network(TCNN) model is defined as follows:

$$v_i(t) = \frac{1}{1+e^{-u_i(t)/\varepsilon}} \qquad (1)$$

$$u_i(t+1) = ku_i(t) + \alpha(\sum_{j=1,j\neq i}^{n} w_{ij}v_j + I_i) - z_i(t)(v_i(t) - I_0) \qquad (2)$$

$$z_i(t+1) = (1-\beta)z_i(t) \qquad (3)$$

where
- v_i = output of neuron i,
- u_i = internal state of neuron i,
- w_{ij} = connection weight from neuron j to neuron i,
- I_i = input bias of neuron i,
- α = positive scaling parameter for inputs,
- k = damping factor of nerve membrane($0 \le k \le 1$),
- $z_i(t)$ = self-feedback connection weight or refractory strength ($z_i(t) \ge 0$),
- ε = steepness parameter of the output function ($\varepsilon > 0$)
- β = damping factor of the time-dependent $z_i(t)$.
- I_0 = positive parameter.

In this model, the variable $z_i(t)$ corresponds to the temperature in the usual stochastic annealing process. So Eq.3 is an exponential cooling schedule for the annealing. The network coincides with the Hopfield network when the value of $z_i(t)$ becomes small enough.

3 The Method to Improve the Original Model

In [2], Chen et al. showed that the parameter β governed the bifurcation speed of the transient chaos. Chen et al. also showed that the parameter α could affect the neurodynamics, that is, the influence of the energy function became too strong to generate the transient chaos when α was too large, and the energy function could not be sufficiently reflected in the neurodyanmaics if α was too small. So in order to make the network have rich dynamics at the beginning, β has to be set to a small value, and α has to be set a suitable value. However, the network uses more steps to reach the saturated state for all neurons if β is small. That is to say, the algorithm will use more CPU time to get a solution to a problem. In addition, it is difficult to find a suitable value of α for different problems. Although Chen et al. have pointed out these problems, until now, no methods are proposed to improve the original transiently chaotic neural network, especially for these parameters.

In order to improve convergent speed and search ability of the original transiently chaotic neural network, we introduce three variables($\alpha(t)$, $\beta(t)$, $\varepsilon(t)$) into TCNN which are all constants(α, β, ε) in the original model. They are updated by:

$$\alpha(t+1) = (1+\lambda)\alpha(t) \quad \text{if } \alpha(t)<0.1 \text{ else } \alpha(t)=0.1 \tag{4}$$

$$\beta(t+1) = (1+\varphi)\beta(t) \quad \text{if } \beta(t)<0.2 \text{ else } \beta(t)=0.2 \tag{5}$$

$$\varepsilon(t+1) = (1-\eta)\varepsilon(t) \quad \text{if } \varepsilon(t)>0.001 \text{ else } \varepsilon(t)=0.001 \tag{6}$$

where λ, φ and η are small positive constants(selected empirically).

At the beginning, $\alpha(0)$ and $\beta(0)$ are set to small values. When $\alpha(t)$ is small, the influence on the energy function is not too strong to generate transient chaos. When $\beta(t)$ is a small value, it can make $z(t)$ decrease slowly at the beginning so that the network has enough time to keep the self-feedback large enough to generate chaos. When $\beta(t)$ becomes large, $z(t)$ decreases quickly. At the same time the influence of the energy function becomes strong when $\alpha(t)$ becomes large gradually. This means the self-feedback signal becomes weak in the motion equation. So using Eqs.4 and 5, we can keep the neural network to have rich chaotic dynamics at the beginning. When both $\alpha(t)$ and $\beta(t)$ become large enough, the chaos disappears quickly. Once bifurcation appears, chaotic signals usually become weak. However, it usually takes many steps to converge to a stable state. That is to say, all neurons use many steps to reach a stable state. In order to help the neural network quickly converge to a stable state for all neurons after the bifurcation with fewer steps, we use Eq.6 to update the variable $\varepsilon(t)$ which is a steepness parameter of the output function(Eq.1). At the beginning, $\varepsilon(t)$ is set to a large value. This means that the steepness of the output function is small. This can help the neural network generate chaos easily. However once $\varepsilon(t)$ becomes small, the steepness of the output function becomes large. That is to say, a small value of the internal state of neurons can make the output of neurons converge to 1 or 0 quickly. Of course this network may get out of hand without keeping these parameters within bounds. A bound must be assigned to each parameter so that these parameters do not change when they reach their bound. In this article, $\alpha(t)$ is bounded on 0.1, $\beta(t)$ on 0.2, and $\varepsilon(t)$ on 0.001, respectively.

4 Simulations on the *N*-Queen Problem

In order to confirm the effectiveness of the proposed method for TCNN, we tested it on *N*-queen problem. The *N*-queen problem is concerned with placing *N*-queens on an *N* by *N* chessboard in such a way that no queen is under attack. The *N*-queen problem has been solved with a variety of artificial neural networks [11,8]. More recently, several chaotic models have been used[6,10,5].

Simulation results are summarized in Table 1. The columns "*N*", "*TCNN1*", "*TCNN2*" and "*Proposed*" represent the number of queens, the results of original TCNN when parameter β was set to a small value 0.001, the results when parameter β

was set to a large value 0.08 in TCNN, and the results of the proposed model. In the proposed model, the parameters are set as follows:

$$k = 0.9; z(0) = 0.08; \beta(0) = 0.002; \alpha(0) = 0.001; \varepsilon(0) = 0.008;$$
$$\lambda = 0.002, \phi = 0.002, \eta = 0.002.$$
(7)

And bounds for $\alpha(t)$, $\beta(t)$, $\varepsilon(t)$ were 0.1, 0.2 and 0.001, respectively. For each instance, 100 simulations were performed by each algorithm with different initial neuron states. The rate of optimal solution and the average steps of each algorithm are presented in Table 1. Note that, in this article, the network is said to be stable and hence terminated if the criterion $\{|V_{ij}(t+1)-V_{ij}(t)|; i,j=1, 2,..., N\} < 5*10^{-5}$ is met.

Table 1. Simulation results on the N-queen problem

N	TCNN1		TCNN2		Proposed	
	Conv.	Steps	Conv.	Steps	Conv.	Steps
20	99%	2145.7	93%	457.2	100%	463.7
50	98%	2354.6	87%	477.9	99%	479.4
100	100%	2511.4	90%	613.6	100%	485.2
200	97%	2587.7	82%	733.4	100%	477.9
300	98%	2713.4	76%	792.1	98%	491.5

From Table 1, the following observations can be made:
1. The proposed model can solve the *N*-queen problem with high global minimum convergent rate, and the steps do not increase when the problem scale becomes large.
2. Although the original model can solve this problem with good solution quality when β is set to a small value, it uses more than 2000 steps to converge to a stable state. When β is set to a large value, the steps are reduced, however the convergent rate is not satisfactory.

So we can conclude that the proposed method is effective to improve the transiently chaotic neural network.

5 Conclusions

We present a method to improve the original transiently chaotic neural network for combinatorial optimization problems. This method makes a transiently chaotic neural network keep rich dynamics, and use fewer steps to converge to a stable state for all neurons after the last bifurcation. This algorithm is tested on the *N*-queen problem. The simulation results show that the proposed model is superior to the original model and another chaotic neural network in light of the optimal convergence rate and average update steps. Moreover this method can be applied to solve other combinatorial optimization problems without tuning any parameters.

References

1. Aihara, K., Takabe, T., Toyoda, M.: Chaotic Neural Networks. Physics Letter A, Vol.144, No.6-7 (1990) 333–340
2. Chen, L., Aihara, K.: Chaotic Simulated Annealing by a Neural Network Model with Transient Chaos. Neural Networks, Vol.8, No.6 (1995) 915–930
3. Hopfield, J.J., Tank, D.W.: Neural Computation of Decisions in Optimization Problems. Biological Cybernetics, Vol.52, No.4 (1985) 141–152
4. Hopfield, J.J., Tank, D.W.: Computing with Neural Circuits: A Model. Science, Vol.233 (1986) 625–633
5. Kwok, T., Smith, K.A.: Experimental Analysis of Chaotic Neural Network Models for Combinatorial Optimization under a Unifying Framework. Neural Networks, Vol.13, No.7 (2000) 731–744
6. Kwok, T., Wang, L., Smith, K.A.: Incorporating Chaos into the Hopfield Neural Network for Combinatorial Optimization. In: Calloas, N., Omaolayole, O. and Wang, L.(eds.): Proceedings World Multiconference on Systemics, Cybernetics and Informatics, Vol.1. Orlando, Florida (1998) 659–665
7. Li, S.Z.: Improving Convergence and Solution Quality of Hofield-Type Neural Networks with Augmented Lagrange Multipliers. IEEE Transactions on Neural Networks, Vol.7, No.6 (1996) 1507–1516
8. Mandziuk, J.: Neural networks for the N-Queens Problem: A Review. Control and Cybernetics, Vol.32, No.2 (2002) 217–248
9. Nozawa, H.: A Neural Network Model as a Globally Coupled Map and Applications Based on Chaos. Chaos, Vol.2, No.3 (1992) 377–386
10. Ohta, M.: Chaotic Neural Networks with Reinforced Self-feedbacks and Its Application to N-queen Problem. Mathematics and Computers in Simulation, Vol.59, No.4 (2002) 305–317
11. Smith, K.A.: Neural Networks for Combinatorial Optimization: A Review of More Than a Decade of Research. INFORMS Journal on Computing, Vol.11, No.1 (1999) 15–34
12. Van den Bout, D.E., Miller, T.K.: Improving the Performance of the Hopfield-Tank Neural Network through Normalization and Annealing. Biological Cybernetics, Vol.62, No.2 (1989) 129–139

A New Neural Network for Nonlinear Constrained Optimization Problems

Zhiqing Meng[1], Chuangyin Dang[2], Gengui Zhou[1], Yihua Zhu[1], and Min Jiang[3]

[1] College of Business and Administration
Zhejiang University of Technology
Hangzhou, Zhejiang 310032, China
mengzhiqing@zjut.edu.cn

[2] Department of Manufacturing Engineering & Engineering Management
City University of Hong Kong
Kowloon, Hong Kong, China
mecdang@cityu.edu.hk

[3] School of Economics and Management
Xidian University
Xi'an, Shanxi 710071, China

Abstract. In this paper we introduce a new neural network for solving nonlinear constrained optimization problems. The energy function for the neural network with its neural dynamics is obtained from an application of the penalty-function method. We show that the system of the neural network is stable and an equilibrium point of the neural dynamics yields an optimal solution for the corresponding nonlinear constrained optimization problem. Based on the relationship between the equilibrium points and the energy function, an algorithm is developed for computing an equilibrium point of the system or an optimal solution to its optimization problem. The efficiency of the algorithm is demonstrated with several numerical examples.

Keywords: Penalty function; Hopfield neural networks; Stability; Equilibrium point

1 Introduction

Hopfield neural networks are very important tools for solving optimization problems [1-7]. Since 1985 a wide variety of Hopfield-like neural networks have been designed for improving the performance of the original model. Hopfield neural networks (HNN) are sorts of feedback neural networks. One of difficulties in this type of neural networks is how to define an energy function and how to make them stable.

The method of exact penalty function is the main tool to solve the problem of nonlinear optimization problems [6-8]. The method can be applied to rebuild the HNN [7]. We use penalty function to construct an energy function for the neural network so that the neural network is stable. By solving the penalty function problem, we can get an equilibrium point of the HNN, which is also a solution of the optimization problem under some mild condition. Therefore, we obtain in this paper a new energy function for Hopfield neural networks, which makes the neural networks have good stability property. We give an algorithm to find out an approximate solution to the corresponding optimization problem, which is also an equilibrium point of the new neural networks system. The testing results show that the algorithm is efficient.

2 A Neural Network for Constrained Optimization Problems

This paper studies a neural network for the following nonlinear constrained optimization problem:

(P)   minimize $f(\mathbf{x})$ subject to $g_i(\mathbf{x}) \leq 0,\ i=1,2,...,m,$

where $f,\ g_i : R^n \to R,\ i \in I = \{1,2,...,m\}$ are differentiable. Let $X = \{\mathbf{x} | g_i(\mathbf{x}) \leq 0,\ i=1,2,...,m\}$.

2.1 A New Neural Network

Assuming that $\mathbf{b}: R^n \times R \to R^n$ is a vector, we obtain a dynamic differentiable system:

$$\frac{d\mathbf{x}}{dt} = B\mathbf{b}(\mathbf{x}(t), t), \qquad (1)$$

where $\mathbf{x}(t) = (x_1(t), x_2(t), \ldots, x_n(t))^T$ is a state vector and $B = \text{diag}(\beta_1, \beta_2, ..., \beta_n)$ with $\beta_i,\ i=1,2,\cdots,n$, being convergence coefficients of the dynamic differentiable system (1).

Definition 2.1.1. If a point $\mathbf{x}^* \in X$ satisfies

$$\mathbf{b}(\mathbf{x}^*, t) = 0, \forall t,$$

$\mathbf{x}^*$ is called an equilibrium point of the dynamic system (1).

Let $p_\varepsilon: R \to R$ be a function given by

$$p_\epsilon(t) = \begin{cases} 0 & if \quad t \leq 0, \\ \frac{1}{3\epsilon} t^{\frac{3}{2}} & if \quad 0 \leq t \leq \epsilon, \\ (t^{\frac{1}{2}} - \frac{2}{3}\epsilon^{\frac{1}{2}}) & if \quad t \geq \epsilon, \end{cases}$$

and $p(t) = \sqrt{\max\{t, 0\}}$. Then, $\lim_{\epsilon \to 0} p_\epsilon(t) = p(t)$. Let

$$F(\mathbf{x}, \rho, \varepsilon) = f(\mathbf{x}) + \rho \sum_{i=1}^{m} p_\varepsilon(g_i(\mathbf{x})).$$

The energy function $E(\mathbf{x})$ for our neural network is defined as

$$E(\mathbf{x}) = F(\mathbf{x}, \rho, \varepsilon) = f(\mathbf{x}) + \rho \sum_{i=1}^{m} p_\varepsilon(g_i(\mathbf{x})), \tag{2}$$

which is differentiable. And the corresponding dynamic differentiable system is given by

$$\frac{d\mathbf{x}}{dt} = -B\nabla E(\mathbf{x}). \tag{3}$$

The new neural network of size n is a fully connected network with n continuous valued units. Let ω_{ij} be the weight of the connection from neuron i to neuron j. Assume that $E(\mathbf{x})$ are twice continuously differentiable. Then we can define the connection coefficients as follows:

$$\omega_{ij} = \frac{\partial E^2(\mathbf{x})}{\partial x_1 \partial x_2}, \quad i, j = 1, 2, \cdots, n. \tag{4}$$

The basic structure of the neural network is as follows.

We define $\mathbf{x} = (x_1, x_2, \cdots, x_n)$ as the input vector of the neural network, $\mathbf{y} = (y_1, y_2, \cdots, y_n)$ as the output vector, and $V(t) = (v_1(t), v_2(t), \cdots, v_n(t))$ as the state vector of neurons. $v_i(t)$ is the state of neuron i at the time t. And this new neural network is a type of Hopfield-like neural network. These kinds of neural networks have two kinds of prominent properties: the state's difference of network corresponds to the negative gradient of the energy function; the connection weights of neurons correspond to the second partial derivatives of the energy function (Hessian matrix).

2.2 Stability Analysis

Consider the following optimization problems:

$$(\text{PI}\rho) \quad \min F(\mathbf{x}, \rho, \varepsilon) \quad s.t. \quad \mathbf{x} \in X.$$

For the stability analysis of (2), we have the following theorems of Lyapunov stability.

Theorem 2.2.1 Let $\mathbf{x}^*$ be an equilibrium point of the dynamic system (3) under the parameter (ρ, ε). If $\mathbf{x} \neq 0$ and $E(\mathbf{x}) \neq 0$, then $\mathbf{x}^*$ is a stable point of the dynamic system (3). And if the weight coefficient matrix $(\omega_{ij})_{n \times n}$ is positive semi-definite, then $\mathbf{x}^*$ is a locally optimal solution to the problem (PIρ).

Proof: According to the assumption, if $\mathbf{x} \neq 0$, we have

$$\frac{dE(\mathbf{x})}{dt} = \sum_{k=1}^{n} \frac{\partial E}{\partial x_k} \frac{dx_k}{dt} = \sum_{k=1}^{n} \frac{\partial E}{\partial x_k}(-\beta_k) \frac{\partial E}{\partial x_k} \leq 0$$

Applying the Lyapunov stability theorem, we obtain that $\mathbf{x}^*$ is a stable point of the dynamic system (3). The second conclusion is obvious. □

Let $F(\mathbf{x},\rho) = f(\mathbf{x}) + \rho \sum_{k=1}^{m} \sqrt{\max\{g_i(\mathbf{x}),0\}}$. Consider the following optimization problem:

$$(P\rho) \quad \min F(\mathbf{x},\rho) \text{ s.t. } \mathbf{x} \in X.$$

Since $\lim_{\varepsilon \to 0} F(\mathbf{x},\rho,\varepsilon) = F(\mathbf{x},\rho)$, we will first study the relationship between $(P\rho)$ and $(PI\rho)$.

Lemma 2.2.1 For any $\mathbf{x} \in X$, $\rho > 0$ and $\varepsilon > 0$, we have

$$0 \leq F(\mathbf{x},\rho) - F(\mathbf{x},\rho,\varepsilon) \leq \frac{2}{3} m\rho\varepsilon^{\frac{1}{2}}. \tag{5}$$

Proof. The definition of $p_\varepsilon(t)$ implies (5). □

Theorem 2.2.2 Let $\{\varepsilon_j\} \to 0$ be a sequence of positive number. Assume that $\mathbf{x}_j$ is a solution to $\min_{\mathbf{x} \in X} F(\mathbf{x},\rho,\varepsilon_j)$ for some $\rho > 0$. Let $\mathbf{x}^*$ be an accumulating point of the sequence $\{\mathbf{x}_j\}$. Then $\mathbf{x}^*$ is an optimal solution to $\min_{\mathbf{x} \in X} F(\mathbf{x},\rho)$.

Definition 2.2.1 A point $\mathbf{x}_\varepsilon$ is ε-feasible or a ε-solution if $g_i(\mathbf{x}_\varepsilon) \leq \varepsilon$ for all $i \in I$.

Theorem 2.2.3 Let $\mathbf{x}^*$ be an optimal solution of (P_ρ) and $\bar{\mathbf{x}} \in X$ an optimal solution of (PI_ρ). Then

$$0 \leq F(\mathbf{x}^*,\rho) - F(\bar{\mathbf{x}},\rho,\epsilon) \leq \frac{2}{3} m\rho\epsilon^{\frac{1}{2}} \tag{6}$$

Proof Lemma 2.2.1 implies this theorem. □

Theorem 2.2.4 Let $\mathbf{x}^*$ be an optimal solution of $(P\rho)$ and $\bar{\mathbf{x}} \in X$ an optimal solution of (PI_ρ). Furthermore, let $\mathbf{x}^*$ be feasible to (P) and $\bar{\mathbf{x}}$ be ε-feasible to (P). Then

$$0 \leq f(\mathbf{x}^*) - f(\bar{\mathbf{x}}) \leq \frac{4}{3} m\rho\epsilon^{\frac{1}{2}}. \tag{7}$$

Proof Since $\bar{\mathbf{x}}$ is ε-feasible to (P), it follows that $\sum_{i \in I} p_\varepsilon(g_i(\bar{\mathbf{x}})) \leq \frac{2}{3} m\varepsilon^{\frac{1}{2}}$. As $\mathbf{x}^*$ is an optimal solution to (P), we have $\sum_{i \in I} p(g_i(\mathbf{x}^*)) = 0$. By Lemma 2.2.1, we get

$$0 \leq (f(\mathbf{x}^*) + \rho \sum_{i \in I} p(g_i(\mathbf{x}^*))) - (f(\bar{\mathbf{x}}) + \rho \sum_{i \in I} p_\varepsilon(g_i(\bar{\mathbf{x}}))) \leq \frac{2}{3} m\rho\epsilon^{\frac{1}{2}},$$

which implies $0 \leq f(\mathbf{x}^*) - f(\bar{\mathbf{x}}) \leq \frac{4}{3} m\rho\epsilon^{\frac{1}{2}}$. □

The following theorem is also obvious.

Theorem 2.2.5 If $\mathbf{x}^*$ is an optimal solution to the problem $(PI\rho)$, then $\mathbf{x}^*$ is an equilibrium point of the dynamic system (3) under the parameter (ρ, ε).

Theorems 2.2.1 and 2.2.5 show that an equilibrium point of the dynamic system yields an approximate optimal solution to the optimization problem $(PI\rho)$. Theorems 2.2.2 and 2.2.3 mean that an approximate solution to $(PI\rho)$ is also an approximate solution to $(P\rho)$ when ε is sufficiently small. Moreover, an approximate solution to (PIρ) also becomes an approximate optimal solution to (P) by Theorem 2.2.4 if the approximate solution is ε-feasible. Therefore, we may obtain an approximate optimal solution to (P) by finding an approximate solution to (PIρ) or an equilibrium point of the dynamic system (3).

3 Applications to Nonlinear Optimization Problems

In order to get an approximate optimal solution to (P) and an equilibrium point of the new neural network system, we propose the following Algorithm I. By Algorithm I, we get an approximate optimal solution to (P) by Theorem 2.2.4, and an equilibrium point of the dynamic system (3) of the neural network.

Algorithm I
Step 1: Given $\mathbf{x}^0, \epsilon > 0, \epsilon_0 > 0, \rho_0 > 0, 0 < \eta < 1$, and $N > 1$.
Let $j = 0$. Construct the energy function (2)
and the dynamic differentiable system (3).
Step 2: Use $\mathbf{x}^j$ as the starting point, and solve $\min_{\mathbf{x} \in X} F(\mathbf{x}, \rho_j, \epsilon_j)$.
Let $\mathbf{x}^j$ be the optimal solution.
Step 3: If $\mathbf{x}^j$ is ϵ-feasible to (P),
then stop and an approximate solution x^j of (P)
and an equilibrium point of dynamics system (3) has been generated.
Otherwise, let $\rho_{j+1} = N\rho_j$ and $\epsilon_{j+1} = \eta\epsilon_j$
and set $j := j+1$ and go to Step 2.

We give an example in the following.
Example 3.1 Consider the problem (P110 in [10]):

$$(P3.1) \quad \min f(\mathbf{x}) = \sum_{j=1}^{n} x_j \ln x_j$$

$$\text{s.t.} \sum_{j=1}^{n} |sin(jt)| x_j \geq nt, \forall t \in T = [0, 1],$$

$$x_j \geq 0, j = 1, 2, ..., n.$$

(P3.1) is an entropy optimization problems with infinitely many linear constraints. As that in [10], we also use the simple discretization method to solve (P3.1). For each problem, we discretize $T = [0, 1]$ into m equal parts and obtain a constraint at $t = i/m, i =, 1, 2, ..., m$. We solve the following problem by Algorithm I.

$$(P3.1)' \min f(\mathbf{x}) = \sum_{j=1}^{n} x_j \ln x_j$$

$$\text{s.t. } g_i(\mathbf{x}) = (i/m)n - \sum_{j=1}^{n} |\sin(ji/m)|x_j), i = 1, 2, ..., m,$$

$$x_j \geq 0, j = 1, 2, ..., n.$$

Table 1.

n	m	No. iter.	ρ_k	e_0	Objective value of Algorithm I	Objective value in [10]
10	10	8	1280	5	4.718230	4.720
10	30	8	1280	5	4.718230	4.721
10	100	8	1280	5	4.719137	4.715
10	300	8	1280	5	4.721301	4.727
10	1000	8	1280	5	4.721945	4.708
30	10	8	1280	15	13.954221	13.950
30	30	9	2560	15	15.283094	15.280
30	100	9	2560	15	15.389501	15.393
30	300	8	1280	15	15.443326	15.454
30	1000	8	1280	15	15.474361	15.467
100	10	14	81920	40	48.341221	48.310
100	30	14	163840	40	53.403638	53.352
100	100	14	163840	40	58.633121	58.453
100	300	14	163840	40	58.553288	58.544
100	1000	14	163840	40	58.698932	58.696

Let $\mathbf{x}^0=(2,2,...,2)$, $\varepsilon=10^{-6}$, $\rho_0=2$, $\eta=0.5, and N=2$. The testing results of Algorithm I is gotten for various different starting ε_0 in Table 1. The solutions generated by Algorithm I are almost same as those in [10]. Table 1 means that we can obtain an approximate optimal solution to (P), and an equilibrium point of the dynamic system (3) of the neural network.

4 Conclusions

In this paper we have studied a new neural network which is a Hopfield-like network when applied to nonlinear optimization problems. An energy function of the neural network with its neural dynamics is constructed based on the method of penalty function. The system of the neural networks has been shown to be stable and its equilibrium point of the neural dynamics also yields an approximate optimal solution for nonlinear constrained optimization problems. An algorithm is given to find out an approximate optimal solution to its optimization problem, which is also an equilibrium point of the system. The numerical example shows that the algorithm is efficient. The convergence rate of algorithm depends on the chosen parameter ρ, ε.

Acknowledgement. This research was partially supported by SRG: 7001364 of City University of Hong Kong

References

1. Hopfield,J.J., Tank,D.W.: Neural computation of decision in optimization problems. Biological Cybernetics 58(1985)67-70. 1985
2. Joya,G., Atencia,M.A., Sandoval,F.: Hopfield neural networks for optimizatiom: study of the different dynamics. Neurocomputing 43(2002)219-237
3. Chen,Y.H., Fang,S.C.: Solving Convex Programming Problems with Equality Constraints by Neural Networks. Computers Math, Applic, 36(7)(1998)41-68
4. Staoshi Matsud: Optimal Hopfield Network for Combinatorial Optimization with Linear Cost Function, IEEE Tans. On Neural Networks, 9(6)(1998)1319-1329.
5. Youshe Xia, Jun Wang: A General Methodology for Designing Globally Convergent Optimization Neural Networks, IEEE Trans. On Neural Networks, 9(6)(1998)1331-1444
6. Zenios, S.A., Pinar,M.C., Dembo,R.S.: A smooth penalty function algorithm for network-structured problems. European J. of Oper. Res., 64(1993)258-277
7. Meng Zhiqing, Hu Qiying, Yang Xiaoqi: A General Model of Non-linear Neural Networks Based on Exact Penalty Function. Automatization (in Chinese), 29(2003)
8. Yang.X.Q.,Meng Z.Q.,Huang,X.X., Pong,G.T.Y.: Smoothing nonlinear penalty functions for constrained optimization, Numerical Functional Analysis Optimization, 24(2003)351-364
9. Lasserre,J.B.:A globally convergent algorithm for exact penalty functions, European Journal of Opterational Research,7 (1981)389-395
10. Shu-Chereng Fang,J,Rajasekera,R. , Tsao,H.S.J.: Entropy Optimization and Mathematical Proggramming. Kluwer,(1997)

Delay PCNN and Its Application for Optimization[*]

Xiaodong Gu[1,2], Liming Zhang[1], and Daoheng Yu[2]

[1]Department of Electronic Engineering, Fudan University, Shanghai,200433, P.R. China
guxiaodong@263.net
[2]Department of Electronics, Peking University, Beijing,100871, P.R. China

Abstract. This paper introduces the DPCNN (Delay Pulse Coupled Neural Network) based on the PCNN and uses the DPCNN to find the shortest path. Cauflield and Kinser introduced the PCNN method to solve the maze[1] and although their method also can be used to find the shortest path, a large quantity of neurons are needed. However, the approach proposed in this paper needed very fewer neurons than proposed by Cauflield and Kinser. Meanwhile, due to the parallel pulse transmission characteristic of the DPCNN, our approach can find the shortest path quickly. The computational complexity of our approach is only related to the length of the shortest path, and independent to the weighted graph complexity and the number of existed paths in the graph.

1 Introduction

Finding the shortest path can be used in routing select [2], web site searching [3], data dimensionality reduction[4], traffic control [5]. Some algorithms of finding the shortest path are developed, such as traditional Dijkstra algorithm. In 1999, Cauflield and Kinser introduced a new approach (C&K approach) for maze running by using PCNN in IEEE Transaction on Neural Networks[1].The C&K approach also can be used in finding the shortest path in graphs. By inspections, except the C&K approach no any other algorithm can both find the shortest path and do so with the minimum effort. However, using the C&K approach to find the shortest path in weighed graphs, a large amount of pulse coupled neurons are needed because per unit length of paths should corresponds one pulse coupled neuron. In this paper, we propose the Delay PCNN (DPCNN) and use it to find the shortest path with a few delay pulse coupled neurons. The DPCNN approach needs very fewer neurons than the C&K approach since only each node in weighed path graph corresponds a delay pulse coupled neuron. On the other hand, our DPCNN approach uses the parallel pulse transmission characteristic of the DPCNN succeeding to the PCNN to find the shortest path non-deterministically and quickly. The computational complexity of our approach is only related to the length of the shortest path, and independent to the weighted path graph complexity and the number of existed paths in the graph.

---

[*] This work was supported by China Postdoctoral Science Foundation (No.2003034282) and National Natural Science Foundation of China (No.60171036 and No.30370392).

In Section 2, the C&K approach is introduced, and its advantages and disadvantages are discussed. Meanwhile, why do the researches in this paper is explained. In Section 3, delay pulse coupled neuron models are developed from pulse coupled neurons and the DPCNN is described. In this paper, the DPCNN is used to find the shortest path. Other applications of the DPCNN could be developed further. In Section 4, the DPCNN algorithm of finding the shortest path is introduced, which both needs very fewer neurons than C&K approach and retains the advantages of the C&K approach completely. In Section 5, analyze the processing of the DPCNN algorithm by an example. Conclusions are obtained in Section 6.

2 C&K Approach and Why Introduce DPCNN Approach

The C&K approach uses the auto-wave generated by the PCNN to run mazes by taking all possible paths to obtain the shortest path. It is a novel non-deterministic method. In the C&K approach, each pixel in the maze corresponds a neuron so the number of neurons increases by the amplification of the maze. If the maze amplifies n times, although it does not change the shortest path, the number of neurons still increases n times. It costs too expensive, especially in hardware implementation. Obviously the C&K approach can be used to find the shortest path in graphs. Using the C&K approach to find the shortest path in graphs, per unit length of paths should corresponds one pulse coupled neuron, so a large amount of pulse coupled neurons are needed. If a PCNN includes too many neurons, it is difficult to implement the network by hardware. However, in order to exploit the potential of artificial neural networks, it would be necessary to implement the network by hardware.

In short, the C&K approach is a novel non-deterministic method. It is its advantages that it could find the shortest path quickly, and could do so with the minimum effort, and is only related to the length of the shortest path, and independent to the path graph complexity. It is its disadvantage that too many neurons are needed to find the shortest path in large mazes or graphs. In this paper, retaining the advantages of the C&K approach, the algorithm of finding the shortest path based on the DPCNN is proposed to hugely reduce the number of neurons.

Aiming at the disadvantage that C&K approach needs too many neurons to find the shortest path, we introduce the DPCNN and use it to find the shortest path with a few neurons. Using the DPCNN approach, each node in the weighed graph corresponds one neuron, whereas using C&K approach, per unit length of paths in the weighed graph should corresponds one pulse coupled neuron. For example, if the length of the path between A and B is 1000 and unit length of paths is 1, and no cross point exists between AB and other paths, using the C&K approach needs 1000 neurons, whereas using the DPCNN approach needs only 2 neurons. When the length of the path between A and B increases to 2000 and unit length of paths is still 1, using the C&K approach needs 2000 neurons, whereas using the DPCNN approach still needs only 2 neurons. If the graph magnifies n times, using the C&K approach, the number of neurons also magnifies n times; whereas using the DPCNN approach, the number of neurons is still as the same as original one before magnification. Therefore, compared with the C&K approach, the DPCNN approach can save neurons greatly.

3 DPCNN

According to the phenomena of synchronous pulse bursts in the cat visual cortex, Eckhorn (1990) introduced the linking field network[6]. In 1993 Johnson developed PCNN[7] from the linking model. PCNN, different from traditional neural networks, can be applied in many fields, such as image processing[8,9], optimization[1].

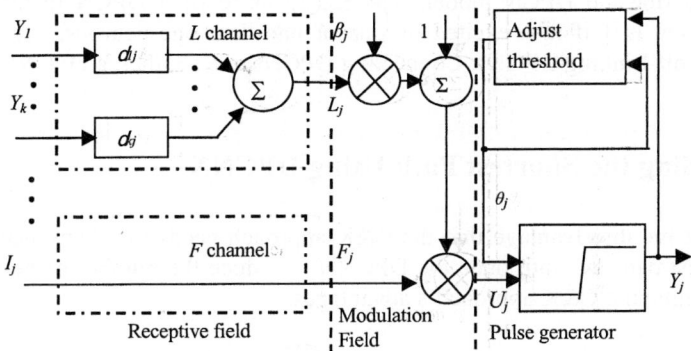

Fig. 1. A Delay Pulse Coupled Neuron (DPCN) model

$$F_j = I_j \tag{1}$$

$$L_j = \text{Step}[\sum_{k \neq j} Y_k(t - d_{kj})] = \begin{cases} 1, & \text{if } \sum_{k \neq j} Y_k(t - d_{kj}) > 0 \\ 0, & \text{else} \end{cases} \tag{2}$$

$$U_j = F_j(1 + \beta_j L_j) \tag{3}$$

$$\frac{d\theta_j(t)}{dt} = -\alpha_j^T + V_j^T Y_j(t), \quad \text{the lower limit of integration is just before the last firing in solution} \tag{4}$$

$$Y_j = \text{Step}(U_j - \theta_j) = \begin{cases} 1, & \text{if } U_j > \theta_j \\ 0, & \text{else} \end{cases} \tag{5}$$

We proposed the DPCNN based on the PCNN. Obtain the DPCNN by introducing delays to the PCNN. Like a Pulse Coupled Neuron (PCN), a Delay Pulse Coupled Neuron (DPCN) consists of three parts: the receptive field, the modulation field, and the pulse generator, see Fig.1. The DPCN receives input signals from other neurons and from external sources through the receptive field. There are two channels in the receptive field. One channel is feeding input (F_j); the other is linking input (L_j). In Fig.1, I_j, an inputs from external sources inputs to F channel. $Y_1,\ldots, Y_k$, output pulses of neuron $1,\ldots,K$ connecting to neuron j respectively, input to the L channel. $d_{1j},\ldots,d_{kj}$ are delays between neuron $1,\ldots,K$ and neuron j respectively. In modulation field, see Fig.1, the linking input L_j is added a constant positive bias firstly. Then it is multiplied by the feeding F_j input and the bias is taken to be unity, see Eq.(3). β_j is the linking strength. The total internal activity U_j is the result of modulation and it is inputted to the pulse generator. If U_j is greater than the threshold θ_j, the neuron output Y_j is 1 (namely the neuron j fires), see Eq.(5). Then Y_j feedbacks to make θ_j rise over U_j immediately so that Y_j turns into 0. Therefore, when U_j is greater than θ_j, neuron j outputs a pulse. The difference between DPCNs and PCNs is in the receptive field. In DPCNs, delays are introduced in L channels or F channels. When DPCNs are used for finding the shortest path, delays are introduced in L channels. The equations from (1) to (5) describe the DPCN model. See Eq.(2), as to each DPCN in this paper, the linking input is 1 if it is excited by one or one than one neurons. We called these DPCNs Unit-Linking DPCNs. Connecting DPCNs one another, a DPCNN appears.

4 Finding the Shortest Path Using DPCNN

Aiming at the disadvantage that the C&K approach needs too many neurons to find the shortest path, we introduce the DPCNN to reduce the number of neurons hugely with retaining the C&K approach's advantages.

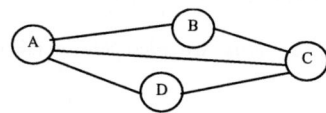

Fig. 2. The weighed graph with 4 nodes corresponding to 4 DPCNs

Using the DPCNN to find the shortest path in weighed graph, each node corresponds to one neuron. If there is a path between two neurons, each neuron's output pulse inputs to the other's L channel, and two symmetric delays between this two neurons equal to the length between them. As an example in Fig.2, there are 4 neurons corresponding to 4 nodes. A, B, C, D indicate respectively 4 neurons as well as 4 nodes. Because delays are symmetric, so $d_{AB}=d_{BA}=|AB|$, $d_{AC}=d_{CA}=|AC|$, $d_{AD}=d_{DA}=|AD|$, $d_{BC}=d_{CB}=|BC|$, $d_{CD}=d_{DC}=|CD|$. The disadvantage of the C&K approach that too many neurons are needed to find the shortest path in large mazes or graphs, is overcome by the DPCNN approach with introducing delays to the Linking channels of the PCNN. Using the DPCNN to find the shortest path in weighed graph, first the starting neuron fires and other neurons do not fire. Next the pulse emitted by starting neuron spreads all over the graph by each possible path, and makes other neurons fire to emit pulses continuously after corresponding delays equal to lengths of corresponding paths. Therefore, the pulse emitted by starting neuron spread over each path like flood. When the end neuron corresponding to the end point fires, the shortest path is obtained. So the computational complexity of DPCNN approach is only related to the length of the shortest path, and independent to the path graph complexity and the number of existed paths in the graph.

In solution process, each neuron only fire once at most, i.e. fired neurons do not emit pulse again, so that the pulse-coupled oscillation is avoided. It guarantees that each path is passed over by the pulse only once, and makes solution easy. That each neuron only fire once also accords with the fact that if each neuron is allowed to fire more than once, the length of the path over which the pulse emitted by the starting neuron at the second time is longer than that at the first time.

In the DPCNN approach, initial values of thresholds of all neurons are equal and they are 1.5. The threshold value of a neuron increase to 100 as soon as it have fired , so that it never fire again. The value of the feeding input of the starting neuron corresponding to the starting point is 2. The values of the feeding inputs of other neuron except the starting neuron are 1.The linking strength (β_j) of each neuron is 1.

The algorithm of finding the shortest path based on DPCNN is described below.

1. Initialization

    Initialize the threshold value, the feeding input value of each neuron. **Fla** is a two-dimension path log sheet and initialize each element of **Fla** 0. At the beginning, the starting neuron fires to emit a pulse and other neurons do not fire.

2. Pulse transmission

    Calculate the internal activity of each neuron and compare it with the corresponding threshold value to obtain the firing state of each neuron. If a neuron fires to emit an output pulse, the output feedbacks make the threshold value rise rapidly so that it never fires again during the processing.

3. Record the path

    If neuron j fires for being excited by the pulse emitted by neuron i, record **Fla**$[i][j]$=1. If neuron j is excited by more than one neurons at the same time, all results should be recorded to **Fla**.

4. If the neuron corresponding to the end point fires, go to step 5; else go to step 2.

5. Obtain the shortest path from **Fla**. End.

If in the 4 step in this algorithm the condition that "*if the neuron corresponding to the end point fires*" is changed to one that "*if all neurons fire*", all shortest paths from the staring point to all other point can be obtained from **Fla**. In a $N*N$ path log sheet **Fla** corresponding to *N*-node graph, if make each node be starting node in turn, using the algorithm described above with the end condition that "*if all neurons fire*" respectively, can obtain N path log sheets. Using these N path log sheets can obtain the shortest path between any two nodes. Obviously, we can get the shortest path between any two nodes from any *N-1* path log sheets of these N path log sheets.

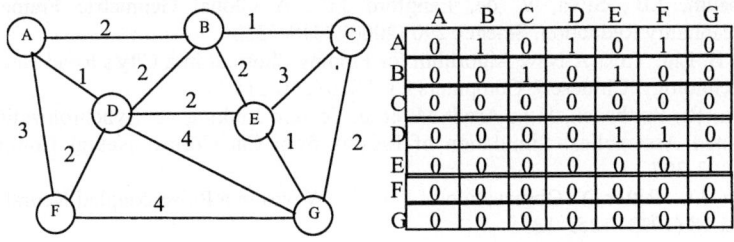

Fig. 3. An example of finding the shortest path

5 Results of Computer Simulations and Discussion

The results of computer simulations show that using DPCNN approach can rapidly find all shortest paths between starting point and end point accurately with a few neurons. An instance is shown in Fig.3. In Fig.3, there are 7 nodes and 13 edges and numbers over the edges indicate lengths between corresponding nodes. A is the starting point. Backtracking from end points to starting point A in the **Fla** in Fig.3, the shortest paths between A and all other nodes can be obtained. The shortest path between A and B is A-B (the length is 2); the shortest path between A and C is A-B-C (the length is 3); the shortest path between A and D is A-D (the length is 1); the shortest path between A and E is A-D-E(the length is 3).

6 Conclusions

In this paper, the DPCNN is introduced and used to find the shortest path in weighed graph to overcome the disadvantage of the C&K approach that too many neurons are needed to find the shortest path in large mazes or weighed graphs. Using the DPCNN approach can rapidly find all shortest paths between starting point and end point accurately with a few neurons. Compared with the C&K approach, the DPCNN approach described in this paper saves a large mount of neurons to find the shortest path with retaining the C&K approach's advantages that it could find the shortest path quickly, and do so with the minimum effort, and is only related to the length of the shortest path, and independent to the path graph complexity.

References

1. Caulfield, H.J., Kinser, J.M.: Finding Shortest Path in the Shortest Time Using PCNN's. IEEE Trans. on Neural Networks 10 (1999) 604-606
2. Ephremides, A., Verdu, S.: Control and Optimization Methods in Communication Network Problems. IEEE Trans. on Auto. Contr. 34 (1989) 930-942
3. Ricca, F., Tonella, P.: Understanding and Restructuring Web Sites with ReWeb, IEEE Multimedia, 8 (2001) 40-51
4. Tenenbaum, J.B., Silva, V. de, Langford J.C.: A Global Geometric Framework for Nonlineaonality Reduction. Science 290 (2000) 2319-2323
5. Yan, H.B. Liu, Y.C.: A New Algorithm for Finding Shortcut in a City's Road Net Based on GIS Technology. Chinese J. Computers 23 (2000) 210-215
6. Eckhorn, R., Reitboeck, H.J., Arndt M. et al: Feature Linking via Synchronization among Distributed Assemblies: Simulation of Results from Cat Cortex. Neural Computation 2 (1990: 293-307
7. Johnson, J.L., Ritter D.: Observation of Periodic Waves in a Pulse-coupled Neural Network. Opt.Lett.18 (1993) 1253-1255
8. Johnson, J.L., Padgett M.L.: PCNN Models and Applications. IEEE Trans. on Neural Networks 10 (1999) 480-498
9. Gu, X.D., Wang H.M., Yu D.H.: Binary Image Restoration Using Pulse Coupled Neural Network. The 8th International Conference on Neural Information Processing, Shanghai, China (2001) 922-927

A New Parallel Improvement Algorithm for Maximum Cut Problem

Guangpu Xia[1], Zheng Tang[1], Jiahai Wang[1], Ronglong Wang[2], Yong Li[1], and Guang'an Xia[3]

[1] Faculty of Engineering, Toyama University, Toyama-shi, Japan 930-8555
xiagp@hi.iis.toyama-u.ac.jp
[2] Faculty of Engineering, Fukui University, Japan 910-8507
[3] Daxing Branch School of Beijing Radio & Television University,
Beijing, China 102600

Abstract. The goal of maximum cut problem is to partition the vertex set of an undirected graph into two parts in order to maximize the cardinality of the set of edges cut by the partition. Enlightened by the elastic net method that was introduced by Durbin and Willshaw for finding shortest route for the Traveling Salesman Problem (TSP), we propose a new parallel algorithm for the maximum cut problem. A large number of instances are simulated to verify the proposed algorithm. The effectiveness of the proposed algorithm is confirmed by the simulation results.

1 Introduction

One of the best known combinatorial optimization graph problem is the maximum cut problem. Let $G = (V, E)$ be an edge-weighted undirected graph, where V is the set of vertices and E is the set of edges. The edge from vertex i to vertex j is represented by $e_{ij} \in E$. $d_{ij} = d_{ji}$ defines weights on edges whose endpoints are vertex i to vertex j. The maximum cut problem is to find a partition of V into two mutually exclusive sets A and B, such that $A \cup B = V$ and $A \cap B = \Phi$ and $\sum_{i \in A, j \in B} d_{ij}$ is maximum. In 1983, Hsu [1] developed a greedy algorithm to approximate the solution to the maximum cut problem on general graphs with arbitrarily weighted edges.

In this paper, we introduce a parallel algorithm analogous to elastic net method for the maximum cut problem. The procedure of the algorithm is similar to that of the Hopfield neural networks. But different to the Hopfield neural networks, we construct the energy function for the maximum cut problem using a nonlinear Gaussian function, which is enlightened by the elastic net method. A large number of randomly generated examples are simulated to verify the proposed algorithm. Simulation results are compared with the ones found by the algorithm of Hsu [1] and Lee et al. [2].

2 Description of the Proposed Algorithm

Let y_{ij} represent whether or not i-vertex ($i=1...N$) should be grouped into vertex subset j ($j =1,2$, which represent vertex set A and vertex set B). For example, the state (y_{i1}=1, y_{i2}=0) indicates that the i-vertex is grouped into vertex set A. The following states (y_{i1}=y_{i2}=0) and (y_{i1}=y_{i2}=1) express no partition and double partition violation, respectively. These partition violation conditions can be expressed by follow:

$$\sum_{i}^{N}(\sum_{j}^{2} y_{ij} -1)^2 = 0 \qquad (1)$$

The maximum cut condition can be expressed by follow:

$$Minimize \; (\sum_{i}^{N}\sum_{k \neq i}^{N}\sum_{j}^{2} d_{ik} y_{ij} y_{kj}) \qquad (2)$$

where d_{ik} is weight on edges whose endpoints are vertex i and vertex k, and is the symmetric matrix.

Enlightened by the elastic net method [4][5], we use a different energy function from that usually used in the Hopfield neural networks[3]. Our energy function is expressed as the following equation:

$$e = -KA \sum_{i}^{N} \phi(C_i, K) + B \sum_{i}^{N}\sum_{k \neq i}^{N}\sum_{j}^{2} d_{ik} y_{ij} y_{kj} \qquad (3)$$

where

$$\phi(C, K) = \exp\{-C^2/(2K^2)\}, \qquad (4)$$

and

$$C_i = \sum_{j=1}^{2} y_{ij} - 1 \qquad (5)$$

which is the i-term of the constraint condition (Eq.(1)) of the maximum cut problem. A and B are positive constants, and the K is a positive scale parameter. From Eq.(3), we can see that the first term of Eq.(3) is smallest if the constraint condition (Eq.(1)) is satisfied, and the second term of Eq.(3) is smallest if the maximum cut is largest.

In order to decrease the energy function (Eq.(3)), so as to find the optimal solution of the maximum cut problem, we can use the gradient descent rule of the Hopfield networks to update the y_{ij}. Takefuji and Lee [6] showed that McCulloch-Pitts neuron model [7] could guarantee an n-variable function converge to a local minimum. The McCulloch-Pitts neuron model used in Takefuji and Lee's theory is represented as :

$$y_{ij} = 1 \; if \; x_{ij} > 0, \; and \; y_{ij} = 0 \; if \; x_{ij} \leq 0 \qquad (6)$$

where x_{ij} is the input of neuron #ij and is updated according the following equation:

$$x_{ij}(t+1) = x_{ij}(t) - \frac{dx_{ij}}{dt} \cdot \Delta t \qquad (7)$$

where $\frac{dx_{ij}}{dt}$ is defined by following motion equation:

$$\frac{dx_{ij}}{dt} = -\frac{\partial e}{\partial y_{ij}} \qquad (8)$$

In the similar procedure with Takefuji and Lee, we use a sigmoid function as the input-output characteristics of the neurons:

$$y_{ij} = 1/\left(1 + e^{(-x_{ij}/T)}\right) \qquad (9)$$

The input x_{ij} to the neuron #ij is updated according the following rule.

$$x_{ij}(t+1) = x_{ij}(t) - \frac{dx_{ij}}{dt} \cdot \Delta t \qquad (10)$$

and the motion equation is defined by follow:

$$\frac{dx_{ij}}{dt} = -K \frac{\partial e}{\partial y_{ij}} \qquad (11)$$

where K is a positive scale parameter which is as same as that in Eq.(3). Using Eq.(3) we have:

$$\frac{\partial e}{\partial y_{ij}} = AC_i \phi \, (C_i, K)/K + B \sum_{k \neq i}^{N} d_{ik} y_{kj} \qquad (12)$$

First, we will prove that our updating rule (Eq.(9) - Eq.(11)) decreases the energy function (Eq.(3)) with the time evolution-collective computational properties.
Consider the derivatives of the energy function e (Eq.(3)) with respect to time t,

$$\frac{de}{dt} = \sum_{ij} \frac{de}{dy_{ij}} \frac{dy_{ij}}{dt} \qquad (13)$$

Using Eq.(11) we have:

$$\frac{de}{dt} = -\frac{1}{K} \sum_{ij} \frac{dy_{ij}}{dt} (\frac{dx_{ij}}{dt}) = -\frac{1}{K} \sum_{ij} \frac{dx_{ij}}{dt} \frac{dy_{ij}}{dx_{ij}} \frac{dx_{ij}}{dt} = -\frac{1}{K} \sum_{ij} (\frac{dy_{ij}}{dx_{ij}}) \left(\frac{dx_{ij}}{dt}\right)^2 \qquad (14)$$

Since y_{ij} is a monotone increasing function of x_{ij} (the sigmoid function), $\frac{dy_{ij}}{dx_{ij}}$ is positive, and K is a positive parameter, each term in Eq.(14) is nonnegative. Therefore

$$\frac{de}{dt} \leq 0 \quad \text{for all } i, j \tag{15}$$

Together with the bound of e, Eq.(15) shows that the time evolution of the system is a motion in state space that seeks out minima in e and comes to stop at such points.

We have proved that for fixed K ($K>0$), the update (Eq.(9) - (11)) would result in a convergence to a local minimum of the energy function e. Now we discuss the behavior of the energy function as the constant K changes. Informally, the first term of Eq.(3) tends to impel the solutions to satisfy the constraints, and the second term of Eq.(3) tries to make the number of removed edges small. Furthermore, it has been proved [5] that because the Gaussian function Eq.(4) is a positive bounded function, at large values of K the energy function is smoothed and there is only one minimum. At small values of K, the energy function contains many local minima, all of which correspond to possible solutions to the problem, and deepest minimum is the optimal solution. Thus, in the same way as the elastic net method, our algorithm proceeds by starting at large K, and gradually reducing K, keeping to a local minimum of e. We would like this minimum that is tracked to remain the global minimum as K becomes small. In our algorithm, when K tends to zero, for e to remain bounded and for every i, $\sum_{j=1}^{2} y_{ij} - 1$ tends to zero, i.e., the constrains must be satisfied. Then the second term (i.e. the cost term) in the expression for e is minimized, and finally, a feasible solution (a local or global minimum) is reached by the second term.

3 Simulation Results

The proposed algorithm was implemented in C++ to a large number of graphs. The parameter values used in the simulations were $A=3$, $B=0.06$, $T=1.42$, the initial value of K was 0.65 and was reduced by 99% every 5 iterations to a final value in the range 0.08-0.09.

The first graph we tested was a graph with 20 vertices and 30 edges which is shown in Fig.1(a). Figure 1(b) shows the results of a simulation on the max cut problem of Fig.1(a). We then gave a typical progressive intermediate solution during the variation of K to illustrate how the proposed algorithm performed in the process of finding good solution.

In order to widely verify the proposed algorithm, we have also tested the algorithm with a large number of randomly generated graphs defined in terms of two parameters, n and p. The parameter n specified the number of vertices in the graph; the parameter p, $0<p<1$, specified the probability that any given pair of vertices constitutes an edge. Integer numbers were given randomly on edges as weight. The range for weight was from -1 to 5. In the experiments, up to 300-vertex graphs with different probability were used to evaluate the proposed algorithm. The simulation result were also compared with that found by the algorithm of Hsu [1] and Lee et al. [2]. For each of instances, 100 simulation runs were performed. Information on the test

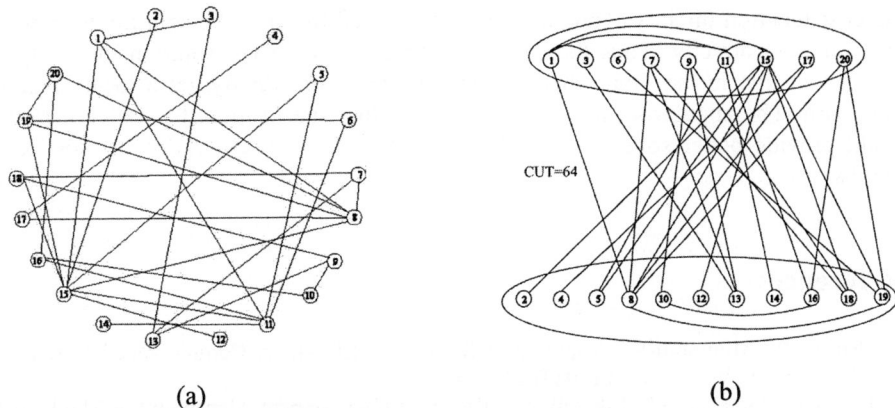

Fig. 1. (a) The max cut problem with 20 vertex and 30 edges. (b) The solution found by proposed algorithm

graphs as well as all results are shown in Table 1. The results that we recorded for each graph are the size of maximum cut produced by the algorithm of Hsu and Lee et al., and by the proposed algorithm, respectively. Table 1 shows that the proposed parallel method could find a better solution than Hsu and Lee et al.'s algorithm in all problems.

Table 1. Comparison of results produced by the algorithm of Hsu, Lee et al., and our algorithm

No. vertex	Probability	No. edges	Lee et al.	Hsu	Propose algorithm
100	0.05	247	467	459	486
100	0.15	742	1145	1111	1178
100	0.25	1235	1765	1747	1781
200	0.05	995	1664	1633	1664
200	0.15	2985	4256	4180	4263
200	0.25	4975	6696	6650	6747
300	0.05	2242	3464	3451	3487
300	0.15	6727	9082	9076	9147
300	0.25	11212	14459	14418	14466

4 Conclusions

Enlightened by the elastic net method, we have demonstrated a successful parallel algorithm of solving the maximum cut problem, and showed its effectiveness by simulation experiments. The algorithm used two different expressions for the constraint condition and the cost function in energy function. The former used a Gaussian function, which is strongly dependent on the scale parameter K, and the latter used the cost function directly. Thus the network worked to force the solution to used the cost function directly. Thus the network worked to force the solution to used the cost function directly. Thus the network worked to force the solution to satisfy the constraints first in the limit where K tends to zero, and finally get a local or global minimum. The simulation results also showed that the proposed parallel algorithm could provide better solutions than Hsu and Lee et al.'s algorithms for most tested graphs.

References

1. Hsu, C.-P.: Minimum-Via Topological Routing. IEEE Trans. Compu.-Aided Des. Integr. Circuits Syst., Vol. 2, No. 4 (1983) 235–246
2. Lee, K.C., Funabiki, N., Takefuji, Y.: A Parallel Improvement Algorithm For The Bipartite Subgraph Problem. IEEE Trans. Neural Networks, Vol. 3, No. 1 (1992) 139–145
3. Hopfield, J.J., Tank, D.W.: "Neural" Computation Of Decisions In Optimization Problems. Bio.Cybern, No. 52 (1985) 142–152
4. Durbin, R., Willshaw, D.: An Analogue Approach Of The Traveling Salesman Problem Using An Elastic Net Method. Nature, 326 (1987) 689–691
5. Durbin, R., Szeliski, R., Yuille, A.: An Analysis Of The Elastic Net Approach To The Traveling Salesman Problem. Neural Computation, Vol. 1 (1989) 348–358
6. Takefuji, Y., Lee, K.C.: Artificial Neural Networks For Four-Coloring Map Problems And K-Colorability Problems. IEEE Trans. Circuits Syst., Vol. 38, No. 3 (1991) 326–333
7. McCulloch, W.S., Pitts, W.H.: A Logical Calculus Of Ideas Immanent In Nervous Activity. Bull. Math. Biophys, Vol. 5 (1943) 115–133

A New Neural Network Algorithm for Clique Vertex-Partition Problem

Jiahai Wang, Xinshun Xu, Zheng Tang, Weixing Bi, Xiaoming Chen, and Yong Li

Faculty of Engineering, Toyama University, Toyama-shi, 930-8555 Japan
wjiahai@hotmail.com

Abstract. In this paper, by adding a nonlinear self-feedback to the maximum neural network (MNN), we propose a new algorithm for the clique vertex-partition problem that introduces richer and more flexible dynamics and can prevent the network from getting stuck at local minima. A large number of instances have been simulated to verify the proposed algorithm.

1 Introduction

A clique of a graph $G=(V, E)$ with a set of vertices V and a set of edges E is a complete subgraph of G where any pair of vertices is connected with an edge in E. A clique vertex-partition problem of a graph G is to partition every vertex in V into a set of disjointed cliques of G. It is known that the problem of partitioning vertices with the minimum number of disjointed cliques in an arbitrary graph is NP-complete [1] [2].

Tseng and Siewiorek [3] proposed an *ad hoc* algorithm for the clique vertex-partition problem in an arbitrary graph. But they neither discussed the time complexity nor the solution quality of their algorithm. Funabiki et al. [1] proposed the first neural network parallel algorithm for solving the clique vertex problem in an arbitrary graph. For solving such combinatorial optimizations, Takefuji et al. [4][5] proposed a maximum neural network (MNN) model and applied it for several NP-complete problems. The MNN always guarantees a valid solution and greatly reduces the search space without a burden on the parameter-tuning. However, the model causes the excessive fixation of the state of the neural network and has a tendency to converge to local minimum easily because it is based on the steepest descent method [6].

In this paper, by adding a nonlinear self-feedback to the MNN, we propose a new algorithm for the clique vertex-partition problem that introduces richer and more flexible dynamics and can prevent the network from getting stuck at local minima. After the nonlinear self-feedback vanishes, the proposed algorithm is then fundamentally reined by the gradient descent dynamics and usually converges to a stable equilibrium point. A large number of instances have been simulated to verify the proposed algorithm.

2 The Proposed Algorithm for Clique Vertex-Partition Problem

The n-vertex m-partition problem can be mapped onto the MNN with $n \times m$ neurons, where it consists of n clusters of m neurons. The ijth neuron has an input U_{ij} and output V_{ij} (where $i=1, \ldots, n, j=1, \ldots, m$). When the output of the ijth neuron is non-zero ($V_{ij}=1$), the ith vertex is partitioned into the jth clique. When the output of ijth neuron is zero ($V_{ij}=0$), the ith vertex is not partitioned into the jth clique. Two constraints must be satisfied in the clique vertex-partition problem. The first constraint is that each vertex must be partitioned into one and only one clique. The second constraint is that any pair of vertices in each clique must be connected with an edge which is given in the original graph. We can formulate the clique vertex-partition problem as the global minimization of the energy function:

$$E = \sum_{j=1}^{m} \sum_{i=1}^{n} \sum_{\substack{k=1 \\ k \neq i}}^{n} (1 - d_{ik}) V_{ij} V_{kj} \qquad (1)$$

where d_{ik} is 1 if the original graph has an edge between the ith vertex and kth vertex, 0 otherwise, and it is always satisfied that $d_{ik} = d_{ki}$ and $d_{ii}=0$. If all the vertex partitions are clique, the energy function E is zero, that is, the second constraint is satisfied. The motion equation of the ijth neuron without the decay term to minimize the energy function is given by:

$$\frac{dU_{ij}}{dt} = -\frac{\partial E}{\partial V_{ij}} = -\sum_{\substack{k=1 \\ k \neq i}}^{n} (1 - d_{ik}) V_{kj} \qquad (2)$$

However, in practice, they are approximated by the first-order Euler method in the form:

$$\Delta U_{ij} = -\sum_{\substack{k=1 \\ k \neq i}}^{n} (1 - d_{ik}) V_{kj} \qquad (3)$$

Therefore, the input U_{ij} for each neuron of the MNN is updated using the following equation based on the first-order Euler method [4][5]:

$$U_{ij}(t+1) = U_{ij}(t) + \Delta U_{ij}(t) \qquad (4)$$

until the network reaches an equilibrium state, and the input/output function of the ijth neuron is given by:

$$\begin{array}{c} \text{If } U_{ij} = \max(U_{i1}, U_{i2}, \ldots U_{im}) \\ \text{then } V_{ij}=1 \text{ else } V_{ij}=0 \end{array} \qquad (5)$$

where the function max( ) returns the first argument with the maximum value. The maximum neuron model always forces one and only one neuron among m neurons to have 1 as the output, that is, each vertex is partitioned into one and only one clique, so the first constraint automatically satisfied.

The advantage of this formulation for the clique vertex-partition problem using the MNN is that the constraint is always automatically satisfied. In this way, the clique vertex-partition problem reduces to obtain a minimum of the energy function without the parameters affecting the global minimum search. At the same time, the maximum neuron model limits the searching space and so reduces the computation load. However, the model causes the excessive fixation of the state of the neural network and has a tendency to converge to local minimum easily because it is based on the steepest descent method [6]. Here, we propose a new parallel algorithm by adding a negative self-feedback to the MNN, as defined below:

$$U_{ij}(t+1) = U_{ij}(t) + \alpha \Delta U_{ij}(t) - z(t)(V_{ij}(t) - \frac{1}{2}) \qquad (6)$$

$$z(t+1) = (1-\beta)z(t) \qquad (7)$$

where α is a positive scaling parameter for inputs, and $z(t)$ is a self-feedback connection weight ($z(t) \geq 0$). The same $z(t)$ value is used for all neurons at t during the updating process. β is a damping factor of the time-dependent $z(t)$ ($0 \leq \beta \leq 1$).

In the proposed algorithm, the updating of input U_{ij} of each neuron is related to the output V_{ij} of itself. The self-feedback term, $-z(t)(V_{ij}(t) - \frac{1}{2})$, becomes negative after the ij-neuron firing at time t (that is, $V_{ij}(t) = 1$). The negative self-feedback inhibits the growth of the input value of neurons at next time, $t+1$. On the other hand, the self-feedback term becomes positive if the ij-neuron is unfired at time t (that is, $V_{ij}(t) = 0$). The positive self-feedback promotes the growth of the input value of neurons at next time, $t+1$. Thus, we can say that the proposed algorithm includes refractory effect, which is one of the characteristics of real biological neurons [7][8]. Each neuron with refractory effect becomes difficult to fire after previous firings. In the expanded maximum neuron model as described by Eq.(2), the input values of all neurons in the same cluster, are compared with each other to extract maximum value. In other words, neurons in the same cluster compete with each other to be fired. Since the self-feedback can inhibit or promote the growth of the input value of neurons in the way mentioned above, it encourages this competition and easily leads to the transition of output state of all neurons. This mechanism contributes to the blocking of continuous firings and encourages other neurons to be eventually fired. Therefore the network can easily escape from a fixed point or a local minimum by changing the output state of neurons in the MNN.

Table 1. Comparison between the algorithm of Funabiki, MNN, and the proposed algorithm for the clique vertex-partition problem

$|V|$	$|E|$	$|C|$	Funabiki		MNN		Proposed algorithm		
			Rate	Step	Rate	Step	Rate	Step	T
8	17	3	90	29	60	10	100	50	0.01
25	162	6	10	190	7	11	100	320	0.01
50	306	17	19	181	9	11	99	350	0.05
50	631	10	11	269	8	12	99	355	0.06
50	939	6	18	184	9	10	97	360	0.06
100	2475	18	24	245	8	11	96	420	0.23
150	5618	26	50	216	8	12	98	421	0.55
200	9968	32	31	227	7	10	97	423	1.02
250	15577	39	55	221	4	11	98	425	1.72
300	22484	46	88	183	2	12	99	424	3.10

In the motion equation (9), the α term means decreasing of the objective function value, which yields a better solution that is closer to a local minimum. The self-feedback term induces a jump from a searching region to other regions in order to avoid getting trapped in a single local minimum; thus the system can generate a new solution.

Given a randomly generated numbers to the initial state of $U_{ij}(0)$ and a sufficiently large initial self-feedback $z(0)$, the competition or the state transition of neurons is drastic when iterating Eq.(5) and Eq.(6). When the self-feedback connection weight $z(t)$ tends toward zero with time evolution according to Eq.(7), the network eventually reduces to the MNN without self-connections which is always allowed to converge valid and good solution. As long as the value of the energy function E is zero, the procedure will terminate.

3 Simulation Results

In order to assess the effectiveness of the proposed algorithm, extensive simulations were implemented in C on AOPEN-PC (PentiumIII, 866 MHz). The parameter values used in the simulations were α=0.013, $z(0)$ =0.08, and β=0.004.

The algorithm of Funabiki and the MNN were also executed for comparison. All methods were executed by 100 simulation runs with different initial state on each graph. As reference [1], under the assumption that the number of clique ($|C|$) is given, ten clique vertex-partition problems in the Table 1 were examined where the number of vertices ($|V|$) and the number of edges ($|E|$) in the graphs were varied from 8 to 300 and from 17 to 22484, respectively. All edges in the graphs were randomly generated.

The number of iteration step was denoted by the number of updating all input U_{ij} and output values V_{ij}. Table 1 summarizes the simulation results where the average number of iteration steps (Step) to converge to solution and the convergence rate (Rate%) are shown. The computation time (T) of proposed algorithm is the average of 100 simulations. The simulation results show that the proposed algorithm converged to the solution by nearly 100% convergence rate in reasonable computation time.

4 Conclusions

By adding a negative self-feedback to the MNN, we proposed a new algorithm for the clique vertex-partition problem that had richer and more flexible dynamics and could prevent the network form getting stuck at local minima. The simulation results showed that the proposed algorithm had superior ability to find a solution of the clique vertex-partition problem over other neural network methods.

References

1. Funabiki, N., Takefuji, Y., Lee, K.C.: A Neural Network Parallel Algorithm for Clique Vertex-Partition Problems. International Journal of Electronics, Vol. 72, No. 3 (1992) 357–372
2. Garey, M.R., Johnson, D.S.: Computers and Intractability: A Guide to the Theory of NP-Completeness. New York: Freeman, W.H.
3. Tseng, C.J., Siewiorek, D.P.: Facet: A Procedure for the Automated Synthesis of Digital Systems. Proceeding of the 20th ACM/IEEE Design Automated Conference (1983) 490–496
4. Takefuji, Y., Wang, J.: Neural Computing for Optimization and Combinatorics. Singapore: World Scientific (1996)
5. Takefuji, Y., Lee, K., Aiso, H.: An Artificial Maximum Neural Network: A Winner-Take-All Neuron Model Forcing the State of the System in a Solution Domain. Biological Cybernetics, Vol. 67 (1992) 243–251
6. Takenaka, Y., Funabiki, N., Higashino, T.: A Proposal Neural Filter: A Constraint Resolution Scheme of Neural Networks for Combinatorial Optimization Problems. IEICE Trans. Fundamentals, Vol. E83-A, No. 9 (2000) 1815–1823
7. Ikeguchi, T., Aihara, K., Itoh, K.: A Novel Chaotic Search for Quadratic Assignment Problems. European Journal of Operational Research, Vol. 139 (2002) 543–556
8. Chen, L., Aihara, K.: Chaotic Simulated Annealing by a Neural Network Model With Transient Chaos. Neural Networks, Vol. 8, No. 6 (1995) 915–930

An Algorithm Based on Hopfield Network Learning for Minimum Vertex Cover Problem

Xiaoming Chen [1,2], Zheng Tang[1], Xinshun Xu[1], Songsong Li[3], Guangpu Xia[1], and Jiahai Wang[1]

[1] Faculty of Engineering, Toyama University,
930-8555, Toyama, Japan
xmchen1@hotmail.com
[2] Faculty of Information, ShenYang University of Technology,
110023, Shenyang, China
[3] Faculty of Engineering, Toyama Prefectural University,
939-0398, Toyama, Japan

Abstract. An efficient algorithm for the minimum vertex cover problem based on Hopfield neural network leaning is presented. The learning algorithm has two phases, the Hopfield network phase and the learning phase. When network gets stuck in local minimum, the learning phase is performed in an attempt to fill up the local minimum valley by modifying parameter in a gradient ascent direction of the energy function. The proposed algorithm is tested on benchmark graphs. The simulation results show that the proposed algorithm is an effective algorithm for the minimum vertex cover problem in terms of the computation time and solution quality.

1 Introduction

Given an undirected graph $G(V, E)$ with a vertex set V and an edge set E, the minimum vertex cover problem is to find a smallest subset $V' \subseteq V$ such that for each edge(a, b) in G $a \in V'$ or $b \in V'$ (or both), V' is said to be a vertex cover of G [1]. The minimum vertex cover problem is a problem of central importance in computer science.

In a landmark paper, Karp has shown that the vertex cover problem is NP-complete [2]. A very simple approximation algorithm based on maximal matching gave an approximation ratio 2 for the general graphs. More recently, parameterized algorithms for the k-vertex cover problem have further drawn researchers' attention, and continuous improvements on the problem have been developed. Buss's algorithm was improved to O(kn+2kk2) by Downey and Fellows. Recently, Frank Dehne et al. have reported that they used fixed parameter tractable algorithm to solve the minimum vertex cover problem on coarse grained parallel machines successfully [3].

In this paper, we introduce an efficient algorithm for the general minimum vertex cover problem based on the Hopfield network learning. The learning algorithm has two phases, the Hopfield network phase and the learning phase. The Hopfield network phase performs gradient descent in state domain, and finds a set of states minimizing the Hopfield network energy. When the Hopfield network gets stuck in local minimum, the learning phase is performed in an attempt to fill up the local minimum valley by modifying parameter in a gradient ascent direction of the energy function. The two phases are repeated until the global minimum or a better solution is obtained. We show here its effectiveness by extensive computational experiments for large random graphs and benchmark graphs.

2 Formulation

For a given undirected graph $G=(V, E)$, V is vertex set and E is edge set. We let $|V|=n$, $|E|=m$. The variables d_{ij} ($i=1,2,...,n$, $j=1,2,...,n$) are binary variables which form the adjacency matrix of graph G. The two values of d_{ij} express the connection exists or not. So the number of vertices of the given graph determines the number of neurons. For example, if there are n vertices in the graph, then n neurons are required. Thus the number of vertices in the cover can be expressed by:

$$E_1 = \sum_i v_i \tag{1}$$

If an edge (i, j) is not covered by a cover, then both v_i and v_j are zero. Then the constrained condition can be written as:

$$E_2 = \sum_i \sum_j d_{ij} \overline{v_i \vee v_j} \tag{2}$$

The objective of the problem is to minimize E_1 with E_2 equal to zero [8]. Thus, the energy function for the minimum vertex cover problem can be described as follow:

$$E = A \sum_i \sum_j d_{ij} \overline{v_i \vee v_j} + B \sum_i v_i \tag{3}$$

Because the last term is a constant, the energy can be written as:

$$E = A \sum_i \sum_j d_{ij} v_i v_j - A \sum_i \sum_j d_{ij} (v_i + v_j) + B \sum_i v_i \tag{4}$$

We can deduce the weights and thresholds of the Hopfield network as:

$$w_{ij} = -2A d_{ij} \qquad h_{ij} = 2 \sum_i d_{ij} - B \tag{5}$$

For the Hopfield neural network, the motion equation is composed of the partial derivation term of the energy function as the gradient descent method.

$$du_i / dt = \sum_{j=1}^{n} w_{ij} v_j + h_i \tag{6}$$

In this paper, we use the following equations to update the neurons.
The internal potential of neuron at time $t+1$ is updated by:

$$u_i(t+1) = u(t) + \frac{du_i(t)}{dt} \tag{7}$$

The neuron state v_i is updated from u_i using the sigmoid function.

3 Gradient Ascent Learning

Hopfield showed that the Hopfield network is guaranteed to converge with energy taking on lower and lower values until it reaches a stable state [5].

In order to help the Hopfield network find a better solution for the minimum vertex cover problem, we proposed a learning method for the Hopfield network (gradient ascent learning) that can help the network escape from local minima. Because energy of the Hopfield neural network is determined by its weights and thresholds, once the Hopfield neural network reaches a stable state, the learning is performed by changing the weights and thresholds of the Hopfield network in gradient ascent direction so that energy at the local minimum increases. Since for a parameter vector $\vec{W}$, the learning requires the parameter to change in the positive gradient direction, we take: (ε is a positive constant.)

$$\Delta \vec{W} = \varepsilon \nabla E(\vec{W}) \tag{8}$$

Then we can obtain: (p and q are small positive constants.)

$$\Delta w_{ij} = p \frac{\partial E}{\partial w_{ij}} = -p v_i v_j \qquad \Delta h_i = q \frac{\partial E}{\partial h_i} = -q v_i \tag{9}$$

Thus the new weights and thresholds become:

$$w'_{ij} = w_{ij} - p v_i v_j \qquad h'_i = h_i - q v_i \tag{10}$$

The change of energy ΔE due to the changes of the weights and thresholds can be described as:

$$\Delta E = -\frac{1}{2}\sum_i \sum_j w'_{ij} v_i v_j - \sum_i I'_i v_i - (-\frac{1}{2}\sum_i \sum_j w_{ij} v_i v_j - \sum_i I_i v_i) = \frac{1}{2} p \sum_i \sum_j v_i^2 v_j^2 + q \sum_i v_i^2 \tag{11}$$

Because p and q are positive constants and v_i is between zero and 1, we can conclude $\Delta E > 0$.

In order to explain the learning method, we use a two-dimensional graph (Fig.1) of the energy function with a local minimum and a global minimum. The energy function value is reflected in the height of the graph. Each position on the energy terrain corresponds to a possible state of the network. For example, if the network is initialized onto the mountainous terrain A, the updating procedure of the Hopfield network makes the state of network move towards a minimum position and reach a steady state B (Fig.1(a)). In equation (11) we have shown that after we change the weights and thresholds the energy will increase. That is to say, the gradient ascent learning makes the previous stable state B becomes a point on the slope of a valley (B'). After updating the Hopfield network with the new weights and thresholds in the Hopfield network updating phase again, point B' goes down the slope of the valley and reaches a new stable state C (Fig.1(b)). Thus, the Hopfield network updating (Phase 1) and the gradient ascent learning (Phase 2) in turn may result in a movement out of a local minimum, and lead the network converge to a global minimum or a new local minimum (Fig.1(c) and (d)).

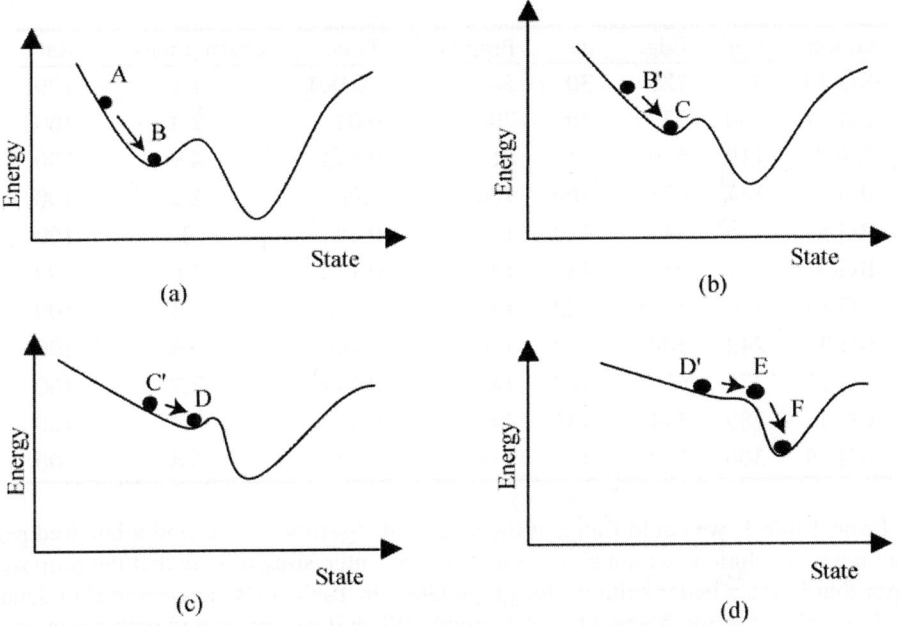

Fig. 1. The conceptual graph of the relation between energy and state transition in the learning process of the Hopfield network with two stable states

4 Simulation Results

In order to assess the effectiveness of the proposed algorithm, extensive simulations were carried out on the randomly generated graphs and the benchmark graphs for the vertex cover problem. For the randomly generated graphs, the solution quality of the proposed algorithm was much better than that of the ratio-2 algorithm, the proposed algorithm competed well with the original Hopfield network and yielded better solutions in most cases. But for the size of this paper, we will not discuss it in detail. The algorithm was implemented in C++ on Pentium 4 1.4G (256M).

The type of graphs we tested was the benchmark graph. Procedures for generating test cases with known answers were described in [4]. In [7] the generating procedures for vertex cover were evaluated experimentally on several approximation algorithms. In [3], some graphs generated by [7] were used to test their algorithm for the vertex cover problem. Here we used some of these graphs in [3] to test the proposed algorithm. The results are shown in Table 1. The algorithm ran 100 times for each graph. In Table 1, |k| refers the size of the optimal solution, and the best solution obtained by the proposed algorithm, the CPU time, learning times and the rate of optimal solutions were given.

Table 1. The simulation results for benchmark graphs

Graph	Ver.	Edg.	|k|	Proposed	Time	Learn_times	Rat
RG.20	50	225	30	30	<0.001	1.1	100
RG.21	100	2370	70	70	0.01	2.3	100
G.408	119	569	81	81	0.113	4.9	100
RG.1	147	675	100	100	0.06	2.2	100
RG.5	175	844	120	120	0.104	3	100
RG.6	192	933	131	131	0.176	3.6	100
RG.14	220	2155	122	122	0.336	5.3	100
RG.3	242	500	120	120	0.106	0.6	100
RG.13	275	675	142	142	0.409	3.7	100
GG.1	289	544	145	144	0.414	3.4	100
RG.15	306	739	148	148	0.945	5.4	100

From Table 1, we could find that the proposed algorithm could find a hundred percent optimal solutions within short time. It is very interesting to note that the proposed algorithm found a better solution for graph GG.1 in Table 1. By analyzing it in detail, we found that this graph was a bipartite graph. When it was generated with a minimum vertex cover of 145 sizes, and the left vertices was also a vertex cover which has smaller size. So we think this graph presented by Frank Dehne was not correct, and the size of the minimum vertex cover should be 144.

In order to gain further insight into the optimization process, in Fig.2 we plotted the variation of energy during the Hopfield network updating phase and the gradient ascent learning phase running on G.408 of Table 2. For this graph the proposed algorithm used four learning times to get the optimal solution which has 81 vertices. From Fig.3, we can see that the Hopfield network updating decreased energy from the initial point A to point A', and generated a local optimum solution with size 89. Then the learning phase raised A' to B and then the second Hopfield network updating decreased B to B', and so did C to C'. But the first three learning did not produce a better solution (A', B' and C'). That is to say the network work did not escape from this local minimum. But the 4th learning produced the optimal solution (E') with size 81 was obtained. We can see clearly that every learning made energy increased and the Hopfield network updating made energy decreased.

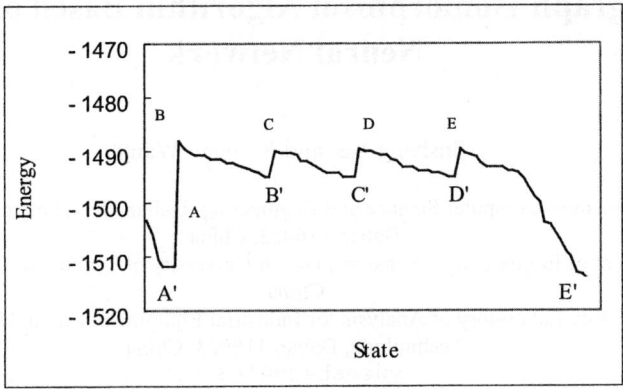

Fig. 2. The graph shows the variation of energy when processing graph G.408 of Table 1

5 Conclusions

We have proposed an efficient algorithm for the minimum vertex cover problem based on the Hopfield neural network and showed its effectiveness by simulation experiments. The simulation results showed that the learning method was effective to help the Hopfield network escape from local minima, and the proposed algorithm was an exactly effective algorithm in terms of computation time and the solution quality.

References

1. Khuri, S., Back, T.: An Evolutionary Heuristic for the Minimum Vertex Cover Problem. KI-94 Workshops (Extended Abstracts), Bonn (1994) 83–84
2. Karp, R.M.: Reducibility Among Combinatorial Problems. Complexity of Computer Computation, Plenum Press, New York (1972) 85–103
3. Dehne, F. et al.: Solving Large FPT Problems on Coarse Grained Parallel Machines. Available: http://www.scs.carleton.ca/~fpt/papers/index.htm
4. Sanchis, L.A.: Generating Hard and Diverse Test Sets for NP-hard Graph Problems. Discrete Applied Mathematics, Vol. 58 (1995) 35–66
5. Hopfield, J.J., Tank, D.W.: Neural Computation of Decisions in Optimization Problems. Bio. Cybern., No. 52 (1985) 141–152
6. Downey, R.G., Fellows, M.R.: Fixed Parameter Tractability and Completeness II: Completeness for W [1]. Theoretical Computer Science, Vol. 141 (1995) 109–131
7. Sanchis, L.A.: Test Case Construction for the Vertex Cover Problem. DIMACS Series in Discrete Mathematics and Theoretical Computer Science, Vol. 15 (1994) 315–326
8. Cao, Q.P., Tang, Z., Wang, R.L.: A Gradient Ascent Learning Algorithm in Weight Domain for Hopfield Neural Network. IEEJ Trans. EIS, Vol. 122, No. 4 (2002) 677–683

A Subgraph Isomorphism Algorithm Based on Hopfield Neural Network

Ensheng Yu[1] and Xicheng Wang[2,3]

[1]Department of Computer Science and Engineering, Dalian University of Technology,
Dalian 116023, China
[2]Department of Engineering Mechanics, Dalian University of Technology, Dalian 116023,
China
[3]State Key Laboratory of Analysis for Industrial Equipment, Dalian University of
Technology, Dalian 116023, China
yuensheng@163.com

Abstract. A subgraph isomorphism algorithm using 2D continuous Hopfield neural network is presented in this paper. Given two graphs G_1 and G_2, the goal is to find a subgraph of G_2 isomorphic to G_1. The rows of the 2D neuron array represent the vertices of G_1, and the columns represent those of G_2. The energy function is defined. The network parameters are deduced from the energy function. The neurons are initialized based on the necessary conditions for subgraph isomorphism. The motion equation is solved using the fourth order Runge-Kutta method. Experimental results show the correctness and validity of the algorithm.

1 Introduction

An undirected graph G consists of a finite set of vertices $V(G)$ and a finite set of edges $E(G)$ joining the vertices. Let p and q denote, respectively, the number of vertices and edges in G, that is, $p=|V(G)|$ and $q=|E(G)|$. v_m is used to denote the mth vertex of G in this paper for simplicity. The adjacency matrix of G, denoted by $A(G)$, is a $p \times p$ matrix $A(G)=(a_{ij})$, where $a_{ij}=1$ if v_i is adjacent to v_j, else $a_{ij}=0$. A subgraph of G is a graph G' such that $V(G') \subseteq V(G)$ and $E(G') \subseteq E(G)$.

Two graphs G_1 and G_2 are isomorphic if there is a one-to-one correspondence between their vertices and another one between their edges that together preserve incidence and adjacency. Two graphs G_1 and G_2, where $|V(G_1)|<|V(G_2)|$, are subgraph isomorphic if there exists a subgraph G_2' of G_2 such that G_1 and G_2' are isomorphic.

The problem of subgraph isomorphism (SI for short) is known to be NP-complete [1]. Some algorithms have been proposed in the past three decades most of which use backtracking method. A well-known algorithm is Ullmann's that is based on backtracking method with an effective refinement procedure to decrease successor states to visit [2]. Another backtracking algorithm VF introduces a set of feasibility rules to cut down the search space [3]. Messmer proposes a non-backtracking algorithm that is devised for SI of an input graph and a set of model graphs [4]. The algorithm presented in this paper is based on 2D continuous Hopfield neural network model.

2 The 2D Continuous Hopfield Model

Artificial neural networks are nonlinear mapping systems whose structure is based on principles observed in the nervous systems of humans and animals. As introduced in [5], Hopfield networks are single-layer networks with feedback connections between neurons. All neurons are globally interconnected and the connections are symmetrical. For SI problem we use the 2D continuous model. For an $m \times n$ neuron array, the motion equation for the neuron at position (i,h) is as follows:

$$\frac{dU_{ih}}{dt} = \sum_{j=1}^{m}\sum_{k=1}^{n} W_{ih,jk} V_{jk} - \frac{U_{ih}}{\tau} + I_{ih} \quad (1)$$

$$V_{ih} = g(U_{ih}) = \frac{1}{1+\exp(-U_{ih}/U_0)} \quad (2)$$

U_{ih}, V_{ih} and I_{ih} are the input, output and bias of the ih-th neuron, respectively; $W_{ih,jk}$ is the weight between the ih-th neuron and the jk-th neuron; and U_0 is a parameter that gives the slope of the function $g(\cdot)$. Let U_0 be 0.001 in the experiment. τ is the value of the time constant, and without loss of generality can be assigned a value of unity. The motion equation decreases the following Liapunov energy function:

$$E = -\frac{1}{2}\sum_{i=1}^{m}\sum_{h=1}^{n}\sum_{j=1}^{m}\sum_{k=1}^{n} W_{ih,jk} V_{ih} V_{jk} - \sum_{i=1}^{m}\sum_{h=1}^{n} I_{ih} V_{ih} \quad (3)$$

That is to say, $dE/dt \leq 0$ is always satisfied. The network will converge to a stable state that is a local minimum of the energy function. If the solution to an optimization problem can be represented by output states of neurons and the optimization problem can be formulated in terms of an energy function with all constraints built in, the network will iterate to minimize the energy function, which results in the solution to the optimization problem.

3 The SI Algorithm Based on Continuous Hopfield Network

3.1 Mapping SI Problem on Hopfield Neural Network

Given two graphs G_1 and G_2, let $p_1=|V(G_1)|$ and $p_2=|V(G_2)|$, where $p_1 < p_2$. A $p_1 \times p_2$ neuron array is constructed. If the network state represents an SI result, it should satisfy the following conditions:

$$\forall i \in \{1,2,\ldots,p_1\} \; \forall j \in \{1,2,\ldots,p_2\} \; V_{ij} \in \{0,1\} \quad (4)$$

$$\forall i \in \{1,2,\ldots,p_1\} \; \sum_{j=1}^{p_2} V_{ij} = 1 \quad (5)$$

$$\forall j \in \{1,2,\ldots,p_2\} \ 0 \leq \sum_{i=1}^{p_1} V_{ij} \leq 1 \qquad (6)$$

If v_i of G_1 can match v_j of G_2, $V_{ij}=1$, else $V_{ij}=0$. Condition (5) denotes v_i can match one and only one v_j. Condition (6) denotes v_j can match at most one v_i. The energy function E is defined to represent the constraints of SI:

$$E = -A\sum_{i=1}^{p_1}\sum_{h=1}^{p_2}\sum_{j=1}^{p_1}\sum_{k=1}^{p_2} C_{ih,jk} V_{ih} V_{jk} + B\sum_{i=1}^{p_1}(\sum_{h=1}^{p_2} V_{ih} - 1)^2 + C\sum_{i \neq j}^{p_1}\sum_{h=1}^{p_2} V_{ih} V_{jh} \qquad (7)$$

A, B and C are weight coefficients. Let $A=B=C=1/2$ in the experiment. B-term and C-term denote the constraints (5) and (6) respectively. $C_{ih,\,jk}$ is the compatibility measure between two corresponding pairs of vertices (v_i,v_h) and (v_j,v_k). According to the nature of SI problem, if v_i is adjacent to v_j, v_h must be adjacent to v_k. So $C_{ih,\,jk}$ is defined:

$$C_{ih,jk} = \begin{cases} 0 & \text{if } a_{ij}^1 = 1 \text{ and } a_{hk}^2 = 0 \\ 1 & \text{otherwise} \end{cases} \qquad (8)$$

$A(G_1) = (a_{ij}^1)$ and $A(G_2) = (a_{ij}^2)$ are the adjacency matrices of G_1 and G_2 respectively. So the A-term of eq. (7) denotes the constraint that any edge in G_1 must be matched to an edge in G_2.

Eq. (7) can be rewritten as follows:

$$E = -A\sum_{i=1}^{p_1}\sum_{h=1}^{p_2}\sum_{j=1}^{p_1}\sum_{k=1}^{p_2} C_{ih,jk} V_{ih} V_{jk} + B\sum_{i=1}^{p_1}\sum_{h=1}^{p_2}\sum_{j=1}^{p_1}\sum_{k=1}^{p_2} \delta_{ij} V_{ih} V_{jk} \qquad (9)$$

$$-2B\sum_{i=1}^{p_1}\sum_{h=1}^{p_2} V_{ih} + Bp_1 + C\sum_{i=1}^{p_1}\sum_{h=1}^{p_2}\sum_{j=1}^{p_1}\sum_{k=1}^{p_2} (1-\delta_{ij})\delta_{hk} V_{ih} V_{jk} =$$

$$-\frac{1}{2}\sum_{i=1}^{p_1}\sum_{h=1}^{p_2}\sum_{j=1}^{p_1}\sum_{k=1}^{p_2} (2AC_{ih,jk} - 2B\delta_{ij} - 2C(1-\delta_{ij})\delta_{hk}) V_{ih} V_{jk} - 2B\sum_{i=1}^{p_1}\sum_{h=1}^{p_2} V_{ih} + Bp_1$$

If $i=j$ then $\delta_{ij}=1$ else $\delta_{ij}=0$. If $h=k$ then $\delta_{hk}=1$ else $\delta_{hk}=0$. Not considering the constant term of eq. (9), we extract the network parameters by comparing eq. (9) with eq. (3):

$$W_{ih,jk} = 2AC_{ih,jk} - 2B\delta_{ij} - 2C(1-\delta_{ij})\delta_{hk} \qquad (10)$$

$$I_{ih} = 2B \qquad (11)$$

From eq. (1), (10) and (11), the motion equation for the ih-th neuron can be got:

$$\frac{dU_{ih}}{dt} = 2A\sum_{j=1}^{p_1}\sum_{k=1}^{p_2} C_{ih,jk} V_{jk} - 2B\sum_{k=1}^{p_2} V_{ik} - 2C\sum_{j \neq i}^{p_1} V_{jh} - \frac{U_{ih}}{\tau} + 2B \qquad (12)$$

3.2 Initialization and Updating of the Neurons

Generally, a random initialization for the neurons is employed, but in [6] Suganthan et al. proposed a biased network initialization scheme and demonstrated that the network performs much better and has faster convergence and relative insensitivity to a number of parameters when the biased initialization is near global minimum. In the network model for SI problem also the biased initialization is employed.

The necessary conditions for SI are used to initialize the neurons. The newly defined necessary condition is that if v_i of G_1 can match v_j of G_2, then all the vertices adjacent to v_i should match the completely different vertices adjacent to v_j. A method based on bipartite graph matching is used to verify this necessary condition. The bipartite graph matching method will be introduced in detail in another paper. If this necessary condition is satisfied, it is easy to see that the following necessary conditions in [7] are satisfied:

1) The degree of v_i must be less than or equal to that of v_j;
2) The sum of degrees of vertices adjacent to v_i must be less than or equal to that for v_j;
3) The maximum degree of vertices adjacent to v_i must be less than or equal to that for v_j.

In addition another necessary condition in [7] is used: the number of edges between vertices adjacent to v_i must be less than or equal to that for v_j.

After using the necessary conditions, a $p_1 \times p_2$ matrix $P=(p_{ih})$ can be got with $p_{ih}=1$ representing v_i can match v_h and $p_{ih}=0$ representing v_i can not match v_h. If there exists a row in which all the elements are equal to 0, it is certain that G_1 isn't isomorphic to any subgraph of G_2, else P is modified as follows:

$$p_{ih} = \frac{p_{ih}}{\sum_{h=1}^{p_2} p_{ih}} + \frac{\eta}{\sum_{h=1}^{p_2} p_{ih}} \quad \forall i \in \{1,2,...,p_1\} \; \forall h \in \{1,2,...,p_2\} \tag{13}$$

Where η is a random number between 0 and 0.2. The aim to use random number is to prevent the network being trapped in an unstable equilibrium.

The resulting matrix P is used to initialize the neurons:

$$\forall i \in \{1,2,...,p_1\} \; \forall h \in \{1,2,...,p_2\} \tag{14}$$
$$V_{ih}^0 = p_{ih}$$
$$U_{ih}^0 = g^{-1}(V_{ih}^0)$$

After initializing the neurons, similar to [8], the fourth order Runge-Kutta method [9] is used to simulate eq. (12):

$$k_1 = hf(U_{ih}^t) \tag{15}$$

$$k_2 = hf(U_{ih}^t + \frac{1}{2}k_1)$$

$$k_3 = hf(U_{ih}^t + \frac{1}{2}k_2)$$

$$k_4 = hf(U_{ih}^t + k_3)$$

$$U_{ih}^{t+1} = U_{ih}^t + \frac{1}{6}(k_1 + 2k_2 + 2k_3 + k_4)$$

Where $f(\cdot)$ is the right-hand side of eq. (12); h is the step size and let it be 0.0002 in the experiment.

3.3 The Algorithm for SI

Now we give the SI algorithm based on continuous Hopfield network.
1) Set the initial states of the neurons using eq. (14);
2) Synchronously update the states using eq. (15);
3) Calculate the new outputs of the neurons using eq. (2);
4) If $\max_{1 \le i \le p_1, 1 \le h \le p_2} |V_{ih}^t - V_{ih}^{t-1}| < \theta$ or reach the maximum iteration times, then output the result, else go to step 2;
5) Verify whether the result of step 4 really denotes an SI between the two input graphs. If not, go to step 1.

Let $\theta = 0.001$ in the experiment. In order to make the ultimate result in accordance with eq. (4), V_{ih} is modified as follows:

$$V_{ih} = \begin{cases} 1 & \text{if } V_{ih} = \max_{1 \le h \le p_2} V_{ih} \\ 0 & \text{otherwise} \end{cases} \quad \forall i \in \{1, 2, ..., p_1\} \; \forall h \in \{1, 2, ..., p_2\} \tag{16}$$

As the energy function has many local minima, though a biased initialization has been used, the network may still be trapped in a local minimum. In order to prevent this happening, in step 5, if the result doesn't denote an SI, the algorithm returns to step 1 to run the network again. This procedure is repeated at most five times.

4 Experimental Results

The random graph server at http://www.ispt.waseda.ac.jp/rgs/ is used to generate many pairs of random graphs that are undirected and connected. Some modification is made for each pair of the generated graphs so that one graph is the subgraph of the other graph. The number of vertices varies from 10 to 200, and the edge density is from 0.1 to 0.8. There are 500 pairs of graphs in all used to test the performance of the algorithm. The Hopfield network is simulated in Linux OS using C programming language. Fig. 1 gives the experimental results of the algorithm.

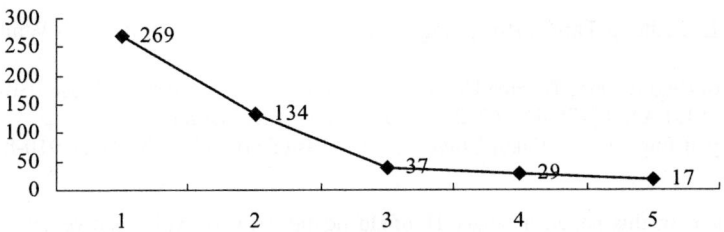

Fig. 1. The experimental results of the algorithm. X-axis is the times of running the Hopfield network. Y-axis is the number of graph pairs that are found SI

From fig. 1 it can be seen that most of the 500 pairs of graphs are found SI in less than 6 times of running the Hopfield network, which shows the correctness and validity of the algorithm.

References

1. Garey, M.R., Johnson, D.S.: Computers and Intractability: A Guide to the Theory of NP Completeness. Freeman & Co., New York (1979)
2. Ullmann, J.R.: An Algorithm for Subgraph Isomorphism. Journal of the Association for Computing Machinery, 23 (1976) 31-42
3. Cordella, L.P., Foggia, P., Sansone, C., Vento, M.: Subgraph Transformations for the Inexact Matching of Attributed Relational Graphs. Computing, Vol. 61 Suppl. 12 (1998) 43-52
4. Messmer, B.T., Bunke, H.: A Decision Tree Approach to Graph and Subgraph Isomorphism Detection. Pattern Recognition, 32 (1999) 1979-1998
5. Hopfield, J.J.: Neurons with Graded Response have Collective Computational Properties like Those of Two-State Neurons. Proc. of the National Academy of Sciences, 81 (1984) 3088-3092
6. Suganthan, P.N., Teoh, E.K., Mital, D.P.: Pattern Recognition by Homomorphic Graph Matching Using Hopfield Networks. Image & Vision Comput. 13 (1995) 45-60
7. Funabiki, N., Kitamichi, J.: A Three-Stage Greedy and Neural-Network Approach for the Subgraph Isomorphism Problem. IEEE International Conference on Systems, Man and Cybernetics, 2 (1998) 1892-1897
8. Lin, W.C., Liao, F.Y., Tsao, C.K., Lingutla, T.: A Hierarchical Multiple-View Approach to Three-Dimensional Object Recognition. IEEE Trans. Neural Networks, 2 (1991) 84-92
9. Burden, R.L., Faires, J.D.: Numerical Analysis. Brooks/Cole Publishing Company (2001)

A Positively Self-Feedbacked Hopfield Neural Network for N-Queens Problem

Yong Li[1], Zheng Tang[1], Ronglong Wang[2], Guangpu Xia[1], and Jiahai Wang[1]

[1] Faculty of Engineering, Toyama University, 3190 Gofuku, Toyama-shi, Japan 930-8555
  TEL/FAX: 81-76-445-6752, liyong@hi.iis.toyama-u.ac.jp
[2] Faculty of Engineering, Fukui University, 3-9-1 Bunkyo, Fukui-shi, Japan 910-8507

Abstract. In this paper, a binary Hopfield neural network with positive self-feedbacks and its collective computational properties are studied. It is proved theoretically and confirmed by simulating the randomly generated Hopfield neural networks with positive self-feedbacks that the emergent collective properties of the original Hopfield neural network also are present in the Hopfield network with positive self-feedbacks. As an example, the network is also applied to the N-Queens problem and results of computer simulations are presented and used to illustrate the computation power of the networks.

1 Introduction

The auto associative memory model proposed by Hopfield[1,2] has attracted considerable interest both as a content address memory (CAM) and, more interestingly, as a method of solving difficult optimization problems [3-5]. The Hopfield neural network contain highly interconnected nonlinear processing elements ("neurons") with specific interconnection strengths between neuron pairs. The output of each neuron is fed back to all other neurons via weights denoted w_{ij}. Each neuron outputs a nonlinearly transform version of the weighted summation of the neurons that are interconnected to it. The nonlinear transformation is characterized by either a hard limiting nonlinear function or a monotonic "sigmoid" function. Hopfield has proven that the network converges when the weights are symmetric with zero diagonal elements (i.e., $w_{ij}=w_{ji}$, $w_{ii}=0$) for the hard limiting nonlinear neuron [1] or nonzero diagonal elements (i.e., $w_{ij} = w_{ji}$, $w_{ii} \neq 0$) for the monotonic "sigmoid" neuron [2] if the nonzero diagonal elements are included in the energy function. Some researchers have studied the properties of Hopfield network with nonzero diagonal elements for solving combinatorial optimization problems. Watanabe et al. [6][7] proposed an oscillatory neuron unit by adding a simple self-feedback, which could also be negative or positive, and an energy value extraction circuit. However, each of them attempts to make some local minima unstable by introducing a negative self-feedback or using an oscillatory neural unit.

2 Hopfield Network with Self-Feedbacks

Unlike the original Hopfield network, the total input to neuron i of the Hopfield network with self-feedbacks takes:

$$x_i = \sum_{j \neq i}^{n} w_{ij} y_j + c_i y_i + h_i \qquad (1)$$

where x_i is the total input, y_i is the output, the element w_{ij} is the symmetric interconnection strength from neuron j to neuron i, h_i is the offset bias and c_i is the self-connection of the neuron i. $c_i = w_{ii}$. Each neuron samples its input at random times. It changes the value of its output or leaves it fixed according to a threshold rule:

$$y_i = \begin{cases} 0 & \text{if } x_i \left(= \sum_{j \neq i}^{n} w_{ij} y_j + c_i y_i + h_i \right) \leq 0 \\ 1 & \text{if } x_i \left(= \sum_{j \neq i}^{n} w_{ij} y_j + c_i y_i + h_i \right) > 0 \end{cases} \qquad (2)$$

We define the energy at time t to be:

$$E = -\frac{1}{2} \sum_i \sum_{j \neq i} w_{ij} y_i(t) y_j(t) - \sum_i h_i y_i(t) \qquad (3)$$

The change in the energy caused by the change in the states of the neurons is:

$$\Delta E = -\frac{1}{2} \sum_i \sum_{j \neq i} w_{ij} [y_i(t+1) y_j(t+1) - y_i(t) y_j(t)] - \sum_i h_i [y_i(t+1) - y_i(t)] \qquad (4)$$

Adding and subtracting $y_i(t+1) y_j(t)$, and simplifying, we get:

$$\Delta E = -\frac{1}{2} \sum_i \sum_{j \neq i} w_{ij} y_i(t+1) \Delta y_j - \frac{1}{2} \sum_i \sum_{j \neq i} w_{ij} \Delta y_i y_j(t) - \sum_i h_i \Delta y_i \qquad (5)$$

Suppose that at time t, the state of the kth neuron is changed. If the change in the state of the ith unit is:

$$\Delta y_i = y_i(t+1) - y_i(t) \qquad (6)$$

we have $\Delta y_k \neq 0$ and $\Delta y_i = 0$ for $i \neq k$. Thus, Eq.(5) can be reduced to:

$$\Delta E = -\frac{1}{2} \sum_{i \neq k} w_{ik} y_i(t+1) \Delta y_k - \frac{1}{2} \sum_{i \neq k} w_{ki} \Delta y_k y_i(t) - h_k \Delta y_k \qquad (7)$$

Because of the facts that $\Delta y_i(t+1) = 0$, i.e., $y_i(t+1) = y_i(t)$ for $i \neq k$ and $w_{ik} = w_{ki}$, using the Eq.(1), Eq.(7) can be rewritten to:

$$\Delta E = -\Delta y_k (x_k(t+1) - c_k y_k(t)) \qquad (8)$$

There are two kinds of changes in state of the kth neuron caused by the updating:
(1) $y_k(t) = 0$, $y_k(t+1) = 1$, $\Delta y_k = 1$
From the threshold rule (Eq.(2)), we have that when $y_k(t+1) = 1$, the value of $x_k(t+1)$ should be positive. Furthermore since Δy_k is positive and $x_k(t+1) > 0$, $y_k(t) = 0$, ΔE is negative.

(2) $y_k(t) = 1$, $y_k(t+1) = 0$, $\Delta y_k = -1$

From the threshold rule (Eq.(2)), we have that the value of $x_k(t+1)$ is negative, and $\Delta y_k < 0$, $y_k(t) = 1$. If c_i is positive or zero, ΔE is negative.

Hence, when the self-feedbacks (c_i) are positive or zero, each change of state by the updating always leads to a reduction in the energy for the network. Note that the theoretical result presented holds for the original Hopfield network since the original one is just one specific case of $c_i = 0$, on the other hand, the negative self-feedback can not guaranteed the network always leads to a reduction in every state change of network. The property can be used in contend address memory (CAM) and any other computational task for which an energy function is essential.

3 Simulation Results for Random Hopfield Network

Experiments were first performed to show the convergence of a Hopfield network with positive self-feedbacks. In simulations, a 50-neuron Hopfield network with positive self-feedbacks $c_i = 1$ (i = 1,2,...50) was chosen. Initial parameters of the network, connection weights and thresholds were randomly generated uniformly between −1.0 and 1.0.

Fig.1 shows the convergence characteristics of the 50-neuron Hopfield network with positive self-feedbacks. From this figure we can see that the Hopfield network with positive self-feedbacks ($c_i = 1$) converges to stable states that do not further change with time. Furthermore, simulations on a randomly generated 50-neuron Hopfield network with different self-feedbacks ($c_i = 1, c_i = -1$ and $c_i = 0$ for $i=1,...50$) were also carried out. Both the Hopfield neural networks with positive self-feedbacks ($c_i = 1$) and without ($c_i = 0$) self-feedbacks seek out minima in $E = -37.86$ and $E = -37.78$,

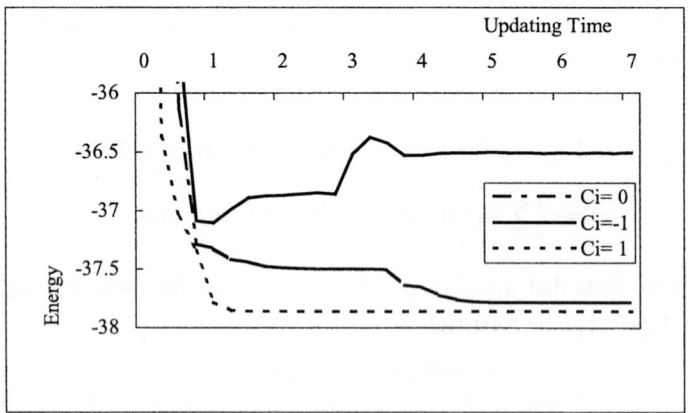

Fig. 1. The convergence characteristic of 50-neuron Hopfield network with different self-feedbacks ($c_i = 1$, -1 and 0, $i = 1,2,...50$).

and come to stop at such point while the network with negative self-feedbacks ($c_i = -1$) does not always lead to a reduction in every state change of the network.

4 Application to N-Queens Problem

N-Queens problem is classic of difficult optimization. The task is given a standard chessboard and N chess queens, to place them on the board so that no queen is on the line of attack of any other queen. The problem can be solved by constructing an appropriate energy function and minimizing the energy function to zero ($E=0$) using an $N \times N$ two-dimensional Hopfield network [8~11].

The objective energy function of the N-Queens problem is given by:

$$E = \frac{A}{2}\left(\sum_{i=1}^{N}\left(\sum_{k=1}^{N}y_{ik}-1\right)^2 + \sum_{j=1}^{N}\left(\sum_{k=1}^{N}y_{kj}-1\right)^2\right) + \frac{B}{2}\sum_{i=1}^{N}\sum_{j=1}^{N}y_{ij}\left(\sum_{\substack{1\le i-k, j-k \le N \\ k \ne 0}}y_{i-k,j-k} + \sum_{\substack{1\le i-k, j+k \le N \\ k \ne 0}}y_{i-k,j+k}\right) \quad (9)$$

where A and B are coefficients, the output $y_{ij}=1$ represents that a queen is placed at i-th orw j-th column on the chessboard, and output $y_{ij}=0$ represents no placement there. The first term becomes zero if one queen is placed in every row. The second term becomes zero if one queen is placed in every column and the third term becomes zero if no more than one queen is placed on any diagonal line.

We can get the total input (x_{ij}) of neuron by using the partial derivation term of the energy function. We have proved that adding a positive self-feedback into the motion equation will not destroy the collective computational properties. For the N-Queens problem, we add a positive self-feedback (C_{ij}) into motion equation and then the total input (x_{ij}) of neuron is given:

$$x_{ij} = -A\left(\sum_{k=1}^{N}y_{ik}-1\right) - A\left(\sum_{k=1}^{N}y_{kj}-1\right) - B\left(\sum_{\substack{1\le i-k, j-k \le N \\ k \ne 0}}y_{i-k,j-k} + \sum_{\substack{1\le i-k, j+k \le N \\ k \ne 0}}y_{i-k,j+k}\right) + C_{ij}y_{ij} \quad (10)$$

Using Eq.(10), networks with positive self-feedbacks for a total 10 chessboard size instances from 10 to 500 queens problem were simulated on a digital computer. 100 simulation runs with different initial states were performed in each of these instances. In the simulations, Eq.(2) were used as the input/output functions of neurons. The self-feedbacks (c_{ij}) were set to 1 for all neurons. The parameters A and B were set to 1. The maximum updating step was set to 1000. Using the same condition, the original Hopfield network ($c_{ij}=0$) was also executed for comparison. The simulation results are shown in Table1. In Table1, the column labeled "conv" is the global convergence times among the 100 simulations and the column labeled "step" is the average numbers of iteration steps required for convergence in the 100 simulations.

The simulation results show that the networks with positive self-feedbacks can almost find optimum solution to all N-Queens problems within short computation times;

while the original Hopfield network can hardly find any optimum solution to the N-Queens problems.

We also compared our results with that found by Takefuji's neural network [8] and the maximum neural network [10]. Table 1 shows the results by the four different networks, where the convergence rates and the average numbers of iteration steps required for the convergence are summarized.

From Table.1 we can see that the Hopfield network with positive self-feedbacks was very effective, and was better than other exiting neural networks in terms of the computation time and the solution quality for the N-Queens problem. Further, the average numbers of iteration steps indicated that the problem size did not strongly reflect the global minimum convergence rate and number of iteration steps.

Table 1. Simulation result

Queens	Proposed network		Hopfield		Takefuji		Maximum NN	
	Conv.	Step	Conv.	Step	Conv.	Step	Conv.	Step
10	91	22	7	518	31	162	26	71
20	100	24	0	--	51	290	47	142
30	100	25	0	--	52	253	53	148
50	100	35	0	--	86	308	78	176
100	100	48	0	--	98	300	99	174
150	100	76	0	--	96	411	95	151
200	100	92	0	--	93	517	95	152
300	100	137	0	--	85	616	95	152
400	100	192	0	--	69	677	87	152
500	100	228	0	--	67	756	86	139

5 Conclusions

In this paper, we have presented theoretical and experimental evidence showing that the Hopfield network with positive self-feedbacks has the same collective computational properties as the original Hopfield network. In order to confirm the practical worth of the proposed network, it was also applied to the N-Queens problem. The simulation results showed that the Hopfield network with positive self-feedbacks was better than the original Hopfield network method and other existing neural network methods for solving the N-Queens problem in terms of the computation time and the solution quality.

References

1. Hopfield, J.J.: Neural Network and Physical Systems with Emergent Collective Computational Abilities. Proc. Natl. Acad. Sci. USA. Vol. 79 (1982) 2554–2558

2. Hopfield, J.J.: Neurons with Graded Response Have Collective Computational Properties Like Those of Two–State Neurons. Proc. Natl. Acad. Sci. USA. Vol. 81 (1984) 3088–3092
3. Hopfield, J.J., Tank, D.W.: 'Neural' Computation of Decisions in Optimization Problems. Bio. Cybern., Vol. 52 (1985) 141–152
4. Hopfield, J.J., Tank, D.W.: Computing with Neural Circuits: A Model. Science, No. 233 (1986) 625–633
5. Tank, D.W., Hopfield, J.J.: Simple Neural Optimization Network: An A/D Converter, Signal Decision Circuit, and Linear Programming Circuit. IEEE Trans, Circuits & Systems, Vol. CAS–33, No. 5 (1986) 533–541
6. Yoshino, K., Watanabe, Y., Kakeshita, T.: Hopfield Neural Network Using Oscillatory Units with Sigmoidal Input-Average out Characteristics. IEICE Trans. Inf. & Syst., Vol. J77–DII (1994) 219–227
7. Watanabe, Y., Yoshino, K., Kakeshita, T.: Solving Combinatorial Optimization Problem Using Oscillatory Neural Network. IEICE Trans. Inf. & Syst., Vol. E 80–D, No. 1 (1997) 72–77
8. Takefuji, Y.: Neural Network Parallel Computing. Kluwer Academic Publishers (1992)
9. Takenaka, Y., Funabiki, N., Nishikawa, S.: Maximum Neural Network Algorithms for N–Queens Problem. J. IPSJ Vol. 37, No. 10 (1996) 1781–1788
10. Takenaka, Y., Funabiki, N., Nishikawa, S.: A Proposal of Competition Resolution Methods on The Maximum Neuron Model through N–Queens Problem. J. IPSJ Vol. 38, No. 11 (1997) 2142–2148
11. Takenaka, Y., Funabiki, N., Higashino, T.: A Proposal of Neuron Filter: A Constraint Resolution Scheme of Neural Networks for Combinatorial Optimization Problems. IEICE Trans. Fundamentals, Vol. E 83–A, No. 9 (2000) 1815–1823

NN-Based GA for Engineering Optimization

Ling Wang[1] and Fang Tang[2]

[1] Department of Automation, Tsinghua University, Beijing 100084, China
wangling@tsinghua.edu.cn
[2] Dept. of Physics, Beijing Univ. of Aeronautics and Astronautics, Beijing, 100083, China
tangfangfang@sohu.com

Abstract. For many engineering optimization problems, there are no explicitly known forms of objective functions in terms of design variables, or it only by complicated analysis or time-consuming simulation to obtain the performance of solution. Aiming at such kind of problems, this paper proposes a neural network (NN)-based genetic algorithm (GA), where the good approximation performance of NN and effective and robust evolutionary searching ability of GA are applied in hybrid sense. That is, NNs are employed in predicting the objective value, while GA is adopted in searching optimal designs based on the predicted performance. Simulation results and comparisons based on a well-known pressure vessel design problem demonstrate the feasibility and effectiveness of the strategy, and much better results are achieved than some existed literature results. In addition, the consistency and statistical quality of the resulted solutions can be improved by applying multiple neural networks.

1 Introduction

Based on the idea of "survival of the fittest", GA is a "generation-evaluation" type of parallel iterative optimization algorithm, which repeats evaluation, selection, crossover and mutation after initialization until the stopping condition is satisfied [1,2]. GA is known as a kind of effective and robust optimization algorithm well suited for discontinuous and multi-modal functions, even in noisy environment. GA does not require the objective function to be continues or even to be available in analytic form and it tends to escape more easily from local optima due to its population-based nature. However, the forms of objective functions of many practical engineering design problems are often not explicitly known in terms of design variables, or sometimes it is not easy to obtain the objective value efficiently [3]. So, it needs complicated analysis or time-consuming simulation to evaluate the performance of design variables. If the GA is applied to such problems, much computational time will be spent in fitness evaluation while less time will be paid for search process, so that the efficiency and quality of the algorithm would be degraded. Recently, meta-modeling techniques, such as regression and NN, have been pursued to approximate the usually unknown input-output function implied by the underlying simulation [3]. Keys *et al* provided some discussions on the performance measure for selection of meta-models [4]. NN is one of the artificial intelligence techniques that play an important role in solving problems with extremely difficult or unknown analytical solution [5]. In this paper, an NN-based GA is proposed for optimization problems without explicitly known form of objective functions, and the feasibility and effectiveness are demonstrated by successfully solving a pressure vessel design problem.

2 NN-Based GA

For practical problems that have no explicitly known forms of objective functions, it requires considerably expensive computational time or some complicated analysis to evaluate the objective value. GA is a "generation-evaluation" process, which only uses fitness value to guide search. If the GA is applied to such kind of problems, it needs an approximation of the fitness value to make the genetic search proceed, and the approximation value also should be given efficiently to guarantee the searching efficiency. NN is a powerful tool for approximation of unknown nonlinear function [5]. Thus, an NN-based GA is proposed which is briefly illustrated in Fig. 1.

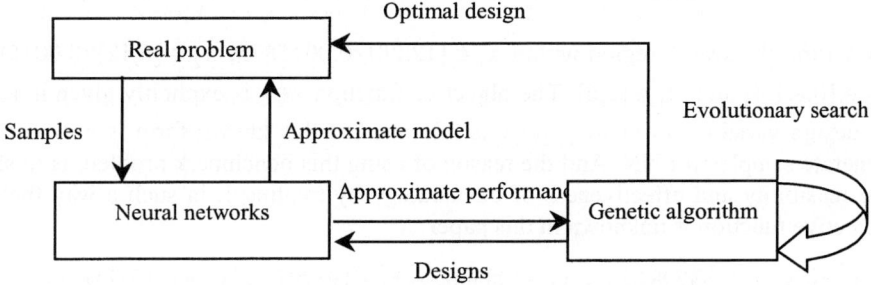

Fig. 1. Framework of NN-based GA

The searching mechanism is briefly described as follows. Firstly, NNs are constructed as approximate models of real problem with certain training algorithms based on a collection of training samples. Then, the GA is employed to explore good designs among solution space. Once the GA generates a new design, the NN will be used to determine its fitness value for the GA to continue its searching process. Until the stopping criterion of the GA is satisfied, the best design resulted by the GA will output and its performance will be determine by detail evaluation based on real problem. To those problems with known form of objective function but it is hard to evaluate the performance, NN still can be established to rapidly provide performance evaluation to enhance the efficiency of genetic search. In a word, the objective value of design can be predicted efficiently by the NN with certain extent of precision degree due to its good approximation performance so as to improve the searching efficiency of GA by without paying much computational time for evaluation. On the other hand, the GA can guarantee good designs due to its effective and robust searching performance.

3 Implementation of NN-Based GA

A pressure vessel design problem is used as example here, which has gained wide study in lots of literatures [7-12]. The design variables include: the thickness of rolled steel plates (x_1, x_2) are integer multiples of 0.0015875m, and the inner radius (x_3) and the length of the cylindrical section (x_4) are continue variables. The objective is to minimize the total cost of manufacturing the pressure vessel, including the cost of material and the cost of forming and welding. The constrained functions $h_1(\cdot)$, $h_2(\cdot)$ and $h_3(\cdot)$ correspond to ASME limits on the geometry.

$$h_1(x_1, x_3) = 0.00049022 x_3 / x_1 - 1 \le 0 \tag{1}$$

$$h_2(x_2, x_3) = 0.000242316 x_3 / x_2 - 1 \le 0 \tag{2}$$

$$h_3(x_3, x_4) = (-\frac{4}{3}\pi x_3^3 + 21.24)/\pi x_3^2 x_4 - 1 \le 0 \tag{3}$$

We limit the search region within $x_1 \in [12,20] \times 0.0015875$, $x_2 \in [6,12] \times 0.0015875$, $x_3 \in [0.8, 1.4]$ and $x_4 \in [2,6]$. The objective function $c(\cdot)$ is explicitly given in terms of design variables in many papers as follows, but this known form is only used to generate samples for NN. And the reason of using this benchmark problem is to show the feasibility and effectiveness of our strategy by treating it in such a way that the objective function is unknown in this paper.

$$c(x_1, x_2, x_3, x_4) = 37982.2 x_1 x_3 x_4 + 108506.3 x_2 x_3^2 + 193207.3 x_1^2 x_4 + 1210711 x_1^2 x_3 \tag{4}$$

NN design. The following BPM algorithm (BP with momentum) [7,9] is used.

$$W(k+1) = W(k) - \lambda \cdot \partial E / \partial W(k) + \beta[W(k) - W(k-1)] \tag{5}$$

where W is the weight matrix, $\lambda \in (0,1)$ is learning rate, β is momentum factor. In following simulation, 50 feasible samples are randomly generated in the limited region. Then, based on the samples, 4-7-1 three-layer NNs are designed by using BPM algorithm with $\lambda = 0.05$, $\beta = 0.4$ and the total training step 10000.

Searching variables. Searching variables $y_1 \in [12,20]$, $y_2 \in [6,12]$, $x_3 \in [0.8, 1.4]$ and $x_4 \in [2,6]$ are encoded as real numbers, and let $x_1 = 0.0015875 \times \text{Int}[y_1]$ and $x_2 = 0.0015875 \times \text{Int}[y_2]$, where $\text{Int}[y_1]$ denotes the maximum integer not larger than y_1. In initialization of GA, P_s (population size=30) solutions are randomly generated in above region, where each searching solution is a combination (y_1, y_2, x_3, x_4).

Fitness function. In order to consider constraints simultaneously, firstly the augmented objective function is defined as follows.

$$C(x_1, x_2, x_3, x_4) = C(X) = c_a(X) + \sum_{i=1}^{3} 100 \cdot [\max(h_i(X), 0)]^2 \tag{6}$$

where $c_a(X)$ is the approximate objective value of design X predicted by NN. Since our goal is to minimize the augmented objective function and the goal of the GA is to maximize the fitness function, so here we simply let the fitness value be the reciprocal of the augmented objective value.

Genetic operators. Proportional selection (e.g. round roulette selection) is applied. For two selected parents, arithmetic crossover is performed with crossover probability p_c (0.9 is used) to generate two new designs. Mutation is designed as $x_{new} = x_{old} + \zeta$, where ζ is a random number subject to normal distribution $N(0,1)$. For the new designs generated by crossover, each design will perform mutation with probability p_m (0.8 is used). Besides, if the new value of the variable is out of the predefined region, the nearer bound value will be used to replace the new value.

Elitist strategy. Top P_s good designs are selected as the new population among the old population and all new designs generated after mutation.

Stopping criterion. If the best design is fixed in 50 successive generations, the algorithm will be stopped and the best design and its performance will be output.

4 Simulations and Comparisons

Firstly, a 4-7-1 NN is designed by using BPM, then above GA based on this single NN is randomly applied ten times. The results are shown in Table 1, where the approximate objective value c_a is predicted by the NN and the true objective value c_t is calculated with the explicit objective function.

Table 1. Ten random searching results of GA based on a single NN

# of simulation	x_1 (m)	x_2 (m)	x_3 (m)	x_4 (m)	c_a	c_t
1	0.0222250	0.0142875	1.1995	3.1398	5068.14	6426.67
2	0.0206375	0.0095250	1.1998	3.1864	5056.44	5365.18
3	0.0190500	0.0111125	1.1998	3.1802	5050.43	5246.72
4	0.0222250	0.0127000	1.1986	3.1114	5092.09	6141.43
5	0.0190500	0.0095250	1.1999	3.1193	5049.94	4942.20
6	0.0222250	0.0127000	1.1992	3.1771	5074.42	6218.12
7	0.0254000	0.0095250	1.1987	3.2011	5096.95	6522.02
8	0.0206375	0.0142875	1.1998	3.1599	5056.17	6081.86
9	0.0190500	0.0142875	1.1999	3.1156	5050.23	5682.77
10	0.0222250	0.0111125	1.1994	3.1995	5070.37	5996.54

From Table 1, firstly it is shown that the predicted best performances are very consist, which means the GA has reached the best design of the predicted objective function. Secondly, it is shown that there still exist certain difference between the predicted and the true performances, which means the approximation performance in sense of the prediction error is still not very good. But even under such prediction condition, good designs can be obtained which strongly shows the robust genetic searching ability even in noisy environment, and it also indicates that for optimization problems the position of the best design is more important than the prediction precision of its performance [4]. So, if the positions of the good designs of the true objective surface and the predicted objective surface by NN have not been greatly changed, the excellence searching ability of the GA can still be well taken advantage.

Next, two NNs are designed and their average output values are used for fitness calculation. Five random searching results of the GA based on double NNs are shown in Table 2. And the statistical performances including mean value and standard derivation of the results of the GA based on a single NN (Table 1) and double NNs (Table 2) are shown in Table 3. Define the null hypothesis $H_0 : \mu_1 = \mu_2$ and the alternative hypothesis $H_1 : \mu_1 > \mu_2$, where μ_i is the expected performances of method i. If

$$t = \frac{\bar{c}_1 - \bar{c}_2}{S_w \sqrt{1/m_1 + 1/m_2}} \geq t_{1-\alpha}(m_1 + m_2 - 2),$$ then μ_1 is significantly larger than μ_2 in

the statistical sense with confidential level $1-\alpha$, where $\bar{c}_i$ is the average performances of method i, S_i is standard deviation, m_i is simulation number, $S_w^2 = [(m_1-1)S_1^2 + (m_2-1)S_2^2]/(m_1+m_2-2)$ and $t_{1-\alpha}(m_1+m_2-2)$ is the $1-\alpha$ quantile of the t-distribution with $m_1 + m_2 - 2$ degrees of freedom. For minimization, we say that method 2 is significantly better than method 1 if the above inequality is satisfied. Setting $\alpha = 0.05$ and by checking the above inequality with the data of Table 3, we get $t = 2.565 > t_{0.95}(13) = 1.771$. So, it is concluded that the performance of GA based on double NNs is significantly better than single NN in statistical sense. In addition, by comparing the statistical standard derivation of true performance, it also can be concluded that the results of GA based on double NNs are much more consist.

Table 2. Five random searching results of GA based on double NNs

# of simulation	x_1 (m)	x_2 (m)	x_3 (m)	x_4 (m)	c_a	c_t
1	0.0190500	0.0111125	1.1992	3.2285	5614.60	5288.84
2	0.0190500	0.0111125	1.1997	3.1898	5603.51	5255.25
3	0.0190500	0.0095250	1.1993	3.2226	5612.04	5036.12
4	0.0190500	0.0111125	1.1999	3.1627	5599.08	5231.07
5	0.0190500	0.0111125	1.2000	3.3061	5613.09	5365.91

Table 3. Comparative results between single model and double models

Model	true performances		approximate performances	
	$\bar{c}_t$	S_t	$\bar{c}_a$	S_a
Double models	5235.44	122.49	5608.46	6.79
Single model	5862.35	530.13	5066.52	17.22

To further show the superior performance of our method, we compare our result with some existed results by different approaches [7-12]. The comparisons are summarized in Table 4, where the unit of design variables in Coello's papers is inch, which is changed into miter here. The superiority of our strategy is obviously shown.

Table 4. Comparison of the results of this paper and some existed literatures

paper	x_1 (m)	x_2 (m)	x_3 (m)	x_4 (m)	Best performance
[7]	0.02858	0.01588	1.212	2.990	8129.80
[8]	0.02858	0.01588	1.481	1.132	7238.83
[9]	0.02858	0.01588	1.481	1.198	7198.20
[10]	0.02540	0.01588	1.316	2.148	7006.90
[11]	0.01905	0.009525	0.986	5.651	5862.27
[12]	0.02064	0.01111	1.069	4.487	6059.94
This paper	0.01905	0.009525	1.1999	3.1193	4942.20

References

1. Wang, L.: Intelligent Optimization Algorithms with Applications. Tsinghua University & Springer Press, Beijing, (2001)
2. Goldberg, D.E.: Genetic Algorithms in Search, Optimization and Machine Learning. Addison-Wesley, Reading, MA, (1989)
3. Banks, J.: Handbook of Simulation. John Wiley & Sons, New York, (1998)
4. Keys, A., Rees, L., Greenwood, A.: Performance Measures for Selection of Metamodels to be Used in Simulation Optimization. Decision Sci., 33 (2002) 31-57
5. Lawrence, J.: Introduction to Neural Networks: Design, Theory, and Applications. California Scientific Software, Nevada, CA, (1994)
6. Jin, Y., Olhofer, M., Sendhoff, B.: A Framework for Evolutionary Optimization with Approximate Fitness Functions. IEEE Trans. EC, 6 (2002) 481-494
7. Sandgren, E.: Nonlinear Integer and Discrete Programming in Mechanical Engineering Systems. J. Mechanical Design, 112 (1990) 223-229
8. Qian, Z., Yu, J., Zhou, J.: A Genetic Algorithm for Solving Mixed Discrete Optimization Problems. Advances in Design Automation, 65 (1993) 499-503
9. Kannan, B.K., Kramer, S.N.: An Augmented Lagrange Multiplier Based Method for Mixed Integer Discrete Continues Optimization and Its Applications to Mechanical Design. J. Mechanical Design, 116 (1994) 405-411
10. Thierauf, G., Jianbo, C.: Parallel Evolution Strategy for Solving Structural Optimization. Engineering Structures, 19 (1997) 318-324
11. Nakayama, H., Arakawa, M., Sasaki, R.: Simulation-based Optimization Using Computational Intelligence. Optimization and Engineering, 3 (2002) 201-214
12. Coello, C.A.C., Montes, E.M.: Constraint-handling in Genetic Algorithms through the Use of Dominance-based Tournament Selection. Advanced Engineering Informatics, 16 (2002) 193-203

Applying GENET to the JSSCSOP

Xin Feng[1], Lixin Tang[1], and Hofung Leung[2]

[1]Department of Systems Engineering, Northeastern University, Shenyang, China
fengxinemail@sohu.com
[2]Department of Computer Science and Engineering, The Chinese University of Hong Kong,
Shatin, N.T., Hong Kong, China

Abstract. GENET is a local search approach with a neural network connectionist architecture for solving constraint satisfaction problems by iterative improvement and incorporates a learning strategy to escape local minima. In this paper, a method within the framework of propagation of posted new constraints and based on the progressive stochastic search of GENET for solving the job shop scheduling constraint satisfaction optimization problem (JSSCSOP) will be presented. The experimental results show that the performance of our method gets competitive when the domain of each variable is not big, even if the size of the problem instances increases.

1 Introduction

The job shop scheduling arose in the manufacturing industries. In the job shop scheduling problem (JSSP), we are given a set of jobs and a set of machines. Each job consists of a set of operations that must be processed in a given order. Furthermore, each operation needs to be processed during an uninterrupted time period of a given length on a given machine. Each machine can handle at most one operation at a time. The purpose is to find a schedule that minimizes the makespan. JSSP have been proven to be strong NP-hard to be solved to optimality [1].

Jain and Meeran [2] summarized the main techniques applied to solve JSSP. In all existing scheduling methods, only the representational assumptions underlying constraint satisfaction problem (CSP) models can naturally accommodate the idiosyncratic constraints that complicate most real world applications. For solving CSP's, iterative repair method has been shown to be a good alternative when compared with the traditional backtracking techniques [3,4]. Wang and Tsang proposed GENET [5], a genetic neural network model for solving general CSP's with binary constraints which based on the iterative repair approach but a heuristic learning rule is used to help the network escaping from local minima. Davenport et al. [6] extend GENET with the illegal and the atmost constraints. Lee J.H.M. et al. [7] presented E-GENET for solving general constraints and proposed several modifications to boost the efficiency. In [8], Lam et al. proposed the progressive stochastic search (PSS) based on GENET. In this paper, we model the JSSP as job shop scheduling constraint satisfaction optimization problem (JSSCSOP) and present a new method within the framework of propagation of posted new constraints and based on the PSS of GENET for solving JSSCSOP.

The paper is structured as follow. The formal definition of the JSSCSOP is given in section 2. The GENET network is briefly introduced in section 3. Section 4 describes the solution methodology. Section 5 shows the results and observations. In section 6, we summarize our contributions.

2 The JSSCSOP

Different from the existing formal description defining job shop scheduling to be a sequencing problem in operation research, we start with the formal description of the job shop scheduling scenario as an assignment problem in constraint satisfaction optimization problem.

Let J be a set of n jobs and M be a set of m machines. Each job J_i consists of a chain of operations $O_{i,1} \prec O_{i,2} \prec \cdots \prec O_{i,m_i}$ $O=\{O_{1,1},\cdots,O_{1,m_1},\cdots,O_{n,1},\cdots,O_{n,m_n}\}$ is the set of total operations to be scheduled. For each operation $O_{i,k} \in O$, a period of $p_{i,k}$ is processing time, $st_{i,k}$ is start time variable, $ft_{i,k}$ is finish time, and $ft_{i,k}= st_{i,k}+ p_{i,k}$. The JSSCSOP can be defined by a quadruple: $P=(ST,T,(C_P+C_C),ct)$, where $ST=\{st_{1,1},\cdots,st_{1,m_1},\cdots,st_{n,1},\cdots,st_{n,m_n}\}$ is the set of start time variables of all operations. $T=\{T_{1,1},\cdots,T_{1,m_1},\cdots,T_{n,1},\cdots,T_{n,m_n}\}$ is the set of associated integer finite domains of variables. C_p is the set of precedence constraints. For all $O_{i,l} \prec O_{i,k}$, if $c_p(\ st_{i,k} \geq st_{i,l}+p_{i,l})$ = $true$, then C_p is satisfied. C_c is the set of machine capacity constraints. For all $O_{i,k}, O_{j,l} \in O$, and $M(O_{i,k})=M(O_{j,l})$, if $c_c(st_{i,k} \geq ft_{j,l}$ ∨ $st_{j,l} \geq ft_{i,k})$ = $true$, then C_c is satisfied. If $st_{i,k}=a_{i,k}$, $a_{i,k} \in T_{i,k}$, then a complete assignment $A=(a_{1,1},\cdots,a_{1,m_1},\cdots,a_{n,m_n})$ is an element of $T_{1,1} \times \cdots \times T_{1,m_1} \times \cdots \times T_{n,m_n}$, corresponding to a schedule. If both C_p and C_c are satisfied simultaneously, the schedule is feasible. Let $CT(P)$ is the set of all feasible schedules, $A_l \in CT(P)$, the optimization objective is $ct_{\min}(P)=\min_{A_l \in CT(P)} ct(A_l)$.

3 Brief Review of GENET

The GENET [6,9] model consists of three components: network architecture which governs the network representation of a CSP; convergence procedure which formulates how the network is updated in the solution searching process and learning strategy which help network to escape local minima.

Consider a CSP (U, D, C), where U is a finite set of variables, D_x is the domain of x ($x \in U$), C is a set of constraints. Each variable $i \in U$ is represented in GENET N by a cluster of *label nodes* $\langle i,j \rangle$, one for each value $j \in D_i$. Each label node $\langle i,j \rangle$ is associated with an output $V_{\langle i,j \rangle}$, which is 1 if value j is assigned to variable i, and 0 otherwise. A label node is *on* if its output is 1; otherwise, it is *off*. A constraint $c \in C$ on variable i_1 and i_2 is represented by weighted connections between incompatible label nodes in clusters i_1 and i_2 respectively. Two label nodes $\langle i_1,j_1 \rangle$ and $\langle i_2,j_2 \rangle$ are connected if $i_1= j_1 \wedge$

$i_2=j_2$ violates c. All connection weights initially set to -1. The input $I_{\langle i,j \rangle}$ to a label node $\langle i,j \rangle$ is:

$$I_{\langle i,j \rangle} = \sum_{\langle k,l \rangle \in A(N,\langle i,j \rangle)} W_{\langle i,j \rangle \langle k,l \rangle} V_{\langle k,l \rangle} \quad (1)$$

Where $A(N, \langle i,j \rangle)$ is the set of all label nodes connected to $\langle i,j \rangle$ and $W_{\langle i,j \rangle \langle k,l \rangle}$ is the connection weight between the label nodes representing the labels $\langle i,j \rangle$ and $\langle k,l \rangle$. A *valid state* S of a GENET network N induces a *valid variable assignment* to the variables of the CSP corresponding to N. Associate an energy $E(N,S)$ with every state S for a network N:

$$E(N,S) = \sum_{(\langle i,j \rangle \langle k,l \rangle) \in N} V_{\langle i,j \rangle} W_{\langle i,j \rangle \langle k,l \rangle} V_{\langle k,l \rangle} \quad (2)$$

When $E(N,S)$ is with negative weights, GENET performs iterative repair by maximizing $E(N,S)$. The $E(N,S)$ of a solution state is always 0. In PSS [8], a list of variables which dictates the sequence of variables to repair must be maintained. When a variable is designated to be repaired, it always has to choose a new value even if its original value should give the best cost value. If $E(N,S)$ is not 0, and no more improvements can be made, the weights on the connections between label nodes which violate a constraint can be adjusted according to the following rule:

$$W^{t+1}_{\langle i,j \rangle \langle k,l \rangle} = W^{t}_{\langle i,j \rangle \langle k,l \rangle} - V_{\langle i,j \rangle} V_{\langle k,l \rangle} \quad (3)$$

Where $W^{t}_{\langle i,j \rangle \langle k,l \rangle}$ is the connection weight between label nodes representing $\langle i,j \rangle$ and $\langle k,l \rangle$ at time t, thus network can escape local minima.

4 Solution Methodology

From the above mentioned, the convergence cycles of GENET network is terminated when $E(N,S)$ is 0. This process only is applicable to obtain a feasible solution for CSP. In this section, we show a transformation for converting CSP into JSSCSOP.

4.1 Initialization

Because most PDRs are very easy to implement and have a low computation burden, we use the well-known EFT, MOR and MWR algorithms to obtain initial feasible solutions respectively. And choose the best one (with the least makespan) to be the upper bound of makespan UB_0. For each $O_{i,k} \in O$, we have $T_{i,k} \subseteq [0, UB_0 - p_{i,k} - 1]$. In the absence of machine capacity constraints, earliest start time constraint are propagated downstream within the job whereas latest start time constraints are propagated upstream. Let $est_{i,k} = \min_{\delta \in T_{i,k}}\{\delta\}$ is the earliest start time, $lst_{i,k} = \max_{\delta \in T_{i,k}}\{\delta\}$ is the latest start time. In J_i, $O_{i,k}, O_{j,l} \in O$, and $c(O_{i,l} \prec O_{i,k}) \in C_p$, after applying the following rules,

$$est_{i,l} = \max\{est_{i,l}, est_{i,k} + p_{i,k}\}, \quad lst_{i,k} = \min\{lst_{i,k}, lst_{i,l} - p_{i,k}\} \quad (4)$$

we can get reduction domain of $T_{i,k}$ and $T_{i,l}$.

4.2 Constraints and Network Connection

In JSSCSOP, there exist maps between variables and variable nodes. In constraint node, we can store penalty values for combinations of value from its relative intermediate nodes. In the existing two types constraint C_p and C_c, C_p can be given a GENET network connection easily, an example in Fig. 1. There are three operations in the same job $J_i \in J$, $O_{i,j-1} \prec O_{i,j} \prec O_{i,j+1}$, if $c_p(\ st_{i,j+1} \geq st_{i,j} + p_{i,j}) = true$, intermediate node a_j^i is 1, otherwise 0.

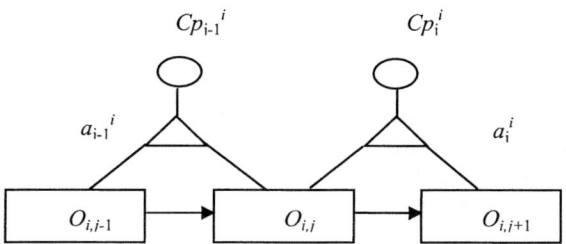

Fig. 1. The representation for precedence constraints in the same job

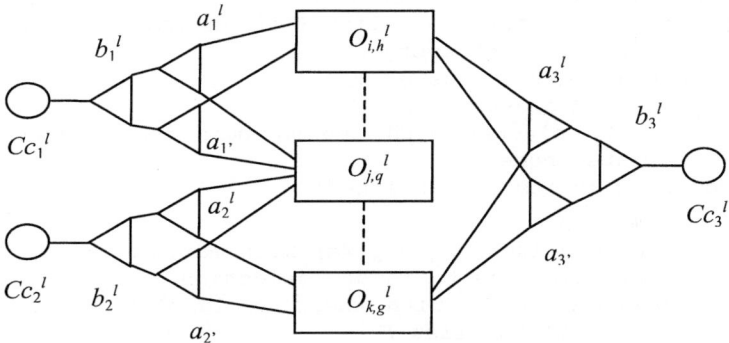

Fig. 2. The representation for machine capacity constraints in the same machine

But C_c is the set of machine capacity constraints which can be represented as disjunctive constraints. For each constraint $C_{ci} \in C_c$, there is a corresponding intermediate node a_i and $a_{i'}$ which holds the initial penalty value for the associated pair under the assignment scheme of C_{ci}. Since we only need one of them satisfied, we add one more intermediate node b_i to store the maximum value between a_i and $a_{i'}$, $b_i^l = \max\{a_i^l, a_{i'}^l\}$. The assignment scheme of the disjunctive constraint would then be based on the value of b_i. An example in Fig. 2. There are three operations will be processed on the same machine l. For Cc_1^l, if $b_1^l = 1$, we know the formula $c_c(st_{i,h} \geq ft_{j,p} \vee st_{j,h} \geq ft_{i,p}) = true$.

Let $ST \leftrightarrow U, T \leftrightarrow D, C_p + C_c \leftrightarrow C$, we can obtain GENET network architecture of JSSCSOP. We use two methods to gather initial complete assignment as the initial state of network. One is current value of each start time variable in the best initial feasible solution of PDR; the other comes from the min-conflict heuristic.

4.3 The Search Strategy

Immediately after the network initialization, all clusters are added to a list F in an arbitrary order, states are changed by asynchronous updating convergence procedure until the list F becomes empty, network is in a feasible solution state. The process is show as follow,

```
Procedure INITIALIZE
    PDRs: EFT, MOR and MWR
    let UB₀ be the best makespan
    for each T_{i,k} do
        T_{i,k}⊆[0,UB₀-p_{i,k}-1]
    end for
    constraint propagation reduce T_{i,k}
Procedure NETWORK CONNECT
    For each W_{(i,j)(k,l)} do
        W_{(i,j)(k,l)} ← -1
    end for
    all clusters in U are initial
        the best initial feasible solution of PDRs
        or the min-conflict heuristic
Procedure CONVERGENCE AND LEARNING
    append all clusters to a list F in an arbitrary order
    while list F is not empty do
      remove and get a cluster u,
            the head cluster
            or the cluster with the minimum input among all on
                label nodes
      denote d as its on label node
      if I_{(u,d)≠0}
          turn on a label node g(≠d) with maximum input,
              breaking tie by random selection
          append clusters with their on label nodes connecting
              to g to the list F
          for all clusters r (≠u)
              denote l_r as its on node
              if l_r is connection to d
                  W_{(u,d)(r,l_r)} ← W_{(u,d)(r,l_r)}-ε    ε is a parameter which can
                      be turned according to the number of variables
              end if
          end for
      end if
    end while
```

Usually we choose the number of variables as the value of ε.

4.4 Posted New Constraints for Optimization

Searching all over the set $CT(P)$ of feasible schedules to find out the optimal solution is time consumed. Here we propagate the posted new constraints of objective function to reduce the search space and the domains of every variable. When a feasible solution is found,

$$P = (ST, T, C_p + C_c + \{ct(P) < UB_0\}, ct)$$ (5)

update the upper bound of makespan from UB_0 to UB_1, the upper bound of each $T_{i,k}$ decreases UB_0-UB_1. Thus updating the clusters would result in new conflicts generate, hence the GENET network try to maximize the energy of network again. Continue to search the better solutions until get to the termination condition.

```
Procedure OPTIMIZE
    INITIALIZE
    NETWORK CONNECT
    While the termination condition is not satisfied do
        if a feasible solution be obtain
            save the feasible solution
            update UB and T
            update NETWORK CONNECT
            CONVERGENCE AND LEARNING
        else
            CONVERGENCE AND LEARNING
        end if
    end while
```

5 Experiments

To test the performance of the method, computational experiment has been carried out on randomly generated problem instances, which were generated by Taillard's random number generator [10]. A JSSCSOP is generated with 3 parameters (n, m, pt), where n is the number of jobs, m is the number of machines, pt is the range of processing time. The combination of parameter levels gives 6 problem scenarios, and for each scenario, 10 different problem instances were randomly generated. Thus totally 60 problem instances were used in the experiment on four algorithms which showed in Table 1. Because algorithms cannot report optimal solutions, the best result of branch and bound (BAB) (during 2000 sec.) is used as the lower bound of the original problem to terminate program.

Table 1. Algorithms

algorithm	Network initialization		Select cluster in the list F	
	feasible solution	min-conflict heuristic	the head	minimum input
G1	√		√	
G2	√			√
G3		√	√	
G4		√		√

All the experiments are performed on a PC with a 2.4 GHz Pentium 4 processor and 512 Mb of memory running WindowsXP. For each problem, 10 runs are presented. All the timings are measured in seconds. All the timing results are the search time only. Fig. 3—8 show the mean time results on different sizes problems.

From the results, the following observations can be made about our Algorithms:

1) All the four algorithms get competitive when the range of processing time is 9. Algorithms can get to the designated lower bound of optimal objective very quickly, even if the size of the problem instance increases. But fail to win for the problems with the processing time range of 99. This is because when the range of processing time is small, the domain of each variable can be small.

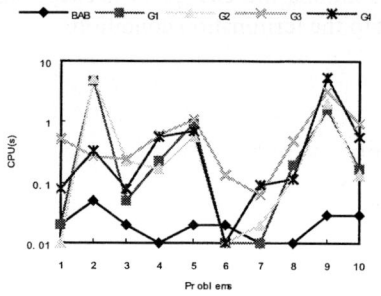

Fig. 3. The mean time results on 5/5/99   **Fig. 4.** The mean time results on 10/5/99

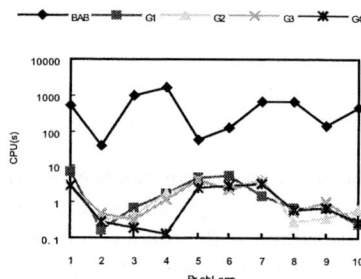 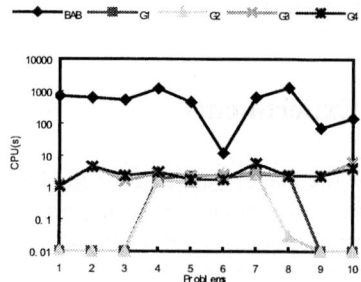

Fig. 5. The mean time results on 10/10/9   **Fig. 6.** The mean time results on 15/10/9

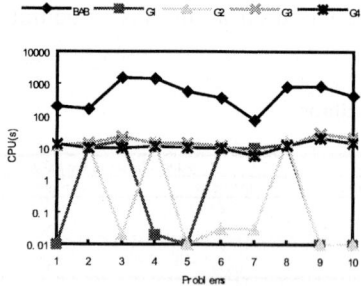 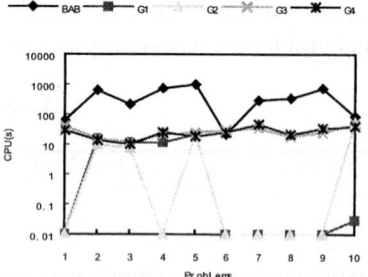

Fig. 7. The mean time results on 20/15/9   **Fig. 8.** The mean time results on 20/20/9

2) G1 and G2 are better than G3 and G4 as the problems size increases. This is consistent with the intuition that it is easy to find a new feasible solution from a set of neighbors of initial feasible solution, especially for large size problems.

6 Concluding Remarks

In this paper, we present a method within the framework of propagation of posted new constraints and based on the PSS of GENET for solving JSSCSOP. The experimental results show that the performance of our method gets competitive when the domain of each variable is not big, even if the size of the problem instance increases. We find out, using such assignment method instead of sequencing methods, this novel method is well suited to solve various large complex scheduling problems in practice when each operation of jobs have fewer possible assignments of values.

Acknowledgements. This research is supported by National Natural Science Foundation of China (Grant No. 60274049 and Grant No. 70171030), Fok Ying Tung Education Foundation and the Excellent Young Faculty Program and the Ministry of Education China.

References

1. Garey, M., Johnson, D.: Complexity Results for Multiprocessor Scheduling under Resource Constraints. SIAM Journal on Computing, 4(4) (1975) 397-411
2. Jain, A. S., Meeran, S.: Deterministic Job Shop Scheduling: Past, Present and Future. European Journal of Operational Research, Vol. 113. (1999) 390-434
3. Backer, B., Furnon, V., Shaw, P., Kilby, P., and Prosser, P.: Solving Vehicle Routing Problems Using Constraint Programming and Meta-heuristics. Journal of Heuristics, 6(4) (2000) 501-523
4. Jussien, N., Lhomme, O.: Local Search With Constraint Propagation and Conflict-based Heuristics. Artificial Intelligence, 139 (2002) 21-45
5. Tsang, E.P.K., Wang, C.J.: A Generic Neural Network Approach for Constraint Satisfaction Problems. In Neural Network Applications, Springer-Verlag, (1992) 12-22
6. Davenport, A., Tsang, E., Wang, C.J., Zhu, K.: Genet: A Connectionist Architecture for Solving Constraint Satisfaction Problems by Iterative Improvement. In Proc. 12th National Conference on Artificial Intelligence, (1994)
7. Lee, J.H.M., Leung, H.F., Won, H.W.: Towards a More Efficient Stochastic Constraint Solver. In Proc. 2th International Conference on Principles and Practice of Constraint Programming, (1996) 338-352
8. Lam, B.C.H. and Leung, H.F.: Progressive Stochastic Search for Solving Constraint Satisfaction Problems. In: Proc. 15th IEEE International Conference on Tools with Artificial Intelligence, (2003)
9. Choi, M.F., Lee, H.M., Stuckey, J.: A Lagrangian Reconstruction of GENET. Artificial Intelligence, Vol. 123. (2000) 1-39
10. Taillard, E.: Benchmarks for Basic Scheduling Problems. European Journal of Operational Research, Vol. 64. (1993) 278-285

Improvements to Bennett's Nearest Point Algorithm for Support Vector Machines

Jianmin Li [1,2], Jianwei Zhang [1], Bo Zhang [2], and Fuzong Lin [2]

[1] TAMS, Faculty of Informatics, University of Hamburg, D - 22527 Hamburg, Germany
{li, zhang}@informatik.uni-hamburg.de
[2] Department of Computer Science and Technology, Tsinghua University, Beijing 100084, P. R. China
{lijianmin, dcszb, linfz}@mail.tsinghua.edu.cn

Abstract. Intuitive geometric interpretation for Support Vector Machines (SVM) provides an alternative way to implement SVM. Although Bennett's nearest point algorithm (NPA) can deal with reduced convex hulls, it has some disadvantages. In the paper, a feasible direction explanation for NPA is proposed so that computation of kernel can be reduced greatly. Besides, the original NPA is extended to handle the arbitrary valid value of μ, therefore a better generalization performance may be obtained.

1 Introduction

Based on statistical learning theory [1], Support Vector Machines (SVM) [2], [3] not only have a concise mathematical form, an excellent generalization ability, but also avoid local optimal solutions and the curse of dimensionality. Therefore, SVM have received more and more attention after being proposed in 1992. They have rapidly developed in theory so that many variants and algorithms have been appearing in recent years. Besides, they have also been applied successfully in many fields, such as handwriting character recognition, object detection, text classification, etc.

Different from many abstract methods, SVM have an intuitive geometric explanation. It is obvious in the separable case in support vector classification (SVC). Maximizing margin, one of the key ideas in SVM, is equivalent to finding the closest points between two convex hulls. Each hull contains samples of one class. Furthermore, it is presented by Bennett [4] and Crisp [5] that in the inseparable case, the solution of soft margin SVM relates to the distance between two reduced/soft convex hulls. A geometric interpretation for support vector regression (SVR) is then proposed by Bennett [6], [7] with a trick of drifting up and down the given samples.

Geometry interpretation not only provides a new angle to understanding SVM but also provides a direct way to implement SVM. An iterative nearest point algorithm to compute the distance between two convex hulls is developed by Keerthi [8]. However, it is only available in the separable case in SVC. A new algorithm suitable to reduced convex hulls is proposed by Bennett [7]. It can be used in both SVC and SVR. Besides the methods dealing with SVM directly, there are some other geometric approaches. For example, guard vectors [9], the superset of support vectors, are extracted from the separable samples by solving a set of linear programming.

However, SVC and SVR are usually solved through their dual problems, the quadratic programming (QP) problems. Due to the requirement of the whole Hessian matrix, many traditional optimization algorithms are not suitable to SVM in large scale problems because the matrix occupies a large block of memory. Most practical methods at present belong to the family of decomposition algorithms in which a much smaller QP on a subset of the variables in the original QP is solved in each iteration. Various decomposition algorithms differ in the size of their working set, the strategy for working set selection, and the method for solving sub-QP. In libsvm, one of the most popular kinds of software, the size of the working set is 2 so that the sub-QP can be solved analytically and a heavy numerical QP overhead can be avoided. It is especially suitable to the large scale problems on the low end platforms, e.g. PC.

In the geometric approach proposed by Bennett, extra optimization software is not necessary either. And since more variables are updated in each iteration, the number of iterations is expected to be reduced. However, it is reported in the paper that the initial implementation is quite slow.

The primary contributions of this paper are to extend and improve the original algorithm. With an interpretation as a feasible direction method, the computation for selecting the working set can be reduced. And the original method is also extended to deal with the arbitrary valid value of coefficient μ.

The paper is organized as follows. In section 2, the geometry explanation of SVM is outlined. Then an interpretation as the feasible direction method is proposed for Bennett's method. Improvement on working set selection and extension to the arbitrary valid value of μ are presented in section 4 and 5. Section 6 is the conclusion.

2 Geometric Interpretation of Support Vector Machines

With a trick [6], where each sample is shifted up and down along y axis so that 2 new samples are constructed and assigned to label +1/−1 according to its moving direction, SVR is converted to SVC on new samples. Therefore, only SVC is involved here.

Assume a training set $\{(x_i, y_i)\}$, $x_i \in I \subseteq R^d$, $y_i \in \{+1, -1\}$, $i = 1, ..., n$. Φ is a nonlinear map from the input space I to a feature space F. $k(x_i, x_j)$ is the inner product in F, i.e. $k(x_i, x_j) = <\Phi(x_i), \Phi(x_j)>$. w is the normal vector of the decision hyperplane in F. A (B) is a matrix consisting of the feature vectors of all the positive (negative) samples respectively. A vector of ones (zeros) with arbitrary dimension is denoted by e (0). The transpose of x (A) is denoted by x' (A'). All the vectors are column vectors.

According to Bennett's presentation, the primal problem in the separate case is:

$$\min_{w,\alpha,\beta} \|w\|^2 - (\alpha - \beta) \tag{1}$$

s.t. $A'w - \alpha e \geq 0, B'w - \beta e \leq 0.$

The corresponding dual problem is as follows:

$$\min_{u,v} \|Au - Bv\|^2 \tag{2}$$

s.t. $e'u = 1, 0 \leq u \leq e, e'v = 1, 0 \leq v \leq e.$

According to the constraints, Au (Bv) is an arbitrary point in the convex hull defined by the positive (negative) samples. Therefore equation (2) is to compute the distance between two convex hulls.

In the inseparable case, slack vectors ξ and η are introduced to measure the error for positive and negative samples. A coefficient μ is used to penalize the sum of the error. The primal and dual problems are as follows:

$$\min_{w,\alpha,\beta} \|w\|^2 - (\alpha - \beta) + \mu(e'\xi + e'\eta) \tag{3}$$
$$\text{s.t.} \quad A'w + \xi - \alpha e \geq 0, \xi \geq 0, B'w - \eta - \beta e \leq 0, \eta \geq 0.$$

$$\min_{u,v} \|Au - Bv\|^2 \tag{4}$$
$$\text{s.t.} \quad e'u = 1, 0 \leq u \leq \mu e, e'v = 1, 0 \leq v \leq \mu e.$$

Au (Bv) is an arbitrary point in the reduced or soft convex hull defined by the positive (negative) samples and the coefficient μ. Therefore equation (4) is to compute the distance between two reduced convex hulls.

According to the equal constraint, it is meaningless to leave μ larger than 1. And equation (2) is in fact a special case of (4) with μ being 1. It should be noticed that even in the separable case, μ can also be smaller than 1[10]. The distance between 2 reduced convex hulls and the corresponding SVM solution is shown in figure 1.

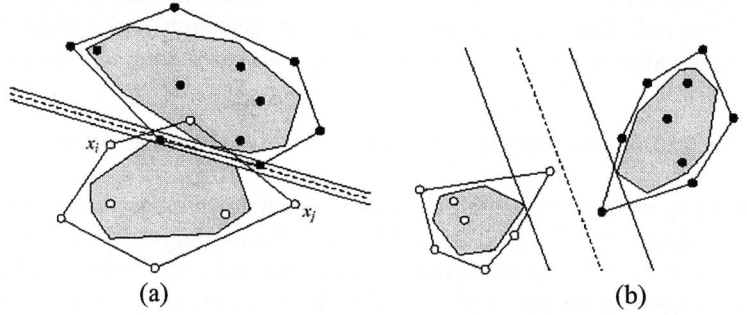

Fig. 1. (a) Soft margin SVMs maximize the margin between reduced convex hulls. (b) The soft margin in the separable case [10].

3 Feasible Direction Interpretation of Bennett's Method

In the nearest points algorithm (NPA) proposed by Bennett, a point in the reduced convex hull $A(\mu)$ which is nearer to the point Bv than the point Au is searched and Au is updated to this newfound point, then a point in the reduced convex hull $B(\mu)$ which is nearer to the point Au than the point Bv is searched and Bv is updated. The procedure is repeated until the optimal condition is satisfied.

Although it is proposed and implemented with a geometric concept, the above method can also be interpreted as a feasible direction method, a family of algorithms in optimization theory. In feasible direction methods for programming problem (5)

$$\min_x \ f(x) \qquad (5)$$

$$\text{s.t.} \quad x \in X.$$

x is updated with equation (6) in each step:

$$x_{k+1} = x_k + \alpha_k d_k \qquad (6)$$

where d_k is a feasible descent direction ($\nabla f(x_k)'d_k<0$), and step length $\alpha_k>0$. d_k and α_k are computed according to some rules so that $f(x_{k+1}) < f(x_k)$, $x_{k+1} \in X$. d_k can also be represented by $x - x_k$, $x \in X$.

The optimal conditions in [7] including corollary 10 for $\mu = 1$, corollary 12 for $\mu = 1/M$, and corollary 14 (kernel version of corollary 12) are all special forms of Theorem 17. In the theorem, $\min\{(c-d)'z | z \in A(\mu)\} \geq (c-d)'c \Leftrightarrow \min\{(c-d)'z - (c-d)'c \mid z \in A(\mu)\} \geq 0 \Leftrightarrow \min\{(c-d)'(z-c) \mid z \in A(\mu)\} \geq 0$. Since both z and c belong to $A(\mu)$, $(z-c)$ is a feasible direction. Besides, $(c-d)$ is $\nabla f(c)$ where $f(z) = \|z-d\|^2$. So the 1st optimal condition means no feasible descent direction at c for z. Similar arguments hold true with the 2nd optimal condition. Furthermore, it can be proven that when u and v are both considered at the same time, if these 2 conditions are satisfied, no feasible descent direction at $(u', v')'$ exists. Therefore, Theorem 17 and relative corollaries are just other descriptions for the cases that no such direction exists.

If the optimal condition is not satisfied, a feasible descent direction is obtained after checking. Then Bennett's method updates the current point with a precise line search along the direction. Consequently, NPA is a kind of feasible direction method.

Furthermore, the method for finding and updating u or v is the same as that in the conditional gradient method since the vector from the current point p to a feasible point with the largest project on ∇f at p is chosen as the direction. However, as ∇f is first projected to the subspace of u or v each time, the NPA here is not a traditional conditional gradient method. It is well-known that the conditional gradient method is convergent when ∇f is Lipschitz continuous, i.e. every limit point of sequence $\{x_k\}$ is a local minimal solution. However, the convergence of Bennett's NPA is not proven.

4 Improvement on the Selecting Working Set

In Bennett's NPA, M ($M = 1/\mu$) components of u or v are selected and updated each time. We also call these components working set in accordance with the decomposition algorithm.

Denote the object function of quadratic problem (4) as $f(u, v)$.

$$f(u,v) = \|Au - Bv\|^2 = \left\|(A \ -B)\binom{u}{v}\right\|^2 = (u' \ v')\binom{A'}{-B'}(A \ -B)\binom{u}{v} \qquad (7)$$

$$= (u' \ v')\begin{pmatrix} A'A & -A'B \\ -B'A & B'B \end{pmatrix}\binom{u}{v} = (u' \ v')Q\binom{u}{v}.$$

Therefore,

$$\nabla f(u,v) = Q\begin{pmatrix} u \\ v \end{pmatrix}. \qquad (8)$$

The next direction d_k is computed through the linear programming (9):

$$\min_{u,v} \; (P\nabla f(u_k, v_k))\left(\begin{pmatrix} u \\ v \end{pmatrix} - \begin{pmatrix} u_k \\ v_k \end{pmatrix}\right) \qquad (9)$$

$$\text{s.t.} \quad e'u = 1, 0 \leq u \leq \mu e, e'v = 1, 0 \leq v \leq \mu e.$$

where P is a project operator. In Bennett's NPA, $\nabla f(u, v)$ is projected to the subspace of u (v) so that the working set only consists of the components of u (v). If the object value is 0 when P is equal to the identity matrix, the distance between 2 reduced convex hulls is obtained. Then the SVM solution can be computed easily.

So, $\nabla f(u, v)$ can be stored and updated whenever u or v is updated. And only M rows of Q are needed to update ∇f since only M components from u (v) are changed. In such a way, much kernel computation can be saved. In the original methods, in order to find the directions for u and v in 2 adjacent steps, all items in Q have to be computed in the worst case. If the memory is not large enough to store Q, the performance of the original method will be bad.

5 Extension to the Arbitrary Valid Value of μ

In NPA, μ is $1/M$ while M is an integer between 1 to the minimum of the amount of positive samples and that of negative samples. With this constraint, any extreme point of a reduced convex hull must be the average of some M feature vectors.

However, as an important coefficient in SVM, it is critical to the performance of decision function. The limitation on its definition domain may be harmful to find the μ with good generalization ability. Fortunately, the limitation can be removed.

In [10], Bern presented a lemma that there is an algorithm with $O(nd)$ arithmetic operations for optimizing a linear function over a reduced convex hull of n points in R^d. The proof is constructional, i.e. it is an algorithm. Assume that the linear function is denoted by a vector w. Compute the projection on w for each point, and then arrange the points according to the projection in decreasing order. The weight for the first $\lfloor 1/\mu \rfloor$ points is μ, the weight of the $\lceil 1/\mu \rceil$th point is $1 - \mu * \lfloor 1/\mu \rfloor$, and that of the others is 0.

Since finding the feasible direction is equivalent to solving a linear programming whose feasible area is a reduced convex hull, the above algorithm can be used. Moreover, although it is originally suitable for input space, it can be extended to feature space even of infinite dimension because only the inner product is used. Assume that w is the weighted sum of l images of vectors in the input space, then $n*l$ kernel operations are needed besides other arithmetic and comparison operations. The computation is almost the same as that of Bennett's NPA.

6   Conclusion

Intuitive geometric interpretation for SVM is not only helpful to understand the concept of SVM but also provides an alternative way to implement SVM. Although Bennett's NPA can deal with reduced convex hulls, it has some disadvantages.

In this paper, a feasible direction explanation for NPA is proposed so that the computation of the kernel can be reduced greatly. Besides, the original NPA is extended to an arbitrary valid value of μ, so a better generalization performance may be obtained.

Since such NPA is not a classical conditional gradient method, ist convergence characteristic should be investigated in the future. And ist practical efficiency also needs to be examined.

Acknowledgements. The work was supported by DAAD PPP D/02/12755 and National Nature Science Foundation of China, Grant No. 60135010 and No. 60321002.

References

1. Vapnik, V. N.: The Nature of Statistical Learning Theory, Springer-Verlag, New York, 2000.
2. Burges, C. J. C.: A Tutorial on Support Vector Machines for Pattern Recognition. Data Mining and Knowledge Discovery, 2(2) (1998) 1-43
3. Smola, A., Schölkopf B.: A Tutorial on Support Vector Regression, NeuroCOLT Technical Report NC-TR-98-030, Royal Holloway College, University of London, UK, 1998
4. Bennet, K. P., Bredensteier, E. J.: Duality and Geometry in SVM Classifiers. In: Pat Langley (ed.): Proceedings of the Seventeenth International Conference on Machine Learning, Morgan Kaufmann, San Francisco (2000) 57-64
5. Crisp, D. J., Burges, C. J. C.: A Geometry Interpretation of µ-SVM Classifiers. In: Solla, S., Leen, T., Muller, K. (eds.): Advances in Neural Information Processing Systems (NIPS 12). MIT Press, Cambridge, MA (2000) 244-251
6. Bi, J., Bennett, K. P.: Duality, Geometry, and Support Vector Regression. In: Dietterich, T., Becker, S., Ghahramani, Z. (eds.): Advances in Neural Information Processing Systems 14 (NIPS 14). MIT Press, Cambridge, MA (2002)593-600
7. Bi, J., Bennett, K. P.: A Geometric Approach to Support Vector Regression. Nerocomputing, 55(1-2) (2003) 79-108
8. Keerthi, S. S., Shevade, S. K., Bhattcharyya C., Murthy K. R. K.: A Fast Iterative Nearest Point Algorithm For Support Vector Machine Classifier Design. IEEE Transactions on Neural Network, 11(1) (2000) 124-136.
9. Yang, M., Ahuja, N.: A Geometric Approach to Train Support Vector Machines. In: Proceedings of IEEE Computer Society Conference on Computer Vision and Pattern Recognition (CVPR 2000). IEEE Computer Society, Los Alamitos, CA (2000) 1430-1437
10. Bern, M., Eppstein, D.: Optimization Over Zonotopes and Training Support Vector Machines. In: Dehne, F., Sack, J.-R., Tamassia R. (eds.): Proceedings of the Seventh International Workshop on Algorithms and Data Structure (WADS 2001). Lecture Notes in Computer Science, Vol. 2125. Springer-Verlag, Heidelberg New York (2001) 111-121

Distance-Based Selection of Potential Support Vectors by Kernel Matrix

Baoqing Li

Department of Computer Science, School of Computing,
National University of Singapore, Singapore 117543
libaoqin@comp.nus.edu.sg
http://www.comp.nus.edu.sg/~libaoqin

Abstract. We follow the idea of decomposing a large data set into smaller groups, and present a novel distance-based method of selecting potential support vectors in each group by means of kernel matrix. Potential support vectors selected in the previous group are passed to the next group for further selection. Quadratic programming is performed only once, on the potential support vectors still retained in the last group, for the construction of an optimal hyperplane. We avoid solving unnecessary quadratic programming problems at intermediate stages, and can take control over the number of selected potential support vectors to cope with the limitations of memory capacity and existing optimizers' capability. Since this distance-based method does not work on data containing outliers and noises, we introduce the idea of separating outliers/noises and the base, by use of the k-nearest neighbor algorithm, to improve generalization ability. Two optimal hyperplanes are constructed on the *base part* and the *outlier/noise part*, respectively, which are then synthesized to derive the optimal hyperplane on the overall data.

1 Introduction

Support vector machines achieve a higher level of accuracy of classification than traditional neural networks such as multilayer perceptrons or radial basis function networks. The underlying principle of a support vector machine is to construct an optimal hyperplane based on the concept of structural risk minimization [1]. That is, not only does it minimize the empirical risk, like a traditional neural network, but simultaneously, it also minimizes the structural complexity by maximizing the margin of the hyperplane, in the hope of greater generalization ability. However, this objective function is optimized by solving a constrained convex quadratic programming problem. When the training data are not linearly separable in the input space, they are mapped via a nonlinear function into a high dimensional feature space, where their corresponding images are linearly separable. Things become more complicated because this quadratic programming problem is now built on this high dimensional feature space. With the help of kernels, fortunately, the quadratic programming problem is reduced back from the high dimensional feature space to the original input space. Even so, it is not still an easy task in general to solve a constrained convex quadratic programming problem, especially when the training data set is extremely large. Several ways to overcome this difficulty have been proposed, for example, chunking [2], decomposition [3], and sequential minimal optimization [4].

In this paper, we follow the idea of decomposing a large data set into smaller groups, each of which is small enough so that the kernel matrix of the data of that group can be stored in memory. In each group, we propose a distance-based method, rather than using the traditional quadratic programming method, for the selection of potential support vectors. The quadratic programming is performed only once on the potential support vectors from *all* the groups. The significance of this proposed method is that

1. we avoid solving unnecessary quadratic programming problems for the selection of potential support vectors at intermediate stages and,
2. we can take control over the number of selected potential support vectors and thus be able to cope with the limitations of memory capacity and existing optimizers' capability.

By intuition, if the distance between two points of different classes is short, they have the potential to become support vectors. Kernels provide a convenient means of distance measure and is the foundation of this paper, which is introduced in Section 2.

However, this distance-based criterion does not work on data with outliers and noises (mislabeled data). Hence the first step should separate outliers/noises (called *outlier/noise part*) and the others (called *base part*) by use of the k-nearest neighbor algorithm, which is described in Section 3.

In Section 4, we select potential support vectors based on distance measure by means of kernel matrix, and construct optimal hyperplanes on both parts (*base part* and *outlier/noise part*). The reason why we do not discard the *outlier/noise part* is that this part still contains certain useful information.

In Section 5, we introduce a generalized form of the hyperplane, which says that the hyperplane is in fact the weighted sum of the training data times their class labels, with more weights given to the data that are more difficult to be correctly classified. Therefore, the overall optimal hyperplane is, as can be induced, the weighted average of the optimal hyperplane on the *base part* and the optimal hyperplane on the *outlier/noise part*.

An illustrated example is given in Section 6.

2 Distance Measure by Kernels

[5] considers kernels, in fact, a measure of distance. Suppose $\mathbf{x}_1$ and $\mathbf{x}_2$ are two training vectors in the input space, they are mapped via a nonlinear function $\varphi(\cdot)$ into a high dimensional feature space, where their corresponding images are $\varphi(\mathbf{x}_1)$ and $\varphi(\mathbf{x}_2)$, respectively. Define kernel K to be the inner product of these two images, $K(\mathbf{x}_1, \mathbf{x}_2) = \varphi^T(\mathbf{x}_1) \cdot \varphi(\mathbf{x}_2)$. The squared Euclidean distance between these two images in the feature space is

$$\|\varphi(\mathbf{x}_1) - \varphi(\mathbf{x}_2)\|^2 = K(\mathbf{x}_1, \mathbf{x}_1) + K(\mathbf{x}_2, \mathbf{x}_2) - 2K(\mathbf{x}_1, \mathbf{x}_2) \ . \quad (1)$$

That is to say, we can easily calculate the distance between two images in the feature space by means of the kernel functions of the two corresponding points in the input space.

This is more clear in the case of radial basis functions as kernels. If kernel K is defined as

$$K(\mathbf{x}_1, \mathbf{x}_2) = \exp\left(-\frac{\|\mathbf{x}_1 - \mathbf{x}_2\|^2}{2\sigma^2}\right),$$

then Equation (1) becomes

$$\|\varphi(\mathbf{x}_1) - \varphi(\mathbf{x}_2)\|^2 = 2 - 2K(\mathbf{x}_1, \mathbf{x}_2).$$

3 Separation of Outliers/Noises and the Base

Intuitively speaking, if two points that belong to two different classes are close enough, they have the potential to become support vectors. We can thus set a threshold of distance, and assume that any two points of different classes, with their distance shorter than this preset threshold, are potential support vectors. This exhaustive search for potential support vectors can be done through kernel matrix without having to solve quadratic a programming problem. The threshold can be adjusted to limit the number of selected potential support vectors to memory capacity.

However, the above criterion does not hold if the training data are noisy (mislabeled) or the outliers of one class stretch to the territory of the other class. Hence our first step should separate outliers/noises and the base.

The nearest neighbor algorithm is an efficient and robust method of clustering. In the simplest 1-nearest neighbor form, a point is classified into the same category as its nearest neighbor. In the k-nearest neighbor form, where $k \geq 2$, the category of a point is determined by the weighted average of its k-nearest neighbors, greater weights are given to its closer neighbors.

We use the 1-nearest neighbor algorithm first to look for suspicious outliers/noises, if the labels of two nearest points do not match, they become suspicious outliers/noises. We then use the k-nearest neighbor algorithm to find out which one of the two is an outlier/noise. The one whose class label is not in accordance with the classification that is determined by using the k-nearest neighbor algorithm is deemed an outlier/noise. Unfortunately, we cannot find out whether it is an outlier or a noise. Sometimes, what's more, outliers/noises occur very close, while the base is distributed sparsely. Such outlier/noise pairs or groups cannot be detected by using the 1-nearest neighbor algorithm, so we have to resort to the k-nearest neighbor algorithm for the detection of possible outliers/noises that are neglected by the 1-nearest neighbor algorithm. More computational resources, of course, will be consumed with a larger k, so we must pick a tradeoff between accuracy and efficiency.

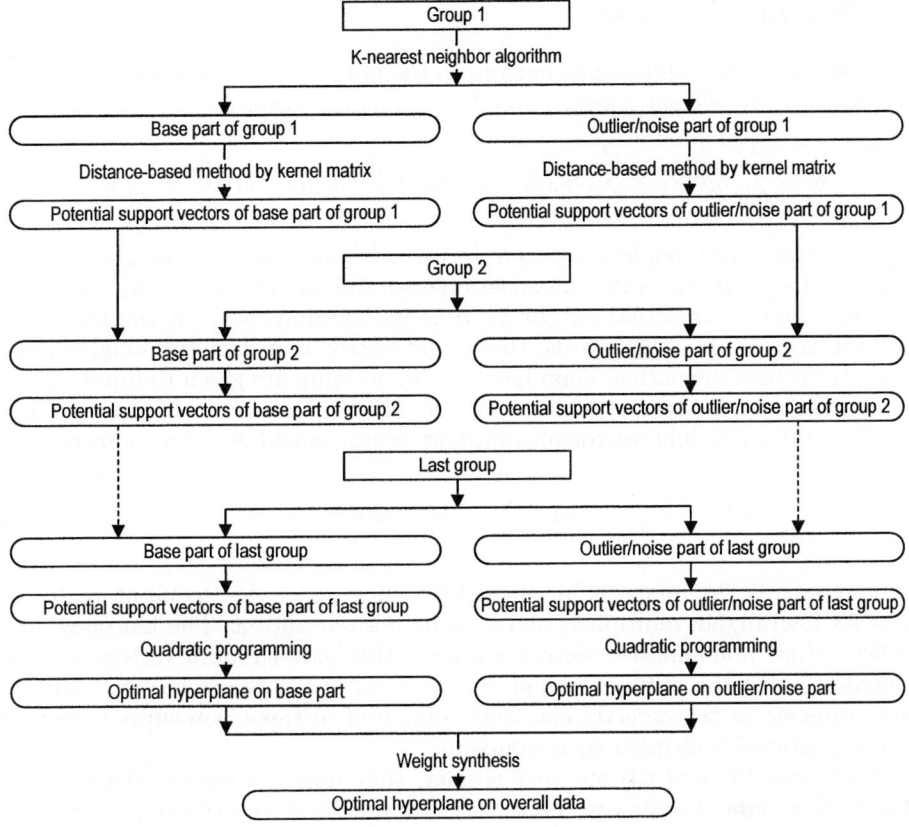

Fig. 1. Procedure for the selection of potential support vectors

4 Procedure for Selection of Potential Support Vectors

As is described in Figure 1, we first decompose the overall data set into smaller groups, e.g., *group 1*, *group 2*, ..., until the *last group*. Each group is small enough to be stored in memory. In each group, say in group 1, data are further divided into the *base part of group 1* and the *outlier/noise part of group 1*, by using the k-nearest neighbor algorithm. In each individual part, distance-based method by means of kernel matrix is used to select potential support vectors. Potential support vectors of both parts of *group 1* are then passed to the corresponding parts of group 2 for further selection. This procedure is repeated until the *last group*. Quadratic programming is performed only once on the potential support vectors still retained in the *last group*, and two optimal hyperplanes on both parts are constructed, respectively. By synthesizing the two hyperplanes (to be discussed in Section 5), an optimal hyperplane on the overall data will be obtained.

5 Weight Synthesis

From Rosenblatt's original perceptron, to the later more complicated multilayer perceptrons, the weight learning rule has a uniform form of

$$\mathbf{w} = \sum_{i=1}^{l} \eta_i e_i \mathbf{x}_i = \sum_{i=1}^{l} \eta_i (d_i - y_i) \mathbf{x}_i = \sum_{i=1}^{l} \eta_i |d_i - y_i| \frac{d_i - y_i}{|d_i - y_i|} \mathbf{x}_i , \qquad (2)$$

provided the initial weight is set to zero, $\mathbf{w}_0 = \mathbf{0}$, where $\mathbf{w}$ is the weight vector to be learned, $e_i = d_i - y_i$ is the classification error due to input vector $\mathbf{x}_i$, its desired output d_i minus its actual output y_i, η_i is its learning rate, l is the number of training vectors. In other words, the weight vector is in fact the weighted sum of input vectors times their class labels, more weights are given to input vectors that produce larger errors (hence are more difficult to be correctly classified.)

The optimal weight vector of a support vector machine is determined by

$$\mathbf{w} = \sum_{i=1}^{l} \alpha_i d_i \varphi(\mathbf{x}_i) , \qquad (3)$$

where $\varphi(\mathbf{x}_i)$ is the image induced in the feature space due to input vector $\mathbf{x}_i$, α_i is its Lagrangian multiplier, and d_i is its desired output. The Lagrange multipliers of all non-support vectors are zero, thus non-support vectors have no contribution to the optimal weight vector. If $\mathbf{x}_i$ is a support vector, then it is more difficult to be correctly classified compared to those non-support vectors, so the Lagrange multiplier α_i is non-zero.

Equations (2) and (3) are very similar, they have a common characteristic that both weight vectors are represented by weighted sum of input vectors (or their corresponding images) times their class labels, more weights are given to the input vectors (or their corresponding images) producing larger errors (more difficult to be correctly classified). This is the reason why we do not discard the *outlier/noise part* completely, because it still contains useful information to contribute to the construction of the overall optimal hyperplane.

If we assume that the *outlier/noise part* contains only outliers, then the optimal hyperplane on the *outlier part* $\mathbf{w}_{outlier}$ should have a positive contribution to the overall optimal hyperplane $\mathbf{w}_{overall}$:

$$\mathbf{w}_{overall} = (1 - \lambda) \mathbf{w}_{base} + \lambda \mathbf{w}_{outlier} .$$

If we assume that the *outlier/noise part* contains only noises, then the optimal hyperplane on the *noise part* $\mathbf{w}_{noise}$ should have a negative (because the training data the whole *noise part* are mislabeled) contribution to the overall optimal hyperplane $\mathbf{w}_{overall}$:

$$\mathbf{w}_{overall} = (1 - \lambda) \mathbf{w}_{base} - \lambda \mathbf{w}_{noise} ,$$

where λ is a positive number much smaller than 1, $0 < \lambda \ll 1$.

The difficulty here is that we can by no means distinguish outliers from noises in the *outlier/noise part*, therefore, weight synthesis is data and task dependent, and is on the basis of trial and error.

6 An Illustrated Example

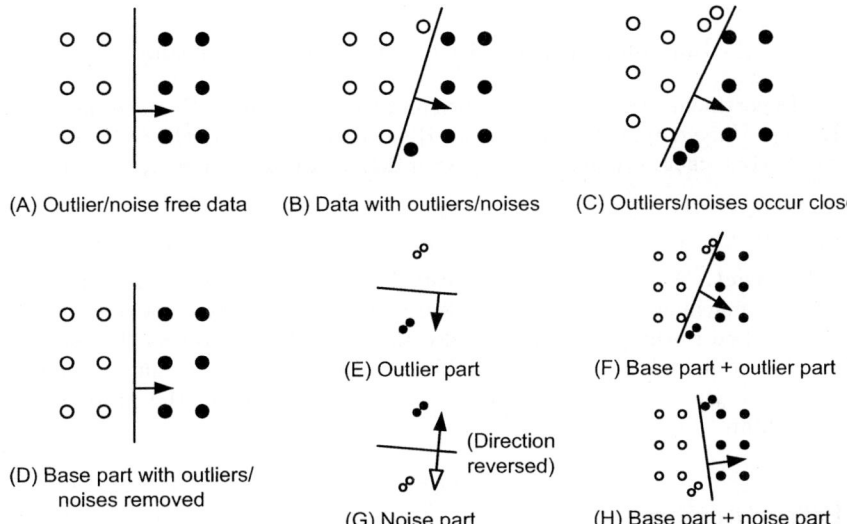

Fig. 2. An illustrated example: (A) If the training data are outlier/noise free, an optimal hyperplane is constructed with the maximum margin; (B) The optimal hyperplane constructed on data with outliers and noises is worse than that constructed on the *base part* in (D), with outliers and noises removed by using the nearest neighbor algorithm; (C) When outliers and noises occur very close, the 1-nearest neighbor algorithm does no good, while the k-nearest neighbor algorithm can help. More computational resources will be consumed with a larger k, and there is a tradeoff between accuracy and efficiency, which is task and data independent; (E) If we assume the *outlier/noise part* contains only outliers, the hyperplane on that part has a positive contribution to the overall hyperplane in (F); (G) If we assume the *outlier/noise part* contains only noises, the hyperplane on that part has a negative contribution to the overall hyperplane in (H), which is again data and task dependent and needs trial and error method

References

1. Vapnik, V.N.: The Nature of Statistical Learning Theory. 2nd edn. Springer (1999)
2. Vapnik, V.N.: Estimation of Dependencies Based on Empirical Data. Springer (1982)
3. Osuna, E.E., Freund, R., Girosi, F.: Support vector machines: Training and applications. A.I.Memo 1602, Massachusetts Institute of Technology (1997)
4. Platt, J.C.: Fast training of support vector machines using sequential minimal optimization. In Scholköpf, B., Burges, C.J.C., Smola, A.J., eds.: Advances in Kernel Methods: Support Vector Learning. The MIT Press (1998)
5. Scholköpf, B.: The kernel trick for distances. In Leen, T., Dietterich, T., Tresp, V., eds.: Advances in Neural Information Processing Systems. Vol. 13. The MIT Press (2000)

A Learning Method for Robust Support Vector Machines

Jun Guo, Norikazu Takahashi, and Tetsuo Nishi

Department of Computer Science and Communication Engineering,
Kyushu University, 6-10-1 Hakozaki, Higashi-ku, Fukuoka, 812-8581, Japan
guojun@kairo.csce.kyushu-u.ac.jp {norikazu, nishi}@csce.kyushu-u.ac.jp

Abstract. We propose an innovative learning algorithm for a support vector machine to be robust. As learning patterns it uses not only the prescribed learning patterns but also their neighbour patterns. The size of the proposed optimization problem to be solved is the same as the original one. Many simulations show the effectiveness of the proposed algorithm.

1 Introduction

One of the important problems on the learning of support vector machines is robustness, i.e., we always want to obtain SVMs to have high ability of generalization. Many methods have so far been proposed[5]-[11]. In this paper we propose a method for a robust learning. Main idea is to adopt many learning points in addition to the original ones. It necessarily yield the large scale optimization problem. So we devise the reduction of the problem size by the use of an averaging method.

We carried out many simulation for several classification problems. The results show very good performance of the proposed method.

2 Preliminaries

Let $\varGamma = \{x_i | i = 1, 2, \cdots, N\}$ be a set of the prescribed learning patterns and we assume that x_i are to be classified into two distinct classes, Class $+1$ (i.e., $d_i = +1$) and Class -1 (i.e., $d_i = -1$). We call these $x_i (i = 1, 2, \cdots, x_N)$ the "*original learning patterns* (the term "learning" will often be omitted)", while we introduce additional learning patterns later.

Let a Kernel function be $K(x, y)$. The function K may be the Gaussian radial basis function or the polynomial function as follows:

$$K(x, y) = e^{-\frac{\|x-y\|^2}{2\sigma^2}} \tag{1}$$

$$K(x, y) = (1 + x^T y)^p \tag{2}$$

The classification of the learning patterns can be usually carried out by solving the following problem:
Original Optimization Problem (Dual)
Maximize the objective function:

$$Q(\alpha) = \sum_{i=1}^{N} \alpha_i - \frac{1}{2} \sum_{i=1}^{N} \sum_{j=1}^{N} \alpha_i \alpha_j d_i d_j K(x_i, x_j) \qquad (3)$$

subject to:

$$0 \leq \alpha_i \quad (i = 1, 2, \cdots, N) \qquad (4)$$

$$\sum_{i=1}^{N} \alpha_i d_i = 0 \qquad (5)$$

Remark 1: The condition (4) can be replaced by $0 \leq \alpha_i \leq C$ for some positive value C.

If σ of the RBF kernel in Eq.(1) is sufficiently small, then all original patterns can be correctly classified. Then, however, we cannot guarantee that patterns x which are very close to an original pattern, for example, x_1, is not necessarily classified into the same class as x_1. That is, it is not robust. To obtain the robust learning, we demand in this paper the following requirement:

Requirement 1: Let one original pattern be x_i and let the nearest original pattern to x_i be x_{*i}. If the patterns x_i and x_{*i} belong to the same class, then we impose that all patterns linearly connecting x_i and x_{*i} are classified into the same class as x_i.

3 Extended Optimization Problem

In this paper we deal with Requirement 1 above. For simplicity of description we focus our discussion on *one* original pattern, say x_1 and let x_{*1} be the nearest original pattern to x_1 and belong to the same class as x_1.

We determine additional $s_1 - 1$ learning patterns x_{1k} $(k = 1, 2, \cdots, s_1 - 1)$:

$$x_{1k} \equiv x_1 + \frac{k}{s_1}(x_{*1} - x_1) \quad (k = 1, 2, \cdots, s_1 - 1) \qquad (6)$$

For convenience we assume that $x_{10} = x_1$ and $x_{1s_1} = x_{*1}$.

For the set of the extended learning patterns Γ_E, which is the union of Γ and $\{x_{1k}\}$ $(k = 1, 2, \cdots, s_1 - 1)$, we can write the optimization problem corresponding to Eqs.(3)-(5) as follows:
Extended Optimization Problem on Γ_E
Maximize the objective function:

$$Q_E(\alpha) = \sum_{i=1}^{N} \alpha_i + \sum_{k=1}^{s_1-1} \alpha_{1k} - \frac{1}{2} \sum_{i=1}^{N} \sum_{j=1}^{N} \alpha_i \alpha_j d_i d_j K(x_i, x_j)$$

$$-\frac{1}{2}\sum_{k=1}^{s_1-1}\sum_{j=1}^{N}\alpha_{1k}\alpha_j d_{1k}d_j K(x_{1k},x_j) - \frac{1}{2}\sum_{i=1}^{N}\sum_{l=1}^{s_1-1}\alpha_i\alpha_{1l}d_i d_{1l}K(x_i,x_{1l})$$

$$-\frac{1}{2}\sum_{k=1}^{s_1-1}\sum_{l=1}^{s_1-1}\alpha_{1k}\alpha_{1l}d_{1k}d_{1l}K(x_{1k},x_{1l}), \text{ (where } d_{1k}=d_1) \tag{7}$$

subject to:

$$0 \leq \alpha_i \ (i=1,2,\cdots,N), \ 0 \leq \alpha_{1k} \ (k=1,2,\cdots,s_1-1) \tag{8}$$

$$\sum_{i=1}^{N}\alpha_i d_i + \sum_{k=1}^{s_1-1}\alpha_{1k}d_{1k} = 0 \tag{9}$$

Let

$$\lambda_{1k}=\frac{k}{s_1}, \quad \mu_{1l}=\frac{k}{s_1} \quad (k,l=0,1,2,\cdots,s_1-1,s_1) \tag{10}$$

Let for convenience

$$\lambda_{10}=\mu_{10}=1, \quad \alpha_{10}=\alpha_1, \quad d_{10}=d_1. \tag{11}$$

We have of course

$$0 \leq \lambda_{1k},\mu_{1l} \leq 1, \quad \lambda_{1k}=\mu_{1k} \quad (k,l=0,1,2,\cdots,s_1-1,s_1) \tag{12}$$

Since $x_1 \approx x_{*1}$, we can approximate $K(x_{1k},\cdot)$ and $K(x_{1k},x_{1l})$ as follows:

$$K(x_{1k},\cdot) = K(\lambda_{1k}x_1 + (1-\lambda_{1k})x_{*1},\cdot) \approx \lambda_{1k}K(x_1,\cdot) + (1-\lambda_{1k})K(x_{*1},\cdot) \tag{13}$$

$$K(x_{1k},x_{1l}) = K(\lambda_{1k}x_1 + (1-\lambda_{1k})x_{*1}, \mu_{1l}x_1 + (1-\mu_{1l})x_{*1})$$
$$\approx \lambda_{1k}\mu_{1l}K(x_1,x_1) + (1-\lambda_{1k})\mu_{1l}K(x_{*1},x_1) + \lambda_{1k}\mu_{1l}K(x_1,x_{*1})$$
$$+(1-\lambda_{1k})(1-\mu_{1l})K(x_{*1},x_{*1}) \tag{14}$$

Substituting Eqs.(13) and (14) into Eq.(7), we have the "*modified objective function*" $\tilde{Q}_E$ as:

$$\tilde{Q}_E(\alpha) = \sum_{i=1}^{N}\alpha_i + \sum_{k=1}^{s_1-1}\alpha_{1k} - \frac{1}{2}\sum_{i=1}^{N}\sum_{j=1}^{N}\alpha_i\alpha_j d_i d_j K(x_i,x_j)$$

$$-\frac{1}{2}\sum_{k=1}^{s_1-1}\sum_{j=1}^{N}\alpha_{1k}\alpha_j d_{1k}d_j\{\lambda_{1k}K(x_1,x_j)+(1-\lambda_{1k})K(x_{*1},x_j)\}$$

$$-\frac{1}{2}\sum_{i=1}^{N}\sum_{l=1}^{s_1-1}\alpha_i\alpha_{1l}d_i d_{1l}\{\mu_{1l}K(x_i,x_1)+(1-\mu_{1l})K(x_i,x_{*1})\}$$

$$-\frac{1}{2}\sum_{k=1}^{s_1-1}\sum_{l=1}^{s_1-1}\alpha_{1k}\alpha_{1l}d_{1k}d_{1l}\{\lambda_{1k}\mu_{1l}K(x_1,x_1)+(1-\lambda_{1k})\mu_{1l}K(x_{*1},x_1)$$

$$+\lambda_{1k}(1-\mu_{1l})K(x_1,x_{*1})+(1-\lambda_{1k})(1-\mu_{1l})K(x_{*1},x_{*1})\} \tag{15}$$

We call the Extended Optimization Problem with the objective function (7) replaced by Eq.(15) "*Linearized Extended Optimization Problem*".

As the result of straightforward but tedious calculation, the third through sixth terms of Eq.(15) can be summarized as:

$$\text{Coefficient of } K(x_1, x_1) = -\frac{1}{2}\left(\sum_{k=0}^{s_1-1} \lambda_{1k}\alpha_{1k}d_{1k}\right)\left(\sum_{l=0}^{s_1-1} \mu_{1l}\alpha_{1l}d_{1l}\right) \quad (16)$$

$$\text{Coefficient of } K(x_1, x_j) = -\frac{1}{2}\left(\sum_{k=0}^{s_1-1} \lambda_{1k}\alpha_{1k}d_{1k}\right)\alpha_j d_j \quad (j \neq 1, *1) \quad (17)$$

$$\text{Coefficient of } K(x_i, x_j) = -\frac{1}{2}\alpha_i\alpha_j d_i d_j \quad (i,j \neq 1, *1) \quad (18)$$

$$\text{Coefficient of } K(x_1, x_{*1}) = -\frac{1}{2}\left(\sum_{k=0}^{s_1-1} \lambda_{1k}\alpha_{1k}d_{1k}\right)\left(\sum_{l=0}^{s_1-1} (1-\mu_{1l})\alpha_{1l}d_{1l}\right) \quad (19)$$

$$\text{Coefficient of } K(x_{*1}, x_{*1}) = -\frac{1}{2}\left(\sum_{k=0}^{s_1-1} (1-\lambda_{1k})\alpha_{1k}d_{1k}\right)\left(\sum_{l=0}^{s_1-1} (1-\mu_{1l})\alpha_{1l}d_{1l}\right) \quad (20)$$

4 Reduction of the Problem Size

Observing Eqs.(16)-(20), we conclude that
Lemma 1: The modified objective function in Eq.(7) is obtained by simply replacing α_1 and α_{*1} with $\tilde{\alpha}_1$ and $\tilde{\alpha}_{*1}$ below:

$$\alpha_1 \rightarrow \tilde{\alpha}_1 \equiv \sum_{k=0}^{s_1-1} \lambda_{1k}\alpha_{1k}, \quad \alpha_{*1} \rightarrow \tilde{\alpha}_{*1} \equiv \sum_{k=0}^{s_1-1} (1-\lambda_{1k})\alpha_{1k} \quad (21)$$

and by keeping other variables α_i ($i \neq 1, *1$) intact.

Let $\tilde{\alpha}_i = \alpha_i$ ($i \neq 1, *1$). Then we see that the vector $[\alpha_1, \alpha_2, \cdots, \alpha_N]$ and $[\tilde{\alpha}_1, \tilde{\alpha}_2, \cdots, \tilde{\alpha}_N]$ can be related by a very simple linear combinations. Thus:
Lemma 2: The Modified Optimum Problem can be rewritten in the same form as the original optimization except for α_i replaced by $\tilde{\alpha}_i$.

Lemma 2 tells us that the augmented variables α_{1k} introduced in the previous section can be approximated by the linear combination of α_1 and α_{*1}. ¿From this fact we can derive the following formulation:

Modified Optimization Problem on Γ
    Let
$$\alpha_{1k} = \lambda_k\alpha_1 + (1-\lambda_k)\alpha_{*1} \quad (22)$$

Maximize the objective function:

$$Q_E(\alpha) = \sum_{i=1}^{N} \alpha_i + \sum_{k=1}^{s_1-1} \alpha_{1k} - \frac{1}{2}\sum_{i=1}^{N}\sum_{j=1}^{N} \alpha_i\alpha_j d_i d_j K(x_i, x_j)$$

$$-\frac{1}{2}\sum_{k=1}^{s_1-1}\sum_{j=1}^{N} \alpha_{1k}\alpha_j d_{1k} d_j K(x_{1k}, x_j) - \frac{1}{2}\sum_{i=1}^{N}\sum_{l=1}^{s_1-1} \alpha_i\alpha_{1l} d_i d_{1l} K(x_i, x_{1l})$$

$$-\frac{1}{2}\sum_{k=1}^{s_1-1}\sum_{l=1}^{s_1-1}\alpha_{1k}\alpha_{1l}d_{1k}d_{1l}K(x_{1k},x_{1l}) \tag{23}$$

subject to:

$$0 \le \alpha_i \ (i=1,2,\cdots,N), \ 0 \le \alpha_{1k} \ (k=1,2,\cdots,s_1-1) \tag{24}$$

$$\sum_{i=1}^{N}\alpha_i d_i + \sum_{k=1}^{s_1-1}\alpha_{1k}d_{1k} = 0 \tag{25}$$

According to Eq.(22) the Modified Optimization Problem on Γ has only N independent variables $\alpha_i (i=1,\cdots,N)$ as the Original Optimization Problem does. By solving the above optimization problem, we obtain the optimum $\tilde{\alpha}_i$ $(i=1,\cdots,N)$. Then $\tilde{\alpha}_{1k} \ge 0$ $(i=1,\cdots,s_1-1)$ can be obtained from Eq.(22).

Remark 2: As easily seen, the Modified Optimization Problem on Γ is the convex quadratic programming.

Remark 3: Until now we discussed about only one original pattern x_1, but we can easily generalize the results into all original patterns x_i $(i=1,\cdots,N)$. See the subsequent simulations.

Remark 4: Since the number of the additional patterns, s_1-1, in the previous results is not important, we may increase it infinitely. Then the summation in some relevant equations can be relpaced by the integration. This may make the calculation for the above optimization problem easy.

5 Simulation Results

We carried out many simulations for several classification problems for the variety of σ in Eq.(1). Figs. 1 and 2 show the comparison of the results of the Original Optimization Problem and the Modified Optimization Problem, respectively, where $\sigma = 0.1$. In these figures the patterns marked by ○ belong to Class $+1$, and those marked by × belong to Class -1. The regions inside the curves show points x satisfying $f(x) < 0$, where $f(x)$ is the discrimination function. We can see from these figures the followings:

1. The Original Optimization Problem can classify the original patterns completely for small value of σ, but the learning is not robust (See Fig.1).
2. The Modified Optimization Problem can classify not only the original patterns but also the extended patterns as described in Requirement 1(See Fig.2).

At the conference we will give many examples showing the effectiveness of our method.

6 Conclusion

We gave a method for the robust learning. Simulation results show that the proposed method works well for some kinds of classification problems, especially complicated classification.

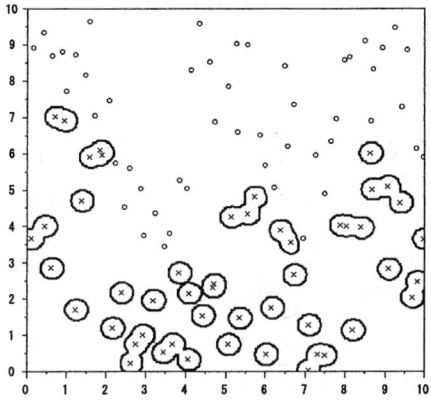
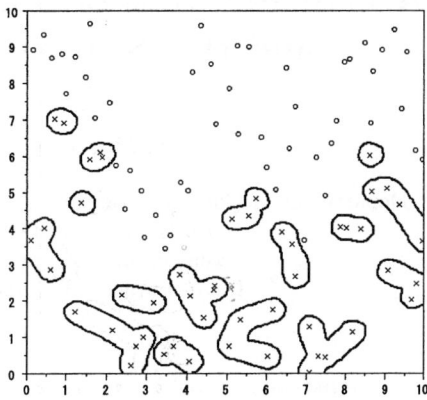

Fig. 1. Results by solving the original optimization problem($\sigma = 0.1$)

Fig. 2. Results by solving the modified optimization problem($\sigma = 0.1$)

References

1. Vapnik, V.N.: The Nature of Statistical Learning Theory. Springer-Verlag, New York (1995)
2. Vapnik, V.N.: Statistical Learning Theory. Wiley, New York (1998)
3. Schölkopf, B., Burges, C.J.C., Smola, A.J.: Advances in Kernel Methods - Support Vector Learning. The MIT Press, Cambridge, Massachusetts, London (1999)
4. Burges, C.J.C.: A Tutorial on Support Vector Machines for Pattern Recognition. Data Mining and Knowledge Discovery 2 (1998) 121-167
5. Buhot, A., Gordon, M.B: Robust Learning and Generalization with Support Vector Machines. J. Phys. A: Math. Gen. 34 (June 2001) 4377-4388
6. Lanckriet, G.R.G., Ghaoui, L.El, Bhattacharyya, C., Jordan, M.I.: A Robust Minimax Approach to Classification. Journal of Machine Learning Research 3 (2002) 555-582
7. Joachims, T.: Estimating the Generalization Performance of an SVM Efficiently. In Proceedings of the International Conference on Machine Learning, San Francisco (2000)
8. Navarrete, P., Ruiz del Solar, J.: On the Generalization of Kernel Machines. Pattern Recognition with Support Vector Machines, Springer (2002) 2388:24-39
9. Shawe-Taylor, J., Cristianini, N.: Robust Bounds on the Generalization from the Margin Distribution. NeuroCOLT Technical Report NC-TR-98-029, Royal Holloway College, University of London, UK (1998)
10. Xia, Y.S., Wang, J.: A One-layer Recurrent Neural Network for Support Vector Machine Learning. IEEE Transactions on Systems, Man and Cybernetics-Part B (2004) 1261-1269
11. Schölkopf, B., Burges, C.J.C., Vapnik, V.N.: Extracting Support Data for a Given Task. In:U.M. Fayyad and R. Uthurusamy (eds.), Proceedings, First International Conference on Knowledge Discovery & Data Mining, AAAI Press, Menlo Park, CA (1995)

A Cascade Method for Reducing Training Time and the Number of Support Vectors

Yi-Min Wen[1,2] and Bao-Liang Lu[1]

[1] Department of Computer Science and Engineering, Shanghai Jiao Tong University,
1954 Hua Shan Rd., Shanghai 200030, China
wenyimin@sjtu.edu.cn, blu@cs.sjtu.edu.cn
[2] Hunan Industry Polytechnic, Changsha 400052, China

Abstract. A novel cascade learning strategy for training support vector machines (SVMs) is proposed to speed up the training of SVMs. The training procedure consists of three steps which are performed in a cascade way. All the subproblems are processed parallelly in each step, and non-support-vector data are filtered out step by step. The simulation results indicate that our method not only speeds up the training procedure while maintaining the generalization accuracy of SVMs but also reduces the number of support vectors.

1 Introduction

In many real-world problems such as text categorization and geography information classification, the sizes of training data sets are usually massive. For example, the Yomiuri News Corpus contains 2,190,512 documents dated 1987-2001. It is necessary to develop efficient methods to deal with these real-world large-scale problems.

Support vector machines (SVMs) [1] have become a popular tool of machine learning. There are two kinds of methods for solving large-scale pattern classification problems. The first method is the incremental learning approach, in which a large-scale problem is divided into many small subproblems that are learned sequently [2]. This approach includes the advanced working set algorithms that use only a subset of the variables as a working set while freezing the others [3], [4]. The shortcoming of this kind of method is that a large number of iterations are required. Therefore, if the training data set is large, the training time will be very long. The second method is the parallel learning method. The basic idea behind this method is to divide a large-scale problem into many subproblems and to parallelly learn these subproblems by many modules. After training, all the trained modules are integrated into a modular system [5], [6], [7]. This kind of method has two main advantages over the existing SVM approaches. 1) It can dramatically reduce training time. 2) It has good scalability and expansibility. However, this method will lead to increasing the number of support vectors.

Based on the essence of support vector (SV) [8], we proposed a novel cascade method for training SVMs for pattern classification. Our method not only speeds up training but also reduces the number of support vectors.

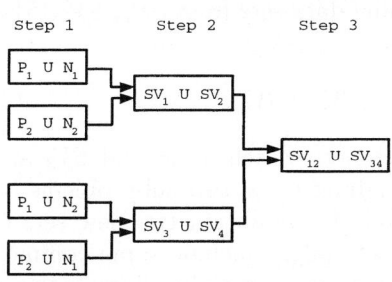 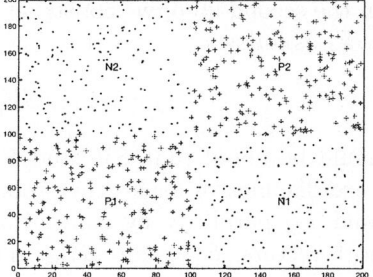

Fig. 1. Illustration of the cascade training method for producing a new smaller training set, $SV_{12} \cup SV_{34}$

Fig. 2. Combined scatter plot of both classes in checkerboard problem. Here, N_1, N_2, P_1 and P_2 denote four subsets

The paper is organized as follows: Section 2 will introduce our cascade method. The experiments are conducted in Section 3. Finally, Section 4 is conclusions.

2 A Cascade Training Method

Because the multi-class classification problem can be transformed into a series of two-class classification problems [5], we only consider two-class classification problems here. We assume that class C_1 includes positive samples and class C_2 includes negative samples. Let $P = \{X_i\}_{i=1}^{L_p}$ be the training set of class C_1 and $N = \{Y_i\}_{i=1}^{L_n}$ be the training set of class C_2. Here, L_p and L_n denote the number of elements in class C_1 and class C_2, respectively. Therefore, the training data set for a two-class problem is given by $T = P \cup N$. The proposed cascade training method has three main steps as illustrated in Fig. 1.

In the first step, the original data sets, P and N, are divided into two subsets by the same ratio r ($0 < r \leq 0.5$), respectively, as follows:

$$P_1 = \{X_i\}_{i=1}^{L_{p_1}}, P_2 = \{X_i\}_{i=L_{p_1}+1}^{L_p}, N_1 = \{Y_i\}_{i=1}^{L_{n_1}}, \text{ and } N_2 = \{Y_i\}_{i=L_{n_1}+1}^{L_n} \quad (1)$$

where $L_{p_1} = \lceil r * L_p \rceil$ and $L_{n_1} = \lceil r * L_n \rceil$.

According to this decomposition, the original two-class problem T is divided into four two-class subproblems as follows:

$$T_1 = P_1 \cup N_1, \ T_2 = P_2 \cup N_2, \ T_3 = P_1 \cup N_2, \text{ and } T_4 = P_2 \cup N_1 \quad (2)$$

An important feature of these subproblems is that there are no common training data between T_1 and T_2, and also between T_3 and T_4. Therefore, these subproblems can be handled by conventional SVM method [4] in a completely parallel way. After training, we can obtain four sets of support vectors, SV_1, SV_2, SV_3, and SV_4, which correspond to T_1, T_2, T_3, and T_4 respectively.

In the second step, we construct two training data sets from SV_1, SV_2, SV_3, and SV_4 as follows:

$$T_{12} = SV_1 \cup SV_2 \text{ and } T_{34} = SV_3 \cup SV_4 \qquad (3)$$

It should be noted that SV_1 and SV_2 are disjunctive each other and SV_3 and SV_4 are disjunctive each other too. After handling these two subproblems on T_{12} and T_{34} parallelly by conventional method [4], we obtain two new sets of support vectors, SV_{12} and SV_{34}. These two sets might include some common support vectors because there might be some common training data between T_{12} and T_{34}. From SV_{12} and SV_{34}, we obtain a new smaller training data set as follows:

$$T_{final} = SV_{12} \cup SV_{34} \qquad (4)$$

Finally, we train a SVM on T_{final} by conventional SVM method [4]. The final SVM will be used in recognition phase and all the trained SVMs obtained in the first and the second steps are discarded and the occupied computing resources are released.

If the subproblems in the first step is too large to solve, we can recursively divide it into four subproblems according to the method mentioned above. In addition, we can extend our method by dividing each class into $k\ (>2)$ subsets and we will discuss this problem later on.

3 Experiments

In order to verify our method, we present three experiments. The first is a artificial problem and the other two are real world problems. We take the tool SVM^{light} [4] for its friendly interface. In all the experiments, "standard method" means that SVMs are trained with the whole training data together by conventional SVM method [4]. The kernel we used is the radial-basis function. All the experiments were performed on a 2.4GHz Pentium 4 PC with 512MB RAM.

3.1 The Checkerboard Experiment

A 2D checkerboard problem is depicted in Fig. 2. The checkerboard divides a 200×200 square into four quadrants. All the points are uniformly distributed in the square. The points labelled by plus are positive samples and the points labelled by dot are negative samples. In this experiment, we randomly generate four training data sets, each of which includes 5000 positive samples and 5000 negative samples. A common test data set, which includes 10000 positive samples and 10000 negative samples, is also generated. Therefore, we get four classification problems, A_1, A_2, A_3, and A_4. Every training data set is divided into four parts, P_1 and P_2, N_1, and N_2.

In this experiment, r is fixed on 0.5. The purpose is to see whether the generalization performance of the SVMs trained by our cascade method is robust

Table 1. Four runnings over four training data sets and one common test data set in the checkerboard experiment, where $\sigma = 31.62$, $c = 1000$, and $r = 0.5$. The row "Averaging" denotes the average results of the four problems, $A_1, A_2, A_3,$ and A_4

Problem	Method	Accuracy		Training	Number
		Training (%)	Test (%)	time(s)	of SV
A_1	Standard method	99.84	99.81	46.39	93
	Cascade method	99.78	99.72	13.08	81
A_2	Standard method	99.89	99.72	38.00	96
	Cascade method	99.85	99.70	15.34	83
A_3	Standard method	99.93	99.84	32.44	88
	Cascade method	99.86	99.75	13.45	79
A_4	Standard method	99.89	99.81	35.50	94
	Cascade method	99.92	99.83	19.87	84
Averaging	Standard method	99.89	99.80	38.08	93
	Cascade method	99.85	99.75	15.44	82

when the training data set varies slightly. From Table 1, we can see that the generalization performance obtained by our method is almost unchanged with the different training data. The last row titled "Averaging" in Table 1 indicates that the generalization accuracy of the SVMs trained by the standard method and the SVMs trained by our method are almost the same, while our method is faster than the standard method and reduces the number of support vectors.

3.2 The Forest Covertype Experiment

In the second experiment, we use the Forest Covertype data set from UCI [9]. This data set is designed to predict the forest cover types in undisturbed forests. The attributes about one instance is 54. In this experiment, we only take all the instances of the sixth and the seventh classes to construct a two-class classification problem A_5 and all the instances of the fourth and the fifth classes to construct another two-class classification problem A_6. In each data set, we randomly take one half of instances for training and the other half for test. The data sets are shown in Table 2.

In this experiment, The training data set are decomposed by r which takes the values of 0.1 and 0.5, respectively. According to Tables 4 and 5, we can see that r does not influence classification accuracy and it only affects the training time. From these two tables, we can conclude that $r = 0.5$ is suitable for our method. Table 6 shows the details of the experiment on problem A_5. We can see that the no-support-vector data are filtered out step by step. If the number of support vectors take only a small proportion of the whole training data set, our cascade method will be much faster than the standard method. From the simulation results shown in Table 4 and 5, we can see that our cascade method can speed up training and reduce the number of support vectors.

3.3 Text Categorization

In the third experiment, we use the Yomiuri News Corpus. Here, we select the data of three classes as shown in Table 3 to construct three two-class classification problems, A_7, A_8, and A_9, by choosing every two classes from these three classes in Table 3. The number of features is 5,000 and r is set to 0.5. From Table 7,we can see that our cascade method outperforms the standard method in training time and reduce the number of support vectors. Because the number of support vectors might take only a small proportion of the whole training data in text categorization, we believe the proposed method is appropriate to dealing with text categorization problems.

Table 2. Distributions of the training and test data used in the second experiment

Problem	Training		Test	
	Positive samples	Negative samples	Positive samples	Negative samples
A_5	8684	10255	8683	10255
A_6	1374	4747	1373	4746

Table 3. The training data and test data used in the third experiment

Category	Data	
	Training	Test
Accidents	34044	8483
Health	35932	7004
By-time	33590	7702

Table 4. Experiment results on problem A_5, where $\sigma = 100$ and $c = 1000$

	Standard method	Cascade method	
		$r = 0.1$	$r = 0.5$
Training time	461.92s	508.04s	268.20s
Number of SV	3943	3827	3778
Training (%)	100	100	100
Test (%)	99.82	99.82	99.82

Table 5. Experiment results on problem A_6, where $\sigma = 100$ and $c = 1000$

	Standard method	Cascade method	
		$r = 0.1$	$r = 0.5$
Training time	27.02s	36.66s	24.69s
Number of SV	1661	1605	1571
Training (%)	100	100	100
Test (%)	96.93	96.93	96.93

4 Conclusions

In this paper we have presented a cascade method for training SVMs. Several experimental results indicate that the proposed method has two attractive features: the first is that it can speed up training while maintaining the generalization accuracy. The second is that the number of support vectors generated by our cascade method is smaller than that of the SVMs trained by the standard method, and this will reduce the time for recognition and simplify the design of classifiers. We believe the proposed method might provide us with a promising approach to deal with large-scale pattern classification problems, such as biological data mining and geography information classification.

Table 6. Details of simulation on problem A_5, where $\sigma = 100$, $c = 1000$, and $r = 0.5$

	Number of training samples	9470	9469	9470	9469
Step 1	Number of SV	3121	1630	2453	2291
	Training time (s)	131.61	56.03	82.95	84.5
	Number of training samples	4751		4744	
Step 2	Number of SV	3772		3770	
	Training time (s)	91.67		91.42	
	Number of training samples	4276			
Step 3	Number of SV	3778			
	Training time (s)	44.92			

Table 7. Simulation results on text categorization, where $\sigma = 2$, $c = 64$, and $r = 0.5$

	Methods	A_7	A_8	A_9
Training accuracy (%)	Standard method	97.74	97.93	96.67
	Cascade method	97.73	97.75	96.67
Test accuracy (%)	Standard method	95.81	96.01	93.62
	Cascade method	95.83	96.02	93.62
Training time (s)	Standard method	12664	7458	18566
	Cascade method	9519	4491	15060
Number of SV	Standard method	10933	9445	12750
	Cascade method	10553	9222	12387

Acknowledgments. This work was supported in part by the National Natural Science Foundation of China via the grant NSFC 60375022. The authors thank Kai-An Wang and Hong Shen for the help on preparing the text categorization data.

References

1. Vapnik, V.N.: Statistical Learning Theory. Wiley Interscience (1998)
2. Syed, N.A., Liu, H., Sung, K.K.: Incremental Learning with Support Vector Machines. In: Proceedings of the Workshop on Support Vector Machines at the International Joint Conference on Artificial Intelligence. Stockholm, Sweden (1999)
3. Osuna, E., Freund, R., Girosi, F.: An Improved Training Algorithm for Support Vector Machines. In: Proceedings of IEEE NNSP'97 (1997) 276-285
4. Joachims, T.: Making Large-scale Support Vector Machine Learning Pratical. In: Schölkopf, B., Burges, C.J., Smola, A.J.(eds.), Advances in Kernel Methods-Support Vector Learning. MIT Press (2000) 169-184
5. Lu, B.L. and Ito, M.: Task Decomposition and Module Combination Based on Class Relations: A Modular Neural Network for Pattern Classification. IEEE Transaction on Neural Networks, Vol.10 (1999) 1244-1256
6. Lu, B.L., Wang, K.A., Utiyama, M., Isahara, H.: A Part-versus-part Method for Massively Parallel Training of Support Vector Machines. In: Proceedings of IJCNN'04. (to appear) Budapest, July 25-29 (2004)

7. Tresp, V.: Scaling Kernel-Based Systems to Large Data Sets. Data Mining and Knowledge Discovery, Vol.5 (2001)
8. Syed, N.A., Liu, H., Sung, K.K.: Handling Concept Drifts in Incremental Learning with Support Vector Machines. In: Proceedings of the ACM SIGKDD International Conference on Knowledge Discovery and Data Mining. San Diego, CA, USA (1999) 317-321
9. Blake, C.L., and Merz, C. J.: UCI. In: ftp://ftp.ics.uci.edu/pub/machine-learning-databases (1998)

Minimal Enclosing Sphere Estimation and Its Application to SVMs Model Selection[*]

Huaqing Li, Shaoyu Wang, and Feihu Qi

Department of Computer Science and Engineering, Shanghai Jiao Tong University,
Shanghai 20030, P.R. China
{waking_lee, wang, fhqi}@sjtu.edu.cn

Abstract. In this paper, we propose a modified sequential minimal optimization (SMO) algorithm for the minimal enclosing sphere estimation (MESE) problem. Being one of the key issues of the VC dimension estimation, the MESE problem has a formulation similar to that of support vector machines (SVMs) training. This allows adoption of the ideas of Platt's SMO algorithm. After careful analysis of the MESE problem, key issues are addressed. Experimental results show the feasibility and effectiveness of the proposed algorithm when applied to SVMs model selection.

1 Introduction

In the success of support vector machines (SVMs) for various pattern classification problems, model selection plays a key role. There are mainly two categories of algorithms for SVMs model selection. Algorithms from the first category estimate the prediction error by testing error on a data set which has not been used for training, while those from the second category estimate the prediction error by theoretical bounds. One of the well known bounds is the VC bound given by Vapnik [1]:

$$R(\alpha) \leq R_{\text{emp}}(\alpha) + \sqrt{\left(\frac{h(\log(2l/h) + 1) - \log(\eta/4)}{l}\right)} \ . \quad (1)$$

where $R(\alpha)$ is the generalization error, $R_{\text{emp}}(\alpha)$ is the training error, h is the VC dimension, l is the size of the training set, η is a user-determined parameter, $0 \leq \eta \leq 1$. With probability $1 - \eta$, the above bound holds.

At present, the cross validation algorithm, which falls into the first category, is one of most popular algorithms employed in literatures [5,3]. Though some theoretical bounds have been explored [2], the use of the VC bound is less reported. However Burges pointed out that despite its looseness, the VC bound can be very predictive for SVMs model selection [1].

---

[*] This work is supported by the National Natural Science Foundation of China (No. 60072029).

The main difficulty of using the VC bound lies in determining the VC dimension. Burges suggested to ease this difficulty by using the following bound on the VC dimension instead of the dimension itself [1]:

$$h \leq \lceil \frac{D_{\max}^2}{M_{\min}^2} \rceil + 1 \ . \tag{2}$$

where $D_{\max}$ is the maximal diameter of a set of gap tolerant classifiers, $M_{\min}$ is the minimal margin of the same set of classifiers.

Then the only thing left is to estimate $D_{\max}$. This problem can be described as follows [1]: Given a training set of data points X_i and a function Φ, which maps the data points from the original space to a high (even infinite) dimensional feature space $\mathcal{H}$, we wish to compute the radius of the smallest sphere in $\mathcal{H}$ which encloses the mapped training data. The corresponding formulation is

$$Minimize \quad R^2 \ . \tag{3}$$

$$subject\ to: \quad R^2 - \|\Phi(X_i) - C\|^2 \geq 0 \quad \forall i \ . \tag{4}$$

where C is the center of the sphere in $\mathcal{H}$. As the problem resembles that of SVMs training, algorithms for the latter can be modified to solve (3).

In [4], Platt proposed a fast, easy to implement algorithm, known as sequential minimal optimization (SMO), for SVMs training. The main virtue of SMO lies in its avoiding the use of a quadratic programming solver by decomposing the original problem into a series of smallest problems which can be analytically solved.

In this paper, the basic ideas of SMO are applied to the minimal enclosing sphere estimation (MESE) problem. Through careful analysis of the problem, Platt's SMO algorithm is appropriately modified and key issues are addressed. By employing the VC bound for SVMs model selection, we show the feasibility and effectiveness of the modified algorithm. The rest of the paper is organized as follows: Section 2 obtains the Lagrangians of (3) and derives the KKT conditions. Details of solving the MESE problem with SMO are discussed in Section 3. In Section 4, SVMs model selection experiments are presented. Conclusions are given in the last section.

2   Reformulation of the MESE Problem

By introducing positive Lagrange multipliers λ_i, the primal Lagrangian of the MESE problem can be obtained [1].

$$Minimize \quad L_P = R^2 - \sum_i \lambda_i (R^2 - \|\Phi(X_i) - C\|^2) \ . \tag{5}$$

This is a convex quadratic programming problem, and we can instead solve the Wolfe dual

$$Maximize \quad L_D = \sum_i \lambda_i K(X_i, X_i) - \sum_i \sum_j \lambda_i \lambda_j K(X_i, X_j) \ . \tag{6}$$

subject to:

$$C = \sum_i \lambda_i \Phi(X_i) \ . \tag{7}$$

$$\sum_i \lambda_i = 1 \ . \tag{8}$$

$$\lambda_i \geq 0 \ . \tag{9}$$

where $K(X_i, X_j)$ is the kernel function.

It's well known that for convex problems, the Karush-Kuhn-Tucker (KKT) conditions are necessary and sufficient for a solution. The KKT conditions of the primal Lagrangian are very simple:

$$\begin{aligned} \lambda_i = 0 &\Rightarrow R^2 - \|\Phi(X_i) - C\|^2 \geq 0 \ , \\ 0 < \lambda_i < 1 &\Rightarrow R^2 - \|\Phi(X_i) - C\|^2 = 0 \ . \end{aligned} \tag{10}$$

3 Solving the MESE Problem with SMO

The basic idea of SMO is to decompose the original large quadratic programming problem into a series of smallest problems which can be solved analytically. Such a strategy avoids using a quadratic programming solver, and makes SMO fast and easy to implement. As to MESE, due to the linear constraint, two Lagrange multipliers are involved in each smallest optimization problem.

3.1 Optimize Two Lagrange Multipliers

Without loss of generality, suppose we are optimizing λ_1 and λ_2 from and old feasible solution: $\lambda_1^{old}, \lambda_2^{old}, \ldots, \lambda_l$. Due to the linear constraint, we have:

$$\lambda_1 + \lambda_2 = \lambda_1^{old} + \lambda_2^{old} = \mu \ . \tag{11}$$

Thus the bounds of λ_1 and λ_2 can be obtained.

$$Low = \max(0, \mu - 1) \leq \lambda_1, \lambda_2 \leq \min(1, \mu) = High \ . \tag{12}$$

Substituting λ_1 with $(\mu - \lambda_2)$, the Wolfe dual can be reformulated as:

$$L_D = \gamma \lambda_2^{\ 2} + \lambda_2[(K_{22} - K_{11}) - 2(V_2^{old} - V_1^{old}) - 2\gamma \lambda_2^{old}] + \text{const} \ . \tag{13}$$

where

$$K_{ij} = K(X_i, X_j) \ . \tag{14}$$

$$\gamma = 2K_{12} - K_{11} - K_{22} \ . \tag{15}$$

$$V_i = \Phi(X_i) \bullet C \ . \tag{16}$$

In most cases, there will be a global maximum and $\gamma < 0$. Hence the optimal λ_2 can be computed as:

$$\lambda_2 = \lambda_2^{old} - \frac{K_{22} - K_{11}}{2\gamma} + \frac{V_2^{old} - V_1^{old}}{\gamma} \ . \tag{17}$$

Since λ_2 is limited, the value obtained from (17) must be clipped before being used to update other parameters.

$$\lambda_2^{clipped} = \begin{cases} Low & \lambda_2 < Low \\ \lambda_2 & Low \leq \lambda_2 \leq High \\ High & \lambda_2 > High \end{cases} \ . \tag{18}$$

Then λ_1 can be updated.

$$\lambda_1 = \lambda_1^{old} - (\lambda_2^{clipped} - \lambda_2^{old}) \ . \tag{19}$$

Under cases when $\gamma = 0$, no global optimum of λ_2 exists. This situation is handled following Platt's strategy [4]: Compute the objective function at both bounds and select the one corresponding to the larger value as λ_2.

3.2 Updating After a Successful Optimization Step

When two multipliers have been successfully optimized, all parameters needed for the next optimization step must be updated. In our implementation, we cached the square of the radius of the estimated enclosing sphere, L2 norm of the center point, and V_i for all examples. The updating rules are given below:

$$\Delta\lambda_1 = \lambda_1 - \lambda_1^{old} \ . \tag{20}$$

$$\Delta\lambda_2 = \lambda_2^{clipped} - \lambda_2^{old} \ . \tag{21}$$

$$\|C\|^2 = \|C_{old}\|^2 + 2 \sum_{i=1,2} \Delta\lambda_i V_i^{old} + K_{11}\Delta\lambda_1^2 + 2K_{12}\Delta\lambda_1\Delta\lambda_2 + K_{22}\Delta\lambda_2^2 \ . \tag{22}$$

$$V_i = V_i^{old} + K_{1i}\Delta\lambda_1 + K_{2i}\Delta\lambda_2 \ . \tag{23}$$

Note that there always exists a Lagrange multiplier larger than zero, provided a successful optimization step has been taken. Suppose $\lambda_1 > 0$, the square of the radius can be updated as (the same thing holds for λ_2):

$$R^2 = \|C\|^2 - 2V_1 + K_{11} \ . \tag{24}$$

3.3 Choosing Two Lagrange Multipliers for Optimization

The heuristics employed to choose two Lagrange multipliers for optimization is very similar to that of Platt's SMO algorithm [4]. However the KKT violations checking is quite different. For MESE, (10) is used for KKT violations checking, which can be further formulated as:

$$\begin{aligned} \lambda_i = 0 &\Rightarrow R^2 - (K_{ii} - 2V_i + \|C\|^2) \geq 0 \ , \\ 0 < \lambda_i < 1 &\Rightarrow R^2 - (K_{ii} - 2V_i + \|C\|^2) = 0 \ . \end{aligned} \quad (25)$$

4 SVMs Model Selection Experiments

In this section, we investigate the use of the VC bound for SVMs model selection in a face recognition problem. One should keep alert that this paper is not dedicated to the face recognition problem itself. Face recognition is performed on the ORL database, which is randomly divided into two equal sets, one for training the other for testing. Feature extraction is done with the Eigenface algorithm.

LIBSVM [3] is employed for SVMs training and testing, as well as model selection with the cross validation algorithm. Model selection with the VC bound is done with a module developed by us with $C++$. The kernel function employed is the *RBF* kernel, which is defined as:

$$K(X_i, X_j) = \exp(-\sigma \|X_i - X_j\|^2) \ . \quad (26)$$

Thereby only two parameters need to be tuned, the penalty parameter c and the kernel parameter σ. While the cross validation algorithm selects both parameters automatically, the VC bound algorithm can only select σ and needs a user-input value of c. In this paper, the parameter ranges are $\log C \in \{-5, \ldots, 15\}$ and $\log \sigma \in \{-15, \ldots, 3\}$, c is set to 2048 for the VC bound algorithm. The standard grid search is performed with twenty samples of c and eighteen samples of σ, both uniformly distributed in log-space.

We also investigate a combination algorithm which works in two steps: (1) Use the VC bound to find an optimal σ'; (2) Use the cross validation algorithm to find the optimal values of c and σ in the ranges of $\log C \in \{-5, \ldots, 15\}$ and $\log \sigma \in \{\sigma' - 1, \sigma', \sigma' + 1\}$ respectively. Experimental results are shown in Table 1.

From Table 1, we can see that the combination algorithm has the best recognition accuracy in all cases, while the other two perform comparatively in most cases. It is interesting to note that the selected parameters by different algorithms may differ widely even when the recognition accuracies are the same. As to computational cost, totally 360 samples need to be examined with the cross validation algorithm to select the best c and σ, while only 18 samples with the VC bound algorithm and 78 samples with the combination algorithm. Taking into account this significant difference, the performances of the combination algorithm and the VC bound algorithm are more appealing than that of the cross validation algorithm. However, the disadvantage of the former two algorithms

Table 1. SVMs model selection results and corresponding recognition accuracy

Number of Eigenfaces	Cross validation			VC bound			Combination		
	log C	log σ	Accuracy	log C	log σ	Accuracy	log C	log σ	Accuracy
20	3	-4	93%	11	0	91%	5	-1	93%
40	5	-6	92%	11	-1	91%	5	-2	92%
60	5	-6	91%	11	-2	91.5%	5	-3	92%
80	5	-6	92%	11	-3	92%	5	-4	93%
100	5	-6	91.5%	11	-3	91.5%	5	-4	92.5%

lies in the requirement of a user-determined c. Empirically we find that when c is large enough, e.g. 2048, the resulted SVMs almost perform well.

Finally it should be pointed out that the recognition accuracies listed in Table 1 are inferior to those reported in many literatures on face recognition with SVMs. This may due to the following reasons: (1) The *RBF* kernel does not fit well for the problem; (2) The best parameters may locate elsewhere than the ranges used; (3) Even though the best parameters locate in the ranges used, they may not be selected due to the relative coarse resolution of the ranges which leads to a grid search of integers in log-space. As this paper is not dedicated to obtaining good face recognition performance, this is not regarded as problematic.

5 Conclusion

In this paper, we modified Platt's SMO algorithm to solve the minimal enclosing sphere estimation (MESE) problem. Though the formulation of MESE is very similar to that of SVMs training, there are a lot of differences in implementation details of the SMO algorithms. The modified SMO algorithm is then used to estimate the VC bound, which is further used for SVMs model selection. Experimental results show that when compared with the popular cross validation algorithm, the VC bound exhibits an appealing potential for SVMs model selection, provided the penalty parameter c is well chosen. Moreover when the two algorithms are appropriately combined, more satisfying performance can be achieved.

References

1. Burges, C.J.: A Tutorial on Support Vector Machines for Pattern Recognition. Data Mining and Knowledge Discovery. **2** (1998) 121–267
2. Chung, K.-M., Kao, W.-C., Sun, T., Wang, L.-L., Lin, C.-J.: Radius Margin Bounds for Support Vector Machines with the RBF Kernel. Neural Computation. **11** (2003) 2643–2681
3. Chang, C.-C., Lin, C.-J.: LIBSVM: A Library for Support Vector Machines. (2002) Online at http://www.csie.ntu.edu.tw/~cjlin/papers/libsvm.pdf

4. Platt, J.: Fast Training of Support Vector Machines Using Sequential Minimal Optimization. In: Scholkopf, B., Burges, C., Smola, A. (eds.): Advances in Kernel Methods: Support Vector Learning. MIT Press (1998) 41–65
5. Staelin, C.: Parameter Selection for Support Vector Machines. (2003) Online at http://www.hpl.hp.com/techreports/2002/HPL-2002-354R1.pdf

Constructing Support Vector Classifiers with Unlabeled Data[*]

Tao Wu and Han-Qing Zhao

Automation Institution, Faculty of Mechatronics & Automation,
National University of Defense Technology,
410073, Changsha, Hunan, China
wutao.nudt@263.net   zhaohanqing@yahoo.com

Abstract. In this paper, a new method is presented to improve the speed and accuracy of SVMs with unlabeled data respectively: one method is to build SVMs with grid points which can be expected to speed SVMs in test phase; another method is to build SVMs with unlabeled data and it was shown that it can improve the accuracy of SVMs when there have a very few labeled data. These two methods are in the frame of quadric programming and no need to increase the computation cost of SVMs greatly, so it is expected to play an important role in some fields for the future.

1 Introduction

Support vector machines (SVMs), a supervised machine learning technique, have been shown to perform well in multiple areas, especially in pattern classification. Here, a supervised machine learning technique requires that all the training data should be labeled.

In this paper, one issue will be concerned: Is it necessary to construct SVMs with labeled data? The answer is absolutely "NO". To prove this, we present two algorithms: one is to construct SVMs with grid points in the feature spaces instead of those labeled training data; the other is to classify the unlabeled data directly with SVM.

The motivation behind this research is two folds:

In many cases, it is found that nearly all the training data come to be support vectors, which lead SVMs to be considerably slower in test phase than some other approaches with similar generalization performance. So if we want to improve the speed of SVMs, it should be better to construct SVMs with some data other than training data.

Recently there has been a growing interest in the use of unlabeled data for enhancing classification accuracy in supervised learning settings. This is called semi-supervised learning. Most studies have shown that there is potential in using unlabeled

---

[*] This work is supported by National Nature Science Foundation of China (No.60234030).

data to enhance the learning process and improve classification ([3]). Finding methods for adding unlabeled data to supervised learning is important because in many real world problems it is easy to collect unlabeled data, but expensive to correctly label it. Constructing SVMs with labeled and unlabeled data will play an important role in semi-supervised learning.

In this paper, we focus on how to construct SVMs with any given data, labeled or unlabeled data. We will show that there is no much difference between the labeled data and unlabeled data for SVMs' construction: we can construct a SVM with labeled or unlabeled data by solving a quadric program.

2 Approximating SVMs with Unlabeled Data

As we know that SVMs are one kind of data driven learning methods and the classifiers are modeled by labeled training data as follows:

$$f(x) = \sum_i \alpha_i y_i k(x, x_i) + b \tag{1}$$

Here, $\{(x_i, y_i)\}_{i=1,\cdots,n}$ are labeled training data, $(\alpha_1, \cdots, \alpha_n; b)$ is the optimal solution of the optimization problem as follows:

$$\max W(\alpha_1, \cdots, \alpha_n; b) = \sum_i \alpha_i - \frac{1}{2} \sum_{i,j} \alpha_i \alpha_j y_i y_j k(x_i, x_j) \tag{2}$$

s.t.

$$\sum_i \alpha_i y_i = 0, \ \alpha_i \geq 0, i = 1, \cdots, n \tag{3}$$

In the expression (1), not all of $\{\alpha_i\}_{i=1,\cdots,n}$ are nonzeros. That means only a little part of training data has contribution to model the classifier. We call this is sparseness, which is one of the attracting features of SVMs. This feature is important to improve the efficiency of SVMs during test phases. However, sparseness can not be often seen in many real examples, which makes SVMs to test a data very slowly.

In order to solve this problem, Burges present a method to reduce the support vectors ([1, 2]). The basic idea of this method is trying to find some data (may not be training data) to represent (1). Obviously, we have to trade off between the correct rate and the number of support vectors and in some case, it is nearly impossible to reduce the support vectors. Does it possible for us to speed SVMs with many support vectors?

SVMs can be seen as a global method to approximate a function because any support vectors should be used in test phase, while in many cases, the values of given data are just correlative with the supports vectors nearby. It is obvious if the chosen kernel function is RBF kernel. So if we can quickly choose some support vectors that are near to the test data, we can obtain the values of these test data very quickly although there are large numbers of support vectors.

To realize this idea, we should align the support vectors. While in (1), support vectors are all from the training data and they are scattered data so it is hard for us to align them. If all the support vectors are grid point of input space, alignment of these data will become enjoyable. So we will present a method to approximate a SVM with grid points, i.e.

$$f(x) = \sum_i \beta_i k(x, \tilde{x}_i) + b' = \sum_i \alpha_i y_i k(x, x_i) + b \tag{4}$$

Here, $\{\tilde{x}_i\}_{i=1,\cdots,n}$ are grid points of feature space.

Obviously, we can get a SVM with grid points as follows: first, constructing a SVM based on the training data; second, reduced the support vectors with method presented by Burges in [1,2]. We call this a two-phase method.

However, we will present a method to solve this problem in one phase: Constructing a SVM with grid points directly by solving an optimization problem.

2.1 Optimization Problem

Suppose that $w = \sum_{i=1}^{n} \alpha_i \phi(\tilde{x}_i)$, here, $\{\tilde{x}_i\}_{i=1,\cdots,n}$ are grid points in the input space, $\phi(\cdot)$ is a nonlinear map introduced by kernel $k(\cdot,\cdot)$, then the optimal plane to classify the training data $\{x_i\}_{i=1,\cdots,l}$ will try to maximize $\|w\|^2$, thus the optimal function com to be:

$$\min_\alpha W(\alpha) = \min w^T \cdot w = \alpha^T H \alpha \tag{5}$$

here,

$$H = (h_{ij}), \quad h_{ij} = k(\tilde{x}_i, \tilde{x}_j)$$

Similarly, the constraint conditions on the label data come to be:

$$y_j(\sum_i \alpha_i k(x_j, \tilde{x}_i) + b) \geq 1, \quad i = 1,\cdots,n, \ j = 1,\cdots,l \tag{6}$$

2.2 Experiment Results

To visualize the result of this algorithm, banana is chosen from UCI benchmark. This is just a toy data and there are two features for every data. To verify our method, first, we use 100 and 200 labeled data to construct traditional SVMs, they are shown on top of Fig. 1. Second, we use our method to approximate them with 25 and 100 grid points. The results are shown on the middle and bottom of Fig. 1.

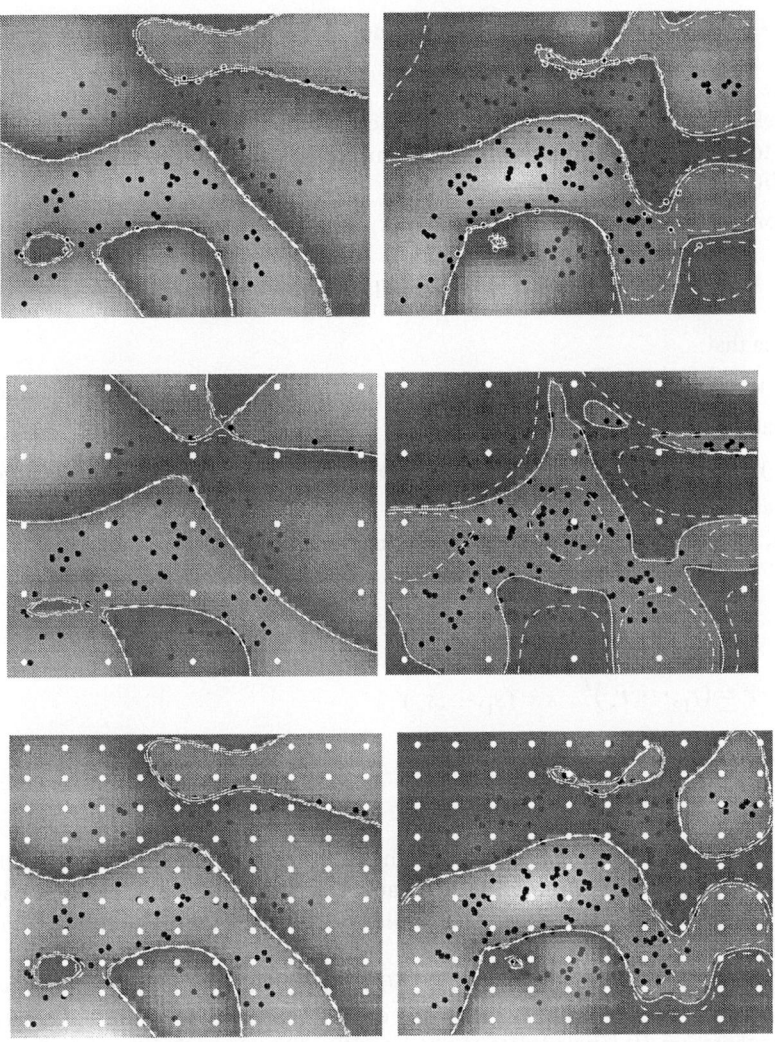

Fig. 1. Building SVMs with banana data

3 Constructing SVMs with Labeled and Unlabeled Data

In section 2, we have shown that SVMs can be represented by unlabeled data, but these unlabeled data have no contribution to the SVM. In this section, we will build SVMs with labeled and unlabeled data. The distribution of unlabeled data will have great effect on the classifier.

3.1 Optimization Problem for Constructing SVMs with Unlabeled Data

There are some optimization models for semi-supervised learning ([4]). While these models need high computation cost. In the fact, the key of semi-supervised problem is how to list constraint conditions for the unlabeled data. We can describe the constraint conditions for unlabeled data as follows:

For any unlabeled data x_i, $\exists\, r_i \geq 0$, $s_i \geq 0$, s.t.

$$w^T \cdot x_i + b + r_i \geq 1 \quad \text{and} \quad w^T \cdot x_i + b - s_i \leq -1 \tag{7}$$

Notice that:

1. If $y_i = 1$, then $w^T \cdot x_i + b_i \geq 1$, so $r_i = 0$, s_i can be any enough large number that (7) can be hold;
2. If $y_i = -1$, then $w^T \cdot x_i + b \leq -1$, so $s_i = 0$, r_i can be any enough large number that (7) can be hold.

Thus, the object function need to be optimized is:

$$\min_{(w,b;r,s)} \phi(w) = \frac{1}{2}(w^T \cdot w) + r^T \cdot s \tag{8}$$

Here, $r = (r_1, \cdots, r_n)^T$, $s = (s_1, \cdots, s_n)^T$.

3.2 Experiments

In this section, we choose two benchmarks from semi-supervised challenge on NIPS'2001. One is "neural image" and another is "horse shoe". "Neural image" have 530 training data and 2710 test data, while "horse shoe" has 400 training data and 800 test data.

In the experiments, we choose first 50 training data for training, but only part of them are labeled. In Table 1 : 2 means every class have only one label data and 10 means there are 10 labeled data (from 1 to 10); 20 means label the data from 1 to 20, and so on. C-SVM means use classical SVMs and S-SVM means semi-supervised SVM presented in this paper. RBF kernel is chosen for all the SVMs and the radius parameter σ is set to 1.

From the result of tables, we can find that the algorithm presented in this section will improve the correct rate of classifiers greatly when the labeled data is very few, but when more data labeled, the effects of unlabeled data turn to be faint.

Table 1. 50 unlabeled data and different number of labeled data for neural image data

#Labeled	2	10	20	30	40	50
C-SVM	942	1457	1702	2066	2047	2034
S-SVM	1908	1744	2068	1674	1937	2059

Table 2. 50 unlabeled data and different number of labeled data for "horse shoe" data

#Labeled	2	10	20	30	40	50
C-SVM	485	476	502	400	373	469
S-SVM	497	537	518	527	533	528

4 Conclusion

In this paper, two methods of constructing SVMs with unlabeled data are presented: One is to use unlabeled data to approximate the SVM based on the labeled data. This method is expected to be another way to speed SVMs during test phase. In this method, the structure of unlabeled data is not used to construct SVMs, but to represent SVMs. Another method of constructing SVMs with unlabeled data structure is also presented in this paper. The innovation of this method is that we build SVMs with unlabeled data by solving a quadric programming problem. The experiments show that this method can improve the accuracy of SVMs greatly when the labeled data are very few.

References

1. Burges, C.J.C.: Simplified Support Vector Decision Rules. In: 13th International Conference of Machine Learning. (1996)71–77
2. Burges, C.J.C., Schölkopf, B.: Improving the Accuracy and Speed of Support Vector Machines. In: Mozer, M., Jordan, M., Petsche, T. (eds.): Advances in Neural Information Processing Systems, Vol. 9. MIT Press, Cambridge MA (1997)
3. Seeger, M.: Learning with Labeled and Unlabeled Data. In: Technical Report. Edinburgh University, UK. (2001). Available from Http://citeseer.ist.psu.edu/seeger01learning.html
4. Fung, G., Mangasarian, O.L.: Semi-Supervised Support Vector Machines for Unlabeled Data Classification. In: Optimization Methods and Software. 15(2001)19-24
5. Bennett, K.P., Demiri, A.Z.: Semi-Supervised Support Vector Machines. In: Advances in Neural Information Processing Systems. (1998)368-374
6. Nigam, K., McCallum, A., Thrun, S., Mitchell, T.: Text Classification from Labeled and Unlabeled Documents Using EM. Machine Learning. (2000)103–134

Nested Buffer SMO Algorithm for Training Support Vector Classifiers

Xiang Wu and Wenkai Lu

State key Laboratory of Intelligent Technology and Systems
Department of Automation, Tsinghua University, Beijing (100084), CHINA
wuxiang00@tsinghua.org.cn, lwkmf@mail.tsinghua.edu.cn

Abstract. This paper presents a new decomposition algorithm for training support vector classifiers. The algorithm uses the analytical quadratic programming (QP) solver proposed in sequential minimal optimization (SMO) as its core solver. The new algorithm is featured by a nested buffer structure, which serves as a working set selection system. This system can achieve faster convergence by imposing restriction on the scope of working set selection. More efficient kernel cache utilization and more economical cache shape are additional benefits, which make the algorithm even faster. Experiments on various problems show that the new algorithm is 1.51 times as fast as LibSVM on average.

1 Introduction

In the past few years, Support Vector Machine (SVM) has gained prominence in the field of machine learning due to its excellent generalization performance. However, the large-scale quadratic programming problem introduced by SVM training cannot be easily solved with standard QP techniques.

Among the methods addressing this difficulty, the most well-studied ones are called "decomposition algorithms", whose basic idea is to decompose the large QP problem into a series of smaller QP sub-problems. That is, in each iteration the algorithm keeps fixed most dimensions of the optimization variable vector, and varies a small subset of dimensions, namely working set, to get maximum reduction of the object function. Decomposition algorithms include Vapnik's chunking algorithm [2], Osuna's fixed working set algorithm [1], Platt's sequential minimal optimization (SMO) [3], Joachims's SVMlight [4] and Keerthi's generalized SMO algorithm [5], etc.

This work continues previous researches on decomposition algorithm. Existing methods are analyzed and a new structure is put forward to achieve better overall efficiency. A nested buffer structure is constructed for fast and efficient working set selection. Shrinking is naturally realized in this selection system. Additionally, a buffer-integrated kernel cache is used in this algorithm. This feature brings the additional benefit of more efficient cache utilization and more economical cache shape, which makes the algorithm even faster.

This paper is organized as follows. First we introduce the SVM training problem we are encountering and the developing history of decomposition algorithms. Second our nested buffer SMO algorithm is described in detail. Then experiments are carried out to test the computation speed of our algorithm. Finally conclusions are made.

2 SVM Training Problem and Decomposition Algorithms

An SVM training problem can be described as follows [7]. Given training examples

$$\{\mathbf{x}_i, y_i\}_{i=1}^{l}, \mathbf{x}_i \in \mathbf{R}^n, y_i \in \{-1,1\}, i = 1,...,l \quad (1)$$

SVM gives a classifying function in the form of

$$\hat{f}(\mathbf{x}) = \text{sgn}\left\{\sum_{i=1}^{l} \alpha_i y_i k(\mathbf{x}_i, \mathbf{x}) - b\right\} \quad (2)$$

where $k(\mathbf{x}_i, \mathbf{x}_j)$ is kernel function, and α_i is the Lagrange multiplier corresponding to training example $\mathbf{x}_i$. $\{\alpha_i\}_{i=1}^{l}$ is the solution of the following QP problem:

$$\min_{\alpha} \ W(\alpha) = \frac{1}{2}\sum_{i,j=1}^{l} \alpha_i \alpha_j y_i y_j k(\mathbf{x}_i, \mathbf{x}_j) - \sum_{i=1}^{l} \alpha_i$$

$$s.t. \quad \sum_{j=1}^{l} \alpha_i y_j = 0 \quad (3)$$

$$0 \leq \alpha_i \leq C, i = 1,...,l$$

The training examples corresponding to non-zero α_i's are called support vectors.

Problem (3) is the QP problem that we are addressing. In this paper several different terms, "α_i's", "variables", "indices", "dimensions" and "examples", all mean the same thing. That is, some dimensions of problem (3)'s solution vector α.

The history of decomposition algorithm can be tracked back to 1979 when SVM was invented. Vapnik described a method to solve the QP problem arising from SVM training, which was called "chunking". At every step the algorithm solves a QP problem that consists of all the identified support vectors and all the variables violating Karush-Kuhn-Tucker (KKT) conditions [2]. In 1997 Osuna et.al. proposed a decomposition algorithm with a fixed-size working set [1]. In 1998 Platt put forward SMO algorithm. The advantage of SMO lies in the fact that its each sub-problem can be done analytically [3]. In 2001 Platt's algorithm was improved by Keerthi, et.al. [5].

In 1998, while introducing his software SVMlight 2.0, Joachims systematically studied working set selection strategy of general decomposition algorithms, and for the first time proposed effective shrinking and kernel cache scheme to speed up decomposition algorithms [4].

3 Nested Buffer SMO Algorithm

3.1 Nested Buffer Structure

Like other SMO based algorithms, nested buffer SMO also belongs to decomposition algorithm and uses two-example analytical solver as its core solver. What makes difference is that it benefits from a new set of peripheral structure. In this paper "buffer" means an artificial subset of variables, or indices. In the process of optimization, we put variables into different buffers according to the current degree of their KKT violation. Buffer 1 (B1) is comprised of the s1 variables that violate KKT conditions the most and therefore most need to be adjusted. Here we use Joachims's feasible direction strategy to define the degree of KKT violation [4]. B2 contains all the items in B1 and other variables that violate KKT conditions to a certain extent, i.e., less violating than those in B1 but more violating than others. B3,...,Bn are formed in the same way. Each Bk contains all the variables whose degree of KKT violation reach a certain threshold. The outer most buffer Bn is the whole training set.

Suppose that in an iteration such a nested structure is established. Then in the next iteration SMO will not choose its working set from all the variables, but will choose from the inner most buffer, B1. Since in the previous iteration the most violating variables are selected and put into B1, it is reasonable to conjecture that the most violating pair of variables is still in B1. (Their violating degrees are usually not subject to severe variation because these values are determined by all the non-zero α_i's.) And therefore it's reasonable to search only B1 for a most violating pair, instead of search the whole training set. This is also because the core solver can lead to a reduction of the object function even if the most violating pair isn't found. The saving in selection time is worth possible decrease in object function's reduction.

After several iterations, we consider that the degree of KKT violation in B1 has fallen to a low level with respect to some indices outside B1. But we still can conjecture that most of violating variables are in B2. Therefore the degree of KKT violation of variables in B2 (including those in B1) is recalculated and those most violating ones are picked to recompose B1. This process of exchanging items with outer buffer according to the reevaluated degree of KKT violation is called a "reconstitution" of B1, or an "update" of B2. Updating B2 every several iterations is economical in two senses: On the one hand, keep B1 always containing important items so that the inner SMO iterations are not wasted. On the other hand, such an update only involves dimensions contained in B2, instead of all the dimensions, most of which still very safely satisfy KKT conditions. And thus computation overhead is saved as much as possible.

In the same way, each Bk is reconstituted each time it detects that the maximum degree of KKT violation in it has fallen down to a low level. Thus indices keep moving and depositing in the nested buffer structure, driven by its center engine, SMO, until all of the variables have satisfied KKT conditions within given tolerance.

3.2 Natural Shrinking and Kernel Cache

Shrinking scheme is to eliminate variables that are likely to be 0 or C as their final value in the middle of iteration process, and thus save the time consumed in examining these variables against KKT conditions in each iteration [4]. Nested buffer SMO naturally realizes shrinking. In fact, buffer itself is a shrinking mechanism. In our algorithm the variables very safely satisfying KKT conditions are set aside in the outermost buffer, while those comparatively "dangerous" ones are held in Bn-1. The shrunk variables will not be checked against KKT conditions until all the variables in Bn-1 have satisfied KKT conditions within a certain tolerance. This tolerance should be determined by the degree of "safety" of the shrunk variables. In this sense buffer Bn-1 realizes the function of shrinking. And the nested buffer system realizes cascade shrinking.

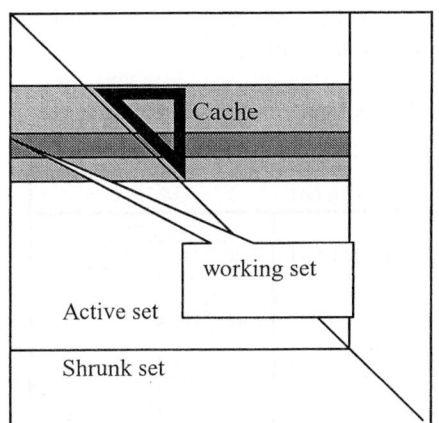
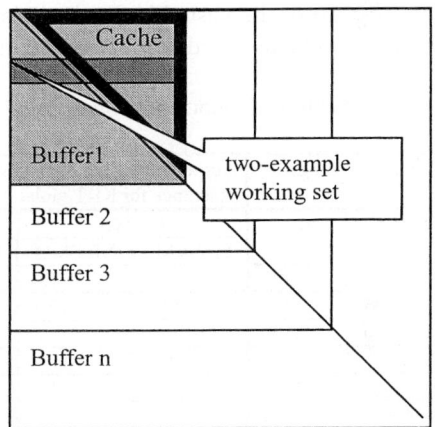

Fig. 1. Illustration of SVMlight and nested buffer SMO's settings for kernel cache The *light colored rectangles* are caches. The *dark rectangles* are the kernel matrix elements accessed in an iteration. The triangles enclosed by bold line are the saved memory by exploiting symmetry.

To reduce the total number of kernel evaluations, i.e., calculation of $k(x_i, x_j)$, a kernel cache is used to store precomputed kernel matrix elements in memory. Nested buffer SMO uses a cache setting different from that of previous decomposition algorithms, and this feature brings additional advantage of more efficient cache utilization and larger actual cache size. In previous algorithms an LRU mode cache stores recently accessed rows of the kernel matrix [4][6][8]. However, when the size of active set (the set of variables not shrunk yet) is large in relation to cache size, the cache can store only several rows of the kernel matrix. What's worse, working set selection strategy often accesses kernel matrix in a random manner. In this case the cache will be updated frequently. This phenomenon is called "thrashing". In this circumstance the algorithm can't benefit from the cache. On the contrary, extra computation time and memory may be spent in keeping an LRU mode cache [8].

Nested buffer SMO naturally circumvented this difficulty by integrating the cache with a certain buffer, Bc. Each index in Bc is allocated a row in the cache. Because the scope of working set selection is restricted to the indices covered by cache, the efficiency of cache utilization is improved. Only when Bc is reconstituted can some cached elements be eliminated, and thus thrashing is seldom incurred. This buffer-integrated cache is easy to realize, cheap in maintenance cost and highly efficient in use. Its high efficiency mainly comes from restriction on the scope of working set selection.

What's more, since the buffer-integrated cache is square in shape, kernel matrix's symmetry can be fully exploited. While caching an m-by-n (m<n) matrix must occupy at least $mn - (1/2)m^2$ float numbers' memory, caching a square matrix of the same number of elements requires storing only half of all the elements. That is to say, given the same available memory size, a square cache can actually hold more kernel matrix elements. Moreover, among all the rectangles of the same area size, square has the largest width, which implies that more indices are covered by the cache (See Figure1).

In one word, the nested buffer is a hierarchical working set selection system with shrinking mechanism and kernel cache naturally integrated with it.

Table 1. Computation time of LibSVM and nested buffer SMO (in seconds)

Common settings		kernel: Gaussian kernel with variance 10; cache size: 10Mb; termination tolerance for KKT violation (on each side): 0.001					
Dataset	Size	C=0.05		C=1		C=20	
		LibSVM	NB SMO	LibSVM	NB SMO	LibSVM	NB SMO
Adult-1a	1605	2.4	1.4	2.0	1.3	2.9	3.8
Adult-4a	4781	20.2	12.2	18.0	12.2	87.1	39.3
Adult-7a	16100	211.0	125.8	224.0	139.5	1583.7	696.1
Adult-a	32561	806.7	517.4	953.6	619.1	7465.9	3673.8
Web-1a	2477	1.2	1.3	1.7	2.0	1.9	2.4
Web-4a	7366	10.7	9.2	24.0	13.9	38.3	19.4
Web-7a	24692	90.5	75.5	161.2	106.3	357.3	203.0
Web-a	49749	354.5	353.4	529.8	502.9	1175.2	1448.3
Total time		1497.2	1096.2	1914.3	1397.2	10712.3	6086.1

4 Experiments and Results

Nested buffer SMO algorithm was tested against LibSVM2.4 [6] on a wide range of benchmark and artificial problems. Both algorithms were written in C++, compiled with Visual C++ 6.0. Both algorithms were run on an unloaded 667MHz Pentium III CPU running Windows 2000. Statistical average on test results indicates that nested buffer SMO is 1.51 times as fast as LibSVM. Due to the limitation of space, here we only list their computation time of experiments on two well-known benchmark datasets, UCI Adult dataset [9] and Web dataset [10]. The common settings of these experiments are listed in the top line of Table 1. Similar experiment settings are also used in [3][4][5][8]. These datasets and the software LibSVM were downloaded from the sites mentioned in the above references. In Table 1 the computation time of LibSVM and nested buffer SMO (abbreviated as NB SMO in the table) are given.

Like other SMO based algorithms, nested buffer SMO uses tolerance of KKT conditions [3] fulfillment as its termination criteria, which guarantees the same precision with that of other algorithms.

Because neither of the two benchmark datasets most commonly used for speed test has more than 50,000 examples, no experiment results on larger datasets are listed. It can be seen from the table that for most problems NB SMO is faster than LibSVM, especially for the most time-consuming ones.

5 Conclusions

In this paper a new improvement on SMO algorithm, called nested buffer SMO, is described. The algorithm constructs a nested buffer system to achieve high efficiency in working set selection, and thus speed up SMO. Another merit of the algorithm is that its cache setting allows for more efficient cache utilization. The cache itself has a more economical shape, too.

Acknowledgement. This work is sponsored by major project of National Natural Science Foundation of China (No. 60390540).

References

1. Osuna E. E., Freund R. and F. Girosi L: Support Vector Machines: Training and Applications. Technical Report, MIT, March (1997)
2. Vladimir N. Vapnik: Estimation of Dependences Based on Empirical Data. Spinger-Verlag, New York (1982)
3. Platt. J.: Sequential Minimal Optimization: A Fast Algorithm for Training Support Vector Machines. Technical Report 98-14, Microsoft Research, Redmond, Washington, April (1998)
4. Joachims T.: Making Large-Scale Support Vector Machine Learning Practical. In: B. Scholkopf, J.C. Burges and A.J. Smola (eds.): Advances in Kernel Methods --- Support Vector Learning, Cambridge MA: MIT Press (1999) 169-184
5. Keerthi, S. S., S. Shevade, C. Bhattacharyya, and K. Murthy: Improvements to Platt's SMO Algorithm for SVM Classifier Design. Neural Computation (2001),13, 637-649
6. Chang, C.-C. and C.-J. Lin: LIBSVM: a Library for Support Vector Machines (Version 2.3). (2001) http://www.csie.ntu.edu.tw/~cjlin/libsvm/
7. Burges, C.J.C.: A Tutorial on Support Vector Machines for Pattern Recognition. Data Mining and Knowledge Discovery (1998), 2(2), 121-167
8. Jianmin, L., Bo, M., Fuzong, L.: A New Strategy for Selecting Working Sets Applied in SMO. Proceedings of 16th International Conference on Pattern Recognition (2002) 427-430
9. http://www.research.microsoft.com/~jplatt/adult.zip (June 2001)
10. http://www.research.microsoft.com/~jplatt/web.zip (June 2001)

Support Vector Classifier with a Fuzzy-Value Class Label

Chan-Yun Yang

Department of Mechanical Engineering, Kuang-Wu Institute of Technology,
No. 151, I-Deh St., Peitou, Taipei, Taiwan 112,
cy.yang@mail.kwit.edu.tw

Abstract. The purpose of this paper is to introduce a concept of fuzzy class memberships to the samples of training set in the support vector classifier. The inclusion of fuzzy values contributed a set of dynamic Lagrangian constraints, which setups a more specific space for searching the optimum, and conducted a more accurate classification performance. The developed model stepped into the sub-structure of the classifier, and involved the complex micro-interactions among the training samples to form a more precise separating hyperplane by fuzzy membership. The micro-interactions also altered the hyperplane and its corresponding margin, and achieved the deep-reaching classification accuracy around the sub-optimal region.

1 Introduction

The purpose of this paper is to introduce a concept of the fuzzy class membership for the samples of training set in the support vector classifier (SVC). The proposition assigned fuzzy values instead of crisp values to the memberships of classes. This inclusion of fuzzy values provided more precise constraints for optimizing results in a more accurate classification performance, especially for much confused data with many outliers in the training stage. In the present study, the outliers are defined as confused samples, which are classified into incorrect partition by a separating scheme. Cao et. al. [1] has been aware of the use of outlier as a stimulus in the mechanism of the SVM by adding a couple of outliers to improve a classical support vector novelty detector. The addition of outliers enhanced the power to balance the monopoly of the single-class. Hence, the separating hyperplane was organized more appropriately riding on the edge. Ke and Zhang [2] discovered that the removal of outliers from the training set generated a smoother hyperplane, leading to a better performance in classification. In editing SVC, the removal of outliers also removes relevant constraints in the optimization search. In contrast with the classical SVC, the modified SVC obtained a much wider margin of the separating hyperplane. Lin and Wang [3] introduced a fuzzy membership s_i as a weighting factor to reduce the influence of slack variables ξ_i in the optimization procedure. The SVM with a fuzzy membership s_i efficiently reduced the effects of outliers, and in turn achieved higher accuracy than that without s_i. Hong and Hwang [4], using the basic idea underlying SVM for multivariate fuzzy models, increased computational efficiency in producing optimal solutions of linear and nonlinear regressions.

Since the introduction of fuzzy concept is to obtain a more precise optimal hyperplane, a key parameter represented by a crisp value should be fuzzified within the training stage to reveal more associated information. For the KKT conditions in the theory of the SVM, the samples near the separating hyperplane have crucial influences in forming a support vector. In fact, it is not sure whether a sample near the hyperplane represents an exact signal or a serious noisy outlier before training. Ke and Zhang [2] discovered this problem, and proposed a two-stage editing algorithm to remove the samples of confused data in the first stage. The edited SVM provided an excellent way to identify the samples of exaggerating influence, but might be oversensitive so as to ignore the effects of these samples. Instead of a complete cancellation, a moderate reduction of the effects of these samples would lead to a precise optimization.

2 Fuzzy Class Label in SVC

According to the notation in the SVM, a dataset S can be expressed as $S = \{(x_i, y_i)\}$, $i = 1, 2, \ldots, n$. Each training sample (x_i, y_i) in S consists of two elements: one is a vector x_i, $x_i \in \Re^d$, representing the location of sample i in the d-dimensional feature space, and the other is a label y_i, $y_i \in \{-1, 1\}$, indicating sample i is in different classes. A proposition to assign the memberships is initiated by reformulating

$$S^* = \{(x_i, y_i^*)\}, \quad i = 1, 2, 3, \ldots, n. \tag{1}$$

In (1), y_i^* denotes a bipolar fuzzy number, representing the potential strength of the exact level of y_i, and can be obtained from a fuzzy membership function:

$$y_i^* = fz(y_i). \tag{2}$$

The fuzzification procedure in $fz(\cdot)$ can be obtained by many algorithms, even as simple as the edited SVM. In the fuzzification, the given label of each sample in the training set is no longer a crisp value. Instead, it became a bipolar fuzzy value:

$$-1 \leq y_i^* \leq 1. \tag{3}$$

With the fuzzy number y_i^*, the canonical hyperplanes of separable case can be rewritten as

$$w \cdot x_i + b \geq +1 \quad \text{for } y_i^* \geq 0, \text{ and} \tag{4}$$

$$w \cdot x_i + b \leq -1 \quad \text{for } y_i^* < 0. \tag{5}$$

Both inequalities (4) and (5) can be merged into an aggregation:

$$y_i^*(w \cdot x_i + b) \geq \mu_i, \quad 0 \leq \mu_i \leq 1, \quad \forall i, \tag{6}$$

where μ_i is a constant assessed by its corresponding y_i^*. For the constraints of optimization, the role of μ_i in (6) is characterized as a fuzzy membership to contract the restricted bounds of the respective constraints of noisy samples. With the constant μ_i obtained earlier, an alternative expression of (6) is given by recovering y_i^* as y_i:

$$y_i(\mathbf{w} \cdot \mathbf{x}_i + b) \geq \mu_i, \quad 0 \leq \mu_i \leq 1, \quad \forall i, \tag{7}$$

and could be expressed in two separated forms:

$$\mathbf{w} \cdot \mathbf{x}_i + b \geq +\mu_i \quad \text{for } y_i = +1, \text{ and} \tag{8}$$
$$\mathbf{w} \cdot \mathbf{x}_i + b \leq -\mu_i \quad \text{for } y_i = -1. \tag{9}$$

Alternatively, the replacement of y_i^* with y_i changes nothing except transferring the bound μ_i to the term $\mathbf{w} \cdot \mathbf{x}_i + b$. The transfer conducted a realization of shrunken bound, from ±1 to ±μ_i, in the inequalities (8) and (9). These two inequalities provide an opportunity to develop a set of dynamic Lagrangian constraints for a more specific space for searching optimum. In the above inequalities, an arbitrary value of μ_i, assigned by the training samples in an additional re-substitution stage before training, can produce a dynamically restricted region of constraints, as shown in Figure 1a. The lower the membership μ_i of a sample, the weaker its relationship with its native class. Through (8) and (9), μ_i of a sample narrows down the restricted region to relax the respective dynamic constraint (Figure 1b), leading to diminishing influence in classification. Hence, with more specific dynamically restricted regions, the optimization problem can be solved more appropriately.

In general, the maximization of margin can be alternatively rewritten as the minimization of $\|\mathbf{w}\|^2$, subject to constraints (7). The Lagrangian of Kuhn-Tucker theorem was introduced to reformulate the problem by minimizing the objective function of

$$L_p(\mathbf{w}, b, \alpha) = \frac{1}{2}\|\mathbf{w}\|^2 - \sum_{i=1}^{n} \alpha_i((y_i(\langle \mathbf{w}, \mathbf{x}_i \rangle + b)) - \mu_i), \tag{10}$$

subject to constraints of

$$\alpha_i \geq 0 \quad \forall i. \tag{11}$$

It should be noted that μ_i is only a constant in (10), so that it should be excluded from the set of independent variables. For optimization, two equations can be obtained by forcing the derivatives with respect to primal variables equal to zero:

$$\partial L_p / \partial \mathbf{w} = 0, \text{ or } \mathbf{w} = \sum_{i=1}^{n} \alpha_i y_i \mathbf{x}_i, \text{ and} \tag{12}$$

$$\partial L_p / \partial b = 0, \text{ or } \sum_{i=1}^{n} \alpha_i y_i = 0. \tag{13}$$

By substituting equations (12) and (13) into (10), the quadratic optimization dual problem becomes

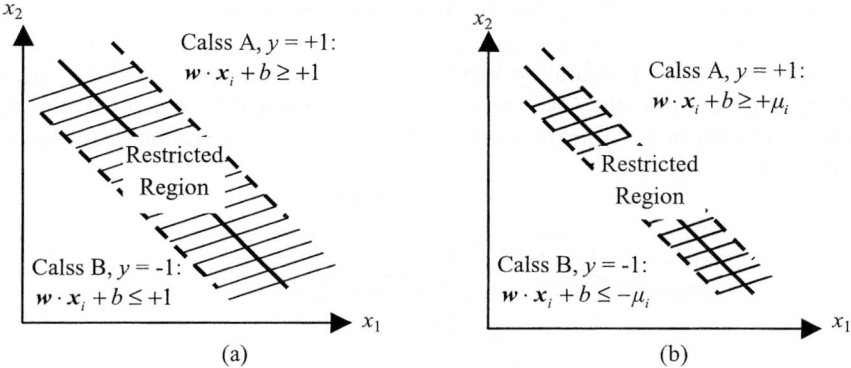

Fig. 1. With fuzzy membership μ_i, the dynamically restricted region becomes narrower

$$\max\ L_D(\alpha) = -\frac{1}{2}\sum_{i=1}^{n}\sum_{j=1}^{n}\alpha_i\alpha_j y_i y_j \langle x_i, x_j \rangle + \sum_{i=1}^{n}\mu_i\alpha_i, \qquad (14)$$

Through L_D, it is shown that the class-membership fuzzified SVC (CMF-SVC) is completely consistent with the theory of classical SVC; the only difference is a scaling factor μ_i.

In order to establish a proper relation to the native class, a metric d_i, expressing the relation strength of outliers which are identified in the resubstitution stage, should be given in advance. In this paper, an Euclidian distance, defined as a distance from a sample location to the separating hyperplane, was chosen as the metric. Based on the metric d_i, a sigmoidal membership function is drawn:

$$\mu_i = fz(y_i) = \begin{cases} 2\exp(-\tau d_i)/(1+\exp(-\tau d_i)), & \tau \geq 0, \text{ for sign}(f(x_i)) \neq y_i \\ 1, & \text{for sign}(f(x_i)) = y_i. \end{cases} \qquad (15)$$

Figure 2 shows the plots of the sigmoidal membership function $fz(\cdot)$. The general characteristic of $fz(\cdot)$ is a two-segment split curve jointing at d_i=0. To a certain end, the value of membership function maintains at full-scale value of 1, indicating that a sample on this side is correctly classified. To the other end, the μ_i of a misclassified outlier i is fuzzified as a fuzzy number which is decreasing with increasing d_i. The decreasing rate depends on the parameter τ in (15).

3   Results and Discussion

Table 1 illustrates the improvement in classification with a linear kernel for several datasets collected from *UCI Repository* [6]. With appropriate parameter setting, some positive effects in enlarging margin and reducing misclassification were found consistently. It is therefore confirmed the proposed CMF-SVC is indeed advantageous. It has to be point out that the „OMC", displayed in column three of Table 1, refers to the minimal misclassification counts that the classical SVC can provide. Although the misclassification reduction in the last column is not significantly large, a slight but valuable improvement from the optimum „OMC" implies that the CMF-SVC has the ability to tune the optimal separating hyperplane more accurately. The ability is the most essential property of the CMF-SVC. However, it should be noted that the enlargement in margin is not linearly correlated to the misclassification reduction as shown in Table 1.

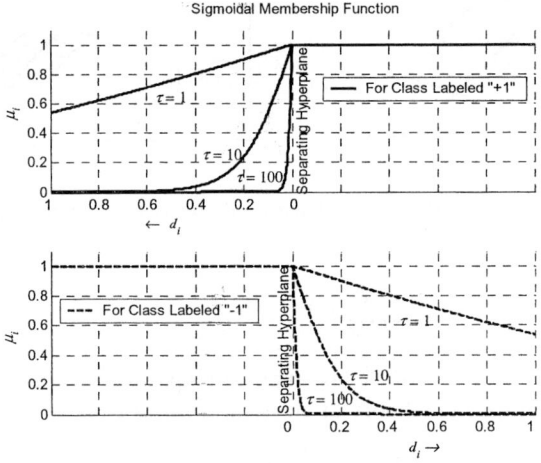

Fig. 2. Sigmoidal membership functions

Table 1. Performance of CMF-SVC

Dataset	Outliers	OMC[1]	τ	Margin Enlargement (%)	Misclassification Reduction (%)
Monks-1	35	146	1	8.33	2.05
Monks-2	68	160	1e4	85.90	8.75
Monks-3	23	80	10	38.02	11.25
Diabetes	92	96	1e2	67.65	11.11

[1]OMC: Original misclassification counts in compared classical model

A visual example for the accuracy improvement is also illustrated in a successive plots in Figure 3, with the known Iris dataset under the membership function $fz(\cdot)$ and the relevant parameter settings $\tau = 1000$. The RBF kernel produced a hyperplane to separate samples in class A and B. From the original and new hyperplanes in Figures 3a and 3b, it can be seen that the new hyperplane is slightly changed due to a moderate enlargement of margin, from which four misclassified samples are moved close to the boundary. With the new hyperplane, four misclassifications were eliminated. It is clear that, if the original hyperplane is near the sub-optimal solution, raising classification accuracy is very difficult. However, with the present method, the fuzzy relation among the training samples could re-evaluate the micro-interactions among these samples to refine the separating hyperplane, change the hyperplane and its corresponding margin moderately, and induce a deep-reaching classification accuracy around the sub-optimal region.

4 Conclusion

In this paper, we presented a crucial approach to fuzzify the class labels of training samples in the support vector scheme. A complete theoretical derivation and general validations on several datasets with a visual example are included in this paper. Experimental results showed that the present approach outperforms the classical method.

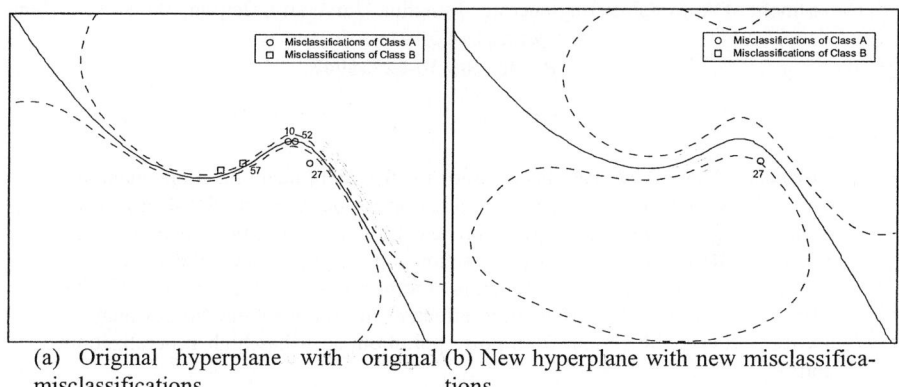

(a) Original hyperplane with original misclassifications

(b) New hyperplane with new misclassifications

Fig. 3. Visual example of margin enlargement and misclassification reduction

Acknowledgement. The author thanks Liou-Chun Chang in KWIT for his helpful comments.

References

1. Cao, L. J., Lee, H. P., Chong, W. K.: Modified Support Vector Novelty Detector Using Training Data With Outliers. Pattern Recognition Letters, Vol. 24 (2003) 2479–2487
2. Ke, H., Zhang, X.: Editing Support Vector Machines. Proceedings of International of Joint Conference on Neural Networks 2001, Vol. 2 (2001) 1464–1467
3. Lin, C. F., Wang, S. D.: Fuzzy Support Vector Machines. IEEE Transactions on Neural Networks, Vol. 13 No. 2 (2002) 464–471
4. Hong, D. H., Hwang, C.: Support Vector Fuzzy Regression Machines. Fuzzy Sets and Systems, Vol. 138 (2003) 271–281
5. Murphy, M.: UCI-Benchmark Repository of Artificial and Real Data Sets. University of California Irvine. http://www.ics.uci.edu/~mlearn (1995)

RBF Kernel Based Support Vector Machine with Universal Approximation and Its Application*

Junping Wang, Quanshi Chen, and Yong Chen

State Key Laboratory of Automobile Safety & Energy Conservation
Department of Automobile Engineering, Tsinghua University, Beijing,100084 P.R.China
fly_wang1110@sina.com
Phone: +86-10-62785947

Abstract. The SVM has been used to the nonlinear function mapping successfully, but the universal approximation property of the SVM has never been proved in theory. This paper proves the universal approximation of the SVM with RBF kernel to arbitrary functions on a compact set and deduces it to the approximation of discrete function. From simulation we can see that the RBF kernel based LS-SVM is more effective in nonlinear function estimation and can prevent the system from noise pollution, so it has high generalization ability.

1 Introduction

Support vector machine (SVM) was originally introduced by Vapnik[1], which is a novel type of learning machine based on Statistical Learning Theory. Compared with the neural network which minimizes the empirical training error and remains a number of weak points such as the existence of many local minima solutions and how to choose the number of hidden units, The SVM aims at minimizing an upper bound of the generalization error through maximizing the margin between the separating hyperplane and the data. This can be regarded as an approximate implementation of the structure risk minimization principle. What makes SVM attractive is the property of condensing information in the training data and providing a sparse representation by using a very small number of data points. So the SVM has been used in many fields such as classification and nonlinear function estimation etc[1]~[3]. Many researchers have presented some algorithms of the SVM and these methods can be used to the nonlinear function estimation. But the universal approximation property of the SVM has never been proved in theory. This paper proves the universal approximation of the SVM with RBF kernel to arbitrary functions on a compact set and deduces it to the approximation of discrete function.

---

* Supported by a grant from the National High Technology Research and Development Program of China (863 Program) (No. 2003AA501100) and China Postdoctoral Science Foundation (No.2003034145).

This paper is organized as follows. In section 2 we discuss the basic Least Square Support Vector Machine (LS-SVM) algorithm for nonlinear function estimation. In section 3, the universal approximation of the SVM with RBF kernel to arbitrary functions on a compact set is proved. Section 4 gives a simulation example of nonlinear function estimation with RBF kernel based LS-SVM.

2   The LS_SVM Algorithm for Nonlinear Function Estimation[1][3]

The basic idea of the SVM is to nonlinearly transform an input space into a feature space which is an element of the Hilbert space, and then to solve a convex optimization problems (typically quadratic programming). The solutions of the SVM are based on the structure risk minimization principle, so it has high generalization ability compared with other methods used in the nonlinear function estimation. In the standard SVM algorithm, the complexity is proportion to the number of training data. The LS-SVM algorithm was applied to huge data sets successfully because the cost function and constraints were different from the standard one.

The LS-SVM model for function estimation in feature space is as follows:

$$y(x) = \omega^T \varphi(x) + b \tag{1}$$

where $x \in R^n, y \in R$, The nonlinear mapping $\varphi(\cdot)$ is similar to the classifier case.

Given a training set $(x_i, y_i), i = 1, 2, \cdots, n$. It is to optimize the cost function

$$\min_{\omega,b,e} J(\omega, e) = \frac{1}{2}\omega^T \omega + \frac{1}{2}\gamma \sum_{i=1}^{n} e_i^2 \tag{2}$$

subject to equality constraints

$$y_i = \omega^T \varphi(x) + b + e_i, i = 1, 2, \cdots, n \tag{3}$$

Construct the Lagrangian

$$L(\omega, b, e; \alpha) = J(\omega, e) - \sum_{i=1}^{n} \alpha_i (\omega^T \varphi(x) + b + e_i - y_i) \tag{4}$$

where α_i are Lagrangian multipliers. The conditions for optimality are given by

$$\begin{cases} \frac{\partial L}{\partial \omega} = 0 \\ \frac{\partial L}{\partial b} = 0 \\ \frac{\partial L}{\partial e_k} = 0 \\ \frac{\partial L}{\partial \alpha_k} = 0 \end{cases} \tag{5}$$

from Eq.(5), we can get

$$\begin{cases} \omega = \sum_{i=1}^{n} \alpha_i \varphi(x) \\ \sum_{i=1}^{n} \alpha_i = 0 \\ \alpha_i = \gamma e_i \\ \omega^T \varphi(x) + b + e_i - y_i = 0 \end{cases} \qquad (6)$$

with solution

$$\begin{bmatrix} 0 & \vec{1}^T \\ \vec{1} & \Omega + \gamma^{-1} I \end{bmatrix} \begin{bmatrix} b \\ \alpha \end{bmatrix} = \begin{bmatrix} 0 \\ y \end{bmatrix} \qquad (7)$$

where $y = [y_1, y_2, \cdots, y_n]^T$, $\vec{1} = [1,1,\cdots,1]^T$, $\alpha = [\alpha_1, \alpha_2, \cdots, \alpha_n]^T$
$\Omega_{il} = \varphi(x_i)^T \varphi(x_l) = K(x_i, x_l). i, l = 1, 2, \cdots, n$

The resulting LS-SVM model for function estimation becomes

$$y(x) = \sum_{i=1}^{n} \alpha_i k(x, x_i) + b \qquad (8)$$

where α_i, b are the solution to the linear system. The kernel $k(\cdot, \cdot)$ is selected as RBF kernel

$$k(x, x_i) = \exp(-\|x - x_i\|^2 / 2\sigma^2) \qquad (9)$$

3 The Universal Approximation Property

It can be proved that RBF Kernel Based SVM is able to approximate any continuous function on compact set with arbitrary accuracy, and this conclusion can be deduced to discrete function.

Theorem: For any continuous real functions g defined on a compact set $U \in R^n$ and any $\varepsilon > 0$, there exists a RBF kernel based SVM f formed by Eq.(8) verifies: $\sup_{x \in U} |f(x) - g(x)| < \varepsilon$.

Lemma (Stone-Weierstrass Theorem): Suppose Z is a set of continuous real functions on compact set U, if it satisfies the conditions below, then the universal closure of Z includes all of continuous functions on U, namely, (Z, d_∞) on $(C[U], d_\infty)$ is compact.

(1). Z is an algebra, that is, set Z is closed to addition, multiplication and scalar multiplication;

(2). Z can isolate each point on U, that is, to every $x, y \in U$, if $x \neq y$, there must exist $f \in Z$ which make $f(x) \neq f(y)$;

(3). Z is not zero at any point on U, to every $x \in U$, there must exist $f \in Z$ which make $f(x) \neq 0$.

Proof: To prove the theorem using Stone-Weierstrass theorem, we should first prove that Y satisfies the three conditions of the lemma. Eq.(8) can be written as

$$y(x) = f(x) = \sum_{i=1}^{n} \alpha_i k(x, x_i) + b = \sum_{i=1}^{n+1} \alpha_i k(x, x_i) = \sum_{i=1}^{n+1} \alpha_i \exp(-\|x - x_i\|^2 / 2\sigma^2)$$

where $\alpha_{i+1} = b$, $x_{i+1} = x$.

(1) (Y, d_∞) is an algebra.

Suppose $f_1, f_2 \in Y$, then f_1, f_2 can be written as:

$$f_1(x) = \sum_{i_1=1}^{n_1+1} \alpha_{i_1} k(x, x_{i_1}) \quad , \quad f_2(x) = \sum_{i_2=1}^{n_2+1} \alpha_{i_2} k(x, x_{i_2})$$

$$f_1(x) + f_2(x) = \sum_{i_1=1}^{n_1+1} \alpha_{i_1} k(x, x_{i_1}) + \sum_{i_2=1}^{n_2+1} \alpha_{i_2} k(x, x_{i_2}) = \sum_{i=1}^{n_1+n_2+2} \alpha_i k(x, x_i) \in Y$$

$$f_1(x) f_2(x) = \left(\sum_{i_1=1}^{n_1+1} \alpha_{i_1} k(x, x_{i_1})\right)\left(\sum_{i_2=1}^{n_2+1} \alpha_{i_2} k(x, x_{i_2})\right) = \sum_{i_1=1}^{n_1+1} \sum_{i_2=1}^{n_2+1} \alpha_{i_1} \alpha_{i_2} k(x, x_{i_1}) k(x, x_{i_2})$$

$$= \sum_{i_1=1}^{n_1+1} \sum_{i_2=1}^{n_2+1} (\alpha_{i_1} \alpha_{i_2}) \exp(-(\|x - x_{i_1}\|^2 + \|x - x_{i_2}\|^2) / 2\sigma^2) \in Y$$

$$cf_1(x) = \sum_{i_1=1}^{n_1+1} (c\alpha_{i_1}) k(x, x_{i_1}) \in Y$$

So (Y, d_∞) is an algebra.

(2) (Y, d_∞) can isolate every point on U

That is, give the parameters of the RBF kernel based SVM defined on U、R, make f has characters as follow: To any given $x^0, y^0 \in U$, when $x^0 \neq y^0$, then $f(x^0) \neq f(y^0)$.

$$f(x_0) - f(y_0) = \sum_{i=1}^{n+1} \alpha_i (\exp(-\|x_0 - x_i\|^2 / 2\sigma^2) - \exp(-\|y_0 - x_i\|^2 / 2\sigma^2))$$

Because the RBF kernel is a non-singular increasing function, while $x_0 \neq y_0$, we can get
$\exp(-\|x_0 - x_i\|^2 / 2\sigma^2) - \exp(-\|y_0 - x_i\|^2 / 2\sigma^2) \neq 0$, so
$f(x_0) - f(y_0) \neq 0$, $f(x_0) \neq f(y_0)$.

(3) (Y, d_∞) is not zero on any point of U

From Eq.(8) we can see that if every $\alpha_i > 0$ ($i = 1, \cdots n+1$) this conclusion can be deduced.

From the three proofed results above we can say that Y is a set of continuous function defined on U, it satisfies the conditions of **Stone-Weierstrass** theorem. The universal closure of Y includes all of continuous function, then, there must exist a $f(x)$ formed as Eq.(8) that can make the theorem hold.

To extend the theorem result to discrete case, we get the deduction as follow:

Deduction: to any $g \in L_2(U)$ and any $\varepsilon > 0$, there must exist a SFNN f formed as Eq.(1) which make $\left[\int_U |f(x) - g(x)|^2 dx\right]^{1/2} < \varepsilon$. Where $U \in R^n$ is a compact set. $L_2(U) \subset \left[g : U \to R \mid \int_U |g(x)|^2 dx < \infty\right]$ and the integral is in the meaning of Lebesgue.

Proof: Because U is compact, so $\int_U dx = V < \infty$. Due to the subset $L_2(U)$ formed by the continuous function defined on U is also compact, therefore, to any $g \in L_2(U)$, there must exist a continuous function on U that makes $\left[\int_U |g(x) - \bar{g}(x)|^2 dx\right]^{1/2} < \varepsilon/2$, From the theorem, there must exit $f \in Y$ that can make $\sup_{x \in U} |f(x) - \bar{g}(x)| < \varepsilon / (2V^{\frac{1}{2}})$ thus:

$$\left[\int_U |f(x) - g(x)|^2 dx\right]^{1/2} < \left[\int_U (\sup_{x \in U} |f(x) - \bar{g}(x)|)^2 dx\right]^{1/2} + \varepsilon/2 < \left(\frac{\varepsilon^2}{2^2 V} V\right)^{1/2} + \frac{\varepsilon}{2} = \varepsilon$$

The deduction is proved.

* A Note: This theorem only proves the universal approximation property of the RBF kernel based SVM. But how to get the parameters of the SVM the LS-SVM algorithm such as Eq.(7) provides an effective method.

4 A Simulation Example

The RBF kernel based LS-SVM method was used to mapping the following nonlinear function.

$$y = 5\sin(5\pi x) + 10\cos^3(10x) \tag{10}$$

x was the input taken in the region $x \in [0,2]$ and y was the output. The noise with the variance $\sigma^2 = 0.25$ was added and a training set with 200 simulation pairs $(x^{(i)}, y^{(i)})$, $i = 1, 2, \cdots, 200$. is given. All the parameters were optimized through the method given in this paper shown as Eq. (7).

The simulation results is shown in Fig.1. In which the solid line shows the original nonlinear function with noisy pollution in input data, the dashed line shows approximation of the function with LS-SVM. From Fig.1 we can see that the SVM is more effective in nonlinear function estimation and can prevent the system from noise pollution, so it has high generalization ability.

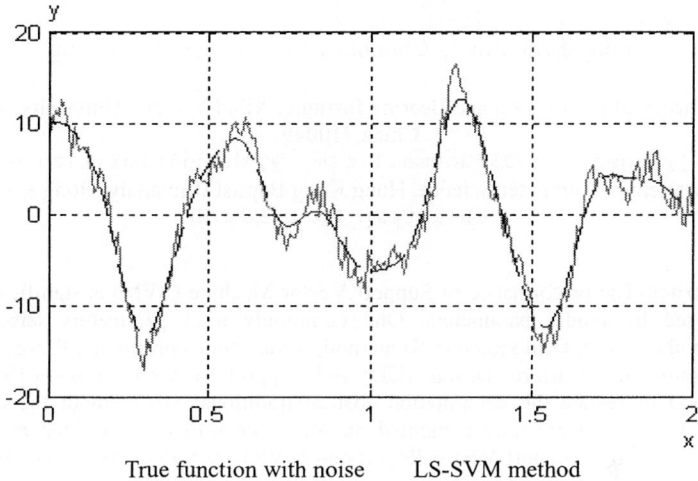

True function with noise   LS-SVM method

Fig. 1. Simulation Results

5 Conclusions

The RBF kernel based SVM has universal approximation property to arbitrary functions on a compact set. From simulation results we can see that the RBF kernel based LS-SVM is more effective in nonlinear function estimation and can prevent the system from noise pollution, so it has high generalization ability.

References

1. Johan A.K. Suykens: Nonlinear Modeling and Support Vector Machines. IEEE Conference on Instrumentation and Measurement Technology. Budapest, Hungary, (2001) pp.287-294, May.
2. Johan A.K. Suykens: Sparse Approximation Using Least Squares Support Vector Machines. IEEE Int. Symposium on Circuit and Systems. Geneva, Switzerland (2000) pp.757-760, May.
3. Feng Rui: Soft Sensor Modeling Based on Support Vector Machine. Information and Control.(2002) vol.31, no.6, pp.567-571.
4. Wang J P, Jing Z L: A Stochastic Fuzzy Neural Network with Universal Approximation and Its Application. Proc. of Int. Conf. On Fuzzy Information Processing Theories and Applications. Beijing: Tsinghua University Press & Springer, (2003) pp. 497-502.

A Practical Parameters Selection Method for SVM

Yongsheng Zhu [1], Chunhung Li [2], and Youyun Zhang [1]

[1] Theory of Lubrication and Bearing Institute, Xi'an Jiaotong University, Xi'an, China,710049
zhu_yongsheng2003@hotmail.com, yyzhang@tlbi.xjtu.edu.cn
[2] Department of Computer Science, Hong Kong Baptist University, Hong Kong,China.
chli@Comp.HKBU.Edu.HK

Abstract. The performance of Support Vector Machine (SVM) is significantly affected by model parameters. One commonly used parameters selection method of SVM, Grid search (GS) method, is very time consuming. Present paper introduces Uniform Design (UD) and Support Vector Regression (SVR) method to reduce the computation cost of traditional GS method: the error bounds of SVM are only computed on some nodes that are selected by UD method, then a Support Vector Regression (SVR) are trained by the computation results. Subsequently, the values of error bound of SVM on other nodes are estimated by the SVR function and the optimized parameters can be selected based on the estimated results. Experiments on seven standard datasets show that parameters selected by proposed method can result in similar test error rate as that obtained by conventional GS method, while the computation cost can be reduced at most from $o(n^m)$ to $o(n)$, where m is the number of parameters, n is the number of levels of each parameter.

1 Introduction

Support vector machine (SVM) is a promising technique for classification and regression developed by Vapnik V.N. et al. [1] .It has been proved to be competitive with the best available learning machines in many applications. But it is also well known that the good performance of SVM highly depends on good parameter settings. In this paper, we focus our attention on parameter selection methods of SVM classifier.

The researches on the parameter selection methods of SVM classifier are still limited. Chapelle O. and Vapnik V.N.[4] developed a gradient decent method based on some error bounds of SVM, such as RM bound [4,5] . However, the RM bound is inaccurate in some cases[6,7], also the initial values of parameters have strong effects on the resultant parameters and the searching efficiency [7]. Till now, grid search (GS) method together with k-fold cross-validate error[2], is the commonly used parameters selection method of SVM. Suppose SVM has m parameters and each parameter is separated into n levels, then total $k \cdot n^m$ SVMs should be trained and tested by GS method. Typically, $m \geq 10$, $n \geq 2$ and $k \geq 5$, the GS method is very time consuming.

The Uniform Design (UD) method [8,9] and Support Vector Regression (SVR)[1] method are introduced to speedup GS method in present paper. UD is one kind of statistical experimental design methods. The object of experimental design is to find some special nodes in a grid of experimental factors so as the real experiments just run on those selected nodes. The experimental results on other nodes are estimated with a regression estimator, which is trained with the real experimental results. Since the number of selected nodes is very small part of total nodes, the experimental time and cost can be saved drastically by UD method. For the similarity between problems of SVM parameters selection and experimental design, UD method is expected to speedup GS method by only computing the error bound on n nodes of total n^m nodes, and then a SVR estimator is trained to estimate the values of error bound of other nodes.

We restrict our attentions on SVM of RBF kernel, which has been widely studied and many reports about its performance are available. For this kind of SVM, two parameters are required to be selected: kernel width σ and error penalty parameter C [1]. RBF kernel is also adopted in SVR.

The rest of paper is organized as follows. Section 2 gives a simple introduction of UD table and settings of parameters of SVR. The experiments method and the results are presented in section 3. Section 4 discusses the results and finally we conclude the paper in section 5.

2 Uniform Design Method and Parameters of SVR

2.1 Uniform Design Table

For limitation of space, this section just provided the most practical part of UD method, the meaning and structure of UD table. The selection method of experimental points(the nodes which are selected from the grid) and other detail theoretic descriptions of UD method please refer to the references listed at website[10].

The selected UD points are listed in a table named Uniform Design table(UD table) in a form of levels' index; one line of UD table presents one UD point. For example, table 1 is a Uniform Design table $U_6(6^4)$ [10](number 6 and superscript 4 in the bracket denote that the table can direct experiment which has at most four factors and each factor has six levels; subscript 6 outside the bracket means six UD points are selected by UD method), the first UD point of table 1 is composed of 5th level of first factor, 4th level of second factor, 6th level of third factor and 2nd level of fourth factor.

Table 1. Uniform Designed table $U_6(6^4)$

	Factor 1	Factor 2	Factor 3	Factor 4
UD point 1	5	4	6	2
UD point 2	4	6	4	6
UD point 3	3	1	3	1
UD point 4	6	3	1	4
UD point 5	1	5	2	3
UD point 6	2	2	5	5

The application of UD method includes two steps: first, run experiments on the selected UD points, and second, evaluate the experimental results of other nodes of the grid by regression method. The most advantage of UD method is time and cost saving by reducing the experimental runs dramatically, such as for table 1, only 6 nodes (UD points) are selected from $6^4 = 1296$ nodes of the grid, hence times of experimental runs are reduced from 1296 to 6.

Many UD tables have been developed for different experiments [10], and this promotes the user's convenience of applying UD method: users can select a suitable UD table for their problem according to the number of factors, number of levels and run times, but need not to know too much theoretic details of UD method.

2.2 Parameters of SVR

Support Vector Regression (SVR) is adopted to deal with the regression problem of proposed parameters selection method. The performance of SVR is also sensitive to its parameters. For SVR with RBF kernel, three parameters(insensitive zone ε, kernel width σ and regularization parameter C) are required to be determined.

Parameters selection of SVR is also a knotty problem and attracts increasing attentions [11-14]. A simple and direct SVR parameters selection methodology [11] is adopted in present paper, in which parameters of SVR are set analytically according to the training samples. Suppose the samples of SVR are $(x_i, y_i), i = 1, \cdots, l$, where x_i is the training sample and y_i is corresponding output, l is the number of the samples, then the parameters of SVR are set as follows[11]:

$$C = \max(|\bar{y} + 3\sigma_y|, |\bar{y} - 3\sigma_y|) . \tag{1}$$

$$\varepsilon = 3\sigma_y \sqrt{\frac{\ln l}{l}} . \tag{2}$$

$$\sigma^d = (0.2 \sim 0.5) * rang(x) . \tag{3}$$

where $\bar{y}$ and σ_y are the mean value and standard deviation of $y_i = 1, \cdots, l$, d is the dimensions of training samples, $rang(x)$ is the scope of attributes of training samples. The performance of the resultant SVR has been proved to be satisfactory in [11] and in our following experiments.

3 Experiments and Results

Seven standard datasets *Banana, Diabets, Image, Ringnorm, Splice, Twonorm,* and *Waveform* of IDA Benchmark repository [16] are adopted to validate the proposed parameters selection method. The data are given in several predefined splits(named "realizations"[16]) into training and testing samples, and the averaged training and testing results over all realizations are expected to be more reasonable in estimating the performance of learning machines.

UD table $U_{37}(37^{12})$ [15] and five-fold cross-validate error bound are adopted in present paper. Considered the computation time, only the parameters of first 10 realizations of each dataset are selected by UD and SVR method, the averaged parameters over those 10 realizations are adopted for all realizations. For the kernel width of SVR, a middle value $\sigma^2 = 0.8$ is adopted, and all samples of SVR are normalized. Finally, both optimized parameters in linear space (the uniformed grid is constructed in $\{C,\sigma\}$ plane) and log space(the uniformed grid is constructed in $\{\ln C, \ln \sigma\}$ plane) are selected separately.

For all seven datasets, two kinds of optimized parameters have been obtained respectively under the criteria of testing error minimization("Optimized Parameters") and RM bound minimization[1]("Near Optimized Parameters") by GS method in 21×21 grid of log space[3]. To eliminate the differences caused by different SVM algorithms, we re-train and re-test SVM with those two kind parameters, and then compare results with that of our method. All results are listed in Table 2, where the columns of "GS method" are the results obtained with the parameters of [3]. The number in bracket is the value of optimized parameter vector (c,σ) of SVM. The smallest error rate of each dataset is expressed in italic.

Table 2. The optimized parameter vector (C,σ) and corresponding test error rate(%)

Name of Databases	UD and SVR method		GS method	
	Linear Space	Log Sapce	Optimized Parameters	Near Optimized Parameters
Banana	(870, 1.54)	(542, 0.73)	(128, 1.4)	(0.5, 0.5)
	10.56 ± 0.52	11.65 ± 0.70	10.53 ± 0.54	*10.45 ± 0.43*
Diabetes	(2550, 265)	(1420, 108.72)	(64, 8)	(0.25, 2)
	23.33 ± 1.65	23.38 ± 1.69	23.97 ± 1.77	24.37 ± 1.80
Image	(246, 2.12)	(166, 3.02)	(16, 1.41)	(16, 0.71)
	3.18 ± 0.62	3.28 ± 0.55	*3.08 ± 0.65*	3.79 ± 0.69
Ringnorm	(3.08, 3.9)	(32.22, 5.71)	(0.25, 2)	(0.25, 2)
	2.06 ± 0.23	2.43 ± 0.32	*1.80 ± 0.16*	*1.80 ± 0.16*
Splice	(1545, 15.2)	(4786, 36.13)	(2, 4)	(0.25, 5.66)
	11.96 ± 0.72	12.10 ± 0.74	12.31 ± 0.66	14.03 ± 0.78
Twonorm	(105, 48.10)	(187.61, 25.82)	(0.25, 2.83)	(0.5, 2.83)
	2.77 ± 0.24	3.15 ± 0.35	*2.46 ± 0.14*	2.56 ± 0.16
Waveform	(29, 9.9)	(162.31, 22.45)	(2, 2.83)	(0.5, 2)
	10.45 ± 0.45	11.12 ± 0.54	*10.25 ± 0.45*	11.43 ± 0.89

---

[1] In [3], several optimized groups of parameters are selected for each kind of dataset; we choose parameters which give the minimal test error rate to compare with ours.

4 Discussion

It is shown in Table 2 that for most datasets, although the optimized parameters obtained by different methods(or in different space) are quite different, the differences among test error rates are relatively small; this proves that there are many "near-optimized parameters" for SVM in a given space, and it is possible to search a group of "near-optimized parameters" in a small domain, therefore the number of UD points need not to be very big. In other experiments, we found 21 UD points can also provide satisfied results[7].

The differences of test error rates between columns of "Linear Space" and "Optimized Parameters" are very small for all datasets, which means that UD and SVR method can find satisfactory parameters for SVM in linear space. In log space, the parameters obtained by UD and SVR method also gives comparable test error rates as "near optimized parameters" which are obtained by GS method (using RM bound as selection criterion), this means that for most real applications, where the selection criterion is always one kind of error bounds and the searching is always carried out in log space, the proposed UD and SVR method are still effective.

For UD and SVR method itself, slightly larger test error is obtained by parameters of log space than that of linear space for each dataset. The reason can be attributed to two facts. First one, searching in uniformed grid of log space corresponds to searching in a non-uniform grid of linear space and the obtained parameters are possibly inaccurate; and second, the estimation error of SVR function can also produce more serious effect on the accuracy of parameters in log space than in linear space. For example, if the estimated parameters of SVR have a deviation Δ from the real optimized parameters, then in linear space, the deviation between obtained parameters and the real optimized parameters is Δ, but in log space, this value will be 2^{Δ}.

As for the computation cost, GS method of [3] searched on 441 nodes, while UD and SVR method only searches on 37 nods. In the case of same error bound and same grid (n^m) are adopted, the computation cost can be reduced from $O(n^m)$ of GS method to $O(n)$ of UD and SVR method, where m is the number of parameters, n is the number of levels of each parameter. The traditional GS method can be speedup at most n^{m-1} times by new method while the performance of resultant SVM can be kept almost unchanged.

5 Conclusion

This paper introduces UD and SVR method to reduce the computation cost of traditional GS method. The proposed method obtains the comparable optimized parameters with the traditional GS method, while the computation cost is reduced from $o(n^m)$ to $O(n)$, where m is the number of parameters, n is the number of levels of each parameter. In experiments, we also found the parameters obtained by new method in linear space possesses slightly lower test error rate than that obtained in log space.

Because UD method has been proved to be useful in experimental design problems of many factors [10], the proposed parameters selection method for SVM is also expected can selected more than 2 parameters effectively. Future works include validating the proposed method in SVM of more than 2 parameters and finding more reasonable parameters selection method for SVR.

Acknowledgment. Thanks are due to the anonymous referees for valuable suggestions. The work is supported by National Natural Science Foundation of China under Grant No.50335030.

References

1. Vapnik, V.N.: The Nature of Statistical Learning Theory. 2nd edn. Springer-Verlag, Berlin Heidelberg New York (1995)
2. Duan, K., Keerthi, S.S., Poo, A.N.: Evaluation of Simple Performance Measures for Tuning SVM Hyperparameters. Neurocomputing, 51 (2003) 41-59
3. Chung, K.M., Kao, W.C., Sun, C.L., Wang, L.L., Lin, C.J.: Radius Margin Bounds for Support Vector Machines with the RBF Kernel. Neural Computation, 15 (2003) 2643-2681
4. Chapelle, O., Vapnik, V.N., Bousquet, O., Mukherjee, S.: Choosing Multiple Parameters for Support Vector Machines. Machine Learning, 46 (2002) 131-159
5. Keerthi, S.S.: Efficient Tuning of SVM Hyperparameters Using Radius/Margin Bound and Iterative Algorithms. IEEE Transactions on Neural Networks, 13 (2002) 1225-1229
6. Schölkopf, B., Mika, S., Burges, C.J.C., Knirsch, P., Müller, K.R., Rätsch, G., Smola, A.J.: Input Space vs. Feature Space in Kernel-Based Methods. IEEE Transactions on Neural Networks, 10 (1999) 1000-1017
7. Zhu, Y.S.: Support Vector Machine and Its Applications in Mechanical Fault Pattern Recognition (in Chinese). Ph.D thesis of Xi'an Jiaotong University. Xi'an, China (2003)
8. Fang, K.T.: The Uniform Design: Application of Number-Theoretic Methods in Experimental Design. Acta Math. Appl. Sin., 3 (1980) 363-372
9. Wang, Y., Fang, K.T.: A Note on Uniform Distribution and Experimental Design. KeXue TongBao (Sci. Bull. China), 26 (1981) 485-489
10. Uniform Design website, http://www.math.hkbu.edu.hk/UniformDesign
11. Cherkassky, V. and Ma, Y.Q.: Practical Selection of SVM Parameters and Noise Estimation for SVM Regression. Neural Network, 17 (2004) 113-126
12. Momma, M., Bennett, K.P.: A Pattern Search Method for Model Selection of Support Vector Regression. In: Kumar, V., Mannila, H., Motwani, R.(eds): Proceedings of the Second Siam International Conference on Data Mining. SIAM, Philadelphia, USA (2002)
13. Cherkassky, V., Ma Y.Q.: Comparison of Model Selection for Regression. Neural Comp, 15 (2003) 1691-1714
14. Chapelle, O., Vapnik, V.N. and Bengio, Y.: Model Selection for Small Sample Regression. Machine Learning, 48 (2002) 9-23
15. Fang, K.T.: Uniform Design and Uniform Design Table (in Chinese). 1st edn. Science Press, Beijin, China (1994)
16. IDA Benchmark Repository, http://ida.first.gmd.de/~raetsch/data/benchmarks.htm.

Ho–Kashyap with Early Stopping Versus Soft Margin SVM for Linear Classifiers – An Application

Fabien Lauer[1], Mohamed Bentoumi[1], Gérard Bloch[1], Gilles Millerioux[1], and Patrice Aknin[2]

[1] Centre de Recherche en Automatique de Nancy (CRAN UMR CNRS 7039) ESSTIN, Rue Jean Lamour, 54519 Vandoeuvre Cedex, France
{fabien.lauer,bentoumi,bloch,millerioux}@esstin.uhp-nancy.fr
[2] Institut National de Recherche sur les Transports et leur Sécurité (INRETS) 2, avenue du Général Malleret-Joinville, 94114 Arcueil Cedex, France
aknin@inrets.fr

Abstract. In a classification problem, hard margin SVMs tend to minimize the generalization error by maximizing the margin. Regularization is obtained with soft margin SVMs which improve performances by relaxing the constraints on the margin maximization. This article shows that comparable performances can be obtained in the linearly separable case with the Ho–Kashyap learning rule associated to early stopping methods. These methods are applied on a non-destructive control application for a 4-class problem of rail defect classification.

1 Introduction

In a classification problem, after the parametrization and variable selection steps, the task is to choose the separating surface form (linear, polynomial ...) and the classifier structure.

In this paper, we discuss linear classification. We compare two learning methods: Ho–Kashyap learning rule [1] and Linear Support Vector Machine (SVM) [2], [3]. For SVM, regularization by soft margin is used to improve the generalization performance. [4] introduced the SVM concepts for Ho–Kashyap classifiers. Regularization is done by adding a parameter in the cost function to control the trade-off between model complexity and the amount of tolerated errors on the training set. We introduce early stopping as another regularization method to avoid overfitting. The learning is stopped before all the training examples are well classified.

We tested these methods in a 4-class linearly separable problem by creating 4 binary sub-classifiers. This application of rail defect classification provided a set of 140 observations which is not enough to split it to a training set and a validation set. Therefore, we used the Leave One Out cross-validation method for the generalization error estimation.

We start in Sect. 2 with some formalism on linear classification, before introducing Ho–Kashyap learning rule (Sect. 2.1) and SVM (Sect. 2.2). Then, in Sect. 3, we apply these methods on the application data and compare the results.

2 Linear Binary Classification

In a binary classification linear problem, the task is to find a separating hyperplane that can separate 2 classes. Let $(x_i, y_i)_{1 \leq i \leq N}$ be a set of training examples with $x_i \in \mathbb{R}^p$ belonging to a class labeled by $y_i \in \{+1, -1\}$. The decision function of a linear classifier is:

$$f(x) = \text{sign}(\langle w, x \rangle + b) \qquad (1)$$

where $\langle .,. \rangle$ stands for dot product and $(w \in \mathbb{R}^p, b \in \mathbb{R})$ are the parameters of the separating hyperplane. If all the training examples are correctly separated, then:

$$y_i (\langle w, x_i \rangle + b) > 0 \quad i = 1, \ldots, N . \qquad (2)$$

2.1 Ho–Kashyap Learning Rule (HK)

Amongst other learning rules for linear classifiers design (perceptron, LMS, linear programming algorithms ... [5]), Ho and Kashyap [1] proposed an iterative gradient descent-based algorithm. Defining a set of N $(p+1)$-dimensional vectors X_i:

$$X_i^t = \begin{cases} (+1, x_i^t) & \text{, if } y_i = +1 \\ (-1, -x_i^t) & \text{, if } y_i = -1 \end{cases} \qquad (3)$$

and a $(p+1)$-dimensional weight vector $W = (b, w^t)^t$ allows to write (2): $\langle W^t, X_i \rangle > 0, i = 1, \ldots, N$. Then defining a $(p+1 \times N)$-matrix $X = [X_1 \, X_2 \ldots X_N]$ gives:

$$W^t X > 0 . \qquad (4)$$

Let B be the "margin" vector with b_i as components. Equation (4) can be rewritten as:

$$W^t X = B^t \qquad (5)$$
$$\text{subject to} \quad b_i > 0 \quad i = 1, \ldots, N .$$

Ho–Kashyap (HK) learning rule solves (5) by minimizing the least squares criterion $J(W, B) = \|W^t X - B^t\|^2$. The margin vector is first initialized to B_0 with all b_i set to small positive values. At each step k, the weight vector W_k is deduced from B_k by:

$$W_k^t = B_k^t X^\dagger \qquad (6)$$

where $X^\dagger = X^t(XX^t)^{-1}$ stands for the pseudo-inverse of X. Then a gradient descent is used to compute a new estimate of the margin vector:

$$B_{k+1}^t = B_k^t - \mu \frac{1}{2} \left(\nabla_B J(W, B) - |\nabla_B J(W, B)| \right) \qquad (7)$$

with μ a positive learning rate.

In order to satisfy the constraints $b_i > 0$, the positive components of $\nabla_B J(W,B)$ are set to 0, thus preventing b_i to decrease and become negative. This is why $\frac{1}{2}(\nabla_B J(W,B) - |\nabla_B J(W,B)|)$ is used instead of $\nabla_B J(W,B)$.

It can be shown [5] that this procedure converges in a finite number of steps $\forall \mu,\ 0 < \mu < 1$, to 0 in the separable case, to a non-zero value otherwise. This makes the tuning of μ not critical.

2.2 Linear SVM

For Linear Support Vector Machine (SVM) binary classifiers, (2) becomes [2]:

$$y_i(\langle w, x_i \rangle + b) \geq 1 \quad i = 1, \ldots, N. \tag{8}$$

We consider now the points that ensure equality in (8). These points belong to the so called *canonical hyperplanes* $H_1 : \langle w, x_i \rangle + b = 1$ and $H_2 : \langle w, x_i \rangle + b = -1$. The distance Δ which separates H_1 and H_2 is equal to $2/\|w\|$ and is called the *margin*. The main difference introduced by a SV classifier is that the optimal separating hyperplane is the one that ensures a maximal margin [2][3], i.e. minimizes $\|w\|$. To build a so called *hard margin SV* classifier, the task is therefore:

$$\begin{aligned} \min \quad & W(w,b) = \tfrac{1}{2}\|w\|^2 \\ \text{subject to} \quad & y_i(\langle w, x_i \rangle + b) \geq 1 \end{aligned} \tag{9}$$

which is equivalent to the maximization problem of the *dual Lagrangian*:

$$\begin{aligned} \max L_{dual} &= \sum_{i=1}^{N} \alpha_i - \tfrac{1}{2} \sum_{i,j=1}^{N} \alpha_i \alpha_j y_i y_j \langle x_i, x_j \rangle \\ \text{subject to } \alpha_i \geq 0,\ i &= 1, \ldots, N \text{ and } \sum_{i=1}^{N} \alpha_i y_i = 0 \end{aligned} \tag{10}$$

where α_i are the Lagrange multipliers. The solution $(\hat{\alpha}_i)$ of (10) allows to determine the couple $(\hat{w}, \hat{b})$:

$$\hat{w} = \sum_{i=1}^{N} \hat{\alpha}_i y_i x_i, \quad \hat{b} = -\frac{1}{2}\langle \hat{w}, x_r + x_s \rangle, \quad \hat{\alpha}_r, \hat{\alpha}_s > 0 \tag{11}$$

where x_r and x_s are two examples for which the corresponding class labels are $y_r = -1$ et $y_s = +1$. The decision function (1) of a SVM classifier is thus given by:

$$f(x) = \text{sign}\left(\sum_{i=1}^{N} \hat{\alpha}_i y_i \langle x_i, x \rangle + \hat{b} \right). \tag{12}$$

From the Karush-Kuhn-Tucker (KKT) conditions [3], we have:

$$\hat{\alpha}_i (y_i [\langle w, x_i \rangle + b] - 1) = 0, \quad i = 1, \ldots, N \tag{13}$$

and therefore only for the points x_i which satisfy $y_i\left[\langle w, x_i\rangle + b\right] = 1$, Lagrange multipliers are non zero: $\hat{\alpha}_i > 0$. These points are called *Support Vectors* (SV).

Most of the time, in practice, the training set contains noise and outliers and a SV classifier calculated from this set can lead to poor generalization. To tackle this problem, slack variables ξ_i which allow errors on the constraints can be introduced. Now, we have the so called *soft margin SVM* problem to solve:

$$\begin{aligned} \min \quad & W(w,b) = \tfrac{1}{2}\|w\|^2 + C\sum_{i=1}^{N}\xi_i \\ \text{subject to} \quad & y_i\left(\langle w, x_i\rangle + b\right) \geq 1 - \xi_i \\ & \xi_i \geq 0 \end{aligned} \tag{14}$$

where C is the regularization parameter which controls the trade-off between training error and model complexity and has to be determined beforehand. Solving the quadratic optimization problem (14) leads to the same dual Lagrangian maximization (10) but subject to:

$$0 \leq \alpha_i \leq C,\ i = 1,\ldots,N \text{ and } \sum_{i=1}^{N} \alpha_i y_i = 0 \ . \tag{15}$$

Two categories of SV can be distinguished: the well classified SV which have $0 < \hat{\alpha}_i < C$, and the misclassified SV which have $\hat{\alpha}_i = C$.

2.3 Generalization and Hyperparameters Tuning

In classification the goal is to minimize the error on future examples which is called the generalization error (GE). The most popular technique for the GE estimation is the *cross-validation* that is independent of the learning machine used. The Leave One Out procedure (LOO) is a cross-validation procedure adapted for a weak data number N and giving an almost unbiased estimation of GE [6]. It consists in dividing the training set in two subsets: a learning subset of $N-1$ examples and a test subset containing only one example. The procedure is repeated N times until all the examples are tested. The estimation of GE is then given by the number of misclassified test examples over N. To lighten calculations, an upper bound of GE can be calculated: k-fold cross-validation which is similar to LOO except that the training set is divided in k subsets. One subset is left for testing and $k-1$ subsets are used for learning. The procedure is thus only repeated k times (typically, $k = 5$ or 10). LOO can be seen as the extreme case of the k-fold cross-validation, where $k = N$.

To avoid overfitting, a certain amount of misclassified training examples can be accepted. In SVM, this is introduced by the soft margin and the regularization parameter C. To tune C, a range of values is scanned and the optimal value is the one corresponding to the minimum of the GE estimation.

For the Ho–Kashyap learning rule, early stopping can make the learning process stop before all the training examples are well classified. Early stopping can be achieved by looking at the GE estimation during the training process and stopping as soon as it is rising. But this method does not yield always to the best minimum of GE (see for instance [7]). In another approach, the training is not stopped but GE is evaluated at all the iterations during the process. Then the lowest GE gives the best classifier. This method can be included in the HK learning to best tune the hyperparameter, here only n, the number of iterations. Indeed, a change in μ yields to a change in n which is automatically tuned.

3 Application

3.1 Context

The application concerns the classification of rail defects signatures. Previous works led to the realization of a suitable double-coils and double-frequencies differential eddy current sensor [8] which can be embarked on a train. After preprocessing, four complex channels (active and reactive parts) are available, which are equivalent to eight real signals.

Tests have been made on a complete subway track. The defects were labeled in 4 classes: switches (ω_1), fishplated joints (ω_2), welded joints (ω_3) and shellings (ω_4). This provided a training set of 140 observations for a 4-class classifier. One observation consists in a window of 500mm width (100 points considering sampling step is 5mm) for each of the 8 signals. The Modified Fourier Descriptors (MFD) [9] result from the 12 first coefficients C_j of the Discrete Fourier Transform (DFT) of the signals of the window by: $d_j = C_j C_{-j} / |C_1 C_{-1}|, j = 1, \ldots, 12$. The number of parameters is thus $p = 96$.

The class-rest approach to the 4-class problem is to split it into 4 binary problems with 4 sub-classifiers dedicated to the separation of one class among the others. Thus, a different subset of variables can be chosen for each. The Orthogonal Forward Regression (OFR) procedure has been applied to rank the parameters with respect to their contribution to each sub-classifier output. Together with the decision criterion introduced in [10], this reduced the input dimensions from $p = 96$ to respectively $p = 15, 15, 8$ and 9. In order to raise ambiguity, the maximum of the 4 sub-classifiers outputs gives the class of the example.

3.2 Results and Comparisons

Table 1 compares the generalization performances (LOO) evaluated with the *Leave One Out* procedure for the Ho–Kashyap learning rule with (HK opt) or without (HK inf) early stopping, SVM with soft margin (SVM $soft$) and SVM with hard margin (SVM $hard$) classifiers. The percentages of well classified examples on the training set are given in (TR set). The results are presented for each sub-classifier as well as for the global classifier before ($Global$ 1) and after ($Global$ 2) raising ambiguity.

The tuning of the hyperparameters is done as described in Sect. 2.3. Since for SVMs, LOO procedure is too much time consuming, 5-fold procedure was used to estimate GE for the tuning of C. Here is an advantage of the HK learning rule: its speed, thanks to which the tuning of n can be made with the LOO estimation of GE which is closer to GE than the 5-fold estimation.

Performances of *HK inf* are similar to the ones of *SVM hard* though it does not maximize the margin. When regularization is used, performances increase in a comparable way for both *SVM soft* and *HK opt*. Different values of μ have been tried for the Ho–Kashyap rule and it showed that it does not really affect the results (by 1 misclassified example in the worst case) but only n in the early stopping procedure.

Table 1. Classification performances

	HK inf		**SVM hard**		**HK opt**		**SVM soft**	
	TR set	LOO	TR set	LOO	TR set	LOO	TR set	LOO
Class ω_1/others	100	94.29	100	90.71	99.29	96.43	99.29	96.43
Class ω_2/others	100	96.43	100	96.43	100	99.29	100	99.29
Class ω_3/others	100	95.71	100	95.00	99.29	97.14	99.29	96.43
Class ω_4/others	100	97.86	100	97.86	100	97.86	100	97.86
Global 1	100	87.14	100	82.86	98.57	92.14	98.57	91.43
Global 2	100	95.71	100	93.57	99.29	97.14	99.29	96.43

4 Conclusion

We reviewed some formalism on linear classification and particularly on linear SVM classifiers. In the particular case of linear classification and on our application, SVM classifiers give very good generalization performances, as expected, by maximizing the margin with only one hyperparameter C to tune. But we showed that a hyperplane trained with a simple learning rule such as Ho–Kashyap can achieve comparable performances with the introduction of early stopping in the learning process and one hyperparameter, the number of iterations n. We used a procedure to tune the hyperparameter C using a simplified cross-validation method, k-fold, to lighten calculation for SVM. Ho–Kashyap learning proved faster and the almost unbiased *LOO* estimation of *GE* could be used for the tuning of n.

References

1. Ho, E., Kashyap, R.L.: An Algorithm for Linear Inequalities and its Applications. IEEE Trans. Electronic Computers **14** (1965) 683-688
2. Burges, C.: A Tutorial on Support Vector Machines for Pattern Recognition. Data Mining and Knowledge Discovery **2** (1998) 121-167
3. Cristianini, N., Shawe-Taylor, J.: An Introduction to Support Vector Machines and Other Kernel-Based Learning Methods. Cambridge University Press (2000)
4. Lęski, J.: Ho–Kashyap Classifier With Generalization Control. Pattern Recognition Letters **24** (2003) 2281-2290
5. Duda, R.O., Hart, P.E., Stork, D.G.: Pattern Classification. 2nd edn. Wiley (2000)

6. Duan, K., Keerthi, S.S., Poo, A.N.: Evaluation of Simple Performance Measures for Tuning SVM Hyperparameters. Neurocomputing **51** (2003) 41-59
7. Prechelt, L.: Early Stopping – But When ? In: Neural Networks: Tricks of the Trade. (1998) 55-69
8. Oukhellou, L., Aknin, P., Perrin, J-P.: Dedicated Sensor and Classifier of Rail Head Defects for Railway Systems. Control Engineering Practice **7** (1999) 57-61
9. Oukhellou, L., Aknin, P.: Modified Fourier Descriptors: a New Parametrization of Eddy Current Signature Applied to the Rail Defect Classification. In: III International Workshop on Advances in Signal Processing for Non Destructive Evaluation of Materials, Québec. (1997)
10. Oukhellou, L., Aknin, P., Stoppiglia, H., Dreyfus, G.: A New Decision Criterion for Feature Selection: Application to the Classification of Non Destructive Testing Signatures. In: European Signal Processing Conference (EUSIPCO), Rhodes, Greece. (1998)

Radar HRR Profiles Recognition Based on SVM with Power-Transformed-Correlation Kernel[*]

Hongwei Liu and Zheng Bao

National Lab of Radar Signal Processing, Xidian University
Xi'an, Shaanxi, 710071, P.R.China
hwliu@xidian.edu.cn

Abstract. Radar automatic target recognition (RATR) based on high-range-resolution (HRR) profiles and support vector machine (SVM) classifier is concerned. The physical mechanism of a RATR performance improvement approach, namely, performing the power transformation to the original HRR signatures, is analyzed based on the properties of HRR profiles. And a novel kernel function, power transformed correlation (PTC) kernel, is designed subsequently for SVM classifiers. The classification performance of SVM and maximum correlation coefficient (MCC) classifier are evaluated based on the measured data.

1 Introduction

The Radar automatic target recognition (RATR) is to identify the unknown target from its radar echoed signatures. Targets high-range-resolution (HRR) profile contains more detail target structure information than that of the low-range-resolution radar echoes, therefore, it plays a very important role in radar ATR community [1-6].

Several literatures [4,6] refer to that the classification performance can be improved significantly by performing power transformation (PT) to the HRR profiles. The reason is explained as the statistical property of HRR profiles after PT is more Gaussian-like, which makes the performance of many classifiers optimal. Departing from radar HRR profile physical properties, we analyses the performance improvement mechanism of PT in RATR in this work, in contrast with that from the viewpoint of statistical properties in literatures.

Support vector machine (SVM) can map features from the original feature space to a high dimensional kernel-induced feature space via kernel function, thus transform the linear non-separable problem to the linear separable problem in high dimensional feature space[7]. One important point in SVM application is the designing of the kernel function. In this work, we design a novel kernel function, power transformed correlation (PTC) kernel, for applying SVM in HRR profiles recognition.

---

[*] This work was partially supported by the National Science Foundation of China under grant 60302009 and the National Defense Advanced Research Foundation of China under grant 413070501.

The remainder of the paper is organized as follows. In Sec. 2, the HRR profiles similarity measurement is discussed. A novel kernel function, PTC kernel, is given in Sec. 3 for SVM application. Followed which in Sec. 4 is the classification example based on measured C band radar data. Our conclusion is summarized in Sec. 5.

2 HRR Profile Similarity Measurement

2.1 Radar HRR Profiles Properties

Generally, the targets size are much larger than the wavelength of the microwave radar, i.e. the radar works in the optics region, the electromagnetism characteristics of targets can be described by the scattering center target model, which is widely used and also proved to be a suitable target model in SAR and ISAR applications. According to this model, a HRR profile is the amplitude of the coherent summations of the complex echoes from target scatterers in each range cell. The m*th* complex returned echo in the n*th* range cell can be written as

$$x_n(m) = \sum_{i=1}^{I_n} \sigma_{n,i} e^{-j\left[\frac{4\pi R_{n,i}(m)}{\lambda} + \theta_{n,i}\right]} \quad (1)$$

where I_n denotes the number of target scatterers in the n*th* range cell, $R_{n,i}(m)$ denotes the distance between radar and the i*th* scatterer in m*th* sampled echo, $\sigma_{n,i}$ and $\theta_{n,i}$ denote the amplitude and initial phase of the i*th* scatterer echo respectively.

If the target orientation changed, its HRR profile will be changed subsequently. Two phenomena are responsible for it. The first is the scatterer's motion through range cell (MTRC). Given target rotation angle larger enough, the scatterers range variation will be larger than a range resolution cell, thus make the HRR profile changed. The target rotation angle limitation to avoid the occurring of MTRC is [1]

$$\Delta \varphi \leq \Delta r / L \quad (2)$$

where $\Delta\varphi$ is the target rotation angle relative to the radar, Δr is the range resolution of radar and L is the target cross length. The second phenomenon is the HRR profile's speckle effect. Because the HRR profile is the coherent summation of the multiple scatterers echoes in one range cell, even the target rotation angle meets the condition in (2), the phase of each scatterer echo will be changed, thus their coherent summation, will be changed subsequently.

If MTRC occurs, it means that the target scattering canter model changed. In this cased, it is required more templates to represent the target HRR profiles. As to the speckle effect, it is required a stable feature extraction approach or a suitable HRR profile similarity measurement to handle it.

2.2 Maximum Correlation Coefficient (MCC) Method

When performing HRR profile based RATR, two issues need to be paid attention to, including time-shift alignment and amplitude normalization. We assume all the HRR profile is 2-norm normalized in this paper. A widely used HRR profiles similarity measurement with time-shift compensated is the maximum correlation coefficient (MCC) method, which is defined as

$$r_{X_1 X_2} = \max_{\tau} \int X_1(t) X_2(t-\tau) dt \qquad (3)$$

where $X_1(t)$ and $X_2(t)$ denote two individual HRR profiles. A larger $r_{X_1 X_2}$ means more similar between the two HRR profiles. The time-shift corresponding to $r_{X_1 X_2}$ is called time-shift compensation factor, which is denotes as τ_{mcc}. It can be proved that the MCC method is equivalent to the Euclidean distance based similarity measurement given time-shift compensated. The link between two approaches is

$$d^2_{X_1 X_2} = 2 - 2 r_{X_1 X_2} = \int (X_1(t) - X_2(t - \tau_{mcc}))^2 dt \qquad (4)$$

where $d_{X_1 X_2}$ is the Euclidean distance between $X_1(t)$ and $X_2(t)$.

2.3 HRR Profile Similarity Measurement Based on Power Transformation (PT)

The speckle of HRR profiles will bring problems to the above similarity measurement methods. Firstly, the speckle may make the HRR profile ill normalized. If there is speckle occurs in one range cell, and results a large amplitude in the range cell, the 2-norm normalization will make the HRR profile in an ill scale, thus decreases the confidence of the similarity measurement of (3) or (4). Secondly, the speckle may make the HRR profile misaligned. The HRR profiles energy distribution along the range dimension play a key role in HRR profile alignment if using MCC method, the typical representation of speckle is the amplitude fluctuation in several range cells, thus destroys the energy distribution structure of HRR profiles, and results a misalignment. Moreover, from (4) it can be seen that the range cells with stronger energy contribute more to the final similarity measurement. Actually, this is unfair in reality, because the radar echo energy of the target components locating at the end of the target, e.g., the airplane's wings and tail, is weak generally, but these echoes represent the target geometry information, which is very important for recognizing different targets.

The PT is proved to be an efficient preprocessing approach in HRR profile based RATR, which can improve the classification significantly. It is defined as

$$Y(t) = X(t)^v, \quad 0 < v \leq 1 \qquad (5)$$

The reason that using PT can improve the classification performance is explained as that the non-normality distributed original HRR profiles will become near normality distribution after PT, thus makes the performance of many classifiers optimal. Instead of from the viewpoint of statistics theory, we will explain the reason from the viewpoint of HRR profiles physical properties below.

Suppose $X_{temp}(t)$ and $X_{test}(t)$ denote a template and a test HRR profile respectively, $Y_{temp}(t)$ and $Y_{test}(t)$ denote their power transformed profiles respectively. Suppose the template and test profiles are aligned, expanding $Y_{temp}(t-\tau_{mcc})$ around $X_{test}(t)$ by a Taylor series up to the first order term

$$Y_{temp}(t-\tau_{mcc}) = X^v_{temp}(t-\tau_{mcc}) \cong X^v_{test}(t) + vX^{v-1}_{test}(t)(X_{temp}(t-\tau_{mcc}) - X_{test}(t)) \qquad (6)$$

Then the Euclidean distance between $Y_{temp}(t-\tau_{mcc})$ and $Y_{test}(t)$ is

$$d^2_{Y_{temp}Y_{test}} = \int (Y_{temp}(t-\tau_{mcc}) - Y_{test}(t))^2 dt \cong \int \left(vX^{v-1}_{test}(t)(X_{temp}(t-\tau_{mcc}) - X_{test}(t))\right)^2 dt \qquad (7)$$

Comparing (7) and (4), the difference is that the contribution of each range cell to the Euclidean distance is weighted by $X^{v-1}_{test}(t)$. The weaker echoes are amplified, and the stronger echoes are compressed. Thus the speckle effect will be decreased in measuring the HRR profiles similarity. Another advantage bring by the PT is that the influence of speckles on HRR profile normalization is decreased, thus can avoid ill normalization. Furthermore, the misalignment caused by the speckle also can be avoided to some extent.

3   Support Vector Machine (SVM) with PTC Kernel

Kernel representations offer a solution by projecting the data into a high dimensional feature space to increase the computational power of the linear leaning machine. A common form of kernel based learning machines can be written as

$$f(X) = \sum_{i=1}^{l} w_i K(X, X_i) + w_0 \qquad (8)$$

where $K(X, X_i)$ is a kernel function that quantifies the similarity between feature vectors X and X_i. The weight w_i quantifies the importance of training feature vector X_i on the classifier. The training procedure includes determining classifier parameters X_i and w. To design a suitable kernel function also is a very important part in kernel machine designing.

Among kinds of kernel learning machines, SVM using the structure risk minimization criterion to train the classification function, which will produce classifier with large margins and lead to better generalization performance. Therefore, SVM is a promising candidate for HRR profiles classification. There is no time-shift alignment issue for the traditional kernel function used in SVM, such as Gaussian RBF kernel, polynomial kernel, Sigmoidal kernel, etc. To be consistent with the HRR profile properties, and use the merit brought by PT, we define a novel kernel function, namely, the power transformed correlation (PTC) kernel

$$K(X(t),X_i(t))=\exp\left(\frac{1-\max_{\tau}\int X^v(t)X_i^v(t-\tau)dt}{\sigma^2}\right) \quad (9)$$

Note that the HRR profiles in above equation are 2-norm normalized after PT. To accelerate the calculation speed, the correlation term in above equation can be implemented via Fast Fourier Transformation (FFT). SVM classifier is a binary classifier, for M-ary classification problem, generally one trains multiple binary SVM classifier, then combine all the output to make a final decision. Two approaches are often used to design each binary classifier, namely, one-*vs*-one and one-*vs*-rest. In what follows, we use the one-*vs*-rest method to build M-ary SVM classifier.

4 Experiment Results

The data used to evaluate the classification performance of SVM with PTC kernel are measured from a C band radar with bandwidth of $400MHz$. Three airplanes, including An-26, Yark-42 and Cessna Citation S/II, are used for classification. The projections of target trajectories onto the ground plane are shown in Fig.1. The measured data of each target are divided to several segments, as shown in Fig.1. To evaluate the generalization performance of the classifiers, we choose test data and training data from different data segments. The 2nd and the 5th segments of Yark-42, the 5th and the 6th segments of An-26, the 6th and the 7th segments of Cessna Citation S/ II are chosen as the training data, all the rest data segments are chosen as test data.

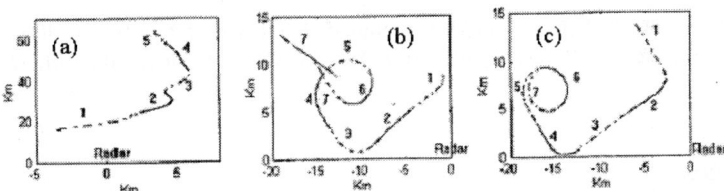

Fig. 1. The projection of target trajectories onto the ground plane (a) Yak-42, (b) An-26 (c) Cessna Citation S/II

Shown in Table 1 are the average classification rates of MCC and SVM classifiers with different parameters. The PT coefficient v is set to as 0.15. It was shown that the PT can improve the classification performance significantly and a better classification performance can be achieved if increasing the template number for the MCC classifier. Obviously, the computation burden will be increased at the same time. The SVM with PT kernel achieves best performance, which uses only 70 support vectors. This means its computation burden is the lowest one among different approaches.

To further compare the classification performance between MCC and RVM, their receiver-operating-characteristic (ROC) curves are shown in Fig.2, which shows that the RVM classifier also has a good ability to reject the false target in addition to its good classification performance.

5 Conclusion

RATR based on HRR profiles is concerned. Departing from the scattering center target model, the physical properties of HRR profiles are analyzed. Based on which we discussed the physical mechanism of power transformation in improving RATR performance. A novel kernel function, PTC kernel, is designed for SVM classifiers. The classification performances of SVM and MCC classifiers are evaluated based on the measured data. The experimental results show that SVM has superior classification performance and lower computation complexity as well.

Table 1. Classification performance comparison between MCC and SVM

	MCC method without PT		MCC method with PT		SVM with PTC kernel
Template number	150	300	150	300	70 (SV number)
Average recognition rate	0.6580	0.7110	0.8820	0.9329	0.9549

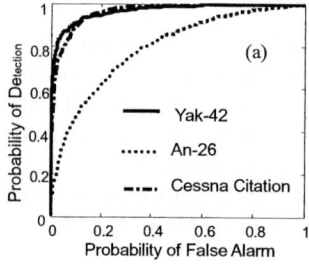 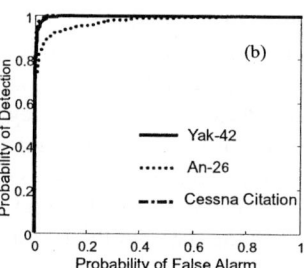

Fig. 2. ROC curves of MCC and SVM, the template number used by MCC is 300, the support vector number of SVM is 70. PT coefficient $v=0.15$, (a) MCC classifier (b) SVM classifier

References

1. Li, H. J., Yang, S. H.: Using Range Profiles as Feature Vectors to Identify Aerospace Objects. IEEE Trans. Antennas and Propagation, Vol.41, No.3 (1993) 261-268
2. Li, H. J., Wang, Y. D., Wang L. H.: Matching Score Properties Between Range Profile of High-Resolution Radar Targets. IEEE Trans. Antennas and Propagation (1996) 444-452
3. Liao, X., Bao, Z., Xing, M.: On the Aspect Sensitivity of High Resolution Range Profiles and Its Reduction Methods, IEEE International Radar Conference (2000) 310–315
4. Williams, R., Westerkamp, J., et al.: Automatic Target Recognition of Time Critical Moving Targets Using 1D High Range Resolution (HRR) Radar, IEEE AES Magazine (2000) 37-43
5. Mitchell, R.A., Westerkamp, J.J.: Robust Statistical Feature Based Aircraft Identification, IEEE Trans. Aerospace and Electronic Systems (1999) 1077-1094
6. Heiden, R., Groen, F. C. A.: The Box-Cox Metric for Nearest Neighbour Classification Improvement, Pattern Recognition (1997) 273-279
7. Burges, C.: A Tutorial on Support Vector Machines for Pattern Recognition, Data Mining and Knowledge Discovery, Vol.2 (1998) 121-167

Hydrocarbon Reservoir Prediction Using Support Vector Machines

Kaifeng Yao[1], Wenkai Lu[1], Shanwen Zhang[2], Huanqin Xiao[2], and Yanda Li[1]

[1] State Key Laboratory of Intelligent Technology and Systems
Dept. of Automation, Tsinghua University
Beijing, 100084, P. R. China
lwkmf@tsinghua.edu.cn
[2] Shengli Oilfield Limited Company, Dongying
Shandong Province, 257100, China

Abstract. Hydrocarbon reservoir prediction using seismic features is a typical classification problem. Numerous methods have been developed for computer-aided reservoir prediction. The prediction accuracy is restricted by the following facts: 1) small amount of samples; 2) small size of features; and 3) the intricate non-linear relation between features and reservoir level. This paper proposes a feature expansion and feature selection method, which maps the features to a higher dimensional feature space and then select proper features, thus mines the 'true' features. The selected features are used for training a linear classifier. Test with seismic data from Guanyinchang district of Sichuan Province and Chengdao district of Shandong Province, the proposed method achieved better prediction result than other methods.

1 Introduction

Seismic information is always used for hydrocarbon reservoir prediction because it's economical to acquire them in a large area. The real reservoir level cannot be determined before sinking a well, thus limits the sample number for training a classification machine. It's hard to get high generalization accuracy with statistical methods for solving problems with small amount of samples. Fuzzy mathematics method has been introduced to determine the fuzzy relations between hydrocarbon accumulation and each seismic feature, and a classifier can be design according to the fuzzy relations [1]. Neural network methods like BP [2] and SOM [3] have also been used for reservoir prediction.

The authors propose the feature expansion and feature selection idea. Expanded features tend to be more linearly related with the classification result. Linear support vector machine (SVM) is applied for the feature selection step. Finally, a simple linear classifier is designed with selected features.

A brief introduction about SVM is given in the next section, and our approach will be introduced in the third section, followed by some experiment examples. The last section gives our conclusion.

2 Support Vector Machine Review

Support vector machine, first introduced by Vapnik, has been widely used in various pattern recognition and regression problems. SVM gives the best generalization accuracy classifier by maximizing the margin between two classes in the feature space [4]-[6]. We present some important ideas and results here.

A linear support vector classifier for a two-class (labeled as 1 and -1) problem is

$$f(x) = sign\left(\sum_n \alpha_n y_n x_n^T x + \beta\right) \qquad (1)$$

Here x_n and y_n are known N samples and labels, x is the test sample vector, α_n and β are parameters solving from an optimized problem, $f(x)$ is the prediction result for sample x. The complexity of solving the SVM problem depends on the training sample number, and is independent of the feature size.

Kernel machines suppose that nonlinear problems in a lower dimensional feature space can always be converted to linear problems in some higher dimensional feature space. The mapping functions to the higher dimensional space are not concerned and need not to be expressed in explicit forms. A kernel function, which denotes the inner product function in the higher dimensional space, covers all the mapping information needed in kernel machines. Using the kernel extension, equation (1) is

$$f(x) = sign\left(\sum_n \alpha_n y_n K(x_n, x) + \beta\right) \qquad (2)$$

$K(u,v)$ is the kernel function representing dot-product of u and v in the mapped space. Frequently used kernel functions include linear, polynomial, radial basis function (RBF), sigmoid function, and spline function, etc. Kernel machines don't care about the explicit mapping function, and some kernel functions don't have explicit mapping form.

SVMs in kernel and linear forms are used in the following feature expansion and feature selection methods.

3 Feature Expansion and Feature Selection Approach

3.1 Feature Expansion

Machines with kernel extension convert nonlinear problems to linear problems in certain new feature spaces. For example, considering a two-class nonlinear classification problem with label function $f(x_1,x_2)=sign(x_1^2+4x_2+2x_1-3x_1x_2+2)$, by extending the feature set (x_1, x_2) to $(x_1, x_2, x_1x_2, x_1^2, x_2^2)$, it's converted to a linear classification problem in the new feature space. This achieves same result as using a 2-degree polynomial kernel function for certain linear classifier (without considering the scale for each new feature).

One of the advantages of using feature expansion instead of kernel machines is that explicitly expanded features can be used for further feature selection, and it's well known that eliminating 'useless' features can improve the performance of a classifier. Any explicit nonlinear functions are acceptable for feature expansion, e.g., spline, RBF, and polynomial etc. Useless features are eliminated in the feature selection step.

3.2 SVM Feature Selection

Linear SVM shown in equation (1) can be simplified as:

$$f(x) = sign(w^T x + \beta) \qquad (3)$$

Apparently, elements of vector x corresponding to large absolute values in w contribute more information for the decision. Thus we can rank the features according to the weight vector w, and eliminate features that contribute less information. A recursive feature elimination (RFE) algorithm is derived [5].

The SVM RFE algorithm:

```
1) Train the linear SVM
2) Compute the weights for the features
3) Eliminate one or more features corresponding to
   small weights
4) Go back to 1) until no features left
```

This algorithm gives a ranked list of features. As shown in next section, we select a feature set that gives good classification performance (for example, low leave-one-out error rate) and with less feature number. Finally, we train a linear SVM based on the selected features. This machine is used for future prediction.

4 Experimental Results

We tested our method on two datasets. One is the Guanyinchang dataset, the other is from Chengdao, Shandong Province.

4.1 Guanyinchang Dataset Experiment

The method is tested with seismic data from Guanyinchang area Sichuan Province, China (see Xiao et al for details). There are 17 labeled samples, and each sample has 7 features. We use former 10 samples for training and later 7 for test. The features are expanded with 3-degree polynomial and then normalized to zero-mean and unit deviation. In figure 1, we evaluate the performance of feature subset by minimizing leave-one-out error number on the training set. Finally, we select 8 and 2 features for further linear classifier design.

The sample distribution in the selected 2-feature space is shown in figure 2. The three classes of samples are nearly linearly separable in this feature space. Two linear classifiers are designed with former 10 samples to separate the three classes, and the generalization error rate is smaller than other methods. In the 8-feature case, the samples are linearly separable, but 2 samples are misclassified because of the small size of training set.

We compared the performance of several types of classifiers in table 1. Xiao used fuzzy mathematics method; Cai used BP neural networks and Xu used SOM neural networks. Our method performs better than these methods, and also better than original SVM method (which is equivalent to our approach without feature selection).

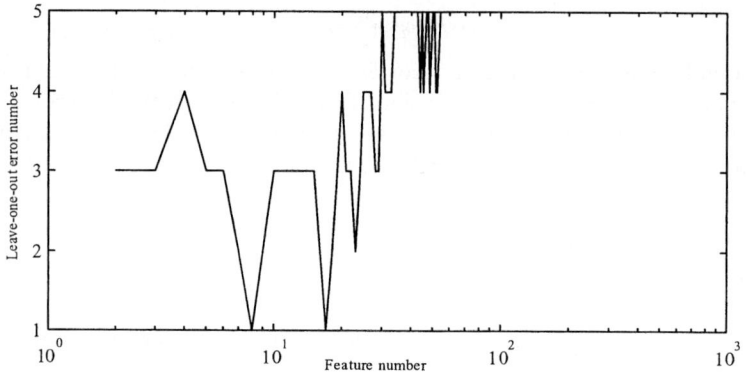

Fig. 1. Relation between the leave-one-out error number and feature size. The error rate reaches minimum at 8 and 11, and trends rising when feature number increases or decreases. We use the following rules to determine the feature number to be used in future classifier: 1) low error rate and 2) less features. We select two cases: 2 features and 8 features in our experiment

Table 1. The prediction results of different types of classifiers. Results with strikethrough lines are misclassified. Cai et al uses 13 training samples; other methods use former 10 samples for training (See [1]-[3]). 'SVM poly' uses 3-degree polynomial kernel function, 'SVM RBF' uses RBF kernel function with $\sigma = 1$. 2 features and 8 features are two cases of linear SVM classifiers based on our method. The final selected feature dimensions are 2 and 8 in these cases

Well #	1	2	3	4	5	6	7	8	9	10	11	12	13	14	15	16	17	Total Errors
Real Val.	0	0	2	2	2	2	1	1	1	2	2	1	2	1	2	0	1	
Xiao	0	~~1~~	~~1~~	2	2	2	1	1	~~2~~	2	2	1	2	1	2	0	~~2~~	4
Cai	0	~~2~~	~~1~~	2	2	2	1	~~2~~	1	2	2	1	2	1	2	0	1	3
Xu	0	~~1~~	~~1~~	2	2	2	1	1	~~2~~	2	~~1~~	1	2	1	~~1~~	0	1	5
SVM Poly	0	0	2	2	2	2	1	1	1	2	~~0~~	1	2	~~2~~	~~0~~	0	~~0~~	4
SVM RBF	0	0	2	2	2	2	1	1	1	2	~~1~~	1	2	~~2~~	~~1~~	0	~~0~~	4
2 features	0	0	2	2	2	2	1	1	1	2	2	~~2~~	2	~~2~~	2	0	1	2
8 features	0	0	2	2	2	2	1	1	1	2	~~1~~	1	2	1	~~1~~	0	1	2

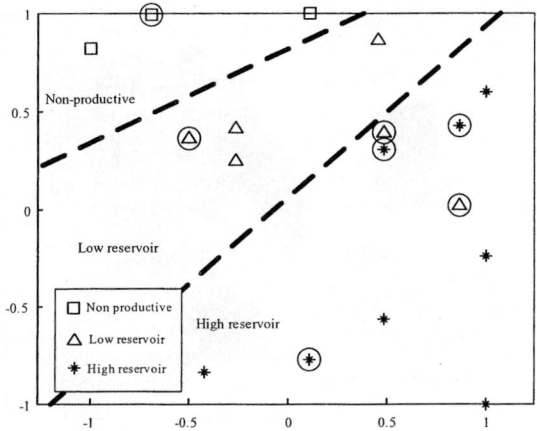

Fig. 2. The sample distribution in the selected 2-D feature space. Rectangles denote non-productive samples; triangles denote low reservoir samples and asterisks are high reservoir samples. Circled samples are 7 test samples. Linear classifiers are designed (as shown in dash lines), which separate the feature space into three areas. We can find that only 2 low reservoir samples are misclassified as high reservoir samples

4.2 Chengdao Dataset Experiment

In this dataset, the 3-D seismic data and 8 labeled wells are known as prior, and the task is to predict the hydrocarbon reservoir in the area. We extract several features from the seismic data, include AR parameters, frequency at maximum power, mean frequency, frequency of accumulated power at 25%, 50% and 75%, frequency of frequency-weighted power at 25%, 50% and 75%, minimum frequency of the log-power. Also we extracted the kurtosis, skewness and the diagonal slice of fourth-order statistics, totally there are 29 features extracted from the seismic data.

In this experiment, we directly use linear SVM for feature selection and training the classifier because the feature dimension is very high compared with the sample number. Finally we selected 8 features for classification, and Figure 3 gives our prediction result in the area. The darker area is predicted to have the higher possibility of reservoir. In this figure, we also draw circles on the well locations. According to the later well discoveries, our prediction result well matches the real result.

5 Conclusion

In this paper, we use SVM for hydrocarbon reservoir prediction. The explicit feature mapping and feature selection method is also introduced. Testing with field data in Guanyinchang area, the error rate dropped about 50% compared with fuzzy mathematic and neural network methods, and also performs better than SVM. The result on Chengdao area is also presented.

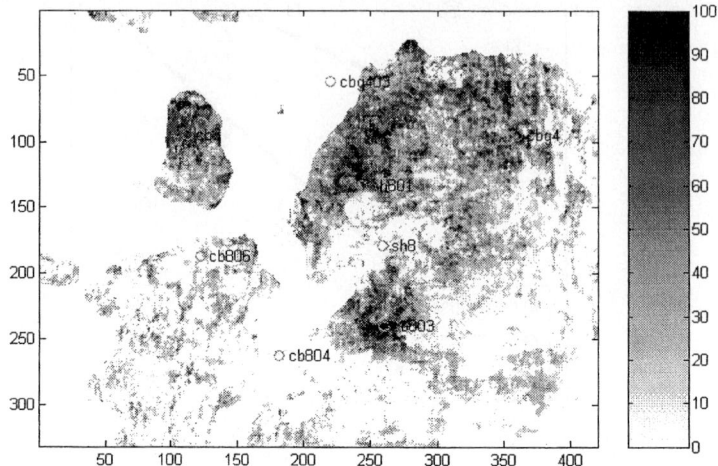

Fig. 3. The prediction result in Chengdao district. The horizontal and vertical axes denote the coordinates on the ground. The value is the probability of the hydrocarbon reservoir. Circles are the 8 well locations. The wells' names are also presented beside the well locations

Acknowledgement. This work is partially sponsored by Chinese National Science and Technology Fund (No. 2001BA605A09), CNPC Innovation Fund and Scientific Foundation for Returned Overseas Chinese Scholars from Ministry of Education.

References

1. Xiao, C.Y., Zhu, B.W.: A Fuzzy Mathematical Method for Predicting Hydrocarbon Accumulation Area by Comprehensively Analyzing Multiple Kinds of Seismic Information. Oil Geophysical Prospecting, 25 (1990) 191–200
2. Cai, Y.D., Gong, J.W., Gan, J.R. Yao, L.S.: Hydrocarbon Reservoir Prediction using Artificial Neural Network Method. Oil Geophysical Prospecting, 28 (1993) 634-638
3. Xu, J.H., Cai, R.: Application of the Supervised SOM Neural Network to Oil and Gas Prediction. Geophysical Prospecting for Petroleum, 37 (1998) 71-76
4. Burges, C.J.C.: Tutorial on Support Vector Machines for Pattern Recognition. Data Mining and Knowledge Discovery, 2 (1998), 121–167
5. Guyon, I., Weston, J., Barnhill, S., Vapnik, V.: Gene Selection for Cancer Classification using Support Vector Machines. Machine Learning, 46 (2002) 389-422
6. Vapnik, V.: The Nature of Statistical Learning Theory. Springer-Verlag, New York (1995)

Toxic Vapor Classification and Concentration Estimation for Space Shuttle and International Space Station

Tao Qian[1], Roger Xu[1], Chiman Kwan[1], Bruce Linnell[2], and Rebecca Young[2]

[1] Intelligent Automation, Inc., 15400 Calhoun Drive, Suite 400, Rockville, MD 20855
{tqian, hgxu, ckwan}@i-a-i.com
[2] NASA Kennedy Space Center (KSC)
Bruce.Linnell-1@ksc.nasa.gov, Rebecca.C.Young@nasa.gov

Abstract. During space walks, the space suits of astronauts may be contaminated by toxic vapors such as hydrazine, which are used for attitude control. Here we present some initial results on vapor classification and concentration estimation by using Support Vector Machine (SVM). The vapor was collected by electronic nose. By collaborating closely with NASA KCS, we achieved great results. For example, for Kam15f (90-second) data set, the classification success rate was 97.5% using SVM as compared to 87% using the linear discriminant method in [1]. Comparative studies were conducted between the SVM classifier and other classifiers such as Back Propagation (BP) Neural Network, Probability Neural Network (PNN), and Learning Vector Quantization (LVQ). In all cases, the SVM classifier showed superior performance over other classifiers. In the concentration estimation part by using SVM, we achieved more than 99% correct estimation of concentration by using the 90th second data samples.

1 Introduction

An electronic nose (e-nose) is an instrument that combines gas sensor arrays and pattern recognition techniques for recognizing both simple and complex odors. Using electronic nose to identify vapor is an effective approach to monitor air contaminants in the Space Shuttle and International Space Station in order to ensure the health and safety of astronauts. The detailed space program applications are:
- Monitoring air contaminant in a closed environment, such as the Space Shuttle and the International Space Station (ISS).
- Monitoring hypergolic propellant contaminants in airlocks.
- Notification of an impending fire which can be disastrous in a closed environment.

Pattern recognition is an important component of the electronic nose system to guarantee accurate vapor identification and concentration estimation. Based on the extracted features of the electronic nose sensors, the pattern recognition system can identify different type of vapors.

Recent theoretical advances and experimental results have drawn considerable attention to the use of kernel functions in data clustering and classification. Among them stands out the SVM. SVMs were first suggested by Vapnik in the 1960s for classification and have recently become an area of intense research owing to developments in the techniques and theory coupled with extensions to regression and density estimation. An SVM is a general architecture that can be applied to pattern recognition and classification, regression estimation and other problems such as speech and target recognition. SVM can be constructed from a simple linear maximum margin classifier that can be trained by solving a convex quadratic programming problem with constraints.

The advantages of SVM include:
- It is a quadratic learning algorithm. Hence, there are no local optima.
- Statistical theory gives bounds on the expected performance of a support machine.
- Performance is better than most learning systems for a wide range of applications including automatic target recognition, image detection, and document classification.
- Although originally designed for 2-class classification, SVMs have been effectively extended to multi-class classification applications. Algorithms, such as One-against-one, DAG, one-against-all, and C&S, have been successfully applied to multi-target recognition.
- There is no over-training problem as compared to conventional learning classifiers such as neural net or fuzzy logic.

The technical activities and contributions of this research are summarized as follows:
- Investigated the performance of the SVM classifier thoroughly. This effort includes proper kernel selection, parameter tuning so that best classification performance can be achieved.
- Compared the SVM classifier with many other existing classifiers. The SVM classifier performs better in most cases if a proper kernel and related parameters are properly chosen.
- Successfully applied the proposed technique to e-nose data classification. Compared with the results in [1], our technique greatly improved the NASA's current classification results. Many good results will be presented in a later section.

The paper is organized as follows. The proposed technical approach will be summarized in Sec. 2. In Sec. 3, we will report e-nose classification and concentration estimation results. Conclusions will be detailed in Sec. 4.

2 Technical Approach and Electronic Nose

2.1 Support Vector Machine

According to references [2-4], the two key elements in the implementation of SVM are the techniques of mathematical programming and kernel functions. The parameters are found by solving a quadratic programming problem with linear equality and inequality constraints, rather than by solving a non-convex, unconstrained optimization problem. The flexibility of kernel functions allows the SVM to search a wide variety of hypothesis spaces. More recently, methods for designing and combining kernels have created a toolkit of options for choosing a kernel in a particular application.

Here the theory of SVM is briefly reviewed in a two-class classification problem, the classes being P, N for $y_i = +1, -1$ respectively. This can easily be extended to k – class classification by constructing k two-class classifiers. Also for multi-class classification problem, one can refer to reference [5]. The geometrical interpretation of Support Vector Machine (SVM) is that the algorithm searches for the optimal separating surface, i.e. the hyperplane. That is, in a sense, equidistant from the two classes. This optimal separating hyperplane has many nice statistical properties as mentioned in Section 1.

2.2 Brief Description of Electronic Nose

The electronic nose (e-nose) consists of an array of non-specific vapor sensors. Popular vapor sensors are metal oxide semiconductor (MOS), surface acoustic wave (SAW), composite polymer (CP), conducting polymer, gas chromatography, mass spectrometry, light spectrum, and electrochemical. A typical time response of a signal sensor is shown in Fig. 1.

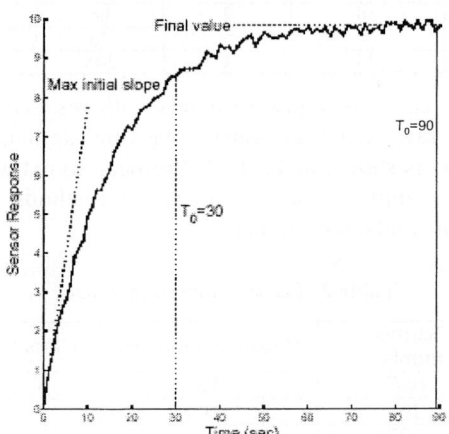

Fig. 1. Typical time response of a single sensor [1]

To the output signals of the sensors, we need to do a very important data preprocessing called feature extraction, which can get the features from signal data, then feed them to the classifier. There are two kinds of features: transient response and steady states. Different transient response features can be extracted from Fig. 1. The sensor measurements at time $T_0 = 30$ and 90 seconds can be used directly as features for classification. We also can do some other preprocessing to the raw signal data to eliminate the noise which is introduced by the sensors and environment.

2.3 Electronic Nose Data Format

NASA supplied us two e-nose data sets. The first data set contains preprocessed data files from Dr. Bruce Linnell at NASA KSC. It has 10 different electronic nose data sets for 8 different situations, which include Airsense, Air2o, Sam1big, Kam15f, Kam15o, Kam20f, Kam20o, and Kam09. The structure of the data provided by NASA is shown in Table 1. For example, the Airsense electronic nose consists of an array of 10 sensors. Therefore, the samples collected by this electronic nose are 10-dimensional. 72 samples are collected, among which there are 8 different vapors. So the classifier will classify 72 samples into 8 classes. Model 30 and Model 90 represent the sensor measurements at time $T_0 = 30$ and 90 seconds as shown in Fig. 1, respectively.

Table 1. The structure of the e-nose data sets

Electronic noses	Model 30			Model 90		
	Sample number	Dimension number	Class number	Sample number	Dimension number	Class number
Airsense	72	10	8	72	10	8
Air3o	944	10	8	1016	10	8
Sam1big	84	5	5	84	5	5
Sam1ahhi	20	5	5	20	5	5
Sam1ahlo	25	5	5	25	5	5
Kam15f	291	38	2	284	38	2
Kam15o	532	38	4	560	38	4
Kam20f	68	38	2	68	38	2
Kam20o	70	38	5	70	38	5
Kam09	596	38	4	548	38	4

The second data set contains 4 raw data sets with respect to Kam15o, Kam15f, Air3f, and Air2o. The raw data files contain the time domain sensor outputs. The structure of the raw data is shown in Table 2. We need to extract features from these raw data sets and also apply some preprocessing methods to remove the noise introduced by the sensors and environment.

Table 2. The structure of raw data

Name	Sample number	Dimension number	Class #	Concentration level
Air2o	1077	10	8	3
Air3F	261	10	3	4
Kam15f	307	38	2	3
Kam15o	647	38	4	3

2.4 Classification Testing Methods

There are two different testing methods which can estimate how well the classifier's general performance is.

Holdout. This method uses half of the data to build the classifier and the rest for testing the classifier. This is the ideal method to calculate the classification success rate when the sample number is much larger comparing with the dimension of the feature of the samples.

RU and Bootstrap. These methods are optimal for small data set classification [6].

RU is an average of two methods. One is the Re-substitution method, which uses all the data sets for training and testing. The other method is called the leave-one-out scheme, which reserves one sample for testing and uses all the rest for training. This leave-one-out scheme repeats for each available sample and takes the average as the final result.

Bootstrap is a scheme that randomly selects training samples from the whole data set, up to the total number of available samples. For example, if the total number of samples is 300, then the training will be performed by randomly selecting 300 samples form the data sets. Some of the samples will be repeated. In the testing part, we simply use all the data samples available.

The final result is the average of the RU and the Bootstrap and the margin of error term is half the difference between them.

3 Classification and Concentration Estimation Results

3.1 SVM Results

In the next few sections, we will use the SVM to perform the classification. Here we briefly describe how we choose some parameters in the SVM for achieving best classification performance.

The SVM has two parameters γ and C need to be selected case by case. For e-nose data, we need to perform a 2-dimension scan to find the optimal parameters for every data set. The flow chart of the scanning process is in Fig. 2.

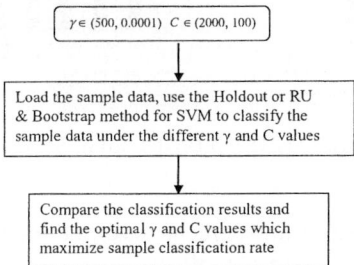

Fig. 2. Flow chart of the 2-dimension parameter scanning program

Classification Results Using Holdout Method. Two types of data sets are available: one is to sample the raw data at 30 seconds and the other one at 90 seconds. For ease of comparison, we plot the bar chart of the classification results generated by the two methods in Fig. 3. It can be seen that SVM improved a few percents for each of the five cases. "all" in the tables means all features (one feature from each sensor in the e-nose) were used for training and testing. Here we also include some recent results performed by NASA by using a nonlinear quadratic estimation technique. For Kam09, the quadratic method achieved 96% correct classification for 30-second data and 100% correct classification for the 90-second data. For Air2o, the quadratic method achieved 97% correct rate for 30-second data and 100% for 90-second data.

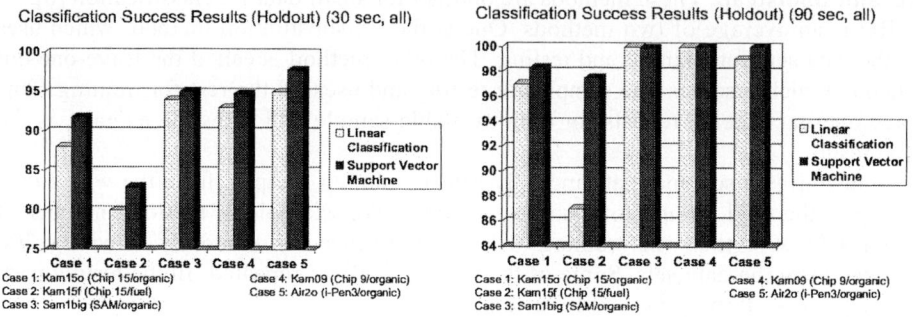

Fig. 3. Performance comparison (holdout) between the SVM and that of [1]

Classification Results Using RU and Bootstrap. The average of RU and the Bootstrap results is calculated, while the margin of error term is half the difference between them. For ease of comparison, bar charts are plotted as shown in Fig. 4. It

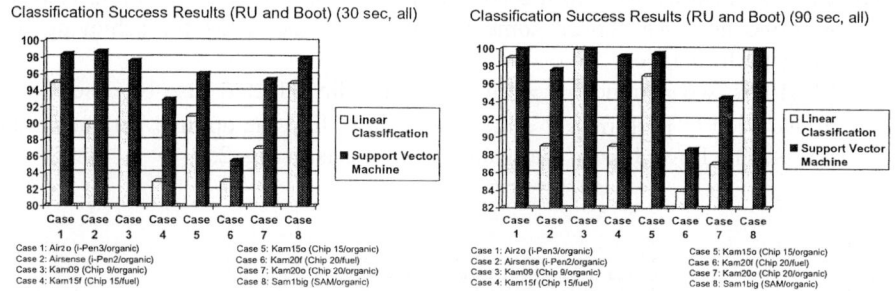

Fig. 4. Performance comparison (RU and Bootstrap) of SVM and the method in [1]

can be seen that SVM improves a few percents over the method in [1]. For example, in the 90-second cases, 8 cases have classification percentage beyond 90 %. This is very encouraging. In addition, we also include some recent results performed by NASA by using a nonlinear quadratic method. For Kam09, the quadratic method achieved 98% correct classification for 30-second data and 100% correct classification for 90-second data. For Air2o, the quadratic method achieved 97% correct classification for 30-second data and 100% correct classification for 90-second data.

3.2 Comparison of PNN, LVQ, and SVM

Besides SVM, we also used other classifiers such as PNN (Table 3) and LVQ (Table 4) to classify the same e-nose data set and compared their performance with that of SVM (Table 5). The following observations are made:
- SVM performs better than PNN and LVQ.
- PNN performs better than LVQ.

Table 3. PNN Classification Success Results (Holdout)

Test name	30 sec, all	90 sec, all
Kam15o (Chip 15/organic)	85.6%±3.2%	97.9%±1.4%
Kam15f (Chip 15/fuel)	79.4%±5.2%	91.9±4.6%
Sam1big (SAM/organic)	93.7%±4.9%	99.7%±2.4%
Kam09 (Chip 9/organic)	84.9%±2%	99.8%±0%
Air2o (i-Pen3/organic)	95.5%±2.2%	100%±0.1%

Table 4. LVQ Classification Success Results (Holdout)

Test name	30 sec, all	90 sec, all
Kam15o (Chip 15/organic)	83.4%±5.3%	96.7%±2.0%
Kam15f (Chip 15/fuel)	74.1%±5.2%	77.8±3.9%
Sam1big (SAM/organic)	93.0%±7.3%	99.7%±2.5%
Kam09 (Chip 9/organic)	82.6%±4.1%	99.5%±0.6%
Air2o (i-Pen3/organic)	96.0%±1.7%	99.9%±0.1%

Table 5. SVM Classification Success Results (Holdout)

Test name	30 sec, all	90 sec, all
Kam15o (Chip 15/organic)	91.8%±2.6%	98.3%±1.2%
Kam15f (Chip 15/fuel)	83%±3.2%	97.5±2.1%
Sam1big (SAM/organic)	95.1%±7%	99.9%±1.2%
Kam09 (Chip 9/organic)	94.8%±2%	100%±0%
Air2o (i-Pen3/organic)	97.8%±1.1%	100%±0%

3.3 Concentration Estimation

The raw data files from NASA include the vapor concentration information. There are several levels of concentration from high, medium, to low. Our objective here is to perform the concentration estimation from these data sets.

For every vapor, NASA performed many experiments under different conditions which include different humidity and different vapor concentrations. The humidity conditions include high, medium, and low, and the concentration conditions include high, medium, and low. So for every vapor, there are 9 different condition combinations in the raw data. For some data sets, they even include another concentration condition: very high. So there are 12 different condition combinations in these files. Some of the data sets do not cover all the conditions, so we just use the data sets which have complete conditions to perform the vapor concentration estimation.

In the vapor concentration estimation, we first construct a SVM to classify the different vapors. Based on the classification result, we then apply a second SVM to classify the different concentrations. For testing, we also need 2 steps to find which vapor it belongs to and which concentration it is by feeding the data into the corresponding SVMs.

We used the holdout method to evaluate the performance of concentration estimation algorithm. The detailed classification and estimation results are shown in Table 6 from a to h for different data sets. For example, Table 6a summarizes the vapor class classification results for data set Air2o. There are 8 classes in Air2o. Table 6b summarizes the concentration estimation results for each class. There are three concentration levels in each class. Similarly, Table 6c and Table 6d summarize results for the 90-second data sets of Air2o. For Air3f data sets, the results are summarized in Table 6e to h. Finally, the combined results for all the cases in Table 6 are shown in Table 7. For 90-second data sets, the overall concentration rate is above 99%.

Table 6. Detailed Concentration Estimation Success Results

Class	1	2	3	4	5	6	7	8
Vapor class classification rate	98.57%	98.24%	99.28%	98.87%	99.55%	99.91%	91.21%	100%

a. Air2o 30 second data set classification result

Concentration\class	1	2	3	4	5	6	7	8
High	97.29%	96.28%	100%	99.84%	100%	100%	87.49%	100%
Low	99.23%	98.12%	100%	98.38%	99.01%	98.19%	89.96%	99.83%
Middle	99.20%	99.39%	99.44%	99.50%	100%±	100%±	94.10%	97.43%

b. Air2o 30 Second data set concentration estimation result

Class	1	2	3	4	5	6	7	8
Vapor class classification rate	98.33%	97.94%	99.82%	99.22%	99.64%	99.37%	90.42%	99.10%

c. Air2o 90 Second data set classification result

Concentration\class	1	2	3	4	5	6	7	8
High	98.25%	97.59%	99.67%	99.36%	100%	100%	97.13%	100%
Low	97.93%	97.96%	100%	98.88%	98.73%	99.79%	100%	100%
Middle	98.93%	99.21%	98.29%	98.40%	100%	100%	97.49%	100%

d. Air2o 90 Second data set concentration estimation result

Class	1	2	3
Vapor class classification rate	99.91%	97.27%	91.21%

e. Air3f 30 sec data set classification result

Concentration\class	1	2	3
High	99.83%	95.85%	92.48%
Low	100%	99.45%	97.90%
Middle	99.83%	94.79%	81.25%
Very High	100%	100%	93.46%

f. Air3f 30 sec data set concentration estimation result

Class	1	2	3
Vapor class classification rate	99.30%	99.82%	99.63%

g. Air3f 90 sec data set classification result

Concentration\class	1	2	3
High	99.65%	100%	99.57%
Low	100%	99.43%	99.55%
Middle	100%	100%	100%
Very High	97.31%	100%	99.67%

h. Air3f 90 sec data set concentration estimation result

Table 7. Combined SVM Concentration Estimation Success Results (Holdout) (Raw Data)

Test name	30 sec, all	90 sec, all
Air3f	96.17%±2.32%	99.6%±1.31%
Air2o (i-Pen3/organic)	98.07%±0.56%	99.16%±0.40%

The result is very encouraging. Comparing with the classification result of the same data (Fig. 3), we can see the concentration estimation result is still very good (especially in 90 second data sets). It shows the second level SVM performed very well. So if we can get a good correct classification rate for the vapors, the SVM guarantees to get accurate vapor concentration estimation.

4 Conclusion

The main goal of this effort was to investigate the classification performance of SVM. Extensive studies for vapor classification, concentration estimation of electronic nose based on data from NASA clearly demonstrated that the performance of SVM is better than other classifiers. We implemented the holdout, RU and Bootstrap algorithms based on the PNN, LVQ, SVM toolboxes, the vapor concentration estimation algorithm based on SVM toolbox, and several preprocessing methods for the raw data of electronic nose which try to eliminate the noise while retaining the difference between classes.

References

1. Linnell, B., Young, R., Buttner, W.: Electronic Nose Vapor Identification for Space Program Applications.
2. Burges, C.: A Tutorial on Support Vector Machines for Pattern Recognition. Data Mining and Knowledge Discovery, 2, Kluwer Academic Publishers, Boston (1998) 121-167
3. Cristianini, N., Shawe-Taylor., J.: An Introduction to Support Vector Machines. Cambridge University Press (2000)
4. Osuna, E., Freund, R., Girosi., F.: Support Vector Machines: Training and Applications. AI Memo 1602, MIT, May (1997)
5. Hsu, C., Lin, C.: A Comparison of Methods for Multiclass Support Vector Machines. IEEE Trans. Neural Networks, Vol. 13, (2002) 415-426
6. Linnell, B.: The Effects of Small Samples on Statistical Pattern Recognition. Ph.D. dissertation, Electrical and Computer Engineering, Dept., North Carolina State University, (2001)

Optimal Watermark Detection Based on Support Vector Machines

Yonggang Fu, Ruimin Shen, and Hongtao Lu

Dept. of Computer Science and Engineering, Shanghai Jiaotong Univ., Shanghai, China
{fyg, rmshen, htlu}@mail.sjtu.edu.cn

Abstract. In this paper, a novel optimal watermark detection scheme based on support vector machine and error correcting codes is proposed. To extract the watermark bits from a possibly corrupted marked image with a lower error probability, we apply both the good generalization ability of support vector machine and the error correction code BCH. Due to the good learning ability of support vector machine, it can learn the relationship between the embedded information and corresponding watermarked image; when the watermarked image is attacked by some intentional or unintentional attacks, the trained support vector machine can recover the right hidden information bits.

1 Introduction

Nowadays, more and more multimedia information including images, videos, audios and documents are reproduced and distributed over Internet and other media. However these attractive properties lead to problems enforcing copyright protection. A potential approach to solve this problem is digital watermarking [1]. One significant merit of digital watermarking over traditional protection methods is to provide a seamless interface so that users are still able to utilize protected multimedia transparently [2].

In order to design robust information hiding scheme, some watermarking algorithm in literature applied error correcting coding(ECC) to improve the bit error rate(BER), such as Bose-Chaudhuri-Hocquenghen (BCH) coding [3,4], Reed-Solomon(R-S) code[5]and Turbo code [6]. Recently, efforts are made to use machine learning technique for watermark embedding and extraction. Neural networks are introduced into watermarking in [7], which makes the watermark detection more robust against common attacks. Genetic algorithm is proposed for selection of the best embedding positions in block based DCT domain watermarking [8]. Hence we can expect that the combination of digital watermarking and machine learning techniques might be a good solution for optimal watermark detection.

In this paper we propose a novel optimal watermark detection scheme making use of support vector machine and BCH coding. This work can be considered as an extension of some existing research [7,9]. In [9], Kutter proposed a spatial domain watermarking scheme for color image. And then Yu et al.[7] improved Kutter's work by applying neural networks. Due to the support vector machine's good learning ability in training process, it can memorize the relationship between the embedded watermark bits and corresponding watermarked image. Applying SVM's good generalization abilities and error correcting ability of BCH coding, optimal watermark detection is reached.

2 Watermark Embedding and Extracting

Support Vector Machine (SVM) is a universal classification algorithm developed by Vapnik and his colleagues [10,11]. The block diagram of proposed watermarking scheme is shown in Fig. 1.

2.1 Watermark Embedding

Since logo watermark is more convincing than the purely binary detection results, the logo image is adopted as the watermark. The total data bits embedded in our proposed watermarking scheme consists of two parts, one is reference mark and the other digital signature (logo). They are denoted as $Rf = r_1 r_2 ..... r_K$ and $S = s_1 s_2 .... s_L$ respectively, where Rf is generated according to secret key $k1$. We apply the BCH code (n, k) to encode the signature S, where n and k represent the length of BCH codeword and the number of bits in each block, respectively, obtaining a bit stream:
$S' = \{s'_i, i = 1,2,...,T\}$, where T is the total bits after BCH encoding, i.e. $T = \dfrac{Ln}{k}$.

Then they are concatenated into a single $W = Rf + S' = w_1 w_2 ..... w_{K+T}$ for embedding. We further suppose W be PN (positive and negative) sequence. The reference mark is embedded only for the training of the support vector machine to extract the data bits.

Let I be a color image with size $M \times N$, defined by $I = [R_\rho, G_\rho, B_\rho]$, where R_ρ, G_ρ, B_ρ are the three image components corresponding to red, green, and blue channel respectively. Suppose $\rho = (i, j)$, $i = 1,2,...,M$ and $j = 1,2,...,N$ be the pixel position for further image accessing. The trained support vector machine is employed to extract the watermark logo S from the tampered image during the extraction stage.

Let $\{\rho_t = (i_t, j_t)\}|_{t=1,2,....K+T}$ be the randomly selected position sequence according to another secret key $k2$ provided by the copyright owner. Here we adopt the

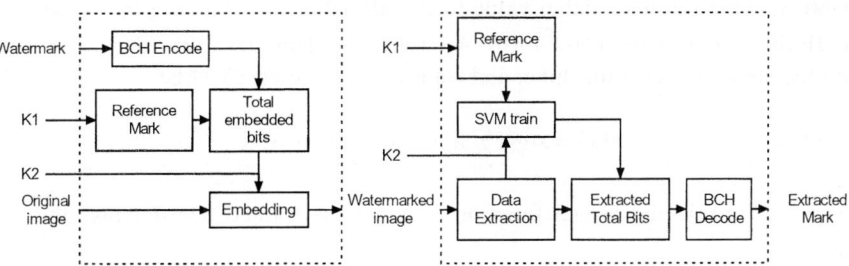

Fig. 1. Watermark embedding and extracting system architecture

conventional random generator according to uniform distribution. To embed the data, the pixels intensity in blue channel is modified as follows:

$$B_{\rho_t} \leftarrow B_{\rho_t} + \alpha w_t, \quad t = 1, 2, \ldots, K+T \tag{1}$$

where α_t is a scaling factor. To embed watermark in the host image as strongly as possible, we vary α_t according to different characteristics of the host image. For different luminance, human eyes have different sensitivity, i.e., the modification in high luminance is less detectable than that low luminance. So the factors α_t what we selected here is $\alpha_t = \beta L_{\rho_t}$, where β is a constant, and

$$L_{\rho_t} = 0.299 R_{\rho_t} + 0.587 G_{\rho_t} + 0.114 B_{\rho_t} \tag{2}$$

is the luminance component of the RGB image at position ρ_t. After the embedding procedure, the corresponding watermarked image denoted as $I' = [R'_\rho, G'_\rho, B'_\rho]$.

2.2 Watermark Extracting

The watermark extracting algorithm can be summarized as following steps:
Step1. Generate the reference mark $Rf = r_1 r_2 \ldots r_K$ and embedding position set $\{\rho_t = (i_t, j_t)\}_{t=1 \ldots K+T}$ according to secret keys $k1$ and $k2$ respectively.
Step2. For each selected position and a slide cross-shaped window with size $2c+1$, the difference between the blue component intensity of central pixel and average intensity of the others within the cross-shaped window is computed $d_{ij} = B'_{ij} - \tilde{B}'_{ij}$, where

$$\tilde{B}'_{\rho=(i,j)} = (\sum_{r=-c}^{c} B'_{i+r,j} + \sum_{r=-c}^{c} B'_{i,j+r} - 2B'_{ij})/(4c) \tag{3}$$

. When the watermarked image is undergone some attacks, since the reference mark and signature data is uniformly embedded into the host image, the relationship between the embedded data bits and watermarked image can be memorized by the training process of support vector machine. Using the good generalization ability of the well trained SVM, we can use this information to extract the hidden data bits. Define training dataset:

$$D = \{d_{i_t-2, j_t}, d_{i_t-1, j_t}, d_{i_t, j_t}, d_{i_t+1, j_t}, d_{i_t+2, j_t}, d_{i_t, j_t-2}, d_{i_t, j_t-1}, d_{i_t, j_t+1}, d_{i_t, j_t+2}, r_t\}_{t=1\ldots K} = \{D_t, r_t\}_{t=1\ldots K} \tag{4}$$

where r_t is the reference label value to the tth pattern for training the SVM and D_t is the tth data element in vector form. Applying the dataset and the data labels, we can train the support vector machine, and suppose the trained SVM be

$$f(x) = sign[\sum_{t=1}^{K} \lambda_t r_t Ker(x, D_t) + b] \tag{5}$$

where Ker is selected kernel function, λ_t s are the trained coefficients, and b is the bias.

Step3. Then watermark extraction dataset (excluding the training dataset) can be extracted from the tampered image in a similar way:

$$E = \{d_{i,-2,j_t}, d_{i,-1,j_t}, d_{i,,j_t}, d_{i,+1,j_t}, d_{i,+2,j_t}, d_{i,,j_t-2}, d_{i,,j_t-1}, d_{i,,j_t+1}, d_{i,,j_t+2}\}_{t=K+1...K+L} = \{D_t\}_{t=K+1,...K+T} \qquad (6)$$

Using the well trained support vector machine and the watermark extraction dataset E, the embedded data bits can be extracted by the output of support vector machine on dataset E, i.e. the extracted data can be denoted as:

$$\overline{w}_i = f(D_{K+i}), \quad i = 1, 2, ..., T \qquad (7)$$

Then the final extracted signature is obtained by BCH decoding. Once all the embedded data bits are extracted, the embedded signature $\tilde{S} = \{\tilde{s}_1 \tilde{s}_2 .....\tilde{s}_L\}$ is reached after BCH soft decision decoding, and then utilized to identify the copyright of owner's intellectual property by comparing S with $\tilde{S}$.

3 Experimental Results

The experimental results with the "Lena" image of $512 \times 512 \times 3 \times 8$ bits are shown in Fig.2 and table 1 respectively. What the signature we adopt is a binary logo image of size 32×25 shown in Fig.2 (b).

Some necessary parameters used in our watermarking scheme are determined by experiments. Firstly, there are several common used SVM kernels, including Linear, RBF and Polynomial. We should select the most suitable one based on many trials on the watermarked image and the attacked ones. From experiments on these three kernels, the Linear and Polynomial can be easily defeated by RBF. As to the kernel parameter σ, all the parameters from 0.1 to 40 with step size 0.2 are tested. When $\sigma \in (5,10)$, the classified results are all acceptable. Hence we adopt RBF kernels SVM with width 8 here. The watermark strength used here is $\beta = 0.3$. As a compromise between the embedded data load and good error correction ability, the BCH code (15, 5) is used in the experiments. Finally, the reference watermark length is decided by SVM performance. It is not necessary to have a huge dataset to train the SVM. Reference mark is a randomly generated bits with length 50 here.

The original Lena image and watermarked version is shown in Fig.2(a) and Fig.2(c) respectively. The Peak signal-to-noise-ratio (PSNR) of the marked image by our method and Yu's method with respect to the original image is about 41.25dB and the watermarked image based on Kutter's method is about 2dB higher. Several types of image processing and attacks are simulated to evaluate the performance of our scheme robustness, including blurring, filtering, mosaic, Jpeg compression, scaling, contrast & luminance enhancement, distortion, and cropping. Due to the limitation of paper space, we exhibit here only two cases of attacks including luminance and contrast enhancement, and distortion attacks for the visual perception.

Fig. 2. (a) Original 512*512; (c) Watermarked; (g) Distorted; (k)Luminance and contrast enhanced. (d)(h)(l) Extracted logo using our method from (c)(g)(k) with bit errors 0, 134,0 respectively; (e)(i)(m) Extracted logo using Kutter's method from (c)(g)(k) with bit errors 22,77,387,598 respectively; (f)(j)(n) Extracted logo using Yu's method from (c)(g)(k) with bit errors 0, 144,40 respectively

When there is no attack, the extracted logo image by our method, Kutter's and Yu's are shown in Fig. 2(d), (e) and (f) respectively. The embedded watermark bits can be exactly extracted with no bit errors by our method and Yu's method, but there are 22 bit errors on the extracted information bits using Kutter's method. In the case of distortion attacks, Fig.2(g) shows the distorted watermarked image with an angle 30° to the left, the extracted logo image by the three method are depicted in Fig.2(h)(i) and (j) respectively. The processed version after luminance and contrast enhancement is shown in Fig.2(k). The watermarked image is attacked by a combination of 75% luminance enhancement and 75% contrast enhancement. Three extracted logo image is shown in Fig.2(l)(m)(n) respectively. The watermark bits can be successfully extracted from the attacked image without any errors by our method, while there are about 40 bit errors in extracted logo by Yu's method, and there are about 498 bits error in extracted logo by Kutter's method, i.e. Kutter's method is completely loss.

Some other quantity comparison results are shown in Table1. From the experimental results in Table1 and Fig.2 against Kutter's and Yu's method, proposed scheme shows superior robustness. The main advantages of proposed scheme come from the good generalization ability of support vector machine. When the watermarked image is tampered, the support vector machine can learn the knowledge and exactly extract the hidden information bits. Furthermore, the BCH coding is helpful for the reliability of the watermarking scheme.

Table 1. Experimental result on different attacks

Attacks	PSNR(dB)	Bit Error Rate (BER)		
		Proposed	Kutter's	Yu's
Attack free	41.25	0	0.0275	0
Jpeg compression (50)	32.55	0.0932	0.2755	0.25
Filtering(3*3)	25.85	0.0102	0.0775	0.0125
Distortion(15°)	20.57	0.1962	0.4351	0.1625
Distortion(50°)	17.48	0.2925	0.5187	0.2702
Scaling (30%)	27.26	0.0615	0.2375	0.2087
Scaling (300%)	37.34	0	0.0275	0
Jitter(1 row+1 column)	28.35	0.1752	0.4738	0.2211
Rotation(15°)	8.75	0.0675	0.3175	0.1412
Cropping (25%)	11.19	0.0675	0.2463	0.2212

4 Conclusions

A novel optimal blind watermarking scheme based on support vector machine and BCH coding is proposed in this paper. The support vector machine can be easily fused with traditional watermarking system to improve the scheme's robustness. By introducing an aiding mark as reference, the support vector machine can be easily trained and the watermark extraction is finished by the trained support vector machine. The experimental result compared with Kutter's and Yu's methods shows outperforming reliability against different type of attacks.

References

1. Swanson, M.D., Kobayashi, M., Tewfik, A.H.: Multimedia data embedding and watermarking technologies. Proceedings of IEEE, 86 (6) (1998) 1064-1087
2. Furon, T. and Duhamel, P.: Copy Protection of Distributed Contents: An Application of Watermarking Technique. In Workshop COST 254: Friendly Exchange through the net, (2000)
3. Huang, J., Elmasry, G.F. and Shi, Y.Q.: Power constrained multiple signaling in digital image watermarking. Proceeding of IEEE workshop on multimedia signal processing (1998) 388-393
4. Huang, J.W., Yun, Q.S.: Reliable information bit hiding. IEEE Transactions on circuits and systems for video technology, 12(10) (2002) 916-920
5. Wu, C.F., Hsieh, W.S.: Image refining technique using watermarking. IEEE Transactions on consumer electronics, 46 (2000) 1-5
6. Pereira, S., Voloshynovskiy, S. and Pun, T.: Effective channel coding for DCT watermarking. Proceeding of IEEE international conference on image processing, 3 (2000) 671-673
7. Yu, P.T., Tsai, H.H., Lin, J.S.: Digital watermarking based on neural networks for color images. Signal Processing, 81 (2001) 663-671
8. Shieh, C.S., Huang, H.C., Wang, F.H., Pan, J.S.: Genetic watermarking based on transform-domain technique. Pattern recognition, 37 (2004) 555-565
9. Kutter, M., Jordan, F., Bossen, F.: Digital signature of color images using amplitude modulation. J.Electronics imaging, 7(2) (1998) 326-332
10. Vapnik, V.: Statistical learning theory. John Wiley, New York (1998)
11. Christopher, J.C.: A tutorial on support vector machines for pattern recognition. Data mining and knowledge discovery, 2 (1998) 121-167

Online LS-SVM Learning for Classification Problems Based on Incremental Chunk

Zhifeng Hao[1], Shu Yu[2], Xiaowei Yang[1], Feng Zhao[1], Rong Hu[1], and Yanchun Liang[3]

[1] Department of of Applied Mathematics, South China University of Technology,
Guangzhou 510640, P. R. China
[2] College of Computer Science and Engineering, South China University of Technology,
Guangzhou 510640, P. R. China
yushu_scut@126.com
[3] College of Computer Science and Technology, Jilin University, Changchun
130012, P. R. China
ycliang@public.cc.jl.cn

Abstract. In this paper an online learning algorithm based on incremental chunk for LS-SVM (Least Square Support Vector Machines) classifiers is proposed. The training of the LS-SVM can be placed in a way of incremental chunk, which avoids computing large-scale matrix inverse but maintaining the precision when training and testing data. This online algorithm is especially useful for the large data set and practical applications where the data come in sequentially. Our experiments with four classification problems in UCI show that compared with LS-SVM, the computational cost of our algorithm is reduced obviously and the accuracy is retained.

1 Introduction

Support vector machine (SVM) [1] is a powerful new tool for data classification and function estimation. The training problem in SVM is equivalent to solve a linearly constrained convex quadratic programming (QP) problem with a number of variables equal to the number of data points. This optimization problem is known to be challenging when the number of data points exceeds a few thousands so that standard QP packages cannot be used even for moderated large data set.

In the previous work, Osuna et al. [2] presented a decomposition algorithm and transformed the original quadratic programming into a series of quadratic programming subproblems. From the view of selecting the working set, Joachims [3] proposed an implementation of the decomposition algorithm based on Osuna's idea, which is called SVM[Light]. Platte [4] has also given a sequential minimal optimization (SMO) algorithm that breaks the large QP problems into a series of smallest possible QP subproblems, which can be solved analytically. Keerthi et al. [5] suggested some improvements to Platte's SMO algorithm for SVM classifier design. As for the convergence of the decomposition algorithm for SVM, Lin [6-7] has given the proofs in detail. Mangasarian et al. [8] proposed the Lagrangian support vector machine (LSVM), for a positive semi-definite nonlinear kernel, a single matrix inversion is required in the space of dimension equal to the number of data points classified. However, the LSVM cannot handle very large nonlinear classification problems efficiently.

Yang et al. [9] presented an extended LSVM (ELSVM) for classifications, which can speed up the convergence of the SVM^{Light} and extend LSVM to the very large nonlinear classification problems.

Suykens et al. [10] proposed a modified version of SVM for classification called least square SVM (LS-SVM), which resulted in a set of linear equations instead of a QP problem. In Chua's work [11], by using the Sherman-Morrison-Woodbury (SMW) matrix identity, LS-SVM can be applied to very large data set with small number of features. But according to [12], the SMW results in numerical instability. The existing LS-SVM algorithm is trained offline in batch way, which is not applicable for real-time classification problems. In these problems data are collected online in real time, groups by groups (or one by one) in a sequence and the training needs to be completed before next data is received. In this case, an online training algorithm would be desirable.

In this paper, an online LS-SVM based on incremental chunk is proposed for the pattern classification problems where input data are supplied in sequence rather than in batch. The algorithm is applied to four UCI data sets. The results show that our algorithm is not only faster than the LS-SVM, but also the training precision and generalization performance is as good as the LS-SVM. It should be noticed that the work of Liu et al. [13] is a specific case of our algorithm.

This paper is organized as follows. In Section 2, we review the offline LS-SVM briefly. The online LS-SVM based on incremental chunk is derived in Section 3. The experimental results of applying the proposed algorithm to four UCI data sets are presented in Section 4. In Section 5, we have some discussions and conclusions.

2   Standard LS-SVM

Consider a training set of N pairs of data points $\{x_k, y_k\}_{k=1}^N$, where $x_k \in R^n$, $y_k \in \{-1,1\}$, solving the LS-SVM is equivalent to determine the following set of linear equations

$$\begin{bmatrix} 0 & Y^T \\ Y & ZZ^T + \gamma^{-1}I \end{bmatrix} \begin{bmatrix} b \\ \alpha \end{bmatrix} = \begin{bmatrix} 0 \\ \vec{1} \end{bmatrix} \quad (1)$$

where

$Z = [\varphi(x_1)^T y_1, \cdots, \varphi(x_N)^T y_N], Y = [y_1, \cdots, y_N]^T, \vec{1} = [1, \cdots, 1]^T, \alpha = [\alpha_1, \cdots, \alpha_k]^T.$

Mercer's condition is applied to the matrix $\Omega = ZZ^T$ with

$$\Omega_{kl} = y_k y_l \varphi(x_k)^T \varphi(x_l) = y_k y_l \psi(x_k, x_l). \quad (2)$$

It is in fact only necessary to invert the $N \times N$ matrix

$$A = \gamma^{-1}I + ZZ^T. \quad (3)$$

Indeed the second row of Eq. (1) gives $bY + A\alpha = \vec{1}$ and together with the first row gives the explicit solution

$$\alpha = A^{-1}(\vec{1} - bY) \text{ where } b = \frac{Y^T A^{-1} \vec{1}}{Y^T A^{-1} Y} \tag{4}$$

Solving (4) requires the inversion of a symmetric and positive definite $N \times N$ matrix, here we use the improved Cholesky decomposition method to decompose matrix A to

$$A = LDL^T \tag{5}$$

where L is a lower triangle matrix whose elements on the diagonal are 1, and D is a diagonal matrix whose elements are non-zero.

3 Online Chunking LS-SVM

Considering that LS-SVM model based on the first N pairs of data has been constructed, and the new data pair $(x_{N+1}, y_{N+1}), \cdots, (x_{N+K}, y_{N+K})$ is fed.

Let

$$\begin{bmatrix} 0 & Y^T \\ Y & ZZ^T + \gamma^{-1}I \end{bmatrix} = A_N, \begin{bmatrix} b \\ \alpha \end{bmatrix} = \alpha_N, \begin{bmatrix} 0 \\ \vec{1} \end{bmatrix} = Y_N \tag{6}$$

Then Eq. (1) changes into

$$A_N \alpha_N = Y_N \Rightarrow \alpha_N = A_N^{-1} Y_N \tag{7}$$

The subscript N means that the current model is based on the first N pairs of data. For $N + K$ pairs of data, one has

$$\alpha_{N+K} = A_{N+K}^{-1} Y_{N+K} \tag{8}$$

where

$$A_{N+K} = \begin{bmatrix} A_N & B^T \\ B & C \end{bmatrix} \tag{9}$$

$$B = \begin{bmatrix} y_{N+1} & \Omega_{N+1,1} & \Omega_{N+1,2} & \cdots & \Omega_{N+1,N} \\ \vdots & \vdots & \vdots & \vdots & \vdots \\ y_{N+K} & \Omega_{N+K,1} & \Omega_{N+K,2} & \cdots & \Omega_{N+K,N} \end{bmatrix} \tag{10}$$

$$C = \begin{bmatrix} \Omega_{N+1,N+1} & \Omega_{N+1,N+2} & \Omega_{N+1,N+3} & \cdots & \Omega_{N+1,N+K} \\ \vdots & \vdots & \vdots & \vdots & \vdots \\ \Omega_{N+K,N+1} & \Omega_{N+K,N+2} & \Omega_{N+K,N+3} & \cdots & \Omega_{N+K,N+K} \end{bmatrix} + \gamma^{-1} I \quad (11)$$

$$Y_{N+K} = [Y_N \; 1 \; 1 \; \cdots \; 1]^T \quad (12)$$

From Ref. [14], we know that the following two lemmas hold:

Lemmas 1. For a matrix $A = \begin{bmatrix} A_{11} & A_{12} \\ A_{21} & A_{22} \end{bmatrix}$, where A_{11}^{-1} and A_{22}^{-1} exist, the following equation is true

$$A^{-1} = \begin{bmatrix} [A_{11} - A_{12} A_{22}^{-1} A_{21}]^{-1} & A_{11}^{-1} A_{12} [A_{21} A_{22}^{-1} A_{12} - A_{22}]^{-1} \\ [A_{21} A_{22}^{-1} A_{12} - A_{22}]^{-1} A_{21} A_{11}^{-1} & [A_{22} - A_{21} A_{11}^{-1} A_{12}]^{-1} \end{bmatrix} \quad (13)$$

Lemmas 2. For matrices A, B, C and D where A^{-1} and C^{-1} exist, the following equations is true

$$(A + BCD)^{-1} = A^{-1} B (C^{-1} + DA^{-1} B)^{-1} DA^{-1} \quad (14)$$

From Lemma 1 and Lemma 2, Theorem 1 can be inferred as follows:

Theorem 1 The matrix A_{N+K}^{-1} in Eq. (8) can be obtained from A_N^{-1} without computing the matrix inverse.

Proof From Eq. (13), one can obtain

$$A_{N+K}^{-1} = \begin{bmatrix} A_N & B^T \\ B & C \end{bmatrix}^{-1} = \begin{bmatrix} [A_N - C^{-1} B^T B]^{-1} & A_N^{-1} B^T [BA_N^{-1} B^T - C]^{-1} \\ [BA_N^{-1} B^T - C]^{-1} BA_N^{-1} & [C - BA_N^{-1} B^T]^{-1} \end{bmatrix} \quad (15)$$

Applying Eq. (14) to the top left submatrix in Eq. (15) yields

$$[A_N - C^{-1} B^T B]^{-1} = A_N^{-1} - A_N^{-1} B^T [-C + BA_N^{-1} B^T]^{-1} BA_N^{-1} \quad (16)$$

It is clear that A_{N+K}^{-1} can be computed from A_N^{-1} without computing the matrix inverse.

End of proof

Using Eqs. (15) and (16), A_{N+K}^{-1} can be computed in an incremental chunk way, which avoids expensive inversion operation. Therefore the corresponding coefficients and bias $\alpha_{N+K} = [b \quad \alpha]^T$ can be computed according to Eq. (8). Then the classifier can work. Given a new input x, the corresponding value $y(x)$ can be estimated by using Eq. (17).

$$y(x) = sign\left[\sum_{k=1}^{N}\alpha_k y_k \psi(x, x_k) + b\right] \quad (17)$$

4 Numerical Results

Our experiments are run on a PC, which utilizes a GenuineIntel~400MHz Pentium II processor with a maximum of 256MB of memory available. In order to test the speed and effectiveness of the presented algorithm, we apply LS-SVM, online LS-SVM (K=1) and online chunk LS-SVM (K>1) to the four same data sets available from the UCI Machine Learning Repository. All features for all experiments are normalized to the range [-1, +1]. A RBF kernel function has been taken with $\sigma = 1$, in this case parameters $\gamma = 1, N = 1$ and $K = 15$ are set. For all the data sets, 90% data are chosen as training data, and the rest are used as testing data. The experimental results are shown in Table 1.

From Table 1, it can be seen that the speed of the online chunk LS-SVM with K>1 is faster than that of K=1. For large data sets, the computing speed of the online chunk LS-SVM with K>1 is improved more than 20% compared with standard LS-SVM. Moreover, it can be inferred from the formulation of online chunk LS-SVM that the solutions of the proposed algorithm would have the same training or testing accuracy as that of the offline LS-SVM, and the experimental results have shown this.

Table 1. Comparison of results obtained by LS-SVM, online LS-SVM (K=1), online LS-SVM (K>1)

Data $l \times n$	Algorithm	Training accuracy (%)	Testing accuracy (%)	Training Time (s)
Tic-Tac-Toe Endgame 958×9	LS-SVM	100	100	48.85
	Online (K=1)	100	100	37.67
	Online (K>1)	100	100	34.15
Image Segmentation 2310×19	LS-SVM	100	66.23	662.01
	Online (K=1)	100	66.23	575.39
	Online (K>1)	100	66.23	541.34
Splice-Juncion Gene Sequences 3190×60	LS-SVM	99.97	95.3	1968.75
	Online (K=1)	99.97	95.3	1629.61
	Online (K>1)	99.97	95.3	1545.85
Sat Image 4435×36	LS-SVM	100	96	7633.13
	Online (K=1)	100	96	5870.66
	Online (K>1)	100	96	5790.27

5 Conclusions

In this paper an online LS-SVM based on incremental chunk for classification problems is proposed. This algorithm provides a preparation for the application of LS-SVM in the online learning problems that have input data supplied in sequence. Experimental results show that the proposed algorithm has the same solution as the offline LS-SVM but lower computational cost. To overcome the shortcoming of losing the sparse feature of LS-SVM, a decreasing way can be added in our algorithm, which is our future wok.

Acknowledgements. This work is supported by the National Natural Science Foundation of China (19901009), Key Project of Ministry of Education of China (02090), Natural Science Foundation of Guangdong Province, China (970472, 000463, 031360), Excellent Young Teachers Program of Ministry of Education of China, "Qian Bai Shi" Talent Young Foundation of Guangdong Province, and Natural Science Foundation of South China University of Technology (E512199, D76010). and FoK Ying Tong Education Foundation (91005).

References

1. Vapnik, V.N.: The Nature of Statistical Learning Theory. Springer Verlag, New York (1995)
2. Osuna, E., Freund, R., Girosi, F.: An Improved Training Algorithm for Support Vector Machines. Neural Networks for Signal Processing-Proceedings of the IEEE. (1997) 276-285
3. Joachims, T.: Making Large-Scale Support Vector Machine Practical. Advances in Kernel Methods-Support Vector Learning. The MIT Press, Cambridge, Massachusetts. (1999) 169-184
4. Platt, J.C.: Fast Training of Support Vector Machines Using Sequential Minimal Optimization. Advances in Kernel Methods-Support Vector Learning. The MIT Press, Cambridge, Massachusetts. (1999) 185-208
5. Keerthi, S.S., Shevade, S.K., Bhattacharyya, C., Murthy, K.R.K.: Improvements to Platt's SMO Algorithm for SVM Classifier Design. Neural Computation. 13(3) (2001) 637-649.
6. Lin, C.J.: On the Convergence of the Decomposition Method for Support Vector Machines. IEEE Transactions on Neural Networks. 12(6) (2001) 1288-1298.
7. Lin, C.J.: Asymptotic Convergence of an SMO Algorithm without any Assumptions. IEEE Transactions on Neural Networks. 13(1) (2002) 248-250
8. Mangasarian, O.L., Musicant, D.R.: Lagrangian Support Vector Machines. Journal of Machine Learning Research. 1 (2001) 161-177
9. Yang, X.W., Shu, L., Hao, Z.F.: An extended Lagrangian Support Vector Machine for classification. Progress in Natural Science. 14(6) (2004) 519-523
10. Suykens, J.A.K., Vandewalle, J.: Least Squares Support Vector Machine Classifiers. Neural Process Letter. 9 (1999) 293-300
11. Chua, K.S.: Efficient computations for large least square support vector machine classifiers. Pattern Recognition Letters. 24 (2003) 75-80
12. Fine, S., Scheinberg, K.: Efficient SVM Training Using Low-Rank Kernel Representations. Journal of Machine Learning Research. 2 (2) (2002) 243-264

13. Liu J.H., Chen, J.P., Jiang, S., Cheng, J.S.: Online LS-SVM for function estimation and classification. Journal of University of Science and Technology Beijing. 10(5) (2003) 73-77
14. Golub, G.H., Van, L.C.F.: Matrix Computations. 3rd edn. The John Hopkins University Press, Baltimore, Maryland. (1996)

A Novel Approach to Clustering Analysis Based on Support Vector Machine

Zhonghua Li, ShaoBai Chen, Rirong Zheng, Jianping Wu, and Zongyuan Mao

College of Automation Science and Eng., South China Univ. of Technology,
Guangzhou 510640, China
honestlee@163.com

Abstract. This paper proposed a novel approach to clustering analysis for a large-scale data based on support vector machines (SVM). For conventional support vector clustering (SVC), data points are mapped by using a Gaussian kernel function to a high dimensional feature space. When mapped back to data space, this sphere is able to separate into several components. However, the dimension of feature space would be very high if a number of data points are mapped. This impairs the efficiency of SVM, and increases its computation. The approach in this paper utilized AIS to compress original data, and new reduced data points are obtained as the input of conventional SVC. Given elevator traffic data, simulation results indicated the applicability of this approach.

1 Introduction

Support vector machine (SVM) is a relatively new approach for classification and regression [1]. Currently, SVM are used in fields such as intrusion detection, software engineering, as well as for many data mining applications.

Different form support vector classification, support vector clustering is unsupervised learning. Data points are mapped from data space to a high dimensional feature space by means of a Gaussian kernel [2]. In feature space, the smallest sphere is searched which can enclose the image of the data. This sphere is then mapped back into data space. Several contours are produced to enclose the data points. Points enclosed by each separate contour are of the same cluster. However, large-scale data points from data space are directly mapped to a high dimensional feature space, which increases the computation and the difficulty of SVC. To reduce the redundancy of data, keeping the feature data and removing the similar data points are necessary.

Artificial immune system (AIS) is another AI method newly developed recently, which interests many researchers. Now it is widely applied in pattern recognition, cluster analysis, function optimization, error detection, etc [3,4]. The theory that redundant similar antibodies are eliminated could be utilized to compress the dataset.

The purpose of this paper is to propose a novel hierarchy clustering approach based on AIS and SVC. The original data points were firstly compressed by means of AIS, and them is fed to SVC. This approach reduced the computation of SVC and improved the efficiency of system.

This paper is organized as follows. In Section 2, the theory of SVC is briefly introduced. Section 3 describes how to design this novel clustering approach. Simulation results and their explanation are listed in Section 4. Finally conclusions are made in Section 5.

2 Support Vector Clustering

Support vector clustering algorithm [2] is introduced briefly. Given $\{x_i\} \in \chi$ be a dataset with N points, Φ is a nonlinear transformation from χ to some high dimension feature space. The smallest enclosing sphere of radius R is looked for. When soft constraints are incorporated by adding a slack variable ξ_j, R should satisfy the following constraint.

$$\|\Phi(x_j) - a\| \leq R^2 + \xi_j, \forall j, \tag{1}$$

where $\|\ \|$ is the Euclidean norm, a is the center of the sphere, and $\xi_j \geq 0$. To solve this problem, the Lagrangian is employed.

$$L = R^2 - \sum_j (R^2 + \xi_j - \|\Phi(x) - a\|)\beta_j - \sum_j \xi_j \mu_j + C \sum_j \xi_j \tag{2}$$

where $\beta_j \geq 0$ and $\mu_j \geq 0$ are both Lagrange multipliers, C is a constant, $C\sum_j \xi_j$ is a penalty term. Using the extreme theorem, we can get:

$$\sum_j \beta_j = 1, \quad a = \sum_j \beta_j \Phi(x_j), \quad \beta_j = C - \mu_j \tag{3}$$

By using the extreme theorem and the KTT complementary conditions, we can eliminate the variables R, a and μ_j, then the Lagrangian is translated into the wolf dual problem, which is a function of the variables β_j:

$$W = \sum_j \Phi(x_j)^2 \beta_j - \sum_{i,j} \beta_i \beta_j \Phi(x_i) \bullet \Phi(x_j), \quad 0 \leq \beta_j \leq C \quad j = 1, ..., N \tag{4}$$

Like the SV method, the dot products $\Phi(x_i) \bullet \Phi(x_j)$ can be represented by an appropriate Mercer Kernel $K(x_i, x_j)$. Gaussian kernel function is typically selected and the Lagrangian W can be expressed respectively:

$$K(x_i, x_j) = e^{-q\|x_i - x_j\|^2} \text{ and } W = \sum_j K(x_j, x_j)\beta_j - \sum_{i,j} \beta_i \beta_j K(x_i, x_j) \tag{5}$$

where q is a width parameter. At each point x, the distance of its image in feature space from the center of the sphere is defined as follows.

$$R^2(x) = \|\Phi(x) - a\|^2 \tag{6}$$

Substituting (3) into (6), we can get:

$$R^2(x) = K(x,x) - 2\sum_j \beta_j K(x_j,x) + \sum_{i,j} \beta_i \beta_j K(x_i,x_j) \qquad (7)$$

And the radius of the sphere is expressed:

$$R = \{R(x_i) \mid x_i \text{ is a SV}\} \qquad (8)$$

In the end, the contour enclosing the points in data space could be defined by:

$$\{x \mid R(x) = R\} \qquad (9)$$

Known form (8), SVs lie on the boundary of each cluster. BSVs are outside, and all other points are inside the clusters.

3 A Novel Hierarchy Clustering Model

This novel hierarchy clustering approach is schematically illustrated in Fig.1. Its procedure could be divided into two steps. The first step is to compress the original data by means of AIS. As an output of AIS, the memory cells are reduced data we need. The second step is support vector clustering with the input of the reduced data.

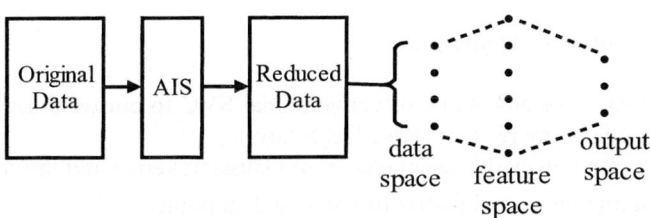

Fig. 1. Structure of hierarchy clustering

3.1 AIS Algorithm

The original traffic flow data are firstly defined as the antigens of AIS, matched by adequate antibodies generated randomly, and antibodies with best affinities are selected to be memory cells. This process is usually named immune memorization [3,4]. The affinity aff_{ij} between antibody i and antigen j and The similarity s_{ij} between antibody i and antibody j are calculated by Euclidean distance. They are:

$$aff_{ij} = sqrt(\sum_{k=1}^{n}(Ab_{ik} - Ag_{jk})^2 \text{ and } s_{ij} = sqrt(\sum_{k=1}^{n}(Ab_{ik} - Ab_{jk})^2 \qquad (10)$$

The normalization of original data is handled by:

$$Ag_j = \{Ag_j - \min\{Ag\}\}/\{\max\{Ag\} - \min\{Ag\}\} \tag{11}$$

The detailed procedure of AIS is described as follows:
Step 1. Inputting data as antigens and normalize them;
Step 2. Randomly initializing N antibodies in $(0,1)$;
Step 3. For each antigen, Computing aff_{ij};

    (1) Choosing m antibodies with the highest affinity to be network cells;
    (2) Cloning network cells. The bigger aff_{ij} is, the more the clone cells are;
    (3) Executing mutation for each clone cell C;
    (4) Calculating the affinity between antigen and C, and choosing clone cells with high affinity to enter memory cell subset M_p;
    (5) Eliminating individuals in M_p whose similarity is less than σ_s.

Step 4. Incorporating M_p into memory cell set M;
Step 5. Eliminating individuals in M whose similarity is less than σ_s;
Step 6. Randomly recruit N individuals as next-generation antibodies together with M. Then return to Step 2.
Step 7. If the network is convergent, M is specified as reduced data.

3.2 Procedure of SVC Algorithm

After AIS, we can take advantage of conventional SVC to cluster those reduce data. Support vector clustering [2] is reviewed as follows.

    1. Determining the width parameter q of Gaussian kernel and the constant C;
    2. Calculating the kernel matrix of reduced data points;
    3. Solving the Lagrangian Multipliers β_j, finding support vectors and outliers;
    4. Calculating the radius of super-plane and $C\sum_j \xi_j$ of distance equation;
    5. Computing the adjacent matrix of reduce data points;
    6. Searching the cluster assignments;
    7. Calculating the performance index of clustering approach.
    8. Mapping back to data space to show the cluster results.

The termination status is that the clusters don't vary within a wide range over q.

4 Simulation Test

Analysis of elevator traffic is crucial for developing a well-performing elevator system [5]. In the simulation, we utilize elevator traffic flow data to verify this algorithm. The traffic flow of some building in Guangzhou is collected in May 2003, which is illustrated in Fig.2.

In the figure, line 1 stands for the number of passengers going upward per 5 minutes, and line 2 shows that of passengers going downward.

A Novel Approach to Clustering Analysis Based on Support Vector Machine   569

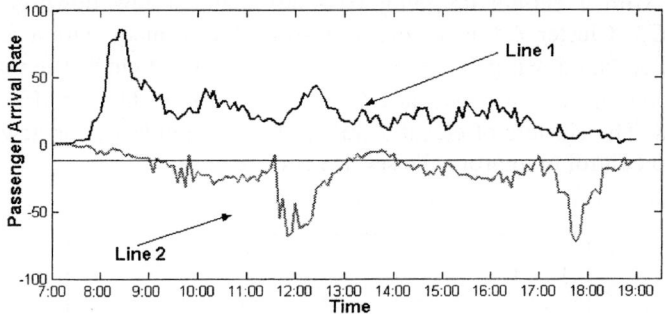

Fig. 2. Traffic flow inside the building

According to Section IV, AIS is firstly used to take feature data from the original input data. The parameters in AIS algorithm are set: σ_s =0.1, rate of natural death $N_{c,max}$ =1.0, m =4, the maximum number of iterations I =15, $N = 20$. The reduced data are gained by using AIS computation.

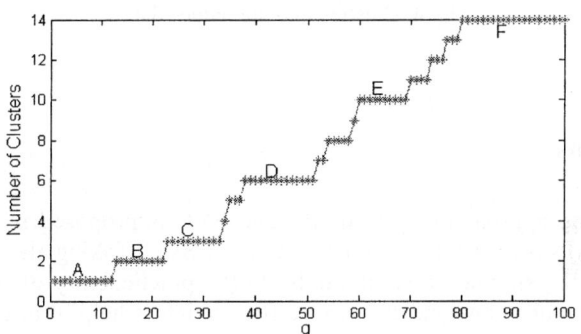

Fig. 3. Number of Clusters over q

For SVC, we employ a policy that q is firstly set as a smaller value and then q is incremented gradually. The number of SVs and the clusters are observed. As q increases, the change of clusters is showed in Fig.3.

In Fig.3, It is clear that the number of clusters shows stability at segment A, B, C, D, E and F. According to the regularity of elevator traffic flow, we choose segment D to observe the distribution of each cluster showed in Fig.4. In the figure, Up-going and down-going are both normalized passenger arrival rate.

Known from Fig.4, the reduced data are classified into 6 clusters. Each cluster has its own clear feature. Cluster $C1$ indicates the up-peak traffic every day, and the number of up-going passengers is much larger than that of down-going passengers. On the contrary, cluster $C2$ stands for the down-peak traffic pattern. Passengers transiting upward are more than those transiting downward. Cluster $C3$ is inter-floor traffic mode with a certain up-going passengers and a few down-going people. Opposite to $C3$, Cluster $C4$ is another inter-floor traffic mode with a few up-going persons and a certain down-going persons. Cluster $C5$ is a light traffic that has a few passengers going upward or downward. Cluster $C6$ is a lunchtime traffic pattern that exists at noon. The clusters of elevator traffic flow are guidance for new strategy of elevator group control to improve the service performance.

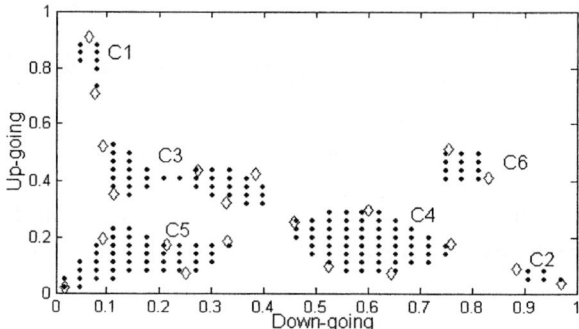

Fig. 4. Clusters into the reduce data

5 Conclusions

A novel clustering approach based on AIS and SVC is proposed in this paper. The introduction of AIS is to reduce the computation of SVC. Taking elevator traffic flow data as an example, simulation results indicate the practicability of this approach. It provides another tool to analyze elevator traffic, which is helpful to guide the design of excellent elevator group system.

References

1. Gunn S.: Support Vector Machines for Classification and Regression. Image Speech and Intelligent Systems Group, University of Southampton (1998)
2. Asa B. H., David H., Hava T. S. and Vapnik V.: Support Vector Clustering. Journal of Machine Learning Research 2 (2001) 125-137
3. Li Z.H., Zhu Y.F., Li C.H. and Mao Z.Y.: Elevator Traffic Flow Analysis Based on Artificial Immune Clustering Algorithm. Journal of South China University of Technology, Vol. 31, 12 (2003) 26-29
4. Castro de L. N., Zuben von F. J.: An Evolutionary Immune System Network for Data Clustering. Proceedings of the Sixth Brazilian Symposium on Neural Networks, Rio de Janeiro (2000) 84-89
5. Barney G. C., Santos dos S. M.: Elevator Traffic Analysis, Design and Control, IEE Peter Peregrinus, London (1985)

Application of Support Vector Machine in Queuing System

Gensheng Hu and Feiqi Deng

College of Automation Science and Engineering, South China University
of Technology, Guangzhou 510640,China
hugs2906@sina.com

Abstract. The solution to performances of queuing system is based on knowing the distributions of customers arrival or service time. Support vector machine (SVM) based on statistical learning theory has been used generally in machine learning because of its good generalization ability. By using SVM we can classify and identity some probability distributions appeared in queuing system and solve the density function regression problem through using support vector regression (SVR). Some other problems needed to be solved are formulated in the end.

1 Introduction

The foundation of support vector machines (SVMs) has been developed by Vaplik [1]. SVMs have become in the last few years one of the most popular approaches to learning from examples. Due to many attractive features, such as good generalization and anti-noise ability, SVMs have been applied in many fields of science and engineering. Many problems have been solved very well based on SVMs in fingerprint classification[2], face identification [3], communication network[4], computer vision[5], text categorization[6], bioinformatics [7], time series prediction[8], etc.

The content researched by queuing theory are deducing the performances of queuing system in the condition of known customer arrival distribution and service time distribution. Since originated by Erlang during his research of telephone communication using probabilistic method in the early 20th century, classical queuing theory has tended to perfection. However, some problems appear when applying theory to practice. Queuing theory assumes that customer arrival and service time submit certain prior distributions. How to discriminate these distributions becomes a major problem in the practice of queuing system. A.G. Konheim deduced some formula for GI/G/1 queuing system[9], but the results expressions are Laplace transform equations and difficult to solve them. In communication network, cells (customers) arrival has different distributions during different time. We want to identify these distributions automatically. The algorithm of SVM based on statistical learning theory solves these problems very well. We can train on some samples drawn from customer arrival or service time making use of SVM. Then we can establish a system for distribution identification automatically. Experiment shows it works very well.

This paper is organized as follows. Following introduction, section 2 introduces the problem of classification and recognition for data distribution of queuing system by using SVM method. The density function regression problem of customer arrival or service time is solved by using SVR in section 3. Section 4 introduces the deducing results of queuing system from samples directly. Section 5 shows some simulations.

2 Classification and Recognition for Data Distribution

In many queuing systems, customer arrival has different distribution during different time. We solve the problem by taking the example of communication system. In communication system, usage frequency during weekday is different from weekend. It is also different during morning, afternoon, before or after midnight. According to experiences, we write distributions obeyed by cells arrival in communication system and estimate distribution parameters by using point estimation method or Bayesian method. Then testify these distributions by using hypothesis testing. So we know a prior the different distributions or the same distribution with different parameters for cells interarrival in communication network.

Given a sample (data) set $\{(x_i, y_i), i = 1,..., N\}$ drawn from these distributions where x_i is a vector consisted of statistical value, $y_i \in \{-1,+1\}$, we train on the data set employing SVM and classify new data based on these results. The detailed procedures are as follows.

Step 1. Gives a sample set $\{(x_i, y_i), i = 1,..., N\}$ drawn from these distributions where x_i is a d-dimensional vector whose entries consist of statistics such as mean, median, skewness, kurtosis, quantiles, etc. $y_i \in \{-1,+1\}$.

Step 2. Solve the following V-SVC problem[10]:

$$\min W(\alpha) = \frac{1}{2} \sum_{i=1}^{N} \sum_{j=1}^{N} \alpha_i \alpha_j y_i y_j K(x_i, x_j) \quad (1)$$

$$\text{subject to } 0 \leq \alpha_i \leq \frac{1}{N}, \quad \sum_{i=1}^{N} \alpha_i y_i = 0, \quad \sum_{i=1}^{N} \alpha_i y_i \geq \nu \quad (2)$$

where $K(.,.)$ is a kernel. It corresponds to an inner product of vectors in higher dimensional feature space if and only if Mercer's conditions are met. See section 3. Here we can adopt the polynomial generator or the Gaussian radial basis function

$$K(x_i, x_j) = (1 + x_i \cdot x_j)^d \text{ or } K(x_i, x_j) = \exp(-\|x_i - x_j\|^2 / 2\sigma^2) \quad (3)$$

The final discriminative function can be shown to take the form

$$f(x) = \text{sgn}(\sum_{i=1}^{N} \alpha_i y_i K(x, x_i) + b) \quad (4)$$

To compute the threshold b, we consider the set of indices for these support vectors S_m. Due to the KKT conditions,

$$b = -\frac{1}{|S_m|} \sum_{x \in S_m} \sum_{j=1}^{N} \alpha_j y_j K(x, x_j) \text{ where } |S_m| \text{ is the cardinality of } S_m. \quad (5)$$

Step 3. Given a test sample randomly drawn from a time interval $[t_1, t_2]$, calculate its summary statistic values mean, median, skewness, kurtosis, quantiles, etc. Substituting these values into (4), and we obtain which distribution the sample belong to immediately.

Step 4. If more than two probability distributions need to be classified, we must use multi-classifier such as one-against-the-rest classifier. We can construct a set of binary classifiers $f^1,..., f^n$, each trained to classify one class from the rest, and combine them by doing the multi-class classification according to the maximal output before applying the sign function, that is, by taking

$$\arg\max_{j=1,...,n} f^j(x) \text{ where } f^j(x) = \sum_{i=1}^{n} y_i \alpha_i^j K(x, x_i) + b^j \quad (6)$$

3 Regression Estimation for Probability Density Function

Regression estimation is the estimation of real-valued functions, rather than just $\{\pm 1\}$ valued ones, as in the case of pattern recognition. Given an i.i.d data set

$$\{(x_i, y_i) \in X \times Y, i = 1,..., N\} \quad (7)$$

where $X \subset R^n, Y \subset R$ is sampled from some unknown probability distribution $P(x, y)$. We transform the input data $x_1,..., x_N$ into a high-dimensional approximately linearly separable feature space using a nonlinear map

$$\Phi : x_i \to \Phi(x_i) \quad (8)$$

Then we do our work in the feature space.

According to Mercer theorem, if K is a kernel satisfying the Mercer conditions, we can construct the mapping Φ into a space H where K acts as a dot product:

$$\langle \Phi(x), \Phi(x_i) \rangle = K(x, x_i) \text{ for almost all } x, x_i \in X. \tag{9}$$

Our goal is to find a function $f \in F$ which is a class of functions $F = \{f | f : X \to Y\}$ with small risk:

$$R[f] = \int_{X \times Y} c(f, x, y) dP(x, y) \tag{10}$$

where c is a loss function.

Because we do not know P, we can not minimize (10) in any case. We can only calculate the empirical risk:

$$R_{emp}[f] = \frac{1}{N} \sum_{i=1}^{N} |y_i - f(x_i)|_\varepsilon \tag{11}$$

here $|y_i - f(x_i)|_\varepsilon = \max(0, |y_i - f(x_i)| - \varepsilon_i)$ is an ε-insensitive loss function.

It would be unwise to attempt to minimize the empirical risk directly. It can be lead to numerical instabilities and bad generalization performance. To avoid this problem is to restrict the class of possible minimizers F of the empirical risk functional $R_{emp}[f]$, such that F becomes a compact set. It is usual to add a capacity control term called regularization term, i.e. $\Omega[f]$, to the original empirical risk function:

$$R_{reg}[f] = R_{emp}[f] + C\Omega[f] \tag{12}$$

where C is a trade-off between minimization of $R_{emp}[f]$ and the smoothness or simplicity which is enforced by small $\Omega[f]$.

The minimization of (12) is equivalent to the following constrained optimization problem:

$$\min \frac{1}{2} \|W\|^2 + C \sum_{i=1}^{N} (\xi_i + \xi_i^*)$$

$$\text{subject to } <W, \Phi(x_i)> + b \leq y_i + \varepsilon_i + \xi_i$$

$$<W, \Phi(x_i)> + b \geq y_i - \varepsilon_i - \xi_i^*$$

$$\xi_i \geq 0, \ \xi_i^* \geq 0, \ i = 1, \ldots, N. \tag{13}$$

Assuming that all trained data is drawn from $X \in [\overline{a}, \overline{b}]$, we add the constraints $F(\overline{a}) = 0$ and $F(\overline{b}) = 1$ for the reason of cumulative probability[11].

We can write the dual optimization problem using Lagrange multiplier

$$\max -\frac{1}{2} \sum_{i,j=1}^{N} (\alpha_i^* - \alpha_i)(\alpha_j^* - \alpha_j) K(x_i, x_j)$$

$$-\sum_{i=1}^{N}(\alpha_i + \alpha_i^*)\varepsilon_i + \sum_{i=1}^{N}(\alpha_i^* - \alpha_i)y_i$$

subject to $\sum_{i=1}^{N}(\alpha_i - \alpha_i^*)(K(x_i,\bar{a}) - K(x_i,\bar{b})) = 0$

$$\sum_{i=1}^{N}(\alpha_i - \alpha_i^*) = 0, \ 0 \leq \alpha_i^{(*)} \leq C \quad (14)$$

The minimizer $f \in F$ of the regularized risk $R_{reg}[f]$ admits a representation of the form: $f(x) = \sum_{i=1}^{N}(\alpha^* - \alpha_i)K(x,x_i) + b$. So probability density function $p(x)$ can be estimated as

$$p(x) = \sum_{i=1}^{N}(\alpha^* - \alpha_i)k(x,x_i) \quad (15)$$

where $k(x,x_i) = \dfrac{\partial d^l}{\partial x^1 \partial x^2 ... \partial x^{d^l}} K(x,x_i)$, $1,2,...,d^l$ being the index of entry of vector x.

4 Deducing Queuing Results from Samples Directly

Due to the complexity of real world, customers arrival or service time don't submit certain distribution exactly. Even if the analytic expressions of these distributions have been written, if they aren't exponential distributions, it is difficult to obtain results of queuing system. SVM based training on random samples, we can deduce the results of queuing system directly. Its basic idea is the same as that used in section 3. The difference is that y_i is a quantile in the sample set (7) in section 3, while in this section y_i is a performance of queuing system, such as queue length or waiting time.

The solving procedures are as follows:

Finding a regression function to estimate the unknown function relation between x_i and y_i:

$$\hat{f}: \hat{f}(x_i) = y_i \quad (16)$$

according to the principle of minimizing its structural risk making use of SVM formulated in section 3.

Utilizing (16), we can predict results of queuing system based on new samples directly. Simulation shows that its accuracy isn't very well. One way to improve its predictive accuracy is to extend the capacity of its sample set which will induce high computational cost and reduce computational speed. We may take feature subset selection method to overcome this problem [12].

5 Experiments

The results of our experiment for data distribution classification are summarized in table 1. We have tested five different distributions whose summary statistic values are mean, median, skewness, kurtosis, quantiles. The data set are drawn from [13]. The results show that Beta and Normal distribution have better classification rate than exponential, Erlang and hyperexponential distribution.

Table 1. Correct identification rate of distribution

Distribution	Success rate (%)		
	100 samples	200 samples	500 samples
Hyperexponential	72	80	90
exponential	83	87	95
Normal	85	90	100
Beta	90	91	100
Erlang	80	85	97
Mean success rate	82	86.6	96.4

We test SVR estimator for probability density function using samples drawn from the hyperexponential density function:

$$f(x) = \sum_{j=1}^{15} \frac{j^2}{1200} e^{-\frac{j}{10}x} \quad x \geq 0 \tag{17}$$

The comparison between the tested density function and its density function regression (we taking Gaussian kernel with $\sigma = 0.5$) is showed in Figure 1.

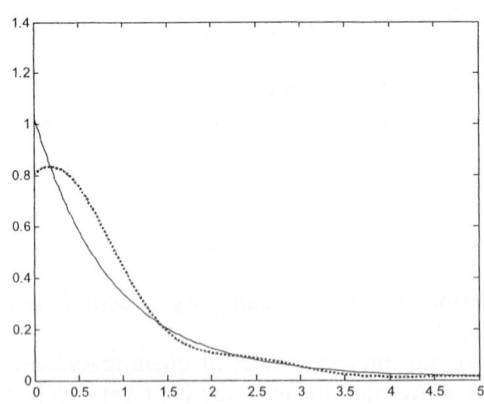

Fig. 1. The tested density function (*solid line*) and its SVR function (*dotted line*)

6 Conclusions

Support vector machine is a new technology appeared in recent years. In this paper, we investigated the application of support vector machine in queuing system. The results show that it has good generalization and recognition under small sample set. However, there are still some problems needed to be solved. The results of density function regression estimation aren't as satisfactory as that of function classification. There are some unbearable gaps between the performances of real world queuing system and the deducing queuing results from samples directly. The solution of all these problems depends on more efficient SVM algorithm.

References

1. Vapnik, V.: The nature of statistical learning theory. Springer, New York (1995)
2. Shah, Shesha, Sastry, P.S.: Fingerprint Classification Using a Feedback-Based Line Detector. IEEE Transactions on Systems, Man, and Cybernetics, Part B: Cybernetics, Vol.34 (2004) 85-94
3. Richman, Michael S.; Parks, Thomas W.; Lee, Hsien-Che: Novel support vector machine-based face detection method. Conference Record of the Asilomar Conference on Signals, Systems and Computers, Vol.1 (1999) 740-744
4. Chen, S.; Samingan, A.K.; Hanzo, L.: Support vector machine multiuser receiver for DS-CDMA signals in multipath channels. IEEE Transactions on Neural Networks, Vol. 12 (2001) 604-611
5. Poggio, Tomaso; Verri, Alessandro: Introduction: Learning and vision at CBCL. International Journal of Computer Vision, Vol.38 (2000) 5-7
6. Leopold, Edda; Kindermann, Jorg.: Text categorization with support vector machines. How to represent texts in input space?. Machine Learning, Vol.46 (2002) 423-444
7. Cho, Sung-Bae: Exploring features and classifiers to classify gene expression profiles of acute leukemia. International Journal of Pattern Recognition and Artificial Intelligence, Vol.16 (2002) 831-844
8. Lee, K.L.; Billings, S.A.: Time series prediction using support vector machines, the orthogonal and the regularized orthogonal least-squares algorithms. International Journal of Systems Science, Vol.33 (2002) 811-821
9. Konheim, A.G.: On elementary solution of the queuing system G/G/1. SIAM Journal on computer, Vol.4 (1975) 540-545
10. Bernhard Schölkopt; Alexander Smola,J.: Learning with kernels. The MIT Press, London, England (2002)
11. Challa, S.; Palaniswami, M.; Shilton, A.: Distributed data fusion using support vector machines. Information Fusion, 2002. Proceedings of the Fifth International Conference on, Vol.2 (2002) 881 – 885
12. Mao, K.Z.: Feature Subset Selection for Support Vector Machines Through Discriminative Function Pruning Analysis. Systems, Man and Cybernetics, Part B, IEEE Transactions on , Vol.34 (2004) 60-67
13. Zhu, J.Y.; Zhang, H.X.; Gao, J.L.; Feng, J.L.: Data distributions automatic identification based on SOM and support vector machines. Proceedings of the First International Conference on Machine Learning and Cybernetics, Vol.1 (2002) 340 – 344

Modelling of Chaotic Systems with Novel Weighted Recurrent Least Squares Support Vector Machines

Jiancheng Sun, Taiyi Zhang, and Haiyuan Liu

Dept. of Information and Communication Eng, Xi'an Jiaotong University.Xi'an, Shaanxi, China
{sunjc, taiyizhang , liuhaiyu}@mailst.xjtu.edu.cn
http://www.xjtu.edu.cn

Abstract. This paper discusses the use of Support Vector Machines(SVM) for dynamic modelling of the chaotic time series. Based on Recurrent Least Squares Support Vector Machines (RLS-SVM), a weighted term is introduced to the cost function to compensate the prediction errors resulting from the positive global Lyapunov exponent in context of the chaotic time series. For demonstrating the effectiveness of our algorithm, the dynamic invariants involves the Lyapunov exponent and the correlation dimension are used for criterions. Finally we apply our method to Santa Fe competition time series. The simulation results shows that the proposed method can capture the dynamics of the chaotic time series effectively.

1 Introduction

Building the model of a dynamical system by the time series analysis has been an important problem in science. Chaotic systems is an important class of the nonlinear dynamical systems in economics, meteorology, chemical processes, biology, hydrodynamics and many other situations. Various techniques for modeling and predicting nonlinear time series are investigated in past years. There are many tradition methods associated with time series analysis such as linear regression and ARIMA models [1]. In addition, some new methods have been proposed to deal with the nonlinear time series analysis such as chaotic attractor[2] and radial basis function[3]. However, in recent years, many researcher addressed the nonlinear time series analysis with the artificial neural networks[4][5].

In this paper, a novel Modified Recurrent Least Squares Support Vector Machines (MRLS-SVM) is developed for realizing the chaotic time series reconstruction. Support Vector Machines(SVM) have become a subject of intensive study and they have been applied successfully to classification and regression[6]. Least squares (LS) versions of SVM greatly simplifies the problem since the solution is characterized by a linear system.[7]. Some iterative operation is necessary for common prediction mission, to deal with this problem, Recurrent Least Squares Support Vector Machines (RLS-SVM) was deduced based on SVM algorithm[8]. In our work, a component including the largest Lyapunov exponent is used in RLS-SVM for weighting the cost function. With this weighting component, we expected the errors results from the later iteration are given less influence since the largest Lyapunov exponent might introduce the exponential distortion in a chaotic time series.

2   Embedding Phase Space of Dynamical System

Deterministic dynamical systems describe the time evolution of a system in some phase space $\Gamma \subset R^m$, for simplicity we will assume that the phase space is a finite dimensional vector space. A state is specified by a vector $x \in R^m$. Then we can describe the dynamics by an explicit system of m first-order ordinary differential equations

$$\frac{d}{dt}x(t) = f(t, x(t)), \quad t \in R \tag{1}$$

A time series can then be thought of as a sequence of observations $s = h\{x\}$ performed with some measurement function h. Since the sequence s (usually scalar) in itself does not properly represent the multidimensional phase space of the dynamical system, we have to employ some technique to unfold the multidimensional structure using the available data. Takens embedding theorem guarantees the reconstruction of a state space representation from a scalar signal alone[9]. Takens states that if the sequence s does indeed consist of scalar measurements of the state of a dynamical system, then under certain genericity assumptions, the time delay embedding provides a one-to-one image of the original set $\{x\}$, provided embedding dimension is large enough. where s are vectors in a new space namely the embedding phase space which are formed from time delayed values of the scalar measurements. For almost all time delay and for some embedding dimension, Takens embedding theorem ensures that there is a smooth map $f : R^m \to R$ such that

$$s_{(n+1)\tau_d} = f(s_{n-(m-1)\tau_d}, \ldots, s_{n-\tau_d}, s_n) = f(s_n) \tag{2}$$

the problem of remodelling becomes equivalent to the problem of estimating the unknown function f in the embedding phase space.

In the chaotic system case, the Lyapunov exponent is proven to be a useful dynamic invariants to characterize the chaotic dynamic system. We can deduce that the system is a chaotic if at least a positive Lyapunov exponent exists. Formally, the Lyapunov spectrum is defined by Wolf *et al.* [10]: given an n-dimensional phase space, the long-term evolution of an infinitesimal sphere is monitored. As the sphere evolves, it will turn into an n-ellipsoid. The i-th one-dimensional Lyapunov exponent is then defined in terms of the length of the resulting ellipsoid's principal axis

$$\lambda_i = \lim_{t \to \infty} \frac{1}{t} \log_2 \frac{p_i(t)}{p_i(0)} \tag{3}$$

The Lyapunov spectrum is then formed by the set $(\lambda_1, \lambda_2, \ldots \lambda_n)$, where λ_i are arranged in decreasing order.

The difference between strange attractors and purely stochastic (random) processes is that the evolution of points in the phase space of a strange attractor has definite structure. The correlation integral provides a measure of the spatial organization of this structure, and is given by

$$C(r) = \lim_{N \to \infty} \frac{1}{N(N-1)} \sum_{i \neq j} \Theta(r - \|s(i) - s(j)\|) \tag{4}$$

where Θ is the Heaviside function. Grassberger and Procaccia found that, for a strange attractor, $C(r) \propto r^v$ for a limited range of r [11]. The power v is called the correlation dimension of the attractor. Thus, we can plot the $\log C(r) - \log r$ graph to identify an attractor.

In addition to above invariants, the Poincar´e map is another parameters to characterize the chaotic system[12]. Since a chaotic system never revisits the same state, it will trace out contours on the Poincar´e map. However, unlike a purely random process, these contours will have definite structure and will graphically indicate the presence of the responsible attractor.

3 Modified Recurrent Least Squares Support Vector Machines and Learning Algorithm

The foundations of Support Vector Machines (SVM) have been developed by Vapnik and possessed popularity promising empirical performance[6]. Suykens and Vandewalle proposed the Recurrent Least Squares Support Vector Machines(RLS-SVM) based on sum squared error(SSE) to deal with some problem in prediction such as iterative operation[8]. Since the chaotic character affect the prediction performance, using the SSE cost function as criterion directly can reduce the prediction precision. In this work, we deal with this problem by consider the influence result from the Lyapunov exponents. In following section, we formulated our algorithm based on the Modified Recurrent Least Squares Support Vector Machines (MRLS-SVM):

1. choose optimal embedding dimension m, time delay τ_d and set $Q \subseteq \{1, \ldots, N\}$,
2. constructed the training data:

$$D = \{((s_{k-(m-1)\tau_d}, \ldots, s_{k-\tau_d}, s_k), s_{k+1}) \in R^m \times R | k \in Q\} \tag{5}$$

3. given initial condition $\hat{s}_i = s_i$ for $i = 1, 2, \ldots m$, the prediction problem is given by:

$$\hat{s}_{(k+1)\tau_d} = f(\hat{s}_{k-(m-1)\tau_d}, \ldots, \hat{s}_{k-\tau_d}, s_k)$$
$$= w^T \varphi(s_k) + b \tag{6}$$

where $\varphi(\cdot): \boldsymbol{R}^m \to \boldsymbol{R}^{n_h}$ is a nonlinear mapping in future space, $w \in \boldsymbol{R}^{n_h}$ is the output weight vector and $b \in \boldsymbol{R}$ is bias term. $\boldsymbol{S}_k = [s_{k-(m-1)\tau_d}; \ldots; s_{k-\tau_d}; s_k]$

4. choose the $\varphi(\cdot)$ and estimate the function $f(\cdot)$ by using training data D
5. predict s by function $f(\cdot)$

In literature [9], Suykens and Vandewalle proposed that the eq.(6) can be convert into optimal problems which can be describe as follows:

$$\min_{w,e} J(w,e) = \frac{1}{2} w^T w + \gamma \frac{1}{2} \sum_{k=m+1}^{N+m} e_k^2 \qquad (7)$$

subject to
$$s_{k+1} - e_{k+1} = w^T \varphi(\boldsymbol{s}_k - \boldsymbol{e}_k) + b \quad k = m+1, \ldots, N+m \qquad (8)$$

where $e_k = s_k - \hat{s}_k$, $\boldsymbol{s}_k = [s_{k-(m-1)\tau_d}; \ldots; s_{k-\tau_d}, s_k]$, $\boldsymbol{e}_k = [e_{k-(m-1)\tau_d}; \ldots e_{k-\tau_d}; e_k]$, and γ is an adjustable constant. The basic idea of mapping function $\varphi(\cdot)$ is to map the data into a high-dimensional feature space, and to do linear regression in this space.

In the above equation, we can see the SSE is used as the cost function. According to the chaotic theory, positive global Lyapunov exponents causes the errors in these predictions to grow exponentially rapidly, and it is conventionally assumed that the prediction error $E(t)$ will grow as

$$E(t) = E(0) e^{\lambda_m t} \qquad (9)$$

where λ_m is the largest Lyapunov exponents. In this paper, we introduce a weighted component to compensate the cost function which described as follows:

$$P_k = \left(e^{\lambda_m \Delta t}\right)^{-(k-m-1)} \qquad (10)$$

where Δt is sampling interval. Rewriting the eq.(7) as follows,

$$\min_{w,e} J(w,e) = \frac{1}{2} w^T w + \gamma \frac{1}{2} \sum_{k=m+1}^{N+m} P_k e_k^2 \qquad (11)$$

To resolve the optimal function eq(11), we define the Lagrangian function

$$L(w,b,e;\alpha) = J(w,e) + \sum_{k=m+1}^{N+m} \alpha_{k-m} \times [s_{k+1} - e_{k+1} - w^T \varphi(\boldsymbol{s}_k - \boldsymbol{e}_k) - b] \qquad (12)$$

where α_i are Lagrange multipliers. SQP has been applied to solve the optimal function eq(11),The resulting recurrent simulation model is described as follows

$$s_{(k+1)\tau_d} = \sum_{l=m+1}^{N+m} \alpha_{l-m} K(\boldsymbol{z}_l, \boldsymbol{s}_k) + b \qquad (13)$$

where $\boldsymbol{z}_l = \boldsymbol{s}_l - \boldsymbol{e}_l$, $\boldsymbol{s}_k = [s_{k-(m-1)\tau_d}; \ldots; s_{k-\tau_d}, s_k]$

The mapping function $\varphi(\cdot)$ can be paraphrased by a kernel function $K(\cdot,\cdot)$ because of the application of Mercer's theorem, which mean that

$$K(x_i, x_j) = \varphi(x_i)^T \varphi(x_j) \tag{14}$$

with RBF kernels one employs

$$K(x_i, x_j) = \exp\left(-\frac{\|x_i - x_j\|}{2\sigma^2}\right) \tag{15}$$

4 Simulation and Results

In the following procedure, we used the laser data of the Santa Fe competition. The data set consists of laser intensity collected from a laboratory experiment[13]

The reconstructed embedding phase space has been discussed in the Sect.2, Cao proposed a method to determine the minimal sufficient embedding dimension which used improved false nearest neighbor method[14]. The time delayed mutual information was suggested by Fraser and Swinney[15] as a effective tool to determine a reasonable τ_d. The time series was divided in two subsets referred to the training and test subset: the training subset consists of 1000 time entries and the test subset consists of 1500 entries.

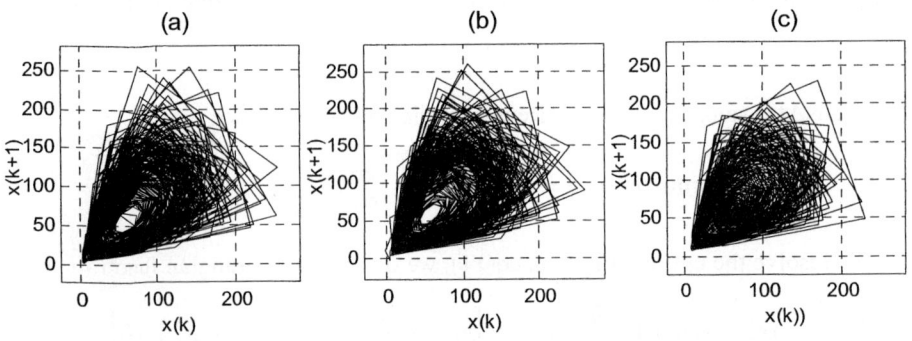

Fig. 1. Poincar'e maps: (a) original system (b) MRLS-SVM reconstruction (c) RLS-SVM reconstruction

We resorted to the estimation of the dynamical invariants of motion from the predicted model. The Poincar'e maps for the Santa Fe time series were drew in Figure 1, where the original attractor (Fig. 1a) is compared with the reconstructed one for MRLS-SVM(Fig. 1b) and RLS-SVM(Fig.1c). The similarity of the Fig.1a and Fig.1b suggests that the two time series represent two distinct trajectories on the same attractor. However, the distinct difference exists in Fig.1a and Fig.1c. Although these plots lead us to believe that the dynamics have been reasonably captured by using MRLS-

5. In figure 3, the largest Lyapunov exponent was also computed using Wolf's ɔdure for different embedding dimensions[12]. Lyapunov exponent of the recon-ted time series seems to follow very closely the spectrum Lyapunov exponent of original time series. In addition, for comparing the performance between RLS-M and MRLS-SVM, we computed the test error of the correlation dimension and argest Lyapunov exponent of the time series in figure 4. this plot illuminate that roposed weighted component can improve the ability to learn the dynamics of the tic system effectively.

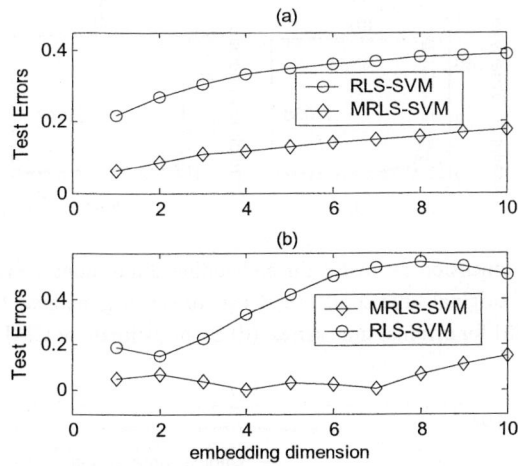

4.Test errors of dynamical invariants of Santa Fe. (a) test errors of Lyapunov exponent. (b) ɔrrors of correlation dimension.

Conclusion

m the simulation results, we can deduce the primary conclusion that is MRLS-M have the ability to capture the dynamics of nonlinear dynamical systems as was ionstrated for the system of Santa Fe competition. This opinion is based on the fact the invariants of the original and generated time series are very similar. We also additional dynamical invariants, the Poincar'e maps, to evaluate the remodeling ormance. In addition, the compare results between the MRLS-SVM and the RLS-M shows that the proposed weighted term can improve the ability to learn the dy-iical invariants of a chaotic dynamical system.

ferences

Brillinger, D . R.: Time series, Data Analysis and Theory, McGraw-hill, New York (1981)
Li K.P., Chen T.L.: Phase Space Prediction Model Based on the Chaotic Attractor. Chin.Phys.Lett., Vol.19, No.7 (2002) 904-907

SVM, for chaotic time series case, the value of correlation dimensi(
Lyapunov exponent is necessary.

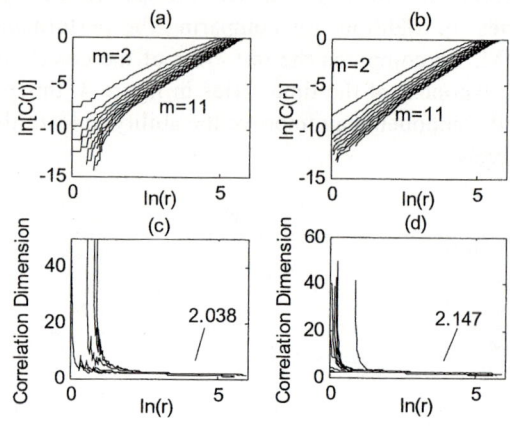

Fig. 2. Correlation dimension estimates for embedding dimensions 2~11.(a)
gral Map (CIM) for original time series. (b) CIM for series generated from
Slope estimate of CIM for original time series. (d) Slope estimate of CIM for

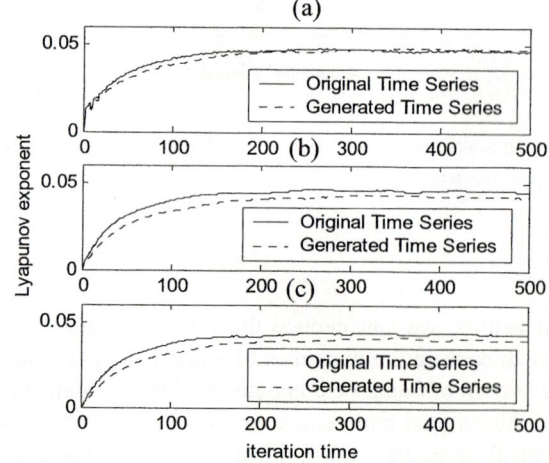

Fig. 3. Estimation of largest Lyapunov exponent for original and generated time
(b)$m=4$,(c)$m=6$

The correlation dimension were computed using Grassberger and P
rithm[13]. In Figure 2, the correlation integral map (CIM) and its slop
both for the original and reconstructed time series. The value of the corr
sion is defined as the slope of the CIM curves for at least 3 consecuti
In this case the correlation dimension for the predicted time series is
region $3<r<5$. The correlation dimension for the time original time seri

3. Casdagli, M.: Nonlinear Prediction of Chaotic Time Series. Physica D 35 (1989) 335-356
4. GanPcay, R.: A Statistical Framework for Testing Chaotic Dynamics Via Lyapunov Exponents, Physica D, 89 (1996) 261-266
5. Lapedes, A., Farber, R.: Nonlinear Signal Processing Using Neural Networks, Los Alamos National Laboratory, LA-UR-87-2662 (1987)
6. Vapnik, V.: The Nature of Statistical Learning Theory, N.Y., Springer (1995)
7. Suykens, J.A.K. (ed.): Least Squares Support Vector Machines, World Scientific (2002)
8. Suykens J.A.K. (ed.): Recurrent Least Squares Support Vector Machines, IEEE Tran. on Circuits and System-I: Fundamental Theory and Applications, Vol. 47, No. 7 (2000) 1109-1114
9. Takens, F.: Detecting Strange Attractors in Fluid Turbulence. In D. Rand and L.S. Young, editors, Dynamical systems and turbulence, Springer-Verlag, Berlin (1981)366-381
10. Wolf A., (ed.): Determining Lyapunov Exponents from a Time Series, Physica D, vol. 16 (1985)285–317
11. Grassberger, P., Procaccia I.: Characterization of Strange Attractors, Physical Review Letters, Vol. 50, No. 5 (1983)346–349
12. Thomsen, J.: Vibrations and Stability. London: McGraw-Hill (1997)
13. Weigend, A. S., Gershenfeld, N. A.: Time Series Prediction: Forecasting the Future and Understanding the Past. Reading, MA: Addison-Wesley (1994)
14. Cao, L.: Practical Method for Determining the Minimum Embedding Dimension of a Scalar Time Series, Physica D (1997) 43-50
15. Fraser, A. M., Swinney H. L.: Independent Coordinates for Strange Attractors from Mutual Information", Phys. Rev. A **33** (1986)1134-1140

Nonlinear System Identification Based on an Improved Support Vector Regression Estimator

Li Zhang and Yugeng Xi

Institute of Automation, Shanghai Jiao Tong University, Shanghai, 200030
zhangli_jasmine@yahoo.com.cn; ygxi@sjtu.edu.cn

Abstract. An improved support vector regression estimator is presented for identifying nonlinear dynamic systems in which the states and the control inputs are separable. The proposed method contains two nonlinear functions with respect to the states and the control inputs respectively. Thus it can be used to estimate the two nonlinear functions in the separable nonlinear dynamic system. The experimental results validate the efficiency of our method.

1 Introduction

So far many methods based on NNs for nonlinear system identification have been proposed [1-3]. If the input of the system satisfies the persistency of excitation (PE) condition and there does not exist disturbance or measure noise, the nonlinear dynamic system identification using NNs could usually obtain good identification results. Generally, it requires a rather large set of hypothesis functions for the nonlinear system identification using NNs. Both the PE condition is not satisfied and there exists noise will lead to the overfitting problem and a large identification (or approximation) error. Therefore choosing a perfected set of hypothesis functions is required, but it is very difficult for complex nonlinear systems. In addition, the learning process of NNs base on the least mean squared error minimization possibly gets into the local minima, which could also lead to a large identification error.

As a new general learning machine, support vector machine (SVM) based on the statistical learning theory [4] has been successfully applied to pattern recognition, regression estimation, signal processing, system identification and others [5-8].

In this paper, an improved support vector regression is presented for identifying the nonlinear dynamic system in which the state variables and the control ones are separable. Our method inherits the good performance of the standard SVR. So it also has very strong nonlinear approximation ability and good generalization performance. It can avoid the overfitting problem and the local minima problem. Simultaneously our method improves the standard SVR based the mixture model of the states and the control inputs. If we know that the state variables and the control ones in a nonlinear system can be separated in advance, we can identify the nonlinear system by using our SVR and obtain a more precise result than that of the standard SVR, because identification model of the improved SVR can eliminate the cross influence between the states and the control inputs essentially. Simulation validates the efficiency of the proposed method.

2 Problem Statement

Here consider a class of discrete nonlinear systems with noise whose state and control variables are separable.

$$y(k+1) = f(\mathbf{y}(k), \mathbf{u}(k)) = g(\mathbf{y}(k)) + h(\mathbf{u}(k)) + e(k+1) \tag{1}$$

where $\mathbf{y}(k) = [y(k), y(k-1), \cdots, y(k-n+1)]^T \in \mathcal{Y} \subset \mathbb{R}^n$ is a state variable, $y(k)$ is the output of the nonlinear system, $\mathbf{u}(k) = [u(k), u(k-1), \cdots, u(k-m+1)]^T \in \mathcal{U} \subset \mathbb{R}^m$ is a control variable, $u(k)$ is the input of the nonlinear system, and $\mathcal{Y}$ and $\mathcal{U}$ are compact sets. $g(\cdot)$ and $h(\cdot)$ are unknown nonlinear functions, $e(k)$ is a disturbance of the system. Here assume that n and m are the known structure orders of the system and $m \leq n$. In order to guarantee the identifiability of a nonlinear system, assume that the outputs of this system are bounded for bounded control input $\mathbf{u}(0) \in \mathcal{U}$ and bounded initial state $\mathbf{y}(0) \in \mathcal{Y}$, and that the nonlinear function $g(\cdot)$ and $h(\cdot)$ are continuous in the bounded compact sets $\mathcal{Y}$ and $\mathcal{U}$, respectively.

If there has a prior knowledge, for example, the state variables can separated from the control ones in a nonlinear system such as Eq. (1), then the standard SVR could not use the prior knowledge. Although we still can perform identification, the identification precision will reduce because of the cross interference between the state variables and the control. To get a satisfied identification result, it requires more training samples. So an improved SVR is proposed to identify nonlinear systems (1).

3 An Improved SVR

It is well known that system identification can be performed using either parallel models or series-parallel ones. Here we adopt series-parallel model to identify the nonlinear dynamic system (1) or predict future values of the outputs using the current outputs and inputs. Given the training sample set $\{(\mathbf{y}(k), \mathbf{u}(k), y(k+1)) \mid \mathbf{y} \in \mathbb{R}^n, \mathbf{u} \in \mathbb{R}^m, y \in \mathbb{R}, k = 1, \cdots, l\}$ generated by Eq. (1), where l is the number of training samples. Firstly, the state variables and the control ones in the training samples are mapped into two feature spaces by two Mercer kernels, respectively. Namely $\mathbf{y} \xrightarrow{\phi} \phi(\mathbf{y})$ and $\mathbf{u} \xrightarrow{\psi} \psi(\mathbf{u})$. Then the linear regression estimation function in these feature spaces could be expressed as

$$\hat{y}(k+1) = \hat{f}(\mathbf{y}(k), \mathbf{u}(k)) = \mathbf{w}_1^T \phi(\mathbf{y}(k)) + \mathbf{w}_2^T \psi(\mathbf{u}(k)) + b \tag{2}$$

By analogy with SVR, we introduce a structure risk to obtain a right balance between the empirical risk minimization and the capacity control of the learning machine. Thus, we have the following primal programming:

$$\min \quad \frac{1}{2}\left(\|\mathbf{w}_1\|^2 + \|\mathbf{w}_2\|^2\right) + C\sum_{k=1}^{l}\left(\xi(k+1) + \xi^*(k+1)\right) \quad (3)$$

s.t. $\quad y(k+1) - \widehat{f}(\mathbf{y}(k), \mathbf{u}(k)) \leq \varepsilon + \xi(k+1)$,

$\widehat{f}(\mathbf{y}(k), \mathbf{u}(k)) - y(k+1) \leq \varepsilon + \xi^*(k+1)$

$\xi(k+1), \xi^*(k+1) \geq 0$, $k = 1, \cdots, l$

where the penalty factor $C > 0$ and the permissible error ε are given by users. Here the roles of the two parameters are identical to that in the standard SVR. The Wolfe dual programming of (3) can be expressed as

$$\max \quad (4)$$
$$-\frac{1}{2}\sum_{k=1}^{l}\sum_{j=1}^{l}(\alpha_k - \alpha_k^*)(\alpha_j - \alpha_j^*)\left[\left(\phi(\mathbf{y}(k)) \cdot \phi(\mathbf{y}(j))\right) + \left(\psi(\mathbf{u}(k)) \cdot \psi(\mathbf{u}(j))\right)\right]$$
$$+\sum_{k=1}^{l}(\alpha_k - \alpha_k^*)\widehat{y}(k+1) - \varepsilon\sum_{k=1}^{l}(\alpha_k + \alpha_k^*)$$
$$\sum_{k=1}^{l}(\alpha_k - \alpha_k^*) = 0$$
$$\alpha_k, \alpha_k^* \in [0, C], \ k = 1, \cdots, l$$

Since ϕ and ψ are two nonlinear mapping functions, we may use two Mercer kernel functions $K_\phi(\cdot, \cdot)$ and $K_\psi(\cdot, \cdot)$ to replace the dot products $\phi(\mathbf{y}(k)) \cdot \phi(\mathbf{y}(k))$ and $(\psi(\mathbf{u}(k)) \cdot \psi(\mathbf{u}(k)))$. The type and parameter values of kernel $K_\phi(\cdot, \cdot)$ can be identical to or different from that of $K_\psi(\cdot, \cdot)$. In doing so, the estimation function of the improved SVR can be written as

$$\widehat{y}(k+1) = \sum_{j=1}^{l}\left(\alpha_j^* - \alpha_j\right)K_\phi\left(\mathbf{y}(j), \mathbf{y}(k)\right) + \sum_{j=1}^{l}\left(\alpha_j^* - \alpha_j\right)K_\psi\left(\mathbf{u}(j), \mathbf{u}(k)\right) + b \quad (5)$$

where the threshold b can be computed by the KKT conditions[8].

4 Simulation

In our experiments, we identified three typical single-input single-output (SISO) nonlinear dynamic systems that could be found in [1]. For the sake of comparison, three methods were used, including the FNN using the static BP algorithm, the standard SVR and the improved SVR presented here. The identification error is defined by $error = \sqrt{\sum_{i=1}^{l_{Test}}(\hat{y}_i - y_i)^2 / l_{Test}}$, where $\hat{y}$ is a estimation of y and l_{Test} is the number of test samples. The noise $e(k)$ added into systems to be identified was independent, zero-mean, and σ^2 variance Gaussian white noise. In all examples, let $\sigma = 0.1$. The training number l of FNNs is 5000, and that of both SVRs is 100. The number of test in the three methods is the same.

Example 1. The system to be identified is described by the difference equation

$$y(k+1) = 0.3y(k) + 0.6y(k-1) + 0.6\sin(\pi u(k)) \qquad (6)$$
$$+ 0.3\sin(3\pi u(k)) + 0.1\sin(5\pi u(k)) + e(k+1)$$

In the training procedure, the amplitude of $u(k)$ was uniformly distributed in the interval $[-1,1]$. While in the test process, the input to the system and the identification model was a sinusoid function $u(k) = \sin(2\pi k / 250)$, $k = 1, \cdots, 500$.

Here, FNN belonged to the class $\mathcal{N}_{3,40,20,1}^3$ [1]. The standard SVR used the Gaussian kernel $K(\mathbf{x}_i, \mathbf{x}_j) = \exp(-\|\mathbf{x}_i - \mathbf{x}_j\|^2 / 2p^2)$ where $p \neq 0 \in \mathbb{R}$ is the kernel parameter. In our method $K_\phi(\cdot,\cdot)$ was the linear kernel $K(\mathbf{x}_i, \mathbf{x}_j) = \mathbf{x}_i^T \mathbf{x}_j$ and $K_\psi(\cdot,\cdot)$ was the Gaussian kernel.

Given a range of values for the Gaussian kernel parameters, we chose a parameter which made the identification error minimum. The results are shown in Table 1. From this example, we can see that the separable-variable SVR has a better identification performance than the multi-layer FNN and the standard SVR.

Example 2. The system to be identified is expressed as

$$y(k+1) = \frac{y(k)y(k-1)[y(k)+2.5]}{1+y^2(k)+y^2(k-1)} + u(k) + e(k+1) \qquad (7)$$

In the training procedure, the amplitude of $u(k)$ was uniformly distributed in the interval $[-2,2]$. While in the test process, the input to the system and the identification model was a sum of two sinusoid functions $u(k) = \sin(2\pi k / 25) + \sin(2\pi k / 10)$, $k = 1, \cdots, 100$.

The multi-layer FNN used a network belonging to $\mathcal{N}_{3,40,20,1}^3$. The standard SVR adopted the Gaussian kernel. In the separable-variable SVR, $K_\phi(\cdot,\cdot)$ was the Gaussian kernel and $K_\psi(\cdot,\cdot)$ was the linear kernel. The results are shown in Table 1.

---

[1] $\mathcal{N}_{3,40,20,1}^3$ which means that the network contains 3 layers and has 3 inputs, 1 output and 2 sets of nodes in the hidden layers, each containing 40, 20 nodes, respectively.

In this example, the identification error obtained by our method is smaller that the other methods.

Example 3. Given the following SISO system

$$y(k+1) = \frac{y(k)}{1+y^2(k)} + u^3(k) + e(k+1) \tag{8}$$

In the training process, let $u(k)$ with the amplitude uniformly distributed over the interval $[-2,2]$. While in the test process, Let $u(k) = \sin(2\pi k/25) + \sin(2\pi k/10)$, $k = 1,\cdots,100$.

The structure of the multi-layer FNN in this example is $\mathcal{N}^3_{2,40,20,1}$. The standard SVR used the Gaussian kernel. In the separable-variable SVR, both $K_\phi(\cdot,\cdot)$ and $K_\psi(\cdot,\cdot)$ are the Gaussian kernel. The experimental results are shown in Table 1.

Table 1. The identification results of three examples

Example	Method	Identification error
1	Multi-layer FNN	0.8286
	Standard SVR	0.4576
	Improved SVR	0.0584
2	Multi-layer FNN	0.2557
	Standard SVR	0.1494
	Improved SVR	0.0959
3	Multi-layer FNN	0.3129
	Standard SVR	0.1564
	Improved SVR	0.0557

5 Conclusion

An improved support vector regression is presented for identifying the nonlinear dynamic system in which the states and the control inputs could be separated. The experimental results on the nonlinear dynamic system described by (1) show that our method has better approximation ability than the standard SVR and FNNs.

So far the training of SVRs is a batch-algorithm and does not exist a recursive algorithm. Therefore the nonlinear system identification using SVRs only could use the series-parallel models. It is a drawback of SVRs compared with NNs. Dynamic identification methods based on the SVR will be the focus of our future work.

References

1. Nerendra, K.S., Parthasarathy, K.: Identification and control of dynamic systems using neural networks. IEEE Transactions on Neural Networks. L (1990) 4-27
2. Polycarpou, M.M., Ioannou, P.A.: Identification and control of nonlinear systems using neural network models: design and stability analysis. Technical report 91-09-01, Department of Electrical Engineering-Systems, University of Southern California (1991)
3. Ioannou, P.A., Sun, J.: Robust adaptive control. Englewood Cliffs, NJ: Prentice-Hall (1996)
4. Vapnik, V.: Statistical Learning Theory. John Wiley and Sons, Inc., New York (1998)
5. Schölkopf, B., Burges, C., Vapnik, V.: Extracting support data for a given task. In Proceedings, First International Conference on Knowledge Discovery & Data Mining. AAAI Press, Menlo Park, CA (1995)
6. Vapnik, V., Golowich, S., Smola, A.: Support vector method for function approximation, regression estimation, and signal processing. In Advances in Neural Information Processing Systems 9 (1997) 281-287
7. Drezet, P.: Support vector machines for system identification. UKACC International Conference on Control'98, 1 (1998) 688-692.
8. Smola, A., Schölkopf, B. A tutorial on support vector regression. NeuroCOLT Technical Report NC-TR-98-030, Royal Holloway College, University of London, UK (1998) Available http://www.kernel-machines.org/

Anomaly Detection Using Support Vector Machines

Shengfeng Tian, Jian Yu, and Chuanhuan Yin

School of Computer and Information Technology, Beijing Jiaotong University
Beijing, 100044, P.R.China
{sftian, jianyu}@center.njtu.edu.cn, xiaoyuehuan@sina.com

Abstract. In anomaly detection, we record the sequences of system calls in normal usage, and detect deviations from them as anomalies. In this paper, one-class support vector machine(SVM) classifiers with string kernels are adopted as the anomaly detector. A sequential learning algorithm for the classifiers is described. Two kinds of kernels are tested with an SVM classifier on the UNM data sets. Results indicate that the string kernel is superior to the RBF kernel for sequence data.

1 Introduction

Intrusion detection has emerged as an important approach to network security. There are two general approaches to intrusion detection: misuse detection and anomaly detection. In anomaly detection, there are many different methods to monitor activities on a computer system. Forrest et al. [1] demonstrated that effective approaches can be obtained by learning program behaviors and detecting deviations from this norm.

Forrest et al.'s approach characterized normal program behaviors in terms of sequences of system calls made by them. They break system calls sequences into substrings of a fixed length N. A separate database of normal behavior for each process of interest is built up. The sequences different from those in the database indicate anomalies in the running process.

Support vector machines, or SVMs, are widely used in pattern recognition problems. But the standard kernel functions can not provide sequence information and the edit distances between two strings calculated by approximate pattern matching algorithms are not valid kernel functions.

In this paper, we revisit the idea of using one-class SVM classifiers for anomaly detection and present a sequential learning algorithm to construct the classifiers. Instead of the traditional kernels, we propose a new string kernel for use in the SVM. Finally an SVM approach with string kernels is applied to anomaly detection using system call sequences.

2 Novelty Detection

In Schölkopf et al. [2], a support vector algorithm was used to characterize the support of a high dimensional distribution. With this algorithm, one can compute a set of contours which encloses the data points. These contours can be considered as normal data boundaries. The data outside the boundaries are interpreted as novelties. In the method suggested by Schölkopf et al. [2], the origin is treated as the only member of the second class after transforming the features via a kernel. They separate the image of one class from the origin. Let $x_1, x_2, \ldots, x_n$ be training examples, $\phi(x)$ be a kernel map which transforms the training examples to a high dimensional feature space. The task is to solve the following quadratic programming problem:

$$\min \tfrac{1}{2} w \cdot w + C \sum_{i=1}^{n} \xi_i - \rho \qquad (1)$$

subject to $w \cdot \phi(x_i) \geq \rho - \xi_i$, $\xi_i > 0 \qquad i=1,\ldots,n$.

Here $C>0$ is the regularization constant and ξ_i are nonzero slack variables. If w and ρ solve the problem, then the decision function

$$f(x) = \text{sign}[w \cdot \phi(x) - \rho] \qquad (2)$$

will be positive for most examples x_i contained in the training set.

The above quadratic programming problem leads to the dual problem

$$\min_{\alpha} \tfrac{1}{2} \sum_{i,j} \alpha_i \alpha_j K(x_i, x_j) \qquad (3)$$

subject to $0 \leq \alpha_i \leq C$, $i = 1,\ldots,n$, and $\sum_i \alpha_i = 1$.

The decision function can be shown to be

$$f(x) = \text{sign}[\sum_{i=1}^{n} \alpha_i K(x_i, x) - \rho] \qquad (4)$$

The offset ρ can be calculated by exploiting that for any $\alpha_i \in (0,C]$, the corresponding example x_i satisfies

$$\rho = \sum_j \alpha_j K(x_j, x_i) \qquad (5)$$

Platt proposed a fast learning algorithm for two-class classification[3]: sequential minimal optimization(SMO), in which only two α_i's are optimized at each iteration. The method consists of a heuristic step for finding the best pair of parameters to optimize and use of an analytic expression to ensure the lagrangian increased monotonically. The idea is adopted in the above problem.

The dual formulation amounts to minimization of

$$W(\alpha) = \tfrac{1}{2}\sum_{i,j} \alpha_i \alpha_j K(x_i, x_j) \qquad (6)$$

respect to α_i and subject to $0 \leq \alpha_i \leq C$, $i = 1,\ldots,n$, and $\sum_i \alpha_i = 1$.
For each α_i, we have

$$\frac{\partial W}{\partial \alpha_i} = 2\sum_j \alpha_j K(x_j, x_i) = 2g_i \qquad (7)$$

Two variables α_i's are selected to be updated according to the following conditions: $g_i > 0$ and $\alpha_i > 0$ or $g_i < 0$ and $\alpha_i < 0$.

If the selected variables are α_i and α_p, the updated values will be $\alpha_i - h$ and $\alpha_p + h$ respectively to satisfy the linear equality constraint. The analytic solution can be obtained from $dW/dh = 0$:

$$h = \frac{g_i - g_p}{K_{ii} + K_{pp} - 2K_{ip}} \qquad (8)$$

where K_{ip} denotes the kernel $K(x_i, x_p)$. The variables α_i and α_p should be selected such that $|g_i - g_p|$ is maximized.

To fulfill the constraints, the flowing bounds should apply to α_i and α_p:
$$L = \max(0, \alpha_i + \alpha_p - C), \quad H = \min(C, \alpha_i + \alpha_p).$$
Then α_i and α_p should be updated as follows:

$$\alpha_i^{new} = \begin{cases} H & \text{if } \alpha_i - h \geq H \\ \alpha_i - h & \text{if } L < \alpha_i - h < H \\ L & \text{if } \alpha_i - h \leq L \end{cases} \qquad (9)$$

$$\alpha_p^{new} = \alpha_p + (\alpha_i - \alpha_i^{new}) \qquad (10)$$

The learning procedure is similar to the SMO. There are two separate choice heuristics: one for α_i and one for α_p. The choice of α_i provides the outer loop of the algorithm. The outer loop iterates over the entire training set, determining an example as α_i which satisfies the inequality conditions. The inner loop iterates over the entire training set, determining an example α_p, which satisfies the inequality conditions and maximizes the value $|g_i - g_p|$. Two Lagrange multipliers are updated according to the above formulas. The learning procedure terminates when the value $|g_i - g_p|$ is smaller than a predefined stopping threshold.

3 String Kernels in SVMs

The kernel function computes the inner product between mapped examples in the feature space. The feature vectors are not calculated explicitly but instead the kernel is computed directly from two examples.

Given a number k≥1, the spectrum kernel concerns the k-length continuous subsequences shared in two strings[4]. The (k,m)-mismatch kernel is similar to spectrum kernel, but at most m mismatches are allowed[5]. In the kernel developed by Lodhi et al.[6], the gaps in the occurrence of the k-length subsequences are allowed.

For intrusion detection, we have three requirements to string kernels. Firstly string kernels are similarity measures between strings which are assessed by number of (possibly non-continuous) matching subsequences shared by two strings, non-continuous occurrences should be penalized. Secondly the feature map is indexed not only by k-length subsequences but all possible subsequences from Σ. Finally the kernel computation is efficient.

To meet the above requirements, we propose a new string kernel. Let Σ be a finite set which we call the alphabet, $x \in \Sigma^*$ denote string defined over the alphabet Σ, $|x|$ the length of x. The neighborhood N(x) generated by x is the set of all subsequences that x contains. Given a number q≥0, the feature map $\Phi(x)$ is defined as

$$\Phi(x) = (\phi_s(x))_{s \in \Sigma^*} \quad (11)$$

$$\phi_s(x) = \begin{cases} \sqrt{q} & \text{s is empty string} \\ 2^{|s|-|x|} & \text{s belongs to N(x)} \\ 0 & \text{otherwise} \end{cases} \quad (12)$$

The kernel K(x,y) is defined as K(x,y) = $\Phi(x) \cdot \Phi(y)$. To ensure the kernel in the interval [0,1], the normalized kernel is defined as

$$K^s(x,y) = [K(x,x)K(y,y)]^{-1/2} K(x,y) \quad (13)$$

The introduction of parameter q in the string kernel is important to classifying performance just as to the function of parameter σ in the RBF kernel.

4 Experiments

In the experiments, the data are taken from the UNM data sets for sendmail which are available at http://www.cs.unm.edu. The traces were obtained at UNM using Sun SPARCstations running SunOs 4.1.1 and 4.1.4. We use "plus" as the normal data set and build up a database of all unique sequences of a given length k=6. For example, suppose we observe the following trace of system calls:

      wait4 open creat link creat write mknod open.

Considering that there are a total of 182 system calls in the SunOS 4.1.x operating system, each system call is assigned an identifying number ranging from 1 to 182. The trace of system calls is changed to a number sequence:

7 5 8 9 8 4 14 5.

For k=6, we get the unique sequences:

7 5 8 9 8 4,   5 8 9 8 4 14,   8 9 8 4 14 5.

In the experiments, the test data sets consist of normal sets including "bounce", "bounce-1", "bounce-2", and "queue", and anomalous sets including "sm-280", "sm-314", "sm-10763", "sm-10801" and "sm-10814". We build up databases of all unique sequences of a given length k=6. The support vector machines are trained with the algorithm derived in Section 2. Two kernel functions are adopted in the algorithm. One is the string kernel $K^s(x,y)$, the another is the standard RBF kernel $K^r(x,y) = \exp[-\|x-y\|^2/\sigma^2]$. The results are shown in Table 1. With the RBF kernel, the best detection performance is achieved when the parameter σ is equal to 100. The minimal difference of novelty numbers between normal data sets and anomalous data sets is 10. With the string kernel, the best detection performance is achieved when the parameter q is equal to 7. The minimal difference of novelty numbers between normal data sets and anomalous data sets is 42. Obviously, the string kernel has the advantage of the RBF kernel in the case of sequence data. The reason why the string kernel is better than the RBF kernel for sequence data is that the sequence information is counted in the string kernel. For example, suppose that there are two sequences as follows:

(1) 1   92   3   94   5   96   (2) 96   1   92   3   94   5.

The two sequences are similar to each other for the string kernel because of the common subsequence "1 92 3 94 5", but are quite different for the RBF kernel.

5 Conclusion

Support vector classifiers using string kernels are suitable to many sequence matching problems such as protein classification, text categorization and intrusion detection. Because command subsequences have different lengths and attackers always insert spurious letters to avoid detection, the string kernel defined in this paper is a good choice in intrusion detection.

Table 1. The detection results of SVMs with parameters C=1, σ=100 and q=7

Data sets	No. of novelties with K^r	No. of novelties with K^s	Normal/Anomalous
bounce	14	2	Normal
bounce-1	11	1	Normal
bounce-2	17	3	Normal
queue	11	1	Normal
sm-280	27	45	Anomalous
sm-314	27	46	Anomalous
sm-10763	28	50	Anomalous
sm-10801	28	50	Anomalous
sm-10814	28	50	Anomalous

References

1. Forrest, S., Hofmeyr, S.A., Somayaji, A.: Intrusion detection using sequences of system calls. Journal of Computer Security, Vol.6 (1998) 151-180
2. Schölkopf, B., Platt, B. J.C., Shawe-Taylor, J., Smola, A.J.: Estimating the support of a high-dimensional distribution. Technical report MSR-TR-99-87, Microsoft Research (1999)
3. Platt, J.C.: Sequential minimal optimization: a fast algorithm for training support vector machines. Technical Report MSR-TR-98-14, Microsoft Research (1998)
4. Leslie, C., Eskin, E. Noble, W.S.: The spectrum kernel: a string kernel for SVM protein classification. In Proceedings of the pacific biocomputing Symposium (2002)
5. Leslie, C., Eskin, E., Weston, J., Noble, W.S.: Mismatch string kernels for SVM protein classification. In: Proceedings of Neural Information Processing Systems (2002) http://www.gs.washington.edu/~noble/papers/leslie_mismatch.pdf
6. Lodhi, H., Saunders, C., Shawe-Taylor, C., Cristianini, N., Watkins, C.: Text classification using string kernels, Journal of Machine Learning Research, Vol.2 (2002) 419-444

Power Plant Boiler Air Preheater Hot Spots Detection System Based on Least Square Support Vector Machines

Liu Han, Liu Ding, Jin Yu, Qi Li, and Yanming Liang

School of Automation & Information Engineering, Xi'an University of Technology, Xi'an 710048, China
{liuhan,liud,liqi}@xaut.edu.cn

Abstract. Air preheater is the important heat exchanger in power plant units. Recombustion accident can be caused by inadequacy combustion of fuel or badly heat-dispersed condition aroused by low air or gas velocity after boiler outage. In the paper, discriminant models of 3 pairs of fire status have been built based on Least Square Support Vector Machines (LS-SVMs) for two kinds of kernel functions. Utilizing polynomial and RBF kernel, the hyperparameters of classifiers were tuned with Leave-one-out(LOO) cross-validation. Receiver Operating Characteristic(ROC) curve comparison shows that LS-SVMs classifiers are able to learn quite well from the raw data samples. Experiment results show that SVMs has good classification and generalization ability and RBF kernel function has more accurate than polynomial kernel function for this problem from the area under the ROC curve (AUC) values of two kernel functions.

1 Introduction

Power plant air preheater is the important heat exchanger using surplus heat of gas from boiler to heating up air. The function of air preheater is to increase temperature of air that system of combustion and coal milling of boiler need as well as to decrease temperature of gas leaving the air preheater and heat loss. In China, rotary air preheater currently has been adopted in boiler of 200MW capacity or above because of characteristic of compact structure, light weight and easily installation. Commonly, clean air preheater is impossible to fire because it is just a heat exchanger. The combustion of air preheater aroused from many factors. But fundamental causation is that superfluous fuels including carbon and oil stockpiled on air preheater heat transfer elements have been oxidated and temperature is rising. When the temperature reach fire point, recombustion is occurred. The conditions of air preheater combustion are composed of superfluous fuel and badly heat-dispersed condition aroused by low air or gas velocity after boiler outage. Inadequately combusted fuel congealed on air preheater heated elements and deposit will be dried and fired on temperature of 350°C. It is difficult to be discovered because of small combustion area during the period of initial stage of combustion. Continuously, combustion keeps on going and heat transfer elements even or full air preheater will be fired when temperature goes on 700°C. The combustion is called air preheater recombustion.

It is proven according to practical experiments that abnormal increase of temperature of gas inlet and outlet duct, air inlet and outlet duct can reflect coincidation of air preheater recombustion. However, once status of abnormal increase of temperature appears, combustion has already come into being a certain extent. So it is necessary to introduce the air preheater hot spots detection system for decreasing harm of combustion minimum. At present, there are air preheater hot spots detection systems to be researched and applied [1][2]. In these systems, thermocouple and infrared sensor are adopted as thermoscope and combustion is judged by comparison of measured temperature value and given alarm threshold. But choice of given alarm threshold strongly depends on experiences and spots situation, which easily results in misinformation. At the same time, artificial neural network has been applied successfully in fire detection [3][4]. However, in many applications neural network has complicated structure and overfull estimated parameters relative to less data samples. As a result, neural network 'overfitting' for data samples appears as well as the generalization ability of neural network is not sufficient. Hot spots detection method based neural network restricts detection accuracy farther rising.

Statistical learning theory (SLT)[5] was introduced by Vapnik in 1995 which focus on machine learning approach in small sample set and support vector machines (SVMs) was a new classification and regression tool based on this theory, which has self-contained basis of statistical learning theory and excellent learning performance. SVMs has already become a new research hotspot in the field of machine learning and has been successfully applied in many fields, such as face identification, handwriting recognition and text classification. In this paper, air preheater hot spots detection system based on support vector machines is proposed. The system combined with engineering experiences for eight years has been gone into many practical applications in boiler of 200MW capacity or above successfully in China or abroad.

2 The Principle of Support Vector Machines and Least Square Support Vector Machines

The principle of empirical risk minimization is essential idea to solve the problem of statistic pattern recognition and statistic machine learning all along. But it is found in practice that excellent generalization ability cannot be achieved to make empirical risk minimization (training error minimum). Especially, excessively small training error may result in poor generalization ability, which is a general case for neural network and called over-fitting. Since postulate of conventional statistic theory is infinite training samples number, in small samples condition the principle of empirical risk minimization based on conventional statistics cannot deduce the principle of expectation risk minimization educed by Bayes Decision Theory. In order to apply conventional statistics in small samples condition, Vapnik built statistics learning theory. In this methodology, excellent generalization ability can be achieved only when both empirical risks and learning machine capacity (VC Dimension) are controlled simultaneously [5].

Support vector machines (SVMs) is based on statistics learning theory. The formulation of SVMs embodies the structure risk minimization (SRM) principle. SVMs has gained wide acceptance due to high generalization ability for a wide range of applications and better performance than other traditional learning machines. In a SVMs, the original input space is mapped into a high dimensional dot product space via a kernel. The new space is called feature space, in which an optimal hyperplane is determined to maximize the generalization ability. The optimal hyperplane can be determined by only few exemplars points called support vectors. SVMs is an approximate implementation of structure risk minimization, which minimizes an upper bound on the error rate of a learning machine on test exemplars instead of minimizing the training error itself as used in neural network-based empirical risk minimization.

Consider a given training set $\{x_i, y_i; i = 1, \cdots l\}$ with input data $x_i \in \mathbb{R}^n$ and corresponding binary class labels $y_i \in \{-1, +1\}$. The input data are mapped to a higher dimensional feature space by nonlinear function $\Phi(\cdot)$ as in the classifier case. The classifier takes the form

$$f(x) = sign[\omega \cdot \Phi(x) + b] \tag{1}$$

Where the term ω denotes weighted vector and the term b denotes a bias term.

In this paper, Vapnik's support vector machines is called standard support vector machines. So this leads to the optimization problem for standard SVMs

$$\min_{\omega,b,\xi} \frac{1}{2}\omega^T \omega + \gamma \sum_{i=1}^{l} \xi_i \tag{2}$$

Subject to

$$\begin{cases} y_i[\omega^T \Phi(x_i) + b] \geq 1 - \xi_i \\ \xi_i \geq 0, \, i = 1, \cdots, l \end{cases} \tag{3}$$

The variables ξ_i are slack variables which are needed in order to allow misclassifications in the set of inequalities and γ is a positive real constant and should be considered as a tuning parameter in the algorithm.

For Least Square Support Vector Machines (LS-SVMs)[6], above formulation has been modified as follows

$$\min_{\omega,b,\xi} \frac{1}{2}\omega^T \omega + \gamma \frac{1}{2} \sum_{i=1}^{l} \xi_i^2 \tag{4}$$

Subject to the equality constraints

$$y_i[\omega^T \Phi(x_i) + b] = 1 - \xi_i \quad i = 1, \cdots, l \tag{5}$$

Important differences with standard SVMs are the equality constraints and the sum squared error term, which greatly simplifies the problem. The solution is obtained after constructing the Lagrangian $L = -\sum_{i=1}^{l} \alpha_i (y_i[\omega^T \Phi(x_i) + b] - 1 + \xi_i) + 1/2 \omega^T \omega + 1/2\gamma \sum_{i=1}^{l} \xi_i^2$ with α_i are the Lagrange multipliers. According to KKT Conditions

[5], the conditions for optimality are given by $\omega = \sum_{i=1}^{l} \alpha_i y_i \Phi(x_i)$, $\sum_{i=1}^{l} \alpha_i y_i = 0$, $\alpha_i = \gamma \cdot \xi_i$ and $y_i[\omega^T \Phi(x_i) + b] - 1 + \xi_i = 0$. Elimination of ω and ξ gives

$$\begin{bmatrix} 0 & Y^T \\ Y & \Omega + \gamma^{-1} I \end{bmatrix} \begin{bmatrix} b \\ \alpha \end{bmatrix} = \begin{bmatrix} 0 \\ \vec{1} \end{bmatrix} \quad (6)$$

with $Y = [y_1; \cdots; y_l]$, $\vec{1} = [1; \cdots; 1]$ and $\alpha = [\alpha_1; \cdots; \alpha_l]$. Mercer's theorem [5] is applied to the matrix Ω with

$$\Omega_{kh} = y_k y_h \Phi(x_k)^T \Phi(x_h) = y_k y_h \cdot K(x_k, x_h), k, h = 1, \cdots, l \quad (7)$$

A positive definite kernel $K(\cdot,\cdot)$ is chosen such that is satisfied Mercer condition. LS-SVMs classifiers with kernel are able to realize strongly nonlinear decision boundaries such as occurring in binary classification problems.

3 Hot Spot Detection System Based on LS-SVMs

3.1 Data Samples

In this research, data samples are acquired from Plant A of 200MW in Shandong Province, Plant B of 300MW in Guizhou Province and Plant C of 300MW in Henan Province in China. Input vector in data samples is five dimensional vectors, which includes 5 temperature points from thermocouples and infrared sensors. In order to exactly decide fire status, we separate air preheater fire status from 3 levels corresponding to 'Level 1' that represents combustion alarms, 'Level 2' that represents combustion pre-alarms and 'Level 3' no alarm, respectively. Table 1 shows number of row data samples from different power plants.

Table 1. Number of data samples of Level 1, Level 2 and Level 3. The rows correspond to the acquisition source, while the columns mention the type of fire status.

	Level 1	Level 2	Level 3	Total
Plant A	13	14	12	39
Plant B	23	8	11	42
Plant C	17	24	8	49
Total	53	46	31	130

3.2 Experiments and Analysis

In this research, three types of fire status need to classified, a multiclass classifier is needed. Three binary LS-SVMs classifiers are constructed and applied to those three pairs, namely Level 1 vs. Level 2, Level 1 vs. Level 3 and Level 2 vs. Level 3. The LS-SVMs classifiers were set up and two kernels, namely polynomial and RBF kernels, were used. For each kernel, the hyperparemeters were tuned to achieve the best leave-one-out(LOO) generalization performance. In this experiment, the hyperparameter γ for the three classifiers with polynomial kernel is optimized at the value of 1. In RBF kernel two hyperparameters are used. Besides γ as a fitting constant, σ controls the width of RBF kernel. Tuned parameters is shown in Table 3. The experiment consists of following steps:

1. the data is divided in a training set(2/3 of data) and a test set(remainder);
2. train the LS-SVMs and use the test set to evaluate the performance;
3. the index of the misclassification is noted.

Table 2 and Table 3 show the results of training and testing of LS-SVMs classifiers with polynomial kernel and RBF kernel respectively.

Table 2. Results of training and testing of LS-SVMs classifiers with polynomial kernel

Index	Level 1 vs. Level 2	Level 1 vs. Level 3	Level 2 vs. Level 3
Number of training data	66	56	52
Number of testing data	33	28	25
γ	1	1	1
Training average error	6.8310	4.1110	5.1023
Training standard deviation	1.2230	1.3129	1.4325
Training ratio of correct classification	89.49%	92.30%	90.11%
Testing average error	4.2746	3.1212	3.9809
Testing standard deviation	1.8513	1.3743	0.7769
Testing ratio of correct classification	87.01%	89.21%	83.77%

Table 3. Results of training and testing of LS-SVMs classifiers with RBF kernel

Index	Level 1 vs. Level 2	Level 1 vs. Level 3	Level 2 vs. Level 3
Number of training data	66	56	52
Number of testing data	33	28	25
γ	0.7576	0.6598	0.4815
σ	3.3248	4.3780	1.7728
Training average error	5.5507	3.0702	4.8309
Training standard deviation	1.0287	1.1760	1.4325
Training ratio of correct classification	91.53%	94.53%	90.77%
Testing average error	3.3675	2.3064	3.9913
Testing standard deviation	1.7610	1.2918	0.7651
Testing ratio of correct classification	89.71%	91.73%	84.01%

The method of Receiver Operating Characteristic(ROC) curves[7] has been used to compare the performance of different classifiers. Each of LS-SVMs classifiers described above will provide predictions for the given test data. The decisions that correspond to the LS-SVMs classifier's output '+1' belong to the positive class. Those that correspond to output '-1' belong to the negative class. By varying a threshold on these sorted decisions we can construct a Receiver Operating Characteristic curves. A ROC curve is a plot of the true positive rate against the false positive rate for the different possible thresholds. Here the ture positive rate is the fraction of the positive instances for which the system predicts 'positive'. The false positive rate is the fraction of the negative instances for which the system erroneously predicts 'positive'. The large the area under the curve(the closer the curve follows the left and top border of the ROC space), the more accurate the test. The expected curve for a system making random predicitons will be a line on the 45-degree diagonal. The evaluation metric we use will be the area under the ROC curve(AUC). The ROC curve for a perfect system has an area of 1. Table 4 summarizes the results obtained from the experiment. It compares the performance of the two kernels used applied to each binary classification problem. As can be seen, for the most of the cases, the LS-SVMs classifiers successfully separate the two classes.

Table 4. Comparision of LS-SVMs classification using polynomial and RBF kernel

	RBF kernel			polynomail kernel		
class pair	L1 vs. L2	L2 vs. L3	L1 vs. L3	L1 vs. L2	L2 vs. L3	L1 vs. L3
AUC	0.8507	0.7954	0.8613	0.8220	0.7902	0.8317
% correct	89.71	84.01	91.73	87.01	83.77	89.21

4 Conclusions

LS-SVMs classifiers have been applied to detection air preheater fire status based on temerature values. Utilizing polynomial and RBF kernel, the hyperparameters of classifiers were tuned with leave-one-out(LOO) cross-validation. ROC curve comparison shows that LS-SVMs classifiers are able to learn quite well from the raw data samples. This shows that the LS-SVMs classifiers generalize well depite that the fact that no feature selection was applied. Future work will focus on classification based on information fusion approach that image, gas physical characteristic like density and so on in air preheater are considered synthetically to get more accurate decisions.

References

1. Liu, H., Liu, D., Li, Q., Shi, W.: Research on Power Plant Boiler Air Preheater Fire Alarm System. Application of Electronic Technique, Vol.24(**6**) (1998) 35-36
2. Yin, G. D.: Hot Point Inspection System of Rotary Air Preheater. Turbine Technology **6** (2003) 137-138
3. Zhang, B. K., Wu, L. B., Wang J. J.: Study on Fuzzy Neural Network for Fire Detection. Journal of Electronics, Vol.22(**4**) (2000) 687-691

4. Wang, X. H., Xiao, J. M., Bao, M. Z.: Ship Fire Detection Based on Fuzzy Neural Network and Genetic Algorithm. Chinese Journal of Scientific Instrumnet, Vol.(3) (2001) 312-314
5. Vapnik, V.: The Nature of Statistical Learning Theory. Springer-Verlag, New York(1999)
6. Suykens, J. A. K., Vandewalle, J.: Least Squares Support Vector Machine Classifiers. Neural Processing Letter, Vol. 9(3) (1999) 293-300
7. Kwokleung, C., Lee, T. W., Pamela, A. S., Michael, H. G., Robert, N. W., Terrence, J. S.:. Comparison of Machine Learning and Traditional Classifiers in Glaucoma Diagnosis. IEEE Trans. on Biomedical Engineering, Vol. 49(9) (2002) 963-974

Support Vector Machine Multiuser Detector for TD-SCDMA Communication System in Multipath Channels

Yonggang Wang[1,2], Licheng Jiao[1], and Dongfang Zhao[2]

[1] Lab. for Intelligent Information Processing, Xidian University, 710071 Xi'an, China
yg.wang@263.net
[2] Air Force Engineering University, 710068 Xi'an, China

Abstract. TD-SCDMA, a part of the third generation mobile communication standards IMT2000 and UMTS is described. A nonlinear multiuser detector based on Support Vector Machine (SVM) is presented and the performance is compared with traditional linear detector. The presented detector can approach optimum multiuser detector than the other linear detectors by using the Midamble of TD-SCDMA. The results of simulation show the feasibility and validity of this detector in multipath channel.

1 Introduction

In the third generation of mobile communication systems code division multiple access (CDMA) will play an important role. Several CDMA based systems have already been standardized by ITU.TD-SCDMA is a special variant of TD-CDMA, which is mainly developed in the Peoples Republic of China, and is now included as a narrowband TDD variant for UMTS. In order to increase the capacity of the system, multiple antennas at the base station and more accurate time synchronization in the uplink have been included in the standard. One of the major problems of CDMA based communication systems is the multiple access interference (MAI) caused by spreading codes that are nonorthogonality due to multipath propagation. MAI can be reduced by multiple antennas and by an appropriate synchronization. Nevertheless, for critical channels, the only way to cope with the problem is multiuser detection (MUD). In the standard for TD-SCDMA linear MUD (joint detector) is proposed. In this paper we derive and investigate a nonlinear MUD based on Support Vector Machine (SVM) [3]. The SVM MUD is a non-blind detector. The algorithm utilizes the midamable of TD-SCDMA system to train the SVM. The presented SVM detector can provides a performance, which is generally close to the optimum MUD than the other linear detectors, while keeping the computational complexity low.

2 TD-SCDMA Communication System

TD-SCDMA is an improved narrowband variant of TD-CDMA. As its name suggests, the TD-SCDMA standard bears two major characteristics: one is to adopt time division duplex (TDD) mode operation for up and downlinks separation; the other is to use synchronous CDMA technology. The use of those techniques in TD-SCDMA offers many benefits. In TD-SCDMA the uplink time synchronization has been improved further. The dominating paths of each user arrive at the same time at the base station. Since orthogonal codes are used, a large portion of the MAI can be avoided. TD-SCDMA makes use of both TDMA and CDMA techniques such that channelization in TD-SCDMA is implemented using both time slots and signature codes to differentiate mobile terminals in a cell.

Table 1. Air Interface Parameters

Parameters	Values
Frame duration T_F	5ms
Burst duration T_{BU}	$675\mu s$
Midamble chips L_m	144
Guard period T_{GP}	$75\mu s$
Data symbols per data block N	22
Symbol duration T_S	$12.5\mu s$
Chips per symbol Q	16
Chip duration T_C	$1/1.28\mu s$
Interleaving depth I_D	4 frames/speech
Coder constraint length L_C	9
Rate R_C	1/3

The frame of TD-SCDMA is composed of 7 main time slots and 2 special time slots allocated around the switching point and used for synchronization purposes. The 7 main time slots can be arbitrarily distributed between uplink and downlink. The only restriction is that uplink and downlink have at least one time slot. This makes TD-SCDMA not only suitable for symmetric voice services, but also for the asymmetric data services (multimedia and internet downloads). The main time slot is composed of two data blocks with $N = 22$ QPSK data symbols spread by a code of length $Q = 16$ which defines the subchannel, a user specific Midamble for channel estimation of length $L_m = 144$ chips, and a guard interval of 16 chips. In table 1 he main air interface parameters of TD-SCDMA are summarized.

3   Signal Model

Due to the midamble and the guard period, the detection of each data block can be considered separately. The optimum receiver for each field of data symbols consists of a bank of channel matched filters, followed by an optimal MUD [1]. Combining the transmit symbols $b_m(k)$ of all subchannels $m = 1, \cdots, M$ at one discrete time instant k into the transmit vector:

$$\mathbf{b}(k) = [b_1(k), b_2(k), \cdots, b_M(k)]^T \tag{1}$$

And similarly defining the vector of received symbols after the matched filters we obtain:

$$\mathbf{r}(k) = \mathbf{P}(k) \begin{bmatrix} \mathbf{b}(k) \\ \mathbf{b}(k-1) \\ \vdots \\ \mathbf{b}(k-L+1) \end{bmatrix} + \mathbf{n}(k) \tag{2}$$

where the $M \times LM$ system matrix is given by

$$\mathbf{P} = \mathbf{S}^T \mathbf{H} \begin{bmatrix} \mathbf{SA} & 0 & \cdots & 0 \\ 0 & \mathbf{SA} & & \vdots \\ \vdots & & \mathbf{SA} & 0 \\ 0 & \cdots & 0 & \mathbf{SA} \end{bmatrix} \tag{3}$$

the signature sequence matrix $\mathbf{S} = [\mathbf{s}_1, \cdots, \mathbf{s}_M]$, and $\mathbf{s}_i = [s_{i,1}, \cdots, s_{i,Q}]^T$ is the signature sequence of the user i. The diagonal signal amplitude matrix $\mathbf{A} = \mathrm{diag}\{A_1, \cdots, A_M\}$. The $M \times LM$ channel impulse response (CIR) matrix $\mathbf{H}$ has the form:

$$\mathbf{H} = \begin{bmatrix} h_0 & h_1 & \cdots & h_{n_h-1} & & & \\ & h_0 & h_1 & \cdots & h_{n_h-1} & & \\ & & \ddots & \ddots & \ddots & \ddots & \\ & & & h_0 & h_1 & \cdots & h_{n_h-1} \end{bmatrix} \tag{4}$$

Orthogonal signature sequences are used in TD-SCDMA, so that the noise vector $\mathbf{n}(k)$ has a variance of $E[\mathbf{n}(k)\mathbf{n}^T(k)] = \sigma_n^2 \mathbf{I}$. We note that the orthogonality of the codes is destroyed by the channel-induced intersymbol interference (ISI). The ISI span depends on the length of the CIR, n_h, related to the length of the chip sequence, Q. For $n_h = 1$, $L = 1$; for $1 < n_h \leq Q$, $L = 2$; for $Q < n_h \leq 2Q$, $L = 3$; and so on.

4 Linear and Optimal Detectors

Joint detection, which is proposed in the standard as MUD, is linear detector. The linear MUD for user i has the form:

$$\hat{b}_i(k) = \mathrm{sign}(y_L(k)) \text{ with } y_L(k) = \mathbf{w}^T \mathbf{r}(k) \tag{5}$$

where $\mathbf{w}$ denotes the detector's weight vector. The most popular solutions are the decorrelating detector and the MMSE solution:

$$\hat{b}_i^{DD} = \text{sign}\left\{(\mathbf{PP}^T)^{-1}\mathbf{p}_i\mathbf{r}(k)\right\} \qquad (6)$$

$$\hat{b}_i^{MMSE} = \text{sign}\left\{(\mathbf{PP}^T + \sigma^2\mathbf{I})^{-1}\mathbf{p}_i\mathbf{r}(k)\right\} \qquad (7)$$

where $\mathbf{p}_i$ denotes the ith column of $\mathbf{P}$.

The linear detector is computationally very simple, and the standard LMS or RLS algorithms can be used to implement adaptively. But a linear MUD only performs adequately in certain situations. Let the N_b possible combinations of $[\mathbf{b}^T(k), \mathbf{b}^T(k-1), \cdots, \mathbf{b}^T(k-L+1)]^T$ be $\mathbf{b}^{(j)} = [\mathbf{b}^{(j)T}(k), \mathbf{b}^{(j)T}(k-1), \cdots, \mathbf{b}^{(j)T}(k-L+1)]^T$ and $b_i^{(j)}$ is the ith element of $\mathbf{b}^{(j)}(k)$. Let us define the set of the N_b noise-free received signal states as:

$$\Re = \left\{\mathbf{r}_j = \mathbf{Pb}^{(j)},\ 1 \leq j \leq N_b\right\} \qquad (8)$$

where $\Re$ can be partitioned into two subsets:

$$\Re_\pm = \left\{\mathbf{r}_j \in \Re : b_i^{(j)} = \pm 1\right\} \qquad (9)$$

If $\Re_+$ and $\Re_-$ are not linearly separable, a linear MUD will exhibit an irreducible error floor even in the noise-free case, as it can only form a hyperplane in the M dimensional received signal space.

Applying the Bayesian classification theory it can be shown that the optimal detector has the form [1]:

$$y_B(i) = f_B(\mathbf{r}(i)) = \sum_{j=1}^{N_b} \beta_j b_i^{(j)} \exp\left(-\frac{\|\mathbf{r}(i) - \mathbf{r}_j\|^2}{2\sigma_n^2}\right) \text{ with } \hat{b}_i(k) = \text{sign}(y_B(k)) \qquad (10)$$

where $b_i^{(j)} \in \{\pm 1\}$ serve as class labels.

5 The Support Vector Machine Detector

The optimal detector (10) requires the knowledge of all the noise-free signal sates, which are unknown to receiver. In practice the receiver can use the midamble of TD-SCDMA to get a block of training samples $\{\mathbf{r}(k), b_i(k)\}_{k=1}^{L_m=144}$. Let us denote the training set of L_m noisy received signal vectors as $\aleph = \{\mathbf{x}_k = \mathbf{r}(k), 1 \leq k \leq L_m\}$ and the set of corresponding class labels as $\mathbf{C} = \{c_k = b_i(k), 1 \leq k \leq L_m\}$. Applying the standard SVM method [2], an SVM detector can be constructed for user i:

$$y_{SVM}(k) = \sum_{j=1}^{L_m} \overline{g}_j c_j K(\mathbf{r}(k), \mathbf{x}_j) + \overline{\eta} \qquad (11)$$

where $K(\cdot)$ is the kernel function of SVM and the set of Lagrangian multipliers $\{\bar{g}_j\}$, denoted in vector form as $\bar{\mathbf{g}} = [\bar{g}_1, \cdots, \bar{g}_{L_m}]^T$, is the solution of the quadratic programming (QP)

$$\bar{\mathbf{g}} = \arg\min_{\bar{\mathbf{g}}} \left\{ \frac{1}{2} \sum_{j=1}^{L_m} \sum_{l=1}^{L_m} g_j g_l c_j c_l K(\mathbf{x}_j, \mathbf{x}_l) - \sum_{j=1}^{L_m} g_j \right\} \tag{12}$$

with the constraints $0 \leq g_j \leq C$ and $\sum_{j=1}^{L_m} g_j c_j = 0$. In this application it was found advantageous to choose the Gaussian kernel function:

$$K(\mathbf{x}_j, \mathbf{x}_l) = \exp\left(-\frac{\|\mathbf{x}_j - \mathbf{x}_l\|^2}{2\rho^2}\right) \tag{13}$$

where the width parameter is related to the root mean square σ_n of the channel noise, an estimate of which can be obtained. The offset constant $\bar{\eta}$ is usually determined from the socalled "margin" SVs, i.e. from those particular $\mathbf{x}_j$ s, for which the corresponding Lagrangian multipliers obey $0 < g_j < C$. Because the optimal decision boundary, defined by $\{\mathbf{r} : f_B(\mathbf{r}) = 0\}$, passes through the origin of the received signal space and possesses certain symmetric properties due to the symmetric structure of $\Re_+$ and $\Re_-$, $\bar{\eta} = 0$ can be used. The user-defined parameter C controls the tradeoff between model complexity and training error. In our application, we will choose it empirically.

The set of SVs, denoted by $\aleph_{SVM}$, is given by those particular $\mathbf{x}_j$ s, which have nonzero Lagrangian multipliers obeying $0 < g_j < C$, where $\aleph_{SVM}$ is usually a small subset of the training data set $\aleph$. These SVs are determined during the training process. Thus the SVM-based MUD requires computing the decision variable

$$y_{SVM}(k) = \sum_{\mathbf{x}_j \in \aleph_{SVM}} \bar{g}_j c_j \exp\left(-\frac{\|\mathbf{r}(i) - \mathbf{r}_j\|^2}{2\rho^2}\right) \text{ with } \hat{b}_i(k) = \text{sign}(y_{SVM}(k)) \tag{14}$$

6 Simulation Results

In this section we compare the performance of different detectors: 1) decorrelating detector, 2) MMSE detector, 3) SVM detector, by extensive simulations of uplink TD-SCDMA communication system in multipath fading channel. The Parameters of simulations are shown in table 1.

In Figure 1 the bit error rate (BER) versus signal-to-noise rate (SNR) is depicted. Note that decorrelating detector (DD) and MMSE detector resulted absolutely unusable. The presented detector, SVM MUD, outperform the others detectors.

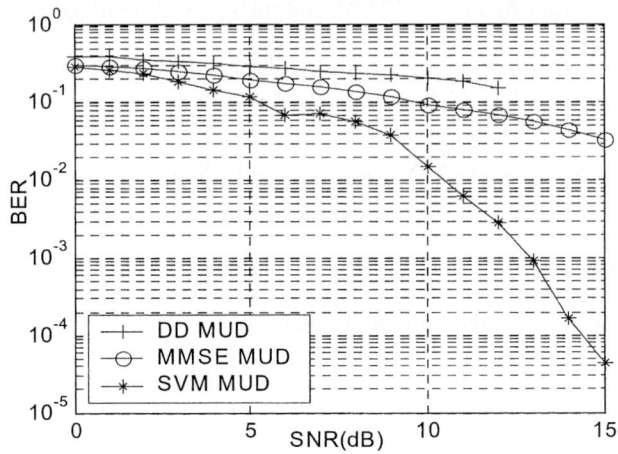

Fig. 1. BER Vs. SNR. Parameters: 8users, 2 codes per user, 3-path fading channel.

7 Conclusions

In this paper, the third generation mobile communication system TD-SCDMA has been introduced. A novel nonlinear multiuser detector based on support vector machine for TD-SCDMA has been proposed. The presented detector resulted in batter performance than traditional linear detector. This, together with the use of other key techniques, such as smart antenna, will make a great increase of the system capacity.

References

1. Verdu, S.: Multiuser Detection. Cambridge Univ. Press, Cambridge UK (1998)
2. Vapnik, V.: The Nature of Statistical Learning Theory. Springer-Verlag, Berlin Heidelberg New York (1995)
3. Chen, S., Samingan, A.K., Hanzo, L.: Support vector machine multiuser receiver for DS-CDMA signals in multipath channels. IEEE Trans. on Neural Networks, vol. 12. IEEE Press, New York (1999) 604-611

Eyes Location by Hierarchical SVM Classifiers

Yunfeng Li and Zongying Ou

CAD & CG Laboratory, School of Mechanical Engineering
Dalian University of Technology, Dalian 116024, P. R. China
yunfengli2004@tom.com, ouzyg@dlut.edu.cn

Abstract. This paper presents a method for eyes location using a two-level hierarchy of SVM (Support Vector Machines) classifiers. On the first level, a two-eye region classifier is obtained by training the SVM using grayscale projections of the two-eye region images. Utilizing this classifier, the region where the two eyes lie can be located by searching the whole face image. On the second level, the left and right eye classifier are obtained by training SVM using grayscale of left and right eye images respectively. Using these two classifiers, the two eyes can be precisely located by searching the output region of the first level. Experimental results show that this method is sufficiently generic and can cope with more various image conditions than exiting techniques.

1 Introduction

Human eyes have always played the most important role during the process of face recognition, eyes location is a prerequisite to almost all automatic face recognition algorithms. During the last decade, numerous attempts have been made to detect and localize eyes, the theories they have been used include image segmentation[1][2][3], grayscale projection[4], edge detection[1][3][5][6], template matching[7] and deformable template matching[5][6], etc. Most of the existing eyes location methods are the combination of one or more of the above theories, those methods emphasize particularly on the use of edge information of eyelids and irises. So the traditional eyes location methods restrict their applications to front face and open eyes. Under general conditions, because of the influences brought by close eyes, spectacles, pose and lighting, etc, eyes patterns become complex and instable, so an effective eyes location algorithm should be robust enough to cope with these problems.

2 Overview of the System

This paper proposes an SVM-based eyes location algorithm which is realized by a two-level classification procedure. On the first level, a two-eye region classifier is obtained by training the SVM using grayscale projections of the two-eye region images. Utilizing this classifier, the region where the two eyes lie is located by searching the whole face image. On the second level, the SVM is trained by using the grayscale of the left and right eye images respectively, left and right eye can be located precisely through searching the left and right part of the exported region from the first level. The diagrammatic description of the operation of the eye location system is shown in Fig. 1.

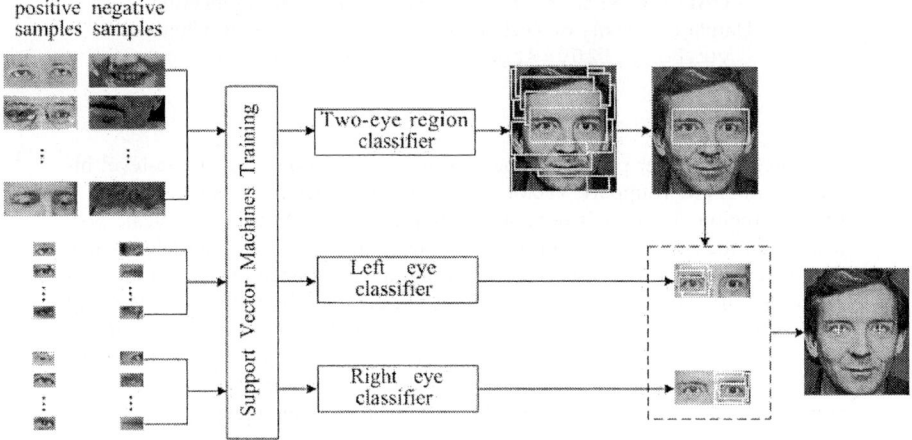

Fig. 1. Diagrammatic description of the operation of the system

3 Support Vector Machines

Support Vector Machines (SVM)[8] perform pattern recognition for two-class problems by finding the decision surface which minimizes the structural risk of the classifier. This is equivalent to determining the separating hyperplane that has maximum distance to the closest samples of the training set. These closest samples are called Support Vectors (SVs).

Considering the problem of separating a set of training samples belong to two separable classes, $(x_1, y_1),...,(x_n, y_n)$, where $x_i \in R^d$, $y_i \in \{-1,+1\}$ with a hyperplane $wx + b = 0$. An optimal hyperplane can be obtained through solving a constrained optimization problem, the decision function can be written as:

$$f(x) = \text{sgn}(\sum_{i=1}^{n} \alpha_i^* y_i k(x_i \cdot x) + b^*) \qquad (1)$$

where α_i^* $(i=1,...,n)$ and b^* are the best solutions to the optimization problem, the kernel $k(x, y)$ is defined by $k(x, y) = \phi(x) \cdot \phi(y)$, ϕ is an implicate nonlinear map which ensures that the training data are linearly separable in new feature space. Every feature target such as two-eye region, left eye or right eye should be uniquely determined in the process of eyes location for each face image, that is in each feature target location process, $f(x)$ must be above zero once and only once, but there may be some cases not accord with this. Because the value of $|w \cdot x + b|$ is directly proportional to the distance from sample x to the hyperplane, the sign of $w \cdot x + b$ should be positive if sample x belongs to positive class and vice versa. As a real number, the bigger the value of $w \cdot x + b$ is, the more likely x belongs to the positive class. So this paper selects the one as feature target that make the value of $w \cdot x + b$ maximal, this lead to the following:

$$x^* = \arg\max(\sum_{i=1}^{n} \alpha_i^* y_i k(x_i \cdot x) + b^*) \qquad (2)$$

4 Eyes Loction

4.1 Two-Eye Region Location

The two-eye region is the part of face image where two eyes and eyebrows lie. The aims of finding the position of two-eye region include narrowing the searching scope for left and right eye respectively, eliminating the disturbance brought by nose and mouth, etc. It also aims to embody the coarse-to-fine policy that can save total search time. On this level, the vectors composed of the vertical and horizontal grayscale projections of the two–eye region images are used for training SVM. Suppose the image of two-eye region $B(i, j)$ has the size of $M \times N$, the horizontal grayscale projection P_h and vertical grayscale projection P_v can be defined as

$$P_h(j) = \sum_{i=1}^{M} B(i, j) \text{ and } P_v(i) = \sum_{j=1}^{N} B(i, j)$$

respectively. If Z stands for the sum of gray values of all the pixels, then the training vector can be represented by the concatenation of $\frac{P_h}{Z} \cdot M$ and $\frac{P_v}{Z} \cdot N$. A two-eye region classifier can be obtained by training the SVM. According to formula (2), using this classifier the two-eye region can be detected and located by searching the whole face image.

4.2 Precise Eyes Location

On this level, the grayscale values of the images are directly used for SVM training, the SVM of left and right eye are trained respectively, thus the classifier of left and right eye can be used for eyes location by searching the left and right part of two-eye region which has been obtained on the first level.

4.3 Support Vector Machines Training

The training samples of SVM are composed of positive sample set and negative sample set. In this paper, there are three positive sample sets that are composed of two-eye region images, left eye images and right eye images respectively. Three negative samples sets corresponding to the positive ones are extracted from the searching fields that do not contain positive samples in each location step, that is the whole face images, the left parts of the two-eye region images and the right parts of the two-eye region images respectively. The negative class is broader and richer, and therefore needs more negative examples to get an accurate definition that separates it from the positive class. In order to select the effective negative examples for targets detection, the method of bootstrapping is used for selecting negative samples, speaking in detail, the false positives are later served as negative examples during the training process. This paper uses SMO (Sequential Minimal Optimization) [9] as training algorithm.

5 Experimental Results

The experiment was performed on the ORL face database[10] which contains 400 facial images corresponding to 40 subjects. There are 10 different images of each subject. We selected 200 samples (5 for each subject) randomly as the training set by which we trained SVM, the remaining 200 samples were used as the test set. The system was implemented by MATLAB language on a computer with an AMD Athlon (1.3GHz) processor and a 256M EMS memory. Fig.2 shows some results of eyes location. Experiments results show an accurate location rate of 90% compared with manual selection and an average speed of 1.9 seconds for one face image.

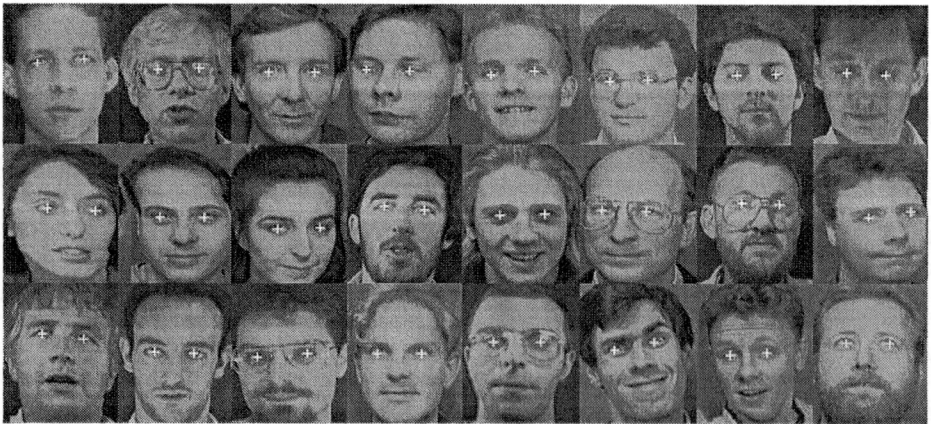

Fig. 2. Some results of eyes location experiment

6 Conclusion

This paper has introduced an approach for eyes location using support vector machines. The system performs the location by means of a two level hierarchy of classifiers. On the first level, the two-eye region classifier detects the two-eye regions in the face images. On the second level, the left eye classifier and the right eye classifier detect their targets independently based on the output of the first level. The two-eye region can be detected more easily compared with one single eye, because it has a large structure and the number of possible candidates is smaller in face image. Besides, the hierarchical detection strategy embodies the time saving policy. The experimental results show that this method is more robust and rapid than existing techniques and can cope with the problems brought by close-eye, side-glance, spectacles and pose during eye location process.

References

1. Rizon, M., Kawaguchi, T.: Automatic eye detection using intensity and edge information. TENCON Proceedings, Vol. 2 (2000) 415–420
2. Tao, L., Kwan, H.K.: Automatic localization of human eyes in complex background. IEEE International Symposium on Circuits and Systems Proceedings, Vol. 5 (2002) 669–672
3. Yang, H., Yuan, B.Z.: Feature extraction in human face recognition system. 5th International Conference on Signal Processing Proceedings, Vol. 2 (2000) 1273–1276
4. Kumar, R.T., Raja, S.K., Ramakrishnan, A.G.: Eye detection using color cues and projection functions. Proceedings of International Conference on Image Processing, Vol. 3 (2002) III-337–III-340
5. Huang, W.M., Mariani, R.: Face detection and precise eyes location. Proceedings of the 15th International Conference on Pattern Recognition, Vol 4 (2000) 722–727
6. Lam, K.M., Yan, H.: An Improved Method for Locating and Extracting the Eye in Human Face Images. Proceedings of the 13th International Conference on Pattern Recognition, Vol. 3 (1996) 411–415
7. Huang, W.J., Yin, B.C., Jiang, C.Y., Miao, J.: A new approach for eye feature extraction using 3D eye template. Proceedings of International Symposium on Intelligent Multimedia, Video and Speech Processing, (2001) 340–343
8. Vapnik, V. N.: The Nature of Statistical Learning Theory. Springer-Verlag, New York(1995)
9. Platt., J.: Sequential Minimal Optimization: A fast algorithm for training support vector machines. Technical Report MSR-TR-98-14, Microsoft Research (1998)
10. The ORL Database of Faces. http://www.uk.research.att.com:pub/data/

Classification of Stellar Spectral Data Using SVM

Fei Xing and Ping Guo

Department of Computer Science
Beijing Normal University, Beijing, 100875, P.R.China
xsoar@163.com pguo@ieee.org

Abstract. In this paper a new technique is developed on stellar spectral classification. Because stellar spectral data sets are usually extremely noisy, wavelet de-noising method is proposed to reduce noise first. Then the support vector machines (SVM) is used for the classification. Experimental results show that in most cases, there will be a better performance using this composite classifier than using SVM with principle component analysis data dimension reduction technique.

1 Introduction

Stellar spectral classification is an important part of automatic recognition of astronomical spectra. Because classifying the spectral data of great bulk manually is a tough job, the technology of automatic and accurate classification on spectral data should be developed.

To automatically recognize stellar spectra, we should build the classifier with training samples first. There are many classification techniques in this research field, among them discriminant analysis is one of the supervised learning classifier building techniques. Quadratic discriminant analysis (QDA) is widely used if sufficient training samples could be supplied [1,2]. Unfortunately, sometimes training samples are usually hard to acquire, and the dimensionality of spectral data is extremely high, thus the estimated covariance matrix will become singular. Linear discriminant analysis (LDA) could be used as one kind of regularization if the total number of samples is larger than the dimension of variables. The covariance matrix, in LDA, is substituted by common covariance matrix. However, in the case of small sample sizes, the common covariance matrix is also singular. To solve the small training sample with high-dimension setting problem, Regularized discriminant analysis (RDA) [3] could be applied. RDA adds the identity matrix as a regularization term to solve the problem in matrix estimation. But parameter optimization of RDA is time consuming. Artificial neural network (ANN) is also a good tool for pattern recognition, it has been successfully used in classification of stellar spectra [4]. Different network models are developed in recent years, the performance of them is data dependent.

Support Vector Machines (SVM) [5] is a new technique for data classification, it has been used successfully in many object recognition applications [6,7,8].

SVM is known to generalize well even in high dimensional spaces under small training sample condition. This characteristic is appropriate for stellar spectral classification where such conditions are typically encountered.

2 Support Vector Machines

SVM was introduced by Vapnik in the late 1960s on the foundation of statistical learning theory [9]. In theory, the SVM classification can be traced back to the classical structural risk minimization (SRM) approach, which determines the classification decision function by minimizing the empirical risk.

SVM uses linear model to implement nonlinear class boundaries through some nonlinear mapping the input vectors **x** into the high-dimensional feature space. The optimal separating hyperplane is determined by giving the largest margin of separation between different classes. For the two-class case, this optimal hyperplane bisects the shortest line between the convex hulls of the two classes. The data are separated by a hyperplane defined by a number of support vectors. The SVM attempts to place a linear boundary between the two different classes, and orient it in such a way that the margin is maximized. The boundary can be expressed as follows:

$$(\mathbf{w} \cdot \mathbf{x}) + b = 0, \qquad \mathbf{w} \in R^N,\ b \in R, \tag{1}$$

where the vector **w** defines the boundary, **x** is the input vector of dimension N and b is a scalar threshold.

The optimal hyperplane is required to satisfy the following constrained minimization as

$$\min\{\frac{1}{2}\|\mathbf{w}\|^2\} \quad \text{with} \quad y_i(\mathbf{w} \cdot \mathbf{x}_i + b) \geq 1,\ i = 1, 2, \ldots, l, \tag{2}$$

where l is the number of training sets.

For a linearly non-separable case, the above formula can be extended by introducing a regularization parameter C as the measurement of violation of the constraints as follows:

$$\min\{\sum_{i=1}^{l} \lambda_i - \frac{1}{2}\sum_{i,j=1}^{l}\lambda_i\lambda_j y_i y_j(\mathbf{x}_i \cdot \mathbf{x}_j)\}$$

$$\text{with} \quad \sum_{i=1}^{l} y_i\lambda_i = 0,\ 0 \leq \lambda_i \leq C,\ i = 1,2,\ldots,l, \tag{3}$$

where the λ_i are the Lagrangian multipliers and are nonzero only for the support vectors. Thus, hyperplane parameters $(\mathbf{w}, b)$ and the classifier function $f(\mathbf{x}; \mathbf{w}, b)$ can be computed by optimization process. The decision function is obtained as follows:

$$f(\mathbf{x}) = \text{sgn}\{\sum_{i=1}^{l} y_i\lambda_i(\mathbf{x} \cdot \mathbf{x}_i) + b\}. \tag{4}$$

In cases where the linear boundary in input spaces will not be enough to separate two classes properly, it is possible to create a hyperplane that allows linear separation in the higher dimension. The method consists in projecting the data in a higher dimension space where they are considered to become linearly separable. The transformation into higher-dimensional feature space is relatively computation-intensive. A kernel can be used to perform this transformation and the dot product in a single step provided the transformation can be replaced by an equivalent kernel function. This helps in reducing the computational load and at the same time retaining the effect of higher-dimensional transformation. The kernel function $K(\mathbf{x}_i, \mathbf{x}_j)$ is defined as follows:

$$K(\mathbf{x}_i, \mathbf{x}_j) = \phi(\mathbf{x}_i) \cdot \phi(\mathbf{x}_j). \tag{5}$$

There are some commonly used kernels:

1. Polynomial: $K(\mathbf{x}_i, \mathbf{x}_j) = [(\mathbf{x}_i \cdot \mathbf{x}_j) + 1]^q$
2. Radial basis: $K(\mathbf{x}_i, \mathbf{x}_j) = \exp(-\|\mathbf{x}_i - \mathbf{x}_j\|^2 / 2\sigma^2)$
3. Sigmoid: $K(\mathbf{x}_i, \mathbf{x}_j) = \tanh(\nu(\mathbf{x}_i \cdot \mathbf{x}_j) + c)$

3 Experiments

The stellar spectra used in our experiments are selected from Astronomical Data Center (ADC). We use 161 stellar spectra contributed by Jacoby *et al.* (1984). Ordered from highest temperature to lowest, the seven main stellar types are O, B, A, F, G, K, and M. The seven main types of stellar spectrum lines are shown in Fig. 1(a).

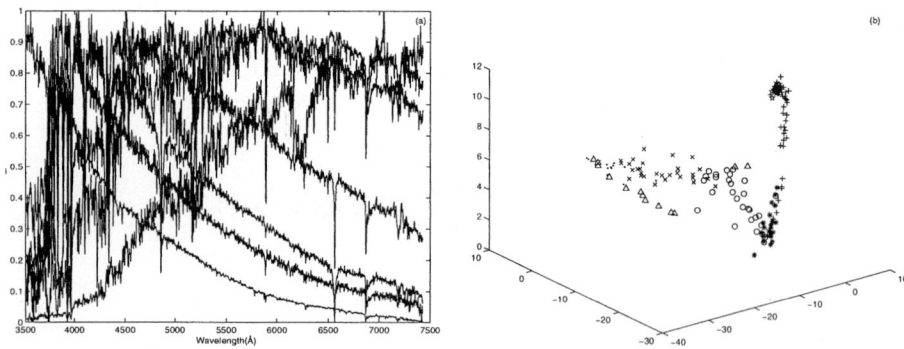

Fig. 1. (*a*) Seven main types of stellar spectra. (*b*) Distribution of stellar spectra in first three principal components space

The bootstrap technique [10] is applied in experiments. 161 samples are divided into two parts, 10 independent random samples drawn from each class are defined by $error = \sqrt{\sum_{i=1}^{l_{Test}}(\hat{y}_i - y_i)^2 / l_{Test}}$, where $\hat{y}$ is a estimation of y and l_{Test} is the number of test samples. The noise $e(k)$ added into systems to be identified was independent, zero-mean, and σ^2 variance Gaussian white noise. In all examples, let $\sigma = 0.1$. The training number l of FNNs is 5000, and that of both SVRs is 100. The number of test in the three methods is the same.

Example 1. The system to be identified is described by the difference equation

$$y(k+1) = 0.3y(k) + 0.6y(k-1) + 0.6\sin(\pi u(k)) \qquad (6)$$
$$+ 0.3\sin(3\pi u(k)) + 0.1\sin(5\pi u(k)) + e(k+1)$$

In the training procedure, the amplitude of $u(k)$ was uniformly distributed in the interval $[-1,1]$. While in the test process, the input to the system and the identification model was a sinusoid function $u(k) = \sin(2\pi k / 250)$, $k = 1, \cdots, 500$.

Here, FNN belonged to the class $\mathcal{N}_{3,40,20,1}^3$ [1]. The standard SVR used the Gaussian kernel $K(\mathbf{x}_i, \mathbf{x}_j) = \exp(-\|\mathbf{x}_i - \mathbf{x}_j\|^2 / 2p^2)$ where $p \neq 0 \in \mathbb{R}$ is the kernel parameter. In our method $K_\phi(\cdot,\cdot)$ was the linear kernel $K(\mathbf{x}_i, \mathbf{x}_j) = \mathbf{x}_i^T \mathbf{x}_j$ and $K_\psi(\cdot,\cdot)$ was the Gaussian kernel.

Given a range of values for the Gaussian kernel parameters, we chose a parameter which made the identification error minimum. The results are shown in Table 1. From this example, we can see that the separable-variable SVR has a better identification performance than the multi-layer FNN and the standard SVR.

Example 2. The system to be identified is expressed as

$$y(k+1) = \frac{y(k)y(k-1)[y(k)+2.5]}{1 + y^2(k) + y^2(k-1)} + u(k) + e(k+1) \qquad (7)$$

In the training procedure, the amplitude of $u(k)$ was uniformly distributed in the interval $[-2,2]$. While in the test process, the input to the system and the identification model was a sum of two sinusoid functions $u(k) = \sin(2\pi k / 25) + \sin(2\pi k / 10)$, $k = 1, \cdots, 100$.

The multi-layer FNN used a network belonging to $\mathcal{N}_{3,40,20,1}^3$. The standard SVR adopted the Gaussian kernel. In the separable-variable SVR, $K_\phi(\cdot,\cdot)$ was the Gaussian kernel and $K_\psi(\cdot,\cdot)$ was the linear kernel. The results are shown in Table 1.

---

[1] $\mathcal{N}_{3,40,20,1}^3$ which means that the network contains 3 layers and has 3 inputs, 1 output and 2 sets of nodes in the hidden layers, each containing 40, 20 nodes, respectively.

In this example, the identification error obtained by our method is smaller that the other methods.

Example 3. Given the following SISO system

$$y(k+1) = \frac{y(k)}{1+y^2(k)} + u^3(k) + e(k+1) \tag{8}$$

In the training process, let $u(k)$ with the amplitude uniformly distributed over the interval $[-2,2]$. While in the test process, Let $u(k) = \sin(2\pi k/25) + \sin(2\pi k/10)$, $k = 1,\cdots,100$.

The structure of the multi-layer FNN in this example is $\mathcal{N}_{2,40,20,1}^3$. The standard SVR used the Gaussian kernel. In the separable-variable SVR, both $K_\phi(\cdot,\cdot)$ and $K_\psi(\cdot,\cdot)$ are the Gaussian kernel. The experimental results are shown in Table 1.

Table 1. The identification results of three examples

Example	Method	Identification error
1	Multi-layer FNN	0.8286
	Standard SVR	0.4576
	Improved SVR	0.0584
2	Multi-layer FNN	0.2557
	Standard SVR	0.1494
	Improved SVR	0.0959
3	Multi-layer FNN	0.3129
	Standard SVR	0.1564
	Improved SVR	0.0557

5 Conclusion

An improved support vector regression is presented for identifying the nonlinear dynamic system in which the states and the control inputs could be separated. The experimental results on the nonlinear dynamic system described by (1) show that our method has better approximation ability than the standard SVR and FNNs.

So far the training of SVRs is a batch-algorithm and does not exist a recursive algorithm. Therefore the nonlinear system identification using SVRs only could use the series-parallel models. It is a drawback of SVRs compared with NNs. Dynamic identification methods based on the SVR will be the focus of our future work.

References

1. Nerendra, K.S., Parthasarathy, K.: Identification and control of dynamic systems using neural networks. IEEE Transactions on Neural Networks. L (1990) 4-27
2. Polycarpou, M.M., Ioannou, P.A.: Identification and control of nonlinear systems using neural network models: design and stability analysis. Technical report 91-09-01, Department of Electrical Engineering-Systems, University of Southern California (1991)
3. Ioannou, P.A., Sun, J.: Robust adaptive control. Englewood Cliffs, NJ: Prentice-Hall (1996)
4. Vapnik, V.: Statistical Learning Theory. John Wiley and Sons, Inc., New York (1998)
5. Schölkopf, B., Burges, C., Vapnik, V.: Extracting support data for a given task. In Proceedings, First International Conference on Knowledge Discovery & Data Mining. AAAI Press, Menlo Park, CA (1995)
6. Vapnik, V., Golowich, S., Smola, A.: Support vector method for function approximation, regression estimation, and signal processing. In Advances in Neural Information Processing Systems 9 (1997) 281-287
7. Drezet, P.: Support vector machines for system identification. UKACC International Conference on Control'98, 1 (1998) 688-692.
8. Smola, A., Schölkopf, B. A tutorial on support vector regression. NeuroCOLT Technical Report NC-TR-98-030, Royal Holloway College, University of London, UK (1998) Available http://www.kernel-machines.org/

Anomaly Detection Using Support Vector Machines

Shengfeng Tian, Jian Yu, and Chuanhuan Yin

School of Computer and Information Technology, Beijing Jiaotong University
Beijing, 100044, P.R.China
{sftian, jianyu}@center.njtu.edu.cn, xiaoyuehuan@sina.com

Abstract. In anomaly detection, we record the sequences of system calls in normal usage, and detect deviations from them as anomalies. In this paper, one-class support vector machine(SVM) classifiers with string kernels are adopted as the anomaly detector. A sequential learning algorithm for the classifiers is described. Two kinds of kernels are tested with an SVM classifier on the UNM data sets. Results indicate that the string kernel is superior to the RBF kernel for sequence data.

1 Introduction

Intrusion detection has emerged as an important approach to network security. There are two general approaches to intrusion detection: misuse detection and anomaly detection. In anomaly detection, there are many different methods to monitor activities on a computer system. Forrest et al. [1] demonstrated that effective approaches can be obtained by learning program behaviors and detecting deviations from this norm.

Forrest et al.'s approach characterized normal program behaviors in terms of sequences of system calls made by them. They break system calls sequences into substrings of a fixed length N. A separate database of normal behavior for each process of interest is built up. The sequences different from those in the database indicate anomalies in the running process.

Support vector machines, or SVMs, are widely used in pattern recognition problems. But the standard kernel functions can not provide sequence information and the edit distances between two strings calculated by approximate pattern matching algorithms are not valid kernel functions.

In this paper, we revisit the idea of using one-class SVM classifiers for anomaly detection and present a sequential learning algorithm to construct the classifiers. Instead of the traditional kernels, we propose a new string kernel for use in the SVM. Finally an SVM approach with string kernels is applied to anomaly detection using system call sequences.

2 Novelty Detection

In Schölkopf et al. [2], a support vector algorithm was used to characterize the support of a high dimensional distribution. With this algorithm, one can compute a set of contours which encloses the data points. These contours can be considered as normal data boundaries. The data outside the boundaries are interpreted as novelties. In the method suggested by Schölkopf et al. [2], the origin is treated as the only member of the second class after transforming the features via a kernel. They separate the image of one class from the origin. Let $x_1, x_2, \ldots, x_n$ be training examples, $\phi(x)$ be a kernel map which transforms the training examples to a high dimensional feature space. The task is to solve the following quadratic programming problem:

$$\min \tfrac{1}{2} w \cdot w + C \sum_{i=1}^{n} \xi_i - \rho \tag{1}$$

subject to $w \cdot \phi(x_i) \geq \rho - \xi_i$, $\xi_i > 0$ $i=1,\ldots,n$.

Here $C>0$ is the regularization constant and ξ_i are nonzero slack variables. If w and ρ solve the problem, then the decision function

$$f(x) = \text{sign}[w \cdot \phi(x) - \rho] \tag{2}$$

will be positive for most examples x_i contained in the training set.

The above quadratic programming problem leads to the dual problem

$$\min_{\alpha} \tfrac{1}{2} \sum_{i,j} \alpha_i \alpha_j K(x_i, x_j) \tag{3}$$

subject to $0 \leq \alpha_i \leq C$, $i = 1,\ldots,n$, and $\Sigma_i \alpha_i = 1$.

The decision function can be shown to be

$$f(x) = \text{sign}[\sum_{i=1}^{n} \alpha_i K(x_i, x) - \rho] \tag{4}$$

The offset ρ can be calculated by exploiting that for any $\alpha_i \in (0,C]$, the corresponding example x_i satisfies

$$\rho = \sum_{j} \alpha_j K(x_j, x_i) \tag{5}$$

Platt proposed a fast learning algorithm for two-class classification[3]: sequential minimal optimization(SMO), in which only two α_i's are optimized at each iteration. The method consists of a heuristic step for finding the best pair of parameters to optimize and use of an analytic expression to ensure the lagrangian increased monotonically. The idea is adopted in the above problem.

The dual formulation amounts to minimization of

$$W(\alpha) = \frac{1}{2}\sum_{i,j}\alpha_i\alpha_j K(x_i, x_j) \qquad (6)$$

respect to α_i and subject to $0 \leq \alpha_i \leq C$, $i = 1,\ldots,n$, and $\sum_i \alpha_i = 1$.
For each α_i, we have

$$\frac{\partial W}{\partial \alpha_i} = 2\sum_j \alpha_j K(x_j, x_i) = 2g_i \qquad (7)$$

Two variables α_i's are selected to be updated according to the following conditions: $g_i > 0$ and $\alpha_i > 0$ or $g_i < 0$ and $\alpha_i < 0$.

If the selected variables are α_i and α_p, the updated values will be $\alpha_i - h$ and $\alpha_p + h$ respectively to satisfy the linear equality constraint. The analytic solution can be obtained from $dW/dh = 0$:

$$h = \frac{g_i - g_p}{K_{ii} + K_{pp} - 2K_{ip}} \qquad (8)$$

where K_{ip} denotes the kernel $K(x_i, x_p)$. The variables α_i and α_p should be selected such that $|g_i - g_p|$ is maximized.

To fulfill the constraints, the flowing bounds should apply to α_i and α_p:
$$L = \max(0, \alpha_i + \alpha_p - C), \quad H = \min(C, \alpha_i + \alpha_p).$$
Then α_i and α_p should be updated as follows:

$$\alpha_i^{new} = \begin{cases} H & \text{if } \alpha_i - h \geq H \\ \alpha_i - h & \text{if } L < \alpha_i - h < H \\ L & \text{if } \alpha_i - h \leq L \end{cases} \qquad (9)$$

$$\alpha_p^{new} = \alpha_p + (\alpha_i - \alpha_i^{new}) \qquad (10)$$

The learning procedure is similar to the SMO. There are two separate choice heuristics: one for α_i and one for α_p. The choice of α_i provides the outer loop of the algorithm. The outer loop iterates over the entire training set, determining an example as α_i which satisfies the inequality conditions. The inner loop iterates over the entire training set, determining an example α_p, which satisfies the inequality conditions and maximizes the value $|g_i - g_p|$. Two Lagrange multipliers are updated according to the above formulas. The learning procedure terminates when the value $|g_i - g_p|$ is smaller than a predefined stopping threshold.

3 String Kernels in SVMs

The kernel function computes the inner product between mapped examples in the feature space. The feature vectors are not calculated explicitly but instead the kernel is computed directly from two examples.

Given a number k≥1, the spectrum kernel concerns the k-length continuous subsequences shared in two strings[4]. The (k,m)-mismatch kernel is similar to spectrum kernel, but at most m mismatches are allowed[5]. In the kernel developed by Lodhi et al.[6], the gaps in the occurrence of the k-length subsequences are allowed.

For intrusion detection, we have three requirements to string kernels. Firstly string kernels are similarity measures between strings which are assessed by number of (possibly non-continuous) matching subsequences shared by two strings, non-continuous occurrences should be penalized. Secondly the feature map is indexed not only by k-length subsequences but all possible subsequences from Σ. Finally the kernel computation is efficient.

To meet the above requirements, we propose a new string kernel. Let Σ be a finite set which we call the alphabet, $x \in \Sigma^*$ denote string defined over the alphabet Σ, $|x|$ the length of x. The neighborhood N(x) generated by x is the set of all subsequences that x contains. Given a number q≥0, the feature map $\Phi(x)$ is defined as

$$\Phi(x) = (\phi_s(x))_{s \in \Sigma^*} \qquad (11)$$

$$\phi_s(x) = \begin{cases} \sqrt{q} & \text{s is empty string} \\ 2^{|s|-|x|} & \text{s belongs to N(x)} \\ 0 & \text{otherwise} \end{cases} \qquad (12)$$

The kernel K(x,y) is defined as $K(x,y) = \Phi(x) \cdot \Phi(y)$. To ensure the kernel in the interval [0,1], the normalized kernel is defined as

$$K^s(x,y) = [K(x,x)K(y,y)]^{-1/2} K(x,y) \qquad (13)$$

The introduction of parameter q in the string kernel is important to classifying performance just as to the function of parameter σ in the RBF kernel.

4 Experiments

In the experiments, the data are taken from the UNM data sets for sendmail which are available at http://www.cs.unm.edu. The traces were obtained at UNM using Sun SPARCstations running SunOs 4.1.1 and 4.1.4. We use "plus" as the normal data set and build up a database of all unique sequences of a given length k=6. For example, suppose we observe the following trace of system calls:
      wait4 open creat link creat write mknod open.

Considering that there are a total of 182 system calls in the SunOS 4.1.x operating system, each system call is assigned an identifying number ranging from 1 to 182. The trace of system calls is changed to a number sequence:

   7 5 8 9 8 4 14 5.

For k=6, we get the unique sequences:

   7 5 8 9 8 4, 5 8 9 8 4 14, 8 9 8 4 14 5.

In the experiments, the test data sets consist of normal sets including "bounce", "bounce-1", "bounce-2", and "queue", and anomalous sets including "sm-280", "sm-314", "sm-10763", "sm-10801" and "sm-10814". We build up databases of all unique sequences of a given length k=6. The support vector machines are trained with the algorithm derived in Section 2. Two kernel functions are adopted in the algorithm. One is the string kernel $K^s(x,y)$, the another is the standard RBF kernel $K^r(x,y) = \exp[-\|x-y\|^2/\sigma^2]$. The results are shown in Table 1. With the RBF kernel, the best detection performance is achieved when the parameter σ is equal to 100. The minimal difference of novelty numbers between normal data sets and anomalous data sets is 10. With the string kernel, the best detection performance is achieved when the parameter q is equal to 7. The minimal difference of novelty numbers between normal data sets and anomalous data sets is 42. Obviously, the string kernel has the advantage of the RBF kernel in the case of sequence data. The reason why the string kernel is better than the RBF kernel for sequence data is that the sequence information is counted in the string kernel. For example, suppose that there are two sequences as follows:

 (1) 1 92 3 94 5 96 (2) 96 1 92 3 94 5.

The two sequences are similar to each other for the string kernel because of the common subsequence "1 92 3 94 5", but are quite different for the RBF kernel.

5 Conclusion

Support vector classifiers using string kernels are suitable to many sequence matching problems such as protein classification, text categorization and intrusion detection. Because command subsequences have different lengths and attackers always insert spurious letters to avoid detection, the string kernel defined in this paper is a good choice in intrusion detection.

Table 1. The detection results of SVMs with parameters C=1, σ=100 and q=7

Data sets	No. of novelties with K^r	No. of novelties with K^s	Normal/Anomalous
bounce	14	2	Normal
bounce-1	11	1	Normal
bounce-2	17	3	Normal
queue	11	1	Normal
sm-280	27	45	Anomalous
sm-314	27	46	Anomalous
sm-10763	28	50	Anomalous
sm-10801	28	50	Anomalous
sm-10814	28	50	Anomalous

References

1. Forrest, S., Hofmeyr, S.A., Somayaji, A.: Intrusion detection using sequences of system calls. Journal of Computer Security, Vol.6 (1998) 151-180
2. Schölkopf, B., Platt, B. J.C., Shawe-Taylor, J., Smola, A.J.: Estimating the support of a high-dimensional distribution. Technical report MSR-TR-99-87, Microsoft Research (1999)
3. Platt, J.C.: Sequential minimal optimization: a fast algorithm for training support vector machines. Technical Report MSR-TR-98-14, Microsoft Research (1998)
4. Leslie, C., Eskin, E. Noble, W.S.: The spectrum kernel: a string kernel for SVM protein classification. In Proceedings of the pacific biocomputing Symposium (2002)
5. Leslie, C., Eskin, E., Weston, J., Noble, W.S.: Mismatch string kernels for SVM protein classification. In: Proceedings of Neural Information Processing Systems (2002) http://www.gs.washington.edu/~noble/papers/leslie_mismatch.pdf
6. Lodhi, H., Saunders, C., Shawe-Taylor, C., Cristianini, N., Watkins, C.: Text classification using string kernels, Journal of Machine Learning Research, Vol.2 (2002) 419-444

Power Plant Boiler Air Preheater Hot Spots Detection System Based on Least Square Support Vector Machines

Liu Han, Liu Ding, Jin Yu, Qi Li, and Yanming Liang

School of Automation & Information Engineering, Xi'an University of Technology, Xi'an
710048, China
{liuhan,liud,liqi}@xaut.edu.cn

Abstract. Air preheater is the important heat exchanger in power plant units. Recombustion accident can be caused by inadequacy combustion of fuel or badly heat-dispersed condition aroused by low air or gas velocity after boiler outage. In the paper, discriminant models of 3 pairs of fire status have been built based on Least Square Support Vector Machines (LS-SVMs) for two kinds of kernel functions. Utilizing polynomial and RBF kernel, the hyperparameters of classifiers were tuned with Leave-one-out(LOO) cross-validation. Receiver Operating Characteristic(ROC) curve comparison shows that LS-SVMs classifiers are able to learn quite well from the raw data samples. Experiment results show that SVMs has good classification and generalization ability and RBF kernel function has more accurate than polynomial kernel function for this problem from the area under the ROC curve (AUC) values of two kernel functions.

1 Introduction

Power plant air preheater is the important heat exchanger using surplus heat of gas from boiler to heating up air. The function of air preheater is to increase temperature of air that system of combustion and coal milling of boiler need as well as to decrease temperature of gas leaving the air preheater and heat loss. In China, rotary air preheater currently has been adopted in boiler of 200MW capacity or above because of characteristic of compact structure, light weight and easily installation. Commonly, clean air preheater is impossible to fire because it is just a heat exchanger. The combustion of air preheater aroused from many factors. But fundamental causation is that superfluous fuels including carbon and oil stockpiled on air preheater heat transfer elements have been oxidated and temperature is rising. When the temperature reach fire point , recombustion is occurred. The conditions of air preheater combustion are composed of superfluous fuel and badly heat-dispersed condition aroused by low air or gas velocity after boiler outage. Inadequately combusted fuel congealed on air preheater heated elements and deposit will be dried and fired on temperature of 350°C. It is difficult to be discovered because of small combustion area during the period of initial stage of combustion. Continuously, combustion keeps on going and heat transfer elements even or full air preheater will be fired when temperature goes on 700°C. The combustion is called air preheater recombustion.

It is proven according to practical experiments that abnormal increase of temperature of gas inlet and outlet duct, air inlet and outlet duct can reflect coindication of air preheater recombustion. However, once status of abnormal increase of temperature appears, combustion has already come into being a certain extent. So it is necessary to introduce the air preheater hot spots detection system for decreasing harm of combustion minimum. At present, there are air preheater hot spots detection systems to be researched and applied [1][2]. In these systems, thermocouple and infrared sensor are adopted as thermoscope and combustion is judged by comparison of measured temperature value and given alarm threshold. But choice of given alarm threshold strongly depends on experiences and spots situation, which easily results in misinformation. At the same time, artificial neural network has been applied successfully in fire detection [3][4]. However, in many applications neural network has complicated structure and overfull estimated parameters relative to less data samples. As a result, neural network 'overfitting' for data samples appears as well as the generalization ability of neural network is not sufficient. Hot spots detection method based neural network restricts detection accuracy farther rising.

Statistical learning theory (SLT)[5] was introduced by Vapnik in 1995 which focus on machine learning approach in small sample set and support vector machines (SVMs) was a new classification and regression tool based on this theory, which has self-contained basis of statistical learning theory and excellent learning performance. SVMs has already become a new research hotspot in the field of machine learning and has been successfully applied in many fields, such as face identification, handwriting recognition and text classification. In this paper, air preheater hot spots detection system based on support vector machines is proposed. The system combined with engineering experiences for eight years has been gone into many practical applications in boiler of 200MW capacity or above successfully in China or abroad.

2 The Principle of Support Vector Machines and Least Square Support Vector Machines

The principle of empirical risk minimization is essential idea to solve the problem of statistic pattern recognition and statistic machine learning all along. But it is found in practice that excellent generalization ability cannot be achieved to make empirical risk minimization (training error minimum). Especially, excessively small training error may result in poor generalization ability, which is a general case for neural network and called over-fitting. Since postulate of conventional statistic theory is infinite training samples number, in small samples condition the principle of empirical risk minimization based on conventional statistics cannot deduce the principle of expectation risk minimization educed by Bayes Decision Theory. In order to apply conventional statistics in small samples condition, Vapnik built statistics learning theory. In this methodology, excellent generalization ability can be achieved only when both empirical risks and learning machine capacity (VC Dimension) are controlled simultaneously [5].

Support vector machines (SVMs) is based on statistics learning theory. The formulation of SVMs embodies the structure risk minimization (SRM) principle. SVMs has gained wide acceptance due to high generalization ability for a wide range of applications and better performance than other traditional learning machines. In a SVMs, the original input space is mapped into a high dimensional dot product space via a kernel. The new space is called feature space, in which an optimal hyperplane is determined to maximize the generalization ability. The optimal hyperplane can be determined by only few exemplars points called support vectors. SVMs is an approximate implementation of structure risk minimization, which minimizes an upper bound on the error rate of a learning machine on test exemplars instead of minimizing the training error itself as used in neural network-based empirical risk minimization.

Consider a given training set $\{x_i, y_i; i=1,\cdots l\}$ with input data $x_i \in \mathbb{R}^n$ and corresponding binary class labels $y_i \in \{-1, +1\}$. The input data are mapped to a higher dimensional feature space by nonlinear function $\Phi(\cdot)$ as in the classifier case. The classifier takes the form

$$f(x) = sign[\omega \cdot \Phi(x) + b] \tag{1}$$

Where the term ω denotes weighted vector and the term b denotes a bias term.

In this paper, Vapnik's support vector machines is called standard support vector machines. So this leads to the optimization problem for standard SVMs

$$\min_{\omega, b, \xi} \frac{1}{2}\omega^T \omega + \gamma \sum_{i=1}^{l} \xi_i \tag{2}$$

Subject to

$$\begin{cases} y_i[\omega^T \Phi(x_i) + b] \geq 1 - \xi_i \\ \xi_i \geq 0, \ i=1,\cdots,l \end{cases} \tag{3}$$

The variables ξ_i are slack variables which are needed in order to allow misclassifications in the set of inequalities and γ is a positive real constant and should be considered as a tuning parameter in the algorithm.

For Least Square Support Vector Machines (LS-SVMs)[6], above formulation has been modified as follows

$$\min_{\omega, b, \xi} \frac{1}{2}\omega^T \omega + \gamma \frac{1}{2} \sum_{i=1}^{l} \xi_i^2 \tag{4}$$

Subject to the equality constraints

$$y_i[\omega^T \Phi(x_i) + b] = 1 - \xi_i \quad i=1,\cdots,l \tag{5}$$

Important differences with standard SVMs are the equality constraints and the sum squared error term, which greatly simplifies the problem. The solution is obtained after constructing the Lagrangian $L = -\sum_{i=1}^{l} \alpha_i (y_i[\omega^T \Phi(x_i) + b] - 1 + \xi_i) + 1/2\omega^T \omega + 1/2\gamma \sum_{i=1}^{l} \xi_i^2$ with α_i are the Lagrange multipliers. According to KKT Conditions

[5], the conditions for optimality are given by $\omega = \sum_{i=1}^{l} \alpha_i y_i \Phi(x_i)$, $\sum_{i=1}^{l} \alpha_i y_i = 0$, $\alpha_i = \gamma \cdot \xi_i$ and $y_i [\omega^T \Phi(x_i) + b] - 1 + \xi_i = 0$. Elimination of ω and ξ gives

$$\begin{bmatrix} 0 & Y^T \\ Y & \Omega + \gamma^{-1} I \end{bmatrix} \begin{bmatrix} b \\ \alpha \end{bmatrix} = \begin{bmatrix} 0 \\ \vec{1} \end{bmatrix} \tag{6}$$

with $Y = [y_1; \cdots; y_i]$, $\vec{1} = [1; \cdots; 1]$ and $\alpha = [\alpha_1; \cdots; \alpha_i]$. Mercer's theorem [5] is applied to the matrix Ω with

$$\Omega_{kh} = y_k y_h \Phi(x_k)^T \Phi(x_h) = y_k y_h \cdot K(x_k, x_h), k, h = 1, \cdots, l \tag{7}$$

A positive definite kernel $K(\cdot, \cdot)$ is chosen such that is satisfied Mercer condition. LS-SVMs classifiers with kernel are able to realize strongly nonlinear decision boundaries such as occurring in binary classification problems.

3 Hot Spot Detection System Based on LS-SVMs

3.1 Data Samples

In this research, data samples are acquired from Plant A of 200MW in Shandong Province, Plant B of 300MW in Guizhou Province and Plant C of 300MW in Henan Province in China. Input vector in data samples is five dimensional vectors, which includes 5 temperature points from thermocouples and infrared sensors. In order to exactly decide fire status, we separate air preheater fire status from 3 levels corresponding to 'Level 1' that represents combustion alarms, 'Level 2' that represents combustion pre-alarms and 'Level 3' no alarm, respectively. Table 1 shows number of row data samples from different power plants.

Table 1. Number of data samples of Level 1, Level 2 and Level 3. The rows correspond to the acquisition source, while the columns mention the type of fire status.

	Level 1	Level 2	Level 3	Total
Plant A	13	14	12	39
Plant B	23	8	11	42
Plant C	17	24	8	49
Total	53	46	31	130

3.2 Experiments and Analysis

In this research, three types of fire status need to classified, a multiclass classifier is needed. Three binary LS-SVMs classifiers are constructed and applied to those three pairs, namely Level 1 vs. Level 2, Level 1 vs. Level 3 and Level 2 vs. Level 3. The LS-SVMs classifiers were set up and two kernels, namely polynomial and RBF kernels, were used. For each kernel, the hyparemeters were tuned to achieve the best leave-one-out(LOO) generalization performance. In this experiment, the hyperparameter γ for the three classifiers with polynomial kernel is optimized at the value of 1. In RBF kernel two hyperparameters are used. Besides γ as a fitting constant, σ controls the width of RBF kernel. Tuned parameters is shown in Table 3. The experiment consists of following steps:

1. the data is divided in a training set(2/3 of data) and a test set(remainder);
2. train the LS-SVMs and use the test set to evaluate the performance;
3. the index of the misclassification is noted.

Table 2 and Table 3 show the results of training and testing of LS-SVMs classifiers with polynomial kernel and RBF kernel respectively.

Table 2. Results of training and testing of LS-SVMs classifiers with polynomial kernel

Index	Level 1 vs. Level 2	Level 1 vs. Level 3	Level 2 vs. Level 3
Number of training data	66	56	52
Number of testing data	33	28	25
γ	1	1	1
Training average error	6.8310	4.1110	5.1023
Training standard deviation	1.2230	1.3129	1.4325
Training ratio of correct classification	89.49%	92.30%	90.11%
Testing average error	4.2746	3.1212	3.9809
Testing standard deviation	1.8513	1.3743	0.7769
Testing ratio of correct classification	87.01%	89.21%	83.77%

Table 3. Results of training and testing of LS-SVMs classifiers with RBF kernel

Index	Level 1 vs. Level 2	Level 1 vs. Level 3	Level 2 vs. Level 3
Number of training data	66	56	52
Number of testing data	33	28	25
γ	0.7576	0.6598	0.4815
σ	3.3248	4.3780	1.7728
Training average error	5.5507	3.0702	4.8309
Training standard deviation	1.0287	1.1760	1.4325
Training ratio of correct classification	91.53%	94.53%	90.77%
Testing average error	3.3675	2.3064	3.9913
Testing standard deviation	1.7610	1.2918	0.7651
Testing ratio of correct classification	89.71%	91.73%	84.01%

The method of Receiver Operating Characteristic(ROC) curves[7] has been used to compare the performance of different classifiers. Each of LS-SVMs classifiers described above will provide predictions for the given test data. The decisions that correspond to the LS-SVMs classifier's output '+1' belong to the positive class. Those that correspond to output '-1' belong to the negative class. By varying a threshold on these sorted decisions we can construct a Receiver Operating Characteristic curves. A ROC curve is a plot of the true positive rate against the false positive rate for the different possible thresholds. Here the ture positive rate is the fraction of the positive instances for which the system predicts 'positive'.The false positive rate is the fraction of the negative instances for which the system erroneously predicts 'positive'. The large the area under the curve(the closer the curve follows the left and top border of the ROC space), the more accurate the test. The expected curve for a system making random predicitons will be a line on the 45-degree diagonal. The evaluation metric we use will be the area under the ROC curve(AUC). The ROC curve for a perfect system has an area of 1. Table 4 summarizes the results obtained from the experiment. It compares the performance of the two kernels used applied to each binary classification problem. As can be seen, for the most of the cases, the LS-SVMs classifiers successfully separate the two classes.

Table 4. Comparision of LS-SVMs classification using polynomial and RBF kernel

	RBF kernel			polynomail kernel		
class pair	L1 vs. L2	L2 vs. L3	L1 vs. L3	L1 vs. L2	L2 vs. L3	L1 vs. L3
AUC	0.8507	0.7954	0.8613	0.8220	0.7902	0.8317
% correct	89.71	84.01	91.73	87.01	83.77	89.21

4 Conclusions

LS-SVMs classifiers have been applied to detection air preheater fire status based on temerature values. Utilizing polynomial and RBF kernel, the hyperparameters of classifiers were tuned with leave-one-out(LOO) cross-validation. ROC curve comparison shows that LS-SVMs classifiers are able to learn quite well from the raw data samples. This shows that the LS-SVMs classifiers generalize well depite that the fact that no feature selection was applied. Future work will focus on classification based on information fusion approach that image, gas physical characteristic like density and so on in air preheater are considered synthetically to get more accurate decisions.

References

1. Liu, H., Liu, D., Li, Q., Shi, W.: Research on Power Plant Boiler Air Preheater Fire Alarm System. Application of Electronic Technique, Vol.24(**6**) (1998) 35-36
2. Yin, G. D.: Hot Point Inspection System of Rotary Air Preheater. Turbine Technology **6** (2003) 137-138
3. Zhang, B. K., Wu, L. B., Wang J. J.: Study on Fuzzy Neural Network for Fire Detection. Journal of Electronics, Vol.22(**4**) (2000) 687-691

4. Wang, X. H., Xiao, J. M., Bao, M. Z.: Ship Fire Detection Based on Fuzzy Neural Network and Genetic Algorithm. Chinese Journal of Scientific Instrumnet, Vol.(**3**) (2001) 312-314
5. Vapnik, V.: The Nature of Statistical Learning Theory. Springer-Verlag, New York(1999)
6. Suykens, J. A. K., Vandewalle, J.: Least Squares Support Vector Machine Classifiers. Neural Processing Letter, Vol. 9(**3**) (1999) 293-300
7. Kwokleung, C., Lee, T. W., Pamela, A. S., Michael, H. G., Robert, N. W., Terrence, J. S.:. Comparison of Machine Learning and Traditional Classifiers in Glaucoma Diagnosis. IEEE Trans. on Biomedical Engineering, Vol. 49(**9**) (2002) 963-974

Support Vector Machine Multiuser Detector for TD-SCDMA Communication System in Multipath Channels

Yonggang Wang[1,2], Licheng Jiao[1], and Dongfang Zhao[2]

[1] Lab. for Intelligent Information Processing, Xidian University, 710071 Xi'an, China
yg.wang@263.net
[2] Air Force Engineering University, 710068 Xi'an, China

Abstract. TD-SCDMA, a part of the third generation mobile communication standards IMT2000 and UMTS is described. A nonlinear multiuser detector based on Support Vector Machine (SVM) is presented and the performance is compared with traditional linear detector. The presented detector can approach optimum multiuser detector than the other linear detectors by using the Midamble of TD-SCDMA. The results of simulation show the feasibility and validity of this detector in multipath channel.

1 Introduction

In the third generation of mobile communication systems code division multiple access (CDMA) will play an important role. Several CDMA based systems have already been standardized by ITU.TD-SCDMA is a special variant of TD-CDMA, which is mainly developed in the Peoples Republic of China, and is now included as a narrowband TDD variant for UMTS. In order to increase the capacity of the system, multiple antennas at the base station and more accurate time synchronization in the uplink have been included in the standard. One of the major problems of CDMA based communication systems is the multiple access interference (MAI) caused by spreading codes that are nonorthogonality due to multipath propagation. MAI can be reduced by multiple antennas and by an appropriate synchronization. Nevertheless, for critical channels, the only way to cope with the problem is multiuser detection (MUD). In the standard for TD-SCDMA linear MUD (joint detector) is proposed. In this paper we derive and investigate a nonlinear MUD based on Support Vector Machine (SVM) [3]. The SVM MUD is a non-blind detector. The algorithm utilizes the midamable of TD-SCDMA system to train the SVM. The presented SVM detector can provides a performance, which is generally close to the optimum MUD than the other linear detectors, while keeping the computational complexity low.

2 TD-SCDMA Communication System

TD-SCDMA is an improved narrowband variant of TD-CDMA. As its name suggests, the TD-SCDMA standard bears two major characteristics: one is to adopt time division duplex (TDD) mode operation for up and downlinks separation; the other is to use synchronous CDMA technology. The use of those techniques in TD-SCDMA offers many benefits. In TD-SCDMA the uplink time synchronization has been improved further. The dominating paths of each user arrive at the same time at the base station. Since orthogonal codes are used, a large portion of the MAI can be avoided. TD-SCDMA makes use of both TDMA and CDMA techniques such that channelization in TD-SCDMA is implemented using both time slots and signature codes to differentiate mobile terminals in a cell.

Table 1. Air Interface Parameters

Parameters	Values
Frame duration T_F	5ms
Burst duration T_{BU}	$675 \mu s$
Midamble chips L_m	144
Guard period T_{GP}	$75 \mu s$
Data symbols per data block N	22
Symbol duration T_S	$12.5 \mu s$
Chips per symbol Q	16
Chip duration T_C	$1/1.28 \mu s$
Interleaving depth I_D	4 frames/speech
Coder constraint length L_C	9
Rate R_C	1/3

The frame of TD-SCDMA is composed of 7 main time slots and 2 special time slots allocated around the switching point and used for synchronization purposes. The 7 main time slots can be arbitrarily distributed between uplink and downlink. The only restriction is that uplink and downlink have at least one time slot. This makes TD-SCDMA not only suitable for symmetric voice services, but also for the asymmetric data services (multimedia and internet downloads). The main time slot is composed of two data blocks with $N = 22$ QPSK data symbols spread by a code of length $Q = 16$ which defines the subchannel, a user specific Midamble for channel estimation of length $L_m = 144$ chips, and a guard interval of 16 chips. In table 1 he main air interface parameters of TD-SCDMA are summarized.

3 Signal Model

Due to the midamble and the guard period, the detection of each data block can be considered separately. The optimum receiver for each field of data symbols consists of a bank of channel matched filters, followed by an optimal MUD [1]. Combining the transmit symbols $b_m(k)$ of all subchannels $m = 1, \cdots, M$ at one discrete time instant k into the transmit vector:

$$\mathbf{b}(k) = [b_1(k), b_2(k), \cdots, b_M(k)]^T \tag{1}$$

And similarly defining the vector of received symbols after the matched filters we obtain:

$$\mathbf{r}(k) = \mathbf{P}(k) \begin{bmatrix} \mathbf{b}(k) \\ \mathbf{b}(k-1) \\ \vdots \\ \mathbf{b}(k-L+1) \end{bmatrix} + \mathbf{n}(k) \tag{2}$$

where the $M \times LM$ system matrix is given by

$$\mathbf{P} = \mathbf{S}^T \mathbf{H} \begin{bmatrix} \mathbf{SA} & 0 & \cdots & 0 \\ 0 & \mathbf{SA} & & \vdots \\ \vdots & & \mathbf{SA} & 0 \\ 0 & \cdots & 0 & \mathbf{SA} \end{bmatrix} \tag{3}$$

the signature sequence matrix $\mathbf{S} = [\mathbf{s}_1, \cdots, \mathbf{s}_M]$, and $\mathbf{s}_i = [s_{i,1}, \cdots, s_{i,Q}]^T$ is the signature sequence of the user i. The diagonal signal amplitude matrix $\mathbf{A} = \mathrm{diag}\{A_1, \cdots, A_M\}$. The $M \times LM$ channel impulse response (CIR) matrix $\mathbf{H}$ has the form:

$$\mathbf{H} = \begin{bmatrix} h_0 & h_1 & \cdots & h_{n_h-1} & & & & \\ & h_0 & h_1 & \cdots & h_{n_h-1} & & & \\ & & \ddots & \ddots & \ddots & \ddots & & \\ & & & h_0 & h_1 & \cdots & h_{n_h-1} \end{bmatrix} \tag{4}$$

Orthogonal signature sequences are used in TD-SCDMA, so that the noise vector $\mathbf{n}(k)$ has a variance of $E[\mathbf{n}(k)\mathbf{n}^T(k)] = \sigma_n^2 \mathbf{I}$. We note that the orthogonality of the codes is destroyed by the channel-induced intersymbol interference (ISI). The ISI span depends on the length of the CIR, n_h, related to the length of the chip sequence, Q. For $n_h = 1, L = 1$; for $1 < n_h \leq Q, L = 2$; for $Q < n_h \leq 2Q, L = 3$; and so on.

4 Linear and Optimal Detectors

Joint detection, which is proposed in the standard as MUD, is linear detector. The linear MUD for user i has the form:

$$\hat{b}_i(k) = \mathrm{sign}(y_L(k)) \text{ with } y_L(k) = \mathbf{w}^T \mathbf{r}(k) \tag{5}$$

where $\mathbf{w}$ denotes the detector's weight vector. The most popular solutions are the decorrelating detector and the MMSE solution:

$$\hat{b}_i^{DD} = \text{sign}\left\{\left(\mathbf{PP}^T\right)^{-1}\mathbf{p}_i\mathbf{r}(k)\right\} \qquad (6)$$

$$\hat{b}_i^{MMSE} = \text{sign}\left\{\left(\mathbf{PP}^T + \sigma^2\mathbf{I}\right)^{-1}\mathbf{p}_i\mathbf{r}(k)\right\} \qquad (7)$$

where $\mathbf{p}_i$ denotes the ith column of $\mathbf{P}$.

The linear detector is computationally very simple, and the standard LMS or RLS algorithms can be used to implement adaptively. But a linear MUD only performs adequately in certain situations. Let the N_b possible combinations of $[\mathbf{b}^T(k), \mathbf{b}^T(k-1), \cdots, \mathbf{b}^T(k-L+1)]^T$ be $\mathbf{b}^{(j)} = [\mathbf{b}^{(j)T}(k), \mathbf{b}^{(j)T}(k-1), \cdots, \mathbf{b}^{(j)T}(k-L+1)]^T$ and $b_i^{(j)}$ is the ith element of $\mathbf{b}^{(j)}(k)$. Let us define the set of the N_b noise-free received signal states as:

$$\Re = \left\{\mathbf{r}_j = \mathbf{P}\mathbf{b}^{(j)},\ 1 \le j \le N_b\right\} \qquad (8)$$

where $\Re$ can be partitioned into two subsets:

$$\Re_\pm = \left\{\mathbf{r}_j \in \Re : b_i^{(j)} = \pm 1\right\} \qquad (9)$$

If $\Re_+$ and $\Re_-$ are not linearly separable, a linear MUD will exhibit an irreducible error floor even in the noise-free case, as it can only form a hyperplane in the M dimensional received signal space.

Applying the Bayesian classification theory it can be shown that the optimal detector has the form [1]:

$$y_B(i) = f_B(\mathbf{r}(i)) = \sum_{j=1}^{N_b} \beta_j b_i^{(j)} \exp\left(-\frac{\|\mathbf{r}(i) - \mathbf{r}_j\|^2}{2\sigma_n^2}\right) \text{ with } \hat{b}_i(k) = \text{sign}(y_B(k)) \qquad (10)$$

where $b_i^{(j)} \in \{\pm 1\}$ serve as class labels.

5 The Support Vector Machine Detector

The optimal detector (10) requires the knowledge of all the noise-free signal sates, which are unknown to receiver. In practice the receiver can use the midamble of TD-SCDMA to get a block of training samples $\{\mathbf{r}(k), b_i(k)\}_{k=1}^{L_m=144}$. Let us denote the training set of L_m noisy received signal vectors as $\aleph = \{\mathbf{x}_k = \mathbf{r}(k), 1 \le k \le L_m\}$ and the set of corresponding class labels as $\mathbf{C} = \{c_k = b_i(k), 1 \le k \le L_m\}$. Applying the standard SVM method [2], an SVM detector can be constructed for user i:

$$y_{SVM}(k) = \sum_{j=1}^{L_m} \overline{g}_j c_j K(\mathbf{r}(k), \mathbf{x}_j) + \overline{\eta} \qquad (11)$$

where $K(\cdot)$ is the kernel function of SVM and the set of Lagrangian multipliers $\{\overline{g}_j\}$, denoted in vector form as $\overline{\mathbf{g}} = [\overline{g}_1, \cdots, \overline{g}_{L_m}]^T$, is the solution of the quadratic programming (QP)

$$\overline{\mathbf{g}} = \arg\min_{\overline{\mathbf{g}}} \left\{ \frac{1}{2} \sum_{j=1}^{L_m} \sum_{l=1}^{L_m} g_j g_l c_j c_l K(\mathbf{x}_j, \mathbf{x}_l) - \sum_{j=1}^{L_m} g_j \right\} \quad (12)$$

with the constraints $0 \leq g_j \leq C$ and $\sum_{j=1}^{L_m} g_j c_j = 0$. In this application it was found advantageous to choose the Gaussian kernel function:

$$K(\mathbf{x}_j, \mathbf{x}_l) = \exp\left(-\frac{\|\mathbf{x}_j - \mathbf{x}_l\|^2}{2\rho^2}\right) \quad (13)$$

where the width parameter is related to the root mean square σ_n of the channel noise, an estimate of which can be obtained. The offset constant $\overline{\eta}$ is usually determined from the socalled "margin" SVs, i.e. from those particular $\mathbf{x}_j$ s, for which the corresponding Lagrangian multipliers obey $0 < g_j < C$. Because the optimal decision boundary, defined by $\{\mathbf{r}: f_B(\mathbf{r}) = 0\}$, passes through the origin of the received signal space and possesses certain symmetric properties due to the symmetric structure of $\mathfrak{R}_+$ and $\mathfrak{R}_-$, $\overline{\eta} = 0$ can be used. The user-defined parameter C controls the tradeoff between model complexity and training error. In our application, we will choose it empirically.

The set of SVs, denoted by $\aleph_{SVM}$, is given by those particular $\mathbf{x}_j$ s, which have non-zero Lagrangian multipliers obeying $0 < g_j < C$, where $\aleph_{SVM}$ is usually a small subset of the training data set $\aleph$. These SVs are determined during the training process. Thus the SVM-based MUD requires computing the decision variable

$$y_{SVM}(k) = \sum_{\mathbf{x}_j \in \aleph_{SVM}} \overline{g}_j c_j \exp\left(-\frac{\|\mathbf{r}(i) - \mathbf{r}_j\|^2}{2\rho^2}\right) \text{ with } \hat{b}_i(k) = \mathrm{sign}(y_{SVM}(k)) \quad (14)$$

6 Simulation Results

In this section we compare the performance of different detectors:1) decorrelating detector, 2) MMSE detector, 3) SVM detector, by extensive simulations of uplink TD-SCDMA communication system in multipath fading channel. The Parameters of simulations are shown in table 1.

In Figure 1 the bit error rate (BER) versus signal-to-noise rate (SNR) is depicted. Note that decorrelating detector (DD) and MMSE detector resulted absolutely unusable. The presented detector, SVM MUD, outperform the others detectors.

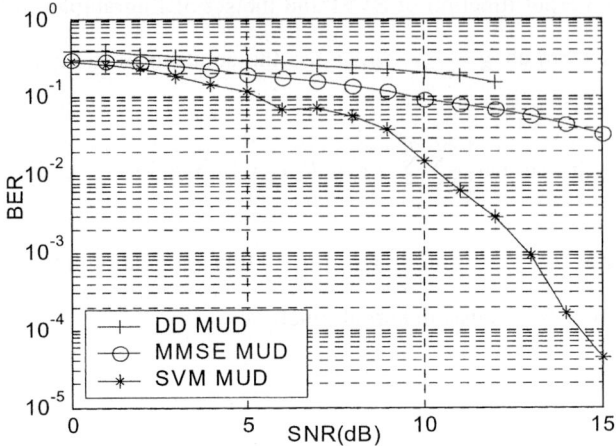

Fig. 1. BER Vs. SNR. Parameters: 8users, 2 codes per user, 3-path fading channel.

7 Conclusions

In this paper, the third generation mobile communication system TD-SCDMA has been introduced. A novel nonlinear multiuser detector based on support vector machine for TD-SCDMA has been proposed. The presented detector resulted in batter performance than traditional linear detector. This, together with the use of other key techniques, such as smart antenna, will make a great increase of the system capacity.

References

1. Verdu, S.: Multiuser Detection. Cambridge Univ. Press, Cambridge UK (1998)
2. Vapnik, V.: The Nature of Statistical Learning Theory. Springer-Verlag, Berlin Heidelberg New York (1995)
3. Chen, S., Samingan, A.K., Hanzo, L.: Support vector machine multiuser receiver for DS-CDMA signals in multipath channels. IEEE Trans. on Neural Networks, vol. 12. IEEE Press, New York (1999) 604-611

Eyes Location by Hierarchical SVM Classifiers

Yunfeng Li and Zongying Ou

CAD & CG Laboratory, School of Mechanical Engineering
Dalian University of Technology, Dalian 116024, P. R. China
yunfengli2004@tom.com, ouzyg@dlut.edu.cn

Abstract. This paper presents a method for eyes location using a two-level hierarchy of SVM (Support Vector Machines) classifiers. On the first level, a two-eye region classifier is obtained by training the SVM using grayscale projections of the two-eye region images. Utilizing this classifier, the region where the two eyes lie can be located by searching the whole face image. On the second level, the left and right eye classifier are obtained by training SVM using grayscale of left and right eye images respectively. Using these two classifiers, the two eyes can be precisely located by searching the output region of the first level. Experimental results show that this method is sufficiently generic and can cope with more various image conditions than exiting techniques.

1 Introduction

Human eyes have always played the most important role during the process of face recognition, eyes location is a prerequisite to almost all automatic face recognition algorithms. During the last decade, numerous attempts have been made to detect and localize eyes, the theories they have been used include image segmentation[1][2][3], grayscale projection[4], edge detection[1][3][5][6], template matching[7] and deformable template matching[5][6], etc. Most of the existing eyes location methods are the combination of one or more of the above theories, those methods emphasize particularly on the use of edge information of eyelids and irises. So the traditional eyes location methods restrict their applications to front face and open eyes. Under general conditions, because of the influences brought by close eyes, spectacles, pose and lighting, etc, eyes patterns become complex and instable, so an effective eyes location algorithm should be robust enough to cope with these problems.

2 Overview of the System

This paper proposes an SVM-based eyes location algorithm which is realized by a two-level classification procedure. On the first level, a two-eye region classifier is obtained by training the SVM using grayscale projections of the two-eye region images. Utilizing this classifier, the region where the two eyes lie is located by searching the whole face image. On the second level, the SVM is trained by using the grayscale of the left and right eye images respectively, left and right eye can be located precisely through searching the left and right part of the exported region from the first level. The diagrammatic description of the operation of the eye location system is shown in Fig. 1.

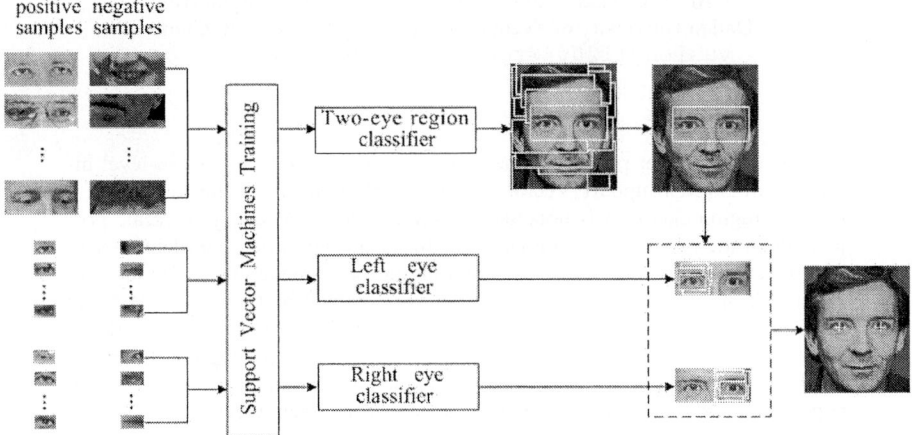

Fig. 1. Diagrammatic description of the operation of the system

3 Support Vector Machines

Support Vector Machines (SVM)[8] perform pattern recognition for two-class problems by finding the decision surface which minimizes the structural risk of the classifier. This is equivalent to determining the separating hyperplane that has maximum distance to the closest samples of the training set. These closest samples are called Support Vectors (SVs).

Considering the problem of separating a set of training samples belong to two separable classes, $(x_1, y_1), ..., (x_n, y_n)$, where $x_i \in R^d$, $y_i \in \{-1, +1\}$ with a hyperplane $wx + b = 0$. An optimal hyperplane can be obtained through solving a constrained optimization problem, the decision function can be written as:

$$f(x) = \text{sgn}(\sum_{i=1}^{n} \alpha_i^* y_i k(x_i \cdot x) + b^*) \qquad (1)$$

where α_i^* $(i = 1, ..., n)$ and b^* are the best solutions to the optimization problem, the kernel $k(x, y)$ is defined by $k(x, y) = \phi(x) \cdot \phi(y)$, ϕ is an implicate nonlinear map which ensures that the training data are linearly separable in new feature space. Every feature target such as two-eye region, left eye or right eye should be uniquely determined in the process of eyes location for each face image, that is in each feature target location process, $f(x)$ must be above zero once and only once, but there may be some cases not accord with this. Because the value of $|w \cdot x + b|$ is directly proportional to the distance from sample x to the hyperplane, the sign of $w \cdot x + b$ should be positive if sample x belongs to positive class and vice versa. As a real number, the bigger the value of $w \cdot x + b$ is, the more likely x belongs to the positive class. So this paper selects the one as feature target that make the value of $w \cdot x + b$ maximal, this lead to the following:

$$x^* = \arg\max(\sum_{i=1}^{n} \alpha_i^* y_i k(x_i \cdot x) + b^*) \qquad (2)$$

4 Eyes Loction

4.1 Two-Eye Region Location

The two-eye region is the part of face image where two eyes and eyebrows lie. The aims of finding the position of two-eye region include narrowing the searching scope for left and right eye respectively, eliminating the disturbance brought by nose and mouth, etc. It also aims to embody the coarse-to-fine policy that can save total search time. On this level, the vectors composed of the vertical and horizontal grayscale projections of the two–eye region images are used for training SVM. Suppose the image of two-eye region $B(i, j)$ has the size of $M \times N$, the horizontal grayscale projection P_h and vertical grayscale projection P_v can be defined as

$$P_h(j) = \sum_{i=1}^{M} B(i, j) \text{ and } P_v(i) = \sum_{j=1}^{N} B(i, j)$$

respectively. If Z stands for the sum of gray values of all the pixels, then the training vector can be represented by the concatenation of $\frac{P_h}{Z} \cdot M$ and $\frac{P_v}{Z} \cdot N$. A two-eye region classifier can be obtained by training the SVM. According to formula (2), using this classifier the two-eye region can be detected and located by searching the whole face image.

4.2 Precise Eyes Location

On this level, the grayscale values of the images are directly used for SVM training, the SVM of left and right eye are trained respectively, thus the classifier of left and right eye can be used for eyes location by searching the left and right part of two-eye region which has been obtained on the first level.

4.3 Support Vector Machines Training

The training samples of SVM are composed of positive sample set and negative sample set. In this paper, there are three positive sample sets that are composed of two-eye region images, left eye images and right eye images respectively. Three negative samples sets corresponding to the positive ones are extracted from the searching fields that do not contain positive samples in each location step, that is the whole face images, the left parts of the two-eye region images and the right parts of the two-eye region images respectively. The negative class is broader and richer, and therefore needs more negative examples to get an accurate definition that separates it from the positive class. In order to select the effective negative examples for targets detection, the method of bootstrapping is used for selecting negative samples, speaking in detail, the false positives are later served as negative examples during the training process. This paper uses SMO (Sequential Minimal Optimization) [9] as training algorithm.

5 Experimental Results

The experiment was performed on the ORL face database[10] which contains 400 facial images corresponding to 40 subjects. There are 10 different images of each subject. We selected 200 samples (5 for each subject) randomly as the training set by which we trained SVM, the remaining 200 samples were used as the test set. The system was implemented by MATLAB language on a computer with an AMD Athlon (1.3GHz) processor and a 256M EMS memory. Fig.2 shows some results of eyes location. Experiments results show an accurate location rate of 90% compared with manual selection and an average speed of 1.9 seconds for one face image.

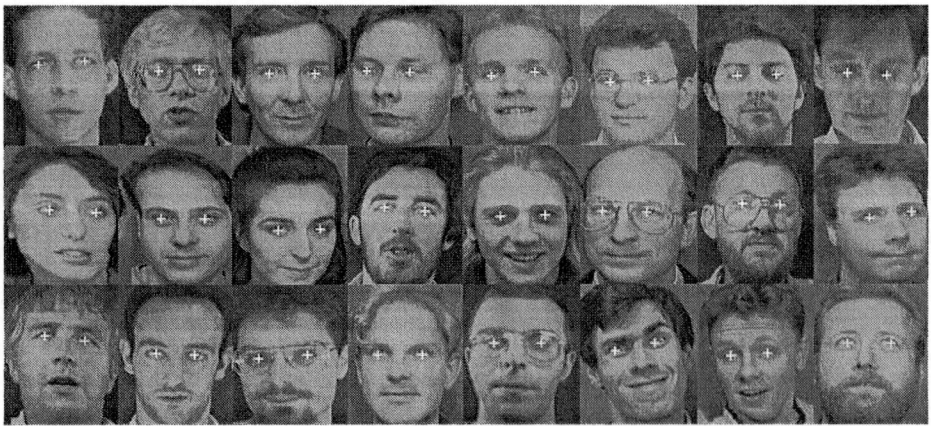

Fig. 2. Some results of eyes location experiment

Table 1. Mean and standard deviation of the classification accuracy of SVM, PCA+SVM and wavelet+SVM

Methods	SVM	wavelet+SVM	PCA+SVM
CCR	81.66%	93.26%	81.30%
STD	3.75	3.08	2.90

used to train the SVM classifier and the remaining samples are used as test samples to calculate correct classification rate (CCR). The experiment is repeated 25 times with random different partition and the mean and standard deviation of the classification accuracy are reported. In the tables presented in this paper, the classification accuracy is reported in percentage.

The SVM is designed to solve two-class problems. For multi-class stellar spectra, a binary tree structure is proposed to solve the multi-class recognition problem. Usually two approaches can be used for this purpose [9,11]: a) The one-against-all strategy to classify between each class and all the remaining. b) The one-against-one strategy to classify between each pair. We adopt the latter one for our multi-class stellar spectral classification, although needing more SVM to be applied, that allows the computing time to be decreased because the complexity of the algorithm depends strongly on the number of training samples.

In the experiment, firstly original data is directly used as the input of SVM. Table 1 shows that direct classification using SVM could achieve 81.66% CCR.

Because the stellar spectral data sets are extremely noisy, the classification rate of directly applying SVM is low. In order to raise the CCR, we propose to adopt wavelet de-noising method to reduce noise first. Wavelet transform, due to its excellent localization property, has become an important tool for de-noising. De-noising by wavelet thresholding was introduced by Donoho and Johnstone [12]. The basic method we use in the experiment involves two steps: 1). Calculate the wavelet coefficients. 2). Identify and zero out wavelet coefficients of the signal which are likely to be noise, remaining wavelet coefficients reserve important high pass features of the signal. Then the SVM is used for the final spectrum recognition (We denote this composite classifier which combines wavelet de-noising and SVM as wavelet+SVM). Table 1 indicates that this method could achieve 93.26% CCR and the smaller standard deviation than direct classification using SVM. According to the hypothesis tests (t-test) applied, the wavelet+SVM method has the mean of the classification accuracy in the validation set significantly ($\alpha=0.05$) larger than direct classification using SVM (p-values equal to 1.97×10^{-8}).

Principal Component Analysis (PCA) [13] is a good tool for dimension reduction, data compression and feature extraction. As a comparison, we use the dimension reduced data with PCA as the input of SVM (We denote this method as PCA+SVM). The distribution of these eigen-spectra in first three principal components (PCs) space is shown in Fig. 1(b). From Fig. 2(a), we can find that the first 10 PCs just have only 0.43% reconstruction error, so we choose them to define a 10-dimensional subspace and map spectra on it to obtain 10-

dimensional vectors. Fig. 2(b) shows a comparison of the original spectrum to the PCA reconstructed spectrum and the wavelet denoised spectrum.

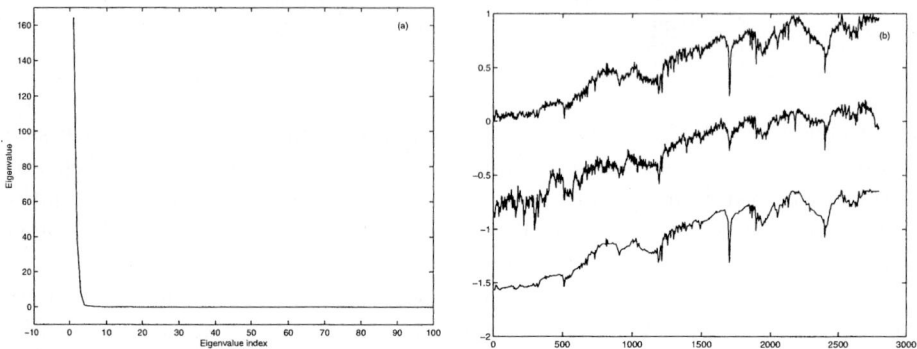

Fig. 2. (*a*) Eigenvalue in decreasing order. (*b*) The de-noising result. *Above line* is original spectrum, *middle line* is with PCA (10 PCs reconstructed spectrum), *bottom line* is with wavelet de-noising

From Table 1, we can see that the performance of PCA+SVM is not very good. It's even worse than direct classification with SVM. According to the hypothesis tests (t-test) applied, the PCA+SVM method has the mean of the classification accuracy in the validation set significantly ($\alpha=0.05$) smaller than wavelet+SVM method (p-values equal to 2.01×10^{-10}). The reason is that SVM can simulate a non-linear projection which can make linearly inseparable data project into a higher dimension space, where the classes are linearly separable. So data dimension has little influence on SVM.

Discriminant analysis is one of the supervised learning classifier building techniques. We also compare QDA and LDA to SVM. The stellar spectrum data are drawn from standard stellar library for evolutionary synthesis and are the same from Ref. [14]. The data set consists of 457 samples and could be divided into 3 classes. The spectrum is of 1221 wavelength points covering the range 9.1 to 160000 nm.

The experiments are conducted as in Ref. [14]. Table 2 shows that the performance of SVM is better than QDA and LDA. The higher CCR and lower standard deviation is achieved.

Table 2. The classification accuracy comparison of QDA, LDA and SVM

Methods	QDA	LDA	SVM
CCR	96.21%	94.88%	99.87%
STD	0.84	0.47	0.18

4 Conclusions

In this paper, a new technique on stellar spectral recognition which combines wavelet and SVM is proposed in this paper. From the experiments we can see that the proposed classifier has a good performance. According to the hypothesis tests (t-test) applied, the classification results of the wavelet+SVM are better than either SVM alone or SVM with PCA data dimension reduction technique. Experiments have been done to demonstrate that the approach offers a very promising technique in automated process of stellar spectra.

Acknowledgements. The research work described in this paper was fully supported by a grant from the National Natural Science Foundation of China (Project No. 60275002) and by National High Technology Research and Development Program of China (863 Program, Project No. 2003AA133060).

References

1. Aeberhard, S., Coomans, D., Vel, O. D.: Comparative Analysis of Statistical Pattern Recognition Methods in High Dimensional Settings. Pattern Recognition, **27** (1994) 1065–1077
2. Webb, A.: Statistical Pattern Recognition. Oxford University Press, London (1999)
3. Friedman, J. H.: Regularized Discriminant Analysis. J. Amer. Statist. Assoc., **84** (1989) 165–175
4. VonHippel, T., Storrie-Lombardi, L. J., Storrie-Lombardi, M.C., Irwin, M. J.: Automated Classification of Stellar Spectra - Part One - Initial Results with Aartificial Neural Networks. R.A.S. Monthly Notices, **269** (1994) 97–104
5. Cortes, C., Vapnik, V.: Support Vector Networks. Machine Learning, **20** (1995) 273–297
6. Dumais, S.: Using SVMs for Text Categorization. IEEE Intelligent Systems, **13** (1998) 21–23
7. Pontil, M., Verri, A.: Support Vector Machines for 3D Object Recognition. IEEE Trans. on Pattern Analysis & Machine Intelligence, **20** (1998) 637–646
8. Drucker, H., Wu, D., Vapnik, V.: Support Vector Machines for Spam Categorization. IEEE Trans. on Neural networks, **10** (1999) 1048–1054
9. Vapnik, V.: The Nature of Statistical Learning Theory. SpringerVerlag, New York (1995)
10. Efron, B., Tibshirani, R.: An Introduction to the Bootstrap. Chaoman & Hall, London (1993)
11. Weston, J., Watkins, C.: Support Vector Machines for Multi-Class Pattern Recognition. Proceedings of the Seventh European Symposium on Artificial Neural Networks, D-Facto, Brussels (1999) 219–224
12. Donoho, D. L.: De-noising by Soft-thresholding. IEEE Trans. on Information Theory, **41** (1995) 613–627
13. Jolliffe, I. T.: Principal Component Analysis. Springer-Verlag, New York (1986)
14. Wang, X., Xing, F., Guo, P.: Comparison of Discriminant Analysis Methods Applied to Stellar Data Classification. Proceedings of SPIE, Vol. 5286. SPIE, Bellingham WA (2003) 758–763

Iris Recognition Using Support Vector Machines

Yong Wang and Jiuqiang Han

School of Electronics and Information Engineering, Xi'an Jiaotong University,
Xi'an, 710049, P. R. China
yongwang@mailei.xjtu.edu.cn, jqhan@mail.xjtu.edu.cn

Abstract. In this work, a new method for iris recognition based on support vector machines was proposed. The recognition consisted of three major components: image preprocessing, feature extraction and classification. Location and normalization methods were employed in image preprocessing. In iris classification and verification, an efficient approach called support vector machines was used. Experimental results show that the proposed method has an emerging performance.

1 Introduction

In recent years, with the development of information and security technology, intelligent personal recognition has become a very important topic. Traditional user authentication schemes are based on passwords, secret codes and/or identification cards or tokens. On one hand, schemes based only on passwords or secret codes can be cracked by intercepting the presentation of such a password, or even by counterfeiting it (via passwords dictionaries or, in some systems, via brutal force attacks). On the other hand, an intruder can attack systems based on identification card or tokens by robbing, copying or simulation them. In that case, biometrics becomes an alternative recognition technique in aspect of protecting ourselves.

Biometrics has also been developed for measuring and analyzing human body characteristics such as fingerprints, retinas, iris, voice patterns, hand¡¡writing and typing patterns, and measurements. And they are being used in a wide range of applications; for example, to increase the security of computer systems, control illegal immigration, check customer signatures at building societies and even to guard the offices. Nevertheless, it is known that, from all of these techniques, iris recognition is the most promising for high security environments [1]. The potential of the human iris for such kind of problems comes from the anatomy of the eye. Some properties of the human iris that enhance its suitability for use in automatic identification include: 1) its inherent isolation and protection from the external environment, being an internal organ of the eye, behind the cornea and the aqueous humor; 2) the possibility of surgically modifying it without high risk of damaging the user's vision; and 3) its physiological response to light, which provides the detection of a dead or plastic iris, avoiding this kind of counterfeit. Also several studies have shown that while the general structure of the iris is genetically determined, the particulars of its minutiae are critically dependent

F. Yin, J. Wang, and C. Guo (Eds.): ISNN 2004, LNCS 3173, pp. 622–628, 2004.
© Springer-Verlag Berlin Heidelberg 2004

on initial conditions in the embryonic mesoderm from which develops. Therefore, there are not ever two irises alike, not even for identical twins [1].

At the moment, two typical prototype iris recognition systems had been developed by Daugman [1] and Wildes et al. [2]. Nevertheless, this biometric technology presents still many open problems. So, more recently, we can find some works in this issue, as given by Boles et al. [3], using wavelet transform, and by Sanchez-Reillo et al. [4], where Gabor filter are used. D. de Martin et al.[5], using dyadic wavelet transform zero-crossing. Here, we develop a new approach using support vector machines (SVMs).

The paper is arranged as follows. Section 2 describes image preprocessing, which involves iris location and iris normalization. Feature extraction uses normalized correlation methods to capture local and global details in an iris. The method is introduced in detail in section 3. Section 4 discusses iris matching based on SVMs. Experiments and results are reported in section 5. The last section gives some concluding remarks.

2 Image Preprocessing

An iris image contains not only the region of interest but also some unuseful part such as eyelid, pupil etc. So, a captured iris image cannot be used directly. After the iris captures and before the feature extraction, two steps are performed: location of the iris inside the image, and normalization of the iris. The procedures are called image preprocessing. The preprocessing is described in the following subsections.

2.1 Iris Localization

A new iris image is shown in Fig. 1(a), it is known that both the inner boundary and the outer boundary of a typical iris can approximately be taken as circle. And gray levels from the outer to the inner are apparently changeable alternatively. The method we employed is gray levels gradient. The overall processing is simple and reliable. First, we scan the iris image through sclera, iris and pupil in two different directions. And iris grayscale signatures are obtained in Fig. 1(b). Detecting the iris inner boundary is to use the gradient of the grayscale signatures. Based on the inner boundary, the iris outer boundary can be found by Daugman[1] method in Eq. (1) through searching small ranges.

$$\max_{(r,x_0,y_0)} \left| G_\sigma(r) * \frac{\partial}{\partial r} \oint_{(r,x_0,y_0)} \frac{I(x,y)}{2\pi r} ds \right| \quad (1)$$

where $I(x,y)$ is original image, $*$ denotes convolution and $G_\sigma(r)$ is a smoothing function such as a Gaussian of scale σ. The complete operator behave in effect as a circular edge detector, blurred at a scale set by σ, that searches iteratively for a maximum contour integral derivative with increasing radius at successively finer scales of analysis through the three parameter space of center coordinates and radius (x_0, y_0, r) defining the path of contour integration.

An example of iris localization is shown in Fig. 1(c). We can see that the iris can be exactly localized using proposed techniques.

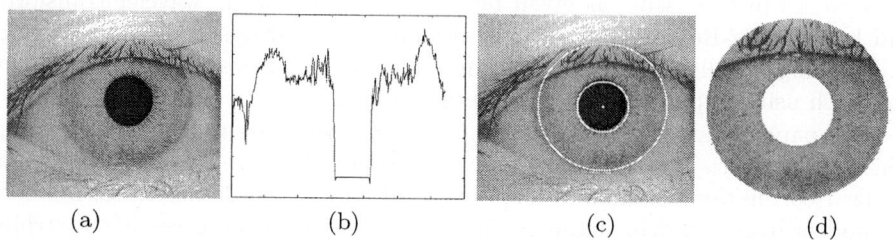

Fig. 1. Iris image preprocessing: (a) Sample iris image; (b) Sample iris signature from the image of (a); (c) Image after iris localization; (d) Iris normalization

2.2 Iris Normalization

The second step of the iris preprocessing is to normalize iris. Iris from different people may be captured in different size, and even for the iris from the same person, the size may change due to the variation of the illumination and other factors. Such elastic deformations in iris texture affect the results of the iris feature extraction and matching. So it is necessary to normalize the located iris. The dimensions of the iris in the image will be scaled to have the same constant diameter regardless of the original size in the images. The distortion of the iris caused by size of inner and outer radius can be realized by image interpolation [6]. The result after iris normalization is shown in Fig. 1(d).

3 Feature Extraction

In order to use the iris pattern for identification. It is important to define a representation that is well adapted for extracting the iris information content from normalized iris image. In this way, we introduce an algorithm for extracting iris features using normalized correlation coefficient. Normalized correlation formulation in Eq. (2) can be calculated.

$$\rho = \sum_{i=1}^{m}\sum_{j=1}^{n} \frac{[p_1(i,j) - \mu_1][p_2(i,j) - \mu_2]}{mn\sigma_1\sigma_2} \quad (2)$$

where $m \times n$ is the size of the two normalized images p_1 and p_2, μ_1 and μ_2 are the gray-level averages of p_1 and p_2, σ_1 and σ_2 are variances of them respectively.

Each pair of iris images can obtain their correlation. With n pairs of iris images, means and variances for ρ can be calculated to form the iris features.

4 Review of the SVMs Form Classification and Verification

In the section, we review the use of SVMs in classification and verification problems.

4.1 SVMs

Let the training set D be $\{(x_i, y_i)\}_{i=1}^{N}$, with each input $x_i \in \Re_m$ and the output label $y_i \in \{\pm 1\}$. The SVMs first maps x to $z = \phi(x)$ in Hilbert space H via a nonlinear map $\phi : \Re_m \to H$. This space H is often called the feature space, and its dimensionality is usually very high (sometimes infinite). Consider the case when the data is linearly separable in H, i.e., there exists a vector $\omega \in H$ and a scalar $b \in \Re$, such that

$$y_i(\langle \omega, \phi(x_i) \rangle + b) \geq 1 \qquad (3)$$

for all elements in the training set. The SVMs may construct a hyperplane $(\langle \omega, \phi(x_i) \rangle + b)$ for which the separation between the positive and negative examples is maximized. Fig. 2 gives the example of classification between two classes using hyperplanes. Then an optimal hyperplane is existed, ω for this "optimal" hyperplane can be found by minimizing $\|\omega\|$, and the resultant solution can be written as $\omega = \sum_{i=1}^{N} \alpha_i y_i \phi(x_i)$ for some $\alpha_i \geq 0$. The vector of α_i's, $\Lambda = (\alpha_1, ..., \alpha_N)$, can be found by solving the following quadratic programming (QP) problem:

$$Maximize\, W(\Lambda) = \Lambda^T 1 - \frac{1}{2}\Lambda^T Q \Lambda \qquad (4)$$

with respect to Λ, subject to the constraints $\Lambda \geq 0$ and $\Lambda^T Y = 0$. Here, $Y^T = (y_1, ..., y_N)$ and Q is a symmetric matrix with elements

$$Q_{ij} = y_i y_j \langle \phi(x_i), \phi(x_j) \rangle \qquad (5)$$

Let $K(x_i, x_j) = \langle \phi(x_i), \phi(x_j) \rangle$ as kernel function. For example, it can be shown that the kernel

$$K(x_i, x_j) = (x_i^T x_j + 1)^d \qquad (6)$$

More generally, it can be shown that any function satisfying Mercer's theorem [9] can be used as kernel, and each such kernel will have an associated map ϕ such that Eq. (6).

During testing, for a test vector, we first compute the activation

$$s = \langle \omega, \phi(x) \rangle + b = \sum_{i=1}^{N} \alpha_i y_i K(x, x_i) + b \qquad (7)$$

with a user-defined threshold s_0, the class label for x is then assigned by the following rule:

$$label = \begin{cases} 1, & s > s_0 \\ -1, & otherwise \end{cases} \qquad (8)$$

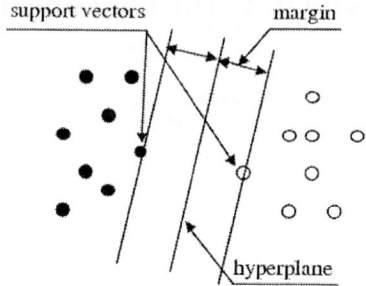

Fig. 2. Classification between two classes using hyperplanes(optimal separating hyperplane with the largest margin identified by the dashed lines, passing the two support vectors)

When the training set is not separable in H, the SVMS algorithm tries to minimize $\|\omega\|$ while at the same time separating the data with minimum number of errors. This is done by minimizing

$$\frac{1}{2}\|\omega\|^2 + C\sum_{i=1}^{N}\xi_i \tag{9}$$

while $\xi_i \geq 0$ satisfying the constraint:

$$y_i\left(\langle \omega, \phi(x_i)\rangle + b\right) \geq 1 - \xi_i \tag{10}$$

Here, C is a regularization parameter controlling the trade-off between model complexity and training error in order to ensure good generalization performance. Again, minimization of (*) can be transformed to a QP problem: maximize Eq. (4) under the constraints $0 \leq \Lambda \leq C_1$ and $\Lambda^T Y = 0$.

4.2 Classification and Verification

All the features obtained should enter a comparison process to determine the user whose iris photograph was taken. This comparison is to be made with the user's template, which will be calculated depending on the comparison algorithm used. The main decision-making tool investigated in the paper is (SVMs). It is based on the principle of structural risk minimization. Here, SVMs is used to classify and verify the iris features.

5 Experimental Results

This section reports some experimental results obtained, detailing the database used, the enrollment process, and final results in classification.

The results have been obtained using a database of eyes of 50 people and 7 photos for each individual. The images were taken in different hours and days.

We select 150 samples (3 photos for each individual) randomly as the training set, from which we calculate iris features and train the SVMs. The remaining 200 samples are used as the test set. In the classification and verification, the kernel function of SVMs used in the experiments is the radial basis function (RBF) defined as $K(x,y) = e^{\frac{-\|x-y\|^2}{2}}$. where x and y denote two samples. The reader is referred to Refs. [10,11] for an evaluation of the relative performances of difference kernel functions.

The results of the experiment are as follows. Verification success rate is up to 97.8%. To evaluate the algorithm, the performance can be measured in terms of three different rates: False Acceptance Rate (FAR): the probability of identifying an intruder as an enrolled user. False Rejection Rate (FRR):probability of rejecting an enrolled user, as if he were an intruder. Error Equalization Rate (EER): the probability of that FAR is equal to FRR.

Fig.3 depicts the plot of ROC and EER according to SVMs classification. Based on the experiments, it is easy to find FRR decreases with FAR increasing. And we can obtain the crossing point between ROC and EER is 2.63%. The experiment results show that SVMs is a promising and effective approach in iris recognition task.

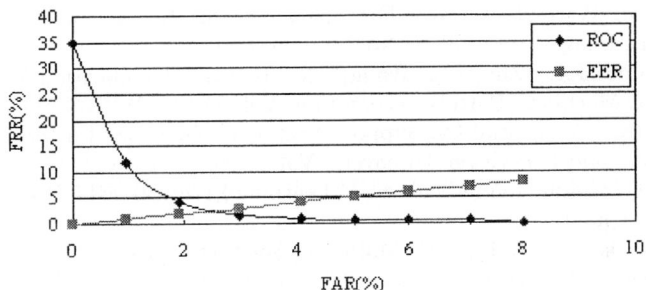

Fig. 3. ROC curve for verification

6 Conclusions

In the paper, we apply SVMs to the task of iris feature classification and verification, using translation-invariant iris feature generated by normalized correlation. RBF kernel are used among SVMs. Experiments are performed on a set of iris images from CASIA iris database. The results show that the SVMs is a promising and effictive classifier in iris recognition.

Acknowledgements. The authors thank National Laboratory of Pattern Recognition, Institute of Automation, Chinese Academy of Sciences for sharing the iris images. This work is funded by research grants from the NSFC (Grant No. 60174030)

References

1. Daugman,J.: High Confidence Visual Recognition of Persons by Test of Statistical Independence. IEEE Transactions on Pattern Analysis and Machine Intelligence, Vol. 15 (1995) 1148–1161
2. Wildes,R., et al.: A System For Automated Iris Recognition.Proc. 2nd IEEE Workshop Applicant. Computer Vision (1994) 121–128
3. Boles,W.W., Boashash,B.: A Human Identification Technique Using Images of the Iris and Wavelet Transform. IEEE Transactions on Signal Processing, Vol. 46, No. 4 (1998) 1185–1188
4. Sanchez-Reillo,R., Sanchez-Avila,C.: Processing of the Human Iris Pattern for Biometric Identification. Proc. Of IPMU 2000. 8th International Conference on Information Processing and Management of Uncertainty in Knowledge Based Systems, Madrid, Spain (2000) 653–656
5. De Martin-Roche,D., Sanchez-Avila,C., Sanchez-Reillo,R.: Iris Recognition For Biometric Identification Using Dyadic Wavelet Transform Zero Crossing. 2001 International Carnahan Conference on Security Technology, London, England (2001) 229–234
6. Ramponi,G.: Warped Distance For Space Variant Linear Image Interpolation. IEEE Trans Image Proc, Vol. 8, No. 5 (1999) 629–639
7. Li,S.T., James,T.K., Zhu,H.L., Wang,Y.N.: Texture Classification Using the Support Vector Machines. Pattern recognition, Vol. 36 (2003) 2883-2893
8. Burges,C.J.C.: A Tutorial On Support Vector Machines for Pattern Recognition. Data Mining and Knowledge Discovery, Vol. 2, No. 2 (1998) 121–167
9. Joachims,T.: Making Large-Scale SVM Learning Practical. MIT Press, Cambridge, MA (1998) Chapter 11
10. Jonsson, K., Kittler, J., Li, Y.P., Matas, J.: Support Vector Machines For Face Authentication. British Machine Vision Conference, T. Pridmore, D.Elliman (Eds.), BMVA Press (1999) 543–553
11. Lehmann, T.M., Gonner, C., Spitzer, K.: Survey: Interpolation Methods In Medical Image Processing. IEEE Trans Med Image, Vol. 18,No. 11(1999) 1049–1075
12. Jain,A.K., Bolle,R., Pankanti,S., et al.: Biometrics: Personal Identification In Networked Society. Kluwer Academic Publishers (1999)

Heuristic Genetic Algorithm-Based Support Vector Classifier for Recognition of Remote Sensing Images

Chunhong Zheng[1], Guiwen Zheng[2], and Licheng Jiao[1]

[1] School of Electronic Engineering, Xidian University, Xi'an 710071, China
[2] 501 branch box of 92 mailbox, Xi'an 710068, China
chzheng@xidian.edu.cn

Abstract. A heuristic genetic algorithm (GA)-based support vector classifier (SVC) for recognition of remote sensing images is presented in this paper. The model parameters of SVC are automatic selected by a heuristic GA to obtain the better performance with high efficiency. Compared with the leave-one-out (loo) method and the trial and error method, this GA-based model parameters selection is simpler and easier to implement. Furthermore, the generalization of the obtained SVC is much improved. Comparative tests conducted on a 2-value remote sensing images demonstrate the better result of the proposed classifier.

1 Introduction

With the rapid development of science and technology, the volume of remote sensing images continues to grow at an enormous rate due to advances in sensor technology for both high spatial and temporal resolution systems. However, most existing information systems for managing and processing remote sensing images further below this development tendency. For a proper exploitation of these data, it is mandatory to develop effective data process techniques able to take advantage of such multisource and multitemporal characteristics of remote sensing image. Therefore, research on the various processing technologies for remote sensing images is still an ongoing hotspot [1-6].

As a key technique for the effective application of remote sensing images, classification of remote sensing images has been extensively study in the past few years. The literature for radar-based classifiers usually categorizes classification algorithms into two different types: supervised and unsupervised [2]. An unsupervised classifier is given raw data, and the classification algorithm separates data into clusters, without any extra information. Each cluster must subsequently be identified as a class by hand. A supervised classifier requires more preparation; it must be given a set of data with corresponding classes (usually called training data) with which it tries to "learn" the character of that particular class. At present, growing interest is being devoted to the supervised classification of remote sensing images, such as Bayesian classifier [2, 3] and neural-network approach [4-6]. But Bayesian classifier requires to building a condition-probability density function model and estimating a prior probability [3]. While the classification techniques based on neural networks need to determine the network parameters by learning form the training data [4-6], and the selected training data have a directly influence on learning time cost and classification accuracy. An unsuitable samples selection will lead to a slow convergence of learning, even though a divergence. Moreover, since the excessive approximation of the neural networks, the generality of the trained neural networks is not favorable [5].

Recently, support vector classifier (SVC) have been proposed as a new learning machine for the cases of high dimensional vectors without no more *prior* information with higher generalization [6-9]. SVC was pioneered by Vapnik in 1992, based on statistical learning theory [9]. One key for effective application of SVC to solve practical problems is how to construct the hyperplane [9]. After introduction of the kernel space theory, the decision function of SVC can be described by supporting vectors by using the dual rule. Such like most available learning machine algorithms, model parameters have a drastic influence on the performance of the SVC [9-12]. So far, most researchers adopted the cross validate method, especially the leave-one-out (loo) method to determine these parameters [10]. This method betrays its localization when the training data is finite. In practice, a trial and error is usually used. This method is not only not to guarantee to obtain the best parameters due to the stochastic search, but also wastes more valuable time with continuous trial and error operation.

In this paper, after the development of a SVC for remote sensing images, the model parameters of the developed SVC are automatic determined by a heuristic GA for better performance and efficiency. Comparative tests on real remote sensing images demonstrate the effectiveness and efficiency of the proposed approach.

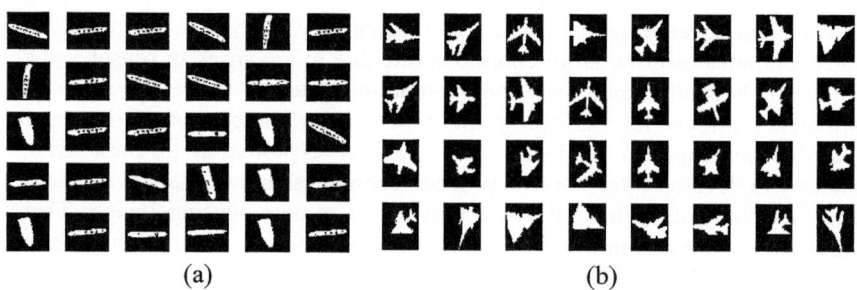

Fig. 1. Examples of remote sensing images. (a) ship, (b) plane

2 Problem Formulation

The obtained remote sensing images consist of complete target object images, partial covered target object images, and various rotary object images. Fig. 1 illustrates an example with ship and plane.

From Fig. 1, it can be seen that the remote sensing images are limited and fragmentary due to complex motion of images. The samples consist of 150 ship images and 160 plane images, which total data is 310 images. 20% of the total samples are selected as training data, the other 80% samples are used as testing data.

Using the bilateral-linear interpolation method, the resolution of the 2-value remote sensing images can be decreased from 128×128 to 25×25. As a result, the pixel of the images can be rearranged as a vector, and the training data can be denoted as $(\mathbf{x}_1, y_1), (\mathbf{x}_2, y_2) \cdots, (\mathbf{x}_l, y_l), \mathbf{x} \in R^n, y \in \{+1, -1\}$.

Assume that the training data can be separated by the following hyperplane

$$y_i(\mathbf{w}^t \cdot \mathbf{x}) + b \geq 1 \qquad i = 1, \ldots, l. \tag{1}$$

The separating hyperplane that has the maximum distance (maximum margin $\gamma = \|w\|^*$) between the hyperplane and the nearest data, is called optimal hyperplane. The optimal hyperplane can be determined by solving the following convex quadratic optimization problem [9]

$$\min \quad \tfrac{1}{2}\|\mathbf{w}\|^2 \tag{2}$$

s.t. $\quad y_i(\mathbf{w}^t \cdot \mathbf{x}) + b \geq 1$.

For nonlinear separable data, the optimal hyperplane can be specified by utilizing the kernel space theory

$$\max \quad W(\alpha) = \sum_{i=1}^{l} \alpha_i - \tfrac{1}{2} \sum_{i,j=0}^{l} \alpha_i \alpha_j y_i y_j K(\mathbf{x}_i \cdot \mathbf{x}_j) \tag{3}$$

s.t. $\quad 0 \leq \alpha_i \leq C, \qquad i = 1, \ldots l$

$$\sum_{i=1}^{l} y_i \alpha_i = 0.$$

where the samples corresponding to $\alpha_i \neq 0 (\alpha^*)$ are termed as support vectors. According to the KTT condition, b in Eqs. (1) and (2) is given [9],

$$b = -\tfrac{1}{2} \sum_{i \in SVs} y_i \alpha_i^* [K(\mathbf{x}_r \cdot \mathbf{x}_i) + K(\mathbf{x}_s \cdot \mathbf{x}_i)]. \tag{4}$$

where $\mathbf{x}_r$ and $\mathbf{x}_s$ are any support vector from each class of samples.

As a result, the decision function of SVC is,

$$D(\mathbf{x}) = \sum_{i \in SVs} \alpha_i^* y_i K(\mathbf{x}_i \cdot \mathbf{x}) + b. \tag{5}$$

The kernel function $K(\mathbf{x}_i, \mathbf{x}_j)$ is generally chosen as Gauss kernel function,

$$K(x_i, x_j) = \exp(-\frac{\|x_i - x_j\|^2}{2 \times \sigma^2}). \tag{6}$$

So that, the model parameter selection of a SVC can be formulated to proper determine the kernel parameter σ and the coefficient C.

3 GA-Based Automatic Model Selection

In this section, a heuristic genetic algorithm (GA) is used to automatic determine the model parameters for SVC effectively. GAs, differing from conventional search techniques, start with an initial set of random solutions, and evolve through successive iterations by genetic operators such as reproduction, crossover and mutation to arrive at better solutions to the problem. After a number of iterations, the algorithms converge to the best solution which hopefully represents the optimal solution [13].

To apply GAs to solve the optimal issue, one has to consider the following issues [13]:

(1). Coding scheme: A real-coded scheme is used to implement the code of the model parameters σ and C. That is to say, each gene of the chromosome stands for one floating-point number corresponding to the parameters. The model parameters σ and C are represented by six bit real float-point number, respectively, which cascades to formulate a twelve bit chromosome.

(2) Produce initial population: Since the real-coded scheme is used, the solution space coincides with the chromosome space. The strategy to produce the initial population is a probabilistic methodology. Considering the diversification of the initial population, the initial values of the designed parameters should be distributed in the solution space as evenly as possible.

(3) Fitness function: The key to solve the automatic model selection of SVC is how to define a proper fitness function to process the GA operation. Based on the theoretical analysis and test validation, the following radius-margin bound is selected as the fitness function,

$$fit = R^2 / \mathbf{num\_total} \cdot \gamma^2 . \qquad (7)$$

where radius $R = 0.5$, γ stands for the maximum margin, that is $\gamma = \|w\|^{-1}$, and num_total represents the number of the training data.

(4) Genetic operation: The used heuristic GA can dynamically adjust the genetic operator to keep the diversification of the population, avoid the premature and speed the convergence. In the primary genetic operation, the basic real-coded GA is performed. With the going of GA operation, the maintenance of elite individual is introduced to avoid the damage of the best individual by crossover and mutation operation, as well as speed of the convergence. Moreover, an adaptive mutation rate technique is also used, which can be expressed as

$$Pm = \frac{\exp(-t/2)}{pop\_size \times \sqrt{L}} . \qquad (8)$$

in which t represents the genetic iteration, pop_size represents the population size, and L stands for the length of the chromosome.

4 Test Results

Based on the available experiences on the 2-value remote sensing images, the model parameters σ and C can be determined as $\sigma \in [0.01 \quad 10000]$, and $C \in [0.001 \quad 1000]$. The mutation rate of the used heuristic GA in the whole iteration is shown in Fig. 2. The best, average, and worst fitness function value with the iteration are shown in Fig.

3, respectively. The model parameters corresponding to the best individual with the iteration is shown in Fig. 4. After the heuristic GA operation, the model parameters σ and C are automatically determined as: $\sigma = 762.1$ and $C = 98.697$.

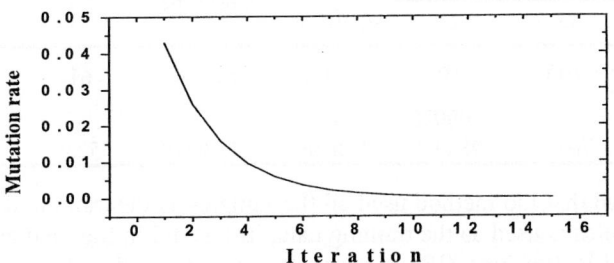

Fig. 2. Mutation rate variation with iteration

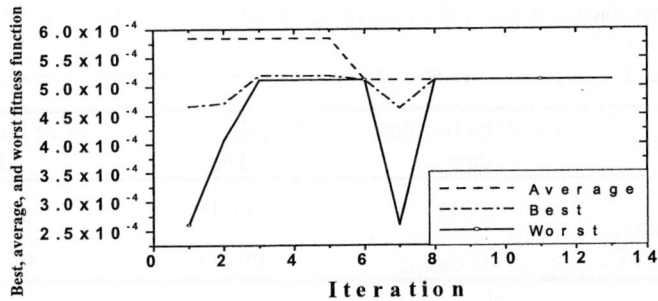

Fig. 3. Best, average, and worst fitness value with iteration

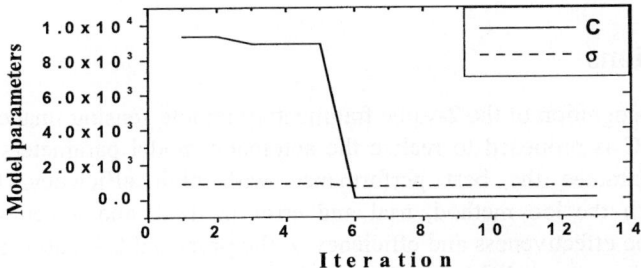

Fig. 4. Model parameters σ and C corresponding to best fitness with iteration

Using the developed SVC with the optimal model parameters obtained by heuristic GA to perform the 5-fold cross validation on the training data, the results are shown in Table 1. For comparison, the results by using the model parameters of SVC determined by trial and error method and loo method are also summarized in Table 1.

Table 1. Comparative results by various model selection methods of SVC

Method	Model parameters		Maximum margin	Cost time	Support vector	Identification Rate(5-fold)
	σ	C				
Trial and error	100	10	0.3091	1.6780	61	97.1%
Loo	1	0.00025	-	-	-	51.290%
GA	762.1	98.697	0.3656	1.5624	52.4	99.6%

It is noticed that loo method used all the samples as the training data, if only 20% of all the samples is used as the training data, just as did in trial and error method and GA, the identification rate (IR) is much lower. From Table 1, it can be seen that heuristic GA-based SVC has the highest identification rate.

Moreover, a comparison of the proposed heuristic GA-based SVC with a neural perceptron was also performed. The results are summarized in Table 2. To validate the effectiveness of the developed method, the testing on the unknown data, (which including 65 ship images and 448 plane images), were also carried out.

Table 2. Comparative results of the GA-based SVC with neural perceptron

Method	IR of the training data	IR of the testing data	IR of the unknown data
Neural perceptron	100%	96.94%	96.02%
GA-based SVC	100%	99.6%	99.42%

From the comparisons, it can be seen that for the training data, the identification rate (IR) of neural perceptron and the GA-based SVC are 100%, while for the testing data and the unknown data, the proposed GA-based SVC has higher identification rate in comparison with the neural perceptron.

5 Conclusions

For the best recognition of the 2-value fragmentary remote sensing images, a heuristic GA-based SVC is proposed to realize the automatic model parameters selection of SVC and guarantee the best performance with high efficiency. Comparisons performed with the loo method, trial and error method, and a neural perceptron demonstrate the effectiveness and efficiency of the proposed GA-based SVC method. Furthermore, the proposed GA-based method realizes an automatic model parameters selection of SVC, which is drastically decrease the costing time of design effective SVC for practical applications.

Acknowledgements. The authors thank the National Nature Science Foundation with 60073053 and 60372047, Specialized Research Fund for the Doctoral Program of Higher Education , and the Younger Researcher Fund of Xidian University.

References

1. Li, J., Narayanan, R.M.: Integrated Spectral and Spatial Information Mining in Remote Sensing Imagery. Vol. 42. IEEE Transactions on Geoscience and Remote Sensing (2004) 673-685
2. Kouskoulas, Y., Ulaby, F.T., Pierce, L.E.: The Bayesian Hierarchical Classifier (BHC) and Its Application to Short Vegetation using Multifrequency Polarimetric SAR. Vol. 42. IEEE Trans. Geoscience and Remote Sensing (2004) 469-477
3. Shkvarko, Y.V.: Unifying Regularization and Bayesian Estimation Methods for Enhanced Imaging with Remotely Sensed Data. Part I: Theory, IEEE Trans. on Geoscience and Remote Sensing (2004)
4. Han, M., Cheng, L., Meng, H.: Application of Four-layer Neural Network on Information Extraction. Vol. 16. Neural Networks (2003) 547-553
5. Bruzzone, L., Prieto, D.F., Serpico, S.B.: A Neural-statistical Approach to Multitemporal and Multisource Remote-sensing Image Classification. Vol. 37. IEEE Trans. Geoscience and Remote Sensing (1999) 1350-1359
6. Shah, S., Sastry, P.S.: Fingerprint Classification using A Feedback-based Line Detector. Vol. 34. IEEE Transactions on Systems, Man and Cybernetics, Part B (2004) 85-94
7. Perez-Cruz, F., Navia-Vazquez, A., Figueiras-Vidal, A.R., Artes-Rodriguez, A.: Empirical Risk Minimization for Support Vector Classifiers. Vol. 14. IEEE Transactions on Neural Networks (2003) 296-303
8. Burges, C. J.C.: A Tutorial on Support Vector Machines for Pattern Recognition. Vol. 2. Data Mining and Knowledge Discovery (1998) 121-167
9. Vapnik, V.: Statistical Learning Theory. John Wiley & Sons (1998)
10. Lee, J.H., Lin, C.J.: Automatic Model Selection for Support Vector Machines. Technical Report, Dept. of Computer Science and Information Engineering, National Taiwan University, Taipei, November (2000)
11. Keerthi, S.S.: Efficient Tuning of SVM Hyperparameters using Radius/Margin Bound and Iterative Algorithms. Vol. 13. IEEE Transactions on Neural Networks (2002) 1225-1229
12. Chapelle, O., Vapnik, V., Bousquet, O., Mukherjee, S.: Choosing Multiple Parameters for Support Vector Machines. Vol. 46. Machine Learning (2002) 131-159
13. Holland, J.H.: Adaptation in Nature and Artificial Systems. MIT Press, Cambridge, MA (1992)

Landmine Feature Extraction and Classification of GPR Data Based on SVM Method

Jing Zhang[1], Liu Qun[1], and Baikunth Nath[2]

[1] Computer Science and Technology Institute
Harbin Engineering University, China, 150001
{zhangjing,liuqun}@hrbeu.edu.cn
[2]Department of Computer Science & Software Engineering
The University of Melbourne, Melbourne, Vic 3010, Australia
bnath@cs.mu.oz.au

Abstract. In this paper, the problem of detecting buried landmine is tackled in the feature extraction and classification. Determining the likelihood set of an unknown pattern (feature vector), extracted from ground penetrating radar data by using SVM method. The advantage of SVM method in feature extraction and classification of image processing is: A classifier works well both on the training samples and on previously unseen samples; In addition, the SVM provides, enable a classification performance improvement based on from high feature dimensions to two or three feature dimensions. Finally, SVM method has a standard theory and a good implementation algorithm.

1 Introduction

Landmines account for over one third of the deaths and injuries on the modern battlefield. One of their main purposes is to maim, creating additional stress on infrastructure. With over 100,000,000 landmines buried and still active throughout the world. In some places, whole areas of arable land cannot be farmed due to the threat of landmines, so they also pose a significant threat to the peace and stability of civilians. In addition, people relief operations are made more difficult and dangerous due to the mining of roads [1].

In order to assist deminers, a range of advanced sensor technologies are being investigated, including metal detectors (MD), Ground Penetrating Radar (GPR), nuclear magnetic resonance , infrared and thermal imaging (TI) sensor etc. We will concentrate on Ground Penetrating Radar (GPR). This ultrawide band radar provides centimeter resolution to locate even small targets [2]. GPR operates by detecting the dielectric contrasts in the soils, which allows it to locate even nonmetallic mines. Unfortunately, this technology can suffer false alarm rates as high as that of metal detectors. As with metal detection, GPR automatic detection and classification algorithms are being developed.

The work described in this paper is situated in the field of data processing and feature extraction applied to GPR signals for the identification of the type of a buried object (mine, stone,...) by eliminating beforehand the signals that do not contain any object signal. In analyzing complicated 3D data sets, it is very common that only a small percentage of the entire data set actually contains interesting features [3]. So the method of feature extraction appeared. Feature extraction has been always mutually studied for exploratory data projection and for classification. Feature extraction for exploratory data projection aims for data visualization by a projection of a high-dimensional space onto two or three-dimensional space, while feature extraction for classification generally requires more than two or three features. Therefore, feature extraction paradigms for exploratory data projection are not commonly employed for classification and vice versa.

GPR data of sample is vector containing the expression high-dimensional space, such as landmine, stone, soil and earth dimensional space etc. An unknown signal is considered as a feature vector ($f = (f_1, \ldots, f_n)$). For each feature vector a decision will be made about the alleged representation of an object, such as classify the feature into only two or three dimensional space.

Many digital processing algorithms that have been developed to facilitate the feature extraction, common methods feature extraction such as contrast stretching, histogram normalization, high pass/low pass/band pass filters, edge enhancement, local contrast enhancement etc [4].

In this paper, we study the GPR data feature extraction and classification by applying of support vector machine (SVM), a machine learning method to divide the features which coming from the original GPR data features (measurements), due to its supposed good performance on extremely scare samples in high-dimensional space.

2 The Method of SVM

The method uses SVM for both feature extraction and for selecting a subset of relevant GPR according to their relative contribution in the feature extraction. This process is done recursively so that a series of GPR subsets and feature extraction models can be obtained in a recursive manner, at different levels of GPR selection [5]. The performance of feature extraction can be evaluated either on an independent test data set or by cross validation on the same data set.

In order to get the feature extraction, there are three steps of the method of SVM. The first step is to get the maximized separate margin; the second one is to minimize the number of misclassification; the last step is the recursive classification and leave-one-out cross validation.

2.1 Maximized Separate Margin

The margin can be defined as the distance between the hyperplane and the samples of the two classes that are closest to the hyperplane of being correctly classified. The detail is as follows:

This optimization problem can be solved by the following dual problem:

$$\max Q(\alpha) = \sum_{i=1}^{n} \alpha_i - \frac{1}{2} \sum_{i,j=1}^{n} \alpha_i \alpha_j y_i y_j (x_i \cdot x_j), \quad (1)$$

subject to

$$\sum_{i=1}^{n} y_i \alpha_i = 0,$$

and

$$0 \leq \alpha_i \leq C, i = 1, \ldots, n.$$

2.2 Minimizing the Number of Misclassification

In order to minimizing the number of training errors, SVM seeks good performance on the training data;
The basic idea of it is described as

$$\min \psi(w, \xi) = \frac{1}{2} \|w\|^2 + C \left(\sum_{i=1}^{n} \xi_i \right), \quad (2)$$

subject to

$$y_i[(w \cdot x_i) + b] - 1 + \xi_i \geq 0, i = 1, \ldots, n,$$

where

$$(x_i, y_i), i = 1, \ldots, n, x_i \in R^d, y_i \in \{+1, -1\}$$

are the training samples which is called support vectors, because only these sample that can support the classification boundary. C represent a constant controlling the trade-off between maximizing the margin and minimizing the errors.
The decision function is

$$f(x) = \text{sgn}\{(w \cdot x) + b\} = \text{sgn}\left\{ \sum_{i=1}^{n} \alpha_i^* y_i (x_i, x) + b^* \right\}. \quad (3)$$

α_i^* is not often non-zero in the final decision of SVM method. $f(x)$ is weighted by the sum of all the features which express the levels of all the GPR and plus a constant term as a threshold. If $f(x) > 0$, then the sample is class 1, otherwise class 2, and the larger the absolute value of $f(x)$ is, the more distinct the sample is from the other class.
The sum of all the features in class 1 and class 2 is

$$m_i^+ = \sum_{x^+ \in \text{class1}} x_i^+, \quad (4)$$

$$m_i^- = \sum_{x^- \in class2} x_i^- . \tag{5}$$

The difference of the two class means in the decision function is:

$$S = \sum_{i=1}^{n} w_i m_i^+ - \sum_{i=1}^{n} w_i m_i^- = \sum_{i=1}^{n} w_i (m_i^+ - m_i^-) . \tag{6}$$

The criterion for ranking the GPR as

$$r_i = w_i (m_i^+ - m_i^-) . \tag{7}$$

2.3 Recursive Classification and Leave-One-Out Cross Validation

Usually, to choose the subset from a feature set can not be finished once. So we use recursive classification. The procedure is, input all the available GPR data as one set and build a model to rank the GPR data, then Select a subset of GPR data and Re-rank the selected GPR data to a new model, and then repeat the selection. That is the recursive classification is: classification - ranking – selection – classification.

If sample size is too small to afford to use an independent test set, we would use the typical choice-cross validation to assess the performance of the classifier. GPR selection depends heavily on the specific samples when sample size is very small, for the cross validation, the sample should be left out as test sample removed from the data set at the very beginning, before any GPR selection procedure.

3 The Feature Extraction and Classification Result

The Core algorithm of SVM is SVMTorch produced by Collobert.R. and Bengio.S [6], but our experience is based on the software package which is the developed SVM algorithm by Xuegong Zhang and Wing H.Wong [7, 8, 9, 10, 11, 12].

For our experimental tests we used ground penetrating radar (GPR) data provided by the DETEC Laboratory of the EPFL [13]. We used data coming from the scanning, with a 1 GHz GPR, of three objects: two non metallic AP mines (type PFM-1 and type PMN) and a stone, all placed horizontally. The scanning was performed by placing the radar antenna on an x-y platform and acquiring a signal at regular positioned at several depths (1, 3, 5 and 10 cm) in sand and in earth. The stone was scanned at 3 cm depth in earth. Out of these data we retrieved a data test set, divided in 6 classes (class 0: PFM-1 in sand, class 1: PMN in sand, class 2: PMN in earth, class 3: Stone in earth, class 4: Sand background, class 5: Earth background). For the mine classes an equal number of samples from each depth were included.

In order to reach the classification accuracy, we use the same SVM scheme of random permutation experiment to get an error rates by chance on randomly assigned classes on the data set. Then the same leave -one-out cross validation procedure is done on every permuted data set. Typically we run 60 permutation experiments with different random seeds each time, and then the estimate the distribution of the error rate on the permuted data, and calculate the permutation p-value for the error to be equal to or less than the error obtained with the true data set. This is a rather strict way to assess the significance of the feature extraction and classification performance. The results indicate that it is possible to discriminate between different buried objects.

We use one image as membership information to show the feature extraction and classification result. Figure 1 shows that there are many features in the original scanning 2D image, include one PMN and one stone in the earth of 5 cm and many other features. Figure 2 represents the results of the feature extraction and classification of it, the image is divided into three classes, PMN in earth, stone in earth and earth background.

4 Conclusions

We study the classification capabilities of the well-known SVM method, which is mostly applied for medical image processing. The landmine image paradigms are evaluated using a SVM feature set. The result of the image shows that the SVM software package has an admirable feature extraction and classification capability, especially for the function of getting unseen samples (samples whose correct classification is known and thus plays the role as supervisors). In addition, the SVM provides, enable a classification performance improvement based on from high feature dimensions to two or three feature dimensions. The SVM method which is used in this study is feasible for feature extraction and classification of landmine image.

Fig. 1. The original scanning 2D image with many features, include one PMN landmine (*black color part in the right*) and one stone (*black color part in the left*) in the earth of 5 cm, and many other features

Fig. 2. The results of the feature extraction and classification image with three classes, PMN in earth (*the white part in the right*), stone in earth (*the white part in the left*) and earth background (*other part in the picture*)

References

1. Hans Chen, Diana D.Hughes, Tan-an Chan: IVE (Image Visualization Environment): A Software Platform for All Three-Dimensional Microscopy Application. Journal of structural Biology (1996) 116.56-60
2. Zoubir A. M., Iskander D. R., I. Chant, and D. Carevic: Detection of Landmines Using Ground-penetrating Radar. SPIE Conf. Detection Remediation Technologies Mines, Minelike Targets, IV, (1999), pp. 1301–1312.
3. Mao J., and Jain A. K.: Artificial Neural Networks for Feature Extraction and Multivariate Data Projection, IEEE Trans. Neural Networks, vol. 6, pp. 296-317, (1995)
4. Sahl H.i, Nyssen E., L.van Kempen, and J.Cornelis: Feature Extraction and Classification Methods for Ultra-sonic and Radar mine Detection. IEEE CESA;98 conference , April 1-4 (1998), Tunis, Tunesia, Volume 4, pp 82-87
5. Xuegong Zhang and Wing H,Wong: Recursive Sample Classification and Gene Selection based on SVM: Method and Software Description. Technical Report, Department of Biostatistics, Harvard School of Public Health (2001)
6. Collobert, R. and Bengio, S. SVMTorch: Support Vector Machines for Llarge-scale regression problems. J. Machine Learn. Res. **1**, 143-160, (2001)
7. Cortes, C. and Vapnik, V: Support-vector Networks, Machine Learning, **20**, pp.273-297 (1995)
8. Joachims, T: Making large-scale SVM learning practical, In Schokopf, B. et al (eds), Advances in Kernel Methods – Support Vector Learning, MIT Press(1999)
9. Vapnik, V.N: The Nature of Statistical Learning Theory (2nd edition), Springer-Verlag, New York (1999a)
10. Vapnik, V. N: An Overview of Statistical Llearning Theory. IEEE Trans. Neural Networks, **10**, pp.988-999 (1999b)
11. Zhang, X: Using Class-center Vectors to Build Support Vector Machines, In: Neural Networks for Signal Processing IX, IEEE Press, (1999) pp.3-11,
12. Zhang, X. and Ke.H: ALL/AML Cancer Classification by Bene Expression Data Using SVM and CSVM approach, In: Genome Informatics 2000, Tokyo: Universal Academy Press, pp. (2000)237-239
13. http://diwww.epfl.ch/lami/detec

Occupant Classification for Smart Airbag Using Stereovision and Support Vector Machines

Hyun-Gu Lee[1], Yong-Guk Kim[2], Min-Soo Jang[1], Sang-Jun Kim[1], Soek-Joo Lee[3], and Gwi-Tae Park[1*]

[1]Dept. of Electrical Engineering, Korea University, Seoul, Korea
{hglee99,gtpark}@korea.ac.kr
[2]School of Computer Engineering, Sejong University, Seoul, Korea
ykim@sejong.ac.kr
[3]Hyundai Autonet Co., Korea

Abstract. Airbag in the cars plays an important role for the safety of occupants. However, Highway Traffic Safety report shows that many occupants are actually killed by wrong deployment of the airbags. For reducing risk caused by airbag, designing a smart airbag is an important issue. The present paper describes an occupant classification system, by which triggering of the airbag deployment can be controlled. The system consists of a pair of stereo cameras and a SVM classifier. Performance of the system shows its feasibility as a vision-based airbag controller.

1 Introduction

Airbags in the cars are crucial for preventing life-threatening and debilitating head and chest injuries by avoiding direct impact contact at the dashboard. Although airbags have saved many lives, the National Highway Traffic Safety Administration (NHTSA) reported that 153 occupants have been killed by the deployment of the airbag itself (2000.4). Particularly, 89 occupants (58.2%) of the death were children or infants in rear-facing infant seats [1]. For solving this problem, NHTSA has recently issued a set of regulations mandating low risk deployment of the airbag.

In this paper, we present a new system that consists of stereovision and occupant classification system using a SVM (Support Vector Machine) classifier. Advantages of using the vision sensor for this application include the ability to capture diverse information about occupant, such as class, pose, distance from occupant to dashboard, and so on. Moreover, it can be installed almost anywhere within the car and it has higher accuracy compared to other sensors based on weight, ultrasound, infrared, and electronic fields [8, 9]. Our vision system is a pair of CCD cameras mounted rear-view mirror position and it is pointed towards the occupant. We first extract a disparity map using a fast stereo algorithm [4, 5], and then down-sampled images of the disparity map are used as input for the SVM classifier [6]. SVM is relatively simple, and yet

---

* The corresponding author

powerful classification technique initially proposed by Vapnik and his colleagues [2, 3]. The well-trained SVM is able to separate between two target classes using a hyperplane. Or it can be extend to be a multiple classifier. Here, we classify the occupants into three classes by using SVM classifier (i.e. occupant: adult or child, child seat and empty seat). In case the result of classification is occupant, an additive process is taken to determine whether the occupant is an adult or child. If the occupant is an adult, the airbag system will deploy the airbag in car crashes, whereas when it turns out as a child seat or empty seat, the airbag system will suppress deployment of the airbag.

In section 2, the details of our stereovision system are described. The SVM classification method is discussed in section 3. In section 4, occupant classification algorithm is explained. Result of experiments is described in section 5. Finally, we summarize our results, and discuss the performance of the whole system in section 6.

2 Stereovision

In this section, we describe our stereovision system and how the system is operated. Figure 1(a) is a schematic illustration of the system. It consists of two CCD cameras. Figure 1(b) shows a pair of stereo images captured by the vision system.

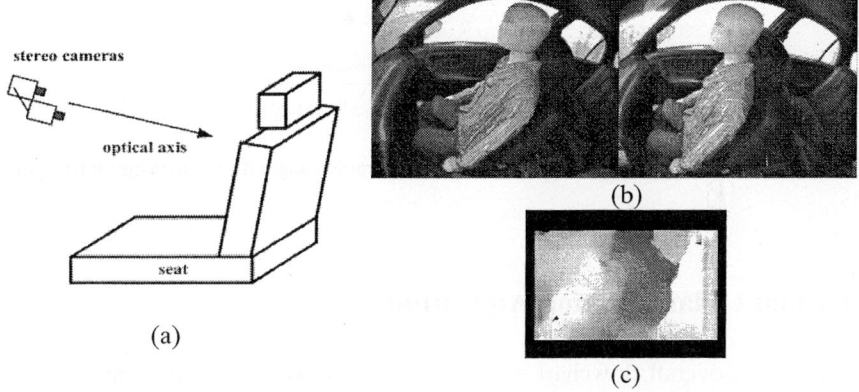

Fig. 1. A stereovision system (a), a pair of stereo images (b), and a disparity map

To obtain a disparity map between two images or from the stereo pair, we adopted a fast SAD (Sum of the Absolute Difference) algorithm [4, 5]. Fig. 1(c) shows a disparity map generated by the fast SAD algorithm. Since the disparity map contains 3D information of an occupant, we can recognize the class of occupant by analyzing the disparity map.

3 Support Vector Machine as a Classifier

The fundamental idea of SVM is to construct a hyperplane as the decision line that separates the positive (+1) classes from the negative (-1) ones with the largest margin [2, 3]. In a binary classification problem, let us consider the training sample $\{(x_i, d_i)\}_{i=1}^{N}$, where x_i is the input pattern for the i-th sample and d_i is the corresponding desired response (target output) with subset $d \in \{-1, +1\}$. The equation of a hyperplane that does the separation is

$$w^T x + b = 0 \tag{1}$$

where x is an input vector, w is an adjustable weight vector, and b is a bias. Fig. 2 shows an optimal hyperplane for the linearly separable case and margin, γ. The aim of the SVM classifier is to maximize the margin. SVM uses diverse kernels in dealing with the input vectors. The standard kernels are polynomial, Gaussian and sigmoid kernels.

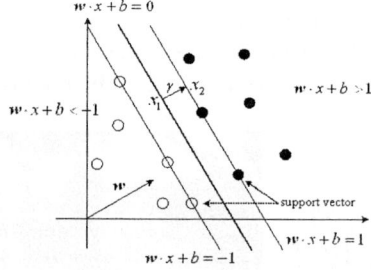

Fig. 2. An optimal hyperplane for the linearly separable case with the maximum margin

4 Occupant Classification Algorithm

Fig. 3 shows an overall flowchart of the occupant classification algorithm. Firstly, a disparity map image is obtained by using a fast SAD, and the background information is removed in the disparity map image. The disparity map image contains 3D information of occupant as well as a dashboard and doors. Since 3D information of a dashboard and doors is unnecessary for occupant classification, we regarded it as background information and removed it. Then, a SVM classifier takes the down-sampled disparity map as an input. In case of an occupant is classified, an additional algorithm determines whether he is a child or an adult.

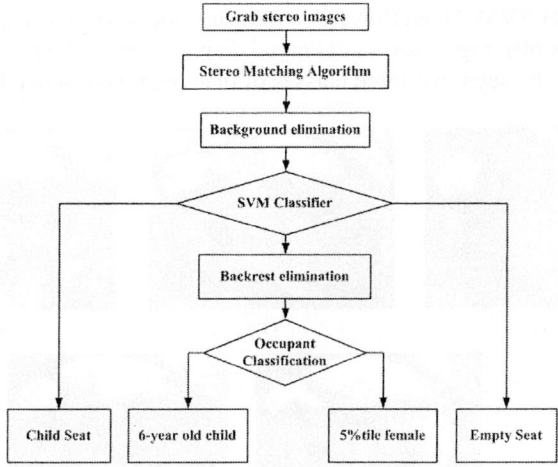

Fig. 3. The flowchart of the occupant classification system

5 Image Database and System Performance

5.1 Stereo Image Database

The performance of the occupant classification algorithm is evaluated using our stereo image database, which consists of many stereo images captured within a car. Table 1 indicates composition of the different classes of occupants. The total of the captured images are 824 and the resolution of an image is 320x240.

Table 1. Composition of the different classes

Class	Occupant	# of iages	
0	Adult	318	6 people, 170cm~180cm
	5th %tile female	166	a realistic looking doll, 140cm
1	6 year-old child	194	a realistic looking doll, 120cm
2	Child seat	124	19 products
3	Empty seat	22	not occupied

In Table 1, we classify car occupants into four classes. Fig. 4 depicts several examples of the occupants such as adult (a, b), child (c), RFIS (Rear Facing Infant Seat) (d) and empty seat (e), respectively. The disparity map is reduced by 1/16 before it is sent to an input of SVM. Therefore, a data set for one disparity map image consists of 4800 values (80x 60). Fig. 5 shows averaged disparity map of each class of occupants. In figure 5, it can be seen that three classes of occupant have separable patterns.

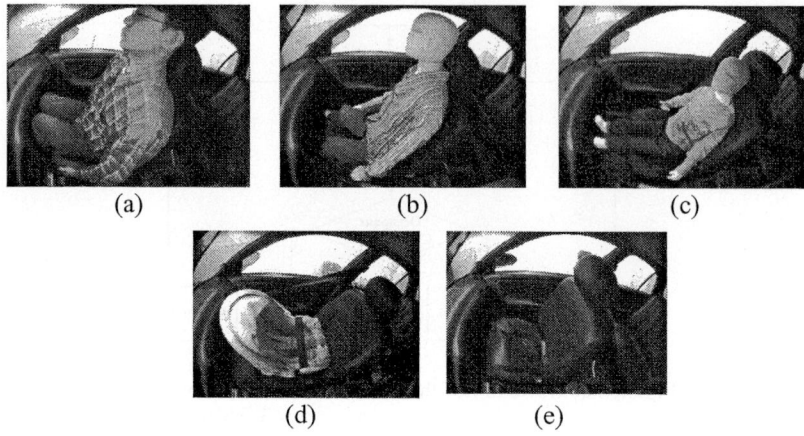

Fig. 4. Images for different occupant class (a) adult (b) 5th %tile female (c) 6-year old child (d) child seat (RFIS) (e) empty seat

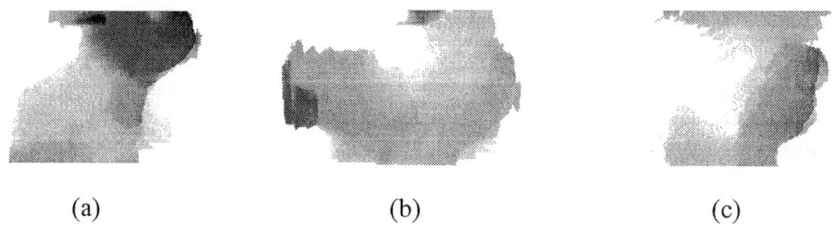

Fig. 5. Averaged disparity map (a) occupant (adult or child) (b) child seat (c) empty seat

5.2 Performance of the System

The total number of data sets within our DB was 824 with 4800 values for each image. We randomly divided the data sets into a training set (50%) and a test set (50%). We used the public domain implementation of SVM, called Lib SVM and two standard kernels (i.e. polynomial kernel and radial basis function kernel) [7]. The experiment was accomplished for 10 times and average results are shown in Table 2. For the polynomial kernel, the correction rate was 85.29%, whereas it was 100.0% for the RBF kernel. However, for the occupant discrimination task, the correction rate of our additional algorithm to distinguish child from adult was 87.2%.

Table 2. Classification rate for each kernels

Kernel	Classification rate
Polynomial	85.29%
RBF	100.0%

6 Conclusions and Discussion

As an initiative to develop a smart airbag system, we have designed a stereovision system combined with a SVM classifier. For the stereovision system, the fast SAD algorithm is adopted to calculate the disparity map of the stereo images. As the disparity map image contains 3-D information of the occupant, it is possible to classify the pose of the occupant by analyzing the disparity map image. SVM is used for classifying occupants, since it is known that the SVM classifier is relatively simple and yet powerful. The stereo image DB was constructed for verifying the performance of the system. Result shows that the performance of the system is satisfactory, suggesting that the vision-based airbag control has a potential.

Acknowledgment. This work was supported by Hyundai Autonet Co.

References

1. http://www.nhtsa.dot.gov
2. Vapnik, V.: The Nature of Statistical Learning Theory. Springer-Verlag, NY, USA (1995) 45-98
3. Haykin, S.: Newral Network. Prentice Hall (1999)
4. Zitnick, C. L., Kanade T.: A Cooperative Algorithm for Stereo Matching and Occlusion Detection. IEEE Trans. on Pattern Analysis and Machine Intelligence, Vol. 22, Issue 7 (2000) 675-684
5. Di Stefano, L., Marchionni, M., Mattoccia, S., Neri, G.: A Fast Area-Based Stereo Matching Algorithm. 15th International Conference on Vision Interface (2002)
6. Haritaoglu, I., Harwood, D., Davis, L. S.: Ghose : A Human Body Part Labeling System Using Silhouettes. 14th international conference on Pattern Recognition, vol. 1 (1998) 77-82
7. http://www.csie.ntu.edu.tw/~cjlin/libsvm/index.html
8. Fultz, W., Griffin, D., Kiselewich, S., Murphy, M., Wu, C., Owechko, Y., Srinivasa, N., Thayer, P.: Selection of a Sensor Suite for an Occupant Position and Recognition System. Proc. Of SAE TOPTEC conference on Vehicle Safety Restraint Systems (1999)
9. Owechkp, Y., Srinivasa, N., Medasani, S., Boscolo, R.: Vision-Based Fusion System for Smart Airbag Applicaions. IEEE, Intelligent Vehicle Symposium, vol. 1 (2002) 245-250

Support Vector Machine Committee for Classification

Bing-Yu Sun[1,2], De-Shuang Huang[1], Lin Guo[1,2], and Zhong-Qiu Zhao[1,2]

[1]Intelligent Computing Lab, Hefei Institute of Intelligent Machines,
Chinese Academy of Sciences
P.O.Box 1130, Hefei, Anhui, 230031, China
[2] Department of Automation, University of Science and Technology of China
bysun@iim.ac.cn

Abstract. In this paper the support vector machine committee is proposed. For a practical pattern recognition problem, usually numerous of features can be used to represent the pattern. SVM committee can utilize these features efficiently and a classifier with better generalization can be obtained. Moreover, a novel aggregation approach of support vector machine committee is also proposed in this paper. The simulating results demonstrate the effectiveness and efficiency of our approach.

1 Introduction

The support vector machine (SVM) is a new and promising classification technique proposed by Vapnik and his group [1]. The working mechanism of the SVM is to learn a separating hyperspace to maximize the margin and produce a good generalization capability. Currently, this technique has been successfully applied in many areas such as face detection, hand-written digit recognition, and so on [2] [3] [4].

In the field of pattern recognition, neural network committee (NNC) is a method that combines many different neural networks (NN) to achieve better generalization performance in comparison with the best performance achievable complying single ANN classifier [5]. With NNC, we can combine different classifiers together and a better result could be obtained. In this paper we will show that the committee strategy can also be applied to SVM and, accordingly, the SVM committee (SVMC) is proposed. Moreover, a novel aggregation method of SVMC is proposed. Traditionally there are several aggregation method of NN committee and the most commonly used one is major voting idea [6]. This approach regards all types of features with same effect on the performance of the SVMC. However, as we known, for a specific recognition problem, individual classifiers relying on different feature sets may attain different success. In order to resolve this problem, a novel weighted voting principle is proposed. In this strategy, each classifier influences the final decision according to not only the performance to the training data, but also its generalization. The experimental results proved our approach is more efficient than major voting method.

The rest of this paper is organized as follows. In Section 2 some basic notions of the SVM is reviewed. The approaches that construct and aggregate the SVMC are described in Section 3, and in Section 4 some simulating results are illustrated. Finally, Section 5 draws several conclusions.

2 Support Vector Machine

An SVM constructs a binary classifier from a set of labeled patterns called training examples. Let $(\mathbf{x}_i, y_i) \in R^l \times \{\pm 1\}, i = 1, \ldots, N$ be such a set of training examples. The SVM finds the optimal separating hyperplane (OSH) $\mathbf{w}^T \cdot \mathbf{x} + b = 0$ that maximizes the margin of the nearest examples from two classes. This OSH is then regarded as the solution to the problem

$$\min J(\mathbf{w}, e) = \frac{1}{2} \mathbf{w} \cdot \mathbf{w} + \frac{1}{2} C \sum_{j=1}^{N} e_j^2 \tag{1}$$

s.t:

$$y_j [\mathbf{w} \cdot \mathbf{x}_j + b] \geq 1 - e_j, \quad j = 1, 2, \ldots, N$$

where C is a constant. Eq. (1) can be solved by constructing a Lagrangian and transformed into the dual:

$$\text{Maximize } W(\alpha) = \sum_{i=1}^{N} \alpha_i - \frac{1}{2} \sum_{i,j=1}^{N} \alpha_i \alpha_j y_i y_j (\mathbf{x}_i \cdot \mathbf{x}_j) \tag{2}$$

s.t:

$$\sum_{i=1}^{N} \alpha_i y_i = 0, \quad 0 \leq \alpha_i \leq C, \text{ for } i = 1, \ldots, N.$$

Then we can have the decision surface of the form:

$$f(\mathbf{x}) = \text{sgn}(\sum_{i=1}^{r} y_i \alpha_i \mathbf{x}_i^s \cdot \mathbf{x} + b) \tag{3}$$

where $\{\mathbf{x}_i^s\}$ is a subset of the training data set. The coefficients α_i and b are the solutions to the quadratic programming problem (2).

To the nonlinearly separable classification problem, we can map the data into another dot product space (called the feature space) F via a nonlinear map $\Phi: R^N \to F$ and perform the above linear algorithm in F. Since the solution has the form:

$$f(\mathbf{x}) = \text{sgn}(\sum_{i=1}^{r} y_i \alpha_i \Phi(\mathbf{x})_i^s \cdot \Phi(\mathbf{x}) + b) \tag{4}$$

it is nonlinear in the original input variables.

According to Mercer's theory, one can choose a kernel function $K(.,.)$ such that

$$K(\mathbf{x}, \mathbf{y}) = \Phi(\mathbf{x}) \cdot \Phi(\mathbf{y}) \tag{5}$$

Then, without considering the mapping explicitly, a nonlinear SVM can be constructed by selecting the proper kernel (for details, see [7])

3 The Construction of Support Vector Machine Committee

3.1 The Architecture of the SVM Committee

Fig.1 illustrates the basic idea of the SVMC. Firstly the feature extraction was done through different methods. Then we have numerous types of features available. For each types of feature a SVM classifier can be got. By aggregating these different SVMs the SVMC can be realized. Note all the types of features should be as independent as possible to each other.

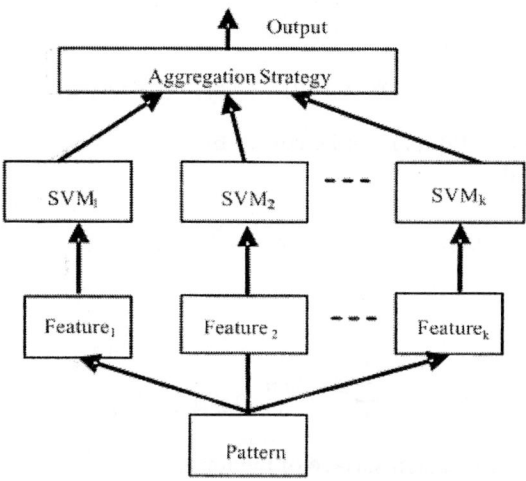

Fig. 1. The architecture of SVM committee

3.2 The Aggregation of the SVMC

Suppose that the task is to use the SVMC comprising K component SVM which is trained through K types of features, and the prediction results of the component SVM are combined through weighted average manner where a weight $\omega_i (i=1,2,...,K)$ satisfying both (6) and (7) is assigned to the i-th component network f_i :

$$0 \leq \omega_i \leq 1 \tag{6}$$

$$\sum_{i=1}^{K} \omega_i = 1 \tag{7}$$

For a given testing vector $\mathbf{x}$, suppose the output of the i-th component SVM is $f_i(\mathbf{x})$ then the output of the committee on $\mathbf{x}$ is:

$$f(\mathbf{x}) = \sum_{i=1}^{K} \omega_i f_i(\mathbf{x}) \tag{8}$$

Suppose we have got the model of the i-th SVM:

$$f_i(\mathbf{x}) = \mathbf{w}_i \cdot \varphi_i(\mathbf{x}) + b_i, \quad i = 1, 2, \ldots, K \tag{9}$$

Then, the output of the SVMC on $\mathbf{x}$ is:

$$f(\mathbf{x}) = \sum_{i=1}^{K} \omega_i \cdot \mathbf{w}_i \cdot \varphi_i(\mathbf{x}) + \sum_{i=1}^{K} \omega_i \cdot b_i \tag{10}$$

Then we can have:

$$f(\mathbf{x}) = \begin{vmatrix} \omega_1 \mathbf{w}_1 \\ \omega_2 \mathbf{w}_2 \\ \vdots \\ \omega_K \mathbf{w}_K \end{vmatrix}^T \cdot \begin{vmatrix} \varphi_1(\mathbf{x}) \\ \varphi_2(\mathbf{x}) \\ \vdots \\ \varphi_3(\mathbf{x}) \end{vmatrix} + \sum_{i=1}^{K} \omega_i \cdot b_i = \overline{\mathbf{w}} \cdot \overline{\varphi}(\mathbf{x}) + \overline{b} \tag{11}$$

where: $\overline{\mathbf{w}} = [\omega_1 \mathbf{w}_1 \quad \omega_2 \mathbf{w}_2 \quad \cdots \quad \omega_K \mathbf{w}_K]$, $\overline{\varphi}(\mathbf{x}) = [\varphi_1(\mathbf{x}) \quad \varphi_2(\mathbf{x}) \quad \cdots \quad \varphi_3(\mathbf{x})]$,

$$\overline{b} = \sum_{i=1}^{K} \omega_i \cdot b_i .$$

From (11) it can be found that the performance of the SVMC depends on the value of ω. The values of $\overline{\mathbf{w}}$ and $\overline{b}$, which are determined by the ω, should ensure that the committee can get better generalization. Just as (1), we can construct following optimization problem:

$$\min J(\overline{\mathbf{w}}, e) = \frac{1}{2} \overline{\mathbf{w}}^T \overline{\mathbf{w}} + \frac{1}{2} \gamma \sum_{j=1}^{N} \xi_j^2 \tag{12}$$

s.t:

$$y_j [\overline{\mathbf{w}}^T \overline{\varphi}(\mathbf{x}_j) + \overline{b}] \geq 1 - \xi_j, \quad j = 1, 2, \ldots, N$$

Using (11) we can eliminate the $\overline{\mathbf{w}}, \overline{b}$, and the corresponding optimization problem can be written as :

$$\min J(\omega, \xi) = \frac{1}{2} \sum_{i=1}^{K} \omega_k^2 \cdot |\mathbf{w}|^2 + \frac{1}{2} \gamma \sum_{j=1}^{N} \xi_j^2 \tag{13}$$

s.t:

$$y_j \cdot [\omega^T \cdot F(\mathbf{x}_j)] \geq 1 - \xi_j, \quad j = 1, 2, \ldots, N$$

where $F(\mathbf{x}_j) = [f_1(\mathbf{x}_j), f_2(\mathbf{x}_j), \ldots, f_K(\mathbf{x}_j)]^T$, $j = 1, 2, \ldots, N$.

Thus this leads to a quadratic programming (QP) problem, and the global optimal value of ω can be easily got. From (13) it can be found that the cost function of SVM committee also consists of a SSE fitting error and a regularization term. The relative importance of these terms is determined by the positive real constant γ. In the case of high noisy data one avoids overfitting problem by taking a smaller γ value.

4 Experimental Results

To evaluate the efficacy of our proposed SVMC and aggregation method, the UCI hand-written digit data was used [8]. This dataset consists of features of handwritten numerals (`0'--`9') extracted from a collection of Dutch utility maps. 200 patterns per class (for a total of 2,000 patterns) have been digitized in binary images. These digits are represented in terms of the following 3 feature sets: A. Profile correlations (dimension: 216) ; B. Fourier coefficients of the character shapes (dimension: 76); C. Pixel averages in 2 x 3 windows (dimension: 240).

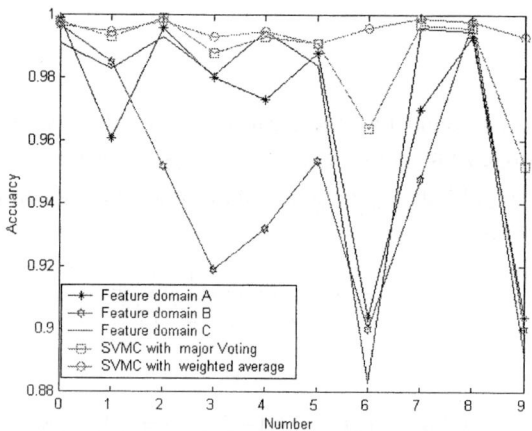

Fig. 2. Classification Accuracies for Each Number (`0'—`9')

From all the patterns 100 patterns per class were selected for training and others for testing. In order to realize the multi-class classification, the one-against-all strategy was selected [9]. During our experiments, the RBF kernel was used and the corresponding parameters were determined by cross-validation method. Fig.2 shows the experimental results. From this table it can be found that the performance of the individual SVM from different feature domains varies greatly. At the same time, the performance of two aggregation methods, major voting and weighted average, is also compared in this figure. Obliviously, weighted average approach is more suitable.

5 Conclusions

In this paper, the SVMC is constructed and applied to do classification problems. For a pattern problem usually several types of features can be used. SVMC can use this feature efficiently and a classifier with improved generalization can be obtained. Moreover, a novel aggregation approach was proposed in this paper. Compared to major voting strategy, our approach can evaluate the relevant importance of each type of the feature and ensure that the SVM committee gets better generalization.. Finally, the simulating results demonstrate the effectiveness and efficiency of our approach. Future research will include how to use this approach to solve more real-world problems.

References

1. Vapnik, V.:The nature of Statistical Learning Theory. Wiley ,New York (1998)
2. Müller, K.-R. , Mika, S., Rätsch, G., Tsuda, K., and Schölkopf, B. :An introduction to Kernel-based Learning Algorithms. IEEE Trans. Neural Network (2001)181–201
3. Osuna, E. , Freund, R. ,and Girosi :Training Support Vector Machines: An Application to Face Detection. In Proc. Computer Vision and Pattern Recognition, San Juan, PR(1997)130–136
4. Pontil, M.,and Verri, A.:Support Vector Machines for 3-D Object Recognition. IEEE Trans. Pattern Anal. Machine Intell(1998) 637–646
5. Caleanu,C.D.:Facial Recognition Using Committee Of Neural Networks", 5th Seminar on Neural Network Applications in Electrical Engineering (2000)
6. Hanaen L.K., Salamon, P. :Neural Network Ensembles. IEEE Trans. Pattern Analysis and Machine Intelligence (1990) 993-1001
7. B. Schoelkopf and A.Smola , Learning with Kernels, MIT Press, Cambridge, MA(2002)
8. Blake, C. : UCI repository of machine learning databases, Department of Information and Computer Science, University of California, Irvine, CA(1998) http://www.ics.uci.edu/~mlearn/MLRepository.html
9. Hsu, C.-W., Lin, C.-J. :A Comparison on Methods for Multi-Class Support Vector Machines.Technical report, Dept. of Computer Science and Information Eng., Nat'l Taiwan Univ(2001)

Automatic Modulation Classification by Support Vector Machines

Zhijin Zhao[1,2], Yunshui Zhou[2], Fei Mei[2], and Jiandong Li[1]

[1] National Key Lab of Integrated Service Network, Xidian University,
Xi'an, Shaan Xi 710071, China
[2] Hangzhou Institute of Electronics Engineering,
Hangzhou, Zhe Jiang 310018, China
Zhaozj03@hziee.edu.cn, zhouys@zjc.zjut.edu.cn,
madrista_1979@hotmail.com

Abstract. Automatic classification of analog and digital modulation signals plays an important role in communication applications such as an intelligent demodulator, interference identification and monitoring, so many investigations have been carried out in the past. Support Vector Machines (SVMs) maps inputs vectors nonlinearly into a high dimensional feature space and constructs the optimum separating hyperplane in space to realize signal classification. In this paper, a new method based on SVM for classifying AM, FM, BFSK, BPSK, USB and LSB is proposed. The classification results for real communication signals using SVMs are given. Compared with radial basis function neural network (RBFNN) method, the method can classify these signals well, and the correct classification rates are above 82%.

1 Introduction

Recently with the diversification of communication styles, a universal demodulator which has a intelligent function becomes needed. One of the functions of a universal demodulator is automatic classification of modulation signals. Namely when the transmitter changes the modulation types adaptively in response to the environment, the demodulator must classify the corresponding modulation types in real time and select the suitable demodulation method.

Many studies using a decision-theoretical or a statistical pattern recognition framework have been carried out [7,8,9]. SVM is a pattern recognition method. It has been used in speech recognition [5], digital recognition and etc. Being different from other learning machines, SVMs [1,2] uses a structural risk minimization (SRM) principle, while others use an empirical risk minimization principle. It uses a kernel function for efficiently performing computations in high dimensional spaces and constructs nonlinear decision function to perform an optimal separating hyperplane in feature space. In this paper a new method based on support vector machines (SVMs) for classifying AM, FM, BFSK, BPSK, USB and LSB is proposed.

2   Support Vector Machines

Support vector machines are based on the structural risk minimization principle and Vapnik-Chervonenkis (VC) dimension from statistical learning theory developed by Vapnik, et al.[1] Traditional techniques for pattern recognition are based on the minimization of empirical risk, that is, on the attempt to optimize performance on the training set, SVMs minimize the structural risk to reach a better performance[2].

We can suppose that S is a set that is made up of points $x_i (i=1,2\cdots,N)$, which belong to R^n, These points are divided into two classes by an objective function y_i,

$$y_i = \begin{cases} 1 & x_i \in S_1 \\ -1 & x_i \in S_2 \end{cases} \tag{1}$$

where S_1 and S_2 belong to different classes. We try to find a hyperplane to separate the two classes, and sort the same class in same side of the hyperplane as much as possible, and make the margin as far as possible. If S can be separated linearly, there may be $w \in R^n$, $b \in R$ to satisfy

$$\left. \begin{array}{ll} w \cdot x_i + b \geq 1 & y_i = 1 \\ w \cdot x_i + b \leq -1 & y_i = -1 \end{array} \right\} \tag{2}$$

Eqn (2) also can be represented by

$$y_i (w \cdot x_i + b) \geq 1 \tag{3}$$

Parameters (w, b) have determined a hyperplane,

$$w \cdot x_i + b = 0 \tag{4}$$

This plane is called the separating hyperplane. The problem of finding the optimal separating hyperplane is converted to an optimal problem as follows.

$$\min \frac{1}{2} \|w\|^2 \quad \text{s.t.} \quad y_i(w \cdot x_i + b) \geq 1 \quad (i=1,2\cdots,N) \tag{5}$$

It is then converted to a dual problem by using Lagrange multiplies,

$$\max \sum_{i=1}^{N} \alpha_i - \frac{1}{2} \sum_{i,j=1}^{N} \alpha_i \alpha_j y_i y_j (x_i \cdot x_j) \quad \text{s.t.} \quad \sum_{i=1}^{N} y_i \alpha_i = 0 \quad \alpha \geq 0 \tag{6}$$

When S cannot be separated linearly, introducing a nonnegative relax factor $\xi = (\xi_1, \cdots \xi_N)$, Eqn (3) can be rewritten as

$$y_i(w \cdot x_i + b) \geq 1 - \xi_i \qquad (7)$$

The optimal problem can be described as

$$\max \sum_{i=1}^{N} \alpha_i - \frac{1}{2} \sum_{i,j=1}^{N} \alpha_i \alpha_j y_i y_j (x_i \cdot x_j) \qquad (8)$$

$$\text{s.t.} \sum_{i=1}^{N} y_i \alpha_i = 0 \quad 0 \leq \alpha_i \leq C$$

Formula (8) is a general form of SVM. When C tends to infinite, formula (8) degenerates into a linear separating problem as formula (6). Replacing $y_i y_j x_i \cdot x_j$ by D_{ij}, the optimal objective function turns to be the maximum $\sum_{i=1}^{N} \alpha_i - \frac{1}{2} \sum_{i,j=1}^{N} \alpha_i D_{ij} \alpha_j$. Obviously, this is a quadratic program. We can solve it by using the sequential minimal optimization (SMO) proposed by Platt [3]. When parameters α_i and b are obtained, the different classes can be distinguished by objective function

$$y = \text{sgn}(w^* \cdot x + b^*) = \text{sgn}(\sum_{i=1}^{N} \alpha_i^* y_i x_i \cdot x + b^*) \qquad (9)$$

In most cases, discrimination is not linear in input space. A higher order function is introduced for mapping a nonlinearly separating problem to a linearly separating problem. Because the optimal problem mentioned above deals with inner product only, a kernel function $K(x_i, x_j)$ can be constructed to substitute the inner product.

Two typical kernel functions are the polynomial kernel function as follows and the Gauss (Radial Basis Function) kernel function, defined by

$$K(x_i, x_j) = [(x_i \cdot x_j) + 1]^d \qquad (10)$$

$$K(x_i, x_j) = \exp(-\frac{|x_i - x_j|^2}{2\sigma^2}) \qquad (11)$$

A kernel function exists when the Mercer condition is satisfied.

When SVM is used for classification, it works like a neural network, which classifies the different classes by inner product between the input vectors and support vectors. Inner product is substituted by kernel function operation.

3 Feature Parameters

In this paper four parameters[6,7] extracted from the signals are used and the brief explanation are given here.

3.1 γ_{max}

γ_{max} is defined by

$$\gamma_{max} = \max|FFT[a_{cn}(i)]|^2 / N_s \qquad (12)$$

where N_S is the number of samples per segment and $a_{cn}(i)$ is the value of the normalized-centered instantaneous amplitude at time instants $t = i/f_s$, $i = 1, \cdots, N_S$, and it is defined by $a_{cn}(i) = a_n(i) - 1$, $a_n(i) = \dfrac{a(i)}{m_a}$, m_a is the average value of instantaneous amplitude evaluated over one segment, i..e.

$$m_a = \frac{1}{N_S} \sum_{i=1}^{N_S} a(i) \qquad (13)$$

Thus, γ_{max} represents the maximum value of the spectral power density of the normalized-centered instantaneous amplitude of the intercepted signal.

3.2 The Spectrum Symmetry P

It is used for measuring the spectrum symmetry around the carrier frequency, and it is based on the spectral powers for lower and upper sidebands of the signal. It is defined as

$$P = \frac{P_L - P_U}{P_L + P_U} \qquad (14)$$

$$P_L = \sum_{i=1}^{f_{cn}} |S(i)|^2 \quad P_U = \sum_{i=1}^{f_{cn}} |S(i + f_{cn} + 1)|^2 \qquad (15)$$

$S(i)$ is the Fourier transform of the signal s, $(f_{cn} + 1)$ is the sample number corresponding to the carrier frequency, f_c and f_{cn} is defined as

$$f_{cn} = (f_c N_S / f_s) - 1 \qquad (16)$$

3.3 σ_{af}

σ_{af} is the standard deviation of the absolute value of the normalized-centered instantaneous frequency f_N, evaluated over the non-weak intervals of a signal segment. And it is defined by

$$\sigma_{af} = \sqrt{\frac{1}{C}[\sum_{a_n(i)>a_t} f_N^2(i)] - [\frac{1}{C}\sum_{a_n(i)>a_t}|f_N(i)|]^2} \qquad (17)$$

where

$$f_N(i) = \frac{f_m(i)}{R_S}, \quad f_m(i) = f(i) - m_f, \quad m_f = \frac{1}{N_S}\sum_{i=1}^{N_S} f(i) \qquad (18)$$

a_t is a threshold for $a(i)$ below which the estimation of the instantaneous frequency is sensitive to the noise, and R_S is the symbol rate of the digital symbol sequence.

3.4 μ_{42}^f

μ_{42}^f is the kurtosis of the normalized-centered instantaneous frequency, defined by

$$\mu_{42}^f = \frac{E[f_N^4(i)]}{\{E[f_N^2(i)]\}^2} \qquad (19)$$

4 Classification Result

For samples of AM, FM, BFSK, BPSK, USB and LSB collected in real environment, experiments have been done by SVM using Gauss and linear kernel functions. For comparison, experiments results using radial basis function neural network (RBFNN) are also given. Through many experiments, the better choice of spread of radial basis functions is 50, and the better choice of biases of RBFNN is 0.016652. The correct classification rates using the same samples applied to train and test are in Table 1. The correct classification rates using the different samples applied to train and test are in Table 2.

From Tables, we can conclude that the SVMs methods can identify correctly these real signals, and the correct classification rate of the SVMs method using linear kernel function is the highest. RBFNN can not effectively classify the modulation signals in real environment. The reason is that SVMs[1,2] uses a structural risk minimization (SRM) principle, while RBFNN uses an empirical risk minimization principle.

5 Conclusion

In this paper, we proposed the method using SVMs for classifying modulation. Results show that better results can be obtained by using linear kernel than Gauss kernel, and the performance of the SVMs method is much better than that of RBFNN's.

Table 1. Comparison of correct rate for various classification methods using same samples

Signal Type	Gauss kernel	Linear kernel	RBFNN
AM	92.68%	97.56%	95.12%
FM	100%	100%	100%
BFSK	100%	100%	95.24%
BPSK	83.78%	100%	91.89%
USB	93.76%	100%	47.54%
LSB	96.3%	100%	61.11%

Table 2. Comparison of correct rate for various classification methods using different samples

Signal Type	Gauss kernel	Linear kernel	RBFNN
AM	90.16%	90.16%	77.87%
FM	90.74%	90.74%	59.26%
BFSK	98.45%	98.45%	100%
BPSK	83.33%	83.33%	67.59%
USB	95.5%	95.5%	14.41%
LSB	82.14%	82.14%	5.36%

References

1. Vapnik, V.N.: Statistical Learning Theory. New York Wiley (1998)
2. Christopher, Burges, C.J.C.: A Tutorial on Support Vector Machines for Pattern Recognition. Data Mining and Knowledge, Vol. 2, No.2 (1998) 121-167
3. Platt, J.: Sequential Minimal Optimization: A Fast Algorithm for Training Support Vector Machines. Microsoft Research Technical Report MSR-TR-98-4 (1998)
4. Taira, S., Murakami, E.: Automatic Classification of Analogue Modulation Signals by Statistical Parameters. MILCOM '99 Conference Proceedings
5. Wang, Z.P., Zhao, L., Zou, C.: Support Vector Machines for Emotion Recognition. Chinese Speech Journal of Southwest University (English Edition) Vol. 19, No.4 Dec. (2003) 307-310
6. Nandi, A.K., Azzouz, E.E.: Automatic Analogue Modulation Recognition. Signal Processing 46 (1995) 211-222
7. Nandi, A.K., Azzouz, E.E.: Modulation Recognition Using Artificial Neural Networks. Signal Processing 56 (1997) 165-175
8. Samir, S., Soliman, Hsue, S.-Z.: Signal Classification Using Statistical Moments. IEEE Transactions on Communications, Vol. 40, No.5 May (1992)
9. Leonardo, M., Reyneri: Unification of Neural and Wavelet Networks and Fuzzy Systems. IEEE Transactions on Neural Network, Vol. 10, No.4 July (1999)

Blind Source Separation Using for Time-Delay Direction Finding

Gaoming Huang[1,2], Luxi Yang[1], and Zhenya He[1]

[1] Department of Radio Engineering, Southeast University, Nanjing China, 210096
[2] Naval University of Engineering, Wuhan China, 430033
redforce@sohu.com lxyang@seu.edu.cn

Abstract. Direction Finding is an important component in Electronic Warfare (EW) system. The direction finding precision may directly affect signal sorting, recognition, location and jamming decision etc. It is an urgent task to improve the direction finding precision. To solve the problems of time-delay direction finding, this paper proposes a novel way mainly based on Blind Source Separation (BSS) to improve time-delay direction finding performance and completely analyzes the algorithm. The implementation of this time-delay direction finding method is therefore simple and the computation burden is light. The direction finding precision by the modified method is much better than what by conventional way. Experiments conducted in this paper show that applying BSS to time-delay direction finding is efficient and feasible.

1 Introduction

In the modern EW signal environments, the signals are denseness and the type of signals is also complicated, which make the task of signal processing in electronic reconnaissance very difficult. As the position of emitters is fixed or the vary speed relative to signals is very slow, then the direction finding become a key technology in current electronic reconnaissance. At the same time, there are more demand for direction finding such as high precision, high resolving ability, multi signal simultaneous processing and real-time direction finding etc. These problems are difficult to be solved by traditional direction finding technology, which urge people to search new methods for them. There are lots kinds of direction finding technologies, which are all face an engineering problem such as non-ideal state of components or error of engineering measurement etc. An effective novel method should been acquired to solve these problems in direction finding, which may be very helpful to signal sorting and recognition in electronic reconnaissance especially to passive location [1].

Time-delay direction finding has been an important method by the high sensitivity, high precision and excellent real-time property. The method of time-delay direction finding can obtain the arrival of azimuth (AOA) of electric wave by the measurement of the difference of the arrival time to each direction finding antenna, which is the time delay. As the bad anti-jamming property and the carrier wave must have determinate modulate, time-delay direction finding has not been a broad application at present. Considering these problems, this paper proposed a novel processing method as: applying BSS method to separate multi mixing signals, then obtaining time delay by correlation, the direction of arrival (DOA) will be obtained by corresponding calculation at last. The structure of this paper is as follows: In the Section 2, we will analyze the issue of time-delay direction finding and pose the problem. In the Section 3, the direction finding algorithm will be analyzed. Some experiments of the algorithm applied in this paper are conducted in the Section 4. Finally a conclusion is given.

2 Problem Formulation

The model of time delay between the signals received by two groups of separate antennas is shown as Fig.1.

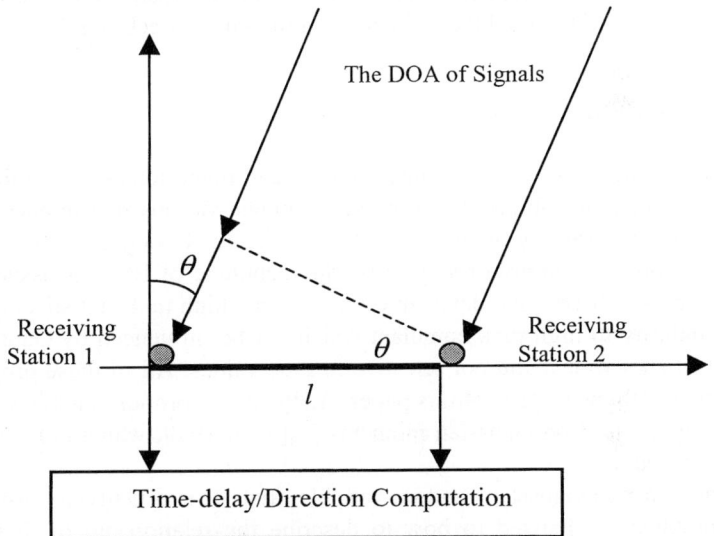

Fig. 1. Signal receiving model

Supposing that each receiving station has m receiving channels, the mutual distance is l. If have $n(n \leq m)$ narrow band signals, $r(\theta_i)$ is the antenna respond to narrow band signal on direction θ_i, then $r(\theta_i) = [1, e^{j\phi_i}, ..., e^{j(m-1)\phi_i}]^T$, where $\phi_i = 2\pi l \sin\theta_i / \lambda$, λ is the wave length. The receiving signal model could be described as:

$$\mathbf{x}(t) = \mathbf{H} \cdot \sum_{i=1}^{n} \mathbf{r}(\theta_i) s_i(t) + \mathbf{n}(t) = \mathbf{A} \cdot \mathbf{s}(t) + \mathbf{n}(t) \tag{1}$$

where $\mathbf{x}(t)$ is the receiving signals of antenna. $\mathbf{A}$ is a $m \times n$ mixing matrix, which is the product of responding function $\mathbf{R} = [\mathbf{r}(\theta_1), ..., \mathbf{r}(\theta_n)]$ and the mixing matrix $\mathbf{H}$ during the signal transmitting process. $\mathbf{s}(t)$ are the source signals including radar signals, interference signals etc. $\mathbf{n}(t)$ are $m \times 1$ dimension noise signals, which include exterior noise, inner noise and electricity noise etc. The signals are assumed as mutually independent and independent with noise in the following analysis. The key problem of direction finding based time-delay is how to eliminate or reduce the affect of noise and mixing signals to time delay computation by correlation, which is the main problem that this paper want to solve.

3 Direction Finding Algorithm

Direction finding algorithm includes three steps: blind separation processing, correlation processing, and azimuth computation. Before all these steps, the receiving signals should be normalized to avoid the influence to following processing.

3.1 Blind Separation Processing

Signals may be affected by outside interferences and inner noises during the transmitting process. The chief object is to remove or reduce various interference and noise, also include the uncertainty by these unexpected signals. Generally we can begin with measure the non-Gaussianity of signals by the technique of BSS. Gaussian moments directly use noisy observation data can estimate some high rank statistical characteristic. It is similarity to high rank cumulant which not be influenced by Gaussian noise, which has a better robust and not sensitive to rude values. All of these properties just are consistent with the request of this paper. A.Hyvärinen proposed a noisy ICA fixed-point algorithm based on Gaussian moments [2][3] in 1999, which is fit for the BSS request of this paper.

Denoting z a non-Gaussian random variable, and n is a Gaussian noise variable σ^2, the problem is changed to how to describe the relationship of $E\{G(z)\}$ and $E\{G(z+n)\}$. We choose $G(\cdot)$ as a density function of a zero-mean Gaussian random variable. The zero-mean and variance c^2 Gaussian density function is defined as:

$$\varphi_c(x) = \frac{1}{c}\varphi(\frac{x}{c}) = \frac{1}{\sqrt{2\pi}c}\exp(-\frac{x^2}{2c^2}) \qquad (2)$$

We can get the equation (3) which defined as a theorem in [3]:

$$E\{\varphi_c(z)\} = E\{\varphi_d(z+n)\} \qquad (3)$$

Thus we maximize, for quasi-whitened data $\tilde{\mathbf{x}}$ the following contrast function:

$$\max_{\|\mathbf{w}\|=1} \left| E\{\varphi_{d(\mathbf{w})}^{(k)}(\mathbf{w}^T\tilde{\mathbf{x}})\} - E\{\varphi_c^{(k)}(v)\} \right|^p \qquad (4)$$

with $d(\mathbf{w}) = \sqrt{c^2 - \mathbf{w}^T \tilde{\mathbf{R}}_{nn} \mathbf{w}}$. It can see that $\varphi_c^{(k)}(x)$ corresponds to a Gaussian density function when $k = 0$. The contrast function is the form of a negative log-density of a super-Gaussian variable when $k = -2$. In fact, $\varphi^{(-2)}(u)$ can be approximated by the function as $G(u) = 0.5 \log \cosh(u)$, which has been widely used in ICA.

We can obtain the following preliminary form of the fixed-point iteration for quasi-whitened data:

$$\mathbf{w}^* = E\{\tilde{\mathbf{x}} \varphi_{d(\mathbf{w})}^{(k+1)}(\mathbf{w}^T \tilde{\mathbf{x}})\} - (\mathbf{I} + \tilde{\mathbf{R}}_{nn}) \mathbf{w} E\{\varphi_{d(\mathbf{w})}^{(k+2)}(\mathbf{w}^T \tilde{\mathbf{x}})\} \tag{5}$$

where $\mathbf{w}^*$, the new value of $\mathbf{w}$, is normalized to unit norm after every iteration. One could adapt c before every step so that $d(\mathbf{w}) = \sqrt{c^2 - \mathbf{w}^T \tilde{\mathbf{R}}_{nn} \mathbf{w}} = 1$. This gives finally the following algorithm with bias removal for quasi-whitened data:

$$\mathbf{w}^* = E\{\tilde{\mathbf{x}} g(\mathbf{w}^T \tilde{\mathbf{x}})\} - (\mathbf{I} + \tilde{\mathbf{R}}_{nn}) \mathbf{w} E\{g'(\mathbf{w}^T \tilde{\mathbf{x}})\} \tag{6}$$

The function $g(\cdot)$ is here the derivative of $G(\cdot)$, and it can be chosen as follows: $g_1(u) = \tanh(u)$, $g_2(u) = u \exp(-u^2/2)$, $g_3(u) = u^3$. These functions include the general non-linear functions used in the fixed-point algorithm. It is clear that adding R_{nn} is the key to remove bias. Using the decorrelation method as noisy-free case, all of the independent components can be obtained at last.

3.2 Correlation Computation

After completing normalization and blind separation processing, the following work is to compute the delay $\Delta \tau$ of the corresponding two antennas. Here we utilize correlation processing. To a same emitter, the receiving time is different as the as the different receiving position. In this case, we must consider the resemble property of the two signals during the time varying. Delaying $\Delta \tau$ to $s_1(n)$ and change it to $s_1(n - \tau)$, then the coefficient $r_{s_1 s_2}(\tau)$ can be written as:

$$r_{s_1 s_2}(\tau) = \sum_{n=-\infty}^{\infty} s_1(n) s_2(n - \tau) \tag{7}$$

When τ varying from $-\infty$ to $+\infty$, $r_{s_1 s_2}(\tau)$ is a function of τ, call $r_{s_1 s_2}(\tau)$ as the coefficient of $s_1(n)$ and $s_2(n - \tau)$, τ is the delay time of $s_2(n - \tau)$. When $|r_{s_1 s_2}(\tau)|$ reach the maximal value at τ_0, then τ_0 is the time difference $\Delta \tau$ of the two signals.

3.3 Azimuth Computation

After $\Delta \tau$ has been obtained, the computation of azimuth is relatively simple. From the model established in the Fig.1, the following equations can be obtained: $\Delta l = l \sin \theta$, $\Delta \tau c = \Delta l = l \sin \theta$, where $c = 3 \times 10^8 m/s$ is the electric wave transmitting speed, the DOA can be obtained as:

$$\theta = \arccos\big((c \cdot \Delta\tau)/l\big) \tag{8}$$

From this equation we can see that the precision of DOA decided by the base line length l and the measure precision $\Delta\tau$. By $\Delta\tau c = \Delta l = l\sin\theta$, $\Delta\tau = (l\sin\theta)/c$ can be obtained. The partial derivative of $\Delta\tau$ can be written as:

$$d(\Delta\tau) = \frac{\partial(\Delta\tau)}{\partial\theta}d\theta + \frac{\partial(\Delta\tau)}{\partial l}dl + \frac{\partial(\Delta\tau)}{\partial c}dc,$$ while l, c are all constant, so the result of partial derivative is $d(\Delta\tau) = \frac{l}{c}\cos\theta d\theta$, then:

$$d\theta = (c \cdot d(\Delta\tau))/(l\cos\theta) \tag{9}$$

From Equation (9) we can know that the precision of direction finding is limited by the base line length l and the measure precision $\Delta\tau$, it also relate to the measure angle. It is obviously when $\theta = 45°$, the effect is the best. So in order to improve direction finding, usually 4 groups of antenna are applied to cover the whole azimuth.

4 Simulations

In order to verify the validity of this direction finding algorithm proposed in this paper, here a series of experiments have been conducted. The background of the experiments are as: Assuming there are two sets of receiving antennas, their distance is 30m, the relative direction of DOA to really north way is 30°. From the establish receiving model in Fig.1 we can know that the delay time is 50ns, if the sampling interval is 1ns, then the delay is 50 sampling intervals. The source signals include one FM noisy jamming and which can be generated using the corresponding steps from [4]. The first step is generating Gaussian noisy voltage, then using a six-apices ellipse filter to form noisy bandwidth. At last taking the outputs from filter to a monofier, the outputs represent the outputs of noisy FM jamming signal. The other signal is a FM signal as: $s_{FM}(t) = \cos(2\pi f_0 t + \phi(t))$, where f_0 is the carrier frequency, $\phi(t)$ is a modulating component. The experiment process is as: First taking a direct correlation to the receiving signals in the interference background (Shown as in Fig.2). From the result as Fig.3, we can see that there are two extremum, which is not the perfect result. Then conduct correlation processing to the separated results (shown as Fig.4) by BSS. The correlation results as Fig.5, the delay time is just 50 sampling intervals. By equation (8) then: $\theta = \arccos(c \cdot \Delta\tau/l) = \arccos\big((3 \times 10^8 \times 50 \times 10^{-9})/30\big) = \arccos(1/2) = 30°$, which is consistent to the assuming conditions.

A lot of experiments have been conducted by the changing the impact of additive noise. The results show that the estimation of delay time after BSS processing is more accuracy and stabilization than direct correlation. The anti-jamming property is also excellent.

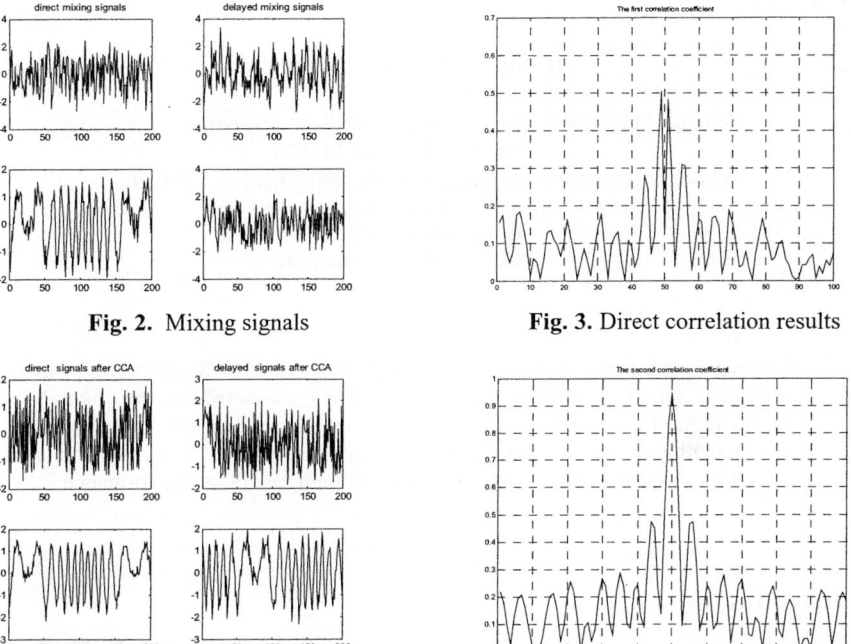

Fig. 2. Mixing signals

Fig. 3. Direct correlation results

Fig. 4. Signal separating results

Fig. 5. The second correlation results

5 Conclusion

This paper proposes applying BSS method to delay-time direction finding, which can effectively overcome the sensitivity to interference of this direction finding method. It can also avoid the localization of time-delay direction finding and improve the precision of it, so as to the location precision. The important contribution of this direction finding method proposed by this paper can improve the applicability of time-delay direction finding.

References

1. Hu, L.Z.: Passive Locating, National Defence Industry Press, Beijing (2004)
2. Hyvärinen, A.: Fast ICA for noisy data using Gaussian moments, ISCAS '99, Proceedings of the 1999 IEEE International Symposium on, Vol.5 (1999) 57 –61
3. Hyvärinen, A.: Gaussian moments for noisy independent component analysis, IEEE Signal Processing Letters, Vol. 6, Issue: 6 (1999) 145 –147
4. Schleher, D.C.: Electronic Warfare in the Information Age, Artech House (1999)

Frequency-Domain Separation Algorithms for Instantaneous Mixtures

Tiemin Mei [1,2] and Fuliang Yin [1]

[1] School of Electronic and Information, Dalian university of technology,
Dalian, P.R. China 116023
[2] School of Information Engineering, Shenyang Institute of Technology,
Shenyang, P. R. China 110168
meitiemin@163.com
flyin@dlut.edu.cn

Abstract. In this paper, we proposed algorithms for the blind separation of instantaneous mixtures in frequency domain. They are based on the joint diagonalization of power spectral density matrices. These algorithms are suitable for both stationary and nonstationary but temporally colored sources. Simulation results show that the new algorithms are of high separation performance.

1 Introduction

For the blind separation of instantaneous mixtures, many algorithms have been proposed in time-domain. These are information theory based algorithms [1], nonlinear neural networks-based algorithms [2,6], and statistics-based (including high and second order statistics) algorithms [3-6]. Some researchers have studied this problem in time-frequency domain [7]. There are some other researchers who have studied this problem in frequency-domain [8]. Frequency-domain algorithms are usually used for the separation of convolutively mixed signals. These algorithms are certainly suitable for the separation of instantaneous mixtures, which is a special case of the convolutive mixtures. As the so-called frequency-domain algorithms concerned, they should be exactly named as time-frequency domain approaches too, because they exploit both the time and frequency domain properties of the sources and the mixing system.

As a special case of the convolutive mixtures, the mixing and unmixing models of instantaneous mixtures in frequency domain are exactly the same (in formula) as those in time domain, this enable us to separate the instantaneous mixtures exactly in frequency domain other than time-frequency domain.

In frequency domain, the temporally colored sources can be looked by analog as complex nonstationary processes. We establish the algorithms on the basis of decorrelation theory of Blind Source Separation (BSS) for nonstationary processes proposed in [2]. Simulation results show that these frequency-domain algorithms are valid and of high performance.

2   Problem Formulation

Assuming $S(t) = [s_1(t), s_2(t), ..., s_n(t)]^T$ ($t=0,1,2,...$), where $s_i(t)$ ($i=1,2,...,n$) are the zero mean, temporally colored and uncorrelated real source signals, that is,

$$E[s_i(t)] = 0, \quad i = 1, 2, ..., n \tag{1}$$

$$r_{s_i s_j}(t_1, t_2) = E[s_i(t_1) s_j(t_2)] = 0, \quad i \ne j \ (i, j = 1, 2, ..., n), \ \forall \ t_1, t_2 \tag{2}$$

Let $X(t) = [x_1(t), x_2(t), ..., x_n(t)]^T$, the mixing model can be expressed as follows.

$$X(t) = AS(t) \tag{3}$$

where A is the unsingular mixing matrix. Let $Y(t) = [y_1(t), y_2(t), ..., y_n(t)]^T$, the output of separation system, and let W be the separation matrix, then the separation model is,

$$Y(t) = WX(t) = WAS(t) \tag{4}$$

Defining the global transform matrix $G = WA$, then

$$Y(t) = GS(t) \tag{5}$$

If we can find out the separation matrix W that makes the global transform matrix G be the product of a diagonal matrix D and a permutation matrix P, then we say that the sources are separated.

The uncorrelation condition (2) can be equivalently expressed in frequency domain as follow.

$$P_{s_i s_j}(\omega) = 0, \ \forall \omega, i \ne j \ (i, j = 1, ..., n) \tag{6}$$

where $P_{s_i s_j}(\omega)$ is the cross power spectral density of source signal. Let $S_F(\omega) = \text{DFT}[S(t)]$, $X_F(\omega) = \text{DFT}[X(t)]$ and $Y_F(\omega) = \text{DFT}[Y(t)]$, where $\text{DFT}[\cdot]$ is the component-wise Discrete Fourier Transform operator; $\omega = \frac{2\pi i}{N}$ ($i = 0, 1, 2, ..., N-1$), N is the length of signals. The mixing and separation model can be expressed in frequency domain as follows:

$$X_F(\omega) = AS_F(\omega) \tag{7}$$

$$Y_F(\omega) = WX_F(\omega) \tag{8}$$

If we take $S_F(\omega)$, $X_F(\omega)$ and $Y_F(\omega)$ as complex stochastic processes, then the frequency-domain mixing and separation model (7) and (8) are exactly the same as the time-domain models (3) and (4). So we can deal with in frequency domain the blind separation problem of instantaneous mixtures the same way as that in time-domain.

3 Algorithm Development

It is proved that if the source signals are subject to the uncorrelation condition (2) and the following condition [2]:

$$\frac{r_{s_i s_i}(t)}{r_{s_j s_j}(t)} \neq \text{Constant}, \quad i \neq j, i, j = 1,...,N \tag{9}$$

then the decorrelation of the output of separation system is a necessary and sufficient condition for the separation of nonstationary stochastic processes, that is,

$$r_{y_i y_j}(t_k) = E[y_i(t_k) y_j(t_k)] = 0 \quad i \neq j; i, j = 1,2,...,n; k = 1,2,...,n \tag{10}$$

In frequency domain, if the source signals are subject to the frequency-domain decorrelation condition (6) and the following condition:

$$\frac{p_{s_i s_i}(\omega)}{p_{s_j s_j}(\omega)} \neq \text{Constant}, \quad i \neq j, i, j = 1,...,N \tag{11}$$

where $p_{s_j s_j}(\omega)$ is the power spectral density of $s_j(t)$. We have the equivalent frequency-domain condition for temporally colored sources as following,

$$p_{y_i y_j}(\omega_k) = E[y_{Fi}(\omega_k) y_{Fj}(\omega_k)] = 0 \quad i \neq j; i, j = 1,2,...,n; k = 1,2,...,n \tag{12}$$

where $y_{Fi}(\omega)$ is the ith component of $Y_F(\omega)$.

Defining the cross power spectral density matrix as $P_F(\omega) = E[Y_F(\omega) Y_F^T(\omega)]$. It is Hermitian and positive definite. The condition (12) is equivalent to the joint diagonalization of $P_F(\omega_i)$ ($i=1,2,...,n$). According to the Hadamard's inequality of positive definite matrix, the following objective function is proposed as that in [2,6],

$$f_i(W) = \frac{1}{2} \log \left[\frac{\|D_F(\omega_i)\|}{\|P_F(\omega_i)\|} \right] \geq 0 \quad i = 1,2,...,n \tag{13}$$

where $D_F(\omega_i) = \text{diag}[P_F(\omega_i)]$, the operator $\text{diag}[\cdot]$ diagonalizes the matrix by setting all the off-diagonal elements to zeros.

The gradient of $f_i(W)$ is,

$$\frac{df_i(W)}{dW} = \text{Re}\left[(D_F^{-1}(\omega_i) P_F(\omega_i) - I) W^{-T}\right] \tag{14}$$

where $\text{Re}[\cdot]$ maintains the real part of the operand, I is the identity matrix. Its natural gradient is [9],

$$\frac{df_i(W)}{dW} W^T W = \text{Re}\left[(D_F^{-1}(\omega_i) P_F(\omega_i) - I) W\right] \tag{15}$$

Dividing $Y_F(\omega)$ into L ($>n$) blocks, the block size is M, the ith block ($i=0,1,…,L-1$) is represented by $Y_F(\omega_i)$. $P_F(\omega_i)$ is estimated from $Y_F(\omega_i)$, that is, $P_F(\omega_i) = Y_F(\omega_i)Y_F^T(\omega_i)$. Optimizing the objective functions alternatively, we obtain the block-based iterative algorithm with natural gradient:

$$W(k+1) = W(k) - \mu \times \text{Re}[D_F^{-1}(\omega_{(\text{mod}(k,L))})P_F(\omega_{(\text{mod}(k,L))}) - I]W(k) \tag{16}$$

where mod(k,L) returns the remainder of $\frac{k}{L}$.

If $P_F(\omega_i)$ is estimated adaptively, we obtain the on-line learning algorithm with natural gradient:

$$W(k+1) = W(k) - \mu \times \text{Re}[D_F^{-1}(\omega_k)P_F(\omega_k) - I]W(k) \tag{17}$$

where $P_F(\omega_k)$ is estimated adaptively with

$$P_F(\omega_k) = \beta P_F(\omega_{k-1}) + (1-\beta)Y_F(\omega_k)Y_F^T(\omega_k), \; (0 < \beta < 1) \tag{18}$$

Complex value computation is needed in the algorithm (16) and (17), so it is very time-consumption. But in fact, the real part and the imaginary part of $S_F(\omega)$, $X_F(\omega)$ and $Y_F(\omega)$ are also subject to the same relationship of (7) and (8), so we can only exploit the real or imaginary part of $Y_F(\omega)$ instead of $Y_F(\omega)$ itself in the algorithms. Discarding the operator Re[·] in (16) and (17), we obtain the real-valued version with the real or imaginary part of $Y_F(\omega)$.

$$W(k+1) = W(k) - \mu[D_F^{-1}(\omega_{(\text{mod}(k,L))})P_F(\omega_{(\text{mod}(k,L))}) - I]W(k) \tag{19}$$

$$W(k+1) = W(k) - \mu[D_F^{-1}(\omega_k)P_F(\omega_k) - I]W(k) \tag{20}$$

where $P_F(\omega_k)$ and $D_F(\omega_k)$ are estimated from the real or imaginary part of $Y_F(\omega)$.

4 Simulations

In this section, we present several simulation results to demonstrate the validity of the proposed algorithms in this paper.

We calculate the Error Index (EI) with the following formula for the performance evaluation:

$$EI = 10\log_{10}\left\{\frac{1}{n}\sum_{i=1}^{n}\left(\sum_{j=1}^{n}\frac{g_{ij}^2}{\max_i g_{ij}^2} - 1\right) + \frac{1}{n}\sum_{j=1}^{n}\left(\sum_{i=1}^{n}\frac{g_{ij}^2}{\max_j g_{ij}^2} - 1\right)\right\} \tag{21}$$

Three speech signals sampled at 16kHz, with length $L=30000$ are used in the experiments. The mixing matrix is as follow:

$$A = \begin{bmatrix} -0.9471 & -1.0559 & -1.2173 \\ -0.3744 & 1.4725 & -0.0412 \\ -1.1859 & 0.0557 & -1.1283 \end{bmatrix} \quad (22)$$

For the block-based algorithm (16) or (19), the mixture signals are transformed to frequency domain with Fast Fourier Transform (FFT) of 32768 samples. We set that $\mu = 0.0003$, block size $M=30$. The FFT of the mixtures, the real and imaginary part of the FFT is exploited respectively in our experiments. The separation results show that they give out very similar convergence behavior. The dynamic curve of the global matrix when the imaginary part is used in separation is shown in Fig.1 (a). The dynamic Error Index curve is shown in Fig.1 (b). After convergence, its averaged Error Index is that EI=48.43 dB.

For comparison, the EI's of P. Comon's algorithm and JADE algorithm (by J.-F. Cardoso) for the same data are 39.57 dB and 39.12 dB respectively (with the programs the authors published on web site). The advantage of our algorithm is obvious.

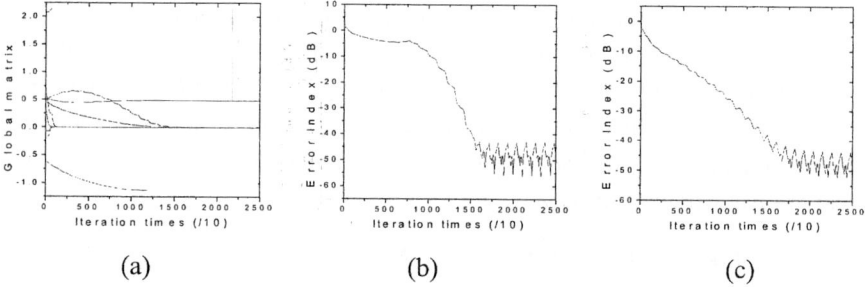

Fig. 1. (a) The block-based algorithm: the convergence curve of global matrix, using imaginary parts of FFTs; (b) the corresponding Error Index curve of (a); (c) the averaged Error Index curve of 100 random mixing matrices.

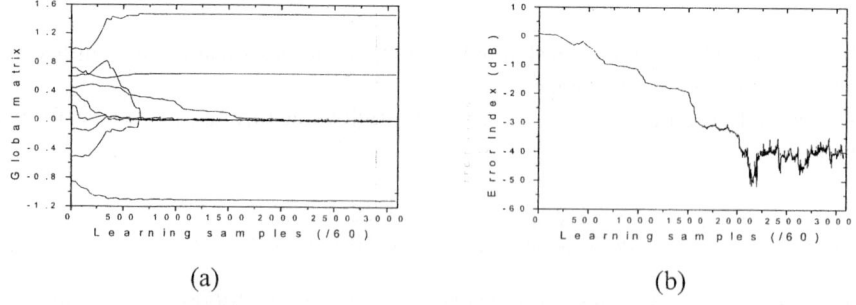

Fig. 2. The on-line learning algorithm: (a) the dynamic global matrix; (b) the dynamic Error Index of (a).

Intensive and extensive experiments show that the algorithm (16) and (19) is very robust, it is valid even when the condition number of the mixing matrix is as high as 7.91e+005. Fig. 1 (c) is the averaged dynamic Error Index of 100 random mixing matrices.

The on-line learning algorithms (17) or (20), their behaviors are very similar too. $\mu = 0.00003$, $\beta = 0.6$, After 6 time runs on the data, the adaptively learning curve of global matrix is shown in Fig.2 (a). The dynamic Error Index curve is shown in Fig.2 (b). After convergence, the averaged Error Index of the last 2000 samples is that EI=40.03 dB. Comparing with the time-domain algorithm in [6] under the same conditions, algorithm (17) or (20) dominates over it at least 4dB as the EI concerned.

5 Conclusions

In this paper, we proposed two BSS algorithms for instantaneous mixture separation in frequency domain. These algorithms are valid for the separation of temporally colored sources. They are suitable for both stationary and nonstationary but temporally colored sources. Simulation results show that they are valid and of high performance.

The research is supported by: National Natural Science Foundations of China (No. 60172072, No. 60372081).

References

1. Bell, J., Sejnowski, T.J.: An Information-Maximization Approach to Blind Separation and Blind Convolution, Neural Computation, vol.7. (1995) 1129–1159
2. Matsuoka, K. (ed.): A Neural Net for Blind Separation of Nonstationary Signals, Neural Networks, vol. 8. (1995) 411–419
3. Comon, P.: Independent Component Analysis, a New Concept?, Signal Processing, vol.36. (1994) 287–314
4. Cardoso, J.F., Souloumiac, A.: Blind Beamforming for Non-Gaussian Signals, IEE Proc., vol. 140. (1993) 362–370
5. Belouchrani, A., Abed-Meraim, K. (ed.): A Blind Source Separation Technique Using Second-order Statistics, IEEE Trans. on SP, vol. 45. (1997) 434–443
6. Choi, S., Cichocki, A., Amari, S.: Equivariant Nonstationary Source Separation, Neural Networks, vol. 15. (2002) 121–130
7. Adel, B., Moeness, G.A.: Blind Source Separation Based on Time-Frequency Signal Representations, IEEE Trans. on SP, vol.46. (1998) 2888–2897
8. Paris, S.: Blind Separation of Convolved Mixtures in the Frequency Domain, Neurocomputing, vol. 22. (1998) 21–34
9. Amari, S.: Natural Gradient Works Efficiently in Learning, Neural Computation, vol. 10. (1998) 251–276

Cumulant-Based Blind Separation of Convolutive Mixtures*

Tiemin Mei[1], Fuliang Yin[1], Jiangtao Xi[2], and Joe F. Chicharo[2]

[1] School of Electronic and Information, Dalian university of technology,
Dalian, P.R. China 116023
meitiemin@163.com
flyin@dlut.edu.cn
[2] School of Electrical, Computer and Telecommunications Engineering,
The University of Wollongong, Australia
jiangtao@uow.edu.au

Abstract. This paper focuses on the blind source separation of convolutive mixtures based on high order cumulants. It is proved that the zero-forcing of pairwise cross-cumulants of the outputs of separation system is a sufficient criterion for the separation of convolutive mixtures. New algorithm is developed based on this criterion. Simulation results are presented to support the validity of the algorithm.

1 Introduction

Blind source separation (BSS) is to restore a set of signal sources from observations that are the mixtures of the signal sources. Separation approaches for instantaneous mixtures are very successful so far, but the separation of convolutive mixtures is still a challenging issue. Many researchers have studied this problem from different aspects. Some researchers investigated this problem on the basis of information theory, such as [1][2]. But the information theory based algorithms depend on the suitable selection of the activation function and the probability distributions of sources. Some frequency domain approaches are also proposed using the available algorithms, which are used for the separation of instantaneous mixtures, but these approaches inevitably encounter the scaling and permutation ambiguities [3][4]. Second and high order statistics are also used for the separation of convolutive mixtures [5-8]. These statistics based algorithms do not have the problems suffered by the above two classes of algorithms, but their disadvantage is that they are not suitable for the separation of mixtures of long reverberating time. In this work, we considered the convolutive mixture model that has been studied in [5,6]. The separation algorithms are based on the backward neural networks.

---

* This work was supported by NNSF of China under Grant No. 60172073 and No. 60372082, Trans-entury Training Program Foundation for Talents by Ministry of Education of China.

The proposed algorithm is simple in computation and doesn't raise the scaling and permutation ambiguities usually suffered by the frequency-domain algorithms [3,4].

2 The Mixing and Separating Models

Assume $S(t)=[s_1(t),s_2(t)]^T$, where $s_1(t)$ and $s_2(t)$ are two zero mean and statistically uncorrelated stochastic processes, that is,

$$E[s_i(t)]=0, \quad i=1,2; \quad E[s_1(t_1)s_2(t_2)]=0, \quad \forall t_1,t_2 \tag{1}$$

We assume $X(t)=[x_1(t),x_2(t)]^T$, where $x_1(t)$ and $x_2(t)$ are the convolutive mixtures of source signals. The FIR filters with equal lengths are used to model the channels. The relations between source signals and observations can be expressed mathematically as follows:

$$x_1(t)=s_1(t)+\sum_{i=0}^{N}a(i)s_2(t-i); \quad x_2(t)=s_2(t)+\sum_{i=0}^{N}b(i)s_1(t-i) \tag{2}$$

Expressing in Z-domain:

$$X_1(z)=S_1(z)+H_{12}(z)S_2(z); \quad X_2(z)=S_2(z)+H_{21}(z)S_1(z) \tag{3}$$

We suppose that $[1-H_{12}(z)H_{21}(z)]$ does not have any zero points on or out of the unit circle in Z-plane, that is, $[1-H_{12}(z)H_{21}(z)]$ is a minimum phase system. We use two FIR filters with equal lengths to model the separation filters.

Let $Y(t)=[y_1(t),y_2(t)]^T$, $Y(t)$ is the output of separation neural networks. The relations between the inputs and the outputs are as follows:

$$y_1(t)=x_1(t)-\sum_{i=0}^{N}c(i)y_2(t-i); \quad y_2(t)=x_2(t)-\sum_{i=0}^{N}d(i)y_1(t-i) \tag{4}$$

Expressing in Z-domain:

$$Y_1(z)=X_1(z)-\hat{H}_{12}(z)Y_2(z); \quad Y_2(z)=X_2(z)-\hat{H}_{21}(z)Y_1(z) \tag{5}$$

Inserting (3) into (5), The relations between the outputs and sources in Z-domain is:

$$Y_1(z)=V_{11}(z)+V_{12}(z); \quad Y_2(z)=V_{21}(z)+V_{22}(z) \tag{6}$$

Where

$$V_{11}(z)=[1-\hat{H}_{12}(z)H_{21}(z)]S_1'(z); \quad V_{12}(z)=[H_{12}(z)-\hat{H}_{12}(z)]S_2'(z) \tag{7}$$

$$V_{21}(z)=[H_{21}(z)-\hat{H}_{21}(z)]S_1'(z); \quad V_{22}(z)=[1-\hat{H}_{21}(z)H_{12}(z)]S_2'(z) \tag{8}$$

$$S_1'(z)=[1-\hat{H}_{21}(z)\hat{H}_{12}(z)]^{-1}S_1(z); \quad S_2'(z)=[1-\hat{H}_{21}(z)\hat{H}_{12}(z)]^{-1}S_2(z); \tag{9}$$

$S_1'(z)$ and $S_2'(z)$ remain the properties of (1).

According to (6), and taking into account the basic assumption (1) and the properties of cumulants, the $(r_1 + r_2)$ order crosscumulant of $y_1(t)$ and $y_2(t-\tau)$ is as follows:

$$c_{y_1 y_2}^{r_1 r_2}(t, t-\tau) = \sum_{l=0}^{r_1} \sum_{k=0}^{r_2} \binom{l}{r_1}\binom{k}{r_2} \mathrm{cum}(v_{11}^l(t), v_{12}^{r_1-l}(t), v_{21}^k(t-\tau), v_{22}^{r_2-k}(t-\tau)) \quad (10)$$

$$= c_{v_{11} v_{21}}^{r_1 r_2}(t, t-\tau) + c_{v_{12} v_{22}}^{r_1 r_2}(t, t-\tau)$$

where $\binom{n}{m} = \dfrac{n!}{m!(n-m)!}$, and

$$c_{v_{11} v_{21}}^{r_1 r_2}(t, t-\tau) \quad (11)$$

$$= \sum_{k_1=0}^{2N} \cdots \sum_{k_{r_1}=0}^{2N} \sum_{l_1=0}^{N} \cdots \sum_{l_{r_2}=0}^{N} \left(\prod_{i=1}^{r_1} w(k_i) \prod_{j=1}^{r_2} p(l_j) \right) c_{s_1}^{r_1+r_2}(t-k_1,\ldots,t-k_{r_1},t-l_1-\tau,\ldots,t-l_{r_2}-\tau)$$

$$w(k) = \delta(k) - \sum_{i=0}^{N} c(i) b(k-i); \quad p(k) = b(k) - d(k)$$

$$c_{v_{12} v_{22}}^{r_1 r_2}(t, t-\tau) \quad (12)$$

$$= \sum_{k_1=0}^{N} \cdots \sum_{k_{r_1}=0}^{N} \sum_{l_1=0}^{2N} \cdots \sum_{l_{r_2}=0}^{2N} \left(\prod_{i=1}^{r_1} q(k_i) \prod_{j=1}^{r_2} u(l_j) \right) c_{s_2}^{r_1+r_2}(t-k_1,\ldots,t-k_{r_1},t-l_1-\tau,\ldots,t-l_{r_2}-\tau)$$

$$u(k) = \delta(k) - \sum_{i=0}^{N} d(i) a(k-i); \quad q(k) = a(k) - c(k)$$

From (11) and (12), because $c_{v_{11} v_{21}}^{r_1 r_2}(t,t-\tau)$ and $c_{v_{12} v_{22}}^{r_1 r_2}(t,t-\tau)$ have very different changing trends, even the two independent sources are of the similar or the same statistics and the mixing system is symmetric, it is impossible that $c_{v_{11} v_{21}}^{r_1 r_2}(t,t-\tau) = -c_{v_{12} v_{22}}^{r_1 r_2}(t,t-\tau)$ for the arbitrary time delay τ.

Thus, let $c_{y_1 y_2}^{r_1 r_2}(t,t-\tau) = 0$ ($\forall t, \tau$ and $r_1, r_2 \geq 1$), there must be that

$$c_{v_{11} v_{21}}^{r_1 r_2}(t,t-\tau) = 0; \quad c_{v_{12} v_{22}}^{r_1 r_2}(t,t-\tau) = 0, \forall t, \tau \quad (13)$$

Because $v_{11}(t)$ and $v_{21}(t)$ are originated from the same source signal $s_1(t)$, the first equation of (13) means that $v_{11}(t) = 0$ or $v_{21}(t) = 0$. $v_{11}(t) = 0$ is impossible when the mixing and separating filters are causal. Thus $v_{21}(t) = 0$ is the unique solution that makes the first equation of (13) holds. Deducing in the same way, we conclude that $v_{12}(t) = 0$ is the unique solution that makes the second equation of (13) holds. In one word, the zero-cumulant condition:

$$c_{y_1 y_2}^{r_1 r_2}(t,t-\tau) = 0 \quad \forall t, \tau \text{ and } r_1, r_2 \geq 1 \quad (14)$$

is a sufficient criterion for the separation of convolutely mixed sources.

From the above discussion, we know that source separation implies that $v_{12}(t) = 0$ and $v_{21}(t) = 0$. Further, in terms of the definitions of $v_{12}(t)$ and $v_{21}(t)$ in (7) and

(8), $v_{12}(t)=0$ and $v_{21}(t)=0$ imply that the separating filters will approximate the mixing filters in frequency domain, respectively. This means that $\hat{H}_{12}(z)=H_{12}(z)$ and $\hat{H}_{21}(z)=H_{21}(z)$ in the non-zero bands of sources, respectively. The outputs of the separation system are as follows when separation is successful:

$$y_1(t)=s_1(t); \qquad y_2(t)=s_2(t) \tag{15}$$

3 Algorithm Development

The backward neural networks is assumed to be strict causal (computable) system, that is, $c(0)=0$ and $d(0)=0$. Inserting the first equation of (4) into (14) and let $r_1=1$, $r_2=r$ and $\tau=1,2,...,N$, we obtain,

$$\mathbf{c}^{lr}_{x_1 y_2} - \mathbf{C}^{lr}_{Y_2 Y_2} \mathbf{W}_1 = \mathbf{c}^{lr}_{y_1 y_2} = 0 \tag{16}$$

Inserting the second equation of (4) into (14) and let $r_1=r$, $r_2=1$ and $\tau=1,2,...,N$, we obtain,

$$\mathbf{c}^{lr}_{x_2 y_1} - \mathbf{C}^{lr}_{Y_1 Y_1} \mathbf{W}_2 = \mathbf{c}^{lr}_{y_2 y_1} = 0 \tag{17}$$

where

$$\mathbf{W}_1 = [c(1),...,c(N)]^T; \qquad \mathbf{W}_2 = [d(1),...,d(N)]^T \tag{18}$$

$$\mathbf{Y}_i = [y_i(t-1),...y_i(t-N)]^T \quad i=1,2; \tag{19}$$

$$\mathbf{c}^{lr}_{x_k Y_l} = [\operatorname{cum}(x_k(t), y_l^r(t-1)),...,\operatorname{cum}(x_k(t), y_l^r(t-N)))]^T \quad k,l \in [1,2], k \neq l. \tag{20}$$

$$\mathbf{c}^{lr}_{y_k Y_l} = [\operatorname{cum}(y_k(t), y_l^r(t-1)),...,\operatorname{cum}(y_k(t), y_l^r(t-N)))]^T \quad k,l \in [1,2], k \neq l. \tag{21}$$

$$\mathbf{C}^{lr}_{Y_k Y_k} = \operatorname{cum}[\mathbf{Y}_k^r \mathbf{Y}_k^T] = [\operatorname{cum}(y_k^r(t-i) y_k(t-j))]_{i,j=1,...,N} \quad k \in [1,2]. \tag{22}$$

Applying Robbins-Monro first-order stochastic approximation methods [6], (16) and (17) give rise to the following adaptive algorithm

$$\mathbf{W}_1(t+1) = \mathbf{W}_1(t) + \mu(t)\mathbf{c}^{lr}_{y_1 y_2} \tag{23}$$

$$\mathbf{W}_2(t+1) = \mathbf{W}_2(t) + \mu(t)\mathbf{c}^{lr}_{y_2 y_1} \tag{24}$$

Special case 1: let r=1 in (23) and (24) and substitute instantaneous values for their ensemble values, we obtain the decorrelation-based algorithm [5,6].

$$\mathbf{W}_1(t+1) = \mathbf{W}_1(t) + \mu_1 y_1(t)\mathbf{Y}_2(t) \tag{25}$$

$$\mathbf{W}_2(t+1) = \mathbf{W}_2(t) + \mu_2 y_2(t)\mathbf{Y}_1(t) \tag{26}$$

Special case 2: let r=3 in (23) and (24), and replace the N by 1 vectors $\mathbf{c}_{y_1 Y_2}^{13}$ and $\mathbf{c}_{y_2 Y_1}^{13}$ with their instantaneous value, we obtain

$$\mathbf{W}_1(t+1) = \mathbf{W}_1(t) + \mu_1 y_1(t) diag\{\mathbf{Y}_2^2(t) - 3\sigma_2^2(t)\}\mathbf{Y}_2(t) \tag{27}$$

$$\mathbf{W}_2(t+1) = \mathbf{W}_2(t) + \mu_2 y_2(t) diag\{\mathbf{Y}_1^2(t) - 3\sigma_1^2(t)\}\mathbf{Y}_1(t) \tag{28}$$

where $\mathbf{Y}_i^r(t) \equiv [y_i^r(t-1),...,y_i^r(t-N)]^T$ $(r=2)$ $(i=1,2)$; $\sigma_i^2(t) = E[\mathbf{Y}_i^2(t)]$ $(i=1,2)$; operator $diag\{\cdot\}$ rearrange a vector into a diagonal matrix whose diagonal entries are the corresponding elements of the vector.

4 Simulations Results

Two Gaussian white noises and two speech signals are used in the experiments. The length of the signals are 30000 samples, sampling frequency is 16kHz. The algorithm (25) and (26) has been studied by [5][6], so we only gave the results of algorithm (27) and (28). The two mixing filters are nonminimum. The largest/smallest pole radius of the mixing filters is 1.66/0.66 and 4.44/0.078, respectively. For speeches, the separation filters approximate the mixing filters over the nonzero frequency bands of sources in frequency domain, refer to Fig.2.

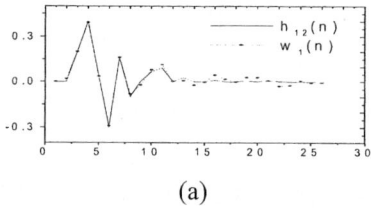
(a)

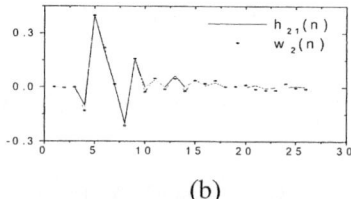
(b)

Fig. 1. Source signals are Gaussian white noises: The solid lines are the mixing filter's impulse responses. The dot lines are separating filter's impulse responses; they approximate the mixing filters in time domain. Before separation, SIR$_1$=4.67dB, SIR$_2$=5.52dB; after separation: SIR$_1$=20.91dB, SIR$_2$=21.54dB.

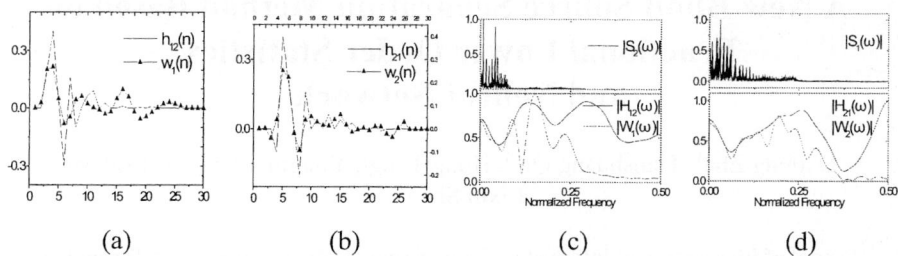

Fig. 2. Source signals are two speeches: (a) and (b) show that separating filters don't approximate mixing filters in time domain; (c) and (d) show that separating filters approximate mixing filters in frequency domain over the non-zero bands of sources. Before separation, SIR_1=8.21dB, SIR_2=5.05dB; after separation: SIR_1=24.64dB, SIR_2=22.71dB.

5 Conclusions

For the given mixing model, we proved that the zero-forcing of pairwise crosscumulants of the outputs of separation system is the sufficient criterion for the separation of convolutive mixtures of independent sources. The proposed algorithm under this criterion is proved to be effective for the separation of convolutive mixtures.

References

1. Bell, A.J., Sejnowski, T.J.: An Information-Maximization Approach to Blind Separation and Blind Deconvolution, Neural Computation, vol.7. (1995) 1129–1159
2. Amari, S., Douglas, S.C., Cichocki, A., Yang, H.H.: Multichannel Blind Deconvolution and Equalization Using the Natural Gradient, in Proc. IEEE Int. Workshop Wireless Communication, Paris, (1997) 101–104
3. Smaragdis, P.: Blind Separation of Convolved Mixtures in the Frequency Domain, Neurocomputing, vol. 22. (1998) 21–34,
4. Parra, L., Spence, C.: Convolutive Blind Separation of Non-stationary Sources, IEEE Trans. Speech and Audio Processing, vol. 8. (2000) 320–327,
5. Van, G.S., Van, C.D.: Signal Separation by Symmetric Adaptive Decorrelation: Stability, Convergence and Uniqueness, IEEE Trans. on SP, vol. 43. (1995) 1602–1612
6. Weinstein, E., Feder, M., Oppenheim, A.V.: Multichannel Signal Separation by Decorrelation, IEEE Trans. Speech and Audio Processing, vol. 1. (1993) 405–413,
7. Kawamoto, M., Matsuoka, K.: A Method of Blind Separation for Convolved Non-stationary Signals, Neurocomputing, vol. 22. (1998) 157–171,
8. Yellin, D., Weinstein, E.: Multichannel Signal Separation: Methods and Analysis, IEEE Trans. Signal Processing, vol. 44. (1996) 106–118

A New Blind Source Separation Method Based on Fractional Lower Order Statistics and Neural Network

Daifeng Zha[1], Tianshuang Qiu[1], Hong Tang[1], Yongmei Sun[1], Sen Li[1], and Lixin Shen[2]

[1]School of Electronic and Information Engineering, Dalian University of Technology,
Dalian, Liaoning Province, P.R. China, 116024
qiutsh@dlut.edu.cn
[2]Dalian Maritime University, Dalian, Liaoning Province, P.R. China, 116024

Abstract. Lower order alpha stable distribution processes can model the impulsive random signals and noises well in physical observation. Conventional blind source separation is based on second order statistics(SOS). In this paper, we propose neural network structures related to multilayer feedforward networks for performing blind source separation based on the fractional lower order statistics (FLOS). The simulation results and analysis show that the proposed networks and algorithms are robust.

1 Introduction

Conventional standard principal component analysis (PCA) and independent component analysis (ICA) are optimal in approximating the input data in the mean-square error sense by describing some second order characteristics[3] of the data. Nonlinear PCA[4][5] method which is related to the higher order statistical techniques is a useful extension of standard PCA. In various PCA methods, the data are represented in an orthonormal basis determined by the second-order statistics (covariance) and higher order statistics of the input data[6][7][8]. Such a representation is adequate for Gaussian data. However, a kind of physical process with suddenly and short endurance high impulse in real world, called lower order alpha stable distribution random process[1][2], has no its second order or higher order statistics. It has no closed form of probability density function so that we can only describe it by the characteristic function as[1][2]:

$$\varphi(t) = \exp\{j\xi t - \delta|t|^{\alpha}[1 + j\beta \operatorname{sgn}(t)\omega(t,\alpha)]\}, -\infty < \xi < \infty, \delta > 0, 0 < \alpha \leq 2, -1 \leq \beta \leq 1 \quad (1)$$

$$\omega(t,\alpha) = \tan\frac{\alpha\pi}{2}(if\ \alpha \neq 1) \quad or \quad \frac{2}{\pi}\log|t|(if\ \alpha = 1)\ )$$

The characteristic exponent α determines the shape of the distribution. Especially, if $\alpha = 2$, it is a Gaussian distribution, and if $\alpha < 2$, it falls into the lower order alpha stable distribution. The dispersion δ plays a role analogous to the variance of the second order process. β is the symmetry parameter and ξ is the location parameter. The typical lower order alpha stable distribution signals are shown in Fig.1.

2 Data Model and Network Structures

Assume that there exist P zero mean source signals $s_i(n), i = 1,2,...P$ that are scalar-valued and mutually statistically independent for each sample value or in practice as independent as possible and the original sources are unobservable, and all that we have are a set of noisy linear mixtures $\mathbf{X}(n) = [x_1(n), x_2(n),...,x_M(n)]^T, n = 1,2,..N$. We can write the signal model in matrix form as $\mathbf{X} = \mathbf{AS} + \mathbf{N}$. Here $\mathbf{X}$ is $M \times N$ observation matrix, $\mathbf{A}$ is $M \times P$ constant mixing matrix with full column rank. $\mathbf{S}$ is $P \times N$ independent source signals matrix, $\mathbf{N}$ denotes $M \times N$ possible additive lower order alpha stable distribution noise matrix. The task of source separation is merely to find the estimation of the sources, knowing only the data $\mathbf{X}$. Consider now a two-layer feedforward neural network structure[3][8] shown in Fig. 2. The inputs of the network are components of the observation matrix $\mathbf{X}$. In the hidden layer there are P neurons, and the output layer consists again of P neurons. Let $\mathbf{V}$ denote the $P \times M$ pre-processing matrix between the inputs and the hidden layer. And $\mathbf{W}^T$, respectively, the $P \times P$ weight matrix between the hidden and output layer.

3 Pre-processing Based on Fractional Order Correlation Matrix

Generally, it is impossible to separate the noise in the input data from the source signals. In practice, if the amount of noise is considerable, the separation results are often fairly poor. Some of the noises can usually be canceled using the standard PCA if the number of mixtures is larger than the number of sources[7]. As lower order alpha stable distribution noise has no second or higher order statistics such as covariance and covariance matrix[1][2], We can define their p th order fractional correlation [1][2] of two p th order random variables η_1, η_2 as $<\eta_1, \eta_2>_p = E\{\eta_1 \eta_2^{<p-1>}\}$ ($0 < p < \alpha \leq 2$, $(\cdot)^{<p>} = |\cdot|^{p-1}(\cdot)^*$), Where * denotes complex conjugation. For $p = 2$, fractional correlation gives usual autocorrelation. Let we define the fractional order correlation matrix $\mathbf{\Gamma}_X$, whose element is shown as

$$< x_i(n), x_j(n) >_p = E\{x_i(n) | x_j(n)|^{p-2} x_j^*(n)\}, \quad 0 < p < \alpha \leq 2 \qquad (2)$$

Then we have $\mathbf{\Gamma}_X = \mathbf{A}\mathbf{\Lambda}\mathbf{A}^H + \sigma \mathbf{I}$. We can interpret $\mathbf{\Lambda}$ as a fractional lower order correlation matrix of the source signals and σ as the additive noise level. A advantage of PCA pre-processing is that it provides a convenient means for estimating the number of the sources or independent components[5][6]. Independent sources lie in "signal" subspace[7] by estimating all of eigenvalues of the matrix $\mathbf{\Gamma}_X$. We can use projection method. Firstly, the data for the PCA processing is made zero mean before they are sent into the PCA networks. Secondly, we project the observation data $\mathbf{X}$ into "signal" subspace via a pre-processing matrix $\mathbf{V}$ that is given as follow:

$$V = D^{-1/2}P^H \atop Z = VX \}, D = diag(\lambda_1, \lambda_2,...,\lambda_P), \lambda_1 \geq \lambda_2 \geq .. \geq \lambda_P, P = [e_1, e_2,...,e_P] \quad (3)$$

where λ_i denotes the ith largest eigenvalue of the data fractional order correlation matrix Γ_X, and the e_i represents the ith corresponding principal eigenvector.

4 Separating Algorithms

During the recent years, many neural network blind separation algorithms have been proposed [4][5][6]. Let us consider the ith output weight vector $W_i, i = 1,2,...P$.. As lower order alpha stable distribution noise has no second order moment, the problem is to learn an weight W_i which maximizes the pth order moment of every elements of $Y(n)$ subject to $W_i W_i^H = I_P$. For each W_i:

$$W_i^{opt} = \arg\max_{W_i} E\{\frac{1}{p}|Z^H(n)W_i|^p\}, (W_i W_i^H = I_P) \quad (4)$$

Define the objective function as

$$J(W_i) = E\{\frac{1}{p}|Z^H(n)W_i|^p\} + \frac{1}{2}\lambda_{ii}(W_i^H W_i - 1) + \frac{1}{2}\sum_{j=1, j \neq i}^{P}\lambda_{ij}W_i^H W_j \quad (5)$$

Here λ_{ij} is the Lagrange multiplier. For each neuron, W_i is orthogonal to the weight vector $W_j, j \neq i$. The estimated gradient of $J(W_i)$ with respect to W_i is

$$\hat{\nabla}J(W_i) = E\{Z(n)|Z^H(n)W_i|^{p-2} conj(Z^H(n)W_i)\} + \sum_{j=1}^{P}\lambda_{ij}W_j \quad (6)$$

At the optimum, the gradients must vanish for $i = 1,2,...P$, and $W_i^H W_j = \delta_{ij}$. These can be taken into account by multiplying (6) by W_j^H from left. We can obtain $\lambda_{ij} = -W_j^H E\{Z(n)|Z^H(n)W_i|^{p-2} conj(Z^H(n)W_i)\}$. Substituting these into (6) we get

$$\hat{\nabla}J(W_i) = [I - \sum_{j=1}^{P} W_j W_j^H] E\{Z(n)|Z^H(n)W_i|^{p-2} conj(Z^H(n)W_i)\} \quad (7)$$

A practical gradient algorithm for optimization problem (4) is now obtained by substitute (7) into $W_i(n+1) = W_i(n) - \mu(n)\hat{\nabla}J(W_i(n))$, where $\mu(n)$ is the gain parameter. The final algorithm is thus

$$W_i(n+1) = W_i(n) - \mu(n)[I - \sum_{j=1}^{P} W_j(n)W_j^H(n)]Z(n)|Z^H(n)W_i(n)|^{p-2} conj(Z^H(n)W_i(n)) \quad (8)$$

$$\mathbf{W}_i(n+1) = \mathbf{W}_i(n) - \mu(n) |y_i(n)|^{p-2} conj(y_i(n))[\mathbf{Z}(n) - \sum_{j=1}^{P} y_j(n)\mathbf{W}_j(n)] \quad (9)$$

Let $g(t) = |t|^{p-2} conj(t)$, then $g(t)$ is appropriate network nonlinear transform function for lower order alpha stable distribution noises. As we know that the instantaneous error of the gradient $\mathbf{I} - \sum_{j=1}^{P} \mathbf{W}_j(n)\mathbf{W}_j^H(n)$ might be zero during the iteration, we modify (9) in order to improve robustness of algorithm as

$$\mathbf{W}_i(n+1) = \mathbf{W}_i(n) - \mu(n)g(y_i(n))[\mathbf{Z}(n) - \sum_{j=1}^{P} g(y_i(n))\mathbf{W}_j(n)] \quad (10)$$

Thus, $\mathbf{W}_1, \mathbf{W}_2, ..., \mathbf{W}_P$ can be obtained by iteration as (10). Let $\mathbf{Y}(n) = [y_1(n), y_2(n), ..., y_P(n)]^T$, and $\mathbf{W} = [\mathbf{W}_1, \mathbf{W}_2, ..., \mathbf{W}_P]$. According to the above derivation by using $g(t) = |t|^{p-2} conj(t)$, for the entire network, the algorithm for learning $\mathbf{W}$ is given as

$$\mathbf{W}(n+1) = \mathbf{W}(n) - \mu(n)[\mathbf{Z}(n) - \mathbf{W}(n)g(\mathbf{Y}(n))]g(\mathbf{Y}^H(n)) \quad (11)$$

5 Performance Analysis

We start from the learning rule (11), and we assume that there exists a square separating matrix $\mathbf{H}^H$ such that $\mathbf{U}(n) = \mathbf{H}^H \mathbf{Z}(n)$. The separating matrix $\mathbf{H}^H$ must be orthogonal (i.e. $\mathbf{H}\mathbf{H}^H = \mathbf{I}_P$). We multiply both sides of (11) by $\mathbf{H}^H$ as

$$\mathbf{H}^H\mathbf{W}(n+1) = \mathbf{H}^H\mathbf{W}(n) + \mu(n)[\mathbf{H}^H\mathbf{Z}(n) - \mathbf{H}^H\mathbf{W}(n)g(\mathbf{W}(n)^H\mathbf{Z}(n))]g(\mathbf{Z}(n)^H\mathbf{W}(n)) \quad (12)$$

$$= \mathbf{H}^H\mathbf{W}(n) + \mu(n)[\mathbf{H}^H\mathbf{Z}(n) - \mathbf{H}^H\mathbf{W}(n)g(\mathbf{W}(n)^H\mathbf{H}\mathbf{H}^H\mathbf{Z}(n))]g(\mathbf{Z}(n)^H\mathbf{H}\mathbf{H}^H\mathbf{W}(n))$$

Define $\mathbf{Q}(n) = \mathbf{H}^H\mathbf{W}(n)$, $\mathbf{W}(n) = (\mathbf{H}^H)^{-1}\mathbf{Q}(n)$, and (12) is written as

$$\mathbf{Q}(n+1) = \mathbf{Q}(n) + \mu(n)[\mathbf{U}(n) - \mathbf{Q}(n)g(\mathbf{Q}(n)^H\mathbf{U}(n))]g(\mathbf{Q}(n)\mathbf{U}(n)^H) \quad (13)$$

Geometrically, the transformation multiplying by the orthogonal matrix $\mathbf{H}^H$ simply means a rotation to a new set of coordinates such that the elements of the input vector expressed in these coordinates are statistically independent. Analogous differential equation of (13) is obtained as matrix form:

$$d\mathbf{Q}/dt = E\{\mathbf{U}g(\mathbf{U}^H\mathbf{Q})\} - E\{g(\mathbf{Q}^H\mathbf{U})g(\mathbf{U}^H\mathbf{Q})\} \quad (14)$$

According to [3], we can easily prove that (14) has a stable solution. For the sake of $\mathbf{Q} = \mathbf{H}^H\mathbf{W}$, thus $\mathbf{W} = (\mathbf{H}^H)^{-1}\mathbf{Q}$ is an asymptotic stable solution of (11). Fig.3 shows the stability and convergence property of the algorithm based on SOS and FLOS. We know the FLOS based algorithm has better stability and convergence than the SOS based algorithm.

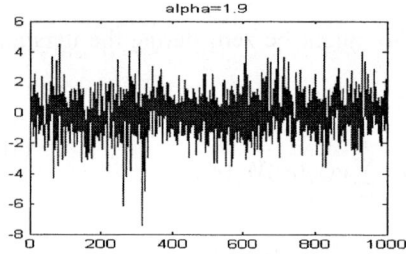 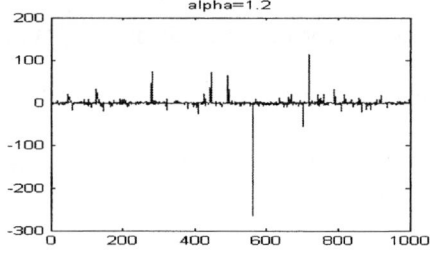

Fig. 1. Alpha stable distribution signals

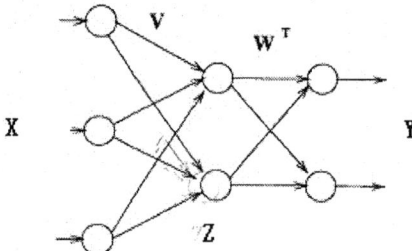 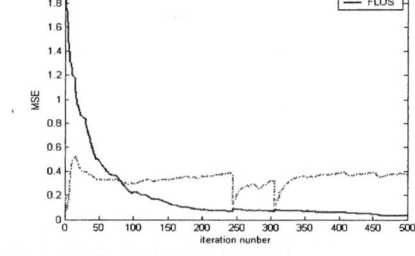

Fig. 2. The structure of neural network

Fig. 3. The stability and convergence performance

6 Experimental Results

Suppose that two sound source signals (one is piano and the other is bird with 10000 samples for each) are sent into a linear microphone array with 5 sensors from different directions. Alpha stable distribution noise ($\alpha = 1.7$) exists in the array at the same time. Two algorithms are used in the experiment, including (1) SOS with nonlinear function $g(t) = \tanh(t)$ and (2) FLOS with $g(t) = |t|^{p-2} conj(t)$. The signal waveforms can be obtained in time domain shown in Fig.4, in which Fig.4(a) and (b) are source signals, Fig4.(c) and (d) are separated signals with SOS based algorithm, and Fig.4(e) and (f) are separated signals with FLOS based algorithm.

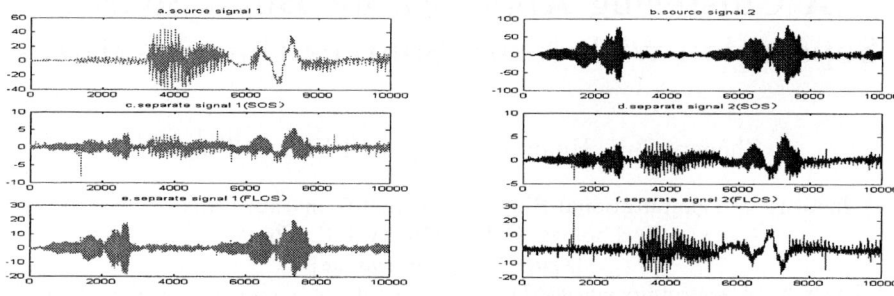

Fig. 4. The source and separated signals

7 Conclusion

The analysis and simulation results show that the proposed networks and BSS algorithms based on the FLOS are more robust than conventional the SOS based algorithm. The separation capability of the former is greatly improved under both Gaussian and fractional lower order stable distribution noise environments.

References

1. Nikias,C. L., Shao, M.: Signal Processing with Alpha-Stable Distributions and Applications. New York: John Wiley & Sons Inc. (1995).
2. Nikias,C. L., Shao, M.: Signal Processing with fractional lower order moments: stable processes and their applications, Proceedings of IEEE, Vol.81, No.7 (1993) 986-1010
3. Karhumen, J., Oja, E., Wang, L., Vigario, R., Joutsensalo, J.: A class of neural networks for independent component analysis. IEEE. Trans. on Neural Networks, vol. 8, No.3 (1997)
4. Yanwu Zhang, Yuanliang Ma: CGHA for principal component extraction in the complex domain. IEEE. Trans. on Neural Networks, vol.8, No.5 (1997)
5. Karhunen, J., Joutsensalo, J.: Nonlinear Generalizations of Principal Component Learning Algorithms. Proceedings of 1993 International Joint Conference on Neural Networks, Vol. 3 (1993)
6. Wang, L., Karhunen, J., Oja, E.:A bigradient optimization approach for robust PCA, MCA, and source separation, Proceedings. IEEE International Conference on Neural Networks, Vol. 4 (1995)
7. Winter, S., Sawada, H.. Makino, S.: Geometrical understanding of the PCA subspace method for overdetermined blind source separation. IEEE. Trans. on Acoustics, Speech, and Signal Processing (2003)
8. Mutihac, R., van Hulle, M.M., PCA and ICA neural implementations for source separation - a comparative study. Proc. of International Joint Conf. Neural Networks, Vol. 1 (2003) 20-24

A Clustering Approach for Blind Source Separation with More Sources than Mixtures

Zhenwei Shi[1], Huanwen Tang[1], and Yiyuan Tang[2,3,4]

[1] Institute of Computational Biology and Bioinformatics, Dalian University of Technology, Dalian 116023, P.R. China
{szw1977@yahoo.com.cn}
[2] Institute of Neuroinformatics, Dalian University of Technology, Dalian 116023, P.R. China
[3] Laboratory of Visual Information Processing, The Chinese Academy of Sciences, Beijing 100101, P.R. China
[4] Key Lab for Mental Health, The Chinese Academy of Sciences, Beijing 100101, P.R. China

Abstract. In this paper, blind source separation is discussed with more sources than mixtures when the sources are sparse. The blind separation technique includes two steps. The first step is to estimate a mixing matrix, and the second is to estimate sources. The mixing matrix can be estimated by using a clustering approach which is described by the generalized exponential mixture model. The generalized exponential mixture model is a powerful uniform framework to learn the mixing matrix for sparse sources. After the mixing matrix is estimated, the sources can be obtained by solving a linear programming problem. The techniques we present here can be extended to the blind separation of more sources than mixtures with a Gaussian noise.

1 Introduction

Independent component analysis (ICA) is a technique that has become increasingly important for a vast range of applications in blind source separation, blind deconvolution, and feature extraction, and so on. The standard formulation of ICA requires at least as many sensors as sources [1,2]. Lewicki et al. [3] proposed a generalized ICA method for learning overcomplete representations of the data that allows for more basis vectors than dimensions in the inputs. Lee et al. [4] demonstrated that three speech signals can be separated given only two mixtures of the three signals using overcomplete representations. Zibulevsky et al. [5] suggested that the mixing matrix and the sources are estimated by using maximum a posteriori approach. In [6], the blind separation technique included two steps. The first step was to estimate a mixing matrix, the second was to estimate sources. The mixing matrix was estimated using K-means clustering algorithm. Motivated by these methods, we present a gradient learning algorithm for generalized exponential mixture model that is able to estimate the mixing matrix.

After the mixing matrix is estimated, the sources are estimated by using linear programming algorithm. Experiments with speech signals demonstrate nice separation results.

2 A Clustering Approach for Blind Source Separation

In blind source separation an sensor signal $x \in R^M$ can be described using an overcomplete basis by the following linear model:

$$x = As, \tag{1}$$

where the columns of the mixing matrix $A \in R^{M \times L}$ where $L > M$ define the overcomplete basis vectors, $s \in R^L$ is the source signal. The source signals are assumed independent such that $p(s) = \prod_{l=1}^{L} p(s_l)$. In addition, each prior $p(s_l)$ is assumed to be sparse typified by the Laplacian distribution [3]. We wish to estimate the mixing matrix A and the source signal s. For a given mixing matrix A, the source signal can be found by maximizing the posterior distribution $p(s|x, A)$ [3]. This can be solved by a standard linear program when the prior is Laplacian [3,6]. Thus, we can estimate the mixing matrix A first.

The phenomenon of data concentration along the directions of the mixing matrix columns can be used in clustering approaches to source separation. In a two-dimensional space, the observations x were generated by a linear mixture of three independent sparse sources (the same three sources of speech signals and mixing matrix as used in Lee et al. [4]), as shown in Fig. 1 (Left) (scatter plot of two mixtures x_1 versus x_2). The three distinguished directions, which correspond to the columns of the mixing matrix A, are visible. In order to determine orientations of data concentration, we projected the data points onto the surface of a unit sphere by normalizing the sensor data vectors at every particular time index t: $x_t = x_t / \|x_t\|$ ($x_t = (x_1(t), x_2(t))^T, t = 1, \ldots, T$). Next, the data points were moved to a half-sphere, e.g., by forcing the sign of the first coordinate $x_1(t)$ to be positive (without this operation each 'line' of data concentration would yield two clusters on opposite sides of the sphere). For each point x_t, the data point $\alpha_t = \sin^{-1}(x_2(t))$ was computed by the second coordinate $x_2(t)$. This is a 1-1 mapping from Cartesian coordinates to polar coordinates, because the data vectors are normalized. Thus, the data $\alpha = \{\alpha_1, \ldots, \alpha_T\}$ also have the centers of three clusters corresponding to the three distinguished directions for two mixtures. The histogram of the data α is presented in Fig. 1 (Right). We can see the histogram of the data α has three cluster distributions and each cluster distribution can be modelled by a unimodal distribution. Thus, the density function generating the data α can be modelled by a linear combination of three unimodal distributions. The coordinates of the centers of the three clusters determine the columns of the mixing matrix A. For example, if the coordinates of the three centers are μ_1, μ_2, μ_3, we can compute the mixing matrix A from

$$A = \begin{pmatrix} \cos(\mu_1) & \cos(\mu_2) & \cos(\mu_3) \\ \sin(\mu_1) & \sin(\mu_2) & \sin(\mu_3) \end{pmatrix}. \tag{2}$$

We can use a linear combination of generalized exponential distributions to describe the density function generating the data α. The generalized exponential distribution is a flexible density model and it can describe uniform, Gaussian, Laplacian and other sub- and super-Gaussian unimodal densities. We therefore write this model as a linear combination of component densities $p(\alpha|k)$ (i.e., the kth cluster density) in the form:

$$p(\alpha) = \sum_{k=1}^{K} p(\alpha|k) P(k), \qquad (3)$$

where the coefficients $P(k)$ are called the mixing parameters satisfying $\sum_{k=1}^{K} P(k) = 1$ ($P(k)$ denotes the prior probability of the data point having been generated from component k of the mixture). Such a model is called a mixture model. We consider that the component densities $p(\alpha|k)$ are modelled as generalized exponential densities, and we call it the generalized exponential mixture model. The generalized exponential density is

$$p(\alpha|k) = \frac{r_k \beta_k^{\frac{1}{r_k}}}{2\Gamma(\frac{1}{r_k})} \exp(-\beta_k |\alpha - \mu_k|^{r_k}), \qquad (4)$$

where μ_k, β_k, r_k are the parameters for the generalized exponential distribution. Clearly $p(\alpha|k)$ is Gaussian when $r_k = 2$, Laplacian when $r_k = 1$, and the uniform distribution is approximated in the limit $r_k \to \infty$ [1]. Thus, we assume that the data α are generated by a generalized exponential mixture distribution, and we can determine cluster centers of the generalized exponential mixture distribution using a specific algorithm (i.e. estimating the cluster centers $\mu_k, k = 1, \ldots, K$). The coordinates of μ_k will determine the columns of the estimated mixing matrix.

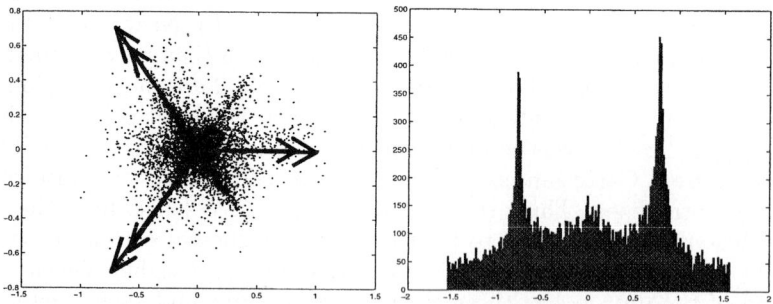

Fig. 1. *Left*: Scatter plot of two mixtures x_1 versus x_2. The basis vectors (the short arrows) show that the three distinguished directions correspond to the columns of the mixing matrix A. After the learning algorithm converges, the learned basis vectors (the long arrows) are close to the true basis vectors (the short arrows). *Right*: The histogram of the data α

3 A Gradient Learning Algorithm for Generalized Exponential Mixture Model

For simplicity, we consider the one-dimensional generalized exponential mixture model. Assume that data $\alpha = \{\alpha_1, \ldots, \alpha_T\}$ are drawn independently and generated by a generalized exponential mixture model. We derive an iterative learning algorithm which performs gradient ascent on the total likelihood of the data. The likelihood of the data is given by the joint density

$$p(\alpha|\Theta) = \prod_{t=1}^{T} p(\alpha_t|\Theta). \qquad (5)$$

The mixture density is

$$p(\alpha_t|\Theta) = \sum_{k=1}^{K} p(\alpha_t|\theta_k, k) P(k), \qquad (6)$$

where $\Theta = (\theta_1, \ldots, \theta_K)$ are the unknown parameters for each $p(\alpha_t|\theta_k, k)$, and that we wish to infer them from the data $\alpha = \{\alpha_1, \ldots, \alpha_T\}$ (the number K is known in advance). The log-likelihood L is then

$$L = \sum_{t=1}^{T} \log p(\alpha_t|\Theta), \qquad (7)$$

and using (6), the gradient for the parameters θ_k is

$$\nabla_{\theta_k} L = \sum_{t=1}^{T} \frac{1}{p(\alpha_t|\Theta)} \nabla_{\theta_k} p(\alpha_t|\Theta) = \sum_{t=1}^{T} \frac{\nabla_{\theta_k}[\sum_{k=1}^{K} p(\alpha_t|\theta_k, k) P(k)]}{p(\alpha_t|\Theta)}$$

$$= \sum_{t=1}^{T} \frac{\nabla_{\theta_k} p(\alpha_t|\theta_k, k) P(k)}{p(\alpha_t|\Theta)}. \qquad (8)$$

Using the Bayes's rule, we obtain

$$P(k|\alpha_t, \Theta) = \frac{p(\alpha_t|\theta_k, k) P(k)}{\sum_{k=1}^{K} p(\alpha_t|\theta_k, k) P(k)}. \qquad (9)$$

Substituting (9) in (8) leads to

$$\nabla_{\theta_k} L = \sum_{t=1}^{T} P(k|\alpha_t, \Theta) \frac{\nabla_{\theta_k} p(\alpha_t|\theta_k, k) P(k)}{p(\alpha_t|\theta_k, k) P(k)}$$

$$= \sum_{t=1}^{T} P(k|\alpha_t, \Theta) \nabla_{\theta_k} \log p(\alpha_t|\theta_k, k). \qquad (10)$$

We assume that the component densities $p(\alpha_t|\theta_k, k)$ are modelled as the generalized exponential, i.e.,

$$p(\alpha_t|\theta_k, k) = \frac{r_k \beta_k^{\frac{1}{r_k}}}{2\Gamma(\frac{1}{r_k})} exp(-\beta_k |\alpha_t - \mu_k|^{r_k}), \quad (11)$$

where $\theta_k = \{\mu_k, \beta_k, r_k\}$ and $\nabla_{\theta_k} L = \{\nabla_{\mu_k} L, \nabla_{\beta_k} L, \nabla_{r_k} L\}$. Gradient ascent is used to estimate the parameters that maximize the log-likelihood. From (10) and (11), we derive an iterative gradient ascent learning algorithm as follows:

$$\triangle \mu_k \propto \sum_{t=1}^{T} P(k|\alpha_t, \Theta)(\beta_k r_k |\alpha_t - \mu_k|^{r_k - 1} sign(\alpha_t - \mu_k)), \quad (12)$$

$$\triangle \beta_k \propto \sum_{t=1}^{T} P(k|\alpha_t, \Theta)(\frac{1}{r_k \beta_k} - |\alpha_t - \mu_k|^{r_k}), \quad (13)$$

$$\triangle r_k \propto \sum_{t=1}^{T} P(k|\alpha_t, \Theta)(\frac{1}{r_k} - \frac{1}{r_k^2}\log\beta_k + \frac{1}{r_k^2}\frac{\Gamma'(\frac{1}{r_k})}{\Gamma(\frac{1}{r_k})} - \beta_k|\alpha_t - \mu_k|^{r_k}\log|\alpha_t - \mu_k|). \quad (14)$$

Practically, the adaptation is stopped once the log-likelihood function stabilizes asymptotically with increasing number of iterations.

4 Simulation Examples

We considered separating three speech sources from two mixtures. The observations x were generated by a linear mixture of the three speech signals used in section 2. Then we used the same method in section 2 to compute the data α. The learning algorithm for the generalized exponential mixture model in section 3 was used for estimated mixing matrix (i.e. estimate the three cluster centers μ_k for α, $K = 3$ here). The parameters were randomly initialized and the learning rate was set to be 0.0005 (typically 40-60 iterations). Fig. 1 (Left) shows the learned basis vectors (the long arrows) and the true basis vectors (the short arrows). After the mixing matrix was estimated, we performed the linear programming algorithm for obtaining the sources [3,6]. The three original signals, two mixtures, and the separated output signals are shown in Fig. 2.

5 Conclusions

Blind source separation is discussed with more sources than mixtures in this paper. If the sources are sparse, the mixing matrix can be estimated by using the generalized exponential mixture models. The coordinates of the cluster

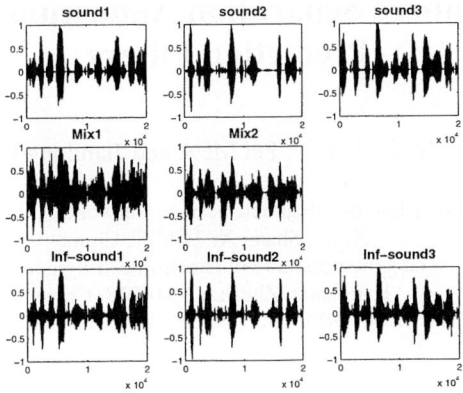

Fig. 2. Three speech experiment: Two mixtures, three original signals and three inferred signals

centers will determine the columns of the estimated mixing matrix. After the mixing matrix is learned, the sources are estimated by solving a linear programming problem. The generalized exponential mixture model is a powerful uniform framework for learning the mixing matrix for sparse sources. The techniques we describe in this paper can be extended to the blind separation of more sources than mixtures with a Gaussian noise [7].

Acknowledgments. The work was supported by NSFC(30170321,90103033), MOE(KP0302) and MOST(2001CCA00700).

References

1. Everson, R., Roberts, S.: Independent Component Analysis: a Flexible Nonlinearity and Decorrelating Manifold Approach. Neural Computation 11 (1999) 1957-1983
2. Shi, Z., Tang, H., Tang, Y.: A New Fixed-Point Algorithm for Independent Component Analysis. Neurocomputing 56 (2004) 467-473
3. Lewicki, M.S., Sejnowski, T.J.: Learning Overcomplete Representations. Neural Computation 12(2) (2000) 337-365
4. Lee, T.W., Lewicki, M.S., Girolami, M., Sejnowski, T.J.: Blind Source Separation of More Sources Than Mixtures Using Overcomplete Representations. IEEE Signal Processing Letters 6(4) (1999) 87-90
5. Zibulevsky, M., Pearlmutter, B.A.: Blind Source Separation by Sparse Decomposition in a Signal Dictionary. Neural Computation 13(4) (2001) 863-882
6. Li, Y., Cichocki, A., Amari, S.: Sparse Component Analysis for Blind Source Separation with Less Sensors Than Sources. 4th International Symposium on Independent Component Analysis and Blind Signal Separation (ICA2003) Japan (2003) 89-94
7. Shi, Z., Tang, H., Liu, W., Tang, Y.: Blind Source Separation of More Sources Than Mixtures Using Generalized Exponential Mixture Models. Neurocomputing (to appear)

A Blind Source Separation Algorithm with Linear Prediction Filters

Zhijin Zhao[1,2], Fei Mei[2], and Jiandong Li[1]

[1] National Key Lab of Integrated Service Network, Xidian University,
Xi'an, Shaan Xi 710071, China
[2] Hangzhou Institute of Electronics Engineering,
Hangzhou, Zhe Jiang 310018, China
Zhaozj03@hziee.edu.cn, madrista_1979@hotmail.com

Abstract. In this paper the theoretical analysis about why to be able to guarantee source separation by blind source separation (BSS) algorithms after linear prediction filtering is presented , and the concept of the linear prediction filter (LPF) application is illustrated. The simulation results verify the derivation of this paper and show that linear prediction analysis can be used in blind source separation for instantaneous mixtures and convolutive mixtures. For the convolutive mixtures, when the separated speech quality is improved, the separation performance is reduced a little.

1 Introduction

Blind source separation (BSS) is an approach to estimate source signals only using the information of mixture signals observed in each input channel and has been widely used in various areas, e.g., noise robust speech recognition, high-quality hands-frees telecommunication systems, discovery of independent sources in biological signals, such as EEG, MEG, and others.

There have been many techniques to blind separate source speech signals[1]. To improve the separated speech quality, linear prediction filter[2-4] has been applied in BSS system. Although some methods[2,4-5] are illustrated effective, there is no explanation why to be able to guarantee source separation by BSS algorithm after linear prediction filtering. In [3] the derivation of the algorithm using linear prediction filter has been presented, but the algorithmic realization is different from its theoretical derivation and the authors themselves have pointed that the exploration of such algorithmic variations is the subject of future work.

In this paper, we present the exploration why to be able to separate source speech signals by the varied algorithm[3]. The concept of the LPF application is illustrated, and then simulation results are presented.

2 Theoretical Analysis

The N signals observed are given by

$$\mathbf{x}(n) = \sum_{k=-\infty}^{\infty} \mathbf{a}(k)\mathbf{s}(n-k) \quad (1)$$

where $\mathbf{s}(n)=[s_1(n),\ldots,s_N(n)]^T$ are statistically independent speech sources, $\mathbf{x}(n)=[x_1(n),\ldots,x_N(n)]^T$ are observation signals, $\mathbf{a}$ is a mixing matrix. Using z-transform we obtain

$$\mathbf{X}[z] = \mathbf{A}[z]\mathbf{S}[z] \quad (2)$$

where

$$\mathbf{A}(z) = \begin{bmatrix} A_{11}(z) & A_{12}(z) & \cdots & A_{1N}(z) \\ A_{21}(z) & A_{22}(z) & \cdots & A_{2N}(z) \\ \vdots & \ddots & \ddots & 0 \\ A_{N1}(z) & \cdots & \cdots & A_{NN}(z) \end{bmatrix}$$

$$X_i(z) = [A_{i1}(z) \quad \cdots \quad A_{iN}(z)]\mathbf{S}[z] \quad i=1,2,\ldots,N \quad (3)$$

$X_i(z)$ and $S(z)$ is z transformation of ith mixture signal $x_i(n)$ and source signals $s(n)$, respectively.

A linear prediction analysis filter (LPAF) $P_i(z)$ is used to each mixture signal $x_i(n)$, the output signal is given by

$$y_i(n) = x_i(n) - \sum_{k=1}^{r} \partial_i(k) x_i(n-k) \quad i=1,2,\ldots,N \quad (4)$$

The filter $P_i(z)$ can be obtained from the mixture signal $x_i(n)$, the output signals $y_i(n)$ is temporally independent residual. In z-domain we have

$$Y_i[z] = P_i[z]X_i[z] \quad i=1,2\ldots,N \quad (5)$$

where $Y_i(z)$ is z transformation of ith output $y_i(n)$. Using matrix notation Eqn(5) can be written as

$$\mathbf{Y}(z) = \begin{bmatrix} Y_1(z) \\ \vdots \\ Y_N(z) \end{bmatrix} = \begin{bmatrix} P_1(z) & 0 & \cdots & 0 \\ 0 & P_2(z) & 0 & 0 \\ \vdots & \ddots & \ddots & 0 \\ 0 & \cdots & 0 & P_N(z) \end{bmatrix} \begin{bmatrix} X_1(z) \\ \vdots \\ X_N(z) \end{bmatrix}$$

Substitution of (2) in above equation gets

$$\mathbf{Y}(z) = \begin{bmatrix} P_1(z) & 0 & \cdots & 0 \\ 0 & P_2(z) & 0 & 0 \\ \vdots & & \ddots & 0 \\ 0 & \cdots & 0 & P_N(z) \end{bmatrix} \mathbf{A}[z] \begin{bmatrix} S_1(z) \\ \vdots \\ S_N(z) \end{bmatrix}$$

$$= \begin{bmatrix} P_1(z)A_{11}(z) & P_1(z)A_{12}(z) & \cdots & P_1(z)A_{1N}(z) \\ P_2(z)A_{21}(z) & P_2(z)A_{22}(z) & \cdots & P_2(z)A_{2N}(z) \\ \vdots & & \ddots & \vdots \\ P_N(z)A_{N1}(z) & \cdots & \cdots & P_N(z)A_{NN}(z) \end{bmatrix} \begin{bmatrix} S_1(z) \\ \vdots \\ S_N(z) \end{bmatrix} \quad (6)$$

According to matrix operation, from Eqn.(6) we can obtain that

$$Y_i(z) = [A_{i1}(z) \quad \cdots \quad A_{iN}(z)] \begin{bmatrix} P_i(z)S_1(z) \\ \vdots \\ P_i(z)S_N(z) \end{bmatrix}$$

$$= [A_{i1}(z) \quad \cdots \quad A_{iN}(z)] P_i(z) \mathbf{S}[z] \quad i=1,2,\ldots,N \quad (7)$$

Comparing Eqn (3) and (7), and using the scaling ambiguity of BSS algorithm, we can obtain the estimation of $\mathbf{A}(z)$ from $\mathbf{Y}(z)$. The original signals can be recovered by Eqn (8), we denote them as $\mathbf{S}^x(z)$, the BSS system is shown in Fig.1.

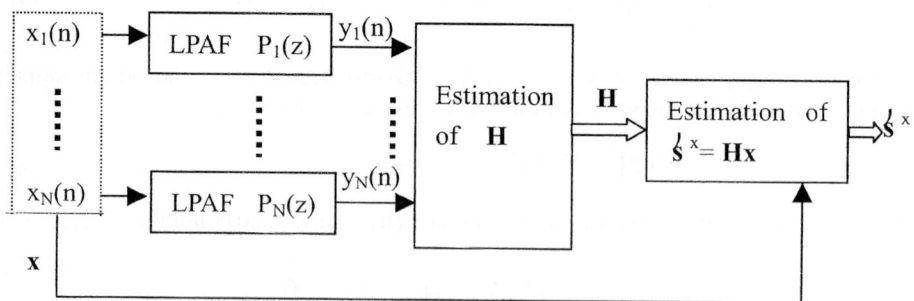

Fig. 1. BSS system

$$\mathbf{S}^x(z) = \mathbf{H}(z)\mathbf{X}(z) \quad (8)$$

Above derivation is suitable for instantaneous mixtures and convolutive mixtures. In the next section we give the simulation results to illustrate its validity.

3 Simulation Results

The source signals and instantaneous mixture signals are shown in Fig.2, the mixing matrix A is given by that

$$A = \begin{bmatrix} 1 & 0.7 & 0.4 \\ 0.6 & 1 & 0.2 \\ 0.8 & 0.5 & 1 \end{bmatrix}$$

Using Fast ICA algorithm and the 13th-order linear prediction analysis filter as filter $P_i(z)$, which is obtained using linear prediction for signals xi(n), to mixtures speech signals with frame size 25ms, the recovered signals $s_i^X(n)$ are shown in Fig.3, and their power spectral densities (PSD) are shown in Fig.4. We can find that $s_i^X(n)$ is the estimation of original signal.

Fig.5 shows two source signals, convolutive mixture signals and separated signals by the BSS system 1 using the BSS algorithm in[6] and 13th-order linear prediction analysis filter to mixture speech signals with frame size 25ms. The source signals are separated approximately.

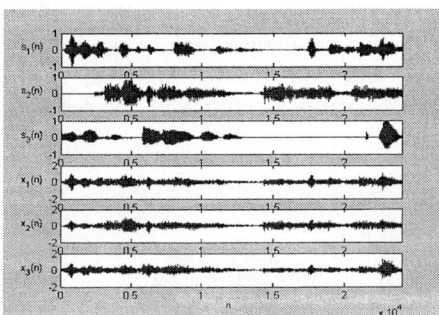

Fig. 2. Source and instantaneous mixture signals

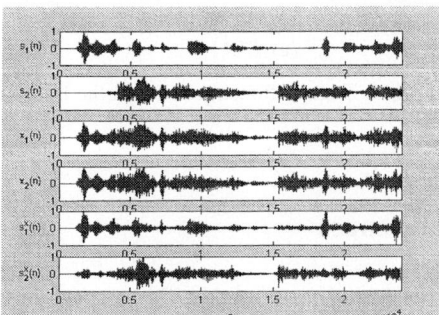

Fig. 3. Separated signals for instantaneous mixtures

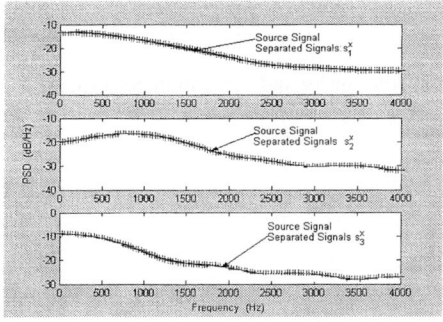

Fig. 4. PSD of Separated signals

Fig. 5. Convolutive mixtures case

To evaluate the affection of the linear prediction filtering on separation performance, we used the signal to interference ratio (SIR), defined as follows:

$$SIR_i = SIR_{Oi} - SIR_{Ii} \quad (9)$$

$$SIR_{Oi} = 10\log\frac{\sum_\omega |Q_{ii}(\omega)S_i(\omega)|^2}{\sum_\omega |Q_{ij}(\omega)S_j(\omega)|^2} \qquad SIR_{Ii} = 10\log\frac{\sum_\omega |A_{ii}(\omega)S_i(\omega)|^2}{\sum_\omega |A_{ij}(\omega)S_j(\omega)|^2}$$

where i j and $\mathbf{Q}(w)=\mathbf{H}(w)\mathbf{A}(w)$. For different frame sizes and different filter orders, SIR_1 and SIR_2 are shown in Fig.6. From Fig.6, we can see that when the frame is equal to 25ms and the order of the linear prediction filter is chose from 10 to 13, the performance is better. The SIR_1 and SIR_2 of the BSS algorithm in [6] without using the linear prediction filter is 8.128dB and 5.0617dB, respectively. So the separation performance is reduced a little, but the quality of the separated speech is improved.

4 Conclusion

We have presented the theoretical analysis of blind source separation algorithm using linear prediction filter and its correct structure. The simulation results verify the derivation in this paper and show that linear prediction analysis can be used in blind source separation for instantaneous mixtures and convolutive mixtures. For the convolutive mixtures, when the separated speech quality is improved, the separation performance is reduced a little.

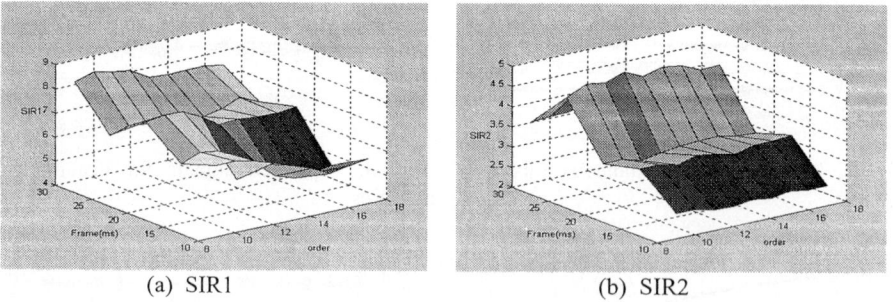

(a) SIR1 (b) SIR2

Fig. 6. The relationship between the frame size, filter order and separation performance

References

1. Aapo, H., Juha, K., Erkki, O.: Independent Component Analysis. John Wiley & Sons, LTD, May (2001)
2. Sun, X., Douglas, S.C.: A Natural Gradient Convolutive Blind Source Separation Algorithm for Speech Mixtures. ICA 2001. San Diego CA Dec (2001) 59-64

3. Douglas, S.C., Sun, X.: Convolutive Blind Separation of Speech Mixtures Using the Natural Gradient. Speech Communication (2003) 65-78
4. Kostas, K., Vicente, Z., Nandi, A.K.: Blind Separation of Acoustic Mixtures Based on Linear Prediction Analysis. ICA2003. Nara, Japan April (2003) 343-348
5. Tsuyoki, N., Hiroshi, S., Kiyohiro, S.: Stable Learning Algorithm for Blind Separation of Temporally Correlated Signals Combining Multistage ICA and Linear Prediction. ICA2003. Nara, Japan April (2003) 337-342
6. Lucas, P., Clay, S.: Convolutive Blind Separation of Non-Stationary Sources. IEEE Trans. SAP, Vol.8, No.3 May (2000) 320-327

A Blind Source Separation Based Micro Gas Sensor Array Modeling Method

Guangfen Wei [1], Zhenan Tang [1], Philip C.H. Chan [2], and Jun Yu [1]

[1] Dept. of Electronic Engineering, Dalian University of Technology, Dalian 116024, China
eegingkowei@tom.com; tangza@dlut.edu.cn; junyu628@21cn.com
[2] Dept. of Electrical & Electronic Engineering,
Hong Kong University of Science & Technology, Hong Kong, China
eepchan@ust.hk

Abstract. Blind Source Separation (BSS) has been a strong method to extract the unknown independent source signals from sensor measurements which are unknown combinations of the source signals. In this paper, a BSS based modeling method is proposed and analyzed for a micro gas sensor array, which is fabricated with surface micromachining technology and is applied to detect the gas mixture of CO and CH_4. Two widely used BSS methods--Independent Component Analysis (ICA) and Nonlinear Principal Component Analysis (NLPCA) are applied to obtain the gas concentration signals. The analyzing results demonstrate that BSS is an efficient way to extract the components which corresponding to the gas concentration signals.

1 Introduction

Gas sensors inherently suffer from imperfections like cross sensitivity, nonlinearity, low stability, and drift. Improving gas sensor performance via better materials, new technologies and advanced design is generally expensive and somewhat limited. Another solution is the intelligent or smart electronic nose design, which combines the gas sensor array and the signal processing tools, allows performance improvements. There have been intensively researches on gas sensor array signal processing methods, including statistic based pattern recognition methods and Artificial Neural Networks [1][2]. Besides the problems of single gas (odor) classification, recognition and quantification, restriction of baseline drift and the effect of interfering gases, gas mixture analysis problems have been studied in the past several years. Pattern recognition method has been proved an efficient gas mixture classification and discrimination method by transforming the gas mixture of some gas concentration combinations to a certain "pattern" [3][4]. Another gas mixture analysis method is the transformation of sensor space to odor space [5]. However, because of the versatility and uncertainty of real sensor models for gas mixture analysis, it is quite a big problem to parameterize the models directly. Thereafter, method of combination of pattern classification and regression is proposed and applied in gas mixture analysis, such as the combination of

Principal Component Analysis (PCA) and ANN [6] and the combination of PCA and principal component regression (PCR) as multicomponent analysis method [7].

Blind source separation (BSS) is the process of extracting unknown independent source signals from sensor measurements which are unknown combinations of the source signals. Initially inspired by signal separation in the human brain, BSS has recently become a very active research area both in statistical signal processing [8] and unsupervised neural learning [9]. The goal of BSS is to recover the waveforms of unobserved signals from observed mixtures without knowing the mixing coefficients. Consequently, source separation cancels sensor crosstalk providing higher accuracy, selectivity, and multisensing capability [10]. There are many approaches to reach the BSS goal, such as Independent Component Analysis (ICA) and nonlinear PCA (NLPCA), which have been applied in lots of areas. Several papers have studied the application of ICA in electronic nose area. Natale, C.D. et al [11] applied ICA into E-nose data for the purpose of counteraction of environmental disturbances and gave a discussion on the connections between ICA and electronic nose. Kermit M. and Tomic O. [12] discussed the application of ICA for the aim of drift restriction and discrimination of specified gases compared with PCA.

Based on the above literatures and works we have done before [13], the connections between the BSS model and the electronic nose data response to gas mixtures are discussed further in this paper. The application of BSS model proposes new insights in the comprehension of the response mechanism of gas sensor array to mixtures.

2 Statement of Blind Source Separation Problem

Consider N samples of the m-dimension observed random vector x, modeled by:

$$x = F(s) + n. \tag{1}$$

where F is an unknown mixing mapping assumed invertible, s is an unknown n-dimensional source vector containing the source signals $s_1, s_2, ..., s_n$, which are assumed to be statistically independent, and n is an additive noise, independent of the sources. The BSS problem is to recover the n unknown actual source signals $s_j(t)$ which have given rise to the observed mixtures. The basic idea in BSS consists in estimating a mapping G, only from the observed data x, such that $y=G(x)$ are statistically independent. The achievements show that BSS problem can be solved under some suitable regularizing constraints and additional prior information [14].

One simplification method is post-Nonlinear mixing approximation, which was introduced by Taleb and Jutten [15], provides the observation x, which is the unknown nonlinear mixture of the unknown statistically independent sources s,

$$x = F(As) + n. \tag{2}$$

where F is an unknown invertible derivable nonlinear function, A is a regular mixing matrix. In a further simplified linear model, the unknown invertible mapping is modeled by a square regular matrix A, $x=As$ [14].

2.1 ICA Approach

ICA is an intensively developed signal processing method in BSS area based on linear mixing model, $x=As$. Most suggested solutions to the ICA problem use the 4th-order cumulant or kurtosis of the signals, defined for a zero-mean random variable v as

$$kurt(v) = E\{v^4\} - 3(E\{v^2\})^2 . \tag{3}$$

For a Gaussian random variable, kurtosis is 0; for densities peaked at 0, it is positive and for flatter densities, negative. The purpose is to find a linear combination of the sphered observations x, say $w^T x$, such that it has maximal or minimal kurtosis under the constraint $\|w\|=1$. To minimize or maximize $kurt(w^T x)$, a neural algorithm based on gradient descent or ascent can be used and w is interpreted as the weight vector of a neuron with input vector x. The objective function can be simplified because the inputs have been sphered.

$$J(W) = E\{(w^T x)^4\} - 3\|w\|^4 + F(\|w\|^2) . \tag{4}$$

where F is a penalty term due to the constraint. To make the learning radically faster and more reliable, the fixed-point iteration algorithms are proposed [16] and applied widely. It is adopted as the IC extraction tool in this paper.

2.2 Nonlinear PCA Approach

The nonlinear PCA approach is based on the definition of nonlinear correlation of random variables y_1 and y_2 as $E\{f(y_1)g(y_2)\}$, and among f and g, at least one is a non linear function, and on a general theorem stating that y_1 and y_2 are independent *iff*

$$E\{f(y_1)g(y_2)\} = E\{f(y_1)\}E\{g(y_2)\} . \tag{5}$$

For all continuous functions f and g that are 0 outside a finite interval. Based on the theorem and the minimum square criterion for PCA, the criterion for NLPCA is

$$J(W) = E\{\left\|x - \sum_{i=1}^{n} g_i(w_i^T x) w_i\right\|^2\} . \tag{6}$$

Because of the nonlinear mixing function, the NLPCA is proposed not only an efficient method to extract the independent components [14], but also based on a post-nonlinear mixing model [17]. It is applied in this paper by contrast with ICA.

3 Modeling Method for Micro Gas Sensor Array

The gas sensor array analyzed in this paper is composed of 4 micro-hotplate (MHP) based sensors with tin oxide film as sensing part. Each sensor is fabricated on a MHP and has the sensing film heated by a polysilicon ring heater surrounding the MHP. A polysilicon resister monitors the operating temperature of MHP. The design and fabrication technology of the MHP has been reported elsewhere [18]. All the sensors are sensitive to CO and CH$_4$ with small different sensitivity. 2 sensors are operated a little more sensitive to CO and the other 2 are operated a little more sensitive to CH$_4$. The response to gas mixtures of one sensor is shown in Fig. 1.

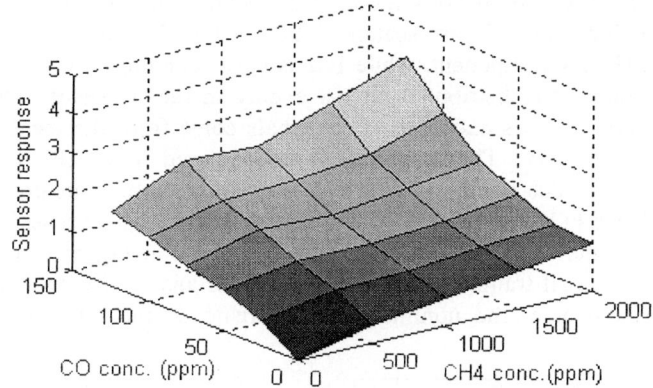

Fig. 1. Micro gas sensor response to CO and CH$_4$ gas mixtures

As can be seen in Fig.1, the signal of each sensor is actually a function of the gas components,

$$x_j(t) = f_j(c_1(t), c_2(t), \ldots, c_n(t)) + n_j, \quad j=1,\ldots,m \tag{7}$$

where $c_i(t)$ is the concentration signal of each gas source, which is to be extracted from the mixture, $f_j(.)$ is the unknown response function of sensor j, which is corresponding to the mixing function, n_j is noise. Linear approximation can be performed in a specified concentration range and the noise-free response of sensors in the array exposed to n compounds can be written as

$$x_j(t) = a_{j1}c_1(t) + a_{j2}c_2(t) + \ldots + a_{jn}c_n(t) = \sum_{i=1}^{n} a_{ji}c_i(t), \quad j=1,\ldots,m \tag{8}$$

where a_{ji} is the sensitivity of sensor j to gas i. This is very the same as the linearly regularized BSS model.

The assumption of independence between the signal sources is the foundation of BSS model. However, the statistics of the detected gas components in some mixtures generally is unknown and some can be obtained from the priori of the analytes. In the experimental period, the BSS model can be applied to model gas sensors, as the ranges and combinations of gas concentrations are controlled subjectively for the purpose of calibration. Therefore, it is a good modeling method for studying the sensors' response mechanism to gas mixtures.

4 Application and Results Discussion

In this paper, mixtures of CO and CH_4 are analyzed with the two above approaches respectively. The concentration range of CO is 0-100ppm, with 25ppm as the interval, while the concentration range of CH_4 is 0-1000ppm, with 500ppm as the interval. The detected concentration points are settled independently. The Independent Components (ICs) and the Nonlinear PCs are scatter plotted in Fig. 2 respectively. IC1 is corresponding to CH_4 gas component, while IC2 is corresponding to CO gas component. The mixture points are distributed almost evenly in the IC space. The differences between the extracted ICs and the source signals come from the measurement noise and less training data. The nonlinear function used in NLPCA algorithm is $y=1/(1+exp(-x))+1$. The nonlinear PCs have almost the same structure as ICs except that the nonlinear PCs are not just corresponding to the gas sources. The difference may come from the specified nonlinear function, the convergent error of the algorithm, noise and small training data set. Because the analyzed compounds are at the low concentration area, the nonlinear characteristic of the NLPCA is not distinguished.

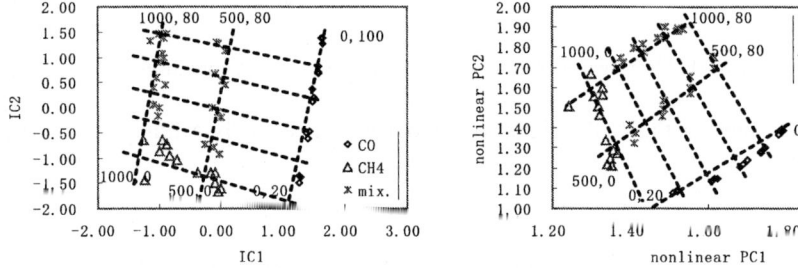

Fig. 2. The scatter plot of ICs and nonlinear PCs obtained from ICA and NLPCA. 1000,80 means that for this compound, CH_4 conc. is 1000ppm and CO conc. is 80ppm

5 Conclusion

Blind Source Separation (BSS) is discussed in this paper to extract the unknown gas source signals from sensor measurements which are unknown combinations of the gas components. ICA and NLPCA method is adopted as the signal processing method for the micro gas sensor array BSS model. The proposed BSS model is verified by analysis of gas mixtures consisting of CO and CH_4. The analyzing results show that the BSS model is effective to extract the independent gas components in mixtures and it provides a new point of view to study the sensing mechanism of gas sensors to mixtures. Further study will be carried out to improve the accuracy and stability.

References

1. Hines, E.L., Llobet, E., Gardner, J.W.: Electronic Noses: A Review of Signal Processing Techniques. IEE proc. Circuits Devices Syst., Vol. 146, No. 6 (1999) 297-310
2. Tanaka, K., Sano, M., Watanabe, H.: Modeling and Control of Carbon Monoxide Concentration Using a Neuro-Fuzzy Technique. IEEE Transactions on Fuzzy Systems, Vol. 3, No.3 (1995) 271-280
3. Boilot, P., Hines, E.L., Gongora, M.A., Folland, R.S.: Electronic Noses Inter-comparison, Data Fusion and Sensor Selection in Discrimination of Standard Fruit Solutions. Sensors and Actuators B, 88 (2003) 80-88
4. Fort, A., Machetti, N., Rocchi, S., Belen, M., Santos, S., Tondi, L., Ulivieri, N., Vignoli, V., Sberveglieri, G.: Tin Oxide Gas Sensing: Comparison among Different Measurement Techniques for Gas Mixture Classification. IEEE Transactions on Instrumentation and Measurement, Vol. 52 (2003) 921-926
5. Pearce, T.C.: Odor to Sensor Space Transformations in Biological and Artificial Noses. Neurocomputing, Vol. 32-33 (2000) 941-952
6. Eklov, T., Lundstrom, I.: Gas Mixture Analysis Using a Distributed Chemical Sensor System. Sensors and Actuators B, 57 (1999) 274-282
7. Capone, S., Siciliano, P., Barsan, N., Weimar, U., Vasanelli, L.: Analysis of CO and CH_4 Gas Mixtures by Using a Micromachined Sensor Array. Sensors and Actuators B, 78 (2001) 40-48
8. Cardoso, J.F.: Blind Signal Separation: Statistical Principles. Proc. IEEE, Vol. 86 (1998) 2009-2025
9. Amari, S., Cichocki, A.: Adaptive Blind Signal Processing-Neural Network Approaches. Proc. IEEE, Vol. 86 (1998) 2026-2047
10. Paraschiv-Ionescu, A., Jutten, C., Bouvier, G.: Source Separation Based Processing for Integrated Hall Sensor Arrays. IEEE Sensors Journal, Vol. 2 No. 6 (2002) 663-673
11. Natale, C.D., Martinelli, E., D'Amico, A.: Counteraction of Environmental Disturbances of Electronic Nose Data by Independent Component Analysis. Sensors and Actuators B, 82 (2002) 158-165
12. Kermit, M., Tomic, O.: Independent Component Analysis Applied on Gas Sensor Array Measurement Data. IEEE Sensors Journal, Vol. 3 No. 2 (2003) 218-228
13. Wei, G.F., Chan, C.H.P., Tang, Z.A., Yu, J.: Independent Component Analysis for Gas Mixture Component Identification and Quantification. Pacific Rim Workshop on Transducers and Micro/nano Technology, Xiamen, China (2002)
14. Hyvarinen, A., Karhunen, J., Oja, E.: Independent Component Analysis. John Wiley & Sons, Inc. (2001)
15. Taleb, A., Jutten, C.: Source Separation in Post-Nonlinear Mixtures. IEEE Transactions on Signal Processing, Vol. 47 No. 10 (1999) 217-229
16. Hyvarinen, A., Oja, E.: A Fast Fixed-Point Algorithm for Independent Component Analysis. Neuron Computation, Vol. 9 (1997) 1483-1492
17. Chalmond, B., Girard, S.C.: Nonlinear Modeling of Scattered Multivariate Data and Its Application to Shape Change. IEEE Trans. Pattern Analysis and Machine Intelligence, Vol.21 No.5 (1999) 422-432
18. Chan, C.H.P., Yan, G.Z., Sheng, L.Y., Sharma, K.R., Tang, Z.A., et al: An Integrated Gas Sensor Technology Using Surface Micro-machining. Sensors and Actuators B, 82 (2002) 277-283

A Novel Denoising Algorithm Based on Feedforward Multilayer Blind Separation*

Xiefeng Cheng[1,2], Ju Liu[2], Jianping Qiao[2], and Yewei Tao[1]

[1] School of Information Science and Engineering, Jinan University
Jinan,250022, P.R. China
{Cheng-x-f-1, Jnutaoyewei}@163.com
[2] School of Information Science and Engineering, Shandong University
Jinan,250100, P.R. China
{Juliu, QianjianPing}@sdu.edu.cn

Abstract. The blind source separation algorithm always does its best to retrieve source signals, so in fact the output is generally a stronger source added by some other weaker sources and noises. The algorithm proposed in this paper assume the mixture of the weaker sources and noises as a new source S_{new}, and the outputs of last separation can be regard as the sources of next separation after proper choice and combination. Then S_{new} can be separated by using blind separation technique repeatedly. The spectrum difference between S_{new} and a selected restored source helps to eliminate the influence of noises and improve the performance of this method. What is more, similitude phase graph is also proposed in this paper, which can show the performance of blind separation algorithm straightly.

1 Introduction

Blind signal processing technique is a hot topic in information processing field which has been widely used in communication, biology and medical applications, image processing, speech processing etc. It aims to restore the source signals as accurate as possible by reasonable assumption and approximation while the priori information of the observed signal is rare.

Blind separation in noisy is difficult because there are a great many unknown terms. At present, most algorithms are under the conditions of non-noise[1]. How to apply the existing method into noisy environment is a problem in the future research.

The rest of this paper is organized as follows. In section one, we analyze the general denoising methods. In section two, a novel denoising algorithm based on feed forward multilayer blind separation is proposed. It works well under different noisy environments which is meaningful for the application of the existing blind separation methods in noisy environments. In section three, a new judgement is proposed to describe the performance of the blind separation algorithm. Simulation results and the analysis are shown in section four and five.

---

* This work was supported by National Natural Science Fundition of China (30000041)

2 Physical Properties of the Noise and the General Denoising Methods

Noise generally refers to the regular or unregular interference mixed into the useful signal. It can be classified into inter-system noise and extra-system noise. For speech signals and image signals the noise generally comes from the outside of the system. After spectrum analysis to large numbers of noisy signals, most noise follows Guassian distribution.

There has been many denoising methods[1-3]:

1. Filter denoising method. Various band-pass and low-pass filters are used to take off the noise.

2. If the background noise under quiet environment can be gotten, the cancellation denoising can be used.

3. Denoising method is introduced in various software applications.

In the experiment we found that the signals $\hat{s}_1$ and $\hat{s}_2$ after blind separation from s_1 and s_2, there is still some residual components of s_2 in $\hat{s}_1$, also, there is still some residual components of s_1 in $\hat{s}_2$. In a sense, blind separation is a technique that separates signals by strengthening a class of signals and weakening the other class through some methods[1-3] such as ICA. Hereby, we propose a novel denoising algorithm based on feed forward multilayer blind separation. Simulation results show that our method can effectively increase the signal-to-noise ratio and similitude coefficient.

3 The Principle

The principle of our method is:

1. Let N is arbitrary noise signal, the mixture model is:

$$y_1 = a_{11} s_1 + a_{12} s_2 + N \quad (1)$$
$$y_2 = a_{21} s_1 + a_{22} s_2 + N$$

After the first BSS (blind signal separation), the separation results are y_{11}^1, y_{21}^1. According to the conclusion of the blind separation algorithm y_{11}^1, y_{21}^1 should mainly conclude one of s_1 or s_2.

$$\begin{cases} y_{11}^1 = s_{11} + s'_{21} + N' \\ y_{21}^1 = s_{21} + s'_{11} + N' \end{cases} \quad (2)$$

Where $s_{11} > \dot{s}_{21}, s_{21} > \dot{s}_{11}$. Let

$$s^1_{21} = s'_{21} + N' \quad (3)$$
$$s^1_{11} = s'_{11} + N'$$

be noise and residual interference signals, thus s^1_{21}, s^1_{11} become new sources:

$$y_{11}^1 = s_1^1 + s'_{21} \qquad (4)$$
$$y_{21}^1 = s_2^1 + s'_{11}$$

2. Because of the permutation indeterminacy of the signals after normal BSS, Full combination of correspondence with the result of first BSS with the feed-forward signal is carried out, that is $\{y_{11}^1, y_1 \bowtie y_{11}^1, y_2 \bowtie y_{21}^1, y_1 \bowtie y_{21}^1, y_2\}$. After the second BSS, we can get the eight output signals. Then computing the similitude coefficient between the eight output signals and y_{11}^1, y_{21}^1 respectively, then classifying the eight signals into two groups according to the similitude coefficient with y_{11}^1 or y_{21}^1. That is, signals whose similitude coefficient with y_{11}^1 is biger will be divided into one group, so we have $\{y_{11}^2, y_{12}^2, y_{13}^2, y_{14}^2\}$. So does y_{21}^1, then we can get $\{y_{21}^2, y_{22}^2, y_{23}^2, y_{24}^2\}$.

3. Go on divide $\{y_{11}^2, y_{11}^1 \bowtie y_{12}^2, y_{11}^1\}...,\{y_{21}^2, y_{21}^1 \bowtie y_{22}^2, y_{21}^1\}...$, and $\{y_{11}^2, y_{21}^2 \bowtie y_{12}^2, y_{22}^2\}...$ into groups to do BSS. In the same way, we can get $y_{11}^3, y_{12}^3 ... y_{1k}^3 ...$ and $y_{21}^3, y_{22}^3 ... y_{2k}^3 ...$. So repeatedly divide y_{1k}^{i-1} and y_{2k}^{i-1} that the feed forward with the y_{1k}^i, y_{2k}^i of this BSS into groups, and it is also divided into groups that the result y_{1k}^i, y_{2k}^i of this BSS to do BSS again. In repeated BSS, because noise and interference s'_{1k}, s'_{2k} in (4) are automatically considered as the new sources, so amplitude of the result will become more and more small. BSS will stop when the mean amplitude of y_{2k}^M or y_{1k}^M is less than a predetermined threshold after the Mth blind separation, then y_{2k}^M or y_{1k}^M is saved as min(y_{2k}^M), min(y_{1k}^M).

4. Regard in y_{1k}^i, y_{2k}^i with maximum mean amplitude as max(y_{1k}^i), max(y_{2k}^i), best $\hat{s}_1, \hat{s}_2$ are estimated by cross spectrum difference of max(y_{1k}^i) and min(y_{2k}^M), max(y_{2k}^i) and min(y_{1k}^M).

5. When weak signals are extracted in a strong noisy environment, the min(y_{2k}^M), min(y_{1k}^M) of the signals may not be gotten. In this case, method of coherence average is adopted. The SNR of the signal after coherence average is M times more than that of the signal without coherence average in an ideal case.

4 Similitude Coefficient and Similitude Phase Diagram

As a quantitative metric of the algorithms' performance, the similitude coefficient is widely used[4]. Let y(n) be the signal after separation, s(n) be the source signal. The similitude coefficient is defined by:

$$\beta = (y_i(n), s_j(n)) = \left|\sum_{n=1}^{N} y_i(n) s_j(n)\right| \Big/ \sqrt{\sum_{n=1}^{N} y_i^2(n) \sum_{n=1}^{N} s_j^2(n)} \quad (5)$$

when $y_i(n) = k s_j(n)$, β_{ij} equals to one. It is clear that the similitude coefficient can not show the magnitude and phase differences between y_i and s_j. So similitude phase diagram is used which can show the differences straightly. The rules of the similitude phase diagram are as follows:

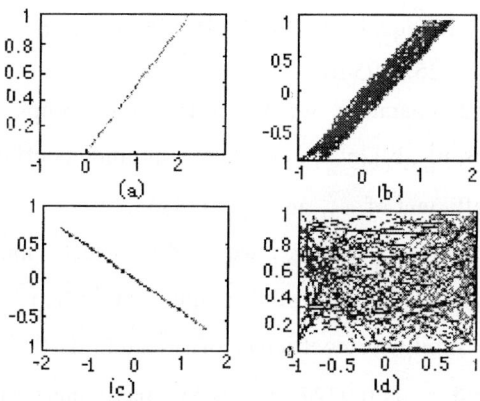

Fig. 1. Examples of similitude phase diagram

Let $f_1(n)$ be a signal. After blind source separation, the magnitude, frequency and phase of $f_1(n)$ is A_1, ω_1 and φ_1 respectively, so does $f_2(n)$.

1. If $A_1 = A_2$, $\varphi_1 = \varphi_2$, $\omega_1 = \omega_2$, so $\beta = 1$ and it is a line of bias with the angle of 45, as shown in fig.1 (a).
2. If $A_1 = kA_2$, $\varphi_1 = \varphi_2$, $\omega_2 = \omega_1$, so $\beta \approx 1$ and it is a line of width d and bias with the angle of 45°, width of $d \propto k$, as shown in fig.1 (b).
3. If $A_1 = A_2$, $\omega_2 = \omega_1$, $\varphi_1 - \varphi_2 = \Delta\varphi$, the angle of the bias will change along with $\Delta\varphi$. When $\Delta\varphi = 180°$, $\beta = 1$, the diagram is shown as fig.1 (c).
4. If $A_1 \neq A_2$, $\omega_2 \neq \omega_1$, $\varphi_1 \neq \varphi_2$, there are n close circles whose filling area has direct proportion with the non-relation of $f_1(n)$ and $f_2(n)$, when all areas are filled, β equals to 0. $f_1(n)$ and $f_2(n)$ are not relative. As shown in fig.1 (d).

5 Simulations

Let s_1, s_2 be tow channel speech signals each of which has 4000 samples, A=[1, 0.56; 0.73, 0.9], N be random noise whose length is 4000, the noisy mixtures can be stated as follows:

$$y_1 = s_1 + 0.56 s_2 + N \qquad (6)$$
$$y_2 = 0.73 s_1 + 0.9 s_2 + N$$

The second-order statistical blind separation algorithm [1] is adopted. It will stop when the mean amplitude is less than 0.1.

The results of the first blind separation are y_{11}^1, y_{21}^1. Similitude coefficient of y_{11}^1 and s_1 is 0.816, SNR1 equals to 22.2634dB; Similitude coefficient of y_{21}^1 and s_2 is 0.8073, SNR2 equals to 25.9395dB.

In the second blind separation, we only select four groups $y_{11}^2, y_{12}^2, y_{21}^2, y_{22}^2$ are from tow input groups{ y_{11}^1, y_1 },{ y_{21}^1, y_2 }, where similitude coefficient of y_{22}^2 and s_1 is 0.8416, similitude coefficient of y_{21}^2 and s_2 is 0.8863.

In the third blind separation, we only select input groups{ y_{21}^2, y_{21}^1 } and{ y_{22}^2, y_{21}^1 },{ y_{11}^2, y_{11}^1 }and{ y_{12}^2, y_{11}^1 }, the outputs are{ y_{21}^3, y_{22}^3 },{ y_{23}^3, y_{24}^3 },{ y_{11}^3, y_{12}^3 } and{ y_{13}^3, y_{14}^3 }, where similitude coefficient of y_{21}^3 and s_1 is 0.9208, where similitude coefficient of y_{23}^3 and s_2 is 0.9327. y_{23}^3 is the maximum term of the mean amplitude.

In the forth blind separation, input groups are from the results of the third time separation and the results of the second separation. The process is stopped when the mean amplitude of y_{21}^4 decreased by the factor of 10 compared with y_{11}^1 in the fourth time blind separation, y_{21}^4 is the result of the blind separation by{ y_{21}^3, y_{24}^3 }.

Finally, the maximum term y_{23}^3 and the minimum term y_{21}^4 are transformed to frequency domain by FFT, spectrum difference in time domain is $\hat{s}_2$, signals $\hat{s}_2$ are shown in fig.2, where similitude coefficient of $\hat{s}_2$ and s_2 is 0.9670, the SNR2 increased to 32.5782dB.

Computation of $\hat{s}_2$ by our method is depicted in fig.2. Similitude phase diagram of y_{21}^1 and s_2 is shown in fig.2 (a) which states that there are not only difference in magnitude but also irregular burr edges between y_{21}^1 and s_2, the irregular burr edges turned out by noise. Similitude phase diagram of $\hat{s}_2$ and s_2 is shown in fig.2 (b) which states that there are only difference in magnitude and no irregular burr edges, the noise is buffed away. $\hat{s}_1$ can also be obtained by this method. The similitude coefficient of $\hat{s}_1$ and s_1 is 0.9556,the SNR1 increased to 30.7640dB. In fig.2, $\hat{s}_2'$ is simulation result of adopting the method of coherence average.

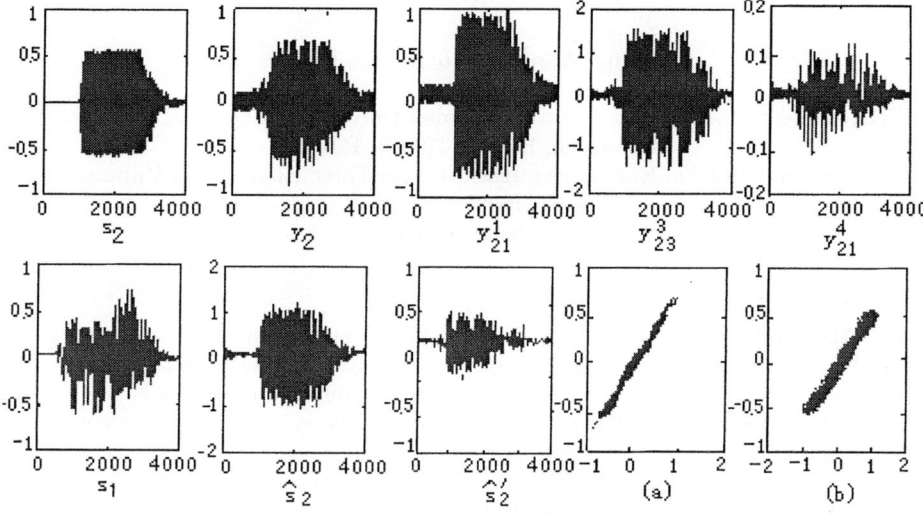

Fig. 2. Computation of $\hat{s}_2$ method

6 Conclusion

The proposed method in this paper makes full use of the different information of between $y_{1k}^{i-1}, y_{2k}^{i-1}$ and y_{1k}^{i}, y_{2k}^{i} in results of multi-blind separation to separate the noise and residual interference signal $s^i{}_{1k}, s^i{}_{2k}$. Simulation results show the effectiveness in denoising. Adjoin proportion coefficient k when to do spectrum difference, SNR can achieve the maximum when k is changed. Noise can be compressed by coherence average too.

In blind signal processing, blind separation of the noisy mixtures is difficult because of too little priori information. At present, most algorithms are under the conditions of non-noise. Therefore, the denoising algorithm based on feed forward multilayer blind separation proposed in this paper is meaningful.

References

1. Liu, J., He, Z.: A Survey of Blind Source Separation and Blind Deconvolution. China. J. Acta Electronica Einica, 30 (2002) 591-597
2. Comon, P.: Independent Component Analysis, A New Concept? Int. J. Signal Processing, 36 (1994) 287-314
3. Wu, S., Feng, C.: Study On Noisy Speech Recognition Methods. China. J. Acta Scientiayum Naturalium Universitatis Pekinensis, 37 (2001) 365-367
4. Qin, H., Xie, S.: Blind Separation Algorithm Based on Covariance Matrix. China. J. Computer Engineeing, 29 (2003) 36-38

The Existence of Spurious Equilibrium in FastICA*

Gang Wang[1,2] and Dewen Hu[1]

[1] Department of Automatic Control, National University of Defense Technology, Changsha, Hunan, 410073, P.R.C.
[2] Telecommunication Engineering Institute, Air Force Engineering University, Xi'an, Shanxi, 710077, P.R.C.
dhu@nudt.edu.cn

Abstract. FastICA is a fast fixed-point algorithm proposed by Hyvärinen for Independent Component Analysis (ICA). For the outstanding performance such as fast convergence and robustness, it is now one of the most popular estimate methods. The existence of spurious equilibrium in FastICA is addressed in this paper, which comes from its applications in Blind Source Separation (BSS). Two issues are involved. The first is on the object function and the second is about the fixed-point algorithm, an approximate Newton's method. Analysis shows the existence of spurious equilibrium, which is the singular point introduced during the algorithm's derivation. Experimental results show the estimates of spurious equilibria, and improvements are proposed by revising the convergence condition.

1 Introduction

FastICA is first proposed by Hyvärinen and Oja in [1] based on the object function of negentropy, a popular nongaussianity measurement. The optimal scheme involved is a fast fixed-point algorithm, an approximate Newton's method. For its simpleness, robustness and fast convergence, FastICA is now one of the most popular algorithms for ICA [2,3].

However in its applications in Blind Source Separation (BSS), estimates from different tests for a certain system are not always the same (up to permutation and scaling), sometimes the estimates are far from the original sources even though the sources fulfill the assumptions of mutual independence and nongaussianity. That is the problem of uniqueness of estimate or the existence of spurious equilibrium. In face representation using FastICA, the independent faces in one test may be different from those of the former even though they been ordered specifically such as by their kurtosis.

As we know the process of nongaussianity-based methods can be separated into two parts, designing of object function and the optimal algorithm [3]. As

---

* Supported by Natural Science Foundation of China (30370416), the Distinguished Young Scholars Fund of China (60225015), Ministry of Science and Technology of China(2001CCA04100) and Ministry of Education of China (TRAPOYT Project).

for the former, Amari showed in [4] that if the sources are mutually independent and nongaussian, popular object functions have no spurious equilibria. In [5] the uniqueness of standard linear ICA was also convinced in mathematic view. In this paper the researches explore where the spurious equilibrium comes from, and concentrate on the rationality of the object function for FastICA and the fixed-point algorithm involved. Improvements are also proposed by revising the convergence condition.

2 Preliminaries

The negentropy is approximated [1,2] using only one nonquadratic function G

$$J(y) = [E\{G(x^T)\} - E\{G(v)\}]^2. \tag{1}$$

When $G(y) = y^4/4$, the kurtosis-based form is obtained

$$J_{\text{kurt}}(y) = (\text{kurt}(y))^2. \tag{2}$$

The stochastic gradient algorithm was presented by Bell and Sejnowski[6]

$$\triangle \mathbf{w} \propto \gamma E\{zg(\mathbf{w}^T z)\}, \mathbf{w} = \mathbf{w}/\|\mathbf{w}\| \tag{3}$$

where z is a sphered form of x, $g(y) = G'(y)$, and practically γ is proposed

$$\gamma = \text{sign}(E\{yg(y) - g'(y)\}). \tag{4}$$

While the original fixed-point algorithm was given [1,2]

$$\mathbf{w} \leftarrow E\{zg(\mathbf{w}^T z)\}, \mathbf{w} \leftarrow \mathbf{w}/\|\mathbf{w}\|. \tag{5}$$

The optima of (5) are estimated at the points where the gradient of Lagrange is zero

$$E\{zg(\mathbf{w}^T z)\} + \lambda \mathbf{w} = 0. \tag{6}$$

Denote $F = E\{zg(\mathbf{w}^T z)\} + \lambda \mathbf{w}$, then the following can be obtained

$$\frac{\partial F}{\partial \mathbf{w}} = E\{zz^T g'(\mathbf{w}^T z)\} + \lambda I \approx E\{zz^T\}E\{g'(\mathbf{w}^T z)\} + \lambda I. \tag{7}$$

According to the Newton's iteration scheme [3]

$$\mathbf{w} \leftarrow \mathbf{w} - [\frac{\partial^2 J_1}{\partial \mathbf{w}^2}]^{-1}\frac{\partial J_1}{\partial \mathbf{w}} \tag{8}$$

herein $F = \partial J_1/\partial \mathbf{w}$, the following is obtained

$$\mathbf{w} \leftarrow \mathbf{w} - [E\{zg(\mathbf{w}^T z)\} + \lambda \mathbf{w}]/[E\{g'(\mathbf{w}^T z)\} + \lambda]. \tag{9}$$

The basic fixed-point iteration is given by multiplying both sides of (9) by $E\{g'(\mathbf{w}^T z)\} + \lambda$

$$\mathbf{w} \leftarrow E\{zg(\mathbf{w}^T z)\} - E\{g'(\mathbf{w}^T z)\mathbf{w}\}, \mathbf{w} \leftarrow \mathbf{w}/\|\mathbf{w}\|. \tag{10}$$

3 The Existence of Spurious Equilibrium

3.1 On the Object Function

We know from (5) that the object function for FastICA (denoted by J_2) is essentially

$$J_2(y) = E\{G(x^T)\} - E\{G(v)\} \tag{11}$$

and the kurtosis-based form is

$$J_{2\text{kurt}}(y) = \text{kurt}(y). \tag{12}$$

The above object functions are different from those for the stochastic gradient. Compare (5) with (3), we can see that λ in (3) has been omitted. Are the spurious equilibria caused by the above relaxing from (3) to (5)? When the convergence condition in FastICA is considered, we can exclude this doubt. If the estimated signal is supergaussian (λ in (4) equals to 1), (11) is equivalent to (1). The iteration converges when $w(t+1) \approx w(t)$ or $||w(t+1) - w(t)|| < \delta$ (δ is the minimum scope for convergence). While the estimated signal is a subgaussian one (λ in (4) equals to -1), (11) is minus to (1). The iteration also converges when $w(t+1) \approx -w(t)$ or $||w(t+1) + w(t)|| < \delta$.

3.2 On the Proximate Newton's Scheme

It is obvious that (8)(9) exist only when $\partial^2 J_1/\partial w^2 \neq 0$, and now let's consider the case of $\partial^2 J_1/\partial w^2 = 0$. Trace back to the Newton's iteration and now first concentrate on the multivariate Taylor series.

Define $J_3(w)$ the Taylor series around the point w_0 thus

$$J_3(w) \approx J_3(w_0) + \frac{\partial J_3(w)^T}{\partial w}|_{w=w_0}(w-w_0) + \frac{1}{2}(w-w_0)^T \frac{\partial^2 J_3(w)^T}{\partial w^2}|_{w=w_0}(w-w_0). \tag{13}$$

Define $J_4(\triangle w) = J_3(w) - J_3(w_0)$ ($\triangle w = w - w_0$) and minimize J_4 with respect to $\triangle w$. Let $\partial J_4/\partial \triangle w = 0$, then

$$\frac{\partial J_3(w)}{\partial w} + \frac{\partial^2 J_3(w)}{\partial w^2} \triangle w = 0. \tag{14}$$

This equation exists when $\partial J_3(w)/\partial w = 0$ and $\triangle w \to 0$. The following can be obtained

$$w = w_0 - [\frac{\partial^2 J_3(w)}{\partial w^2}]^{-1} \frac{\partial^2 J_3(w)}{\partial w^2}|_{w=w_0}. \tag{15}$$

As we know (15) is ill-conditioned when $\partial^2 J_3(w)/\partial w^2 \to 0$. From (9) to (10), it is obvious that the case of ill-condition has not been considered. That is to say the singular point can also be the solution for (10) which will cause spurious equilibrium practically.

In a 2-dimention system of BSS, spurious equilibria in a kurtosis-based fixed-point scheme are considered. Denote $w_1 = (w_1, w_2)$ the demixing vector, k_{z_1} and

k_{z_2} the kurtosis of z_1 and z_2 ($z = (z_1, z_2)$) respectively. Let $w_2 = \pm\sqrt{1-w_1^2}$, then

$$J_{2kurt} \propto w_1^4 k_{z_1} + (1-w_1^2)^2 k_{z_2}. \tag{16}$$

And when $\partial^2 J_{2kurt}/\partial w_1^2 = 0$ the following spurious equilibria ($w = (w_1, w_2)$) can be obtained

$$w_1 = \pm\sqrt{\frac{k_{z_2}}{3(k_{z_1}+k_{z_2})}}, w_2 = \pm\sqrt{\frac{3k_{z_1}+2k_{z_2}}{3(k_{z_1}+k_{z_2})}}, or \mp\sqrt{\frac{3k_{z_1}+2k_{z_2}}{3(k_{z_1}+k_{z_2})}}. \tag{17}$$

While $w_1 = \pm\sqrt{1-w_2^2}$, other spurious equilibria symmetrical to (17) can also be obtained.

4 Experimental Results and Improvements

Follow the above 2-dimention system. To make the sources as independent as possible, s_1 and s_2 are mined from two signals (a square signal and a sine one) by FastICA. For simplify they are both of 0-mean, 1-variance, and the former is a square signal, later a quasi-sawtooth one as showed in Fig.1.(a). The kurtosises are 1 and 1.9340 respectively. The mixing matrix applied is A=$\begin{bmatrix} 0.8 & 0.6 \\ 0.6 & -0.8 \end{bmatrix}$, and the mixtures x = As are depicted in Fig.1.(b) (herein we can see that x_1 and x_2 are both sphered signals).

In applying FastICA, the active function is selected $g(x) = x^3$ (corresponding to the kurtosis-based form), and the deflation scheme is applied for multiple components. From y = $W^T x = W^T vAs$, W = (w_1, w_2), define $Q^T = W^T vA$, Q = (q_1, q_2), and in which $q_1 = (q_1, q_2)$ ($q_1^2 + q_2^2 = 1$). Then whether element w_1 or w_2 in W is spurious equilibrium can be justified by element q_1 or q_2 in Q. Only when Q is a permutation (or nearly permutation), are s recovered from the mixtures, otherwise the two convergence points for w or q are spurious equilibria and s_1 s_2 spurious independent components. Signals in Fig.2.(a) are the estimates of spurious equilibria while Q=$\begin{bmatrix} -0.3472 & 0.9378 \\ 0.9378 & 0.3472 \end{bmatrix}$ and the kurtosises are 1.5761 and 1.5159 respectively. It matches well the analysis in Part 3 ($\partial^2 J_{2kurt}/\partial w_1^2 = 0$ and (17)). Change the mixing matrix A we can also get these two groups (for Q and estimated components) up to scaling and permutation.

The relation between the object function of J_{2kurt} (denote J) and q_1 is depicted in Fig.2.(b), in which "△" and "▽" correspond to independent components, while "*" and "+" two spurious equilibria (the same symbols denote the two permutations for one estimate). In Table 1. five initial points for w are given respectively for the given mixing matrix as above, where w* corresponds to the independent sources and w** the spurious ones.

From the above we can see that the spurious equilibria are far from the equilibria fulfilling $\partial J_3(w)/\partial w = 0$. That is to say, to get rid of the spurious independent components, we can only revise the convergence condition for the algorithm.

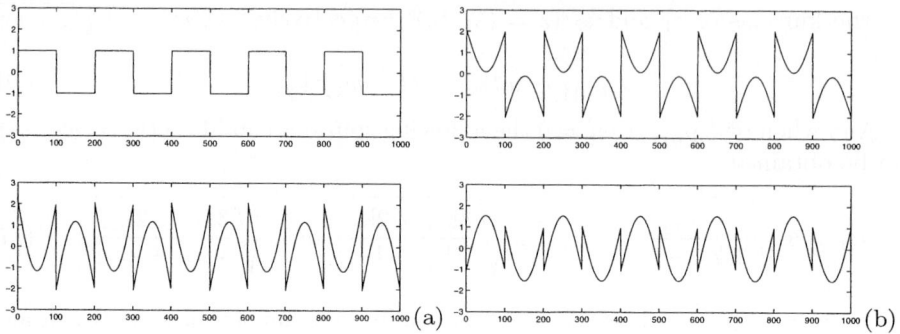

Fig. 1. (a). Original sources; (b). Two mixtures.

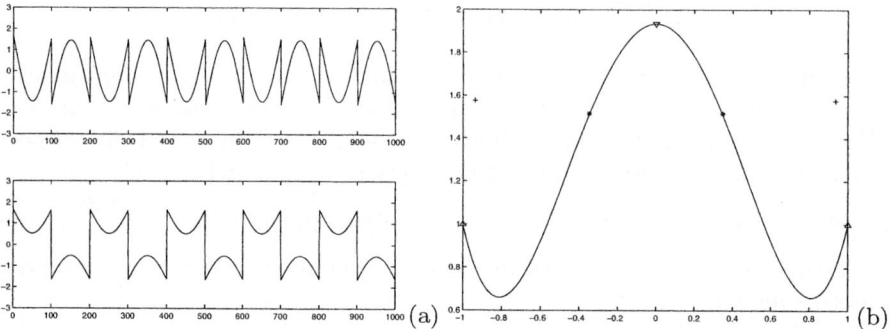

Fig. 2. (a). Spurious independent components; (b). $J_{2\text{kurt}}$ and the estimate of q_1

For example assume a small disturbance ε for w_0 estimated by FastICA, and normalize $w_0 \pm \varepsilon$ by $w_0 \pm \varepsilon \leftarrow w_0 \pm \varepsilon/||w_0 \pm \varepsilon||$. Then if $J_3(w_0 + \varepsilon) \cdot J_3(w_0 - \varepsilon) < 0$, rerun FastICA, otherwise algorithm converges and w_0 is the final estimate corresponding to one independent source. As for the small disturbance ε, the minimum scope for convergence in FastICA should be considered, for example the relation of $||\varepsilon|| > \delta$ must be satisfied.

However why FastICA converges at equilibrium but not spurious equilibrium in most cases? There are at least two causes. First the elements in demixing matrix are initialized randomly between -0.5 and 0.5, and secondly it tends to the edges($\exists\ q_i \to \pm 1, 1 \leq i \leq m$) but not the singular points for the normalization($w \leftarrow w/||w||$) after each iteration [7].

5 Conclusion

This paper concentrates on the existence of spurious equilibria in FastICA, and the analysis is based on the object function and the fixed-point algorithm. It is showed that the object function for FastICA is a relaxation form of conventional approximation of negentropy, and fortunately its affects are compensated by

Table 1. Initial points for w

w^*	w^{**}
(0.31797, 0.16023)	(−0.12163, 0.36001)
(0.34622, 0.16023)	(−0.19724, 0.041674)
(−0.19538, −0.31035)	(−0.054904, 0.4381)
(−0.0034476, 0.39977)	(0.33812, −0.48036)
(−0.15881, 0.034079)	(0.33182, 0.0028129)

applying flexible convergence assumption. For the fixed-point iteration, it's convinced that besides the independent sources, spurious equilibria that correspond to the singular points can also be obtained. Experimental results confirm the analysis. Improvements are also proposed for FastICA to get rid of the spurious equilibria when a new restriction is added into the convergence condition.

References

1. Hyvärinen, A.,Oja, E.: A Fast Fixed-Point Algorithm for Independent Component Analysis. Neural Computation, Vol.9 (1997) 1483-1492.
2. Hyvärinen, A.: Fast and Robust Fixed-Point Algorithms for Independent Component Analysis. IEEE Transactions on Neural Networks Vol.10 (1999) 626-634.
3. Hyvärinen, A., Karhunen, J., Oja,E.: Independent Component Analysis. John Wiley, New York (2001).
4. Cichocki, A., Amari, S.: Adaptive Blind Signal and Image Processing: Learning Algorithms and Applications. John Wiley, New York (2003).
5. Theis, F.J.,: Mathematics in Independent Component Analysis. Erscheinungsjahr (2002) 30-33.
6. Amari, S., Cichocki, A., Yang, H.H.: A New Learning Algorithm for Blind Signal Separation. In Advances in Neural Information Processing Systems, NIPS-1995, Vol.8 (1996) 757-763.
7. Hyvärinen, A.: Independent Component Analysis by Minimization of Mutual Information. Technique Report A46, Helsinki University of Technology, Laboratory of Computer and Information Science (1997).

Hardware Implementation of Pulsed Neural Networks Based on Delta-Sigma Modulator for ICA

Hirohisa Hotta, Yoshimitsu Murahashi, Shinji Doki, and Shigeru Okuma

Graduate School of Nagoya University, Furocho chikusaku nagoyashi, Japan
{hotta,murahasi,doki,okuma}@okuma.nuee.nagoya-u.ac.jp

Abstract. In order to ride on the strength of parallel operation which is feature of Neural Network (NN), It should be implemented on hardware. Considering hardware implementation of NN, it is preferable that the circuit scale of each neuron is compact. We have been proposed Pulsed NN based on Delta-Sigma Modulation (DSM-PNN). The Delta-Sigma Modulator (DSM), which is a key of our proposed DSM-PNN, is well known as a method to convert the input signal into 1-bit pulse stream. DSM-PNN copes with both operating accuracy and compact circuit scale, Because PNN has an advantage of whose circuit scale is compact, and a accuracy of operation. In this paper, an outline of DSM-PNN and how to realize a learning rule of NN are explained by taking ICA as example.

1 Introduction

There are several features of Neural Network (NN) such as learning, nonlinear-operation and parallel-operation. Especially, parallel-operation is paid our attention as a key role of realizing high-speed signal processing. Considering applications in the field of engineering, hardware implementation is preferable to realize the parallel operation of NN. Hardware design, however, takes a lot of time and effort, thus most of NN have been realized in software. Recently, the disadvantage of hardware design has been reduced due to the enhancement of hardware design environment using FPGA. For example, progress of development environment using HDL, improvement of circuit scale, speed, price of FPGA and so on.

Considering hardware implementation of NN, the circuit scale and operation accuracy are important. When it comes to circuit scale, effective pulsed neuron model using 1-bit connection have been proposed. In the pulsed neuron model using 1-bit, the data represented by multiple-bit is converted to the data represented by single-bit pulse stream. This model can reduce the circuit scale because connection area becomes small. The conventional pulsed neuron models using 1-bit connection, however, has poor operation accuracy and is hard to adapt to application which requires high operation accuracy, for example, control and signal processing. In order to realize the precise 1-bit operation, we focus attention on the Delta-Sigma Modulation (DSM) which can generate 1bit-pulse stream accuracy and proposed pulsed NN based on DSM ([1])([2]). The advantage of DSM can be summarized into three points.

- The 1bit-pulse stream generated by DSM can keep to its accuracy even if it is operated repeatedly. Because The quantization noise of DSM is shaped into high frequency band and suppressed in the signal band.
- DSM can be realized by using simple digital circuit elements such as comparator and adder, whose circuit scale is small.
- The quantization noise of DSM can be estimated easily, because the relation between operation speed and accuracy is formulated.

Therefore, using DSM, compact and precise Pulsed NN can be realized. In this paper, firstly, DSM-PNN is introduced. Secondly, the method how the algorithm, here take Independent Component Analysis (ICA) as an example, can be realized using DSM-PNN is explained. Lastly, DSM-PNN realizing ICA is implemented in FPGA, and is checked operations.

2 DSM-PNN

Fig.1 shows 3-Layer NNs as example of DSM-PNN. DSM-PNN consists of Weight Multiplier (WM), Multi-input Adder (MA), and DSM. In this section, the circuit which multiply the weight value with input signal (WM) and the circuit that adds weighted input signals (MA) are explained. Additionally, multiplier which multiplies 1-bit pulse stream and another one (Single-bit Multiplier:SBM), which is necessary to realize several learning algorithms, is explained.

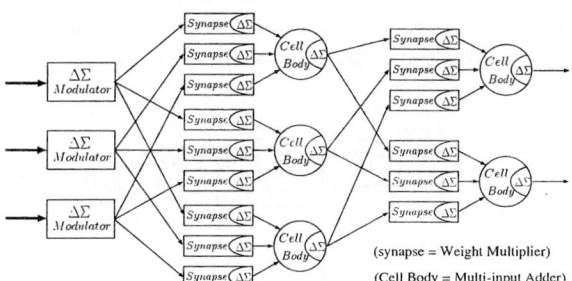

Fig. 1. DSM-PNN

2.1 Weight Multiplier (WM)

In this paper, it is called WM that the element multiplies input signal x represented by 1-bit pulse stream with synaptic weight W represented by multi-bit signal. It consists of selector and DSM. Fig.2.(a) shows WM in the case in which W is 2. x is multiplied with W by selector, and the result, which is multi-bit pulse stream, is converted to 1-bit pulse stream by DSM.

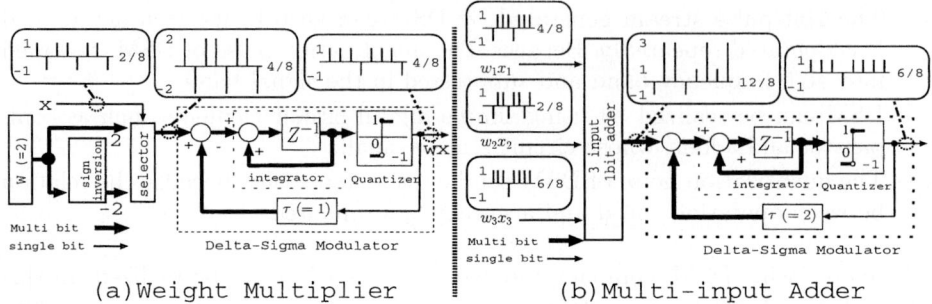

Fig. 2. Weight multiplier and Multi-input Adder

2.2 Multi-input Adder (MA)

In this paper, the element, which adds the weighted signals from synapse circuit, is called MA. It consists of 1-bit adder with n-dimensional inputs and DSM. Fig.2.(b) shows MA that has 3-dimensional inputs. 3-dimentional inputs are added by 1-bit adder, and the result, which is represented by multi-bit pulse stream, is converted into 1-bit pulse stream by DSM.

2.3 Single-Bit Multiplier (SBM)

In this paper, it is called SBM that the element multiplies input signal x represented by 1-bit pulse stream with input signal y represented by 1-bit pulse stream. It consists of WM, DSM and other simple circuits. Fig.3 shows SBM in the case in which $x = 2/8$, $y = 4/8$ and $\tau = 8$. y is converted into multi-bit signal by LPF, and multiplied with x by WM.

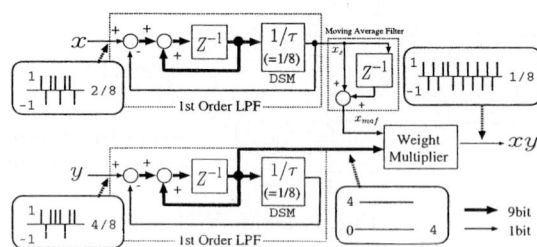

Fig. 3. Single-bit multiplier

3 Realization of ICA for DSM-PNN

In this section, the method of realizing ICA, which is an algorithm to separate mixture signals, using DSM-PNN is proposed. Here, we treat the following problem. When two independent signals $s=[s_1,s_2]$ are mixed by unknown mixture matrix A. The mixture signals $x=[x_1,x_2]$ is given by

$$x = As \tag{1}$$

The aim is to recover s_1, s_2 from x_1, x_2. Another way of saying, the aim is to find out the estimate matrix W which satisfy the following formula.

$$s = Wx \tag{2}$$

Several algorithms of ICA has been proposed. Here, we adopt natural gradient approach proposed by Amari, Cichocki and Yang([3]). The estimate matrix W can be obtained by following formulas.

$$\Delta w_{11} = \alpha \{w_{11} - y_1^4 w_{11} - y_1^3 y_2 w_{21}\} \tag{3}$$

$$\Delta w_{12} = \alpha \{w_{12} - y_1^4 w_{12} - y_1^3 y_2 w_{22}\} \tag{4}$$

$$\Delta w_{21} = \alpha \{w_{21} - y_2^4 w_{21} - y_1 y_2^3 w_{11}\} \tag{5}$$

$$\Delta w_{22} = \alpha \{w_{22} - y_2^4 w_{22} - y_1 y_2^3 w_{12}\} \tag{6}$$

Fig.4. shows DSM-PNN for ICA. It consists of synaptic circuit, Multi-input Adder, and Biquadrator.

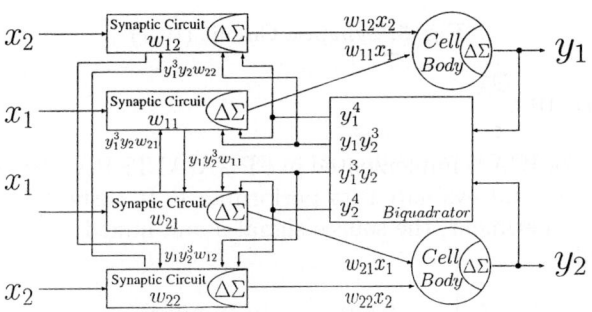

Fig. 4. Pulsed neuron model of ICA

3.1 Biquadrator

In this paper, the circuit which calculates $y_1^k y_2^{(4-k)}$ where $k = 0, 1, 3, 4$ from y_1, y_2 is called "Biquadrator". Fig.5. shows this circuit. It consists of combination of proposed SBM.

3.2 Synaptic Circuit

The synaptic circuits calculate the weight multiplication of forward pass and update W. It consists of DSM, Weight Multiplier, and Multi-input Adder.

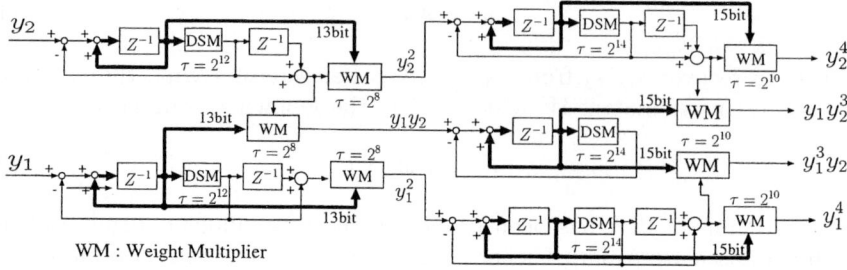

Fig. 5. Biquadrator

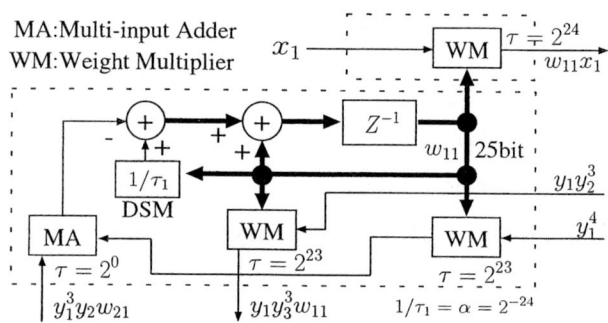

Fig. 6. Synaptic Circuit (W_{11})

4 Experiment

The DSM-PNN for ICA is implemented in FPGA(ALTERA EPF20K200EFC484 -2X 200,000 gates) and evaluated its performance. It costs 2356 LCs to implement it. In this experiment, the source signal s and mixed signal x are given by following formula.

$$\begin{bmatrix} x_1(t) \\ x_2(t) \end{bmatrix} = \begin{bmatrix} 0.6 & 0.4 \\ 0.4 & 0.6 \end{bmatrix} \begin{bmatrix} \sin(2 \cdot 60\pi t) \\ sgn(sin(2 \cdot 50\pi t)) \end{bmatrix} \quad (7)$$

And, initial value of W is given by

$$\begin{bmatrix} w_{11} & w_{12} \\ w_{21} & w_{22} \end{bmatrix} = \begin{bmatrix} 8.24 \times 10^{-1} & -3.13 \times 10^{-2} \\ 2.23 \times 10^{-1} & 6.64 \times 10^{-2} \end{bmatrix} \quad (8)$$

Fig.7 shows source signal s, mixed signal x, and separated signal y filtered by LPF. For this result, the correct performance of DSM-PNN for ICA is shown. Besides, the convergence time, which is defined as the time until weight coefficient reduce the error to less than 3 percents, of DSM-PNN is compared to that of PC (Pentium4:3GHz, DDR-SDRAM:1GB). PC calculates ICA which is described by C language on the same learning rate and initial parameters. As a result, the convergence time of DSM-PNN was 7.8 second, and that of PC was 98.8 second. For this result, DSM-PNN whose base clock is 33.3MHz is 12.6 times superior to PC whose base clock is 3GHz. It is shown that DSM-PNN can operate ICA fast.

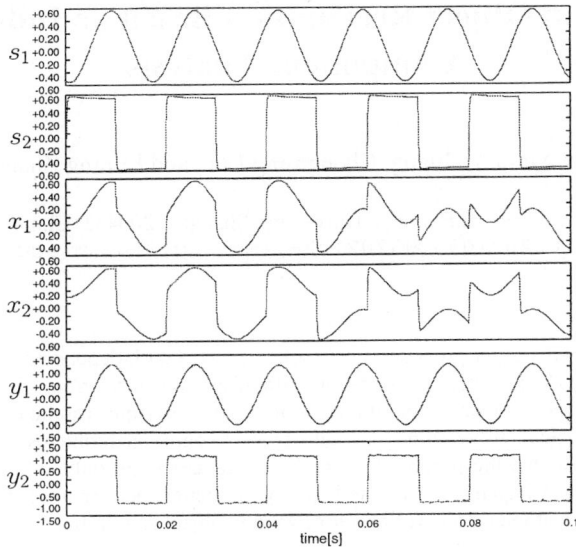

Fig. 7. Sources, mixture signals, and output signals

5 Conclusion

In this paper, we proposed DSM-PNN for ICA, and describe how to implement in FPGA. In the experiment, DSM-PNN succeeds to separate mixture signals. And, comparing with PC, it is shown that DSM-PNN can operate fast.

References

1. Murahashi, Y., Hotta, H., Doki, S., Okuma, S.: Hardware Realization of Novel Pulsed Neural Networks Based on Delta-Sigma Modulation with GHA Learning Rule. The 2002 IEEE Asia-Pacific Conference on Circuits and Systems
2. Murahashi, Y., Hotta, H., Doki, S., Okuma, S.: Pulsed Neural Networks Based on Delta-Sigma Modulation with GHA Learning Rule and Their Hardware Implementation. The IEICE Transactions on Information and Systems, PT.2 Vol.J87-D-2 NO.2 (2004) 705-715 (Japanese)
3. Amari, S.: Natural gradien learning works efficiently in learning Neural Computation (1997)

Study on Object Recognition Based on Independent Component Analysis*

Xuming Huang, Chengming Liu , and Liming Zhang

Dept. E.E, Fudan University, Shanghai 200433, China
{022021035,032021036,lmzhang}@fudan.edu.cn

Abstract. This paper proposes a new scheme based on Independent Component Analysis(ICA) for object recognition with affine transformation. For different skewed shapes of recognized object, an invariant descriptor can be extracted by ICA, and it can solve some skewed object recognition problems. Simulation results show that the proposed method can recognize not only skewed objects but also misshapen objects, and it has a better performance than other traditional methods, such as Fourier method in object recognition.

1 Introduction

Objects recognition based on affine transform is a typical problem of rigid object recognition, and has important applications in computer vision. The fundamental difficulty in recognizing an object from 3 D space is that the appearance of shape depends on the observing angle. In general, affine transform can be considered as a depiction of the same object obtained from different angles. The ability to quickly recognize the same object under any affine transform is very important for the practical applications in military, industrial automation, etc.

Recently, some methods have been proposed for the object recognition under affine transform. They can be divided into two groups: (i) region-based and (ii) contour based techniques. Affine moment invariance for object recognition is a region-based method[1]. Although the moment-based methods can be applied to gray-scale or binary images, they are sensitive to variations in background shading or object illumination. In addition, only a few of low order moments can be used in practical application because the high order moments are sensitive to noise. Methods based on affine curves, such as affine arc length curve, enclosed area curve and curvature scale space[2,3], and Fourier descriptors, belong to the contour-based method. However, those methods are sensitive to the starting point. Oirrak et al proposed a new affine invariant descriptors using Fourier[4,5]. Applying it to the object recognition, one can accurately recognize the complicated objects under low optical distortion. Unfortunately, the recognition correct rate decreases greatly when the distortion is large.

---

* This research was supported by a grant from the National Science Foundation (NSF 60171036, 30370392), China.

In this paper, we propose a new affine-invariant descriptor based on ICA. It is robust to noise and independent of contour sampling order because the ICA method uses statistical information. The fast algorithm for ICA proposed by Hyvärinen[6,7] is adopted in this paper.

2 Object Recognition Based on ICA

Suppose object's contour X be sampled as N discrete point. Each point is denoted a coordinate vector. Let $X'(l')$ be a linear affine transform from $X(l)$ with affine matrix Z and translation vector B, that can be written as

$$X'(l') = ZX(l) + B \tag{1}$$

Here, l and l' are different because of the difference sampling on the contour X and X'. It's obvious that $B = 0$ when X and X' are set in their centroid coordinate, respectively, so in the following, we take $B = 0$ for analytic convenience.

If $X(l)$ is sampled on the contour X by random, $X(l)$ is a random vector. Consider that X' and X are the linear combination of the same source S with A' and A, then $X'(l') = A'S'(l')$ and $X(l) = AS(l)$. By using ICA, we can adapt matrix W' and W to obtain $S'(l')$ and $S(l)$, and A' and A, that satisfy

$$W'X'(l') = W'A'S'(l') \approx S'(l'), A' = W'^{-1}, WX(l) = WAS(l) \approx S(l), A = W^{-1} \tag{2}$$

According to property of ICA by [6], $S'(l')$ and $S(l)$ is almost the same on the statistical characteristic, only the order of their elements and sampling are difference.

For the unknown sample order of the contour data, we couldn't compare $S'(l')$ and $S(l)$ directly, we use $S'(l')$ and $S(l)$ to compute Centroid-Contour Distance(CCD), respectively. The CCD of $S(l)$ is defined as:

$$d_s(l) = \sqrt{S_x(l)^2 + S_y(l)^2}, l = 1, 2, \ldots, N \tag{3}$$

Here $S_x(l)$ and $S_y(l)$ are the components of $S(l)$ on x axis and y axis. Then sort the $d_s(l)$, we get $d_{sort}(t)$, which satisfies:

$$d_{sort}(t_1) < d_{sort}(t_2) < \ldots < d_{sort}(t_N), t_1 < t_2 < \ldots < t_N \tag{4}$$

It is obvious that whatever the order of samples or components is the curve $d_{sort}(t)$ is affine invariant. In view of different curve length of actual object contour, before recognition we should resample the sorted curve $d_{sort}(t)$ to assure the same length of every sorted CCD curve, and the re-sampled $d_{sort}(t)$ is also an affine invariant curve. So in the following, $d_{sort}(t)$ is the representation of the re-sampled and sorted CCD curve for expression convenience.

3 Simulation and Applications

In this section, some simulation results confirm our method proposed in section 2, and some applications show that the proposed method is better than Fourier method[5] that is well known as an effective method recently.

Example 1: Fig.1(a) is an airplane's contour, which is extracted by any image processing method, and its affine transform shape is shown in Fig.1(b). Fig.2 shows the independent variables for Fig.1(a) and (b) on x and y axis by random sampling.

From Fig.2, we only can see the random variables, but not the shape. Recompose the two random variables we can get two shapes, shown in Fig.3, and the re-sampled and sorted CCD curve of $S(l)$ and $S'(l')$ are the same, it's an affine invariant curve, and is shown in Fig.4, the length of curve is fixed on 256 points.

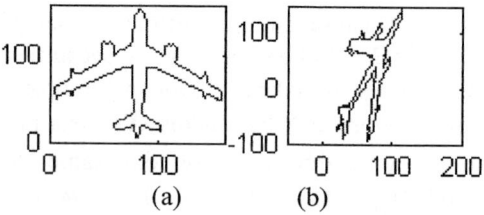

Fig. 1. (a) Airplane's counter; (b) skewed shape of (a).

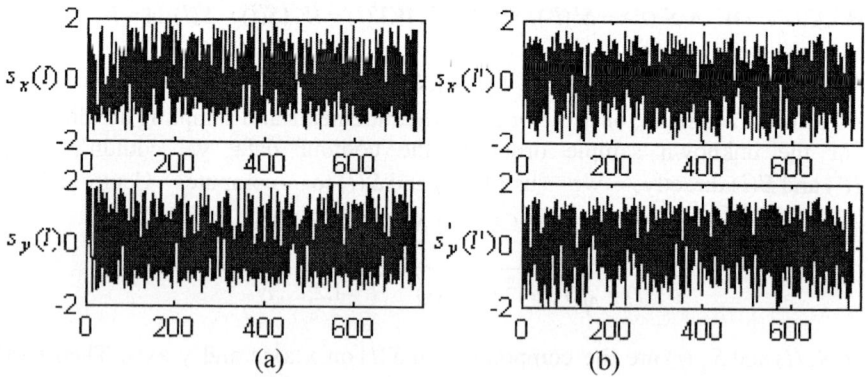

Fig. 2. The ICA results, S and S' for X and X' by random sampling. (a) Two random variables of S; (b) Two random variables of S'.

To test the robustness of the proposed method, we add random noise on the sampled shape, the sorted CCD curve will differ with the affine invariant curve, but the difference is slight. Fig.5 shows the correlative coefficient between noised CCD curve and the affine invariant curve in Fig.4 under different noise magnitude.

From Fig.5 we could see that the correlative coefficient's change is in the range of 0.2% even the noise magnitude is reach 10pixels.

Example 2: To test the discrimination ability of our proposed method, we use real airplane images. Fig.6(a) shows ten known airplane models that represent ten different objects. From different view angle, we will get different affine contour images. One kind of the affine models corresponding to 10 distorted images, with 40° skewed on horizontal and 10° on vertical direction, are shown in Fig.6(b), and we take these as test images. Table 1 shows the maximal correlative coefficient between current test image and each model (the recognition result is shown in the brackets) by using our method and the Fourier method[5]. This result shows that our method has better performance than the Fourier method in the case of high skewed shape recognition. (One airplane can not be correctly recognized for proposed method and four airplanes for Fourier method.) For slighter distortion our method also could recognize accurately.

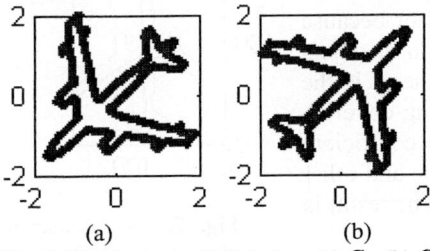
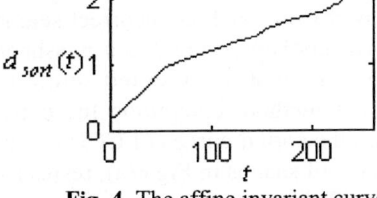

Fig. 3. The recomposed shapes. (a) S. (b) S'.

Fig. 4. The affine invariant curve of S and it's all affine shapes.

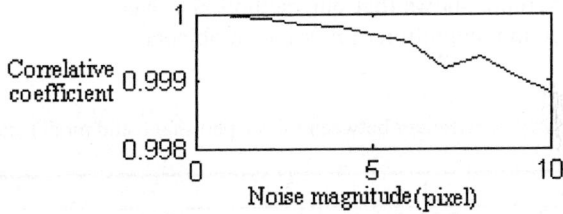

Fig. 5. The correlative coefficient under different noise magnitude.

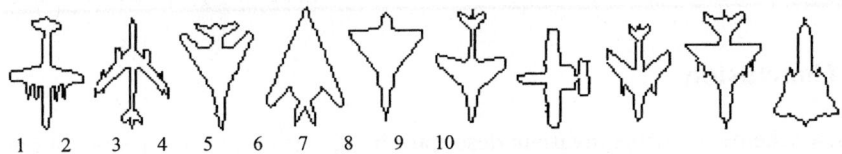

Fig. 6. (a) Ten known airplane models.

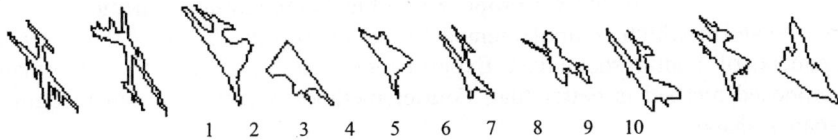

Fig. 6. (b) The skewed images correspond to the model images in Fig.6(a) respectively.

Table 1. The recognized results for each test image in Fig.6(b) using our method and Fourier method respectively.

Test image	1	2	3	4	5
Our Method	0.9995(1)	0.9986(2)	0.9976(3)	0.9978(5)	0.9971(5)
Fourier Method	0.9955(1)	0.9281(2)	0.9280(3)	0.9746(4)	0.8922(4)
Test image	6	7	8	9	10
Our Method	0.9987(6)	0.9998(7)	0.9980(8)	0.9988(9)	0.9987(10)
Fourier Method	0.9784(6)	0.9447(4)	0.9571(8)	0.9115(4)	0.9261(6)

Example 3: In practical object recognition applications, it is difficult to perfectly extract object's contour because of many reasons such as incorrect segmentation or object is partial misshapen. Fig.7 is a misshapen plane, it is the 2# plane in Fig.6(a) whose left aerofoil is being covered. Using our method, computing the correlative coefficient between the sorted curves of CCD of this shape and other ten curves of shapes in Fig.6(a), respectively. The result is shown in Table 2.

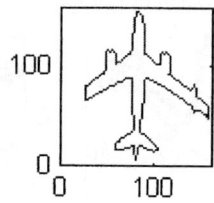

Fig. 7. A misshapen plane corresponds to 2# in Fig. 6(a)

The result in Table 2 shows that the misshapen plane is best similar with the 2# plane, and the recognition result is correct. This example shows that our method also has a well performance in recognition of misshapen objects

Table 2. Correlative coefficient between misshapen shape and model shape in Fig.6(a).

Plane index	1	2	3	4	5
Correlative coefficient	0.9898	0.9984	0.9798	0.9818	0.9744
Plane index	6	7	8	9	10
Correlative coefficient	0.9956	0.9793	0.9959	0.9928	0.9877

4 Conclusion

A new scheme for affine invariant descriptor by ICA is proposed, in which we assume some random series, random sampled from contour of an object and its any skewed shapes, can be considered as a linear combination of affine invariant independent components. Based on this we propose an objects recognition scheme for different affine transform which is implemented by computing the CCD of the independent components of random shape data. Experimental results show that the performance of the proposed method is better than Fourier methods, and it also can recognize the misshapen shapes.

References

1. Zhao, A., Chen, J.: Affine curve moment invariants for shape recognition. Pattern Recognition, vol. 30, no.6 (1997) 895-901
2. Arbter, K., Snyder, W. E., Burkhardt, H., Hirzinger, G.: Application of Affine-Invariant Fourier Descriptors to Recogniton of 3-D Objects. IEEE Trans. PAMI, vol. 12, no.7 (1990) 640-647
3. Abbasi, S., Mokhtarian, F.: Affine-Similar Shape Retrieval: Application to Multiview 3-D Object Recognition. IEEE Trans. IP, Vol. 10, no.1 (2001) 131-139
4. Oirrak, A. E., Daoudi, M., Aboutajdine, D.: Estimation of general 2D affine motion using Fourier descriptors. Pattern Recognition, 35 (2002) 223-228
5. Oirrak, A. E., Daoudi, M., Aboutajdine, D.: Affine invariant descriptors using Fourier series. Pattern Recognition Letters, 23 (2002) 1109-1118
6. Hyvärinen, A.: Fast and robust fixed-point algorithms for independent component analysis. IEEE Trans. NN, 10(3) (1999) 626-634
7. Hyvärinen, A., Oja, E.: A fast fixed-point algorithm for independent component analysis. Neural Computation, 9 (1997) 1483-1492

Application of ICA Method for Detecting Functional MRI Activation Data

Minfen Shen[1], Weiling Xu[1], Jinyao Yang[2], and Patch Beadle[3]

[1] Key Lab. of Guangdong, Shantou University, Guangdong 515063, China
mfshen@stu.edu.cn
[2] Ultrasonic Institute of Shantou, Shantou, Guangdong, China
[3] School of System Engineering, Portsmouth University, Portsmouth, U.K.

Abstract. Functional magnetic resonance imaging (fMRI) is a widely used method for many applications, but there is still much debate on the preferred technique for analyzing these functional activation image. As an effective way for signal processing, Independent Component Analysis (ICA) is used for detection of fMRI signal. Several experiments with real data were also carried out using FastICA algorithm. ICA procedure was applied to decompose the independent components that are a very useful way to restrain the impact caused by noise. The results indicate that ICA significantly reduces the physiological baseline fluctuation component and random noise component. The true functional activities by computing the linear correlation coefficients between decomposed time-series of fMRI signals and stimulating reference function were successfully detected.

1 Introduction

Functional magnetic resonance imaging has emerged as a useful and noninvasive method for studying the functional activity of human brain. We have found that it is possible to indirectly detect the changes in blood-oxygenation levels which are the result of neuronal activation based on the fMRI technique [1]. Nowadays the research of fMRI mainly concentrates on the fields such as cognitive neuroscience, clinical medicine [2].

The purpose of fMRI signal processing is to detect the brain functional activation area. Recent studies of fMRI signal processing concentrate on noise reduction, image registration and functional activity detection. The software package, such as SPM and AFNI, are widely used in fMRI signal analysis. Generally, image pre-processing often utilize Gauss smoothing to reduce noise. This type of global smoothing considers edges undesirably. Bayesian processing and Wiener smoothing overcome the disadvantage of global smoothing, and smooth the image only using local correlated neighborhood [3], which can retain local information of the image. Functional activated area is identified by the time-series of voxel individually, so the motion artifact caused by head movements slightly will results in false detection of activity. In order to enhance the reliability of functional activity detection, we apply image registration to correct the movements during the collection of fMRI signals [4].

Commonly, we consider human head as a rigid body, so the six parameters (3 of translation, 3 of rotation) that describe rigid body movement can be applied to image registration. We applied projection transform to adjust the rigid body movements in this work.

Independent Component Analysis (ICA) technology developed as a latest multi-channel signal processing method in fields of blind source decomposition technique [5]. ICA decomposes multi-channel observed signals into several independent components using various arithmetics according to the principle of statistic independent. In this paper, ICA is used for decomposing noise after introducing a model for explaining how the event-related true signals which reflect the saturation of blood oxygenation. The model is employed to detect the fMRI activation areas.

2 Principles and Methods

2.1 Principle of ICA

ICA is a new method for decompose the statistical independent components from the multidimensional mixture of observed data, and belongs to the class of general linear models. The framework of ICA is shown in Fig1. In the framework of ICA, s is the signal vector with dimension m, x is the observed mixed vector with dimension n, A is the unknown mixing matrix, W is the blind decompose matrix, and y is the decomposed vector with dimension m.

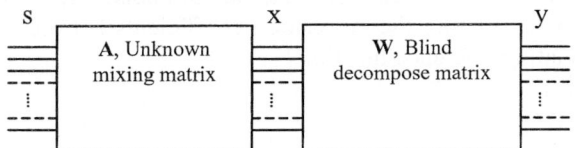

Fig. 1. The framework of ICA

$$x = As \tag{1}$$

The problem now become to find out a matrix W which is the inverse of A:

$$y = Wx = WAs \tag{2}$$

The observed mixed vector x can be decomposed into unmixing independent components y by finding an unmixed matrix W. We can compute the unmixed matrix W based on the maximizing log-likelihood rule. Fast infomax-based ICA algorithm is used to perform ICA decompose [6]. In order to simplify the ICA algorithm, first we center the observed signals x by removing the means of them. Then we perform the procedure of whitening x by a linear system.

2.2 The Model of fMRI Signals

Assuming that the time-series of individual voxel in fMRI signals are consisted of two independent components, the event-related signals (s_1), which is reflect the saturation of blood oxygenation, and the noise (s_2) comprising physiological baseline fluctuation component and random noise component. When we consider an individual voxel time-series x_1, we can approximately count that x_2, which is the maximal correlation coefficient with x_1 among the eight-connected neighborhood, as the same source of x_1. Then the observation signals can be defined as

$$x = [x_1, x_2]^T \tag{3}$$

In this formula x_1 and x_2 are both the linear combination of the signals s_1 and s_2. Utilizing the decomposition technique of ICA, we can get the blind decomposed signals y_1 and y_2. Simulate signal and ICA decomposed result are shown in Fig.2.

Finally, the correlation between the voxel time-series and the stimulating reference function shows the influence of experiment stimulation to the changing of this voxel's blood oxygenation saturation. If the correlation is high, this voxel belongs to the functional activity, and vice versa. We can get the "true" signals by using ICA technique to restrain psychological noise and random noise, and detect the functional activation area by calculating the correlation coefficients between decomposed true signals and stimulating reference function.

3 fMRI Signals Acquisition

All experimental data were obtained using a 1.5-T scanner and selected fast Echo Planar Imaging (EPI) technique, which uses traverse scan with TR=3s. Firstly, we will do locating scan to the observed person, and the imaging matrix is 256×256. The time-series of the experiment stimulation is: after 30s' experiment preparation, visual stimulation will be started, and the duration is 30s. Then, it needs 30s' interval. Five trials were repeated. The imaging matrix of EPI is 64×64.

4 Results

The fMRI image will be grouped according to different locating layer. Removing the parts of 30s' experiment preparation, the data stored in every layer are 64×64×100. Remove the noise existed in every frame, 64×64 matrix size, by self-adapting Wiener filter. Then, take the first frame as basic reference, and map other 99 frames by projection transform. According to the model mentioned above, individual voxel time-series in fMRI signal are simplified as a noise-free ICA model composed by two independent components, s_1 and s_2, observed signal $x = As$. The actual fMRI signal is consisted of a main component s1 and multi-noise components, which can be defined more approximately by noise ICA model: $x = As + n$, in which n indicates that the number of independent noise components is more than 1. Because the energy of s_1 component is much higher than it of noise components in the activated voxel, we still can get a better-decomposed result when the multi-noise components are simplified as s_2. Fig.3 shows the decomposed results of time-series of an activated voxel utilizing FastICA technique.

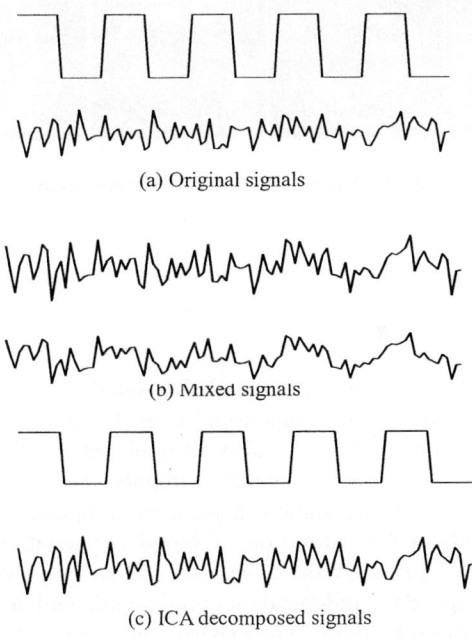

Fig. 2. Simulate signals and ICA decompose results

After restraining psychological noise and random noise caused by apparatus utilizing ICA technique, we can calculate the correlation coefficients, c1 and c2, between decomposed results and stimulating reference function. The voxel will be identified as activated voxel if c of which is greater, and vice versa.

For reducing the computational burden of functional activated area detection, we can use automatic threshold to discriminate background and brain area so that we do ICA separation only to the voxel of brain area. If only one main component can be decomposed from the voxel time-series, then we can determinate that this voxel does not belong to the activated area. At last, we project the activated area detection matrix, 64×64 size, to the locating image mapping with bicubic interpolation; and show the result of activated area detection using 64-class pseudocolor. Fig.4 shows the experiment result.

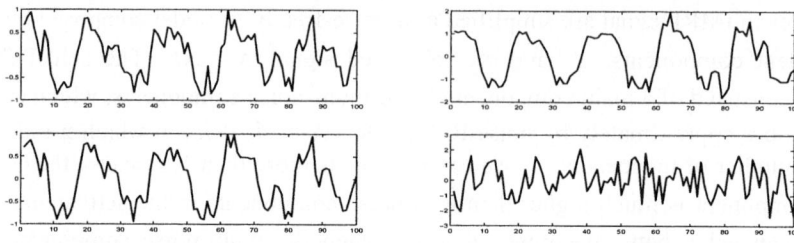

Fig. 3. The decomposed results of time-series of an activated voxel utilizing FastICA technique

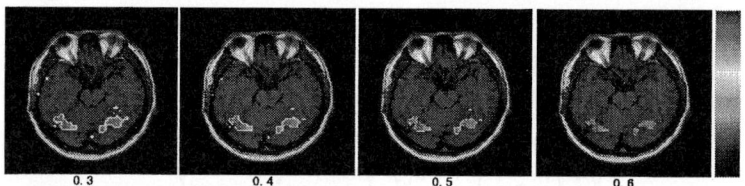

Fig. 4. The experiment result of functional activation areas with red color

5 Conclusion

ICA technique was used to investigate the fMRI signal. Both outline of the data's pre-processing and the indication of functional activated area in the locating image were analyzed. The proposed method is valid for nonlinear analysis since the method does not involve the character of the brain's impulse response. The results have shown that fMRI signals comprise some independent components, such as the event-related signal which reflect the saturation of blood oxygenation, the physiological noise, the random noise caused by the apparatus and so on. The fastICA algorithm was employed to decompose the independent components, which is a very useful way to restrain the impact caused by noise. The experimental result also indicated that ICA can help to effectively detect the true functional activity by computing the linear correlation coefficients between decomposed time-series of fMRI signals and stimulating reference function.

Acknowledgements. The research was supported by the Natural Science Foundation of China (60271023), the Natural Science Foundation of Guangdong (021264 and 32025) and the Key Grant of the Education Ministry of China (02110).

References

1. Xuejun Sun, Maili Liu, Caohui Ye: Progress of Functional Magnetic Resonance Imaging in Brain Research. China Journal of Neurosci, Vol. 3. Science Press, Beijing (2001) 270-272

2. Enzhong Li, Xuchu Weng: Asymmetry of Brain Function Activity: fMRI Study Under Language and Music Stimulation. Chinese Medical Journal, Vol. 02. Chinese Medical Association Publishing House, Beijing (2000) 154-158
3. T.Kim, L.Al-Dayeh, and M.Singh: FMRI Artifacts Reduction Using Bayesian Image Processing. IEEE Transactions on Nuclear Science, Vol. 46, IEEE, Piscataway NJ (1999) 2134-2140
4. Netsch Thomas, Roesch Peter, Weese Juergen: Grey Value-based 3-D Registration of Functional MRI Time-series: Comparison of Interpolation Order and Similarity Measure. Proceedings of SPIE - The International Society for Optical Engineering, Vol. 3979. Society of Photo-Optical Instrumentation Engineers, Bellingham WA (2000) 1148-1159
5. Comon Pierre: Independent Component Analysis: A New Concept. Signal Processing, Vol. 36. Elsevier Science Publishers, Amsterdam (1994) 387-314
6. Hyvarinen A.: Fast and Robust Fixed-point Algorithms for Independent Component Analysis. IEEE Trans Neural Networks , Vol. 10. IEEE, Piscataway NJ (1999) 626-634
7. Friston K.J., Jezzard P., and Turner R.: Analysis of Functional MRI Time-series. Human Brain Mapping, Vol. 1. (1994) 153-171
8. Huang, X., Woolsey, G.A.: Image Denoising Using Wiener Filtering and Wavelet Thresholding. IEEE International Conference on Multi-Media and Expo, (2000) 1759-1762

Single-Trial Estimation of Multi-channel VEP Signals Using Independent Component Analysis

Lisha Sun[1], Minfen Shen[1], Weiling Xu[1], and Francis Chan[2]

[1] Key Lab. of Guangdong, Shantou University, Guangdong 515063, China
lssun@stu.edu.cn
[2] Dept. of Electronic Engineering, Hong Kong University, Hong Kong

Abstract. Single-trial estimation of visual evoked potential (VEP) became a very interesting and challenge problem in biomedical signal processing at present. A method based on the independent component analysis (ICA) was proposed for single-trial detection of multi-channel VEP signals contaminated with background electroencephalograph (EEG). It achieved the satisfied results to show that the presented ICA method can be as an effective tool for single-trial VEP signal estimation in low SNR situation.

1 Introduction

VEPs represent an objective non-invasive method for theoretical research of visual information processing, as well as a diagnostic tool for early detection of some functional brain disorders [1]. VEPs can also be regarded as an objective testing of visual parameters in many clinical applications. In comparison to the modern brain imaging techniques, the VEPs examination is a low price method with very high temporal resolution [2].

Estimation of VEP signal is still a problem in biomedical signal processing so far. We have to estimate the significance VEPs from the EEG with a very low SNR. Generally, the usual methods for detecting the VEP signals adopt the coherent averaging method. Averaging methods improve the SNR by increasing the stimulating times, however, it ignores variation among every experimental VEP response. It results in affecting directly detection of the changing VEPs for each stimulus [3, 4, 5]. To overcome this limitation, this paper uses ICA technique for the purpose of single-trial estimation of multi-channel VEPs. The algorithm for blind separation of the signals is based on the infomax-ICA [6, 7]. Some simulations and real EEG data set analysis were also implemented and discussed.

2 Methods

2.1 Fast ICA Algorithm

The algorithm of the FastICA is briefly provided in the following discussions. After data whitening, the FastICA algorithm takes following form when the fourth-order statistics kurtosis is used:

(1) Choose randomly an initial vector w(0) and normalize it to have a unit norm.
(2) Compute the next estimation of a ICA basis vector after whitening using the following fixed-point iteration rule:

$$w(k) = Y[Y^T w(k-1)] - 3w(k-1) \quad (1)$$

Normalize w(k) by dividing it by its norm, that is

$$w(k) = w(k) / \|w(k)\| \quad (2)$$

(3) If w(k) unconverged, go back to (2) until |wT(k)w(k-1)| is sufficiently close to 1.

The above procedure is the fixed-point rule for estimating one ICA basis vector. The other ICA basis vectors can be estimated sequentially if necessary by projecting a new initial basis vector w(0) onto the subspace which is orthogonal to the subspace spanned by the previously found ICA basis vectors, and following then the same procedure.

3 Simulation Results

Fig.1 shows 3 independent sources s1-s3. After the linear mixture, the output of the linear mixtures was shown in Fig. 2. They look as if they were completely noise, but actually there are some quite structured underlying source signals hidden in these observed signals. What we need to do is to find the original signals from the 3 mixtures x1, x2 and x3. This is the blind source separation (BSS) problem which means that we know very little if anything about the original sources. Based on the proposed ICA procedure, the problem of the independent sources blind separation can be solved with the FastICA algorithm. The corresponding results of the blind separation were illustrated in Fig. 3. The mixture matrix is as

$$A = \begin{bmatrix} 0.1 & 0.09 & 1.1 \\ 0.12 & 0.1 & 1.2 \\ 0.15 & 0.15 & 1.3 \end{bmatrix} \quad (3)$$

The mixture matrix via ICA after whitening is also given as

$$A' = \begin{bmatrix} 0.8900 & -0.4470 & -0.0902 \\ -0.4448 & -0.8071 & -0.3882 \\ 0.1007 & 0.3856 & -0.9171 \end{bmatrix} \quad (4)$$

Finally, the separating matrix has the following form:

$$W = \begin{bmatrix} 0.8900 & -0.4448 & 0.1007 \\ -0.4470 & -0.8071 & 0.3856 \\ -0.0902 & -0.3882 & -0.9171 \end{bmatrix} \quad (5)$$

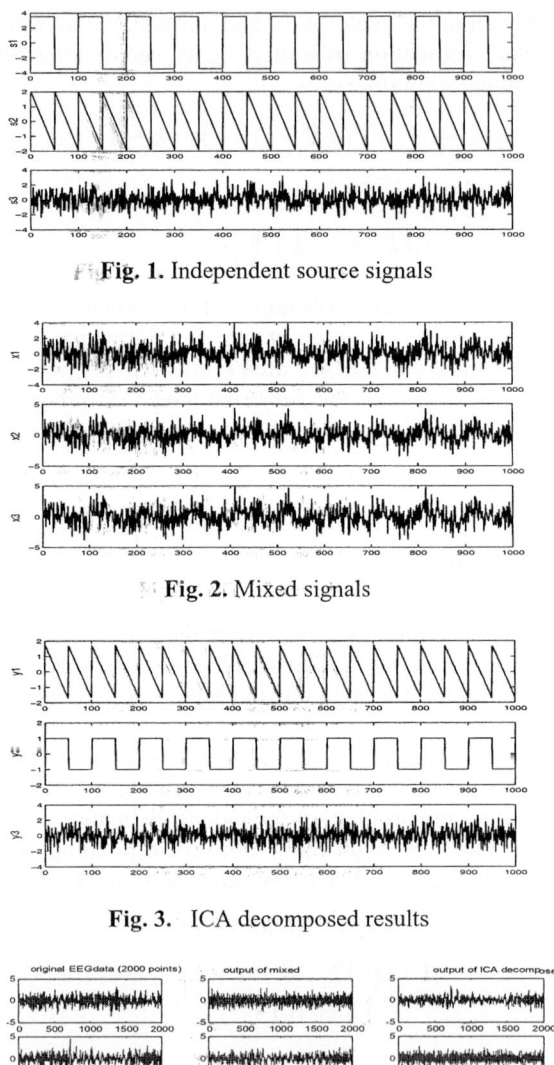

Fig. 1. Independent source signals

Fig. 2. Mixed signals

Fig. 3. ICA decomposed results

Fig. 4. Simulating the EEG data of fact with 2000 data points

The VEP measures the electrical response of the brain's primary visual cortex to a visual stimulus. Most clinical VEP recordings involve placing a patient in front of a black and white checkerboard pattern displayed on a video monitor. The stimulation module was connected to the PC in order to synchronize the acquisition with the stimulus. In general, clinic EEG data either include super-Gaussian's signals (for example, ERPs), or sub-Gaussian signals(for example, EOG and working frequency disturb etc.). We validated feasibility of the extended Informax ICA algorithm, and also did a simulating experimentation about the EEG data of fact, the resulting was shown in Fig. 4. The result shows the extended ICA algorithm can greatly realize the blind source separation. Meanwhile the signals include both super-Gaussian and sub-Gaussian's signals. However, it is noted that this result cannot illuminate to directly obtain the visual evoked potential by ICA decomposing because we came down to multi-channel data.

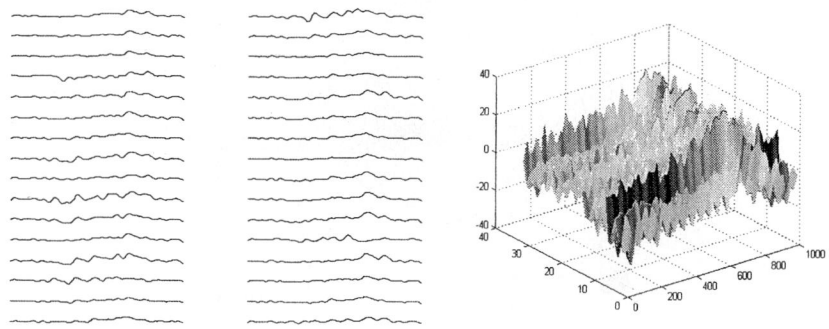

Fig. 5. 32 recordings of the evoked EEGs. Left: 2-dimension. Right: 3-dimension

This experimental data is clinical evoked EEG data. The check alternate black/white to white/black at a rate of approximately twice per second. The electrode systems was used according to the international 10-20 standard. The channel number of recordings is 128, repeated 200-trial with sampling frequency at 250Hz. We symmetrically chosen 32 channels from the whole data, and chosen one of the trials. The observing sources of VEPs recordings were shown in Fig. 5.

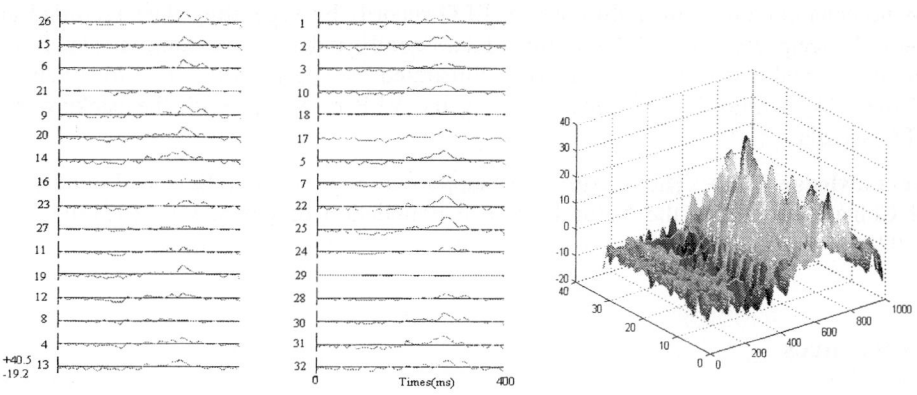

Fig. 6. The results of real VEP responses after ICA decomposed. Left: 2-dimension. Right: 3-dimension

The same procedure above using ICA was used for the detection of single-trial VEP signals in Fig.5. Then the correct or estimated VEP signals were successfully obtained, as shown in Fig. 6. Moreover, the 32 channel VEP responses were used to constructe a topographic map which provides the dynamic information of the potential distribution on the scalp at the specified time. Fig. 7 gives an example at the time of 0.274 second.

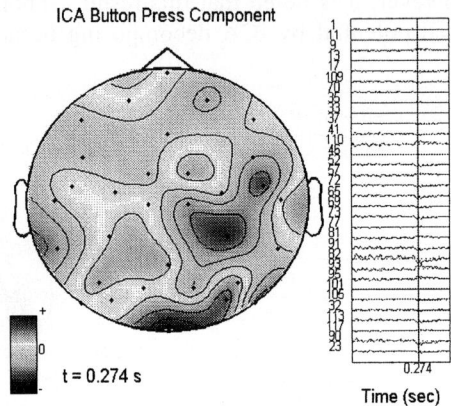

Fig. 7. Latent period map

4 Conclusions

ICA method was applied in this contribution to solve the problem of estimating the single-trial VEP signals. The FastICA algorithm was used to perform the blind separation of the original signals. The result in this paper showed that the ICA method provided a new and useful window fro dealing with many brain and non-brain phenomena contained in multi-channel EEG records by separating data recorded at multiple scalp electrodes into a sum of temporally independent components. Both simulation and real EEG signal analysis illustrated the applicability and effectiveness of the ICA method for real-time detecting the VEP responses with the background EEG signals.

Acknowledgements. The research was supported by the Natural Science Foundation of China (60271023), the Natural Science Foundation of Guangdong (021264 and 32025).

References

1. Regan D.: Human Brain Electrophysiology, Evoked Potentials and Evoked Magnetic Field in Science and Medicine. Elsevier Science Publishing, New York, 1989
2. Livingstone M. and Hubel D.: Segregation of Form, Color, Movement and Depth: Anatomy, Physiology, and Perception. Science, Vol. 240. (1988) 740-749
3. Karjalaninen P. A., Kaipio J. and et al.: Subspace Regularization Method for the Sing-trial Estimation of Evoked Potentials. IEEE Transactions on Biomedical Engineering, Vol. 18. IEEE, Piscataway NJ (1999) 849-860

4. Nitish V. T.: Adaptive Filtering of Evoked Potential, IEEE Transactions on Biomedical Engineering, Vol. 34. IEEE, Piscataway NJ (1987) 6-12
5. Comon P.: Independent Component Analysis, A new Concept. Signal Processing, Vol. 36. Elsevier Science Publishers, Amsterdam (1994) 287-314
6. Bell A. J.,Sejnowski TJ: An Information-maximization Approach to Blind Separation and Blind Deconvolution. Neural Computation, Vol. 7. MIT Press, Cambridge (1995) 1129-1159
7. Lee T. W. and et al: Independent Component Analysis Using an Extended Infomax Algorithm for Mixed Subgaussian and Supergaussian. Neural Compution, Vol. 11. MIT Press, Cambridge (1999) 409-433
8. Tootell R., Hadjikhani N., Mendola J., Marrett S. and Dale A.: From Retinotopy to Recognition: fMRI in Human Visual Cortex. Trends Cogn Sci Vol. 2. (1998) 174-183
9. Derrington A. and Lennie P.: Spatial and Temporal Contrast Sensitivities of Neurones in Lateral Geniculate Nucleus of Macaque. J Physiol Lond, Vol. 375. (1984) 219-240
10. Sclar G. Maunsell J. and Lennie P.: Coding of Image Contrast in Central Visual Pathways of the Macaque Monkey. Vision Res, Vol. 30. (1990) 1-10
11. Fung K. Chan F. et al: A Tracting Evoked Potential Estimator. Signal Processing, Vol. 36. (1994) 287-314

Data Hiding in Independent Components of Video

Jiande Sun, Ju Liu, and Huibo Hu

School of Information Science and Engineering, Shandong University
Jinan 250100, Shandong, China
{jd_sun, juliu, huibo_hu}@sdu.edu.cn

Abstract. Independent component analysis (ICA) is a recently developed statistical technique which often characterizes the data in a natural way. Digital watermarking is the main technique for copyright protection of multimedia digital products. In this paper, a novel blind video watermarking scheme is proposed, in which ICA is applied to extract video independent components (ICs), and a watermark is embedded into the ICA domain by using a 4-neighboring-mean-based method. The simulation shows that the scheme is robust to MPEG-2 compression and able to temporally resynchronize.

1 Introduction

After the invention of digital video, its copyright protection issues have become important as it is possible to make unlimited copies of digital video without quality loss. Video watermark is a proposed method of video copyright protection. Watermarking of digital video has taken increasingly more significance lately. To be useful, video watermark should be perceptually invisible, blind detection, temporal resynchronization, robust to MPEG compression, etc. Many video watermark algorithms have been proposed, most of which have embedded watermark into the extracted frame feature [1], [2] or block-based motion feature [3], [4], [5]. However, these features are all based on frames, i.e. they are not the real features of video. It results they are weak in terms of robustness.

Independent Components Analysis (ICA) is a novel signal processing and data analysis method. Using ICA, people can recover or extract the source signals only from the observations according the stochastic property of the input signals. It has been one of the most important methods of blind source separation and received increasing attentions in pattern recognition, data compression, image analyzing and so on [6], [7], for the ICA process derives features that best present the data via a set of components that are as statistically independent as possible and characterizes the data in a natural way.

In this paper, the FastICA algorithm [8] is used to extract the independent components (ICs) of video. The video ICs are watermarked by modifying their wavelet coefficients according to the 4-neighboring-mean-based algorithm [9]. Simulations show that the watermark can be detected blindly. In addition, the scheme is robust to MPEG-2 compression and can re-synchronizing temporally. ICA and video ICs extraction are presented in section 2. Section 3 describes the proposed scheme. And simulations in Section 4 show its robustness to MPEG compression and the ability to temporal resynchronization. Finally, there is a conclusion.

2 ICA and Video ICs Extraction

2.1 Problem Formulation

Assume that there are m sensors and n source signals in the mixture system. The relationship between sources and observations is:

$$x = As \qquad (1)$$

where $x = [x_1, x_2, \cdots, x_m]^T$ are m mixtures, $s = [s_1, s_2, \cdots, s_n]^T$ are n mutually independent unknown sources and A is a mixing matrix. ICA is to estimate the source signal s or a de-mixing matrix C only from the observation signal x according to the statistical characteristic of s. So the estimation of source signal $\hat{s}$ is gotten through $\hat{s} = Cx$.

The FastICA algorithm developed by Hyvainen and Oja [8] is used to extract the video ICs in this paper.

2.2 FastICA Algorithm

The FastICA algorithm used in this paper is a fixed-point algorithm for independent component analysis (ICA), which provides good decomposition results efficiently. It pre-whitens the observation by performing Principal Component Analysis (PCA). The observed signal x is transformed to $v = Tx$, whose components are mutually uncorrelated and all have unit variance.

The objective function of FastICA by kurtosis is:

$$kurt(D^T v) = E\{(D^T v)^4\} - 3[E\{(D^T v)^2\}]^2 = E\{(D^T v)^4\} - 3\|D\|^4 \qquad (2)$$

The de-mixing matrix learning algorithm is:

$$D(k) = E\{v(D(k-1)v)^3\} - 3D(k-1) \qquad (3)$$

where k is the number of iteration. The de-mixing matrix is $C = D^T T$.

2.3 Video ICs

In [10], Hateren and Ruderman show their results of performing independent component analysis on video sequences of natural scenes, which are qualitatively similar to spatiotemporal properties of simple cells in primary visual cortex. They divides a whole video into video blocks just like what is shown in the middle of Fig. 1.

And the video blocks are concatenated to be the video vectors, which are the observed signals x of ICA model (1). The ICs, i.e. s of (1), are the spatiotemporal constituents of video blocks.

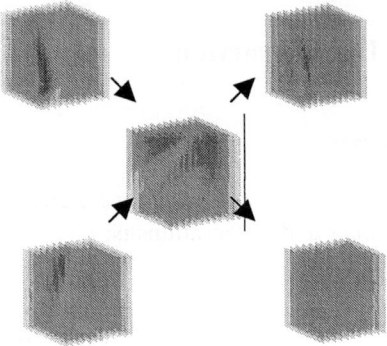

Fig. 1. The description of video block, its ICs and ICFs in [10] is shown here. The independent components (ICs) are on the left, the independent component filters (ICFs) are on the right and the middle are the original video blocks

3 Watermark Embedding and Detection

3.1 Watermark Embedding Algorithm

Before watermark embedding, where the data will be hidden should be determined. The extracted video ICs are considered as independent videos, whose frames are the slices just like the sub-images shown in Fig.2. The slices of the same order in their respective ICs are collected, among which the one that has the maximum variance is selected to be embedded watermark.

 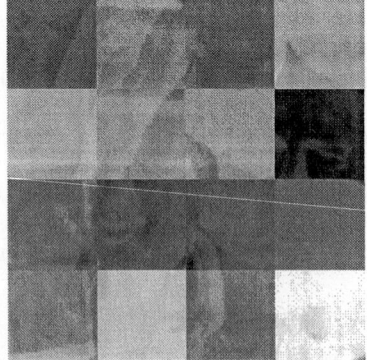

Fig. 2. The first frame of the experimental video is on the left. And on the right, there are the slices of the 16 extracted video ICs, which are the same order in ICs. These ICs can be regarded as some independent video, whose frames are this kind of slices

Pseudo-random 0,1 sequences are used as watermarks W_n embedded into the wavelet domain of the selected slices. If the coefficient $p_{i,j}$ will be watermarked, the mean of its four neighboring coefficients is $mp_{i,j} = \frac{1}{4}(p_{i,j-1} + p_{i-1,j} + p_{i,j+1} + p_{i+1,j})$. And these four coefficients are not watermarked. The embedding algorithm is following.

$$p'_{i,j} = p_{i,j} + \alpha(p_{i,j} - mp_{i,j}), F_n(m) = 1 \begin{cases} p_{i,j} > mp_{i,j}, W(m) = 0 \\ p_{i,j} < mp_{i,j}, W(m) = 1 \end{cases} \quad (4)$$

$$p'_{i,j} = p_{i,j} - \alpha(mp_{i,j} - p_{i,j}), F_n(m) = 0 \begin{cases} p_{i,j} > mp_{i,j}, W(m) = 1 \\ p_{i,j} < mp_{i,j}, W(m) = 0 \end{cases} \quad (5)$$

where $m = 1, 2, \cdots, r$, r is the length of watermark sequence, n is the slice order in the video ICs, α is a weight. F_n is a symbol to determine the existence of watermark.

3.2 Watermark Detecting Algorithm

Before detecting watermark, the received video should be resynchronized temporally. The detecting algorithm is:

$$W_n^*(m) = 0 \begin{cases} p_{i,j}^* > mp_{i,j}^*, F_n(m) = 0 \\ p_{i,j}^* < mp_{i,j}^*, F_n(m) = 1 \end{cases} \quad (6)$$

$$W_n^*(m) = 1 \begin{cases} p_{i,j}^* > mp_{i,j}^*, F_n(m) = 1 \\ p_{i,j}^* < mp_{i,j}^*, F_n(m) = 0 \end{cases} \quad (7)$$

where $p_{i,j}^*$ and $mp_{i,j}^*$ are the watermarked coefficient and its neighboring mean of the received video, respectively. W_n^* is the extracted watermark of the slice n. The similarity between the extracted watermark and the original one is defined:

$$Sim(W_n, W_n^*) = 1 - \frac{\sum_{m=1}^{r} XOR[W_n(m), W_n^*(m)]}{r} \quad (8)$$

4 Simulations

The experimental video consists of 16 256×256 frames. And the whole video is divided into 16 video blocks of the same size. We select 16 numbers from 100 to 250 as the seed to generate pseudo- random 0,1 sequences, which will be used as the watermark. Every to-be-watermarked slice is watermarked with different sequences.

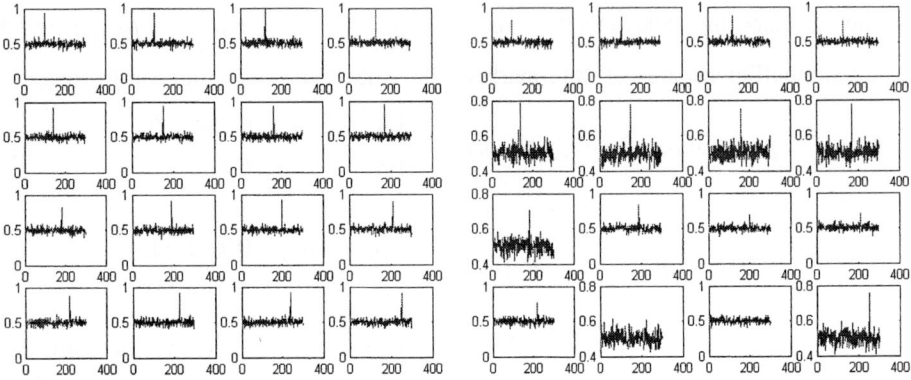

Fig. 3. The similarity between the original symbol and the extracted one of each watermarked slice with (right) and without (left) MPEG-2 compression is listed. Obviously the similarities are affected by the compression, but the detection peaks are still outstanding at the very states. X-axis denotes the state of pseudo-random sequences, while the y-axis is the similarity

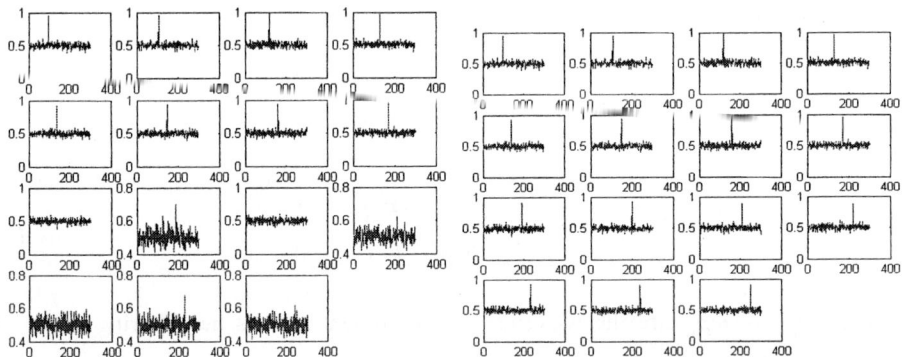

Fig. 4. The similarities listed here are the detection results of frame dropping experiment. The 9th frames are dropped at the receive terminal. The detection results without resynchronization are list on the left, in which the detection results of the first 8 frames are correct, and the detection results are damaged from the 9th frame. That shows that the first dropped frame is the 9th frame. Maybe there are many dropped frames, but the 9th is the first one is confirmed. The detection results after temporal synchronization are shown on the right. If there are many frames dropped, the resynchronization can be achieved after finding all the dropped frames. X-axis denotes the state of pseudo-random sequences, while the y-axis is the similarity

5 Conclusion

In this paper, a novel blind video watermarking scheme is presented, which is based on the video independent component extraction by FastICA. The watermark is embedded into the slices of the obtained video ICs according to the 4-neighboring-mean algorithm. The simulation shows the feasibility of this scheme. The results demonstrate its robustness to MPEG-2 compression and the ability to temporal resynchronization are good. The robustness to other attacks, such as collusion, and the tradeoff between the robustness and invisibility are the next research focus. And new application of ICA in video analysis is also interesting.

References

1. Bhardwaji, A., Pandey, T.P., Gupta, S.: Joint Indexing and Watermarking of Video Using Color Information, IEEE 4th Workshop on Multimedia Signal Processing (2001) 333 – 338
2. Liu, H., Chen, N., Huang J., Huang X., Shi, Y.Q.: A Robust DWT-Based Video Watermarking Algorithm, IEEE International Symposium on Circuits and Systems, Vol. 3 (2002) 631 – 634
3. Zhang, J., Maitre, H., Li, J., Zhang, L.: Embedding Watermark in MPEG Video Sequence, IEEE 4th Workshop on Multimedia Signal Processing (2001) 535 – 540
4. Bodo, Y., Laurent, N., Dugelay, J.: Watermarking Video, Hierarchical Embedding in Motion Vectors, Proceedings of International Conference on Image Processing, Vol. 2 (2003) 739 – 742
5. Zhu, Z., Jiang, G., Yu, M., Wu, X.: New Algorithm for Video Watermarking, 6th International Conference on Signal Processing, Vol.1 (2002) 760 - 763
6. Hurri, J., Hyvarinen, A., Karhunen, J., Oja, E.: Image Feature Extraction Using Independent Component Analysis, Proc. IEEE Nordic Signal Processing Symposium, Espoo Finland (1996)
7. Larsen, J., Hansen, L. K., Kolenda, T., Nielsen, F. A.: Independent Component Analysis in Multimedia Modeling, 4th International Symposium on Independent Component Analysis and Blind Signal Separation (ICA2003), Japan (2003) 687-695
8. Hyvrinen, A., Oja, E.: A Fast Fixed-Point Algorithm for Independent Component Analysis, Neural Computation, Vol. 9, No. 7 (1997) 1483-1492
9. Hong, I., Kim, I., Han, S.S.: A Blind Watermarking Technique Using Wavelet Transform, Proceedings of IEEE International Symposium on Industrial Electronics, Vol. 3 (2001) 1946 – 1950
10. van Hateren, J.H., Ruderman, D.L.: Independent Component Analysis of Natural Image Sequences Yields Spatio-Temporal Filters Similar to Simple Cells in Primary Visual Cortex, Proceedings of the Royal Society of London B, 265(1412) (1998) 2315-2320

Multisensor Data Fusion Based on Independent Component Analysis for Fault Diagnosis of Rotor

Xiaojiang Ma[1] and Zhihua Hao[1,2]

[1] Institute of Vibration Engineering, Dalian University of Technology, Dalian,116023, P.R. China
mxjiang@dlut.edu.cn
[2] Tangshan College,Tangshan,063000, P.R. China
haozhihua@yahoo.com.cn

Abstract. Independent Component Analysis is applied to multi-channel vibration measurements to fuse the information of several sensors, and provide a robust and reliable fault diagnosis routine. Independent components are obtained from the measurement data set with FastICA algorithm, and their AR modeling estimates are calculated with BURG method. A probabilistic neural network is applied to the AR modeling parameters to perform the fault classification. Similar classification is applied directly to vibration measurements. Based on the results with real measurement data from the rotor test rig, it is shown that data fusion with ICA enhances the fault diagnostics routine.

1 Introduction

Vibration monitoring has been used in rotating machines fault detection for decades [1-3]. In [3], It is claimed that vibration method is the most effective methods of assessing the overall health of a rotor system. Vibration data is often measured with multiple sensors mounted on different parts of the machines. For each machine there are typically several vibration signals being monitored in addition to some static parameters like load, speed. The examination of data can be tedious and sensitive to errors. Also, fault related machine vibration is usually corrupted with structural machine vibration and noise from interfering machinery. Further, dependent on the sensor position, large deviation on noise may occur in measurements.

Due to these problems intelligent compression of the multisensor measurements data may aid in data management for fault diagnosis purpose. Independent component analysis (ICA) may be used to combine the information of several sensors into a few "virtual sensors" that allows for smaller data dimensionality and less dependence on the position of the ensemble of sensors, whereas the relevant diagnostic information may still be retained. This means that distributing an ensemble of sensors over a machine casing would then allow for extraction of the diagnostic information of the fault source, regardless the exact position of each sensor.

In this article, ICA is studied to provide a robust fault diagnostic routine for a rotor system. Ypma & al. have utilized similar approach with application to fault diagnostics of a submersible pump in [4]. Sanna & al. have applied ICA and SVM with application to fault detection of an induction motor in [5]. In our study, resulting independent components are further processed with autoregressive (AR) modeling and probabilistic neural network based classification to obtain the fault condition of a rotor system.

2 Independent Component Analysis

2.1 Basis of ICA

The ICA (also named blind source separation) aims at extracting independent signals from their linear mixtures or extracting independent features from signals having complex structure. What makes the difference between ICA and other multivariate statistical methods is that it looks for components that are both statistically independent, and nongaussian. For example in PCA, the redundancy is measured by correlations between data elements, while in ICA the idea of independent is used. Statistical independence means that the value of any one of the components gives no information of the other components.

Consider a situation where there are a number of signal emitted by a rotating machine vibration and interfering vibration sources. Assume further that there are several vibration sensors. The sensors are in different positions so that each records a mixture of the original source signals with slightly different weights. Lets denote i th recorded mixture with x_i and j th original source with s_j. The phenomenon can be described with an equation $X = AS$, where elements of X are x_i and elements of S are s_j. The elements a_{ij} of the matrix A are constants coefficients that give the mixing weights that are assumed to be unknown. A is called a mixing matrix. A blind source separation problem is to separation original sources from observed mixtures of the sources, while blind means that we know very little about the original sources and about the mixture. It can be safely assumed that the mixing coefficients are different enough to make matrix A invertible. Thus there exists a matrix $W = A^-$ that reveals the original sources $S = WX$. After this the problem is, how to estimate the coefficient matrix W. A simple solution to the problem can be found by considering the statistical independence of different linear combinations of X.

In [6], problem of finding independent components is formulated with the concept of mutual information. First differential entropy H of a random vector $y = (y_1, y_2, \cdots, y_n)$ with density $f(\cdot)$ is defined as follows:

$$H(y) = -\int f(y) \log f(y) dy. \tag{1}$$

A Gaussian variable has the largest entropy among all variables of equal variance. Differential entropy can be normalized to get the definition of negentropy:

$$J(y) = H(y_{gauss}) - H(y). \qquad (2)$$

where y_{gauss} is a Gaussian random vector of the same covariance matrix as y. Negentropy is zero for Gaussian variable and always nonnegative. Mutual information I of random variable $y_i, i = 1,...,n$ can be formulated using negentropy J, and constraining the variables to be uncorrelated:

$$I(y_1, y_2, ..., y_n) = J(y) - \sum_i J(y_i). \qquad (3)$$

Since mutual information is the information theoretic measure of the independence of random variable, it is natural to use it as a criterion for finding the ICA transform. Thus the matrix W is determined so that the mutual information of the linear combinations of mixed variables X is minimized.

The FastICA algorithm[6] is a computationally highly efficient method for performing the estimation of independent components. In this paper, a FastICA MATLAB-package developed at Laboratory of Computer and Information Science in the Helsinki University of Technology was used.

2.2 Separating the Independent Components of Fault Signal on Rotor

Two machine fault conditions which occur in rotating machine are misalignment and foundation loose. Vibration data were available from a small test rig which had been used to simulate these two conditions that occur simultaneously. The rig was consisted of a rotating shaft with a rotor driven by an electric motor held in place by two bearing block. Vibrations were measured using two accelerometers mounted in the vertical directions on the two bearing block. The analysis frequency is 1000 Hz, rotating speed is 3000r/min (working frequency is 50 Hz). The length of vibration signals is 1024 points. The waves of measured vibration signals were shown in Fig.1(a), where symbol "A" denote A channel measured data, symbol "B" denote B channel measured data. Two independent components separated by ICA were shown in Fig.1(b). We used the deflation approach with cubic nonlinearity. We could not judge how about the effect of the ICs alone. Then the amplitude spectrums of ICs (shown in Fig.2(a)) were calculated. From the spectrum of IC1, we can find a spectrum line in 100 Hz, it is the 2X of rotating frequency. This is the obvious feature of misalignment fault. From the spectrum of IC2, we can find that obvious spectrum lines are 2X, 3X and 6X of rotating frequency respectively. These features correspond to the foundation loose fault.

From above, we conclude that the independent source signals (misalignment and foundation loose) were obtained from the double channel compound signals by the ICA. It is further shown that ICA is a feasible method in mechanical fault diagnosis.

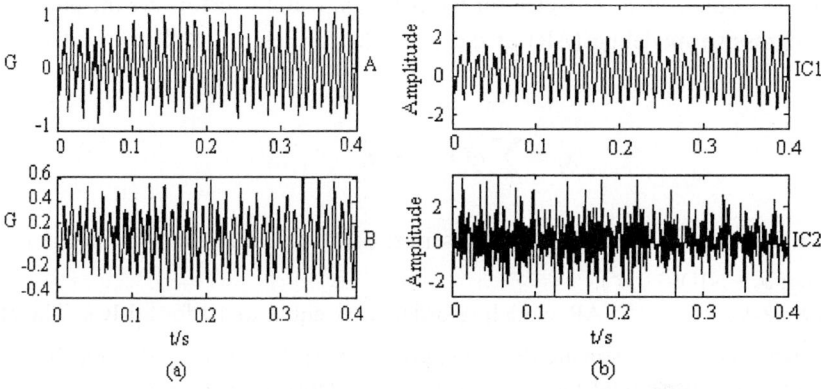

Fig. 1. (a) Is the waves of measured vibration signals; (b) is the waves of independent components

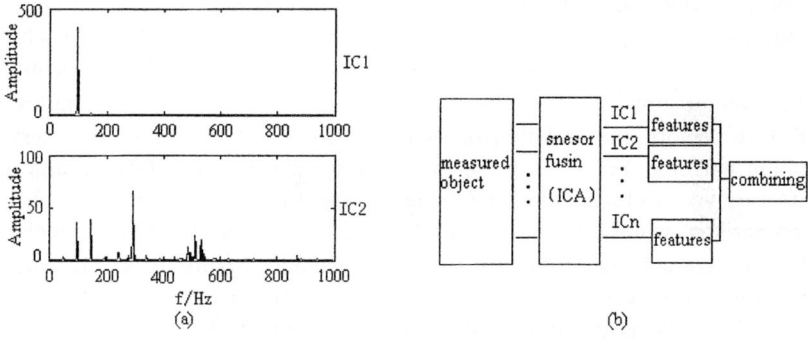

Fig. 2. (a) Is the amplitude spectrums of independent components; (b) is the illustration of sensors fusion based on ICA

3 Data Fusion Based ICA on Rotor Fault

The process of multisensor data fusion was shown in Fig.2(b). For a plant tested, the original dataset was constructed through M measurement sensors mounted on different parts of the machine. The FastICA algorithm was applied for the multidimensional dataset to obtain N independent components ($M \geq N$). One side, the process can compress the measurements into a small amount of channel combinations-statistically independent components of the measurements-that could clearly indicate faults in the machine; on the other hand, these independent components allows for less dependence on the position of the ensemble of sensors, whereas the relevant diagnostic information may still be retained.

Then the features were extracted for the ICs. The feature vectors are the parameters of autoregressive modeling (AR) of the ICs. The formula of $AR(p)$ is defined as follows:

$$x_t = \sum_{i=1}^{p} \varphi_i x_{t-i} + a_t . \tag{4}$$

Where x_t (t=1,2,...,n) are measured time series; φ_i (i=1,2,..., p) are parameters of AR modeling; $a_t \sim$ NID(0, σ^2).

In the next section, the AR modeling order p is equal to 10 for all ICs. The Burg method was applied to estimate the parameters φ_i (i=1,2,...,10) of AR modeling. The parameters were combined into a feature vector whose dimensionality was equal to $10 \times N$, where N denotes the number of ICs.

In order to illustrate the effect of this data fusion routine, two fault conditions, e.g. misalignment single fault, foundation loose and misalignment compound fault, were simulated from the test rig above mentioned. Vibrations were measured using two accelerometers. Each fault condition was measured to generate 16 samples. For these two conditions, there are 32 samples. We assume that there are two ICs for each fault condition sample. Then the feature vectors were formed in the way above mentioned. The feature vector dimensionality was equal to 10×2 for each of the samples. For the 32 feature vectors, the probabilistic neural network (PNN) was used to serve as classification machine. For each of the two fault signals, the classification algorithm took 12 feature vectors for training and the remaining 4 feature vectors for testing.

For comparison, without using the data fusion, the parameters of AR modeling φ_i (i=1,2,...,10) were calculated directly for the measured vibration data. Training and testing feature vectors were formed in the same way as earlier.

The comparing of classification results were shown in table 1.

Table 1. Comparing of classification results

samples	Compound fault				Single fault			
	1	2	3	4	1	2	3	4
Using ICA	0	1	1	1	0	0	0	0
	1	0	0	0	1	1	1	1
Without using ICA	1	1	1	0	0	1	0	1
	0	0	0	1	1	0	1	0

From the Table 1, we can see that only the first sample of compound fault was wrongly classified to the single fault after data fusion using ICA, all of the other samples were classified correctly. However, without using ICA, three samples (the fourth of compound fault, the second and fourth of single fault) of testing data were wrongly classified. This could indicate that faults are more easily detected from fused vibration measurements than pure vibration signals.

4 Conclusion

ICA was applied to multisensor vibration measurements to fuse the measurement information of several channels, and provide robust routine for the rotor fault detection. ICs were found from the measurements set with a FastICA algorithm and their features of AR modeling were estimated with Burg method. The PNN was applied to the feature vector sets to perform the fault diagnosis. Similar classification was applied directly to the features of AR modeling of vibration measurements. Results gained with the data fusion methods were compared to the results gained without data fusion. The results show that the data fusion methods enhance the fault diagnostic routine.

References

1. Betta, G., Liguori, C., Paolillo, A., Pietrosanto, A.: A DSP-based FFT analyzer for the fault diagnosis of rotating machine based on vibration analysis. Proc. IEEE Conf. on Instrumentation and Measurement Technology. Budabest, Hungary, (2001) 572-577
2. Marcal, R.F.M., Negreiros, M., Susin, A.A., Kovaleski, J.L.: Detecting faults in rotating machines. IEEE Instrumentation & Measurement Magazine, 3(4), (2000) 24-26
3. Laggan, P.A.: Vibration monitoring. Proc. IEE Colloquium on Understanding your Condition Monitoring, (1999) 1-11
4. Ypma, A., Pajunen, P.: Rotating machine vibration analysis with second-order independent component analysis. Proc. of the Workshop on ICA and Signal Separation, Aussois, France, (1999) 37-42
5. Pöyhönen, S., Jover, P., Hyötyniemi, H.: Independent component analysis of vibrations for fault diagnosis of an induction motor. IASTED International Conference on CIRCUITS, SIGNALS, and SYSTEMS. Cancun, Mexico, (2003) 203-208
6. Hyvärinen, A.: Fast and robust fixed-point algorithms for independent component analysis. IEEE Trans. On Neural Networks, 10(3), (1999) 626-634

Substructural Damage Detection Using Neural Networks and ICA

Fuzheng Qu, Dali Zou, and Xin Wang

School of Mechanical Engineering, Dalian University of Technology, Dalian China, 116023
{fzqu,crane}@dlut.edu.cn

Abstract. Frequency response function (FRF) data are sensitive indicators of structural physical integrity and thus can be used to detect damage. This paper deals with structural damage detection by using back propagation neural networks (BPNN). Features extracted from FRFs by applying Independent component analysis (ICA) are used as input data to NN. The Latin hypercube sampling (LHS) is adapted for efficient generation of the patterns for training the neural network. Substructural identification technique is also used to cope with complex structure. A truss is presented to demonstrate the effectiveness of the present method with good accuracy, even if the measurements are incomplete and full of noise.

1 Introduction

Once a structure is damaged, its dynamic properties and responses will also change. These changes can be used to detect structural damage. Generally speaking, there are three types of measured data. They are vibration response-time histories, measured FRFs and modal parameters. According to these three types, there are also three damage-identification techniques. Over the last decade, although the modal parameter-based methods are used very popular, the methods are difficult to apply to large structures or slight damaged structures for their low insensitivity to damage and large analysis errors. Since FRFs can provide much more information in a desired frequency range and are readily available in an industrial environment, the FRF-based methods can be taken as promising methods to detect large and complex structural damage [1].

For the amount of FRF data is very large, it is not practical to use all these data directly. How to extract the features from FRFs about structural state efficiently without losing much useful information must be considered. Principal component analysis (PCA) as a well-known method for feature extraction has been used to solve this problem [1]. PCA decorrelates the input data using second-order statistics and thereby generates compressed data with minimum mean-squared reprojection error. As a matter of fact, when a structure is damaged, the measured data inevitably include higher-order statistic information, so using PCA may omit some useful information. Recently the applications of ICA originally developed for blind source separation are developed very quickly [2]. ICA minimizes both second-order and higher-order dependencies in the input, thus using ICA here can extract more useful information about the state of structure.

Neural networks have unique capability to be trained to recognize given patterns and to classify other untrained patterns. Hence, neural networks have already been used for estimating the structural parameters with proper training [1,3]. In this study, the neural network-based approach is extended to identify the damage of a complex structural system with many unknowns. Substructural identification [3] is employed to overcome the issues. The features extracted from FRFs using ICA are utilized as input to neural networks, and LHS technique [4] is employed for efficient generation of training patterns. A truss example is presented to demonstrate the effectiveness of the proposed method.

2 Independent Component Analysis (ICA)

Let S be the matrix of unknown source signals and X be the matrix of observed mixtures, If A is the unknown mixing matrix, then the mixing model is written as

$$X = AS \tag{1}$$

It is assumed the source signals are independent of each other. In order to extract the independent components, ICA algorithms try to find an un-mixing matrix W such that

$$U = WX = WAS \tag{2}$$

is an estimation of the independent source signals.

Among the large amount of ICA algorithms, the FastICA is the most popular used. In this algorithm, non-Gaussianity (independent) is measured using approximations to negentropy (J) which is more robust than kurtosis based measures and fast to compute. The approximation has the following form

$$J(u_i) = [E\{G(u_i^T x_t)\} - E\{G(v)\}] \tag{3}$$

where u_i is a vector, comprising one of the rows of the matrix U. v is a standardized Gaussian variable. G is a non-quadratic function. It can be shown that maximizing any function of this form will also maximize independence [5]. To simplify the FastICA algorithm, two preprocessing steps are applied [6]. First, the data is centered by subtracting the mean of each column of the data matrix X. Second, whitening the data matrix by projecting the data onto it's principal component directions i.e. $X \rightarrow KX$, where K is a pre-whitening matrix can be obtained by using PCA. The use of PCA whitening also has the property of reducing the dimension of X.

3 Substructural Damage Detection

For a structure with many unknown parameters, it is not practical to identify all the unknowns at the same time, because most of identification techniques require expensive computation that would be prohibitive. In the present study, the local identification is proposed to subdivide a structure into several substructures. The substructure to be identified is called as the internal substructure, while the others as the external substructures. Since the parameters to be estimated are limited to an internal substructure, it is expected that the numerical problems such as divergence or falling into local minima may be avoided. Another advantage is that this approach requires measurement data only on the substructure of interest, instead of on the whole structure [3].

4 Illustrative Numerical Example

4.1 Truss Structure

A numerical example is a two-span planar truss, which contains 55 truss elements, 24 nodes and 44 nodal DOFs as shown in Fig.1. Values for the material and geometric properties are as follows: the elastic modulus is 200 GPa; the cross-sectional area is 0.03 m2; and the mass density is 8000kg/m3. Given only the elastic modulus will decrease, if the structure is damaged. Instead of identifying the whole elastic modulus simultaneously, the structure was subdivided into three substructures and only the internal substructure will be identified. It was assumed that the unknown elastic modulus scaling factors for the elements in the internal substructure were between 0.3 and 1. The 5 y-direction FRFs were assumed to be measured corresponding to 5 test nodes in the internal substructure. Assuming the element 3, 6 and 9 is damaged; the corresponding elastic modulus scaling factors is 0.5, 0.8 and 0.9.

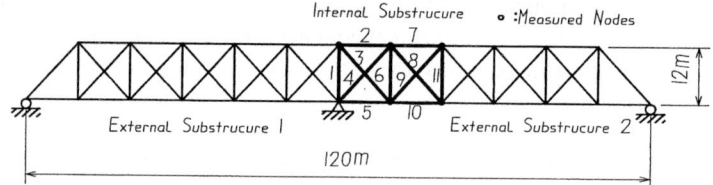

Fig. 1. Substructuring for localized damage identification

4.2 Training Patterns

In order to get an accurate identification using NN, several identification iterations should be processed. Each training patterns include two parts. The generation of first part uses the LHS, 8 sampling points between 0.3 and 1 are selected with an equal interval of 0.1 for each elastic modulus scaling factor in the internal substructures. Consequently, the number of the possible sample cases is 8, which is not be enough to represent the system adequately. Hence, 100 iterations of LHS process are taken, and 800 training patters are sampled. The second part is depended on the result of last NN identification (The original identification result is assumed that no element is damaged) and given only one element is damaged; the elastic modulus of the damage element is between 0.3 and 1 as above. Thus, the number of second possible sample cases is 8×11=88. So the total number of the training patterns is 800+88=888.

4.3 Features Extraction by ICA

The number of spectral lines in each FRF is 1001 which from 0 Hz to 200Hz with an interval of 0.2 Hz. So a 5005×888 matrix is formed. Using PCA for data whitening, select the principal components until the reconstruction quality is over 85%. Then using ICA for further feature extraction. Those extracted independent components can be used as the input of NN.

4.4 Training Processes

Although many types of artificial neural network are used in practice, a four-layer back-propagation (BP) type network was selected for the present example. The output vector of the BPNN is just the sampled values in the internal substructure. The number of the neurons in the input layers varies depending on the results of ICA. The numbers of the neurons in the first hidden, the second hidden and the output layers are taken to be 150,50,11. Every training process took about 5000 epochs to learn the pattern representation using the standard BP algorithm.

4.5 Damage Detection

Two initial test cases are carried out. One is free of noise, the other is contaminated by 10% noise. Just as shown in Fig.2 and Fig. 3, the identifications at the both cases are very closed to the real damage. As NN has the ability to resist the noise, the measurement errors have little effect to the damage identification. For the large and middle damage (50% and 20%), one iteration is enough to have a good identification. But for the slight damage (<10%), more iterations should be done to get the correct result.

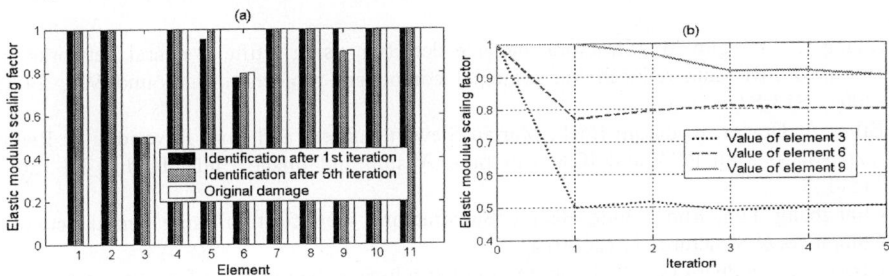

Fig. 2. (a) Identification using 5 noise-free FRFs. (b)Identification process of element 3,6 and 9

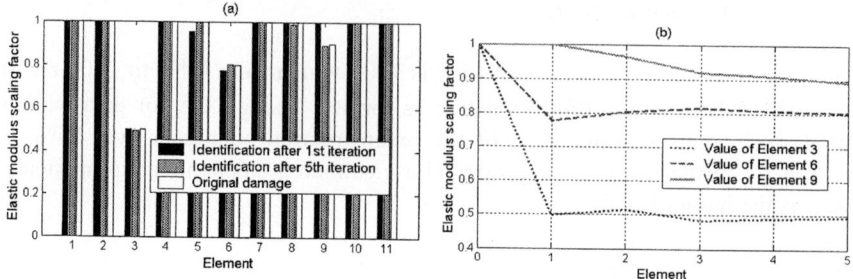

Fig. 3. (a) Identification using 5 noise-polluted FRFs (b)Identification process of element 3,6 and 9

5 Conclusion

A neural network–based damage identification using FRF data is presented for a complex structural system, particularly in the case of noisy and incomplete measurements obtained from a few sensors. With the help of ICA, the FRF data can be compressed and the features about the state of structure can also be extracted to identify the damage. The substructural identification improved the ability to identify the damage of large structure efficiently. The validation of the method is demonstrated by a truss numerical simulation, and the identification is also satisfying.

References

1. Zhang C., Imregun M.: Structural damage detection using artificial neural networks and measured FRF data reduced via principal component projection. J. Sound Vib. 242(5) (2001). 813-827
2. Bruce A. Draper, Kyungim Baek, Marian Stewart Bartlett, and Ross Beveridge J.: Recognizing faces with PCA and ICA. Computer Vision and Image Understanding 91 (2003) 115–137
3. Chungbang Yun, Eun Young Bahng: Substructural identification using neural networks. Computers & Structures 77 (2000) 41-52
4. Olsson A., Sandberg G., Dahlblom O.: On Latin hypercube sampling for structural reliability analysis. Structural safety 25 (2003) 47-68
5. Hyvärinen A.: The fixed-point algorithm and maximum likelihood estimation for independent component analysis. Neural Processing Letters 10 (1999) 1-5.
6. Cao L.J., Chua K.S., Chong W.K., Lee H.P., Gu Q.M.: A comparison of PCA, KPCA and ICA for dimensionality reduction in support vector machine. Neurocompting 55 (2003) 321-336

Speech Segregation Using Constrained ICA

Qiu-Hua Lin[1], Yong-Rui Zheng[1], Fuliang Yin[1], and Hua-Lou Liang[2]

[1] School of Electronic and Information Engineering,
Dalian University of Technology, Dalian 116023, China
qhlin@dlut.edu.cn
[2] School of Health Information Sciences,
The University of Texas at Houston, Houston, TX 77030, USA

Abstract. In natural environment, speech often occurs concurrently with acoustic interference. How to effectively extract speech remains a great challenge. This paper describes a novel constrained Independent Component Analysis (ICA) approach, the ICA with reference (ICA-R), to speech segregation. Different from the traditional ICA which recovers simultaneously all the source signals, the ICA-R extracts only some desired source signals from the mixtures of source signals by incorporating some *a priori* information into the separation process. We show how the ICA-R can be applied to separate a target speech signal from interfering sounds by exploiting a proper reference signal, which is based on the different characteristic between speech signal and its environmental noises, i.e., the speech signal has pitch and its harmonic frequencies whereas the noises usually do not. Results of computer experiments demonstrate the efficiency of the proposed method.

1 Introduction

Independent Component Analysis (ICA) aims to recover a set of unknown mutually independent source signals from their observed linear mixtures without knowing the mixing coefficients [1]. It has been applied to communications, biomedical engineering, etc. [2]-[4]. The problem of speech segregation has also received attention from researchers investigating ICA [5]-[7]. A recent variation, constrained ICA, also referred to the ICA with reference (ICA-R) [8], was proposed to allow *a priori* information about the desired source signals to be provided in the separation process. This differs from most approaches to ICA where the algorithms recover simultaneously all the source signals; the ICA-R extracts only the desired source signal which is the closest one, in some sense, to a constructed reference signal based on prior knowledge. Clearly the reference signal does not need to be exactly the same as the desired source signal.

The aim of this paper is to present an application of ICA-R to extract a speech signal from its noisy linear mixtures by using a proper reference signal derived from *a priori* information about the different characteristics between speech signal and its environmental noises, e.g., the speech pitch has fundamental and harmonic frequencies whereas the noises usually do not. The main advantage of the proposed method is that the speech signal is significantly enhanced by including prior information into the ICA separation process, which would greatly facilitate many applications, including automatic speech recognition and speaker identification.

2 ICA with Reference (ICA-R)

Suppose that there exist M independent source signals $\mathbf{s}(t) = [s_1(t), \ldots, s_M(t)]^T$ and N observed mixtures of the source signals $\mathbf{x}(t) = [x_1(t), \ldots, x_N(t)]^T$ (usually $N \geq M$). The linear ICA assumes that these mixtures are instantaneous, and noiseless, i.e.,

$$\mathbf{x}(t) = \mathbf{As}(t) \tag{1}$$

where $\mathbf{A}$ is a $N \times M$ mixing matrix containing the mixing coefficients. The goal of the classical ICA is to find an $M \times N$ demixing matrix $\mathbf{W}$ such that M output signals

$$\mathbf{y}(t) = \mathbf{Wx}(t) = \mathbf{WAs}(t) = \mathbf{PDs}(t) \tag{2}$$

where $\mathbf{P} \in \mathbf{R}^{M \times M}$ is a permutation matrix and $\mathbf{D} \in \mathbf{R}^{M \times M}$ is a diagonal scaling matrix.

Instead of separating all M number of independent sources from N mixed signals, ICA-R extracts L ($L<M$) number of desired independent sources from N mixed signals by incorporating some *a priori* information to the ICA learning algorithm as reference signals. These reference signals, denoted by $r(t) = [r_1(t), \ldots, r_L(t)]^T$, carry some information of the desired sources but not identical to the corresponding desired signals. In the following, we briefly describe the one-unit ICA-R. It finds one weight vector $\mathbf{w}$, i.e., one row of the demixing matrix $\mathbf{W}$, so that the output signal $y(t) = \mathbf{w}^T \mathbf{x}(t)$ equals to $s^*(t)$, which is the desired one of all source signals, by using $r(t)$ as the reference signal. For simplicity, the time index t is omitted in the equations below.

A flexible and reliable approximation of the negentropy $J(y)$ introduced by Hyvärinen in [9] is defined as the contrast function of one-unit ICA:

$$J(y) \approx \rho[E\{G(y)\} - E\{G(v)\}]^2 \tag{3}$$

where ρ is a positive constant, v is a Gaussian variable having zero mean and unit variance, $G(\cdot)$ can be any nonquadratic function.

The closeness between the estimated output y and the reference signal r is measured by norm $\varepsilon(y,r)$ which has a minimal value when y corresponds to the desired source s^*. A threshold ξ can be used to distinguish this desired source s^* from other source signals such that the formula $g(\mathbf{w}) = \varepsilon(y,r) - \xi \leq 0$ is satisfied only when $y = s^*$ among all source signals. Treating $g(\mathbf{w})$ as feasible constraint to the contrast function in (3), the problem of ICA-R can be modeled in the framework of constrained independent component analysis [8] as follows:

$$\text{maximize} \quad J(y) \approx \rho[E\{G(y)\} - E\{G(v)\}]^2 \quad (4)$$
$$\text{subject to} \quad g(\mathbf{w}) \leq 0, h(\mathbf{w}) = E\{y^2\} - 1 = 0.$$

The equality constraint $h(\mathbf{w})$ is included to ensure that the contrast function $J(y)$ and the weight vector $\mathbf{w}$ are bounded. In [8], a Newton-like learning algorithm is derived by finding the maximum of an augmented Lagrangian function corresponding to (4):

$$\mathbf{w}_{k+1} = \mathbf{w}_k - \eta \mathbf{R}_{\mathbf{xx}}^{-1} L'_{\mathbf{w}_k} / \delta(\mathbf{w}_k) \quad (5)$$

where k denotes the iteration index, η is the learning rate, $\mathbf{R}_{\mathbf{xx}}$ is the covariance matrix of the input mixtures $\mathbf{x}$,

$$L'_{\mathbf{w}_k} = \rho E\{\mathbf{x}G'_y(y)\} - 0.5\mu E\{\mathbf{x}g'_y(\mathbf{w}_k)\} - \lambda E\{\mathbf{x}y\},$$

$$\delta(\mathbf{w}_k) = \rho E\{G''_{y^2}(y)\} - 0.5\mu E\{g''_{y^2}(\mathbf{w}_k)\} - \lambda$$

where $G'_y(y)$ and $G''_{y^2}(y)$ are the first and the second derivatives of $G(y)$ with respect to y, and $g'_y(\mathbf{w}_k)$ and $g''_{y^2}(\mathbf{w}_k)$ are those of $g(\mathbf{w}_k)$. μ and λ are the Lagrange multipliers learned by the gradient-ascent method:

$$\mu_{k+1} = \max\{0, \mu_k + \gamma g(\mathbf{w}_k)\}, \quad (6)$$

$$\lambda_{k+1} = \lambda_k + \gamma h(\mathbf{w}_k). \quad (7)$$

where γ is the scalar penalty parameter.

3 Reference Signal for Extracting a Speech Signal

A proper and available reference signal is important for ICA-R to extract the desired signal. Thus the most important thing for the proposed method is to construct a proper reference signal before we extract a speech signal from its noisy linear mixtures with ICA-R.

Examining the speech signal and its environmental noises such as random noise and cocktail party noise, we may easily find their distinct characteristics, e.g., the speech has pitch and its harmonic frequencies whereas the noises usually do not. As an example, Fig. 1 shows four different source signals in Fig. 1(a)-(d): speech signal, random noise, noise bursts, and cocktail party noise, which are the speech and intrusions used in [10] (sampled at 16KHz and 28591 samples), as well as their corresponding power density spectra (Fig. 1 right panel). It is evident from the power spectrum of the speech in Fig. 1(f) that the speech signal has the regular distribution of the pitch and its harmonic frequencies. Such an observation, however, is not pronounced for random noise, noise bursts, and cocktail party noise in Fig. 1(g)-(i).

Since the periodic rectangular pulses with pitch frequency of the speech in Fig.1 (a), as shown in Fig. 1(e) (only 8000 samples of 28591 samples for identification), also have fundamental frequency and its harmonic frequencies (see Fig. 1(j)), which is similar to that of the speech signal, we can generate periodic rectangular pulses, whose frequency is approximately equal to that of the speech signal pitch, as the reference signal r to extract the speech signal from its noisy mixtures in the proposed method.

There are a variety of methods available to estimate the pitch frequency of the speech signal [11]. The accurate estimation of pitch frequency of the speech is not crucial for the ICA-R: simply using one of methods is sufficient. We first pick one of standard methods to estimate approximately the speech pitch, and then construct the reference signal base on the estimated pitch frequency.

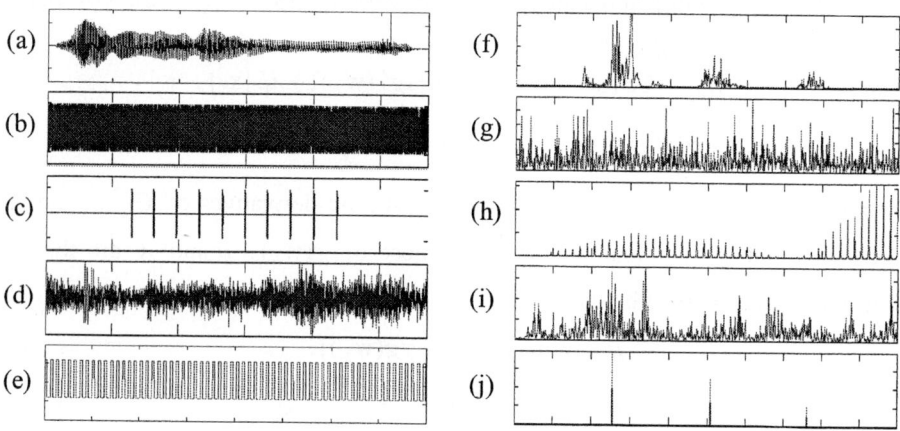

Fig. 1. Comparison of the power density spectrum of five different source signals. (a) Speech signal. (b) Random noise. (c) Noise bursts. (d) Cocktail party noise. (e) Periodic rectangular pulses with the pitch frequency of the speech signal in (a). (f)-(j) Power density spectrum of (a)-(e), respectively.

4 Experimental Results

To illustrate the performance of the proposed method, we carried out extensive computer experiments with the synthesized mixtures of the speech signals and intrusions used in [10]. Fig. 2 shows one example of extracting the speech signal in Fig. 1(a) from its linear noisy mixtures of the four source signals in Fig. 1(a)-(d), as shown in Fig. 2(a)-(d), with the following mixing matrix:

$$A = \begin{bmatrix} 0.9355 & 0.0579 & 0.1389 & 0.2722 \\ 0.9169 & 0.3529 & 0.2028 & 0.1988 \\ 0.4103 & 0.8132 & 0.1987 & 0.0153 \\ 0.8936 & 0.0099 & 0.6038 & 0.7468 \end{bmatrix} \quad (8)$$

The reference signal generated is the periodic rectangular pulses having the pitch frequency of the speech signal in Fig. 1(a), as shown in Fig. 1(e). The learning equations (5)–(7) of ICA-R are then used to extract the speech signal from the four noisy mixtures in Fig. 2(a)-(d). In the ICA-R algorithm, $G(y) = \exp(-y^2/2)$ is used since the speech signal is super-Gaussian signal. The closeness between the ICA-R output signal y and the corresponding reference signal r is defined as the negative of the correlation $\varepsilon(y,r) = -E\{y\,r\}$, i.e., $g(\mathbf{w}) = -E\{y\,r\} - \xi$. The threshold ξ is initialized with a small value to avoid the ICA-R algorithm going to a local optimum, and then is gradually increased to converge at the global maximum.

As a result, the desired weight vector $\mathbf{w}$ is first derived, and the desired speech signal is then produced. Fig. 2(e) shows the satisfactory result. We can see that the speech signal is extracted from the noisy mixtures with high accuracy.

To quantify the performance of the proposed method, we computed the signal-to-noise ratio (SNR) index for the speech signal in Fig. 1(a) in the extracted signals by ICA-R in Fig. 2(e), which is defined as follows:

$$SNR = 10\log_{10} \frac{\sum_{t=1}^{T}[(\mathbf{w}^T\mathbf{A})_i s_i(t)]^2}{\sum_{j=1}^{M}\sum_{t=1}^{T}[(\mathbf{w}^T\mathbf{A})_j s_j(t)]^2}, \quad i \neq j \tag{9}$$

where $\mathbf{w}$ is the desired weight vector derived from the ICA-R algorithm, $\mathbf{A}$ is the mixing matrix in (8), $(\mathbf{w}^T\mathbf{A})_i$ denotes the ith column of $\mathbf{w}^T\mathbf{A}$, $s_i(t)$ is the speech signal to be extracted, and $s_j(t)$ denotes the noises. Besides, M is the number of the source signals $s(t)$, and T is the data length of them, e.g., $M=4$, $T=28591$ for the above example. The computed SNR is 21.80dB, which also shows a satisfactory result.

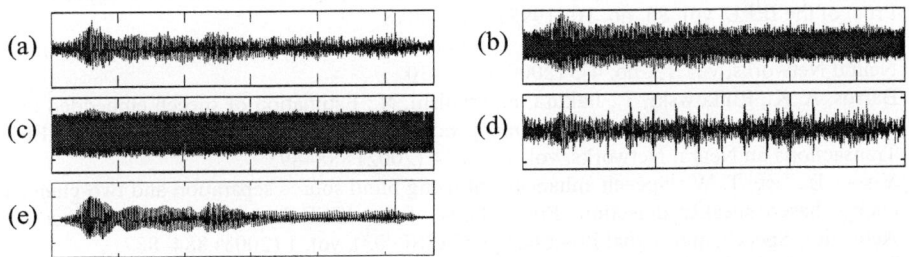

Fig. 2. One example of extracting a speech signal from its noisy mixtures by using ICA-R. (a)-(d) Four noisy mixtures of (a)-(d) in Fig. 1. (e) Extracted speech signal by the ICA-R with the reference signal in Fig. 1(e).

5 Conclusion

The ICA-R has been shown to be applicable to the extraction of a speech signal in its noisy mixtures by using a proper reference signal constructed by the periodic rectangular pulses with the similar frequency to that of the speech signal pitch. Experiment results on synthesized data demonstrate the effectiveness of the proposed method.

Since ICA-R extracts only one desired signal, its computational requirements for learning is substantially reduced compared with the traditional ICA [8].

The proposed method can serve as a speech enhancement scheme when the environmental noises are approximately additive. The proposed method may also be extended to cancellation of other noise signals in their noisy mixtures if the reference signal for a desired signal is not readily available compared to that for other noise signals. In such a case, it is needed first to extract a specific noise by one-unit ICA-R with a reference signal, or to extract some specific noises by multi-unit ICA-R [8] with multiple reference signals, then to reduce these noises by post-processing such as adaptive noise cancellation.

Acknowledgement. This work was supported by the National Natural Science Foundation of China under Grant No. 60172073 and No. 60372082, Trans-Century Training Program Foundation for the Talents by the Ministry of Education of China, and Foundation for University Key Teachers by the Ministry of Education of China.

References

1. Comon, P.: Independent component analysis, a new concept?. Signal Processing, vol. 36, no. 3 (1994) 287–314
2. Cardoso, J. F.: Blind signal separation: statistical principles. Proc. of the IEEE, vol. 86, no. 10 (1998) 2009–2025
3. Amari, S., Cichocki, A.: Adaptive blind signal processing - Neural network approaches. Proc. of the IEEE, vol. 86, no. 10 (1998) 2026–2048
4. Hyvärinen, A., Oja, E.: Independent component analysis: algorithms and applications. Neural Networks, vol. 13, no. 4-5 (2000) 411–430
5. Barros, A. K., Rutkowski, T., Itakura, F., Ohnishi, N.: Estimation of speech embedded in a reverberant and noisy environment by independent component analysis and wavelets. IEEE Transactions on Neural Networks, vol. 13, no. 4 (2002) 888–893
6. Visser, E., Lee, T. W.: Speech enhancement using blind source separation and two-channel energy based speaker detection. Proceedings of the IEEE International Conference on Acoustics, Speech, and Signal Processing (ICASSP '03), vol. 1 (2003) 884–887
7. Lee, J. H., Jung, H. Y., Lee, T. W., Lee, S. Y.: Speech enhancement with MAP estimation and ICA-based speech features. Electronics Letters, vol. 36, no. 17 (2000) 1506–1507
8. Lu, W., Rajapakse, J. C.: ICA with reference. Proc. Third Int. Conf. on ICA and Blind Source Separation (ICA 2001) (2001) 120–125
9. Hyvärinen, A.: New approximations of differential entropy for independent component analysis and projection pursuit. Advances in Neural Information Processing Systems, vol. 10 (1998) 273–279
10. Cooke, M. P.: Modeling Auditory Processing and Organization. Cambridge, U.K.: Cambridge Univ. Press (1993)
11. Tabrikian, J., Dubnov, S., Dickalov, Y.: Maximum a-posteriori probability pitch tracking in noisy environments using harmonic model. IEEE Transactions on Speech and Audio Processing, vol. 12, no. 1 (2004) 76–87

Adaptive RLS Implementation of Non-negative PCA Algorithm for Blind Source Separation*

Xiao-Long Zhu[1], Xian-Da Zhang[1], and Ying Jia[2]

[1] Department of Automation, Tsinghua University, Beijing 100084, China
{xlzhu_dau, zxd-dau}@mail.tsinghua.edu.cn
[2] Intel China Research Center, Beijing 100020, China
ying.jia@intel.com

Abstract. Recently, Plumbley and Oja presented a non-negative PCA algorithm, which performs blind separation of non-negative well-grounded sources using a rectification nonlinearity. The algorithm is based on an ordinary gradient learning, and the observations require prewhitening by eigenvalue decomposition. In this paper, we apply our previously developed natural-gradient-based RLS algorithm to optimize the non-negative PCA criterion. In addition, we propose here an RLS-type preprocessing algorithm, which can whiten the data in terms of covariance matrix. These two algorithms are adaptive, and can work in a cascade mode. The validity of the new implementation is confirmed through computer simulations.

1 Introduction

The blind source separation (BSS) problem consists of recovering mutually independent but otherwise unobserved source signals from their linear mixtures without knowing the mixing coefficients (for a review, see e.g. [1]). This kind of blind technique has significant potential applications in various fields, such as wireless telecommunications, radar, audio and acoustics, image enhancement, biomedical signal processing, etc.

The instantaneous noise-free linear mixing model in BSS is largely a solved problem under the usual assumptions of independent nongaussian sources and full rank mixing matrix. However, with some prior information on the sources, like positivity, constant modulus or finite alphabet, new simplified algorithms may become possible. Following this clue, Plumbley and Oja [2], [3] presented recently a non-negative principle component analysis (PCA) algorithm, which is special case of the nonlinear PCA algorithm [4], but using a rectification nonlinearity.

The original non-negative PCA algorithm is based on an ordinary gradient learning, and it does not make use of the structure information, i.e., orthonormality, of the separating matrix; therefore it converges slowly. On the other

---

* This work was supported in part by the Chinese Postdoctoral Science Foundation and in part by the Foundation of Intel China Research Center.

hand, the non-negative PCA algorithm requires the observations to be whitened in terms of covariance matrix, and this stage is primarily performed off-line via eigenvalue decomposition, so strictly speaking, the whole separation process cannot be thought to be adaptive.

It is widely recognized that the recursive-least-squares (RLS) algorithm is better than the least-mean-square (LMS)-type one in convergence rate and tracking capability [5], [6], and the natural gradient learning works more efficiently than the ordinary gradient one [7], [8]. We proposed in this paper a natural-gradient-based RLS implementation of the non-negative PCA algorithm, which combines our previously presented RLS algorithm [9] and a newly developed adaptive RLS-type whitening approach. The validity of the new implementation is confirmed by computer simulations.

2 Problem Statement

Consider the instantaneous noise-free linear mixing model:

$$\mathbf{x}_t = \mathbf{A}\mathbf{s}_t \tag{1}$$

where $\mathbf{A}$ is a nonsingular $n \times n$ real mixing matrix, $\mathbf{x}_t = [x_{1,t}, \cdots, x_{n,t}]^T$ is the available vector of observations, and $\mathbf{s}_t = [s_{1,t}, \cdots, s_{n,t}]^T$ is a vector of real independent stationary sources, all but perhaps one of them nongaussian. The goal is to reconstruct the original sources given just the observations. To this end, one usually adjusts an $n \times n$ matrix $\mathbf{B}$, called the 'separating matrix', such that the output vector

$$\mathbf{y}_t = \mathbf{B}\mathbf{x}_t \tag{2}$$

is a scaled and permuted version of the source vector.

Among many higher-order contrast functions, the following nonlinear PCA criterion is shown to perform BSS [4], [6], [9]

$$J(\mathbf{W}) = E\left\{\|\mathbf{v}_t - \mathbf{W}^T\mathbf{g}(\mathbf{W}\mathbf{v}_t)\|^2\right\} \tag{3}$$

where $\mathbf{v}_t = \mathbf{U}\mathbf{x}_t$ is the whitened vector with an $n \times n$ whitening matrix $\mathbf{U}$, $\mathbf{y}_t = \mathbf{W}\mathbf{v}_t$, and $\mathbf{g}(\mathbf{y}_t) = [g_1(y_{1,t}), \cdots, g_n(y_{n,t})]^T$ denotes a vector of nonlinearly-modified output signals. In (3), $\mathbf{W}$ is constraint to be orthonormal such that $\mathbf{W}^T\mathbf{W} = \mathbf{W}\mathbf{W}^T = \mathbf{I}$. The total separating matrix is $\mathbf{B} = \mathbf{W}\mathbf{U}$.

In many real-world applications, e.g. in the analysis of images, text, or air quality, the sources are always non-negative [10]. With such a prior information, Plumbley and Oja proposed the 'non-negative PCA' algorithm, as [2], [3]

$$\mathbf{W}_t = \mathbf{W}_{t-1} + \eta_t \mathbf{z}_t \left[\mathbf{v}_t^T - \mathbf{z}_t^T \mathbf{W}_{t-1}\right] \tag{4}$$

where $\eta_t > 0$ is a learning rate, and the nonlinearity in $z_{i,t} = g_i(y_{i,t}) = \max(y_{i,t}, 0)$ is a rectification function (In this case, (3) is referred to as the non-negative PCA criterion). (4) is a special case of the nonlinear PCA algorithm [4], and it performs blind separation of non-negative well-grounded sources [3]. A source s_i is called non-negative if $\Pr(s_i < 0) = 0$, and such a source will be called well-grounded if $\Pr(s_i < \delta) > 0$ for any $\delta > 0$, i.e., s_i has nonzero probability density function all the way down to zero.

Most of the existing algorithms for BSS assume the observations to be zero mean, or transformed to be so, and the whitening matrix $\mathbf{U}$ is determined by $\mathbf{U}E\left\{\mathbf{x}_t\mathbf{x}_t^T\right\}\mathbf{U}^T = \mathbf{I}$. Differently, the non-negative PCA algorithm does not remove the mean $\bar{\mathbf{x}}_t$ from the data in order to keep information on the non-negativity of the sources. That is to say, it is the covariance matrix instead of the correlation matrix of $\mathbf{v}_t$ that is the identity matrix via prewhitening. Given a sequence of real data vectors $\mathbf{x}_t$, let the eigenvalue decomposition of the covariance matrix $\Sigma_\mathbf{x} = E\left\{(\mathbf{x}_t - \bar{\mathbf{x}}_t)(\mathbf{x}_t - \bar{\mathbf{x}}_t)^T\right\}$ be $\Sigma_\mathbf{x} = \mathbf{E}\mathbf{D}\mathbf{E}^T$, then for any orthonormal matrix $\mathbf{T}$, the whitening matrix [2], [3]

$$\mathbf{U} = \mathbf{T}\mathbf{D}^{-0.5}\mathbf{E}^T \qquad (5)$$

ensures $\Sigma_\mathbf{v} = \mathbf{U}\Sigma_\mathbf{x}\mathbf{U}^T = \mathbf{I}$.

3 Adaptive RLS Implementation of Non-negative PCA

The whitening algorithm (5) relies on the eigenvalue decomposition, and it is difficult to be used in real time due to heavy computations. To prewhiten the observations on-line, we apply the following cost function:

$$J_2(\mathbf{U}) = \mathrm{trace}(\mathbf{U}\Sigma_\mathbf{x}\mathbf{U}^T) - \log\det(\mathbf{U}\Sigma_\mathbf{x}\mathbf{U}^T) - n \qquad (6)$$

which reaches the global minimum of zero if and only if $\mathbf{U}\Sigma_\mathbf{x}\mathbf{U}^T = \mathbf{I}$ [11]. For wide-sense stationary signals, we use the sample average

$$\widehat{\Sigma}_\mathbf{x} = \frac{1}{t}\sum_{i=1}^{t}(\mathbf{x}_i - \bar{\mathbf{x}}_i)(\mathbf{x}_i - \bar{\mathbf{x}}_i)^T \qquad (7)$$

to approximate the covariance matrix. Setting the gradient of $J_2(\mathbf{U})$ equal to zero and applying the matrix inverse lemma, we obtain an RLS-type whitening algorithm, as

$$\mathbf{U}_t = \frac{t}{t-1}\left[\mathbf{U}_{t-1} - \frac{\widetilde{\mathbf{v}}_t\widetilde{\mathbf{v}}_t^T}{t-1+\widetilde{\mathbf{v}}_t^T\widetilde{\mathbf{v}}_t}\mathbf{U}_{t-1}\right] \qquad (8)$$

where $\widetilde{\mathbf{v}}_t = \mathbf{U}_{t-1}(\mathbf{x}_t - \bar{\mathbf{x}}_t)$. Although this algorithm is similar in form to the one proposed in [12], a time-varying forgetting factor accounts for its efficiency in the sense of faster convergence speed and better steady-state accuracy.

With the whitened vector $\mathbf{v}_t = \mathbf{U}_{t-1}\mathbf{x}_t$ in hand, we can use the natural gradient with orthogonality constraint [8]: $\widetilde{\nabla}J(\mathbf{W}) = \mathbf{W}\mathbf{W}^T \cdot \nabla J(\mathbf{W}) - \mathbf{W}\cdot[\nabla J(\mathbf{W})]^T\cdot\mathbf{W}$, where $\nabla J(\mathbf{W})$ is the ordinary gradient of $J(\mathbf{W})$ with respect to

W, to optimize the non-negative PCA criterion. Somewhat tedious but straightforward manipulation yields

$$\widetilde{\nabla} J(\mathbf{W}_t) = 2 \sum_{i=1}^{t} \lambda^{t-i} \left\{ -\mathbf{W}_t \mathbf{W}_t^T \mathbf{z}_i \mathbf{v}_i^T + \mathbf{W}_t \mathbf{v}_i \mathbf{z}_i^T \mathbf{W}_t \right\} \quad (9)$$

where λ is a forgetting factor ($0 \leq \lambda < 1$). According to (9), the optimal weight matrix is

$$\mathbf{W}_{opt,t} = \left[\sum_{i=1}^{t} \lambda^{t-i} \mathbf{y}_i \mathbf{z}_i^T \right]^{-1} \left[\sum_{i=1}^{t} \lambda^{t-i} \mathbf{z}_i \mathbf{v}_i^T \right] \quad (10)$$

and it can be computed iteratively exploiting the RLS-type algorithm [9]:

$$\begin{aligned} \mathbf{y}_t &= \mathbf{W}_{t-1} \mathbf{v}_t, \quad \mathbf{z}_t = \max(\mathbf{y}_t, 0) \\ \mathbf{Q}_t &= \mathbf{P}_{t-1} / (\lambda + \mathbf{z}_t^T \mathbf{P}_{t-1} \mathbf{y}_t) \\ \mathbf{P}_t &= \frac{1}{\lambda} \left[\mathbf{I} - \mathbf{Q}_t \mathbf{y}_t \mathbf{z}_t^T \right] \mathbf{P}_{t-1} \\ \mathbf{W}_t &= \mathbf{W}_{t-1} + \left[\mathbf{P}_t \mathbf{z}_t \mathbf{v}_t^T - \mathbf{Q}_t \mathbf{y}_t \mathbf{z}_t^T \mathbf{W}_{t-1} \right]. \end{aligned} \quad (11)$$

Taking $\lambda = 0$ in (9), we get the stochastic natural-gradient-based non-negative PCA algorithm [9]:

$$\mathbf{W}_t = \mathbf{W}_{t-1} + \eta_t \left[\mathbf{z}_t \mathbf{v}_t^T - \mathbf{y}_t \mathbf{z}_t^T \mathbf{W}_{t-1} \right] \quad (12)$$

which is closely related to above RLS-type rule (11). In fact, it holds at the separation points that $\mathbf{P}_t \approx \mathbf{Q}_t \approx \alpha(\lambda) \cdot \mathbf{I}$ with $\alpha(\lambda)$ being a very small value [12].

It should be stressed that the non-negative PCA algorithm (4) is based on the ordinary gradient, the separating matrix will converge to an orthonormal matrix, but it may be not true during the learning process. In contrast, both the LMS-type algorithm (12) and the RLS-type one (11) are based on the natural gradient (9), they make use of the Stiefel manifold structure of the object function (3), and hence can be expected to perform BSS more satisfactorily.

4 Computer Simulations

In order to verify the effectiveness of the proposed RLS implementation of the non-negative PCA algorithm, we consider separation of the four source signals used in [3]: uniformly distributed in $[0, \sqrt{12}]$ such that each one is non-negative and well-grounded. In simulations, the mixing matrix **A** is fixed, but the elements of which are randomly assigned in the range $[-1, +1]$.

We run the adaptive algorithm (8) to prewhiten the observations and apply the whitening error

$$E_{wht} = \frac{1}{n^2} \left\| \mathbf{I} - \mathbf{U} \Sigma_\mathbf{x} \mathbf{U}^T \right\|_F^2 \quad (13)$$

where $\|\cdot\|_F$ denotes the Frobenius norm, to measure its performance. Fig. 1 plots the whitening error averaged over 500 independent runs (in each run, both the mixing matrix and the sources are randomly generated). Also plotted is the average curve of the existing RLS whitening algorithm [12] with a forgetting factor 0.993. Clearly, (8) performs more efficiently due to usage of the time-varying forgetting factor, and it can achieve almost perfect whitening (after about 2000 iterations), which is of particular importance in BSS or otherwise the orthonormality of the separating matrix would not hold any more and the separation would degrade.

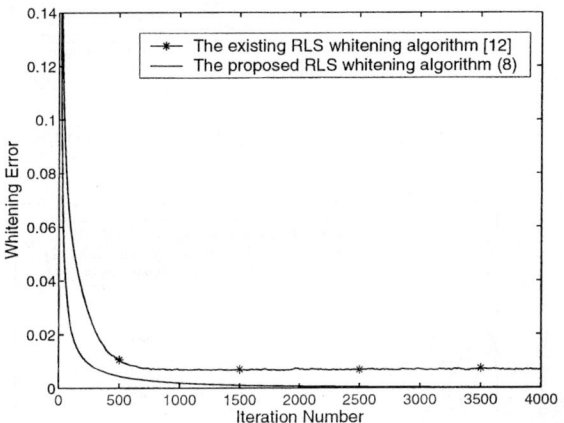

Fig. 1. The whitening errors averaged over 500 independent runs

To evaluate the performance of the BSS algorithm, we use the cross-talking error [6], [9]

$$E_{ct} = \frac{1}{n}\sum_{i=1}^{n}\left(\sum_{j=1}^{n}\frac{|c_{ij}|}{\max_k |c_{ik}|} - 1\right) + \frac{1}{n}\sum_{j=1}^{n}\left(\sum_{i=1}^{n}\frac{|c_{ij}|}{\max_k |c_{kj}|} - 1\right) \qquad (14)$$

and the permutation error [2], [3]

$$E_{perm} = \frac{1}{n^2}\left\|\mathbf{I} - \mathrm{abs}(\mathbf{C}^T)\mathrm{abs}(\mathbf{C})\right\|_F^2 \qquad (15)$$

where $\mathbf{C} = \mathbf{WUA} = \{c_{ij}\}$, and $\mathrm{abs}(\mathbf{C})$ is the matrix of absolute values of the elements of $\mathbf{C}$. Both errors vanish when perfect separation is achieved.

As comparison, we run the non-negative PCA algorithm (4) and the natural-gradient-based one (12) at the same time. All the three algorithms employ the newly proposed whitening algorithm (8) to obtain the whitened vector $\mathbf{v}_t$. The identical initial matrices $\mathbf{W}_0 = diag(sign(\mathbf{v}_1))$ are chosen such that all of the outputs are non-negative at the beginning. To understand the behaviors of the

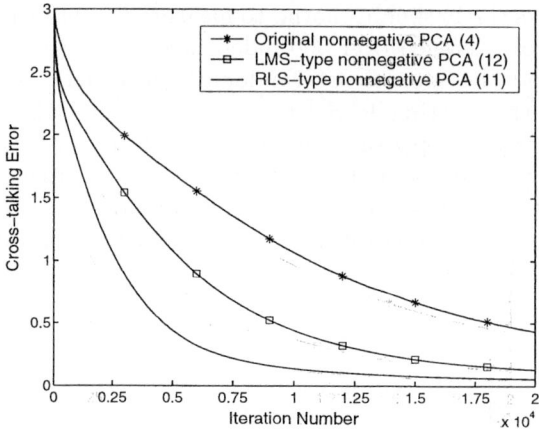

Fig. 2. The cross-talking errors averaged over 500 independent runs

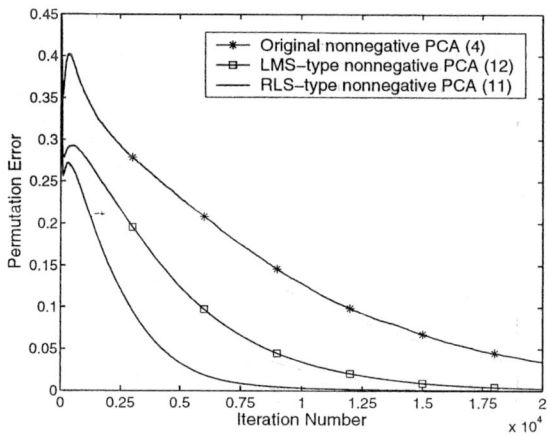

Fig. 3. The permutation errors averaged over 500 independent runs

basic algorithms as closely as possible, we use constant learning rates $\eta_t = 0.01$ and constant forgetting factor $\lambda = 0.98$. Another 500 independent runs are executed, and the average cross-talking errors and permutation errors are depicted in Fig. 2 and Fig. 3, respectively. Obviously, the non-negative PCA algorithm (11) works most satisfactorily, which agrees with what we expect, since it is based on the natural gradient learning (9), and at the same time it is an RLS-type algorithm. From these two figures, we can also see that the natural-gradient-based LMS algorithm (12) outperforms the non-negative PCA algorithm (4) which applies an ordinary gradient learning.

5 Conclusions and Discussions

In this paper, we first developed an RLS-type whitening algorithm, and then presented an adaptive natural-gradient-based RLS implementation of the non-negative PCA algorithm, by incorporating our previously proposed algorithm for prewhitened BSS [9]. Computer simulations showed that our new implementation performs satisfactorily.

It should be pointed that the proposed prewhitening algorithm (8) can be used widely in other BSS algorithms and other signal processing problems, where stationary signals need to be whitened in the sense of covariance matrix.

Finally, for the non-negative BSS problem, only a rotation matrix is required to achieve separation of the sources provided the observations are prewhitened, and a Givens matrix works for the case of two sources [2], [3]. If there are more than two sources, a Householder matrix does. Therefore, the non-negative PCA criterion (3) can be rewritten as

$$J_3(\mathbf{h}) = E\left\{\left\|\mathbf{v}_t - \left(\mathbf{I} - \frac{2\mathbf{h}\mathbf{h}^T}{\mathbf{h}^T\mathbf{h}}\right)\mathbf{g}\left((\mathbf{I} - \frac{2\mathbf{h}\mathbf{h}^T}{\mathbf{h}^T\mathbf{h}})\mathbf{v}_t\right)\right\|^2\right\} \tag{16}$$

from which new simplified and more efficient batch or adaptive algorithms may be derived.

References

1. Hyvarinen, A., Karhunen, J., Oja, E.: Independent Component Analysis. Wiley, New York (2001)
2. Plumbley, M.: Algorithms for Nonnegative Independent Component Analysis. IEEE Trans. Neural Networks 14 (2003) 534-543
3. Plumbley, M., Oja, E.: A Nonnegative PCA Algorithms for Independent Component Analysis. IEEE Trans. Neural Networks 15 (2004) 66-76
4. Oja, E.: The Nonlinear PCA Learning in Independent Component Analysis. Neurocomputing 17 (1997) 25-46
5. Haykin, S.: Adaptive Filter Theory. 3rd edn. Prentice-Hall, Englewood Cliffs New Jersey (1996)
6. Karhunen, J., Pajunen, P., Oja, E.: The Nonlinear PCA Criterion in Blind Source Separation: Relations with Other Approaches. Neurocomputing 22 (1998) 5-20
7. Amari, S.: Natural Gradient Works Efficiently in Learning. Neural Computation 10 (1998) 251-276
8. Amari, S.: Natural Gradient Learning for Over- and Under-Complete Bases in ICA. Neural Computation 11 (1999) 1875-1883
9. Zhu, X.L., Zhang, X.D.: Adaptive RLS Algorithm for Blind Source Separation Using a Natural Gradient. IEEE Signal Processing Letters 9 (2002) 432-435
10. Lee, D.D., Seung, H.S.: Learning the Parts of Objects by Nonnegative Matrix Factorization. Nature 401 (1999) 788-791
11. Cardoso, J.F., Laheld, H.: Equivariant Adaptive Source Separation. IEEE Trans. Signal Processing 44 (1996) 3017-3029
12. Zhu, X.L., Zhang, X.D., Ye, J.M.: Natural Gradient-Based Recursive Least-Squares Algorithm for Adaptive Blind Source Separation. Science in China Ser. F 47 (2004) 55-65

Progressive Principal Component Analysis

Jun Liu[1], Songcan Chen[1], and Zhi-Hua Zhou[2]

[1] Department of Computer Science and Engineering,
Nanjing University of Aeronautics and Astronautics,
Nanjing 210016, China
maurice_nt@sohu.com, s.chen@nuaa.edu.cn
[2] National Laboratory for Novel Software Technology,
Nanjing University, Nanjing 210093, China
zhouzh@nju.edu.cn

Abstract. Principal Component Analysis (PCA) is a feature extraction approach directly based on a whole vector pattern and acquires a set of projections that can realize the best reconstruction for an original data in the mean squared error sense. In this paper, the progressive PCA (PrPCA) is proposed, which could progressively extract features from a set of given data with large dimensionality and the extracted features are subsequently applied to pattern recognition. Experiments on the FERET database show its face recognition performance is better than those based on both $E(PC)^2A$ and FLDA.

1 Introduction

The traditional PCA is an effective approach of extracting features and dimensionality reduction and has partially successfully been applied to pattern recognition such as face recognition with one training (face) pattern per person [1]. It operates directly on a whole vector pattern to extract so-needed global features for subsequent recognition by using a set of found projectors from a given training pattern set. The extracted features can maximally preserve or reconstruct original pattern information in the mean squared error sense. However, when the dimensionality of given pattern is very large, for example, 1000 or larger, extracting features directly from these large dimensional patterns *once* exists some processing difficulty such as computational complexity for large scale covariance matrix constructed by the training set. With the divide-and-conquer technique, the subpattern-based PCA (SpPCA) [2] eliminates such a difficulty via first dividing the whole pattern into K equally-sized subpatterns {Sp1, Sp2, ..., SpK} as shown Fig. 1 and then separately performing PCA on each training subpattern set of sharing the same (feature) components so as to obtain a classification performance gain and robustness to partially missing subpatterns. However, SpPCA only performs independently PCAs in different divisions and thus disregards useful contextual information among different subpattern sets, which produces unfavorable influence on classification performance. In this paper, a novel PCA (PrPCA) is developed to avoid the problem but still to preserve its robustness to partially missing information. It operates progressively or subpattern-by-subpattern on given pattern to extract features from a set of large dimensional patterns as shown Fig.1.

2 Proposed PrPCA

Except for extracting features in a *progressive* mode, the idea behind the proposed PrPCA is almost identical to the SpPCA so that PrPCA can keep both its simplicity and robustness but still possesses better recognition accuracy as shown in Table 1. The following is a formulation for the PrPCA.

PrPCA consists of two steps. The first step is a partition step, i.e., an original whole pattern is partitioned into a set of equally-sized (actually, not limited to this) subpatterns in non-overlapping ways and then all those subpatterns sharing the same original feature components are respectively collected from the training set to compose corresponding training subpattern sets. In the second step, PCA is performed progressively on these subpattern sets as shown Fig.1.

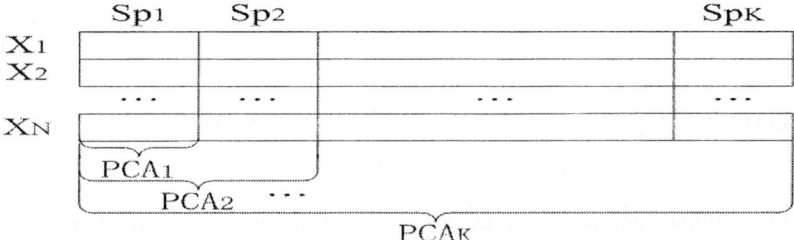

Fig. 1. Progressive feature extraction of PrPCA

More specifically, we are given a set of training patterns with large dimensionality:

$$X = \{X_1, X_2, \cdots, X_N\} \quad (1)$$

with each column vector X_i ($i=1, 2, \ldots, N$) having m dimensions. Now according to the first step, an original whole pattern is first partitioned into K d-dimensional subpatterns in a nonoverlapping way and reshaped into a d-by-K matrix

$$X_i = (X_{i1}, X_{i2}, \cdots, X_{iK}) \quad (2)$$

with

$$X_{ij} = (x_{i((j-1)d+1)}, \cdots, x_{i(jd)})^T \quad (3)$$

being the jth subpattern of X_i and $i=1,2,\ldots, N$ and $j=1,2,\ldots,K$. Then according to the second step, we perform PCA progressively from left to right as shown in Fig. 1, i.e., PCA$_1$ performs on the first subpattern set $\{X_{11}, X_{21}, \ldots, X_{N1}\}$ and then obtains corresponding reduced subpatterns $\{P_{11}, P_{21}, \ldots, P_{N1}\}$ which are, in turn, used to augment the subpattern dimensionality of the next set to construct new progressive training set

$$\left\{ \begin{bmatrix} P_{11} \\ X_{12} \end{bmatrix}, \begin{bmatrix} P_{21} \\ X_{22} \end{bmatrix}, \cdots, \begin{bmatrix} P_{N1} \\ X_{N2} \end{bmatrix} \right\} \quad (4)$$

on which PCA$_2$ is performed, ..., and so on. Finally we will terminate to some PCA$_k$ ($k \leq K$) depending on the predefined recognition accuracy. Here for completeness of description, we give a concise introduction to the PCA as follows:

Without loss of generality, let us abuse a notation of the training set

$$X = \{X_1, X_2, \cdots, X_N\} \tag{5}$$

and perform PCA on it. Define a covariance matrix for the X,

$$C = \frac{1}{N} \sum_{i=1}^{N} (X_i - \overline{X})(X_i - \overline{X})^T \tag{6}$$

where

$$\overline{X} = \frac{1}{N} \sum_{i=1}^{N} X_i \tag{7}$$

is the sample mean. PCA can find a set of optimal projection vectors

$$\Phi = (\varphi_1, \varphi_2, \cdots, \varphi_l) \tag{8}$$

to ensure the minimal reconstruction error by solving the eigenvalue-eigenvector system

$$C\Phi = \Phi \Lambda \tag{9}$$

under the constraints that

$$\Phi^T \Phi = I \tag{10}$$

where I is an identity matrix and

$$\Lambda = diag(\lambda_1, \lambda_2, \cdots, \lambda_l) \tag{11}$$

a diagonal matrix composed by the first l largest non-negative eigenvalues of C in a descending order and thus their corresponding first l eigenvectors compose the Φ. l is the smallest value that is determined by the criterion

$$\sum_{i=1}^{l} \lambda_i \bigg/ \sum_{i=1}^{N} \lambda_i \geq \theta \tag{12}$$

Here θ is a user-predetermined parameter. In this way, the final obtained PCA$_k$ ($k \leq K$) projections can be used to extract features for any pattern and subsequently those extracted features are in turn used to pattern recognition with 1-nearest neighbor (1NN). To verify the feasibility of the PrPCA, we use a real FERET face database [3] to carry out the following experiments in the next section.

3 Experimental Results

3.1 Dataset

The FERET database comprises 400 gray-level frontal view face images from 200 persons, with the size of 256x384. There are 71 females and 129 males, each of whom has two images (*fa* and *fb*) with different race, different gender, different age, different expression, different illumination, different occlusion, different scale, etc. The *fa* images are used for training while the *fb* images for testing. In our experiments, all faces are normalized to satisfy some constraints so that each face could be appropriately cropped. Those constraints include that the line between the two eyes is parallel to the horizontal axis, the inter-ocular distance (distance between the two eyes) is set to a fixed value, and the size of the image is fixed. Here the eyes are manually located and after a series of rotating and resizing, the cropped image size is 60×60 pixels and the inter-ocular distance is 28 pixels. In the experiments, we use two partition ways for the whole pattern: sequent (seq) and random (rand) and investigate their influence on recognition results.

3.2 Results and Conclusions

We compare the obtained classification results with those of $E(PC)^2A$ (best 85.5%)[1][Appendix] and FLDA(best 86.5%) [4].

Table 1 shows the recognition accuracies (RAs) of both PrPCAs and SpPCAs for different partitions with changeable sizes of blocks and the testing patterns including differently missing blocks.

Table 1. Recognition accuracies (RA) (%) and percentages of kept blocks (POKB). Notice: *a*: θ are all set to 0.99999 in this row to obtain the RAs and POKBs; *b*: θ values corresponding to achieve the best RAs.

Block size	3×3		3×5		5×5	
	RA	POKB	RA	POKB	RA	POKB
PrPCA(seq)	86.5	93.75	86.5	92.5	86.5	92.36
PrPCA(rand)	89	41.3	88	37.5	88	38.9
SpPCA[a]	85	100	85	100	85	100
SpPCA	87(0.9)[b]	100	87.5(0.8)[b]	100	87(0.8)[b]	100

Fig.2 shows three randomly missing block cases able to achieve corresponding best RAs with different block-size partitions.

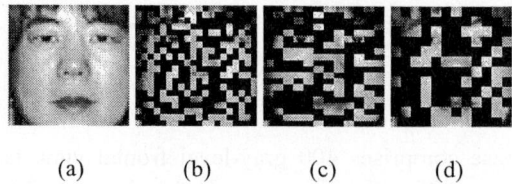

(a) (b) (c) (d)

Fig. 2. Randomly kept block percentages and corresponding best recognition accuracies (%). (a) Original face; (b) 41.3 (3×3) and 89; (c) 37.5 (3×5) and 88; (d) 38.9 (5×5) and 88. Note: the dark blocks represent discarded or missing parts.

From the table 1, we can observe that the RAs (88-89%) of PrPCA with random partitions are higher than the RAs (86.5% and 87.5%) of both the counterparts with sequential partitions, the SpPCAs and FLDA[4]. At the same time, the tests on the SpPCA indicate the RAs (85%s) with preserving all original information are not necessarily better than the ones (87-87.5%) with partial original information preserved (θ values are respectively set to 0.9, 0.8 and 0.8 as listed in Table 1). Furthermore, in order to test the robustness of the proposed PrPCA, we adopt a randomly discarding image block way for the faces to be recognized (as shown in Fig. 2) to examine its RAs and find that it can still achieve the best RA of 88% in the random partition even when blocks of 61.1% in faces are missed (equivalently, 38.9% blocks are kept.). These results confirm feasibility and effectiveness of the PrPCA and finally the idea used here can be applied to other similar problems.

Acknowledgements. This work was supported by the National Natural Science Foundation of China under the Grant No. 60271017, the National Outstanding Youth Foundation of China under the Grant No. 60325207, the Jiangsu Science Foundation under the Grant No. BK2002092, the Jiangsu Science Foundation Key Project, the *QingLan* Project Foundation of Jiangsu Province, and the Returnee Foundation of China Scholarship Council. Portions of the research in this paper use the FERET database of facial images collected under the FERET program.

References

1. Chen, S.C., Zhang, D.Q., Zhou, Z.-H.: Enhanced $(PC)^2A$ for face recognition with one training image per person. Pattern Recognition Letters, 2004, in press
2. Chen, S.C., Zhu, Y. L.: Subpattern-based principle component analysis. Pattern Recognition 37 (2004) 1081-1083
3. Phillips, P.J., Wechsler, H., Huang, J., Rauss, P.J.: The FERET database and evaluation procedure for face-recognition algorithms. Image and Vision Computing 16 (1998) 295-306
4. Chen, S.C., Liu, J., Zhou, Z.-H.: Making FLDA applicable to face recognition with one sample per person. Pattern Recognition 37 (2004) 1553-1555

Appendix: E(PC)²A

The aim of E(PC)²A is to augment the sample size by constructing the original face's n-order projections. Let $I(x,y)$ be an intensity image of size $N_1 \times N_2$, where $x \in [1, N_1]$, $y \in [1, N_2]$, and $I(x,y) \in [0,1]$, we can define the second-order projection of the original image as

$$P_2(x, y) = \frac{V_2(x) H_2(y)}{J_{mean}} \quad (13)$$

where J_{mean} is the mean value of $J(x,y)$ which is defined as the square of $I(x,y)$, that is, $J(x,y) = I(x,y)^2$, and V_2 and H_2 are defined respectively as:

$$V_2(x) = \frac{1}{N_2} \sum_{y=1}^{N_2} J(x, y) \quad (14)$$

$$H_2(y) = \frac{1}{N_1} \sum_{x=1}^{N_1} J(x, y) \quad (15)$$

Then, through combining the original image with its first and second-order projections, a new projection-combined image, i.e. $I_2(x, y)$ as shown in Eq.A.4, can be obtained, where α and β are parameters used to control the bias of the projections.

$$I_2(x, y) = \frac{I(x, y) + \alpha P_1(x, y) + \beta P_2(x, y)}{1 + \alpha + \beta} \quad (16)$$

Extracting Target Information in Multispectral Images Using a Modified KPCA Approach

Zhan-Li Sun[1,2] and De-Shuang Huang[1]

[1] Intelligent Computing Group, Hefei Institute of Intelligent Machines, Chinese Academy of Sciences, P.O.Box 1130, Hefei, Anhui, 230031, China
[2] Department of Automation, University of Science and Technology of China, Hefei, Anhui, 230026,China
{sun_zhl}@iim.ac.cn

Abstract. In this paper, a modified kernel principal component analysis (KPCA) approach is proposed to extract target information in multispectral images. The advantage of this method is to be able to extract target information with fairly small computation complexity compared to the standard KPCA when a large number of input samples need to be processed. Finally, some experimental results demonstrate that our proposed approach is effective and efficient for analyzing and interpreting the multispectral images.

1 Introduction

Extracting target information included in the images is one of the most important tasks for the scene interpretation of multispectral images. In the past, target information was usually derived from the second or third principal component (PC) images obtained after the PCA for multispectral images [1]. This method, however, can only guarantee that those linear features can be extracted while the nonlinear features may be lost. The kernel principal component analysis (KPCA) [2] can be used to extract nonlinear features in images by the kernel trick. Compared to other nonlinear feature extraction approaches, e.g., Hebbian networks [3], autoassociative multilayer perceptrons [4], etc, the KPCA has the advantage of no nonlinear optimization being addressed; and it is essentially involved in only linear algebraic operation as simple as the PCA.

Unfortunately, neither the KPCA nor its new algorithms [5,6] proposed recently are difficult to directly process multispectral images with the single KPCA when a large number of samples need to be processed (Note that this point can be verified by the Matlab codes provided by the authors.). Therefore, in this paper, a modified KPCA approach, in which the kernel matrix computation is in batch or block model, is proposed to tackle the above problem. It can efficiently extract target information without storing and computing the big kernel matrix that is encountered in the standard KPCA

The remainder of this paper is organized as follows. The fundamental principle of extracting target information in multispectral images by the modified KPCA is introduced in Section 2. Experimental results and related discussions are presented in Section 3. Finally, Section 4 concludes some conclusive remarks.

2 Main Results

Let us briefly introduce the problem background at first. Assume that an observation data matrix X is obtained from p multispectral images $X_j, (j = 1, \cdots p)$, while each column of X, $x_i(i = 1, 2, \cdots, q)$, is usually referred to as an observation. In the following, we will extract target information from all observations in X by a modified KPCA approach. The main steps for the method are briefly introduced as follows.

Since the size of the observations in the data matrix X is too large, it is infeasible to perform the KPCA directly use all observations due to the limitation of compute memory size. Therefore, we firstly divide the matrix X into several patches, e.g., choose l observations from X in sequence as an input subset $S_i(i = 1, 2, \cdots, c)$ that forms a data matrix for the KPCA, where $c = q/l$. Subsequently, we can define a $l \times l$ kernel matrix K_{s_i} for every set $S_i(i = 1, 2, \cdots, c)$ with elements

$$K_{ij} = k(x_i, x_j) \tag{1}$$

where x_i and x_j are the sample vectors in the set $S_i(i = 1, 2, \cdots, c)$, and $k(\cdot)$ is a kernel function. Radial basis function and polynomial function can be usually used as kernel functions.

After all kernel matrices $K_{s_i}(i = 1, 2, \cdots, c)$ are computed, we can derive the mean kernel matrix

$$K = (K_{s_1} + K_{s_2} + \cdots + K_{s_c}) \tag{2}$$

Correspondingly, the kernel matrix $\widetilde{K}$ of centered data in the Hilbert space can be obtained by

$$\widetilde{K} = K - A_l K - K A_l + A_l K A_l \tag{3}$$

where $(A_l)_{ij} = 1/l$. From eqn. (3), we can obtain the eigenvalue decomposition of the matrix $\widetilde{K}$ as follows:

$$l\lambda\alpha = \widetilde{K}\alpha \tag{4}$$

where λ is an eigenvalue of $\widetilde{K}$ and α an eigenvector expansion coefficient. The nth eigenvector expansion coefficient α^n should satisfy the constraint:

$$\lambda_n(\alpha^n, \alpha^n) = 1 \tag{5}$$

where λ^n is the nth eigenvalue of $\widetilde{K}$.

At last, a suitable number of eigenvectors is chosen according to requirement. Let us denote S the mean of all sets $S_i(i = 1, 2, \cdots, c)$

$$S = (S_1 + S_2 + \cdots, S_c)/c \tag{6}$$

Then the nth principal component (PC) image is given approximatively by projecting input vector matrix X on the direction of the nth eigenvector:

$$(V^n \cdot \Phi(x)) = \sum_{t=1}^{l} \alpha_t^n k(x_t, x_i), \ i = 1, 2, \cdots, q \tag{7}$$

where $\phi(\cdot)$ is a nonlinear mapping, V^n is the nth eigenvector, x_i denotes an observation vector of X, $x_t(t=1,2,\cdots,l)$ is the sample vector in S.

Consequently, the procedure of nonlinear feature extraction by the modified KPCA is summarized as follows:

Step 1. Given the number of observations included in subsets, then devide the data matrix X into subsets $S_i(i=1,2,\cdots,c)$; compute kernel matrices $K_{s_i}(i=1,2,\cdots,c)$ for every set $S_i(i=1,2,\cdots,c)$ using eqn. (1), respectively.

Step 2. Compute the mean kernel matrices K and $\widetilde{K}$ by the eqns. (2) and (3).

Step 3. Compute α_n according to eqns. (4) and (5).

Step 4. Given the number of the principal components (PCs), project all observations on the direction of the PCs using eqns. (6) and (7).

Step 5. Evaluate the PC images by some performance indices, adjust the number of subset S_i and σ^2 of radial basis function, then go to Step 1.

3 Experiment Results

In this section, we will present the experimental results by the modified KPCA approach for multispectral images, and related performance indices and discussions are also given here. The corresponding six original multispectral images used in our experiments are obtained from George Washington University, as depicted in Fig.1, respectively. Note that each image contains 192×192 pixels. Thus the observation data matrix X obtained from the six multispectral images is a $6 - by - 192 \times 192$, i.e., $p = 6, q = 192 \times 192$.

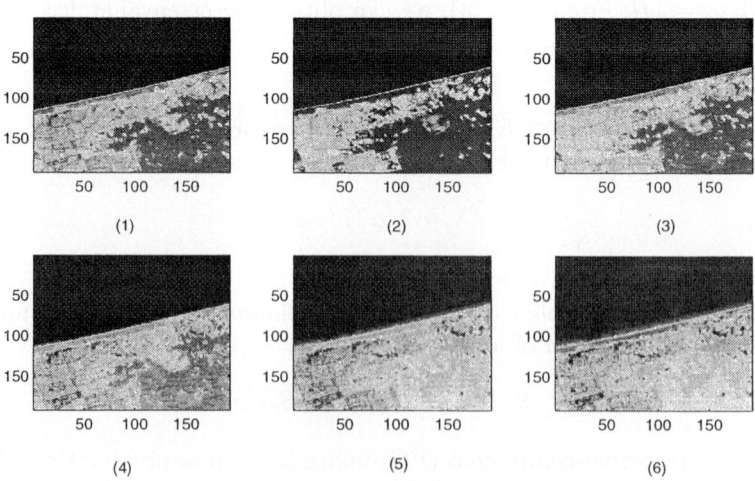

Fig. 1. The six original multispectral images

3.1 Experimental Results

Firstly, in order to be easily compared to later experimental results by the modified KPCA, we use the PCA to analyze the multispectral images. In order to save space, here we don't give the PC images obtained after the PCA. Note that the number of the PC images obtained must be less than or equal to the one of the original multispectral images.

Further, we use the modified KPCA to extract the target information. After a continuous adjustment, the final parameters in experiments by our method are given in Table 1, where σ^2 is the shape parameter of radial basis function. As a result, the twelve PC images obtained after our proposed method are shown in Fig.2.

Table 1. The final parameters in experiments by our method

l	c	σ^2
12	3072	31623

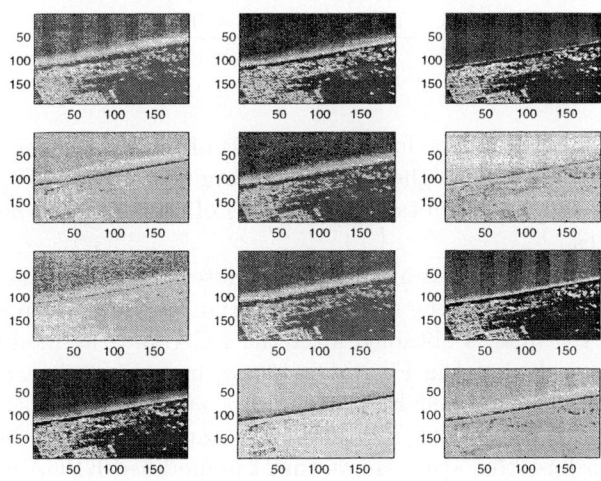

Fig. 2. The twelve PC images obtained by the modified KPCA approach

Note that more PC images are obtained than original images by the modified KPCA, which is very significant because target information is usually retained in the last few PC images. From Fig.2, it can be found that more target information is displayed in the PC images compared to the original multispectral images, especially in the upper part of these PC images.

3.2 Performance Analysis and Related Discussion

In the following, the signal to noise ratios and the normalized variances will be chosen as the performance indices to evaluate the resulted PC images.

(1). Signal to noise ratio:

The definition of SNR introduced in literature [7] is formulated as follows:

$$SNR = \delta_i^2 / \delta_n^2 \tag{8}$$

where $\delta_i^2, i = 1, 2, \cdots, n$ are the variances in a decreasing order obtained after the PCA and the modified KPCA. As a result, the SNRs comparison corresponding to the PCA and the modified KPCA (MKPCA) is shown in Table 2, respectively.

Table 2. The SNRs comparison corresponding to the PCA and the modified KPCA

PCA	1^{st}PC	2^{nd}PC	3^{rd}PC	4^{th}PC	5^{th}PC	6^{th}PC
SNR	5084	48.604	18.004	9.1421	1.878	1
MKPCA	1^{st}PC	2^{nd}PC	3^{rd}PC	4^{th}PC	5^{th}PC	6^{th}PC
$SNR(\times e+012)$	3.3726	2.1559	1.9113	1.8159	1.7703	1.5911
MKPCA	7^{th}PC	8^{th}PC	9^{th}PC	10^{th}PC	11^{th}PC	12^{th}PC
$SNR(\times e+012)$	1.435	1.2691	1.0057	0.92176	0.90525	

From the Table 2, it can be found that SNRs of the PC images obtained after the modified KPCA are far higher than ones obtained after the PCA. Therefore, high quality PC images are obtained by means of our proposed method.

(2). Variances

Since information included in an image is determined to some extent by the variance amount, therefore, the normalized variances comparison corresponding to the resulted PC images obtained after the PCA and the modified KPCA is also given here, as depicted in Fig.3. Obviously, from Fig.3, it can be seen that the variances for the modified KPCA decomposition decrease slowly than the one for the PCA. Since target information is usually retained in the last few PC images, therefore, this again shows that the modified KPCA method can be used to efficiently extract the target information in multispectral images.

3.3 Conclusions

In this paper, we extract the target information included in multispectral images by a modified KPCA. The experimental results demonstrated that our proposed method was efficient and feasible for this task. Future research work will include how to interpret the scene of spectral images processed.

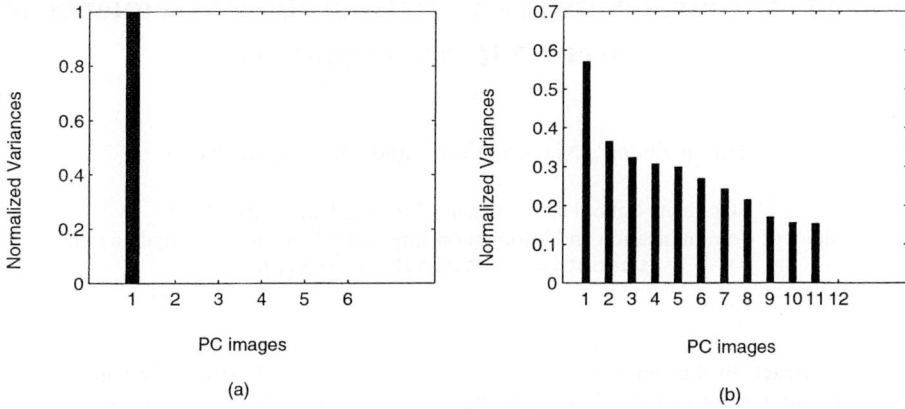

Fig. 3. The normalized variances comparison corresponding to the resulted PC images obtained after the PCA and the modified KPCA. (a) PCA; (b) the modified KPCA

Acknowledgments. The authors gratefully acknowledge Professor Harold Szu at Geroge Washington University for providing the spectral images.

References

1. Chitroub, S., Houacine, A., Sansal, B.: Unsupervised Learning Rules for POLSAR Images Analysis. Neural Networks for Signal Processing, Proceedings of the 2002 12th IEEE Workshop on (2002) 567-576
2. Schölkopf, B., Smola, A., Müller, K.-R.: Nonlinear Component Analysis as a Kernel Eigenvalue. Neural Comput., no. 10 (1998) 1299-1319
3. Oja, E.: A Simplified Neuron Model as A Principal Component Analyzer. J.Math. Biology, no.15 (1982) 267-273
4. Diamantaras, K.I., Kung, S.Y.: Principal Component Neural Networks. New York: Wiley (1996)
5. Kim, K.I., Franz, M.O., Schölkopf, B.: Image Modeling Based on Kernel Principal Component Analysis, http://www.kyb.tuebingen.mpg.de/publication.html
6. Schölkopf, B., Mika, S., Burges, C.J.C., Knirsch, P., Müller, K.-R., Smola, A.J.: Input Space VS. Feature Space in Kernel-Based Methods. IEEE Trans. Neural Networks, no. 5 (1999) 1000-1017
7. Ready, P., Wintz, P.: Information Extraction, SNR Improvement, and Data Compression in Multispectral Imagery. IEEE Trans. Communications, Vol. 21, no. 10 (1973) 1123-1131

Use PCA Neural Network to Extract the PN Sequence in Lower SNR DS/SS Signals

Tianqi Zhang[1], Xiaokang Lin[1], and Zhengzhong Zhou[2]

[1] Graduate School at Shen zhen, Tsinghua University 518055
[2] Shool of Communication and Information Eng., UEST of China Chengdu 610054
zhangtianqi@tsinghua.org.cn

Abstract. In this paper, we firstly propose an approach of discrete Karhunen-Loeve transformation to blind estimation of the PN (Pseudo Noise) sequence in lower SNR DS/SS signals. As the K-L approach is based on the decomposition of autocorrelation matrix, it has computational defects when the signal vectors became longer. In order to overcome the defects of K-L approach, we choose the PCA (Principal Components Analysis) neural networks to extract the PN sequence. Theoretical analysis and experimental results are provided to show that the approach can work well on lower SNR input DS/SS signals. The proposed method can be extended to the case of DS/CDMA (Direct Sequence Code Division Multiple Access) too.

1 Introduction

Since the spectrum density of direct sequence spread spectrum (DS/SS, DS) signals are always lower than that of the noises in communication channel, the DS signals have obvious capability of anti-jamming and lower probability interception. However, if some signal parameters are known, such as period and chip rate of the PN (Pseudo Noise) sequence, blind estimation of the PN sequence from DS signals directly is possible [4][5], which can be useful in communication management and military communication scout. A method of time-delayed autocorrelation was proposed to de-spread DS signal without the PN sequence [1]. Because some spectral correlation computations were required, it was difficult to carry out in real-time. Furthermore, it only did de-spread DS signal without the PN sequence, but it didn't utilize or analyze any structure information of the PN sequence. We used PCA (Principal Component Analysis) NN (Neural Networks) to get the PN sequence blind estimation of the DS signal in [4]. Several days ago, we incidentally found a letter [5] using the method which had an analogy to the method in [4] to discuss the interesting problem of PN sequence estimation. But the methods both in [4] and [5] did need to estimate the synchronous point between the symbol waveform and the observation windows. It is very hard to estimate the synchronous point from lower SNR DS signals yet.

In this paper, we first express the PN sequence blind estimation in Karhunen-Loeve transformation. Furthermore, we implement the estimation by using the PCA NN to extract the principal components of DS signal. We overcome difficulties about computational memory size and speed of the K-L method in case of longer received signal vectors. The approaches in this paper don't need to search the synchronous point between symbol waveform and PN sequence waveform, either. Therefore the approaches remedy the defects of method in [4] and [5]. Getting PN sequence, we can receive DS signal on our own, and realize blind de-spread of DS signal at last.

In this paper, we assume that the PN sequence repetition period is exactly equal to one symbol period, and one symbol is synchronously modulated by one period of PN sequence according to the fact of engineering. For long code DS signals, we have another paper to discuss it.

2 Signal Model

The base band DS signal $x(t)$ corrupted by the white Gaussian noise $n(t)$ with the zero mean and σ_n^2 variance can be expressed as [1]

$$x(t) = s(t - T_x) + n(t) \tag{1}$$

Where $s(t) = d(t)p(t)$ is the DS signal, $p(t) = \sum_{j=-\infty}^{\infty} p_j q(t - jT_c)$, $p_j \in \{\pm 1\}$ is the PN sequence, $d(t) = \sum_{k=-\infty}^{\infty} m_k q(t - kT_0)$, $m_k \in \{\pm 1\}$ is uniformly distributed with $E[m_k m_l] = \delta(k - l)$, $\delta(\cdot)$ is the Dirac function, $q(t)$ denotes a pulse chip with period of T (T may be T_0 or T_c here). Where $T_0 = NT_c$, N is the length of PN sequence, T_0 is the period of PN sequence, T_c is the chip duration, T_x is the random time delay and uniformly distributed on the $[0, T_0]$.

It is well known that the PN sequence and synchronization are required to de-spread the received DS signals. But we only have the knowledge of T_0 and T_c.

3 Subspace Analysis Based on K-L Transformation

The received DS signal is sampled and divided into non-overlapping temporal windows, the duration of which is T_0. Then one of the received signal vector is

$$\mathbf{X}(k) = \mathbf{s}(k) + \mathbf{n}(k), \quad k = 1, 2, 3, \cdots \tag{2}$$

Where $\mathbf{s}(k)$ is the k-th vector of useful signal, $\mathbf{n}(k)$ is the white Gaussian noise vector. The dimension of vector $\mathbf{X}(k)$ is $N = T_0 / T_c$. If the random time-delay is T_x, $0 \leq T_x < T_0$, $\mathbf{s}(k)$ may contain two consecutive symbol bits, each modulated by a period of PN sequence, i.e.

$$\mathbf{s}(k) = m_k \mathbf{p}_1 + m_{k+1} \mathbf{p}_2 \tag{3}$$

Where m_k and m_{k+1} are the two consecutive symbol bits, $\mathbf{p}_1$ ($\mathbf{p}_2$) is the right (left) part of the PN sequence waveform.

According to K-L transformation, we definite the $\mathbf{p}_1$ and $\mathbf{p}_2$ as

$$\mathbf{u}_i^T \mathbf{u}_j = \delta(i-j) \; , \; i,j=1,2 \qquad (4)$$

Where $\mathbf{u}_1$ and $\mathbf{u}_2$ are ortho-normal vectors. From $\mathbf{u}_1$ and $\mathbf{u}_2$, we have

$$X(k) = m_k \|\mathbf{p}_1\| \mathbf{u}_1 + m_{k+1} \|\mathbf{p}_2\| \mathbf{u}_2 + \mathbf{n}(k) \qquad (5)$$

The autocorrelation matrix of $X(k)$ may be estimated as

$$\hat{\mathbf{R}}_X(M) = \frac{1}{M} \sum_{i=1}^{M} X(i) X^T(i) \qquad (6)$$

Assume $s(k)$, $n(k)$ are mutually independent, substitute Eq.(5) into Eq.(6) yields

$$\hat{\mathbf{R}}_X(\infty) = \mathbf{U}_s \mathbf{\Lambda}_s \mathbf{U}_s^T + \mathbf{U}_n \mathbf{\Lambda}_n \mathbf{U}_n^T = \sigma_n^2 \left\{ \left(SNR \cdot \frac{T_0 - T_x}{T_c} \right) \cdot \mathbf{u}_1 \mathbf{u}_1^T + \left(SNR \cdot \frac{T_x}{T_c} \right) \cdot \mathbf{u}_2 \mathbf{u}_2^T + \mathbf{I} \right\} \qquad (7)$$

Where $\mathbf{I}$ is an identity matrix of dimension $N \times N$, the expectation of m_k is zero. The variance of m_k is σ_m^2, the symbol is uncorrelated from each other. The energy of PN sequence is $E_p \approx T_c \|\mathbf{p}\|^2$, the variance of $s(k)$ is $\sigma_s^2 = \sigma_m^2 E_p / T_0$, $SNR = \sigma_s^2 / \sigma_n^2$.

The row vectors of $\mathbf{U}_s$ and $\mathbf{U}_n$ are corresponding to the eigenvectors of eigenvalue $\lambda_{R1} = [1 + SNR \cdot (T_0 - T_x)/T_c] \sigma_n^2$, $\lambda_{R2} = (1 + SNR \cdot T_x / T_c) \sigma_n^2$ and σ_n^2, and exist $\lambda_{R1} \geq \lambda_{R2} > \sigma_n^2$. It is clear that the eigenvalues of $\hat{\mathbf{R}}_X(\infty)$ are dependent on T_x. When $T_x \neq 0$, the biggest eigenvalue is λ_{R1}, the sign of the corresponding eigenvector $\mathbf{p}_1 = \text{sign}(\mathbf{u}_1)$. The second biggest eigenvalue is λ_{R2} and the sign of the corresponding eigenvector $\mathbf{p}_2 = \text{sign}(\mathbf{u}_2)$. We can recover a period PN sequence from $\mathbf{p} = \mathbf{p}_2 + \mathbf{p}_1 = \text{sign}(\mathbf{u}_2) + \text{sign}(\mathbf{u}_1)$. When $T_x = 0$, λ_{R1} and $\mathbf{p}_1 = \text{sign}(\mathbf{u}_1)$ which denote a period of PN sequence.

Because the accumulation of $\hat{\mathbf{R}}_X$ estimation by Eq. (6) is a de-noise process, we can estimate the PN sequence by decomposition of $\hat{\mathbf{R}}_X$ even when SNR is lower. However, the memory size and computational speed will become problems when N becomes bigger. In the following context, we will propose to use the PCA NN to solve these problems.

4 Implementation of the Neural Networks

As in Fig.1, a two-layer PCA NN is used to estimate the PN sequence in DS signal blindly [3]. The number of input neurons is given by $N = T_0 / T_c$.

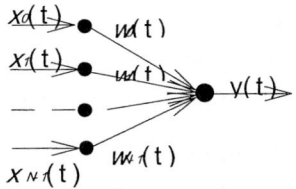

Fig. 1. Neural Networks

In the first case, assume $T_x = 0$, then we can divide the received signal into non-overlapping temporal windows, the duration of which is $T_0 = NT_c$. Assume one of the vectors is

$$\mathbf{X}(t) = \mathbf{X}(k) = \left[x(t), x(t-T_c), \cdots, x[t-(N-1)T_c]\right]^T = \left[x_0(t), x_1(t), \cdots, x_{N-1}(t)\right]^T \quad (8)$$

Where $\{x_i(t) = x(t - iT_c), i = 0,1,\cdots, N-1\}$ are sampled by one point per chip. The synaptic weight vector is

$$\mathbf{w}(t) = \left[w_0(t), w_1(t), \cdots, w_{N-1}(t)\right]^T \quad (9)$$

Where the sign of $\{w_i(t), i = 0,1,\cdots, N-1\}$ denotes the i-th bit of estimated PN sequence. The output layer of NN has only one neuron, its output is

$$y(t) = \sum_{i=0}^{N-1} w_i(t) x_i(t) = \mathbf{w}^T(t)\mathbf{X}(t) = \mathbf{X}^T(t)\mathbf{w}(t) \quad (10)$$

The original Hebbian algorithm of the NN is $\mathbf{w}(t+1) = \mathbf{w}(t) + \beta y(t)\mathbf{X}(t)$. However, we will use the algorithm

$$\mathbf{w}(t+1) = \mathbf{w}(t) + \beta y(t)\left[\mathbf{X}(t) - \mathbf{w}^T(t)\mathbf{X}(t)\mathbf{w}(t)\right] \quad (11)$$

Where β is a positive step-size parameter, it can be fixed or time-varied. In order to achieve good convergence performance, we express β as $\beta = 1/d_{t+1}$, and $d_{t+1} = Bd_t + y^2(t)$.

According to the theory of Liung's statistical analysis [2], if the step size $\beta \to 0$, the learning curves of Eq.(11) are the same as the dynamic curves of the following differential equation

$$d\mathbf{w}(t)/dt = y(t)\mathbf{X}(t) - y^2(t)\mathbf{w}(t) = \mathbf{X}(t)\mathbf{X}^T(t)\mathbf{w}(t) - \mathbf{w}^T(t)\mathbf{X}(t)\mathbf{X}^T(t)\mathbf{w}(t)\mathbf{w}(t) \quad (12)$$

As $\mathbf{X}(t)$ is statistically independent of $\mathbf{X}(t+k), (k>1, k=2,3,4,\cdots)$, and $\mathbf{X}(t)$ is independent of $\mathbf{w}(t)$. Statistically averaging two side of Eq.(12), yields

$$d\mathbf{w}(t)/dt = \mathbf{R}\mathbf{w}(t) - \mathbf{w}^T(t)\mathbf{R}\mathbf{w}(t)\mathbf{w}(t) \quad (13)$$

From Eq.(13), we can find out the following conclusions:

(a) Let $\mathbf{w}_f$ denote the asymptotically stable equilibrium point of Eq.(13) with a domain of attraction $D(\mathbf{w}_f)$; then the vector $\mathbf{w}(t)$ enters a compact subset A of the domain of attraction $D(\mathbf{w}_f)$ infinitely often. That means $\lim_{t\to\infty}\mathbf{w}(t)=\mathbf{w}_f$, with $P\{\mathbf{w}(t)\in A, \text{infinitely\_often}\}=1$. (b) Let λ_i ($\lambda_0\geq\lambda_1\geq\cdots\geq\lambda_{N-1}\geq 0$) and $\mathbf{C}_i$ denote the eigenvalues and its corresponding orthogonal normalized eigenvectors of the correlation matrix $\mathbf{R}_X=E[\mathbf{XX}^T]$ respectively, the result must be: $\mathbf{w}_f=\pm\mathbf{C}_0$.

The analysis and proof of (a) and (b) are detailed in [3]. We discuss the principal component analysis algorithm, and give the following points:
(1) The NN has two symmetrical attractors. (2) The probability that $\mathbf{w}(t)$ converge to $\pm\mathbf{C}_0$ is 1. (3) When the NN has converged at its stable points, we will obtain $\lambda_0(\infty)=\lambda_0$, and $\lim_{t\to\infty}\mathbf{w}(t)=\mathbf{w}_f=\pm\mathbf{C}_0$. The PN sequence is $\hat{\mathbf{P}}=\text{sign}[\mathbf{w}_f]=\pm\mathbf{p}$.

In the second case, we assume $T_x\neq 0$. According to the result of subspace analysis based on K-L transformation, we'll have to extract the first and second principal component before realizing the whole PN sequence estimation. When we extract the second principal component, we can use the NN in Fig.1 again, but its input signals are changed to $\mathbf{X}'(t)$: $\mathbf{X}'(t)=\mathbf{X}(t)-y(t)\mathbf{w}(t)$. When we input $\mathbf{X}'(t)$ to the NN in Fig.1, $\mathbf{w}(t)$ will converge to the second principal component vector of $\mathbf{X}(t)$.

5 Simulations and Conclusions

Simulations: In the figures is $SNR=-20\log 10(\sigma_n)$, because the useful signals in experiment are square waves with amplitude of ± 1. From it we get learning curves and performance curves when $Tx/To=1/5$.

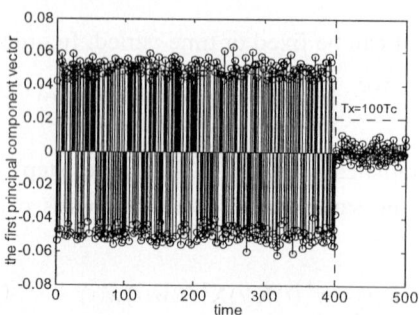
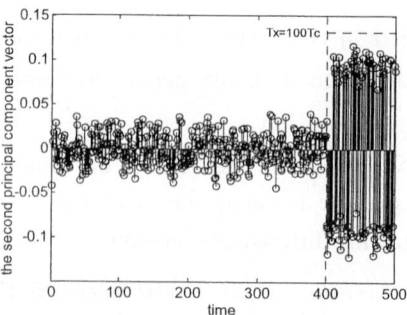

Fig. 2. First principal component vector   Fig.3. Second principal component vector

Fig.2 and Fig.3 denote the first and second principal component vector with $N=500bit$ at $Tx/To=1/5$ respectively. From them, we can estimate the Tx parameter of the received signals.

Fig.4 shows the learning curves of the NN, where $N=500bit$ (truncation of m sequence), $Tx=100Tc$. The curves from left to right correspond to the noise standard deviation $\sigma_n = 1.0, 2.0, \cdots, 10.0$. We can see the convergence property of the NN is very good. Fig.5 denotes the performance curves. It shows the time taken for the NN to perfectly estimate the PN sequence for lengths of *N=100bit* , *200bit* and *500bit* at *Tx/To=1/5*. Under the same condition, the longer of the PN sequence, the performance is better.

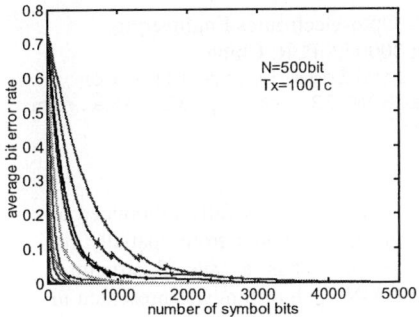
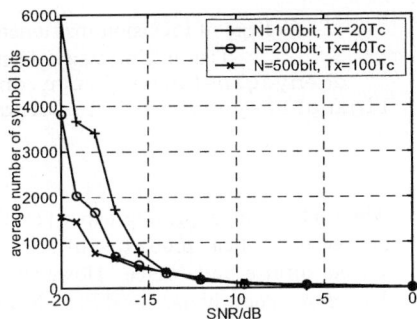

Fig. 4. Learning curves (N=500bit)   **Fig. 5.** Performance curves

Conclusions: Under common circumstances, if we have known PN sequence, we can obtain -20dB ~ -30dB of *SNR* threshold when we de-spread the received DS signals. In [1] Gardner used the method of "blind de-spreading" to achieve –15dB of the *SNR* threshold, but on the same condition, we can realize threshold of $SNR = -20.0dB$ easily, hence the performance of the methods in this paper is better. Besides this, the NN has higher speed than the K-L transformation method, it can solve the difficult problem of PN sequence blind estimation perfectly, furthermore, it can be used in management, interception and interference of DS/SS communications , and be extended to the case of DS-CDMA systems.

References

1. French, C.A., Gardener, W.A.: Spread-Spectrum Despreading without the Code. IEEE Trans.Com. Vol. 34 (1986) 404-407
2. Ljung, L.: Analysis of Recursive Stochastic Algorithms. IEEE Trans. On AC. Vol. 22 (1977) 551-575
3. Haykin, S.: Neural Networks-A Comprehensive Foundation. Prentice Hall PTR, Upper Saddle River, NJ, USA (1999)
4. Chang, T.Q., Guo, Z.X.: A Neural Networks to Estimation the PN Sequence of DS/SS Signals. the Ninth Telemetry and Telecommand Technology Annual Meeting of China, Haikou, China (1996) 535-537
5. Dminique, F., Reed, J.H.: Simple PN Code Sequence Estimation and Synchronization Technique Using the Constrained Hebbian Rule. Electronics Letters, Vol. 33 (1997) 37-38

Hierarchical PCA-NN for Retrieving the Optical Properties of Two-Layer Tissue Model

Yaqin Chen, Ling Lin, Gang Li, Jianming Gao, and Qilian Yu

College of Precision Instrument & Opto-electronics Engineering,
Tianjin University, Tianjin 300072, P. R. China
chenyaqin@twtmail.tju.cn, linling815@vip.sina.com,
ligang59@eyou.com, jianminggao80@163.com, y_ql@sina.com

Abstract. In the preceding paper [1], PCA-NN was successfully introduced to deduce the optical properties of semi-infinite tissue model from spatially resolved diffuse reflectance. However, tissue often has a layered structure. Therefore, a new hierarchical PCA-NN (HPCA-NN) algorithm was presented in this paper for extracting the optical properties of multi-layer tissue model from the spatially resolved reflectance. For simplicity, we concentrated on the two-layer model that simulated a skin layer with thickness of 5 mm and the semi-infinite underlying muscle layer. The results showed that the method can achieve high predictive accuracy with the rms errors (RMSEs) < 1% for the top-layer optical properties and the RMSEs < 5% for the bottom-layer optical properties. All the results were based on Monte Carlo simulations.

1 Introduction

In recent years, great efforts have been devoted to the noninvasive determination of tissue optical properties, namely, absorption coefficient μ_a and reduced scattering coefficient μ_s', from spatially resolved diffuse reflectance [2]. Most of the algorithms for solving this nonlinear inverse problem employ an assumption that tissue is homogeneous. However, biological tissue usually exhibits a complicated layer structure, for example, skin, stomach, esophagus, and brain. So the study of deducing the optical properties with consideration of layered tissue structure is of great importance for medical and clinical applications.

At present, few reports pay attention to the determination of the optical properties of simply two-layer tissue model, in which the bottom layer has infinitely thickness [3]. In the literature, the solution of diffusion approximation for two-layer model was often used to estimate the two-layer optical properties by fitting to the spatially resolved reflectance. Nevertheless, the accuracy of the estimated optical properties could not meet the demand for actual applications due to the intrinsic limits of diffusion theory. In addition, traditional fitting methods such as nonlinear least-squares fitting are difficult to perform in real-time and robustness. Consequently, it is urgent to develop a robust way with better real-time performance for deducing the optical properties based on a more exact model of light propagation in tissue. The most popular accurate model is Monte Carlo (MC) simulations in spite of its expensive computation.

In our previous work, PCA-NN has been proved to be a more efficient and robust way for retrieving the optical properties of semi-infinite tissue model from spatially resolved diffuse reflectance calculated by MC simulations [1]. Although the PCA-NN method can be extended to multi-layer tissue model for simultaneously estimating the optical properties, the prediction performance would not be good for all the optical properties with a finite training set. Considering that the spatially resolved reflectance can characterize each layer sequentially [4], we proposed a new hierarchical PCA-NN (HPCA-NN) algorithm for retrieving the optical properties of multi-layer tissues. As the preliminary study, this paper focused on the two-layer tissue model. The choice of optical properties for the two-layer model, as listed in Table 1, was consistent with skin and underlying muscle, and the top-layer thickness was chosen at 5 mm. Moreover, the following discussions were based on the absolute spatially resolved diffuse reflectance, denoted by $R_d(r)$.

Table 1. Optical properties of two-layer tissue model [5]

Parameter	Definition	Value
μ_{a1} / mm^{-1}	Top-layer absorption coefficient	0.022 – 0.03
μ_{s1}' / mm^{-1}	Top-layer reduced scattering coefficient	1.3 – 1.6
μ_{a2} / mm^{-1}	Bottom-layer absorption coefficient	0.018 – 0.026
μ_{s2}' / mm^{-1}	Bottom-layer reduced scattering coefficient	0.4 – 0.7

2 Methods

2.1 Monte Carlo Simulations

In this paper, three data sets of $R_d(r)$, including a training data set A for semi-infinite tissue model, a training set B and a test set C for two-layer tissue model, were respectively generated by MC method [1]. The training set A consisted of 15 × 16 matrix of reflectance. The values of μ_a and μ_s' in this matrix were incremented in steps of 0.002 and 0.1 mm^{-1}, respectively, within $0.002 < \mu_a < 0.03$ mm^{-1} and $0.3 < \mu_s' < 1.8$ mm^{-1}. The training set B included 400 profiles covering the ranges of μ_{a1}, μ_{s1}', μ_{a2} and μ_{s2}' (Table 1, Value column) on the combinations of a 5 × 4 × 5 × 4 matrix. The test set C consisted of 100 simulations whose optical properties were randomly selected in the ranges of μ_{a1}, μ_{s1}', μ_{a2} and μ_{s2}'.

2.2 HPCA-NN Algorithm

The PCA-NN method for determining the optical properties of semi-infinite tissue model from spatially resolved diffuse reflectance has been described thoroughly [1]. Based on this method, a new hierarchical PCA-NN (HPCA-NN) approach was presented for extracting the optical properties of two-layer tissue model. In the HPCA-

NN algorithm, the two-layer optical properties are estimated sequentially from the spatially resolved reflectance $R_d(r)$ that result from photons propagated mainly within the top layer or the bottom layer. As depicted in Fig. 1, the top-layer optical properties μ_{a1} and μ_{s1}' are firstly deduced from the reflectance values $R_d(r)$ at $r < D1$ by calibrating the outputs of the first PCA-NN, which is established on semi-infinite tissue model. Then, with the inputting of the estimated μ_{a1} and μ_{s1}', the second PCA-NN established on two-layer tissue model is used to deduce the bottom-layer optical properties μ_{a2} and μ_{s2}' from $R_d(r)$ at $r < D2$. For the particular two-layer tissue model discussed in this paper, we chose $D1 = 5$ mm and $D2 = 35$ mm.

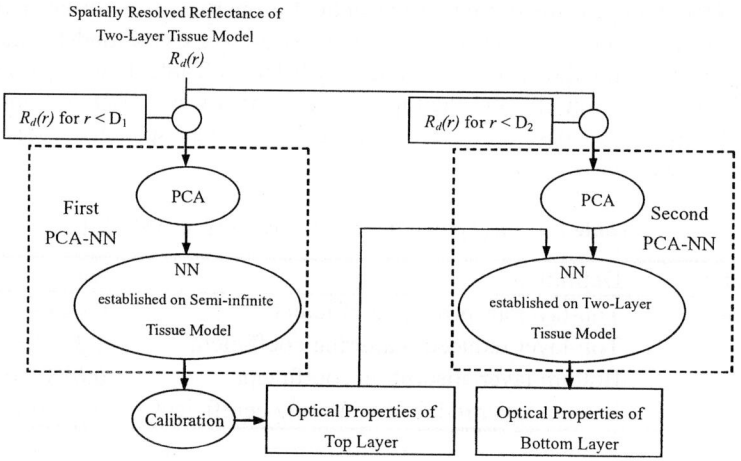

Fig. 1. Schematic representation of the HPCA-NN algorithm

In this paper, a two-layer backpropagation NN architecture was chosen for both PCA-NNs. The first PCA-NN was trained with the training data set A to make the association between the reflectance data and the corresponding optical properties of semi-infinite tissue model μ_a and μ_s'. Since the first three principal components of the set A accounted for almost the total data variance, the first NN was constructed with 3 inputs, 5 hidden units and 2 outputs. Different to the first PCA-NN, the second PCA-NN was trained with the training set B and the associated top-layer optical properties μ_{a1} and μ_{s1}' as inputs, and the corresponding bottom-layer optical properties μ_{a2} and μ_{s2}' as outputs. Considering that the number of samples in the set B was limited, we constructed the second NN with 10 inputs, 8 hidden nodes and 2 outputs. This implied that the first eight principal components of the training set B explaining more than 95% of the data variance were introduced into the second NN.

Furthermore, the calibration unit in the HPCA-NN algorithm was prerequisite to the determination of μ_{a1} and μ_{s1}' since the first PCA-NN was established on semi-infinite tissue model. In this paper, the calibration factors were respectively determined by fitting the outputs of the first PCA-NN to the true μ_{a1} and μ_{s1}' when inputting the training set B.

3 Results and Discussion

As described above, the HPCA-NN algorithm was based on the sequential estimation of the optical properties of two layers. The first PCA-NN after training predicted values with the rms errors (RMSEs) of 2.0% for μ_a and 0.15% for μ_s' on the training set A, and 1.1% for μ_{a1} and 0.19% for μ_{s1}' on the training set B. However, owing to the set B, which consisted of the reflectance profiles of two-layer tissue model, different to the training set A of semi-infinite model, the estimated μ_{a1} and μ_{s1}' for the set B respectively distributed around the regression line y = x + 0.0006 with the coefficient of determination R^2 = 0.992 and the line y = x + 0.0039 with R^2 = 0.999, as shown in Fig. 2. Thus, the calibration factors were accordingly determined to be 0.0006 for μ_{a1} and 0.0039 for μ_{s1}'. Fig.3 shows the estimated μ_{a1} and μ_{s1}' for the test set C, which were determined by calibrating the predicted values of the first PCA-NN, with the prediction RMSEs of 0.77% for μ_{a1} and 0.14% for μ_{s1}'.

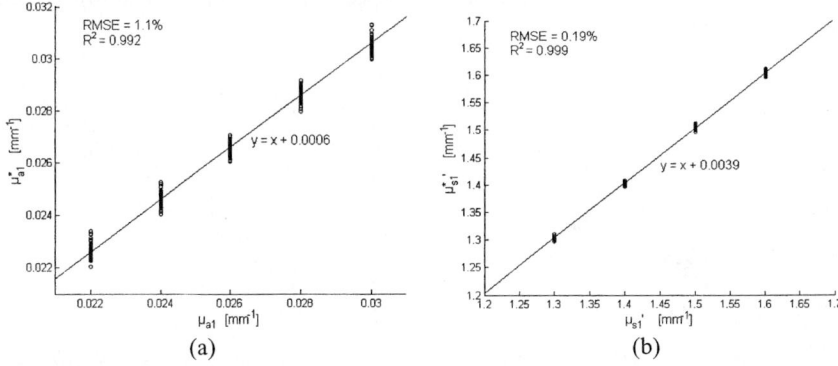

Fig. 2. Estimated absorption and reduced scattering coefficients of the top layer (μ_{a1}^* and $\mu_{s1}'^*$) for the training set B determined by the first PCA-NN versus the true μ_{a1} and μ_{s1}' used in the Monte Carlo simulations are respectively shown in (a) and (b). The diagonal lines are the regression lines.

For the second PCA-NN in the HPCA-NN trained with the training set B, the training results showed that the prediction RMSEs in μ_{a2} and μ_{s2}' were 1.9% and 2.2%, respectively. In the training procedure, we made an assumption that the true μ_{a1} and μ_{s1}' of the set B were known $a\ prior$ and input to the second PCA-NN. Under the similar assumption, the test results for the test set C indicated that the RMSEs in μ_{a2} and μ_{s2}' were respectively 2.5% and 3.7%. However, in practice, the top-layer optical properties are usually unknown. Therefore, we employed the estimated μ_{a1} and μ_{s1}', which were determined by the foregoing process of the HPCA-NN algorithm, to predict the bottom-layer optical properties μ_{a2} and μ_{s2}'. Fig. 4 shows the predicted μ_{a2} and μ_{s2}' for the test set C, which were determined by the second PCA-NN with the inputting of the estimated μ_{a1} and μ_{s1}', with the predictive accuracies of 4.0% for μ_{a2} and 4.6% for μ_{s2}'.

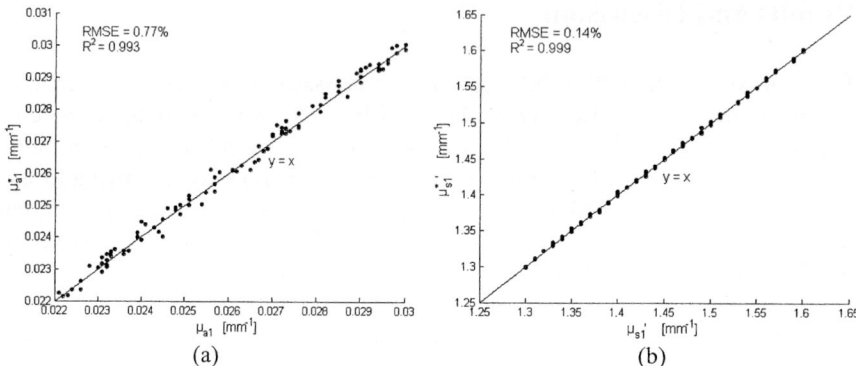

Fig. 3. Estimated absorption and reduced scattering coefficients of the top layer (μ_{a1}^* and $\mu_{s1}'^*$) for the test set C determined by calibrating the outputs of the first PCA-NN versus the true μ_{a1} and μ_{s1}' used in the Monte Carlo simulations are respectively shown in (a) and (b). The diagonal lines are the lines of equality.

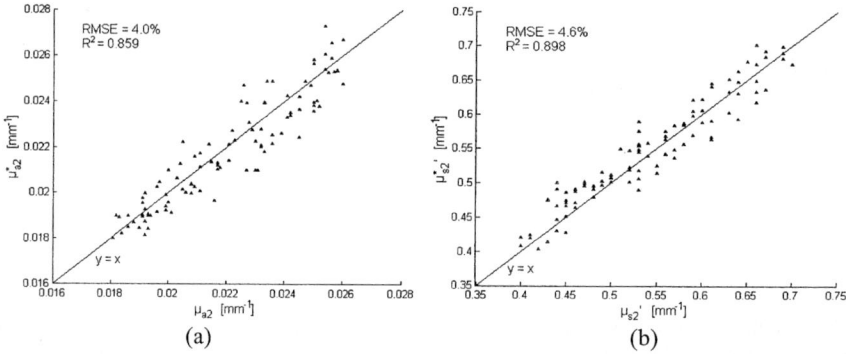

Fig. 4. Estimated absorption and reduced scattering coefficients of the bottom layer (μ_{a2}^* and $\mu_{s2}'^*$) for the test set C determined by the second PCA-NN with the inputting of the estimated μ_{a1} and μ_{s1}' versus the true μ_{a2} and μ_{s2}' used in the Monte Carlo simulations are respectively shown in (a) and (b). The diagonal lines are the lines of equality.

For comparison with the HPCA-NN method, we also built a PCA-NN established on two-layer tissue model, which had the similar two-layer backpropagation NN structure with 8 inputs, 8 hidden neurons and 4 outputs, to simultaneously estimate the optical properties of two layers. This PCA-NN was trained with the training set B to make the relationship between the reflectance data and the corresponding μ_{a1}, μ_{s1}', μ_{a2} and μ_{s2}'. From the results for the reflectance $R_d(r)$ at $r < 25$, 30, 35, 40 and 45mm, the best predictive accuracy was achieved from $R_d(r)$ at $r < 25$ mm with the RMSEs of 2.2% for μ_{a1}, 0.58% for μ_{s1}', 11% for μ_{a2}, and 8.4% for μ_{s2}' on the training set B, and 2.5% for μ_{a1}, 0.69% for μ_{s1}', 9.6% for μ_{a2}, and 12% for μ_{s2}' on the test set C. But compared with the accuracy of HPCA-NN, the prediction performance of PCA-NN was worse due to the limited training samples in the set B for the simultaneous estimation of four parameters μ_{a1}, μ_{s1}', μ_{a2} and μ_{s2}'.

4 Conclusions

In this paper, we presented a hierarchical PCA-NN algorithm for determining the optical properties of two-layer tissue model. The basic idea of HPCA-NN is to sequentially estimate the optical properties of two layers by constructing two PCA-NNs whose outputs and inputs are interconnected. From the preliminary results, we concluded that the HPCA-NN method is a promising technique for retrieving the optical properties of two layers from the absolute spatially resolved reflectance with high predictive accuracy of 0.77% for μ_{a1}, 0.14% for μ_{s1}', 4.0% for μ_{a2} and 4.6% for μ_{s2}'. And the algorithm can achieve good real-time performance due to the employment of neural network techniques.

Acknowledgements. This research was supported by the National Natural Science Foundation of China (No. 60174032), and '211 Engineering' of Tianjin University.

References

1. Chen, Y.-Q., Lin, L., Li, G., Yu, Q.-l.: Determination of Tissue Optical Properties from Spatially Resolved Relative Diffuse Reflectance by PCA-NN. IEEE Int. Conf. Neural Networks & Signal Processing, Vol. 1 (2003) 369–372
2. Farrell, T. J., Patterson, M. S., Essenpries, M.: Influence of Layered Tissue Architecture on Estimates of Tissue Optical Properties Obtained from Spatially Resolved Diffuse Reflectometry. Applied Optics, Vol. 37 (1998) 1958–1972
3. Alexandrakis, G., Farrell, T. J., Patterson, M. S.: Accuracy of the Diffusion Approximation in Determining the Optical Properties of a Two-layer Turbid Medium. Applied Optics, Vol. 37 (1998) 7401–7409
4. Fawzi, Y. S., Youssef, A.-B. M., El-Batanony, M. H., Kadah, Y. M.: Determination of the Optical Properties of a Two-layer Tissue Model by Detecting Photons Migrating at Progressively Increasing Depths. Applied Optics, Vol. 42 (2003) 6398–6411
5. Lin, L., Niwayama, M., Shiga, T., Kudo, N., Takahashi, M., Yamamoto, K.: Two-layered Phantom Experiments for Characterizing the Influence of a Fat Layer on Measurement of Muscle Oxygenation Using NIRS. Proc. SPIE, Vol. 3257 (1998) 156–166

features. No restrictions are implied for the type of local features. However, the features must represent simple spatial characteristics.

PCA is designed in such a way that the original feature set is represented by a number of effective "features" and yet retains most of intrinsic information contained in the data. Here, a single-layer feed-forward NN is used to perform PCA. Its structure is given in Fig. 2.

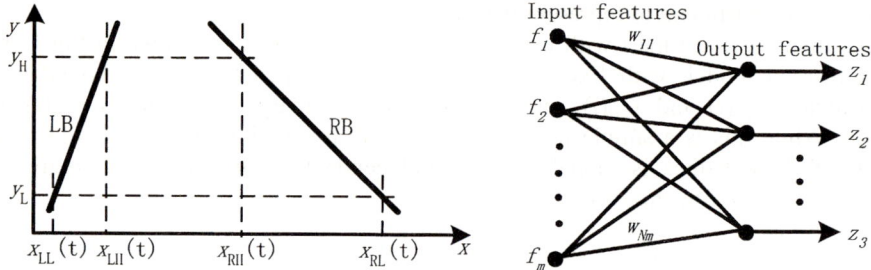

Fig. 1. Straight-road model

Fig. 2. Feedforward PCA NN with a single layer of nodes

The PCA is trained using the Generalized Hebbian Algorithm(GHA).This is an unsupervised learning algorithm based on a Hebbian learning rule. The GHA implementation in the training stage and output vector Z are computed as following:

$$\Delta w_{j,i}(n) = \eta z_i(n) f_i(n) - \eta z_j(n) \sum_{k=1}^{j} w_{k,i}(n) z_k(n). \tag{3}$$

$$z_j(n) = \sum_{i=1}^{m} w_{j,i}(n) f_i(n) \quad i=1,...,m, \; j=1,...,N. \tag{4}$$

$$w_{j,i}(n+1) = w_{j,i}(n) + \Delta w_{j,i}(n). \tag{5}$$

$$Z = wF. \tag{6}$$

In our approach, the input vector is the 6 dimensional feature vector F and the output Z is taken to 3 dimensions, where w is the matrix of the PCA coefficients. For fast, the proposed method uses only one spatial feature described by the vertical edge extraction mask of Prewitt since vertical edge is more pronounced than horizontal one.

4 Conclusions

In this paper, we presented a hierarchical PCA-NN algorithm for determining the optical properties of two-layer tissue model. The basic idea of HPCA-NN is to sequentially estimate the optical properties of two layers by constructing two PCA-NNs whose outputs and inputs are interconnected. From the preliminary results, we concluded that the HPCA-NN method is a promising technique for retrieving the optical properties of two layers from the absolute spatially resolved reflectance with high predictive accuracy of 0.77% for μ_{a1}, 0.14% for μ_{s1}', 4.0% for μ_{a2} and 4.6% for μ_{s2}'. And the algorithm can achieve good real-time performance due to the employment of neural network techniques.

Acknowledgements. This research was supported by the National Natural Science Foundation of China (No. 60174032), and '211 Engineering' of Tianjin University.

References

1. Chen, Y.-Q., Lin, L., Li, G., Yu, Q.-l.: Determination of Tissue Optical Properties from Spatially Resolved Relative Diffuse Reflectance by PCA-NN. IEEE Int. Conf. Neural Networks & Signal Processing, Vol. 1 (2003) 369–372
2. Farrell, T. J., Patterson, M. S., Essenpries, M.: Influence of Layered Tissue Architecture on Estimates of Tissue Optical Properties Obtained from Spatially Resolved Diffuse Reflectometry. Applied Optics, Vol. 37 (1998) 1958–1972
3. Alexandrakis, G., Farrell, T. J., Patterson, M. S.: Accuracy of the Diffusion Approximation in Determining the Optical Properties of a Two-layer Turbid Medium. Applied Optics, Vol. 37 (1998) 7401–7409
4. Fawzi, Y. S., Youssef, A.-B. M., El-Batanony, M. H., Kadah, Y. M.: Determination of the Optical Properties of a Two-layer Tissue Model by Detecting Photons Migrating at Progressively Increasing Depths. Applied Optics, Vol. 42 (2003) 6398–6411
5. Lin, L., Niwayama, M., Shiga, T., Kudo, N., Takahashi, M., Yamamoto, K.: Two-layered Phantom Experiments for Characterizing the Influence of a Fat Layer on Measurement of Muscle Oxygenation Using NIRS. Proc. SPIE, Vol. 3257 (1998) 156–166

Principal Component Analysis Neural Network Based Probabilistic Tracking of Unpaved Road

Qing Li[1,2], Nannig Zheng[1], Lin Ma[1], and Hong Cheng[1]

[1]The Institute of Artificial Intelligence and Robotics, Xi'an Jiaotong University, Xi'an 710049, China
{qli,lma,hcheng,qli}@aiar.xjtu.edu.cn,
nnzheng@mail.xjtu.edu.cn,
[2]The Engineering Institute, Air force Engineering University, Xi'an 710038, China

Abstract. Based on principal component analysis neural network, within a probabilistic framework, we introduce a new Monte Carlo tracking technique for autonomous navigation of land vehicle on unpaved road. The use of straight-road model and particle filter allows us to handle blurry road boundaries, and the use of color space transform, local spatial features and principal component analysis facilitates adaptation to the road conditions. Experimental results verify the algorithm in different conditions.

1 Introduction

There is increasing interest in autonomous navigation of land vehicles. The building of autonomous vehicles is a complex and challenging task with huge potentials in both civil and military application domains. One of the key problems in an autonomous land vehicle (ALV) is the tracking of the road. This is difficult task since the environment influences the road condition, the road follows a varying path, and the vehicle undergoes complex dynamic motion.

Some successful algorithms to track road are presented in the literatures [1][2]. However, there are some challenging problems related to tracking road. The difficulties come from: obtaining and maintaining precise geometrical road model; the complex algorithms required to search for position and match road features, and excessive computation.

In this paper, based on principal component analysis neural network, within a probabilistic framework, we introduce a new Monte Carlo tracking technique for autonomous navigation of land vehicle on unpaved road. The use of straight-road model and particle filter allows us to handle blurry road boundaries and the use of color space transform, local spatial features and principal component analysis facilitates adaptation to the road conditions.

The state representation of road is discussed in section 2 and principal component analysis of road is presented in section 3. Particle filter is described in section 4. Some experimental results are shown in section 5 and the paper is concluded in section 6.

2 Road Model

In this section, we introduce an efficient representation scheme for the important aspects of the road state and thereby define the road tracking problem as a prediction filter. To begin with, we need to represent the road shape and position compactly with appropriate parameters and define the set of parameters as a state. For unpaved roads, two lines are sufficient to represent the road shape.

An image $I(t)$ at time t with image plane coordinates (x,y) is shown in Fig.1. The straight-road model, with fixed y_H and y_L, is represented by the left boundary x-coordinates $x_{LL}(t)$ and $x_{LH}(t)$, and by the right boundary x-coordinates $x_{RH}(t)$ and $x_{RL}(t)$. While this is a simple model, it is suitable for tracking most unpaved roads. Each of these values is modeled as an independent second-order auto-regressive dynamics. We define the state at time t as

$$x_t = (x_{LL}(t), x_{LH}(t), x_{RH}(t), x_{RL}(t)). \tag{1}$$

The dynamics then reads

$$x_{t+1} = Ax_t + Bx_{t-1} + Cv_t, \; v_t \sim N(0, \Sigma). \tag{2}$$

Matrices A, B, C and Σ defining this dynamics are unknown and heuristically determined or could be learned from a set of representative sequences where correct tracks have been obtained in some way. For the time being we use an ad-hoc model composed of four independent constant velocity dynamics, $x_{LL}(t)$ and $x_{LH}(t)$, and $x_{RH}(t)$ and $x_{RL}(t)$ with respective standard deviations 1 pixel/frame, 0.5 pixel/frame, 0.5 pixel/frame and 1 pixel/frame.

3 Principal Component Analysis of Unpaved Road

The Hue-Saturation-Value(HSV) color space is used in order to decouple chromatic information from shading effects. Pixels with large and small intensities are not included in the training data set, because hue and saturation become unstable in this range. This paper exploits the colors of the images, but also their local spatial characteristics. The HSV color components of each pixel are considered as the first three features. The entire feature set is completed by additional spatial features that are extracted from neighboring pixels[3].

Let $N(r,c)$ denote the local neighboring region of pixel(r,c). Therefore, the color of each pixel (r,c) can be associated with local image characteristics extracted from the region $N(r,c)$. These characteristics can be considered as local spatial features of the image and can be helpful for the tracking road process. That is, using the values of the colors of $N(r,c)$, f_k, $k=4,5,\ldots$, $K+3$ local features can be defined as image spatial features. No restrictions are implied for the type of local features. However, the features must represent simple spatial characteristics.

PCA is designed in such a way that the original feature set is represented by a number of effective "features" and yet retains most of intrinsic information contained in the data. Here, a single-layer feed-forward NN is used to perform PCA. Its structure is given in Fig. 2.

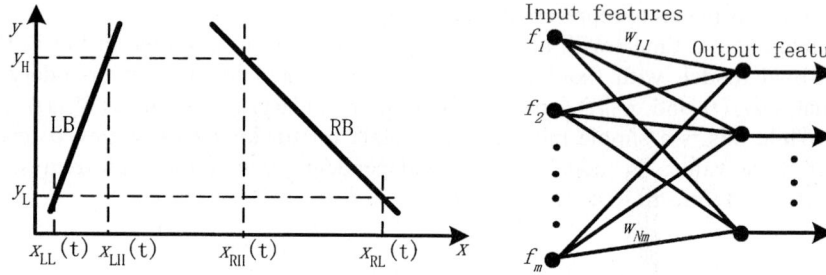

Fig. 1. Straight-road model

Fig. 2. Feedforward PCA NN with a single layer of nodes

The PCA is trained using the Generalized Hebbian Algorithm(GHA). This is an unsupervised learning algorithm based on a Hebbian learning rule. The GHA implementation in the training stage and output vector Z are computed as following:

$$\Delta w_{j,i}(n) = \eta z_i(n) f_i(n) - \eta z_j(n) \sum_{k=1}^{j} w_{k,i}(n) z_k(n). \tag{3}$$

$$z_j(n) = \sum_{i=1}^{m} w_{j,i}(n) f_i(n) \quad i = 1,...,m, \, j = 1,...,N. \tag{4}$$

$$w_{j,i}(n+1) = w_{j,i}(n) + \Delta w_{j,i}(n). \tag{5}$$

$$Z = wF. \tag{6}$$

In our approach, the input vector is the 6 dimensional feature vector F and the output Z is taken to 3 dimensions, where w is the matrix of the PCA coefficients. For fast, the proposed method uses only one spatial feature described by the vertical edge extraction mask of Prewitt since vertical edge is more pronounced than horizontal one.

4 Probabilistic Tracking

In visual tracking problems, the likelihood is non-linear, and often multi-modal, with respect to the hidden state. As a result, the Kalman filter and its approximations are usually not suitable. However, the recursion can be used within a sequential Monte Carlo frame-work. Sequential Monte Carlo techniques for filtering time series and their use in the specific context of visual tracking[4][5] have been described at length in the literature.

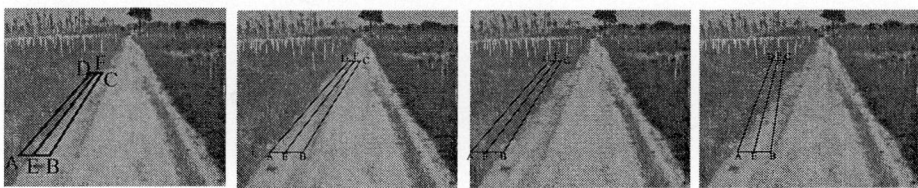

Fig. 3. The different positions of boundary window

4.1 Statistical Model

A trapezoid ABCD represents boundary window, and the line EF divides the trapezoid ABCD into two smaller trapezoids AEFD and EBCF. Point E and point F are the midpoints of line segment AB and DC, respectively. The length of segment AB and DC is constant. The statistical characteristics of the sum of principal component vary with different position of EF. Define $z=z_1+z_2+z_3$ and $s=s_1/s_2+s_1/s_3$. The standard deviation of z in ABCD is defined as s_1, one in AEFD as s_2 and one in EBCF as s_3. Thus four cases appear:

s_1 is large, but s_2 and s_3 are small when line EF is just the boundary of the road(See the first one in Fig. 3); s_1 and s_2 and s_3 are small when trapezoid ABCD is on off-road (See the third one in Fig. 3); s_1 and s_2 and s_3 are small when trapezoid ABCD is on road(See the second one in Fig. 3); s_1 and s_2 and s_3 are large when one part of trapezoid ABCD is on off-road and other part of trapezoid ABCD is on road(See the last one in Fig. 3). Thus s increases when line EF approaches boundary of unpaved road.

4.2 Estimation by Particle Filtering

In order to estimate the shape of the road ahead of the vehicle, we have chosen a particle filter, the CONDENSATION algorithm[4]. The term particle filter refers to a mechanism for estimating a probability distribution over the state space x(*t*) given observations from a stream of images. The distribution is approximated by a set of "particles" pairs{x,π}, where, as above, x is a state vector, and π is a weight that reflects the plausibility of x as a representation of the true state of the system. Importantly, the method places no assumptions on the distributions involved, and it is this power to represent arbitrary, multi-modal distributions that proves useful when tracking in the presence of the clutter that often confounds uni-modal methods such as the Kalman filter.

Input $x(t)$, the algorithm may be summarized as follows[5]:
Current particle set: $\{x_t^m\}$, $m=1\ldots M$; Prediction: for $m=1\ldots M$, draw $\tilde{x}_{t+1}^m$ from second-order automatic regressive dynamics; Weighting: for $m=1\ldots M$ compute $\pi_{i+1}^m = K\exp(\lambda S(t))$, with K is used as normalization; Selection: for m=1…M, sample index α(m) from discrete probability $\{\pi_{i+1}^k\}_k$ over $\{1\ldots M\}$, and set $x_{t+1}^m = \tilde{x}_{t+1}^{\alpha(m)}$.

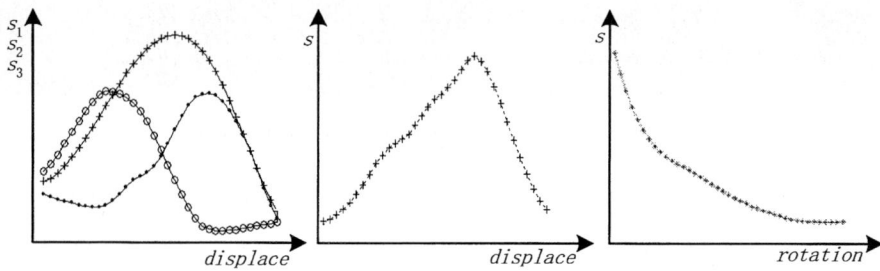

Fig. 4. (a) Change of s_1 and s_2 and s_3 when ABCD displaces horizontally from the third one in Fig.3 to the second one in Fig.3; (b) Change of s when ABCD displaces horizontally from the third one in Fig.3 to the second one in Fig.3; (c) Change of s when ABCD rotates from the first one in Fig.3 to the last one in Fig. 3

5 Experimental Results

The tracking road algorithm was tested on a variety of unpaved road image sequences in rural country. The implementation used a set of 100 particles to estimate *x*. Our estimate for the state is obtained from the mean of the particle set, and, if desired, a confidence measure in road geometry can be derived from the variance of the particle set. Particle filter initialization was facilitated by the manual boundary specification for a single initial image frame. One frame of video sequences tracked is shown in Figs.5. We cannot judge system accuracy, but it is possible to assess robustness to different road by observing the algorithm's performance on video sequences. We have found that algorithm functions robustly under many conditions. While the straight-road model does not mark the curved road boundaries precisely, tracking is maintained and the information is sufficient to navigate the vehicle. For sharp right-angled corners or the curved sampled often, the straight-road model can be represented by a piecewise straight-road model. The computational complexity of the algorithm is related to number of particle and size of boundary window.

6 Conclusions

This paper has presented a road tracking algorithm designed for unpaved roads. The recursive prediction of road model makes it possible to process only a small neighborhood of the image without losing important information. The algorithm has shown robustness to variations in road properties. Owing to the under-lying estimation scheme, the algorithm is able to initialize and recover automatically in the few cases where track is lost. The straight-road models are also less sensitive when weighing particle since there are fewer parameters to the model. Using particle filter within probabilistic framework we further improve its robustness. We are currently working to improve real time performance.

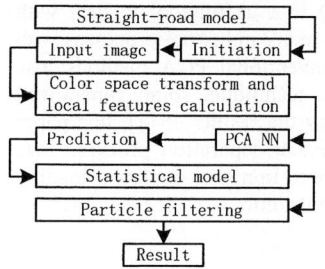

Fig. 5. Implementation of tracking

Fig. 6. One frame of video sequences tracked

Acknowledgment. This work was supported by the National Natural Science Foundation of China under Grant 60024301.

References

1. Turk, M.A., Morgenthaler, D.C., Gremban, K.D.: VITS: A Vision System for Autonomous Land Vehicle Navigation. IEEE Transaction of Pattern Analysis and Machine Intelligence, 1988,10 (3) 342-361
2. Jeong, H., Oh, Y., Park, J.H., Koo, B.S.: Vision-Based Adaptive and Recursive Tracking of Unpaved Roads. Pattern Recognition Letters 23 (2002) 73-82
3. Papamarkos, N., Antonis, E.A., Strouthopoulos, C.P.: Adaptive Color Reduction. IEEE Transactions on System, Man, and Cybernetics-Part B: Cybernetics, 32 (1) (2002) 44-56
4. Isard, M., Blake, A.: Condensation-Conditional Density Propagation for Visual Tracking. Int. J. Computer Vision, 29 (1) (1998) 5-28
5. Perez, P., Hue, C., Vermaak, J., Gangnet, M.: Color-Based Probabilistic Tracking. In: Heyden A. (eds.): ECCV. Lecture Notes in Computer Science, Vol. 2350. Springer-Verlag, Berlin Heidelberg New York (2002) 661-675

Chemical Separation Process Monitoring Based on Nonlinear Principal Component Analysis[*]

Fei Liu and Zhonggai Zhao

Institute of Automation, Southern Yangtze University
Wuxi, 214036, P. R. China
fliu@thmz.com

Abstract. Principal component analysis (PCA) is a useful tool to deal with linear relationship among process variables. For many industrial processes with variables containing nonlinear relationship, conventional PCA methods lose their power. Instead, applying neural network technique, some generalized linear PCA methods are presented. Motivated by the results of [1], this paper discusses monitoring and diagnosis for a chemical separation process. Two neural networks are employed, one of which is used to model nonlinear loading functions, and another to map principal components onto corrected data set.

1 Introduction

As a statistical analysis tool, PCA can exploit linear relationship among process variables and reduce data dimensionality. For linear or weak nonlinear raw data, PCA technique can greatly reduce data dimensionality and extract useful information from a large amount of process data. This provides great convenient for field engineer or manipulator to monitor industrial process. However, for strong nonlinear raw data, conventional PCA cannot achieve its aim. By replacing linear loading vectors in PCA, principal curve and principal surface is presented [2], on which raw data projects and then produces nonlinear principal components (PCs). An iterative algorithm to find principal curve is also presented, but the method cannot produce a nonlinear principal component model in sense of a principal loading. In [3], two neural networks are introduced to overcome above weakness. One is used to model relationship between nonlinear principal components and raw data set, and another is used to resume raw data set from nonlinear PCs. But this method results in large calculation. In the meantime, also applying neural network, a novel method extends PCA to nonlinear cases by optimizing network inputs [1]. The converged inputs are nonlinear PCs and the obtained network also can reconstruct raw data set from nonlinear PCs.

This paper discusses monitoring and diagnosis for an industrial chemical separation process, in which the relationships among variables can not be only generalized in linear form. If linear PCs model is used, there need a large number of PCs. This will bring field workers a lot of PCs monitoring plot, which does not take best advantage

---

[*] Supported by the Open Project Program of National Lab. of Industrial Control Technology.

of PCA and makes on-line monitoring difficulty. Motivated by existing results [1], this paper also uses two neural networks. In order to avoid converging at local minimum points, the training method of neural networks integrates decreasing learning rate and momentum disturbance term [5].

2 Statistical Monitoring

Assuming a $n \times m$ data set X, linear PCA follows: $X = t_1 p_1^T + t_2 p_2^T + \cdots + t_m p_m^T$, where $t_1, t_2 \cdots t_m$ represent principal components and $p_1, p_2 \cdots p_m$ are loading vectors corresponding to principal components. If there are linear relationships in X, then $X = t_1 p_1^T + t_2 p_2^T + \cdots + t_l p_l^T + E$, where E is the residual and $l \ll m$. For nonlinear PCA, $X = F(T) + E$, where $T = [t_1, t_2 \cdots t_l]$ is nonlinear principal score matrix, $F(\cdot)$ is the nonlinear loading function representing principal components model. For the ith sample, Squared Prediction Error (SPE) of the model is $SPE_i = e_i e_i^T$, where $e_i = (X_i - \hat{X}_i) = (X_i - F(T))$. If process is in control, E is normally distributed. Then the SPE can be well approximated by so-called $g \cdot \chi_h^2$ distribution, where g is a constant and h is effective degrees of freedom of the χ-squared distribution [4]. The mean and variance of the $g \cdot \chi_h^2$ distribution ($\mu = gh, \sigma^2 = 2g^2 h$) equate to the sample mean (m) and variance (v). Based on the distribution, SPE control limit at significance α can be calculated by: $SPE_\alpha = (v/2m)\chi^2_{2m^2/v,\alpha}$, where $\chi^2_{2m^2/v,\alpha}$ is the critical value of χ-squared variance with $2m^2/v$ at significance α. That is to say SPE_i will not beyond control limit if process runs well. At same time, the scatter character of T also belongs to same „normal" cluster. According to these features, SPE plot and score plot are often used to monitor industrial processes. If the SPE value of current sample is beyond control limit, or principal component lies far away from „normal" population, perhaps something is going wrong with the process.

Contribution of process variables to SPE also tells useful information about process. For the kth variable in the pth sample, contributes to SPE like following formula: $de_{pk} = \left\| q_{pk} - \hat{q}_{pk} \right\|^2$, where $\hat{q}_{pk}$ is the prediction of q_{pk}.

3 Neural Network Inputs Optimizing and Algorithm

To model unknown nonlinear loading function $F(\cdot)$, set up a three-layer neural network characterized by what the nodes of hidden-layer are more than that of inputs. While inputs are unknown, the goal of network training is to minimize error between the raw data and outputs of network. Similar to weigh adjusting of network, inputs of network are also modified using errors back propagated from output layer. Hidden nodes use sigmoid functions and input and output nodes use linear functions.

The detail algorithm is as following: Assume Q is the training data set, Z and U are the outputs and inputs of network, respectively. The objective function to be minimized in network training is $E = \|Z - Q\|^2$, the steepest descent direction is $\Delta U = -\eta \frac{\partial E}{\partial U} = \eta(Z-Q)\frac{\partial Z}{\partial U}$, $\Delta W = -\eta \frac{\partial E}{\partial W} = \eta(Z-Q)\frac{\partial Z}{\partial W}$, where $Z = W\sigma(B+VU)$, and W, V respectively represent network weights of output layer and input layer, $\sigma(\cdot)$ is a sigmoid function, B is bias of nodes, η is learning rate of network. The main difference between traditional back-propagation network and above network is that weights and inputs of network are modified in the meanwhile time. In order to quicken training, decreasing learning rate and momentum disturbance term are introduced [5]. That is:

$$\Delta W(k+1) = \eta(1-m_c)(Z-Q)\frac{\partial Z}{\partial W} + m_c \Delta W(k), \qquad (1)$$

$$\Delta U(k+1) = (1-m_c)\eta(Z-Q)\frac{\partial Z}{\partial U} + m_c \Delta U(k), \qquad (2)$$

where $\eta(k) = \eta(0)/(1+\frac{k}{r})$, r is a constant which is determined by experience, m_c is the coefficient of momentum and it is usually around 0.9.

The number of input nodes is first supposed to equal to the number of linear PCs, and then to decrease or increase according to training results until the error is under a threshold. In practice, it is best to get the least inputs nodes. For an optimizing network with f input nodes, a hidden nodes, and n output nodes trained on m samples of data, the number of hidden nodes is limited by following inequality and equality: $(fa + fn) + (f + a + n) + fm \ll mn$, $a = \sqrt{f+n} + \alpha$, $\alpha \in (1,10)$.

4 Chemical Separation Process Online Monitoring

4.1 Process Description

A schematic view of actual separation process is briefly described as following. Two series of raw flow are fed into mixed tank, and then into raffinate tower through heat exchanger. In raffinate tower, the top level of tower is achieved by manipulating reflux flow via reflux flow pump. To maintain quality of product close to a desired set point or between a specified range, the temperature differences between the 5th and 21st plate is manipulated by temperature and flow controllers. The bottom material is pumped and returned to tower through a heating stove burning oil and gas. In industrial practice, this process is suspected to be nonlinear, a series of real data under operating condition is showed in Fig.1, in which fifteen process variables (measured values of temperature, pressure, level and flow rate) are normalized and included.

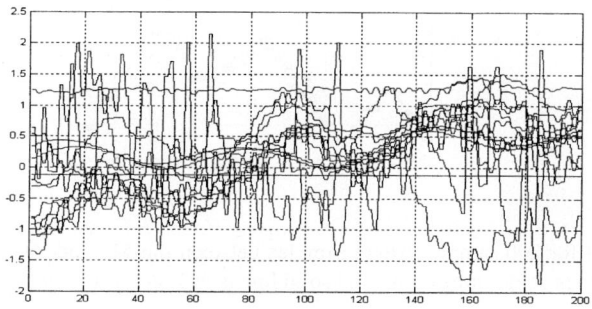

Fig. 1. Raw data for neural network training

4.2 Nonlinear PCs Model

As shown in Fig.1, 200 samples out of real raw data under operating condition are used to build nonlinear PCs model. Consider 4 linear PCs model, which can explain 84.92% of the variance of real data, following setup uses a neural network to extract nonlinear PCs.

Firstly, let the number of input nodes be 4 and hidden nodes 15, network is trained with raw data. For a specified error threshold of training error, we adjust the number of input nodes until it is the smallest. For above specific case, two input nodes can meet the aim by trial. Thus, a 2-15-15 three-layer network is built which can reconstruct original data from PCs. Secondly, look for nonlinear loading functions, which can map the original data to PCs. Another three-layer back-propagation network is then used to model the relationship between raw data and PCs.

4.3 Statistical Online Monitoring

Based on above off-line nonlinear PCs model, the SPE control limit SPE_α can be calculated by other real data under operating condition, and then the normal region, in which the scores of monitored samples should lie, is decided. For on-line monitoring, in every time interval, the score and SPE of current sample data are calculated by above model and then is depicted respectively in score plot and SPE plot. By means of these plots, one (e.g. field worker) can judge whether the process is out of normal control region. It is well known that monitoring methods rely on score plots. Monitoring a process by nonlinear scores only needs 3 plots, while 9 plots by using linear PCs. Obviously, nonlinear PCs eases monitoring workload greatly. Following gives some results related to chemical separation process monitoring and diagnosis.

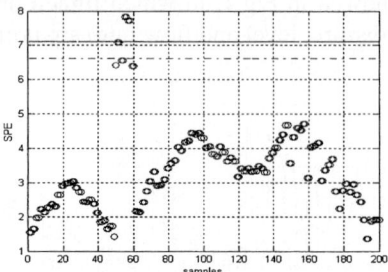

Fig. 2. SPE plot for 200 samples where samples between the 51st and 60th exceed the control limit. Dotted line is 95% control limit and solid line is 99% control limit

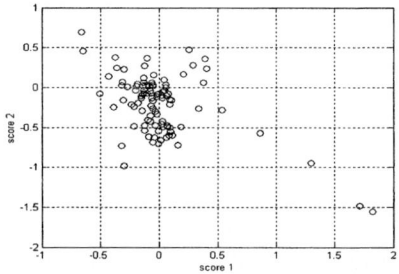

Fig. 3. Score-score plot. There are a lot of points out of the cluster

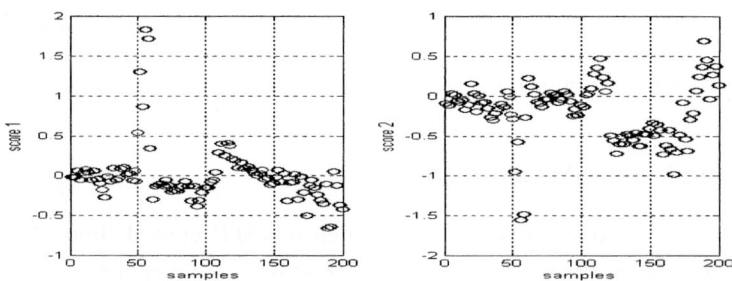

Fig. 4. Score plot. From the 50th to 60th samples, the scores are far away from the scores of the other samples

According to above Fig. 2 to Fig. 4, a lot of sample dots lie far out the normal cluster. Especially, from time interval 50th to 60th, SPE goes beyond the control limit. Combining with other corresponding plots, it is clear that some faults may happen in process.

To detect fault reason or which variables are abnormal, the plot of contribution to SPE is used. In Fig.5, it is obvious that the contribution of the 7th variable to SPE is bigger than that of other variables, which implies that something involved in the 7th variable may be fault.

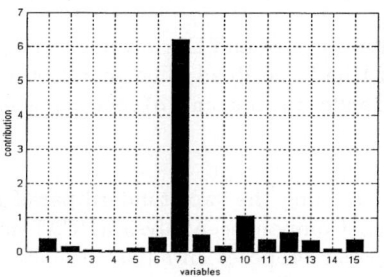

Fig. 5. Contribution plot

5 Conclusions

Two neural networks are introduced in nonlinear PCA for chemical separation process monitoring and diagnosis. One aims to extract nonlinear PCs and develop the model for reconstructing data, and another to capture nonlinear loading functions. Based on these trained networks, two indices, SPE and T plot, are selected as monitoring indictor of the process. Furthermore, the contribution plot is used to detect fault variables. The practice results show that nonlinear PCA is a very useful and convenient statistical analysis tool.

References

1. Tan, S., Mavrovouniotis, M. L.: Reducing Data Dimensionality Through Optimizing Neural Network Inputs. AIChE J. 41 (1995) 1471-1480
2. Hastie, T., Stuetzle, W.: Principal Curves. Journal of the American Statistical Association 84 (1989)502-516
3. Dong, D., McAvoy, T. J.: Nonlinear Principal Analysis-Based on Principal Curves and Neural Networks. Computers and Chemical Engineering 30 (1996) 65-78
4. Box, G. E. P.: Some Theorems on Quadratic Forms Applied in the Study of Analysis of Variance Problems: Effect of Inequality of Variance in One-Way Classification. The Annals of Mathematical Statistics 25 (1954) 290-302
5. Christian, D., John, M.: Note on Learning Rate Schedules for Stochastic Optimization. In: Lippmann, R. P., Moody, J. E., Touretzky, D. S. (Eds.): Neural Information Processing Systems (1991) 832-838

An Adjusted Gaussian Skin-Color Model Based on Principal Component Analysis

Zhi-Gang Fan and Bao-Liang Lu

Department of Computer Science and Engineering, Shanghai Jiao Tong University,
1954 Hua Shan Road, Shanghai 200030, China
zgfan@sjtu.edu.cn, blu@cs.sjtu.edu.cn

Abstract. By combining the two standard paradigms of unsupervised learning, Principal Component Analysis (PCA) and Gaussian density estimation, this paper proposes an adjusted Gaussian skin-color model for skin-color detection. This method is more robust than the standard Gaussian model because it can weaken the bias caused by noise and enhance the fitness of the mathematical model. The experiments show that this method works well for the real-world images with complex backgrounds.

1 Introduction

Automatic detection of human faces is a very difficult task. Face detection in color images that begin with skin color modeling has been the topic of extensive research for the several past decades. In recent years, surveys on face detection have been made [8], [9], [4]. Some adaptive and unsupervised methods for skin-color modeling have been introduced in resent years [7], [3], [1]. We propose an adjusted Gaussian skin-color model based on Principal Component Analysis (PCA) for skin-color detection. We use single Gaussian model and don't use Gaussian mixture models because we find that, in our dataset, the distribution of skin-color data is unimodal. Some researchers have the same point of view [7]. We show that our model significantly outperforms the standard Gaussian model.

2 Adjusted Gaussian Skin-Color Model

2.1 Gaussian Model and PCA

The multivariate Gaussian density is a very important and unique density function in statistical theory. It may be viewed as the central idea in second-order statistics. The distribution of an n-dimensional random vector x is Gaussian if its probability density function has the form:

$$p_x(x) = \frac{1}{(2\pi)^{n/2} |C_x|^{1/2}} \exp(-\frac{1}{2}(x-m_x)^T C_x^{-1}(x-m_x)) \tag{1}$$

where m_x is its mean, and C_x is the covariance matrix of x.

Principal component analysis (PCA) is a popular method for features extraction and pattern representation. It can find the projection directions that maximize the total scatter across all variables and remove the mutual correlation between the elements. Denote the training set of m samples by $X = (x_1, x_2, ..., x_m) \subset R^{n \times m}$, where x is an n-dimensional vector. Define the covariance matrix as follows:

$$C_x = \frac{1}{m} \sum_{i=1}^{m} (x_i - m_x)(x_i - m_x)^T \qquad (2)$$

Then, the eigenvalues and eigenvectors of the covariance C_x are calculated. Let $A = (a_1, a_2, ..., a_r) \subset R^{n \times r} (r \leq n)$ be the r eigenvectors corresponding to the r largest eigenvalues $\{\lambda_1, \lambda_2, ..., \lambda_r\}$. Thus, for the original random vector x which belongs to set $X = (x_1, x_2, ..., x_m) \subset R^{n \times m}$, its corresponding eigenfeature y, belonging to the set $Y = (y_1, y_2, ..., y_m) \subset R^{r \times m}$, can be obtained by projecting x into the eigenfeature space as follows:

$$y = A^T (x - m_x) \qquad (3)$$

The transformed vector y is a random vector with zero mean and its covariance matrix D is related to that of x by the following equation:

$$D = A^T C_x A \qquad (4)$$

where D is a diagonal matrix having the eigenvalues of C_x along its diagonal. According to equation (4), we can obtain the following equation:

$$C_x = (A^T)^{-1} D A^{-1} \qquad (5)$$

Then, we can obtain the following equation:

$$C_x^{-1} = A D^{-1} A^T \qquad (6)$$

From equations (3) and (6), we have

$$\begin{aligned}
(x - m_x)^T C_x^{-1} (x - m_x) &= (x - m_x)^T (A D^{-1} A^T)(x - m_x) \\
&= ((x - m_x)^T A) D^{-1} (A^T (x - m_x)) \\
&= y^T D^{-1} y \\
&= \sum_{i=1}^{r} \frac{y_i^2}{\lambda_i}
\end{aligned} \qquad (7)$$

By using the right part of equation (7), we can change equation (1) into the following form:

$$p_x(x) = \frac{1}{(2\pi)^{n/2} \prod_{i=1}^{r} \lambda_i^{1/2}} \exp(-\frac{1}{2} \sum_{i=1}^{r} \frac{y_i^2}{\lambda_i}) \qquad (8)$$

2.2 Adjusted Model

Through PCA, the Gaussian density function can be changed into the form as equation (8). When $r = n$, the equation (8) is equal to the original one, equation (1), but it can transform the coordinate axes into the principal axes of Gaussian model. Thus, not only we can calculate more feasible, but also, the more important, we can easily evaluate the fitness of the Gaussian model along the principal axes directions. In practice, the Gaussian skin-color model always has considerable bias because of the noise in the sample set. Fig.1 shows such situation, in which a histogram of the skin-color samples projected on the second principal axes of the Gaussian skin-color model is illustrated. We denote the histogram as $f(x)$. The mean is located on the zero point. The bias can obviously be seen because the main peak of $f(x)$ floats away from the mean. The reason for the bias is that the noise is inevitably mixed into the sample

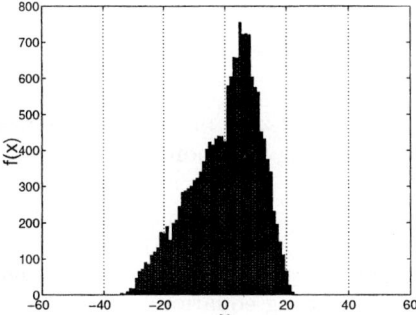

Fig. 1. The histogram $f(x)$

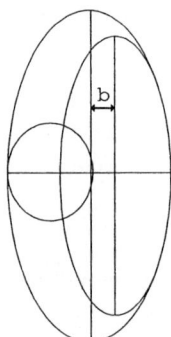

Fig. 2. The bias of the Gaussian model affected by noise

set when sampling. Usually, the Gaussian model is disturbed by the noise on one side such as the situation shown in Fig.2, in which the two ellipses denote two Gaussian models, the small circle denotes the noise, and the parameter b denotes the bias. The small ellipse is disturbed by the circle on the left side, so, it is replaced by the big ellipse which produces the bias b. To adjust the bias, we must move the vertical principal axes to right.

Because of the big bias, we must adjust the model to weaken the bias and enhance the fitness. According to equation (8), we can effectively adjust the Gaussian model along the principal axes direction. We can adjust different principal axes independently at the same time disregarding relationship among them because these axes are uncorrelated. On the ith principal axes, we define the bias as two components, the bias on mean as b_i and the bias on variance as c_i. Thus, we can define the adjusted Gaussian skin-color model in the form:

$$p_x(x) = \frac{1}{(2\pi)^{n/2} \prod_{i=1}^{r}(c_i \lambda_i)^{1/2}} \exp(-\frac{1}{2} \sum_{i=1}^{r} \frac{(y_i - b_i)^2}{c_i \lambda_i}) \qquad (9)$$

2.3 Measurement of the Biases b_i and c_i

On the ith principal axes, we can measure the biases b_i and c_i according to the histogram of the training samples projected on this principal axes. The histogram $f(x)$ shown in Fig.1 is just this type of histogram. Obviously, the function $f(x)$ is a noisy function and should be smoothed. We use a low-pass filter $g(x)$ to smooth $f(x)$ through convolution operation:

$$h(x) = f(x) * g(x) \qquad (10)$$

Assume function $h(x)$ reaches its maximum value at point x_m: $\max(f(x)) = f(x_m)$. Specifically, we begin the convolution operation from the point x_m, and go on along the two sides of this point respectively. The smoothed histogram $h(x)$ is shown in Fig.3. In order to measure the biases, we have to locate the

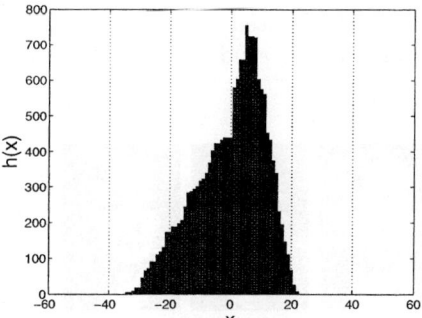
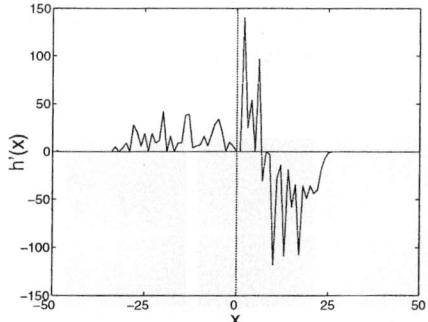

Fig. 3. The smoothed histogram $h(x)$ **Fig. 4.** The $h'(x)$

main peak of the histogram. Differentiating the function $h(x)$ we can obtain its derivative $h'(x)$ and produce reliable peaks at the inflection points. Fig.4 shows $h'(x)$. The two mutually perpendicular lines indicate the location of zero point. We can obviously see the high peak part is on the right side of the mean point. Assume that the high peak part is in the interval (x_1, x_2). For this task, selecting the interval (x_1, x_2), careful consideration should be made. In practice, the interval (x_1, x_2) should include the point x_m and cannot include the deep valleys parts whose bottoms reach the zero line and two sides are high peaks.

High peak part of $h'(x)$ in the interval (x_1, x_2) indicates the main peak of $h(x)$. Thus, to adjust the biases, we should move mean point to the central point of the interval (x_1, x_2). Assume $h(x) > 0$ in the interval (x_n, x_p). On the ith principal axes, we can measure the biases through the following formulas :

$$d = (x_2 - x_1)/2 \qquad (11)$$
$$d_1 = x_p - x_n \qquad (12)$$
$$b_i = d + x_1 \qquad (13)$$
$$c_i = 2d/d_1 \qquad (14)$$

However, in practice, on some principal axes the biases are too small to be adjusted necessarily. Thus, on the ith principal axes, we change the formulas (13), and (14) to the following forms:

$$b_i = \begin{cases} d + x_1 & |d + x_1| > Q_i \\ 0 & \text{otherwise} \end{cases} \quad (15)$$

$$c_i = \begin{cases} 2d/d_1 & |d + x_1| > Q_i \\ 1 & \text{otherwise} \end{cases} \quad (16)$$

where Q_i is a user defined constant threshold on the ith principal axes.

3 Implementation and Experiments

We use the standard Gaussian skin-color model and adjusted Gaussian skin-color model respectively in our experiments for comparison study. Fig.5 shows the

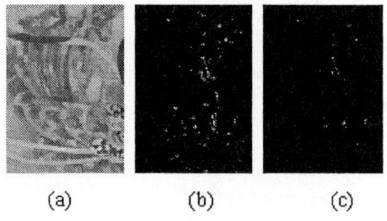
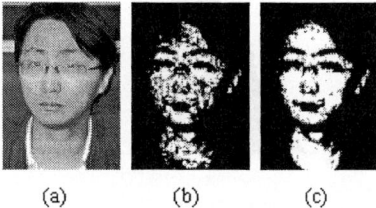

Fig. 5. (a) A non-face image; (b) Binary image after segmentation based on standard Gaussian model; (c) Binary image after segmentation based on adjusted Gaussian model

Fig. 6. (a) A face image; (b) Binary image after segmentation based on standard Gaussian model; (c) Binary image after segmentation based on adjusted Gaussian model

detection result on non-face images. The segmentation (Fig.5(b)) based on standard Gaussian model shows that much non-skin color is misjudged as skin color. In contrast, the misjudgment rate is very low in the segmentation (Fig.5(c)) by using our adjusted Gaussian model. Fig.6 shows the detection result on face images. The segmentation (Fig.6(c)) obtained through adjusted Gaussian model is much better than the segmentation (Fig.6(b)) processed by using standard Gaussian model.

Table 1 shows the number of skin-color pixels detected by adjusted model is much larger than that detected by the standard model. Through Table 1, we can see that the adjusted Gaussian skin-color model outperforms the standard Gaussian skin-color model.

Table 1. Result of experiments on face images

Methods	No. all pixels in test images	No. the pixels be detected	Ratio between detected pixels and all pixels (%)
Standard model	2665872	425007	15.94%
Adjusted model	2665872	669042	25.10%

4 Conclusions

As an improvement to standard Gaussian skin-color model, we analyze the adjusted Gaussian skin-color model based on PCA. The experimental results indicate that in face images, our model can detect much more skin-color than the standard model; while in non-face images, very lower misjudgment rate is obtained with our model. Furthermore, the flexibility of our adjusted model can be used in many related color-detection system and applied to real-time application.

Acknowledgements. The authors thank Bin Huang for his help and discussion during preparing this paper. This work was partially supported by the National Natural Science Foundation of China via the grant NSFC 60375022.

References

1. Bergasa, L. M., Mazo, M., Gardel, A., Sotelo, M. A., Boquete, L.: Unsupervised and Adaptive Gaussian Skin-Color Model, *Image and Vision Computing*, vol. 18, (2000) 987-1003
2. Greenspan, H., Goldberger, J., Eshet, I.: Mixture Model for Face Color Modeling and Segmentation, *Pattern Recognition Letters*, Vol. 22, (2001) 1525-1536
3. Soriano, M., Martinkauppi, B., Huovinenb, S., Laaksonenc, M.: Adaptive skin color modeling using the skin locus for selecting training pixels, *Pattern Recognition*, Vol. 36, (2003) 681-690
4. Vezhnevets, V., Sazonov, V., Andreeva, A.: A Survey on Pixel-Based Skin Color Detection Techniques, In *Proc. Graphicon-2003*, Moscow, Russia (2003) 85-92
5. Moghaddam, B., Pentland, A.: Probabilistic visual learning for object representation, *IEEE Trans. Pattern Anal. and Machine Intell.*, vol. 19, no. 7, (1997) 696-710
6. Turk, M., Pentland, A.: Eigenfaces for Recognition, *Journal of Cognitive Neuroscience*, vol. 3, no. 1, (1991) 71-86
7. Hsu, R. L., Mottaleb, M. A., Jain, A. K.: Face detection in color images, *IEEE Trans. Pattern Anal. and Machine Intell.*, vol. 24, no. 5, (2002) 696-706
8. Yang, M. H., Kriegman, D., Ahuja, N.: Detecting Faces in Images: A Survey, *IEEE Trans. Pattern Anal. and Machine Intell.*, vol. 24, no. 1, (2002) 34-58
9. Hjelm, E., Low, B. K.: Face Detection: A Survey, *Computer Vision and Image Understanding*, vol. 83, no. 3, (2001) 236-274

Convergence Analysis for Oja+ MCA Learning Algorithm

Jiancheng Lv, Mao Ye, and Zhang Yi

Computational Intelligence Laboratory, School of Computer Science and Engineering
University of Electronic Science and Technology of China, Chegndu 610054, P. R. China
zhangyi@uestc.edu.cn, jeson_cha@yahoo.com
http://cilab.uestc.edu.cn

Abstract. The convergence of Oja+'s MCA learning algorithm was proven in past by using a deterministic continuous-time dynamical system with restrictive condition that the learning rate must converge to zero. This paper gives a new proof for the convergence of the Oja+'s MCA algorithm via a corresponding deterministic discrete-time (DDT) dynamical system. This approach allows the learning rate to be some constant. In this paper, the fixed points of the DDT system are determined and an invariant set is obtained. Based on the invariant set, the convergence is proven.

1 Introduction

MCA neural networks can extract smallest principal component from input data without calculating the correlation matrix in advance. Many MCA neural networks have been proposed, such as the OJAn's algorithm[3], the Luo-Unbehauen-Cichocki's algorithm[5], the Feng-Bao-Jiao's learning algorthhm[4] and EXIN's algorithm[2]. Since the algorithms are described by discrete stochastic equations, it is very difficult to study the MCA learning algorithm directly[6]. Traditionally, the convergence of MCA learning algorithms are proved via some deterministic continuous time (DCT) system[1]. This method requies the learning rate must converge to zero. However, in practice, the learning rates are often taken as constants.

This paper discusses the convergence of the well known Oja+ MCA learning algorithm. A new proof is given by studying its corresponding deterministic discrete time system (DDT). This approach allows the learning rate to be some constant. Thus, the results are more applicable.

2 Oja+'s MCA Algorithm

Consider the linear single neuron network

$$y(k) = w^T(k)x(k), \quad k = 0, 1, 2, \cdots, \tag{1}$$

where $x(k)$ is the input vector, $y(k)$ is the output vector, and $w(k)$ is the weight vector. The Oja+ MCA learning algorithm is described as follows[1,8]

$$w(k+1) = w(k) - \eta[y(k)x(k) - \mu(y^2(k) + 1 - \|w(k)\|^2)w(k)] \quad (2)$$

where $k \geq 0$, $\eta > 0$ is the learning rate, $\mu > 0$ is some constant. $\|*\|$ represents the Euclidean norm. Taking the conditional expectation $E\{w(k+1)/w(0), x(i), i < n\}$ to (2) and using the condition expected value as the next iterate, a corresponding DDT system is given as

$$w(k+1) = w(k) - \eta[Cw(k) - w^T(k)Cw(k)w(k) + \mu(\|w(k)\|^2 - 1)w(k)] \quad (3)$$

for $k \geq 0$, where $C = E[x(k)x^T(k)]$. Next, we will study the convergence of (3) which will shed some lights on the convergence of (2).

Suppose $\lambda_1 \geq \lambda_2 \cdots \geq \lambda_n > 0$ be all the eigenvalues of the matrix C and the unit eigenvectors $v_i, i = 1, \cdots, n$ associated with the eigenvalue λ_i construct an orthonormal base in R^n. Clearly, all the eigenvectors v_i as well as the zero vector are equilibria of (3). Assume the multiplicity of the smallest eigenvalue (denoted by σ) of C is $m(1 \leq m \leq n)$, and denote $V_\sigma = \{v_{n-m+1}, \cdots, v_n\}$, Therefore, for $k \geq 0$, $w(k) \in R^n$ can be represented as

$$w(k) = \sum_{i=1}^{n} z_i(k)v_k = \sum_{i=1}^{n-m} z_i(k)v_k + \sum_{i=n-m+1}^{n} z_i(k)v_i \quad (4)$$

where $z_i(k)(i = 1, \cdots, n)$ are some constants.

Clearly, from (3) and (4), it holds that

$$z_i(k+1) = [1 + \eta(\mu - \lambda_i - w^T(k)(\mu I - C)w(k))]z_i(k), \quad (5)$$

for $k \geq 0$.

3 Convergence Analysis

Theorem 1. *The equilibrium point v_n is locally asymptotical stable if*

$$\eta < \frac{1}{\mu - \lambda_n}. \quad (6)$$

Other equilibrium points of (3) are unstable.

Proof: From (3), it follows that

$$\frac{\partial G}{\partial w} = I + \eta[(\mu I - C) - w^T(k)(\mu I - C)w(k) - 2(\mu I - C)w(k)w^T(k)]. \quad (7)$$

For the equilibrium 0, it holds that

$$\left.\frac{\partial G}{\partial w}\right|_0 = I + \eta(\mu I - C) = J_0. \quad (8)$$

The eigenvalues of J_0 are $\rho = \eta(\mu - \lambda_i) + 1 > 1$ for $i = 1, 2, \cdots, n$. According to the Lyapunov theory, the equilibrium point 0 is unstable.

For the equilibrium $v_j (j = 1, \cdots, n)$, it follows that

$$\left. \frac{\partial G}{\partial w} \right|_{v_j} = I + \eta[(\mu I - C) - (\mu - \lambda_j)I - 2(\mu I - C)v_j v_j^T] = J_j. \tag{9}$$

By some simple computations, the eigenvalues of J_j are given as

$$\begin{cases} \alpha = \eta(\lambda_j - \lambda_i) + 1, & \text{if } i \neq j \\ \alpha = -2\eta(\mu - \lambda_j) + 1, & \text{if } i = j. \end{cases} \tag{10}$$

If $j \neq n$, it holds that $|1 + \eta(\lambda_j - \lambda_n)| > 1$, clearly, v_j is unstable. Suppose that $j = n$, if $\eta < \dfrac{1}{\mu - \lambda_n}$, it follows that

$$\begin{cases} |\eta(\lambda_n - \lambda_i) + 1| < 1, & \text{if } i \neq n \\ |1 - 2\eta(\mu - \lambda_n)| < 1, & \text{if } i = n. \end{cases} \tag{11}$$

Thus, v_n is asymptotical stable. The proof is completed.

It is easy to see that some trajectories of (3) may diverge. Next, an invariant set is given to guarantee that the trajectories starting from the invariant set will be bounded.

Theorem 2. *Denote*

$$S = \left\{ w \mid w \in R^n, w^T(\mu I - C)w \leq \frac{1}{\eta} \right\}, \tag{12}$$

if $\mu > \lambda_1$ and

$$\eta \leq \frac{0.8899}{\mu}, \tag{13}$$

then, S is an invariant set of (3).

Proof (sketch): From (3) and (4), it follows that

$$w^T(k+1)(\mu I - C)w(k+1)$$
$$= \sum_{i=1}^n (\mu - \lambda_i)[1 + \eta(\mu - \lambda_i - w^T(k)(\mu I - C)w(k)]^2 z_i^2(k)$$
$$\leq \sum_{i=1}^n (\mu - \lambda_i)[1 + \eta(\mu - \lambda_n - w^T(k)(\mu I - C)w(k)]^2 z_i^2(k)$$
$$= [1 + \eta(\mu - \lambda_n - w^T(k)(\mu I - C)w(k)]^2 w^T(k)(\mu I - C)w(k)$$
$$\leq \max_{0 \leq s \leq \mu - \lambda_n + 1/\eta} \left\{ [1 + \eta(\mu - \lambda_n - s)]^2 \cdot s \right\}. \tag{14}$$

It can be calculated out that

$$w^T(k+1)(\mu I - C)w(k+1) \leq \frac{4}{27\eta}(1 + \eta(\mu - \lambda_n))^3. \tag{15}$$

Since $\eta \leq 0.8899/\mu$ and $\mu < \lambda_i$, then

$$\frac{4}{27\eta}(1 + \eta(\mu - \lambda_n))^3 \leq \frac{1}{\eta}. \tag{16}$$

Thus,

$$w^T(k+1)(\mu I - C)w(k+1) \leq \frac{1}{\eta}, \tag{17}$$

i.e., $w(k+1) \in S$. Therefore, S is an invariant set. The proof is completed.

Since S is an invariant set, each trajectory starting from a point in S will remain in it for ever. Moreover, the next theorem shows a more strong result. That is any trajectory starting from the invariant set will converge to the unit eigenvector associated with the smallest eigenvalue of the correlation matrix.

Theorem 3. *If*

$$\eta \leq \frac{0.8899}{\mu}, \tag{18}$$

where $\mu > \lambda_i$, then the invariant S is an attractive set of equilibrium v_n.

The proof for the above theorem is omitted here due to the space limitation.

4 Conclusions

The dynamic behavior of Oja+'s MCA learning algorithm has been studied. The convergence of the corresponding DDT system of Oja+'s MCA learning algorithm has been proved. It shows that the attractor of the DDT system is the unit eigenvector associated with the smallest eigenvalue of the correlation matrix. Its attractive set S has been determined. Unlike DCT method, the DDT approach allows the learning rate to be constant.

References

1. Oja, E.: Principal Components, Minor Components, and Linear Neural Network. IEEE Trans. Neural Networks Vol. **5** (1992) 927-935
2. Cirrincione, G., Cirrincione, M., Hérault, J., and Huffel, S. V.: The MCA EXIN Neuron for the Minor Component Analysis. IEEE Trans. Neural Networks Vol. **13** No.1 (2002) 160-187
3. Xu, L., Oja, E. and Suen, C.: Modified Hebbian Learning for Curve and Surface Fitting. Neural Networks Vol. **13** (1992) 441-459
4. Feng, D. Z., Bao, Z. and Jiao, L. C.: Total Least Mean Squares Algorithm. IEEE Trans. Signal Processing Vol. **46** (1998) 2122-2130

5. Luo, F., Unbehauen, R. and Cichocki, A.: A Minor Component Analysis Algorithm. Neural Networks Vol. **19** No. **2** (1997) 291-197
6. Zhang, Q. F.: On the Discrete-Time Dynamics of a PCA Learning Algorithm. Neurocomputing Vol. **55** (2003) 761-769
7. Zuffiria, P. J.: On the Discrete-Time Dynamics of the Basic Hebbian Neural-Network Node. Neural Computation Vol. **11** No. **2** (2000) 529-533
8. Fiori, S., Piazza, F.: Neural MCA for Robust Beamforming. Proc. of International Symposium on Circuits and Systems(ISCAS'2000) Vol. **III** (2000) 614-617

On the Discrete Time Dynamics of the MCA Neural Networks

Mao Ye and Zhang Yi

Computational Intelligence Laboratory, School of Computer Science and Engineering,
University of Electronic Science and Technology of China
maoye@sina100.com, zhangyi@uestc.edu.cn

Abstract. Minor component analysis(MCA) by neural network is endowed with a stochastic discrete-time(SDT) weight vector learning algorithm. It is very difficult to study such algorithm directly. Previous theoretical results on the SDT algorithm are based on its deterministic continuous-time(DCT) ODE asymptotic approximation. However, in general, they are not equivalent at all. Since the behavior of the conditional expectation of the weight vector can be studied by the deterministic discrete-time(DDT) algorithm, it is reasonable to study the SDT algorithm by its DDT algorithm indirectly. By studying the DDT algorithms, we can prove that some of previous MCA neural networks are globally convergent if the learning rate is a variable.

1 Introduction

MCA neural network is a statistical method used to extract the eigenvector associated to the smallest eigenvalue of the covariance matrix of the input data on-line, for instances[2,5,7,8,9]. It is important for many applications, such as beamforming[4], curve and surface fitting[9], adaptive signal processing[6].

All of these MCA neural networks are described by stochastic discrete-time (SDT) systems. And many previous proofs on the global convergence of the SDT algorithm are based on the continuous-time ODE asymptotic approximation [3, 5,9]. However, By the fundamental stochastic approximation theorem, one key assumption will not be satisfied in general. That is the learning rate of the algorithm should converge to zero, this condition is not always satisfied in practice due to the round-off limitation and tracking requirements. Thus the convergence of continuous-time differential equations does not imply the convergence of corresponding discrete ones when the learning rate does not converge to zero.

Up to now, the rigorous mathematical literature about MCA is very poor. The divergence of some MCA learning algorithms with constant learning rate is reported in [1]. However, for the variable learning rate, the discrete-time dynamics of previous MCA learning algorithms have not been studied yet. In this paper, the discrete-time dynamics of such learning algorithms is studied. With conditions on the learning rate, a few MCA learning algorithms will be globally convergent.

2 Analysis of MCA Neural Networks

Consider a linear neuron with the input $x(k) = [x_1(k), x_2(k), \cdots, x_n(k)]^T \in R^n$, the weight vector $w(k) \in R^n$ and the output $y(k) = w^T(k)x(k)$ at time k. $x(k)(k = 0, 1, \cdots)$ is a zero-mean discrete-time stochastic process. Such process is constructed as a sequence $x(0), x(1), \cdots$ of independent and identically distributed samples upon a distribution of a random variable.

Let the covariance matrix $C = E[x(k)x^T(k)]$ and $\lambda_1, \lambda_2, \cdots, \lambda_n$ be all of eigenvalues of C ordered by $\lambda_1 \geq \lambda_2 \geq \cdots > \lambda_n > 0$. Since C is a symmetric matrix, then there exists an orthonormal basis of R^n composed by eigenvectors of C. Suppose that $\{v_i | i = 1, \cdots, n\}$ is an orthonormal basis in R^n and each v_i is a unit eigenvector of C corresponding to the eigenvalue λ_i.

Assume $\eta(k) > 0$ is the learning rate and $\| * \|$ stands for the Euclidean norm, using the notations in [2], the existing MCA learning algorithms can be summarized as follows,

(1). The classical OJA learning algorithm[7] based on the Hebbian learning law:
$$w(k+1) = w(k) - \eta(k)y(k)\left[x(k) - y(k)w(k)\right], \tag{1}$$

(2). The OJA+ learning algorithm[4,7]:
$$w(k+1) = w(k) - \eta(k)\left[y(k)x(k) - y^2(k)w(k) + \mu\left(\|w(k)\|^2 - 1\right)w(k)\right], \tag{2}$$

where μ is a constant. The condition $\mu > \lambda_1$ is required for the stability reason [4].

(3). The EXIN[2], OJAn[9] and Luo-Unbehauen-Cichocki's[5] learning algorithm:
$$w(k+1) = w(k) - \eta(k)F(\|w(k)\|^2) \times \left\{y(k)x(k) - \frac{y^2(k)}{w^T(k)w(k)}w(k)\right\} \tag{3}$$

and
$$F(u) = \begin{cases} 1/u & \text{EXIN} \\ 1 & \text{OJAn} \\ u & \text{LUO,} \end{cases}$$

(4). The Feng-Bao-Jiao's[3] learning algorithm:
$$w(k+1) = w(k) - \eta(k)[w^T(k)w(k)y(k)x(k) - w(k)]. \tag{4}$$

Taking the conditional expectation $E\{w(k+1)|w(0), x(i), i < k\}$ to the OJA and OJA+ learning algorithms and identifying the conditional expected value as the next iterate. OJA DDT is
$$w(k+1) = w(k) - \eta(k)\left[Cw(k) - w(k)^T Cw(k)w(k)\right], \tag{5}$$

and OJA+ DDT is
$$w(k+1) = w(k) - \eta(k)[Cw(k) - w^T(k)Cw(k)w(k)$$
$$+ \mu(\|w(k)\|^2 - 1)w(k)]. \tag{6}$$

Theorem 1. *The vector v_n is the unstable fixed point of OJA DDT. And OJA+ DDT is not globally convergent.*

Proof. Linearizing (5) at v_n, and let $e(k) = w(k) - v_n$, after some basic computations, it follows that

$$e(k+1) = \left[I - \eta(k)\left(C - \lambda_n I - 2\lambda_n v_n v_n^T\right)\right] e(k) + o(\|e(k)\|^2).$$

Let $G(k) = \left[I - \eta(k)\left(C - \lambda_n I - 2\lambda_n v_n v_n^T\right)\right]$, the eigenvalues of matrix $G(k)$ are the following, $\gamma_n = 1 + 2\lambda_n \eta(k)$, and $\gamma_j = 1 - \eta(k)(\lambda_j - \lambda_n)$, $j \neq n$. By Lyapunov theory, since $\gamma_n > 1$, v_n is unstable.

Let $\overline{C} = -(C - \mu I)$ in OJA+ DDT, it follows that

$$w(k+1) = w(k) + \eta(k)[\overline{C}w(k) - w(k)^T \overline{C} w(k) w(k)]. \tag{7}$$

If $\mu > \lambda_1$, actually, algorithm (7) is the DDT of classical OJA PCA learning algorithm which extracts the largest eigenvector associated with the eigenvalue $\mu - \lambda_n$. By the proof in [10], (7) is not globally convergent. □

Taking the conditional expectation to (3), it follows that

$$w(k+1) = w(k) - \eta(k)F(\|w(k)\|^2) \times \left\{Cw(k) - \frac{w^T(k)Cw(k)}{w^T(k)w(k)}w(k)\right\}. \tag{8}$$

Assume V_{λ_n} is the eigensubspace of the smallest eigenvalue λ_n, and $V_{\lambda_n} = \text{span}\{v_n\}$.

Theorem 2. *(8) is globally convergent if $w(0)$ is not orthogonal to V_{λ_n} and $\eta(k)$ satisfies the following conditions,*

$$c_0 < \lambda_1 \eta(k) F(\|w(k)\|^2) \leq \gamma \quad \text{if} \quad \|w(k)\|^2 \leq 1,$$
$$c_0 \|w(k)\| < \lambda_1 \eta(k) F(\|w(k)\|^2) \|w(k)\| \leq \gamma \quad \text{if} \quad \|w(k)\|^2 \geq 1,$$

where $\gamma < 1$ for all $k \geq 0$ and c_0 is a small positive constant.

Proof. Define $\theta(k) = \left[1 - \frac{\eta(k)F(\|w(k)\|^2)(\lambda_{n-1} - \lambda_n)}{1 + \eta(k)\lambda_1 F(\|w(k)\|^2)}\right]^2$. Clearly, $0 < \theta(k) < \left[1 - \frac{c_0(\lambda_{n-1} - \lambda_n)}{\lambda_1(1+\gamma)}\right]^2 < 1$, for all $k > 0$. Since the vector set $\{v_1, v_2, \cdots, v_n\}$ forms an orthogonal basis of R^n, for each $k \geq 0$, $w(k)$ can be represented as

$$w(k) = z_n(k)v_n + \sum_{j=1}^{n-1} \epsilon_j(k) v_j \tag{9}$$

where $z_n(k)$ and $\epsilon_j(k)$ are some constants. Because $w(0)$ is not orthogonal to V_{λ_n}, $z_n(0) \neq 0$.

Substitute (9) into (8), it follows that

$$z_n(k+1) = \left[1 - \eta(k)F(\|w(k)\|^2)\left(\lambda_n - \beta(k)\right)\right] z_n(k), \tag{10}$$

and
$$\epsilon_j(k+1) = \left[1 - \eta(k)F(\|w(k)\|^2)\left(\lambda_j - \beta(k)\right)\right]\epsilon_j(k), \tag{11}$$
$1 \leq j \leq n-1$ for $k \geq 0$, where
$$\beta(k) = \frac{\sum_{i=1}^{n-1} \lambda_i \epsilon_i^2(k) + \lambda_n z_n^2(k)}{\|w(k)\|^2}.$$

It is easy to see that, $\lambda_1 \geq \beta(k) \geq \lambda_n$ for $k \geq 0$.

Since $1 - \eta(k)F(\|w(k)\|^2)(\lambda_n - \beta(k)) > 1 > 0$, for $k \geq 0$, then it follows from (10) that $z_n(k) > 0$ for all $k \geq 0$ if $z_n(0) > 0$, and $z_n(k) < 0$ for all $k \geq 0$ if $z_n(0) < 0$. Without loss of generality, we assume that $z_n(0) > 0$, thus, $z_n(k) > 0$ for all $k \geq 0$.

By (10), we have
$$z_n(k+1) = \left[1 - \eta(k)F(\|w(k)\|^2)(\lambda_n - \beta(k))\right] \cdot z_n(k) \geq z_n(k)$$
for $k \geq 0$. This shows that $\{z_n(k)\}$ are monotone increasing sequences.

For each $j (1 \leq j \leq n-1)$, it holds that $1 - \eta(k)F(\|w(k)\|^2)(\lambda_j - \beta(k)) \geq 1 - \gamma > 0$ for $k \geq 0$. Then, from (10) and (11), it follows that

$$\left[\frac{\epsilon_j(k+1)}{z_n(k+1)}\right]^2 = \left[\frac{1 - \eta(k)F(\|w(k)\|^2)(\lambda_j - \beta(k))}{1 - \eta(k)F(\|w(k)\|^2)(\lambda_n - \beta(k))}\right]^2 \cdot \left[\frac{\epsilon_j(k)}{z_n(k)}\right]^2$$
$$\leq \left[1 - \frac{\eta(k)F(\|w(k)\|^2)(\lambda_j - \lambda_n)}{1 + \eta(k)F(\|w(k)\|^2)(\beta(k) - \lambda_n)}\right]^2 \cdot \left[\frac{\epsilon_j(k)}{z_n(k)}\right]^2$$
$$\leq \left[1 - \frac{\eta(k)F(\|w(k)\|^2)(\lambda_j - \lambda_n)}{1 + \eta(k)\lambda_1 F(\|w(k)\|^2)}\right]^2 \cdot \left[\frac{\epsilon_j(k)}{z_n(k)}\right]^2$$
$$\leq \prod_{i=0}^{k+1} \theta(i) \cdot \left[\frac{\epsilon_j(0)}{z_n(0)}\right]^2 \to 0, \text{ as } k \to +\infty, \quad (j = 1, 2, \cdots, n-1).$$

It follows that
$$\lim_{k \to +\infty} \beta(k) = \lim_{k \to +\infty} \frac{\lambda_n z_n^2(k) + \sum_{j=1}^{n-1} \lambda_j \epsilon_j^2(k)}{z_n^2(k) + \sum_{j=1}^{n-1} \epsilon_j^2(k)} = \lambda_n.$$

By the properties of $\theta(i)$, there must exist constants $M > 0, \alpha > 0$, such that
$$\beta(k) - \lambda_n \leq Me^{-\alpha k} \text{ for } k \geq 0. \tag{12}$$

Using (10), it follows that
$$z_n(k+1) - z_n(k) = \eta(k)F(\|w(k)\|^2)(\beta(k) - \lambda_n)z_n(k)$$

for $k \geq 0$. By the conditions of Theorem 2 and the estimate of (12), it holds that $z_n(k) - z_n(0) \leq \frac{\gamma}{\lambda_1} \sum_{i=1}^{k} M e^{-\alpha i}$. Thus, we have $z_n(k)$ is bounded. Since $z_n(k)$ is monotone increasing, so

$$\lim_{k \to +\infty} z_n(k) = z_n. \tag{13}$$

Then, using (13),

$$\epsilon_j^2(k) = \left[\frac{\epsilon_j(k)}{z_n(k)}\right]^2 \cdot z_n^2(k) \to 0 \text{ as } k \to +\infty. \tag{14}$$

From (9), together with (13) and (14), it follows that

$$\lim_{k \to +\infty} w(k) = z_n v_n \in V_{\lambda_n}.$$

This shows that $w(k)$ converges to the smallest eigenvector direction. □

Remark 1. By the proof, we know the norm of weight vector $w(k)$ is increasing. Thus EXIN DDT is globally convergent when the learning rate is a constant.

Taking the conditional expectation to (4), it follows that

$$w(k+1) = w(k) - \eta(k)[\|w(k)\|^2 C w(k) - w(k)], \quad \text{for } k \geq 0. \tag{15}$$

Theorem 3. *Feng DDT (15) is bounded, and the weight vector converges to the v_n direction, if $w(0)$ is not orthogonal to V_{λ_n}, $c_0 < \eta(k)\|w(k)\|^2 \lambda_1 < \gamma$, $0 < \eta(k) < \gamma$ for all $k \geq 0$ where $\gamma < 1$ and c_0 is a small positive constant.*

Proof. Substitute (9) into (15), it follows that

$$z_n(k+1) = \left[1 - \eta(k)\left(\|w(k)\|^2 \lambda_n - 1\right)\right] z_n(k), \tag{16}$$

and

$$\epsilon_j(k+1) = \left[1 - \eta(k)\left(\|w(k)\|^2 \lambda_j - 1\right)\right] \epsilon_j(k), \tag{17}$$

$1 \leq j \leq n-1$ for $k \geq 0$.

If the conditions of Theorem 3 are satisfied, by the similar proof in Theorem 2, we have that

$$\lim_{k \to +\infty} \left|\frac{\epsilon_j(k)}{z_n(k)}\right|^2 = 0. \tag{18}$$

Next, we will show $w(k)$ is bounded. By (16), (17), it follows that

$$\|w(k+1)\| \leq \left[1 + \eta(k)\left(1 - \|w(k)\|^2 \lambda_n\right)\right] \|w(k)\|. \tag{19}$$

If $\|w(k)\|^2 \geq \dfrac{1}{\lambda_n}$, we have that $\|w(k+1)\| \leq \|w(k)\|$. When $0 \leq \|w(k)\|^2 \leq \dfrac{1}{\lambda_n}$, we have

$$\|w(k+1)\| \leq \left[1 + \eta(k)\left(1 - \|w(k)\|^2 \lambda_n\right)\right] \|w(k)\|$$
$$\leq [1 + \eta(k)] \|w(k)\| \leq 2\|w(k)\| \leq \dfrac{2}{\lambda_n}.$$

Thus, $\|w(k)\|$ is bounded. By (18), it follows that

$$\lim_{k \to +\infty} \epsilon_j^2(k) = \lim_{k \to +\infty} z_n^2(k) \left|\dfrac{\epsilon_j(k)}{z_n(k)}\right|^2 = 0 \qquad (20)$$

for $1 \leq j \leq n-1$, i.e., the weight vector converges to the v_n direction. □

Remark 2. If $\eta(k) = \eta_0$ for $k > 0$, η_0 is very small, and satisfies the conditions of Theorem 3, (16) can be approximated by its ODE equation with error $O(\eta_0)$. By defining Lyapunov function as in [3], combining with Theorem 3, we can prove that $w(k)$ of Feng DDT will converge to $\dfrac{v_n}{\sqrt{\lambda_n}} + O(\eta_0)$. This result also partially illustrates the reason of oscillation of algorithm (4).

3 Conclusions

The discrete-time dynamics of MCA neural networks is studied, rigorous mathematical analysis is given which does not change the discrete-time dynamical system to the corresponding ODE system. We find that some of previous MCA neural networks are globally convergent, and theoretically confirm that algorithm EXIN is the best MCA learning algorithm which is claimed in [2]. However, if the learning rate is a constant, by Remark 1, the convergence rate of EXIN learning algorithm is decreasing. Future study is to derive a learning algorithm which is globally convergent with a constant learning rate and a invariant convergence rate.

References

1. Anisse, T., Gianalvo, C.: Against the convergence of the minor component analysis neurons. IEEE Trans. Neural Network 10 (1999) 207-210
2. Cirrincione, G., Cirrincione, M., Herault, J., Huffel, S.V.: The MCA EXIN Neuron for the minor component anlysis. IEEE Trans. Neural Network 13 (2002) 160-187
3. Feng,D.Z., Bao,Z., Jiao, L.C.: Total least mean squares algorithm. IEEE Trans. Signal Processing 46 (1998) 2122-2130
4. Fiori,S., Piazza,F.: Neural MCA for robust beamforming. Proc. of International Symposium on Circuits and Systems(ISCAS'2000) III (2000) 614-617
5. Luo, F., Unbehauen, R., Cichocki, A.: A minor component analysis algorithm. Neural Networks 10(1997) 291-297

6. Mathew, G., Reddy, V.: Developement and analysis of a neural network approach to Pisarenko's harmonic retrieval method. IEEE Trans. Signal Processing 42(1994) 663-667
7. Oja, E.: Principal components, Minor components, and linear neural networks. Neural Networks 5(1992) 927-935
8. Oja, E., Wang, L.: Robust fitting by non-linear neural units. Neural networks 9(1996) 435-444
9. Xu, L., Oja,E., Suen,C.: Modified Hebbian learning for curve and surface fitting. Neural Networks 5(1992) 441-457
10. Zuffiria, P.J.: On the discrete time dynamics of the basic hebbian nerual network node. IEEE Trans. Neural Network 13(2002) 1342-1352

The Cook Projection Index Estimation Using the Wavelet Kernel Function

Wei Lin[1,2], Tian Zheng[1,2,3], Fan He[1], and Xian-bin Wen[2]

[1] Department of Applied Mathematics, Northwestern Polytechnical University, Xi'an, 710072, China
[2] Department of Computer Science & Technology, Northwestern Polytechnical University, Xi'an, 710072, China
[3] Key Laboratory of Education Ministry for Image Processing and Intelligent Control, Huazhong University of Science & Technology, Wuhan 430074, China
linwei@nwpu.edu.cn

Abstract. The key procedure of exploratory projection pursuit is to optimize a criterion function, which is called the projection pursuit index. The cook family index estimated by the wavelet kernel function is given in this paper. And the asymptotic unbiasedness and the convergence property of the projection index are proved. Also, as the fast computing of this kind projection index, it is suited for the processing of a large data. Some results of projection index based on the wavelet kernel estimation are compared with that of the Gauss kernel estimation.

1 Introduction

In remote sensing application, multispectral data and multichannel data, especially the polarimetric data are often used in the terrain classification. The high-dimensional data processing is quite important. Till now, several processing methods for the high-dimensional data are proposed, such as the principle component analysis (PCA), the independent component analysis (ICA), and the projection pursuit (PP). Although there are much difference in them, they are all based on linear projections, and PP has been rapidly gaining prominence in application [1-3]. In application, the computation of the projection indices is very important for PP. In general, explicit computation of a projection index is impossible though an index may be computed analytically for certain restricted situations. Some approximation methods, which are based on polynomial base functions, have been produced. For example, index based on expansions in terms of orthogonal polynomials, the Legendre index and the Hermite index [1-2]. In this paper, a new approximation technology for projection index is proposed, that is, wavelet kernel density functions is constructed to approximate index. The wavelet kernel density functions not only possess the feature kernel density function, but also compute simply and fast. That is indicated by experimentation.

This paper is organized as follows. The wavelet kernel function estimation of the cook projection index is given in section 2. Section 3 gave the property of the asymptotic unbiasedness and the convergence in mean squared and proved in appendix. Some experimentation results are given to compare with that of the Gauss kernel-based projection index in section 4. The property is proved in Appendix.

2 The Basic Theory

2.1 Projection Index

The projection index is a measure, which gauge the interestingness of the projection data. In general, the projection index may be divided into two kinds. The first kind is called the density-form projection index, which needs to compute the probability density function of projection data, such as the Cook family projection index, entropy index, etc. The other kind is called the non-density-form projection index, which needn't to compute the probability density function of projection data, for example, moments index, etc. Whereas the non-density-form projection index represents the data incompletely as it is based on the feature of data, the density-form projection index often is taken to be a measure. The Cook family projection index is one of them. It is given by

$$\text{PI}(y) = \int_{\mathbb{R}} \{f(y) - g(y)\}^2 g(y) dy. \tag{1}$$

Where $y = TZ$, and T is a general transformation on $\mathbb{R}$. $Z = \alpha^T X$, and X is the p-dimensional vector, which is centered and sphere. α indicates a projection direction vector, which satisfies $\alpha^T \alpha = 1$. $f(y)$ denotes the probability density function (PDF) of y. $g(y)$ is the PDF of the $\varphi(z)$ by T transformation, here $\varphi(z)$ is PDF of the standard normal distribution.

In a later part, we discuss the cook family projection index.

2.2 The Estimation of Projection Index Using the Wavelet Kernel Function

The computation of the density-form projection index is complicated. In most situations, it is not represents in an analytic form. So it always approximated using a polynomial function to replace the $f(y)$. The Legendre-based functions, the Hermite-based functions and the kernel-based function are often used to approach the $f(y)$. As the wavelet kernel function posses the feature like the kernel density function, meanwhile it posses the localization of frequency. Hence, we adopt the wavelet kernel function to estimate the projection index.

Definition 1. Assume that X is a p-dimensional random vector, $\alpha^T = (\alpha_1, \alpha_2, \cdots \alpha_p)$ indicates a p-dimensional projection direction, $Y = \alpha^T X$ is a one-dimensional data. $f(y)$ denotes the PDF of Y. $\hat{f}(y)$ is the estimation of the $f(y)$ using the wavelet kernel function. Then $\text{PI}(\hat{f}(y))$ (short form as $\widehat{\text{PI}}(y)$), which is the wavelet estimation of the projection index $\text{PI}(f(y))$, is given as follows:

$$\widehat{\mathrm{PI}}(y) = \int_{\mathbb{R}} \{\frac{1}{n}\sum_{i=1}^{n}\mathbb{K}_m(y,Y_i) - g(y)\}^2 g(y)dy. \qquad (2)$$

Where $\mathbb{K}_m(y,Y_i)$ is the reproducing kernel of V_m space. $\mathbb{K}(y,x) = \sum_k \phi(y-k)\phi(x-k)$. The scale function $\phi(x)$, that produces the reproducing kernel, is r-regular, and satisfies Zak's transformation conditions. For a detail expression consult literature [4], page 221.

3 Property of the Wavelet Estimation of the Projection Index

Two lemmas with density function are given before the properties of the wavelet estimation of the projection index.

Lemma 1. Let $f(y)$ be a one-dimensional density function and A be a set with $\mathbb{A} = \{y \in \mathbb{R} \mid f(y) > M\}$. M is an enough large positive. Then $\mathbb{A}$ is a zero-measure set.

Lemma 2. Assume $f(y), g(y)$ are continuous density functions, then there exists a constant M, such that

$$\int_{\mathbb{R}} f(y)g(y)dy \leq M. \qquad (3)$$

The proofs of lemmas are simply. As the length of paper is required, they are omitted. Thereinafter we'll descript the properties of the asymptotic unbiasedness and the convergence in mean squared of the estimation of wavelet kernel function of the cook projection index.

Theorem 1. (Asymptotic unbiasedness) Assumption $f(y)$ is a continuous one-dimensional probability density function, let $\hat{f}_m(y)$ denote to be the wavelet estimation of $f(y)$. The estimation of the cook family projection index $\widehat{\mathrm{PI}}(y)$ is an asymptotic unbiased estimation of $\mathrm{PI}(y)$. Where $g(y)$ is a known probability density function.

Theorem 2. (Convergence in mean squared) Supposed $f(y)$ and $g(y)$ both are one-dimensional continuous probability density function on the compact support $\mathbb{T} = (-T, T)$, $\hat{f}_m(y)$ is the wavelet estimation of $f(y)$, for the cook family projection index

$$\mathrm{PI}(y) = \int_{\mathbb{T}} (f(y) - g(y))^2 g(y) dy. \qquad (4)$$

The wavelet estimation can be represented by

$$\widehat{\mathrm{PI}}(y) = \int_{\mathbb{T}} (\hat{f}_m(y) - g(y))^2 g(y) dy. \qquad (5)$$

Then (i) if y belongs to the compact support set $\mathbb{T}$

$$E|\widehat{\mathrm{PI}}_m(y) - \mathrm{PI}(y)|^2 \to 0, \; m \to \infty, \text{and } m = O(\log n). \qquad (6)$$

(ii) if $f(y)$ belongs to the Sobolev smoothness space W^s, $s > \lambda + \frac{1}{2}$, and $m \approx \dfrac{\log 2}{2\lambda + 1} \cdot \log n$, then

$$E\,|\widehat{PI}_m(y) - PI(y)|^2 = O(n^{-\frac{2\lambda}{2\lambda+1}}). \tag{7}$$

The proofs of theorem 1 and theorem 2 are given in appendix A and B.

4 Experimental Results and Analysis

We adopted sample data form different mixture models to compute the cook family projection index. The mixture models produce by normal, lognormal, Rayleigh distribution. For the briefness, $g(y)$ replace by $\varphi(y)$.

Three methods are used for computing the cook family projection index. The results are show in Table 1. Where $f(y) = \pi_1 f_1(y) + (1 - \pi_1) f_2(y)$

Table 1. The values of the cook family projection index by three methods

Mixture type $\pi_1 = 0.3$	Sample size	True value	Kernel density	Wavelet kernel
Mix normal $f_1(y) \sim N(-5,1)$ $f_2(y) \sim N(5,1)$	500	0.0922	0.0912	0.0912
	1000	0.0905	0.0910	0.0910
	1500	0.0903	0.0899	0.0899
	2000	0.0899	0.0897	0.0897
Mix_rayleigh $f_1(y) \sim Rayl(1)$ $f_2(y) \sim Rayl(3)$	500	0.0945	0.0925	0.0925
	1000	0.0934	0.0941	0.0941
	1500	0.0924	0.0910	0.0910
	2000	0.0905	0.0896	0.0898
Mix lognormal $f_1(y) \sim LN(-5,1)$ $f_2(y) \sim LN(5,1)$	500	0.0927	0.0923	0.0925
	1000	0.0918	0.0915	0.0915
	1500	0.0902	0.0904	0.0903
	2000	0.0890	0.0889	0.0890

From table 1, the cook family projection index can be approximate by the kernel density and wavelet kernel density function very well. But the kernel density estimation of the index needs to calculate the different optimized bandwidth for various numbers of samples. And the values of projection index using wavelet kernel function are almost of no difference for different scales.

5 Conclusions

The estimation of the cook family projection index with wavelet kernel function is given in this paper. The property of this estimation, and the computing of projection index are discussed. As the localization of wavelet transformation in time and frequency area, it is suitable for large mount of computing, especially under the situation of the image processing. The later work about the projection direction will be considered, under the condition of the wavelet kernel-based projection index.

Acknowledgment. This work is supported in part by the National Natural Science Foundation of China (60375003) and the Doctorate Foundation of Northwestern Polytechnical University (CX200327).

References

1. Hall, P.: On Polynomial-based Projection Indices for Exploratory Projection Pursuit. Ann. Statistics. 17 (1989) 589-605
2. Cook, D., Buja, A., and Cabrera, J.: Projection Pursuit Indices Based on Expansions with Orthonormal Functions. Journal of Computational and Graphical Statistics. 2 (1993) 225-250
3. Klinke, S. and Cook, D.: Kernel-based Projection Pursuit Indices in XGobi. Statistical Software Newsletter in Computational Statistics and Data Analysis. 25 (1997) 363-369
4. Vidakovic, B. (ed.): Statistics Modeling by Wavelets. John Wiley & Sons publication, New York (1999)

Appendix A. The Proof of the Theorem 1

Proof: $|\widehat{EPI}_m(y) - PI(y)| = |E\int_\mathbb{R}(\widehat{f}_m(y) - g(y))^2 g(y)dy - \int_\mathbb{R}(f(y) - g(y))^2 g(y)dy|$

$\leq E\int_\mathbb{R}|(\widehat{f}_m(y) - f(y))(\widehat{f}_m(y) + f(y) - 2g(y))g(y)|dy$

$\leq E\left(\int_\mathbb{R}|(\widehat{f}_m(y) - f(y))\sqrt{g(y)}|^2 dy\right)^{\frac{1}{2}}\left(\int_\mathbb{R}|(\widehat{f}_m(y) + f(y) - 2g(y))\sqrt{g(y)}|^2 dy\right)^{\frac{1}{2}}$

Hereinafter proof the expression $\int_\mathbb{R}|(\widehat{f}_m(y) + f(y) - 2g(y))|^2 g(y)dy$ is bounded.

When there least exists one of $\widehat{f}_m(y), f(y), g(y)$, which is non-bounded, suppose that $\mathbb{A}_1 = \{y \in \mathbb{R} | \widehat{f}_m(y) > M\}$, $\mathbb{A}_2 = \{y \in \mathbb{R} | f(y) > M\}$, $\mathbb{A}_3 = \{y \in \mathbb{R} | g(y) > M\}$. Based on **Lemma 1**, $\mathbb{A} = \mathbb{A}_1 \cup \mathbb{A}_2 \cup \mathbb{A}_3$ is a zero-measure set.

So, $\int_\mathbb{R}|(\widehat{f}_m(y) + f(y) - 2g(y))|^2 g(y)dy$

$= \int_{\mathbb{R}-\mathbb{A}}|(\widehat{f}_m(y) + f(y) - 2g(y))|^2 g(y)dy + \int_\mathbb{A}|(\widehat{f}_m(y) + f(y) - 2g(y))|^2 g(y)dy$

$= \int_{\mathbb{R}-\mathbb{A}}|(\widehat{f}_m(y) + f(y) - 2g(y))|^2 g(y)dy \leq \int_{\mathbb{R}-\mathbb{A}} 16M^2 g(y)dy = 16M^2$.

And $|\widehat{f}_m(y) - f(y)| = O(2^{-\lambda m})$. For a proof and comprehensive discussion consult [4], pages 221-222.

Hence $|\widehat{EPI}_m(y) - PI(y)| = ME\left(O(2^{-\lambda m})\int_\mathbb{R} g(y)dy\right)^{\frac{1}{2}} = O(2^{-\lambda m})$.

If $m \to \infty$, $|\widehat{EPI}_m(y) - PI(y)| = O(2^{-\lambda m}) \to 0$, i.e. $\lim_{m \to \infty}|\widehat{EPI}_m(y) - PI(y)| = 0$. #

Appendix B. The Proof of the Theorem 2

Proof: $|\widehat{\text{PI}}_m(y) - \text{PI}(y)| = |\int_{\mathbb{T}} (\widehat{f}_m(y) - g(y))^2 g(y)dy - \int_{\mathbb{T}} (f(y) - g(y))^2 g(y)dy|$

$\leq \int_{\mathbb{T}} |(\widehat{f}_m(y) - f(y))(\widehat{f}_m(y) + f(y) - 2g(y))g(y)| dy$

$\leq \left(\int_{\mathbb{T}} |(\widehat{f}_m(y) - f(y))\sqrt{g(y)}|^2 dy\right)^{\frac{1}{2}} \left(\int_{\mathbb{T}} |(\widehat{f}_m(y) + f(y) - 2g(y))\sqrt{g(y)}|^2 dy\right)^{\frac{1}{2}}$.

For the compact support set $\mathbb{T}$, based on the proof of the Theorem 1, we have that $\int_{\mathbb{R}} |(\widehat{f}_m(y) + f(y) - 2g(y))|^2 g(y)dy$ is bounded, i.e. there exists a constant $M > 0$, such that $\int_{\mathbb{T}} |(\widehat{f}_m(y) + f(y) - 2g(y))|^2 g(y)dy \leq 16M^2$.

The same proof as above that $\int_{\mathbb{T}} |\widehat{f}_m(y) - f(y)|^2 g(y)dy$ is also bounded.

Then $E|\widehat{\text{PI}}_m(y) - \text{PI}(y)|^2$

$\leq E\left(\int_{\mathbb{T}} |(\widehat{f}_m(y) - f(y))|^2 g(y)dy\right)\left(\int_{\mathbb{T}} |(\widehat{f}_m(y) + f(y) - 2g(y))|^2 g(y)dy\right)$

$\leq 16M^2 E\left(\int_{\mathbb{T}} |\widehat{f}_m(y) - f(y)|^2 g(y)dy\right) = 16M^2 \int_{\mathbb{T}} E|\widehat{f}_m(y) - f(y)|^2 g(y)dy$.

(i) The theorem 7.2.2(i) in literature [4], for arbitrarily $\varepsilon > 0$

$E|\widehat{f}_m(y) - f(y)|^2 < \varepsilon/M^2$, $m \to \infty, m = O(\log n)$,

Thereby $E|\widehat{\text{PI}}_{j_0}(y) - \text{PI}(y)|^2 \leq \varepsilon \int_{\mathbb{T}} g(y)dy = \varepsilon$,

So $E|\widehat{\text{PI}}_m(y) - \text{PI}(y)|^2 \to 0$, if $m \to \infty$, and $m = O(\log n)$.

(ii) The theorem 7.2.2(ii) in literature [4],

$E|\widehat{\text{PI}}_m(y) - \text{PI}(y)|^2 \leq M^2 \int_{\mathbb{T}} E|\widehat{f}_m(y) - f(y)|^2 g(y)dy = O(n^{-\frac{2\lambda}{2\lambda+1}})$. #

Automatic Cluster Number Determination via BYY Harmony Learning

Xuelei Hu and Lei Xu

Department of Computer Science and Engineering,
The Chinese University of Hong Kong, Shatin, NT, Hong Kong
{xlhu, lxu}@cse.cuhk.edu.hk

Abstract. Selection of the number of clusters is a crucial problem in clustering. Conventionally, it was effected via cost function based criteria such as AIC and MDL. In this paper we empirically investigate automatic selection of the number of clusters via BYY harmony empirical learning. Results of experiments show that the true number of clusters can be automatically obtained during BYY harmony empirical learning. It is superior to conventional methods in that it needs much less computational cost.

1 Introduction

Clustering is a typical unsupervised learning technique for finding groups or clusters in multivariate data. When the optimum number of clusters is not known a priori, model selection techniques can be used to determine the number of clusters based on mixture models [1,2]. It is conducted via a two-phase style implementation that first obtains a set of candidate models under the maximum likelihood (ML) principle, usually in help of expectation-maximization (EM) algorithm [3], and then select the 'optimal' model according to one of model selection criteria, such as Akaike's information criterion (AIC) [4], the minimum description length (MDL) criterion [5] which formally coincides with the Bayesian inference criterion (BIC) [6]. A major shortcoming of this implementation is that a whole set of candidate models must be obtained first.

Bayesian Ying-Yang (BYY) harmony learning is capable of performing not only the two-phase style model selection via a set of new model selection criteria but also automatic model selection during parameter learning [2,7]. As a typical example, BYY harmony learning has been used on Gaussian mixture with several new results. Started from 1995 [8,9,10] and further summarized in [7], the selection of the number of clusters can be made either by criteria or automatically via adaptive algorithms from BYY harmony learning. Among them, the simplest example is the hard-cut EM algorithm [8,11]. This paper further empirically investigates automatic cluster number determination via implementing the hard-cut EM algorithm. Comparing with the conventional two-phase implementation by the criteria AIC and MDL, experiments have shown that the same or similar optimal model can be automatically obtained by using the hard-cut EM algorithm with much less computing cost.

2 Gaussian Mixture Based Clustering and Model Selection on Number of Clusters

Gaussian mixture model based clustering assumes that the data are distributed according to a mixture of Gaussian distributions, denoted by

$$p(x) = \sum_{l=1}^{k} \alpha_l G(x|m_l, \Sigma_l) \qquad (1)$$

with $\alpha_l \geq 0, l = 1, ..., k$, and $\sum_{l=1}^{k} \alpha_l = 1$, where and throughout this paper, $G(x|m, \Sigma)$ denotes a Gaussian density with mean vector m and covariance matrix Σ. Provided with a set of observations $\{x_t\}_{t=1}^{n}$ the task of model based clustering is to estimate the number of components k and the model parameters $\theta_k = \{m_1, ..., m_k, \Sigma_1, ..., \Sigma_k, \alpha_1, ..., \alpha_k\}$. Given k, one most popular method to estimate θ is maximum likelihood, i.e. estimating θ_k by maximizing the log likelihood function

$$L(\theta_k) = \ln \prod_{i=1}^{n} p(x_i) = \sum_{i=1}^{n} \ln \sum_{l=1}^{k} \alpha_l G(x_i|m_l, \Sigma_l), \qquad (2)$$

which can be effectively implemented by the expectation-maximization (EM) algorithm [3].

One remaining problem is how to select the number of components, which is usually conducted via a two-phase style implementation. In the first phase, we define a range of values of k from k_{min} to k_{max} which is assumed to contain the optimal k. At each specific k, we estimate the parameters θ_k by maximizing Eq. (2). In the second phase, we obtain an estimate of the number $\hat{k}$ according to

$$\hat{k} = \arg\min_{k} \{J(\hat{\theta}_k, k), k = k_{min}, ..., k_{max}\}, \qquad (3)$$

where $J(\hat{\theta}_k, k)$ is some model selection criterion.

Two frequently used model selection criteria are AIC and MDL. Generally, these two model selection criteria take the form [12]

$$J(\hat{\theta}_k, k) = -2L(\hat{\theta}_k) + A(n)D(k) \qquad (4)$$

where $L(\hat{\theta}_k)$ is the log likelihood Eq. (2), $D(k)$ is the number of independent parameters in k-component mixture, $A(n)$ is a function with respect to the number of observations. According to [1], generally we have $D(k) = (k-1) + k(d+d(d+1)/2)$ where d is the dimension of x. If an identical spherical covariance $\Sigma_l = \sigma^2 I$ is considered we simply have $D(k) = k + kd$. Moreover, different approaches lead to different choices of $A(n)$. We have $A(n) = 2$ for AIC [4] and $A(n) = \ln n$ for MDL [5].

Instead of selecting k from a set of candidate models, BYY harmony learning can make automatic cluster number selection in parallel with parameter learning [7]. In the special case of Gaussian mixture model, BYY harmony learning in a situation without considering any regularization, called empirical learning, estimates the parameters by maximizing the harmony measure Eq. (38) in [7] with $z_x = 1$, also denoted by

$$H(\theta_k, k) = \frac{1}{n} \sum_{i=1}^{n} \sum_{l=1}^{k} P(l|x_i) \ln(\alpha_l G(x_i|m_l, \Sigma_l)). \quad (5)$$

Explained in [7] under the name of the least complexity nature, this maximization will push some α_l towards to zero if the corresponding component $G(x_i|m_l, \Sigma_l)$ is extra and thus can be discarded for modelling a given set of observations. Thus, as long as k is initialized at a value large enough, an appropriate k will be determined automatically in parallel with parameter learning that implements Eq. (5).

A direct and simplest way for implementing Eq. (5) is the hard-cut EM algorithm which was firstly proposed in [8,11]. It alternates two steps. First, by maximizing $H(\theta_k, k)$ with respective a free $P(l|x_i)$, resulting in

$$P(l|x_i) = \begin{cases} 1, \, l = \arg\min_l \ln(\alpha_l G(x_i|m_l, \Sigma_l)); \\ 0, \, \text{otherwise}. \end{cases} \quad (6)$$

Second, the rest parameters are updated also via Eq. (5) with $P(l|x_i)$ fixed as above. Specifically, we have

Step 1 For $i = 1, ..., n$ and $l = 1, ..., k$ calculate $\hat{P}(l|x_i)$ according to Eq. (6)
Step 2 For $l = 1, ..., k$, update parameters by $\hat{\alpha}_l = \frac{1}{n} \sum_{i=1}^{n} \hat{P}(l|x_i)$, $\hat{m}_l = \frac{1}{n\hat{\alpha}_l} \sum_{i=1}^{n} \hat{P}(l|x_i) x_i$, and $\hat{\Sigma}_l = \frac{1}{n\hat{\alpha}_l} \sum_{i=1}^{n} \hat{P}(l|x_i)(x_i - \hat{m}_l)(x_i - \hat{m}_l)^T$.

Several other implementations of Eq. (5) were proposed in [7]. One is implementing Eq. (5) via the gradient ascending approach either in a batch way or adaptively per sample (see Tab.2 in [7]), in help of $\alpha_l = e^{c_l}/\sum_{j=1}^{k} e^{c_j}$ for satisfying its constraint (see Tab.2(D) in [7]). The other way is inserting $P_b(l|x_i) = \alpha_l G(x_i|m_l, \Sigma_l)/\sum_{l=1}^{k} \alpha_l G(x_i|m_l, \Sigma_l)$ (e.g., see Eq. (40) in [7]) into Eq. (5) and then updating the rest parameters, which was further investigated in [13]. Another is letting $P(l|x_i)$ in Step 2 of the above hard-cut EM being replaced by a combination $P(l|x_i) + \gamma P_b(l|x_i)$ with γ gradually decreasing from a large value to zero (see Eq. (42) and Eq. (43) in [7]).

3 Simulation Results, Comparison, and Discussion

We illustrate the experimental performance of BYY harmony empirical learning for automatic selection of the number of clusters simply via the above hard-cut EM algorithm. Also, we show an experimental comparison with two-phase style methods via the EM algorithm for ML learning and the criteria AIC and MDL for model selection. The initial parameter estimates for the hard-cut EM algorithm and the EM algorithm were obtained by k-mean algorithm and the initial estimates of α_l were set equally to $1/k$.

3.1 Automatic Model Selection

Data points of size 1000 were randomly generated from a 5-component bivariate Gaussian mixture distribution with equal mixing priors, and equal spherical covariance matrix as shown in Fig. 1(a). We used a Gaussian mixture with equal spherical covariance matrix and initially set $k = 10$. The initial model is shown in Fig. 1(b). Shown in Fig. 1(c) and (d) are the results of the hard-cut EM algorithm after 10 iterations and after 17 iterations. We observe that extra components were removed during learning by pushing corresponding α_l into zero. The algorithm converges to the true model and selects the correct number of clusters.

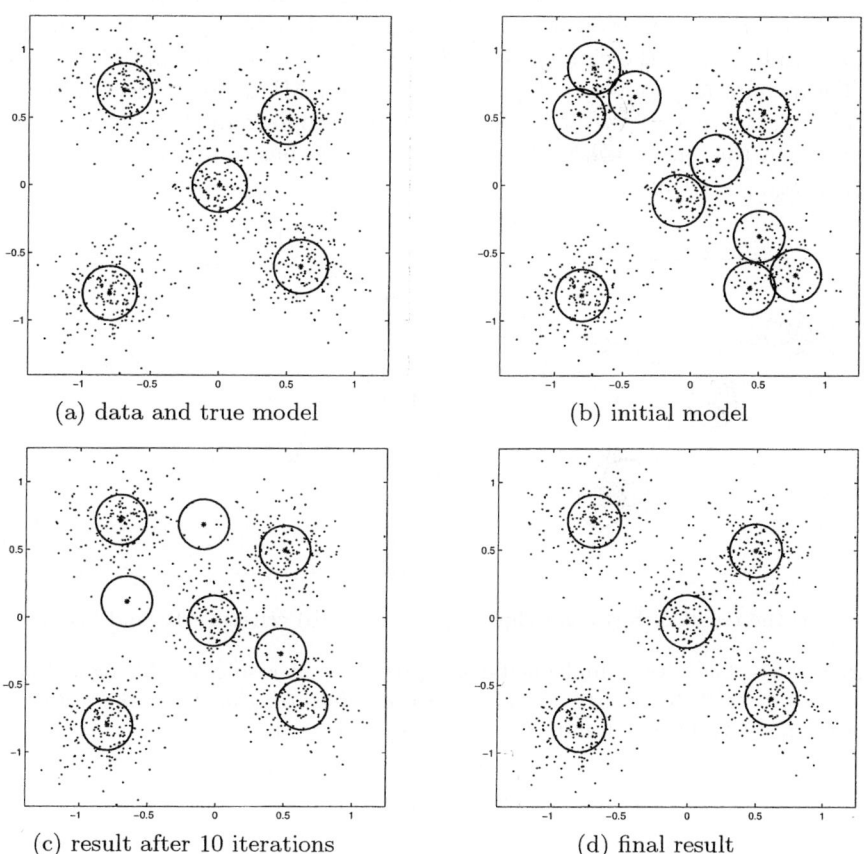

Fig. 1. Automatic selection of the number of clusters via BYY harmony empirical learning. (a) data points and the true model with $k = 5$, (b) initial model with $k = 10$, (c) the model after 10 iterations with $k = 8$, (d) the final learned model after 17 iterations with $k = 5$

3.2 Comparison with EM Algorithm

We implemented the EM algorithm on the same data set with the same initialization as shown in Fig. 1(b). The final result is shown in Fig. 2(a). The EM algorithm cannot select the number of clusters because the maximized log likelihood function is a non-decreasing function of k. Moreover, the EM algorithm required 382 iterations before convergence. After many experiments, which cannot be shown here due to space limitation, we find that the hard-cut EM algorithm converges faster than EM algorithm about 10 times.

3.3 Comparison with Conventional Methods

We further show the results of two-phase style methods by using criteria: AIC and MDL on the same data set. From $k_{min} = 1$ to $k_{max} = 10$, we used EM algorithm to obtain the candidate models and then used AIC and MDL to select k. The results are shown in Fig. 2(b). AIC and MDL both selected the right number of clusters. However it took much more time than the automatic model selection by the hard-cut EM algorithm because we has to perform EM algorithm $k_{max} - k_{min} + 1$ times.

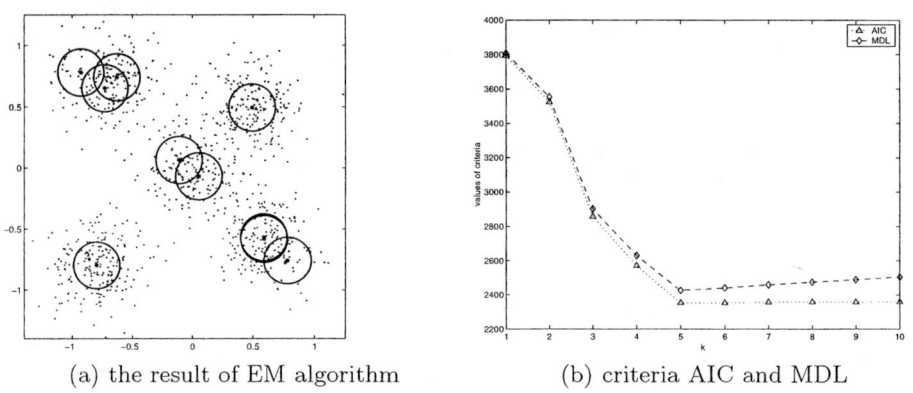

(a) the result of EM algorithm (b) criteria AIC and MDL

Fig. 2. Two-phase style implementation. (a) the final result of EM algorithm with $k = 10$ after 382 iterations, (b) the curves of values of $J(\hat{\theta}_k, k)$ of the criteria AIC (dotted line with triangle) and MDL (dashed line with diamond)

4 Conclusion

We have empirically investigated automatic selection of the number of clusters via BYY harmony empirical learning and compared it with the conventional ML learning based model selection criteria: AIC and MDL. We have observed that the BYY harmony empirical learning based automatic model selection is more advantageous in term of needing much less computing cost.

Acknowledgement. The first author would like to thank Dr. Kai Chun Chiu, Dr. Zhiyong Liu, and Dr. Changyin Sun for helpful discussions.

References

1. Bozdogan, H.: Mixture-Model Cluster Analysis Using Model Selection Criteria and a New Informational Measure of Complexity. In: Proc. of the First US/Japan Conference on the Frontiers of Statistical Modeling: An Informational Approach. Volume 2., Dordrecht, the Netherlands, Kluwer Academic Publishers (1994) 69–113
2. Xu, L.: BYY Harmony Learning, Structural RPCL, and Topological Self-Organizing on Mixture Models. Neural Networks **15** (2002) 1125–1151
3. Dempster, A., Laird, N., D.Rubin: Maximum Likelihood Estimation from Incomplete Data via the EM Algorithm. J. Royal Statistical Soc. B **39** (1977) 1–38
4. Akaike, H.: A New Look at Statistical Model Identification. IEEE Transactions on Automatic Control **19** (1974) 716–723
5. Rissanen, J.: Modeling by Shortest Data Description. Automatica **14** (1978) 465–471
6. Schwarz, G.: Estimating the Dimension of a Model. The Annals of Statistics **6** (1978) 461–464
7. Xu, L.: Best Harmony, Unified RPCL and Automated Model Selection for Unsupervised and Supervised Learning on Gaussian Mixtures, Three-Layer Nets and ME-RBF-SVM Models. International Journal of Neural Systems **11** (2001) 43–69
8. Xu, L.: Bayesian-Kullback Coupled Ying-Yang Machines: Unified Learings and New Results on Vector Quantization. In: Proc. of ICONIP95, Beijing, China (1995) 977–988
9. Xu, L.: Cluster Number Selection, Adaptive EM Algorithms and Competitive Learnings, Invited Talk. In: Proc. of IEEE Intl Conf. on NNSP95. Volume 2., Nanjing, China (1995) 1499–1502
10. Xu, L.: New Advarces on the Ying-Yang Machine, Invited Talk. In: Proc. of 1995 Intl Symp. on Artificial Neural Networks, Taiwan (1995) IS07–12
11. Xu, L.: Bayesian Ying-Yang Machine, Clustering and Number of Clusters. Pattern Recognition Letters **18** (1997) 1167–1178
12. Sclove, S.L.: Some Aspects of Model-Selection Criteria. In: Proc. of the First US/Japan Conference on the Frontiers of Statistical Modeling: An Informational Approach. Volume 2., Dordrecht, the Netherlands, Kluwer Academic Publishers (1994) 37–67
13. Ma, J., Wang, T., Xu, L.: A Gradient BYY Harmony Learning Rule on Gaussian Mixture with Automated Model Selection. in press, Neurocomputing (2004)

Unsupervised Learning for Hierarchical Clustering Using Statistical Information

Masaru Okamoto, Nan Bu, and Toshio Tsuji

Department of Artificial Complex System Engineering
Hiroshima University
Kagamiyama 1-4-1, Higashi-Hiroshima, Hiroshima, 739-8527 JAPAN
{okamoto, bu, tsuji}@bsys.hiroshima-u.ac.jp
http://www.bsys.hiroshima-u.ac.jp

Abstract. This paper proposes a novel hierarchical clustering method that can classify given data without specified knowledge of the number of classes. In this method, at each node of a hierarchical classification tree, log-linearized Gaussian mixture networks [2] are utilized as classifiers to divide data into two subclasses based on statistical information, which are then classified into secondary subclasses and so on. Also, unnecessary structure of the tree can be avoided by training in a cross-validation manner. Validity of the proposed method is demonstrated with classification experiments on artificial data.

1 Introduction

Recently, there have been growing interests in using bioelectric signals such as electromyogram (EMG) to conduct man-machine interface. In order to discriminate an operater's intentions from bioelectric signals efficiently, several attempts have been made so far [1], [2]. Generally, such pattern discrimination is performed by estimating the relationship between the bioelectric signals as feature vectors and the corresponding intentions as class labels. However, difference between classes in the bioelectric signals of elderly or handicapped people is ambiguous, and this relates to poor reliability of the class labels available. To overcome this problem, clustering analysis has been widely adopted, in which a collection of patterns is organized into clusters based on similarity.

The previously proposed clustering analysis techniques can be dichotomized as either k-means algorithm or hierarchical clustering. The k-means algorithm identifies a partition of the input space. On the other hand, the hierarchical clustering performs a nested series of partitions and finally performs a grouping with a suitable number of classes. Also, in order to determine the number of class automatically, a clustering algorithm using self organizing maps (SOM) [5] has been proposed [6]. In this method, estimation of the number of classes is carried out based on the number of the data belonging to each node of SOM. However, when parameters in this method were not set up appropriately, such method may fail to perform satisfying clustering for complicated data.

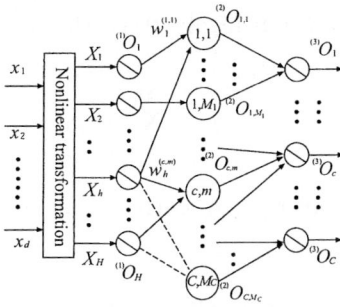

Fig. 1. Structure of LLGMN

In this paper, a novel hierarchical clustering method is proposed. In this method, a probabilistic NN that is derived from the Gaussian mixture model (GMM), called a log-linearized Gaussian mixture network (LLGMN) [2], is utilized for partition at each non-terminal node. The proposed method can estimate the number of terminal nodes corresponding to the number of classes according to the statistical information obtained solely from the training data.

2 LLGMN

2.1 Structure of LLGMN

The structure of LLGMN is shown in Fig.1. First, an input vector $\boldsymbol{x} \in \Re^D$ is converted into a modified vector $\boldsymbol{X}$ as follows:

$$\boldsymbol{X} = [1, \boldsymbol{x}^T, x_1^2, x_1x_2, \cdots, x_1x_D, x_2^2, x_2x_3, \cdots, x_2x_D, \cdots, x_D^2]^T . \tag{1}$$

The first layer consists of H units corresponding to the dimension of $\boldsymbol{X}$, and an identity function is used for an activation function of each unit. The variable $^{(1)}O_h$ ($h = 1, \cdots, H$) denotes the output of the hth unit in the first layer. Each unit in the second layer receives the output $^{(1)}O_h$ weighted by coefficients $w_h^{(c,m)}$ ($c = 1, \cdots, C; m = 1, \cdots, M_c$). C denotes the number of classes and M_c is the number of components belonging to class c. The relationship between the input $^{(2)}I_{c,m}$ and the output $^{(2)}O_{c,m}$ of unit $\{c, m\}$ in the second layer can be defined as

$$^{(2)}I_{c,m} = \sum_{h=1}^{H} {}^{(1)}O_h w_h^{(c,m)} , \tag{2}$$

$$^{(2)}O_{c,m} = \frac{\exp[^{(2)}I_{c,m}]}{\sum_{c'=1}^{C} \sum_{m'=1}^{M_{c'}} \exp[^{(2)}I_{c',m'}]} , \tag{3}$$

where $w_h^{(C,M_C)} = 0$. The relationship between the input $^{(3)}I_c$ and the output is described as,

$$^{(3)}O_c = {}^{(3)}I_c = \sum_{m=1}^{M_c} {}^{(2)}O_{c,m} . \quad (4)$$

The output of the third layer $^{(3)}O_c$ corresponds to the posterior probability $P(c|x)$ of class c.

2.2 Supervised Learning Algorithm [2]

Consider a training set $\{x^{(n)}, T^{(n)}\}$ $(n = 1, \cdots, N)$, where $T^{(n)} = \{T_1^{(n)}, \cdots, T_C^{(n)}\}$. If the input vector $x^{(n)}$ belongs to class c, $T_c^{(n)} = 1$, and $T_{c'}^{(n)} = 0$ for all of the other class c'. An energy function according to the minimum log-likelihood training criterion can be derived as:

$$J_{SV} = -\sum_{n=1}^{N}\sum_{c=1}^{C} T_c^{(n)} \log{}^{(3)}O_c^{(n)} . \quad (5)$$

In the training process, modification of the LLGMN's weight $w_h^{(c,m)}$ is defined as:

$$\Delta w_h^{(c,m)} = -\eta \sum_{n=1}^{N} \frac{\partial J_{SV}^n}{\partial w_h^{(c,m)}} , \quad (6)$$

$$\frac{\partial J_{SV}^n}{\partial w_h^{(c,m)}} = ({}^{(2)}O_{c,m}^{(n)} - \frac{{}^{(2)}O_{c,m}^{(n)}}{{}^{(3)}O_c^{(n)}} T_c^{(n)}) X_h^{(n)} , \quad (7)$$

where $\eta > 0$ is the learning rate.

3 Hierarchical Clustering

The divisive clustering starts from a single cluster, and terminates when a termination criterion has been satisfied, so that the training data are divided into the appropriate number of clusters. At each non-terminal node, LLGMN is used to achieve binary splits. Even for data of complicated distributions, interpretable clustering can be made after a nested series of binary splits. In this section, after the description of the proposed unsupervised learning algorithm of LLGMN, division validation according to the statistical properties of the training data and pruning law are explained.

3.1 Unsupervised Learning Algorithm

Given the number of classes, C, the entropy used as cost function is defined as:

$$J_{SO} = -\sum_{n=1}^{N}\sum_{c=1}^{C} {}^{(3)}O_c^{(n)} \log {}^{(3)}O_c^{(n)} , \quad (8)$$

where N is the number of total data. The proposed unsupervised learning algorithm modifies weights by minimizing Eq. (8), However, for some initial weights, the LLGMN may be trained to cluster all training data into one class, and cost function, J_{SO}, may converge to such a local minimum. Therefore, in the proposed method, the initialization of the weights is carried out to prevent the LLGMN to converge to one of the local minima, and the number of classes is restricted to two.

Let us consider that LLGMN clusters data into two classes: C_1 and C_2. First, $\boldsymbol{x}_1$ and $\boldsymbol{x}_2$ are chosen for the initialization of weights from the total training data set $\boldsymbol{A}$ according to the following equation,

$$(\boldsymbol{x}_1, \boldsymbol{x}_2) = \operatorname*{argmax}_{\boldsymbol{x}^{(i)}, \boldsymbol{x}^{(j)} \in \boldsymbol{A}} (||\boldsymbol{x}^{(i)} - \boldsymbol{x}^{(j)}||) . \tag{9}$$

Then the set $\boldsymbol{B}$, which means the set of the utilized data, is set as $\{\boldsymbol{x}_1, \boldsymbol{x}_2\}$. Assuming that $\boldsymbol{x}_1$ and $\boldsymbol{x}_2$ are labeled with C_1 and C_2, respectively. Training of LLGMN is performed using the supervised learning rule [2] in order to classify $\boldsymbol{x}_1$ and $\boldsymbol{x}_2$ into C_1 and C_2 respectively. Then, with the initialized weights, unsupervised learning of the LLGMN is performed using the set $\boldsymbol{B}$. The mean values of $\bar{\boldsymbol{x}}_{C_1}$ and $\bar{\boldsymbol{x}}_{C_2}$ are calculated using the training data clustered into C_1 and C_2, respectively. One datum $\boldsymbol{x} \in \boldsymbol{A} - \boldsymbol{B}$, from which the distance to either $\bar{\boldsymbol{x}}_{C_1}$ or $\bar{\boldsymbol{x}}_{C_2}$ is the smallest, is added into the set $\boldsymbol{B}$. Then, modification of the weight $\Delta w_h^{(c,m)}$ is defined as:

$$\Delta w_h^{(c,m)} = -\eta \frac{\partial J_{SO}^{(n)}}{\partial w_h^{(c,m)}} , \tag{10}$$

$$\frac{\partial J_{SO}^{(n)}}{\partial w_h^{(c,m)}} = -(J_{SO} - \log {}^{(3)}O_c^{(n)}) {}^{(2)}O_{c,m}^{(n)} X_h^{(n)} . \tag{11}$$

After training with a pre-defined number of times, another training datum is selected from the set $\boldsymbol{A} - \boldsymbol{B}$ and added into the set $\boldsymbol{B}$. This step of training repeats, until all the training data is added into $\boldsymbol{B}$, that is to say, $\boldsymbol{B} = \boldsymbol{A}$.

3.2 Division Validation

With the proposed method, unnecessary splits may occur when the hierarchy of the tree becomes too deep. In this method, cross-validation is adopted and the posterior probabilities of the validation data is utilized to determine whether to split a node or not. First, the validation data is prepared and the entropy $H(\boldsymbol{x})$ is defined as:

$$H(\boldsymbol{x}) = -\sum_{c=1}^{C} {}^{(3)}O_c^{(n)} \log {}^{(3)}O_c^{(n)} . \tag{12}$$

Then, the average value H_E of $H(\boldsymbol{x})$ is utilized as the termination criterion.

$$H_E = \frac{1}{|\boldsymbol{N}_c|} \sum_{\boldsymbol{x}^{(n)} \in \boldsymbol{N}_c} H(\boldsymbol{x}^{(n)}) , \tag{13}$$

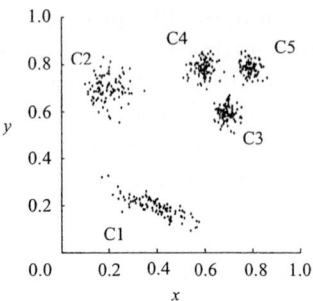

Fig. 2. Examples of artificial data

where N_c stands for the set of validation data belonging to the node under consideration, and $|N_c|$ is the number of validation data in N_c. If H_E is higher than a threshold H_T, splitting of the corresponding node is terminated. On the other hand, if all validation data of the node in consideration are clustered into one class, outliers may exist in the training data and the division of this node must be terminated. Also, for occasions when there is only one training data in a node, further split of this node must be terminated, since division is impossible. With this validation, a classification tree can be constructed based on the statistical properties of the training data, and can cluster complicated data into a proper number of classes.

3.3 Pruning Law

In the proposed method, outliers are always classified into some terminal nodes (clusters) separated from other major clusters. Especially, when the hierarchy of the tree grows too large, the influence of outliers becomes prominent because of a decrease of the number of training data in each node. After the classification tree is constructed, pruning is conducted to improve the clustering efficiency. The number of training data left in each terminal node is utilized as a decision index of pruning. If the ratio of the number of training data in a terminal node to the total training data number is lower than threshold α_T, this node and its counter are merged into their father node. With this pruning law, excessive splits may be prevented, and the number of clustering may not increase corresponding to the number of outlier data.

3.4 Experiments

Numerical simulations were carried out in order to verify the proposed method. The feature data is illustrated in Fig. 2: There are 2-dimensional data $x \in \Re^2$, and generated from five classes, C_i ($i = 1, 2, \cdots, 5$). Each class consists of one normal distribution. The number of training data for each class is 100, and the number of validation data for each class is 200. The LLGMN includes seven units in the first layer, two units in the second layer corresponding to the total number of components, and two units in the third layer. To construct the classification tree, threshold of entropy H_T is set as 0.2, threshold of pruning α_t as 0.01, learning rate η as 0.01, and training times in each addition of training data as 100. The classification tree starts from the root node, where training data are divided into two nodes at each non-terminal node, and finally, a hierarchical tree is constructed from five terminal nodes. To validate the generalization ability, 300 samples for each class that are not used in training process were clusterd, and the discrimination rate for 20 independent trials was $98.5 \pm 0.64\%$. It can be found that the proposed method can estimate the number of classes and achieve high classfication rate.

4 Conclusion

In this paper, to deal with the discrimination problem of ambiguous teacher signals, a hierarchical clustering method was proposed. In this method, entoropy of the LLGMN's outputs at each node are used as the termination criterion, and unnecessary splits in the structure of the classification tree can be avoided, so that the proposed method can make an interpretable and reasonable partition of the training data according solely to its statistical characteristics.

In future works, we would like to carry out discrimination experiments on various data and to examine the influence of the parameters to the clustering result. Furthermore, we would like to establish an improved method that determines the value of theresholds such as α_T automatically.

References

1. Hiraiwa, A., Shimohara, K., Tokunaga, Y.: EMG Pattern Analysis and Classification by Neural Network. IEEE International Conference on Syst., Man and Cybern., (1989) 1113–1115
2. Tsuji, T., Fukuda, O., Ichinobe, H., Kaneko, M.: A Log-linearized Gaussian Mixture Network and its Application to EEG Pattern Classification. IEEE Trans. on System, Man and Cybernetics-Part C: Applications and Reviews, Vol. 29., No. 1. (1999) 60–72
3. Anderberg, M.R.: Cluster Analysis for Applications. Academic Press, New York (1974)
4. Ward, J.H.: Hierarchical Grouping to Optimize an Objective Function. Journal of the American Statistical Association, Vol. 58., No. 301. (1963) 235–244
5. Kohonen, T.: Self-organization and Associative Memory. Third Edition, Springer-Verlag, Berlin (1994)
6. Terashima, M., Shiratani, F., Yamamoto, K.: Unsupervised Cluster Segmentation Method Using Data Density Histogram on Self-organizing Feature Map. IEICE Transactions on Information and Systems, PT. 2, Vol. J79., No. 7., (1996) 1280–1290 (in Japanese)

Document Clustering Algorithm Based on Tree-Structured Growing Self-Organizing Feature Map*

Xiaoshen Zheng[1], Wenling Liu[1], Pilian He[2], and Weidi Dai[2]

[1] College of Marine Science and Engineering, Tianjin University of Science and Technology, 300450, Tianjin, China
[2] Department of Computer Science and Technology, Tianjin University, 300072, Tianjin, China
zhengxiaoshen@163.com

Abstract. Document clustering is widely studied in text mining. In this paper, document clustering algorithm based on Tree-Structured Growing Self-organizing Feature Map (TGSOM) is presented as an extended version of the clustering algorithm of Self-organizing Map (SOM) in neural network, which has a dynamic tree-structure generated during the training process. TGSOM's growth speed can be controlled through the function of the Spread Factor (SF), and the precision of clustering results is different because of the difference value of SF. The user can get the hierarchical clustering results through changing the size of SF in different steps during clustering.

1 Introduction

In reality, a substantial portion of the available information is stored in document databases, which consist of large collections of documents from various sources, such as news articles, research papers, books, digital libraries, e-mail messages and Web pages. Document databases are rapidly growing due to the increasing amount of information available in electronic forms, such as electronic publications, e-mail, CD-ROMs and the Word-Wide Web (which can also be viewed as a huge, interconnected, dynamic document databases). So clustering analysis is very important in text mining as a means of data mining.

In this paper, we put forward a document clustering algorithm based on Tree-Structured Growing Self-organizing Feature Map (TGSOM), which has a dynamic structure generated and less processing time during the training process. The user can get the hierarchical clustering results through changing the size of SF in different steps during clustering.

The rest of this paper is organized as follows: Section 2 analyzes the neural network development. Section 3 describes the network structure and training method of TGSOM. Section 4 validates the clustering performance of TGSOM through experiments. Finally, section 5 is for conclusion remarks.

---

* Supported by the technology development Foundation of Tianjin (No: 043600411)

2   Neural Network Development

Up to now, the research of neural network has gained great improvement and put forward some models. Kohonen, a Finland scholar, raised the Self-organizing Map SOM[1] model. Clustering is performed that several units compete for the current object with SOM. SOM is believed to resemble processing that can occur in the brain and widely apply in the information processing community, and it is regarded as a useful method of clustering analysis in data mining and text mining.

When the SOM is used, the number of nerve cell on competes layers should be designated in advance, which limit of network structure reduces the convergence speed of models. So some emluator trainings about different value of nerve cell should do to confirm the network structure adapting to special application. Because of the clustering relation between documents is unknown when SOM is used in text mining, the network structure should be generated dynamically during the training.

Then D. Alahakon puts forward the Growing Self-organizing Map (GSOM)[2], which includes four neural cell forming square structure on compete levels initialization, and the neural cells of dynamically growing keep the regular planar structure as initialization. However, new node can only grow on the edge of the network, not the right location, and 1-3 nodes need to grow once but only one is effective, which result in some redundancy nodes and lower efficiency.

Based on the dynamically growing ideas of GSOM, we bring forward the TGSOM, which has a dynamic structure generated and less processing time during the training process. The user can get the hierarchical clustering results through changing the value of the Spread Factor (SF) in different steps during clustering. The main difference between GSOM and TGSOM as following: the TGSOM model adopts the flexible tree-structure and the new nodes can grow on needing location, the TGSOM model can grow only one new node once and few redundancy nodes.

3   Tree-Structured Growing Self-Organizing Feature Map (TGSOM)

The efficiency of TGSOM is over than that of GSOM, and the tree-structure of TGSOM is similarity to that of SCONN[3]. But the training results of TGSOM is beyond to that of SCONN because TGSOM's growth speed can be controlled through the function of SF and the hierarchical clustering results can be got through changing the value of SF in different steps during clustering.

3.1   Network Structure

TGSOM is regarded as an improvement of SOM model, which network structure shown in Figure1, including input layer and output layer.
In Fig.1 *(a)* is the initialize state of a root node, *(b)* is the state of a 8 nodes, which forms the planar tree-structure. The input layer nodes X1-Xn are achieved complete interlinkage, and the weight of nodes is self-organized during the TGSOM growing.

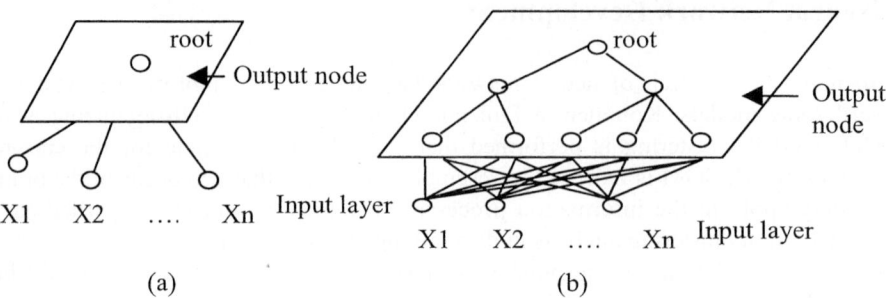

Fig. 1. TGSOM network structure

3.2 Basic Concepts of TGSOM

Definition 1: For input vector *v*, the nearest node to *v* in compete layer is called best matching node (bmn).
Definition 2: The distance between input vector *v* and bmn is called the error of bmn and *v*, marked as *E*, where *D* is the dimensions of *v*.

$$E = \sum_{k=1}^{D} (v_k - w_{bmn,k})^2 \tag{1}$$

3.3 Training Method of TGSOM

The training process of TGSOM method is following:
(1) Initialization
   (a) Given the random value in the range of [0,1] to root node vector weight;
   (b) Calculate the Growth Threshold (GT) based on the user requirement.
(2) Training
   (a) Random (or ordering) selecting the training sample *v* from the vector set *V*.
   (b) Looking for the bmn from the present network node in TGSOM.
   (c) Calculating the error *E* between bmn and *v* according to the formula (1). If *E<GT* then turn (d) to adjust operation; or else turn(e)to growing.
   (d) Adjusting the weight of bmn neighborhood.

$$w_j(k+1) = \begin{cases} w_j(k), j \notin N_{k+1} \\ w_j(k) + LR(k) \times (v_k - w_j(k)), j \in N_{k+1} \end{cases} \tag{2}$$

Where the variable *LR(k)* will reduce when *k* increasing; $w_j(k+1)$ and $w_j(k)$ are the weight of node *j* before and after adjusting respectively; N_{k+1} is the neighborhood of bmn at the k+1st training.
   (e) Growing the new child node of bmn, and $w_{child} = v$.
   (f) *LR(K+1)=LR(k)*α*, where *α* is the adjust factor of *LR*, and 0<*α*<1.
   (g) Repeat (a)-(f) until the whole samples in vector set *V* have been trained.

(3) Repeating (2) until the growing nodes achieve saturation.
(4) Smoothness
   Reducing *LR* to adjust the weight of node.
   The tree network structure is got when training of TGSOM completed.

3.4 Selecting the Parameter of GT

In the algorithm of TGSOM, the new node is growing when $E>GT$, where E is got from the formula (1). We introduce SF into the TGSOM model. $GT=D*f(SF)$, $0<SF<1, 0<f(SF)<1$, given $f(SF)=(1-SF)^2$, namely, $GT=D*(1-SF)^2$. As a result, if the smaller of SF, namely the bigger of GT, the cruder clustering can be achieved, or else the refined clustering can be achieved.

According to the above, the SF is the random value of the closed interval [0,1]. During document clustering, at first the SF is smaller, the crude document clustering can be gained, then the accurate results may get when the value of SF is bigger.

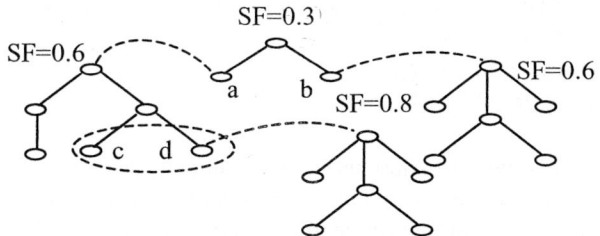

Fig. 2. Hierarchical clustering processing corresponding to different SF

In fig.2, *SF=0.3* the crude clustering is got, then, the documents which included in the nodes *a* and *b* in the above results are clustered second respectively when *SF=0.6*. Last, the documents which include nodes *d* are clustered third when *SF=0.8*.

4 Experiment and Result

We use the front page (CIRB011) of NTCIR-3[2] Chinese standard data collections as test database. Part of documents of the collections are used in our experiment, including spo (sport), int (international politics), pol (national politics), art, ent (entertainment), eco (economics), fin (finance), sto(stock), tec(science and technology), med (medical), edu (education) etc.

---

[2] The NTCIR was co-sponcered by Japan Society for Promotion of Science (JSPS) as part of JSPS "Research for Future" Program" and National Center for Science Information Systems (NACSIS) since 1997, by JSPS and Research Center for Information Resources at National Institute of Informatics. Participation is invited from anyone interested in research on information access technologies. NTCIR Workshops are periodical event which are held once par one and half years.

We mix the test documents into a document set and select the former 50 documents, then get the clustering results that show in figure 3 when SF is 0.4 and 0.6 respectively.

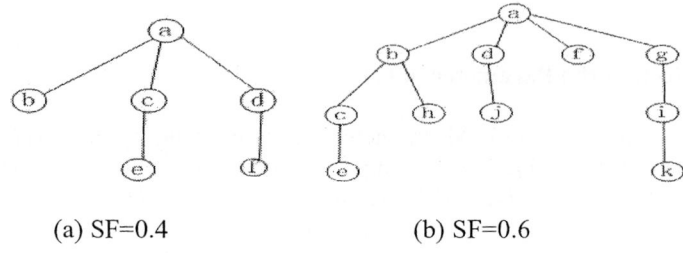

(a) SF=0.4   (b) SF=0.6

Fig. 3. The clustering results when SF is 0.4 and 0.6

When *SF=0.4*, the clustering results show in Fig.3 *(a)*(the letters corresponding to the nodes, and the sequence number corresponding to the number of document set).
  b: 1,2,4,5,6,7,10,11,18,23,28,29,30,32,33,36,37,45,46,48,49,50
  c: 3,8,9,13,19,20,35,39
  d: 12,17,21,22,24,34,38,42,44
  e: 14,15,16,25,26,27,31,43,47
  f: 40,41

From above results, we can know these 50 documents are clustered into 5 clusters, in which eco, tec and pol have been clustered triumphantly, (b: eco; c: tec; d: pol), but the other three classes cannot be separated completely because of the smaller value of SF.

In this experiment, the clustering processing needs 2 training periods and 50 seconds. However, at the same experiment condition, we get the similar clustering with the GSOM algorithm, the clustering processing needs 5 training periods and 50 minutes. So it can be seen that the TGSOM algorithm's efficiency is higher than that of GSOM obviously with the documents increase and clustering efficiency improvement.

When *SF=0.6*, the clustering results show in Fig.3 *(b)*.
  b: 1,4,30
  c: 2,5,6,7,11,18,23,28,29,32,33,37,45,46,48,49,50
  d: 3,8,9,13,19,35,39
  e: 10,36
  f: 12,17,21,22,24,34,38,42,44
  g: 14,15,16,47
  h: 20
  i: 25,31,43
  j: 26,27
  k: 40,41

From the above results, we can know these 50 documents are clustered into 10 clusters, including classes and their sub-class.

Table 1. The number of node comparison about TGSOM and GSOM

Clustering results	TGSOM			GSOM		
	Value of SF	Number of node	Valid nodes	Value of SF	Number of node	Valid nodes
1	0.4	6	5	0.1	26	13
2	0.6	11	10	0.85	145	20

From the table, the number of node in TGSOM is less than that of GSOM, which shows that the growing speed of TGSOM is exceeded that of GSOM obviously.

From above experiments, we know that the value of SF can control the growth of network and the precision of document clustering.

5 Conclusion

The TGSOM algorithm applies the idea of self-organization and grows its network structure dynamically, which overcome the limit of changeless network structure of SOM. At the same time, the TGSOM algorithm has the less number of nodes and higher training efficiency than that of SOM and GSOM. On the other hand, the TGSOM algorithm applies the value of SF to control the growth of network and the precision of document clustering, the user can get the hierarchical clustering results through changing the size of SF in different steps during clustering. In a word, the documents clustering algorithm based on TGSOM is an effective and scalable clustering one.

References

1. Kohonen, T.: Self-organizing Maps. Springer, Berlin (1995)
2. Alahakoon, D., Halgamuge, S.K.: Dynamic Self-organizing Maps with Controlled Growth for Knowledge Discovery. IEEE Trans. on Neural Networks. 11 (2000) 601-614
3. Choi, D., Park, S.: Self-creating and Organizing Neural Networks. IEEE Trans. on Neural Networks. 5 (1994) 561-575
4. Chen, K.H, Chen, H.H.: Overview of CLIR Task at the Third NTCIR Workshop. NTCIR Workshop 3 Meeting (2002) 1-13

Improved SOM Clustering for Software Component Catalogue

Zhuo Wang, Daxin Liu, and Xiaoning Feng

Computer Science and Technology College, Harbin Engineering University, P.R.China,
150001
wangzhuo0812@sina.com

Abstract. The faceted catalogue is one of most popular methods among software component catalogues. However, the traditional faceted catalogue needs experts to establish and maintain glossary. To solve the problem, SOM clustering is applied in software catalogue to automatically form the glossary. By the improved SOM clustering, faceted descript values are clustered and the centers are taken as faceted terms. In SOM algorithms, distances of components are reduced or extended according to component connectors. So the components in this catalogue are not isolated but have relations with each other. And the problem that clustering result is disturbed by the input order, is also solved in the improved SOM clustering.

1 Introduction

The clustering algorithms attempt to organize unlabelled input vectors into clusters or 'natural groups' such that points within a cluster are more similar to each other than points belonged to different clusters [1]. There are five types of clustering methods: hierarchical clustering, partitioning clustering, density-based clustering, grid-based clustering, model-based clustering [2]. Each type has its advantages and disadvantages. Self-Organizing Map (SOM), proposed by Kohonen, has been widely used in many industrial areas such as pattern recognition, biological modeling, data compression signal processing, and data mining [3]. Attempts have been made to cluster data by using SOM. SOM is used to develop clustering algorithms. The number of output neurons is equal to the desired number of clusters. SOM is, however, conceptually different from clustering [1].

Software reuse is being pursued as a promising way of increasing productivity, assuring quality and meeting deadlines in software development [4]. Software reuse is concerned with the technological and organizational issues arising from the usage of already existing software components to build new applications [5]. Software development costs would be greatly reduced if programmers could easily reuse components of preexisting programs. However, to make software reuse operational the software developers need to be provided with large libraries containing reusable software components. Furthermore, the user of such a library must be assisted in locating components that are functionally close to the required component. Clustering the stored software components into groups of semantically similar components, i.e. software components that exhibit similar behavior, performs this task [6].

Among of software component catalogue, the faceted categorize is introduced by many software component repository system. However, the glossary of the faceted categorize is established and maintained by manual work at present. This problem restricts the flexibility of the faceted term description and the search condition. On the other hand, the traditional component catalogue considers a component is an independent object. In fact, the interoperability of components is very important to system integration. Therefore, a faceted categorize based on clustering (Sfaceted categorize) is proposed in this paper. By Sfaceted categorize, the glossary is automatically established and gradually improved, and the component connector is considered one of the factors when components are classified.

This paper is organized as follows. Section 2 summarizes the importance of the component connectors, and presents classify of component connectors. Section 3 describes the component denotation. Section 4 includes the SOM structure and the Advance SOM clustering training process. Section 5 describes some empirical results and analysis. Section t summarizes the advantage of this Faceted Catalogue.

2 Sfaceted Catalogue

In the traditional faceted catalogue, the glossary is defined in advance. So users must select terms to describe components. However, the glossary of the Sfceted catalogue is empty initially. The user can freely input several paragraphs as the facet value to describe components. The method only defines the form of the paragraph, but doesn't restrict the content and the words. The similarity of facet values is computed. The facet values are clustered by an improved SOM clustering algorithm. The centers of the classes are regarded as the facet terms. And other values are the synonyms of the terms to form the glossary. The similarity between the synonym and the term must be more than a parameter. If the similarity is too small, the synonym will be deleted from the glossary. The new value settles in the clustering determined by the advanced SOM clustering. After a new term added, the centers of clustering need to be adjusted. Because the structure of terms of facet is a tree structure, the clustering algorithm must be iterated. Moreover, in the Sfaceted catalogue, a software component is not insolated, but an independent number of some systems that has relationship with each other. To achieve the intention, the distance of components can be shortened or extended in clustering.

2.1 Classification of Component Connectors

Although components have become the predominant focus of researchers and practitioners, they only address one aspect of large-scale development. Another important aspect, particularly magnified by the emergence of the Internet and the growing need for distribution, is interaction among components. Component interaction is embodied in the notion of software component connectors. Connectors manifest themselves in a software system as shared variable accesses. In large, and especially distributed systems, connectors becomes key determinants of system properties, such as performance, resource utilization, global rates of flow, scalability, reliability, security, and so forth.

In Sfaceted categorize, the component connector affects the components' ascription at the clustering border. At the clustering border the component C belongs to the clustering in which there are the most components connected to C. The following is the six classes of the component connectors.

- Data sharing: the components access the same data through database or other data structures.
- Instructions linking: the components instruct each other, or the third object instructs the components.
- Networking protocols: the components transfer information by networking protocol.
- Business collaboration: functions of the components complete a business activity by collaboration.
- Message transferring: one or more components send message to start up other components.
- Service channel: this connector is for some special components. The function of these components is to provide service to other components or software. The connector is established between the service supporter and the user.

2.2 SOM Structure

Self-organizing Map proposed by Kohonen is a family of self-organizing neural networks widely used for clustering and visualization. SOM training belongs to the class of competitive learning algorithms. Similar to K-Means, the reference clustering in SOM is randomly initialized. SOM follows a winner-take-part competitive learning process. For each incremental input, the network identifies the winner clustering that is most similar to the input, and updates the weights of the winner as well as the winner's neighbors in order to incorporate the input. Here, an improved SOM clustering is provided to automatically produce Sfaceted glossary. The SOM consists of a regular, two-dimensional, grid of map units. Each unit i is represented by a prototype vector $\lambda_i = \{X_{1i}, X_{2i}, X_{3i}, X_{4i}, X_{5i}, X_{6i}, X_{7i}\}$. λ_i is a vector denotation for each facet of a component. X_{1i} is the vector of the component terms for one facet. X_{2i} is the clustering ID in which there are most components that have Data sharing connector with the components. X_{3i} is the clustering ID in which there are most components that have Instructions linking connector with the components. X_{4i} is the clustering ID in which there are most components that have Networking protocols connector with the components. X_{5i} is the clustering ID in which there are most components that have Business collaboration connector with the components. X_{6i} is the clustering ID in which there are most components that have Message transferring connector with the components. X_{7i} is the clustering ID in which there are most components that have Service channel connector with the components. The units are connected to adjacent ones by a neighborhood relation. The number of map units, which typically varies from a few dozen up to several thousand, determines the accuracy and generalization capability of the SOM. Here experts according the clustering define the number of map units. During training, the SOM forms an elastic net that folds onto the „cloud" formed by the input data. Data points lying hear each other in the input space are mapped onto nearby map units. Thus, the SOM can be interpreted as a topology-preserving mapping from input space onto the 2-D grid of map units, as shown in Fig. 1.

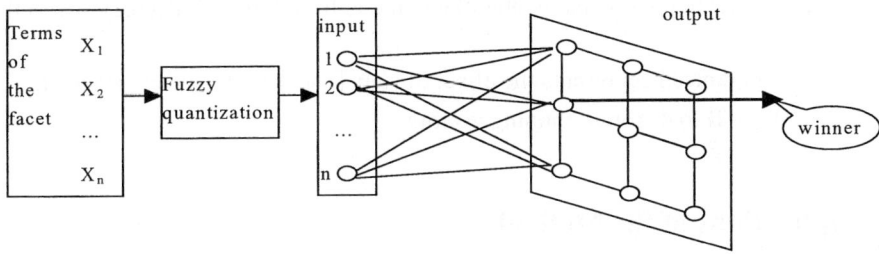

Fig. 1. SOM Structure

2.3 SOM Training

1.The vector that denotes a software component is presented is presented to the network as input neuron. Each output neuron expresses a clustering. Initialize the weights ω_{ij} ($i = 1, 2, ..., N, j = 1, 2, ..., M$) of each neuron N_{ij} with random values or use a predefined map we obtained in another training. The experts define the values by experience of project practice in order to improve training rate and accurateness.

2.Input a new mode vector to the network.

3.Compute the distance between the input modes and each output neuron.

$$d_j = \left[(x_1(t) - \omega_{j1}(t))^2 + \sum_{i=2}^{7} (x_i(t) - \eta * \omega_{ji}(t))^2 \right]^{\frac{1}{2}} \quad (1)$$

η: the effect of each input neuron, the first input neuron is the main factor of clustering, the other input neurons is the less important factor. So the range of η is from 0 to 1.

$x_i(t)$: The output computed by the input neuron i at the step t.

$\omega_{ji}(t)$: The weight between the input i and the output neuron j at the step t.

4.Select the winning neuron N_{max} that stores in the weights the closest pattern to the input b_{ij}.

5.Update the weights of the selected neuron and its neighbors:

$$\omega_{ji}(t+1) = \omega_{ji}(t) + \eta(t)\lambda(t)(x_i(t) - w_{ji}(t))$$
$$j \in NE_j^*(t), \qquad 1 \le i \le n \tag{2}$$

$$\lambda(t) = \frac{x_i(t-1) - \omega_{ji}(t-1)}{x_i(t) - \omega_{ji}(t)} \tag{3}$$

$\eta(t)$: The time regulation parameter, the range is from 0 to 1. It decreases with the time increasing.

$\lambda(t)$: The parameter represents the effect of step $(t-1)$ to the weight at step t.

6. Repeat steps 2) to 4) for all training vectors.

3 Applications of the Method

Five facts are defined in the experiment, including „Function", „Operation Object", „Environment", „Form", and „Performance". „Function" is the business function of components. „Operation Object" is the business entity, which components operate on. „Environment" is the hardware and software environment in which components run. „Form" includes the component standard that components following and the program language of components. „Performance" includes the resource that components use and the time efficiency of components.

In the experiment, 120 components in the repository need clustering. These components come from several real application systems. They are J2EE components, CORBA components, and .NET components. Their functions include making ERP system, graphic displaying, computing and the other common utility functions. We present the number of term produced automatically and the number of term defined by experts for each facet. We count the error clustering component number for each facet. We compute the error rate of component clustering (error component number/ total component number). These experiment data are shown in table 1. The clustering results of the Environment facet and the Performance facet are better, because the semantic range of the terms of these facets is small. The clustering result of the Function facet is worst, because the description of a component function is various. But the error rate is in the acceptable range.

4 Conclusion

With the development of software reuse, many artificial intelligence technologies are applied in this area to improve the automatic grade. Software component catalogue is one of the important problems in software reuse. The faceted catalogue is a popular application adopted by many components repository systems. However, glossary of facets needs to be established and maintained by men. And it restricts the application of the faceted catalogue. SOM clustering is a useful algorithm applied in exploratory pattern-analysis, grouping, decision-making, and machine-learning situations. By the improved SOM clustering, the faceted catalogue can produce a glossary automatically. According to the feature of the software component, the faceted values are clustering and the cluster centers is regarded as the faceted terms. When computing the distance between the input modes and each output neuron we use a parameter to adjust the effect of each weight. When the weights are updated, the effect of the previous weight modification is considered. The advanced SOM clustering has resolved the problem that the clustering result is affected by the input sequence. By the experiment, Sfaceted catalogue is proved that it is not affected by the input irrelevant word. When the number of input faceted values is far more than the faceted terms, Sfaceted catalogue is an effective and convenient catalogue.

Table 1. Experiment data of each facet by SFaceted catalogue

	produced term	Expert term	defining	error component	error rate
Function	27	38		16	13.3%
Operation Object	8	10		11	9.2%
Environment	6	7		8	6.7%
Form	9	6		7	5.8%
Performance	5	5		3	2.5%

References

1. Pal, N.R., Bezadek, J.C., Tsao, E.C.K.: Generalized Clustering Networks and Kohonen's Self-organizing Scheme. IEEE.Tans. Neural Networks, 4(4) (1993) 549–557
2. Han, J., Kamber, M.: Data Mining: Concepts and Techniques. Morgan-Kaufman, Sn Francisco (2000)
3. Kohonen, T.: Self-organizing Maps. Springer-Verlag, Berlin, Germany (1997)
4. Constantopoulos, P., Doerr, M.: Component Classification in the Software Information Base. Obejct-Oriented Software Composition, Prentice-Hal (1995) 177–200
5. Merkl, D.: Self-organizing Maps and Software Reuse. Computational Intelligence in Software Engineering, World Scientific (1998)
6. Merkl, D.: Content-Based Software Classification by Self-organization. Proc. of the IEEE International Conference on Neural Networks (1995) 1086-1091

Classification by Multilayer Feedforward Ensembles[*]

Mercedes Fernández-Redondo, Carlos Hernández-Espinosa,
and Joaquín Torres-Sospedra

Universidad Jaume I. Dept. de Ingeniería y Ciencia de los Computadores. Avda Vicente Sos Baynat s/n. 12071 Castellon. Spain.
{redondo, espinosa}@icc.uji.es

Abstract. As shown in the bibliography, training an ensemble of networks is an interesting way to improve the performance with respect to a single network. However there are several methods to construct the ensemble and there are no complete results showing which one could be the most appropriate. In this paper we present a comparison of eleven different methods. We have trained ensembles of 3, 9, 20 and 40 networks to show results in a wide spectrum of values. The results show that the improvement in performance above 9 networks in the ensemble depends on the method but it is usually marginal. Also, the best method is called "Decorrelated" and uses a penalty term in the usual Backpropagation function to decorrelate the network outputs in the ensemble.

1 Introduction

The most important property of a neural network (NN) is the generalization capability. The ability to correctly respond to inputs which were not used in the training set.
One technique to increase the generalization capability with respect to a single NN consist on training an ensemble of NN, i.e., to train a set of NNs with different weight initialization or properties and combine the outputs of the different networks in a suitable manner to give a single output.

It is clear from the bibliography that this procedure in general increases the generalization capability [1,2].

The two key factors to design an ensemble are how to train the individual networks and how to combine the different outputs to give a single output.

Among the methods of combining the outputs, the two most popular are *voting* and *output averaging* [3]. In this paper we will normally use *output averaging* because it has no problems of ties and gives a reasonable performance.

In the other aspect, nowadays, there are several different methods in the bibliography to train the individual networks and construct the ensemble [1-3], [5-10].

However, there is a lack of comparison among the different methods and it is not clear which one can provide better results.

In this paper, we present a comparison among eleven different methods.

---

[*] This research was supported by the project MAPACI TIC2002-02273 of CICYT in Spain.

2 Theory

In this section we briefly review the different ensemble methods.

Simple Ensemble. A simple ensemble can be constructed by training different networks with the same training set, but different random weight initialization.

Bagging. This ensemble method is described in reference [5]. It consists on generating different datasets drawn at random with replacement from the original training set. After that, we train the different networks in the ensemble with these different datasets, we use one dataset for each network.

Bagging with Noise (BagNoise). It was proposed in [2], we use in this case datasets of size 10·N (number of training points) generated in the same way of Bagging, where N is the number of training points of the initial training set. Also we introduce a random noise in every selected training point drawn from a normal distribution.

Boosting. This ensemble method is reviewed in [3]. It is conceived for a ensemble of only three networks. The first network is trained with the whole training set. After this training, we pass all patterns through the first network and we use a subset of them, which has 50% of patterns incorrectly classified and 50% classified correctly. With this new training set we train the second network. After that, the original patterns are presented to both networks. If the two networks disagree in the classification, we add the training pattern to the third training set.

CVC. It is reviewed in [1]. In k-fold cross-validation, the training set is divided into k subsets. Then, k-1 subsets are used to train the network and results are tested on the subset that was left out. Similarly, by changing the subset that is left out of the training process, one can construct k classifiers. This is the technique used in this method.

Adaboost. We have implemented the algorithm "Adaboost.M1" in the reference [6]. In the algorithm the successive networks are trained with a training set selected at random, but the probability of selecting a pattern changes depending on the correct classification of the pattern and on the performance of the last trained network. The algorithm is complex and the full description should be looked for in the reference.

Decorrelated (Deco). This ensemble method was proposed in [7]. It consists on introducing a penalty added to the usual error function. The term for network j is:

$$Penalty = \lambda \cdot d(i,j)(y - f_i) \cdot (y - f_j) \qquad (1)$$

Where λ determines the strength of the penalty term and should be found by trial and error, y is the target of the training pattern and f_i and f_j are the outputs of networks number i and j in the ensemble. The term $d(i,j)$ is in equation 2.

$$d(i,j) = \begin{cases} 1, & \text{if } i = j-1 \\ 0, & \text{otherwise} \end{cases} \qquad (2)$$

Decorrelated2 (Deco2). It was proposed also in reference [7]. It is basically the same method of "Decorrelated" but with a different term $d(i,j)$ in the penalty:

$$d(i,j) = \begin{cases} 1, & \text{if } i = j-1 \text{ and } i \text{ is even} \\ 0, & \text{otherwise} \end{cases} \quad (3)$$

Evol. This ensemble method was proposed in [8]. In each iteration (presentation of a training pattern), it is calculated the output of the ensemble for the input pattern by voting. If the output is correctly classified we continue with the next iteration. Otherwise, the network with an erroneous output and lower MSE is trained in this pattern until the output of the network is correct. This procedure is repeated for several networks until the vote of the ensemble correctly classifies the pattern.

Cels. It was proposed in [9]. This method also uses a penalty term added to the usual error function. In this case the penalty term for network number i is in equation 4.

$$Penalty = \lambda \cdot (f_i - y) \cdot \sum_{j \neq i} (f_j - y) \quad (4)$$

Where y is the target and f_i and f_j the outputs of networks i and j.

Ola. This ensemble method was proposed in [10]. In this method, first, several datasets are generated by using bagging. Every network is trained in one of this datasets and in *virtual data*. The *virtual data* for network i is generated by selecting randomly samples for the original training set and perturbing the sample with a random noise drawn from a normal distribution with small variance. The target for this new virtual sample is calculated by the output of the ensemble without network number i for this sample. For a full description of the procedure see the reference.

3 Experimental Results

We have applied the eleven ensemble methods to ten different classification problems. They are from the UCI repository of machine learning databases. Their names are Cardiac Arrhythmia Database (Aritm), Dermatology Database (Derma), Protein Location Sites (Ecoli), Solar Flares Database (Flare), Image Segmentation Database (Image), Johns Hopkins University Ionosphere Database (Ionos), Pima Indians Diabetes (Pima), Haberman's survival data (Survi), Vowel Recognition (Vowel) and Wiscosin Breast Cancer Database (Wdbc).

We have constructed ensembles of a wide number of networks, in particular 3, 9, 20 and 40 networks in the ensemble. In this case we can test the results in a wide set of situations.

We trained the ensembles of 3, 9, 10 and 40 networks. We repeated this process of training an ensemble ten times for different partitions of data in training, cross-validation and test sets. With this procedure we can obtain a mean performance of the ensemble for each database (the mean of the ten ensembles) and an error in the performance calculated by standard error theory. The results of the performance are in table 1 and 2 for the case of ensembles of three networks and in table 3 and 4 for the case of nine, we omit the results of 20 and 40 networks by the lack of space by the improvement of increasing the number of networks is in general not important.

Table 1. Results for the ensemble of three networks

	ARITM	DERMA	ECOLI	FLARE	IMAGEN
Single Net.	75.6 ± 0.7	96.7 ± 0.4	84.4 ± 0.7	82.1 ± 0.3	96.3 ± 0.2
Adaboost	71.8 ± 1.8	98.0 ± 0.5	85.9 ± 1.2	81.7 ± 0.6	96.8 ± 0.2
Bagging	74.7 ± 1.6	97.5 ± 0.6	86.3 ± 1.1	81.9 ± 0.6	96.6 ± 0.3
Bag_Noise	75.5 ± 1.1	97.6 ± 0.7	87.5 ± 1.0	82.2 ± 0.4	93.4 ± 0.4
Boosting	74.4 ± 1.2	97.3 ± 0.6	86.8 ± 0.6	81.7 ± 0.4	95.0 ± 0.4
Cels_m	73.4 ± 1.3	97.7 ± 0.6	86.2 ± 0.8	81.2 ± 0.5	96.82 ± 0.15
CVC	74.0 ± 1.0	97.3 ± 0.7	86.8 ± 0.8	82.7 ± 0.5	96.4 ± 0.2
Decorrelated	74.9 ± 1.3	97.2 ± 0.7	86.6 ± 0.6	81.7 ± 0.4	96.7 ± 0.3
Decorrelated2	73.9 ± 1.0	97.6 ± 0.7	87.2 ± 0.9	81.6 ± 0.4	96.7 ± 0.3
Evol	65.4 ± 1.4	57 ± 5	57 ± 5	80.7 ± 0.7	77 ± 5
Ola	74.7 ± 1.4	91.4 ± 1.5	82.4 ± 1.4	81.1 ± 0.4	95.6 ± 0.3
Simple Ens.	73.4 ± 1.0	97.2 ± 0.7	86.6 ± 0.8	81.8 ± 0.5	96.5 ± 0.2

Table 2. Results for the ensemble of three networks (continuation of Table 1)

	IONOS	PIMA	SURVI	VOWEL	WDBC
Single Net.	87.9 ± 0.7	76.7 ± 0.6	74.2 ± 0.8	83.4 ± 0.6	97.4 ± 0.3
Adaboost	88.3 ± 1.3	75.7 ± 1.0	75.4 ± 1.6	88.43 ± 0.9	95.7 ± 0.6
Bagging	90.7 ± 0.9	76.9 ± 0.8	74.2 ± 1.1	87.4 ± 0.7	96.9 ± 0.4
Bag_Noise	92.4 ± 0.9	76.2 ± 1.0	74.6 ± 0.7	84.4 ± 1.0	96.3 ± 0.6
Boosting	88.9 ± 1.4	75.7 ± 0.7	74.1 ± 1.0	85.7 ± 0.7	97.0 ± 0.4
Cels_m	91.9 ± 1.0	76.0 ± 1.4	73.4 ± 1.3	91.1 ± 0.7	97.0 ±0.4
CVC	87.7 ± 1.3	76.0 ± 1.1	74.1 ± 1.4	89.0 ± 1.0	97.4 ± 0.3
Decorrelated	90.9 ± 0.9	76.4 ± 1.2	74.6 ± 1.5	91.5 ± 0.6	97.0 ± 0.5
Decorrelated2	90.6 ± 1.0	75.7 ± 1.1	74.3 ± 1.4	90.3 ± 0.4	97.0 ± 0.5
Evol	83.4 ± 1.9	66.3 ± 1.2	74.3 ± 0.6	77.5 ± 1.7	94.4 ± 0.9
Ola	90.7 ± 1.4	69.2 ± 1.6	75.2 ± 0.9	83.2 ± 1.1	94.2 ± 0.7
Simple Ens.	91.1 ± 1.1	75.9 ± 1.2	74.3 ± 1.3	88.0 ± 0.9	96.9 ± 0.5

By comparing the results of table 1, 2, 3 and 4 with the results of a single network we can see that there the improvement by the use of the ensemble methods depends clearly on the problem. For example in databases Aritm, Flare, Pima and Wdbc there is not a clear improvement.

In the rest of databases there is an improvement, perhaps the most important one is in database Vowel.

There is, however, one exception in performance in the method Evol. This method did not work well in our experiments. In the original reference the method was tested in the database Heart. The results for a single network were 60%, for a simple ensemble 61.42% and for Evol 67.14%. We have performed some experiments with this database and our results for a simple network are 82.0 ± 0.9, clearly different.

Now, we can compare the results of tables 1, 2, 3 and 4 for an ensemble of different number of networks. We can see that the results are in general similar and the improvement of training an increasing number of networks, for example 20 and 40, is in general marginal. Taking into account the computational cost, we can say that the best alternative for an application is an ensemble of three or nine networks.

We have also calculated the percentage of error reduction of the ensemble with respect to a single network. We have used equation 5 for this calculation.

Table 3. Results for the Ensemble of nine networks

	ARITM	DERMA	ECOLI	FLARE	IMAGEN
Adaboost	73.2 ± 1.6	97.3 ± 0.5	84.7 ± 1.4	81.1 ± 0.7	97.3 ± 0.3
Bagging	75.9 ± 1.7	97.7 ± 0.6	87.2 ± 1.0	82.4 ± 0.6	96.7 ± 0.3
Bag_Noise	75.4 ± 1.2	97.0 ± 0.7	87.2 ± 0.8	82.4 ± 0.5	93.4 ± 0.3
Cels_m	74.8 ± 1.3	97.3 ± 0.6	86.2 ± 0.8	81.7 ± 0.4	96.6 ± 0.2
CVC	74.8 ± 1.3	97.6 ± 0.6	87.1 ± 1.0	81.9 ± 0.6	96.6 ± 0.2
Decorrelated	76.1 ± 1.0	97.6 ± 0.7	87.2 ± 0.7	81.6 ± 0.6	96.9 ± 0.2
Decorrelated2	73.9 ± 1.1	97.6 ± 0.7	87.8 ± 0.7	81.7 ± 0.4	96.84 ± 0.18
Evol	65.9 ± 1.9	54 ± 6	57 ± 5	80.6 ± 0.8	67 ± 4
Ola	72.5 ± 1.0	86.7 ± 1.7	83.5 ± 1.3	80.8 ± 0.4	96.1 ± 0.2
Simple Ens	73.8 ± 1.1	97.5 ± 0.7	86.9 ± 0.8	81.6 ± 0.4	96.7 ± 0.3

Table 4. Results for the ensemble of nine networks (continuation of Table 3)

	IONOS	PIMA	SURVI	VOWEL	WDBC
Adaboost	89.4 ± 0.8	75.5 ± 0.9	74.3 ± 1.4	94.8 ± 0.7	95.7 ± 0.7
Bagging	90.1 ± 1.1	76.6 ± 0.9	74.4 ± 1.5	90.8 ± 0.7	97.3 ± 0.4
Bag_Noise	93.3 ± 0.6	75.9 ± 0.9	74.8 ± 0.7	85.7 ± 0.9	95.9 ± 0.5
Cels_m	91.9 ± 1.0	75.9 ± 1.4	73.4 ± 1.2	92.7 ± 0.7	96.8 ± 0.5
CVC	89.6 ± 1.2	76.9 ± 1.1	75.2 ± 1.5	90.9 ± 0.7	96.5 ± 0.5
Decorrelated	90.7 ± 1.0	76.0 ± 1.1	73.9 ± 1.3	92.8 ± 0.7	97.0 ± 0.5
Decorrelated2	90.4 ± 1.0	76.0 ± 1.0	73.8 ± 1.3	92.6 ± 0.5	97.0 ± 0.5
Evol	77 ± 3	66.1 ± 0.7	74.8 ± 0.7	61 ± 4	87.2 ± 1.6
Ola	90.9 ± 1.7	73.8 ± 0.8	74.8 ± 0.8	88.1 ± 0.8	95.5 ± 0.6
Simple Ens	90.3 ± 1.1	75.9 ± 1.2	74.2 ± 1.3	91.0 ± 0.5	96.9 ± 0.5

$$PorError_{reduction} = 100 \cdot \frac{PorError_{single\ network} - PorError_{ensemble}}{PorError_{single\ network}} \quad (5)$$

The value of the percentage of error reduction ranges from 0%, where there is no improvement by the use of a particular ensemble method with respect to a single network, to 100%. There can also be negative values, which means that the performance of the ensemble is worse than the performance of the single network.

This new measurement is relative and can be used to compare more clearly the different methods. Furthermore we can calculate the mean performance of error reduction across all databases this value is in table 3 for ensembles of 3, 9, 20 and 40 networks.

Table 5. Mean percentage of error reduction for the different ensembles

	Ensemble 3 Nets	Ensemble 9 Nets	Ensemble 20 Nets	Ensemble 40 Nets
Adaboost	1.33	4.26	9.38	12.21
Bagging	6.86	12.12	13.36	12.63
Bag_Noise	-3.08	-5.08	-3.26	-3.05
Boosting	-0.67			
Cels_m	9.98	9.18	10.86	14.43
CVC	6.18	7.76	10.12	6.48
Decorrelated	9.34	12.09	12.61	12.35
Decorrelated2	9.09	11.06	12.16	12.10
Evol	-218.23	-297.01	-375.36	-404.81
Ola	-33.11	-36.43	-52.53	-47.39
Simple Ens	5.58	8.39	8.09	9.72

According to this global measurement *Ola*, *Evol* and *BagNoise* performs worse than the *Simple Ensemble*. The best methods are *Bagging*, *Decorrelated* and *Decorrelated2*. In total there are only four methods which perform better than the *Simple Ensemble*.

Also in this table, we can see the effect of increasing the number of networks in the ensemble. There are two methods (*Adaboost* and *Cels*) where the performance seems to increase slightly with the number of networks in the ensemble. But other methods like *Bagging*, *CVC*, *Decorrelated*, *Decorrelated2* and *Simple Ensemble* does not increase the performance beyond 9 or 20 networks in the ensemble. The reason can be that the new networks are correlated to the first ones or that the combination method (the average) does not exploit well the increase in the number of networks.

4 Conclusions

In this paper we have presented experimental results of eleven different methods to construct an ensemble of networks, using ten different databases. We trained ensembles of 3, 9, 20 and 40 networks in the ensemble. The results showed that in general the improvement by the use of the ensemble methods depends clearly on the database, in some databases there is an improvement but in other there is not improvement at all. Also the improvement in performance from three or nine networks in the ensemble to a higher number of networks depends on the method. Taking into account the computational cost, an ensemble of nine networks may be the best alternative for most of the methods. Finally, we have obtained the mean percentage of error reduction over all databases. According to the results of this measurement the best methods are "Decorrelated", "Bagging" and "Cels" and there are only four or five methods which perform better than the "Simple Ensemble".

References

1. Tumer, K., Ghosh, J.: Error Correlation and Error Reduction in Ensemble Classifiers. Connection Science. **8** no. 3 & 4, (1996) 385-404
2. Raviv, Y., Intrator, N.: Bootstrapping with Noise: An Effective Regularization Technique. Connection Science. **8** no. 3 & 4, (1996) 355-372
3. Drucker, H., Cortes, C., Jackel, D., et al.: Boosting and Other Ensemble Methods. Neural Computation. **6** (1994) 1289-1301
4. Verikas, A., Lipnickas, A., et al.: Soft Combination of Neural Classifiers: A Comparative Study. Pattern Recognition Letters. **20** (1999) 429-444
5. Breiman, L.: Bagging Predictors. Machine Learning. **24** (1996) 123-140
6. Freund, Y., Schapire, R.: Experiments with a New Boosting Algorithm. Proceedings of the Thirteenth International Conference on Machine Learning. (1996) 148-156
7. Rosen, B.: Ensemble Learning Using Decorrelated Neural Networks. Connection Science. **8** no. 3 & 4, (1996) 373-383
8. Auda, G., Kamel, M.: EVOL: Ensembles Voting On-Line. Proc. of the World Congress on Computational Intelligence. (1998) 1356-1360
9. Liu, Y., Yao, X.: A Cooperative Ensemble Learning System. Proc. of the World Congress on Computational Intelligence. (1998) 2202-2207
10. Jang, M., Cho, S.: Ensemble Learning Using Observational Learning Theory. Proceedings of the International Joint Conference. on Neural Networks. **2** (1999) 1281-1286

Robust Face Recognition from a Single Training Image per Person with Kernel-Based SOM-Face

Xiaoyang Tan [1,2], Songcan Chen [2,3], Zhihua Zhou [1], and Fuyan Zhang [1]

[1] National Laboratory for Novel Software Technology
Nanjing University, Nanjing 210093, China
[2] Department of Computer Science and Engineering
Nanjing University of Aeronautics & Astronautics, Nanjing 210016, China
[3] National Laboratory of Pattern Recognition
Institution of Automation, Chinese Academy of Sciences, Beijing 100080, China
{txy123,s.chen}@nuaa.edu.cn {zhouzh,fyzhang}@nju.edu.cn

Abstract. In this paper, a kernel-based SOM-face method is proposed to recognize expression variant faces under the situation of only one training image per person. Based on the localization of the face, an unsupervised kernel-SOM learning procedure is carried out to capture the common local features and the non-Euclidean structure of the image data, so that a compact and robust representation of the face can be obtained. Experiments on the FERET face database show that the Kernel-based SOM-face method can obtain higher recognition performance than the regular SOM-face method.

1 Introduction

One of the factors that have strongly affected the performance of face recognition is the face representation model. For example, the eigenface, as a classical representation model for face recognition, tries to find a linear mapping that mostly keeps the variation between the face images [1]. However, due to its linearity in nature, this PCA-based representation can not always capture the non-linear structure of face images. Fisherface is also widely used, which aims to extract the most discriminant features from the face image [2]. However, like other LDA-based methods, Fisherface suffers from the *small sample problem* and will fail in the situation where there is only one training image per person available [3-5]. The limitations discussed above suggest additional research needed on the representation of face image.

Recently, several researchers have tried to model the inherent nonlinear structure of the complex image data using nonlinear methods, such as neural networks [6], support vector machines [7], and kernel methods [8]. In a previous work [5], an SOM-based face representation model called "SOM-face" (see Fig.1) has been proposed to deal with both the nonlinear problem and the small sample problem in face recognition. In this paper, motivated by the success of kernel methods in pattern regression, a kernel-based SOM-face method is proposed. This method generalizes the strength of the kernel method and SOM network while at the same time overcomes some of the shortcomings of regular SOM-face method.

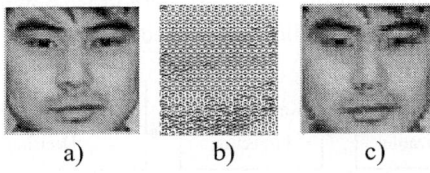

a)   b)   c)

Fig. 1. Example of an original image, its projection and the reconstructed image a) Original face image. b) The distribution of image in the topological space. c) "SOM face" reconstructed

2 Kernel SOM

The basic idea of Self-Organizing Map [9] is to find and adapt the winner neuron and its topological neighbors according to the current input vector so as to reveal the hidden statistical structures of the input space. However, the commonly used distance measure in the regular SOM algorithm is the Euclidean norm, thus the non-Euclidean neighborhood structure in the input space can hardly be revealed.

Recently, Pan et al. [10] and Andras [11] independently proposed the kernel-SOM algorithm, whose key feature is that the updates of the high dimensional weight vectors are made indirectly by updating the low dimensional vectors in the original space, thus the method can be intrepreted as a way to induce different non-Euclidean distance measures for the original space using different kernel fuctions. Common cases for the kernel fuctions are the polynomial, RBF and logarithmic kernels, etc.

In particular, for the RBF kernel ($K(x, y)=\exp(-\|x-y\|^2/2\sigma^2)$) and logarithmic kernel ($K(x, y)=\log((1+\|x-y\|^2)/\sigma^2)$), we have the following kernel-SOM updating rules respectively:

$$\Delta w(t) = \eta(t)h(t)\frac{e^{-\|x-w(t)\|^2/2\sigma^2}}{\sigma^2}(x-w(t)) \triangleq \eta(t)h(t)\rho_{d,\sigma}(t)(x-w(t)) \quad (1)$$

$$\Delta w(t) = \eta(t)h(t)\frac{1}{\sigma^2+\|x-w(t)\|^2}(x-w(t)) \triangleq \eta(t)h(t)\rho'_{d,\sigma}(t)(x-w(t)) \quad (2)$$

By comparing the new updating rules (1) or (2) with the regular SOM updating rules $w(t)=\eta(t)h(t)(x-w(t))$, it can be found that a scale factor $\rho_{d,\sigma}(t)$ is added in the kernel version, which is affected by the distance d between the input vector x and the winner neuron in the input space under certain kernel paremeter σ at time t. The value of the scale factor decreases as d increases. This means that the attraction of the winner neuron and its neighborhood to the outlier or noisy data will be lessened.

3 The Proposed Method

A high-level block diagram of the proposed method is shown in Fig. 2. The details of the method are described in the following subsections.

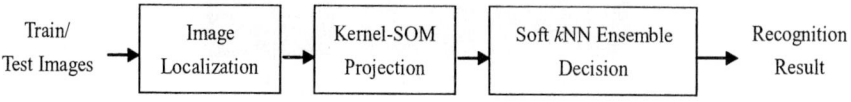

Fig. 2. Block diagram of the proposed method for face recognition

3.1 Localizing Face Image

Since local features are relatively less sensitive to the occlusion and variation in face (such as expression, pose, and illumination) than global features [12], we pay much attention to the local representation of face. One of the simplest ways to localize the face image is to divide the image into M different non-overlapping local sub-blocks with equal sizes, each of which potentially represents specific local information of the image. As a result of the process, a set of sub-block vectors (SBV) is obtained.

In this way, the information of face image is distributed and represented by several low dimensional local features instead of only one high dimensional vector. This helps relax the small sample problem. Furthermore, the sub-pattern dividing process can also help increase the diversity of the SOM-faces, which is useful for the soft kNN ensemble in the later step.

3.2 The Kernel-SOM Projection

There are two main problems to be solved then, i.e. 1) Since localization process more likely results in many identical sub-blocks from different face classes, that is, some sub-blocks may belong to or be shared by several different classes at the same time, it usually causes the so-called one-to-many mapping problem. 2) In the situation of classification of data with a large number of classes, it is indeed difficult to find a hard class boundary for each class. The first problem makes most supervised learning methods such as radial basis function network (RBF) and multi-layer perceptron network (MLP) fail in the sense of their one-to-one mapping characteristic.

In our previous work [5], we found Self-Organizing Maps (SOM) is a suitable option for those problems, because the neurons of SOM may have multiple class labels at the same time, thus providing the capability of *one-to-many* mapping, on the other hand, the topological preservation property of SOM make it possible to represent the content of each class in a nonlinear way [9]. In this work, we replace the regular SOM with kernel-SOM. The use of kernel method helps find a simpler class boundary structure, which, accordingly, leads to a more robust classification.

The training of kernel-SOM is similar to that of the regular SOM. Its training process is divided into two phases as recommended by [9], that is, an ordering phase and a fine-adjustment phase. We notice that the scale factor $\rho_{d,\sigma}(t)$ in (1) or (2) may hurt the ordering performance because of its very small values in the initial phase. Thus we do not make any modification to the regular learning rule in the ordering phase, and the reference vectors are adjusted according to (1) or (2) only in the second phase.

3.3 Soft Nearest Neighbor Decision

Since the SOM makes similar input patterns clustered to adjacent neurons, the relationship between spatially adjacent neurons could be used to improve the classification performance. This leads to a soft *k*NN ensemble decision scheme. A separate soft *k*NN classifier is constructed for each sub-block of the face image to calculate the confidence value for its membership in every class according to:

$$c_{jk} = \log(\tau+1) / \log(d_{jk}+1) \quad (3)$$

where $d_{j,k}$ is distance between the *j*-th neuron of the face and its *k*-th nearest neighbors, and ô is the minimum among all the distances. This defines a confidence value C_{jk} for the *j*-th sub-block's membership in *k*-th class, that is, the higher the confidence value for a class, the more likely a sub-block will belong to that class.

Then, the label of the test image can be obtained through majority voting, as follows,

$$Label = \arg\max_k (\sum_{j=1}^{M} c_{jk}), k = 1,\ldots C \quad (4)$$

where *M* is the total number of sub-blocks of a face and *C* is the number of face class.

4 Experiments

The experimental face database used in this work comprises 400 gray-level frontal view face images from 200 persons, with the size of 256×384. Each person has two images (*fa* and *fb*, used as training gallery and probes respectively) with different facial expressions. All the images are randomly selected from the FERET face database [13]. Before the recognition process, the raw images are normalized 60×60 pixels and the inter-ocular distance is 28 pixels.

The details of the experiments are given below. In the localization phase, a block size of 3×3 is used. Then two kind of kernel SOM, i.e. RBF-kernel SOM and log-kernel SOM are trained respectively with 100 updates in the first phase and 400 updates in the second one, with σ=2. The initial weights of all neurons are set to the greatest eigenvectors of the training data, and the learning parameter and the neighborhood widths of the neurons converge exponentially to 0 with the time of training.

Table 1 presents the performance of the proposed method with reference to other template-based approaches, such as nearest neighbor (1-NN), eigenface, and $E(PC)^2A$[4] concerning the *top 1 match rate*. Table 1 reveals that the proposed method achieves higher recognition accuracy than other approaches in dealing with the *one image per person problem*, such as eigenface and $E(PC)^2A$.

Table 1. Comparison of recognition accuracies (%) for six approaches

Method	Accuracy
1-NN	84.0
Eigenface	83.0
$E(PC)^2A$	85.5
Regular SOM-face	87.5
RBF-kernel-SOM-face	88.5
LOG-kernel-SOM-face	89.5

Next, we study the influence of the k value. Experimental results are presented in Fig. 3. It can be observed that when the k value is small (e.g. $k<30$), the regular SOM-face method performs better than the two kernel-SOM-face methods. However, the overall performance is not so good (<86.5%). When k gradually increases to the range between 60 and 120, the performances of all the three compared methods increase as well, ranging from 86% to 89.5%. Among them, the log-kernel SOM-face performs the best, next the RBF-kernel SOM-face, and both the kernel SOM-face methods perform better than the regular SOM-face method.

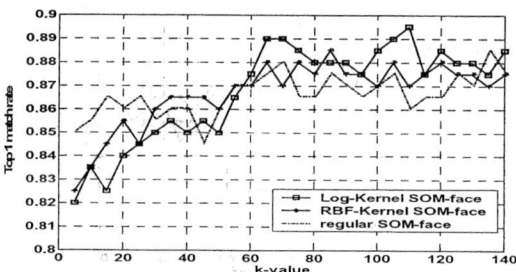

Fig. 3. Top 1 matching rate as a function of k-value.

5 Conclusion

In this paper, a novel face representation and recognition approach is presented, where faces are first localized, then represented by their kernel-SOM-based proximities with non-Euclidean distance measure embedded in. Experimental results show the superiority of using kernel-SOM-face to using regular SOM-face in scenarios where only one training image per person is available.

Acknowledgement. This work was supported by the National Natural Science Foundation of China under the Grant No. 60271017, the National Outstanding Youth Foundation of China under the Grant No. 60325237, the Jiangsu Science Foundation under the Grant No. BK2002092, and the Jiangsu Science Foundation Key Project. Portions of the research in this paper use the FERET database of facial images collected under the FERET program.

References

1. Turk, M., Pentland, A.: Eigenfaces for recognition. Journal of Cognitive Neuroscience 3 (1991) 71-86
2. Belhumeur, P., Hespanha, J., Kriegman, D.: Eigenfaces vs. Fisherfaces: recognition using class specific linear projection. IEEE Trans. Pattern Analysis and Machine Intelligence 19 (1997) 711-720
3. Wu, J., Zhou, Z.-H.: Face recognition with one training image per person. Pattern Recognition Letters 23 (2002) 1711-1719
4. Chen, S.C., Zhang, D.Q, Zhou, Z.-H.: Enhanced $(PC)^2A$ for face recognition with one training image per person. Pattern Recognition Letters, in press
5. Tan, X.Y, Chen, S.C., Zhou Z.-H., Zhang, F.Y.: Recognizing seriously occluded, expression variant faces from single training image per person with SOM-based kNN ensemble. Technical Report, Computer Science Department, Nanjing University, China, Dec. 2003
6. Raytchev, B., Murase, H.: Unsupervised face recognition by associative chaining. Pattern Recognition 36 (2003) 245-257
7. Pang, S., Kim, D., Bang, S.Y.: Membership authentication in the dynamic group by face classification using SVM ensemble. Pattern Recognition Letters 24 (2003) 215-225
8. Lu, J., Plataniotis, K.N., Venetsanopoulos, A.N.: Face recognition using kernel direct discriminant analysis algorithms. IEEE Trans. Neural Networks 14 (2003) 117-126
9. Kohonen, T., Self-Organizing Map. 2nd edition. Springer-Verlag, Berlin (1997)
10. Pan, Z.S., Chen, S.C., Zhang, D.Q.: A kernel-based SOM classification in input space. Acta Electronica Sinica 32 (2004) 227-231 (in Chinese)
11. Andras, P.: Kernel-kohonen networks. International Journal of Neural Systems, 12 (2002) 117-135
12. Pentland, A., Moghaddam, B., Starner, T.: View-based and modular eigenspaces for face recognition. In: Proceedings of the IEEE International Conference on Computer Vision and Pattern Recognition, Seattle, WA, (1994) 84-91
13. Phillips, P.J., Wechsler, H., Huang, J., Rauss, P.J.: The FERET database and evaluation procedure for face recognition algorithms. Image and Vision Computing 16 (1998) 295-306

Unsupervised Feature Selection for Multi-class Object Detection Using Convolutional Neural Networks

Masakazu Matsugu[1] and Pierre Cardon[2]

[1] Canon Inc. HVS Research Dept. 5-1, Morinosato-Wakamiya, Atsugi, 243-0193 Japan
matsugu.masakazu@canon.co.jp
[2] EU-Japan Center for Industial Cooperation, 13-3, Ichibancho, Tokyo, 102-0082 Japan

Abstract. Convolutional Neural Networks (CNN) have proven to be useful tools for object detection and object recognition. They act like feature extractor and classifier at the same time. In this study we present an unsupervised feature selection procedure for constructing a training set for the CNN and analyze in detail the learnt receptive fields. We then introduce, for the first time, a figural alphabet to be used for low-level feature detection with CNN. This alphabet turned out to be useful in detecting a vocabulary set of intermediate level features and considerably reduces the complexity of the CNN. Moreover we propose an optimal high-level feature selection procedure and apply this to the challenging problem of car detection. We demonstrate promising results for multi-class object detection using obtained figural alphabet to detect considerably different categories of objects (e.g., faces and cars).

1 Introduction

In this work, we address the problem of selecting optimal local features for multi-class object detection [10]. A crucial aspect for object detection is the choice of an optimal set of features. In [2] and [11], an interest point operator and a k-means clustering algorithm are used to extract and regroup high-level features for estimating the parameters of the underlying probabilistic model. In [5], the image entropy is adopted to select interesting areas in the image and a Self-Organizing Map (SOM) [6] to organize the big amount of extracted high-level features, then a clustering algorithm is used to regroup similar units in the SOM to an automatic determined number of macro-classes. In [9], sub-optimal features are selected, for training the convolutional neural networks (CNN), by trial and error and extracted manually.

CNN [8] as well as *neocognitrons* [3] have been used for face detection and recognition [9], [7]. In [9], a variant of back-propagation algorithm is proposed to teach each layer separately (sequential BP: SBP) so that the extracted features are controlled, and also some specific parts of the face can be detected. The first two layers are trained with intermediate-level features (e.g. eye-corners), while the subsequent layers are trained with more complex, high-level features (e.g. eyes, faces...). This requires to select a training set of features. By selecting a limited set of features for a specific object, we may expect to find a restricted yet useful set of receptive fields as in neurophysiological studies [1], [4].

In this work, we present an unsupervised feature extracting and clustering procedure, using an interest operator combined with a SOM (Section 2). This method combines the advantages of both [11] and [5] by selecting a limited number of features and regrouping them using a topographic vector quantizer (SOM); acting like a vector quantizer and introducing a topographic relation at the same time.

The obtained feature classes are self-organized, low-and intermediate-level features that are used to train the 2 first layers of the CNN and obtain a minimum set of 4 alphabetical receptive fields by back-propagation (Section 3). This alphabet considerably reduces the complexity of the network by decreasing the number of parameters and can be used for detection of different object classes (e.g. faces, cars,...). We also introduce a method to select optimal high-level features and illustrate it with the car detection problem (subsection 3.3).

2 Unsupervised Local Feature Extraction in CNN

We use a modified architecture of CNN [9], which inherits classical architecture with shared weights, local receptive field and subsampling layers. The whole network is described in the lower part of Fig.1.

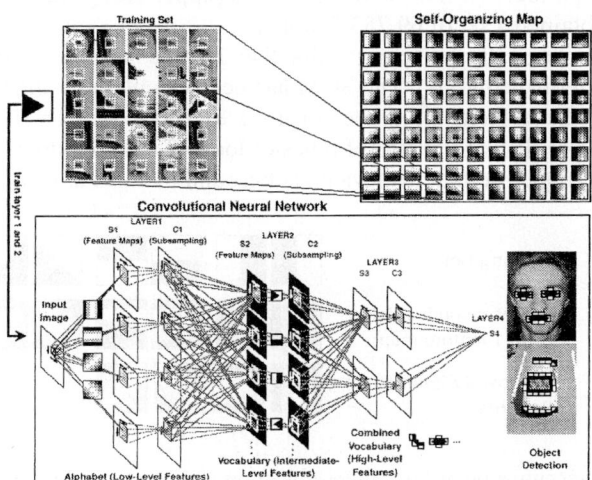

Fig. 1. CNN model (lower half) for multi-class object detection and alphabetical local features (upper right) obtained from SOM with training set (upper left)

Some specific local fragments of image extracted a priori, by using the proposed method in this study, are used to train the first two layers of the CNN. First, we train the CNN to recognize only one feature (one output plane in S2). A sequential back-propagation algorithm [9] is used for learning and weights are updated after each training pattern (fragments of images) is presented. A fixed number of 100 epochs has been used. For each training set, a different number of cell-planes in layer S1 have been tested. The network has essentially four distinct sets of layers: S1-C1, S2-C2, S3-C3, S4 (S_k: the kth feature detecting layer; C_j: the jth feature pooling layer for subsampling). Layers S3-C3 and above are concerned with object specific feature detection. In order to limit the number of features to object-relevant features, an interest point operator is used. This operator selects corner-like features in the image.

Having selected a restricted number of points we extract features around these points. These features are used as learning set for the SOM well suited for classifying and visualizing our feature set. It turned out that the illumination has a big influence on the classification of our features, so we have rescaled the feature set between -1 and 1 before applying the SOM. Each unit of the SOM defines a training set for the CNN. Once lower-level alphabetical feature detectors are formed, higher level feature detectors can be obtained from BP with connections between neurons below intermediate layers fixed.

3 Results

3.1 Unsupervised Selection of Low Level Features

Since we are interested in low-level features to train the first two layers of the CNN, we have chosen to extract small (7 x 7) features. With a database of 904 (size: 208 x 256) images (300 faces (frontal view), 304 cars (upper view) and 300 various types of images), we obtained a set of 69,753 features.

We start by manually selecting units that have a simple character (horizontal, vertical and diagonal contrast). The SOM has been calculated using the SOM Toolbox in Matlab. We have fixed the umber of units to 100, based on the assumption that there are not more than 100 different types of local (7 x 7) features in an object image appropriately cropped so that irrelevant background features are cut out.

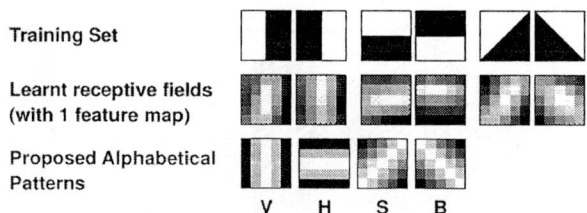

Fig. 2. Learnt receptive fields for low-level features and corresponding alphabetical patterns

For each cluster we only consider the 300 features, which are the closest to the SOM-unit, in terms of Euclidean distance. 200 features are used for training, 50 features for validation and the last 50 units for testing. The results have been obtained with a test set of 50 features and optimal receptive fields have been selected by cross-validation. We see that for such simple features, only one cell-plane in S1 is sufficient to obtain good detection results. We also notice that the learnt receptive fields (Fig.2) have a regular pattern. Based upon these patterns we propose a set of 4 alphabetical patterns *V*, *H*, *S*, *B* (hereafter, represents vertical, horizontal, slash, backslash, respectively) described in Fig.2.

3.2 Intermediate Level Features as Vocabulary

We observe that some feature clusters in the SOM have a more complex aspect as shown in Fig.3. We claim that these more complex features can be detected using the simple receptive fields, described in the previous section.

Considering for example the feature described in Fig.3, we see that this eye-corner type feature can be decomposed into 2 local alphabetical features. After training the CNN to detect this type of feature, we obtain the results with FAR: 0%, 4%, 8%, and 10% for *H, V, S*, and *B*, respectively.

The usefulness of our alphabetical set appears when we want to detect several high-level features with a small number of receptive fields, using synergies between the features. Let us consider the features used to detect a complete eye or a mouth [9]. They can be decomposed to 2 horizontal, 2 slash and 2 back-slash components (Fig.4).

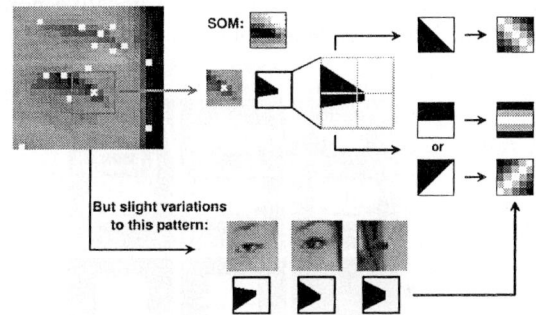

Fig. 3. Intermediate-level feature (eye-corner) decomposed in low-level features

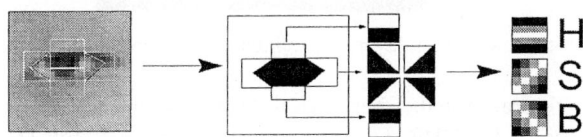

Fig. 4. High-level feature (i.e., eye) for face-detection, decomposed into low-level features

With a limited set of three fixed receptive fields *H, S* and *B* it turned out that we reach a detection rate of eye-corner comparable to that of using six learnt receptive fields. Our alphabetical set, being close to the optimal set of weights, therefore outperforms the learnt weights. We can extend these results for different types of complex features and construct a vocabulary set that can be recognized with *H, V, S,* and *B*. For illustration purposes, we have tested our alphabet with images from which features have been extracted. It turned out that we could detect, in the S2 layer, eye- and mouth-corners as well as the side mirrors of a car, using only three receptive fields (*H, S* and *B*).

3.3 Extraction of High Level Features Using Alphabetical Features

A last question to be answered is which vocabulary we should use, in other words, what features are important to detect a specific object. To find these features we apply classical BP (hereafter referred as GBP: global BP), not the proposed SBP, to the entire CNN with connections below S3 layer (S1-C1-S2-C2) fixed, and analyze the output of Layer3 (high-level features). The GBP converges to a local minimum, therefore the algorithm will tend to extract sub-optimal features to minimize the detection error.

To examine the validity of our scheme, we applied our method to a training set of images of bright-colored cars with significant variance in shape, illumination, size and orientation. The size of the images used for learning was 156 x 112, and 90 images were used for training and 10 images for validation. We aim to find characteristic high-level features for the detection of this type of cars and for this particular view. In addition, we need to tailor our model to be able to distinguish between cars and other rectangular objects. So, we have included a set of negative non-car examples, with similar rectangular shape but which were not cars.

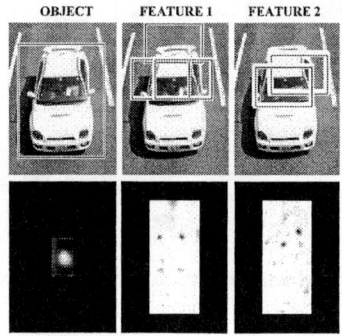

Fig. 5. (a) Car detection in the top layer, (b),(c)intermediate detection results for feature1 and 2

We used our figural alphabet between the input layer and S1 (fixed), and obtained other receptive fields by GBP. Using the four receptive fields H, V, S, and B, the number of cell-planes in S1-C1, S2-C2, S3-C3 and S4 were respectively 4, 3, 2 and 1. Respective sizes of the receptive fields were 5 x 5(Input-S1), 9 x 9(C1-S2), 25 x 45(C2-S3), and 79 x 29(C3-S4). After a fixed number of 500 epochs, we selected the receptive fields by cross-validation. With these receptive fields, we obtained a detection rate of 72% (test set of 50 cars) and 13 false alarms for 100 non-car images.

Having discovered the important features for our object detection problem, we obtain object specific vocabulary to select to construct these high-level features. We can use SBP as in [9] to train the higher level layers in the CNN: to train layer by layer with the selected vocabulary features. By doing so we limit the computing time which can be very long in the case of GBP and obtain very feature specific detectors in each layer. Typical detection result is shown in Fig.5.

4 Summary

We have proposed an automatic feature extraction procedure combining, for the first time, an interest point operator [11] with a SOM [5] to extract a training set of features for the CNN. By training the CNN with this training set, we found figural alphabets of 4 simple receptive fields obtained by SBP [9] are good enough to detect frontal view of cars as well as faces. We have shown that in spite of the simplicity of this alphabet it gives remarkable results, comparable and sometimes better than the learnt receptive fields with average detection rate over 95% for different types of features (see subsection 3.2). After obtaining alphabetical feature detectors in the S1 and S2 layer of CNN, we applied GBP to the S3 and S4 layers of CNN, with lower level weights fixed, to obtain higher level feature detectors (e.g., cars and faces), thereby obtaining sub-optimal vocabulary set. The optimality was examined in terms of cross-validation. In summary, we showed that the proposed method can be used to extract useful, generic local features for multi-class object detection (e.g., face and car detection) in the framework of convolutional neural networks.

References

1. Blackmore, C., Cooper, G. E.: Development of the Brain Depends on the Visual Environment. Nature, 228 (1970) 477-478
2. Burl, M., Leung, T., Perona, P.: Face Localization via Shape Statistics. In: Intl. Workshop on Automatic Face and Gesture Recognition (1995)
3. Fukushima, K.: Neocognitron: A Self-organizing Neural Network Model for a Mechanism of Pattern Recognition Unaffected in Shift Position. Biol. Cybern., 36 (1980) 193-202
4. Hubel, D., Wiesel, T.: Receptive Fields, Binocular Interaction and Functional Architecture in the Cat's Visual Cortex. Journal of Physiology, 160 (1962) 106-154
5. Ikeda, H., Kashimura, H., Kato, N., Shimizu, M.: A Novel Autonomous Feature Clustering Model for Image Recognition. In: Proc. of the 8th International Conference on Neural Information Processing (2001)
6. Kohonen, T.: Self-organizing Maps. Springer-Verlag, Berlin (1985)
7. Lawrence, S., Giles, G. L., Tsoi, A. C., Back, A. D.: Face Recognition: A Convolutional Neural Network Approach. IEEE Transactions on Neural Networks, 8 (1995) 98-113
8. Lecun Y., Bengio, Y.: Convolutional Networks for Images, Speech, and Time-Series. In: Arbib, M. (ed.): Handbook of Neural Networks and Brain Sciences, MIT Press, Cambridge (1995) 255-258
9. Matsugu, M., Mori, K., Ishii, M., Mitarai, Y.: Convolutional Spiking Neural Network Model for Robust Face Detection. In: Proc. of the 9th International Conference on Neural Information Processing (2002) 660-664
10. Papageorgiou, C. P., Oren, M., Poggio, T.: A General Framework of Object Detection. In: Proc. of International Conference on Computer Vision (1998) 555-562
11. Weber, M., Welling, M., Perona, P.: Unsupervised Learning of Models for Recognition. In: Proc. of the 6th European Conference on Computer Vision (2000)

Recurrent Network as a Nonlinear Line Attractor for Skin Color Association

Ming-Jung Seow and Vijayan K. Asari

Department of Electrical and Computer Engineering
Old Dominion University
Norfolk, VA 23529
{mseow, vasari}@odu.edu

Abstract. A novel learning algorithm for pattern association in a recurrent neural network is proposed in this paper. Unlike the conventional model in which the memory is stored in an attractive fixed point at discrete location in state space, the dynamics of the proposed learning algorithm represents memory as a line of attraction. The region of convergence at the line of attraction is defined by the statistical characteristics of training data. The performance of the learning algorithm is compared with Bayesian model in experiments on skin color segmentation.

1 Introduction

One of the major goals of both biological neural network modeling and artificial neural network research is to discover better learning rules to yield networks that can learn more difficult tasks, such as the tasks that the brain can handle [1-3]. A fundamental problem in associative memory concerns how memory is encoded by the neural network learning algorithm [4].

An associative memory model can be formulated as an input – output system. The input to the system can be an N dimensional vector $x \in \mathbb{R}^N$ called the memory stimuli, and the output can be an L dimensional vector $y \in \mathbb{R}^L$ called the memory response. The relationship between the stimuli and the response is given by $y = f(x)$, where $f : \mathbb{R}^N \to \mathbb{R}^L$ is the associative mapping of the memory. Each input – output pair or memory association (x, y) is said to be stored and recalled in the memory.

In this paper, we propose a novel learning algorithm for a recurrent neural network suitable for multiple-valued pattern association using the concept of nonlinear line attractor. In most models of associative memory, memories are stored as attracting fixed points at discrete locations in state space [5]. Fixed-point attractor may not be suitable for patterns, which exhibit similar characteristics [2, 6-8]. As a consequence, it may be more appropriate to represent the data using a nonlinear line attractor network [9-10]. Our experiments with human skin color demonstrate that the proposed learning rule can learn and characterize skin color precisely as a nonlinear line attractor.

2   Learning Model

The relationship of each neuron with respect to every other neuron can be expressed as a *k*-order polynomial for stimulus-response pair (x^s, y^s) corresponding to the sth pattern and is given by

$$y_i^s = \frac{1}{n}\sum_{j=1}^{n}\left(w_{(k,ij)}^s\left(x_j^s\right)^k + w_{(k-1,ij)}^s\left(x_j^s\right)^{k-1} + \cdots + w_{(1,ij)}^s x_j^s + w_{(0,ij)}^s\right) \quad \text{for } 1 \leq i \leq n \tag{1}$$

The resultant memory w^s_m of the mth order term in a n neuron network where $0 \leq m \leq k$ can be expressed as

$$w_m^s = \begin{pmatrix} w_{(m,11)}^s & \cdots & w_{(m,1n)}^s \\ \vdots & \ddots & \vdots \\ w_{(m,n1)}^s & \cdots & w_{(m,nn)}^s \end{pmatrix} \tag{2}$$

To combine w^s to form a memory matrix w; we need to utilize statistical methods. The least squares estimation approach to this problem can determine the best fit line when the error involved is the sum of the squares of the differences between the expected outputs and the approximated outputs. Hence, the weight matrix must be found that minimizes the total least squares error

$$E\left(w_{(0,ij)}, w_{(1,ij)}, \cdots w_{(k,ij)}\right) = \sum_{s=1}^{P}\left(y_i^s - P_k\left(x_j^s\right)\right)^2 \quad \text{for } 1 \leq i, j \leq N \tag{3}$$

where

$$P_k\left(x_j\right) = w_{(k,ij)}\left(x_j^s\right)^k + w_{(k-1,ij)}\left(x_j^s\right)^{k-1} + \cdots + w_{(1,ij)} x_j^s + w_{(0,ij)} = \sum_{z=0}^{k} w_{(z,ij)}\left(x_j^s\right)^z \tag{4}$$

A necessary condition for the coefficients $w_{(0,ij)}, w_{(1,ij)}, \ldots, w_{(k,ij)}$, to minimize the total error E is that

$$dE / dw_{(d,ij)} = 0 \quad \text{for each } d = 0, 1, \cdots, k \tag{5}$$

It can then be simplified to a set of normal equations as

$$w_{(0,ij)}\sum_{s=1}^{P}\left(x_j^s\right)^0 + w_{(1,ij)}\sum_{s=1}^{P}\left(x_j^s\right)^1 + \cdots + w_{(k,ij)}\sum_{s=1}^{P}\left(x_j^s\right)^k = \sum_{s=1}^{P} y_i^s\left(x_j^s\right)^0$$

$$w_{(0,ij)}\sum_{s=1}^{P}\left(x_j^s\right)^1 + w_{(1,ij)}\sum_{s=1}^{P}\left(x_j^s\right)^2 + \cdots + w_{(k,ij)}\sum_{s=1}^{P}\left(x_j^s\right)^{k+1} = \sum_{s=1}^{P} y_i^s\left(x_j^s\right)^1$$

$$\vdots$$

$$w_{(0,ij)}\sum_{s=1}^{P}\left(x_j^s\right)^k + w_{(1,ij)}\sum_{s=1}^{P}\left(x_j^s\right)^{K+1} + \cdots + w_{(k,ij)}\sum_{s=1}^{P}\left(x_j^s\right)^{2k} = \sum_{s=1}^{P} y_i^s\left(x_j^s\right)^k$$

The coefficients $w_{(0,ij)}, w_{(1,ij)},...,w_{(k,ij)}$ can hence be obtained by solving the above linear equations. Figure 1 shows the weight graph illustrating the concept of training based on the above theory. A weight graph is a graphical representation of the relationship between the ith neuron and the jth neuron for P patterns. Utilization of the weight graph can help visualize the behavior of one neuron pair.

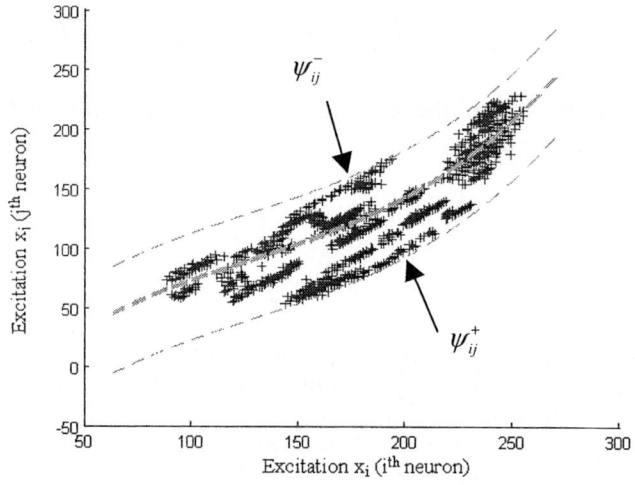

Fig. 1. Weight graph

The network architecture is a single layer fully connected recurrent neural network. The net output of the network can be computed as

$$Net_i = \frac{1}{n}\sum_{j=1}^{n}\left[w_{(k,ij)}x_j^k + w_{(k-1,ij)}x_j^{k-1} + \cdots + w_{(1,ij)}x_j + w_{(0,ij)}\right] \quad 1 \leq i \leq n \quad (6)$$

Net_i can be thresholded considering the region of distance of the weight components in the weight graph shown in Figure 1. That is, in order to consider each pattern, we need to find the region where the threshold can encapsulate each pattern. Mathematically the threshold function can be expressed as

$$o_i^{new} = f(net_i) = \begin{cases} o_i^{old} & \text{if } \tau_i^- \leq \left[o_i^{old} - net_i\right] \leq \tau_i^+ \\ net_i & \text{otherwise} \end{cases} \quad \text{for } 1 \leq i \leq n \quad (7)$$

where

$$\tau_i^- = \varphi \frac{1}{n}\sum_{j=1}^{n}\psi_{ij}^- \quad \text{for } 1 \leq i \leq n \quad (8)$$

$$\tau_i^+ = \varphi \frac{1}{n}\sum_{j=1}^{n}\psi_{ij}^+ \quad \text{for } 1 \leq i \leq n \quad (9)$$

φ is a constant such that $0 \leq \varphi \leq 1$ \hfill (10)

$$\psi_{ij}^- = \min_{\forall P}\left\{\left(w_{(k,ij)}\left(x_j^s\right)^k + w_{(k-1,ij)}\left(x_j^s\right)^{k-1} + \cdots + w_{(1,ij)}x_j^s + w_{(0,ij)}\right) - x_i^s\right\} \quad (11)$$

$$\psi_{ij}^+ = \max_{\forall P}\left\{\left(w_{(k,ij)}\left(x_j^s\right)^k + w_{(k-1,ij)}\left(x_j^s\right)^{k-1} + \cdots + w_{(1,ij)}x_j^s + w_{(0,ij)}\right) - x_i^s\right\} \quad (12)$$

The window width between τ_i^- and τ_i^+ decides if the O_i^{old} should be updated or not. That is, τ_i^- and τ_i^+ calculate the mean of the maximum distances from the approximated function.

3 Human Skin Color Modeling

Figure 2 shows an image and the skin color occurrences in that image in the Red, Green, and Blue (RGB) space (256×256×256). It can be observed that the skin colors are forming a nonlinear pipeline in the RGB space. That is, it is possible to describe the skin color mathematically using the nonlinear line attractor network.

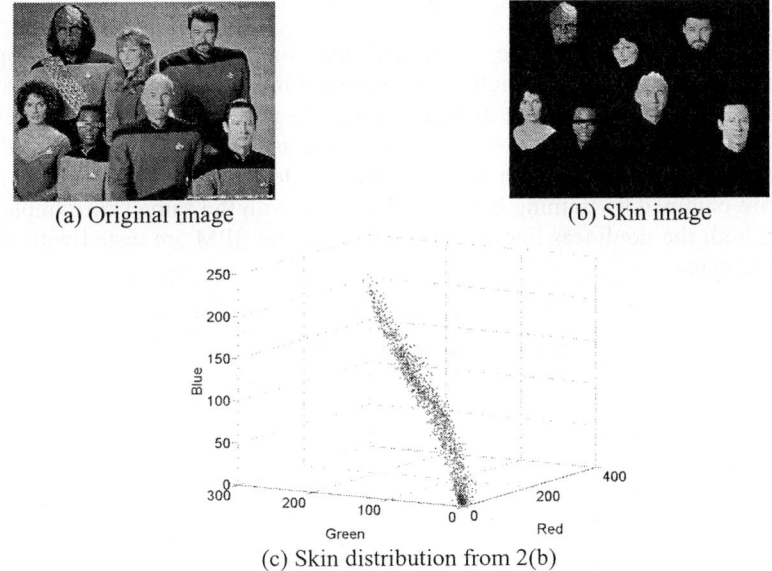

(a) Original image

(b) Skin image

(c) Skin distribution from 2(b)

Fig. 2. Skin occurrence as 3D scatter graph

The training set consists of 8,000 skin samples collected from a Sony EVI-D30 surveillance camera. The collected samples include skins of people belonging to different races. The collected skin samples are used for training the nonlinear line attractor network. Figure 3 shows the decision surfaces of the proposed algorithm. It can be seen that the network is able to distinguish skin and non-skin regions even though the network is trained only with skin samples.

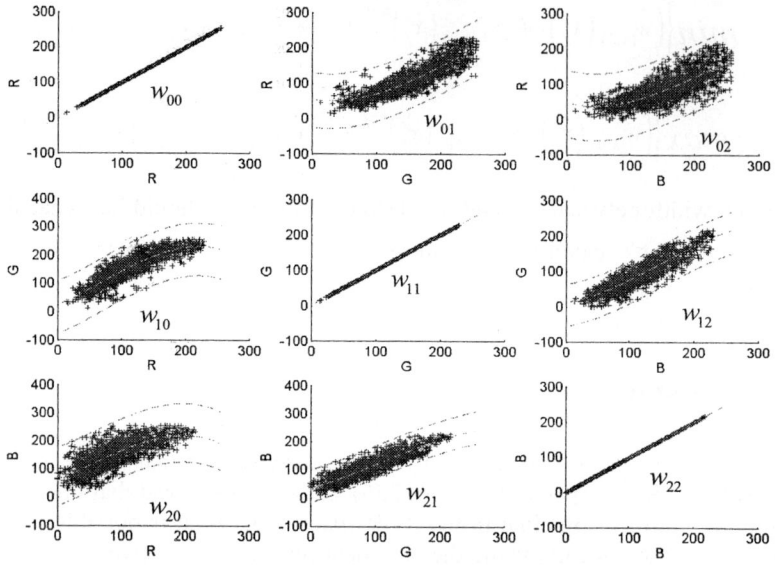

Fig. 3. Memory of the attractor network

Classifier performance can be quantified by computing the receiver operating characteristic (ROC) curve, which shows the relationship between correct detections and false detections. It is shown in Figure 4 that the performance of the skin classifier trained using the proposed learning algorithm is better compared to the skin probability mask (SPM), which is a classical model for skin detection [11]. These results are obtained by training both the classifiers with the same skin samples. After training, both the nonlinear line attractor network and SPM are tested with the same set of test images.

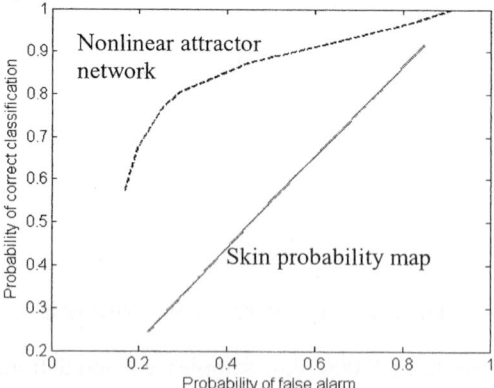

Fig. 4. Average ROC curves over 10 images

4 Conclusion

A novel learning algorithm for the training of multiple-valued neural network utilizing the concept of nonlinear line attractor has been presented. The technique is based on a mathematical model of statistical nature of the patterns under consideration. The method of least squares estimation has been utilized to approximate the relationship between the relative magnitudes of the output of a neuron with respect to the output of another neuron. The threshold function has been represented by the statistical characteristics of the input patterns. Experiments conducted on several benchmark problems have shown that the network can learn and distinguish human skin regions successfully.

References

1. Bengio, Y., Bengio, S., Cloutier, J.: Learning a Synaptic Learning Rule. Proc. IEEE Int. Conf. Neural Networks, Vol. 2. Seattle Washington (1991) 969-974
2. Seow, M.J., Asari, K.V.: Associative Memory Using Ratio Rule for Multi-Valued Pattern Association. Proc. IEEE Int. Conf. Neural Networks, Vol. 4. Portland Oregon (2003) 2518-2522
3. Seow, M.J., Asari, K.V.: Learning Using Distance Based Training Algorithm for Pattern Recognition. Pat. Recogn. Let. 25 (2004) 189-196
4. Haykin, S.: Neural Network: A Comprehensive Foundation. Macmillan, New York (2001)
5. Müezzinoglu, M.K., Güzelis, C., Zurada, J.M.: A New Design Method for the Complex-Valued Multistate Hopfield Associative Memory. IEEE Trans. Neural Networks. 14 (2003) 891-899
6. Brody, C.D., Kepecs, R.A.: Basic Mechanisms for Graded Persistent Activity: Discrete Attractors, Continuous Attractors, and Dynamic Representations. Curr. Opin. Neurobiol. 13 (2003) 204-211
7. Stringer, S.M., Trappenberg, T.P., Rolls, E.T., de Araujo, I.E.T.: Self-Organizing Continuous Attractor Networks and Path Integration: One-Dimensional Models of Head Direction Cells. Network: Comput. Neural Syst. 13 (2002) 217-242
8. Seung, H.S.: Continuous Attractors and Oculomotor Control. Neural Networks. 11 (1998) 1253-1258
9. Seow, M.J., Valaparla, D.P., Asari, K.V.: L2-Norm Approximation Based Learning in Recurrent Neural Networks for Expression Invariant Face Recognition. Proc. IEEE Int. Conf. System, Man & Cybernetics. Washington DC (2003) 3512-3517
10. Seow, M.J., Asari, K.V.: Associative Memory Based on Ratio Learning for Real Time Skin Region Extraction. Proc. IEEE Int. Work. Applied Imagery and Pattern Recognition. Washington DC (2003) 151-156
11. Brand, J.D., Mason, J.S.D.: A Skin Probability Map and Its Use in Face Detection. Proc. Int. Conf. Image Processing, Vol. 1. Thessaloniki Greece (2001) 1034-1037

A Bayesian Classifier by Using the Adaptive Construct Algorithm of the RBF Networks

Minghu Jiang[1,3], Dafan Liu[1], Beixing Deng[2], and Georges Gielen[3]

[1]Lab of Computational Linguistics, Dept. of Chinese Language,
Tsinghua Univ., Beijing, 100084, China.
jiang.mh@tsinghua.edu.cn
[2]Dept. of Electronics Eng., Tsinghua Univ., Beijing, 100084, China.
[3]Dept. of Electrical Eng., MICAS, K.U.Leuven, B3001 Heverlee, Belgium

Abstract. In paper we propose a Bayesian classifier for multiclass problem by using the merging RBF networks. The estimation of probability density function (PDF) with a Gaussian mixture model is used to update the expectation maximization algorithm. The centers and variances of RBF networks are gradually updated to merge the basis unites by the supervised gradient descent of the error energy function. The algorithms are used to construct the RBF networks and to reduce the number of basis units. The experimental results show the validity of our method which gives a smaller number of basis units and obviously outperforms the conventional RBF learning technique.

Keywords: Gaussian mixture model (GMM), radial basis function (RBF), expectation-maximization (EM), maximum-likelihood (ML)

1 Introduction

GMM density estimates can be used to adaptive RBF classifier. The key problem in designing of an RBF network is proper choice of the number and location of centers. Too small a number of these units may result in poor approximation and classification accuracy, whereas too large a number of these units can cause overfitting of the data and poor generalization performance [1]. Several methods have been proposed to determine the sizes and construction of RBF networks [2], [3]. According to Bayesian criterion to minimize the probability of misclassification, we select the highest posterior probability as the class of input data, the GMM is used to realize the PDF estimation of sub-RBF network for each class, and the RBF network of the common basis units is used to respond over all classes. The centers and variances of RBF networks are updated to merge the basis unites by the supervised iterative training.

2 Modeling of the Conditional Density

Assume that the PDF estimation is the GMM output [4], which is the linear weight combination of M_k Gaussian densities:

$$p(x^{(n)}/C_k) = y_k(x^{(n)}) = \sum_{j=1}^{M_k} P(j/C_k)p(x^{(n)}|\theta_j,C_k). \quad (1)$$

Where $P(j/C_k)$ is mixing coefficients of the jth component, the prior probability of the data can been generated from the jth basis function associated with class k, $\theta_{k,j} = \{\mu_{k,j}, \sigma^2_{k,j}\}$ is initialed by using the K-means clustering algorithm. $j=1,2,\ldots, M_k$, and $n=1,2,\ldots, N_k$, $N = \sum_{k=1}^{K} N_k$ (M_k and N_k are respectively the numbers of basis units and training data). To ensure the mixture is a true PDF, these priors must satisfy:

$$\sum_{j=1}^{M_k} P(j/C_k) = 1, \ 0 \le P(j/C_k) \le 1, \ k=1,2,\ldots,K. \quad (2)$$

According to Bayesian theorem, we have:

$$P(C_k/x^{(n)}) = \frac{P(C_k)p(x^{(n)}/C_k)}{\sum_{k'} P(C_{k'})p(x^{(n)}/C_{k'})} = \frac{P(C_k)p(x^{(n)}/C_k)}{p(x^{(n)})} = \gamma_k y_k(x^{(n)}). \quad (3)$$

$$\text{Class of } x^{(n)} = \arg\max_k \{P(C_k/x^{(n)})\} = \arg\max_k \{y_k(x^{(n)})\}. \quad (4)$$

The posterior of the jth basis unit can be estimated by Bayesian theorem:

$$P(j/x^{(n)}, C_k) = \frac{p(x^{(n)}|\theta_j, C_k)P(j/C_k)}{p(x^{(n)}/C_k)}, \text{ for } k=1,2,\ldots,K. \quad (5)$$

Where $\gamma_k = P(C_k)/p(x^{(n)})$, $p(x^{(n)}|\theta_j,C_k)$ is the PDF of the jth component with class k. It is important that the shape and number of the basis units are compatible with the output space distribution of the data [5]. We select Gaussian function as basis units, for simplicity, the covariance matrix of Gaussian function is assumed to be a diagonal one. The conditional density will peak when the training pattern $x^{(n)}$ is near the center vector $\mu_{k,j}$. The approach for determining the parameters of GMM from a given data set is to use ML estimation. $\theta_{k,j}$ can be estimated by using the EM algorithm to maximize the likelihood function of Eq. (1) [5], it equates to minimize the log-likelihood error:

$$\hat{E}_{MLE} = \arg\min_{\theta_j} \{-\sum_{n=1}^{N_k} \ln[\sum_{j=1}^{M_k} P(j/C_k)p(x^{(n)}|\theta_j,C_k)]\}. \quad (6)$$

Taking the partial derivative of Eq. (6) with respect to $\mu_{k,j}$ and $\sigma^2_{k,j}$, respectively, and setting them equal to zero, we have the parameters updated as:

$$\mu_{k,j} = \frac{\sum_{n=1}^{N_k} P(j/x^{(n)}, C_k) x^{(n)}}{\sum_{n=1}^{N_k} P(j/x^{(n)}, C_k)} \text{ and } \sigma_{k,j}^2 = \frac{\sum_{n=1}^{N_k} P(j/x^{(n)}, C_k) \left\| x^{(n)} - \mu_{k,j} \right\|^2}{L \sum_{n=1}^{N_k} P(j/x^{(n)}, C_k)}. \quad (7)$$

From Eqs. (2) and (6), we introduce a Langrange multiplier into Eq. (6), given [1]:

$$\hat{E}_{MLE} = \arg\min_{\theta_j} \{ -\sum_{n=1}^{N_k} \ln[\sum_{j=1}^{M_k} P(j/C_k) p(x^{(n)} | \theta_j, C_k)] + \lambda [\sum_{j}^{M_k} P(j/C_k) - 1] \}. \quad (8)$$

Taking the partial derivative of Eq. (8) with respect to $P(j/C_k)$, setting it equal to zero, after simple derivation, we have the priors:

$$P(j/C_k) = \frac{1}{N_k} \sum_{n=1}^{N_k} P(j/x^{(n)}, C_k) = \frac{1}{N_k} \sum_{n \in C_k} \frac{p(x^{(n)} | \theta_j, C_k) P(j/C_k)}{p(x^{(n)}/C_k)}. \quad (9)$$

$$P(j) = \sum_k P(C_k) P(j/C_k) = \frac{1}{N} \sum_k \sum_{n \in C_k} \frac{p(x^{(n)} | \theta_j, C_k) P(j/C_k)}{p(x^{(n)}/C_k)}. \quad (10)$$

Eq. (1) can be expressed by a sub-RBF network which shown as:

$$y_k(x^{(n)}) = \sum_{j=1}^{M_K} w_{k,j} \vartheta_j(x^{(n)}) = \sum_{j=1}^{M_K} w_{k,j} \exp(-\frac{\left\| x^{(n)} - \mu_{k,j} \right\|^2}{2\sigma_{k,j}^2}). \quad (11)$$

Where $w_{k,j} = \frac{P(j/C_k)}{(2\pi\sigma_{k,j}^2)^{L/2}} > 0$, $\vartheta_j(x^{(n)})$ denotes the Gaussian basis function of the jth center. For the sub-RBF network of each class, the centers are shifted and variances are expanded to merge the basis unites by the gradient of the desirable approximating PDF estimation. We start with a little large number of basis units, the EM algorithm is divided into E-step and M-step, and $\theta_{k,j}$ are updated by the unsupervised learning algorithm. E-step (use the parameter values to determine the posterior probability):

$$P^{(t+1)}(j/x^{(n)}, C_k) = \frac{p^{(t)}(x^{(n)} | \theta_j, C_k) P^{(t)}(j/C_k)}{p^{(t)}(x^{(n)}/C_k)}. \quad (12)$$

M-step (re-estimating the parameters of the linear models and obtain the new ML estimate from the current estimate):

$$P^{(t+1)}(j/C_k) = \frac{t}{t+1} P^{(t)}(j/C_k) + \frac{1}{t+1} P^{(t)}(j/x^{(n)}, C_k). \quad (13)$$

$$\mu_{k,j}^{(t+1)} = \mu_{k,j}^{(t)} + \eta_\mu^{(t)} \sum_{n=1}^{N_k} (x^{(n)} - \mu_{k,j}^{(t)}) \, P^{(t)}(j/x^{(n)}, C_k). \tag{14}$$

$$(\sigma_{k,j}^2)^{(t+1)} = (\sigma_{k,j}^2)^{(t)} + \eta_\sigma^{(t)} \sum_{n=1}^{N_k} (\|x^{(n)} - \mu_{k,j}^{(t)}\|^2 - (\sigma_{k,j}^2)^{(t)}) \, P^{(t)}(j/x^{(n)}, C_k). \tag{15}$$

Where $\eta_\mu^{(t)}$ and $\eta_\sigma^{(t)}$ are the learning rates. In the following part, we describe a method to construct adaptively the number of basis units for too many hidden nodes.

3 A Bayesian Classifier of the Merging RBF Networks

Because a Bayesian classifier of the RBF networks exist too many hidden nodes, and this may lead to lower performance and over-fitting. We need sufficiently utilize the information of K initial networks to reduce the their size. For any hidden units, the generated response of the hidden layer is strongest when the input vector is closest to their centers, the weight updates are calculated for these strong responses of basis units. For these very near centers, we should merge them according to the supervised training of output layer, for these centers that are very weak response to input data should be canceled. Assume that E is the error function, the target output $d_k^{(i)}$ denotes the desired posteriors. From Eq. (3), we have:

$$p(C_k/x^{(n)}) = \tilde{y}_k(x^{(n)}) = \gamma_k y_k(x^{(n)}) = \sum_{j=1}^M \tilde{w}_{k,j} \vartheta_j(x^{(n)}) = \sum_{j=1}^M \tilde{w}_{k,j} \exp(-\frac{\|x^{(n)} - \mu_j\|^2}{2\sigma_j^2}). \tag{16}$$

Due to many basis unites by adding all classes into Eq. (16), in order to decrease its number, $\theta_{k,j}$ are updated to merge the basis unites by the supervised learning:

$$\mu_j^{(t+1)} = \mu_j^{(t)} - \eta_\mu^{(t)} \frac{\partial E^{(t)}}{\partial \mu_j^{(t)}} = \mu_j^{(t)} + \eta_\mu^{(t)} \sum_k \sum_{n \in C_k} \frac{2\tilde{w}_{k,j}(d_k^{(n)} - \tilde{y}_k(x^{(n)}))(x^{(n)} - \mu_j^{(t)}) \vartheta_j(x^{(n)})}{N(\sigma_j^2)^{(t)}}. \tag{17}$$

$$(\sigma_j^2)^{(t+1)} = (\sigma_{k,j}^2)^{(t)} + \eta_\sigma^{(t)} \sum_k \sum_{n \in C_k} \frac{\tilde{w}_{k,j}(d_k^{(n)} - \tilde{y}_k(x^{(n)}))(x^{(n)} - \mu_j^{(t)})^2 \vartheta_j(x^{(n)})}{N(\sigma_{k,j}^4)^{(t)}}. \tag{18}$$

$$\eta_\mu^{(t+1)} = \begin{cases} g\eta_\mu^{(t)} & E^{(t)} < E^{(t-1)} \\ h\eta_\mu^{(t)} & E^{(t)} > E^{(t-1)} \\ \eta_\mu^{(t)} & E^{(t)} = E^{(t-1)} \end{cases}, \quad \eta_\sigma^{(t+1)} = \begin{cases} g\eta_\sigma^{(t)} & E^{(t)} < E^{(t-1)} \\ h\eta_\sigma^{(t)} & E^{(t)} > E^{(t-1)} \\ \eta_\sigma^{(t)} & E^{(t)} = E^{(t-1)} \end{cases}, \; g>1, \; h<1. \tag{19}$$

Eqs. (13)-(15) enable the two neighboring centers to be closer and their variances to be extended. For the class k, when two neighboring basis functions have almost the same variance and very near center, the two basis functions are merged into one, i.e.:

$$\left\|\frac{\mu_i - \mu_j}{\sigma_i}\right\|^2 < \beta \text{ and } \left\|\frac{\mu_i - \mu_j}{\sigma_j}\right\|^2 < \beta. \qquad (20)$$

Where β is a threshold, then the new center and variance are determined as:

$$\mu_{new} = (\mu_i + \mu_j)/2 \, , \, \sigma_{new} = \sqrt{\sigma_i \sigma_j} \, . \qquad (21)$$

Weight of the output layer is updated on iterative step $(t+1)$ as:

$$\tilde{w}_{k,j}^{(t+1)} = \tilde{w}_{k,j}^{(t)} - \eta_w \frac{\partial E^{(t)}}{\partial w_{k,j}^{(t)}} = \tilde{w}_{k,j}^{(t)} + \eta_w (d_k^{(n)} - \tilde{y}_k(x^{(n)}))\vartheta_j(x^{(n)}). \qquad (22)$$

Where η_w is learning step length. For the ith basis unit, if $\tilde{w}_{k,i} < 1/N$ holds for all k, shows the significance and contribution of this unit are sufficiently small, then the ith unit is cancelled, retraining it until converges. The training is divided into two phases: The first phase is the training of each sub-RBF network to estimate the PDF of each class by a GMM updated with the EM algorithm. Then using the supervised gradient descent of the posterior probability merges the similar basis units and removes the redundant units. The second phase is the supervised gradient descent of the error function to realize the training of center mergence.

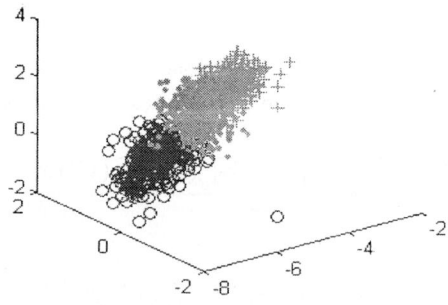

Fig. 1. Classification for Gamma-Ray Bursts, three classes

4 Experimental Results

We demonstrate the validity of the proposed method by experimental data of the Gamma-Ray Burst, the data is got from BATSE of Compton Observatory Science Support Center in NASA. The dataset includes more than 1000 data, we use the three attributes: pulse lighteness, intensity and length. The 500 sets are used for training, and the other is used for testing. We get three classes by K-means clustering algorithm, their distributions are shown in Fig.1. The parameters of each sub-RBF network are calculated from about 40 hidden units for each class, then the hidden units and common basis units are merged by using iterative gradient descent algorithm to maximize fit to the training data with η=0.001. It has been observed that about 23 hidden units are reduced most without degrading network performance. A summary of the results was shown in Table I. The classification accuracy of the test set in the best instance for the conventional RBF networks is from 92.4% to 88.2%, and the classification accuracy of the test set for our mergence algorithm of common basis units is 93.3%. The experimental results show the our algorithm can provide more sufficient accuracy estimation for multiclass problems, show that the supervised training of center mergence can remove redundant centers, this leads to a smaller RBF network which outperforms the RBF networks trained in the conventional approach.

Table 1. Classification for Gamma-Ray Bursts

Number of hidden units	10	15	20	25	30	35	40
Error Energy	0.1233	0.08	0.0767	0.0767	0.08	0.0933	0.0833

5 Conclusions

A Bayesian classifier by using the adaptive construct algorithm of the RBF Networks provides an optimal construct of the parameters and size of the RBF networks and to reach the minimum misclassification, the model of common basis units is well fits input data for multi-class problems. The experimental simulations show the validity of proposed method which gives a smaller number of basis units and obviously outperforms the conventional RBF learning technique.

References

1. Miller, D. J., Uyar, H. S.: Combined Learning and Use for a Mixture Model Equivalent to the RBF Classifier. Neural Computation, 10 (1998) 281-293
2. Chen, S., Wu, Y., Luk, B. L.: Combined Genetic Algorithm Optimization and Regularized Orthogonal Least Squares Learning for Radial Basis Function. IEEE Transactions on Neural Networks. 10 (1999) 1239-1243
3. Alba, J. L., Docio, L., Docampo, D., et al: Growing Gaussian Mixtures Network for Classification Application. Signal Processing. 76 (1999) 43-60
4. Utsugi, A.: Density Estimation by Mixture Models with Smoothing Priors. Neural Computation. 10 (1998) 2115-2135
5. Lotlikar R., Kothar, R.: Bayes-Optimality Motivated Linear and Multilayered Perceptron-Based Dimensionality Reduction. IEEE Transactions on Neural Networks. 11 (2000) 452-463

An Improved Analytical Center Machine for Classification

FanZi Zeng[1], DongSheng Li[2], and ZhengDing Qiu[1]

[1] Beijing Jiaotong University, Institute of Information and Science, Beijing, 100044
[2] Dongjian Hydropower Plant, Hunan, 423400
zeng_f@126.com

Abstract. Analytical center machine (ACM) has remarkable generalization performance based on analytical center of version space and outperforms SVM. From the analysis of geometry of machine learning and principle of ACM, it is showed that some training patterns are redundant to the definition of version space. Redundant patterns push ACM classifier away from analytical center of the prime version space so that the generalization performance degrades, and slow down the classifier and reduce the efficiency of storage. Thus, an incremental algorithm is proposed to remove redundant patterns and embed into the frame of ACM that yields a redundancy-free, accurate analytical center machine for classification called RFA-ACM. Experiments with Heart, Thyroid, and Banana datasets demonstrate the validity of RFA-ACM.

1 Introduction

Support vector machines (SVMs) [1-3] have been recently introduced as a new tool for machine learning application due to its outstanding performance in generalization on unseen data. In fact, SVM classifier corresponds to the center of the largest inscribed hypersphere in version space, thus when the version space is elongated or asymmetric, SVMs is not very effective. Recently, a classifier called ACM is proposed by T.B.Trafalis[4], which corresponds to the analytical center of the version space and outperforms SVMS. From the analysis of geometry of learning machine and the principle of ACM classifier, it reveals some constraints consistent with training data are redundant. Redundant constraints put ACM classifier away from the analytical center of prime version space so that the generalization performance degrades, and reduce the speed of classification and the efficiency of storage. To address the problem, an algorithm to remove the redundant constraints is proposed and combined with the ACM classifier algorithm, which yields a redundancy-free, accurate ACM classifier called RFA-ACM. Experiments on three datasets: Heart, Thyroid, and Banana demonstrate the validity of RFA-ACM.

2 Geometry of Linear Learning Machine and ACM Classifier

Let training set $T = \{(x_j, y_j)_{j=1}^m\} \subset \Re^d \times \{+1, -1\}$. For perceptron algorithm, the objective is to solve the following feasibility problem:

$$y_j \cdot (x_j \bullet w + b) > 0 \ \forall \ j = 1, \cdots, m \tag{1}$$

where $\bullet$ represents the inner product. Let $\tilde{w} = \begin{bmatrix} w \\ b \end{bmatrix} \in \Re^{d+1}$, $\tilde{x}_j = \begin{bmatrix} x_j \\ 1 \end{bmatrix} \in \Re^{d+1}$, slightly rewrite (1) as follows

$$\tilde{S}_j = y_j \tilde{x}_j \bullet \tilde{w} > 0 \ \forall \ j = 1, \cdots, m \tag{2}$$

We consider the augmented weight space $\tilde{w}$, in which each pattern $(\tilde{x}_j, y_j)$ is represented as a hyperplane passing through the origin. From inequality (2), the first inequality reduces the augmented weight space $\tilde{w}$ in $\Re^{d+1}$ to a half space; the second inequality further reduces the half space into a cone. The following inequalities further reduce the cone or keep the cone unchangeable. Those training patterns consistent with inequalities keeping the cone unchangeable are redundant to the definition of the feasible region. After all the hyperplanes have been defined, a cone in $\Re^{d+1}$ will describe the set of feasible solutions in the augmented weight space.

Any feasible solution $\tilde{w}$ will separate the training set, but generalization will be rather poor in most instances. Thus, the concept of analytical center of the feasible region is introduced to select the optimal hyperplane.

We introduce the concept of logarithmic barrier function to describe the analytical center of the feasible region. Define the slack variable $\tilde{S}_j \in \Re$ as a measure of how close or how far away the current solution is from the constraint j. If the current solution violates constraint j, $\tilde{S}_j$ will be negative. An interior point of the feasible region is defined as a point, for which all slacks $\tilde{S}_j$ are positive. For the set of constraints, which currently describe the feasible region, the slacks can be expressed as follows:

$$\tilde{S}_j = y_j \tilde{x}_j \bullet \tilde{w} > 0 \ \forall \ j = 1, \cdots, m \tag{3}$$

Furthermore, A logarithmic function $\Phi : \Re^+ \to \Re$ is introduced, which goes to infinity when the current solution approaches the boundary of the feasible region. More specifically,

$$\Phi(\tilde{S}) = -\ln(\tilde{S}) \cdot \tag{4}$$

For m training patterns, the logarithmic function Φ becomes as follows

$$\Phi(\tilde{S}) = -\sum_j^m \ln(\tilde{S}_j), \tilde{S}' = (\tilde{S}_1, \cdots, \tilde{S}_m) \cdot \tag{5}$$

Where $\tilde{S}'$ represents the transpose of $\tilde{S}$. The minimization of the logarithmic barrier function $\Phi(\tilde{S})$ constitutes an excellent means to calculate a point, i.e. a weight vector $\tilde{w}^*$, which is far away from all constraints and is optimal in the sense that it minimizes logarithmic average of classification error.

To retrieve weight vector $\tilde{w}^*$, we employ the concept of analytical center and have to solve the minimization problem (5). Unfortunately, the minimization of logarithmic barrier function can only be calculated, if the feasible region is compact, which is decidedly not the case for a cone in $\Re^{d+1}$. So a hypersphere in $\Re^{d+1}$ with radius R around the origin is added as constraint to compactify the feasible region. Therefore our goal is to calculate the minimum of $\Phi(\tilde{w})$ while making sure at the same time that the optimal weight vector $\tilde{w}^*$ satisfies the spherical constraint. Following the above, the optimal weight vector, which corresponds to the ACM classifier, can be computed by solving the optimization problem:

$$\min \quad \Phi(\tilde{w}) = -\sum_{j}^{m} \ln(y_j \tilde{x}_j \bullet \tilde{w}) \tag{6}$$

$$\text{s.t.} \quad h(\tilde{w}) = \frac{1}{2}\tilde{w}'\tilde{w} - 1 = 0$$

If the linear inseparable case is encountered, the kernel strategy is used to map the training dataset into feature space, in which training dataset is linear separable.

3 Incremental Algorithm of Removing Redundant Constraints and RFA-ACM Classifier

In order to facilitate the analysis of incremental algorithm of removing redundant constraints and RFA-ACM classifier, two definitions are introduced firstly.

Definition 1 Given the bounded polytope $R=\{x\in\Re^n\,|\,Ax\leq b, A\in\Re^{N\times n}, b\in\Re^n\}$, $N>2$, the i-th inequality is redundant if and only if $R=R_i=\{x\in\Re^n\,|\,A_j'x\leq b_j, j=1,\cdots,i-1,i+1,\cdots,N\}$, i.e. $\nexists x\in R_i: A_i'x > b_i$.

Definition 2 Feasible region defined by (3) is prime, if there are no redundancies in the feasible region.

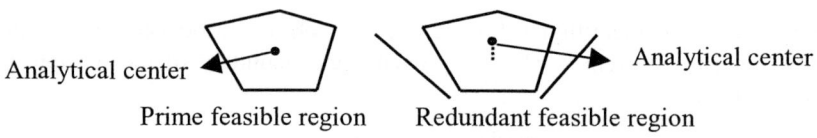

Analytical center   Analytical center

Prime feasible region   Redundant feasible region

Fig. 1. The affect of redundancies on the analytical cente

Fig. 1 illustrates the affects that the redundant constraints have on the analytical center. In the case, the analytical center moves upward and away from the two redundant inequalities. This is due to the presence of the logarithmic terms associated with redundancies in the objective function (6)[5]. Moreover, the theory of sparseness bound[6] shows that sparse feature can attain lower bound of generalization compared to dense feature. Therefore, redundant constraints degrade the generalization performance, and intuitively slow down the classifier, reduce the efficiency of storage. Thus, removal of redundant constraints is crucial to the performance of the classifier based on analytical center.

From the definition 1, we have the redundancies removal algorithm 1 of the polytope R as follows.

Redundancies removal algorithm 1
Input: bounded polytope R
Output: prime polytope (redundancies free) of R: $\Phi = \{x \in \Re^n \mid Cx \leq d\}$

Step 1 Let $i \leftarrow 1, C \leftarrow [], d \leftarrow []$
Step 2 If $\max_{x \in R_i} A_i'x > b_i$, then $C \leftarrow \begin{bmatrix} C \\ A_i \end{bmatrix}, d \leftarrow \begin{bmatrix} d \\ b_i \end{bmatrix}$

Step 3 If $i<N$ then $i=i+1$, go to step 2, else stop.

As to unbounded holytope, such as cone, because the linear programming $\max_{x \in R_i} A_i'x$ in step 2 has no solution for each possible value of i, algorithm 1 is not applied to the removal of redundancies for the unbounded holytope. Thus, we discuss the removal of redundancies for the unbounded holytope as follows.

Given cone $\Omega = \{x \in \Re^n \mid a_i'x \geq 0, i \in I, I = \{1, \cdots, m\}\}$, $\Omega_k = \{x \in \Re^n \mid a_i'x \geq 0, i \in I/k\}$, I/k represents the indices I except k.

Proposition 1 If $x^* \in \Omega_k$ and $a_k'x^* < 0$, then constraint k is definitely not redundant.

Proof: we use apagoge. Suppose the constraint k is redundant, according the definition 1, we have $\Omega_k = \Omega$, this is equivalent to the following linear inequality system no solution for all $i \in I/k$,

$$a_k'x < 0$$
$$a_i'x \geq 0, \text{ for all } i \in I/k$$

This contradicts the condition of proposition 1: $x^* \in \Omega_k, a_k'x^* < 0$, thus constraint k is definitely not redundant.

Proposition 2 If $x^* \in \Omega_k$ and $a_k'x^* \geq 0$, then constraint k may be redundant, may be not.

Proof: we use apagoge. Suppose the constraint k is definitely redundant.

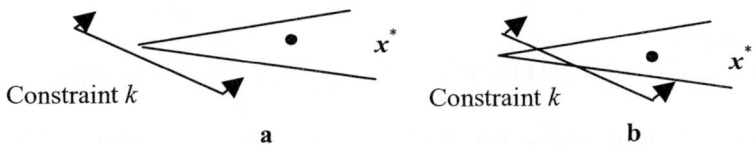

Fig. 2. When $a_k'x^* \geq 0$, the possible distribution of constraint k

From Fig. 2, we can see that when $x^* \in \Omega_k$ and $a_k'x^* \geq 0$, constraint k in Fig.2-a is redundant, but constraint k in Fig.2-b is not. This means that Figure 2 gives counterexample of proposition 2, i.e. that If $x^* \in \Omega_k$ and $a_k'x^* \geq 0$, the constraint k is definitely redundant is not correct. Thus, proposition 2 is correct.

In order to make the cone to be a bounded holytope, an extra constraint must be added. If we have a point x^* in cone, an equality constraint $(x^*)'x \leq \|x^*\|^2$, which passes through x^* with norm x^* and with the distance $\|x \cdot\|^2$ to the origin, can meet the goal. Then $\Omega_k = \{x \in \Re^n \mid a_i'x \geq 0, (x^*)'x \leq \|x^*\|^2, i \in I/k\}$ becomes the bounded holytope. Therefore, we have the incremental algorithm of removal of redundancies for the cone as follows.

Incremental algorithm 2
Input: cone $\Omega = \{x \in \Re^n \mid a_i'x \geq 0, i \in I, I = \{1, \cdots, m\}\}$
Output: redundancies-free cone Θ
Step 1 Take two constraints to form the cone $\Theta = \{x \in \Re^n \mid a_i'x \geq 0, i \in \tilde{I}, \tilde{I} = \{1,2\}\}$, let $k=2$
Step 2 Arbitrarily take a point in the cone $x^* \in \Theta$, namely $x^* \in \text{int}(\Theta)$
Step 3 Take constraint $k=k+1$, if $k>m$, then stop, else go to step 4
Step 4 If $a_k'x^* < 0$, then constraint k is not redundant, and let $\tilde{I} \leftarrow \begin{bmatrix} \tilde{I} \\ k \end{bmatrix}$, go to step 3, else go to step 5
Step 5 If $\min_{x \in \Theta} a_k'x < 0$, then go to step 3, else let $\tilde{I} \leftarrow \begin{bmatrix} \tilde{I} \\ k \end{bmatrix}$ and go to step 6
Step 6 Arbitrarily takes a point in the cone $x^* \in \Theta$, go to step 3.

Now we embed the algorithm 2 into the algorithm of ACM classifier, which yields a redundancies-free, accurate ACM classifier called RFA-ACM. The algorithm of RFA-ACM is completely described as follows.

Incremental algorithm for RFA-ACM
Input: cone $\Omega = \{x \in \Re^n \mid a_i'x \geq 0, i \in I, I = \{1, \cdots, m\}\}$
Output: analytical center of prime feasible region, which corresponding to RFA-ACM
Step 1 Take two patterns to form the cone $\Theta = \{x \in \Re^n \mid a_i'x \geq 0, i \in \tilde{I}, \tilde{I} = \{1,2\}\}$, let $k=2$.
Step 2 Calculate the analytical center of Θ, i.e. solve the minimization problem (6) over Θ. Denote the corresponding analytical center as $x^* \in \Theta$.
Step 3 Take pattern $k=k+1$, if $k>m$, then stop, else go to step 4.
Step 4 If $a_k'x^* < 0$, then constraint k is not redundant, and let $\tilde{I} \leftarrow \begin{bmatrix} \tilde{I} \\ k \end{bmatrix}$, go to step 6, else go to step 5.
Step 5 If $\min_{x \in \Theta} a_k'x < 0$, then go to step 3, else let $\tilde{I} \leftarrow \begin{bmatrix} \tilde{I} \\ k \end{bmatrix}$ and go to step 6.
Step 6 Calculate the analytical center of Θ, i.e. solve the minimization problem (6) over Θ. Denote the corresponding analytical center as $x^* \in \Theta$, go to step 3.

4 Experiments and Results

Experiments on three datasets: Heart, Thyroid, Banana, which are from http://www.first.gmd.de/~raetsch/, are conducted to test the generalization performance of RFA-ACM, and the comparison of generalization performance among SVM [7], ACM [4] and RFA-ACM are conducted. Datasets are randomly partitioned as follows: The datasets Heart and Thyroid exhibit a training set to test set ratio of 60%:40%, while the dataset Banana is partitioned based on a ratio of 10%:90%. The kernel function used is $k(x_i, x_j) = \exp(\|x_i - x_j\|^2 / 2\sigma^2)$, the value of σ is shown as the fourth column in Table 1. Table 1. shows the generalization errors for SVM [8], ACM [4], as well as for RFA-ACM and demonstrates that RFA-ACM effectively improves the generalization performance.

Table 1. Generalization error with SVM, ACM, RFA-ACM on benchmark data

Dataset	SVM [7]	ACM [4]	RFA-ACM	σ
Heart	25.40	21.87	20.10	10.00
Thyroid	5.30	4.91	4.63	3.00
Banana	16.20	14.73	12.87	0.50

5 Conclusion

Analytical center machine based on analytical center of version space (ACM) has remarkable generalization performance and outperforms SVM. Redundant constraints to the description of version space push ACM classifier away from analytical center of the prime version space so that the generalization performance degrades, and slow down the classifier and reduce the efficiency of storage. Therefore, an incremental algorithm to remove redundant constraints is proposed to address the above problems and embed into the frame of ACM that yields a redundancy-free, accurate analytical center machine for classification called RFA-ACM. Experiments with Heart, Thyroid, and Banana datasets demonstrate the validity of RFA-ACM.

References

1. Schölkopf, B., Burges, C.J.C., Smola, A.J.: Advances in Kernel Methods: Support Vector Learning. The MIT Press, Cambridge Massachusetts (1999)
2. Smola, A., Bartlett, P., Sch¨olkopf, B., Schuurmans, D.: Advances in Large Margin Classifiers. The MIT Press, Cambridge Massachusetts (2000)
3. Vapnik, V.: The Nature of Statistical Learning Theory. Springer Verlag (1995)
4. Trafalis, B.T., Malysche, M.A.: An Analytic Center Machine. Machine Learning, Vol. 46 (1/3) (2002) 203–223
5. Caron, J.R., Greenberg, J.H., Holder, G.A.: Analytic Centers and Repelling Inequalities. http://www-math.cudenver.edu/ccm/reports/
6. Graepel, T.: PAC-Bayesian Pattern Classification with Kernels Theory, Algorithm, and an Application to the Game of Go. PhD Dissertation (2002)
7. Herbrich, R., Graepel, T., Campbell, C.: Robust Bayes Point Machines. Proceedings of ESANN 2000 (2000)

Analysis of Fault Tolerance of a Combining Classifier

Hai Zhao and Bao-Liang Lu

Department of Computer Science and Engineering, Shanghai Jiao Tong University,
1954 Huashan Rd., Shanghai 200030, China
{zhaohai,blu}@cs.sjtu.edu.cn

Abstract. This paper mainly analyses the fault tolerant capability of a combining classifier that uses a K-voting strategy for integrating binary classifiers. From the point view of fault tolerance, we discuss the influence of the failure of binary classifiers on the final output of the combining classifier, and present a theoretical analysis of combination performance under three fault models. The results provide a theoretical base for fault detection of the combining classifier.

1 Introduction

The original purpose of this research arises from two research directions which are different but have some relations.

The first direction is neural network research with fault tolerance. In engineering, the needs of reliability has converted pure hardware concern to hardware-software concern. As a classification model with fault tolerance in essential, the fault tolerance of neural network models has been extensively studied during the past several years [1][2]. The main research approach of fault tolerance is based on the practical fault model or theoretical fault model, e.g., researchers study the shortcut failure of single point with fault tolerance analysis, and then improve the performance or effect through the revised algorithm.

The second direction is concerned with the binarization of multi-class problems which is to be more and more important in the last decade [3][4]. The common ground of those researches is that they studied the integrated output ability of a certain base classifier model, or discussed the total classification error caused by the output combination of some special base classifiers.

Our study focuses on a combining classifier in which component classifier are integrated into a modular classifier with a voting strategy called K-voting in this paper. This combining classifier can also be regarded as a min-max modular neural network[5], which only decompose a K-class classification problem into $\binom{K}{2}$ two-class subproblems and no any further decomposition is performed on two-class subproblems. Throughout this paper, this combining classifier is called K-voting classifier for short.

The basic fault unit for analysis is the binary classification module. Under three fault models, we give the quantitative expression of the ultimate classification effect based on the failure probabilities of binary classifiers.

The rest of the paper is organized as follows: In Sections 2 and 3 we briefly introduce K-voting classifier and mathematical notation. Three different fault models will be presented in Section 4. The experimental results and comment on theoretical and experimental results are presented in Section 5. Conclusions of our work and the current line of research are outlined in Section 6.

2 The Combining Procedure of K-Voting Classifiers

Now, we give a simple introduction to combination procedure for K-voting classifier.

we use one-against-one method for task decomposition. Suppose a K-class classification problem is considered, by using one-against-one decomposition method, we get $\binom{K}{2}$ independent two-class problems. The training samples of every two-class problem come from two different classes. K-voting combination rule is defined as: if the outputs of *K-1* binary classifiers in all support the same class ID, then the final classification result is just this class, otherwise, the result is unknown.

3 Mathematical Notation

After binarization, a binary classifier is noted as X_{ij}, which means it has learned from examples in class i and class j. We also take the notation X_{ij} as a stochastic variable with the two-point probability distribution.

For a given testing example S, we denote its class ID as *V(S)* The non-fault output probability p_{ij} of a binary classifier X_{ij} is constrained by

$$p_{ij} = p_{ji} \text{ for } 0 < i, j \leq K, \text{and } i \neq j \tag{1}$$

and X_{ij}'s random guess probabilities for a sample from one non-ij class:

$$P(X_{ij} = i | V(S) = i') = q_{ij} \text{ and } P(X_{ij} = j | V(S) = i') = 1 - q_{ij}, \forall i' \neq i \text{ and } i' \neq j \tag{2}$$

The distribution of classes in the number of testing data:

$$\alpha_i \text{ for } 0 < i \leq K, \text{and } \sum_{i=1}^{K} \alpha_i = 1 \tag{3}$$

The correct rate, incorrect rate, and unknown rate obtained by the original K-voting classifier are denoted by

$$P_{OA}, P_{OE}, \text{ and } P_{OF}, \tag{4}$$

respectively. Note that analysis in this paper is based on a simplified assumption: If all occurrence of fault states are regarded as stochastic variants, then these variants are independent with each other.

4 Fault Models

In this section, we will discuss three different fault models and analyse the performance of K-voting classifier under those assumptions.

4.1 Complete Failure Model

We consider the fault model as the output of each binary classifier is complete failure, i.e., fault state is an undefined output state of binary classifier.

Firstly, suppose that the original classification procedure is a faultless procedure, i.e., all binary classifiers can always output the correct results.

In K-voting combination, as to ultimate output class ID i, only those binary classifiers will have the effect on the final output: $X_{mn}, m = i$ or $n = i$. That is, only under such condition $X_{ij} = i, \forall j, 1 \leq j \leq K, i \neq j$, the output can be i.

All the other binary classifiers will have nothing with the final output and fault occurrence is independent, so to the ith class testing sample, the probability of effective output of K-voting classifier will be $\prod_{j=1,j\neq i}^{K} p_{ij}$. After weighting, we get the available output probability of all classes: $P'_{A1} = \sum_{i=1}^{K}(\alpha_i \prod_{j=1,j\neq i}^{K} p_{ij})$.

Finally, according to (4), we get the correct rate, incorrect rate and unknown rate in practical case as follows:

$$P_{A1} = P_{OA} \sum_{i=1}^{K}(\alpha_i \prod_{j=1,j\neq i}^{K} p_{ij}) \quad P_{E1} = P_{OE} \sum_{i=1}^{K}(\alpha_i \prod_{j=1,j\neq i}^{K} p_{ij}) \quad (5)$$

$$P_{F1} = P_{OF} \sum_{i=1}^{K}(\alpha_i \prod_{j=1,j\neq i}^{K} p_{ij}) + (1 - \sum_{i=1}^{K}(\alpha_i \prod_{j=1,j\neq i}^{K} p_{ij})) \quad (6)$$

4.2 Complete Inverse Fault Model

Suppose that the outputs of all binary classifiers can be inversed to irrational output under a certain probability, i.e., to output the rational result under a certain probability, and in other cases, the output is the reverse of the rational judgment according to classification algorithm.

As the same in Section 4.1, we also consider the output performance of K-voting combination while all original outputs of binary classifiers are always correct. because of the independent assumption, at this time, the correct rate of K-voting classifier is given by

$$P'_{A2} = P_{OA} \sum_{i=1}^{K}(\alpha_i \prod_{j=1,j\neq i}^{K} p_{ij}) \quad (7)$$

Because of the fault of inverse, the probability that those binary classifiers can not handle the classification is changed. At this time, for non-i,j classes, the output probability of class ID i by the binary classifier X_{ij} should be revised as:

$$q'_{ij} = q_{ij}p_{ij} + (1 - q_{ij})(1 - p_{ij}) \tag{8}$$

Consider the probability of misclassification of all the samples from class i,

$$\sum_{j=1,j\neq i}^{K}(1-p_{ij})\prod_{m=1,m\neq i,m\neq j}^{K} q'_{jm}$$

After weighting, the probability of misclassification for all samples will be:

$$P'_{E2} = \sum_{i=1}^{K}(\alpha_i \sum_{j=1,j\neq i}^{K}(1-p_{ij})\prod_{m=1,m\neq i,m\neq j}^{K} q'_{jm}) \tag{9}$$

According to (7) and (9), we may get the unknown rate of K-voting classifier:

$$P'_{F2} = 1 - \sum_{i=1}^{K}(\alpha_i(\prod_{j=1,j\neq i}^{K} p_{ij} + \sum_{j=1,j\neq i}^{K}(1-p_{ij})\prod_{m=1,m\neq i,m\neq j}^{K} q'_{jm})) \tag{10}$$

In practice, the unknown rate is very low, so we can omit the output probability of the case that the unknown output is inverted to the correct output or misclassification. But, we can not exclude the case that the inversion causes the final output from the incorrect to the correct. According to Section 4.1, we can get the probability that the correct is inversed from the incorrect as follows:

$$\sum_{i=1}^{K}(\alpha_i \sum_{j=1,j\neq i}^{K} q'_{ij} \prod_{m=1,m\neq i,m\neq j}^{K}(1-p_{jm}))$$

Ultimately, we get the actual correct rate, incorrect rate, and unknown rate as follows:

$$P_{A2} = P_{OA}\sum_{i=1}^{K}(\alpha_i \prod_{j=1,j\neq i}^{K} p_{ij}) + P_{OE}\sum_{i=1}^{K}(\alpha_i \sum_{j=1,j\neq i}^{K} q'_{ij} \prod_{m=1,m\neq i,j}^{K}(1-p_{jm})) \tag{11}$$

$$P_{E2} = P_{OE}\sum_{i=1}^{K}(\alpha_i \prod_{j=1,j\neq i}^{K} p_{ij}) + P_{OA}\sum_{i=1}^{K}(\alpha_i \sum_{j=1,j\neq i}^{K}(1-p_{ij}) \prod_{m=1,m\neq i,j}^{K} q'_{jm}) \tag{12}$$

$$P_{F2} = P_{OF} + P_{OA}(1 - \sum_{i=1}^{K}(\alpha_i(\prod_{j=1,j\neq i}^{K} p_{ij} + \sum_{j=1,j\neq i}^{K}(1-p_{ij})\prod_{m=1,m\neq i,m\neq j}^{K} q'_{jm})))$$

$$+P_{OE}(1 - \sum_{i=1}^{K}(\alpha_i(\sum_{j=1,j\neq i}^{K} p_{ij} + \sum_{j=1,j\neq i}^{K} q'_{ij} \prod_{m=1,m\neq i,m\neq j}^{K}(1-p_{jm})))) \tag{13}$$

Because we have omitted the probability of the unknown output inverse to the correct or the incorrect, (11) and (12) will be underestimated, and (13) will be overestimated.

4.3 Pseudo-Correct Output Model

In this subsection, we assume that the binary classifier may not give out the normal output according to rational judgment, but give a random guess according to a certain probability when the binary classifier is in its fault state. Suppose the probability of the output of class ID i under the fault situation of binary classifier X_{ij} is given by:

$$r_{ij} \text{ for } 0 < i,j \leq k \text{ and } i \neq j \tag{14}$$

Consider that the outputs of all the binary classifiers are always correct. Then the actual output caused by fault will be

$$p'_{ij} = p_{ij} + (1 - p_{ij})(\frac{\alpha_i}{\alpha_i + \alpha_j}r_{ij} + \frac{\alpha_j}{\alpha_i + \alpha_j}(1 - r_{ij})), 0 < i,j \leq k, i \neq j \tag{15}$$

Therefore, we may convert the pseudo-correct output model to the complete inverse fault model. As mentioned in Section 4.2, according to (8), we obtain:

$$q'_{ij} = q_{ij}p'_{ij} + (1 - q_{ij})(1 - p'_{ij}) \tag{16}$$

Finally, by substituting the p_{ij} with p'_{ij} in (11), (12), and (13), we will obtain the ultimate classification performance of the classifier in this fault model.

5 Computer Simulations and Discussion

We carry out simulations on a practical ten-class classification task. The output of each binary classifier will be disturbed before K-voting combination as a simulation of fault occurrence. The correct rate, the incorrect rate, and the unknown rate is 65.6198%, 28.5537%, and 5.8264% under non-fault condition, respectively. For page limitation, only comparison between the theoretical and the practical performance under pseudo-correct fault model is presented. The experimental results are shown in Table 1. For convenience, non-fault probability of each binary classifier is set to the same value and each r_{ij} is set to 0.5.

The simulation result basically coincides with our theoretical analysis. Also, as we mentioned before, the correct rate and incorrect rate do be underestimated and unknown rate does be overestimated. Under any fault model, we find that:

a) K-voting combination is highly sensitive to the fault of binary classifier.
b) The fault of a K-voting classifier is largely unknown output, instead of the misclassification.

6 Conclusions and Future Work

We have proposed a mathematical model for analysis of the fault tolerance of binarization classification procedure with a quantitative approach in this paper. We also give a better understand of the binarization procedure of the K-voting

Table 1. Simulation results on pseudo-correct model

Reliability	Correct rate(%)		Incorrect rate (%)		Unknown rate(%)	
	Actual	Theoretical	Actual	Theoretical	Actual	Theoretical
50%	5.5372	3.695282	5.4545	1.607959	89.0083	94.696659
75%	20.4545	17.262967	11.6116	7.511781	67.9339	75.225152
90%	41.0331	39.288998	19.8347	17.096155	39.1322	43.614747
95%	51.8595	50.942594	23.4298	22.167083	24.7107	26.890222
99%	62.6446	62.411657	27.3554	27.157713	10.0000	10.430530

classifier. The analysis results give a theoretical base of the influence caused by the fault of binary classifier on the ultimate combination classification result.

In fact, our analysis procedure has no relation with the features of binary classifiers or even the performance of binary classifiers. Therefore, our method has the common sense to some degree. In addition, the fault model we presented, especially the third one, has some value in practice application.

The further work may focus on the direction of fault detection. Through observation on the classification performance before and after fault occurrence, to locate the binary classifier with fault is possible under some prior fault model.

Acknowledgements. The authors would like to thank Mr Yi-Ming Wen, Mr Bin Huang and Miss Lu Zhou for their helpful advice. This research was partially supported by the National Natural Science Foundation of China via the grant NSFC 60375022.

References

1. Bolt, G., Austin J., Morgan, G.: Fault-tolerant Multilayer Perceptron Networks. Department of Computer Science, York University, Hestington, York, UK, TechRep:YCS180 (1992)
2. Hammadi, N.C., Ito, H.: A Learning Algorithm for Fault Tolerant Feed Forward Neural Networks. IEICE Trans. on Information and Systems, E80-D(1) (1997) 21-27
3. Fürnkranz,J.: Round Robin Classification. The Journal of Machine Learning Research, Vol.2 (2002) 721-747
4. Allwein, E.L., Schapire, R. E., Singer, Y.: Reducing Multiclass to Binary: A Unifying Approach for Margin Classifiers. Journal of Machine Learning Research, Vol.1 (2000) 113-141
5. Lu, B.L., Ito, M.: Task Decomposition and Module Combination Based on Class Relations: a Modular Neural Network for Pattern Classification. IEEE Transactions on Neural Networks, Vol.10 (1999) 1244-1256

Training Multilayer Perceptron with Multiple Classifier Systems

Hui Zhu, Jiafeng Liu, Xianglong Tang, and Jianhuan Huang

Pattern Recognition Center of Harbin Institute of Technology, Harbin, China, 15001
hzhu@hope.hit.edu.cn

Abstract. An idea of training multilayer perceptron (MLP) with multiple classifier systems is introduced in this paper. Instead of crisp class membership i.e. {0, 1}, the desired output of training samples is assigned by multiple classifier systems. Trained with these samples, the network is more reliable and processes better outlier rejection ability. The effectiveness of this idea is confirmed by a series of experiments based on bank check handwritten numeral recognition. Experimental results show that for some recognition applications where high reliability is needed, MLP trained with multiple classifier systems label samples is more qualified than that of MLP trained with crisp label samples.

1 Introduction

Because of the excellent discrimination ability, Multilayer perceptron (MLP) has been widely applied in character recognition [1][2]. For certain applications such as bank check recognition, a high reject rate will be acceptable and the error rate should be kept as low as possible [3]. Unfortunately, the output of MLP can't be used as a useful confidence measurement. MLP with thresholding criteria on the outputs is not adequate for pattern verification and its performance will be poorer for patterns do not belong to any of classes defined in the application, i.e. outliers [4].

Besides poor outlier rejection ability, the solutions of MLP are difficult to be interpreted and understood [5]. Trained with crisp class membership samples, strong discriminant ability as well as poor reliability can be obtained. The reason of poor reliability is that the outputs of other classes are all equally coped with and set to be zero which should not be the real situation. Examples presented by Chiang show that handwritten character classes are not crisp sets, they should be treated as fuzzy sets [6]. Considering these situations, Paul Gader et al. state that training neural network-based character recognizers using fuzzy class memberships instead of crisp class memberships can result in more useful character recognition modules for handwritten word recognition [7]. They use fuzzy k-nearest neighbor algorithm to assign desired outputs to training samples and a higher word recognition rate is obtained. In order to improve the rejection ability and reliability of a single neural network classifier on confusion and unrecognizable patterns, knowledge of five experts is gathered for assigning class membership to training samples. Training with these samples, a more reliable network can be obtained [3].

Instead of human experts, an idea of assigning class membership to MLP training samples with multiple classifier systems (MCS) is proposed in this paper. Some classifiers such as exemplar classifiers, probabilistic classifiers are used to form MCS for assigning class membership to training samples. Experiments based on bank check handwritten numerals recognition are carried out to confirm the effectiveness of this method. Some Renminbi symbols in bank check are gathered to test its outlier rejection ability. We compare the effectiveness of this method with fuzzy k-nearest neighbor algorithm. The results show that MLP trained with MCS assign class membership training samples possesses excellent outlier rejection ability. The neural network is more reliable and it presents some similar performance as the corresponding MCS does. The paper is organized as follows. Section 2 gives the idea of assigning class membership to training samples with MCS. Experimental details are presented in section 3. Section 4 concludes this work.

2 Assign Training Sample Class Memberships with Multiple Classifier Systems

Let $D=\{D_1, D_2, \ldots, D_L\}$ be L Classifiers. For an input pattern vector x and class label $\Omega = \{1, 2, \ldots, c\}$ of $C-class$ problem, denote the output of the ith classifier as $D_i(x) = [d_{i,1}(x), d_{i,2}(x), \ldots, d_{i,c}(x)]$ where $d_{i,j}(x)$ is the output of classifier D_i for jth class on pattern x.

Because different classifiers always use different features or different models, the outputs value of them are usually in different range which makes it hard to combine them directly. A transformation function T should be applied to transform these outputs into the same scale. Let $T(D_i(x)) = \{t_{i,1}(x), t_{i,2}(x), \ldots, t_{i,c}(x)\}$ be the transformed outputs of D_i on pattern x. An effective output transformation function T should satisfy the criteria in [8]. Some useful output transformation function can be obtained in the literature [8][9][10].

Let $D(x) = [d_1(x), d_2(x), \ldots, d_c(x)]$ be fused output on input pattern x, $D(x) = F(T(D_1(x)), T(D_2(x)), \ldots, T(D_L(x)))$ where F is fusion rules. Details of fusion rules on measure level information such as Minimum, Maximum, Average, Product, Probabilistic product, Fuzzy integral, Dempster-Shafer combination, Decision templates etc. can be obtained in [11].

For a training pattern x from class $i \in \Omega$, let $O(x) = [o_1(x), o_2(x), \ldots, o_c(x)]$ be class membership of pattern x. A crisp membership will be $o_i(x) = 1.0$ and $o_j(x) = 0.0, \forall j \neq i \in \Omega$. MCS will assigns class membership to training samples to be $o_i(x) = 1.0$ and $o_j(x) = d_j(x), \forall j \neq i \in \Omega$ where $d_j(x)$ is the fused output.

3 Experiments

Handwritten numeral in bank check is gathered for setting up experiments with 7176 training samples and 22729 testing samples. Symbol of Renminbi (RMB) is gathered to test the outlier rejection ability of MLP. Some of samples are shown in fig.1.

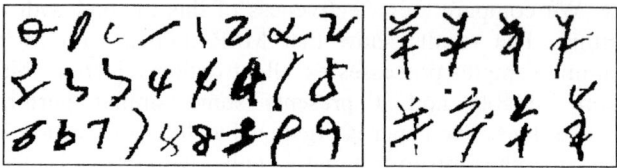

Fig. 1. Experimental samples (a) handwritten numeral; (b) RMB symbols

Three classifiers including modified quadratic discriminant classifier (MQDC), learning vector quantization classifier (LVQ) and support vector machine (SVM) are adopted to form MCS. The detail of MQDC and LVQ classifier can be found in [12]where MQDC is denoted by MQDF2. The SVM adopted in this experiment using polynomial with degree 3 as kernel. There are 200 prototypes for each class of LVQ.

Three kinds of feature including directional element feature (DEF) [13], transition feature (TRSF) [14] and bar feature (BARF) [7] are adopted in this experiment. The original dimension of DEF is 196 and is translated into 64 with Karhunen-Loève transform. The dimension of TRSF and BARF is 100 and 120 respectively.

We denote SVM, MQDC and LVQ with DEF, BARF and TRSF by S1, S2, S3, MQ1, MQ2, MQ3, L1, L2, L3 respectively. These nine classifiers build the entire classifier pool. We use three ($L=3$) classifiers selected from it to build MCS for assigning class membership to training samples.

The output of SVM is the number of support discriminators. Minimum output normalization method is adopted to normalize its output.

$$t_{S,j}(x) = \frac{d_{S,j}(x) - \min\{d_{S,1}(x),......,d_{S,c}(x)\}}{\sum_{i=1}^{c}(d_{S,i}(x) - \min\{d_{S,1}(x),......,d_{S,c}(x)\})} \quad (1)$$

To satisfy the criteria that the larger the outputs are, the more likely the corresponding class is the pattern, we transform outputs of MQDC and LVQ by two steps. First, they are normalized with equation (1) and let $d'_{MQ,L,j}(x)$ be the intermediate results. The final output normalization results can be obtained by equation (2).

$$t_{MQ,L,j}(x) = \frac{1 - d'_{MQ,L,j}(x)}{\sum_{i=1}^{c}[1 - d'_{MQ,L,i}]} \quad (2)$$

Average rule is adopted to combine multiple classifiers' outputs in this experiment. The method can be expressed as follows:

$$D(x) = \frac{\sum_{i=1}^{L} T(D_i(x))}{L} \qquad (3)$$

The MLP is based on TRSF with 100 input nodes, 150 hidden nodes and 10 output nodes. It is trained by BP with crisp, MCS assign and fuzzy $k-nearest$ neighbor algorithm assign samples respectively. k is set to be 20 for the fuzzy algorithm. We denote those MLP by MLP_CRISP, MLP_FUZZY and MLP_MCS respectively.

We test recognition rate of each classifier in the classifier pool and the results is shown in table 1.

Table 1. Recognition rate of single classifier in the classifier pool (%)

	S1	S2	S3	MQ1	MQ2	MQ3	L1	L2	L3
tra.	93.4	91.6	93	92.6	96.7	92	83.5	88	90.9
tes.	92.3	95.7	96.8	78.0	95.6	90.1	85.6	89	93.3

The recognition rate of MLP_CRISP, MLP_FUZZY and MLP_MCS are shown in table 2. The corresponding MCS recognition rates are shown in bracket.

Table 2. Recognition rate of MLP trained with different label training samples(%)

	Crisp	Fuzzy	L1L2L3	MQ1MQ2MQ3	S1S2S3	L3MQ2S3	L3MQ3S3
tra.	94.1	93.8	93.7	93.2	94.3	93.4	93.9
			(93.14)	(96.66)	(96.3)	(93.2)	(93.1)
tes.	96.2	96.1	95.9	95.5	95.9	95.3	95.5
			(95.8)	(91.3)	(97.5)	(96.9)	(96.8)

MLP_CRISP possesses the highest recognition rate. Because only very little number of class membership assigned by fuzzy k-nearest neighbor algorithm is non-zero and the value is very near to zero, performance of MLP_FUZZY is very similar to MLP_CRISP. The recognition rate of MLP_MCS is much lower than that of MLP_CRISP and MLP_FUZZY, but the difference among them is very small. There is no obvious relationship between recognition rate of MLP_MCS and that of the corresponding MCS. But it is interesting that the recognition rate of MLP trained with MCS which comprises of the same classifier with different features is a little higher than that of MLP trained with MCS which comprises of different classifier with the same or different features.

We use the thresholding criteria that a pattern will be rejected whenever its highest output does not reach a given threshold to test the reliability of MLP and 0.99 is selected as the threshold. Fig.2 shows the result of recognition rate versus rejection rate on handwritten numeral testing dataset.

From fig.2 we can get that the reliability of MLP_CRISP is very poor, the output of MLP_CRISP can't be used as an effective confidence measurement. MLP_FUZZY is more reliable than MLP_CRISP. MLP trained with MCS which comprises of different classifier with the same or different features presents the best reliability as well as a high rejection rate. It should be mentioned that the MCS should be designed carefully because there exists MLP trained with certain MCS whose reliability is even lower than MLP_CRISP.

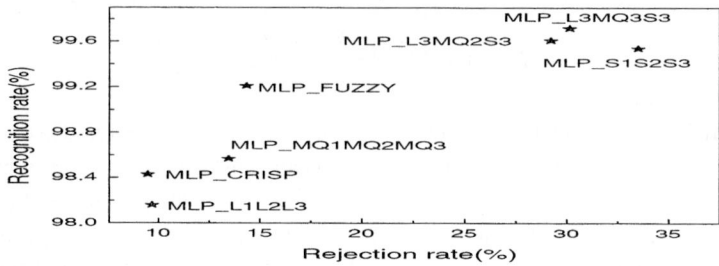

Fig. 2. Recognition rate vs. rejection rate

We test the outlier rejection ability of MLP on 1102 RMB symbols, the results are shown in table 3. It shows that MLP trained with certain MCS possesses a high outlier rejection rate, it is more reliable than MLP_CRISP. But also the MCS should be built carefully to achieve high reliable networks.

Table 3. Outlier rejection rate of MLP (%)

	Crisp	Fuzzy	L1L2L3	MQ1MQ2MQ3	S1S2S3	L3MQ2S3	L3MQ3S3
RMB	62.52	71.42	62.52	59.62	97.01	91.29	86.84

Finally we investigate the solutions of MLP_CRISP and MLP_MCS. Fig.3 shows the solution class sequence arranged according to outputs of each recognition model in decrease order on handwritten numeral "0". The left of Fig.3 is the image of sample numeral "0".

```
MLP_CRISP:      0 4 6 8 9 1 2 3 5 7
MLP_L3MQ3S3: 0 8 4 2 6 9 1 7 3 5
L3MQ3S3:           0 8 4 2 6 9 3 5 7 1
```

Fig. 3. Solution sequence of MLP_CRISP and MLP_MCS

Because crisp class membership treats each output except the desired one equally i.e. set to be zero, so the solution of MLP_CRISP is hard to predict and interpreted. Trained with MCS, MLP presents a similar solution as the corresponding MCS does. It suggests that the solution of MLP is influenced by input features as well as by its supervisors.

4 Conclusions

An idea of training MLP with MCS assign class membership training samples is proposed in this paper. The class membership of training samples is set to be the combination measure level outputs of MCS. A series of experiments based on bank check handwritten numeral recognition is carried out to test the effectiveness of MLP_MCS. The results show that although the recognition rate of MLP_MCS is a litter lower than that of MLP_CRISP, MLP_MCS is more reliable. MLP_MCS presents excellent outlier rejection ability and its output can be used as an effective confidence measurement. Another interesting result is that MLP_MCS presents a similar solution results as the corresponding MCS does. It suggests that the solution of MLP is influenced by input features as well as by its supervisors.

But it should be mentioned that the MCS should be carefully build because there exists MLP trained with certain MCS which is less reliable than MLP_CRISP.

Although most of discussions are based on character recognition, the idea of training MLP with MCS assign class membership training samples can be applied in any applications where a reliable MLP is needed.

References

1. Cho, S.B.: Neural-network Classifiers for Recognizing Totally Unconstrained Handwritten Numerals. IEEE Trans. Neural Networks, Vol.3, No.1 (1997) 43~53
2. Cun, Y.Le., Matan, O., Boser, B., Denker, J.S., Henderson, D., Howard, R.E., Hubbard, W., Jacket, L.D., Baird, H.S.: Handwritten Zip Code Recognition with Multilayer Networks. Proceedings of the 10th International Conference on Pattern Recognition 2 (1990) 35~40
3. Wu, X.J., Suen, C.Y.: Injection of Human Knowledge into The Rejection Criterion of a Neural Network Classifier. Joint 9th IFSA World Congress and 20th NAFIPS International Conference 1 (2001) 499~505
4. Gori, M., Scarselli, F.: Are Multilayer Perceptrons Adequate for Pattern Recognition and Verification?. IEEE Trans. Pattern Anal. and Mach. Intell., Vol.20, No.11 (1998) 1121-1132
5. Lippmann, R.P.: Pattern Classification Using Neural Networks. IEEE Communications Magazine, Vol.27, No.11 (1989) 47 - 50, 59-64
6. Chiang, J.-H.: A Hybrid Neural Network Model in Handwritten Word Recognition. Neural Networks, Vol.11, No.2 (1998) 337-346
7. Gader, P., Mohamed, M., Chiang, J.-H.: Comparison of Crisp and Fuzzy Character Neural Networks in Handwritten Word Recognition. IEEE Trans. Fuzzy Systems, Vol.3, No.3 (1995) 357-363
8. Huang, Y.S., Suen, C.Y.: A Method of Combining Multiple Classifiers-A Neural Network Approach. Proceedings of the 12th IAPR International Conference on Pattern Recognition- Conference B: Computer Vision & Image Processing 2 (1994) 9-13, 473-475
9. Hakan Altınçay, Mübeccel Demirekler: Undesirable Effects of Output Normalization in Multiple Classifier Systems. Pattern Recognition Letters 24 (2003) 1163-1170
10. Duin, R.P.W., Tax, D.M.J.: Classifier Conditional Posterior Probabilities. Advances in Pattern Recognition, Proceedings of Joint IAPR International Workshops SSPR '98 and SPR '98, Sydney, NSW, Australia, Lecture Notes in Computer Science 1451 (1998) 611-619
11. Kuncheva, L.I., Bezdek, J.C., Duin, R.P.W.: Decision Templates for Multiple Classifier Fusion: An Experimental Comparison. Pattern Recognition 34 (2001) 299-314
12. Liu, C.-L., Sako, H., Fujisawa, H.: Performance Evaluation of Pattern Classifiers for Handwritten Character Recogniton. International Journal on Document Analysis and Recognition 4 (2002) 191-204
13. Kato, N., Suzuki, M., Omachi, S., Aso, H., Nemoto, Y.: A Handwritten Character Recognition System Using Directional Element Feature And Asymmetric Mahalanobis Distance. IEEE Trans. Pattern Anal. and Mach. Intell., Vol.21, No.3 (1999) 258~262
14. Gader, P.D., Mohamed, M., Chiang, J.-H.: Handwritten Word Recognition with Character and Inter-Character Neural Networks. IEEE Trans. Systems, Man and Cybernetics-Part B, Vol.27, No.1 (1997) 158-164

Learning the Supervised NLDR Mapping for Classification

Zhonglin Lin[1], Shifeng Weng[1], Changshui Zhang[1],
Naijiang Lu[2], and Zhimin Xia[2]

[1] State Key Laboratory of Intelligent Technology and Systems
Department of Automation, Tsinghua University, Beijing 100084, P. R. China
{linzl02,wengsf00}@mails.tsinghua.edu.cn
zcs@mail.tsinghua.edu.cn
[2] Shanghai Cogent Biometrics Identification Technology Co., Ltd.

Abstract. In researches and experiments, we often work with large volumes of high-dimensional data and regularly confront the problem of dimensionality reduction. Some Non-Linear Dimensionality Reduction (NLDR) methods have been developed for unsupervised datasets. As for supervised datasets, there is a newly developed method called SIsomap, which is capable of discovering structures that underlie complex natural observations. However, SIsomap is limited from the fact that it doesn't provide an explicit mapping from original space to embedded space, and thus can't be applied to classification. To solve this problem, we apply neural network to learning that mapping and then to classify. We test our method on a real world dataset. To prove the effectiveness of our method, we also compare it with a related classification method, Extended Isomap. Experiments show that our proposed method has satisfactory performance.

1 Introduction

In recent years, some new nonlinear dimensionality reduction technologies have been developed, such as Isomap [1] and LLE [2]. They are capable of discovering the structure of unsupervised dataset. Isomap uses geodesic distance (i.e., distance metrics along the manifolds) but not Euclidean distance to reflect intrinsic similarity between any two points. The classical multidimensional scaling method is then applied to finding a set of low dimensional points with similar pairwise distance. Although these methods demonstrate excellent results to find the embedding manifolds that best describe the data points with minimum reconstruction error, they are not learning method for supervised learning and thus may not be optimal from the classification viewpoint.

Extended Isomap tries to apply Isomap to pattern classification problems. The core idea of the algorithm is the definition of feature vector: each data point's vector of geodesic distance to any other points is defined as the feature vector of that data point. Then, classification is done on these feature vectors.

How to utilize supervised information during dimensionality reduction is a key point of supervised learning. Weng et, al. proposed a novel algorithm named

Supervised Isometric Mapping (SIsomap) for supervised learning [4]. SIsomap has many advantages in finding the true structure of a supervised dataset and in making the true structure visual in low dimensional space. SIsomap utilizes supervised information effectively which prompts us to develop it further for classification. As the application of SIsomap for classification is limited because of the absence of explicit dimensionality reduction mapping, we need a mapping learner to work here. Next, we can design classifier in lower dimensional feature space. Many classification techniques can be used if the explicit mapping is learned. In this paper, we take BP network as a mapping learner and classifier, and get satisfactory results.

2 Related Work

2.1 Isomap

Isomap is is a well-known nonlinear dimensionality reduction method, i.e., uncovering the natural parameter space of a nonlinear manifold. It retains the pairwise distance on manifold between points in the low dimensional space. So, Isomap can obtain isometric embedding of manifold.

Isomap includes the following procedures [1]:

- Construct the neighborhood graph G over all data points.
- Compute the shortest path between all pairs of points.
- Construct d-dimensional embedding using standard MDS [5].

The first two steps are the key to Isomap: measuring the local metric by Euclidean Distance and the global metric by Geodesic Distance. After these two steps have been completed, a suitable unsupervised nonlinear dissimilarity metric will be defined. In the last step, Isomap preserves the Geodesic Distance between samples.

We can see that in Isomap the pairwise Euclidean distance in classical MDS is replaced by the geodesic distance along the manifold. It's the essential idea of Isomap. We can also see that Isomap only retains such distance, but does not use any supervised information, i.e., it does not adjust distances according to supervised information. So, it is not appropriate for supervised learning.

2.2 Extended Isomap

Extended Isomap [3] is an algorithm for classification. The first two steps of Extended Isomap are similar to original Isomap method. The main algorithmic feature of Extended Isomap is that each data point's vector of geodesic distance to any other points is defined as the feature vector of that data point. Next, classification is done in this newly defined space. For example, Linear Discriminant Analysis is used in [3]. This kind of feature definition is the core idea of Extended Isomap.

2.3 SIsomap

SIsomap [4] is a novel algorithm of dimensionality reduction for supervised learning. It defines the supervised local metric as the projection of the unsupervised local metric on the local gradient direction of supervised information.

SIsomap assumes that the supervised information $f(x)$ is first-order differentiable function. Note that the gradient of $f(x)$, denoted by $f'(x)$, indicates a direction along which the supervised information changes fastest. In the supervised sense, the part of the original difference along this direction is more important than the total original difference between the two samples. Thus the gradient direction of $f(x)$, denoted as $P(x) = f'(x)/\|f(x)\|$, is the direction which is especially interested in. SIsomap defines the supervised local metric as follows (Fig.1 gives an illustration.):

$$d^{(S)}(x, x') = \left|(P(x))^T (x - x')\right|, \qquad (1)$$

where x' is an adjacent sample of x, and the superscript (S) means "supervised".

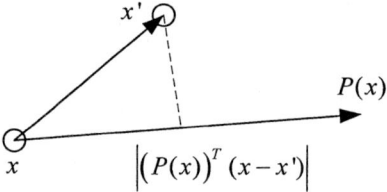

Fig. 1. Illustration of supervised local metric $d^{(S)}(x, x')$

Having defined the supervised local metric between any two neighboring samples, the global metric can be formulated as follows:

$$d^{(SI)}(x_i, x_j) = \inf_r \int_r |(P(x))^T dr|, \qquad (2)$$

where r is any path connecting x_i and x_j along the manifold, and the superscript (SI) indicates "supervised" and "isometric".

The remaining steps of SIsomap are computing the global metric by propagating the local metric, and employing MDS to reduce the dimensionality by preserving $d^{(SI)}$. These steps are the same as in Isomap.

Two versions of SIsomap are proposed in [4]. One version(SIsomap1) is designed for cases in which the supervised information is one-dimensional continuous attributes. The other version(SIsomap2) is designed for the case in which the datasets have two-class label attributes. SIsomap2 makes the data of same class closer and at the same time makes the data from different classes farther. So, it is a NLDR technique for two-class data problems and it is the base of our learning.

3 Learning the SIsomap2 Mapping

SIsomap2 is presented for the second case, in which the supervised information $f(x)$ are two-class label attributes. Things are different for the interface region and the non-interface region. First, a patch $S(x)$ is defined: $S(x) \equiv N(x,k) \cup \{x\}$, where $N(x,k)$ is the set of the k nearest neighbors of x.

For a patch $S(x)$ located in the interface region, there are samples from the two classes. It is impossible to differentiate $f(x)$, because it is neither differentiable nor continuous. So a log function of posterior probability ratio is introduced as new supervised information:

$$g(x) = \ln \frac{p(w_1|x)}{p(w_0|x)}, \qquad (3)$$

where $p(w_k|x)$ is the posterior probability of class w_k, given x. Obviously, $g(x)$ is well consistent with $f(x)$. Comparing $g(x)$ with $f(x)$, it is more reasonable to assume $g(x)$ is continuous and first-order differentiable.

For a patch in the non-interface region, samples' s labels are same. That means there are no differences in supervised information. Since all the directions of this local patch are of equal importance, the Euclidean Distance is used to measure local metric.

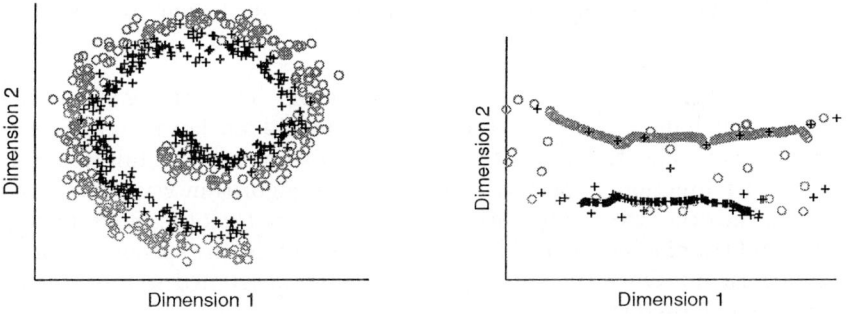

Fig. 2. Double-spiral dataset (left) and SIsomap's result on it (right). In both figures, circle represents data from one class and plus sign represents the other

Sisomap2 was tested on double-spiral dataset in [4]. Fig.2 shows the original double-spiral dataset and its supervised learning result. The two figures show that although the distribution of data might be complex in original space, the structure in embedded space found by SIsomap has excellent discriminability. This result may lead to good performance of classification and inspires us to learn such a mapping. Then, starting from these low dimensional data, we can design classifier with excellent performance. Neural network is an appropriate tool for both learning mapping and classification.

4 Experimental Results

Our method was used to classify the breast cancer Wisconsin dataset available from [6]. This dataset contains 699 patterns with nine attributes: clump thickness, uniformity of cell size, and etc. So the original input space is 9-dimensional. The class output includes two classes: benign and malignant. We removed the 16 instances with missing values from the dataset to construct a new dataset with 683 instances. We randomly selected 500 instances and applied SIsomap to obtaining each instance's new expression with three dimensions. Fig.3 shows one of such results. Then, a BP network was used to learn such a mapping.

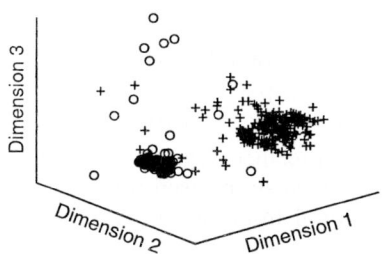

Fig. 3. Result of breast cancer dataset by SIsomap2 in 3-dimensional space. Circle represents benign samples and plus sign represents malignant samples respectively

We also setup a BP network for classification. The settings of network's architecture included: 2 layers, three neurons in hidden layer and one neuron in output layer. The transition functions of two layers were tan-sigmoid and linear transfer function respectively. The training algorithm was batch gradient descent. In the other two comparative experiments, we took the same strategy for choosing and training network. That meant we didn't do any special optimization for any algorithm. Once the network was set up and trained well, classification could be done subsequently.

When classifying by Extended Isomap, we used BP network instead of Fisher Linear Discriminant method for classification. Since training often failed because of the high dimensionality of feature vector (In Extended Isomap, the defined feature vector has the dimension of one less than the number of sample. That was several hundreds in our experiment.), we firstly employed PCA on all feature vectors to get compact expressions and then employed BP network method to classify. In our experiment, we set the percentage of reserved energy of PCA as 95%.

In our experiment, we repeated every experiment 10 times, and got the average and variance of error rate. The training set was randomly selected every time, and the remain data constituted test set. The results of our algorithm (SIsomap+BP), using BP network directly (BP) and Extended Isomap (EIsomap) are listed as below.

Table 1. Comparison of three methods' performance on error rate and variance

Method	Error Rate	Variance
SIsomap+BP	2.64%	2.49e-004
BP	3.54%	1.81e-004
EIsomap	3.39%	2.57e-004

The experiments show that our method outperforms the other two. The main reason is that SIsomap makes the best of supervised information in dimensionality reduction and BP network learns the mapping well.

5 Conclusion and Discussion

SIsomap is limited from the fact that it doesn't provide an explicit mapping from original space to embedded space. In this paper, we applied neural network to learning that mapping and then to classification. We tested our method on breast cancer dataset and got excellent result.

Our method has additional merit. The dimensionality reduction results from SIsomap are often visual in 2 or 3 dimensional space. These visualization may provide helpful instruction in designing appropriate classifier. For example, if the feature space is linear discriminative, Fisher Linear Classifier can be used. Neural network can be taken as a general choice. This paper used BP neural network as a choice of classifier.

References

1. Tenenbaum, J.B., Silvam, V.D., Langford, J.C.: A Global Geometric Framework for Nonlinear Dimensionality Reduction. Science, Vol. 290 (2000) 2319-2323
2. Roweis, S., Saul, L.: Nonlinear Dimensionality Reduction by Locally Linear Embedding. Science, Vol. 290 (2000) 2323-2326
3. Yang, M.H.,: Face Recognition Using Extended Isomap. Proceedings of International Conference on Image Processing, Vol. 2 (2002) 117-120
4. Weng, S.F., Zhang, C.S., Lin, Z. L.: Exploring the Structure of Supervised Data. Submitted to ECML/PKDD (2004)
5. Borg, I., Groenen, P.: Modern Multidimensional Scaling: Theory and Application. Springer-Verlag, Berlin Heidelberg New York (1997)
6. Blake, C., Merz, C.: UCI Repository of Machine Learning Databases [http://www.ics.uci.edu/~mlearn/MLRepository.html]. Irvine, CA: University of California, Department of Information and Computer Science. (1998)

Comparison with Two Classification Algorithms of Remote Sensing Image Based on Neural Network

Yumin Chen[1], Youchuan Wan[2], Jianya Gong[1], and Jin Chen[1]

[1] State Key Laboratory for Information Engineering in Surveying, Mapping and Remote Sensing, Wuhan University, China, 430079
{ycymlucky, gongjiaya, chenjin215}@hotmail.com
[2] School of Remote Sensing and Information Engineering, Wuhan University, China,430079
wych@public.wh.hb.cn

Abstract. The traditional approaches of classification are always unfavorable in the description of information distribution. This paper describes the BP neural network approach and the Kohonen neural network approach to the classification of remote sensing images. Two algorithms have their own traits and can be good used in the classification. A qualitative comparison demonstrates that both original images and the classified maps are visually well matched. A further quantitative analysis indicates that the accuracy of BP algorithm is better than the result of the Kohonen neural network

1 Introduction

The traditional approaches of classification are visual interpretation and threshold method. However, these approaches are based on the spectrum of one type of target at some channels and ignore it at other channels, and it is difficult that many thresholds are adjusted at the same time to appropriate value. The artificial neural network has the basic characters of human brains. The peculiarity of neural network is massive parallel computing, nonlinear dynamics of consecutive time, real time processing and self-organization, self-learning. In this paper, describes the BP neural network algorithm and the Kohonen algorithm to the classification of remote sensing images.

2 The BP Neural Network Using in Classification of the Images

The BP neural network algorithm devises two steps. The feed forward step yields activation values for every input pattern. Since it is not possible to compute directly the error of the hidden-layer neurons, the output error is back-propagation to the hidden neurons through the weighted connections.

In the remote sensing image processing, the training data is the RGB value of each pixel, which is known the desired output pattern. The output layer represents the number of classes desired presented as a linear array, in which the value of each element is standardized between 0 and 1 (Fig.1). Training the network consists in presenting all the patterns one by one in input, updating the weights according to the committed error and verifying if the output error is below a chose threshold.

```
Input layer           Output layer
(R₁, G₁, B₁)<—> No.1<—>(1, 0,..., 0)
(R₂, G₂, B₂)<—> No.2 <—>(0, 1,..., 0)
      ......
(Rₙ, Gₙ, Bₙ)<—> No.n <—>(0, 0,..., 1)
```

Fig. 1. The input layer and output layer of BP algorithm

The process of algorithm can be found in books about neural network and doesn't describe concretely now. Here, Application of the gradient descent method yields the following iterative weight update rule:

$$W_{ij}(n+1) = W_{ij}(n) + \eta d_i^k X_j^{k-1} + \alpha \Delta W_{ij}(n) \tag{1}$$

Where W_{ij} is the weight of the connection from node I to node j, η the learning factor, α the momentum factor, and d_i^k the node error. For output node I, d_i^k is then given by:

$$d_i^k = X_i^k (1-X_i^k)(X_i^k - Y_i) \tag{2}$$

The node error at an arbitrary hidden node I is:

$$d_i^k = X_i^k (1-X_i^k) \sum W_{ij} X_j^{k-1} \tag{3}$$

3 The Kohonen Neural Network Using in the Images

Kohonen neural networks (KNNs) are a type of self-organizing network and use an unsupervised learning algorithm. The Kohonen unit with the largest activation is declared "the winner". The important concept of KNNs is that not only is the winning unit modified, but also are the units in a neighborhood around the winning unit.

In remote sensing image processing, the input array is the RGB value of each pixel in remote sensing image and the output value is the winning node N_{j*} in the output matrix, which is the classification result of the pixel (Fig.2).

```
Input layer                              output layer
                                         ⎡ 0, 0, ..., 0 ⎤
(Rᵢ, Gᵢ, Bᵢ) <———> Result is j* <———>    ⎢ 0, j*, ..., 0 ⎥
                                         ⎣ 0, 0, ..., 0 ⎦
```

Fig. 2. The input layer and output layer of BP algorithm

In the learning step, Adjust the weight W_{ij} connected between the output node N_{j*} and neighborhood Nc:

$$W_{ij}(t+1) = W_{ij}(t) + \eta(t)(x_i(t) - W_{ij}(t)) \quad i \in \{1,2,...,n\} \ j \in Nc_{j*}(t) \tag{4}$$

Where $\eta(t)$ is the learning factor, $\eta(t) = 0.9(1-t/1000)$, $0 < \eta(t) < 1$. $Nc_{j*}(t)$ is reduced as the t increase. When t increased enough, $Nc_{j*}(t) = \{N_{j*}\}$, it's just left the winning Node N_{j*}, the algorithm is finished.

4 Experimental Results

In the experiments, a Landsat-TM image, over the area of Hubei province near the Honghu city, taken on August 2000 is used. The data consists of three bands (7(2.08-2.35um), 4(0.76-0.9um), 3(0.63-0.69um)). The seventh band uses blue color; the fourth band uses green color, and the third band uses red color. The image size is 512 * 512 pixels. The aim of the classification was to distinguish between the 8 categories: S_1—grassplot; S_2—paddy field; S_3—cropland; S_4—woodland; S_5—bare soil; S_6—soil; S_7—lake; S_8—river. The spectrum of vegetation (S_1, S_2, S_3, S_4) or water body (S_7, S_8) is resembling, so it is very difficult to differentiate between them using the traditional approaches.

In this study, the BP neural network consists of 4 input nodes, 8 output nodes, and 13 nodes in hidden layer. The parameter selection is as follows: η and α in equ.1 are set to 0.02 and 0.9. From study area 760 sample patterns are pick up randomly. The summed squared error (SSE)< 0.002, the BP neural network method used 83s, on 5267000 iteration and it's overall accuracy is 86.209%, kappa coefficient is 0.644. The resultant classified image is shown in Table1 and Fig.3 (b).

In the Kohonen algorithm, It is select 8×8 matrix as the output matrix, then the connective weight matrix is W[24×8] (Each output node connect with 3 input nodes, so the weight matrix is the structure of 3×8×8, expressed in 2D space is 24×8 matrix). The mediate coefficient of studying speed is $\eta(t) = 0.9(1-t/1000)$. The neighborhood $Nc_{j*}(t)$ is exchanged by the iteration number t. The rule is defined as the neighborhood Nc is 7×7 matrix at first which center at the winner node j^*, then the neighborhood Nc reduced to 6×6 matrix, 5×5 matrix as the t increased. At last, the neighborhood Nc just left the winner node j^*. The algorithm used 96s, the error matrix is showed in Table2 and it's overall accuracy is 85.636%, kappa coefficient is 0.636. The resultant classified image is shown in Fig.3 (c).

Table 1. The error matrix of the BP algorithm

Test \ Truth	S_1	S_2	S_3	S_4	S_5	S_6	S_7	S_8	Total
S_1	83	7	8	3	0	0	0	0	101
S_2	7	95	6	4	0	0	0	0	112
S_3	8	6	75	5	0	0	0	0	94
S_4	2	2	1	88	0	0	0	0	93
S_5	0	0	0	0	73	6	3	2	84
S_6	0	0	0	0	5	82	2	3	92
S_7	0	0	0	0	0	1	85	11	97
S_8	0	0	0	0	2	1	10	74	87
total	100	110	90	100	80	90	100	90	760

Overall Accuracy is 86.209%, Kappa = 0.644

Table 2. The error matrix of the Kohonen neural network algorithm

Test \ Truth	S_1	S_2	S_3	S_4	S_5	S_6	S_7	S_8	Total
S_1	86	6	8	2	0	0	0	0	102
S_2	7	95	7	7	0	0	0	0	116
S_3	4	8	73	6	0	0	0	0	91
S_4	3	1	2	85	0	0	0	0	91
S_5	0	0	0	0	70	6	4	2	82
S_6	0	0	0	0	7	81	3	3	94
S_7	0	0	0	0	1	2	85	9	97
S_8	0	0	0	0	2	1	8	76	87
total	100	110	90	100	80	90	100	90	760

Overall Accuracy is 85.636%, Kappa = 0.636

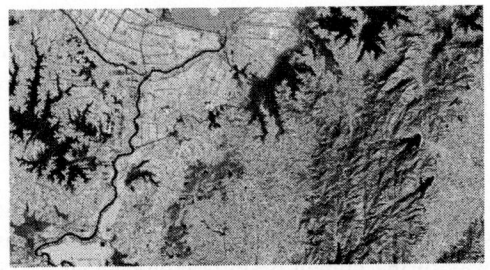

(a) The original Landsat-TM image

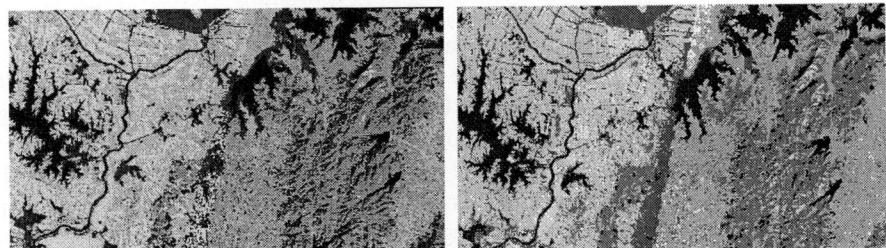

(b) The classified image by BP algorithm   (c) The classified image by Kohonen algorithm

Fig. 3. The result in the experiment of the neural network algorithm

Comparing with these two neural network algorithms, we can find that the overall accuracy of two has not very much difference. The overall accuracy of BP algorithm is 0.573% higher than the KNNs. As the resembling spectrum of vegetation (S1, S2, S3, S4), KNNs accuracy is higher in the S1 spectrum, but the accuracy of BP algorithm is higher in other spectrum S2, S3, S4. It maybe in the KNNs color space the green color has more difference to other color, so it can easy to classify the grassplot (green color). The BP algorithm is a supervised learning algorithm and the output error is back-propagation to the hidden neurons through the weighted connections. The BP algorithm has good stability and can't get fluctuate greatly. So BP algorithm has higher accuracy in many instances than KNNs.

5 Conclusions

The BP algorithm and KNNs are all come from artificial intelligence approach. The contacts between units are referred to as connections of synapses and are usually weighted. So these two kinds of algorithms have a lot of resemblance. However, there are also some differences between them, such as blow:

   a. The thinking of these two neural network algorithms is different. The BP algorithm is a back-propagation neural network. The output neuron's error is given by the absolute difference between the activation value and the desired value. All weighted connections are updated according to the error associated to each neuron immediately after feeding a pattern to the network. KNNs are a type of self-organizing network that recognizes the statistical characteristics of input datasets. It's more like the human brain, which can study and memory the knowledge. The more important is not only the winning unit modified, but also are the units in a neighborhood around the winning unit.

   b. BP neural network is a supervised learning algorithm, which judges whether the concept is correct according to learned knowledge or suggestion information, and then updates the knowledge database. Kohonen neural network is an unsupervised learning algorithm. The study way of the human brain is independent, i.e. it has the ability of self-organization and self-adaptation. In the complicated and non-steady environment it can adjust one's own thinking and ideas. This is the function reached directly depending on the stimulation from the external world.

   c. The study and memory ability of the human brain has not only very strong robustness, but also elasticity and plasticity. Human brain can remember the knowledge learnt very firmly, study new knowledge constantly and forget some knowledge that is not commonly used or unimportant. There are two ways to reflect the input from the external world——"form bottom to top" and "from top to bottom". The former means that the human brain can correctly distinguish and grasp the different objects and relations (spatial and temporal, logic relation, etc.). The latter involves the ability of people concentrating the attention and neglecting something. BP neural network use the way of "form top to bottom" to implement classification, and use the way of "form bottom to top" to train the back-propagation of error. Then Kohonen neural network train sample and classify images mainly through the way of "form top to bottom".

   d. The two algorithms have their own advantage and disadvantage in practice. BP algorithm has many hidden layer nodes, which give more adjusting parameters. But if there are too many hidden nodes, it will affect the calculation speed. It is the difficult problem to decision the number of hidden layers and hidden nodes. Kohonen algorithm has only input layer and output layer, no hidden layer in the middle. It is more important to decide the output layer, which has no united formula, and if there is more output layer point, it will be more adjustable, but add the amount of calculation. The key problem of the two algorithms is how to adjust the connective weight matrix. Because of the supervised learning, BP algorithm need more similar of the activation value and the desired value, so adjusting formula of weight matrix can be easy defined. However, Kohonen is an unsupervised learning algorithm and according to the rule of repression in the neural network. There is no united formula to define the neighborhood, the reduce function of the neighborhood and the winning rule of neural cell, which need to define according to the practice. Therefore, Kohonen algorithm is more similar to the thinking mode of the human brain in theory, but Kohonen algorithm has many uncertain parameters. If the parameters aren't chosen appropriately, the accuracy of Kohonen algorithm is less than BP algorithms. That is the reason why BP algorithm is more and more widely used in practice.

References

1. Chungen, Y., Lasse, R., Zhongyang, L.: Methods to Improve Prediction Performance of ANN Models. Simulation Modelling Practice and Theory. 11(2003) 211–222
2. Kimes, D.S., Nelson, R.F.: Attributes of Networks for Extracting Continuous Vegetation Variables. Int. J. Remote Sensing. 19(1998) 2639–2663
3. Rangsanseri, Y., Thitimajshima, P., Promcharoen, S.: A Study of Neural Network Cassification of JERS-1/OPS Iages. In: 19th Asian Conference on Remote Sensing (1998)
4. Mannan, B., Roy, J., Ray, A.K.: Fuzzy ARTMAP Supervised Classification of Multi-Spectral Remotely-Sensed Images. Int. J. Remote Sensing. 19(1998) 767–774
5. Rongyi, Y., Shenchu, X.: Analysis on Design of Kohonen Network System Based on Classification of Complex Signals. Photonics and Technology, Vol. 8., BeiJing (2002)
6. Hu, Y., Wu, G.: Modified RLS Training Algorithm for Multi-Layer Feed-Forward Neural Networks. System Engineering and Electronics Technology, Vol. 22., BeiJing (2000) 77–80
7. Aoyama, T., Hanxi, Z., Yoshihara, L.: Forecasting of the Chaos by Iterations Including Multi-Layer Neural Network. Proceedings of the International Joint Conference on Neural Networks (2000) 467–471
8. Fiset, R., Cavayas, F.: Automatic Comparison of a Topographic Map with Remotely Sensed Images in a Map Updating Perspective: The Road Network Case. Int. J. Remote Sensing. 18(1997) 991–1006

Some Experiments with Ensembles of Neural Networks for Classification of Hyperspectral Images[*]

Carlos Hernández-Espinosa, Mercedes Fernández-Redondo, and
Joaquín Torres-Sospedra

Universidad Jaume I. Dept. de Ingeniería y Ciencia de los Computadores. Avda Vicente Sos Baynat s/n. 12071 Castellon. Spain.
{espinosa, redondo}@icc.uji.es

Abstract. A hyperspectral image is used in remote sensing to identify different type of coverts on the Earth surface. It is composed of pixels and each pixel consist of spectral bands of the electromagnetic reflected spectrum. Neural networks and ensemble techniques have been applied to remote sensing images with a low number of spectral band per pixel (less than 20). In this paper we apply different ensemble methods of Multilayer Feedforward networks to images of 224 spectral bands per pixel, where the classification problem is clearly different. We conclude that in general there is an improvement by the use of an ensemble. For databases with low number of classes and pixels the improvement is lower and similar for all ensemble methods. However, for databases with a high number of classes and pixels the improvement depends strongly on the ensemble method.

1 Introduction

A hyperspectral image is used in remote sensing (RS) to identify different type of coverts of the Earth surface. One image is formed of pixels of spatial resolution like a normal image, but in this case each pixel is composed of spectral bands of the electromagnetic spectrum registered by a sensor.

There is a division between multispectral and hyperspectral images in RS, if the number of spectral bands of each pixel in the image is less than 20 the image is called multispectral, otherwise (more than 20 bands) the image is called hyperspectral. The limit is 20 bands, but usually a hiperspectral image has more than 200 bands, as it is the case of the images captured by the sensor AVIRIS which are used in this research.

One of the problems of processing RS images is the supervised classification of pixels. This problems consist on classifying the different pixels into a set of different surface covering, given a known classification of part of the pixels.

The problem of classification of RS images has traditionally been performed by classical statistical methods. However, recently other techniques like neural networks (NN), in particular Multilayer Feedforward (MF) have been applied [1].

---

[*] This research was supported by the project MAPACI TIC2002-02273 of CICYT in Spain.

Beside that, it is well known that one technique to increase the performance with respect to a single NN is the design of an ensemble of NNs, i.e., a set of NNs with different initialization or properties in training and combine the different outputs in a suitable and appropriate manner.

This technique has also been applied in the classification of remote sensing images. For example in [2], it is used a simple ensemble of MF networks with the fuzzy integral as combination method. Finally in [3], an ensemble of NNs is used for the estimation of chlorophyll.

However, in the experiments cited above multispectral images (with less than 20 spectrall bands) are used and it is rare in the bibliography the utilization of hyperspectral images.

Obviously the problem of classification is different when using a multispectral or a hyperspectral image. In the case of a multispectral image, we will have a NN with less than 20 inputs, which is a normal number of inputs. However, in the case of a hyperspectral image we will have big NNs with around 220 inputs. The results can not be extrapolated for one case to the other.

In this paper we present experiments of eight different methods of constructing the ensembles of MF networks and with four hyperspectral images as data.

The output combination method employed was in all cases *output averaging*, other methods will be tried in future research.

2 Theory

In this section we briefly review the different ensemble methods.

2.1 Simple Ensemble

A simple ensemble can be constructed by training different networks with the same training set, but with different random weight initialization.

2.2 Bagging

This ensemble method is described in reference [4]. It consists on generating different datasets drawn at random with replacement from the original training set. After that, we train the different networks in the ensemble with these different datasets.

2.3 Boosting

This ensemble method is reviewed in [5]. It is conceived for a ensemble of only three networks. The first network is trained with the whole training set. After this training, we pass all patterns through the first network and construct the new training set with 50% of patterns incorrectly classified and 50% correctly classified. With this new training set we train the second network. After that, the original patterns are presented to both networks. If the two networks disagree in the classification, we add the training pattern to the third training set. Otherwise we discard the pattern.

2.4 CVC

It is reviewed in [6]. In *k-fold cross-validation*, the training set is divided into k subsets. Then, k-1 subsets are used to train the network. Similarly, by changing the subset that is left out of the training process, one can construct k classifiers, each of which is trained on a slightly different training set. This is the technique used in this method.

2.5 Adaboost

We have implemented the algorithm denominated "Adaboost.M1" in the reference [7]. In the algorithm the successive networks are trained with a training set selected at random from the original training set, but the probability of selecting a pattern changes depending on the correct classification of the pattern and on the performance of the last trained network.

2.6 Decorrelated (Deco)

This ensemble method was proposed in [8]. It consists on introducing a penalty term added to the error function. The penalty term for network number *j* in the ensemble is:

$$Penalty = \lambda \cdot d(i,j)(y - f_i) \cdot (y - f_j) \tag{1}$$

Where λ is a parameter, *y* is the target and f_i and f_j are the outputs of networks *i* and *j* in the ensemble. The term *d(i,j)* is 1 for *i=j-1* and 0 otherwise.

2.7 Decorrelated2 (Deco2)

It was proposed also in reference [8]. It is basically the same method of "Decorrelated" but with a different term *d(i,j)*. In this case *d(i,j)* is 1 when *i=j-1* and *i* is even.

3 Experimental Results

The four hyperspectral images are extracted from two scenes obtained from the *AVIRIS imaging spectrometer*, we describe the scenes in the following paragraphs.

Indian Pines 1992 Data: This data consist of a 145x145 pixels by 224 bands of reflectance data with about two-thirds agriculture and on-third forest or other natural perennial vegetation. Since the scene is taken in June some of the crops present, corn, soybeans, are in the early stages of growth with less than 5% coverage. The ground truth available is designated in sixteen classes and is not all mutually exclusive. From this scene following other experiments [9] and with the intention of comparing the results with the technique of *support vector machines*, we have used two images: the full scene (denominated PINES in our experiments) for which there is a ground truth covering 49% of the scene and it is divided among 16 classes ranging in size from 20 to 2468 pixels, and a subset of the full scene (denominated SUB_PINES in our experiments) consisting of pixels [27 – 94] x [31 – 116]. For this subscene there is a ground truth for over 75% and it is comprised of the three row crops, Corn-notill, Soybean-notill, Soybean-mintill, and Grass-Trees. Following other works we have reduced the number of bands to 200 by removing the region of water absorption.

Salinas 1998 Data: This scene was acquired on October 9, 1998, just south of the city of Greenfield in the Salinas Valley in California. This data includes bare soils (with five subcategories), vegetables (broccoli with two subcategories, romaine lettuce with 4 subcategories, celery and corn_senesced and green weeds) and vineyards fields (with three subcategories). For a more detailed description of the subcategories see reference [9]. From this scene two images are extracted. The first one (denominated Sal_A in our experiments) comprising 86 x 83 pixels which include the six classes. The second image (denominated Sal_C in our experiments) comprising 217 x 512 pixels which includes the 16 classes described above.

In table 1, there is a brief description of the databases, the columns "Ninput" and "Noutput" are the number of inputs and number of classes in the image respectively. And columns "Ntrain", "Ncross", and "Ntest" are the number of pixels included in the training set, cross-validation set and testing set respectively. Also we have a column headed "Nhidden" which contains the optimal number of hidden units for a Multilayer Feedforward network.

Table 1. General characteristics of the images and networks

Database	Ninput	Nhidden	Noutput	Ntrain	Ncross	Ntest
PINES	200	50	16	6633	1658	2075
SUB_PINES	200	15	4	2812	703	878
SAL_A	224	4	6	3423	855	1070
SAL_C	224	36	16	34644	8660	10825

We trained ensembles of three and nine networks. We kept the number of networks in the ensemble low because of the computational cost. We repeated the process of training an ensemble two times with different partitions of data in training, cross-validation and test sets. In this way, we can obtain a mean performance of the ensemble for each database (the mean of the two trials) and an error in the performance calculated by standard error theory. The results of the performance are in table 2 for the case of ensembles of three networks and in table 3 for the case of nine. We have also included the mean performance of a single network for comparison.

The results of table 2 show that in general there is an improvement by the use of an ensemble except in the case of *Boosting*. The improvement depends on the method and database. The database with lower improvement is SUB_PINES. In the case of database SAL_A the improvement of the ensemble is more or less regular for all ensemble methods. Finally, in databases PINES and SAL_C the improvement is low for same methods and high for others, it seems that the methods which modify the training set (*Adaboost, Bagging* and *CVC*) are the best in the case of database SAL_C, and the methods with penalty in the error function (*Decorrelated* and *Decorrelated2*) and the *Simple Ensemble* are the best in database PINES.

As a conclusion, it seems that we can get an increased performance in images of a higher number of pixels and classes, like PINES and SAL_C, but there is no a clear candidate among the different ensemble methods.

Table 2. Results for the ensemble of three networks

	PINES	SUB_PINES	SAL_C	SAL_A
Single Network	91.0 ± 0.2	96.27 ± 0.16	86.03 ± 0.15	99.07 ± 0.19
Adaboost	91.42 ± 0.10	96.0 ± 0.3	95.1 ± 0.2	99.48 ± 0.14
Bagging	92.77 ± 0.10	95.9 ± 0.3	95.9 ± 0.4	99.57 ± 0.14
Boosting	90.5 ± 0.7	95.05 ± 0.06	86.1 ± 0.7	98.0 ± 0.2
CVC	91.5 ± 0.7	96.0 ± 0.5	94.799 ± 0.018	99.48 ± 0.05
Decorrelated	93.3 ± 0.7	96.30 ± 0.17	86.5 ± 0.2	99.39 ± 0.14
Decorrelated2	93.5 ± 0.3	96.7 ± 0.3	86.4 ± 0.2	99.39 ± 0.14
Simple Ensemble	93.63 ± 0.19	96.2 ± 0.4	86.6 ± 0.3	99.43 ± 0.09

Table 3. Results for the ensemble of nine networks

	PINES	SUB_PINES	SAL_C	SAL_A
Single Network	91.0 ± 0.2	96.27 ± 0.16	86.03 ± 0.15	99.07 ± 0.19
Adaboost	92.53 ± 0.10	96.46 ± 0.00	95.90 ± 0.18	99.57 ± 0.04
Bagging	93.54 ± 0.3	96.0 ± 0.3	96.3 ± 0.2	99.67 ± 0.14
CVC	93.3 ± 0.3	96.5 ± 0.6	96.4 ± 0.3	99.62 ± 0.09
Decorrelated	93.7 ± 0.7	96.5 ± 0.3	86.5 ± 0.2	99.48 ± 0.05
Decorrelated2	94.0 ± 0.3	96.8 ± 0.5	86.5 ± 0.3	99.48 ± 0.14
Simple Ensemble	94.53 ± 0.07	96.2 ± 0.5	86.6 ± 0.2	99.48 ± 0.14

By comparing the results of tables 2 and 3, we can see that there is a general improvement by increasing the number of networks in the ensemble. The method which has the highest increasing in performance is *CVC*. In the rest the improvement is usually less than 1%. However, as a trade off the computational cost is greater, which a important factor to take into account. For example the training time of a neural networks for database PINES was six days in a Pentium 4 processor at 2,4Ghz.

As mentioned before these four images have been used in the reference [9] and we reproduce in table 4 the results of classification with *support vector machines*.

Table 4. Results of using support vector machines (SVM), comparison with other methods

	PINES	SUB_PINES	SAL_C	SAL_A
SVM	87.3	95.9	89	99.5
Single NN	91.0 ± 0.2	96.27 ± 0.16	86.03 ± 0.15	99.07 ± 0.19
Best Ensemble of 9 NNs	94.53 ± 0.07	96.8 ± 0.5	96.4 ± 0.3	99.67 ± 0.14

As shown in table 4 a single NN is a useful alternative to a *SVM*, it performs better in databases PINES and SUB_PINES and worse in SAL_C and SAL_A. We have also included the best results of an ensemble of 9 NNs, as we can see if we select the ensemble methods appropriately we can outperform a single NN and a SVM. The improvement seems to be more important in images with a higher number of pixels and classes, and therefore more difficult to classify.

4 Conclusions

In this paper we have presented experimental results of eight method of constructing an ensemble of Multilayer Feedforward networks in the application area of hyperspectral image classification. For this experiments we have used a total of four images extracted from two scenes. The results show that in general there is an improvement by the use of an ensemble except in the case of *Boosting*. The improvement depends on the method and database. In databases with a low number of classes and pixels like SUB_PINES and SAL_A (where the general performance of a single network is high) the improvement of the ensemble is lower and more or less regular for all ensemble methods. But, for databases with higher number of pixels and classes like PINES and SAL_C the improvement is low for same methods and high for others, it seems that the methods which modify the training set (*Adaboost*, *Bagging* and *CVC*) are the best in the case of database SAL_C, and the methods with penalty in the error function (*Decorrelated* and *Decorrelated2*) and the *Simple Ensemble* are the best in database PINES. It can be an interesting research to try both alternatives in new application images. Furthermore, we have reproduced the results of *support vector machines* for these images and we have seem that a neural network is a interesting alternative, specially in the case of constructing an appropriate ensemble with several networks.

References

1. Blamire, P.A.: The Influence of Relative Image Sample Size in Training Artificial Neural Networks. International Journal of Remote Sensing. **17** (1996) 223-230
2. Kumar, A.S, Basu, S.K., Majumdar, K.L.: Robust Classification of Multispectral Data Using Multiple Neural Networks and Fuzzy Integral. IEEE Trans. on Geoscience and Remote Sensing. **35** no. 3, (1997) 787-790
3. Slade, W.H., Miller, R.L., Ressom, H., Natarajan, P.: Ensemble Neural Network for Satellite-Derived Estimation of Chlorophyll. Proceeding of the International Joint Conference on Neural Networks. (2003) 547-552
4. Breiman, L.: Bagging Predictors. Machine Learning. **24** (1996) 123-140
5. Drucker, H., Cortes, C., Jackel, D., et al.: Boosting and Other Ensemble Methods. Neural Computation. **6** (1994) 1289-1301
6. Tumer, K., Ghosh, J.: Error correlation and error reduction in ensemble classifiers. Connection Science. **8** no. 3 & 4, (1996) 385-404
7. Freund, Y., Schapire, R.: Experiments with a New Boosting Algorithm. Proceedings of the Thirteenth International Conference on Machine Learning. (1996) 148-156
8. Rosen, B.: Ensemble Learning Using Decorrelated Neural Networks. Connection Science. **8** no. 3 & 4, (1996) 373-383
9. Gualtieri, J.A., Chettri, S.R., Cromp, R.F., Johnson, L.F.: Support Vector Machine Classifiers as Applied to AVIRIS Data. Summaries of the Eight JPL Airborne Science Workshop. (1999) 1-11

SAR Image Recognition Using Synergetic Neural Networks Based on Immune Clonal Programming

Shuiping Gou and Licheng Jiao

National Key Laboratory for Radar Signal Processing and
Institute of Intelligent Information Processing
Xidian University, Xi'an, 710071, China
shuipinggou@163.com

Abstract. A method for SAR image recognition algorithm is proposed, which makes use of the global optimal search ability and the quick local search ability of Immune Clonal Programming (ICP) [1] to obtain the prototype vectors in Synergetic Neural Networks (SNN) [2]. As a result, the recognition performance of SNN is improved. Moreover, a study has been made of multi-class recognition using SNN, a bottleneck problem of SNN, and the strategy of One-Against-One [3] is introduced in this paper. Simulation result shows the recognition accuracy rate of SNN is satisfied.

1 Introduction

High-dimension and nonlinear problem is described as a group of low-dimension nonlinear equations in Synergetic. Haken generalized Synergetics theory to the field of pattern recognition and put forward a new theory of neural network, synergetic neural network (SNN), which used in pattern recognition in 1990s [2]. Recently, real image recognition based on synergetic neural network has already gained more and more attention of researchers. In 1996, possibility of Cellular Neural Networks (CNN) implementing synergetic image processing models, which is proposed by Knneth [4]. But the essential of NN is based on steepest-decent and it has defect such as easily-getting into local minimum, poor search performance in large-scale and multi-peak spaces, and so on [6]. Hence, SNN based on evolutionary learning has become hotspot gradually in the field of intelligent computation and its application has achieved successfully in some areas.

In addition, SAR has used widely military and civil areas in recently years. Object recognition of SAR image based on object model is a international research focus on the whole [6,7]. In this paper, we adopt SNN based on ICP to recognize SAR image and remote sense images.

The organization of this paper is as follows. In section 2, we briefly describe the basic principle and dynamics equation of SNN. In section 3, image recognition algorithm using SNN based on ICP is discussed. Simulation results and their analysis are given in section 4. Finally, a conclusion is drawn in section 5.

2 Synergetic Neural Networks

The basic principle of SNN is that the pattern recognition procedure is consistent with the dynamic process [2]. q, a pattern remained to be recognized, is constructed by a dynamic process which translates q into one of prototype pattern vectors v_k through status $q(t)$, namely, this prototype pattern is closest to $q(0)$. The process is described in the following dynamic equation:

$$\dot{q} = \sum_{k=1}^{M} \lambda_k (v_k^+ q) v_k - B \sum_{k \neq k'} v_k (v_{k'}^+ q)^2 (v_k^+ q) - Cq(q^+ q) + F(t) \;, \tag{1}$$

where q is the status vector of input pattern with initial value q_0, λ_k is attention parameter. If and only if λ_k is positive, patterns can be recognized. v_k is prototype vector and v_k^+ is its adjoint vector that satisfies $(v_k^+, v_{k'}) = v_k^+ v_{k'} = \delta_{kk'}$. The second term on the right of equation (1) plays a patterns recognition role in defining the potential surface. The second term defines the competition among prototypes and controls the location of ridges in the potential surface. B and C are the constant.

Based on the idea of synergetics the pattern recognition procedure can be viewed as the competition process of many order parameters, whose dynamic equation is as follows:

$$\dot{\xi} = \lambda_k \xi_k - B \sum_{k' \neq k} \xi_{k'}^2 \xi_k - C(\sum_{k'=1}^{M} \xi_{k'}^2) \xi_k \;, \tag{2}$$

Where ξ_k satisfies initial condition:

$$\xi_k(0) = v_k^+ \cdot q(0) \;. \tag{3}$$

Haken has proved that when $\lambda_k = C > 0$, the largest initial order parameter will win and the network will then converge. In this case, the SNN can meet some real-time needs for it avoids iteration [2].

3 Algorithm of SNN Based on Immune Clonal Programing

The ICP runs on the basis of memory units and ensures to converge more quickly to the global optimal solution than common evolutionary algorithms. The essence of immune clone is that clonal operator can produce a variation population around the optimal of parents according to their affinity, and then the searching area is enlarged which is helpful for avoiding prematurity in course of evolution. In Artificial Intelligent Computation [8], antibody, antigen, the affinity between antibody and antigen are similar to the definition of the objective function and restrictive condition. And the ICP is according to the affinity function $f(*)$, a point $a_i(k) \in A(k)$ in the solution space will be divided into q_i different points $a'_k(k) \in A'(k)$, by using clonal operator, a new antibody population is attained after performing clonal mutation and clonal selection. Hence, the algorithm includes three operating steps: clone, clonal mutation, and clonal selection.

In SNN, learning of networks problem can be reduced to how to get prototype pattern vector v_k and adjoint vector v_k^+ [2,4]. In this paper, we adopt ICP method to obtain prototype vector. Then the prototype vectors are coded with real number, namely, a series of the prototype vectors are arranged according to the certain order, which composes of a individual among solution population. On the other hand, numbers of accurate recognition sample in the training sets are treated as fitness function f.

The prototype vectors of SNN are feature vectors extracted from image while the optimal process is based on ICP. Then, image recognition algorithm is as follows:

Step1 Initialize the prototype vector V and get initialization population V_k ($k = 1,2,\cdots,n$). Set iteration number L and the size of population.

Step2 Iteration: Calculate the corresponding adjoint vectors to recognize image of training sets, then select better individual in sequence to form new population according to affinity.
Clone the population: $V'(K) = \Theta(V(K)) = [\Theta(v_1(k)), \Theta(v_2(k)), \cdots, \Theta(v_N(k))]^T$
where N is the size of clone, which is too big and the algorithm is timeconsuming while N is too small and clone operation may be invalidation.

Step3 Clone Mutation: $V''(K) = T_m^c(V'(k))$, where mutation uses Gauss number. In order to retain original population antibody information, clone mutation is not acted as original population but acted on cloned antibody.

Step4 Calculate the affinity of $V''(k) : \{f(V''(K))\}$

Step5 Clonal selection: $U''(K+1) = T_s^c(U''(k))$. Select the best individual with better affinity than its parents to get new parent population, and then update population to implement information exchange.
$K=K+1$.

Step6 If K is larger L or affinity is satisfied, the algorithm is over. Otherwise return step2.

4 Experiment Results and Analysis

Experiment 1: Image collection is composed of 1064 binary value remote sense images, containing 608 plane images and 456 marine ones with the size of 128×128. A part of them are shown in Fig.1.

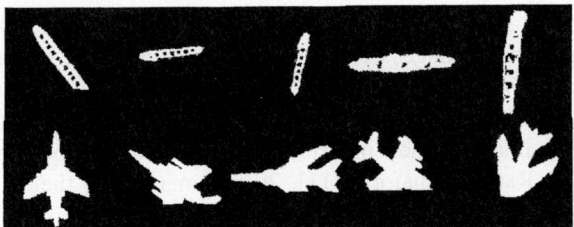

Fig. 1. Part of plane and marine images

Then shape features based on Hu [9] invariant moments are extracted from all images. The dimension of feature vectors is 7. We randomly select 50 as training data and the rest as test data from each class. In this experiment, some parameters of the method presented are defined as: the size of initial population is 25, mutation probability 0.1 and the clonal scale 8. The termination condition is specified the recognition accuracy rate of training samples be 100%, otherwise the iterative number 50. The average recognition results of 20 times are shown in Table 1, in which substitutes SVM for Support Vector Machines method [10] and GA for Genetic Algorithm and the string 'ARR' replaces the average recognition rate.

Table 1. Comparison of recognition results by several methods

	SVM	SNN (GA)	SNN (ICP)
Scale of training samples	1/3	1/10	1/10
Training time (s)	296.5	87.63	46.14
Test time (s)	11.21	0.0321	0.0316
ARR of training samples (%)	100	100	100
ARR of test samples (%)	94.5	93.76	99.0

From Table 1 it is obvious that the SNN method has much shorter training time and much quicker recognition speed than the SVM. On the other hand, the number of training data of the SNN is not much more than the SVM while recognition accuracy rate is raised yet. This is because the prototype vectors in SNN represent different patterns and input pattern type is less than feature dimension of prototype pattern vectors and the correlation between patterns is small. In synergetics, orthonormality of prototype vectors to eliminate the correlation between patterns, which can construct more rational classification space, is avoided by calculating the order parameters, namely, the inner product of state vector q and adjoint vectors.

In addition, SNN based on ICP has two advantages over SNN based on GA, i.e. higher recognition accuracy rate and shorter training time, since the ICP approach combines the global optimal search with powerful local search, which makes iteration number of the algorithm decrease and get optimal solution of problem utmost in the limited iteration times. But the GA requires infinite iteration obtain optimal value, which is impossible and the result is search ability is limited.

Experiment 2: Data sets are SAR images of Washington D.C, USA at http://www.sandia.gov/radar/images consisting of 12 bridges signed with their cropped sub-images shown in Fig.2.

Each bridge and a pseudo-object cropped are rotated, scaled and offset center and then we obtain 60 sub-images. Their invariant features and 25-dimension feature vectors are obtained with Radon transform [11]. 20 sub-images are then treated as training sets and 40 ones as test data for each class objects. In this experiment, one-against-one (OAO) [3] is adopted for SNN. Other parameters for the method are the size of initial population 10, mutation probability 0.1 and the clonal scale 5. The average recognition results of 20 runs are shown in Table 2.

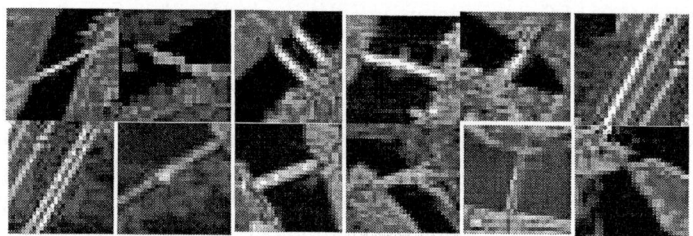

Fig. 2. Bridge sub-images cropped of the SAR image

20 sub-images are then treated as training sets and 40 ones as test data for each class object. In this experiment, one-against-one (OAO) [3], a common method for multi-class recognition problem, is adopted for SNN. Other parameters of the method are set to the size of initial population is 10, mutation probability 0.1 and the clonal scale 5. The average recognition results of 20 times are shown in Table 2.

Table 2. Comparison of results using different

	SVM (OAO) [12]	SNN (ICP)	SNN (ICP+OAO)
Training time (s)	751.37	131.75	379.46
Test time (s)	86.83	1.187	1.187
Recognition rate	93.65%	75.165	91.08%

From Table 2 we can see that multi-class recognition problem in SNN is solvable resorting to the strategy of the OAO. But the OAO is time-consuming. Fortunately, SNN based on ICP has lesser training time and its test recognition speed is very quick than the SVM. It is important for time to recognize SAR image. However, recognition accuracy rate of SNN is not much as the SVM. This is due to the fact that pattern classification using equation (3) in SNN requires the correlation between pattern vectors be very low, which is very difficult for analogue object recognition to ensure. So the reconstruction of order parameters in SNN to modify the equation (3) deserves our research further.

5 Conclusions and Discussion

In this paper, we have presented an approach SNN based on immune clonal programming to optimize prototype vectors, which implemented SAR image and remote sense images recognition. In addition, we argue that the existing SNN algorithm worse multi-class recognition performance and introduce the one-against-one technique, which classification accuracy is satisfactory. Moreover, optimization of attention parameters, construction of order parameters and optimal learning of adjoint vector deserve us deep research in SNN.

References

1. Du, H.F.: Immune Clonal Computing and Aritificial Immune Networks. Xidian University Postdoctoral Research Report (2003) 68–80
2. Haken, H.: Synergetic Computers and Recognition–a Top-down Approach to Neural Nets. Berlin: Springer-Verlag (1991)
3. Weston, J., Watkins, C.: Multi-class Support Vector Machines. Technical Report CSD-TR-98-04, Royal Holloway University of London (1998)
4. Kenneth, R.C., Chua, L.O.: A Synergetics Approach to Image Processing in Cellular Neural Networks. IEEE Int'l Symposium on Circuits and System, Circuit & Systems Connecting the World Proceedings, Piscataway, NJ, USA: IEEE (1996) 134–137
5. Shutton, R.S.: Two Problems with Backpropagation and Other Steepest-decent Learning Procedures for Networks, In Proceedings of 8th Annual Conference of the Cognitive Science Soiety, Erlbaum, Hillsdale, NJ (1986) 823–831
6. Zhao, Q., Jose, C.P., Victor, I.B.: Synthetic Aperture Radar Automatic Target Recognition with Three Strategies of Learning and Representation. Optical Engineering, 5 (2000) 1230–1244
7. Abhijit, M., Arthur, V.F.: Multi-class SAR-ATR using Shift-invariant Correlation Filters, Pattern Recognition, 4 (1993) 619–626
8. Jiao, L.C., Du, H.F.: Development and Prospect of the Artificial Immune System. Chinese Journal of Electronic, 10 (2003) 1540–1548
9. Hu, M.K.: Visual Pattern Recognition by Moment Invariants. IRE Transactions on Information Theory. 2 (1962) 179–187
10. Zhang, Y.N.: Intelligent Object Recognition Based on Support Vector Machines. Northwestern Polytechnical University Press, Xi'an (2002)
11. Hou, B., Liu, F., Jiao, L.C.: Linear Feature Detection Based on Ridgelet. Science in China, Ser. E., 4 (2003) 141–152
12. Zhang, X.R.: Research on Feature Extraction and Target Recognition in Remote Sensing Images, Master dissertation, Xidian University, 1 (2003)

Speaker Identification Using Reduced RBF Networks Array*

Han Lian, Zheng Wang, Jianjun Wang, and Liming Zhang

Dept. E.E, Fudan University, Shanghai 200433, China
{022021020, 022021022, wangjj,
lmzhang}@fudan.edu.cn

Abstract. An efficient reduced RBF networks array for speaker recognition is proposed in this paper. The reduced method combines the advantages of IOC and ROLS algorithms. Not only the number but also the positions of data centers are adapted in training progress, which optimizes the structure of network very well. Moreover, the experiments on a closed set, text-independent speaker recognition system show that, better robustness and simpler networks can be achieved through this improved algorithm in comparison with classical RBFN.

1 Introduction

The purpose of speaker identification is to determine a speaker's identity from the speech utterances. Essentially, speaker identification is a pattern recognition problem of a speech signal[1]. It consists of two processes: training and recognition. This paper focuses on the close-set, text-independent speaker identification task.

One of the two most important aspects of a speaker recognition system is the identification of discriminating features representing the specific characteristics of the voice of the speakers. Another important aspect is the choice of classifier[2]. These two are critical to the overall performance of any speaker recognition system. The block diagram of the system is given in Fig.1. Neural networks have played an active role in pattern recognition and many types of NNs, such as multilayer perceptrons (MLP), time delay neural networks (TDNN), radial basis function (RBF) networks, learning vector quantization (LVQ) and other forms of neural networks, have been applied in speaker identification[3]. There are two reasons for us to choose the RBF as the type of network. The first one is because of the successful use of GMM[4] in speaker identification. The RBF network is a three-layer NN, which has the same underlying structure as the GMM when Gaussian function is selected as the type of basis function in the RBF network. The second reason is the efficient training algorithm of the RBF network, which makes the RBF network easy to use.

---

* This research is supported by the National Science Foundation (NSF 60171036, 30370392), China.

The key point in design of radial basis function networks is to specify the number and the locations of the centers[5]. A common training approach is to first choose the network centers using an unsupervised technique to reflect, in some way, the distribution of the input training data. The approaches can be: randomly chosen input vectors from training data; vectors obtained from applying clustering algorithm, such as k-means clustering, to the network input training data (IC) or to both the input and output data (IOC), in which the output vector is also involved in the clustering process. Since the structure of centers designed in the IOC way is more effective, we choose IOC to get the initial locations of the centers.

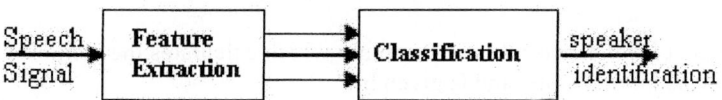

Fig. 1. Diagram of speaker identification

A satisfactory network performance is frequently achieved provided the network has a sufficient number of centers. However, it is difficult to know in advance how many network centers are required. Thus, the number of centers used is usually larger than necessary, especially in the application of utterance features. In this paper, the recursive orthogonal least-squares (ROLS)[6] algorithm is applied to the selection of suitable network centers from an initial set using the information available after training.

Mel's frequency cepstral coefficients (MFCC)[7] are extracted as the features for speaker characteristics. The system is composed of some binary classifiers, while the binary partitioned approach has been shown as an efficient solution for reducing training time[8]. The experiments on speaker recognition system show that, through improved adaptive RBF neural network array, better robustness and simpler networks can be achieved in comparison with classical RBFN.

The paper is organized as follows. Section 2 gives a brief introduction to RBF networks and shows the reduced method for the classifier. Section 3 shows results of speaker identification experiments, followed by conclusions in Section 4.

2 Reduced RBF Networks

2.1 RBF Networks Architecture

An RBFN with N inputs, M hidden units and one output is shown in Fig.2. Each hidden unit is a Gaussian kernel function with output related to the distance between the input vectors and the centroid of the basis function. Using the Gaussian kernel function, RBFN is capable of forming an arbitrarily close approximation to any continuous function.

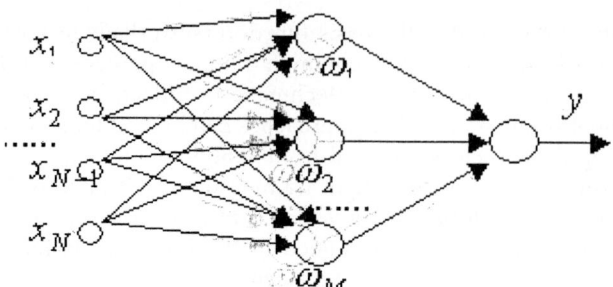

Fig. 2. RBFN structure

The output layer forms a linear combiner which calculates the weighted sum of the outputs of the hidden units, and is given by,

$$y = \sum_{i=1}^{M} w_i \phi_{ci}(x) + b \tag{1}$$

where $x \in R^n$ is the input vector, $y \in R$ is the output node, w_i is the weight from the ith RBF center, b is the output bias, and (2) are the Gaussian kernel functions,

$$\phi_{c_i}(x) = \exp(-\parallel x - c_i \parallel^2 / r_i) \tag{2}$$

where $c_i \in R^n$ is the ith center, and r_i are the function widths. Once the centers have been fixed, the optimal linear weights can be determined straightforwardly by using a linear least squares algorithm, or taking the pseudo inverse.

2.2 Reduced Structure of RBFN

The IOC method is applied to construct the initial locations and number of RBFN centers. In IOC method, clustering is applied to the augmented vectors which are obtained by concatenating the output vector to the weighted input vector and the resulting cluster codebook vectors are rescaled and projected into the input space to obtain the centers. Therefore, the locations of centers are influenced not only by the input sample spread but also by the output sample deviations.

Then after an initial RBFN is developed, the centers are sequentially removed while minimizing the effect on the network output error, chosen by the backward selection method of ROLS. It is a method for optimally reducing the size of a trained RBFN by removing one center at each stage so as to retain the highest achievable accuracy with the remaining centers. Generally, as each center is removed the network complexity is reduced at the expense of accuracy. In this paper, Akaike's final prediction error (FPE) criterion[6] is used as the particular metric of compromise between network complexity and accuracy.

Finally, gradient descent algorithm is applied to tune the reduced structure of network centers, and using pseudo inverse the optimal weights in the final network are achieved, which means much less time for learning as compared to using the integrated structure of centers.

3 Experiments

3.1 Database

The database consists of speech utterances randomly read from Chinese poems by 12 male and 8 female students under normal laboratory conditions. For every speaker, there are 20 segments of utterances, and each segment is about 2 seconds, including silence of about 1 second. The speech signal was sampled at 16kHz with 16 bits A-D conversion.

3.2 Feature Extraction

The digital speech data is pre-emphasized with $1-0.97Z^{-1}$, then segmented into frames of 410 sample points with an overlap of 250 sample points. Twelve orders of MFCC of each frame are adopted as the speakers' features.

3.3 Subnets Training

The structure of an RBF subnet is as follows: The number of nodes in the first layer is 12 and in the output layer it is one, and the number of nodes in the hidden layer will be automatically determined in training. Hence, the architecture of each subnet in the RBFN array is flexible enough to fit the separability between the two training patterns.

For N speakers, there are $N(N-1)/2$ subnets by using binary partitioned approach. For speaker i, there is an total error distance $\varepsilon_n = \sum (d - f(x))$. Assume that $k=argmin(\varepsilon_n)$, and the speaker k is selected for speaker identification.

3.4 Experiment Results

Based on the database of 20 speakers, we construct a RBFN array formed by 190 binary partitioned subnets using reduced RBF methods. Each subnet performs a classifier for two patterns. We choose randomly 64 or 128 samples from features to form the initial centers of network. During the training process, the number of centers is gradually decreasing. After training, the histogram of the number of nodes of all subnets in the hidden layer is shown in Fig.3. The mean of number of hidden nodes is 42.5, decreasing by 34% from 64, and 62.94, decreasing by 51% from 128.

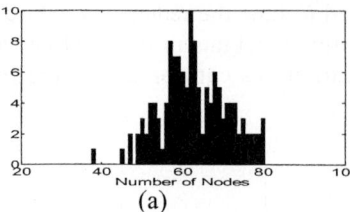 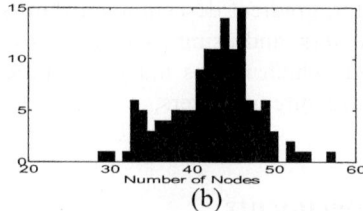

Fig. 3. Histogram of number of hidden nodes (a).The number of initial nodes is 128; (b).The number of initial nodes is 64

The performances of IC-RBF, IOC-RBF, and reduced RBFN are compared with the same MFCC and the same initial hidden unit's number 64 or 128. Here, IC-RBF means the RBFN array constructed by IC method, IOC-RBF is constructed by IOC method, which are both called classical RBF, and reduced RBFN is presented in this paper. Training with 5, 10, 15 segments of each speaker respectively, and testing with the residual, the identification results of 20 speakers are shown in Table.1.

Table 1. Comparison of identification rate of different RBF

Training Number	IC - RBF		IOC - RBF		Reduced RBF	
	64	128	64	128	64	128
5	78.24%	88.24%	80.29%	90.00%	82.65%	92.06%
10	92.35%	82.06%	93.53%	82.65%	96.18%	83.82%
15	90.29%	94.12%	91.18%	95.59%	91.76%	96.47%

From the results, it is evident that reduced structure of RBFN performs a better identification accuracy over classical RBFN. Note that in spite of the expense of accuracy when we reduce the complexity of the network, we can get a much smaller RBF network than before, as be shown in Fig.3. Thus, it becomes easier to perform a tuning to all the residual centroids with gradient descent algorithm, which make the centers more effective. Moreover, less memory of store and faster calculation for identification are benefited from the reduction of RBFN.

4 Conclusion

An adaptively reduced RBFN array system is proposed for speaker recognition, and the text-independent identification performance reaches 96.47% in a 20-speaker's database. Experimental results show that improvement in the accuracy of recognition is achieved using the reduced RBFN array as compared with the classical RBFN array. Results also show that by adopting reduction processing, the architecture of network gets smaller and more effective.

References

1. Xicai, Y., Datian, Y., Chongxun, Z.: Neural networks for improved text independent speaker identification. IEEE Engineering in Medicine and Biology Magazine, Vol. 21 (2002) 53-58
2. Mak, M.W., Allen, W.G., Sexton, G.G.: Speaker identification using radial basis functions. 3th International Conf on Artificial Neural Networks. IEEE Press (1993) 138-142
3. Kevin, R.F., Richard, J.M., Khaled, T.A.: Speaker recognition using neural networks and conventional classifiers. IEEE Trans on Speech and Audio processing, Vol. 2 (1994) 194-205
4. Reynodls, D., Rose, R.: Robust text-independent speaker identification using Gaussian mixture speaker models. IEEE Trans on Speech and Audio processing, Vol. 3 (1995) 72-83
5. Uykan, Z., Guzelis, C., Celebi, M.E.: Analysis of input-ouput clustering for determining centers of RBFN. IEEE Trans on Neural Networks, Vol. 11 (2000) 851-858
6. Gomm, J.B., Ding, L.Y.: Selecting radial basis function network centers with recursive orthogonal squares training. IEEE Trans on Neural Networks, Vol. 11 (2000) 306-314
7. Yang, X., Chi, H.: Digit Speech Signal Processing. Beijing. Publishing House of Electronic Industry (1995) 48-90
8. Rudasi, L., Zahorian, S.A.: Text-independent speaker identification using binary-pair partitioned neural networks. International Joint Conf on Neural Networks IJCNN. IEEE press (1992) 679-684

Underwater Acoustic Targets Classification Using Welch Spectrum Estimation and Neural Networks

Chunyu Kang, Xinhua Zhang, Anqing Zhang, and Hongwen Lin

Research Center of Signal and Information, DaLian Navy Academy, 116018, DaLian, China
dlkangcy@sohu.com

Abstract. In this paper we analyze the frequency spectrum signature of passive sonar target radiated noise, propose a kind of method to extract the object feature based on Welch power spectral density estimation. Then we extract the feature of thousands of stylebooks in many operating conditions and present a dynamic programming method for feature selection in order to get better feature vectors. Finally, the classification experiment for these noise stylebooks is done using the BP Neural Network (BPNN) and the K-Nearest Neighbor (KNN) methods. Results of experiment show that the methods are practical and effective. The method can be applied to the underwater acoustic targets classification.

1 Introduction

The central technology of underwater acoustic equipments and intelligence weapon systems is automatic classification for underwater acoustic targets. Both feature extraction and target classification are always two main problems need to be solved. Classification of underwater acoustic targets is a classic pattern recognition problem that need for high credibility and simultaneity[1]. Target classification results relies on the result of feature extraction and feature selection, but different classification results come from different classifier, even for the same feature vectors. The K-Nearest Neighbor (KNN) and BP Neural Network (BPNN) is conventional machine learning theory. It is an important method that the continuous spectrum and line spectrum of signal is used to automatically recognize and classify the targets in sonar, radar, speech recognition and so on. The power spectral density of the passive sonar signal is a representative superposition of continuous spectrum and line spectrum. It has very important meaning that the feature extraction based on continuous and line spectrum is used to underwater acoustic targets classification.

Many specialists and scholars have been done much study on this subject [1-4]. This paper analyze the frequency spectrum signature of the passive sonar target radiated noise , propose a kind of method to extract the object signature based on Welch power spectral density estimation. And then the feature vectors are selected using a dynamic programming method. Finally, the classification experiment is done using the BPNN and the KNN method.

2 Method of Welch Power Spectrum Estimation [5]

There are many power spectrum estimation methods such as Bartlett, Nuttall, Welch method and so on. Through comparing, the Welch power spectral density method is used to extract the feature of passive sonar target radiated noise.

The Welch method is also named weight overlap average method. The basic thought is to truncate the time signal $x_N(n)$ (N is the length of signal) using the window function $w(n)$ and to estimate the spectrum of the truncated data segment $x_N^i(n)$, then moving the window function and repeating them until the signal $x_N(n)$ through all windows. At last, the Welch power spectrum estimation of the signal $x_N(n)$ is the average of the power spectrum estimation $\hat{P}_{PER}^i(\omega)$ of all data segments.

3 Feature Extraction Based on the Welch Power Spectral Density

By comparing the Welch power spectrum of the radiated noise of lots of passive sonar targets, the Welch spectrum form of three different passive sonar targets (I, II, III) is approximately like Fig.1. And by comparing the spectrum of three type targets in different operating conditions, some conclusions are obtained [6].

- The spectrum form of the same kind of targets in different operating conditions is general identical. It shows that the Welch spectrum has certain stability to figure the radiated noise.
- There are relatively evident differences in spectrum form of different kind of targets, which shows that the Welch spectrum can token the difference and provide separability of different kind of targets.
- The Welch spectrum energy of class I mainly focuses on 0 ~ 2KHz, and the variety is relatively smooth on 2 ~ 4KHz. The spectrum variety of class II is relatively smooth and the main energy of class III is on 0 ~ 1KHz.
- The energy of the radiated noise of three kinds of targets basically focuses on low frequency and attenuates relatively quickly on high frequency.

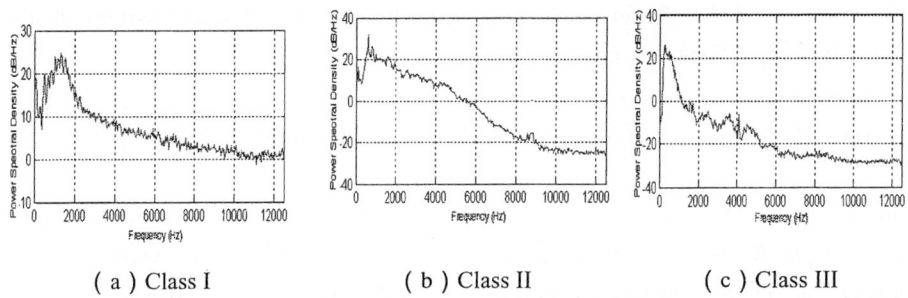

(a) Class I (b) Class II (c) Class III

Fig. 1. Welch power spectral density of three kinds of targets

By analyzing lots of spectrum of different kinds of targets, the radiated noise are filtered through a IIR band-pass filter (20Hz ~ 6KHz) before the feature extraction.

4 Feature Selection Based on the Dynamic Programming Method

4.1 Mechanism of Feature Selection [1]

The dimension of pattern feature space decides the structure complexity and recognition performance in a sense. In fact, when the feature dimension increases the train samples must be doubly increased in order to retain the recognition performance. Selecting the high quality and small amount feature vectors from multitudinous character information under certain sample information conditions is expected in project applications.

Suppose $\Omega_C = \{W, \sigma_\omega, P\}$ is the class probability space of samples, where W is the pattern class aggregate of the recognition system, namely $W = \{\omega_1, \omega_2, \cdots \omega_C\}$, C is the pattern class number, σ_ω is a σ algebra that the subset of W creates, P is a probability measure under σ_ω. The prior probability of all kinds of classes is $p(\omega_i), (i = 1, 2, \cdots C)$. And suppose $\Omega_F = \{X, \beta_x, p(X)\}$ is the feature probability space of samples, where $X = [x_1, x_2, \cdots x_F]^T$ is the F dimensions feature vectors of samples, β_x is the Borel field of X, $p(X)$ is the feature probability density function under β_x.

The original indeterminacy of the pattern class

$$H(\Omega_C) = -\sum_{i=1}^{C} p(\omega_i) \log p(\omega_i) \tag{1}$$

is fixed to the known pattern class problem. But the posterior entropy

$$H(\Omega_C | \Omega_F) = -\sum_{i=1}^{C} \int_X p(X, \omega_i) \log p(\omega_i | X) dX \tag{2}$$

can be changed by feature selection according to the known pattern feature information Ω_F. The change is

$$I(\Omega_C, \Omega_F) = H(\Omega_C) - H(\Omega_C | \Omega_F) = \sum_i \sum_j p(\omega_i, X_j) \log \left| \frac{p(\omega_i, X_j)}{p(\omega_i) p(X_j)} \right|, \tag{3}$$

namely the mutual information between the pattern feature and class feature space, where $X_j \in R^F$ is the number j selection of X. It shows the whole relativity between the feature vectors of samples and other type targets. When no mutual information of type space among feature vectors, it has most simple form as eq. (4).

$$I_s(\Omega_C, \Omega_F) = \sum_{i=1}^{C} \sum_{k=1}^{F} p(\omega_i, x_k) \log \left| \frac{p(\omega_i, x_k)}{p(\omega_i) p(x_k)} \right| \tag{4}$$

The purpose of feature selection is to make the maximal mutual information as eq. (3). If eq. (3) is directly used as a performance index of feature selection the calculation is quite complex. And that, eq. (4) is only the special form of eq. (3) and does not meet the common feature distributing condition. In fact, it has the relation with eq. (3) as follows.

$$I(\Omega_C, \Omega_F) \leq I_s(\Omega_C, \Omega_F) \tag{5}$$

It is a good approximation of mutual information $I(\Omega_C, \Omega_F)$ if $I_s(\Omega_C, \Omega_F)$ subtracts a positive modified item. Thus the feature selection performance index which is simple and good approximation of mutual information $I(\Omega_C, \Omega_F)$ can be used, shown in eq. (6).

$$V(M) = I_s(\Omega_C, \Omega_M) - \alpha \frac{1}{M} \sum_{\substack{x_i, x_j \in \Omega_M \\ x_i \neq x_j}} I(x_i, x_j) \tag{6}$$

Where α is the feature relativity coefficient, currently choosing 0.1 ~ 1.0, $I(x, y)$ is the mutual information between feature x and feature y, namely $I(x, y) = \int p(x, y) \log \left| \frac{p(x, y)}{p(x) p(y)} \right| dxdy$, where $p(x)$ is the probability density function of feature x and $p(y)$ is the probability density function of feature y, $p(x, y)$ is the joint probability density function of feature x and feature y.

4.2 Feature Selection Arithmetic [7]

Suppose the original feature aggregate is $O = (x_1, x_2, \cdots x_F)$, where F is the total feature dimensions that can be selected. The state variable $s_k \subset O$ is the selected feature aggregate on k phase, $F_k \subset O$ is the aggregate that can be selected on k phase, $d_k(s_k) \in F_k$ is the decision-making variable. So the state variable is $s_k = T_{k-1}(s_{k-1}, d_{k-1}(s_{k-1}))$ on k phase, where T_{k-1} is the state transforming function on s_{k-1} state. $P_{k,n-1}(s_k)$ is the all allowed strategy aggregate from k to $n-1$ phase. The decision-making performance function eq. (6) from original state s_0 to state s_{n+1} according to strategy $p_{0,n}$ is eq. (7).

$$V_{0,n}(s_0, p_{0,n}) = \sum_{x_j \in S_{n+1}} \sum_{i=1}^{C} p(\omega_i, x_j) \log \left| \frac{p(\omega_i, x_j)}{p(\omega_i) p(x_j)} \right| - \alpha \frac{1}{N_s} \sum_{\substack{x_i, x_j \in S_{n+1} \\ x_i \neq x_j}} I(x_i, x_j) \quad (7)$$

$$= V_{0,n-1}(s_0, p_{0,n-1}) + v_n(s_n, d_n)$$

Where

$$v_n(s_n, d_n) = \sum_{i=1}^{C} p(\omega_i, d_n) \log \left| \frac{p(\omega_i, d_n)}{p(\omega_i) p(d_n)} \right| - \alpha \frac{1}{N_s} \sum_{x_j \in S_n} I(x_j, d_n) \quad (8)$$

is the performance index on $n+1$ phase and state s_n using decision-making d_n, N_s is the element number of selected feature aggregate.

According to dynamic programming theory, the full condition that permission strategy $p_{0,n-1}^* = (d_0^*, d_1^*, \cdots, d_{n-1}^*)$ is the optimal strategy is

$$V_{0,n-1}(s_0, p_{0,n-1}^*) = \max_{p_{0,k-1} \in P_{0,k-1}(s_0)} \left| V_{0,k-1}(s_0, p_{0,k-1}) + \max_{p_{n,k-1} \in P_{k,n-1}(s_k^p)} V_{k,n-1}(s_k^p, p_{k,n-1}) \right| \quad (9)$$

to any $0 < k < n-1$ and s_0, where $s_k^p = T_{k-1}(s_{k-1}, d_{k-1})$ is the k phase state that is confirmed by original state s_0 and sub-strategy $p_{0,k-1}$.

5 Neural Network Classifier [8]

A neural network can be entirely specified by three basic components: the neuron (node) model or its input/output equation, the network topology, and the training or learning rule. In this paper, the BP neuron model which uses tangent sigmoid transfer function is adopted. Two-Layer feed-forward neural network is used for classification. And the node numbers of the input are the dimensions of feature vector, the hidden layer has 30 nodes and the output layer has three nodes which correspond with the target type. The training rule is gradient descent with momentum back-propagation which can quickly converge.

6 Experiments and Conclusions

All of the experimental data come from real noise signal radiated from three types targets (I, II, III) which work in various operating, ocean and weather conditions. The total samples are 4506. The train and test samples are selected according to 1:10.
If the feature selection isn't adopted (the feature dimensions are 256) the collectivity right recognition probability of BPNN is 94.84% and the right probability of KNN method is 92.88%. Table 1. is the right recognition probability of each type target.

Table 1. Right recognition probability of each type target

Method \ Class	I	II	III
BPNN	86.48%	97.49%	85.33%
KNN	82.40%	96.58%	81.33%

If the feature selection is adopted (the feature dimensions are 114) the collectivity right recognition probability of BPNN is 95.64% and the right probability of KNN method is 93.21%. Table 2. is the right recognition probability of each type target.

Table 2. Right recognition probability of each type target

Method \ Class	I	II	III
BPNN	86.22%	98.40%	86.67%
KNN	82.65%	96.62%	82.00%

According to the recognition experiment results, the conclusions are obtained.
– The feature vector based on Welch power spectral density has a good classification effect to passive sonar target classification or recognition, and can provide evidence for the Identification Friend or Foe in future war. It shows that the Welch feature vectors have separability among types.
– After feature selection, the feature dimensions reduce and recognition results are improved. It is said that the feature selection using dynamic programming method is feasible.
– Different classifier has different classification results. It shows that the classifier acts as an important part for target recognition. Therefore, the classifier should be studied next.

References

1. Zhang, X.H.: Recognition of Underwater Targets Based on the Theories of Intelligent Information Processing. PhD Thesis, ZheJiang University China (1996) 1-42
2. Zeng, Q.J., Wang, F., Huang, G.J.: Technique of Passive Sonar Target Recognition Based on Continuous Spectrum Feature Extraction. Journal of Shanghai JiaoTong University, Vol.36. Shanghai (2002) 382-386
3. Wu, G.Q., Li, J., Chen, Y.M.: Ship Radiated Noise Recognition (I) the Overall Framework Analysis and Extraction of Line Spectrum. Journal of Acoustics,Vol.23.Pekin(1998)394-400
4. Jing, Z.H., Xiang, D.Q., Wang, Y.Y.: A Simulation Experiment of Underwater Targets Identification Based on Line Spectrum Features. Journal of System Simulation, Vol.12. Pekin (2000) 642-644
5. Hu, G.S.: Digital Signal Processing-Theory, Arithmetic and Realization. TsingHua University Press, Pekin (1998) 320-338
6. Wang, B.G., Dong, D.Q., Xie, S.Y.: Object Through Signature Simulation Based on Power Spectrum Density Estimation. Journal of System Simulation, Vol.14. Pekin (2002) 25-33
7. Zhang, X.H.: Dynamic Programming Method for Feature Selection. Journal of Automatic, Vol.24. Pekin (1998) 675-680
8. Jiang, Z.L.: Introduction to Artificial Neural Networks. Higher Education Press, Pekin (2001) 39-52

Automatic Digital Modulation Recognition Using Wavelet Transform and Neural Networks

Zhilu Wu, Guanghui Ren, Xuexia Wang, and Yaqin Zhao

Institute of Electronics and Information Technology, Harbin Institute of Technology,
Harbin 150001, China
{wuzhilu, rgh, wangxuexia, yaqinzhao}@hit.edu.cn

Abstract. This paper presents an efficient digital modulation classification method based on wavelet transform and artificial neural networks (ANN). The method performs feature extraction via the discrete wavelet transform of the underlying digital signals because of the usefulness of the wavelet in de-noising and in compressing the digital signals. The features extracted from wavelet coefficients are then presented to the ANN for pattern recognition and classification. In addition, a less nodes output player and error back propagation learning with momentum are used to speed up the training process and improve the convergence of the ANN. Experimental results and performance evaluation of the method are given and it is found that the benefits of the developed method are that its structure is simple and it performs well at low signal to noise ratio (SNR) with high overall success rates.

1 Introduction

Recent developments in the field of ANN have made them a powerful tool to pattern recognition and classification. ANN is a system that is deliberately constructed to make use of some organizational principles resembling those of the human brain. They represent the promising new generation of information processing systems. Neural networks are good at tasks such as pattern matching and classification, function approximation, optimization and data clustering, while traditional computers, because of their architecture, are inefficient at these tasks [1].

Automatic modulation recognition of digital communication signals is an important signal processing problem in communications and its related fields. Many papers have been written on the topic of automatic modulation recognition using neural networks methods [2]-[4]. In these papers, key features are extracted from the instantaneous amplitude, the instantaneous phase and the instantaneous frequency of the intercepted signal. These features are normalized, and then analyzed by a trained multi-layer perceptron (MLP), which have a high accuracy. However, a degradation of performance at higher signal noise ratios (SNR) will appear when the network is trained on signals with lower SNR.

One of the primary requirements for neural network classifiers to have good generalization capability is the availability of a compact set of features that capture all the major attributes of the intercepted signals in a relatively small number of the components. The wavelet transform is hence an obvious candidate for the extraction of features used for pattern recognition and classification [5]. Wavelet transform is widely used in many fields of signal processing, especially image compression, speech processing, computer vision. For communication signal processing, it has been used in automatic digital modulation recognition and other tasks.

It has been demonstrated before that wavelets and neural networks can be successfully combined for the pattern recognition and classification. In recent research work, one method is the so-called Classification Wavelet network (CWN), in which the wavelet transform is applied in the first hidden layer of the network to extract compact features from input signals, and is followed by further layers to perform classification [6]. Another method, which is used in our work, is that the features were extracted from the approximation coefficients of Discrete Wavelet Transform (DWT) of the communication signal after certain level of decomposition, and the extracted features are then fed to the neural network subsystem for classification.

This paper presents a new method in automatic digital modulation recognition. In the next Section, we introduce the fast wavelet transform and the pre-processing in which the input key features are extracted from the wavelet coefficients of the wavelet decompositions at multi-levels. In Section 3, the modified structure of ANN is described and the BP learning with momentum algorithm is introduced. Section 4 gives the simulation results and the evaluation performance. Finally, a conclusion is made in Section 5.

2 Feature Extraction Using Wavelet Coefficients

As we know, feature extraction is the key to pattern recognition. How to define a representation that is well adapted for extracting the information content of digital communication signals is an important problem in modulation recognition. Intuitively, two available sources of information involved with time and frequency domains are inherent in communication signals. When the signal includes important structures that belong to different scale, it is often helpful to decompose the signal into a set of 'detail components' of various sizes. DWT analyzes signals at different frequency bands with different resolutions by decomposing a signal into coarse approximation and detail information, so it is widely used in pattern recognition and classification.

DWT employs two sets of function, known as the scaling functions and wavelet functions, which can be viewed as low-pass and high- pass filters, as shown in Fig. 1. The original signal, S, passes through two completely filters, namely low-pass and high-pass filters, and emerges as two signals: approximation, A, and detail, D, coefficients. The approximation coefficients are the high-scale, low-frequency components of the signal, while the detail coefficients are the low-scale, high-frequency components. The DWT, or filtering, process can be repeated until only one approximation coefficient is found.

In our research, the feature extraction structure is performed using the wavelet coefficients of the wavelet decompositions at 9 levels, as shown in Fig. 2. Wavelet decomposition is obtained using the Daubechies 3-coefficients filters. The computed coefficients in the sequences d_1, d_2, ..., d_9 form the wavelet decomposition of the measured signal S. The wavelet coefficients in the sequence d_m can be interpreted as the details of the signal S at coarser and coarser resolutions as m is increased. It means that d_1 represents the highest frequency region, d_2 the next lower frequency region, and so on.

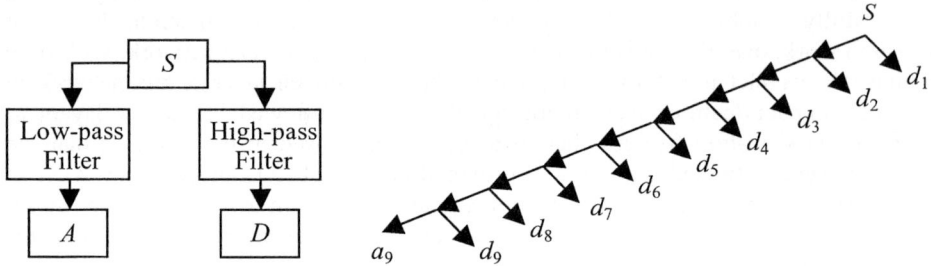

Fig. 1. DWT decomposition structure   **Fig. 2.** Wavelet coefficients of the DWT at 9 levels

Since the parameter assessment using all wavelet coefficients will often turn out to be tedious or leads to inaccurate results, a preprocessing routine that computes robust features correlated to the process parameters of interest is highly desirable. From the procedure described in previous subsection, sequences d_1, d_2, ..., d_9 are determined from a set of representative signals. The input feature vectors to a signal interpretation procedure can now be computed from these sequences, which consists of simply determining every component t_m of the feature vector $(t_1, t_2, ..., t_9)$ through the Euclidean norm $t_m=\|d_m\|_2$ of each such sequence d_m. This means that each feature t_m is determined as the square root of the energy of the wavelet coefficients in the corresponding cluster d_m. Consequently, the number of features for a signal S is equal to the number of sequences determined with the method described in the previous subsection. The physical meaning of each feature t_m is equal to that of a wavelet coefficient, namely it represents time and frequency information of the regarded signals S. A single feature especially describes a certain frequency rang (a scale m), which is equal to that described by the wavelet coefficients underlying this feature. And so constructed feature vectors are robust to noise in the corresponding signals S, which has been proofed in recent research work [7].

3   ANN Classification System

In recent years, we have seen the application of ANN to many pattern recognition problems, including communication signal modulation recognition. Because the convergence characters depend on the complexity of the structure including the number of the layers and the output nodes, in this paper we choose a simple structure of ANN with a 5-node input layer, a single hidden layer with 25 nodes and only a 3-node output layer instead of a 7-node output layer for classifying 7 signals: π/4OQPSK, OQPSK, GMSK, 16QAM, 4PSK, 2PSK, and 4FSK. Inputs to this network are the feature vectors extracted from the wavelet coefficients of the received signals. At the end of its training, the net performs a binary classification on each given input pattern. The value of each neuron in the output layer are designated as '1' and '0', which forms the output vector from '001' to '111' to expressing different signals.

Among the many neural networks learning algorithms, the error back propagation (BP) algorithm is considered the most useful learning algorithm. However, in training and learning phase of the ANN, a network using standard BP learning algorithm may get stuck in a shallow local minimum. In order to improve the convergence characters, the algorithm of BP learning with momentum is also employed in this paper to slide through such a minimum. The momentum allows the network to respond not only to the local minimum but also to the recent trends in the error surface. In the learning steps, for every epoch, the updating of the network weights is based on the rule of $W_n^{(i)} = W_n^{(i-1)} + \Delta W_n^{(i)}$, where $n = 1, 2$ or 3, $I = 1, 2, \ldots, K$, K is the maximum number of epochs, W_n are the updates in the weights.

4 Experiments

In this experiment, the simulation is carried in MATLAB environments. We select an ANN with a 25-node hidden layer using error back propagation learning with momentum, and choose Daubechies wavelets class-3 with the decomposition performed up to level 9 in the filtering process for features extraction.

In this paper, 7 types of band-limited digitally modulated signals, as mentioned in Section 3, are used as the training signal, in which the SNR is equal to 20dB of Gaussian noise. We measure the performance of the developed recognition system by using different types of the received signals while SNR equals to 20dB, 15dB, 10dB, and 8dB, respectively. The results of the performance evaluation of the proposed method for recognizing the 7 signals based on 400 realizations for each modulation type at different SNRs are displayed in Table 1.

Table 1. The accuracy rates for each signal at different SNRs

Signals	20dB	15dB	10dB	8dB
π/4OQPSK	99.75%	100%	95.5%	93.5%
OQPSK	90.25%	96.75%	93.75%	94.25%
GMSK	100%	100%	99.5%	100%
16QAM	98.5%	96%	99.5%	94.5%
4PSK	100%	98%	90%	100%
2PSK	100%	100%	100%	100%
4FSK	98.25%	93%	93.25%	97.25%

The simulation results show that using the same testing data that corrupted by Gaussian noise sequences at SNR=20dB, the ANN classifier based on DWT shows high accuracy. While the random noise level increase from 20dB to 8dB, the ANN classifier still can give high accurate results. The developed method exhibits a satisfying performance at the SNR even as low as 8 dB, especially for 2PSK and GMSK signals. All experimental results show the correctness of the method proposed and the feasibility of the recognition system.

5 Conclusions

In this paper we discuss an efficient digital modulation classification method based on DWT and ANN. A new method in feature extraction for signal recognition is introduced, in which Daubechies wavelets at order 3 (db3) has been used with the decomposition performed at level 9. The de-noised and compressed feature vectors obtained from DWT are used as the inputs to neural network classifier. The ANN classifier consists of feed forward neural network using back propagation learning rule with momentum to train the network.

In our research, computer simulations of different types of band-limited digital modulated signals corrupted by Gaussian noise sequences with different SNRs have been carried out to train ANN and measure the performance of the system. The high recognition rates from table 1 indicate the robustness of the developed modulation classification method based on DWT and ANN. However, there are 9.75% incorrect classifications. The misclassification percent might reduce with a larger train set and different wavelets family in feature extraction. Our further work will continue in this direction.

References

1. Dabbous, T.E., Shafie, K.A.: A Dynamic Model for Neural Networks. 16th National Radio Science Conference (NRSC'99), Ain Shams University, Cairo, Egypt (1999) 1-10
2. Nandi, A.K., Azzouz, E.E.: Algorithms for Automatic Modulation Recognition of Communication Signals. IEEE Transactions on Communications, Vol. 46 (1998) 431–436
3. Kavalov, D.: Improved Noise Characteristics of a Saw Artificial Neural Network RF Signal Processor for Modulation Recognition. 2001 IEEE Ultrasonics Symposium (2001) 19-22
4. Kremer, S.C., Shiels, J.: A Testbed for Automatic Modulation Recognition Using Artificial Neural Networks. CCECE'97 (1997) 67-70
5. Habibi, A.: Introduction to Wavelets. IEEE (1995) 879-885
6. Hoffman, A.J., Tollig, C.J.A.: The Application of Classification Wavelet Networks to the Recognition of Transient Signals. IEEE (1999) 407-410
7. Stefan, P., Kamarthi, S.V.: Feature Extraction from Wavelet Coefficients for Pattern Recognition Tasks. IEEE Transactions on Pattern Analysis and Machine Intelligence, Vol. 21 (1999)

Local Face Recognition Based on the Combination of ICA and NFL

Yisong Ye, Yan Wu, Mingliang Sun, and Mingxi Jin

Department of Computer Science and Engineering, Tongji University, Shanghai 200092, China
yeyisxp@hotmail.com

Abstract. This paper presents a local face recognition algorithm that is based on independent component analysis (ICA) and the nearest feature line (NFL). First, we separate a face image into several facial components. Then, we extract feature through combination of principal component analysis (PCA) and ICA; in the step of recognition, we first get each part of distance by NFL, then we calculate the synthetical distance by combining different parts. Compared with holistic image representation, this method has many advantages, such as a much higher recognition rate, more stable and flexible in practice.

1 Introduction

Currently, the main technologies of face recognition can be categorized into three kinds: template-based method, geometry characteristic - based method and model-based method1. As one kind of template-based method, Eigenface method [2] can extract features by using 2nd-order covariance of face image matrix. But central-limit theorem limits the energy difference of second-order matrix, and some important feature of human face may be contained in the high-order relationships among the image pixels [3], and thus, it is most likely to obtain feature parameters with much better sorted performance by utilizing a certain high-order statistic method. Independent component analysis (ICA) [4] is a high-order analysis method based on signal statistic characteristics. So, it may be used to extract face features.

A number of current face recognition algorithms are based on the whole face for recognition. In many practical occasions, however, it is necessary to recognize a person by local characteristics, such as recognition of the escaped criminals wearing sunglasses or covering the face. Therefore, synthesizing the statistic and geometric characteristics of human face, we propose a recognition algorithms based on local facial space analysis. Firstly, by using manually aligned eye position, we segment a face into two parts according to the geometric characteristics of human face, removing hair style and other useless information, then processing principal component analysis (PCA) and ICA for respective part, and calculating corresponding nearest feature line (NFL) distance, ultimately processing comprehensive recognition by setting reasonable coefficient of weight.

2   ICA for Local Face Recognition

To traditional holistic recognition approach based on PCA/ICA/LDA, such external unavoidable factors as illumination and posture will cause great changes of gray-scale image data, and eventually will decrease the stability of recognition. In order to eliminate the effect of illumination, Wang and Tan [5] proposed 2nd-order PCA approach, also some other algorithms based on PCA are proposed by simply truncating some maximal eigenvectors which can optimize extraction; Lien [6] proposed a recognition approach based on optical flow for the sake of eliminating the influence of expression and posture, Liu [7] proposed Eigenflow approach, etc. These approaches generally have the flaws of bad practicability or complicated calculation. At the same time, in statistics, some local face can tolerant a certain environmental change and even the same environmental factor can result in different effect to the different area. So the local ICA approach can enhance the stability of recognition. In the recognition step, this approach can adapt well to different practical conditions by adjusting the coefficient of weight flexibly. For example, to the sample of wearing sunglasses, we can obtain the comprehensive recognition by reducing the weight coefficient of eyes area or only making use of the lower part of the face.

In addition, considering that the actual face recognition system only makes use of five-sense image, therefore, we clipped the whole face image and discard the much useless information such as hairstyle in this paper.

For simplifying, manually positioning of human eyes is introduced to complete the automatic reduction and upper-lower segment (illustrated in Fig. 1.) which produces four pattern space: original image, clipped image, upper part of the face (upper face) and lower part of the face (lower face), and then acquires the basis of four types of space by ICA training in respective pattern space.

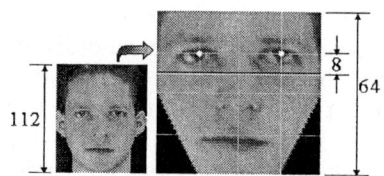

Fig. 1. A typical example for face clipping and segment

For a certain input pattern F^i (i-sample identification), new space vector-group coordinate $d^i = (d_1^i, d_2^i, d_3^i, d_4^i)$ is obtained by implementing respective ICA spatial projection. Suppose the weight vector $\alpha = (\alpha_1, \alpha_2, \alpha_3, \alpha_4)$, and then get the syntactical distance D^i:

$$D^i = d^i \cdot \alpha^T = \sum_{j=1}^{4} \alpha_j d_j^i \tag{1}$$

Finally, complete the final recognition step by applying certain classifier. In this paper, we adopt the NFL distance.

3 Recognition Based on NFL

Suppose we are given L classes of patterns, each contains Nc samples $\{x_1^c, x_2^c, \cdots, x_{Nc}^c\}$. For the test sample y and any x_i^c, x_j^c in the $\{x_1^c, x_2^c, \cdots, x_{Nc}^c\}$, define distance:

$$d_{NFL}(\overline{x_i^c x_j^c}, y) = \|y - p\| \qquad (2)$$

Where $p = x_i^c + \mu(x_j^c - x_i^c)$, and $\mu = \dfrac{(y - x_i^c) \cdot (x_j^c - x_i^c)}{\|x_j^c - x_i^c\|^2}$

The final judgment is measured by:

$$d_{NFL}(\overline{x_{i*}^{c*} x_{j*}^{c*}}, y) = \min_{1 \le c \le L} \min_{1 \le i \le Nc} d_{NFL}(\overline{x_i^c x_j^c}, y) \qquad (3)$$

4 Experiments and Results

4.1 Experiment Process

We use ORL face database as training and testing samples. We put all 40 patterns as training pattern, and select 5 images randomly for each pattern as training samples, that's 200 images in all, the resident 5 images for each pattern as testing samples.

Training Steps:
1. Clip and segment each image integrally, and constitute a column vector in accordance with the order of row to form training set $\{F_i^1, F_i^2, \cdots, F_i^M\}$ (M=200 is the number of samples, i=1,2,3,4 is the identification of each pattern.) Calculate the zero mean of matrix $<Fi>$;
2. First make K-L transformation to $<Fi>$ according to the algorithm of eigenface [2], then sort the eigenvector and eigenvalue. Choose the first M' principal component as Xi;
3. Do ICA training to Xi according to Bell and Sejnowski's InfoMax algorithm [8], and get the independent component subspace in token of X. InfoMax has several parameters; for this study, the block size was 50, the initial learning rate was 0.0005 and, after 1000 iterations, it was reduced every 200 epochs to 0.0003, 0.0002, and 0.0001. As to the permutation and other parameters are exactly the

same as those used in [3]. To avoid over fitting, we check output covariance. If nicely scaled, the covariance should approximate 3.5*I.
4. Project X_i to independent component subspace, and get the ICA coefficient in token of the image.

Recognition Steps:
1. Get zero-mean training vector group according to Step 1 described in the Training steps, then project them to principle space of training sample respectively, finally get the eigen coefficient of principle space Y_i;
2. Project Y_i into independent component subspace, and get ICA feature parameters of the independent component space;
3. Get comprehensive parameters by the linear combination of the empirical weight for ICA feature parameter of each piece, then recognize with NFL approach.

4.2 Experiment Results

Fig. 2 illustrates each part's principle component and independent component basis image respectively. We find that independent component basis image could depict the detailed facial information much better, as well as retaining holistic outline. It also can be found that the recognition rate of ICA is superior to that of PCA. Contemporarily, ICA can achieve very high recognition rate with less independent components, which also verifies that the characteristic encoding efficiency of ICA is higher than that of PCA and which is more obviously especially in the space of low statistic complexity9.

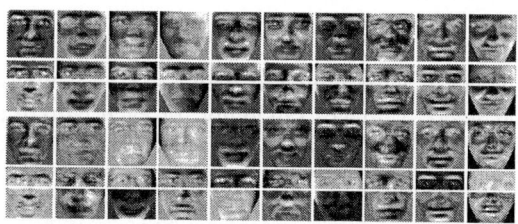

Fig. 2. Basis images of principal component (the first three lines) and independent component (the last three lines) for the local facial image with 10 independent components

Fig. 3 shows that the recognition rate of combination of upper and lower part of the face is apparently superior to respective recognition rate, and it can reach 92.5% with 30 independent components, meanwhile the whole curve change stably with the gradually increasing number of independent components, so it is with other combinations. Table 1 only lists the average recognition rate of partial combinations in correspondence with ICA basis range [10,200]. We can see from Table 1 that, the recognition rate of original human face in ORL set is apparently superior to that of the clipped face, and it is mainly because that the same samples in ORL set have an ideal information such as constant hairstyle which take up large area of image. But, the face recognition system in practice usually extracts facial part only for recognition. Table 1 shows that the average recognition rate can reach 92.85% easily by combination recognition of upper and lower part of the face. It is proved by large numbers of experiments, that ICA based on local face not only obviously improves the recognition performance and robustness, but also achieves better characteristic encoding efficiency. Table 1 also lists classification results comparing with other typical classification algorithms. We can see that, the NFL algorithm is much more efficient than others.

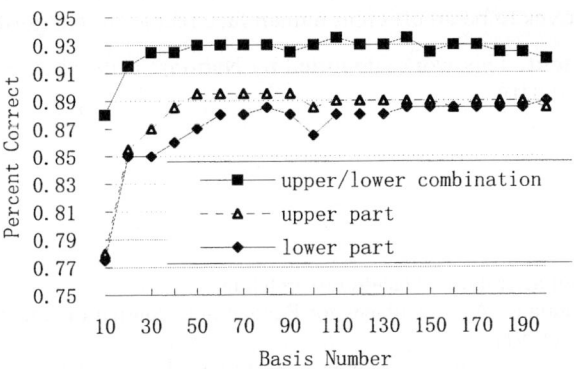

Fig. 3. The average recognition performance obtained for the upper-part face, the lower-part face and the combination of both

Table 1. Corresponding to various partial combinations, the average correct recognition rate of different classifier in ICA basis range [10,200]

	Combination weight (#1:#2:#3:#4)	NFL	Euclidean distance	Cosine angle [10]	Similarity 11
Original face (#1)	/	0.9645	0.9265	0.9585	0.9575
Clipped face (#2)	/	0.9173	0.8908	0.8850	0.8518
Upper part (#3)	/	0.8820	0.8015	0.8265	0.8140
Lower part (#4)	/	0.8718	0.8130	0.7988	0.7565
#12 combination	0.90:0.10:0:0	0.9700	/	/	/
#34 combination	0:0:0.60:0.40	0.9253	/	/	/
#234 combination	0:0.08:0.55:0.37	0.9285	/	/	/

Documentation [12] listed the most optimal recognition rate of the other face recognition approach with the same experiment conditions. Experiments in this paper obtain the most optimal recognition rate by 97.50% via synthesizing all the local images, which is superior to that of holistic Eigenface, ICA, Fisherface, and slightly better than that of Kernel Eigenface(96%), Kernel Fisherface(97%).

5 Conclusion

This paper presents a combined face recognition algorithm of local independent component. Local independent component algorithm takes both statistic and geometric characters into account, gets rid of limitation of holistic independent component algorithm that is sensitive to illumination and expression, and can achieve better characteristic encoding efficiency. Comparing with traditional holistic recognition algorithm, large numbers of experiments prove that local independent component algorithm is much more preponderant in recognition rate, robustness and flexibility. It proves to be an efficient human face recognition algorithm.

Acknowledgement. This work supported by National Natural Science Foundation of China (No. 60135010).

References

1. http://www.jdl.ac.cn/project/faceId/research.htm
2. Turk, M., Pentland, A.: Eigenfaces for Recognition. Journal of Cognitive Neuro-science, Vol. 3, No. 1 (1991) 71-86
3. Bartlett, M.S., Movellan, J.R., Sejnowski, T.J.: Face Recognition by ICA. IEEE Trans. on Neural Networks, Vol. 13, No. 6 (2002) 1450-1463
4. Comon, P.: Independent Components of Analysis-A New Concept. Signal Processing, Vol. 36 (1994) 287-314
5. Wang, L., Tan, T.K.: Experimental Results of Face Description Based on the 2nd-order Eigenface Method, ISO/IEC JTC1/SC21/WG11/M6001, Geneva, May (2000)
6. Lien, J.J., Zlochower, A., Cohn, J.F., Kanade, T.: Automated Facial Expression Recognition. Proceedings of the Third IEEE International Conference on Automatic Face and Gesture-Recognition, Nara Japan, April (1998) 390-395
7. Liu, X.M., Chen, T., Kumar, B.K.V.: Face Authentication for Multiple Subjects Using Eigenflow. Pattern Recognition, Vol. 36, No. 2 (2003) 313-328
8. Bell, A.J., Sejnowski, T.J.: An Information-Maximization Approach to Blind Separation and Blind Deconvolution. Neural Computation, Vol. 7, No. 6 (1995) 1129-1159
9. Draper, B.A., Baek, K., Bartlett, M.S., Beveridge, J.R.: Recognizing Faces with PCA and ICA. CVIU, Vol. 91, No. 1-2 (2003) 115-137
10. Bian, Z.Q., Zhang, X.G.: Pattern Recognition (in Chinese). Tsinghua University Press, Oct. (2002) 4-6
11. Su, G.D., Zhang, C.P.: MMP-PCA Face Recognition Method. ELECTROLICS LETTERS, Vol. 38, No. 25, Dec. (2002) 1654-1657
12. Kwak, N., Chong, H.C., Ahuja, N.: Face Recognition Using Feature Extraction Based on Independent Component Analysis. Image Processing, Vol. 2 (2002) 22-25

Facial Expression Recognition Using Kernel Discriminant Plane

Wenming Zheng[1,2], Xiaoyan Zhou[1,2], Cairong Zou[2], and Li Zhao[1,2]

[1] The Research Center for Science of Learning, Southeast University, Nanjing, Jiangsu, 210096, China
[2] The Engineering Research Center of Information Processing and Application, Southeast University, Nanjing, Jiangsu 210096, China
{wenming_zheng, xiaoyan_zhou, cairong, zhaoli}@seu.edu.cn

Abstract. Facial expression recognition (FER) based on a new nonlinear feature extraction method, called kernel discriminant plane (KDP), is proposed in this paper. KDP is a nonlinear extension of the Sammon's optimal discriminant plane (ODP) via the kernel trick. The recognition procedure is divided into two steps: (1) we select 34 fiducial points manually from each facial image and use the coordinates of these points as the input data of the facial image; (2) we construct a multiple binary classifier for the classification purpose of this task. The better performance of the proposed method is confirmed by the Japanese Female Facial Expression (JAFFE) database.

1 Introduction

Automatic facial expression recognition (FER) could be traced back to the preliminary work of Suwa et al. [1] in 1978 and gained much popularities starting with the pioneering work of Mase and Pentland [2] in the nineties. More recently, facial expression analysis has become a very hot topic in computer vision and pattern recognition, and there are various approaches have been proposed to this goal. For literature surveys, see [3][4][5]. This paper is an extension of the previous work done by Lyons et al. [6] on FER using a nonlinear feature extraction method, called kernel discriminant plane (KDP). In literature [6], Lyons et al. proposed to use the Fisher's linear discriminant analysis (FLD) method for FER. However, for a binary classification task, FLD can get only one discriminant vector. Moreover, FLD works well only for linear patterns and might fail for nonlinear patterns such as facial expression patterns, whose distribution is a nonlinear manifold [7] in the facial expression space.

Motivated by the kernel based machine learning method [9], we extend Sammon's optimal discriminant plane (ODP) [8] from linear domain to a nonlinear one via the kernel trick [10][11][12], and therefore propose the kernel discriminant plane (KDP) method. Then we perform the FER task based on the KDP method and the previous work of Lyons et al.. The main advantage of KDP for FER task is that it can extract two nonlinear features in the input space and therefore may perform better than FLD when it comes to the FER task.

2 KDP Formulations

Suppose that $\mathbf{X}_1 = \{\mathbf{x}_1^j\}_{j=1,\cdots,N_1}$ and $\mathbf{X}_2 = \{\mathbf{x}_2^j\}_{j=1,\cdots,N_2}$ are the observations in R^n from two different classes. Let $\Phi: R^n \to F$ be a nonlinear mapping that maps the observations to a high (even infinite) dimensional feature space F. The between-class scatter matrix $\mathbf{S}_B^\Phi$ and the total-scatter matrix $\mathbf{S}_T^\Phi$ in F are defined as:

$$\mathbf{S}_B^\Phi = \frac{N_1 N_2}{N}(\mathbf{u}_1^\Phi - \mathbf{u}_2^\Phi)(\mathbf{u}_1^\Phi - \mathbf{u}_2^\Phi)^T \tag{1}$$

$$\mathbf{S}_T^\Phi = \sum_{i=1}^{2}\sum_{j=1}^{N_i}(\Phi(\mathbf{x}_i^j) - \mathbf{u}^\Phi)(\Phi(\mathbf{x}_i^j) - \mathbf{u}^\Phi)^T \tag{2}$$

where $\mathbf{x}^T$ represents the transpose of $\mathbf{x}$, $\Phi(\mathbf{x}_i^j)$ the j th sample in the i th class, $N = N_1 + N_2$, $\mathbf{u}_i^\Phi$ the mean of the i th class samples and $\mathbf{u}^\Phi$ the mean of all samples:

$$\mathbf{u}_i^\Phi = \frac{1}{N_i}\sum_{j=1}^{N_i}\Phi(\mathbf{x}_i^j), \quad \mathbf{u}^\Phi = \frac{1}{N}\sum_{i=1}^{c}\sum_{j=1}^{N_i}\Phi(\mathbf{x}_i^j) \tag{3}$$

According to literature [8], the first discriminant vector ω_1 of Sammon's ODP in the feature space F is the kernel Fisher's discriminant vector (KFD) [12], and the second one is the unit eigenvector that maximizes $J(\omega)$ under the orthogonal constraint $(\omega_1^\Phi)^T \omega_2^\Phi = 0$, where $J(\omega) = \frac{\omega^T \mathbf{S}_B^\Phi \omega}{\omega^T \mathbf{S}_T^\Phi \omega}$, or $J(\omega) = \frac{\omega^T(\mathbf{u}_1^\Phi - \mathbf{u}_2^\Phi)(\mathbf{u}_1^\Phi - \mathbf{u}_2^\Phi)^T \omega}{\omega^T \mathbf{S}_T^\Phi \omega}$.

According to literature [13], we obtain that any solution ω can be expressed as:

$$\omega = \Phi(X)(\mathbf{I} - \mathbf{M})\alpha \tag{4}$$

where $\Phi(X) = [\Phi(\mathbf{x}_1^1) \quad \Phi(\mathbf{x}_1^2) \quad \cdots \quad \Phi(\mathbf{x}_2^{N_2})]$, $\alpha = [\alpha_{11} \quad \alpha_{12} \quad \cdots \quad \alpha_{2N_2}]^T$, $\mathbf{I}$ is the identity matrix, and $\mathbf{M}$ a $N \times N$ matrix with all terms equal to $1/N$. Thus, $\mathbf{u}_1^\Phi = \Phi(X)\mathbf{w}_1$, $\mathbf{u}_2^\Phi = \Phi(X)\mathbf{w}_2$, $\mathbf{u}^\Phi = \Phi(X)\mathbf{1}_N$, $\mathbf{S}_T^\Phi = \Phi(X)(\mathbf{I} - \mathbf{M})(\mathbf{I} - \mathbf{M})^T(\Phi(X))^T$ where $\mathbf{1}_N$ is a vector with all terms equal to $1/N$ and

$$\mathbf{w}_1 = [w_{11} \quad w_{12} \quad \cdots \quad w_{1N}]^T, w_{1i} = \begin{cases} 1/N_1, i \le N_1; \\ 0, i > N_1 \end{cases} \tag{5}$$

$$\mathbf{w}_2 = [w_{21} \quad w_{22} \quad \cdots \quad w_{2N}]^T, w_{2i} = \begin{cases} 0, i \le N_1; \\ 1/N_2, i > N_1 \end{cases}$$

Let $\mathbf{K}$ be an $N \times N$ block matrix whose elements are given by $(\mathbf{K}_{pq})_{p=1,2;q=1,2}$, where $\mathbf{K}_{pq} = (k_{ij})_{i=1,\cdots,N_p;j=1,\cdots,N_q}$ is a $N_p \times N_q$ matrix. Suppose that the element $(k_{ij})_{pq}$ of $\mathbf{K}_{pq}$ can be expressed as the dot product of $\Phi(\mathbf{x}_p^i)$ and $\Phi(\mathbf{x}_q^j)$, i.e., $(k_{ij})_{pq} = \Phi^T(\mathbf{x}_p^i)\Phi(\mathbf{x}_q^j)$. Then we obtain that

$$\mathbf{K} = (\Phi(X))^T \Phi(X) \quad (6)$$

2.1 The First Discriminant Vector of KDP

Let ω_1 be the first discriminant vector of KDP. From equation (4), we obtain that to find ω_1 maximizing $J(\omega)$ is equivalent to find coefficient vector α_1 that maximizes:

$$J_1(\alpha) = \frac{\alpha^T (\mathbf{L}_1 - \mathbf{L}_2)(\mathbf{L}_1 - \mathbf{L}_2)^T \alpha}{\alpha^T \mathbf{T} \alpha} \quad (7)$$

where $\mathbf{L}_i = (\mathbf{I} - \mathbf{M})^T \mathbf{K} \mathbf{w}_i$, $\mathbf{T} = (\mathbf{I} - \mathbf{M})^T \mathbf{K}(\mathbf{I} - \mathbf{M})(\mathbf{I} - \mathbf{M})^T \mathbf{K}(\mathbf{I} - \mathbf{M})$.

Let $\mathbf{T}(0)$ represent the null space of $\mathbf{T}$ and $\overline{\mathbf{T}(0)}$ the orthogonal complement of $\mathbf{T}(0)$. Suppose that ξ_i ($i=1,\cdots,r$) form a basis of $\overline{\mathbf{T}(0)}$. Let

$$\mathbf{Q} = [\xi_1 \quad \cdots \quad \xi_r] \quad (8)$$

then there exists $\boldsymbol{\beta}_1 \in R^{r \times 1}$ satisfying $\boldsymbol{\alpha}_1 = \mathbf{Q}\boldsymbol{\beta}_1$. Thus, to find $\boldsymbol{\alpha}_1$ maximizing $J_1(\boldsymbol{\alpha})$ is equivalent to find $\boldsymbol{\beta}_1$ that maximizes:

$$J_2(\boldsymbol{\beta}) = \frac{\boldsymbol{\beta}^T \mathbf{Q}^T (\mathbf{L}_1 - \mathbf{L}_2)(\mathbf{L}_1 - \mathbf{L}_2)^T \mathbf{Q} \boldsymbol{\beta}}{\boldsymbol{\beta}^T \mathbf{Q}^T \mathbf{T} \mathbf{Q} \boldsymbol{\beta}} = \frac{(\boldsymbol{\beta}^T \tilde{\ })^2}{\boldsymbol{\beta}^T \tilde{\mathbf{T}} \boldsymbol{\beta}} \quad (9)$$

where $\tilde{\ } = \tilde{\mathbf{L}}_1 - \tilde{\mathbf{L}}_2$, $\tilde{\mathbf{L}}_i = \mathbf{Q}^T \mathbf{L}_i$ ($i=1,2$), $\tilde{\mathbf{T}} = \mathbf{Q}^T \mathbf{T} \mathbf{Q}$.

From literature [8], we obtain that the direction of $\boldsymbol{\beta}_1$ can be expressed as: $\boldsymbol{\beta}_1 = \tilde{\mathbf{T}}^{-1}\tilde{\ }$.

It follows that the direction of ω_1 can be formulated by

$$\omega_1 = \Phi(X)(\mathbf{I} - \mathbf{M})\boldsymbol{\alpha}_1 = \Phi(X)(\mathbf{I} - \mathbf{M})\mathbf{Q}\boldsymbol{\beta}_1 \quad (10)$$

2.2 The Second Discriminant Vector of KDP

Let $\omega_2 = \Phi(X)(\mathbf{I} - \mathbf{M})\mathbf{Q}\boldsymbol{\beta}_2$ be the second discriminant vector. From Eq. (10), we have

$$\omega_1^T \omega_2 = \boldsymbol{\beta}_1^T \mathbf{R} \boldsymbol{\beta}_2 = 0 \quad (11)$$

where $\mathbf{R} = \mathbf{Q}^T (\mathbf{I} - \mathbf{M})^T \mathbf{K} (\mathbf{I} - \mathbf{M}) \mathbf{Q}$. Let $C = \frac{(\boldsymbol{\beta}_2^T \tilde{\ })^2}{\boldsymbol{\beta}_2^T \tilde{\mathbf{T}} \boldsymbol{\beta}_2} - \lambda \boldsymbol{\beta}_1^T \mathbf{R} \boldsymbol{\beta}_2$. Setting the partial derivative of C with respect to $\boldsymbol{\beta}_2$ equal to zero, we obtain that $2\tilde{\rho} - 2\rho^2 \tilde{\mathbf{T}} \boldsymbol{\beta}_2 - \lambda \mathbf{R} \boldsymbol{\beta}_1 = 0$, where $\rho = \frac{\boldsymbol{\beta}_2^T \tilde{\ }}{\boldsymbol{\beta}_2^T \tilde{\mathbf{T}} \boldsymbol{\beta}_2}$. Therefore, we obtain that

$$\beta_2 = \frac{1}{\rho}\tilde{T}^{-1}\left[\tilde{} - \frac{\lambda}{2\rho}R\beta_1\right] \quad (12)$$

From equation (11), we obtain $\beta_1^T R\beta_2 = \frac{1}{\rho}\beta_1^T R\tilde{T}^{-1}\left[\tilde{} - \frac{\lambda}{2\rho}R\beta_1\right] = 0$. Thus

$$\frac{\lambda}{2\rho} = \frac{\beta_1^T R\tilde{T}^{-1}\tilde{}}{\beta_1^T R\tilde{T}^{-1}R\beta_1} \quad (13)$$

Combination equation (12) and (13), we obtain that the direction of β_2 is:

$$\beta_2 = \left[\tilde{T}^{-1} - \frac{\tilde{}^T \tilde{T}^{-1} R\tilde{T}^{-1}}{\tilde{}^T \tilde{T}^{-1} R\tilde{T}^{-1} R\tilde{T}^{-1}}\tilde{T}^{-1}R\tilde{T}^{-1}\right]\tilde{} \quad (14)$$

The direction of the second discriminant vector of KDP is given by

$$\omega_2 = \Phi(X)(I-M)Q\beta_2 \quad (15)$$

Thus, the coefficient β_i is divided by $\sqrt{\beta_i^T Q^T (I-M)^T K(I-M)Q\beta_i}$ in order to normalize the corresponding vector ω_i in F: $(\omega_i)^T \omega_i = 1$. For a test point t, the projection can be computed by $\omega_i^T(\Phi(t)-u^\Phi) = \beta_i^T Q^T (I-M)^T (\kappa - K1_N)$, where $\kappa = [\kappa_1 \ \kappa_2 \ \cdots \ \kappa_N]^T$, $\kappa_{(i-1)N_1+j} = k(x_i^j, t)$.

3 Experiments

In this experiment, the Japanese Female Facial Expression (JAFFE) database, which has been previously used in literatures [6][13][14], will be used to re-conduct the FER task based on KDP. The JAFFE database contains 213 images of 7 categories of facial expressions (neutral, happiness, sadness, surprise, anger, disgust and fear) posed by 10 Japanese female models. The original images are all sized 256×256 pixels with a 256-level gray scale. For each facial image, we locate 34 fiducial points manually by referring to the literatures [6][13][14] and use the coordinates of these points to form a vector of 68 elements as the input data. Let $\{(u_i, v_i) | i=1,\cdots,34\}$ denote the 34 fiducial point positions of a given facial image. Then, a 68×1 vector $x = [u_1 \ \cdots \ u_{34} \ v_1 \ \cdots \ v_{34}]^T$ is generated corresponding to this facial image. Figure 1 illustrates the 34 fiducial point positions located in this paper.

We use the "leave-one-image-out" strategy and "leave-one-subject-out" strategy, respectively, to test the performance of the proposed method. In the "leave-one-image-out" strategy, only one facial image is used as the testing data and the other 212 facial images as the training data. The experiment is repeated for all the possible 213 trials and the recognition results are averaged. In the "leave-one-subject-out" strategy, we select one subject of images as the testing data and the other nine as the training data.

This is repeated for all the ten possible trials and the experimental results are averaged. For each trial, the polynomial kernel function $k(\mathbf{x},\mathbf{y}) = (<\mathbf{x},\mathbf{y}>)^d$ and the gaussian kernel function $k(\mathbf{x},\mathbf{y}) = \exp(-\|\mathbf{x}-\mathbf{y}\|^2 / \sigma)$ are used, respectively.

Considering that the experiment is a multiple class recognition problem, we adopt the one-against-one strategy [15] to conduct the experiment, respectively. Table 1 and Table 2 show the experimental results. From Table 1 and Table 2, we can see that both KFD and KDP achieve better than their respective linear version. Moreover, we also see that the KDP achieves the best recognition rate among the methods, which can reach as high as 99.06% classification rate for "leave-one-image-out" strategy and 83.57% for "leave-one-subject-out" strategy.

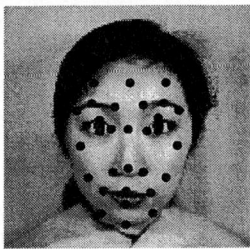

Fig. 1. The 34 fiducial points

Table 1. Experimental results using "leave-one-image-out" strategy

Method	Classification Rate
FLD	95.31%
ODP	98.59%
KFD (σ = 1e5)	97.65%
KDP (σ = 1e4)	**99.06%**

Table 2. Experimental results using "leave-one-subject-out" strategy

Method	Classification Rate
FLD	72.3%
ODP	82.16%
KFD (σ = 1e4)	77%
KDP (σ = 1e4)	**83.57%**

4 Conclusion

We propose the kernel version of Sammon's optimal discriminant plane in this paper and apply it for facial expression recognition task. Although the selection of the kernel function is still an open problem, our experimental results show that KDP with gaussian kernel can achieve the recognition rate as high as 99.06% using the "leave-one-image-out" strategy and 83.57% using the "leave-one-subject-out" strategy, which is much higher recognition rate than the results using the FLD method: the recognition rate of FLD is 95.31% using "leave-one-image-out" strategy and 72.3% using "leave-one-subject-out" strategy in this paper (the previous results reported by Lyons et al. [6] using "leave-one-subject-out" strategy is 72% using the same database).

References

1. M. Suwa, N. Sugie, K. Fujimora, "A preliminary note on Pattern Recognition of Human Emotional Expression", Proceedings of the Fourth International Joint Conference on Pattern Recognition, IEEE Computer Society (1978) 408-410.
2. K. Mase, A. Pentland, "Recognition of Facial Expression From Optimal Flow", IEICE Transaction (E), Vol. 74, (1991) 3474-3483.
3. B. Fasel, and J. Luettin, "Automatic Facial Expression Analysis: A Survey", Pattern Recognition, Vol. 36, (2003) 259-275.
4. M. Pantic, and Leon J. M. Rothkrantz, "Automatic Analysis of Facial Expression: The State of the Art", IEEE Trans. On Pattern Analysis and Machine Intelligence, Vol. 22, (2000) 1424-1445.
5. M. Pantic, and Leon J. M. Rothkrantz, "Toward an Affect-Sensitive Multimodal Human-Computer Interaction", Proceedings of the IEEE, Vol.91, (2003) 1370-1390.
6. Michael J. Lyons, Julien Budynek, and Shigeru Akamatsu, "Automatic Classification of Single Facial Images", IEEE Trans. On Pattern Analysis and Machine Intelligence, Vol.21, (1999) 1357-1362.
7. Y. Chang, C. Hu, and M. Turk, "Manifold of facial expression", Proc. of IEEE International Workshop on Analysis and Modeling of Faces and Gestures (2003) 28-35.
8. J. W. Sammon, Jr., "An optimal discriminant plane," IEEE Trans. Computer, C-19, (1970) 826-829.
9. Alberto Ruiz, Pedro E. López-de-Teruel, "Nonlinear kernel-based statistical pattern analysis," IEEE Trans. on Neural Networks, vol. 12, (2001) 16-32.
10. V. N. Vapnik, "The Nature of Statistical Learning Theory: Springer (1995).
11. B. Schölkopf, A. J. Smola, and K.-R. Müller, "Nonlinear component analysis as a kernel eigenvalue problem," Neural Computation, vol.10, MIT Press (1998) 1299-1319.
12. S. Mika, G. Rätsch, J. Weston, B. Schölkopf, and K.-R. Müller, "Fisher discriminant analysis with kernels," in Neural Networks for Signal Processing IX, Y.-H. Hu, J. Larsen, E. Wilson, and S. Douglas, Eds. Piscataway, IEEE, (1999) 41-48.
13. Michael J. Lyons, Shigeru Akamatsu, Miyuki Kamachi and Jiro Gyoba, "Coding Facial Expression with Gabor Wavelets", In Proceedings of Third IEEE International Conference on Automatic Face and Gesture Recognition, IEEE Computer Society (1998) 200-205.
14. Z. Zhang, M. Lyons, M Schuster, and S. Akamatsu, "Comparison between Geometry-based and Gabor-wavelets-based Facial Expression Recognition using multi-layer perception", In Proceedings of Third IEEE International Conference on Automatic Face and Gesture Recognition, IEEE Computer Society (1998) 454-459.
15. Guodong Guo, Stan Z. Li, and Kapluk Chan, "Face Recognition by Support Vector Machines", In Fourth IEEE International Conference on Automatic Face and Gesture Recognition, IEEE Computer Society (2000) 196-201.

Application of CMAC-Based Networks on Medical Image Classification[*]

Weidong Xu, Shunren Xia, and Hua Xie

Key laboratory of Biomedical Engineering of Ministry of Education, Zhejiang University,
Hangzhou, China, 310027
temco@zju.edu.cn

Abstract. Three CMAC-based neural networks (CMAC, FCMAC, ANFIS) are respectively introduced to deal with four medical image classification problems. According to our experiments, the features and applicability of these three classifiers are discussed. At the same time, a series of optimization methods based on wavelet transform, genetic algorithm, and self-organization competition neural network are applied on FCMAC and ANFIS in order to increase convergence speed and reduce memory requirement.

1 Introduction

Nowadays, in many computer-aided diagnosis and content-based image retrieval systems, artificial neural network (ANN) technology is widely used, mainly in medical image classification, where the most important factor of networks is not the approaching precision, but the generalization ability and convergence speed.
Cerebellar model articulation controller (CMAC) is a table-lookup local network [1]. Different from multi-layer perceptrons (MLP, a global approaching network), CMAC has simpler calculation, higher convergence speed and better generalization ability, non-existing local minimum, so that it has been widely applied in control and classification fields. In this paper, three CMAC-based networks are applied on four image classification problems, and then their features and applicability are discussed.

2 CMAC

Of the three networks, the first one is Albus CMAC [1]. It is an association neural network based on cerebella model, which can respond quickly to any input (no local minimum), and gives similar output if similar input given (generalization ability), like conditioned reflex.

The structure of CMAC is demonstrated as Fig.1. S domain has N_s inputs, each input S_n ($n \in (0, N_s-1)$) has N_d dimension signals, and each dimension signal X_i is quantized in Q_i levels ($i \in (0, N_d-1)$) and mapped to C_g ($C_g > 1$) quantizing perceptrons

---

[*] Supported by NSFC (No. 60272029) and NSF of Zhejiang Province (No. M603227)

in M domain. If the two nearest quantization levels have C_g-1 overlapped perceptrons, each signal X_i in S domain needs $P_i=Q_i+C_g$-1 perceptrons in M domain.

When a N_d-dimension input is mapped to M domain, a C_g*N_d mapping matrix ($M_{i,j}$, $i\in (0, N_d$-1$), j\in (0, C_g$-1$)$) of integer is gained. Each mapping column vector V_j ($j\in (0, C_g$-1$)$), can be comprehended as a N_d-place integer, which denotes the physical address of weight coefficients in A domain. The number of all possible addresses is:

$$A_d = P_0 \prod_{i=1}^{N_d-1} \frac{P_i}{C_g} = (Q_0+C_g-1) \prod_{i=1}^{N_d-1} \frac{Q_i+C_g-1}{C_g} \quad (1)$$

Hash-coding technology is usually applied on CMAC, to map the addresses in A domain to the addresses A_j in A_p domain (many-to-one mapping). If the memory space in A_p domain is not very small, the overlapping phenomenon that different addresses in A domain are mapped to the same address in A_p domain can be ignored. Finally, the output of CMAC is just the sum of the C_g weight coefficients in A_p domain:

$$F(S_n) = \sum_{j=0}^{C_g-1} W_{Aj} \quad (2)$$

CMAC usually adopts least mean square(LMS) method as learning algorithm. If the expected output of sample S_n is $d(S_n)$, $W(n+1)$ will be defined as:

$$W_{Aj}(n+1) = W_{Aj}(n) - \frac{\eta}{C_g} \frac{\partial E}{\partial W} = W_{Aj}(n) - \frac{\eta}{C_g} \frac{\partial E}{\partial e} \frac{\partial e}{\partial F} \frac{\partial F}{\partial W}$$
$$= W_{Aj}(n) + \frac{\eta e}{C_g} = W_{Aj}(n) + \frac{\eta}{C_g}(d(S_n) - F(S_n)) \quad (3)$$

3 FCMAC and ANFIS

The second CMAC-based neural network is Fuzzy CMAC [2][3], which was developed in 1990s. Fuzzy theory can easily make use of knowledge, but ratiocination process isn't clear and convenient; CMAC has adaptive learning ability, but can hardly apply knowledge. Fuzzy CMAC combines their advantages, and overcomes their defects.

The structure of FCMAC is demonstrated as Fig.2. It is divided into 5 layers. Different from CMAC, it doesn't quantize inputs. Instead, X_i will be transformed to a set of fuzzy membership grade coefficients. This process is fuzzification, and this layer is fuzzified layer. Membership functions have the same form, but different fuzzy centers $Z_{i,k}$ and fuzzy widths $\sigma_{i,k}$ ($i\in (0, N_d$-1$)$, $k\in (0, M_i$-1$)$, M_i denotes the number of fuzzy labels $F_{i,k}$). After that, each X_i has a M_i-dimension fuzzy membership vector.

The third layer is fuzzy association layer. A usual method is multiplication cross of fuzzy labels of different dimension signals, i.e. multiplication of fuzzy membership grade coefficients. Thus its outputs Y_m ($m\in (0, A_y$-1$)$) are A_y cross products of fuzzy membership grade coefficients, called relevance grade coefficients of fuzzy rules.

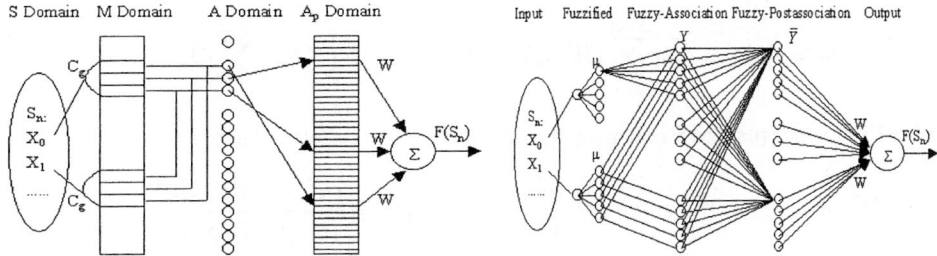

Fig. 1. CMAC **Fig. 2.** FCMAC

The fourth layer is fuzzy postassociation layer, used to normalize all relevance grade coefficients. The final output of FCMAC doesn't need to be defuzzified, since this normalization process can be considered as a barycenter-based defuzzification.

$$A_y = \prod_{i=0}^{N_d-1} M_i \qquad (4)$$

$$Y_m = \prod_{i=0}^{N_d-1} \mu_{F_i}, \quad \overline{Y}_m = Y_m \bigg/ \sum_{m=0}^{A_y-1} Y_m \qquad (5)$$

The final layer is output layer, where there are Ay weight coefficients Wm. The output of FCMAC is defined as the sum of Ay products of each normalized relevance grade coefficient and its corresponding weight coefficient:

$$F(S_n) = \sum_{m=0}^{A_y-1} (\overline{Y}_m W_m) = \sum_{m=0}^{A_y-1} ((Y_m W_m) \bigg/ \sum_{m=0}^{A_y-1} Y_m) \qquad (6)$$

FCMAC also adopts LMS learning algorithm. W(n+1) is defined as:

$$\begin{aligned}W_m(n+1) &= W_m(n) - \eta \frac{\partial E}{\partial W} = W_m(n) - \eta \frac{\partial E}{\partial e} \frac{\partial e}{\partial F} \frac{\partial F}{\partial W} \\ &= W_m(n) + \eta e \overline{Y}_m = W_m(n) + \eta \overline{Y}_m (d(S_n) - F(S_n))\end{aligned} \qquad (7)$$

The third CMAC-based neural network adaptive network based fuzzy inference system (ANFIS) was proposed in 1993 [4], which is very similar to FCMAC, and the only difference is that ANFIS is based on Takagi-Sugeno fuzzy model, while FCMAC on Mamdani fuzzy model. ANFIS is divided into 6 layers. The anterior 4 layers are the same as those of FCMAC, which won't be repeated here. The fifth layer of ANFIS is result layer, used to compute the result of each fuzzy rule. Different from FCMAC, the result of each fuzzy rule in ANFIS is not the product of each relevance grade coefficient and its corresponding weight coefficient, but the product of each relevance grade coefficient and its corresponding polynomial Pm, composed by Nd dimension input signals Xi and Nd+1 corresponding result weight coefficients Wi,m. The final layer is output layer, which is used to add up all fuzzy rule results O_m:

$$O_m = \overline{Y}_m P_m = \overline{Y}_m(\sum_{i=0}^{N_d-1}(W_{i,m}X_i) + W_{N_d,m}), \quad F(S_n) = \sum_{m=0}^{A_y-1} O_m \quad (8)$$

ANFIS also adopts LMS learning algorithm. $W(n+1)$ is defined as:

$$\begin{aligned} W_{i,m}(n+1) &= W_{i,m}(n) - \eta\frac{\partial E}{\partial W} = W_{i,m}(n) - \eta\frac{\partial E}{\partial e}\frac{\partial e}{\partial F}\frac{\partial F}{\partial O}\frac{\partial O}{\partial W} \\ &= W_{i,m}(n) + \eta e\overline{Y}_m X_i = W_{i,m}(n) + \eta\overline{Y}_m X_i(d(S_n) - F(S_n)) \\ W_{N_d,m}(n+1) &= W_{N_d,m}(n) + \eta\overline{Y}_m(d(S_n) - F(S_n)) \end{aligned} \quad (9)$$

4 Optimization

4.1 Dimensionality Reduction

The most troublesome problem of CMAC is the high memory requirement. CMAC (A_d) uses hash-coding, but FCMAC and ANFIS can't. The memory requirement of FCMAC (A_y) and ANFIS ($A_y*(N_d+1)$) seems much lower than that of CMAC ($A_d>>A_y$, for $Q_i>>M_i$), however, when dimensionality turns bigger, it is also terrible. To reduce it, the most effective way is to reduce the input dimensionality.

The first dimensionality reduction method is called problem decomposition. For example, in the human chromosome classification, a simple CMAC can be designed to classify all the chromosomes into 3 groups (long, middle and short), with the length. And then, in each group, another CMAC can be designed to classify the chromosomes into 3 groups (side-centromere, middle-centromere and others), with the centromere position. At last, a FCMAC or ANFIS can be designed to classify the chromosomes in 9 groups, with other signals. It's like a cascade connection of networks, but cannot be applied in all circumstances.

The second method is called wave analysis. At first, the N_d dimension input signals should be arranged, making correlated signals gathered and uncorrelated ones apart. Then, normalize all the signals, making the N_d dimension signals looks like a normalized discrete wave with N_d points. After that, a wavelet transform is carried out on this wave. From the decomposed wavelet domain, N_c wave feature parameters can be gained ($N_c<N_d$), which can primarily describe the N_d-point wave, and are regarded as a new N_c-dimension input. In this way, the input dimensionality is reduced, but some information will be lost during the mapping from N_d signals to N_c parameters.

4.2 Other Improvements

After the training, the weight coefficients of FCMAC or ANFIS have been adjusted, but the condition parameters $C_{i,k}$, including fuzzy centers $Z_{i,k}$ and fuzzy widths $\sigma_{i,k}$, haven't been updated, making convergence speed and approaching precision down. A usual condition parameter training method is LMS method. The error is back propagated (BP) from the output layer to the fuzzified layer, and condition parameters are accordingly updated to reduce the error (upper: for FCMAC; lower: for ANFIS).

$$C_{i,k}(n+1) = C_{i,k}(n) - \eta \frac{\partial E}{\partial e}\frac{\partial e}{\partial F}\frac{\partial F}{\partial Y}\frac{\partial Y}{\partial \mu}\frac{\partial \mu}{\partial C} = C_{i,k}(n) + \eta e W_m \frac{\partial Y}{\partial \mu}\frac{\partial \mu}{\partial C}$$

$$C_{i,k}(n+1) = C_{i,k}(n) - \eta \frac{\partial E}{\partial e}\frac{\partial e}{\partial F}\frac{\partial F}{\partial O}\frac{\partial O}{\partial Y}\frac{\partial Y}{\partial \mu}\frac{\partial \mu}{\partial C} = C_{i,k}(n) + \eta e P_m \frac{\partial Y}{\partial \mu}\frac{\partial \mu}{\partial C}$$
(10)

Another condition parameter training method is based on GA (genetic algorithm) [3]. The initial condition parameters are transformed into binary chromosome firstly. Then genetic algorithm is iterated until the learning result is satisfying. After the chromosome is transformed back, these parameters not only can be regarded as the final ones, but also can continue to be trained with LMS-BP method above.

Fuzzy label number reduction is used to improve the performance of Fuzzy neural networks. A good reduction method is based on self-organization competition neural network (SOCNN) [5], which can classify a series of vectors into several groups adaptively. Since excessive fuzzy label number reduction will reduce the approaching precision, the initial fuzzy label number must be big enough (7 or 9). Train the networks, and the updated condition parameters of each dimension are inputted into one SOCNN, whose input dimensionality is the same as fuzzy label number of this dimension. After the training, such Nd SOCNN will give the practical type Ti ($<=Mi$) of condition parameters of each dimension. Replace Mi with Ti, and the fuzzy label number of each dimension is reduced, leaving approaching precision unreduced.

Learning rate is a key factor of NN training. Here, learning rate R is firstly defined as a comparatively big value ($R_0 \in (0, 2)$). If the divergence does not take place, the best learning rate R_b has been found, otherwise R is decreased with a certain searching step, and repeat the above process. After that, a descending function is defined to adjust the learning rate during the training. At the beginning, the learning rate R can be very big, to improve the convergence speed; after much training, the learning rate R should approach the R_b, in order to ensure the convergence.

5 Experiments

In order to compare three CMAC-based networks, four medical image classification problems are presented. They are cell classification (P1, input: the gray-histogram and color of a cell, 3 dimensions; output: existence of the karyon, 1 dimension; 200 samples), gastroscope classification (P2, input: the shape, gray histogram and color of a gastroscope image, 8 dimensions; output: existence of gastric ulcer, 1 dimension; 100 samples), human chromosome classification (P3, input: the length, texture and centromere position of a chromosome, 6 dimensions; output: number of chromosome, 24 dimensions; 500 samples), and mammogram microcalcification classification (P4, input: the gray, contrast and shape of a point cluster, 10 dimensions; output: existence of microcalcification cluster, 1 dimension; 200 samples).

The optimization methods described here have been applied in the practical experiments. To reduce the input dimensionality, problem decomposition method has been applied on P3 (from 6 to 4), while wave analysis method has been applied on P2 (from 8 to 5) and P4 (from 10 to 5). From Tab. 1, it can be found out that CMAC has highest convergence speed and simplest computation, but can hardly deal with complex classification problems like P3, P4; FCMAC and ANFIS can deal with both simple and complex classification problems, but the learning time is much longer.

6 Conclusion

Three CMAC-based neural networks have been introduced. For simple classification problem, CMAC classifier is a better choice due to its high convergence speed and short learning time. As for complex classification problem, FCMAC or ANFIS classifier is a more considerable choice, because its nonlinear approaching precision is much higher than CMAC. But if the input dimensionality is too high, FCMAC or ANFIS can hardly be realized; so input dimensionality reduction must be carried out. The best dimensionality of FCMAC or ANFIS is 3~5 dimensions. Besides, if FCMAC or ANFIS is used, a series of optimization methods should be applied on it, with which the learning time and approaching precision could be improved.

Table 1. Experiment results of three classifiers on four classification problems (including memory space(MS), training times(TT), RMS error(RMSE) and learning time(LT))

P1	MS	TT	RMSE	LT	P2	MS	TT	RMSE	LT
CMAC	1024	10000	0.0006	2.7s	CMAC	8192	100000	0.0011	58s
FCMAC	125	200000	0.0057	11.3s	FCMAC	3000	100000	0.0121	127s
ANFIS	500	100000	0.0033	17.2s	ANFIS	18000	200000	0.0053	414s
P3	MS	TT	RMSE	LT	P4	MS	TT	RMSE	LT
CMAC	8192	200000	0.0427	57s	CMAC	8192	100000	0.0514	61s
FCMAC	4500	200000	0.0083	113s	FCMAC	3000	100000	0.0105	132s
ANFIS	22500	200000	0.0051	374s	ANFIS	18000	100000	0.0073	233s

References

1. Albus, J. S.: Data Storage in the Cerebellar Model Articulation Controller (CMAC). Journal of Dynamic Systems, Measurement, and Control, Trans. of ASME, Vol. 97. ASME, New York (1975) 228~233
2. Brown, M., Harris, C. J.: Neurofuzzy Adaptive Modeling and Control. Prentice Hall (1997)
3. Yu, H., Yu, J. S.: A Novel Fuzzy CMAC Network and Its Application. Control and Instruments in Chemical Industry, Vol. 30. Lanzhou (2003) 28~32
4. Jang, J. -S. R.: ANFIS: Adaptive-Network-Based Fuzzy Inference System. IEEE Trans. on System, Man, and Cybernetics, Vol. 23. IEEE (1993) 665~685
5. Cong, S., Li, G. D.: The Decrease of Fuzzy Label Number Using Self-Organization Competition Network. International ICSC/IFAC Symposium on Neural Computation (NC'98), Vienna University of Technology, Austria (1998) 106-112

Seismic Pattern Recognition of Nuclear Explosion Based on Generalization Learning Algorithm of BP Network and Genetic Algorithm

Daizhi Liu, Renming Wang, Xihai Li, and Zhigang Liu

Section 602, Xi'an Research Institute of High Technology, Hongqing Town, Baqiao,
Xi'an 710025, People's Republic of China
daizhiliu@163.com

Abstract. During the pattern recognition using BP neural network, the generalization performance often becomes poor. To improve the generalization performance of BP Network, a novel BP network generalization learning algorithm based on suboptimal criterion of fitting error of random assistant samples is presented. And we apply this algorithm to the classification of underground nuclear explosion earthquake events and natural earthquake events. Experimental results indicate that this method is effective and can improve the identification rate of underground nuclear explosions and natural earthquakes.

1 Introduction

In September 1996, a treaty banning underground nuclear test is subscribed in Geneva, and it decides to use 6 kinds of technologies to monitor compliance with this treaty. Among these technologies, the seismic method is chosen firstly.The procedure for using seismic means to monitor nuclear explosions is usually divided into three parts: (1)detection of signals;(2)location of the source of the signals; and (3)identification of the source as either earthquake or explosion.Part(1)and (2) are the common works in earthquake monitoring system, identification of the source is the critical[1], and it always is the key research of many countries[1][2][3][4][5][6]. Artificial neural network (ANN) is a kind of physical simulation of human brain, and one of its applications is the classification. However, the theory of ANN is not consummate, such as the network training is slow and frequently converges locally and so on. This limits the application of ANN greatly. In this paper, we discuss the generalization performance of BP neural network, and introduce a novel BP network generalization learning algorithm based on suboptimal criterion of fitting error of random assistant samples to improve generalization performance, and we apply this algorithm to the classification of underground nuclear explosions and natural earthquakes.

2 Generalization Learning Algorithm of BP Network Based on Genetic Algorithm

For BP network, under under-determined status, there will be many solutions satisfying the training accuracy at the same time, and starting from the random initial point on the error hyperplane (i.e. initial network weights), BP algorithm advances step by step along the descending direction of error gradient, and approaches one of the solutions finally. Therefore, in loose under-determined status, the more the under-determined solutions, the larger the range of solutions, and the bigger the probability of poor generalization performance of training of BP network. The paper[7] demonstrates this phenomenon by a curve fitting experiment. Essentially, the degree of the over-fitting phenomenon is determined by the initial weights of network, and once initial weights are determined, network weights will descend along the gradient on the error plane and approach one of the fixed under-determined solutions, furthermore, with the improvement of the training accuracy, the over-fitting phenomenon will become more obvious. If we use genetic algorithm (GA) to train network weights, on condition that training accuracy is satisfied, there will be many individuals satisfying the training accuracy. These individuals represent different under-determined solutions of network in a certain sense. Based on these under-determined solutions, we can adjust the fitness function in order to make the whole population converge along the direction of under-determined solutions with optimal generalization ability. The key is how to choose the individuals with good generalization ability. One method is to carry out generalization experiment with another group of known samples and check the generalization performance of individuals (i.e. check the recognition rate), and then choose and evolve them. However, the number of training samples is always limited in practice, and the method has the same effect as the method in which the check samples are also used to be training samples, so it has little practical value. In this paper, a new learning GA algorithm based on the suboptimal criterion of fitting error of random assistant samples is presented.

2.1 BP Neural Network Learning Algorithm Based on GA

The steps of BP Neural Network Learning Algorithm based on GA are as follows:

1)Chromosome encoding. Each weight of neural network is encoded as a string in a certain order,The specific form of code string is given by

$$\underbrace{W_{00}^1 W_{01}^1 \cdots \theta_0^1 \cdots W_{k0}^1 W_{k1}^1 \cdots \theta_k^1}_{\text{parameters of the first hidden layer}} \cdots \underbrace{W_{00}^N W_{01}^N \cdots \theta_0^N \cdots W_{k0}^N W_{k1}^N \cdots \theta_k^N}_{\text{parameters of the output layer}}$$

Each gene in the code string represents a specific connecting weight and threshold. Element W_{ji}^n represents the connecting weight between node i of layer n-1 and node j of layer n, and element θ_j^n represents the threshold of node j of layer n-1.

2)Define evaluation function. The output mean square error E of network is used to evaluate fitness. The smaller E, the higher the fitness of the network. The evaluation function is defined as follows:

$$fit(X) = 1/E = L/\sum_{l=1}^{L}\sum_{j=1}^{J}(t_{ij} - O_{ij})^2$$

where E represents the mean square error generated by training samples, L, the number of training samples and J, the number of output node.

3)Initialize population. In BP algorithm, the initial weights are generated in interval [-1,1]. Since GA couldn't adjust weights itself, in order to produce feasible solutions as much as possible in population and avoid that too small initial weights lead to too slow convergence speed, the initial values of population are randomly generated in interval [-10,10].

4)Genetic operators. In general cases, the crossover probability P_c and mutation probability P_m are obtained through experience, so for different problems, repeated experiments are needed to determine their values. In order to overcome the uncertainty of these values, the concept of adaptive genetic algorithm (AGA) is presented to realize variable mutation probability.

2.2 Suboptimal Criterion of Fitting Error of Random Assistant Samples

Let the number of training samples set X be l, the largest possible range of X be $\Re_x$, the expected output set be $T = \{t_1, t_2, \cdots, t_l\}$, and the actual network output be $Y = \{O_1, O_2, \cdots, O_l\}$. The assist sample set is generated as follows: choose q points randomly denoted by $A = \{a_1, a_2, \cdots, a_q\}$ within the space $\Re_x$, and no any assumption is made for the target output of the assistant samples. Suppose the actual network output set of the assistant samples is $B = \{b_1, b_2, \cdots, b_q\}$ where B is q points in the output space.

During the training, when network weights have evolved to a certain degree, there will be many under-determined network solutions satisfying the network training accuracy. Therefore the network output of each under-determined solution is identical to network target output, while for assistant sample points, difference will inevitably exist between output set B of each network individual (if the differences do not exist, we can ensure that each under-determined solution is actually the same solution), and the over-fitting phenomenon of some network individuals must be very serious, while other individuals will be relatively good. Our goal is try to eliminate those individuals with bad fitting effect from the population by selection operator. Then, how to evaluate the output fitting effect of network individuals? Actually we can evaluate it according to the minimum square fitting error(MSFE) E_{mse} of the common output set $B+Y$, We increase the fitness of of network individuals with little fitting error to increase the probability to be preserved and inherited. By this way, the whole population will converge in the direction of best generalization ability and satisfactory training accuracy.

Suppose input space is an n-dimensional one, the MSFE E_{mse} of output set $B+Y$ is defined as summation of the MSFE of each one-dimensional coordinate. Since the computation of the MSFE is very complicated, the multi-dimensional linear interpolation error E_{app} of the random assistant sample points is used to replace E_{mse}, which can decrease the complexity of the evaluation function greatly. Suppose the dimension of input space is n, the linear interpolation error E_{app} of assistant sample point a_i can be computed as follows:

(1) On the j-th coordinate of input space, find out two training sample points xd and xu which have the shortest distance from point $a_i(a_{i1}, a_{i2}, \cdots, a_{in})$, and their coordinates are denoted by $(xd_1, xd_2, \cdots, xd_n)$ and $(xu_1, xu_2, \cdots, xu_n)$ where $xd_j < a_{ij} < xu_j$. Suppose the expected network output of xd and xu are td and tu respectively, the linear interpolation error of the assistant point is given by

$$E_{ai} = \sum_{j=1}^{n} [bi - td + \frac{(td - tu) * (a_{ij} - xd_j)}{xu_j - xd_j}]^2 \quad (1)$$

(2) Compute the mean square interpolation error of q assistant points

$$E_{app} = \frac{1}{q}\sum_{i=1}^{q} = \frac{1}{q}\sum_{i}^{q}\sum_{j=1}^{n}[bi - td + \frac{(td - tu) * (a_{ij} - xd_j)}{xu_j - xd_j}]^2 \quad (2)$$

since q points used in each generation in the evolution are generated randomly, after some time of evolution, the assistant sample points will spread to the whole valid input space. Because evolution is progressing in the direction of decreasing E_{app}, the result of evolution is to decrease the minimum square fitting error E_{mse} as much as possible.

2.3 Generalization Learning Algorithm

The key step of generalization learning algorithm is the definition of the evaluation function. The fitting evaluation function should characterize both the output mean square error of network and the fitting error of assistant samples, and it can be denoted by

$$fit(X) = \begin{cases} 1/E = p/\sum_p\sum_j(t_{pj} - O_{pj})^2, E > e_{min} \\ 1/e_{min} + 1/E_{app}, E < e_{min} \end{cases} \quad (3)$$

where E represents the actual output square error of the network, e_{min} output accuracy of training samples, E_{app}, the linear interpolation fitting error of assistant samples. Assistant samples are generated randomly in each generation, which guarantee the continuous improvement of the generalization ability of the evolutionary population.

3 Seismic Pattern Recognition of Nuclear Explosion Based on Generalization Learning Genetic Algorithm

In this section, we will apply the generalization learning algorithm to the classification of underground nuclear explosions and natural earthquakes. There are

100 nuclear explosion samples and natural earthquake samples respectively in our database, 50 nuclear explosion samples and 50 earthquake samples are chosen as the training samples, the rest as the testing samples. The neural network structure is 11-5-1, which is obtained by CC algorithm [7],and we use generalization learning Genetic Algorithm (GA) and adapted GA algorithm [8][9]to train the network weights. The size of population is 500, and the number of random assistant samples is 100, the selection operator is generated by random league matches, the size of random league matches is 3. The network training accuracy is $e_{min} = 0.01$. After evolution, the curve of optimal individual fitness is shown in Fig.1.

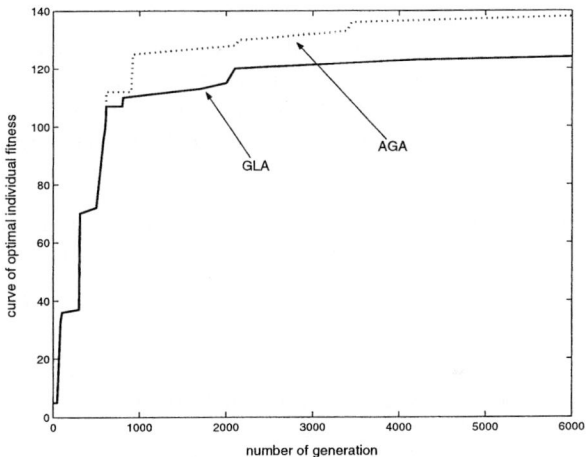

Fig. 1. Curve of optimal individual fitness(GLA denotes generalization learning algorithm)

From Fig.1, we can find that with the increasing of generations, the training accuracy of optimal individual of the population is higher. When the generation number is 500, the fitness of optimal individual of the population reaches $1/e_{min} = 100$, the network output mean square error of the optimal individual is 0.01, and at this time, generalization learning GA algorithm starts to use random assistant samples to improve generalization performance, but AGA algorithm still to improve the network output accuracy. After evolution, the recognition results of the two kinds of algorithm are listed in Table 1.

From Table 1, we find that the generalization performance of generalization learning GA algorithm is obviously better than AGA algorithm, and the testing recognition rate is higher.

Table 1. Recognition results of two algorithms (TR denotes training recognition rate, CR denotes testing recognition rate, E denotes earthquake recognition rate, N denotes nuclear explosion recognition rate, and T denotes total mean recognition rate, GLA denotes generalization learning GA algorithm)

	TR(%)			CR(%)		
	E	N	T	E	N	T
AGA	100	100	100	96	92	94
GLA	98	98	98	96	94	95

4 Concluding Remarks

In this paper, we introduce a method of improving generalization performance of BP network, and apply this algorithm to the recognition of underground nuclear explosions and natural earthquakes. Experimental results show that this algorithm is effective as the classifier of seismic pattern recognition pathology.

Acknowledgment. The authors should express our thanks to the National Natural Science Foundation for its support(No. 40274044).

References

1. Weiwen Xiao: Review on Nuclear Explosion Detection by Seismic Methods (in Chinese), Detection of Nuclear Explosions (1986) 115-130
2. Yuxiu Wang: Recognition and Reconnaissance of Nuclear Explosions (in Chinese), Detection of Nuclear Explosions (1986) 190-205
3. Zhongliang Wu, et.al.: Research on Seismic Recognition of Nuclear Explosions from 1990 (in Chinese), World Earthquake Exploration Abroad, Vol.4 (1994) 1-5
4. Ke Zhao, Daizhi Liu, et.al.: Feature Extraction on Time-Frequency Plane of Seismic Pattern Recognition of Nuclear Explosion (in Chinese), Nuclear Electronics and Detection Technology, Vol.20 (2000) 272-278
5. Daizhi Liu, Ke Zhao, et.al.: Attractor Analysis in Seismic Pattern Recognition of Nuclear Explosion (in Chinese), Acta Electronica Sinica, Vol.25 (1997) 122-125
6. Daizhi Liu, Ke Zhao, et.al.: Fractal Analysis with Applications to Seismological Pattern Recognition of Underground Nuclear Explosions, Signal Processing (2000) 1849-1861
7. Renming Wang: On Survey System for Nuclear Explosion and Its Key Technologies (in Chinese), The Second Artillery Institute of Engineering, Xi'an (2003)
8. Fahlmanse: Fast-Learning Variationon Back-Propagation. Anempirical Study. Morgan Kaufmann Publisher, SanMateo CA (1988) 38-51
9. Paoy: Adaptive Pattern Recognition and Neural Networks. Addison-Wesley Publishing Company, NewYork (1989) 1-5

A Remote Sensing Image Classification Method Using Color and Texture Feature

Wen Cao, Tian-Qiang Peng, and Bi-Cheng Li

Information Engineering University,
No. 306, P.O. BOX 1001, Zhengzhou, Henan, 450002, China.
Monopolist_cw@Hotmail.com

Abstract. In this paper, we present an effective remote sensing image classification method using color and texture feature based on D-S evidence theory and neural networks. In our method, the multiresolution Gabor filtering and the color components in PCA color space are applied. Firstly, PCA techniques are applied to RGB values of the original image. We apply components of the image besides the first principal component to train and classify the image using B-P neural network, then, we obtain a classification result. secondly, the texture images can be classified in multiple scales and orientations using the Gabor filtering, then, we obtain the second classification result. Finally, the two classification results of the B-P neural network are fused with evidence theory. The fused result is regarded as the final classification result of the original image. The experimental results show that the new method is efficient and improves the classification accuracy largely.

1 Introduction

Remote sensing images are major sources of data and information that are used in various fields such as environmental studies, forest management, and urban change detection [1]. Classification is a common and powerful information extraction method, which is used in remote sensing. There are many classification methods that have their own advantages and drawbacks. Classification of remote sensing images has traditionally been performed by classical statistical methods (e.g., Bayesian and k-nearest-neighbor classifier).

A non-parametric classification method, neural networks, becomes important for real data processing. Neural networks have the properties of parallel processing ability, adaptive capability for multi-spectral images, good generalization and not requiring the prior knowledge of the probability distribution of the data, so when compared with statistical classification methods, neural network methods show huge superiority [2].

The method based on Neural Networks using color and texture feature has been analyzed separately. To increase the efficiency of the image classification and the image interpretation, spectral information and spatial information are then required to collect together. So, a classification method based on Neural Networks is proposed, in this paper, which color and texture feature are utilized via D-S evidence theory. Firstly, there exists relativity among RGB bands in RGB color space, simultaneously, principal component analysis (PCA) is used to create uncorrelated spectral bands [3]. The principal components are utilized to study and train using B-P neural network instead of RGB values of the original image, then, a classification result using color information is obtained. Secondly, the texture information is one of the spatial information, so Gabor texture features of the neighbors around each pixel are extracted from the original image [4]. These features are utilized to study and train using B-P neural network, then, a classification result using spatial information is obtained [5]. Finally, the two classification results are fused with a new D-S evidence theory combination rule [6].

2   Principal Component Analysis

Let $X = [\vec{X}_1, \vec{X}_1, \ldots, \vec{X}_N]$ be a set of RGB values of pixels that belong to some different kinds of regions collected, where N is the number of training samples.

The mean vector $\vec{M}$ is computed as $\vec{M} = \sum_{i=1}^{N} \vec{X}_i$ and we get Φ by subtracting $\vec{M}$ from $\vec{X}_i$'s: $\Phi = [\vec{\Phi}_1, \vec{\Phi}_1, \ldots, \vec{\Phi}_N]$, where $\vec{\Phi}_i = \vec{X}_i - \vec{M}$. Then,

$$S_N \Psi = \Psi \Lambda \tag{1}$$

where S_N is the covariance matrix of Φ, $\Psi = [\vec{\Psi}_1, \vec{\Psi}_2, \vec{\Psi}_3]^T$ and $\Lambda = [\lambda_1, \lambda_2, \lambda_3]$ ($\lambda_1 \geq \lambda_2 \geq \lambda_3$) represent the eigenvalues and eigenvectors, respectively.

Two eigenvectors, $\vec{\Psi}_2$ and $\vec{\Psi}_3$, that correspond to smallest eigenvalues, λ_2 and λ_3, represent two directions with smallest spread of $\vec{X}_i$'s. So, a PCA color space is defined by the two axes, $\vec{\Psi}_2$ and $\vec{\Psi}_3$. W_{pca} is a linear projection matrix of which two vectors are $\vec{\Psi}_2$ and $\vec{\Psi}_3$, then the RGB values of the samples are converted into the principal components using followed equation:

$$Y_i = W_{pca}^T X_i \tag{2}$$

3   Gabor Filters

Gabor filters constitute another widely used descriptor of texture for segmentation and classification [4]. In a given window ($9*9$) of the texture, the given image $I(x, y)$ is convoluted with the Gabor function $G_{mn}(x,y)$ for scale m and orientation n to obtain the signal response:

$$w_{mn}(x,y) = \sum_s \sum_t I(x-s, y-t) G_{mn}(s,t) \qquad (3)$$

These magnitude coefficients represent the energy content at different scale and orientation of the image. The mean and the standard deviation offer an effective descriptor in remote sensing application. The feature vector is created using μ_{mn} and σ_{mn} as the feature components. So, Four scales and 4 orientations are used and the feature vector is given by $f_{G(x,y)} = (\mu_{00}, \sigma_{00}, \mu_{01}, \sigma_{01}, \cdots, \mu_{33}, \sigma_{33})$.

4 D-S Evidence Theory

Evidence theory is a reasoning method dealing with uncertainty problems [6], [7]. It is established on a non empty set Θ, where, Θ is called a frame of discernment and consists of an exhaustive set of elements mutually excluding. Any supposition A in question should belong to the power set of $\Theta: 2^{\Theta} = \{A | A \subseteq \Theta\}$. A basic probability assignment function (BPAF) is a function $m: 2^{\Theta} \to [0,1]$ which satisfies the following conditions:

$$m(\Phi) = 0 \qquad \sum_{A \subset \Theta} m(A) = 1 \qquad (4)$$

$m(A)$ is a measure of belief attributed exactly to the A and to none of the proper subsets of the A. Φ denotes the empty set. The elements of Θ that have a non zero mass are called focal elements and the union of all focal elements is called the core of the m-function. Evidence is composed of a body of evidence ($A, m(A)$) and given a body of evidence, a belief function Bel and a plausibility function Pl in the set 2^{Θ} are defined as:

$$Bel(A) = \sum_{B \subset A} m(B) \qquad \forall A \subset \Theta \qquad (5)$$

$$Pl(A) = 1 - Bel(A^C) \qquad \forall A \subset \Theta \qquad (6)$$

$Bel(A)$ measures the total belief that the object is in A and $Pl(A)$ measure the total belief that can move into A. The interval [$Bel(A), Pl(A)$] is the range of belief in A and measures the uncertainty degree in A.

D-S evidence theory provides a useful combination rule with which evidence from different sources can be combined. The combination rule is as follows:

$$m(\Phi)=0 \quad m(A)=\frac{1}{1-k}\sum_{A\cap B_j\cap C_l\cdots=A}m_1(A)\bullet m_2(B_j)\bullet m_3(C_l)\bullet\cdots \forall A\subset\Theta \qquad (7)$$

$$k=\sum_{A_i\cap B_j\cap C_l\cap\cdots=\Phi}m_1(A_i)\bullet m_2(B_j)\bullet m_3(C_l)\bullet\cdots \qquad (8)$$

assures the degree of conflict between evidence sources. The coefficient $1/(1-k)$ is called a normalization factor and is used to avoid non zero mass from being assigned to the empty set Φ after combination.

5 The New Classification Method

The paper proposes an effective remote sensing image classification method using color and texture feature based on D-S evidence theory and neural networks.

Firstly, we choose some samples and obtain a linear projection matrix W_{pca}, then, the RGB values of the samples are converted into the two principal components besides the first principal component. The set Y of two uncorrelated principal components is trained using BP neural network, then, the RGB values of each pixel of the original image are converted into the two principal components using former (3) equation. In this way, a classification result of the B-P neural network is obtained using color feature of image. Where, the neural network classifier has 2 nodes in the input layer, 20 nodes in the hidden layer and 4 nodes in the output layer.

There exists correlation among RGB bands of each pixel. PCA transform allows us to use orthogonal uncorrelated components for image description. That is useful for independent image processing each of three components. This transform is less sensitive to noise compared to the nonlinear transforms and keeps metric of the original color space. As we know, the first principal component mostly reflects the brightness information, So it is not appropriate for the input of classifier, and two principal components are regarded as evidences of image classification instead of RGB values. It can produce more accurate classification result based on a small amount of training samples than the RGB values of the original image.

Secondly, the color image is converted into the gray image.

Then, the texture images can be classified in multiple scales and orientations using the multiresolution Gabor filtering. Through studying and training the feature vector, the second classification result of the B-P neural network is obtained using texture feature. Where, the neural network classifier has 16 nodes in the input layer, 20 nodes in the hidden layer and 4 nodes in the output layer.

Finally, the two classification results (decisions) of the B-P neural network are fused with evidence theory. The fused result is regarded as the final classification result of the original image.

6 Experimental Results

For the experiments, we use a remote sensing image of Ning Bo area, and a sub image containing an airfield is selected to test the classification effect of the method. We divide the image into four categories: water, vegetation, soil, and and 25 sample for each category are choose. Fig. 1 shows the original sub image and the classification results of different classification methods. (a) is the original sub image; (c) is the classification results of the image directly by a neural network. (d) and (e) are classification results of the image after a PCA transform and a Gabor filtering separately by a neural network; (f) shows the combined classification results of (d) and (e) with evidence theory.

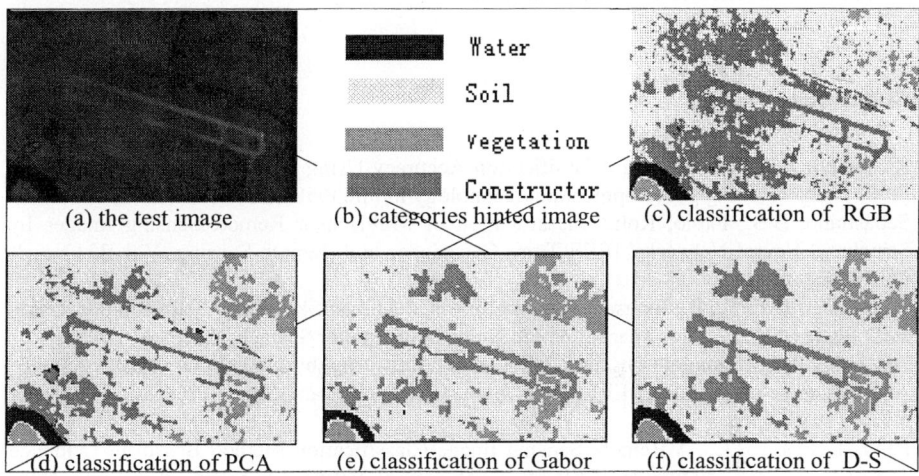

Fig. 1. The test image and the classification results

Airport belongs to the class of constructor by human visualization and its surroundings is mainly soil, so the desired classification result should distinguish airport from soil obviously meaning a clear airport contour. Using early neural network methods to classify, it is difficult to attain the desired classification result. It can be seen form (c) that the contour of airport is not clear and there are lots of error classifications in its edge especially at the top of airport course. The reason for above phenomenon is that the spectrum feature vector of neighbor pixels at the edge of airport is similar to that of pixels belonging to the class of constructor, i.e. the phenomenon of "same spectrum with different land matters" leads to the error classification. Compare to (c), (d) shows a better classification results with less correlation after the PCA transform. (e) utilizes the spatial information of the remote sensing image by a Gabor filtering, and get a more smoothed classification results than (c), however, lost of details information is lost. It can be observed that the airport contour in (f) is clearer than in Fig. 1(d) and Fig. 1(e) and the error classification phenomena is reduced thereby improving the classification effect and the classification precision of the whole image.

7 Conclusion

In order to make full use of the information included in a remote sensing image, an effective remote sensing image classification method using color and texture feature based on D-S evidence theory and neural network is presented in this paper. Firstly, PCA technique are applied to RGB values of the original image, then the two components of the PCA transform are sent to a BP neural network classifier. Secondly, Gabor texture features of the neighbors in the N*N square around a pixel are extracted from the original image, and the texture features are classified by a BP neural network. Finally, the two classification results (decisions) of the B-P neural network are fused with evidence theory. Experimental results show that the new method is efficient and improves the classification accuracy largely.

References

1. Alesheikh, A.Ali.: Improving Classification Accuracy Using Knowledge Based Approach. (2003) http://www.gisdevelopment.net/technology/ip/mi03058.htm
2. Sebastiano, B.S., Fabio, Roli.: Classification of Multisensor Remote-Sensing Images by Structured Neural Networks. IEEE Trans.Geoscience and Remote Sensing, Vol. 33. No. 3. (1995) 562–578
3. Juneho, Y., Jiyoung, P., Jongsun, K., JongMoo, C.: Robust Skin Color Segmentation Using a 2D Plane of RGB Color Space. Lecture Notes in Computer Science, (2003) 413–420
4. Daugman, J.G.: Complete Discrete 2-D Gabor Transforms by Neural Networks for Image Analysis and Compression. IEEE Trans. Acoustics, Speech and Signal Processing, Vol. 36. No. 7. (1988) 1169–1179
5. Peng, T.Q., Li, B.C.: A Remote Sensing Image Classification Method Based on Evidence Theory and Neural Networks. IEEE Int. Conf. Neural Network & Signal Processing, Nanjing, China, Vol.II, (2003) 240–244
6. Dempster, A.P.: Upper and Lower Probabilities Induced by a Multi-valued Mapping. Ann. Mathematical Statistics, Vol. 38. (1967) 325–339
7. Shafer, G.A.: Mathematical Theory of Evidence. Princeton, N J: Princeton U P (1976)

Remote Sensing Image Classification Based on Evidence Theory and Neural Networks

Gang Chen, Bi-Cheng Li, and Zhi-Gang Guo

Information Engineering University,
No. 306, P.O. BOX 1001, Zhengzhou, Henan, 450002, China
Maplechen111@Hotmail.com

Abstract. An effective method based on evidence theory and neural networks to classify the remote sensing images is brought forward in the paper. Firstly, with the spatial information in consideration, the original image is smoothed with a modified gradient inverse weighting smoothing method, then the classification of the original and smoothed images is performed separately using a B-P neural network. Finally, result comes out after fusing the two classification results with D-S evidence theory. Experimental results demonstrate that the proposed method is effective and can improve the classification accuracy.

1 Introduction

The automatic classification technique of remote sensing images is a branch of pattern recognition techniques in remote sensing field and has an extensive use in military and civilian areas [1]. It aims to the identification of remote sensing images, i.e. recognizing and classifying ground cover information in remote sensing images thereby distinguishing the corresponding ground truth and extracting the required information [2].

Artificial neural networks have been widely used in pattern recognition area and also widely used for the classification of remote sensing images in recent years [3][4]. Neural networks are nonparametric classifiers. They have the ability of parallel-processing and adaptive property. And no prior knowledge of the probability distribution of the data is needed when they are used for classification. When compared with statistical classification methods, neural network methods show great superiority. Early neural network methods for classifying remote sensing images are on the basis of pixel-by-pixel, i.e. judging which class a certain pixel belongs to only according to the value of the pixel without its neighboring pixels' values in consideration [5]. However, the classification method only using the value of a single pixel usually cannot obtain the desired result [6]. Li etc. presented that there exists certain correlation between neighboring pixels in remote sensing images, and the neighbors can provide useful spatial information for classification [7]. Many researchers have applied the spatial information to the classification methods based on neural networks. In general, the neighbors in the N*N square around a pixel are first obtained and then the value of each pixel in the range is regarded as the input of neural networks. Bischof and some other researchers classified remote sensing images with N=9 and their classification result is much better than that obtained by using the value of a single pixel as input, however, it needs considerable computation.

In order to utilize the spatial information to improve classification precision, to accelerate the training speed of neural networks as well as to decrease the complexity of computation at the same time, Peng etc. proposed a smoothed image-aided classification method of remote sensing image which results in good performance [8]. Compared with [8], the paper considers more about the correlations between a pixel and its neighbors. Firstly, a smoothed image is obtained using a modified gradient inverse weighting smoothing method. And then, the original image and the smoothed one are separately classified with the BP neural network. Finally, the two classification results are regarded as two pieces of evidence and fused with a new D-S evidence theory combination rule [9]. Further, the influence of the reliability of the classifiers to the final fusion results is taken into account.

2 Base of Evidence Theory and a New Combination Rule

2.1 Base of Evidence Theory

The theory of evidence was first proposed by Dempster in early 1967 and then extended by Shafer as a mathematical framework for the representation of uncertainty. Let $m(A)$ be the basic probability assignment function assigned to any supposition A on the frame of discernment Θ, then the combination rule is as follows:

$$m(\Phi) = 0$$

$$m(A) = \frac{1}{1-K} \sum_{A_{i1} \cap A_{i2} \cap \cdots \cap A_{in} = A} m_1(A_{i1}) m_2(A_{i2}) \cdots m_n(A_{in}), \tag{1}$$

$$(\forall A \subset \Theta, A \neq \Phi)$$

where

$$K = \sum_{A_{i1} \cap A_{i2} \cap \cdots \cap A_{in} = \Phi} m_1(A_{i1}) m_2(A_{i2}) \cdots m_n(A_{in}). \tag{2}$$

It measures the degree of conflict among evidence sources.

2.2 A New Combination Rule

When evidence sources are highly conflicting, the traditional combination rule (1) will generate unreasonable result. To combine the conflicting evidence sources, Li presented a new combination rule [9], which assigns the evidences' conflicting probability to every proposition according to its average supported degree. The new combination rule improves the reliability and rationality of combination results. Although evidences conflict with one another highly, results obtained are still good enough.

3 Classification of Remote Sensing Images

We propose an effective classification method of remote sensing images based on evidence theory and neural networks in this paper. The processing is as Fig. 1.

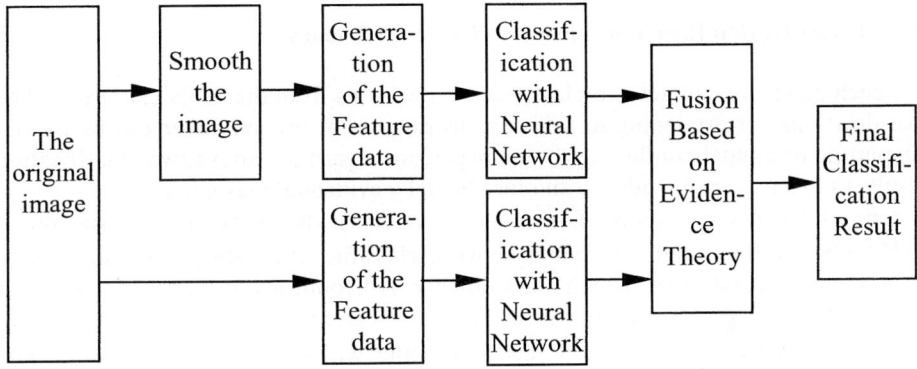

Fig. 1. Realization of remote sensing image classification based on evidence theory

3.1 Acquisition of the Smoothed Image

There are spatial correlations among pixels of remote sensing image, and the correlations are demonstrated by the successiveness of the distribution of ground cover. Given a pixel (not including the ones at the boundary of classes) belonging to some class, the class type of its neighboring pixels should be the same as its own. Bibliography [8] replaced the central pixel with the average value of the pixels besides the central pixel in the square, so the spatial information of the central pixel was included in the average value. But this model simply thought that every neighbor had the same influence on the central one. And the difference between them was not considered. In general, a neighboring pixel similar to the central one has a strong influence on it and should be assigned a larger weight, on the contrary, a relatively smaller weight should be assigned to a dissimilar one. So a modified gradient inverse weighting smoothing method is used to smooth the three bands of the remote sensing image separately in this paper. The difference between our modified method and the conventional gradient inverse weighting method [10] is that we assign no weight to the central pixel, i.e. the value of the smoothed pixel is determined totally by its neighbors.

3.2 Neural Network Classifier

Two neural network classifiers are designed in this paper, each neural network classifier has 3 nodes in the input layer, 20 nodes in the hidden layer and 4 nodes in the output layer. Every node of the output layer represents one class type. The inputs of the classifier are the RGB bands of the remote sensing image, and the outputs are the degrees of the current input pixel belong to each class type. Some sample pixels are selected from the original image and the smoothed one to train the two neural network classifiers separately. Then, the two classifiers are used to classify the original remote sensing image and the smoothed one.

3.3 Classification Based on Fusion of Evidence Theory

For each pixel, we can get two classification decisions from the ANN classifiers. The two decisions corresponding to different distribution of ground cover can be mutual supporting or mutual conflicting. In the paper, we regard the above two classification results as two pieces of evidence and fuse them by evidential reasoning.

One of the key problems of fusion with evidence theory is the construction of BPA. In [8], the outputs of neural network classifier after being normalized were used as the belief degree to each class directly. A hypothesis of that model was that the belief degree of the classifier to the whole discernment frame was zero, namely, $m(\Theta) = 0$. However, it is know that a classifier will never has a recognition rate of 100%, so the outputs of the classifier are not completely believable. Assigning the unbelievable part to the whole discernment frame should be a reasonable solution. Let $T_1, T_2, \cdots, T_M$ be the different ground cover classes, which is M in total, and $\Theta = \{T_1, T_2, \cdots T_M\}$ be the frame of discernment. For a pixel, the actual outputs using the original image as the inputs of the BP network are normalized as $y_1^{(1)}, y_2^{(1)}, \cdots y_M^{(1)}$, and the actual outputs using the smoothed image as the inputs of the BP network are normalized as $y_1^{(2)}, y_2^{(2)}, \cdots y_M^{(2)}$. For the same pixel, the basic probability assignment function to two pieces of evidence m_1, m_2 are as follows:

$$m_1 : m_1(T_1) = \varepsilon_r^{(1)} y_1^{(1)}, \quad m_1(T_2) = \varepsilon_r^{(1)} y_2^{(1)}, \cdots, m_1(T_M) = \varepsilon_r^{(1)} y_M^{(1)} \quad (3)$$
$$m_1(\Phi) = 0, \quad m_1(\Theta) = 1 - \varepsilon_r^{(1)}$$

$$m_2 : m_2(T_1) = \varepsilon_r^{(2)} y_1^{(2)}, \quad m_2(T_2) = \varepsilon_r^{(2)} y_2^{(2)}, \cdots, m_2(T_M) = \varepsilon_r^{(2)} y_M^{(2)} \quad (4)$$
$$m_2(\Phi) = 0, \quad m_2(\Theta) = 1 - \varepsilon_r^{(2)}$$

where, $\varepsilon_r^{(1)}$, $\varepsilon_r^{(2)}$ are the recognition rates of the two classifiers from a test set.

Evidence m_1 and evidence m_2 are combined using the new combination rule of D-S evidence theory [9] to produce a piece of new evidence. According to the rule that the determined target class should own the maximum, we can get the fusion classification results.

4   Experimental Results and Performance Comparison

The test image in the paper is a SPOT image of Ning Bo area. By human visualization, the image is divided into four categories: water, vegetation, soil, and building. We choose 20 samples for each category to train the two neural network classifiers. Then a sub-image is selected from the test image to get the classification rate (average classification rate of the four categories) of the two classifiers, which are 82.4% and 78.6% in our experiment.

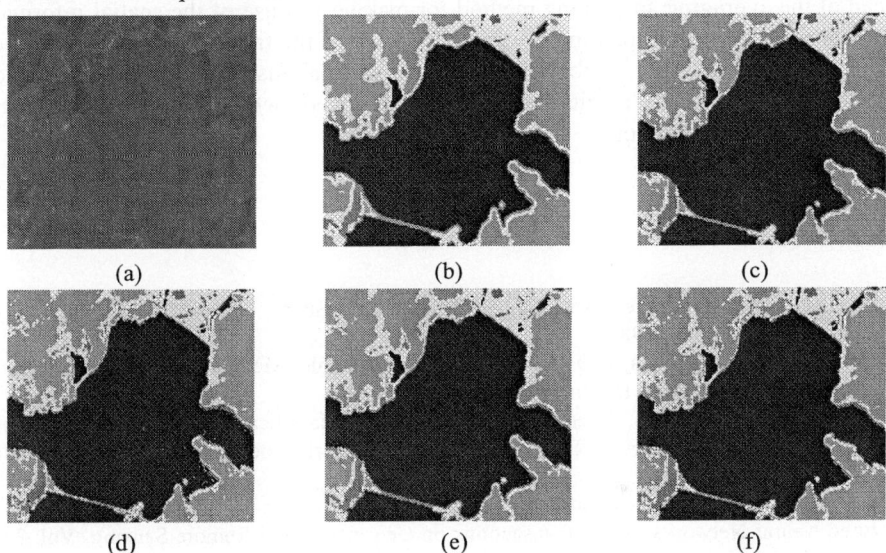

Fig. 2. The tested image and classification results

In Fig. 2, (a) shows the original test image, (d) is the classification results of the original test image after a neural network classifier, and classification results of the averaging smoothed image and the smoothed image generated by the modified gradient inverse weighting smoothing method after a ANN classifier are showed in (b) and (c), (e) and (f) are the classification results of the method used in [8] and the method used in the paper.

It is obvious that there is more detailed information in (c) than in (b) with better classification effect on the borders of different classes, and the same conclusion can be achieved from (f). For a better comparison between the method proposed in the paper and the method used in [8], we manually selected 800 test samples, which include 400 samples belonging to water, 200 samples belonging to vegetation, 100 of soil and the same number of building. By using the method proposed in this paper, an average classification rate of 93.5% is got which is superior to the classification rate 90.25% resulting from the method proposed in [8], and the experimental results are in accordance with our visual effects.

5   Conclusion

It is difficult to conquer the phenomenon of "same spectrum from different land matters" by using traditional methods based on neural networks and the error classifications often exist on the borders of different classes, which thereby degrades the classification precision. On the basis of [8], we propose an improved classification method based on evidence theory and neural networks. In our method, three improved aspects were proposed. First, a modified gradient inverse weighting smoothing method is used instead of the averaging smoothing method for making full use of the spatial information. Second, a new combination rule takes the place of the traditional one. Third, the influence of the reliability of the classifiers on the final fusion results is taken into account. The experimental results show that the improved method can provide a better precision of the classification.

References

1. Li, J.G.: Computer Pattern Recognition Technology. Shang Hai Jiao Tong University Publisher, Shang Hai (1986).
2. Sun, J.B., Shu, N., Guan, Z.Q.: Remote Sensing Principle, Method and Application. Bei Jing Mapping Publisher, Bei Jing (1997).
3. Chen, S.W., Chen, C.F., Chen, M.S.: Neural-Fussy Classification for Segmentation of Remotely Sensed Images. IEEE Transactions on Signal Processing, Vol. 45, No. 11. (1997) 2639-2654.
4. Serpico, S.B., Roli, F.: Classification of Multisensor Remote-Sensing Images by Structured Neural Networks. IEEE Transactions on Geoscience and Remote Sensing, Vol. 33, No. 3. (1995) 562-578.
5. Bischof, H., Schneider, W., Pinz, A.J.: Multispectral Classification of Landsat-Images Using Neural Networks. IEEE Transactions on Geoscience and Remote Sensing, Vol. 30, No. 3. (1992) 482-490.
6. Bruzzone, L., Conese, C., Masell, F., Roll, F.: Multisourse Classification of Complex Rural Areas by Statistical and Neural-Network Approaches. Photogrammetric Engineering & Remote Sensing, Vol. 63, No. 5. (1997) 523-533.
7. Li, Q., Wang, Z.Z.: An Integration of Remote Sensing Information Classification Methods Based on ANN and Experience Knowledge. Automation Transactions, Vol. 26, No. 2. (2000) 232-239.
8. Peng, T.Q., Li, B.C.: A Remote Sensing Image Classification Method Based on Evidence Theory and Neural Networks. Journal of Data Acquisition & Processing, Vol. 18, No. 2. (2003) 170-174.
9. Li, B. C.: Effective Combination Rule of Evidence Theory. Proceeding of SPIE, Vol. 4554. (2002) 237-240.
10. Wang, D.C.C., Vagnucci, A.H.: Gradient Inverse Weighting Smoothing Schema and The Evaluation of Its Performance. Computer Graphics and Image Processing, 15th. (1981).

Neural Networks in Detection and Identification of Littoral Oil Pollution by Remote Sensing *

Bin Lin [1], Jubai An[1], Carl Brown[2], and Hande Zhang [3]

[1]Dalian Maritime University, P.R. China 116026
linbin@newmail.dlmu.edu.cn
jubaian@dlmu.edu.cn
[2]Environmental Technology Center, Environment Canada, Ottawa, Canada
Brown.Carl@ec.gc.ca
[3]Aerial Department of China Marine Supervision, Qing Dao, P.R. China
Hande99@sohu.com

Abstract. In order to differentiate classes of oil-spills on water surface, a neural network (NN) approach is applied for spectral data analysis and identification of airborne laser fluorosensor in this paper. The target to be detected may be one of the following: seawater, lube, diesel, etc. The primary requirement for airborne sensors is to identify the substances targeted by the laser beam. Pearson Correlation Coefficient (PCC) method is one of the most current approaches. This paper outlines the NN model for the identification of the spilled oils, and makes a comparison with PCC in an effort to increase the level of confidence in the identification results. The results of ground tests using known targets show an increased confidence in the results when using the NN Model compared to that of PCC. It is believed that the NN model would play a significant role in the ocean oil-spill identification in the future.

1 Introduction

Pollution in marine environment, especially oil spills, is a great danger to living organisms, to seawater and even human. The ever-increasing number of oil spills, due to leaks from pipelines or accidental and illegal ship discharges makes the pollution more serious. Researches have been conducted to control the littoral oil pollution.

1.1 Background

One of the most important tasks is to detect discriminate different oil pollution on sea/ice. Remote sensing (RS) is becoming an increasingly important tool for the effective detection of oil-spill. Among different types of RS methods, the airborne laser fluorosensor RS is now considered as the most effective method in detecting oil spills by means of the characteristic fluorescence emission of the oil.

---

* Funded by the National Natural Science Foundation of China. (NSFC Project 40346028)

Fingerprints spectra of a variety of substances are shown in Figure1. The target for each laser pulse may be any of the following: crude oil, heavy oil, diesel oil, lube, jet fuel, seawater, stone, sand. The former five substances may be foreground materials, and the latter three may be background materials. The spectrum is in the form of 64-channel spectra that covering the spectral range from 330 to 650 nm.

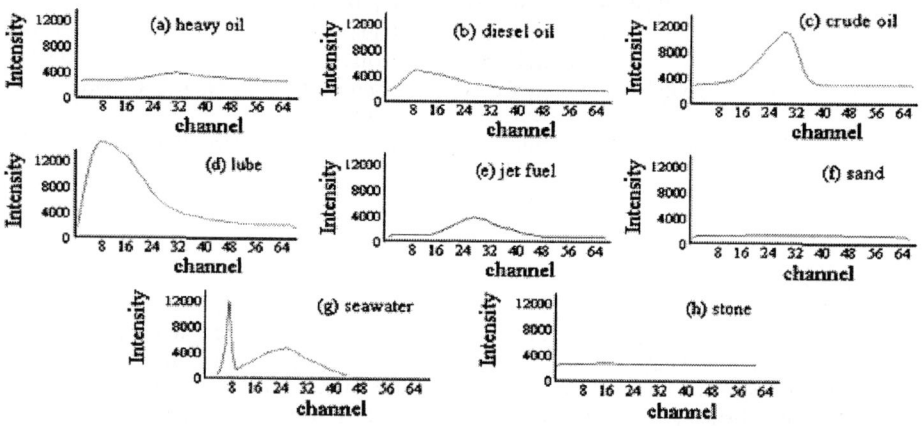

Fig. 1. Fingerprints spectra of eight pure substances by laser fluorescence RS

The achievement of real-time detection and identification of oils and related petroleum products in the marine environment by airborne laser fluorosensors, requires computationally efficient algorithms. Our goal is to identify the results of each laser shot as some combination for the oils or other substances. This requirement necessitates an efficient means of performing the classification.

1.2 Current Approaches to Oil-Spill Classification

Presently, the most common algorithm used to determine the presence of oil contamination with laser fluorosensors is based on the Pearson Correlation Coefficient (PCC). Standard reference fluorescence spectra (fingerprints) of the above "pure" substances are stored in the airborne system computer. Correlation coefficients are calculated for the "live" spectrum versus the above "fingerprints". When the value of the correlation versus the "fingerprints" is greater than a certain level and is greater than the correlation with the seawater spectrum, the "live" spectrum is identified as being of that substance.

Although PCC is useful to classify the oils, there are still some problems: there may be more than one value of the correlation coefficients versus the "fingerprints" which is above a given level and is greater than the correlation versus the seawater spectrum, and that some of the correlation coefficients for a target are very similar in magnitude. These problems lead to confusion about the nature of the substance producing the "live" spectrum and reduce the confidence in the identification. Some uncertainties, and unknown and complex factors between the "live" spectrum and the substances targeted by laser beam may be the reasons for concern.

Thus, the modeling of the relationships between the "live" spectrum and targeted substance by precise mathematical models becomes increasingly difficult.

Recently, Neural Network (NN) that serves as powerful computational frameworks has gained much popularity in fields of RS. This paper outlines our preliminary NN model to allow the oil-spill classification to be performed.

2 A Neural Network (NN) Identification Model

Figure 2 shows the procedure of the oil detection system. It mainly consists of two parts: (a) data acquisition and pre-processing; (b) classifier using neural network.

2.1 Data Acquisition and Preprocessing

In order to perform the NN training and simulating we require training spectra data together with testing spectra data. All the data are acquired from ground simulation experiment, provided by Emergencies Science Division, Environment Canada.

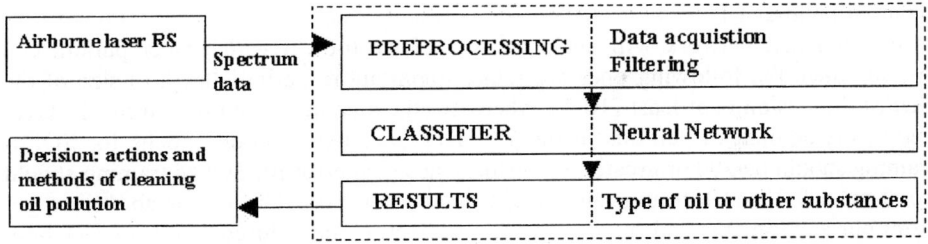

Fig. 2. The procedure of the oil detection system

The selection of spectra is an important job for the performance of network. We tried to filtrate certain amount of typical spectra of oils or other substances and thrown off the ones that have widely varying magnitudes. It will take the network very long time to train if there are many bad sample spectra in the training data set, and thus will result in bad classification outcome.

2.2 A Classifier Model

Self-organizing feature maps (SOFM) Network can be used in problems of pattern recognition and classification, like our application of oil-spill classification.

Suppose that we want the network to recognize the following prototype vectors: $\{ P_1, P_2, ... P_8 \}$. Table 1 shows the vectors and their corresponding target substances.

Table 1. The target substances and the corresponding prototype vectors

Target substance	Heavy oil	Diesel oil	Crude oil	Lube	Jet fuel	Seawater	Stone	Sand
Prototype vectors	P_1	P_2	P_3	P_4	P_5	P_6	P_7	P_8

The input data to the NN are in form of vectors or matrix. Since the target substances may be any of the above eight substances, we could define a matrix P (64×8) to be the desired prototype vectors:

$$P = [P_1, P_2, P_3, P_4, P_5, P_6, P_7, P_8]^T.$$

Actually each row vector ($P_1, P_2, ..., P_8$) is a spectrum data of a specific kind of pure substance. P can be trained by SOFM NN, and goal vectors is absent because SOFM NN apply the clustering function without tutor learning. The vectors M (64×m, m is the amount of testing spectra) to be classified are testing spectra data of sample pixels, acquired from readable binary files to input the network.

There are four steps in the training process: a) assemble the sample data set; b) create the network object; c) train the network; d) simulate the network response to new inputs. After initializing proper parameters, a two-dimensional self-organizing map will learn the topology of its inputs' space. In the progress of training and simulation, the negative distances between each neuron's weight vector and the input vector are calculated to get the weighted inputs. The weighted inputs are also the net inputs. The net inputs compete so that only the neuron with the most positive net input will output a 1.

We find that network will performs well if we train it with proper parameters. Among them the following ones are rather important to be initialized: a) size of the SOFM dimension is at least [10 3], where 10=the amount of neurons in the 1st layer and 3=the amount of neurons in the 2nd layer; b) maximal train epochs for SOFM training should be 200 or greater; c) performance goal error for SOFM training should be at most 0.01; d) learning rate should be at most 0.05. The values above are got from our experiments and the classification effect is quite closed to our expectation. So it is crucial not to set the parameters beyond the scope to get good outcome.

3 Results Analysis

The accuracy of the algorithm for identification is our main concern. Both the NN and PCC Model have been tested with the spectra of sampled pixels from ground tests with known targets. The comparison of test results is shown in Table 2.

Table 2 shows that the correct identification rate of NN (75%) is much higher than that of PCC model (45%). Although a test with sampled pixel, it still could indicate that NN model is an algorithm with higher accuracy. After analyzing we conclude that NN model only has difficulty in identifying stone and sand because the spectra of these two substances have too much similarity (Figure 1). However, PCC model has such confusing results as crude oil and jet oil, diesel and crude oil, which have more differentia in their spectra.

Table 2. Results of comparison of NN and PCC on the spectra of sampled pixels

Pixel NO.	Expected Result	Results of identification		Pixel NO.	Expected Result	Results of identification	
		PCC	NN			PCC	NN
14	Heavy oil	Heavy oil	Heavy oil	610	Jet oil	Crude oil or jet oil	Jet oil
18	Lube	Lube	Lube	612	Sand	Stone	Stone
36	Sand	Stone	Stone	683	Crude oil	Diesel or Crude oil	Crude oil
44	Crude oil	Crude oil	Crude oil	754	Sand	Stone	Sand
75	Diesel	Diesel or Crude oil	Diesel	815	Crude oil	Crude oil	Crude oil
96	Seawater	Seawater	Seawater	886	Seawater	Seawater	Seawater
167	Jet oil	Crude oil or jet oil	Jet oil	907	Heavy oil	Jet oil	Jet oil
468	Heavy oil	Jet oil	Heavy oil	918	Lube	Lube	Lube
469	Crude oil	Crude oil	Crude oil	1119	Crude oil	Crude oil	Crude oil

The laser sensor hardware together with software will be installed on the airplane flying on the sea to conduct the oil detection task. Based on the requirement of real-time identification oils in the marine environment, a big problem that must be considered is the computation time for conducting the identification process. Based on the experiences of the PCC Algorithm of the LEAF system developed by the Emergencies Science and Technology Division of Environment Canada, this NN algorithm is able to provide a real-time identification of a 64-band spectrum of 8 oils at a sampling rate of 100Hz using a Pentium IV computer.

In conclusion, NN model is effective and advanced as classifier for littoral oil-spill in that NN algorithm can extract the internal features of parameter by self-organizing and thus its accuracy is much higher than PCC.

The results of this study demonstrate that NN is a very useful tool in spectra data analysis and identifying for detecting littoral oil-spills by airborne laser fluorosensor RS. Testing is still going on to assess and trying to improve the NN model.

4 Future Work

The preliminary results from our experimental NN give a positive indication, but an amount of research remains before it can function as an effective and timely oil-spill detection system. An effective NN-based approach to such pollutant detection must be highly adaptive, which means not only for oil detection but also for the detection of other marine pollution. It is believed that such problem could be solved using the same NN model mentioned above as long as we have the sample spectra data of these pollutants, like chlorophyll. Thus, another important factor affecting the detection system is to acquiring the sample spectra data as training data to NN.

In addition, calculating the concentration of pollutants is another significant problem. Laser RS may work with other RS methods, like thermal Infrared Radiation

(IR), visible camera, Ultraviolet (UV), and satellite radar, to get enough information so as to calculating the concentration. Neural networks are still important tool in fusing the detecting results of different RS sensors.

5 Conclusion

Neural networks provide a number of advantages in the detection of oil pollution. It is proved that the training has developed a network that not only fits the training data, but also fits real-world data that the network will process operationally. Also a software package for post-flight analysis for the airborne detection of oil spills by laser-induced fluorescence, which employs both the NN and PCC Model, has been completed. The early results of our tests of these technologies show significant promise, and our future work will involve the refinement of this approach and the development of a full-scale system. The NN model would play a significant role in the field of ocean pollutant detection in the future.

Acknowledgements. We want to thank Environmental Technology Center, Environment Canada, Ottawa, for providing the spectrum data by airborne laser fluorosensor remote sensing that our research needed.

References

1. Wang, N.L., Pan, X.G.: Application of Matlab/NNTool in Neural Network System. Computer Simulation 4 (2004) 125-128
2. Lin, B., An, J.B.: Study on Detection Method of Spilled Oil at Sea by ANN of Laser Remote Sensing. Marine Environmental Science 1 (2004) 47-49
3. Keiner, E.L.: Estimating Oceanic Chlorophyll Concentrations with Neural Networks. Remote Sensing 1 (1999) 189-194

Identifying Pronunciation-Translated Names from Chinese Texts Based on Support Vector Machines

Lishuang Li, Chunrong Chen, Degen Huang, and Yuansheng Yang

Department of Computer Science and Engineering, Dalian University of Technology,
116024 Dalian, China
lils@dlut.edu.cn

Abstract. This paper presents a method of automatic recognition of pronunciation-translated names (P-Names) based on support vector machines (SVMs): extracting the character itself, character-based part-of-speech (POS) tag, frequency information of a character in P-Name table and context information as the attributes of feature vectors, a training set is established. The machine learning models of automatic identification of P-Names based on support vector machines are obtained using polynomial kernel functions. The testing results show that this method is efficient for identifying pronunciation-translated names.

1 Introduction

Unlike English, written Chinese does not delimit words by spaces and there is no clue to tell where the word boundaries are. So it is necessary to segment Chinese texts before processing. Pronunciation-translated names (P-Names) are the Chinese characters translated from foreign names according to their pronunciation. The presence of P-Names brings more ambiguities to Chinese word segmentation since each character in a P-Name can be used as a common character. Therefore, recognition of P-Names helps to reduce ambiguities in word segmentation and improve the performance of Chinese information retrieval since many unknown words are P-Names, especially for international Chinese news.

One way to perform this task effectively is to rely on statistics derived from a large corpus in which the P-Names are annotated, unfortunately, annotated data is difficult to obtain.

In this paper, we propose a method based on support vector machines to identify P-Names. It needs less training samples, and has better generalization performance.

2 Support Vector Machines (SVMs)

Support vector machines first introduced by Vapnik [1] are a supervised machine learning algorithm for binary classification. SVMs have advantages over conventional statistical learning algorithm: SVMs have high generalization performance independent of dimension of feature vectors; SVMs can carry out their learning with all combinations of given features without increasing computational complexity by introducing the kernel function. SVMs have been applied to many pattern recognition problems. In the field of natural language processing, they are applied to text categorization [2], chunking identification [3], Chinese unknown word identification [4], etc.

2.1 Optimal Hyperplane

Given training examples

$$S = \{(x_1, y_1), (x_2, y_2), ..., (x_\ell, y_\ell)\}, x_i \in R^n, y_i \in \{-1, +1\} \quad (1)$$

x_i is a feature vector (n dimension) of the i-th sample. y_i is the class (positive(+1) or negative(-1) class) label of the i-th sample. ℓ is the number of the given training samples. SVMs find an "optimal" hyperplane: $(w \cdot x + b) = 0$ to separate the training data into two classes. The optimal hyperplane can be found by solving the following quadratic programming problem (we leave the details to Vapnik [5]):

$$\max \quad \sum_{i=1}^{\ell} \alpha_i - \frac{1}{2} \sum_{i,j=1}^{\ell} \alpha_i y_i \alpha_j y_j K(x_i \cdot x_j) \quad (2)$$

$$\text{subject to} \quad \sum_{i=1}^{\ell} y_i \alpha_i = 0, \quad 0 \leq \alpha_i \leq c, i = 1, 2, ..., \ell$$

The function $K(x_i, x_j) = \phi(x_i) \cdot \phi(x_j)$ is called kernel function. Given a test example, its label y is decided by the following function:

$$f(x) = \text{sgn}[\sum_{x_i \in SV} \alpha_i y_i K(x_i \cdot x) + b] \quad (3)$$

2.2 Multi-class Classifiers

Basically, SVMs are binary classifiers, thus we must extend SVMs to multi-class classifiers in order to solve multi-class discrimination problems. There are two popular methods to extend a binary classification task to that of K classes: *one class vs. all others* and *pairwise*. Here, we employ the simple *pairwise* method. This idea is to build $K \times (K-1)/2$ classifiers considering all pairs of classes, and final decision is given by their weighted voting.

3 Identifying P-Names from Chinese Texts Using SVMs

Firstly, we segment and assign part-of-speech (POS) tags to words in the texts using a Chinese lexical analyzer. Secondly, we break segmented words into characters and assign each character its features. Lastly, a machine learning model SVMs-based to identify the P-Names is set up by choosing proper kernel functions.

3.1 P-Name Chunk Tags

We use the *Inside/Outside* representation for proper chunks:
I Current token is inside of a chunk.
O Current token is outside of any chunk.
B Current token is the beginning of a chunk.

A chunk is considered as a P-Name word in this case. Every character in the training set is given a tag classification of B, I or O, that is, $y_i \in \{B, I, O\}$.

3.2 Feature Extraction for P-Names

Since P-Names are identified from the segmented texts, the mistakes of word segmentation can result in error identification of P-Names. So we must break words into characters and extract features for every character. Table 1 summarizes types of features and their values.

Table 1. Summary of features and their values

Type of feature	Value
POS	n-B, v-I, p-S
Frequency information of a character in P-Name table	Y or N
Character	surface form of the character itself
Previous BIO tag	B-character, I-character, O-character

The POS tag from the output of lexical analysis is subcategorized to include the position of the character in the word. The list of POS tags is shown in Table 2.

Table 2. POS tags in a word

POS tag	Description of the position of the character in a word
<POS>-S	one-character word
<POS>-B	first character in a multi-character word
<POS>-I	intermediate character in a multi-character word
<POS>-E	last character in a multi-character word

If the frequency of a character in P-Name table is larger than a threshold, the value is assigned to Y, otherwise assigned to N.

The "character" is surface form of the character in the word.

We also use previous BIO-tags as features.

Whether a character is inside a P-Name or not, it depends on the context of the character. Therefore, we use contextual information of two previous and two successive characters of the current character as features.

3.3 Choosing Kernel Functions

Here, we choose polynomial kernel functions: $K(x, x_i) = [(x \cdot x_i) + 1]^d$ to build an optimal separating hyperplane by testing d=1, 2, 3 respectively.

Table 3. Results for person names extraction

	Recall	Precision	F-measure
d=1	88.37	82.87	85.53
d=2	91.99	86.20	89.01
d=3	92.72	85.82	89.14

Table 4. Results for place names extraction

	Recall	Precision	F-measure
d=1	87.89	77.42	82.32
d=2	90.96	79.90	85.07
d=3	93.14	77.17	84.40

4 Experiments and Results

We used more than one-month news of year 1998 from the People's Daily as the training and testing corpus (divided into 2 parts randomly with a size ratio for training/testing of 3/1) to conduct an open test experiment. It contains about 8755 P-Names (5591 place names and 3164 person names). The person names and place names of P-Names are trained and tested respectively. The results of our experiments in recall, precision and F-measure are given in Table 3 and Table 4.

5 Conclusions and Future Work

We identified P-Names from Chinese texts using support vector machines: the features of character-based POS, the frequency information of a character in P-Name table, character itself and contextual information were extracted. The results show that SVMs can avoid overfitting in the high dimension feature space and can achieve better F-measure with small training sets.

To increase the precision, we can add more features related to P-Names, such as character types, the information of a character inside P-Names used as a common character. We also can experiment with different context window size, different chunking representation, and combine with rule-based model.

In the field of Chinese information processing, neural network is another effective method and has been applied to word segmentation, POS tagging, shallow parsing [6], etc, it is probably an efficient approach for recognition of pronunciation-translated names, but to our knowledge, no related research about P-Names is reported up to now. Application neural network to recognition of P-Names and comparison with SVMs are on the way.

References

1. Vapnik, V.N.: The Nature of Statistical Learning Theory. Springer-Verlag, Berlin (1995)
2. Joachims, T.: Text Categorization with Support Vector Machines: Learning with Many Relevant Features. In: Nedellec, C., Rouveirol, C. (eds.): Machine Learning. Lecture Notes in Computer Science, Vol. 1398. Springer-Verlag, Heidelberg (1998) 137–142
3. Li, H., Zhu, J., Yao, T.:SVM Based Chinese Text Chunking. Journal of Chinese Information Processing. 2 (2004) 1-7
4. Goh, C.L., Asahara, M., Matsumoto, Y.: Chinese Unknown Word Identification Using Character-Based Tagging and Chunking. In: ACL 2003: 41st Annual Meeting of the Association for Computational Linguistics, Interactive Poster/Demo Sessions, Companion Volume of the Proceesings (2003) 197-200
5. Vapnik, V.N.: Statistical Learning Theory. John Wiley & Sons, New York (1998)
6. Xi, C., Sun, M.: Automatic Prediction of Chinese Phrase Boundary Location with Neural Networks. Journal of Chinese Information Processing. 2 (2002) 20-26

FPGA Implementation of Feature Extraction and Neural Network Classifier for Handwritten Digit Recognition*

Dongsheng Shen, Lianwen Jin, and Xiaobin Ma

School of electronics and information, South China Univ. of Tech. Guangzhou 510640
{dsshen,eelwjin}@scut.edu.cn

Abstract. FPGA (Field Programmable Gate Arrays) implementation of an off-line handwritten digit recognition system based on elastic meshing directional feature extraction and integrated neural network classifier is proposed in this paper. Elastic meshing directional feature extraction is used to extract the feature of normalized 32*16 handwritten digit images. Integrated neural network classifier with BP (back-propagation) learning algorithm is designed as classifier. The pipeline technology and multi-buffer technology are used in the FPGA implementation of elastic meshing directional feature extraction. FPGA implementation architecture of neural network computing unit and integrated neural network classifier is proposed in this paper. Experiment shows that compared with software-based implementation, FPGA-based system can greatly speed up off-line handwritten digit recognition and is suitable for application in some real-time situations where high process speed and portability are required.

1 Introduction

Generally, pattern recognition system is implemented using software technology. However, the speed of software-based implementation is low, and software-based implementation relies on computer and is not suitable for using in some environments where high portability is needed. FPGA-based implementation is a good solution to the problem. FPGA uses basic hardware units such as gate; register to implement algorithm and FPGA has parallel execution ability. Therefore, FPGA can greatly speed up some algorithms, especially parallel algorithm. In addition, FPGA-based implementation is suitable for using in the environments where portability is needed. In recent years, FPGA is used to implement many real-time pattern recognition systems [1][2][3]. This paper proposes an approach to implement an off-line handwritten digit recognition system in FPGA. The system is based on elastic meshing directional feature extraction [4] and integrated BP neural network classifier [5]. Experiment shows that compared with conventional software-based implementation, FPGA-based implementation can greatly speed up off-line handwritten digit recognition system and is suitable for real-time application.

---

*   Project sponsored by:NSFC(No.60275005), GDNSF(No.2003C50101, 011611).

2 Introduction of an Offline Handwritten Digit Recognition System

2.1 Elastic Meshing Directional Feature Extraction [4]

Figure 1 is the block diagram of elastic meshing directional feature extraction [4].

During the process of extracting elastic meshing directional feature, four edge detection templates (shown in Figure 2) are used to decompose the whole binary character image into four directional sub-images, namely, horizontal, vertical, right up diagonal and left up diagonal sub-images.

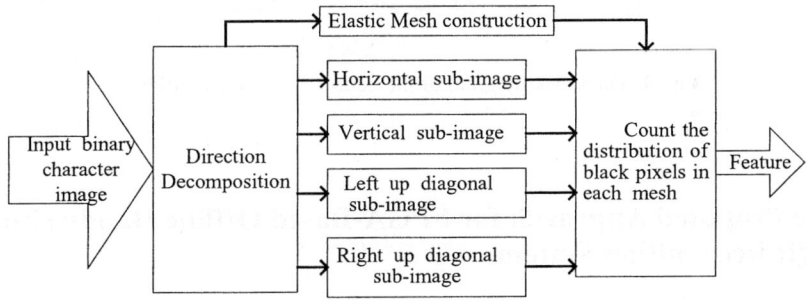

Fig. 1. The block diagram of elastic meshing directional feature extraction

-1	0	1
-1	0	1
-1	0	1

-1	-1	-1
0	0	0
1	1	1

0	1	1
-1	0	1
-1	-1	0

-1	-1	0
-1	0	1
0	1	1

Fig. 2. Four edge detection templates used to decompose binary image

After a character is decomposed into four directional sub-images, elastic mesh described above are applied into four sub-images respectively, and the distribution of black pixels in each mesh is calculated as the feature. Detail of the feature extraction can be found in the [4].

2.2 Integrated BP Neural Network Classifier [5]

We use four three-layer BP neural networks to recognize the components of horizontal feature, vertical feature, right up diagonal feature, left up diagonal feature respectively. Outputs of four networks input to a BP neural network and final recognition result is got. Figure 3 is the block diagram of the BP integrated neural network classifier [5].

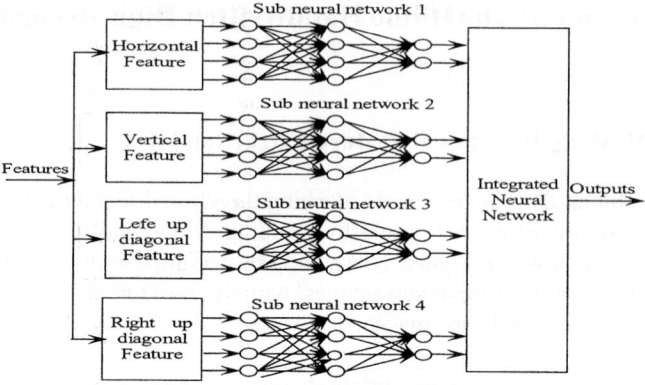

Fig. 3. The block diagram of the neural network classifier

3 The Proposed Approach for FPGA-Based Offline Handwritten Digit Recognition System

3.1 FPGA-Based Elastic Meshing Directional Feature Extraction

The block diagram for FPGA-based elastic meshing directional feature extraction is showed in figure 4.

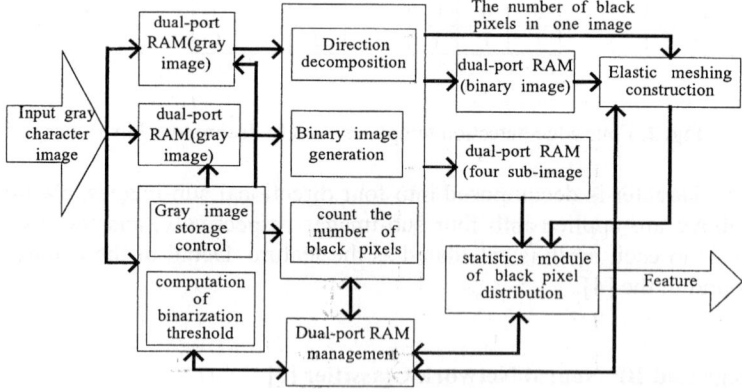

Fig. 4. The block diagram of FPGA implementation of elastic meshing directional feature extraction

The whole system consists of gray image storage module, binarization module, direction decomposition module, elastic mesh construction module, statistics module of black pixel distribution, dual-port RAM state management module and some dual-port RAM. Detail of the whole system is described as follows.

The pipeline technology is adopted for the FPGA-based implementation of feature extraction. The whole process consists of four sub-processes: (1) Gray image storage control (compute binary threshold simultaneously); (2) Direction decomposition (generate binary image from gray image and count the number of black pixels in one image simultaneously); (3) Elastic mesh construction; (4) Statistics of black pixel distribution in each mesh. Each process is implemented using an independent function module. Accordingly, an input gray image goes through four sub-processes in order. Feature is finally outputted. Because each sub-process runs simultaneously, data buffers are adopted by each function module, otherwise the function module may overlap results while going on transaction on next image. Multi-buffer technology is adopted in this paper. Data in one buffer is derived the same input gray image. Data in different buffers is derived form different input gray images. One function module generates the data, and after a time interval, the data is completely processed by the successive function module. The time interval varies small. The number of buffers needed between function modules is determined by how much new data generated during the time interval. There are two kinds of buffers, dual-port RAM and registers. Dual-port RAM is used to store data, which occupy large memory space. Registers are used to store some values such as the number of black pixels in one image, binary threshold and elastic mesh

In order to make use of the space of dual-port RAM well, each dual-port RAM is regarded as a cyclic queue. Buffer in dual-port RAM is equal to node in cyclic queue. Two pointers "front" and "rear" are used to indicate the position of the queue's hand and tail respectively. Their initial value is set to zero. In a cyclic queue, "front" pointer and "rear" pointer point at the same position when the cyclic queue is empty or full. Therefore, a Boolean variable is set to discriminate between the cyclic queue's empty state and full state. It is stored in the "dual-port RAM state management" module.

As the input is gray scale image, a pre-processing module is used to convert the gray image into binary image. A binary algorithm based on arithmetic mean of gray value [6] is used. It can achieve good threshold effect and is easy to be implemented in FPGA.

Several shift registers are used in direction decomposition module to decompose the binary image into four sub-images. The size of input image is 32*16 lattice, namely one row contain 16 pixels. The size of edge detection templates is 3*3 lattice. Therefore, the length of shift registers is 16*3=48 bits. The area that four edge detection templates applied consists of the registers' bits 0, 1,2, 16,17,18,32,33 and 34. The pixel in the register bit 0 is the pixel under being decomposed. The pipeline technique is adopted in the direction decomposition module. The direction decomposition process consists of four processes, which are given below:

(1) Take the new pixel from dual-port RAM that store gray image according to order of binary image's row and convert the new pixel into binary pixel; (2) The bits in the registers shift left one bit and save the incoming bit in the registers bit 0; (3) Template operation; (4) Save result in dual-port RAM that stores four sub-images.

For the pixels of the first row and the second row, the edge detection templates cannot be applied. Thus, they do not belong to any sub-images. However, the binary values of these pixels should be putted into the shift registers for the successful process of the following pixels.

We use 4*4 elastic mesh in elastic mesh construction module. The number of black pixel in the binary image is first computed. Then the number is used to divide the image into four parts in horizontal direction. Moreover, the number is used to divide the image into four parts in vertical direction.

Four independent function modules are used in statistics module of black pixel distribution to process four sub-images respectively. They run simultaneously to count the distribution of black pixels in each mesh. The result is output as the feature.

The reasonable arrangement of data storage structure can enhance data exchange speed between function modules. In the dual-port RAM that stores binary image, one word has eight bits, corresponding to eight pixels of binary image. Then, elastic mesh construction module can read eight pixels' binary value once. In the dual-port RAM that stores four sub-images, one word has four bits. Each bit corresponds to one bit of four sub-images. Therefore, statistics module of black pixel distribution in each mesh can read data necessary once, and then processes four sub-images simultaneously.

3.2 Implementation of a Three-Layer BP Neural Network Using Mixed Parallel Computing Unit

A mixed parallel computing unit [7] is proposed in Figure 5 to implement the basic three-layer BP neural network. All links between the input layer and middle layer of neural network is processed in parallel (link parallelism); at the same time, neurons of output layer is also processed in parallel (neuron parallelism).

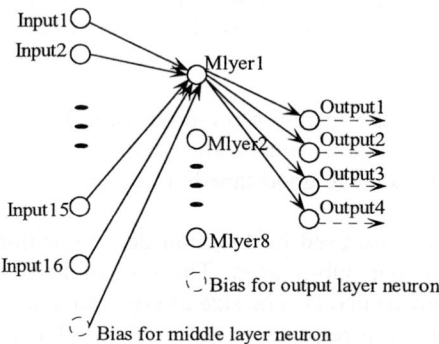

Fig. 5. Mixed parallel computing unit for three-layer BP neural network

There are two types of computing model: one for link parallelism and another for neuron parallelism. The neuron can get bias by a predetermined look up table (LUT).

Computing model between input layer and middle layer is showed in Figure 6.

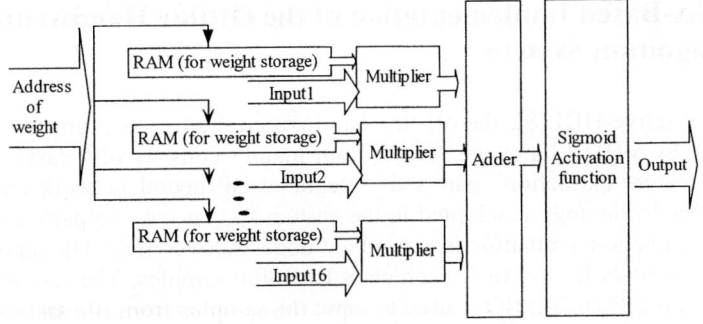

Fig. 6. Computing model between input layer and middle layer

RAM is used for link weights. Adder has multiple-input and single-output in middle layer for all links. Besides, a LUT based sigmoid function is attached for function mapping.

The computing model for output layer neuron is shown in Figure 7. Figure 7(a) is a diagram for four parallel output neurons, and Figure 7(b) is the details for each neuron. Figure 7(a) indicates that the computing results of middle layer are transferred to next layer by broadcasting. Therefore, all conditions for output neurons computing can be met at the same time.

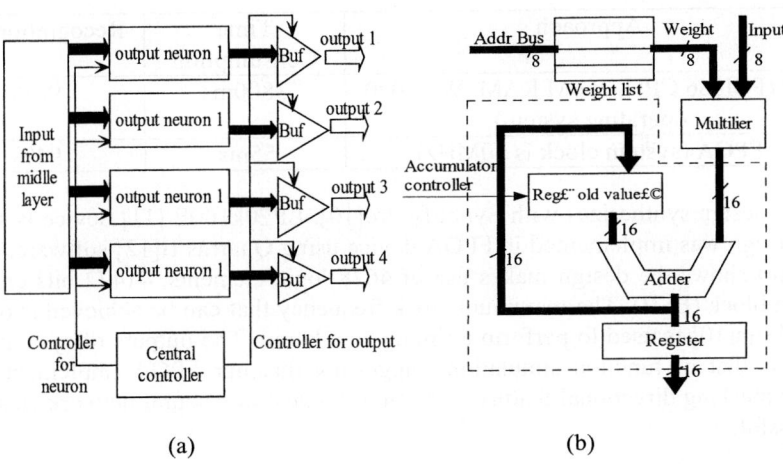

Fig. 7. Structure of computing model for output neuron

Integrated BP neural network classifier is implemented using one mixed parallel computing unit described above in time division multiplex approach.

4 FPGA-Based Implementation of the Offline Handwritten Digit Recognition System

Under the Active-HDL[8], the off-line handwritten digit recognition system was implemented by VHDL language. The system mainly consists of "elastic meshing directional feature extraction" part and "integrated BP neural network classifier" part. The pipeline technology is adopted in the system so that the two parts can run simultaneously. Function simulation is carried out under Active-HDL. The inputs of system are Arabic digitals from 0 to 9, each class have 400 samples. The size of each inputting sample is 32*16. TextIO is used to input the samples from file and save output to file. Moreover, we use a PC (PIII866 cpu,256M RAM,Win 2000 OS) to do the same simulation . Experiment results are given in table 1.

From table 1, it can be seen that the speed of FPGA-based implementation is about 70 times faster than that of PC-based implementation at least. Weights parameters of neural network are expressed using floating-point number in software-based implementation. However, in order to economize hardware resources, weights of neural network are expressed using fixed-point number in FPGA-based implementation. Therefore, recognition rate in FPGA-based implementation is a little lower than that of software-based implementation. More study on the selection of weight accuracies can be found in paper [9].

Table 1. Experiments result of function simulation of FPGA system and PC-based system

Approach	Time Consumption	Recognition rate
PC (PIII866 CPU, 256M RAM, Win2000 operating system)	3800ms	96%
FPGA (system clock is 50MHz)	55ms	94%

The design synthesizes with Synplify Pro[10]. EP20k160E [11] device is selected. The design was implemented in FPGA device using Quartus II[12] software. Quartus II report shows the design makes use of 4628 logic elements, 41472 bits embedded system block (ESB). The maximum clock frequency that can be achieved is 65 MHz. ModelSim[10] is used to perform a timing simulation. The outputs of system are the same as that of function simulation, suggesting that our FPGA implementation of elastic meshing directional feature extraction & integrated neural network classifier is successful.

5 Conclusions

Compared with software-based implementation, FPGA-based implementation can greatly speed up off-line handwritten digit recognition and is suitable for using in some real time application where high process speed and portability are required.

References

1. Athanas, P., Abbott, L., Cherbaka, M., Pudipeddi, B., Paar, K.: A Custom Computing Solution to Automated Visual Inspection of Silicon Wafers. Southeastcon '97, 'Engineering new New Century', Proceedings, IEEE (1997) 315–319
2. Jean, J., Liang, X., Drozd, B., Tomko, K.: Accelerating an IR Automatic Target Recognition Application with FPGAs. Field-Programmable Custom Computing Machines, FCCM '99 Proceedings, Seventh Annual IEEE Symposium on (1999) 290 –291
3. Miteran, J., Bailly, R., Gorria, P.: Real Time Image Segmentation Using FPGA and Parallel Processor. TENCON '96, Proceedings, 1996 IEEE TENCON. Digital Signal Processing Applications, Vol. 1 (1996) 233-236
4. Jin, L., Xu, B.: Directional Cellular Feature Extraction with Elastic Meshing for Handwritten Chinese Character Recognition. Journal of Circuits and Systems, Vol. 2 (1997) 7-12
5. Luo, X., Dong, S., Jin, L., Xu, H.: An Integrated Neural Network Based Classifier for Handwritten Digit Recognition. Computer Engineering, Vol. 28 No.8 (2002) 69-71
6. Gao, Y., Zhang, L., Wu, G.: An Algorithm for Threshold Based on Arithmetic Mean of Gray. Journal of Image and Graphics Value, Vol. 4 (1999) 524-528
7. Ma, X., Jin, L., Shen, D., Yin, J.: A Mixed Parallel Neural Network Computing Unit Implementation in FPGA. IEEE Int. Conf. Neural Networks & Signal Processing, Nanjing, China (2003)
8. Wang, Y., Zhang, Z.: VHDL Programming and Simulation. Posts & Telecom Press (2000)
9. Piche, S.W.: The Selection of Weight Accuracies for Madalines Neural Networks. IEEE Transactions on ,Vol. 6, Issue 2 (1995) 432- 445
10. Wang, C., Xue, X., Zhong, X.: The Way of Using FPGA/CPLD Design Tool Xilinx ISE 5.X. Posts & Telecom Press (2003) 85-160
11. Wang, J., Yang, J.: Digial System Design and Verilog HDL. 1st edition. Publishing House of Electronics Industry, Beijing (2002) 51-62
12. Chen, X., Teng, L.: Starting VHDL and Its Application. Posts & Telecom Press (2000) 165-192

Replicator Neural Networks for Outlier Modeling in Segmental Speech Recognition

László Tóth and Gábor Gosztolya

Research Group on Artificial Intelligence
H-6720 Szeged, Aradi vértanúk tere 1., Hungary
{tothl, ggabor}@inf.u-szeged.hu

Abstract. This paper deals with outlier modeling within a very special framework: a segment-based speech recognizer. The recognizer is built on a neural net that, besides classifying speech segments, has to identify outliers as well. One possibility is to artificially generate outlier samples, but this is tedious, error-prone and significantly increases the training time. This study examines the alternative of applying a replicator neural net for this task, originally proposed for outlier modeling in data mining. Our findings show that with a replicator net the recognizer is capable of a very similar performance, but this time without the need for a large amount of outlier data.

1 Introduction – Neural Nets in Speech Recognition

Speech recognition does not naturally fit into the usual pattern classification scheme where the items to be classified are represented by a fixed number of features. Rather, speech is a continuous stream of information where the possible utterances vary in length and their number is practically unlimited. A possible solution is to trace the problem back to the recognition of some properly chosen (fixed-size) building blocks. During recognition these building blocks have to be found, identified, and the information they provide needs to be combined. The most successful solution, Hidden Markov Modeling (HMM) [6], does exactly this. The traditional HMM methodology seeks to model the distribution of the building units by means of Gaussian mixtures. Applying neural nets for this task instead became very popular in the mid-nineties after it became widely known that Artificial Neural Nets (ANN) approximate the class posteriors. It was claimed that ANNs are more flexible, attain a better performance due to their discriminative nature, and require an order of magnitude fewer parameters. Since then ANNs have become a widely accepted alternative to Gaussian-based modeling in the speech community under the name "HMM/ANN hybrids" [1].

In addition to this, many authors have criticized two other important aspects of hidden Markov modeling, namely the choice of units and the way their probability scores are aggregated. Traditionally, the building units are small uniform ($\sim$30ms) signal chunks, and the probabilities assigned to them are combined by multiplication. Both of these simplifications are unrealistic from a perceptual point of view. Several alternatives have been proposed, and one of them is the so-called segmental modeling approach where the building units are longer, variable-length signal intervals [8]. This technology offers

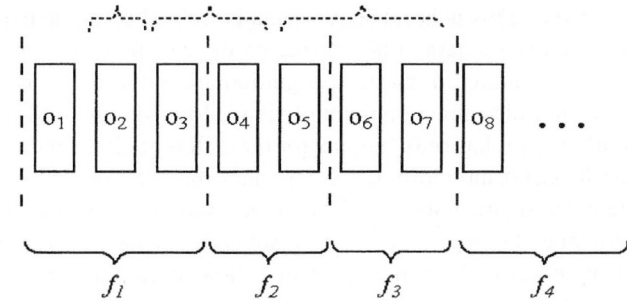

Fig. 1. An illustration of the connection between observation vectors and phonetic segments in speech recognition. The dotted brackets on top denote examples of anti-phone segments

better recognition results, but also introduces special problems. One of these, the problem of outlier segments, is in the focus of this paper. Section 2 really describes the problem in a nutshell, and Section 3 presents one possible solution, replicator neural networks (RNN). In Section 4 we discuss the experiments we preformed to justify the use of RNNs, then we round off with some remarks in Section 5.

2 Segmental Modeling and the Problem of Outliers

The preprocessed speech signal arrives at the input of a speech recognizer in the form of a vector series $o_1, ..., o_T$ (see Fig. 1.). The goal of recognition is both to find the correct phonetic segmentation of the signal (denoted by dashed lines in Fig. 1) and to correctly identify the segments in the form of phonetic labels $f_1, ..., f_N$. Let us assume for the moment that the positions of the segment boundaries have been determined, so the task is only to identify the segments. In the standard frame-based approach the ANN is responsible for supplying label-probabilities to each vector (data frame). The probability associated with the whole segment is then obtained by multiplying the frame-based values. In segmental models the features vectors of a segment are first transformed into a (fixed-dimensional) segmental feature set, and this forms the input data for the ANN. Here, the neural net is responsible for identifying the *whole segment in one*, not merely its individual feature vectors. It was reported by several authors that in segment classification this approach beats the standard frame-based technique even with a surprisingly simple segmental feature set [2][3][7][9].

Thus far, however, we assumed the segment boundaries were given. In real life this this is not so, of course. The solution is to evaluate every reasonable set of segment boundaries, thus incorporating the task of segmental classification in a search process. During this search the recognizer will be confronted with vector subseries that do not correspond to real phonetic segments, that is, they overlap real phonetic boundaries. The frame-based system will automatically handle these "anti-phones": if the frame-based probability of a class label f is high on some part of the segment, but is low on another one, then after multiplication the whole segment will get a small probability of being the phone f. The ANN of the segmental system, however, cannot automatically cope with segments like these. This is because it is usually trained on a manually segmented database, which naturally contains only examples of real phonetic segments. So the anti-phone segments encountered during recognition are not seen during training, and behave as outliers from a classification point of view. Moreover, the ANN has no way of reporting these outliers as it has only outputs for the different class labels. One possible solution might be to insert an additional component into the model that assesses the correctness of the given segmentation [10]. Another option is to introduce an additional class into the segmental classifier that will correspond to the outlier segments [3][9]. In the latter case we are confronted by the problem of how to train the outlier class. In our earlier work we generated outlier examples by taking quasi-random segments from the training corpus so that they overlap real segments or are incorporated within them (e.g. the intervals bounded by the dotted brackets in Fig. 1) [9]. On average, six such anti-phone samples were created per phone, thus seriously increasing the amount of training data needed. In practice, however, we found that these examples are not representative enough in the sense that the recognizer still behaves unexpectedly in many cases (i.e. it accepts obvious outliers as phones). It would have been possible to generate even more outlier examples, but we preferred to avoid this option for several reasons. These are the following:

- Apart from obvious cases (e.g. segments that heavily overlap a real boundary) it is not a trivial matter to see how the anti-phone segments should be generated. It might be, for example, that the segments generated according to Fig. 1 could still sound like one phone so, perceptually, they are not really anti-phones. In addition, the manual segmentation of the training corpus may also contain mistakenly positioned boundaries.
- Generating even more anti-phones per segment would cause the training data to be overwhelmed by one class which, as we observed, has a detrimental effect on the learning process.
- One characteristic of speech recognition is that the training databases are enormous. Even the training corpora that are considered small contain hundreds of thousands of phone instances. Creating dozens of outlier examples for each of these really did not sound appealing, especially regarding the training time.

This is why we were looking for a method that allows 1-class learning, that is learning from positive examples (in our case phonetic segments) only. Unfortunately, standard perceptron-based neural nets are not suitable for this task mainly because their responses are not localized. A network with radial basis functions (RBFN) would have been a possible choice, but we did not want to give up our well-tried multilayer perceptron network. Instead, we were looking for some simple extension to our current system. This is where replicator neural networks come in the picture.

3 Replicator Neural Networks

The basic idea behind a Replicator Neural Net (RNN) [4][5] is simple enough: the input data is also used as the desired output data. Consequently, by minimizing the mean square error during training we force the net to reconstruct its training patterns with the smallest

5 Conclusions

This paper investigated the feasibility of using a replicator neural network to assess the outlyingness of hypothesized segments in a segmental speech recognizer. This was motivated by the hope that, by doing this, a relatively simple and efficient model could replace the tedious process of generating and training outlier samples for a traditional MLP. The experiments justified our belief that RNNs indeed have the potential for this task as they yielded a performance similar to our usual methodology. Now, further studies are required to understand under what conditions they may behave worse (as in the second feature set) than our standard system, and whether they can be made to outperform it. Hence we plan to conduct more experiments in the future to precisely identify what these factors are.

References

1. Bourlard, H. A., Morgan, N.: Connectionist Speech Recognition – A Hybrid Approach. Kluwer Academic (1994)
2. Clarkson, P., Moreno, P. J.: On the Use of Support Vector Machines for Phonetic Classification. Proceedings of ICASSP'99 (1999) 585-588
3. Glass, J. R.: A Probabilistic Framework for Feature-Based Speech Recognition. Proceedings of ICSLP'96 (1996) 2277-2280
4. Hawkins, S., He, H. X., Williams, G. J., Baxter,R. A.: Outlier Detection Using Replicator Neural Networks. Proc. DaWak'02 (2002)
5. Hecht-Nielsen, R.: Replicator Neural Networks for Universal Optimal Source Coding. Science, Vol. 269. (1995) 1860-1863
6. Huang, X. D., Acero, A., Hon, H-W.: Spoken Language Processing. Prentice Hall (2001)
7. Kocsor, A., Tóth, L., Kuba Jr., A., Kovács, K., Jelasity, M., Gyimóthy, T., Csirik, J.: A Comparative Study of Several Feature Space Transformation and Learning Methods for Phoneme Classification. International Journal of Speech Technology, Vol. 3, Number 3/4 (2000) 263-276
8. Ostendorf, M., Digalakis, V., Kimball, O. A.: From HMMs to Segment Models: A Unified View of Stochastic Modeling for Speech Recognition. IEEE Trans. ASSP, Vol. 4. (1996) 360-378
9. Tóth, L., Kocsor, A., Kovács, K.: A Discriminative Segmental Speech Model and its Application to Hungarian Number Recognition. Proc. TSD'2000 (2000) 307-313
10. Verhasselt, J., Illina, I., Martens, J. P., Gong, Y., Haton, J. P.: Assessing the Importance of the Segmentation Probability in Segment-Based Speech Recognition. Speech Communication, Vol. 24, No. 1 (1998) 51-72

A Novel Approach to Stellar Recognition by Combining EKF and RBF Net

Ling Bai[1,2] and Ping Guo[1]

[1] Department of Computer Science
Beijing Normal University, Beijing, 100875, P.R. China
[2] National Laboratory of Pattern Recognition
Institute of Automation Chinese Academy of Sciences, Beijing, 100080, P.R. China
mandybailing@163.com; pguo@ieee.org

Abstract. A new approach to stellar recognition is proposed. It uses extended Kalman filter as a feature selector and pre-classifier for stellar spectra data, while radial basis function neural networks is adopted for classification. Experiments with real-world data set show that the performance of the proposed technique is quite well, and the correct classification rate can reach as high as 93%. It is shown that the result using EKF is better than that using principle component analysis data dimension reduction technique.

1 Introduction

With the huge sky survey telescope being built, more than 1×10^7 spectra of faint celestial objects will be collected. The scientific exploitation of these will require powerful, robust, and automated classification tools tailored to the specific survey.

Stellar spectra are extremely noisy and voluminous. Consequently, the acceptable method of classification must be both computationally efficient and robust to noise. As we know, neural networks have already had many applications on astronomical data classification [1]. But the correct classification rate (CCR) is very low if the spectra data has not been pre-processed. The most traditional feature extraction method is to use principle component analysis (PCA) [3] data dimension reduction technique. PCA is usually used directly for ill-conditioned data by extracting the latent variables. The number of latent variables is lower than the number of samples [4]. However, as we know, PCA is only efficient when the raw data can be separated linearly. Therefore, we should develop new methods to deal with the pre-process problem.

Extended Kalman filter (EKF) addresses the general problem of trying to estimate the state $\mathbf{x} \in \Re^n$ of a discrete-time controlled process that is governed by a linear stochastic difference equation [5]. For many systems, EKF has proven to be a useful method of obtaining good estimates of the system state. In this paper, EKF is applied as a pre-processor in the stellar recognition system.

2 Background

2.1 Extended Kalman Filter

Here, we briefly review the essential of EKF [5]. The process is governed by the non-linear stochastic difference equation:

$$\mathbf{x}_k = f(\mathbf{x}_{k-1}, \mathbf{u}_k, \mathbf{w}_{k-1}) \qquad (1)$$

with a measurement $\mathbf{y} \in \Re^l$, that is

$$\mathbf{y}_k = h(\mathbf{x}_k, \mathbf{v}_k), \qquad (2)$$

where the random variables $\mathbf{w}_k$ and $\mathbf{v}_k$ represent the process and measurement noise.

To estimate a process with non-linear difference and measurement relationships, we begin by writing new governing equations that linearize an estimate about Eq. 1 and Eq. 2,

$$\mathbf{x}_k \approx \tilde{\mathbf{x}}_k + \mathbf{A}(\mathbf{x}_{k-1} - \hat{\mathbf{x}}_{k-1}) + \mathbf{W}\mathbf{w}_{k-1} \qquad (3)$$

$$\mathbf{y}_k \approx \tilde{\mathbf{y}}_k + \mathbf{H}(\mathbf{x}_k - \tilde{\mathbf{x}}_k) + \mathbf{V}\mathbf{v}_k \qquad (4)$$

where

- $\mathbf{x}_k$ and $\mathbf{y}_k$ are the actual state and measurement vectors.
- $\tilde{\mathbf{x}}_k$ and $\tilde{\mathbf{y}}_k$ are the approximate state and measurement vectors from Eq. 1 and Eq. 2.
- $\hat{\mathbf{x}}_k$ is a posteriori estimate of the state at step k.
- the random variables $\mathbf{w}_k$ and $\mathbf{v}_k$ represent the process and measurement noise.
- $\mathbf{A}$ is the Jacobian matrix of partial derivatives of f with respect to $\mathbf{x}$.
- $\mathbf{W}$ is the Jacobian matrix of partial derivatives of f with respect to $\mathbf{w}$.
- $\mathbf{H}$ is the Jacobian matrix of partial derivatives of h with respect to $\mathbf{x}$.
- $\mathbf{V}$ is the Jacobian matrix of partial derivatives of h with respect to $\mathbf{v}$.

The complete set of EKF equations is shown below in Table 1.

Table 1. EKF equations

EKF time update equations:	EKF measurement update equations:
$\hat{\mathbf{x}}_k^- = f(\hat{\mathbf{x}}_{k-1}, \mathbf{u}_k, 0)$	$\mathbf{K}_k = \mathbf{P}_k^- \mathbf{H}_k^T (\mathbf{H}_k \mathbf{P}_k^- \mathbf{H}_k^T + \mathbf{V}_k \mathbf{R}_k \mathbf{V}_k^T)^{-1}$
$\mathbf{P}_k^- = \mathbf{A}_k \mathbf{P}_{k-1} \mathbf{A}_k^T + \mathbf{W}_k \mathbf{Q}_{k-1} \mathbf{W}_k^T$	$\hat{\mathbf{x}}_k = \hat{\mathbf{x}}_k^- + \mathbf{K}_k(\mathbf{y}_k - h(\hat{\mathbf{x}}_k^-, 0))$
	$\mathbf{P}_k = (\mathbf{I} - \mathbf{K}_k \mathbf{H}_k) \mathbf{P}_k^-$

$\mathbf{A}_k$ and $\mathbf{W}_k$ are the process Jacobians at step k, and $\mathbf{Q}_k$ is the process noise covariance at step k. $\mathbf{H}_k$ and $\mathbf{V}_k$ are the measurement Jacobians at step k, and $\mathbf{R}_k$ is the measurement noise covariance at step k.

2.2 RBF Network

A RBF network consists of the l-dimension input $\mathbf{x}$ being passed directly to a hidden layer. Suppose there are c neurons in the hidden layer. Each of the c neurons in the hidden layer applies an activation function. The outputs of the network consist of sums of the weighted hidden layer neurons [6].

When input spectra are already projected to a feature space, a RBF network classifier is invoked for the final classification.

The k-th component of RBF neural network output looks like [7]:

$$z_k(\mathbf{x}) = \sum_{j=1}^{c} \mathbf{w}_{kj}\varphi_j(\|\mathbf{x} - \mu_j\|) + \mathbf{w}_{k0}$$

$$= \sum_{j=0}^{l} \mathbf{w}_{kj}\varphi_j(\|\mathbf{x} - \mu_j\|)$$

$$z_k(x) = \mathbf{W}_k \varphi(\mathbf{x}) \tag{5}$$

where $\mathbf{x}$ is a l-dimension input vector, $\mathbf{w}_{k0}$ is a set of bias constants, $\varphi_0(\|\mathbf{x} - \mu_j\|) \equiv 1$ and radial basis functions are chosen as of Gaussian type

$$\varphi_j(\|\mathbf{x} - \mu_j\|) = \exp[-\frac{1}{2\gamma_j^2}\|\mathbf{x} - \mu_j\|^2]. \tag{6}$$

The parameter μ_j is the center and γ_j is the standard deviation of the Gaussian function.

3 Classification Experiment and Discussion

There are seven main spectral types of stars in the order of decreasing temperature, namely: O – He II absorption; B – He I absorption; A – H absorption; F – Ca II absorption; G – strong metallic lines; K – bands developing; and M – very red. The stellar spectra used in our experiments are selected from Astronomical Data Center (ADC). Among them, 161 stellar spectra contributed by Jacoby *et al* (1984) are used in our work together with 96 observed by Pickles *et al* (1985). The spectra taken from the above libraries have resolutions of 0.14 and 0.5 nm, respectively. All spectra are digitized and linearly interpolated to the wavelength range of 360 $\sim$ 742 nm with a step of 0.5 nm.

In the experiments, the leave-one-out cross validation technique is used to select learning samples. $n_j = 16$ training samples are randomly drawn from each class, 15 for training and one for testing. Finally the average CCR of RBF classifier is reported.

First, EKF in Table 1 are used for de-noising and feature extraction. To recursive calculate parameters in EKF, from our experience, we should make some assumption as follows:

1. The process control parameter u_k is ignored in our experiment.
2. f in Eq. 1 is chosen as a linear function.
3. The variance of the process and measurement noise $\mathbf{w}_k$, $\mathbf{v}_k$ can be considered approximately as a very small value.
4. h in Eq. 2 is chosen as Gaussian function.
5. An identity matrix is chosen as the initial value of $\mathbf{P}$ matrix, and zero is chosen as the initial value of $\hat{\mathbf{x}}_k$.

After iteratively computation using equations in Table 1, we have both the estimate of input $\mathbf{x}$ and the estimate of matrix $\mathbf{U}$. Fig. 1 shows some de-noising results. From the figure we can see that the spectra lines become smooth, it shows that EKF can reduce noise by some content. The output of EKF can easily be

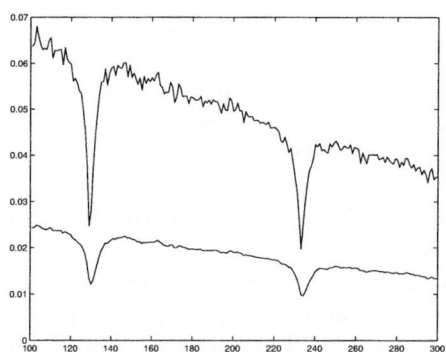

Fig. 1. The de-noising result. *Above line* is without, *bottom line* is with EKF

calculated via the estimate matrix $\mathbf{U}$. The three dimension distribution in feature space of EKF is shown in Fig. 2(a). From the figure we can see that, when the output of EKF is projected to $\mathbf{Y}$ space, some classes are well separated. They will be recognized quite exactly if using RBF network for final classification.

Then, the feature vectors are used as the input of a RBF network. In the experiments, we design the structure of RBF network with 35 hidden neurons. For the network output vector $\mathbf{z}$, we use one-of-k encoding method.

The structure of a RBF network is very important for raising CCR. With different numbers of RBF hidden neurons, the RBF network as a final classifier performs differently in the classification of samples. If a proper neuron number is selected, we can build a better classifier in recognition system. In this study, we also use 70-node structure RBF network to classify the stellar data. From experiments we find that the CCR is nearly the same as the 35-node structure RBF network. The RBF network performance is considerable stable when the number of node is over 35.

As a comparison, we also use 7-neuron structure RBF network to classify the data. The center μ_i is estimated by the mean value of 15 samples in each class, and γ_i is corresponding to deviation of μ_i. From the experiments we can see that

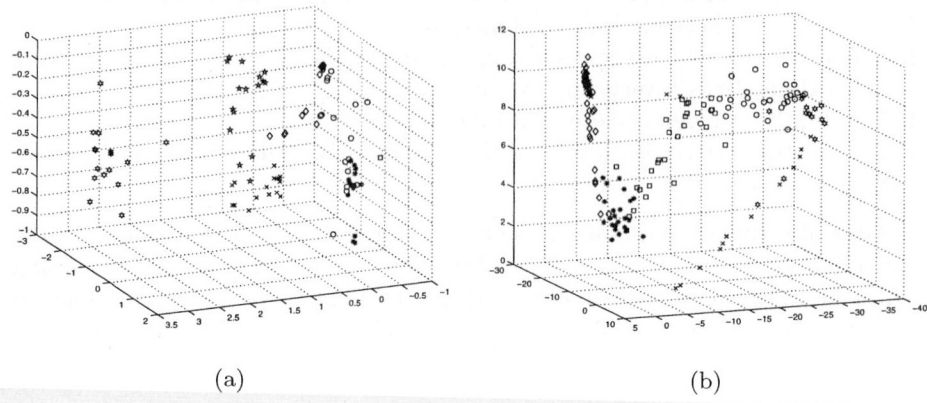

Fig. 2. (*a*) The 3-D distribution of feature data in **Y** space. (*b*) The 3-D view of stellar data in the first three principle component space

the results when we use 35-neuron are much better than the results when we use 7-neuron RBF network.

Similarly, 105-neuron structure RBF network is also adopted as a comparison. This is equivalent to exactly interpolation. In this case the classification accuracy is degraded. The result illustrates that hidden neurons not only transfer the characteristic of the feature vectors, but also transfer the noise of the feature vectors to output. Table 2 show the comparison of different structure RBF network.

PCA is a good tool for dimension reduction, data compression and feature extraction. PCA forms a set of linearly independent basis vectors with which to describe the data, and can be useful as a classification system by using only the most significant few components. In our study, we can also directly use PCA to extract principal component for such ill-conditioned problem [3].

The distribution of the raw stellar spectra for the first three principal components is shown in Fig. 2(b). Some classes are overlapped and can not be separated from each other linearly. Compared with Fig. 2(a), we can see that EKF as a data pre-process technique can separate classes quite well, if RBF net is adopted, high CCR can be obtained easily, but PCA is a little worse in this case. PCA is a linear transformation of the data, so any linear classification model which uses a fit to the first few principal components is likely to be too simplistic for

Table 2. Comparison of RBF net with different number of hidden neurons

Neurons	c=7	c=35	c=70	c=105
CCR	82%	93%	92%	78%

multiparameter classification. EKF, on the other hand, can be made arbitrarily nonlinear, and has a convenient means for investigating degrees of complexity and nonlinearity (through the use of different f and h function). Table 3 shows the comparison of CCR between PCA and EKF. It is obvious that the use of EKF as a pre-processor is much better than that the use of PCA.

Table 3. Comparison of CCR of EKF vs. PCA

Methods	EKF	PCA
CCR	93%	71%

4 Conclusions

A composite classifier which combines EKF and RBF neural network is applied to stellar spectral recognition in this paper. First, EKF is employed for denoising and pre-classifying, then RBF network is used for final classification. Experiments show that EKF is efficient for this non-linear separating problem. The proposed classifier gives excellent classification results, we believe that it is a very promising technique in spectra recognition.

Acknowledgments. The research work described in this paper was fully supported by a grant from the National Natural Science Foundation of China (Project No. 60275002) and by National High Technology Research and Development Program of China (863 Program, Project No. 2003AA133060).

References

1. Gupta, R. A., Gulati, R., Gothoskar, P., Khobragade, S.: Stellar Spectral Classification Using Automated Schemes. Astrophysical Journal, **426** (1994) 340-344
2. Qin, D. M., Guo, P., Hu, Z. Y., Zhao, Y. H.: Automated Classification of Celestial Spectra Based on Statistical Mixture Modeling with RBF Neural Networks. Chinese J. of Astro. & Astrophy., **3** (2003) 277-286
3. Jolliffe, I.T.: Principal Component Analysis. Springer-Verlag, New York (1986)
4. Martens, H., Naes, T.: Multivariate Calibration. Wiley, New York (1991)
5. Links at the Web site: http://www.cs.unc.edu/~welch/kalman Links.html
6. Dan, S.: Training Radial Basis Neural Networks with the Extended Kalman Filter. Neurocomputing, **48** (2002) 455-475
7. Bishop, C. M.: Neural Networks for Pattern Recognition. Oxford University Press, London (1995)

Pattern Recognition Based on Stability of Discrete Time Cellular Neural Networks

Zhigang Zeng[1,2], De Shuang Huang[3], and Zengfu Wang[1]

[1] Department of Automation, University of Science and Technology of China
Hefei, Anhui, 230026, China
zhigangzeng@iim.ac.cn
[2] Department of Mathematics, Hubei Normal University,
Huangshi, Hubei, 435002, China
[3] Intelligent Computing Lab, Hefei Institute of Intelligent Machines,
Chinese Academy of Sciences, P. O. Box 1130, Hefei Anhui 230031, China

Abstract. In this paper, some sufficient conditions are obtained to guarantee that discrete time cellular neural networks (DTCNNs) can have some stable memory patterns. These conditions can be directly derived from the structure of the neural networks. Moreover, the method of how to estimate of the attracting domain of such stable memory patterns is also described in this paper. In addition, a new design algorithm for DTCNNs is developed based on stability theory (not based on the well-known perceptron training algorithm), and the convergence of the design algorithm can be guaranteed by some stability theorems. Finally, the simulating results demonstrate the validity and feasibility of our proposed approach.

1 Introduction

Cellular neural networks (CNNs), first introduced in [1], are of great interest due to the fact that such networks are the easiest to be implemented in hardware. CNNs include generally the class of feedback neural networks with local interconnections, and they are also suitable for very large-scale integration (VLSI) implementations of associative memories. The goal of associative memories is to store a set of desired patterns as stable memories such that a stored pattern can be retrieved when the initial pattern contains sufficient information about that pattern. In practice, the desired memory patterns in associative memories are usually represented by bipolar vectors (or binary vectors). For instances, the literatures [2-8] directly regarded CNNs as associative memories, Literature [5] discussed the synthesis of CNNs with space-invariant cloning templates; literature [7] developed a design algorithm based on the eigenstructure method; literature [2] considered the realization of associative memories via the class of (zero-input) CNNs introduced in [1], and developed a new synthesis procedure (design algorithm) for CNNs with space-invariant cloning templates based on the well-known perceptron training algorithm. This paper also provides a new design algorithm for DTCNNs based on stability theory [9-12](not based on the well-known perceptron training algorithm), and the convergence of the design algorithm can be guaranteed by some stability theorems.

This paper consists of the following sections. Section 2 describes some preliminaries. The main results are stated in Sections 3. A new design method of DTCNNs for pattern memory is presented in Sections 4. Simulation results of one illustrative example are given in Section 5. Finally, concluding remarks are made in Section 6.

2 Preliminaries

In this paper, we always assume that $t \in \mathcal{N}$, where $\mathcal{N}$ is the set of all natural number.

Consider DTCNNs

$$\Delta x_i(t) = -c_i x_i(t) + \sum_{j=1}^{n} a_{ij} f(x_j(t)) + \sum_{j=1}^{n} b_{ij} f(x_j(t - \tau_{ij}(t))) + u_i, \quad (1)$$

where $\Delta x_i(t) = x_i(t+1) - x_i(t), i = 1, 2, \cdots, n$, $x = (x_1, x_2, \cdots, x_n)^T \in \Re^n$, is the state vector, $c_i > 0$, $A = (a_{ij}) \in \Re^{n \times n}$ and $B = (b_{ij}) \in \Re^{n \times n}$ are connection weight matrices, $0 \le \tau_{ij}(t) \le \tau$ is time delay, and $\tau_{ij}(t)$ and τ are nonnegative integers, for $\forall r \in \Re$,

$$f(r) = \frac{1}{2}(|r+1| - |r-1|), \quad (2)$$

$u = (u_1, \cdots, u_n)^T \in \Re^n$ is the external input vector.

The initial value problem for DTCNNs (1) requires the knowledge of initial data $\{x(-\tau), \cdots, x(0)\}$. This vector is called initial string in [11]. For every initial string, there exists a unique solution $\{x(t)\}_{t \ge -\tau}$ of (1) that can be calculated by the explicit recurrence formula $x_i(t+1) = (1-c_i)x_i(t) + \sum_{j=1}^{n} a_{ij} f(x_j(t)) + \sum_{j=1}^{n} b_{ij} f(x_j(t - \tau_{ij}(t))) + u_i$.

Define the saturation region $\Omega^{(s)} = \{\Pi_{i=1}^{n}(-\infty, -1)^{\delta^{(i)}} \times (1, +\infty)^{1-\delta^{(i)}}, \delta^{(j)} = 1 \text{ or } 0, j = 1, 2, \cdots, n\}$. Hence, $\Omega^{(s)}$ is made up of 2^n elements.

A vector α will be called as a (stable) memory vector (or simply, a memory) of DTCNNs if $\alpha = f(\beta)$, where β is an asymptotically stable equilibrium point of DTCNNs (1).

Use $\mathcal{B}^n$ to denote the set of n-dimensional bipolar vectors, i.e., $\mathcal{B}^n = \{x \in \Re^n, x_i = 1, \text{ or } -1, i = 1, 2, \cdots, n\}$. Hence, $\mathcal{B}^n$ is made up of 2^n elements. For any $(s_1, s_2, \cdots, s_n)^T \in \mathcal{B}^n$ define an operator:

$$\mathcal{L}(s_i) = \begin{cases} (1, \infty), & s_i = 1, \\ (-\infty, -1), & s_i = -1. \end{cases}$$

Consequently, $(s_1, s_2, \cdots, s_n)^T$ and $\prod_{i=1}^{n} \mathcal{L}(s_i) = \mathcal{L}(s_1) \times \mathcal{L}(s_2) \times \cdots \times \mathcal{L}(s_n)$ are one-to-one correspondence.

Design Problem: Given m ($m \leq 2^n$) vectors $\alpha^1, \alpha^2, \cdots, \alpha^m \in \mathcal{B}^n$, choose the connection weight between the i-th and j-th neurons a_{ij}, b_{ij} and u such that $\alpha^1, \alpha^2, \cdots, \alpha^m \in \mathcal{B}^n$ are stable memory vectors of DTCNNs.

Definition 1. The equilibrium point x^* of (1) is said to be locally asymptotically stable in region $\mathcal{D}$, if for $\forall \varepsilon > 0$ there exists $T > t_0$ such that for $\forall t \geq T$, $\|x(t; t_0) - x^*\| \leq \varepsilon$, where $x(t; t_0)$ is the solution of DTCNNs (1) with any initial string $\{x(t)\}_{t_0 - \tau \leq t \leq t_0}$, $x(s) \in \mathcal{D}, s \in [t_0 - \tau, t_0]$, and $\mathcal{D}$ is said to be a locally asymptotically attracting set of the equilibrium point x^*. When $\mathcal{D} = \Re^n$, x^* is said to be globally asymptotically stable.

Definition 2. The point x^* is said to be an isolated equilibrium point of (1), if x^* is an equilibrium point of (1) and there exists $\delta > 0$ such that for $\forall \tilde{x} \in \{x \mid 0 < \|x - x^*\| < \delta, x \in \Re^n\}$, $\tilde{x}$ is not an equilibrium point of (1).

3 Stability Theory for Pattern Memory

Theorem 1. Choose $(s_1, s_2, \cdots, s_n)^T \in \mathcal{B}$. DTCNNs (1) have neither more nor less than one isolated equilibrium point located $\Omega = \prod_{i=1}^n \mathcal{L}(s_i)$, if and only if for $\forall i \in \{1, 2, \cdots, n\}$, $\sum_{j=1}^n (a_{ij} + b_{ij}) s_j + u_j) s_i > c_i$.

Theorem 2. For $\forall (s_1, s_2, \cdots, s_n)^T \in \mathcal{B}$, denote $\Omega = \prod_{i=1}^n \mathcal{L}(s_i)$. If $x^* \in \Omega$ is an equilibrium point of (1) and $c_i \in (0, 1), i = 1, 2, \cdots, n$, then x^* is locally asymptotically stable, and Ω is locally asymptotically attracting region of x^*.

Proof. If $x(s) \in \prod_{i=1}^n \mathcal{L}(s_i), s \in [t - \tau, t]$, then from (1) and (2),

$$\Delta x_i(t) = -c_i x_i(t) + \sum_{j=1}^n (a_{ij} + b_{ij}) s_j + u_i. \tag{3}$$

Obviously, since $c_i \in (0, 1), i = 1, 2, \cdots, n$, if (3) has an equilibrium point, then this equilibrium point is globally asymptotically stable. In addition, from (3), $x_i(t+1) = (1 - c_i) x_i(t) + \sum_{j=1}^n (a_{ij} + b_{ij}) s_j + u_i$. Since $x^* \in \Omega$ and $x_i^* = (\sum_{j=1}^n (a_{ij} + b_{ij}) s_j + u_i)/c_i$, if $x(s) \in \Omega, s \in [t - \tau, t]$ then from (3), $x_i(t+1) = (1 - c_i) x_i(t) + \sum_{j=1}^n (a_{ij} + b_{ij}) s_j + u_i = (1 - c_i) x_i(t) + c_i x_i^*$. Hence $x_i(t+1) s_i > 0$; i.e., $x(t+1) \in \Omega$. They imply that x^* is locally asymptotically stable, and Ω is locally asymptotically attracting region of x^*.

Consider DTCNNs with two cells

$$\begin{cases} \Delta x_1(t) = -c_1 x_1(t) + \sum_{j=1}^2 a_{1j} f(x_j(t)) + \sum_{j=1}^2 b_{1j} f(x_j(t - \tau_{1j}(t))) + u_1, \\ \Delta x_2(t) = -c_2 x_2(t) + \sum_{j=1}^2 a_{2j} f(x_j(t)) + \sum_{j=1}^2 b_{2j} f(x_j(t - \tau_{2j}(t))) + u_2. \end{cases} \tag{4}$$

Denote saturation region $\Omega^{(s)} = \bigcup_{i=1}^4 \Omega_i^{(s)}$, where $\Omega_1^{(s)} = (-\infty, -1) \times (1, \infty)$, $\Omega_2^{(s)} = (1, \infty) \times (1, \infty)$, $\Omega_3^{(s)} = (-\infty, -1) \times (-\infty, -1)$, $\Omega_4^{(s)} = (1, \infty) \times (-\infty, -1)$.

Corollary 1. DTCNNs (4) have an isolated equilibrium point located in $\Omega_1^{(s)}$ or $\Omega_2^{(s)}$ or $\Omega_3^{(s)}$ or $\Omega_4^{(s)}$, if and only if respectively one of the following inequalities holds,

$$a_{11} - a_{12} + b_{11} - b_{12} - u_1 > c_1, -a_{21} + a_{22} - b_{21} + b_{22} + u_2 > c_2;$$
$$a_{11} + a_{12} + b_{11} + b_{12} + u_1 > c_1, a_{21} + a_{22} + b_{21} + b_{22} + u_2 > c_2;$$
$$a_{11} + a_{12} + b_{11} + b_{12} - u_1 > c_1, a_{21} + a_{22} + b_{21} + b_{22} - u_2 > c_2;$$
$$a_{11} - a_{12} + b_{11} - b_{12} + u_1 > c_1, -a_{21} + a_{22} - b_{21} + b_{22} - u_2 > c_2.$$

Proof. Choose $(s_1, s_2) = (-1, 1)$ or $(s_1, s_2) = (1, 1)$ or $(s_1, s_2) = (-1, -1)$ or $(s_1, s_2) = (1, -1)$, according to Theorem 1, Corollary 1 holds.

Theorem 3. Let $d_{ij} = a_{ij} + b_{ij}$. If $c_1, c_2 \in (0, 1)$ and one of the following cases holds,

$$\begin{cases} d_{11} - d_{12} - u_1 > c_1, -d_{21} + d_{22} + u_2 > c_2, \\ d_{11} + d_{12} + u_1 < c_1, d_{11} - d_{12} + u_1 < c_1, d_{11} + d_{12} - u_1 < c_1; \end{cases} \quad (5)$$

$$\begin{cases} d_{11} + d_{12} + u_1 > c_1, d_{21} + d_{22} + u_2 > c_2, \\ d_{11} - d_{12} + u_1 < c_1, d_{11} + d_{12} - u_1 < c_1, d_{11} - d_{12} - u_1 < c_1; \end{cases} \quad (6)$$

$$\begin{cases} d_{11} + d_{12} - u_1 > c_1, d_{21} + d_{22} - u_2 > c_2, \\ d_{11} + d_{12} + u_1 < c_1, d_{11} - d_{12} + u_1 < c_1, d_{11} - d_{12} - u_1 < c_1; \end{cases} \quad (7)$$

$$\begin{cases} d_{11} - d_{12} + u_1 > c_1, -d_{21} + d_{22} - u_2 > c_2, \\ d_{11} + d_{12} + u_1 < c_1, d_{11} - d_{12} - u_1 < c_1, d_{11} + d_{12} - u_1 < c_1, \end{cases} \quad (8)$$

then DTCNNs (4) have neither more nor less than one isolated equilibrium point located in the saturation region $\Omega^{(s)}$, which is locally asymptotically stable. In addition, if (5) holds, then the equilibrium point is located in $\Omega_1^{(s)}$; if (6) holds, then the equilibrium point is located in $\Omega_2^{(s)}$; if (7) holds, then the equilibrium point is located in $\Omega_3^{(s)}$; if (8) holds, then the equilibrium point is located in $\Omega_4^{(s)}$.

Proof. According to Corollary 1 and Theorem 2, Theorem 3 holds.

Remark. It is similar to Theorem 3 that some sufficient conditions can be obtained to guarantee that (4) has two or three or four memory patterns.

Corollary 2. If $a_{11} = 1.2, a_{12} = 0.3, a_{22} = 1, a_{21} = 0, u_1 = 0.3, u_2 = 0.3, c_1 = c_2 = 0.9, b_{ij} = 0$, then DTCNNs (4) have neither more nor less than one isolated equilibrium point located in the saturation region $\Omega^{(s)}$, which is locally exponentially stable and is located in $\Omega_2^{(s)}$.

Proof. According to Theorem 3, Corollary 2 holds.

4 A New Method of Design

Step 1. Desired patterns are denoted by vector that belong to $\mathcal{B}^n$; i.e., we obtained m vectors $\beta^{(1)}, \beta^{(2)}, \cdots, \beta^{(m)} \in \mathcal{B}^n$, where m is the number of the desired patterns, n is the number of neurons of designed DTCNNs and is even.

Step 2. Denote $\beta^{(i)} = (s_1^{(i)}, s_2^{(i)}, \cdots, s_n^{(i)})^T$, $\alpha_1^{(i)} = (s_1^{(i)}, s_2^{(i)})^T$, $\alpha_2^{(i)} = (s_3^{(i)}, s_4^{(i)})^T, \cdots, \alpha_k^{(i)} = (s_{2 \times k-1}^{(i)}, s_{2 \times k}^{(i)})^T, \cdots, \alpha_{n/2}^{(i)} = (s_{n-1}^{(i)}, s_n^{(i)})^T$.

Step 3. According Theorem 3, design $n/2$ connection weight matrixes $A_{11}, A_{22}, \cdots, A_{n/2,n/2}, B_{11}, B_{22}, \cdots, B_{n/2,n/2}, C_{11}, C_{22}, \cdots, C_{n/2,n/2}$ and 2 dimensional vector $\bar{u}_k$ such that $\alpha_k^{(i)}, (i = 1, 2, \cdots, m)$ are stable memory vectors of DTCNNs (4).

Step 4. Desired connection weight matrices are $A = diag\{A_{11}, \cdots, A_{n/2,n/2}\}$, $B = diag\{B_{11}, \cdots, B_{n/2,n/2}\}$, $C = diag\{C_{11}, \cdots, C_{n/2,n/2}\}$, and the desired external input vector is $u = (u_1^T, \cdots, u_{n/2}^T)^T$. Then $\beta_k^{(i)}, (i = 1, \cdots, m)$ are stable memory of DTCNNs (1).

5 Example

Example. The same example as introduced in [2] is used here. Consider DTCNNs with the structure of 12 neurons ($n = 12$). The objective is to store the four patterns shown in Fig. 1 as stable memories (black $= -1$ and white $= 1$). Denote $\beta_1, \beta_2, \beta_3, \beta_4$ be desired patterns depicted respectively in Fig. 1. According to

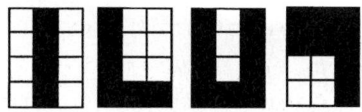

Fig. 1. The desired pattern

the method of design described in section 4, choose A, B, C and u such that $\Delta y(t) = -Cy(t) + Af(y(t)) + Bf(y(t - \tau(t))) + u$ has neither more nor less than four stable memory patterns $\beta_1, \beta_2, \beta_3, \beta_4$. Fig. 2 depicted simulation results of cells with several random initial strings. Fig. 3 describes the memorial process of β_2 with one random initial string when $t = 3, 6, 9, 12$.

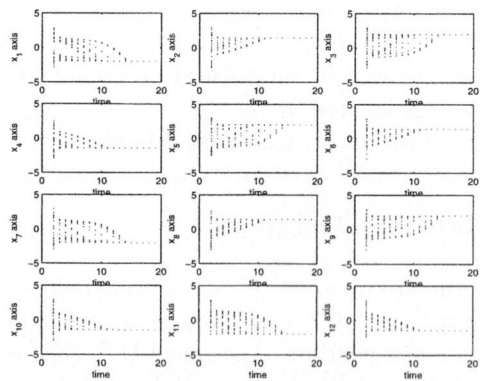

Fig. 2. Transient behavior of cell in DTCNNs

Fig. 3. The memorial process of β_2 with one random initial string

6 Concluding Remarks

In this paper, a new design algorithm for DTCNNs was developed based on stability theory (not based on the well-known perceptron training algorithm), and the convergence of the design algorithm can be guaranteed by some stability results. These results can guarantee that the desired patterns are stable memory patterns of the designed DTCNNs. And the method of how to estimate the attracting domain of such stable memory patterns is also described in this paper. These results can be directly derived from the structure of the neural networks. Hence, it is very convenient for ones to design DTCNNs for storing desired memory patterns efficiently. Finally, the simulating results showed the validity and feasibility of our proposed approach. Future research works will include how to use our proposed approach to solve more practical problems, and extend our approach to a more general case.

References

1. Chua, L. O., Yang, L.: Cellular Neural Networks: Theory. IEEE Trans. Circuits Syst., **35** (1988) 1257-1272
2. Lu, Z. J., Liu, D. R.: A New Synthesis Procedure for A Class of Cellular Neural Networks with Space-invariant Cloning Template. IEEE Trans. Circuits Syst., (1998) 1601-1605
3. Brucoli, M., Carnimeo, L., Grassi G.: Discrete-time Cellular Neural Networks for Associative Memories with Learning and Forgetting Capabilities. IEEE Trans. Circuits Syst. I, **42** (1995) 396-399
4. Chua, L. O., Roska, T.: The CNN Paradigm. IEEE Trans. Circuits Syst. I, **40** (1993) 147-156
5. Liu, D. R.: Cloning Template Design of Cellular Neural Networks for Associative Memories. IEEE Trans. Circuits Syst. I, **44** (1997) 646-650
6. Liu, D. R., Lu, Z.: A New Synthesis Approach for Feedback Neural Networks Based on the Perceptron Training Algorithm. IEEE Trans. Neural Networks, **8** (1997) 1468-1482
7. Michel, A. N., Farrell J. A.: Associative Memories via Artificial Neural Networks. IEEE Contr. Syst. Mag., **10** (1990) 6-17
8. Seiler, G., Schuler, A. J., Nossek J. A.: Design of Robust Cellular Neural Networks. IEEE Trans. Circuits Syst. I, **40** (1993) 358-364
9. Liao, X. X., Wang, J.: Algebraic Criteria for Global Exponential Stability of Cellular Neural Networks with Multiple Time Delays. IEEE Trans. Circuits and Systems I, **50** (2003) 268-275

10. Zeng, Z. G., Wang, J., and Liao, X. X.: Global Exponential Stability of A General Class of Recurrent Neural Networks with Time-varying Delays. IEEE Trans. Circuits and Systems Part I, **50** (2003) 1353-1358
11. Mohamad, S., Gopalsamy, K.: Exponential Stability of Continuous-time and Discrete-time Cellular Neural Networks with Delays. Applied Mathematics and Computation, **135** (2003) 17-38
12. Nikita, E. B., Danil, V. P.: Stability Analysis of Discrete-Time Recurrent Neural Networks. IEEE Trans. Neural Networks, **13** (2002) 292-303

The Recognition of the Vehicle License Plate Based on Modular Networks*

Guangrong Ji[1], Hongjie Yi[1], Bo Qin[1], and Hua Xu[2]

[1] College of Information Science and Engineering, Ocean University of China, Qingdao, 266003, China
{grji,qinbo}@mail.ouc.edu.cn , yihongjie@hotmail.com
[2] Institute of Laser Shandong Academy of Sciences, Jining, 272017, China
xuhua2338375@163.com

Abstract. This paper presents a novel method of real-time recognizing the vehicle license based on the integrated modularized neural network. Some locations in the license plate contain Arabic numeral only and the others contain Arabic numeral, English letter or Chinese character. According to the complexity, the two kinds of network structures, ω_i / ω_j and $\omega_i / \overline{\omega_i}$, are organized and the outputs are integrated under the given fuzzy rules to perform the recognition tasks. The results of the experiments have shown that this method can recognize the characters quickly and accurately.

1 Introduction

The recognition of vehicle license plates is one of the most important tasks in intelligent traffic field, which is one of the applications in computer vision and pattern recognition. It can be applied to the management of crossings, carports, charging in intersections, highways and Customs etc. So the realization of real-time auto-identifying vehicle license plates is very important to the computerization of traffic management. Currently, some applications of the intelligent traffic management system have already been carried out in some developed countries, but the recognition of vehicle license plates is difficult, because of the multiform characters and the different illumination environment.

The recognition system can be divided into several steps, picture collection, pretreatment, location, character division and recognition. The pretreatment usually does some picture enhancement, recover and transform in order to enhance its characteristics. So it can be located accurately. We had brought forward a location method based on RGB information and modular neural networks [1]. The method can locate license plates of vehicles under the difference of the illumination environment, even distinguish the type of the vehicles. The character dividing system divides the plates into several single characters including Chinese characters, English letters and Arabic numerals. Then each character is standardized before recognitions.

---

* The National 863 Natural Science Foundation of P. R. China (2001AA636030) and the Natural Science Foundation of Shandong Province, P. R. China (Y2000G03), fully supported this research.

So far, many scholars had done widespread and far-reaching researches in this field, and brought forward some new methods, such as simply template matching, outline matching, projection series characteristics matching, outline projection matching and template matching based on Hausdorff distances. For standardized characters all the methods above will get a desirable result. But, usually the picture contains noise, so the characters divided from the picture have some flaws. A single method cannot obtain a perfect result. Zhi Chai [2] had brought forward a multi-recognizer fusion method, it can improve the recognizing precision to a certain extent. But it cannot recognize the blurry picture or the anamorphic character efficiently. Xiaojing Liu [3] had brought forward a projection and coefficient transform method. This method had a clear mathematics model, but it is too slow to use in real–time system. The Neural Network is widely used in pattern recognition for its parallel processing and robust ability. A single neural network cannot achieve a satisfactory recognition because of the multiform characters to be recognized in the vehicle license plates. According to the real-time requirement in the recognition of the vehicle license plate, we present a novel method of recognizing the license based on the integrated modularized neural network in this paper. Through the division of the target, we trained many modularized neural networks separately, and then integrate them under the fuzzy theory. The results of the experiments have shown that this method can recognize the characters quickly and accurately, which is robust.

2 The Theory of Modular Neural Network

"Dividing and integrating" had been proved a very useful way to large-scale problem. Recently years it had been a hotspot of system design [4] [5]. Baoliang Lu [6] had brought forward a task dividing and module integrating neural network. He divide K-class problem into many simple 2-class problems, train neural network for every 2-class problem, and then integrate these neural networks to get the solution of the K-class problem. The experiment had proved it's effective.

The K-class problem training set is as follows:

$$T_i = \{(X_l, y_l^{(i)})\}, \quad i = 1, 2, \ldots, K, \quad X_l \in R^d, \quad y_l^{(i)} \in R^1 \tag{1}$$

$X_l \in R^d$ is input vector, $Y_l \in R^K$ is output vector.

A K-class problem can be divided into many 2-class problems. Its training set is as follows:

$$T^{(u,v)}{}_{ij} = \{(X_l^{(iu)}, 1-\varepsilon)\}_{l=1}^{U} \cup \{(X_l^{(jv)}, \varepsilon)\}_{l=1}^{V}, \quad u = 1, 2, \ldots, \leq N_i, \tag{2}$$
$$v = 1, 2, \ldots, \leq N_j, \quad i = 1, 2, \ldots, K-1, \quad j = i+1$$

$\chi_l^{(iu)} \in C_i$, $\chi_l^{(jv)} \in C_j$. So we can divide the training set for every network, and integrate these networks for the original problem.

3 The Feature of Vehicle License Plate

Nowadays, in China the vehicle license plate contains 7 characters, including Chinese characters, English letters and Arabic numerals. The pictures below (fig.1) are some vehicle license plates, which show the forms above. According to the different characteristics, we use different methods so as to get an optimized result.

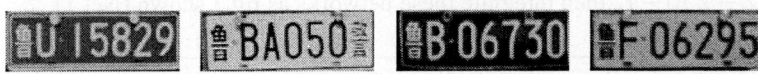

Fig. 1. Current vehicle license plate in China. The first character is a Chinese character standing for the province. The second character is a letter standing for the city. The third character is a letter or an Arabic numeral. The fourth to sixth characters are Arabic numerals. The last character is an Arabic numeral or a Chinese character. If it is a Chinese character it stands for the special use of the vehicle.

4 Training Set Making

Firstly, we collect enough pictures of vehicle license plate; then locate the license plate using modularized networks and divide it into separate characters; finally, do some pre-process and make the characters standardized. The training set is shown as fig.2

鲁 苏 甫 赣 鄂 桂 甘 晋 蒙 陕 吉 冀 闽 贵 粤

川 青 藏 宁 琼 京 津 豫 沪 云 辽 黑 湘 皖 新

A B C D E F G H J K L M N P Q R S T U V W

1 2 3 4 5 6 7 8 9 警

Fig. 2. The characters in the license plate we had got

5 Net Making and Training

As we had discussed above, we use different structures of the net to deal with different characters.

5.1 For Small Sample Space, We Use ω_i / ω_j Classifier

The forth to sixth characters are Arabic numerals. According to Baoliang Lu's [6] method, we divide $C_{10}^2 = 45$ training set as follows:

T_{01} T_{02} T_{03} T_{04} T_{05} T_{06} T_{07} T_{08} T_{09}
T_{12} T_{13} T_{14} T_{15} T_{16} T_{17} T_{18} T_{19}
....
T_{89}

Train the 45 BP networks separately. The topology structure of every network is $512 - 256 - 1$, then integrate these networks as fig.3 shown (Net 1).

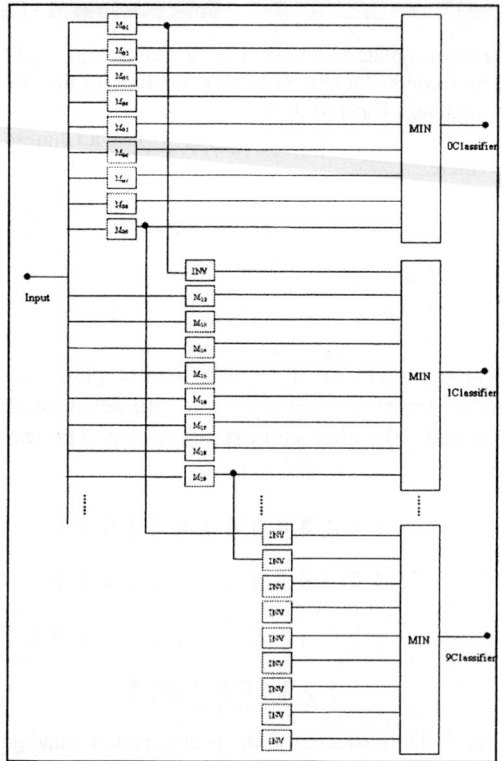

Fig. 3. The integration topology of ω_i / ω_j classifier. In the picture every small rectangle stands for a separately neural networks, the large rectangle stands for integrate rules.

The last character is Arabic numeral or Chinese character, but there are only a few special Chinese characters. A few Chinese character classifiers are added in this net (Net 2).

5.2 For Large Sample Space, We Use $\omega_i/\omega_{\bar{i}}$ Classifier

The second and third characters are letters or Arabic numerals. The sample space is very large. If still using the method above, the training set will be $C_{36}^2 = 630$. It makes the network too complexity to realize. So we use $\omega_i/\omega_{\bar{i}}$ classifier, for the large sample space. The 36 training sets are as below:

T_A T_B T_C T_D T_E T_F T_G T_H T_I T_J T_K T_L T_M T_N T_O T_P T_Q T_R T_S T_T T_U T_V T_W T_X T_Y T_Z T_0 T_1 T_2 T_3 T_4 T_5 T_6 T_7 T_8 T_9

Train the 36 BP networks separately. The topology structure of every network is $512 - 256 - 1$, then integrate these networks as fig.4 shown (Net 3).

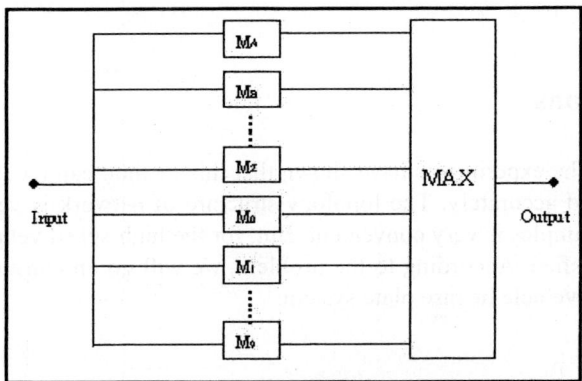

Fig. 4. The integration topology of $\omega_i/\omega_{\bar{i}}$ classifier. In the picture every small rectangle stands for a separately neural networks, the large rectangle stands for integrate rules.

The first character is Chinese character. There are 30 samples. We use the similar net (Net 4).

6 Experiments and Result Analysis

Pick a picture randomly, using it test the trained network. The procedure and result of the experiments are shown as fig.5

The first character "lu" and the second the third characters "U" and "I" send to NET3 and NET 4. While the input is "lu", the output of "lu" classifier is maximum, the other classifier is smaller, so we recognize the character as "lu". The forth to sixth characters and the last character send to NET 1 and NET 2. While the input is the character, the output of the very classifier is maximum. After logic operates we can get the output. Finally, integrate these networks, we can get the result as above. The system is simulated on the personal computer (P4 2.4G, 512M). It is used about 0.7s to recognize a segmented plate. Totally we use this system to recognize 107 vehicle plates, the accuracy achieves about 96.9%.

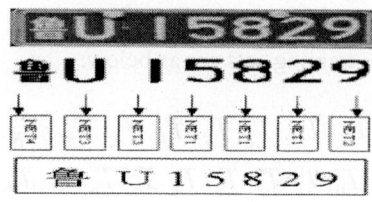

Fig. 5. The procedure of the experiments

7 Conclusions

The results of the experiments have shown that this method can recognize the characters quickly and accurately. The topology structure of network is very clear. Adding and reducing samples is very convenient. But, for the high-speed vehicles, the correctness is not satisfied. According to the problem, we will go on consummating the recognition of the vehicle license plate system.

References

1. Guangrong Ji, Fang JI: The Location of the Vehicle License Plates Based on Modularized Networks. Proceedings of The 12th National Conference on Neural Networks, Beijing (2002) 655-660
2. Zhi Chai, Qingzhou Tao, Yanmei Yu, Xiaohai He: A Fast and Practicable Way to Recognize the Characters of Vehicle License Plate. Journal of Sichuan University (Natural Science Edition) Jun.2002 Vol.39 No.3 465-468
3. Xiaojing liu, Yu Cheng: A Research of Vehicle License Plate Automatic Recognition. Journal of Nanjing University of Aeronautics & Astronautics Vol 30,No 5 573-577
4. Zhao Q.F: A Study on Co-evolutionary Learning of Neural Networks. Lecture Notes in Artificial Intelligence, NO.1141, Springer-Verlag, Heidelberg (1997)
5. Zhao Q.F: A Combination of IEA and CEA. Lecture Notes in Artificial Intelligence, NO.1141, Springer-Verlag, Heidelberg (1997)
6. B. L. Lu, Y. Bai, H. Kita and Y. Nishikawa: A Multisieving Neural-network Architecture That Decomposes Learning Tasks Automatically. in Proc. IEEE Conf. Neural Networks, Orlando, FL, June 28-July 2. (1994) 1319-1324

Neural Network Associator for Images and Their Chinese Characters

Zhao Zhang

Missouri Western State College, St. Joseph MO 64507, USA,
zhang@mwsc.edu

Abstract. A neural network with two layers of processing elements (PEs) is investigated to associate a group of 48x48 gray scale images with the corresponding 16x16 Chinese character bitmap images. The input layer has 2304 (48x48) PEs, and the output layer has 256 (16x16) PEs. In order to reduce the size of the weight matrix, each input PE only connects to two output PEs, for a total of 4608 connections. Each output PE connects to 18 input PEs (4608/256=18). The connections are assigned randomly, and remain unchanged throughout this the investigation. The momentum based back propagation learning algorithm is used to train the weight matrix. The system is capable of perfectly learning input-output associations when inputs are orthogonal to each other. However, the investigation finds that close to perfect learning still occur for a random set of images.

1 The Neural Network Associator

This investigation try to associate the 48x48 gray scale images with their Chinese characters. A two layer neural network is used. One layer is the input layer, and the other is the output layer. The input layer has 2304 (48x48) PEs, each PE is connected to a single pixel of the input image. The output layer has 256 (16x16) PEs, one PE for each pixel of the output image.

For a fully connected neural network, the size of the weight matrix will be 2304x256. The matrix is too large to efficiently train and test this neural network. To reduce the size of the weight matrix, each input PE only connects to two output PEs (other than connects to 256 PEs if fully connected). So the total number of connections between the two layers is 4608. For each output PE, it is connected to 18 input PEs (4608/256=18). For a fully connected network, each output PE needs to connect to 2304 input PEs. The connections are established randomly, and remain unchanged for this investigation. Figure 1 shows the structure of the neural network.

There are eight exemplars to train and test the neural network, each exemplar contains a 48x48 gray scale image and a 16x16 bitmap image of its corresponding Chinese character. The eight input pictures are pictures of cat, car, flower, bridge, dog, ship, shoes, and tree. The format of the associated Chinese characters is 16x16 bitmap image. A handwritten font is used to create the bitmap images.

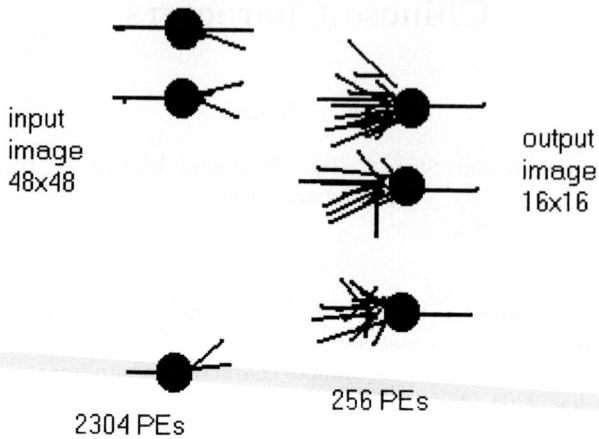

Fig. 1. The structure of the neural network system. The total connections between input and output PEs are 4608. Each output PE randomly connects to 18 input PEs

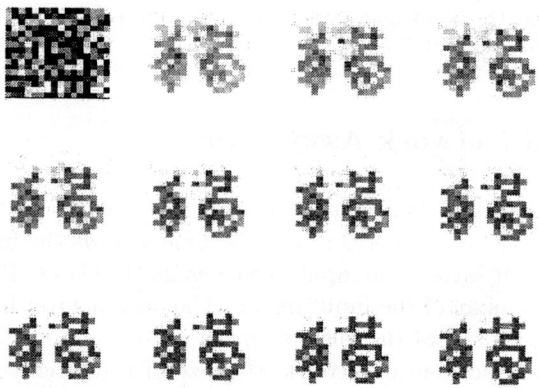

Fig. 2. The output bitmap image during the training process, with 30 epochs intervals between each result

Initially, the weight matrix is a random weight matrix. During the training process, the weights are updated based on squared error (the total error energy) and the local first-order gradients in the back propagation plane. It is called one run or one epoch after the training process go through all eight exemplars.

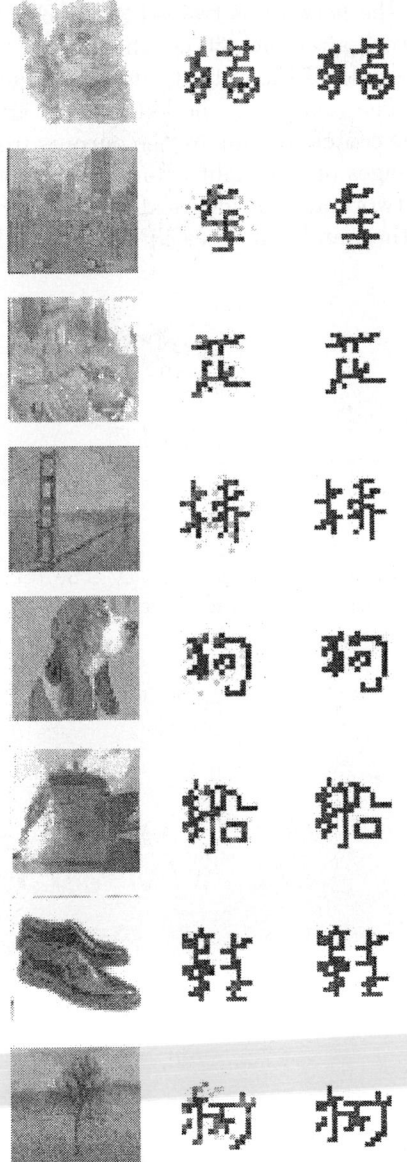

Fig. 3. The results of the trained neural network for all eight exemplars. The images on the left column are the 48x48 pictures, the bitmap images on the center column are the outputs from the trained neural network. The bitmap images on the right column are the desired outputs.

2 Results and Discussion

Figure 1 shows the output results during the learning process. The first output at the upper left corner is the result of the untrained neural network. The next image is the result after the network is trained 30 epochs. The figure shows the result of the neural network after each 30 epochs interval.

Figure 3 shows the results of the output, after the neural network has been trained for 360 epochs. The images on the left column are the 48x48 pictures, the bitmap images on the center column are the outputs from the trained neural network. The bitmap images on the right column are the desired outputs. The outputs of the neural network are not identical to the desired outputs, but they are very close to each other, and can be easily recognized by someone who can read Chinese.

Exploring Various Features to Optimize Hot Topic Retrieval on WEB

Lan You[1], Xuanjing Huang[1], Lide Wu[1], Hao Yu[2], Jun Wang[2], and Fumihito Nishino[2]

[1] Department of Computer Science, Fudan University, Shanghai, 200433
{lan_you,xjhuang,ldwu}@fudan.edu.cn
[2] FUJITSU R&D Center Co., Ltd
{yu,jwang}@frdc.fujitsu.com, nishino@jp.fujitsu.com

Abstract. BBS is an electrical forum on Web where people discuss many topics. So it's a challenging problem to retrieve hot topics from it. There are various features of hot topics. Though count of posts on BBS about topic is a simple and effective feature for hotness of topic, it is shown in the paper that a better result can be obtained if irrelevant posts are filtered out before counting; and a much better result can be obtained if more features are combined using BPNN.

1 Introduction

BBS is a popular platform on Web where people talk about various topics everyday. There are too many topics for people even to browse the topic list. Most people only care those hot topics. As the scale of BBS is becoming larger and larger, it is rather a hard job for human to find hot topics in these tons of information. We explored features of hot topics and utilized them to do hot topic retrieval on BBS.

Our work is similar to topic detection, which is one of the major tasks of TDT [1]. Some topic detection systems [2] [3] use term-based algorithms to discover topics in a corpus of news archive. The corpus consists of temporally ordered news stories, which can be viewed as a stream-like structure. Some other systems [4] [5] also use the temporal information in a form of time window. In James Allan's work [6], they exploit time by incorporating a time penalty in their thresholding model. However, all these works are different from ours in that messages on BBS have a tree-like structure instead of a stream-like one, and relationship between messages is more complicated than that between documents in news archive.

The most similar work we found is Hiromi OZAKU's automatic recommendation of hot topics in discussion-type newsgroups [7]. Their system, called HISHO, find hot topics by counting number of articles and other conditions in a news spool. This enumerative method is also taken by many existing BBS to select hot topics. But we think it is not sufficient because the hotness of topics on BBS is influenced by many features. All these features should be integrated. We chose Neural Net based algorithm to do this integration because of its flexibility and capability of characterization.

In this paper, we explore three types of features of hot topics: numeral features, authority features and temporal feature. Each of them has some impact on the hotness of topic. To fully utilize these features is not just a simple work like counting. In section 2, we will discuss our relevance-based method for extracting useful features from the metadata. In the same section, we also suggest two Neural Net based algorithms for combining them. Section 3 presents the results of our experiments, which will show the influence of each feature on topic hotness, the comparison between the relevance-based method and the enumerative method for measuring features and the effect of the combination of features.

2 Using Features on BBS

2.1 Features of Hot Topics

2.1.1 Numeral Features

There are some intuitive features of topics which can be described by their number. This kind of features includes the number of threads (*#threads*), the number of replies (*#replies*) and the number of authors (*#authors*) within a topic. It's quite intuitive that the larger these numbers, the hotter the topic.

But besides those substantial articles, there are still many garbage messages on BBS. So we use a relevance-based method to exclude such useless information. In this method, we consider the relevance of the message content. Messages are classified into three kinds of classes: domain-relevant classes, irrelevant class and no-content class. Messages relevant to a certain domain are classified into domain-relevant class, and messages irrelevant to all the domains are classified into irrelevant class. No-content class contains those messages only having trivial words.

The classification is done using a Naive Bayes classifier. The following formula describes the principle of this classifier.

$$c = \arg\max_{c_j \in C} p(c_j) \prod_{w_i \text{ in message}} p(w_i | c_j), \tag{1}$$

where c is the target class, C is the collection of classes, c_j is class j, w_i is the ith word in the vocabulary, which contains all the words in the training set. So $P(c_j)$ is the prior probability of class c_j, $P(w_i|c_j)$ is the condition probability. They can be calculated by these formulas:

$$P(c_j) = \frac{\text{number of messages in each class}}{\text{total number of training set}}, \quad P(w_i | c_j) = \frac{n_i + 1}{n + N}, \tag{2}$$

where n_i is number of w_i occurring in the messages in class c_j, n is the total number of words in the messages in class c_j, and N is the number of words in the vocabulary.

Based on the classification, we define topic domain as the domain to which most messages in the topic are relevant. Then we can get two more numbers for a topic:

$$\begin{aligned}\text{\#relevant replies} &= \text{the number of replies that are relevant to its topic domain}, \\ \text{\#relevant authors} &= \text{the number of authors whose messages are relevant to its topic domain}.\end{aligned} \tag{3}$$

2.1.2 Authority Features

2.1.2.1 Authority of Posters
A message posed by a famous poster is always very influential so that many people will follow it. We define a famous poster as one that has posted a lot of substantial messages that led to a few of hot discussion. Thus we can use the reply relationship between messages. This reply relationship is like the link relationship between web pages. So we use an extended HITS algorithm [8] to analyze poster authority. We compute posters' authority scores recursively with the following formula:

$$a(i) = \sum_{j \to i} w(i,j) * h(j), \quad h(j) = \sum_{j \to i} w(i,j) * a(i), \quad (4)$$

where *a(i)*, *h(i)* are poster *i*'s authority and hub score respectively, and *w(i,j)* is the weight of the reply relationship of poster *j* to poster *i*. For each topic, we calculate its *average authority score* and *average author rank percentage* as follow:

$$average\ authority\ score = \frac{\sum_{i \in Topic} a(i)}{\#\,relevant\ authors},$$

$$average\ author\ rank\ percentage = \frac{\sum_{i \in Topic} \frac{rank(i)}{|authors\ in\ the\ board|}}{\#\,relevant\ authors}, \quad (5)$$

where *a(i)*, *rank(i)* and *#relevant authors* has been defined before, *Topic* is the set of relevant authors in the topic, *rank(i)* is the rank of poster *i* on a certain message board according to his authority score, and |*authors in the board*| is the number of authors appearing on the certain message board.

2.1.2.2 Authority of the Message's Source
There are some messages cited from other authoritative medias (like Routers). These messages always arise hot discussion. To measure message's authority, we maintain a list of authoritative medias. The source authority of a message is defined as:

$$source\_authority(m) = \begin{cases} 1 & \text{if the message } m \text{ is cited from other authoritative media} \\ 0 & \text{else} \end{cases}. \quad (6)$$

Defining *M(topic)* the set of messages in a topic, we get the topic source authority:

$$topic\ source\ authority = \sum_{m \in M(topic)} source\_authority(m). \quad (7)$$

2.1.3 Temporal Feature
This feature is the average interval between every two messages in a topic. If there are a lot of messages talking about one topic in a short time, it is likely to be a hot topic. The hot discussion always lasts about 30 percent of the whole topic lifecycle. So we decided to use interval information as such: Let T to be the hour which has the maximum messages. Then compute the *average interval* in the period from T_{min} (10 percent of lifecycle before T) to T_{max} (10 percent of lifecycle after T). The formula is:

$$Average\ Interval = \frac{T_{max} - T_{min}}{\#\,of\ messages\ posted\ between\ T_{min}\ and\ T_{max}}. \quad (8)$$

2.2 Combination of Features

2.2.1 Hotness Classification
To judge the hotness of a topic can be treated as a problem of classification. The above features can be used to classify topics into hot and general ones. We use two Neural Net based algorithms to do linear combination and non-linear combination.

2.2.2 Linear Combination
The linear combination of the features can be described by the following formula:

$$Hotness\ score = \sum_{i=1}^{n} w_i t_i , \qquad (9)$$

where n is the number of features, t_i is the value of feature i, and w_i is the weight of feature i. A threshold is set here to realize the hotness classification. We use a single layer Perceptron to train the weights.

2.2.3 Non-linear Combination
The following formula is non-linear combination of features:

$$Hotness\ score = f(t_1, t_2, \cdots, t_n) , \qquad (10)$$

Since we don't know the form of non-linear function f, we use a BPNN (Back-Propagation Neural Net) with one hidden layer. The activation function taken on hidden nodes is the bipolar sigmoid function.

The BPNN we used only has one output node Y. A topic is represented as a n-dimension vector. After feeding it throughout the network, a hotness score is given by the output node for each topic. A threshold is set at the output node Y. Those topics that get a score above the threshold are classified as hot.

3 Experiments

3.1 Experiments Set-Up

We selected the BBS of Yahoo Finance (http://biz.yahoo.com/co/). It has a relatively complete list of stocks, and a large amount of visitors. We chose messages from IBM, SUN, Microsoft, Dell four boards during 12 specific time span. Clustering was done over them. After thread clustering, 1174 candidate topics are generated, among which 957 are randomly chosen as training data and the remaining 217 are chosen as testing data. Then several assessors read these topics and judged their hotness. There are 128 hot topics in the training data, and 18 in the testing data.

3.2 Evaluation on Hot Topic Retrieval by Utilizing Features

3.2.1 Retrieval by Single Feature
The retrieval results when only using single features are shown in table 1.

It seems the performance of *#replies* is rather good with 0.643 best f-score so that it is reasonable for most existing BBS to use this feature to determine the hotness of a thread. But the result of our relevance-based method is better than that of enumerative method in that the average precision and best f-score of *#relevant replies* and *#relevant author* is higher than those of *#replies* and *#authors*.

Table 1. Retrieval Results When Only Using Numerical Features. Average precision is the average of the precision value obtained after each hot topic is retrieved. Best f-score is the value of the highest f-score among all the results according to different thresholds

	F1	F2	F3	F4	F5	F6	F7	F8	F9
Average precision	0.365	0.723	0.765	0.735	0.844	0.131	0.260	0.243	0.284
Best f-score	0.359	0.619	0.643	0.643	0.811	0.238	0.313	0.214	0.320

#threads	F1	#replies	F4	Average author rank percentage	F7
#authors	F2	#relevant replies	F5	Topic source authority	F8
#relevant authors	F3	Average authority score	F6	Average interval	F9

And *average author rank percentage* is better than *average authority score*, though they both are measurements of poster's authority. This is because *average author rank percentage* considers the domain-specific authority, which means a famous poster in a certain domain may be not so famous in another domain.

Though the performance of *average interval* is not so well when only using this feature, we can see later that it does work in the combination methods.

3.2.2 Retrieval by Combination of Features

In the combination method, we don't use all the features because of information redundancy. We omit *#replies* and *#authors* and *average authority score*. We suggest three kinds of combination in table 2.

Table 3 is the results of combination methods when using BPNN and Perceptron.

Table 2. Description of Combination Methods

Combination method	Involved features
Combination 1	#threads + #relevant replies + #relevant authors
Combination 2	#threads + #relevant replies + #relevant authors + average author rank percentage + topic source authority
Combination 3	#threads + #relevant replies + #relevant authors + average author rank percentage + topic source authority + average interval

Table 3. Retrieval Results of Combination Methods

	combination1		combination2		combination3	
	BPNN	Perceptron	BPNN	Perceptron	BPNN	Perceptron
Average precision	0.938	0.900	0.943	0.919	0.958	0.928
Best f-score	0.882	0.824	0.923	0.857	0.941	0.865

We can see that the retrieval performance improves when adding more features into the combination. Though the performance of single authority features and single temporal feature is poor (see table 1), they do work in the combination. When adding the authority features, the average precision and best f-score rise about 2% and 4% using Perceptron, and about 0.5% and 5% using BPNN. And when adding the temporal feature, the average precision and best f-score rises about 1% and 1% using Perceptron, and about 1.5% and 2% using BPNN. And on average, the average precision using BPNN is about 3.5% higher than using Perceptron, and the best f-score is about 8% higher. This is because the feature space is more likely to be non-linearly separable. So BPNN works better.

The advantage of BPNN can also be proved by the following precision-recall graphs. They are comparison of BPNN and Perceptron in each combination method. The curves of BPNN are always better than those of Perceptron.

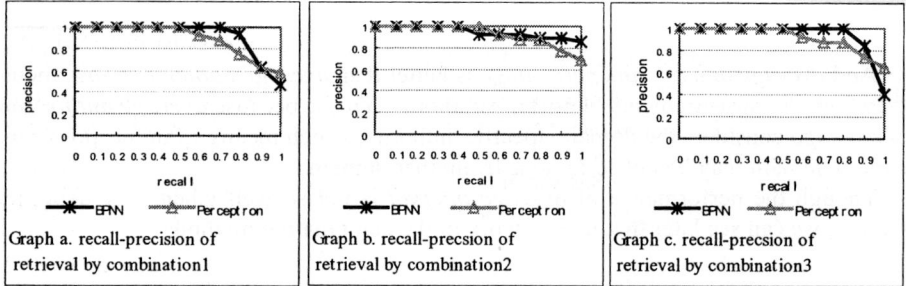

Graph a. recall-precision of retrieval by combination1

Graph b. recall-precsion of retrieval by combination2

Graph c. recall-precsion of retrieval by combination3

Fig. 1. Recall-Precision Graphs of results using combination methods

Table 4. Comparison of Retrieval Results by Combination Method and Best Single Features

	Combination3/BPNN	#replies	#relevant
Average precision	0.958	0.735	0.844
Best f-score	0.941	0.643	0.811

In table 4, we compare the best combination method (combination3 using BPNN) and the best single features (*#replies* and *#relevant replies*). The average precision and the best f-score of combination method are much better than single feature.

4 Conclusion

We explore the features of hot topics on BBS. The influence of single features and their combination on the hotness of topic is discussed in our experiments. It seems our relevance-based method works better than the simple enumerative method to measure the *#replies* and *#authors*. This insures our hot topic retrieval algorithm works better than the scheme used by most existing BBS. Considering the complicated structure of BBS, we do hot topic retrieval using combination of all types of features. Each type of features shows its influence on the combination and non-linear combination of features is better than linear combination and much better than single feature.

References

1. TDT: The 2003 Topic Detection and Tracking (TDT2003) Task Definition and Evaluation. http://www.nist.gov/speech/tests/tdt/tdt2003/index.htm. (Jun. 2003)
2. Walls, F., Jin, H., Sista, S., Schwartz, R.: Topic Detection in Broadcast News. Proceedings of the DARPA Broadcast News Workshop. Morgan Kaufmann Publishers (1999) 193-198
3. Dharanipragada, S., Franz, M., McCarley, J.S., Roukos, S., Ward, T.: Story Segmentation and Topic Detection in the Broadcast News Domain. Proceedings of the DARPA Broadcast News Workshop. Morgan Kauffmann Publishers (1999) 65-68
4. Yang, Y., Pierce, T., Carbonell, J.: A Study on Retrospective and On-Line Event Detection. Proceedings of ACM SIGIR Conference on Research and Development in Information Retrieval (1998) 28-36
5. Bun, K.K., Ishizuka, M.: Topic Extraction from News Archive Using TF*PDF Algorithm. Proceedings of the 3rd International Conference on Web Information Systems Engineering. WISE (2002) 73-82
6. Allan, J., Papka, R., Lavrenko, V.: Online New Event Detection and Tracking. Proceedings of ACM SIGIR Conference on Research and Development in Information Retrieval (1998) 37-45
7. Ozaku, H., Utiyama, M., Murata, M., Uchimoto, K., Ishara, H.: Automatic Recommendation of Hot Topics in Discussion-Type Newsgroups. Proceedings of the 5th International Workshop on Information Retrieval with Asian Languages (2000) 203-204
8. Kleinberg, J.: Authoritative Sources in a Hyperlinked Environment. Proceedings of the 9th Annual ACM-SIAM Symposium on Discrete Algorithms. ACM Press (1998) 668-677

Author Index

Adachi, Masaharu   I-217, I-395
Aguirre, Carlos   I-55
Aknin, Patrice   I-524
An, Dexi   II-555
An, Jubai   I-977
Asari, Vijayan K.   I-870

Bai, Ling   I-1002
Bao, Zheng   I-531
Beadle, Patch   I-726, II-350, II-356
Bentoumi, Mohamed   I-524
Bi, Guo   II-595
Bi, Weixing   I-338, I-425
Bi, Yingwei   II-405
Bian, Zheng-Zhong   II-453
Bloch, Gérard   I-524
Bo, Liefeng   I-264
Brown, Carl   I-977
Bu, Nan   I-834

Cai, Feng   II-418, II-430, II-920
Cai, Minglun   II-418
Campos, Doris   I-55
Cao, Guang-yi   II-150
Cao, Jinde   I-78, II-976
Cao, Li   II-218
Cao, Wen   I-965
Cao, Wenming   II-786
Cao, Xiu   II-188
Cao, Yijia   II-188
Cao, Yukun   II-688
Cao, Zongjie   II-399
Cardon, Pierre   I-864
Chai, Lin   II-381
Chan, Francis   I-732
Chan, Philip C.H.   I-696
Chau, Kwokwing   II-970
Chaudhari, Narendra S.   I-362, II-494
Chen, Chunrong   I-983
Chen, Gang   I-971
Chen, Guanrong   II-627
Chen, Huaidong   II-399
Chen, Jin   I-906, II-567, II-595
Chen, Jinmiao   I-362, II-494
Chen, Junying   I-317

Chen, Nian-Yi   I-389
Chen, Peng   II-574
Chen, Qing   II-755
Chen, Quanshi   I-512
Chen, ShaoBai   I-565
Chen, Shuang   I-199
Chen, Songcan   I-180, I-768, I-858
Chen, Tianping   I-38, I-144
Chen, Xiaoming   I-401, I-425, I-430
Chen, Xiaoping   I-311
Chen, Xingfu   II-601
Chen, Xiyuan   II-805
Chen, Yaqin   I-786
Chen, Yen-Wei   I-186
Chen, Yong   I-102, I-120, I-512, II-639
Chen, Yuehui   I-211
Chen, Yumin   I-906
Chen, Zhe   II-607
Cheng, Guojun   II-110
Cheng, Hong   I-792
Cheng, Xiefeng   I-702
Cheung, Albert   II-627
Chi, Wei   II-920
Chicharo, Joe F.   I-672
Choi, Hyung-Jun   I-239, II-988
Chou, PenChen   II-83
Chu, Fulei   II-767
Chu, Xiaodong   II-581
Cui, Pingyuan   II-129

Dai, Meng   II-681
Dai, Weidi   I-840, II-424
Dai, Xianhua   II-519
Dang, Chuangyin   I-406
Deng, Beixing   I-876
Deng, Bing   II-324
Deng, Feiqi   I-571
Ding, Haiyan   II-459
Ding, Liu   I-598
Ding, Zijia   II-549
Doki, Shinji   I-714
Dominguez, David   I-14, I-20
Dong, Jiwen   I-211
Du, Cheng   II-411
Du, Hai   II-163

Du, Haifeng I-299
Du, Minghui II-269
Du, Ruxu II-169
Duan, Shukai II-813

Elhabian, Tarek II-248, II-281
Ertunc, Metin II-254

Fan, Chunling II-749
Fan, Jian II-35
Fan, Quanyi II-866
Fan, Zhi-Gang I-804
Fang, Qun II-381
Fang, Yonghui I-276
Feng, Chun-Bo I-32
Feng, Guori I-235
Feng, Xiaoning I-846
Feng, Xin I-454
Feng, Zhipeng II-767
Fernández-Redondo, Mercedes I-223, I-229, I-852, I-912
Fu, JianGuo II-701
Fu, Xiao-wei II-150
Fu, Yonggang I-552

Gao, Feng II-749
Gao, Hangshan I-126
Gao, Jianfeng I-311
Gao, Jianming I-786
Gao, Peng II-904
Gielen, Georges I-305, I-876
Gong, Jianya I-906
Gong, Shengguang II-110
Gosztolya, Gábor I-996
Gou, Shuiping I-918
Gu, Wenjin II-117
Gu, Xiaodong I-26, I-413
Guan, Huanxin I-96
Guan, Xiaohong II-657
Guan, Xinping II-71, II-169, II-743
Gui, Jinsong II-799
Guo, Chengan I-370
Guo, Chuangxin II-188
Guo, Dongming II-737
Guo, Jun I-474
Guo, Lin I-648
Guo, Ping I-616, I-1002, II-369, II-393
Guo, Shide II-293
Guo, Shu-Mei II-375
Guo, Ying II-405

Guo, Zhenhe II-675
Guo, Zhi-Gang I-971

Hacker, Roger I-383
Han, Gyu-Sik II-988
Han, Jiuqiang I-622
Han, Liu I-598
Han, Min II-200, II-713
Hao, HanYong II-537
Hao, Pengwei I-168
Hao, Ping II-737
Hao, Zhifeng I-558
Hao, Zhihua I-744
He, Daihai I-138
He, Fan I-822
He, Hanlin I-44
He, Jingrui II-931
He, Jun II-595
He, Junfeng II-138
He, Mi I-138
He, Pilian I-840, II-424
He, Wuhong I-299
He, XiaoFu II-681
He, You II-886
He, Zhenya I-660
Hernández-Espinosa, Carlos I-223, I-229, I-852, I-912
Horzyk, Adrian I-150
Hotta, Hirohisa I-714
Hou, Yuexian II-424
Hsu, Chin-Yuan II-375
Hu, Bao-Gang I-7
Hu, Chang II-218
Hu, Dewen I-199, I-708, II-25, II-525
Hu, Gensheng I-571
Hu, Huibo I-738
Hu, Lingyun I-286
Hu, Mengdi II-786
Hu, Qinglei II-42
Hu, Qiying II-1000
Hu, Rong I-558, II-561
Hu, Shiyan II-669
Hu, Xuelei I-828
Hu, Yunan II-129
Hua, Yu II-982
Huang, Chengkui II-779
Huang, Degen I-983
Huang, De-Shuang I-114, I-648, I-774, I-1008, II-476, II-513
Huang, Gaoming I-660

Huang, Guangbin II-755
Huang, Hong-Zhong II-820
Huang, Houkuan II-707
Huang, Jianhuan I-894
Huang, Jie II-59
Huang, JiFeng II-681
Huang, Liangwei II-381
Huang, Xin II-476, II-513
Huang, Xuanjing I-1025
Huang, Xuming I-720
Huang, Yan-Xin II-337
Huang, Ying II-448
Huang, Yong II-886
Hwang, TsenJar II-83

Jang, Min-Soo I-642
Jang, Sung-Whan II-65
Ji, Ce I-96
Ji, Guangrong I-1015
Ji, Ye I-332
Jia, Ying I-761
Jiang, Chun-di II-19
Jiang, Hairong II-163
Jiang, Min I-406, II-1000
Jiang, Minghu I-305, I-876
Jiang, Qiuhao I-108
Jiang, Quanyuan II-188
Jiang, Yadong II-892
Jiang, Yuan I-356
Jiang, YuGang II-369
Jiao, Licheng I-258, I-264, I-299, I-605, I-629, I-918, II-236, II-275
Jin, Chun II-904
Jin, Lianwen I-988
Jin, Mingxi I-941
Jin, Wuyin I-138
Jin, Xiaoyi I-299
Jin, Yuqiang II-129
Jin, Zhihua II-749

Kanae, Shunshoku II-507
Kang, Chunyu I-930
Kang, Haigui II-799
Kang, Renke II-737
Kang, Zhiyu II-381
Kavak, Adnan II-254
Kawabata, Shuma II-436
Kawamoto, Masashi II-488
Kim, Hyun-Ki I-156
Kim, Sang-Jun I-642

Kim, Yong-Guk I-642
Kim, Yong-Soo I-174
Kim, Yongkab I-162
Kirk, James S. II-695
Kneupner, Klaus II-761
Kong, Fan-Sheng I-323
Koroutchev, Kostadin I-14, I-20
Kotani, Makoto I-395
Kuh, Anthony I-370
Kuhlenkötter, Bernd II-761
Kwan, Chiman I-543

Lai, Hsin-Hsi II-898
Lan, Xiang I-144
Lauer, Fabien I-524
Lee, Chang-Shing II-375
Lee, Dae-Won I-239, I-350, II-287
Lee, Dong-Yoon II-65
Lee, Hyo-Seok II-988
Lee, Hyuk-Soon II-287
Lee, Hyun-Gu I-642
Lee, Jaewook I-239, I-350, II-287, II-988
Lee, Shu Tak II-387
Lee, Soek-Joo I-642
Lei, Xusheng II-1
Leung, Hofung I-454
Li, Baoqing I-468
Li, Bi-Cheng I-965, I-971
Li, Bo I-332, II-701
Li, Chongrong II-931
Li, Chuandong I-61, I-102
Li, Chung Lam II-387
Li, Chunhung I-518
Li, DongSheng I-882
Li, Fucai II-567, II-595
Li, Gang I-786, II-362
Li, Guo-Zheng I-389
Li, Hai-Bin II-820
Li, Haiyang II-531
Li, Hongzhai II-393
Li, Huaqing I-487
Li, Ji II-549
Li, Jiandong I-654, I-690
Li, Jianmin I-462
Li, Kaili II-224
Li, Li II-581
Li, Linglai II-543
Li, Lishuang I-983
Li, Muguo II-163
Li, Ping II-181

Li, Qi   I-598
Li, Qing   I-792
Li, Ruqiang   II-567
Li, Sen   I-678
Li, Shaoyuan   II-175, II-755, II-848
Li, Shouju   II-792
Li, Shuchen   II-181
Li, Songsong   I-430
Li, Ting   II-501
Li, Wuzhao   II-619
Li, Xi   II-150, II-470
Li, Xiaobing   II-405
Li, Xiaoli   II-169, II-743
Li, Xiaoou   II-77, II-212
Li, Xiaoxia   II-362
Li, Xihai   I-959, II-925
Li, Xueming   I-132
Li, Xunming   I-32
Li, Yan   II-743
Li, Yanda   I-537, II-448
Li, Yong   I-401, I-419, I-425, I-442
Li, Yuan   II-962
Li, Yunfeng   I-611, II-688
Li, Zhancheng   II-350, II-356
Li, Zhengxue   I-235
Li, Zhonghua   I-565
Li, Zifeng   II-1007
Lian, Han   I-924
Lian, Shiguo   II-627
Liang, Hua-Lou   I-755
Liang, Xinying   II-574
Liang, Yanchun   I-558, II-104, II-337
Liang, Yanming   I-598
Liao, Xiaofeng   I-61, I-102, I-120, I-132, II-633, II-639, II-645, II-688
Liao, Xiaoxin   I-44
Lin, Bin   I-977
Lin, Chih-Chung   I-377
Lin, Fuzong   I-462
Lin, Hongwen   I-930
Lin, JiaJun   II-681
Lin, Ling   I-786, II-362
Lin, Qiu-Hua   I-755
Lin, Wei   I-822
Lin, Xiaokang   I-780
Lin, Yang-Cheng   II-898
Lin, Yiping   I-67
Lin, Zhonglin   I-900
Linnell, Bruce   I-543
Liu, Bing   I-270

Liu, Chengming   I-720
Liu, Chunsheng   II-826
Liu, Dafan   I-876
Liu, Daizhi   I-959, II-925
Liu, Daxin   I-846
Liu, Fang   I-293, II-236, II-275
Liu, Fei   I-798, II-123
Liu, Guangyuan   I-276
Liu, Haiyuan   I-578
Liu, Hongbo   I-332
Liu, Hongwei   I-531
Liu, Jiafeng   I-894
Liu, Jian   II-35
Liu, Jinxin   II-848
Liu, Ju   I-702, I-738, II-311, II-317
Liu, Jun   I-768
Liu, Lijun   I-282
Liu, Ming   II-91
Liu, Peng   II-91
Liu, Qun   I-636
Liu, Weixiang   II-470
Liu, Wenhong   II-501
Liu, Wenling   I-840
Liu, Wenqi   II-144
Liu, Xinlu   II-904
Liu, Yaqiu   II-42
Liu, Yi   II-156, II-874
Liu, Yingxi   II-792
Liu, Yu   I-317
Liu, Yue   II-962
Liu, Yuliang   II-362
Liu, Yutian   II-581
Liu, Zengrong   I-67
Liu, Zhendong   II-163
Liu, Zhengkai   II-675
Liu, Zhigang   I-959
Long, Weijiang   I-244
Lu, Bao-Liang   I-480, I-804, I-888
Lu, Hongtao   I-552, II-651
Lu, Jun   I-389
Lu, Naijiang   I-900
Lu, Wei   II-144, II-651
Lu, Wen-Cong   I-389
Lu, Wenkai   I-500, I-537
Lu, Wenlian   I-38
Lu, Yinghua   II-104
Luo, Siwei   II-482
Luo, Zhiwei   II-7
Lv, Jiancheng   I-810

Ma, Fei II-35
Ma, Guangfu II-42
Ma, Lin I-792
Ma, Lixin II-436
Ma, Qing II-1007
Ma, Runnian I-126
Ma, Xiaobin I-988
Ma, Xiaojiang I-744
Ma, Xiuli I-258
Ma, Ye II-701
Mao, Weihua I-90
Mao, Zongyuan I-565
Matsugu, Masakazu I-864
Mei, Fei I-654, I-690
Mei, Tiemin I-666, I-672
Meng, Guang II-613
Meng, Zhiqing I-406, II-1000
Millerioux, Gilles I-524
Miyajima, Hiromi II-436
Murahashi, Yoshimitsu I-714
Murray-Smith, Roderick II-52

Nakao, Zensho I-186
Nan, Dong I-235
Nath, Baikunth I-636
Nishi, Tetsuo I-474
Nishino, Fumihito I-1025

Oh, Sung-Kwun I-156, I-162, I-174, II-65
Okamoto, Masaru I-834
Okuma, Shigeru I-714
Ortiz-Gómez, Mamen I-229
Otten, Lambert I-383
Ou, Zongying I-611
Ouyang, Gaoxiang II-169

Pan, Daru II-269
Pan, Leilei I-383
Pan, Qielu II-7
Pang, Guibing II-262
Park, Byoung-Jun I-174
Park, Daehee I-162
Park, Gwi-Tae I-642
Park, Ho-Sung I-156, I-162
Pascual, Pedro I-55
Pedrycz, Witold I-156, I-162, I-174
Peng, Jinye II-663
Peng, Tian-Qiang I-965

Qi, Feihu I-487

Qi, Ronggang II-1007
Qian, Tao I-543
Qiao, Jianping I-702
Qin, Bo I-1015
Qiu, Tianshuang I-678, II-299, II-405, II-501, II-607
Qiu, Yuhui I-276
Qiu, ZhengDing I-882, II-994
Qu, Changwen II-886
Qu, Fuzheng I-750
Qu, Jianbo II-779

Ren, Bing II-854
Ren, Guanghui I-936
Ren, XuChun II-841
Rodríguez, Francisco B. I-14, I-20
Roh, Seok-Beom II-65

Saeki, Noboru II-488
Sakane, Akira II-488
Sato, Norihisa I-395
Sbarbaro, Daniel II-52
Seow, Ming-Jung I-870
Serrano, Eduardo I-14, I-20, I-55
Shang, Ruiqiang II-242
Shao, Chao II-707
Shao, Cheng II-860
Shao, Dingrong II-248, II-281
Shao, Zhiqiong I-235
Shen, Dongsheng I-988
Shen, Lixin I-678
Shen, Minfen I-726, I-732, II-350, II-356
Shen, Ruimin I-552
Shen, Ruiming II-651
Shen, Xianfeng II-832
Shen, Zhang-Quan I-323
Shi, Aiguo II-418, II-430, II-920
Shi, Haixiang II-230
Shi, Weiren II-880
Shi, Xiangquan II-344
Shi, Zhenwei I-684
Shi, Zhewen I-317
Shi, Zhiwei II-200
Shiba, Kenji II-488
Shigei, Noritaka II-436
Shu, Jiwu II-719
Song, Baoquan II-525
Song, Xigeng II-767
Song, Xin I-344
Song, Yanhui II-549

Song, Yibin II-224
Su, Feng II-886
Su, Guangda II-411
Su, Hang II-587
Su, Jianbo I-293, II-1, II-7
Su, Jing II-453
Sulistiyo I-186
Sun, Bing-Yu I-648
Sun, Changyin I-32
Sun, Daoheng II-826
Sun, Fuchun II-13
Sun, Hequan II-799
Sun, Hui II-587
Sun, Jiancheng I-578
Sun, Jiande I-738
Sun, Lisha I-732
Sun, Mingliang I-941
Sun, Shiliang II-950
Sun, Weitao II-719
Sun, Yongmei I-678
Sun, Youxian II-194, II-912
Sun, Yuanyuan II-581
Sun, Zengqi I-286, II-537
Sun, Zhan-Li I-774
Sun, Zonghai II-194, II-912

Tadeusiewicz, Ryszard I-150
Tai, Huiling II-892
Takahashi, Norikazu I-474
Tan, Gang II-442
Tan, Guozhen II-937
Tan, Xiaoyang I-858
Tan, Ying II-675
Tanaka, Yoshiyuki II-488
Tang, Fang I-448
Tang, Guoping II-645
Tang, Hao II-866
Tang, Haoyang I-61
Tang, Hong I-678
Tang, Huanwen I-684
Tang, Lixin I-454
Tang, Xianglong I-894
Tang, Yiyuan I-332, I-344
Tang, Zhenan I-696
Tang, Zheng I-338, I-401, I-419, I-425, I-430, I-442
Tao, Haihong II-236, II-275
Tao, Jun II-848
Tao, Ran II-324
Tao, Yewei I-702

Tian, Fengchun I-383
Tian, Shengfeng I-592
Tian, Xue II-713
Tong, Hanghang II-931
Torres-Sospedra, Joaquín I-223, I-229, I-852, I-912
Tóth, László I-996
Tsaih, Ray I-377
Tsuji, Toshio I-834, II-488

Valdes, Arturo II-52

Wada, Kiyoshi II-507
Wan, Anhua I-90
Wan, Chunru I-270
Wan, Yong II-311
Wan, Youchuan I-906
Wang, Bin II-331
Wang, Dan II-59
Wang, Gang I-49, I-708
Wang, Guangxing II-242
Wang, Haijun I-72
Wang, Han I-344
Wang, Hong-bin II-19
Wang, Hong-Qiang II-476
Wang, Hongrui II-19, II-206
Wang, Hongyu II-607
Wang, Jiahai I-401, I-419, I-425, I-430, I-442
Wang, Jianjun I-1, I-924
Wang, Jie II-943
Wang, Jing II-163, II-994
Wang, Jinkuan I-344
Wang, Jun I-1025
Wang, Junping I-512
Wang, Lidan II-813
Wang, Lin I-84
Wang, Ling I-264, I-448, II-236, II-275
Wang, Lipo I-270, II-230
Wang, Peijin II-224
Wang, Renming I-959
Wang, Ronglong I-419, I-442
Wang, Shaoyu I-487
Wang, Shoujue II-786
Wang, Wei II-200, II-453, II-657
Wang, Wenhui II-555
Wang, Xi-Huai II-956
Wang, Xia II-206
Wang, Xiao Tong II-701
Wang, Xicheng I-436

Wang, Xin   I-750, II-175, II-848
Wang, Xiong   II-97
Wang, Xiukun   I-332
Wang, Xudong   II-601
Wang, Xuening   II-25
Wang, Xuexia   I-936
Wang, Xugang   I-338
Wang, Yan   II-337, II-362
Wang, Yang   II-832
Wang, Yong   I-622
Wang, Yonggang   I-605
Wang, Yongxue   II-854
Wang, Yuan   II-962
Wang, Yukun   II-269
Wang, Yuwen   II-399
Wang, Zengfu   I-114, I-1008
Wang, Zhen   II-549
Wang, Zheng   I-924
Wang, Zhengzhi   II-525
Wang, Zhiquan   II-627
Wang, Zhongjie   II-175, II-848
Wang, Zhongsheng   I-44
Wang, Zhuo   I-846
Wei, Guangfen   I-696
Wei, Lixin   II-206
Wei, Xiaopeng   I-72
Wen, Jianming   II-866
Wen, Xian-bin   I-822
Wen, Yi-Min   I-480
Wen, Yingyou   II-242
Weng, Shifeng   I-205, I-900
Wong, Fai   I-205
Wong, Kwan-Yee Kenneth   II-731
Wong, Shu-Fai   II-731
Wu, Gengfeng   II-35, II-962
Wu, Hao   II-13
Wu, Jianping   I-565
Wu, Lide   I-1025
Wu, Qinghui   II-156, II-860
Wu, Tao   I-494
Wu, Wei   I-235, I-282
Wu, Weidong   II-826
Wu, Xiang   I-500
Wu, Xing   II-567
Wu, Yan   I-941
Wu, Ying   I-138
Wu, Zhilu   I-936

Xi, Jiangtao   I-672
Xi, Youmin   I-126

Xi, Yugeng   I-586, II-755
Xia, Guang'an   I-419
Xia, Guangpu   I-401, I-419, I-430, I-442
Xia, Shunren   I-953
Xia, Zhimin   I-900
Xiang, Tao   I-132
Xiao, Di   II-633
Xiao, Huanqin   I-537
Xiao, Jian-Mei   II-956
Xie, Guangzhong   II-892
Xie, Hongmei   II-663
Xie, Hua   I-953
Xie, Ji-Gang   II-994
Xing, Fei   I-616
Xing, Hong-Jie   I-7
Xiong, Wenjun   I-108
Xiong, Zhangliang   II-344
Xiong, Zhihua   II-97
Xu, Bowen   II-866
Xu, Hongji   II-311, II-317
Xu, Hua   I-1015
Xu, Jianhua   I-252
Xu, Jianxue   I-138
Xu, Lei   I-828
Xu, Roger   I-543
Xu, Shiguo   II-713
Xu, Weidong   I-953
Xu, Weijun   I-1, II-725
Xu, Weiling   I-726, I-732
Xu, Wenji   II-262
Xu, Xin   II-25
Xu, Xinhe   II-181
Xu, Xinshun   I-401, I-425, I-430
Xu, Xu   II-104
Xu, Yinfeng   II-725
Xu, Yongmao   II-97, II-773
Xu, Yuesheng   I-235
Xu, ZongBen   I-1
Xue, Jin   II-399

Yamamoto, Takayoshi   II-574
Yan, Dongwei   II-943
Yan, Huajun   II-465
Yan, Lan-Feng   II-453
Yang, Bo   I-211, II-71
Yang, Chan-Yun   I-506
Yang, Jialin   II-832
Yang, Jianhua   II-144
Yang, Jie   I-235, I-389
Yang, Jinyao   I-726, II-350

Yang, Luxi   I-660
Yang, Shangming   I-193
Yang, Simon X.   I-383, II-880
Yang, XianHui   II-874
Yang, Xiaowei   I-558
Yang, Xuhua   II-194, II-912
Yang, Yongqing   II-619, II-976
Yang, Yu-Jiu   I-7
Yang, Yuansheng   I-983
Yang, Zeying   II-779
Yang, Zi-Jiang   II-507
Yao, Jin   II-832
Yao, Kaifeng   I-537
Yao, Man   II-601
Yao, ZhenHan   II-841
Ye, Datian   II-459
Ye, Hao   II-561
Ye, Mao   I-810, I-815
Ye, Yisong   I-941
Yeh, Chung-Hsing   II-898
Yi, Hongjie   I-1015
Yi, Zhang   I-810, I-815, II-465
Yigit, Halil   II-254
Yin, Chuanhuan   I-592
Yin, Fuliang   I-666, I-672, I-755, II-299
You, Lan   I-1025
Young, Rebecca   I-543
Yu, Daoheng   I-26, I-413, II-293
Yu, Ensheng   I-436
Yu, Guoqiang   II-950
Yu, Hao   I-1025
Yu, Jian   I-168, I-592
Yu, Jin   I-598
Yu, Jinyong   II-117
Yu, Jun   I-696
Yu, Qilian   I-786
Yu, Shu   I-558
Yu, Wen   II-77, II-212
Yuan, Jianping   II-381
Yuan, Kun   I-78
Yuan, Wenjiang   II-937
Yuan, Yanbo   II-269
Yuan, Yuman   II-117
Yue, Heng   II-175

Zeng, Fanming   II-110
Zeng, FanZi   I-882
Zeng, Jin   II-880
Zeng, Zhigang   I-114, I-1008
Zha, Daifeng   I-678

Zhai, Xiaobing   II-262
Zhang, Anqing   I-930, II-501
Zhang, Bin   II-925
Zhang, Bo   I-462, II-248, II-281
Zhang, Bofeng   II-962
Zhang, Changshui   I-205, I-900, II-950
Zhang, Daoqiang   I-180
Zhang, Dianjun   II-156
Zhang, Fuyan   I-858
Zhang, GuangZheng   II-513
Zhang, Guicai   II-595
Zhang, Hande   I-977
Zhang, Hao   II-13
Zhang, Hongyu   II-854
Zhang, Huaguang   I-49, I-96
Zhang, Jianwei   I-462
Zhang, Jianzhong   II-531
Zhang, Jie   II-97, II-773, II-874
Zhang, Jing   I-636
Zhang, Jingfen   II-613
Zhang, Li   I-586
Zhang, Liming   I-26, I-413, I-720, I-924, II-331
Zhang, Liqing   I-235, II-442
Zhang, Ming   II-587
Zhang, Naimin   I-235
Zhang, Qiang   I-72
Zhang, Qun   II-163
Zhang, Shanwen   I-537
Zhang, Taiyi   I-578
Zhang, Tianqi   I-780
Zhang, Weidong   II-110
Zhang, Wenxiu   I-244
Zhang, Xian-Da   I-761
Zhang, Xiang   II-761
Zhang, Xiangliang   II-657
Zhang, Xinhua   I-930
Zhang, Xiuling   II-169
Zhang, Xuegong   I-252
Zhang, Xuxiu   II-501
Zhang, Yanglan   II-982
Zhang, Yi   I-193
Zhang, Yonghui   II-156, II-860
Zhang, Yongsheng   II-418, II-430, II-920
Zhang, Youyun   I-518
Zhang, Yu   II-537
Zhang, Yunfeng   II-362
Zhang, Yuzheng   II-350
Zhang, Zengke   II-138
Zhang, Zhao   I-1021

Zhang, Zhengwei II-561
Zhang, Zhiqiang II-169
Zhao, Chun I-90
Zhao, Deyou II-613
Zhao, Dongfang I-605
Zhao, Feng I-558
Zhao, Hai I-888
Zhao, Han-Qing I-494
Zhao, Jian II-663
Zhao, Jianli II-242
Zhao, Jianye II-293
Zhao, Ke II-925
Zhao, Li I-947
Zhao, Lianwei II-482
Zhao, Ming-Yang II-820
Zhao, Shi Jian II-773
Zhao, Xing-Ming II-476
Zhao, Yaqin I-936
Zhao, Zhijin I-654, I-690
Zhao, Zhong-Qiu I-648
Zhao, Zhonggai I-798
Zheng, Chunhong I-629
Zheng, Guiwen I-629
Zheng, Nannig I-792
Zheng, Nanning II-470
Zheng, Qin I-317
Zheng, Rirong I-565
Zheng, Tian I-822
Zheng, Weimin II-719
Zheng, Wenming I-947
Zheng, Xiaoshen I-840
Zheng, Xufei I-276

Zheng, Yong-Rui I-755
Zhou, Bo II-418, II-430, II-920
Zhou, Chun-Guang II-337
Zhou, Donghua II-91, II-543, II-555
Zhou, Dongsheng I-72
Zhou, Fuchang II-595
Zhou, Gengui I-406
Zhou, Hao II-331
Zhou, Hong II-305
Zhou, Jin I-144
Zhou, Jinjin II-262
Zhou, Mingquan II-663
Zhou, Tsing II-639
Zhou, Xiaoyan I-947
Zhou, Xin II-663
Zhou, Yunshui I-654
Zhou, Zhengzhong I-780
Zhou, Zhi-Hua I-180, I-356, I-768, I-858
Zhou, Zongtan I-199, II-525
Zhu, Daqi II-619, II-976
Zhu, Hui I-894
Zhu, Ruijun II-299
Zhu, Xiao-Long I-761
Zhu, Xin-jian II-150
Zhu, Yihua I-406
Zhu, Yongsheng I-518
Zong, Ziliang I-338
Zou, Cairong I-947
Zou, Dali I-750
Zou, Jiyan II-587
Zou, Shu-Xue II-337
Zurada, Jacek M. II-695

Lecture Notes in Computer Science

For information about Vols. 1–3058

please contact your bookseller or Springer

Vol. 3177: Z.R. Yang, H. Yin, R. Everson (Eds.), Intelligent Data Engineering and Automated Learning – IDEAL 2004. VXIII, 852 pages. 2004.

Vol. 3174: F. Yin, J. Wang, C. Guo (Eds.), Advances in Neural Networks - ISNN 2004, Part II. XXXV, 1021 pages. 2004.

Vol. 3173: F. Yin, J. Wang, C. Guo (Eds.), Advances in Neural Networks - ISNN 2004, Part I. XXXV, 1041 pages. 2004.

Vol. 3172: M. Dorigo, M. Birattari, C. Blum, L. M.Gambardella, F. Mondada, T. Stützle (Eds.), Ant Colony, Optimization and Swarm Intelligence. XII, 434 pages. 2004.

Vol. 3158: I. Nikolaidis, M. Barbeau, E. Kranakis (Eds.), Ad-Hoc, Mobile, and Wireless Networks. IX, 344 pages. 2004.

Vol. 3157: C. Zhang, H. W. Guesgen, W.K. Yeap (Eds.), PRICAI 2004: Trends in Artificial Intelligence. XX, 1023 pages. 2004. (Subseries LNAI).

Vol. 3156: M. Joye, J.-J. Quisquater (Eds.), Cryptographic Hardware and Embedded Systems - CHES 2004. XIII, 455 pages. 2004.

Vol. 3153: J. Fiala, V. Koubek, J. Kratochvíl (Eds.), Mathematical Foundations of Computer Science 2004. XIV, 902 pages. 2004.

Vol. 3152: M. Franklin (Ed.), Advances in Cryptology – CRYPTO 2004. XI, 579 pages. 2004.

Vol. 3150: G.-Z. Yang, T. Jiang (Eds.), Medical Imaging and Virtual Reality. XII, 378 pages. 2004.

Vol. 3148: R. Giacobazzi (Ed.), Static Analysis. XI, 393 pages. 2004.

Vol. 3146: P. Érdi, A. Esposito, M. Marinaro, S. Scarpetta (Eds.), Computational Neuroscience: Cortical Dynamics. XI, 161 pages. 2004.

Vol. 3144: M. Papatriantafilou, P. Hunel (Eds.), Principles of Distributed Systems. XI, 246 pages. 2004.

Vol. 3143: W. Liu, Y. Shi, Q. Li (Eds.), Advances in Web-Based Learning – ICWL 2004. XIV, 459 pages. 2004.

Vol. 3142: J. Diaz, J. Karhumäki, A. Lepistö, D. Sannella (Eds.), Automata, Languages and Programming. XIX, 1253 pages. 2004.

Vol. 3140: N. Koch, P. Fraternali, M. Wirsing (Eds.), Web Engineering. XXI, 623 pages. 2004.

Vol. 3139: F. Iida, R. Pfeifer, L. Steels, Y. Kuniyoshi (Eds.), Embodied Artificial Intelligence. IX, 331 pages. 2004. (Subseries LNAI).

Vol. 3138: A. Fred, T. Caelli, R.P.W. Duin, A. Campilho, D.d. Ridder (Eds.), Structural, Syntactic, and Statistical Pattern Recognition. XXII, 1168 pages. 2004.

Vol. 3136: F. Meziane, E. Métais (Eds.), Natural Language Processing and Information Systems. XII, 436 pages. 2004.

Vol. 3134: C. Zannier, H. Erdogmus, L. Lindstrom (Eds.), Extreme Programming and Agile Methods - XP/Agile Universe 2004. XIV, 233 pages. 2004.

Vol. 3133: A.D. Pimentel, S. Vassiliadis (Eds.), Computer Systems: Architectures, Modeling, and Simulation. XIII, 562 pages. 2004.

Vol. 3131: V. Torra, Y. Narukawa (Eds.), Modeling Decisions for Artificial Intelligence. XI, 327 pages. 2004. (Subseries LNAI).

Vol. 3130: A. Syropoulos, K. Berry, Y. Haralambous, B. Hughes, S. Peter, J. Plaice (Eds.), TEX, XML, and Digital Typography. VIII, 265 pages. 2004.

Vol. 3129: Q. Li, G. Wang, L. Feng (Eds.), Advances in Web-Age Information Management. XVII, 753 pages. 2004.

Vol. 3128: D. Asonov (Ed.), Querying Databases Privately. IX, 115 pages. 2004.

Vol. 3127: K.E. Wolff, H.D. Pfeiffer, H.S. Delugach (Eds.), Conceptual Structures at Work. XI, 403 pages. 2004. (Subseries LNAI).

Vol. 3126: P. Dini, P. Lorenz, J.N.d. Souza (Eds.), Service Assurance with Partial and Intermittent Resources. XI, 312 pages. 2004.

Vol. 3125: D. Kozen (Ed.), Mathematics of Program Construction. X, 401 pages. 2004.

Vol. 3124: J.N. de Souza, P. Dini, P. Lorenz (Eds.), Telecommunications and Networking - ICT 2004. XXVI, 1390 pages. 2004.

Vol. 3123: A. Belz, R. Evans, P. Piwek (Eds.), Natural Language Generation. X, 219 pages. 2004. (Subseries LNAI).

Vol. 3121: S. Nikoletseas, J.D.P. Rolim (Eds.), Algorithmic Aspects of Wireless Sensor Networks. X, 201 pages. 2004.

Vol. 3120: J. Shawe-Taylor, Y. Singer (Eds.), Learning Theory. X, 648 pages. 2004. (Subseries LNAI).

Vol. 3118: K. Miesenberger, J. Klaus, W. Zagler, D. Burger (Eds.), Computer Helping People with Special Needs. XXIII, 1191 pages. 2004.

Vol. 3116: C. Rattray, S. Maharaj, C. Shankland (Eds.), Algebraic Methodology and Software Technology. XI, 569 pages. 2004.

Vol. 3114: R. Alur, D.A. Peled (Eds.), Computer Aided Verification. XII, 536 pages. 2004.

Vol. 3113: J. Karhumäki, H. Maurer, G. Paun, G. Rozenberg (Eds.), Theory Is Forever. X, 283 pages. 2004.

Vol. 3112: H. Williams, L. MacKinnon (Eds.), Key Technologies for Data Management. XII, 265 pages. 2004.

Vol. 3111: T. Hagerup, J. Katajainen (Eds.), Algorithm Theory - SWAT 2004. XI, 506 pages. 2004.

Vol. 3110: A. Juels (Ed.), Financial Cryptography. XI, 281 pages. 2004.

Vol. 3109: S.C. Sahinalp, S. Muthukrishnan, U. Dogrusoz (Eds.), Combinatorial Pattern Matching. XII, 486 pages. 2004.

Vol. 3108: H. Wang, J. Pieprzyk, V. Varadharajan (Eds.), Information Security and Privacy. XII, 494 pages. 2004.

Vol. 3107: J. Bosch, C. Krueger (Eds.), Software Reuse: Methods, Techniques and Tools. XI, 339 pages. 2004.

Vol. 3106: K.-Y. Chwa, J.I. Munro (Eds.), Computing and Combinatorics. XIII, 474 pages. 2004.

Vol. 3105: S. Göbel, U. Spierling, A. Hoffmann, I. Iurgel, O. Schneider, J. Dechau, A. Feix (Eds.), Technologies for Interactive Digital Storytelling and Entertainment. XVI, 304 pages. 2004.

Vol. 3104: R. Kralovic, O. Sykora (Eds.), Structural Information and Communication Complexity. X, 303 pages. 2004.

Vol. 3103: K. Deb, e. al. (Eds.), Genetic and Evolutionary Computation – GECCO 2004. XLIX, 1439 pages. 2004.

Vol. 3102: K. Deb, e. al. (Eds.), Genetic and Evolutionary Computation – GECCO 2004. L, 1445 pages. 2004.

Vol. 3101: M. Masoodian, S. Jones, B. Rogers (Eds.), Computer Human Interaction. XIV, 694 pages. 2004.

Vol. 3100: J.F. Peters, A. Skowron, J.W. Grzymała-Busse, B. Kostek, R.W. Świniarski, M.S. Szczuka (Eds.), Transactions on Rough Sets I. X, 405 pages. 2004.

Vol. 3099: J. Cortadella, W. Reisig (Eds.), Applications and Theory of Petri Nets 2004. XI, 505 pages. 2004.

Vol. 3098: J. Desel, W. Reisig, G. Rozenberg (Eds.), Lectures on Concurrency and Petri Nets. VIII, 849 pages. 2004.

Vol. 3097: D. Basin, M. Rusinowitch (Eds.), Automated Reasoning. XII, 493 pages. 2004. (Subseries LNAI).

Vol. 3096: G. Melnik, H. Holz (Eds.), Advances in Learning Software Organizations. X, 173 pages. 2004.

Vol. 3095: C. Bussler, D. Fensel, M.E. Orlowska, J. Yang (Eds.), Web Services, E-Business, and the Semantic Web. X, 147 pages. 2004.

Vol. 3094: A. Nürnberger, M. Detyniecki (Eds.), Adaptive Multimedia Retrieval. VIII, 229 pages. 2004.

Vol. 3093: S.K. Katsikas, S. Gritzalis, J. Lopez (Eds.), Public Key Infrastructure. XIII, 380 pages. 2004.

Vol. 3092: J. Eckstein, H. Baumeister (Eds.), Extreme Programming and Agile Processes in Software Engineering. XVI, 358 pages. 2004.

Vol. 3091: V. van Oostrom (Ed.), Rewriting Techniques and Applications. X, 313 pages. 2004.

Vol. 3089: M. Jakobsson, M. Yung, J. Zhou (Eds.), Applied Cryptography and Network Security. XIV, 510 pages. 2004.

Vol. 3087: D. Maltoni, A.K. Jain (Eds.), Biometric Authentication. XIII, 343 pages. 2004.

Vol. 3086: M. Odersky (Ed.), ECOOP 2004 – Object-Oriented Programming. XIII, 611 pages. 2004.

Vol. 3085: S. Berardi, M. Coppo, F. Damiani (Eds.), Types for Proofs and Programs. X, 409 pages. 2004.

Vol. 3084: A. Persson, J. Stirna (Eds.), Advanced Information Systems Engineering. XIV, 596 pages. 2004.

Vol. 3083: W. Emmerich, A.L. Wolf (Eds.), Component Deployment. X, 249 pages. 2004.

Vol. 3080: J. Desel, B. Pernici, M. Weske (Eds.), Business Process Management. X, 307 pages. 2004.

Vol. 3079: Z. Mammeri, P. Lorenz (Eds.), High Speed Networks and Multimedia Communications. XVIII, 1103 pages. 2004.

Vol. 3078: S. Cotin, D.N. Metaxas (Eds.), Medical Simulation. XVI, 296 pages. 2004.

Vol. 3077: F. Roli, J. Kittler, T. Windeatt (Eds.), Multiple Classifier Systems. XII, 386 pages. 2004.

Vol. 3076: D. Buell (Ed.), Algorithmic Number Theory. XI, 451 pages. 2004.

Vol. 3075: W. Lenski, Logic versus Approximation. VIII, 205 pages. 2004.

Vol. 3074: B. Kuijpers, P. Revesz (Eds.), Constraint Databases and Applications. XII, 181 pages. 2004.

Vol. 3073: H. Chen, R. Moore, D.D. Zeng, J. Leavitt (Eds.), Intelligence and Security Informatics. XV, 536 pages. 2004.

Vol. 3072: D. Zhang, A.K. Jain (Eds.), Biometric Authentication. XVII, 800 pages. 2004.

Vol. 3071: A. Omicini, P. Petta, J. Pitt (Eds.), Engineering Societies in the Agents World. XIII, 409 pages. 2004. (Subseries LNAI).

Vol. 3070: L. Rutkowski, J. Siekmann, R. Tadeusiewicz, L.A. Zadeh (Eds.), Artificial Intelligence and Soft Computing - ICAISC 2004. XXV, 1208 pages. 2004. (Subseries LNAI).

Vol. 3068: E. André, L. Dybkjær, W. Minker, P. Heisterkamp (Eds.), Affective Dialogue Systems. XII, 324 pages. 2004. (Subseries LNAI).

Vol. 3067: M. Dastani, J. Dix, A. El Fallah-Seghrouchni (Eds.), Programming Multi-Agent Systems. X, 221 pages. 2004. (Subseries LNAI).

Vol. 3066: S. Tsumoto, R. Słowiński, J. Komorowski, J.W. Grzymała-Busse (Eds.), Rough Sets and Current Trends in Computing. XX, 853 pages. 2004. (Subseries LNAI).

Vol. 3065: A. Lomuscio, D. Nute (Eds.), Deontic Logic in Computer Science. X, 275 pages. 2004. (Subseries LNAI).

Vol. 3064: D. Bienstock, G. Nemhauser (Eds.), Integer Programming and Combinatorial Optimization. XI, 445 pages. 2004.

Vol. 3063: A. Llamosí, A. Strohmeier (Eds.), Reliable Software Technologies - Ada-Europe 2004. XIII, 333 pages. 2004.

Vol. 3062: J.L. Pfaltz, M. Nagl, B. Böhlen (Eds.), Applications of Graph Transformations with Industrial Relevance. XV, 500 pages. 2004.

Vol. 3061: F.F. Ramos, H. Unger, V. Larios (Eds.), Advanced Distributed Systems. VIII, 285 pages. 2004.

Vol. 3060: A.Y. Tawfik, S.D. Goodwin (Eds.), Advances in Artificial Intelligence. XIII, 582 pages. 2004. (Subseries LNAI).

Vol. 3059: C.C. Ribeiro, S.L. Martins (Eds.), Experimental and Efficient Algorithms. X, 586 pages. 2004.